Computational Fluid Dynamics 2004

C. Groth · D. W. Zingg (Editors)

Computational Fluid Dynamics 2004

Proceedings
of the Third International Conference
on Computational Fluid Dynamics, ICCFD3,
Toronto, 12–16 July 2004

With 627 Figures and 47 Tables

Springer

Clinton Groth
David W. Zingg
University of Toronto
Institute of Aerospace Studies
4925 Dufferin Street
Toronto, Ontario M3H 5T6
Canada
e-mail: groth@utias.utoronto.ca
dwz@oddjob.utias.utoronto.ca

Library of Congress Control Number: 2006928440

ISBN-10 3-540-31800-3 Springer Berlin Heidelberg New York
ISBN-13 978-3-540-31800-2 Springer Berlin Heidelberg New York

Springer is a part of Springer Science+Business Media

springer.com

Typesetting by the editors, final processing by LE-TeX GbR, Leipzig
Production: LE-TEX Jelonek, Schmidt & Vöckler GbR, Leipzig
Cover design: *design & production*, Heidelberg

Printed on acid-free paper 54/3100/YL 5 4 3 2 1 0

Preface

This book contains the Proceedings of the Third International Conference on Computational Fluid Dynamics (ICCFD), held in Toronto, Ontario, Canada from July 12 through 16, 2004. The ICCFD series has evolved into the leading international conference series for scientists, mathematicians, and engineers interested in the computation of fluid flow. Invited keynote lectures were given by renowned researchers in the areas of detached-eddy simulation, micro-flow simulations, control of shocks, multiscale models of the circulatory system, and high-end computing in aerospace.

Abstracts were received from 23 countries. The executive committee, consisting of C. Bruneau, J.-J. Chattot, D. Kwak, N. Satofuka, K. Srinivas, and myself, was responsible for selection of papers. Each of the members had a separate subcommittee to carry out the evaluation. As a result of this careful peer review process, 123 papers were selected for oral presentation and a further 67 for poster presentation. The conference was attended by 172 delegates from 22 different countries.

Thanks are due to our sponsors, NASA and the Institute for Aerospace Research of the National Research Council of Canada. In particular, the generous grant from NASA is a key factor in the success of this conference series. I would also like to thank the staff at the Westin Harbour Castle, all who participated in the organization of the conference, including the review process, and the students of the CFD group at the University of Toronto Institute for Aerospace Studies for the tremendous help they provided toward the success of this conference.

These Proceedings contain a snapshot of the field of computational fluid dynamics as of 2004. They present a vibrant field with strong capabilities in many areas of application and a myriad of opportunities for future contributions to science and engineering.

Toronto, Canada
September 2004

David W. Zingg
Conference Chair

Contents

Part III Adaptive Meshing

Part IV Algorithms

Part XVII Optimization

Part XVIII Parallel Algorithms

Part XX Technical Notes

Part I

Invited Lectures

Topics in Detached-Eddy Simulation

Philippe R. Spalart

Boeing Commercial Airplanes, P.O. Box 3707, Seattle, WA 98124, USA
philippe.r.spalart@boeing.com

The paper re-visits the motivation of DES, and then touches on its diffusion in CFD codes; grid concerns including both users' mis-conceptions and actual DES issues; the use of DES as a pure LES with wall modeling; and possible long-term improvements.

1 Introduction

The DES approach to high-Reynolds-number separated flows is seven years old [1], although the first true results appeared only five years ago [2]. Its best description is in [3], and a broader review in [4]. The central motivation is the observation that Large-Eddy Simulation (LES) is powerful in regions of massive separation and other free shear flows such as jets, but much too costly in the large areas of thin boundary layers (BL's) which cover aircraft and vehicles. Therefore, affordable CFD approaches need to treat these with Reynolds-Averaged modeling. No theoretical rebuttal has been made by LES proponents of this pessimistic statement, which has had an influence at least in Europe. Even as a "grand challenge" and with generous assumptions, the estimated readiness date of pure LES for a wing remains at the year 2045.

On the other hand, progress in Reynolds-Averaged Navier-Stokes (RANS) models outside thin shear flows has remained very modest, whether in terms of the numerical practicality of the models, or their accuracy. The two dominant models are 12 years old. This field being idea-limited, a "readiness date" cannot be projected. The pessimistic view is that a general RANS model with certain engineering accuracy is out of reach of human intelligence. However, keeping the other sources of error in CFD below engineering accuracy will never be certain either, considering the users' training and their need for rapid answers. In any case, RANS has its place, especially for attached flows which place low demands both on the models' physics and the users' competence.

The LES cost estimates of 1997 [1] can be confronted with recent findings. Even forceful studies such as LESFOIL found that in 2002 the limit on the spanwise domain size for LES of an airfoil was near 1% of its chord [5, 6] which is insufficient when the BL thickness δ exceeds 8%. Over the trailing edge, even the best Reynolds stresses were not very close to experiment. Now, extrapolating to the wing considered in [1, 4], its turbulent domain is 2,000 times larger, and the time interval much longer. Since the ONERA LES used 6 million points per side of the airfoil and per % of chord, without a turbulent leading edge or lower surface, the extrapolation to a wing leads to well over 10^{10} points, which is consistent with the estimate of 10^{11} in [4]. Recall that it is a weak function of Reynolds number (because under the assumptions made for wall modeling, only the slow thinning of the BL influences the cost).

The very resilient issues with LES and RANS led to a consensus in many circles that RANS/LES hybrid methods are the only ones with a chance in external separated flows, and to the creation of other hybrids, in particular LNS and SAS [7, 8].

These other, more recent hybrids have not yet spread outside the groups that created them but DES has, and is offered in vendor CFD codes including Cobalt, CFD++, STAR-CD, Acusolve and Fluent [9, 10]. Their capability to resolve "LES content", with short waves and high frequencies, needs to be verified; Cobalt results have been impressive. It is not clear what proportion of users can make an accurate use of DES, because significant additional decisions must be made for the grid and time step, relative to RANS CFD, and a substantial increase in cost must be accepted. It appears that the vendors provide publications and consultation to new DES users, rather than comprehensive sets of instructions. A manual for DES grid design is found at http://techreports.larc.nasa.gov/ltrs/PDF/2001/cr/NASA-2001-cr211032.pdf. However, no manual can be a substitute for the combination of experience, intuition for separation and turbulence, and effort in visualizing solutions. Also essential is a willingness to apply grid refinement, and to admit that CFD is not yet able to produce an accurate solution when the problem is simply too challenging. Examples would be: an aircraft with Active Flow Control through a tiny suction slot; a complete car; and a prediction of noise over the entire audible range. Yet, these tasks are in high demand.

2 Grid Issues

2.1 Expectations for grid count

In some studies, DES is compared with LES on the same flow, and is expected to provide similar accuracy on a *coarser* grid than LES. This is most often incorrect. If a pure LES is possible, the flow cannot contain extended turbulent BL's. The BL's, probably, are simply laminar, so that DES does not provide its fundamental additional capability over LES. The difficulty resides in the region of massive separation, and there is no reason why DES would accept a coarser grid than LES does. The DES SGS model is one among many plausible ones. Therefore, it is fair to compare DES and LES on the same grid, and to count that DES can also treat the flow at high Reynolds numbers with turbulent BL's, *without* a dramatic reduction of the grid spacing [3].

Another error is apparent in some LES studies of flows with turbulent BL's. These used pure SGS models such as Smagorinsky's or its dynamic derivative in those BL's, with grids much too coarse to resolve the BL eddies. Such simulations are, effectively, DES with an inappropriate RANS model, one which has no credentials to simulate an entire BL. The eddy viscosity is, furthermore, grid-dependent even in the attached BL, which would result in a continued drift of the separation point if the grid were refined.

Grid refinement in unsteady three-dimensional simulations is very demanding. A refinement that doubles the total number of points is questionable. An unquestionable refinement consists in doubling the number of points in every direction, and the number of time steps. This raises the cost by a factor of 16, and is rarely achieved. A fair compromise is to use a factor of $\sqrt{2}$ in each direction, especially if this is done twice [3]. With second-order numerics and a given equation, such a refinement reduces the error by half. The situation in DES and LES is not as simple, because the differential equation depends on the grid spacing, and the order of accuracy depends on which quantity is considered. Nevertheless, if an LES were grossly under-resolved, it is very unlikely that a refinement by $\sqrt{2}$ would fail to reveal it.

Refining the grid in only one or two directions is not consistent, unless the coarse run has given strong evidence that only these directions were under-resolved. In [11], the spanwise grid spacing Δz around a cylinder was left the same while the x-y grid was refined. Furthermore, Δz was already larger than the x-y spacings in the sensitive region, so that the capability to resolve eddies was unchanged. The refinement was illusory. In contrast, Morton *et al.* applied systematic refinement via a parameter in the grid generation [10].

The common approach is to learn about the flow from simulations on relatively coarse grids and to generate grids that are finer by the factor $\sqrt{2}$ in the more sensitive regions, but not everywhere [3]. The grid is optimized, based on flow visualizations. The grid count does not quite rise by $\sqrt{8}$ but, in the user's judgment, the quality of the resolution did improve by $\sqrt{2}$. The neatest package would come from re-running the coarse simulation on a grid obtained by uniformly de-refining the optimized fine grid.

2.2 Grey Area, Ambiguous Grids and "Grid-Induced Separation"

Concurrently with its encouraging results on airfoils, thin wings, and cylinders, weaknesses of DES were discovered, notably by Caruelle, Deck, and Menter [12, 13]. It was always recognized that the location of separation will always be controlled by the RANS model, so that perfection is not expected, no matter how fine the grid. The primary new concern is that, starting from a valid RANS solution (Type I in fig.1), gradually refining the grid alters the solution in obscure ways. In the extreme, it leads to a serious problem, called "Grid-Induced Separation" (GIS) by Menter [13]. It was not anticipated in [1] that simulations would encounter this with grids intended for the RANS mode, but the evidence is here. The reduced grid spacing begins

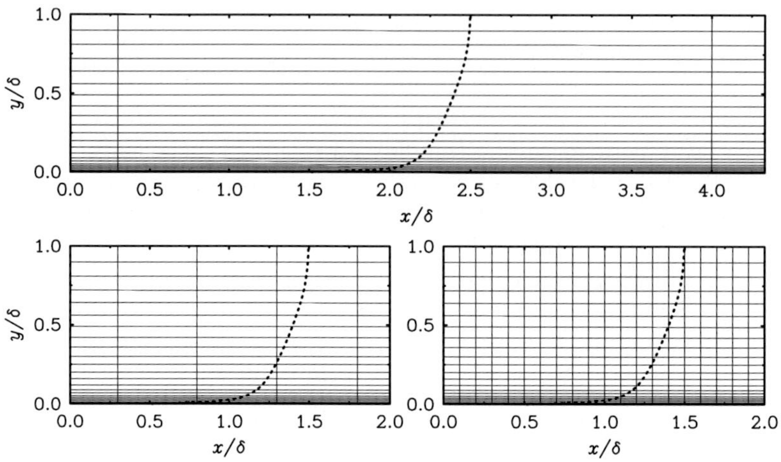

Fig. 1. DES grids in a boundary layer. Top, Type I, natural DES; left, Type II, ambiguous spacing; right, Type III, LES. See section 2.2 and 3. - - -, mean velocity. δ is BL thickness. Assume $\Delta z \approx \Delta x$.

to lower the eddy viscosity, in the direction of LES, but not enough to allow "LES content" (eddies) to develop. The grid is "ambiguous" (Type II) and the DES equations fail to recognize that pure RANS behavior was intended. The solution then is essentially a RANS with too weak an eddy viscosity; the strongest symptom is premature separation. Furthermore, DES fails to give much of a signal of this failure. Sometimes, GIS results from refining the grid to a spacing that is un-necessarily fine, and regions in which the wall-parallel spacing is very fine in both directions cannot be extensive, simply because the grid count would be extremely large. For instance, an efficient grid adaptation at the foot of a shock wave would refine the spacing only normal to the shock, and therefore not cause GIS. In that sense, two-dimensional exercises as in [13] over-state the GIS issue. DES, of course, is never two-dimensional. Nevertheless, the ideal hybrid method would never produce GIS, even with substantial thickening of the BL. Some effort was applied against GIS but without much success, at least if the modifications are required to preserve the simplicity of the DES equations and avoid zonal divisions.

A related danger is that such an un-intended drop of eddy viscosity can fortuitously improve these near-RANS predictions, because turbulence models fail somewhat more often by producing an excess than a deficit of eddy viscosity. See, for instance, the difference between Menter's BSL and SST models: SST is the favorite, and always returns a lower eddy viscosity and therefore earlier separation. Also observe that all simple models produce far too much eddy viscosity inside vortices. It is much preferable not to attribute to DES an improvement which is not deserved, and would overshoot if the grid were refined further (but still short of LES mode).

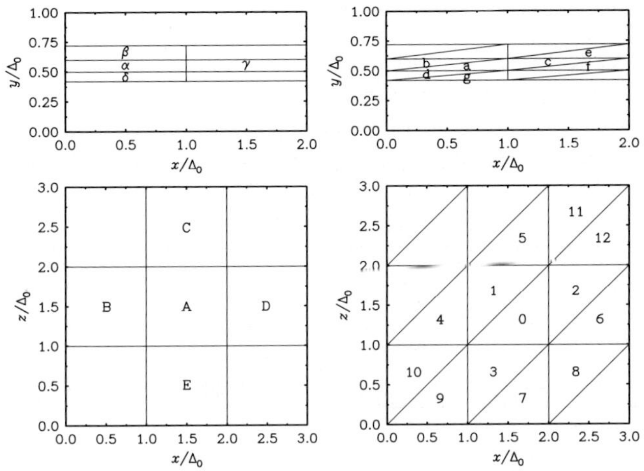

Fig. 2. Possible grid arrangements in a BL. Upper frames, side views; lower frames, top views. Left frames, structured grids; right frames, regular prisms or tetrahedra.

2.3 Definition of Δ in unstructured grids

The length scale Δ controls the eddy viscosity in the LES regions. The definition of DES [1] included $\Delta \equiv \max(\Delta x, \Delta y, \Delta z)$, for structured grids. A clear statement for unstructured grids is essential now that such codes are running DES. Figure 2 illustrates the challenge in a BL. The cell names are placed at the cell centroids. The target for Δ is Δ_0, at least at first sight. This is natural in the structured hexahedral grids on the left, where the distances taken will be from α to γ and from A to B, for instance. On the right, the grids are regular, but considered as unstructured. A common grid type is with prisms, which look like the upper-left frame from the side, and the lower-right frame from the top. Considering cell 0, the natural procedure will be to calculate the distance only to 1-3, the cells which share a face with 0. Then, Δ will be only $\Delta_0\sqrt{5}/3$, or about $0.75\Delta_0$. If cells that share only a corner are included, cells 4-7 give Δ_0, but cells 10-11 drive Δ to $\Delta_0\sqrt{20}/3$, which is much larger. The cell diameter is $\sqrt{2}\Delta_0$ (or a little more, because of the cell thickness in 3D). For consistency with the cubic cells of the 3D simulations used to set the value of C_{DES}, the diameter would be divided by $\sqrt{3}$. Therefore, the diameter measure finally gives $\sqrt{2/3}\Delta_0$, which appears satisfactory.

A more subtle question is: what is the legitimate value for Δ in this prism grid? It has twice the degrees of freedom of the hexahedral grid, but is the effective resolution better by $\sqrt{2}$? The diameter of the cells is a plausible measure, since it is the longest distance over which some derivative of the function is assumed to be uniform in the re-construction scheme. The diameter of the triangles is the same as that of the squares. On the other hand, the average distance to the points (1-3) used to calculate the gradient is also a plausible measure, and is not far from $\Delta_0/\sqrt{2}$. In the end, the effect is weak for prisms, and C_{DES} has never been a sensitive constant.

The effect could be stronger for other cell shapes, such as tetrahedra, and arrangements. These will not be analyzed here, but code writers should be aware. In no situation should the scale Δ reduce to a distance similar to that from a to g in the upper-right frame. In general, using for Δ the cell diameter, divided by $\sqrt{3}$, appears to be the safest approach.

3 Use of DES as a pure LES

3.1 Analysis of the channel-flow DES study by Nikitin *et al.*

The paper, NNWSS in short, addressed the simplest of candidates for LES among wall-bounded flows [14]. The objective was to understand the behavior of DES in a thick BL with a grid fine enough for LES mode inside it (Type III). This is an "un-natural" use of DES, but some geometries will impose it, and it also is the only possible solution to the issue that no RANS model will ever give a perfect prediction of separation and reattachment. This type of DES can also be viewed as an LES with "wall modeling". This makes it a candidate for atmospheric BL simulations, for instance; it has unlimited Reynolds-number capability, and the S-A model can treat rough surfaces.

In the NNWSS work, DES functioned as expected, as an unsteady RANS very near the wall and an LES in the center of the channel. This has not been achieved by other RANS-LES hybrids, nor by SAS [8]. The velocity profiles revealed the expected "modeled log layer" very near the wall, and "resolved log layer" part-way up the BL. These two layers matched in slope (the Kármán constant κ), but not in level (the intercept C). Some observers took this as a substantial failure of DES, thus failing to appreciate several crucial aspects of that work.

The study was conducted under very tight constraints. The formulas of DES were used without any adjustments for this new rôle, and fit in a small space; simplicity and clarity are tangible advantages for any model, as confirmed by the agreement between three different codes. The grids had identical spacings in the wall-parallel directions, therefore indifferent to the flow direction; the spacing strategy in the wall-normal direction was systematic. A Reynolds number Re_τ of 80,000, far out of reach of DNS, was reached on modest computers. Substantial grid refinement was conducted, and simply lowered the RANS/LES interface without disturbing either of the log layers.

This should be compared with the countless channel "LES" studies which hardly exceed the Reynolds numbers accessible to DNS, even with massive computing resources, or to the studies that involve highly complex sub-grid-scale (SGS) and wall models, or two zones. These would be difficult to extend to general geometries; often, they are not even clearly defined in the publication. They contain numerous disposable constants. The height of this practice is when the Kármán "constant" used in the model is subjected to large variations. Not only is this against any theory, but it would destroy the accuracy at high Reynolds numbers. In most of the studies NNWSS can be compared with, grids are finer in the spanwise direction than in the flow direction, which is practical only in the simplest of flows. Some SGS models have stability problems, and require averaging in the wall-parallel or other homogeneous directions; this is inconsistent with the idea that a dynamic model responds to local conditions better than an algebraic model such as Smagorinsky's. It also makes them far from ready for non-trivial geometries. NNWSS had no such issues. DES provides a dynamic SGS model, in a sense, but makes no claims that it has better physics than an algebraic model; simply, it is done to unify the LES treatment with the RANS treatment.

A full and simple solution to the log-layer mismatch in DES has not been found, and flow visualizations examined after the paper went to press were disappointing in that the near-wall structures were much weaker and more elongated than expected [15]. An expedient solution is to re-set the resolved log layer by offsetting the modeled log layer; this is achieved by changing the c_{v1} constant of the S-A model from 7.1 to 4 [14], and restores the skin-friction coefficient C_f to a correct level, from a level that is about 15% too low. Piomelli *et al.* addressed the deeper problem by intensifying the resolved turbulence with random forcing, applying a "backscatter model" [15]. Unfortunately, this requires explicit intervention and extra parameters, which is a serious obstacle to routine use. The method loses its readiness for general geometries and grids, and the user burden is higher.

3.2 Switch from RANS to LES mode within an attached boundary layer

RANS is best where the BL is thinnest while, at least in a research exercise such as LESFOIL, LES could be the final answer to separation prediction. This is because it would reduce the rôle of empiricism, increasingly as the grid is refined. DES allows LES to be initiated after the BL has transitioned and thickened sufficiently, but well before separation. This would make a "DESFOIL" very competitive, although delicate; it must not allow any ambiguous-grid situation, and LES content must be created deliberately. An abrupt change in grid spacing near 35% chord, from Type I to Type III, will prevent GIS, and the challenge is to generate mature LES content within as short a space as possible. There is a single solution field, only with special measures locally to "trip" the BL. It is known that raw random numbers do not meet this challenge: the "turbulence" takes many BL thicknesses to recover. A very useful alternative is the recycling method of Lund *et al* [16]. It has been very effective for BL's without pressure gradient, and can be greatly simplified; its extension to pressure gradients appears manageable in 2D.

4 Long-term improvements

4.1 Optimization of the RANS model

The remaining rôle of the model is in BL's, and shallow separation bubbles. This motivated the introduction of the SST model in DES [17], but there would be no sense in going beyond two equations. The essence of DES is to employ simple RANS models, tuned for thin shear layers. It would be logical to re-calibrate the model for these flows only, ignoring the free shear flows which normally are in LES mode. This would give the same number of adjustable parameters for a smaller class of flows. The obstacle is that achieving a level of accuracy that is conclusively superior will require excellent accuracy in experiment and calculations for a large number of flows. Recent findings on wind-tunnel wall effects, for instance, are poor omens of this [18]. Also note how the value of the Kármán constant has been challenged by recent experiments, so that the accepted range of $[0.40, 0.41]$ has given way to $[0.38, 0.436]$, which is wide. This is the most fundamental of constants in a BL model, so that perfection for a RANS model, even in BL's only, is as elusive as ever.

4.2 Solution to ambiguous grids

A proposal derives from the observation that GIS occurs in RANS-type grids that still have BL character, that is, shallow cells (figure 1). The DES length-scale limiter starts controlling $\tilde{d}$ when Δ is less than $d/0.65$, whereas the wall-normal spacing Δn is usually less than $d/10$. Therefore, there is a range of situations for which the cell aspect ratio AR could be a tool. Δ would be multiplied by a function of AR that equals 1 for AR near 1, so that it is passive in normal LES grids, and exceeds 1 for higher values of AR. There are concerns over regions away from the wall where the grid may have high aspect ratio, either fortuitously or because of adaptation to a thin flow feature. It was tested with only moderate success by J. Forsythe (personal communication, 2003), who will at this meeting present an alternate proposal, based on a function of d and $C_{DES}\Delta$ that is not simply their minimum, but instead overshoots $C_{DES}\Delta$ when they are nearly equal. This must be viewed as a partial solution, just like the use of AR, because it contains a parameter that limits by how much $\tilde{d}$ can exceed $C_{DES}\Delta$. Therefore, further refinement will defeat them, unless the user explores the solution and raises the limit again, which is somewhat against the spirit of DES.

Menter *et al.* use the F_2 function to disable the limiter inside a BL, for the SST version of DES [13]. Only sudden separation can drive the length scale $2k^{3/2}/\epsilon$ small enough, relative to the wall distance d, for LES mode to begin. This approach favors the RANS mode of DES. It seems to increase the possibilities for multiple solutions. A version for S-A with $F_2(r)$ will be tested.

4.3 Other challenges

Automatic grid adaptation is a major goal for industrial CFD codes. Adaptation in steady solutions is taking a more mathematical turn, with adjoint methods in particular, and emphasizes anisotropic refinement to shear layers and shock waves. Adaptation in DES has started, but is isotropic away from walls and remains more empirical, typically being based on mean vorticity [19].

Another legitimate need of engineering practice is error estimation. The drag of an airliner does not need to be known "as closely as possible"; it needs to be known to better than 1%. Scientific journals also ask for the numerical uncertainty to be "accurately quantified", and a true answer is usually impossible. This is already very difficult for the numerical errors, more difficult for the LES errors, and near-impossible for the RANS errors. Numerical errors may eventually be estimated from a single solution. LES errors can be scoped by vigorous grid refinement. RANS errors can be scoped by switching models, but not reliably. It is possible for a new flow type to make all the models err in the same direction, so that a test between models is not instructive. The free vortex seems to be a clear example of that. In all cases, real-life geometries with a very different level of sensitivity in different regions pose much more difficult problems than simple geometries.

5 Summary

DES has been rather successful and well-understood, and has not required any essential modification since its creation in 1997. However, perfection is not expected from any method in an endeavor as complex as the numerical prediction of turbulence, especially since the numerical power at the engineers' disposal remains marginal for many "real-life" problems, and utterly insufficient for the rest. Therefore, RANS-LES hybrids will be helpful for many years, but user training and judgment will be essential as will experience sharing via publications. Not only is the approach imperfect, but it can be mis-used; in that sense, robustness almost becomes a liability. Fully solving the issue of ambiguous grids is a priority, but has proven to be a resilient difficulty. The RANS component also may be improved, with the usual emphasis on separation. Another welcome change would be a numerically efficient system to control laminar regions; a magnificent one would be to predict transition, within the Navier-Stokes solution and even in unstructured grids.

Acknowledgements

The author is highly grateful for the extensive simulations and numerous fruitful discussions he owes to Prof. M. Strelets and his group, to Prof. K. Squires, and to Dr. J. Forsythe.

References

1. P. R. Spalart, W.-H. Jou, M. Strelets, S. R. Allmaras: Comments on the feasibility of LES for wings, and on a hybrid RANS/LES approach. First AFOSR International Conference on DNS/LES, Aug. 4-8 1997, Ruston, Louisiana.
2. M. Shur, P. R. Spalart, M. Strelets, A. Travin: Detached-eddy simulation of an airfoil at high angle of attack. 4th Int. Symp. Eng. Turb. Modelling and Measurements, May 24-26 1999, Corsica. Elsevier.
3. A. Travin, M. Shur, M. Strelets, P. Spalart: Detached- Eddy Simulations past a Circular Cylinder. Flow, Turb. Comb. **63**, 293 (2000).
4. P. Spalart: Strategies for turbulence modelling and simulations. Int. J. Heat Fluid Flow. **21**, 252 (2000).
5. C. P. Mellen, J. Frölich, W. Rodi: Lessons from the European LESFOIL project on LES of flow around an airfoil. AIAA J. **41**, 4:573-581 (2003).
6. I. Mary, P. Sagaut: Large eddy simulation of flow around an airfoil near stall. AIAA J. **40**, 6:1139-1145 (2002).
7. P. Batten, U. Goldberg, S. Chakravarthy: LNS – an approach towards embedded LES. AIAA-2002-0427.
8. F. R. Menter, M. Kuntz, R. Bender: A scale-adaptive simulation model for turbulent flow predictions. AIAA 2003-0767.
9. R. Allen, F. Mendonça, D. Kirkham: RANS and DES turbulence model predictions of noise on the M219 cavity at M=0.85. Colloquium EUROMECH 449, Dec. 9-12 2003, Chamonix, France.
10. S. A. Morton, J. R. Forsythe, K. D. Squires, K. E. Wurtzler: Assessment of unstructured grids for detached-eddy simulation of high Reynolds number separated flows. 8th ISGG Conf., Honolulu, June 2002.
11. M. Breuer, N. Jovičić, K. Mazaev: Comparison of DES, RANS and LES for the separated flow around a flat plate at high incidence. Int. J. Num. Meth. in Fluids. **41**:357-388 (2003).
12. S. Deck, E. Garnier, P. Guillen: Turbulence modelling applied to space launcher configurations. J. Turbulence **3** (2002).
13. F. R. Menter, M. Kuntz, L. Durand: Adaptation of eddy viscosity turbulence models to unsteady separated flow behind vehicles. Symp. “The aerodynamics of heavy vehicles: trucks, buses and trains”. Monterey, USA, Dec. 2-6 2002.
14. N. V. Nikitin, F. Nicoud, B. Wasistho, K. D. Squires, P. R. Spalart: An Approach to Wall Modeling in Large-Eddy Simulations. Phys. Fluids **12**, 7 (2000).
15. U. Piomelli, E. Balaras, H. Pasinato, K. D. Squires, P. R. Spalart: The inner-outer layer interface in large-eddy simulations with wall-layer models. Int. J. Heat Fluid Flow **24**:538-550 (2003).
16. T. S. Lund, X. Wu, K. D. Squires: Generation of turbulent inflow data for spatially-developing boundary layer simulations. J. Comp. Phys. **140**:233-258 (1998).
17. M. Strelets: Detached Eddy Simulation of massively separated flows. AIAA-2001-0879.
18. A. Garbaruk, M. Shur, M. Strelets, P. R. Spalart: Numerical study of wind-tunnel wall effects on transonic airfoil flows. AIAA J. **41** 6:1046-1054 (2003).
19. S. A. Morton, M. B. Steenman, R. M. Cummings, J. R. Forsythe: DES grid resolution issues for vortical flows on a delta wing and an F-18C. AIAA-2003-1103.

High-End Computing Challenges in Aerospace Design and Engineering

F. Ronald Bailey

Advanced Management Technology Inc., M.S. 258-6, NASA Ames Research Center, Moffett Field, CA 94035-1000. fbailey@mail.arc.nasa.gov

Abstract
High-End Computing (HEC) has had significant impact on aerospace design and engineering and is poised to make even more in the future. In this paper we describe four aerospace design and engineering challenges: Digital Flight, Launch Simulation, Liquid Rocket Engine Fuel System and Digital Astronaut. The paper discusses modeling capabilities needed for each challenge and presents projections of future near and far-term HEC computing requirements. NASA's HEC Project Columbia is described and programming strategies presented that are necessary to achieve high real performance

1 Introduction

High-End Computing (HEC) has had a major impact on design and engineering in the aerospace industry. HEC simulation is routinely used to improve understanding of complex physics phenomenon and thus lead to improved design solutions. HEC is enabling the use of CFD to significantly reduce wind-tunnel testing in vehicle design and to provide data that cannot be obtained by wind-tunnel experiments. None of these would be possible without the more than five orders-of-magnitude increase in HEC performance over the past three decades, which, in turn, has motivated the development of models of increasing fidelity and complexity. As a result, HEC applications have significantly reduced cost, lowered risk and improved performance. However, there is still huge potential for even greater benefits from expanded application of HEC in aerospace and it is timely to look at what significant advances in simulation can be initiated now. Here we explore four challenging and potentially fruitful areas for advancement.

The first challenge is Digital Flight that simulates aircraft dynamic flight and advances the application of CFD to broader areas of the flight envelope. The second challenge is space vehicle Launch Simulation in which high-fidelity modeling of the mission profile is used to improve mission planning and design as well as provide better assessments of risk. The third challenge is Liquid Rocket Engine Fuel System Simulation that exemplifies the use of high-fidelity modeling of complex systems for development of space transportation systems. Finally, the fourth challenge is the Digital Astronaut that models the human body's response to a prolonged space environment.

In the following we describe each challenge and give examples of recent pioneering developments in CFD modeling that provide motivation to pursue further development. We propose the solution of model problems as one means of advancing our solution capability in areas of physics modeling, algorithms and programming. These model problems are meant to be solved in the next few years and we have chosen NASA's Project Columbia, offering a peak performance of 60 TeraFLOPS, as a candidate HEC platform. The problems have been sized to be solved on a 12 TeraFLOPS platform. This is 20 percent of Project Columbia and is considered the practical available portion of a system that is shared by all of NASA. Solution times are estimated based on CPU hours used by known codes on smaller, precursor problems and we implicitly assume that computational efficiency remains constant as we move to our model problems. Next, we estimate HEC platform performance needed to solve a complete problem with the fidelity and turnaround necessary to meet the practical demands of design and engineering. Finally, we give a short description of Columbia and discuss programming challenges necessary to achieve good performance.

2 Digital Flight

One of the most pressing needs in aerospace vehicle design is the accurate prediction of stability and control characteristics throughout the flight envelope. Accurate data is especially critical for automatic control systems. Heretofore, the prediction of stability and control parameters has depended on expensive wind-tunnel and flight tests. In many cases the predictions were inadequate and vehicles exhibited unexpected stability and control problems discovered in flight test - sometimes with catastrophic consequences and at the cost of human lives. We are now on the threshold of Digital Flight: the ability to predict aerodynamic stability and control parameters over the entire flight envelope and to simulate dynamic flight behavior using CFD. The payoff is substantial. Digital Flight will lead to a much better understanding of flow characteristics and to improved designs. Design cycle time will be reduced and flight control laws improved. Wind-tunnel and flight tests will be reduced leading to decreased cost. Moreover, project risk will be reduced and safety increased.

The important issues in stability and control involve separated flows. Longitudinal, directional and lateral instability are all characterized by massively separated vortical and wake flows. Resolving these requires very fine grids to capture the large gradients present. Because placing fine grids everywhere is not practical, adaptive grid refinement techniques are being developed that refine the grid in regions of large-gradients as the solution evolves. Figure 1 shows the application of adaptive grid refinement applied to vortical flow simulation [1]. Figure 1a shows the chine vortex extending beyond the aircraft tail which results from refinement of the initial grid. Figure 1b shows the result of the final refined grid that correctly predicts the chine vortex bursting ahead of the vertical tail.

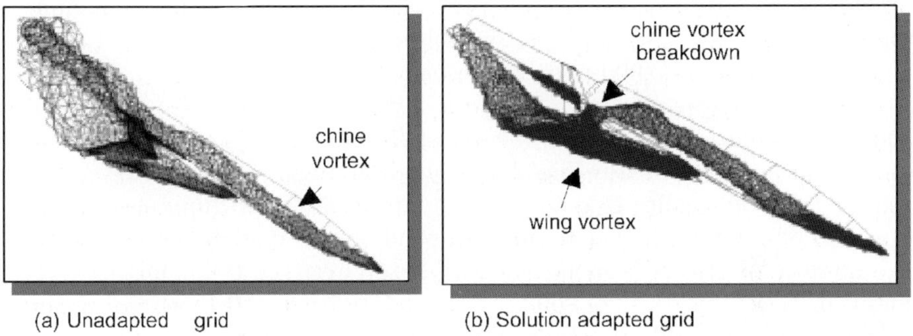

Fig. 1. Solution Adaptive Grid for Vortex Breakdown [1]

Reynolds-averaged Navier-Stokes (RANS) models are generally sufficient for quantitative predictions of mean flow characteristics as long the flow is not dominated by massive separation. Prediction of massively separated flows is much improved by use of Detached-Eddy Simulation (DES) [2]. DES employs a Reynolds-averaged turbulence model near the wall and Large-Eddy Simulation in separated regions. It combines the efficiency of RANS near the wall with the capability of LES to resolve geometry-dependent, unsteady, three-dimensional turbulent portions of the flow. Figure 2 compares experimental lift and pitching moment coefficients vs. angle-of-attack for the F/A-18E with RANS and two turbulence models (SA Baseline and SST Baseline) and time-averaged calculations (DES Baseline and DES Adapted) [3]. DES with adapted grid show good agreement over the entire angle-of-attack range.

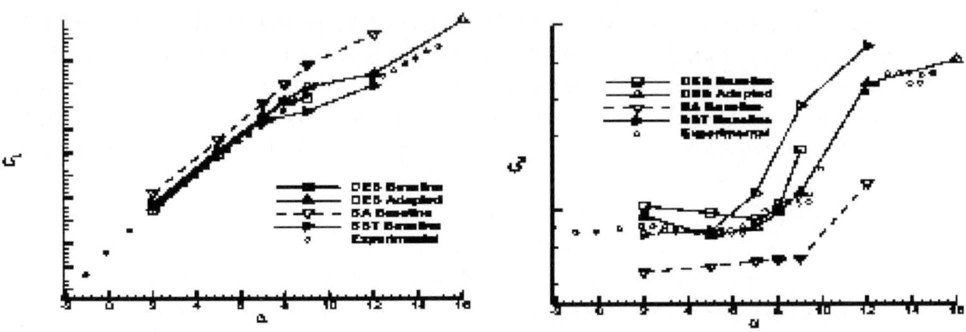

Fig. 2. Comparison of experimental and calculated CL and CM [3].

Now consider a Digital Flight model problem that solves unsteady RANS to simulate a six degree-of-freedom vehicle flight trajectory and includes solution adaptive grids and DES, but ignores aeroelastic effects. Quantities such as wind angles, velocity vector, forces, moments, control surface deflections, thrust, etc. are captured at points along the trajectory. These quantities populate a data base used for control law design and analysis and for input to piloted flight simulators. Data is typically captured at 50-80 Hz and the data base may contain 300,000 trajectory points.

Experience from the Drag Prediction Workshop II [4,5] suggests that to achieve adequate quantitative results of lift, drag and moment for performance purposes at steady flight and modest angle of attack using RANS requires a good quality grid with grid adaptation and a minimum of 25 million grid points for a half configuration wind-tunnel model. For Digital Flight the number of grid cells must be doubled to cover the full configuration domain, increased in areas of solution grid adaptation and increased again for higher flight Reynolds numbers. We estimate that 60 million grid points is the minimum needed to investigate solution adequacy over a broad range of flight conditions.

To estimate computation time requirements we observe that for well known codes such as OVERFLOW-D [6, 7] about 5000 time steps [8] are required for flow to travel one body length in an RANS simulation. For a typical fighter configuration maneuvering at 0.8 Mach number the flow travels about 15 body lengths in one second. The addition of DES has been observed to decrease the time step (increase solution time) by factors of 5-10 [3]. Based on the computing times reported for OVERFLOW-D [9] on a SGI Origin 3000 and an estimated 10 times increase in computations due to DES, we estimate one second of flight requires a solution time of 12 hours on a 12 TeraFLOPS platform. At an 80 Hz sample rate it takes nine minutes to obtain a database data at one trajectory point

The practical application of Digital Flight to vehicle design requires computational result production match design flow time. For example, a typical database of 300,000 trajectory points requires about six months of wind-tunnel tests (including model construction, test preparation, etc.). To match this rate our model simulation needs to calculate a trajectory point per minute and this would require a 100 TeraFLOPS platform. For practical applications the platform requirement may be larger due to uncertainties in grid resolution and modeling of separation physics. Therefore, we estimate a HEC platform in the range of 100-500 TeraFLOPS peak performance would be needed for practical design and engineering application.

3 High-Fidelity Launch

As NASA space missions and systems become more complex associated cost and risk become growing concerns. Advances in HEC will enable the high-fidelity simulation of airframe, propulsion system and auxiliary systems throughout launch and accent. High-accuracy launch simulations will make it possible to plan and evaluate the readiness to launch with much lower uncertainty. Integrated, high-fidelity modeling will enable simulation of failures and associated vehicle response. Rapid turnaround will enable analysis of multiple, dynamic configurations even at the conceptual design level. As HEC capability advances, running statistical significant numbers of simulations will enable risk assessment inputs to better represent reality in investigations such as propulsion system failure as well as the ability to recover from failures.

Simulation of a flight mission to orbit can be divided into a launch phase and an ascent phase. The launch phase requires a software procedure with the capability to simulate lift-off in the launch pad environment including exhaust heat radiation, acoustic, debris and local weather effects. The ascent phase requires a software procedure that can simulate flight to orbit including dynamics of booster and auxiliary tank separation. Ascent simulation is essentially Digital Flight applied to multiple rigid bodies and their dynamic separation. The governing equations may be Euler or RANS at lower altitudes and Boltzman equations in the low density upper atmosphere. A high-fidelity CFD vehicle flight model capable of six degree-of-freedom (6-DOF), multiple body flight is a key component for both launch and ascent phases.

Recent simulation of the shuttle Columbia (STS-107) debris trajectories [10] is representative of current capability in complex 6-DOF modeling. OVERFLOW-D [6, 7] solving the RANS equations and CART3D [11, 12], solving the Euler equations, were used to perform unsteady, moving-body, CFD simulations of the entire shuttle/debris flow field and the aerodynamic forces and moments acting on the debris. Both are capable of 6-DOF simulation of and rigid-body relative motion among an arbitrary number of bodies. Over 40 OVERFLOW-D and 400 CART3D 6-DOF simulation were preformed in the investigation of foam debris shed from the region of the left bipod-ramp of STS-107. The analysis provided an estimate of the debris trajectory, impact velocity and foam size that was instrumental in establishing the possibility for a piece of foam debris to cause massive damage to the Shuttle Orbiter wing RCC panels and T-seals. Figure 3 shows a CART3D simulated trajectory that closely resemble the strike location observed on film.

Since we treated Digital Flight in the previous section, we concentrate on launch simulation. A simulation model would be designed to treat the whole launch environment until the vehicle has cleared the launch tower. The model would integrate 6 DOF multiple-body motion, debris impact, propulsion system vibration and exhaust, acoustics due to exhaust, fuel accumulation in the exhaust plume, exhaust chemistry including fuel burning, thermal stress on the vehicle structure and finally weather at the launch site. The model would be very complex and integrate data from propulsion simulation, meso-

Fig. 3. CART3D 6-DOF debris trajectory [10].

scale weather prediction and experiment. Thus, it is necessary to consider developing the model in stages in which each stage adds new or increased fidelity.

Consider a first stage conceptual model that treats vehicle motion, debris motion and exhaust plume effects in the launch environment, including the presence of the launch facilities. We assumed that viscous effects are not significant and that vehicle motion and exhaust blast waves can be modeled by time accurate Euler equations. Simplified plume heating and chemical reaction models are also assumed. From exhaust blast wave speed and launch tower structure details we determine time and length scales. Thus, we pick a time step based exhaust plume pressure wave velocity and a spatial resolution of, say, one foot. Utilizing CART3D and a 100 million point grid this launch simulation could be completed in about a day on a 12 TeraFLOPS platform or one sixth of Project Columbia.

For practical design and engineering application the simulation would be more complex, perhaps requiring 5-10 times more computation, need to be completed in about a day, and require a HEC platform capable of 60-120 TeraFLOPS.

4 Liquid Rocket Engine Fuel Subsystem

The performance and reliability of rocket engines is critical to space transportation missions. A critical design challenge is rocket engine turbomachinery, which is the most expensive component in terms of development and operations and is the major cause of engine failures. We currently lack a practical capability to predict transient, 3-dimensional, environments internal to turbopumps and therefore must rely on the costly test-fail-fix cycle.

Given advanced simulation capabilities in turbomachinery we can reduce development and operations cost, reduce development time, quantify design margins and increase safety and robustness.

The goal of Liquid Rocket Engine Fuel System Simulation is to provide a high-fidelity framework for the design and analysis of the fuel/oxidizer supply subsystem for a liquid rocket propulsion system including unsteady turbopump flow analysis. The system is basically made up of multiple pumps and feed lines and its simulation will provide the basis for determining and addressing root causes of transient flow/cavitation induced vibration that result in structural damage such as turbine blade cracks and breakage.

The first major challenge in developing a system simulation capability is to model flow through turbopumps. High-performance turbopump design is currently a semi-empirical process that experience has shown misses many important features of turbopump flows, thus CFD simulation can add greatly to improved design. Especially valuable is information such as transient flow phenomena at start-up, and non-uniform flows that impact vibration and structural integrity. Challenges to developing CFD models are significant. Rocket turbopumps have complex geometries including full and partial blades, tip leakage and an exit boundary to a diffuser. Their flows include a number of complex flow phenomena including boundary layer transition, turbulent boundary layer separation, and wakes tip vortices as well as influences of three-dimensional and Reynolds number effects. Modeling cavitation is perhaps the greatest challenge as no models have advanced to the point of producing quantitative results for engineering.

Recent advancement in turbopump simulation has been demonstrated by an unsteady computation for the SSME turbopump impeller/diffuser using the INS3D code developed at NASA Ames Research Center [13]. To resolve the unsteady interaction between the rotating and stationary parts an overset grid was used. The entire configuration including inlet guide vanes, impeller blades and diffuser blades was constructed using 34.3 million grid points in 114 zones. Instantaneous snapshots of particle traces and pressure surfaces from these computations are shown in Figure 4. In this simulation one impeller rotation requires approximately 36 hours using 480 SGI Origin 3000 CPUs (600 MHz) or the equivalent of about 18,000 CPU hours.

The next step to advancing liquid fuel subsystem simulation is modeling a multi-stage turbopump. We consider a model problem consisting of a six stage turbo-pump. We ignore cavitation, upstream and downstream manifolds and ducting, and coupled shroud and hub cavity flows. A six stage turbo-pump is estimated to require 150 million grid points and about 1 million Origin CPU hours to simulate 10 revolutions, which is sufficient time for start-up transient flow disturbances to die out. We estimate an INS3D simulation on a 12 TeraFLOPS HEC platform would be completed in about four days. This turnaround is considered reasonable for a flow analysis and for simulation development.

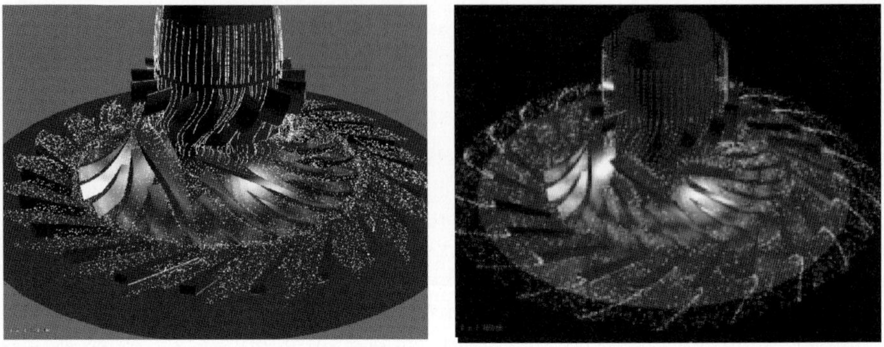

Fig. 4. Particle traces and pressure surfaces for unsteady turbopump computations (first rotation and end of third rotation) [13].

Advancing the model to include the items ignored is estimated to require a 2- to 5-fold increase in HEC performance. A typical fuel subsystem with four pumps and piping may require an addition five-fold increase. Finally, for practical design analysis purposes a two day turnaround is acceptable. Thus for a complete design simulation we estimate that a 240 - 600 TeraFLOPS HEC platform is required.

5 Digital Astronaut Circulatory System

The Digital Astronaut is a new NASA effort aimed at an integrated modeling and database system that enables the efficient construction and utilization of a class of quantitative models of the whole human body in order to simulate the function of a normal human being during and after a space voyage. This system will include HEC enabled computer simulation models that will allow detailed study of the effects of weightless space flight as well as the effects of the altered spacecraft environment on astronauts. The Digital Astronaut will be capable of appropriate structural integration, spanning the required multiple levels of biological organization, from the whole body through the organs, tissues, and cells to the genes and proteins. In addition, the Digital Astronaut will be capable of integrating multiple coupled physiological subsystems and components of biological networks (circulatory, respiratory, musculoskeletal, etc.) into a consistent whole-body model. Finally the Digital Astronaut will provides a data integration function for integrating space and analogue related empirical data, phenomenological observations and experimental studies with theoretical principles. The resulting system will help biomedical researchers understand the human effects of space flight, to use this knowledge to improve medical care for space voyagers and to design appropriate countermeasures that reduce the biomedical risks of space flight.

Simulating the effects of space travel on the human circulatory system is a key element of the Digital Astronaut. The altered cardiac output due to deconditioning during space flight and readaption on return impacts the blood circulation in the human body. This is particularly evident in the brain where altered blood supply impacts oxygen supply to certain parts. Analysis of this condition requires the capability to simulate blood flow in arteries and capillaries. Hemodynamic modeling challenges are significant [14]. Vascular networks exhibit anatomically complex geometry and their three-dimensional reconstruction requires techniques using magnetic resonance imaging, magnetic resonance angiogram and computed tomography to obtain accurate anatomical vasculature. Blood is a non-Newtonian fluid where red blood cell aggregation at low shear rates which makes the apparent blood viscosity increase and a higher shear rates red blood cell deformation which makes apparent viscosity decrease. Blood vessels exhibit distensible wall motion due to heart pulse and relative diameter change up to 20 percent requiring a structural deformation model for the arterial walls. For simulation of major arteries to be computationally manageable minor arteries such as arterioles, venules and capillaries need to be truncated and an auto-regulation model is required for modeling the outflow boundaries conditions. Finally, the effects of varying gravity including those on deformable wall motion and the resulting human circulatory flow patterns must be modeled.

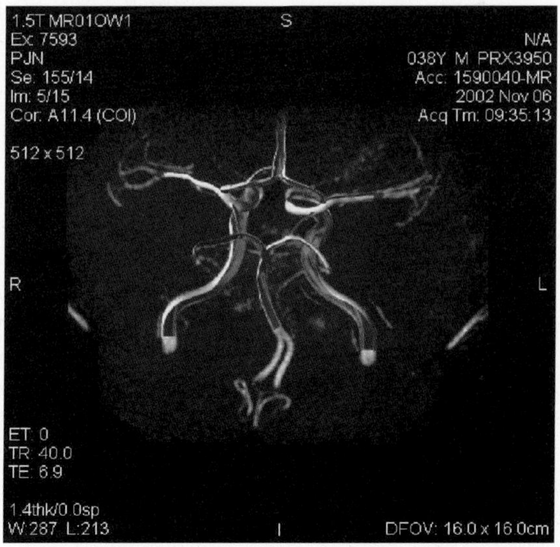

Fig. 5. Circle of Willis Simulation Superimposed on MRA [14].

A pioneering example of simulation of circulatory blood flow in the brain is illustrated in Figure 5 by the blood flow distribution through a realistic

Circle of Willis configuration superimposed on an MRA image [14]. Results were obtained using an MPI-OpenMP hybrid version of the INS3D code [15] and they represent the first simulation of blood circulation using non-Newtonian flow models within deformable walls. This pioneering simulation required 3000 SGI Origin 3000 CPU (600MHz) hours for 1 million grid points.

Circulatory blood flow simulation is in its infancy with many challenges remaining in geometry and physics modeling. A challenging next step would be simulating circulatory blood flow in the large vessels of the heart. We estimate 10 million grid point heart simulation with INS3D can be completed in less than eight hours on a 12 TeraFLOPS platform.

For analysis and counter measure development, two days to simulate the complete brain-heart circulatory system is a reasonable target and we estimate a HEC platform capable of 100 - 200 TeraFLOPS would be required for a 100 million grid simulation.

6 Project Columbia

Project Columbia has been initiated by NASA to provide an unprecedented HEC capability to solve large-scale computational aerospace science and engineering problems. The system, with a peak performance in excess of 60 TeraFLOPS, is being installed in the NASA Advanced Supercomputing (NAS) facility at Ames Research Center and is scheduled for full operation by the end of 2004. The system is configured as a cluster of 20 SGI Altix 3700 computers each with 512 processors and 1024 GB of memory. The total system has 10,240 Intel Madison processors, 20TB of memory and utilizes the Linux operating system. Two communication fabrics connect the Altix systems. An Infiniband switch fabric provides low latency MPI communication and a 10 Gigabyte Ethernet switch fabric provides user access and I/O communications. Each SGI Altix uses the NumaLink communication scheme to implement a non-uniform memory access architecture resulting in a single system image. Processors share a single address space and each processor is provided with low latency access to global memory. Columbia can be configured into a capability portion and a capacity portion. Four Altix systems will be linked via advanced NumaLink to allow MPI to use global shared memory constructs to significantly reduce inter processor communication latency. This 2048 processor subsystem will provide a powerful 12 TeraFLOPS "capability" platform for pioneering more finely-grained applications. The remaining 16 Altix systems then provide a powerful 48 TeraFLOPS "capacity" platform for the bulk of NASA's large-scale science and engineering applications.

7 Programming Strategies for Columbia

The turnaround time and peak performance estimated for pioneering design and engineering applications are summarized in Table 1. We have estimated the required peak performance, but the actual application code performance obtained on Columbia will depend to a large degree on problem characteristics, such as number of grid points and zones, algorithm used and the programming approach. Our goal is to achieve no less than about 80 percent linear scalability for Euler Codes, 40 percent for RANS codes, and 20 percent processor efficiency overall to be consistent with our experience on the SGI Origin 3000 systems at NAS. In some cases, such at Digital Fight, there is obvious coarse grained parallelism and good scalability potential since several trajectories can be run in parallel to populate a stability and control data base. Some cases, like simulating the human circulatory system, good scalability for large processor count may not be so obvious and innovative programming strategies may be necessary to achieve good scalability.

Table 1. Turnaround Time and HEC Performance Estimates for Aero Challenge Pioneer and Design & Engineering Applications

Aero HEC Challenge	Pioneer Problem		Design & Engr. Problem	
	Solution Time	Platform Peak Perf. (TeraFLOPS)	Solution Time	Platform Peak Perf. (TeraFLOPS)
Digital Flight	9 min.	12	1 min.	100-500
High Fid. Launch	1 day	12	1 day	60-120
Rocket Fuel System	4 days	12	2 days	240-600
Digital Astronaut	8 hrs	12	2 days	100-200

Columbia's architecture provides flexibility in programming approaches for increased scalability. It supports both MPI and OpenMP programming. Parallel programming among the Altix systems is implemented by MPI while programming within each Altix can be implemented by MPI or OpenMP. Columbia is also well suited for hybrid programming [16]. For example, within each Altix system MPI is used to implement coarse grained parallelism among Altix systems and their nodes while shared-memory OpenMP is used to implement parallelism within each node. Hybrid programming has the advantage of increasing the number of processors available to work on the problem while requiring loop-level programming to span only a small number of processors. Note, that connecting four Altix systems by NumaLink to form a capability system will allow MPI to use global shared memory constructs to dramatically reduce latency across the 2048 processors.

Multi-Level Parallelism (MLP) is another hybrid programming approach that takes advantage of the Altix single system image design. MLP is an extension of the Cray shared memory programming model of the 1980-90s [17]. It was developed for NUMA architectures that permit shared memory access to global data such as the SGI Origin series and SGI Altix. MLP replaces MPI with UNIX/LINUX forked processes and has a total of three routines in the library. Because communication is through shared memory loads/stores for all tasks, latencies are on the order of hundreds of nanoseconds rather than several microseconds. Lower communication latency improves the performance of coarse parallelism. Fast dynamic load balancing is supported by the dynamic creation of virtual nodes and the shared memory interface. Thus, MLP provides high levels of scaling efficiency for large processor counts.

Hybrid programming is well suited to CFD codes where the grid domain is portioned into zones. During iteration each zone's solution is updated independently and then the global solution is updated by exchange of zonal boundary data. MPI/MLP tasks are initiated to perform the zonal iteration on groups of zones assigned to each node in a manner that attempts to balance the workload among tasks. As each zone is processed the task takes advantage of loop-level parallelism using OpenMP directives. MLP scalability experiments with highly optimized OVERFLOW [18] code conducted on a 1024 SGI Origin 3000 platform showed scalability to nearly 1000 processors.

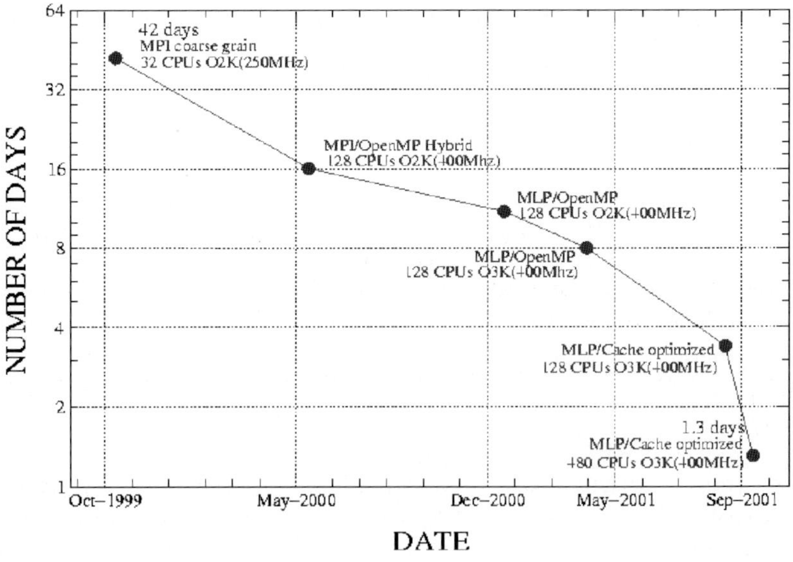

Fig. 6. Runtime Improvement for INS3D Turbopump (34 million grid points, 114 zones) on SGI Origin 3000.

All HEC platform processors suffer from the fact that processor performance is improving at a much greater rate than that of memory. Large cache memories have been introduced to bridge this gap and Columbia's processors have six MB of cache (nine MB on the capability system). To get the most benefit from cache the programming strategy needs to map the problem so that cached operands flow uninterrupted during the execution of fine-grained loops.

Figure 6 illustrates the benefits of hybrid programming and cache optimization by showing the history of runtime improvement in INS3D turbopump simulations. The initial simulation required 42 days using 32 CPUs (250 MHz) on a SGI Origin 2000 using MPI and was reduced to just over a day using 480 CPUs (400 MHz) on a SGI Origin 3000 using MLP with cache use optimized.

8 Conclusion

Recent CFD development have demonstrated the feasibility of embarking on new HEC challenges in aerospace design and engineering that will decrease costs, improve performance and reduce risk. NASA's Columbia Project will provide a 60 TeraFLOPS HEC platform with the capability needed to foster the development of new models to meet these challenges. While a lot of work lies ahead in developing accurate physics models, algorithms and programming strategies to achieve adequate real, sustained performance the payoff will be truly significant.

References

1. S. Z. Pirzadeh: A Solution-Adaptive Unstructured Grid Method by Grid Subdivision and Local Remeshing. AIAA Journal of Aircraft, **37**(5):818-824 (2000).
2. P. R. Spalart, W-H Jou, M. Strelets and S. R. Allmaras: Comments on the Feasibility of LES for Wings, and on RAHS/LES Approach. **Advances in DNS/LES**, 1st AFOSR Int. Conf. on DNS/LES, Greyden Press, Columbus OH, August 1997.
3. J. R. Forsythe and S. H. Woodson: Unsteady CFD Calculations of Abrupt Wing Stall Using Detached-Eddy Simulation. AIAA Paper 2003-0594, January 2003, Reno, Nevada.
4. K. R. Laflin, O. Brodersen, M. Rakowitz, J. C. Vassberg, E. N. Tinoco, R. A. Wahls, J. H. Morrison, and J. Godard: Summary of Data from the Second AIAA CFD Drag Prediction Workshop. AIAA Paper 2004-0555, Jan. 2004. Reno, NV.
5. M. J. Hemsch and J. H. Morrison: Statistical Analysis of CFD Solutions from 2nd Drag Prediction Workshop. AIAA Paper 2004-0566, Jan. 2004. Reno, NV.
6. R. Meakin: Adaptive Spatial Partitioning and Refinement for Overset Structured Grids. Computational Methods Applied Mechanical Engineering, **189**:1077-1117, 2000.

7. R. Meakin: Automatic Off-Body Grid Generation for Domains of Arbitrary Size. AIAA Paper 2001-2536. June 2001. Anaheim, CA.
8. R. Meakin. Private Communication.
9. T. A. Nygaard and R. L. Meakin: An Aerodynamic Analysis of a Spinning Missile with Dithering Canards. AIAA Paper 2002-2799. June 2002. St. Louis, MO.
10. R. J. Gomez, D. Vicker, S. E. Rogers, M. J. Aftosmis, W. M. Chan, R. Meakin, and S. Murman: STS-107 Investigation Ascent CFD Support. AIAA Paper 2004-2226, June 2004. Portland, OR.
11. M. J. Aftosmis, M. J. Berger and J. E. Melton: Robust and Efficient Cartesian Mesh Generation for Component-Based Geometry. AIAA Journal **36**(6):952-960. June 1998.
12. M. J. Aftosmis, M. J. Berger and G. Adomavicius: A Parallel Multilevel Method for Adaptively Refined Cartesian Grids with Embedded Boundaries. AIAA Paper 2000-0808, Jan. 2000. Reno, NV.
13. C. Kiris, W. Chan, and D. Kwak: A Three-Dimensional Parallel Time-Accurate Turbopump Simulation Procedure Using Overset Grid Systems. **Computational Fluid Dynamics 2002**, pp. 673-684. Springer-Verlag, Berlin. 2002
14. C. S. Kim, C. Kiris, D. Kwak, and T. David: Numerical Models of Human Circulatory System under Altered Gravity: Brain Circulation. AIAA Paper 2004-1092, Jan 2004. Reno, Nevada.
15. C. Kiris and D. Kwak: Aspects of Unsteady Incompressible Flow Simulations. Computers & Fluids, **31**: 627-638, 2002.
16. L. Oliker, X. Li, P. Husbands, and R. Biswas: Effects of Ordering Strategies and Programming Paradigms on Sparse Matrix Computations. SIAM Review **44**(3):373-393, 2002.
17. J. R. Taft: Achieving 60 GFLOP/s on the Production Code OVERFLOW-MLP. Parallel Computing, **27**(4):521-536. 2001.
18. D. C. Jespersen, T. H. Pulliam, and P. G. Buning: Recent Enhancements to OVERFLOW. AIAA Paper 97-0644, Jan 1997. Reno, NV.

Control of Shocks in CFD

Claude Bardos and Olivier Pironneau

Laboratoire Jacques-Louis Lions (LJLL), Université Paris VII and Université Paris VI-IUF `Olivier.Pironneau@upmc.fr`

Summary. Shock Control is a type of inverse problem for which the most suited solution method seems to be least square and optimal control algorithms (see [8]). As control theory assumes differentiability, there are mathematical difficulties when the modelling uses a system of conservation laws like the shallow water or Euler equations (see [15] for example). In this paper we study the differentiated equations of some equations of Fluid Dynamics and show that Calculus of Variation can be applied and control problem can be solved numerically more or less as usual, provided derivatives are understood as distributions.

Key words: Inverse problems, optimal control, shocks

1 Introduction

Consider a system of conservation laws in $\mathcal{R}^n$

$$\partial_t u + \nabla \cdot F(u) = 0 \;\; u(0) = u_0 \tag{1}$$

To find the initial condition u_0 that brings the shock in a given position on could consider an ideal response at time T, u_d and minimize

$$\min_{u_0 \in \mathcal{U}} E(u_0) = \frac{1}{2}\int_D |u(T) - u_d|^2 \;\; : \;\; \text{subject to (1)} \tag{2}$$

Calculus of Variations gives

$$\delta E = \int_{\mathcal{R}^n} (u(T) - u_d)\delta u(T) \qquad \partial_t \delta u + \nabla \cdot ((F'_u)^T \delta u) = o(|\delta u|) \tag{3}$$

and then introduce an adjoint state u^* solution of

$$\partial_t u^* + F'_u \nabla \cdot u^* = 0, \;\; u^*(T) = I_D(u - u_d) \tag{4}$$

and show by an integration by parts that

$$\delta E = \int_{\mathcal{R}^n} u^*(0) \cdot \delta u(0) + o(\|\delta u(0)\|) \tag{5}$$

The validity of this Calculus is usually taken for granted but it so happens that in the presence of shocks it is not correct.

Indeed take the simple case where $\mathcal{U}$ is the set of indicator functions of spheres of center $x(a)$, a being a scalar parameter. Then $\delta u = u'_a \delta a$ but u being discontinuous u' has a Dirac singularity because the shock position depends on a (see below). So right from the beginning the calculus above is wrong because $(u(T) - u_d)\delta u(T)$, for instance, is the product of a discontinuous function by a Dirac measure.

Differentiability of hyperbolic systems with shocks has been studied first by Majda [14] for the stability with respect to small perturbations. Godlewski et al[10] also showed differentiablity in one dimension by another method. Then Bressan made an extensive study of the problem in his book [3]. Counter enxemples can also be found in [4] when shocks meet. For practical purpose Giles [9] showed most of the properties for the adjoint state equations that we will rederive here using distribution theory.

The problem is not new obviously! Already in the eighties Pierre Perrier at Dassault Industries had put the question to us; later Hafez and Jameson also wondered if the usual set of variations for the Euler equation state variables were enough to allow for shock displacements.

One important application, beside shock stability, is for the optimization of supersonic business jets for which the sonic boom must be kept below an admissible level. It is an optimal shape design and it has been studied by Alonzo et al in [5] and Mohammadi et al in [6] and [7]. The problem then is

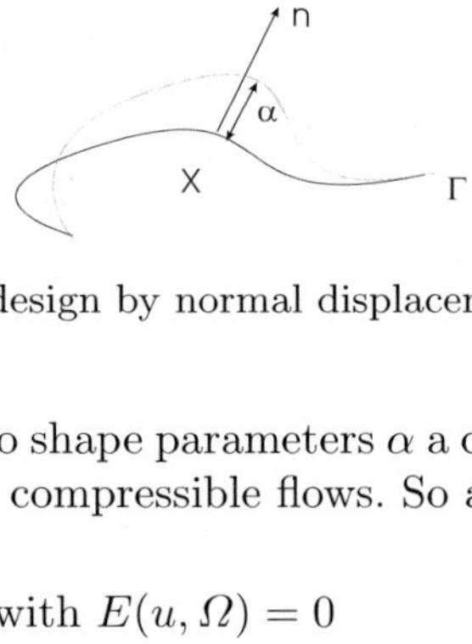

Fig. 1. Optimal shape design by normal displacement of a reference shape

to minimize with respect to shape parameters α a criteria involving u solution of the Euler equations for compressible flows. So a simple calculation gives:

$$J(u) := \int_{\Omega(\alpha)} f(u)\mathrm{d}x \text{ with } E(u,\Omega) = 0$$

$$\partial\Omega' = \{x + \alpha n(x) \in \partial\Omega\} \Rightarrow \ \delta J = \int_{\partial\Omega} f(u)\alpha d\gamma + \int_{\Omega} f'_u(u)\delta u + o(|\alpha|)$$

with, for some χ, $\mathbf{E}'(\mathbf{u},\mathbf{\Omega}) \cdot \delta\mathbf{u} = \chi\alpha$. Then with $\mathbf{E}'^T p = f'(u)$

$$\delta J = \int_{\partial\Omega} (f(u) + p\chi)\alpha d\gamma + o(|\alpha|) \tag{6}$$

The problem is to justify this calculus i.e. to show that $o(|\alpha|)$ is small and (both problems are connected) to find what is χ. On could say that these are mathematical questions while in practice one needs only to differentiate the discretized system and no such difficulties arise in finite dimension; however the following shows that the difficulties are also there for discrete systems and shows its face in the convergence towards the continuous system when the mesh size tends to zero.

To illustrate the difficulty, consider the problem of finding the motion of the shock due to a change in the incidence angle for a NACA airfoil. Automatic Differentiation (see [12]) gives the answer as shown on figure 1. There the Dirac mass in the derivatives is clearly visible.

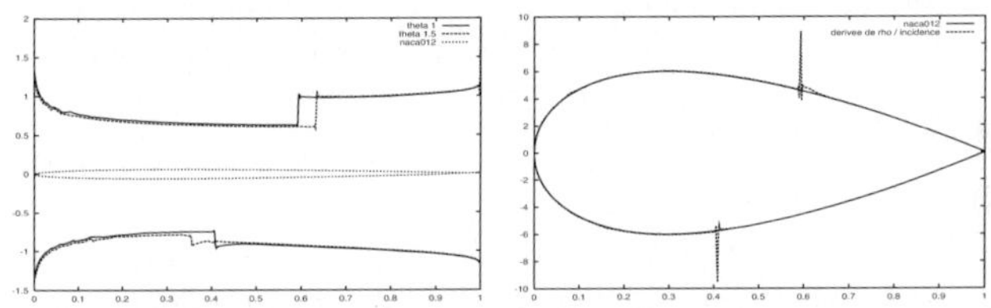

Fig. 2. Optimal shape design by normal displacement of a reference shape and sensitivity of density with respect to incidence angle. (from Dicesare et al [11]).

2 Derivatives with Distribution Theory

We summarize here the results published in [1]. More details can be found in [2].

For a given value of the parameter a (incidence for instance) we denote by C^1_S be the space of z-continuously differentiable functions outside a shock surface S. Now we assume that everything depends in a smooth and differentiable way of a parameter a. Then for any point $z(a) \in S(a)$ the normal variation of z with respect to a has an intrinsic value $z' \cdot n$ and

Proposition 1. *The a-derivative of $v \in C^1_S$ in the sense of distribution is*

$$v' = v'_a - [v]z' \cdot n\ \delta_S \tag{7}$$

where v'_a is the pointwise a-derivative.

For example, let $H(z)$ be the Heavyside function, ($H = 1$ if $z > 0$, 0 otherwise), and consider

$$v(z) = v^-(z, a) + (v^+(z, a) - v^-(z, a))H(z - z(a)) \tag{8}$$

Then obviously

$$v' = v^{-\prime} + [v'_a]H(z - z(a)) - [v]z'(a)\delta(z - z(a)) = v'_a - [v]z'\delta(z - z(a)) \tag{9}$$

A technical lemma is needed

Lemma 1. *Let w be vector valued but tangent to $S(a)$. Let $\nabla_S\cdot$ the surface divergence:*

$$\nabla\cdot(w\delta_S)=\delta_S\nabla_S\cdot w \tag{10}$$

Corollary 1.

$$\nabla\cdot\mathbf{v}=0\ \Rightarrow\ \nabla\cdot\mathbf{v}'=0\ \Rightarrow\ [\mathbf{v}\cdot n]'=0 \tag{11}$$

Notation When f is smooth left and right of S, let

$$\bar f=f^{\pm}\text{ in }\Omega^{\pm}\text{ and }\bar f=\frac{1}{2}(f^{+}+f^{-})\text{ on }S(a)$$

Proposition 2. *For $f,g\in C^1_S$:*

$$(fg)'=f'\bar g+\bar f g' \tag{12}$$

Proof: Derivatives of fg, f and g may have Dirac masses on $S(a)$, but by $[fg]=\bar f[g]+[f]\bar g$, so we have equality of the Dirac masses on S and by the usual rule of derivatives of product we also have equality outside S.

2.1 Application to Burgers' equation

Consider the standard Burger equation

$$\partial_t u+\partial_x\frac{u^2}{2}=0\text{ in }Q:=\mathcal{R}\times(0,+\infty),\ \ u(x,0)=u^0(x,a) \tag{13}$$

with compatible (entropy condition) initial data, u^0, discontinuous at $x=x(t,a)$; the trick is to notice that it is the divergence of $\boldsymbol{v}=(u,u^2/2)^T$ with respect to the variable $z=\{t,,x\}$:

$$\nabla_{t,x}\boldsymbol{v}=\partial_t u+\partial_x\frac{u^2}{2}=0 \tag{14}$$

When $\boldsymbol{v}$ is discontinuous then the div will generate a Dirac measure. Recall that

$$\nabla\cdot\mathbf{v}=\nabla_x\cdot\mathbf{v}+[\mathbf{v}]_S\cdot n \tag{15}$$

and writing that the div is zero as a distribution theory will require that the Dirac part is also zero and that is precisely the Rankine-Hugoniot (RK) condition

$$\dot x:=\partial_t x=\bar u:=\frac{1}{2}(u^{+}(x(t,a),t,a)+u^{-}(x(t,a),t,a)). \tag{16}$$

an equation for the velocity of the shock.

Assume that there is only one shock, then (see [3, 13]) sensitivity analysis shows that

$$\partial_t u'_a + \partial_x(u u'_a) = 0 \ \text{ in } \Omega\backslash S \qquad \dot{x}' = \bar{u}'_a + x'\partial_x \bar{u} \ \text{ on } S \tag{17}$$

where x' is the derivative of x with respect to a, is a complete system to define x'. The proof is obvious it suffices to differentiate Burger's equation and the RK condition. To generalize this kind of results to systems however is not easy (see Majda [14]). So let us show that the result can be obtained with distribution derivatives, in a rather straightforward manner.

Proposition 3. *The derivative of u solution of Burger's equation satisfies*

$$\partial_t u' + \partial_x(\bar{u}u') = 0 \quad u'(0) = u'_0 \tag{18}$$

read in the sense of distribution theory

This is because of Corollary 1 and of Proposition 2. Notice that it gives also

$$\partial_t u'_a + \partial_x(u u'_a) + \delta_S(-[u]\bar{u}'_a + d_t([u]x')) = 0, \ u'(x,0) = u^{0'} \tag{19}$$

Remark 1. Numerical linearization of Burger's equation inevitably leads to a discrete form of (18), which of course makes sense only in variation form. Notice that a space-time variational form would be preferable.

2.2 Application to Control

Consider the problem

$$\min_a \{J := \frac{1}{2}\int_R |u(T) - u_d|^2 \ : \partial_t u + \partial\frac{u^2}{2} = 0 \ u(0) = u^0(a)\} \tag{20}$$

according to the new results calculus of variation is now

$$J' = \int_R \overline{(u - u_d)}|_T u'(T) \text{ with } \partial_t u' + \partial_x(\overline{u}u') = 0, \ u'(0) = u^{0'} \tag{21}$$

Define the adjoint

$$\partial_t p + \bar{u}\partial_x p = 0, \ p(T) = \overline{(u - u_d)}|_T \tag{22}$$

Use integration by parts

$$0 = -\int_{R\times(0,T)} u'(\partial_t p + \bar{u}\partial_x p) = -\int_R (pu')|_0^T \tag{23}$$

$$\Rightarrow \ J' = \int_R p(T)u'(T) = \int_R p(0)u^{0'} \tag{24}$$

Because of the bar in $\bar{u}$ the adjoint contains a stand alone equation at the shock:

$$\frac{d}{dt}p(x(t),t) = 0, \; p(x(T),T) = \overline{(u-u_d)}|_T \tag{25}$$

which is the condition which defines p in the triangle between the two characteristics issued from $T, x(T)$.

Notice that p is continuous at the shock, but discontinuous on these two characteristics issued from $T, x(T)$. As an illustration consider

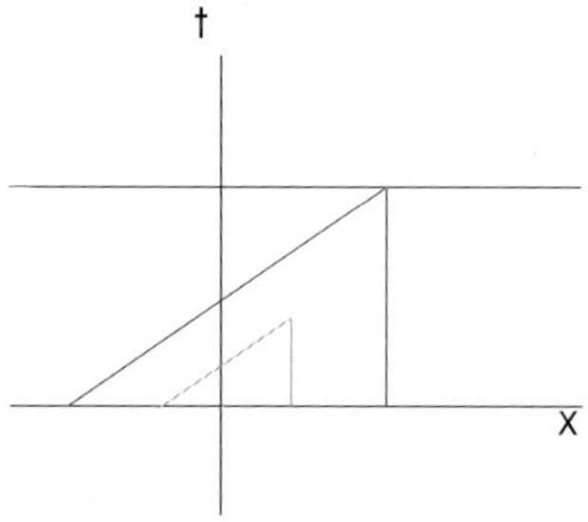

Fig. 3. Characteristics for the adjoint equation, going back from T to 0 and leaving a triangle in the shade where the value of the solution is determined by its value on the shock.

$$\partial_t u + \partial_x(\frac{u^2}{2}) = 0 \; u(x,0) = 1 - H(x-a) \tag{26}$$

for which there is an analytical solution: $u = 1 - H(x - (a + \frac{t}{2})$. So the derivative of

$$J := \frac{1}{2}\int_{-1}^{\infty} u(T)^2 \text{ is } J' = \frac{1}{2}\frac{d}{da}\int_{-1}^{a+\frac{T}{2}} = \frac{1}{2} \tag{27}$$

The theory implies

$$J' = \int_{-1}^{\infty} \bar{u}(T)u'(T) = \int p(x,0)u'(x,0) = p(0,0) \tag{28}$$

where

$$\partial_t p + \bar{u}\partial_x p = 0 \; p(T) = \bar{u}(T) = \frac{1}{2} \;\Rightarrow\; p(0,0) = \frac{1}{2} \tag{29}$$

The numerical computation of p is not easy because the numerical scheme has to hit exactly the point $T, x(T)$. So one way is to widen the shock at time 0 so that there are several points at the shock at time T. 1 Another way is to avoid the adjoint altogether and use u' directly. This is possible only if the number of parameters (like a) is small. Then to avoid computing Dirac

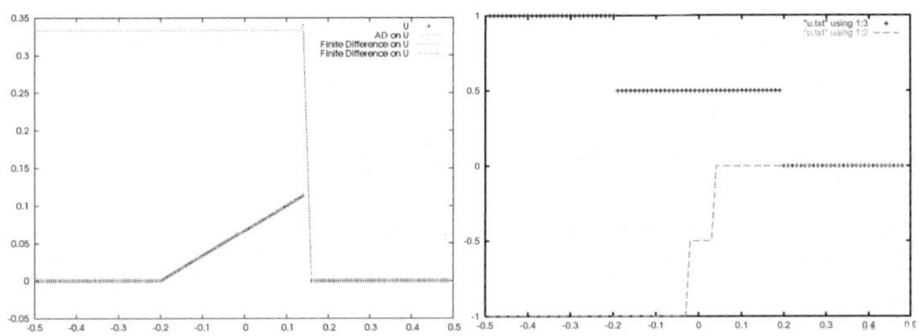

Fig. 4. Sensitivity by AD compared with finite difference shows perfect agreement: u' by AD, u' by $\frac{u(a+da)-u(a)}{da}$ and U' at $t = 0.3$.

masses the best is to integrate u' in x so as to transform the Dirac masses into discontinuities:

$$\partial_t u' + \partial_x(\bar{u}u') = 0 \;\Rightarrow\; \partial_t U' + \bar{u}\partial_x U' = 0 \text{ with } U' = \int^x u'(y,t)\mathrm{d}y \quad (30)$$

Then the shock displacement is proportional to the jump $[U']$, 0.11 in the following example when $u^0 = (1 + \frac{a}{3})I(-x)$

Remark 2. Automatic differentiation (like in aldol-C) works on U'. Chi-Wang Shu's Discontinuous Galerkin scheme applied to U' gives the following results for Burger's equation with x-periodic boundary conditions on $C = [-1, 1]$

$$\partial_t u + \partial_x F(u) = 0,\; u(x,0) = u_0(x) \text{ in } C \times (0,T) \quad (31)$$

reformulated as

$$\partial_t \phi + F(\partial_x \phi) = 0,\; \phi(x,0) = \int^x u(x,0) \quad (32)$$

In our example (see figure 4)$F(u) = \frac{1}{2}(u+1)^2,\; u_0(x) = -\cos(a + \pi x)$

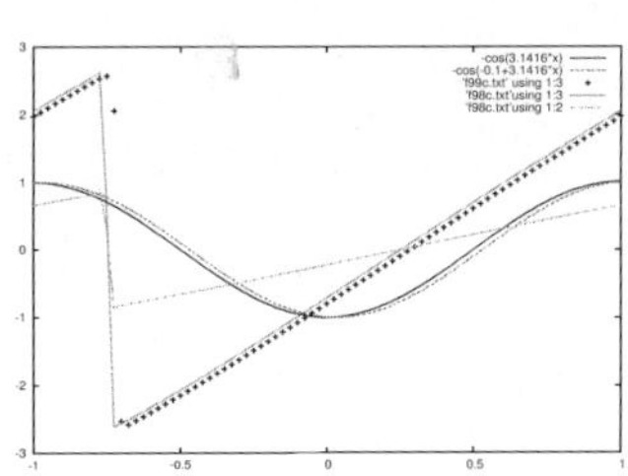

Fig. 5. Sensitivity by AD compared with finite difference shows perfect agreement for a finite volume scheme based on the primitive variable U.

3 Euler's Equations

The method extends to systems. Consider for example the energy equation of the Euler system for inviscid compressible flows:

$$\partial_t e + \partial_x(u(\gamma e - (\gamma-1)\frac{\rho u^2}{2})) = 0 \; p = (\gamma-1)(e - \frac{\rho u^2}{2}) \tag{33}$$

Then the trick is to decompose all non-linear terms into binary products, here with u and $w := u^2$

$$v = \rho u \;\Rightarrow\; v' = u'\bar\rho + \bar u \rho' \;\Rightarrow\; u' = \frac{v'}{\bar\rho} - \frac{\bar u}{\bar\rho}\rho'$$

$$\Rightarrow\; w' = 2\bar u u' = 2\frac{\bar u}{\bar\rho}v' - 2\frac{\bar u^2}{\bar\rho}\rho' \partial_t\rho' + \partial_x v' = 0$$

$$\partial_t v' + \partial_x(-\frac{3-\gamma}{2}\frac{\bar u \bar v}{\bar\rho}\rho' + \frac{3-\gamma}{2}(\bar u + \frac{\bar v}{\bar\rho})v' + (\gamma-1)e') = 0$$

$$\partial_t e' + \partial_x((-\gamma\frac{\bar u \bar e}{\bar\rho} + (\gamma-1)\frac{\bar u^2 \bar v}{\bar\rho})\rho'$$

$$+(\gamma\frac{\bar e}{\bar\rho} - \frac{\gamma-1}{2}(\overline{u^2} + 2\frac{\bar v \bar u}{\bar\rho}))v' + \gamma\bar u e') = 0 \tag{34}$$

This last system contains the linearized RK conditions.

3.1 A control problem with the shallow water system

Consider a simple example with

$$\begin{aligned}\partial_t u + \frac{1}{2}\partial_x(u^2 + 2\rho)\partial_t\rho + \partial_x(\rho u) = 0\\ r(0) = (0.5 + a)I_{x<0} + 0.5u(0) = 0\end{aligned} \tag{35}$$

It gives

$$\partial_t u' + \partial_x(\bar u u' + \rho') \qquad \partial_t\rho' + \partial_x(\bar\rho u' + \bar u\rho') = 0 \tag{36}$$

and after integration in x,

$$\partial_t U' + \bar u\partial_x U' + \partial_x R' = 0 \qquad \partial_t R' + \bar u\partial_x R' + \bar\rho\partial_x U' = 0 \tag{37}$$

The Numerical Scheme for U', R' is essentially the same as for u, ρ because the Jacobians are the same:

$$\begin{aligned}\partial_t W + A\partial_x W = 0,\; A = Q\Lambda Q^{-1}\; V = Q^{-1}W \;\Rightarrow\\ \partial_t V + \Lambda\partial_x V + Q^{-1}(\partial_t Q + A\partial_x Q)V = 0\end{aligned} \tag{38}$$

In the case of the Shallow water equation,

$$A = \begin{pmatrix} u & \rho \\ 1 & u \end{pmatrix} \; \Lambda = \begin{pmatrix} u+\sqrt{\rho} & 0 \\ 0 & u-\sqrt{\rho} \end{pmatrix} \; Q = \begin{pmatrix} \sqrt{\rho} & -\sqrt{\rho} \\ 1 & 1 \end{pmatrix}$$

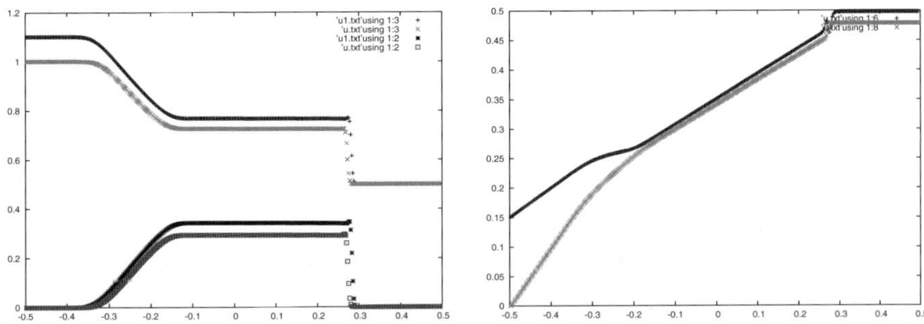

Fig. 6. At $T = 0.3$ shock moves by 0.012 with a shock strength of 0.25, while $[R'] = 0.03$ giving, with $\delta a = 0.1$, a change in shock pos=0.012. Right: green R by AD, red, R by numerical scheme.

$$Q^{-1} = \begin{pmatrix} \frac{1}{2\sqrt{\rho}} & \frac{1}{2} \\ -\frac{1}{2\sqrt{\rho}} & \frac{1}{2} \end{pmatrix} \quad W = \begin{pmatrix} \rho \\ u \end{pmatrix} = \begin{pmatrix} \sqrt{\rho}(V_1 - V_2) \\ V_1 + V_2 \end{pmatrix} \quad V = \begin{pmatrix} \frac{W_1}{2\rho} + \frac{W_2}{2} \\ -\frac{W_1}{2\rho} + \frac{W_2}{2} \end{pmatrix}$$
$$Q^{-1}(\partial_t Q + A\partial_x Q)V = \frac{V_1 - V_2}{4}\partial_x \begin{pmatrix} -u + 2\sqrt{\rho} \\ u + 2\sqrt{\rho} \end{pmatrix} \tag{39}$$

In the results presented we have use the method of characteristics applied to the system in V.

As before, when the number of unknown is small (3 in our case) it is better not to use the adjoint state because it is numerically too difficult to compute. We have used R', U' directly to compute the derivative of E and a descent method with fixed stepsize (not the best of course but the easiest to validate the method). The results are shown on figure 7.

4 Multi-dimensional Systems

The same theory applies to multi-dimensional systems. To use the same strategy we need a Hamilton-Jacobi formulation. For the Euler system for $W(x,t) \in \mathcal{R}^p$, notice that it is the divergence of a long vector:

$$\partial_t W + \partial_x F_1(W) + \partial_y F_2(W) = 0 \Leftrightarrow \nabla_{t,x,y} \cdot \begin{pmatrix} W^T \\ F_1(W)^T \\ F_2(W)^T \end{pmatrix} = 0. \tag{40}$$

Therefore

$$\big(W \; F_1(W) \; F_2(W)\big) = \big(\partial_x c - \partial_y b \;\; \partial_y a - \partial_t c \;\; \partial_t b - \partial_x a\big) \tag{41}$$

with $a(x,t),\ b(x,t),\ c(x,t) \in \mathcal{R}^p$

$$\partial_t c + F_1(\partial_x c - \partial_y b) - \partial_y a = 0,$$

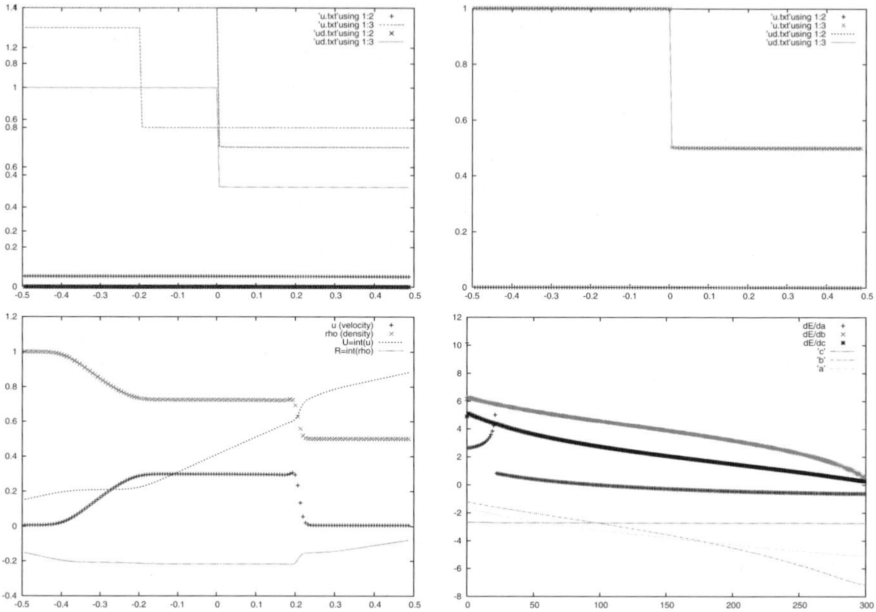

Fig. 7. Top left: Initial (u^0 in red and ρ^0 in green) and target (u_d^0 in blue and ρ_d^0 in magenta) velocities and densities. Top right: Computed (in red and green) and target (in blue and magenta) velocities and densities. Bottom left : Solution (u, ρ, U, R) at time T=0.3. Bottom right: Convergence: Gradients (in red, green and blue) and parameters (in magenta, light blue and yellow) with respect to a, b, c versus iteration number (all use log scales).

[pironneau-fig5]

$$\partial_t b - F_2(\partial_x c - \partial_y b) - \partial_x a = 0 \tag{42}$$

Now change variable from $c \to c - \partial_y \int^t a,\ b \to b - \partial_x \int^t a$, and notice that this change leaves $\partial_x c - \partial_y b$ unchanged; therefore with $V = (c, -b) \in \mathcal{R}^{p\times p}$ and $\mathcal{F} = (F_1, F_2)\ :\ \mathcal{R}^p \to \mathcal{R}^{p\times p}$ the system is rewritten as

$$\partial_t V + \mathcal{F}(\nabla\cdot V) = 0,\ \ V(0) = \nabla\cdot W^0 \tag{43}$$

4.1 Application to the shallow water equations

Consider

$$\begin{aligned} &\partial_t \rho + \nabla\cdot v = 0 \\ &\partial_t v + \nabla\cdot(\frac{v\otimes v}{\rho} + \frac{\rho^2}{2}) = 0 \end{aligned} \tag{44}$$

Then this leads to

$$\nabla \cdot V = W = \begin{pmatrix} \rho \\ v_1 \\ v_2 \end{pmatrix} \quad \mathcal{F}(W) = \begin{pmatrix} W_2 & W_3 \\ \frac{W_2^2}{W_1} + \frac{W_1^2}{2} & \frac{W_2 W_3}{W_1} \\ \frac{W_2 W_3}{W_1} & \frac{W_3^2}{W_1} + \frac{W_1^2}{2} \end{pmatrix} \tag{45}$$

The linearized equation is

$$\partial_t V' + \overline{\mathcal{F}'(W)} \nabla \cdot V' = 0 \tag{46}$$

and for the shallow water equations $\nabla \cdot V' = W' = (\rho', v_1', v_2')$:

$$F'(W)W' = \left| \begin{pmatrix} 0 & 1 & 0 \\ W_1 - (\frac{W_2}{W_1})^2 & 2\frac{W_2}{W_1} & 0 \\ -\frac{W_2 W_3}{W_1^2} & \frac{W_3}{W_1} & \frac{W_2}{W_1} \end{pmatrix} \begin{pmatrix} W_1' \\ W_2' \\ W_3' \end{pmatrix} \begin{pmatrix} 0 & 0 & 1 \\ -\frac{W_2 W_3}{W_1^2} & \frac{W_3}{W_1} & \frac{W_2}{W_1} \\ W_1 - \frac{W_3^2}{W_1^2} & 0 & 2\frac{W_3}{W_1} \end{pmatrix} \begin{pmatrix} W_1' \\ W_2' \\ W_3' \end{pmatrix} \right| \tag{47}$$

4.2 Numerical Examples

For a given velocity field u let $\Sigma = \{x, t) \;:\; u(x,t) \cdot n < 0, \; x \in \Gamma\}$ and consider

$$\partial_t \rho + \nabla \cdot (\rho u) = 0, \;\; \rho(0) = \rho^0, \;\; \rho = \rho_\Gamma \text{ on } \Sigma \tag{48}$$

Then the same transformation gives

$$\partial_t \boldsymbol{V} + \mathcal{F}(\rho) = 0, \;\; \nabla \cdot \boldsymbol{V} = \rho, \;\; \mathcal{F}(\rho) = \rho \boldsymbol{u}. \tag{49}$$

For a flow in $x_1 > 0$, an upwind F.D. explicit scheme is then

$$\boldsymbol{V}_{i,j}^{m+1} = \boldsymbol{V}_{i,j}^{m} - \frac{\delta t}{h}((V_1^m{}_{i,j} - V_1^m{}_{i-1,j} + V_2^m{}_{i,j+1} - V_2^m{}_{i,j+1})\boldsymbol{u}_{i,j} \tag{50}$$

5 Conclusion

Optimal control of systems with shocks can be handled numerically more or less like ordinary systems except that there are two additional difficulties

- When the optimization criteria involves u explicitly in a boundary integral, the adjoint equation has a special pointwise behavior on the shock with a boundary condition involving $\bar{u}$ and this is a serious numerical difficulty. If it cannot be avoided (by reformulating the problem) it can be handled either by special numerically switching schemes as in Chatot [16] for sonic points, or by widening the shock over several grid points. Artificial viscosity is also a solution but only if the optimality conditions are derived from the discretized system and not by discretizing the continuous optimality conditions.

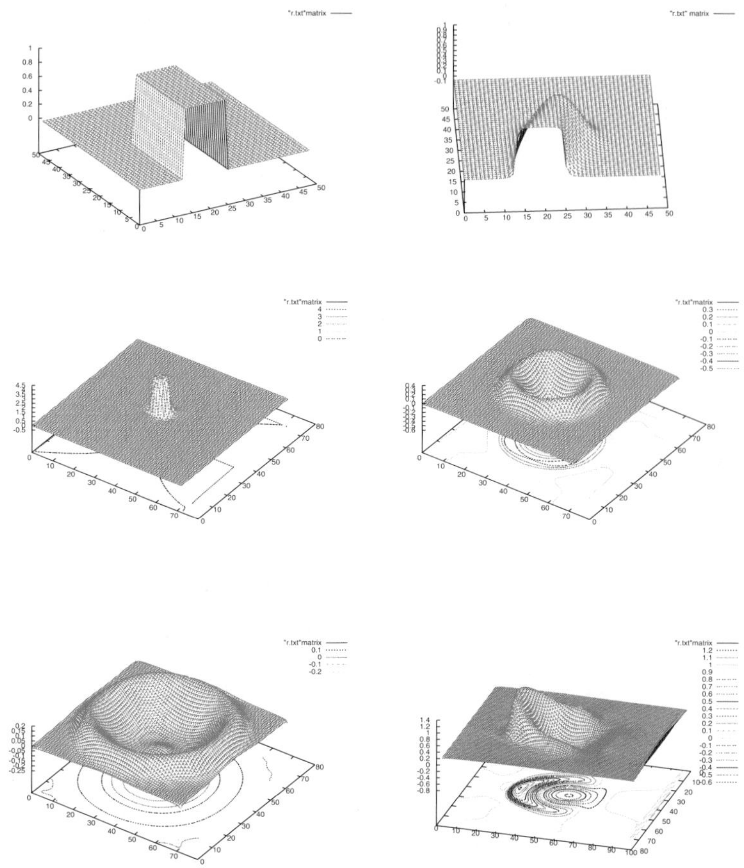

Fig. 8. Top: convection of a passive scalar entering from one boundary a) when the velocity is aligned with the flow, b) when it is not. Middle and bottom left: Flow due to a stone falling in a lake and obtained like above and below by integrating the Hamilton-Jaboci formulation of the Navier-Stokes equations. Bottom right: same but the stone has an initial velocity.

– The linearized equation has Dirac singularities. Normally this equation is used theoretically but not numerically. If one wishes to avoid the adjoint equation or one uses automatic differentiation in direct mode then this linearized system should be used in its Hamilton-Jacobi form because there Diracs become jump discontinuities.

Finally for the wave drag shape optimization of airfoils one sees that it would be better to minimize the integral of the pressure smoothed locally near the shock so as to avoid the above difficulty; but then more analysis needs to be done to check that the solution converges to the discontinuous one when the smoothing factor tends to zero.

References

1. C. Bardos and O. Pironneau : A formalism for the differentiation of conservation laws. C.R.Acad. Sci. Paris, t. , Série I 335(10):839-845 (2002).
2. C. Bardos and O. Pironneau : Data Assimilation in the Presence of Shocks. (to appear).
3. Bressan A. and Marson A.: A Variational Calculus for Discontinuous Solutions of Systems of Conservation Laws. Comm. in PDE. Vol 20 (9 and 10) , 1491-1552, (1995).
4. Campos-Pinto M., Cohen A., Dahmen W. and DeVore R.: On the Stability of Nonlinear Conservation Laws in the Hausdorff Metric, IGPM Report, RWTH Aachen (2004).
5. Alonso JJ, Kroo IM, Jameson A.. Advanced algoritms for design and optimization of QSP. *AIAA Pap.* 2002-0144. Reno, NV. (2002).
6. Mohammadi, B, Pironneau, O.: Control of Shocks and Sonic Booms. ECCOMAS proceedings, P. Neittanmakki ed. Jyvaskyla, Finland (2004).
7. Mohammadi, B, Pironneau, O.: Applied shape optimization for fluids. Oxford University Press, Oxford (2000).
8. Lions, J.-L. : *Optimal Control of Distributed Systems.* Springer Verlag, (1970).
9. Giles M.A. and N.A. Pierce. Analytic adjoint solutions for the quasi-one-dimensional euler equations. *Journal of Fluid Mechanics*, 426:327–345, 2001.
10. Godlewski E. and Olazabal M. and P.A. Raviart. On the linearization of hyperbolic systems of conservation laws. Application to stability. In *Équations aux dérivées partielles et applications*, pages 549–570. Gauthier-Villars, Éd. Sci. Méd. Elsevier, Paris, 1998.
11. Di Cesare N. and O. Pironneau. Shock sensitivity analysis. *Computational Fluid Dynamics Journal*, 9(2), 2000.
12. Griewank, A. *Evaluating Derivatives, Principles and Techniques of Algorithmic Differentiation.* Vol 19, Frontiers in Applied Mathematics. SIAM. (2000).
13. Tai-Ping Liu. Hyperbolic and Viscous Conservation Laws. CBMS-NSF Regional Conference Series in Applied Mathematics, 72. (SIAM), Philadelphia, PA, 2000.
14. Majda A.: The stability of multidimensional shock fronts Memoirs of the AMS January 1983 Volume 41 & 43 number 275.
15. Daeszcu D.N. and Navon I. M. Adaptative observations in the context of 4D-Var data assimilation. Meteorlogy and Atmospheric Physics, 85, 205-226 (2004).
16. Chattot, J-J. Springer (2000).

A Domain Decomposition Framework for Fluid-Structure Interaction Problems

Simone Deparis[1], Marco Discacciati[1], and Alfio Quarteroni[1,2]

[1] Institute of Analysis and Scientific Computing, CMCS-IACS, EPF Lausanne
simone.deparis@epfl.ch, marco.discacciati@epfl.ch
[2] MOX, Politecnico di Milano alfio.quarteroni@epfl.ch

Summary. In this note we review some classical algorithms for fluid-structure interaction problems and we propose an alternative viewpoint mutuated from the domain decomposition theory. This approach yields preconditioned Richardson iterations on the Steklov-Poincaré nonlinear equation at the fluid-structure interface.

Key words: Fluid-structure interaction, domain decomposition, finite element approximation, blood flow

1 Introduction

In this work, we address the numerical solution of fluid-structure interaction problems, in the case where implicit time advancing schemes are used. This yields at each time step a coupled system which is highly nonlinear, since the fluid domain depends on the unknown displacement of the structure. Standard strategies for solving this nonlinear problems are fixed point based methods such as Block-Gauss-Seidel (BGS) iterations (cf. [11, 15, 16]) or block Newton methods with exact or inexact jacobian (cf. [4, 7, 8, 9, 12]).

In this paper we revise these approaches, then we reformulate the fluid-structure interaction problem in a domain decomposition framework. On the associated interface problem we propose several splitting algorithms which are mutuated from subdomain iterative procedures. In particular we introduce several preconditioners that are obtained from the Steklov-Poincaré nonlinear operators associated with the fluid and the structure problems.

We consider a numerical example originated from the modeling of the interaction between blood-flow and vessel wall in a cylindric tract of an artery.

2 Fluid-structure interaction

In order to address each problem in its natural setting, we choose to consider the fluid in an ALE (Arbitrary Lagrangian Eulerian) formulation (cf. [4, 16]) and the structure in a pure Lagrangian framework.

The system under investigation occupies a moving domain Ω_t in its actual configuration. It is made of a deformable structure Ω_t^{s} (such as an arterial wall, a pipe-line, ...) surrounding a fluid under motion (blood, oil, ...) in the complement Ω_t^{f} of Ω_t^{s} in Ω_t (see Fig. 1).

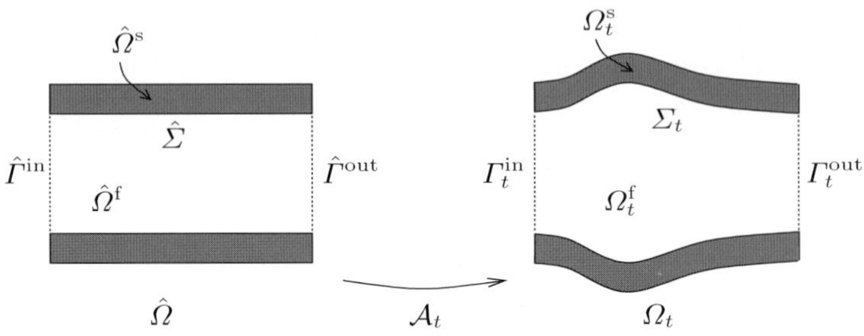

Fig. 1. ALE mapping between the initial configuration and the configuration at time t.

We assume the fluid to be Newtonian, viscous, homogeneous and incompressible. Its behavior is described by its velocity and pressure. The elastic solid under large displacements is described by its velocity and its stress tensor. The classical conservation laws of the continuum mechanics govern the evolution of these unknowns.

We denote by Γ_t^{in} and Γ_t^{out} the inflow and outflow sections of the fluid domain, by $\boldsymbol{n}_{\mathrm{f}}$ the fluid domain's outward normal on $\partial\Omega_t^{\mathrm{f}}$ and by $\boldsymbol{n}_{\mathrm{s}}$ the one of the structure on the reference boundary $\partial\hat{\Omega}^{\mathrm{s}}$. The boundary conditions on the fluid inlet and outlet can be either natural or essential (i.e., of Neumann or Dirichlet type, respectively), while on the interface we impose that the fluid and structure velocities match and so do the normal stresses. For simplicity, we assume zero body forces on both the structure and the fluid and that the boundary conditions on the remaining part of the structure boundary are of Dirichlet or of Neumann type.

The problem consists in finding the time evolution of the configuration Ω_t^{f}, as well as the velocity $\boldsymbol{u}$ and pressure p for the fluid and the displacement $\boldsymbol{d}$ of the structure. We define the ALE mapping

$$\forall t\,,\ \mathcal{A}_t : \hat{\Omega}^{\mathrm{f}} \to \Omega_t^{\mathrm{f}},$$

i.e. a map that retrieves at each time the current configuration of the computational domain Ω_t^{f}. Note in particular that on the reference interface $\hat{\Sigma}$, $\boldsymbol{n}_{\mathrm{f}} \circ \mathcal{A}_t = -\boldsymbol{n}_{\mathrm{s}}$. We denote by $\hat{\boldsymbol{x}}$ the coordinates on the reference configuration $\hat{\Omega}^{\mathrm{f}}$ and by $\boldsymbol{w} = \frac{d\mathcal{A}_t}{dt}$ the domain velocity.

For simplicity, we denote in short by Fluid(...) and Str(...) the fluid and structure problems, respectively. Precisely, for given vector functions $\boldsymbol{u}_{\mathrm{in}}$, $\boldsymbol{g}_{\mathrm{f}}$ and $\boldsymbol{f}_{\mathrm{f}}$, Fluid$(\boldsymbol{u}, p, \mathcal{A}_t; \boldsymbol{u}_{\mathrm{in}}, \boldsymbol{g}_{\mathrm{f}}, \boldsymbol{f}_{\mathrm{f}})$ means that we consider the following problem whose solution is $\boldsymbol{u}$, p, $\mathcal{A}_t$:

$$\text{Fluid}(\boldsymbol{u}, p, \mathcal{A}_t; \boldsymbol{u}_{\text{in}}, \boldsymbol{g}_{\text{f}}, \boldsymbol{f}_{\text{f}}) : \begin{cases} \boldsymbol{\Delta} \mathcal{A}_t = 0 \text{ in } \hat{\Omega}^{\text{f}}, \\ \mathcal{A}_t = 0 \text{ on } \partial \hat{\Omega}^{\text{f}} \setminus \hat{\Sigma}, \\ \Omega_t^{\text{f}} = \mathcal{A}_t(\hat{\Omega}^{\text{f}}), \\ \rho_{\text{f}} \left(\left. \dfrac{\partial \boldsymbol{u}}{\partial t} \right|_{\hat{\boldsymbol{x}}} + (\boldsymbol{u} - \boldsymbol{w}) \cdot \boldsymbol{\nabla} \boldsymbol{u} \right) \\ \qquad = \operatorname{div}(2\mu \boldsymbol{\epsilon}(\boldsymbol{u})) - \boldsymbol{\nabla} p + \boldsymbol{f}_{\text{f}} \text{ in } \Omega_t^{\text{f}}, \\ \operatorname{div} \boldsymbol{u} = 0 \text{ in } \Omega_t^{\text{f}}, \\ \boldsymbol{u} = \boldsymbol{u}_{\text{in}} \text{ on } \Gamma_t^{\text{in}}, \\ \boldsymbol{\sigma}_{\text{f}}(\boldsymbol{u}, p) \cdot \boldsymbol{n}_{\text{f}} = \boldsymbol{g}_{\text{f}} \text{ on } \Gamma_t^{\text{out}}, \end{cases} \tag{1}$$

where ρ_{f} is the fluid density, μ its viscosity, $\boldsymbol{\epsilon}(\boldsymbol{u}) = \frac{(\boldsymbol{\nabla u} + (\boldsymbol{\nabla u})^T)}{2}$ is the strain rate tensor and $\boldsymbol{\sigma}_{\text{f}}(\boldsymbol{u}, p) = -pId + 2\mu\boldsymbol{\epsilon}(\boldsymbol{u})$ the Cauchy stress tensor (Id is the identity matrix). Note that (1) does not univocally define a solution $(\boldsymbol{u}, p, \mathcal{A}_t)$ as no boundary data are prescribed on the interface Σ_t.

Similarly, for given vector functions $\boldsymbol{g}_{\text{s}}$, $\boldsymbol{f}_{\text{s}}$, $\text{Str}(\boldsymbol{d}; \boldsymbol{g}_{\text{s}}, \boldsymbol{f}_{\text{s}})$ means that we consider the following problem whose solution is $\boldsymbol{d}$:

$$\text{Str}(\boldsymbol{d}; \boldsymbol{g}_{\text{s}}, \boldsymbol{f}_{\text{s}}) : \begin{cases} \rho_{\text{s}} \dfrac{\partial^2 \boldsymbol{d}}{\partial t^2} = \operatorname{div}(\boldsymbol{\sigma}_{\text{s}}(\boldsymbol{d})) - \gamma \boldsymbol{d} + \boldsymbol{f}_{\text{s}} \text{ in } \hat{\Omega}^{\text{s}}, \\ \boldsymbol{\sigma}_{\text{s}}(\boldsymbol{d}) \cdot \boldsymbol{n}_{\text{s}} = \boldsymbol{g}_{\text{s}} \text{ on } \partial \hat{\Omega}^{\text{s}} \setminus \hat{\Sigma}, \end{cases} \tag{2}$$

where $\boldsymbol{\sigma}_{\text{s}}(\boldsymbol{d})$ is the first Piola–Kirchoff stress tensor, γ is a coefficient accounting for possible viscoelastic effects, while $\boldsymbol{g}_{\text{s}}$ represents the normal traction on external boundaries. Appropriate models have to be chosen for the structure depending on the specific problem at hand. The reader may refer, e.g., to [1, 3, 10, 18].

Similarly to what we have noticed for (1), problem (2) can not define univocally the unknown $\boldsymbol{d}$ because a boundary value on $\hat{\Sigma}$ is missing.

When coupling the two problems together, the "missing" boundary conditions are indeed supplemented by suitable matching conditions on the reference interface $\hat{\Sigma}$. More precisely, if we denote by λ the interface variable corresponding to the displacement $\boldsymbol{d}$ on $\hat{\Sigma}$, at any time the coupling conditions on the reference interface $\hat{\Sigma}$ are

$$\begin{aligned} \mathcal{A}_t &= \lambda, \\ \boldsymbol{u} \circ \mathcal{A}_t &= \dot{\boldsymbol{d}}_{\hat{\Sigma}}, \\ (\boldsymbol{\sigma}_{\text{f}}(\boldsymbol{u}, p) \cdot \boldsymbol{n}_{\text{f}}) \circ \mathcal{A}_t + \boldsymbol{\sigma}_{\text{s}}(\boldsymbol{d}) \cdot \boldsymbol{n}_{\text{s}} &= 0, \end{aligned} \tag{3}$$

where $\dot{\boldsymbol{d}}_{\hat{\Sigma}}$ denotes the temporal derivative of $\boldsymbol{d}_{|\hat{\Sigma}}$.

The system of equations (1)-(3) identifies our coupled fluid-structure problem.

2.1 Fluid and structure operators

We suppose the problem to be discretized in time. When the solution is available at time t^n, we look for the solution at the new time level $t^{n+1} = t^n + \delta t$. When no ambiguity occurs, all the quantities will be referred to at time $t = t^{n+1}$.

In view of domain decomposition formulation we introduce two interface operators S_{f} and S_{s} as follows.

S_{f} is the Dirichlet–to–Neumann map in Ω_t^{f},

$$S_{\mathrm{f}} : \lambda \mapsto \text{find } (\boldsymbol{u}, p, \mathcal{A}_t) : \begin{cases} \mathrm{Fluid}(\boldsymbol{u}, p, \mathcal{A}_t; \boldsymbol{u}_{\mathrm{in}}, \boldsymbol{g}_{\mathrm{f}}, \boldsymbol{f}_{\mathrm{f}}) \\ \boldsymbol{u}_{|\Sigma_t} \circ \mathcal{A}_t = (\lambda - \boldsymbol{d}^n_{|\hat{\Sigma}})/\delta t \end{cases}$$
$$\mapsto \sigma_{\mathrm{f}} = (\boldsymbol{\sigma}_{\mathrm{f}}(\boldsymbol{u}, p) \cdot \boldsymbol{n}_{\mathrm{f}})_{|\Sigma_t} \circ \mathcal{A}_t,$$

while S_{s} is the Dirichlet–to–Neumann map in $\hat{\Omega}^{\mathrm{s}}$,

$$S_{\mathrm{s}} : \lambda \;\mapsto\; \text{find } \boldsymbol{d} : \begin{cases} \mathrm{Str}(\boldsymbol{d}; \boldsymbol{g}_{\mathrm{s}}, \boldsymbol{f}_{\mathrm{s}}) \\ \boldsymbol{d}_{|\hat{\Sigma}} = \lambda \end{cases} \quad \mapsto \sigma_{\mathrm{s}} = \boldsymbol{\sigma}_{\mathrm{s}}(\boldsymbol{d}) \cdot \boldsymbol{n}_{\mathrm{s}}.$$

Let us introduce the corresponding homogeneous operators that will serve as building blocks in the construction of our preconditioners. For any given λ, we define the homogeneous Dirichlet–to–Neumann maps in Ω_t^{f} and $\hat{\Omega}^{\mathrm{s}}$, respectively, by

$$\begin{aligned} \bar{S}_{\mathrm{f},\lambda}(\delta\lambda) &= S_{\mathrm{f}}(\lambda + \delta\lambda) - S_{\mathrm{f}}(\lambda) \\ \bar{S}_{\mathrm{s},\lambda}(\delta\lambda) &= S_{\mathrm{s}}(\lambda + \delta\lambda) - S_{\mathrm{s}}(\lambda). \end{aligned} \tag{4}$$

Next we formally introduce the tangent operators for the fluid and structure problems. First of all, let Fluid'_λ (respectively, Str'_λ) denote the linearized fluid (respectively, structure) problem around λ (the reader may refer to [7, 8] for a precise definition), where we consider homogeneous boundary conditions on $\partial\Omega_t^{\mathrm{f}} \setminus \Sigma_t$ and on $\partial\hat{\Omega}^{\mathrm{s}} \setminus \hat{\Sigma}$ for the fluid and the structure, respectively, and null body forces for both problems.

The definitions of the tangent operators are as follows.

For the fluid:

(i) Dirichlet–to–Neumann tangent map in Ω_t^{f} about a given point λ. For any increment $\delta\lambda$, $\delta\sigma_{\mathrm{f}} = S'_{\mathrm{f}}(\lambda) \cdot \delta\lambda$ is computed as follows:

$$S'_{\mathrm{f}}(\lambda) : \delta\lambda \mapsto \text{find } (\delta\boldsymbol{u}, \delta p, \delta\mathcal{A}_t) : \begin{cases} \mathrm{Fluid}'_\lambda(\delta\boldsymbol{u}, \delta p, \delta\mathcal{A}_t) \\ \delta\boldsymbol{u}_{|\Sigma_t} \circ \delta\mathcal{A}_t = \frac{\delta\lambda}{\delta t} \end{cases}$$
$$\mapsto \delta\sigma_{\mathrm{f}} = (\boldsymbol{\sigma}_{\mathrm{f}}(\delta\boldsymbol{u}, \delta p) \cdot \boldsymbol{n}_{\mathrm{f}})_{|\Sigma_t} \circ \delta\mathcal{A}_t.$$

(ii) Neumann–to–Dirichlet tangent map about a given point λ. For any increment $\delta\sigma_{\mathrm{f}}$, $\delta\lambda = \left(S'_{\mathrm{f}}(\lambda)\right)^{-1} \delta\sigma_{\mathrm{f}}$ is computed as follows:

$$(S_{\mathrm{f}}'(\lambda))^{-1} : \delta\sigma_{\mathrm{f}} \mapsto \text{find } (\delta\boldsymbol{u}, \delta p, \delta\mathcal{A}_t) : \begin{cases} \mathrm{Fluid}_\lambda'(\delta\boldsymbol{u}, \delta p, \delta\mathcal{A}_t) \\ (\boldsymbol{\sigma}_{\mathrm{f}}(\delta\boldsymbol{u}, \delta p) \cdot \boldsymbol{n}_{\mathrm{f}})_{|\Sigma_t} \circ \delta\mathcal{A}_t = \delta\sigma_{\mathrm{f}} \end{cases}$$
$$\mapsto \delta\lambda = \delta t\, \delta\boldsymbol{u}_{|\Sigma_t} \circ \delta\mathcal{A}_t.$$

For the structure:

(*iii*) Dirichlet–to–Neumann tangent map about a given point λ. For any increment $\delta\lambda$, $\delta\sigma_{\mathrm{s}} = S_{\mathrm{s}}'(\lambda) \cdot \delta\lambda$ is computed as follows:

$$S_{\mathrm{s}}'(\lambda) : \delta\lambda \mapsto \text{find } \delta\boldsymbol{d} : \begin{cases} \mathrm{Str}_\lambda'(\delta\boldsymbol{d}) \\ \delta\boldsymbol{d}_{|\hat{\Sigma}} = \delta\lambda \end{cases} \mapsto \delta\sigma_{\mathrm{s}} = \boldsymbol{\sigma}_{\mathrm{s}}(\delta\boldsymbol{d}) \cdot \boldsymbol{n}_{\mathrm{s}} \text{ on } \hat{\Sigma}.$$

(*iv*) Neumann–to–Dirichlet tangent map about a given point λ. For any increment $\delta\sigma_{\mathrm{s}}$, $\delta\lambda = (S_{\mathrm{s}}'(\lambda))^{-1}\,\delta\sigma_{\mathrm{s}}$ is computed as follows:

$$(S_{\mathrm{s}}'(\lambda))^{-1} : \delta\sigma_{\mathrm{s}} \mapsto \text{ find } \delta\boldsymbol{d} : \begin{cases} \mathrm{Str}_\lambda'(\delta\boldsymbol{d}) \\ \boldsymbol{\sigma}_{\mathrm{s}}(\delta\boldsymbol{d}) \cdot \boldsymbol{n}_{\mathrm{s}} = \delta\sigma_{\mathrm{s}} \end{cases} \mapsto \delta\lambda = \delta\boldsymbol{d}_{|\hat{\Sigma}}\,.$$

Remark 1 (special cases). (a) In some cases it might occur that the model adopted for the structure has a lower dimension than that of the fluid (see, e.g., [18], where a two-dimensional model is used for the fluid whereas a one-dimensional model is used for the structure). In such cases the structure domain reduces to the interface Σ_t. The methods presented here still apply; however, in that case the application of the operator S_{s} implies simply a computation of the residue.

(b) When the fluid or the structure are linear like in the test case addressed in Sect. 5, the tangent maps are equivalent to the homogeneous operators, precisely

$$\bar{S}_{\mathrm{f},\lambda}(\delta\lambda) = S_{\mathrm{f}}'(\lambda) \cdot \delta\lambda \qquad \text{or} \qquad \bar{S}_{\mathrm{s},\lambda}(\delta\lambda) = S_{\mathrm{s}}'(\lambda) \cdot \delta\lambda.$$

3 The problem at the interface

With the help of the maps S_{f} and S_{s} previously introduced, there are at least three ways to express the coupled problem in terms of the interface variable λ. From the mathematical side they are all equivalent. However, they can inspire different iterative methods.

3.1 Fixed point formulation

The most obvious way to reformulate the coupled problem (1)-(3) is through the following fixed point formulation: find an interface displacement λ such that

$$S_{\mathrm{s}}^{-1}(-S_{\mathrm{f}}(\lambda)) = \lambda. \tag{5}$$

The explanation is as follows: the displacement λ generates the stress $S_{\mathrm{f}}(\lambda)$ on the interface. The stress applied to the structure is then $-S_{\mathrm{f}}(\lambda)$ and its corresponding displacement has to reproduce λ.

When treating fluid-structure interaction problems, this is the usual way to consider the coupling. The existing algorithms are conceived to solve equation (5). See for example [8, 11, 16].

Fixed point iterations

A standard algorithm to solve problem (5) is based on relaxed fixed point iterations. One iteration of the fixed point algorithm reads: for a given λ^k, do

$$\begin{aligned} \sigma_{\mathrm{f}}^k &= S_{\mathrm{f}}(\lambda^k), \\ \bar{\lambda}^k &= S_{\mathrm{s}}^{-1}(-\sigma_{\mathrm{f}}^k), \\ \lambda^{k+1} &= \lambda^k + \omega^k(\bar{\lambda}^k - \lambda^k), \end{aligned} \tag{6}$$

where all equalities are valid on $\hat{\Sigma}$. The choice of the relaxation parameter ω^k is crucial for the convergence of the method (see [2] for a recent analysis). With this purpose, in [14] the authors apply a generalization to the vector case of the Aitken acceleration technique. This allows a dynamic (and automatic) choice of the relaxation parameter, according to the formula

$$\omega^k = -\frac{\left(\bar{\lambda}^k - \lambda^k - \bar{\lambda}^{k-1} + \lambda^{k-1}\right) \cdot \left(\lambda^k - \lambda^{k-1}\right)}{\left\|\bar{\lambda}^k - \lambda^k - \bar{\lambda}^{k-1} + \lambda^{k-1}\right\|^2}.$$

In fact, as pointed out in [4], this choice of ω^k is the one that minimizes the norm of

$$\left\|\left(\lambda^k - \lambda^{k-1}\right) + \omega\left(\bar{\lambda}^k - \lambda^k - \bar{\lambda}^{k-1} + \lambda^{k-1}\right)\right\|.$$

The main advantage of this method is that the fluid and structure problems are solved independently. In fact, each step k of the algorithm (6) implies:

1. apply S_{f} to a given displacement λ^k, that is solve the fluid problem in Ω_t^{f} with boundary condition $\boldsymbol{u}_{|\Sigma_t} \circ \mathcal{A}_t = (\lambda^k - \boldsymbol{d}^n_{|\hat{\Sigma}})/\delta t$ on $\hat{\Sigma}$; then compute the stress $\sigma_{\mathrm{f}}^k = (\boldsymbol{\sigma}_{\mathrm{f}}(\boldsymbol{u}^k, p^k) \cdot \boldsymbol{n}_{\mathrm{f}})_{|\Sigma_t} \circ \mathcal{A}_t$ on the interface $\hat{\Sigma}$;
2. apply the inverse of S_{s}^{-1} to $-\sigma_{\mathrm{f}}^k$, that is solve the structure problem in $\hat{\Omega}^{\mathrm{s}}$ with boundary condition $\sigma_{\mathrm{s}}(\boldsymbol{d}^k) \cdot \boldsymbol{n}_{\mathrm{s}} = -\sigma_{\mathrm{f}}^k$ on $\hat{\Sigma}$; then compute the correction $\overline{\lambda}^k$ of the displacement at the iterate k.

The main drawback of this method is that it may result in a slow convergence rate.

3.2 Rootfinding formulation

A formulation of the interface problem slightly different from (5) (cf. [7, 8, 9, 12]) is: find an interface displacement λ such that

$$\Phi(\lambda) = 0, \quad \text{where } \Phi(\lambda) = S_{\mathrm{s}}^{-1}(-S_{\mathrm{f}}(\lambda)) - \lambda. \tag{7}$$

Newton algorithm

Let $J(\lambda)$ denote the Jacobian of $S_{\mathrm{s}}^{-1}(-S_{\mathrm{f}}(\lambda))$ in λ. Given λ^0, for $k \geq 0$, a step of the Newton algorithm associated to problem (7) reads:

$$\begin{aligned} &\sigma_{\mathrm{f}}^k = S_{\mathrm{f}}(\lambda^k), \\ &\bar{\lambda}^k = S_{\mathrm{s}}^{-1}(-\sigma_{\mathrm{f}}^k), \\ &(J(\lambda^k) - Id)\mu^k = -(\bar{\lambda}^k - \lambda^k), \\ &\lambda^{k+1} = \lambda^k + \omega^k \mu^k. \end{aligned} \tag{8}$$

When $J(\lambda)$ represents the exact Jacobian, ω^k can be taken equal to one. Otherwise, it can be computed by a line search technique (see, e.g., [17]).

Note that the Jacobian of $S_{\mathrm{s}}^{-1}(-S_{\mathrm{f}}(\lambda^k))$ in λ^k has the following expression

$$J(\lambda^k) = -\left[S_{\mathrm{s}}'\left(S_{\mathrm{s}}^{-1}(-S_{\mathrm{f}}(\lambda^k))\right)\right]^{-1} \cdot S_{\mathrm{f}}'(\lambda^k) = -\left[S_{\mathrm{s}}'\left(\bar{\lambda}^k\right)\right]^{-1} \cdot S_{\mathrm{f}}'(\lambda^k). \tag{9}$$

We point out that algorithm (8) requires an additional step with respect to (6): in fact we have to solve the linear system with matrix $J(\lambda^k) - Id$. Indeed, while the computation of $\left[S_{\mathrm{s}}'\left(\bar{\lambda}^k\right)\right]^{-1} \cdot \delta\sigma$ (for any given $\delta\sigma$) does only require the derivative with respect to the state variable at the interface, the computation of $S_{\mathrm{f}}'(\lambda^k) \cdot \delta\lambda$ implies also shape derivatives, since a variation in λ determines a variation of the fluid domain. This is a nontrivial task. In the literature, several approaches have been proposed to solve exactly the tangent problem [7], or else to approximate it by either simpler models for the fluid [5, 8], or through finite differences schemes [9, 13, 20]. However, the lack of *a priori* criteria for selecting optimal finite difference infinitesimal steps may lead to a reduction of the overall convergence speed [8].

3.3 Domain decomposition (or Steklov-Poincaré) formulation

The computational domain is naturally split into the fluid domain Ω_t^{f} and the structure Ω_t^{s}. It will be clear from the context, whether we impose Dirichlet or Neumann boundary conditions on the interface Σ_t, while on $\partial\Omega_t^{\mathrm{f}} \setminus \Sigma_t$ and $\partial\Omega_t^{\mathrm{s}} \setminus \Sigma_t$ the problems have always the same kind of boundary conditions. We refer to [19] for a general setting of domain decomposition methods.

The Steklov–Poincaré interface equation associated to the differential problem is: find λ such that

$$S_{\mathrm{f}}(\lambda) + S_{\mathrm{s}}(\lambda) = 0. \tag{10}$$

Note that the dependence on the data is hidden in the definition of the operators S_{f} and S_{s}. This is necessary since the problem at hand is nonlinear.

Remark that equation (10) can formally be retrieved by applying S_{s} to both sides of (5).

Preconditioned (nonlinear) Richardson method

Since the Steklov–Poincaré problem (10) is nonlinear, the Richardson method must be interpreted in a slightly different way than what is done in the literature for the linear case (see, e.g., [19]). Given λ^0, for $k \geq 0$, the iterative method reads:

$$\begin{aligned} \sigma_{\mathrm{f}}^k &= S_{\mathrm{f}}(\lambda^k), \\ \sigma_{\mathrm{s}}^k &= S_{\mathrm{s}}(\lambda^k), \\ \sigma^k &= -\left(\sigma_{\mathrm{f}}^k + \sigma_{\mathrm{s}}^k\right), \\ \mu^k &= P^{-1}\sigma^k, \\ \lambda^{k+1} &= \lambda^k + \omega^k \mu^k, \end{aligned} \tag{11}$$

with appropriate choice of the scalar ω^k. Every equation should still be intended on $\hat{\Sigma}$. The preconditioner P, that must be chosen appropriately, maps the interface variable onto the space of normal stresses, say $P : \Lambda \to Y$. Note that if $(S_{\mathrm{f}} + S_{\mathrm{s}})$ is affine, if P is taken equal to $\bar{S}_{\mathrm{f}} + \bar{S}_{\mathrm{s}}$ and $\omega^k = 1$, then the Richardson method converges in one iteration. It is also possible to choose a preconditioner which depends on the iterate λ^k or more generally on the iteration step k. In these cases we will denote it by P_k.

A general strategy to compute the relaxation parameter ω^k is given by :

$$\omega^k = -\frac{\left(\mu^k - \mu^{k-1}\right) \cdot \left(\lambda^k - \lambda^{k-1}\right)}{\|\mu^k - \mu^{k-1}\|^2}. \tag{12}$$

This value of ω^k is the one that minimizes the norm

$$\left\|\left(\lambda^k - \lambda^{k-1}\right) + \omega\left(\mu^k - \mu^{k-1}\right)\right\|,$$

over all possible values of ω. This criterium generalizes to the vector case the Aitken extrapolation technique (see [4]).

At each step k, algorithm (11) requires to solve independently the fluid and the structure problems and to apply a preconditioner. Precisely,

1. apply S_{f} to λ^k, that is solve the fluid problem as already illustrated for algorithm (6);
2. apply S_{s} to λ^k, that is solve the structure problem with boundary condition $\boldsymbol{d}^k_{|\hat{\Sigma}} = \lambda^k$ on $\hat{\Sigma}$ and compute the stress $\sigma_{\mathrm{s}}^k = \sigma_{\mathrm{s}}(\boldsymbol{d}^k) \cdot \boldsymbol{n}_{\mathrm{s}}$ on $\hat{\Sigma}$;
3. apply the preconditioner P^{-1} to the total stress σ^k on $\hat{\Sigma}$.

Note that steps 1. and 2. can be performed in parallel.

Remark 2. If no preconditioner is used, then at the differential level P should be intended as being the projection operator $\mathcal{I}$ from the space of displacements Λ to the space of stresses Y, so that

$$\lambda^{k+1} = \lambda^k + \omega^k \mathcal{I}^{-1} \sigma^k \in \Lambda. \tag{13}$$

At the algebraic level, this remark can be omitted since in that case we are always dealing with vectors of $\mathbb{R}^n$.

The crucial issue is how we can set up a preconditioner (more precisely, a scaling operator) in order for the iterative method to converge as quickly as possible. We address this problem in the next subsections.

4 Preconditioners for the domain decomposition formulation

In this section we discuss some classical choices of the preconditioner for the Richardson method applied to the domain decomposition approach. We also compare the proposed preconditioners to the fixed point and Newton strategies that we have illustrated in Sects. 3.1 and 3.2.

4.1 Dirichlet–Neumann and Neumann–Neumann preconditioners

We define a generic linear preconditioner (more precisely, its inverse):

$$P_k^{-1} = \alpha_\mathrm{f}^k \, S_\mathrm{f}'(\lambda^k)^{-1} + \alpha_\mathrm{s}^k \, S_\mathrm{s}'(\lambda^k)^{-1}, \tag{14}$$

for two scalars α_f^k and α_s^k. Another possibility is to use the operators $\bar{S}_{\mathrm{f},\lambda^k}$ and $\bar{S}_{\mathrm{s},\lambda^k}$ instead of $S_\mathrm{f}'(\lambda^k)$ and $S_\mathrm{s}'(\lambda^k)$, respectively:

$$P_k^{-1} = \alpha_\mathrm{f}^k \, \bar{S}_{\mathrm{f},\lambda^k}^{-1} + \alpha_\mathrm{s}^k \, \bar{S}_{\mathrm{s},\lambda^k}^{-1}. \tag{15}$$

In the special case in which both S_f and S_s are linear, since the tangent problems are equivalent to the homogeneous problems, (14) and (15) coincide. Otherwise, (15) is a nonlinear operator and can be considered as an approximation of (14) to be used in order to avoid the solution of the linearized problems.

From (15) we retrieve the following special cases:

– If $\alpha_\mathrm{f}^k = 1$ and $\alpha_\mathrm{s}^k = 0$, then

$$P_k^{-1} = P_{DN}^{-1} = \bar{S}_{\mathrm{f},\lambda^k}^{-1}.$$

P_{DN} is called a *Dirichlet-Neumann preconditioner* and

$$P_{DN}^{-1}(\sigma^k) = \bar{S}_{\mathrm{f},\lambda^k}^{-1} \left(-S_\mathrm{f}(\lambda^k) - S_\mathrm{s}(\lambda^k) \right);$$

– If $\alpha_{\rm f}^k = 0$ and $\alpha_{\rm s}^k = 1$, then

$$P_k^{-1} = P_{ND}^{-1} = \bar{S}_{{\rm s},\lambda^k}^{-1}.$$

This is called a *Neumann-Dirichlet preconditioner* and

$$P_{ND}^{-1}(\sigma^k) = \bar{S}_{{\rm s},\lambda^k}^{-1} \left(-S_{\rm f}(\lambda^k) - S_{\rm s}(\lambda^k)\right);$$

– If $\alpha_{\rm f}^k + \alpha_{\rm s}^k = 1$, then

$$P_k^{-1} = P_{NN}^{-1} = \alpha_{\rm f}^k \bar{S}_{{\rm f},\lambda^k}^{-1} + \alpha_{\rm s}^k \bar{S}_{{\rm s},\lambda^k}^{-1}$$

which is called a *Neumann-Neumann preconditioner.*

In the Dirichlet–Neumann (or the Neumann–Dirichlet) case the computational effort of a Richardson step may be reduced to the solution of only one Dirichlet problem in one subdomain and one Neumann problem in the other.

For both cases (14), (15), it is possible to choose the parameters $\alpha_{\rm f}^k$, $\alpha_{\rm s}^k$ and ω^k dynamically in the following way. We define $\omega_{\rm f}^k = \omega^k \alpha_{\rm f}^k$ and $\omega_{\rm s}^k = \omega^k \alpha_{\rm s}^k$ and we look for $\omega_{\rm f}^k$ and $\omega_{\rm s}^k$ that minimize

$$\left\| (\lambda^k - \lambda^{k-1}) + \omega_{\rm f} \left(\mu_{\rm f}^k - \mu_{\rm f}^{k-1}\right) + \omega_{\rm s} \left(\mu_{\rm s}^k - \mu_{\rm s}^{k-1}\right) \right\|,$$

over all possible values of $\omega_{\rm f}$ and $\omega_{\rm s}$, which corresponds to solving the linear system

$$A^T A \begin{pmatrix} \omega_{\rm f}^k \\ \omega_{\rm s}^k \end{pmatrix} = -A^T \left(\lambda^k - \lambda^{k-1}\right), \tag{16}$$

where A is the two column matrix

$$A = \left(\left(\mu_{\rm f}^k - \mu_{\rm f}^{k-1}\right); \left(\mu_{\rm s}^k - \mu_{\rm s}^{k-1}\right) \right).$$

Again, this can be regarded as a generalized Aitken criterium (see [4]). Finally, we set

$$\lambda^{k+1} = \lambda^k + \omega_{\rm f}^k \mu_{\rm f}^k + \omega_{\rm s}^k \mu_{\rm s}^k.$$

Note that this automatic choice generalizes the one carried out in (12).

Remark 3. In the linear case we obtain the simple expressions

$$P_{DN}^{-1}(\sigma^k) = S_{\rm f}^{-1}(-S_{\rm s}(\lambda^k)) - \lambda^k;$$

$$P_{ND}^{-1}(\sigma^k) = S_{\rm s}^{-1}(-S_{\rm f}(\lambda^k)) - \lambda^k.$$

The convergence of the preconditioned Richardson algorithm (11) with P_{DN}^{-1} or P_{ND}^{-1} is proved, thanks to the equivalence with fixed point iterations, in [2]. The proof can be extended to P_{NN}^{-1} using the classical theory of domain decomposition (cf. [19]). A complete proof for the nonlinear case will be given in a future work.

Comparison with fixed point iterations

The previously introduced fixed point iterations (6) can be regarded as special instances of the general preconditioned domain decomposition algorithm (11) for a suitable choice of the preconditioner.

In fact, let us consider the special case when the structure model is linear (but not necessarily the one of the fluid). Then, if we choose $\alpha_{\mathrm{f}}^k = 0$ and $\alpha_{\mathrm{s}}^k = 1$, the algorithm (11) is equivalent to the fixed-point algorithm (6) (see section 3.1). Indeed, from (11)

$$\mu^k = (S'_{\mathrm{s}}(\lambda^k))^{-1}\left(-S_{\mathrm{f}}(\lambda^k) - S_{\mathrm{s}}(\lambda^k)\right) = S_{\mathrm{s}}^{-1}\left(-S_{\mathrm{f}}(\lambda^k)\right) - \lambda^k,$$

hence $\lambda^{k+1} = \lambda^k + \omega^k(\bar{\lambda}^k - \lambda^k)$, which coincides with the last equality of (6).

4.2 The Robin-Robin preconditioner

A further possibility is offered by the following preconditioner

$$P_{RR} = \frac{1}{\gamma_{\mathrm{f}} + \gamma_{\mathrm{s}}}\left(\gamma_{\mathrm{f}}\mathcal{I} + S'_{\mathrm{f}}(\lambda^k)\right)\mathcal{I}\left(\gamma_{\mathrm{s}}\mathcal{I} + S'_{\mathrm{s}}(\lambda^k)\right),$$

where $\mathcal{I}$ is the projection operator from Λ to Y (see Remark 2), while γ_{f} and γ_{s} are positive parameters which can be chosen using a suitable error minimization strategy (see [6]). We call P_{RR} a *Robin-Robin preconditioner.*

4.3 The Newton algorithm on the Steklov-Poincaré equation

The genuine Newton algorithm applied to the Steklov-Poincaré problem (10) is retrieved by using the algorithm (11) (with $\omega^k = 1$) and choosing P at the step k as

$$P_k = S'_{\mathrm{f}}(\lambda^k) + S'_{\mathrm{s}}(\lambda^k). \tag{17}$$

Note that to invert P_k one must use a (preconditioned) iterative method and may approximate the tangent problems to accelerate the computations.

Comparison with the Newton algorithm (8) on problem (7)

The Richardson algorithm (11) for the Steklov-Poincaré formulation (10) with preconditioner given by (14) (with $\alpha_{\mathrm{f}}^k = \alpha_{\mathrm{s}}^k = 1$) is not equivalent to the Newton algorithm (8) applied to problem (7). In fact, the Newton algorithm (8) could be regarded as a Richardson method (11), choosing however a nonlinear preconditioner defined as

$$P_k(\mu) = S_{\mathrm{s}}\left(S'_{\mathrm{s}}(\bar{\lambda}^k)^{-1} \cdot \left(S'_{\mathrm{f}}(\lambda^k) + S'_{\mathrm{s}}(\bar{\lambda}^k)\right) \cdot \mu \right), \tag{18}$$

where $\bar{\lambda}^k = S_{\mathrm{s}}^{-1}(-S_{\mathrm{f}}(\lambda^k))$.

In this case, for $\sigma^k = -(S_\mathrm{f}(\lambda^k) + S_\mathrm{s}(\lambda^k))$, we would obtain

$$\begin{aligned} P_k^{-1}(\sigma^k) &= \left(S_\mathrm{f}'(\lambda^k) + S_\mathrm{s}'(\bar\lambda^k) \right)^{-1} \cdot S_\mathrm{s}'(\bar\lambda^k) \cdot S_\mathrm{s}^{-1}(-S_\mathrm{f}(\lambda^k) - S_\mathrm{s}(\lambda^k)) \\ &= \left(\left[S_\mathrm{s}'(\bar\lambda^k) \right]^{-1} \cdot S_\mathrm{f}'(\lambda^k) + Id \right)^{-1} \left(S_\mathrm{s}^{-1}(-S_\mathrm{f}(\lambda^k)) - \lambda^k \right). \end{aligned}$$

We see that this is equivalent to (8). In fact (9) is equal to the first bracket in the last line.

Remark 4. Note that if (only) the structure is linear, the preconditioner defined in (18) is also linear and becomes

$$P_k = S_\mathrm{f}'(\lambda^k) + S_\mathrm{s}'(\lambda^k),$$

which is exactly (17). This is a Newton method applied to (7) or (10). However, we would like to remark that the domain decomposition approach allows us to set up a completely parallel solver. In fact, the fluid and the structure subproblems can be computed simultaneously (and independently) for both the residue computation (operators S_f and S_s) and the application of the preconditioner (operators S_f' and/or S_s').

Table 1. Comparison among fixed point, rootfinding and Steklov-Poincaré approaches.

Fixed point	Rootfinding	Steklov-Poincaré
$S_\mathrm{s}^{-1}(-S_\mathrm{f}(\lambda)) = \lambda$	$S_\mathrm{s}^{-1}(-S_\mathrm{f}(\lambda)) - \lambda = 0$	$S_\mathrm{f}(\lambda) + S_\mathrm{s}(\lambda) = 0$
Fixed point iter.	Newton iter.	Prec. Richardson iter.
$\sigma_\mathrm{f}^k = S_\mathrm{f}(\lambda^k)$ $\overline{\lambda}^k = S_\mathrm{s}^{-1}(-\sigma_\mathrm{f}^k)$ $\lambda^{k+1} = \lambda^k + \omega^k(\overline{\lambda}^k - \lambda^k)$	$\sigma_\mathrm{f}^k = S_\mathrm{f}(\lambda^k)$ $\overline{\lambda}^k = S_\mathrm{s}^{-1}(-\sigma_\mathrm{f}^k)$ $(J(\lambda^k) - Id)\mu^k = -(\overline{\lambda}^k - \lambda^k)$ $\lambda^{k+1} = \lambda^k + \omega^k\mu^k$	$\sigma^k = S_\mathrm{f}(\lambda^k) + S_\mathrm{s}(\lambda^k)$ $\mu^k = P^{-1}(-\sigma^k)$ $\lambda^{k+1} = \lambda^k + \omega^k\mu^k$
1 flow solve 1 structure solve	1 flow solve 1 structure solve 1 Jacobian solve	1 flow solve 1 structure solve 1 precond. solve

5 A linear test case

We have considered a simple model for the fluid and the structure that has been proposed by Causin, Gerbeau and Nobile in [2]. Due to the linearity of the problem, we could not test the effectiveness of the tangential operators as preconditioners; we leave this test to a future work where we shall consider a general nonlinear context. However, even in this simple case the domain decomposition approach that we advocate can provide better convergence results than those obtained using the classical fixed point approach.

The domain under consideration is a rectangle Ω 6 cm long and 1 cm wide as represented in Fig. 1. The one-dimensional wall $\hat{\Sigma}$ is on the top of the domain and coincides with the interface. We consider a pressure wave entering the computational domain from the left and a symmetry condition on the x-axis. We consider only small displacements and neglect the domain deformation in the fluid domain. For this reason we will drop the dependence on t of both Ω_t^{f} and Σ_t.

The inviscid incompressible fluid is modeled as follows:

$$\mathrm{Fluid}(\boldsymbol{u}, p; p_{\mathrm{in}}, 0) : \begin{cases} \rho_{\mathrm{f}} \dfrac{\partial \boldsymbol{u}}{\partial t} + \nabla p = 0 \text{ in } \Omega^{\mathrm{f}},\ \forall t, \\ \mathrm{div}\boldsymbol{u} = 0 \text{ in } \Omega^{\mathrm{f}}, \\ p = p_{\mathrm{in}}(t) \text{ on } \Gamma^{\mathrm{in}}, \\ p = 0 \text{ on } \Gamma^{\mathrm{out}}, \\ \frac{\partial p}{\partial n} = 0 \text{ on the } x\text{-axis}, \end{cases} \tag{19}$$

where $\rho_{\mathrm{f}} = 1g/cm^2$ is the blood density.

The structure is described by a string model on $\hat{\Sigma}$ with homogeneous Dirichlet values as boundary conditions at the end points of $\hat{\Sigma}$:

$$\mathrm{Str}(d; 0) : \left\{ \rho_{\mathrm{s}} h_{\mathrm{s}} \frac{\partial^2 d}{\partial t^2} - kGh_{\mathrm{s}} \frac{\partial^2 d}{\partial x^2} + \frac{Eh_{\mathrm{s}}}{1-\nu^2} \frac{d}{R_0^2} = f_{\mathrm{s}}, \right. \tag{20}$$

where we set the wall density $\rho_{\mathrm{s}} h_{\mathrm{s}} = 1.1g/cm^2$, the Young modulus $E = 7.5 \cdot 10^5 dynes/cm^2$, the Poisson coefficient $\nu = 0.5$ and the shear modulus $kG = 2.5 \cdot 10^5 dynes/cm^2$.

The coupling conditions on the interface $\hat{\Sigma}$ impose the continuity of the normal velocity:

$$\boldsymbol{u} \cdot \boldsymbol{n}_{\mathrm{f}} (= u_2) = \frac{\partial d}{\partial t} \tag{21}$$

as well as the equilibrium of the stresses

$$f_{\mathrm{s}} = -p. \tag{22}$$

In view of the momentum equation restricted to $\hat{\Sigma}$, (21) becomes

$$\frac{\partial p}{\partial n} = \rho_{\mathrm{f}} \frac{\partial^2 d}{\partial t^2}. \tag{23}$$

We consider the interface variable $\lambda = \partial p / \partial n$; the associated Steklov-Poincaré interface problem expresses the equation (22) and takes the following form

$$S_{\mathrm{f}}(\lambda)+S_{\mathrm{s}}(\lambda)=\chi \quad \text{on } \hat{\Sigma}, \tag{24}$$

where $S_{\mathrm{f}}(\lambda)$ and $S_{\mathrm{s}}(\lambda)$ are linear Steklov-Poincaré operators, while χ is a right-hand side accounting for the boundary data. (We warn the reader that all along the previous sections the right hand side was indeed incorporated in the definition of the nonlinear Steklov-Poincaré operators).

Using the notations of Sect. 2.1, the operators in (24) may be written explicitly as

$$S_{\mathrm{f}}:\lambda \mapsto \text{ find } (\boldsymbol{u}(\lambda),p(\lambda)): \begin{cases} \mathrm{Fluid}(\boldsymbol{u}(\lambda),p(\lambda);0,0) \\ \dfrac{\partial p(\lambda)}{\partial n}=\lambda \text{ on } \hat{\Sigma} \end{cases}$$
$$\mapsto S_{\mathrm{f}}(\lambda)=p(\lambda)_{|\hat{\Sigma}}, \tag{25}$$

and

$$\begin{aligned} S_{\mathrm{s}}:\lambda \mapsto \text{ find } d(\lambda): \frac{\partial^2 d(\lambda)}{\partial t^2}&=\frac{\lambda}{\rho_{\mathrm{f}}} \\ \mapsto S_{\mathrm{s}}(\lambda)=\rho_{\mathrm{s}}h_{\mathrm{s}}\frac{\partial^2 d(\lambda)}{\partial t^2}&-kGh_{\mathrm{s}}\frac{\partial^2 d(\lambda)}{\partial x^2}+\frac{Eh_{\mathrm{s}}}{1-\nu^2}\frac{d(\lambda)}{R_0^2} \quad \text{on } \hat{\Sigma}. \end{aligned} \tag{26}$$

For the numerical approximation, the fluid equations (19) reduce to the Laplace equation for the pressure

$$\mathrm{Fluid}(p;p_{\mathrm{in}},0): \quad \Delta p=0 \text{ in } \Omega^{\mathrm{f}},$$

with the same boundary conditions as in (19). Then, both the pressure p and the displacement d are discretized using the Galerkin method with $\mathbb{P}1$ (piecewise linear) finite elements. Moreover, the Newmark method is used for the time-discretization of the structure problem.

In (26), d can be computed as the solution of an ordinary partial differential equation. At the discrete level, d^{n+1} can be retrieved by finite differences from d^n and $\dot{d}^n$ computed at the previous time step by the Newmark method.

We have applied the Richardson method (11) considering the following preconditioners:

a1) $P^{-1}=P_{DN}^{-1}=\bar{S}_{\mathrm{f}}^{-1}$, i.e. the fluid problem acts as preconditioner;
b1) $P^{-1}=P_{ND}^{-1}=\bar{S}_{\mathrm{s}}^{-1}$, i.e. the structure problem acts as preconditioner;
c1) $P^{-1}=P_{NN}^{-1}=\alpha_{\mathrm{f}}^k\bar{S}_{\mathrm{f}}^{-1}+\alpha_{\mathrm{s}}^k\bar{S}_{\mathrm{s}}^{-1}$, i.e. both problems enter in the preconditioner.

Since the operators S_{f} and S_{s} are linear, they coincide with S_{f}' and S_{s}', respectively.

The acceleration parameters ω^k for preconditioners a1) and b1) and α_{f}^k, α_{s}^k for c1) have been computed using the Aitken extrapolation techniques (12) and (16), respectively.

We have considered several time steps δt and space discretization parameters h in order to test the three above preconditioners. The results are reported in table 2.

As we can see both preconditioners P_{DN}^{-1} and P_{ND}^{-1} give satisfactory convergence results and we remark that in both cases the relaxation parameters ω^k computed using the Aitken approach belong to the range of stability indicated in [2]. However, the Neumann-Neumann type preconditioner c1) allows us to improve the convergence rate of the algorithm, sensibly reducing the number of iterations required to satisfy a convergence test with tolerance 10^{-5}.

Table 2. Number of iterations for the choice of the preconditioners defined in a1), b1), c1) and mean value of the acceleration parameters. The values $h = 0.15$ and $\delta t = 1.e\text{-}5$ have been used for the upper and the lower table, respectively.

Precond.	P_{DN}^{-1}		P_{ND}^{-1}		P_{NN}^{-1}		
δt	ω_{mean}	iter.	ω_{mean}	iter.	$(\omega\alpha_{\mathrm{f}})_{mean}$	$(\omega\alpha_{\mathrm{s}})_{mean}$	iter.
1.e-4	0.26	45	0.33	101	0.16	0.10	33
1.e-5	0.26	44	0.29	87	0.16	0.10	33
1.e-6	0.26	44	0.29	87	0.17	0.09	33

Precond.	P_{DN}^{-1}		P_{ND}^{-1}		P_{NN}^{-1}		
h	ω_{mean}	iter.	ω_{mean}	iter.	$(\omega\alpha_{\mathrm{f}})_{mean}$	$(\omega\alpha_{\mathrm{s}})_{mean}$	iter.
0.3	0.35	33	0.29	86	0.23	0.07	27
0.15	0.26	44	0.29	87	0.16	0.10	33
0.1	0.22	54	0.30	86	0.13	0.11	36
0.05	0.16	75	0.29	85	0.09	0.14	43
0.015	0.09	131	0.36	107	0.04	0.19	55

In Fig. 2 we have represented the relaxation parameters $\omega^k\alpha_{\mathrm{f}}^k$ and $\omega^k\alpha_{\mathrm{s}}^k$ (obtained for fixed $h = 0.15$ and $\delta t = 1.e\text{-}5$) which characterize the preconditioner P_{NN}^{-1} defined in c1) and whose mean values have been reported in table 2. Notice that they sensibly vary at each iteration.

Finally, we have tested the Richardson method (11) with preconditioner P_{NN}^{-1} as in c1) with respect to different values of the wall density $\rho_{\mathrm{s}} h_{\mathrm{s}}$, i.e. we have multiplied and divided by 10 the actual density. We have set $h = 0.15$ and $\delta t = 1.e\text{-}5$; as shown in table 3, the Aitken acceleration (16) can effectively compute the relaxation parameters assuring good convergence results.

In a second stage, we have slightly changed our framework by considering as interface variable the stresses on $\hat{\Sigma}$: $\sigma = -p (= \sigma_{\mathrm{s}})$ on $\hat{\Sigma}$ (rather than the displacement λ given by (23)). This yields a generalization of the so-called FETI method, which in the case of standard elliptic boundary value problems can be seen as the dual approach to (24) (see [21]). In fact, in this case we have to solve the interface equation

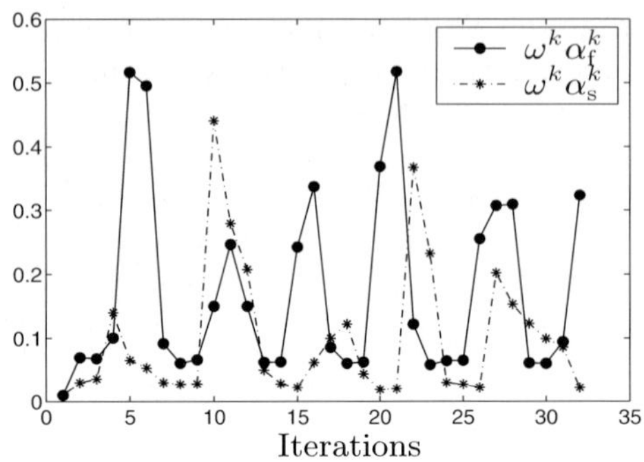

Fig. 2. Relaxation parameters $\omega^k \alpha_\mathrm{f}^k$ and $\omega^k \alpha_\mathrm{s}^k$ computed using (16).

Table 3. Number of iterations with respect to different wall densities $\rho_\mathrm{s} h_\mathrm{s}$ for the preconditioner c1) corresponding to the values $h = 0.15$, $\delta t = 1.e\text{-}5$.

wall density	$(\omega\alpha_\mathrm{f})_{mean}$	$(\omega\alpha_\mathrm{s})_{mean}$	iter.
$0.1 \cdot \rho_\mathrm{s} h_\mathrm{s}$	0.60	0.005	15
$1 \cdot \rho_\mathrm{s} h_\mathrm{s}$	0.16	0.10	33
$10 \cdot \rho_\mathrm{s} h_\mathrm{s}$	0.02	0.60	25

$$S_\mathrm{f}^{-1}(\sigma) + S_\mathrm{s}^{-1}(\sigma) = \widetilde{\chi} \tag{27}$$

defined through the inverses of the Steklov-Poincaré operators (25) and (26). A similar framework can be set up for the nonlinear coupled problem (1)-(3) as well.

Here the preconditioners become:

a2) $P^{-1} = \bar{S}_\mathrm{f}$;
b2) $P^{-1} = \bar{S}_\mathrm{s}$;
c2) $P^{-1} = \alpha_\mathrm{f}^k \bar{S}_\mathrm{f} + \alpha_\mathrm{s}^k \bar{S}_\mathrm{s}$

and, again, the Aitken strategy has been used to accelerate the convergence.

We have considered several time steps δt and space discretization parameters h in order to test the effectiveness of the preconditioners based on the FETI approach. The results are reported in table 4.

Also in this case we can see that the combination of both the fluid and the structure operators gives a better preconditioner than each one taken separately.

6 Conclusions

In this paper we have considered a general fluid-structure interaction problem. For its numerical solution we have analyzed three different approaches,

Table 4. Number of iterations for the choice of the preconditioners a2), b2), c2) and mean value of the acceleration parameters. The values $h = 0.15$ and $\delta t = 1.e\text{-}5$ have been used for the upper and the lower table, respectively.

Precond.	$\bar{S}_\mathrm{f}$		$\bar{S}_\mathrm{s}$		$\alpha_\mathrm{f}^k \bar{S}_\mathrm{f} + \alpha_\mathrm{s}^k \bar{S}_\mathrm{s}$		
δt	ω_{mean}	iter.	ω_{mean}	iter.	$(\omega\alpha_\mathrm{f})_{mean}$	$(\omega\alpha_\mathrm{s})_{mean}$	iter.
1.*e*-4	0.19	61	0.42	54	0.10	0.17	36
1.*e*-5	0.18	59	0.36	63	0.10	0.17	37
1.*e*-6	0.18	60	0.36	64	0.10	0.17	37

Precond.	$\bar{S}_\mathrm{f}$		$\bar{S}_\mathrm{s}$		$\alpha_\mathrm{f}^k \bar{S}_\mathrm{f} + \alpha_\mathrm{s}^k \bar{S}_\mathrm{s}$		
h	ω_{mean}	iter.	ω_{mean}	iter.	$(\omega\alpha_\mathrm{f})_{mean}$	$(\omega\alpha_\mathrm{s})_{mean}$	iter.
0.3	0.17	56	0.50	45	0.08	0.27	31
0.15	0.18	59	0.36	63	0.10	0.17	37
0.1	0.18	61	0.31	77	0.11	0.13	40
0.05	0.19	63	0.22	109	0.12	0.08	46
0.015	0.20	64	0.13	191	0.15	0.03	56

one based on a fixed-point iteration, another on Newton's method, a further on domain decomposition iterations. In particular, we have focused on the domain decomposition reformulation and we have proposed a general Richardson algorithm to solve the associated Steklov-Poincaré interface equation, with several choices of preconditioner.

The analogies with the two other approaches are discussed, and a numerical test on a simplified linear model for blood-flow simulation in an arterial vessel is investigated. The results show in particular that the so-called Neumann-Neumann preconditioner has the potential of handling efficiently significant variations of numerical discretization parameters as well as physical parameters.

Acknowledgements

This research has been supported by the Swiss National Science Foundation (Project 20-101800), by the European Community's RTN Project HPRN-CT-2002-00270 HaeMOdel, by the Italian reseach project Cofin2003-MIUR "Numerical Modeling for Scientific Computing and Advanced Applications" and by INDAM.

References

1. S. Anicic and A. Léger. Formulation bidimensionnelle exacte du modèle de coque 3D de Kirchhoff-Love. *C. R. Acad. Sci. Paris Sér. I Math.*, 329(8):741–746, 1999.

2. P. Causin, J.-F. Gerbeau, and F. Nobile. Added-mass effect in the design of partitioned algorithms for fluid-structure problems. Technical Report 5084, INRIA, 2004.
3. D. Chapelle and K. Bathe. *The Finite Element Analysis of Shells - Fundamentals.* Springer Verlag, 2003.
4. S. Deparis. *Numerical analysis of axisymmetric flows and methods for fluid-structure interaction arising in blood flow simulation.* PhD thesis, École Polytechnique Fédérale de Lausanne (EPFL), 2004.
5. S. Deparis, J.-F. Gerbeau, and X. Vasseur. GMRES preconditioning and accelerated Quasi-Newton algorithm and application to fluid structure interaction. In preparation, 2004.
6. M. Discacciati. An operator-splitting approach to nonoverlapping domain decomposition. Technical Report 14, EPFL, 2004. Submitted to Numer. Math.
7. M. Fernández and M. Moubachir. A Newton method using exact Jacobians for solving fluid-structure coupling. Technical Report 5085, INRIA, 2004.
8. J.-F. Gerbeau and M. Vidrascu. A quasi-Newton algorithm based on a reduced model for fluid structure problems in blood flows. *M2AN*, 37(4):631–648, 2003.
9. M. Heil. An efficient solver for the fully-coupled solution of large-displacement fluid-structure interaction problems. *Comput. Methods Appl. Mech. Engrg.*, 2003. In press.
10. P. Le Tallec. *Introduction à la Dynamique des Structures.* Ellipse, Paris, 2000.
11. P. Le Tallec and J. Mouro. Fluid structure interaction with large structural displacements. *Comput. Methods Appl. Mech. Engrg.*, 190(24-25):3039–3067, 2001.
12. H. Matthies and J. Steindorf. Numerical efficiency of different partitioned methods for fluid-structure interaction. *Z. Angew. Math. Mech.*, 2(80):557–558, 2000.
13. H. Matthies and J. Steindorf. Partitioned strong coupling algorithms for fluid-structure interaction. *Computer & Structures*, 81:805–812, 2003.
14. D. P. Mok and W. A. Wall. Partitioned analysis schemes for the transient interaction of incompressible flows and nonlinear flexible structures. In K. S. W. Wall, editor, *Trends in Computational Structural Mechanics.* K.U. Bletzinger,CIMNE, Barcelona, 2001.
15. D. P. Mok, W. A. Wall, and E. Ramm. Accelerated iterative substructuring schemes for instationary fluid-structure interaction. In K. Bathe, editor, *Computational Fluid and Solid Mechanics*, pages 1325–1328. Elsevier, 2001.
16. F. Nobile. *Numerical approximation of fluid-structure interaction problems with application to haemodynamics.* Thesis n. 2548, École Polytechnique Fédérale de Lausanne (EPFL), 2001.
17. A. Quarteroni, R. Sacco, and F. Saleri. *Numerical Mathematics.* Springer, New York – Berlin – Heidelberg, 2000.
18. A. Quarteroni, M. Tuveri, and A. Veneziani. Computational vascular fluid dynamics: problems, models and methods. *Comp. Vis. Science*, 2:163–197, 2000.
19. A. Quarteroni and A. Valli. *Domain Decomposition Methods for Partial Differential Equations.* Oxford Science Publications, Oxford, 1999.
20. T. Tezduyar. Finite element methods for fluid dynamics with moving boundaries and interfaces. *Arch. Comput. Methods Engrg.*, 8:83–130, 2001.
21. A. Toselli and O. Widlund. *Domain Decomposition Methods.* Springer, New York, 2004. In press.

Micro Flow Simulation Using Kinetic and Continuum Approaches

Koji Morinishi[1]

Kyoto Institute of Technology, Matsugasaki, Sakyo-ku, Kyoto, 606-8585 Japan
morinisi@kit.ac.jp

Summary. Computational method based on a kinetic model Boltzmann equation has been developed for micro scale low speed flows. Navier-Stokes method with the first order slip boundary condition is also developed for comparison. The results obtained with both the methods are compared with those of the direct simulation Monte Carlo method and experiments for supersonic flows. Numerical results for low speed flows over a micro circular cylinder and a micro sphere are also obtained with the methods, while it is difficult to obtain the low speed flow results with the direct simulation Monte Carlo method. Results of the Navier-Stokes equations with the slip boundary condition generally agree with those of the kinetic model Boltzmann equation if the Knudsen number is less than 0.1. A kinetic/continuum hybrid method has also been developed. The hybrid method may be a promising tool for analyzing whole flow regimes from free molecule to continuum flows.

1 Introduction

Numerical simulation for gas flows about micro devices is one of the recent new frontiers of computational fluid dynamics (CFD). It may be an essential tool for understanding the fluid behavior around micro-electro-mechanical systems (MEMS). The Knudsen number ($Kn = \lambda/L$, the ratio of molecular mean free path λ to a characteristic flow length L) of the flows in MEMS is typically between 0.01 and 10, where conventional Navier-Stokes approach with no-slip boundary condition is not valid completely. The Navier-Stokes approach with slip boundary condition may be valid for the slip flow regime ($0.01 < Kn < 0.1$), while Boltzmann (kinetic) approach may be preferable for the transition flow regime ($0.1 < Kn < 10$).

One of the numerical methods widely used for the Boltzmann equation is the direct simulation Monte Carlo (DSMC) method [1]. The DSMC method is a powerful simulation tool for hypersonic rarefied gas flows, where the inherent statistical scatter of DSMC easily disappear into large changes of flow quantities in hypersonic flows. The method, however, becomes a poor simulation tool for the flows in micro scales because the micro flows are low speed flows, where huge sample size is required to reduce the statistical scatter to a level of small changes of flow quantities in the low speed flows [2, 3]. For example, Fig. 1 shows the density contours obtained with the DSMC method at a million time steps for a flow over a micro circular cylinder at a free stream

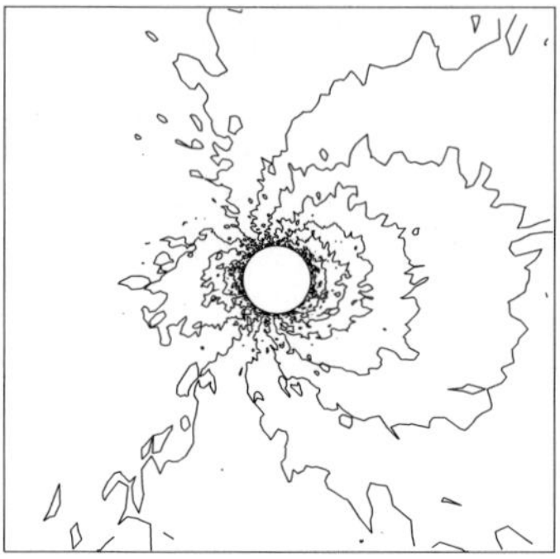

Fig. 1. Density contours obtained with DSMC method at $M_\infty = 0.1$ and $Kn = 0.1$

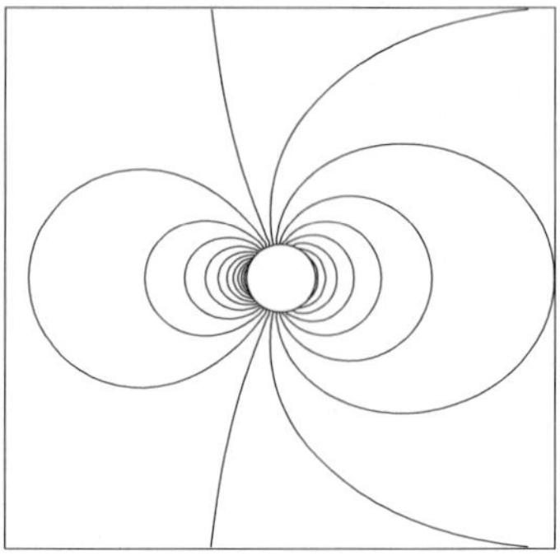

Fig. 2. Density contours obtained with kinetic method at $M_\infty = 0.1$ and $Kn = 0.1$

Mach number of 0.1 and a Knudsen number of 0.1. Even at a million time steps the statistical scatter is still so large compared to the small changes of flow quantities that any meaningful results can not be obtained.

In contrast to the DSMC method, the CFD method based on a kinetic model Boltzmann equation [4, 5] is free from the statistical scatter. The results obtained with the method for the same flow conditions are plotted in Fig. 2. The method is definitely superior to the DSMC method for the subsonic flow simulation, though it is rather expensive for hypersonic flow simulation [6]. Efficiency of the method may be further improved by the implementation of parallel computing [7]. Moreover a Boltzmann/Navier-Stokes hybrid method, an universal tool for analyzing whole flow regimes from free molecule to continuum flows, may easily be constructed, since the same CFD methods for the Navier-Stokes equations are adopted for the kinetic model Boltzmann equation.

In this study, fundamental potentialities of the kinetic model Boltzmann method are demonstrated for simulating low speed gas flows in micro scales. Implementation of the compressible Navier-Stokes method with the first order slip boundary condition is also made for comparison. The reliability of those methods are first examined for supersonic flows so that direct comparison of the numerical results can be made with those of the DSMC method. Typical simulation of low speed micro flows is carried out for a circular cylinder and a sphere. Fundamental potentialities of the kinetic/continuum hybrid method are also demonstrated briefly.

2 Kinetic Approach

The motion of gas molecules at any Knudsen number is governed by the well known Boltzmann equation. In the transition regime from continuum to free molecular flow, especially, it is naturally desirable to obtain the solution of the Boltzmann equation instead of the Navier-Stokes equation. Because of its complex collision integral term, however, the solution requires an exceedingly formidable task except for few simple problems. In this paper, instead of the full Boltzmann equation, we use a kinetic model equation, which correctly resembles the lower 13 moments of the Boltzmann equations. The numerical algorithm described can essentially be applied to the Boltzmann equation as well as any other kinetic model equations.

2.1 Kinetic Model Equation

The kinetic model Boltzmann equation in nondimensional form without any external force may be written as follows:

$$\frac{\partial f}{\partial t} + \boldsymbol{c} \cdot \frac{\partial f}{\partial \boldsymbol{x}} = \nu(f_0 - f) \tag{1}$$

where f is the velocity distribution function which depends on the time t, the physical space $\boldsymbol{x}$, and the molecular velocity $\boldsymbol{c}$. The distribution function f_0 of the BGK model [8], which is the most fundamental model, is the local equilibrium distribution function f_e:

$$f_e = \frac{n}{(\pi T)^{3/2}} \exp\left(-\frac{\boldsymbol{C}^2}{T}\right). \tag{2}$$

where n is the number density and T the temperature The peculiar velocity $\boldsymbol{C}$ is defined with the molecular velocity $\boldsymbol{c}$ and the macroscopic flow velocity $\boldsymbol{u}$ as:

$$\boldsymbol{C} = \boldsymbol{c} - \boldsymbol{u} \tag{3}$$

For a higher model equation which correctly resembles the lower 13 moments, the distribution function f_0 is given with the local equilibrium distribution function f_e as

$$f_0 = f_e \left[1 + \frac{2}{5pT}(1 - Pr)\boldsymbol{C} \cdot \boldsymbol{q} \left(\frac{2\boldsymbol{C}^2}{T} - 5\right)\right] \tag{4}$$

where p is the pressure, $\boldsymbol{q}$ the heat flux vector, and Pr the Prandtl number (=2/3 for a monatomic gas).

The macroscopic flow quantities are obtained from the following moments of the distribution function.

$$
\begin{aligned}
n &= \int f d\boldsymbol{c} \\
n\boldsymbol{u} &= \int \boldsymbol{c} f d\boldsymbol{c} \\
\frac{3}{2} nT &= \int \boldsymbol{C}^2 f d\boldsymbol{c} \\
\mathbf{q} &= \int \boldsymbol{C}\boldsymbol{C}^2 f d\boldsymbol{c}
\end{aligned}
\tag{5}
$$

The pressure p is obtained from the equation of state:

$$p = nT \tag{6}$$

The collision frequency ν is usually defined as:

$$\nu = \frac{8nT^{1-s}}{5\sqrt{\pi}Kn} \tag{7}$$

where Kn is the reference Knudsen number based on the reference length L and the molecular mean free path at reference state λ_∞ which is defined as

$$\lambda_\infty = \frac{16\mu_\infty}{5mn_\infty\sqrt{2\pi RT_\infty}} \tag{8}$$

where μ is the viscosity coefficient, m the mass of a single molecule, and R the gas constant. The viscosity coefficient is assumed to depend on the temperature as:

$$\mu \propto T^s \tag{9}$$

where the Maxwell molecules correspond to the power s of 1 and the hard sphere molecules to 0.5.

For two dimensional flow problems, the following reduced distribution function g and h are conveniently introduced so that the operational count and storage of the computation can be reduced.

$$
\begin{aligned}
g(t, x, y, c_x, c_y) &= \int f dc_z \\
h(t, x, y, c_x, c_y) &= \int c_z^2 f dc_z
\end{aligned}
\tag{10}
$$

For nondimensionalizing the basic equations, a reference number density n_∞, a reference temperature T_∞, a reference velocity C_∞, and a reference length L are introduced. The reference velocity C_∞ is the most probable molecular thermal speed which is defined as:

$$C_\infty = \sqrt{2RT_\infty} \tag{11}$$

2.2 Numerical Procedure

For obtaining numerical solution of the kinetic model Boltzmann equation, any conventional numerical methods, such as finite difference, finite volume, and finite element methods can be adopted. In this study an upwind gridless method [9, 10] is adopted for estimation of the convective terms of the equation.

The spatial derivatives of the distribution function f at any computational point i can be evaluated with the following linear combination form in the cloud of neighboring points $C(i)$.

$$\frac{\partial f}{\partial \boldsymbol{x}} = \sum_{k \subset C(i)} \boldsymbol{a}_{ik} f_{ik} \tag{12}$$

where the subscript k denotes the index of the point which belongs to the cloud $C(i)$. The sum is obtained over all member points of $C(i)$ except the point i itself. The function values f_{ik} are evaluated at the midpoint between points i and k, which may be obtained with simple arithmetical averages.

The coefficients $\boldsymbol{a}_{ik}$ are once obtained at the beginning of computation and stored. Several methods can be used for obtaining the coefficients. For an example, the x-component of the coefficients $\boldsymbol{a}_{ik}$ can be obtained with solving the following system of equations using QR or singular value decompositions.

$$\sum_{k \subset C(i)} a_{ik} f_{ik}^{(m)} = d^{(m)} \tag{13}$$

The components of $f^{(m)}$ and $d^{(m)}$ are given with:

$$f^{(m)} \in \left(1\,,\; x\,,y\,,z\,,\; x^2\,,\; y^2\,,\; z^2\,,\; xy\,,\cdots\right) \tag{14}$$

and

$$d^{(m)} \in (0\,,\,1\,,\,0\,,\,0\,,\,0\,,\,0\,,\,0\,,\,0\,,\cdots)\;. \tag{15}$$

The convective terms of the kinetic model equation can be evaluated with the gridless method as:

$$\begin{aligned} \boldsymbol{c} \cdot \left.\frac{\partial \boldsymbol{q}}{\partial \boldsymbol{x}}\right|_i &= \sum_{k \subset C(i)} (\boldsymbol{a}_{ik} \cdot \boldsymbol{c}) \boldsymbol{q}_{ik} \\ &= \sum_{k \subset C(i)} \boldsymbol{F}_{ik} \end{aligned} \tag{16}$$

where $\boldsymbol{q}$ denotes $\boldsymbol{q} = (g, h)^T$ for the two dimensional problems and $\boldsymbol{q} = f$ for the three dimensional problems. The numerical flux $\boldsymbol{F}$ are estimated as

$$\boldsymbol{F}_{ik} = \frac{1}{2}\left\{A_{ik}(\boldsymbol{q}_{ik}^{+} + \boldsymbol{q}_{ik}^{-}) - |A_{ik}|(\boldsymbol{q}_{ik}^{+} - \boldsymbol{q}_{ik}^{-})\right\} \tag{17}$$

where A_{ik} is defined as

$$A_{ik} = \boldsymbol{a}_{ik} \cdot \boldsymbol{c} \tag{18}$$

The third order method may be obtained, for example, if $\mathbf{q}_{ik}^{-}$ are evaluated with the following reconstruction [11].

$$\mathbf{q}_{ik}^{-} = \mathbf{q}_i + \frac{1}{2}\left(\omega_0 \delta\mathbf{q}_{ik}^{-} + \omega_1 \delta\mathbf{q}_{ik}\right) \tag{19}$$

Here $\delta\mathbf{q}_{ik}^{-}$ are obtained with:

$$\delta\mathbf{q}_{ik}^{-} = 2\nabla\mathbf{q}_i \cdot \mathbf{r}_{ik} - \delta\mathbf{q}_{ik} \tag{20}$$

where $\mathbf{r}_{ik}$ is the vector from the point i to k and $\delta\mathbf{q}_{ik}$ are defined as:

$$\delta\mathbf{q}_{ik} = \mathbf{q}_k - \mathbf{q}_i \ . \tag{21}$$

The weights ω_0 and ω_1 for the third order linear method are:

$$\omega_0 = \frac{1}{3}\ , \quad \omega_1 = \frac{2}{3}\ . \tag{22}$$

After evaluating the convective and collision terms, following implicit Euler method is used for the temporal discretization of the kinetic model equation.

$$\left(\frac{1}{\Delta t_i} + \nu + \sum_{k \subset C(i)} A_{ik}^{+}\right) \Delta\boldsymbol{q}_i + \sum_{k \subset C(i)} A_{ik}^{-} \Delta\boldsymbol{q}_k = RHS_i \tag{23}$$

where RHS are the evaluation of the convective and collision terms and $A^{\pm}$ are defined as follows

$$A^{\pm} = \frac{1}{2}(A \pm |A|) \tag{24}$$

The solution of this linear system of equation is obtained with a LU-SGS procedure [12] as:

$$\Delta\boldsymbol{q}_i^{*} = D_i^{-1}\left(RHS_i - \sum_{k \subset L(i)} A_{ik}^{-} \Delta\boldsymbol{q}_k\right) \tag{25}$$

$$\Delta\boldsymbol{q}_i = \Delta\boldsymbol{q}_i^{*} - D_i^{-1} \sum_{k \subset U(i)} A_{ik}^{-} \Delta\boldsymbol{q}_k \tag{26}$$

where $C(i) = L(i) \bigcup U(i)$ and D_i are defined with

$$D_i = \left(\frac{1}{\Delta t_i} + \nu + \frac{1}{2} \sum_{k \subset C(i)} |A_{ik}|\right) \tag{27}$$

The macroscopic flow quantities, for example, the number density, can be obtained with numerical quadrature as

$$n = \int\int\int f dc_x dc_y dc_z = \sum\sum\sum wf \tag{28}$$

where w are the weights of quadrature. Simple equally spaced trapezoidal rule is used for high speed flow cases ($Ma \geq 0.3$). For low speed flow cases ($Ma < 0.3$) the weights are once determined at the beginning of computation and stored so that the lower moments of the equilibrium distribution function f_e are exactly obtained.

2.3 Boundary Conditions

Perfect diffuse reflection is assumed for the interaction of molecules with the wall surface. That is, molecules which strike the surface are subsequently emitted with fully accommodating to the wall temperature and velocity. The distribution function for the molecules reflecting from the wall surface is given as

$$f_w = \frac{n_w}{(\pi T_w)^{3/2}} \exp\left(-\frac{\mathbf{C}^2}{T_w}\right) \qquad c_n > 0 \tag{29}$$

The density of molecules diffusing from the surface is determined from the following mass balance condition

$$\int_{c_n>0} c_n f d\mathbf{c} = -\int_{c_n<0} c_n f d\mathbf{c} \tag{30}$$

Substituting (29) into the left hand side of (30) and applying the numerical quadrature to both the integrals, the density can be obtained as

$$n_w = -\frac{1}{\Theta}\sum_{c_n<0} w c_n f \tag{31}$$

At the beginning of computation, Θ are once obtained from the following equation and stored at each point on the wall surface.

$$\Theta = \sum_{c_n>0} \frac{w c_n}{(\pi T_w)^{3/2}} \exp\left(-\frac{\mathbf{C}^2}{T_w}\right) \tag{32}$$

At inlet and outlet boundaries, the local equilibrium distribution functions are specified for incoming molecules. For outgoing molecules, simple extrapolation of the distribution function is used, which may not affect numerical results because the distribution functions on inner computational points are updated with the upwind gridless solver.

3 Continuum Approach

The basic equations of continuum approach is the compressible Navier-Stokes equations which may be written in following form.

$$\frac{\partial \boldsymbol{q}}{\partial t} + \boldsymbol{\nabla} \cdot \boldsymbol{F} = \boldsymbol{\nabla} \cdot \boldsymbol{R} \tag{33}$$

where $\boldsymbol{q}$ is the conservative vector, $\boldsymbol{F}$ the convective fluxes, and $\boldsymbol{R}$ the viscous fluxes. The conservative vector and the flux terms are given with:

$$\boldsymbol{q} = \begin{pmatrix} \rho \\ \boldsymbol{u} \\ e \end{pmatrix} \quad , \quad \boldsymbol{F} = \begin{pmatrix} \rho \boldsymbol{u} \\ \rho \boldsymbol{u}\boldsymbol{u} + p\boldsymbol{I} \\ \boldsymbol{u}(e+p) \end{pmatrix} \quad , \quad \boldsymbol{R} = \begin{pmatrix} 0 \\ \boldsymbol{\tau} \\ \boldsymbol{u} \cdot \tau - \boldsymbol{q} \end{pmatrix} \tag{34}$$

where e is the total energy per unit volume, τ the viscous stress tensor, and q the heat flux vector. The total energy e is defined for a perfect gas as:

$$e = \frac{p}{\gamma - 1} + \frac{1}{2}\rho \boldsymbol{u}^2 \tag{35}$$

where γ is the ratio of specific heats.

The Navier-Stokes equations are solved using the upwind gridless and LU-SGS methods [9, 10].

At the wall surface, the first order slip conditions are implemented as:

$$u_s - u_w = \frac{2 - \sigma_v}{\sigma_v} Kn \frac{\mu}{\rho \sqrt{T_w}} \frac{\partial u}{\partial n} + \frac{3}{4}\sqrt{\frac{\pi}{2}} Kn \frac{\mu}{\rho T_w} \frac{\partial T}{\partial s} \tag{36}$$

$$T_s - T_w = \frac{2 - \sigma_t}{\sigma_t} \frac{2\gamma}{\gamma + 1} \frac{Kn}{Pr} \frac{\mu}{\rho \sqrt{T_w}} \frac{\partial T}{\partial n} \tag{37}$$

where n and s denote the normal and tangential directions to the surface, respectively. The momentum accommodation coefficient σ_v and the energy accommodation coefficient σ_t are set to unity in all the computation presented here.

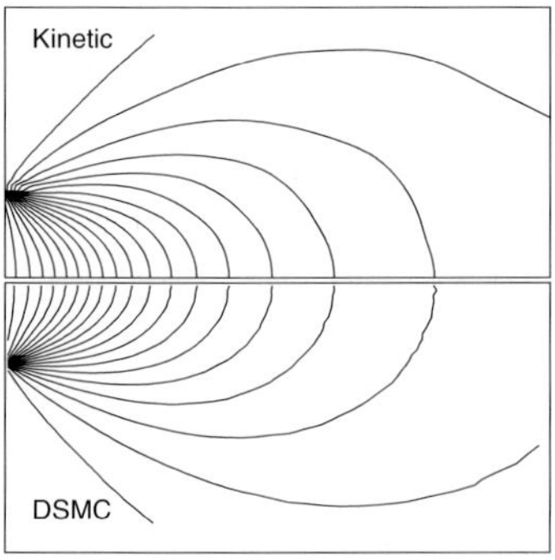

Fig. 3. Density contours obtained with kinetic model and DSMC

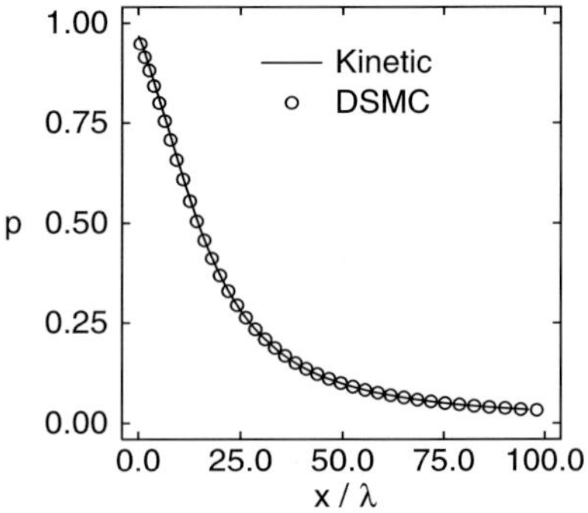

Fig. 4. Pressure distributions along the center line obtained with kinetic model and DSMC

4 Application to Supersonic Flows

Before applying the kinetic model Boltzmann equation method to low speed micro scale flows, several simulation of supersonic flows are carried out so that comparisons of the numerical results with those of the DSMC method can be made directly.

4.1 Sonic Expansion into a Vacuum

The first test case is the two-dimensional sonic expansion of monatomic gas from a nozzle lip into a vacuum. The computational domain is a rectangle with $0 \le x \le 100\lambda_\infty$ and $-45\lambda_\infty \le y \le 45\lambda_\infty$. The extent of nozzle is $30\lambda_\infty$, which is located at $x = 0$ and $-15\lambda_\infty \le y \le 15\lambda_\infty$. The comparison of the density contours obtained with the kinetic model Boltzmann equation method and the DSMC method is made in Fig. 3. While the computations are carried out for the whole computational domain with the hard sphere collision model, the results of the kinetic model Boltzmann equation method are plotted for the upper half domain and the DSMC results for the lower half domain. The pressure distributions along the center line are compared in Fig. 4. Both the methods exactly predict the same flow field.

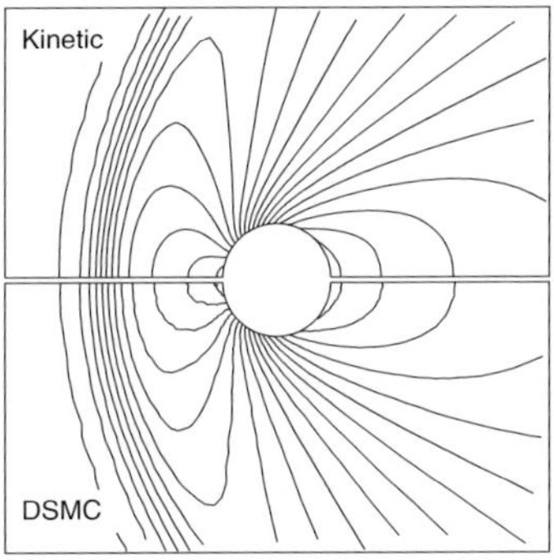

Fig. 5. Density contours obtained with kinetic model and DSMC

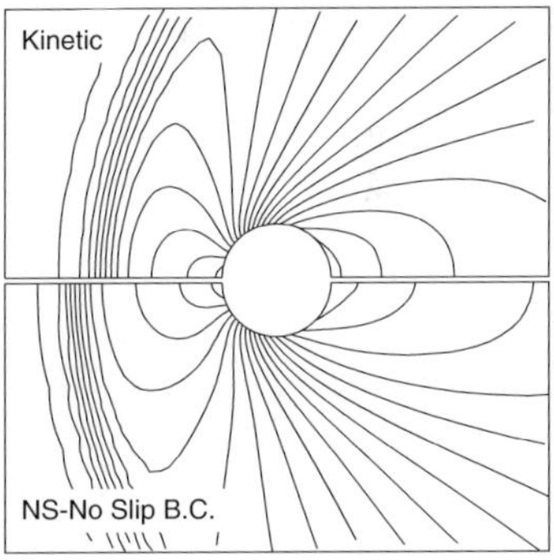

Fig. 6. Density contours obtained with kinetic model and Navier-Stokes with no-slip

4.2 Supersonic Flows over a Circular Cylinder

Supersonic flows over a circular cylinder are computed at a free stream Mach number of 2.0 and several Knudsen numbers from 0.01 to 1.0. Comparison of the density contours obtained with the kinetic model Boltzmann equation method and the DSMC method is made in Fig. 5 at a free stream Knudsen

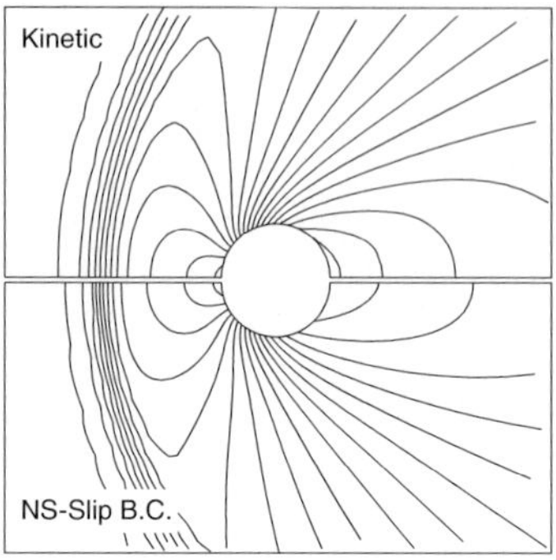

Fig. 7. Density contours obtained with kinetic model and Navier-Stokes with slip

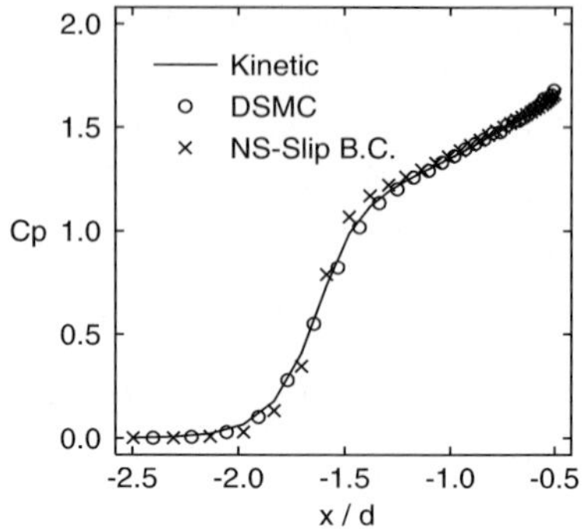

Fig. 8. Pressure distributions along the front stagnation streamline

number of 0.1. While both the computations are carried out for the whole computational domain with 160×48 cells and the hard sphere collision model, the kinetic model results are plotted for the upper half domain and the DSMC results for the lower half domain. Very good agreement is generally observed over the whole flow field.

Figures 6 and 7 show the comparison of the density contours obtained with the kinetic equation and the Navier-Stokes equations. The results with the no-slip boundary conditions are plotted in Fig. 6 and the results with the slip boundary conditions in Fig. 7. Large discrepancy is observed over the whole flow field between the two results in Fig. 6. The results of the Navier-Stokes equations with the slip boundary conditions well agree with those of the kinetic equation in the front region including the position of bow shock, while discrepancy is still observed in the wake. The density in the wake is lower than the free stream density. This large rarefaction effect leads the Navier-Stokes equations to fail to predict the wake flow field. The slip boundary conditions well remedy the continuum solution in the front flow field at the Knudsen number. The fact is also found in the comparisons of the pressure distributions along the front stagnation streamline as shown in Fig. 8. Very good agreement between the kinetic model results and the DSMC results is again observed. The results of the Navier-Stokes equations with the slip boundary conditions are also well compared with the other two results.

Figures 9 and 10 show the comparison of the density contours obtained at a free stream Knudsen number of 1.0. The comparison between the kinetic model results and the DSMC results is generally good, while the Navier-Stokes equations with slip boundary conditions fail to predict the front flow field as well as the wake. The density contours obtained at a free stream Knudsen number of 0.01 are plotted in Fig. 11. As expected, the comparison

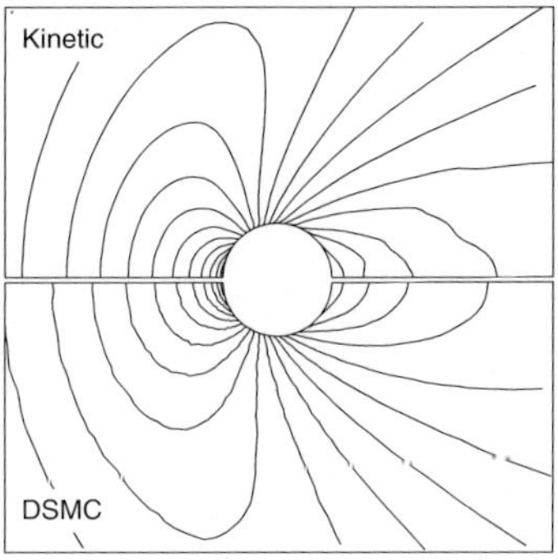

Fig. 9. Density contours obtained with kinetic model and DSMC

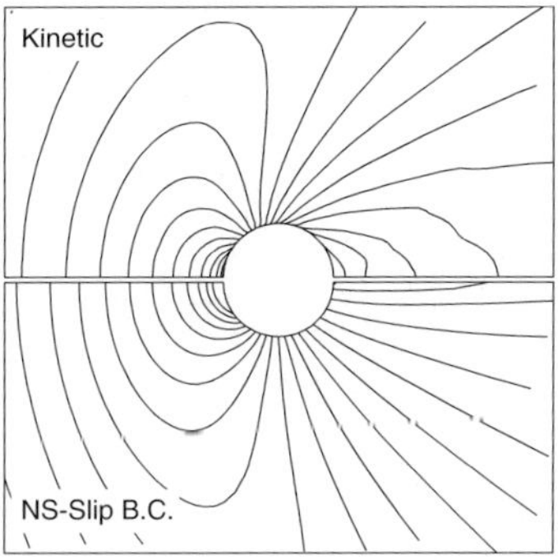

Fig. 10. Density contours obtained with kinetic model and Navier-Stokes with slip

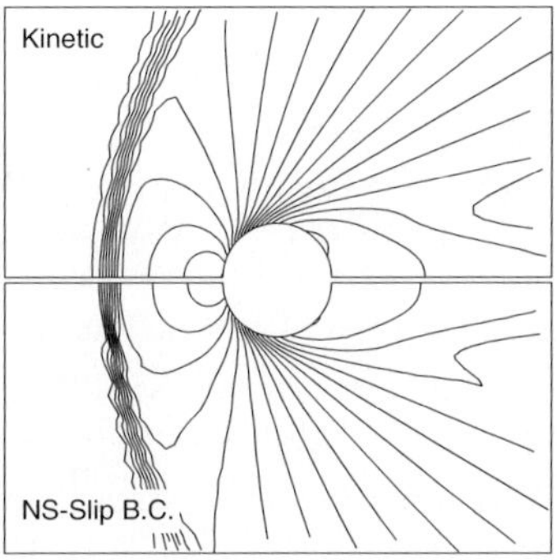

Fig. 11. Density contours obtained with kinetic model and Navier-Stokes with slip

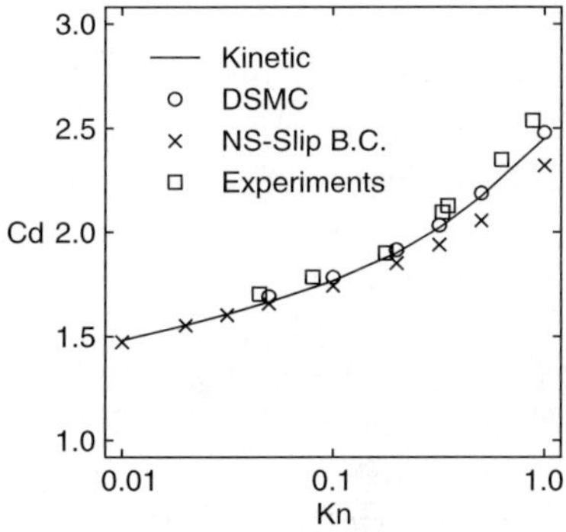

Fig. 12. Drag coefficient as a function of Knudsen number

between the kinetic results and the continuum results at the low Knudsen number is quite good in the whole flow field.

Figure 12 shows drag coefficients as a function of the Knudsen number. In addition to the three numerical results, the experiments of Maslach and Schaaf [13] are plotted with square symbols. Very good agreement between the kinetic model results and the DSMC results is again observed. The results of the Navier-Stokes equations with the slip boundary conditions well agree with the kinetic model results if the Knudsen number is less than 0.1.

5 Application to Low Speed Micro Scale Flows

Since it is difficult that meaningful results are obtained for low speed flows with the DSMC method, numerical results of the kinetic model Boltzmann

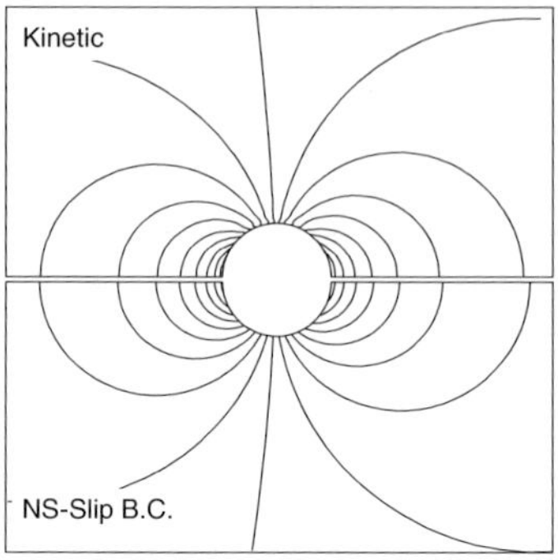

Fig. 13. Pressure contours obtained at $M_\infty = 0.01$ and $kn = 0.1$

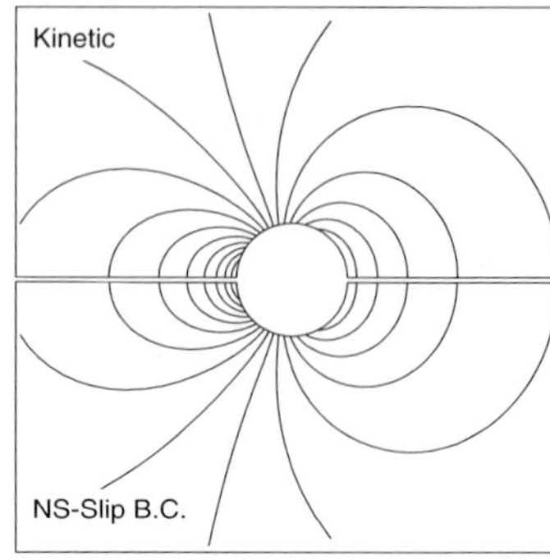

Fig. 14. Pressure contours obtained at $M_\infty = 0.01$ and $kn = 0.01$

equation method are compared with those of the Navier-Stokes equations with the slip boundary condition.

5.1 Low Speed Flows over a Micro Circular Cylinder

Flows over a micro circular cylinder are considered as the first test case for low speed flows. A series of numerical results are obtained at free stream Mach numbers of 0.1 and 0.01 and Knudsen numbers from 0.01 to 1.0. Comparison of the pressure contours obtained with the kinetic model Boltzmann equation method and the Navier-Stokes equations with the slip boundary condition at a free stream Mach number of 0.01 and a Knudsen number of 0.1 is made in Fig. 13. For the low speed flow case, since the maximum change of density in the whole flow field is about $\pm 0.1\%$ and rarefaction effect is not large even in the wake, there is good general agreement between the two pressure contours. As expected, better comparison can be observed for a lower Knudsen number case. Figure 14 shows the comparisons of the pressure contours obtained at a free stream Mach number of 0.01 and a Knudsen number of 0.01. Very good agreement is obtained for the pressure contours.

The drag coefficients are plotted as a function of the Knudsen number in Fig. 15. The comparison between the kinetic prediction and the Navier-Stokes prediction is generally good if the Knudsen number is less than 0.1. Discrepancy more than 10%, however, is found between the two results at the Knudsen number of 1.0. Figure 16 shows the drag coefficients as a function of the Reynolds number. The Reynolds number Re is related to the Knudsen number Kn and Mach number M_∞ as follows:

$$Re = \sqrt{\frac{\gamma\pi}{2}} \frac{M_\infty}{Kn} \tag{38}$$

A low speed (low Mach number) and high Knudsen number flow is a very low Reynolds number flow. For the low Reynolds number flows, the numerical

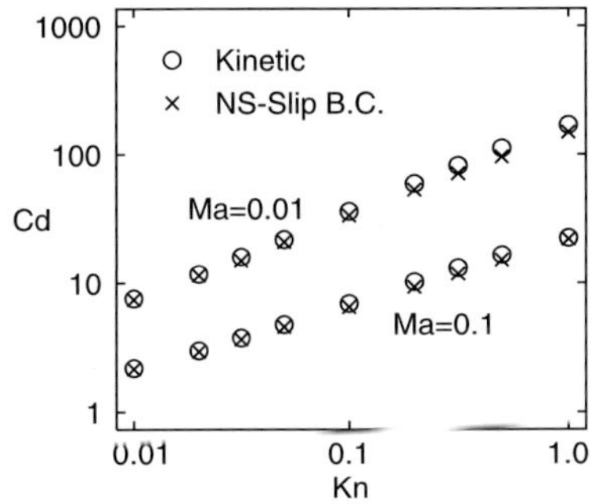

Fig. 15. Drag coefficient as a function of Knudsen number

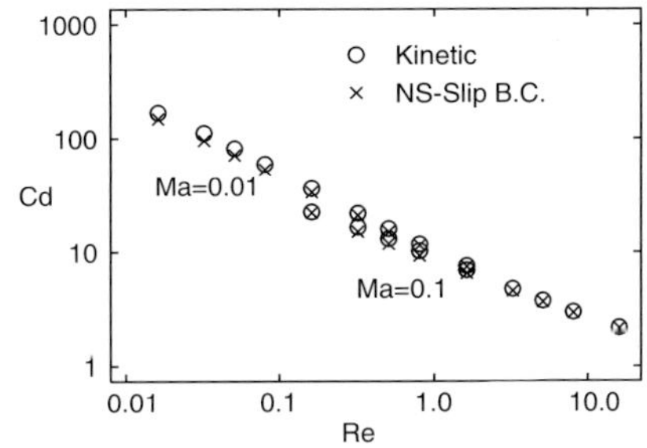

Fig. 16. Drag coefficient as a function of Reynolds number

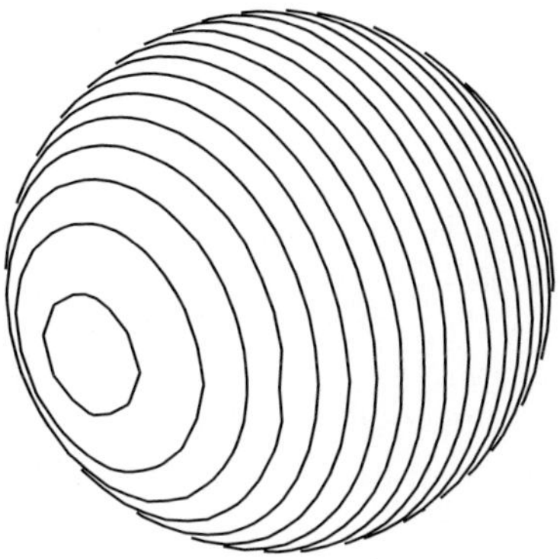

Fig. 17. Pressure contours on a micro sphere

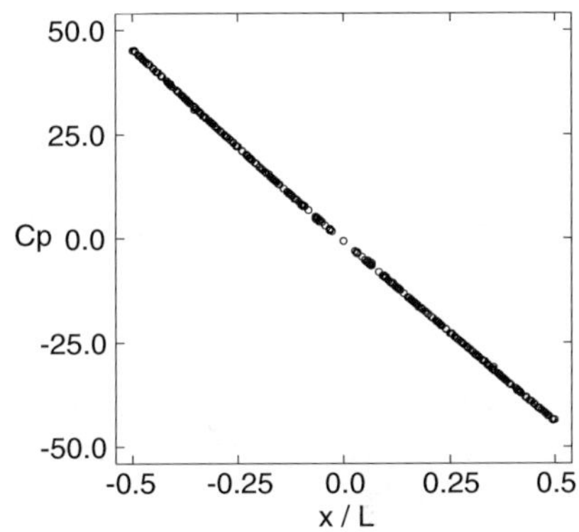

Fig. 18. Pressure distribution as a function x on a micro sphere

predictions of drag coefficient strongly depend on the radius of computational domain. For example, the outer boundary should be located at a distance of more than 200 diameters of the circular cylinder for the case of $M_\infty = 0.01$ and $kn = 0.01$.

5.2 Low Speed Flows over a Micro Sphere

A flow past a micro sphere is one of the most fundamental problems in micro scale flows. A series of numerical results are again obtained at free stream Mach numbers of 0.1 and 0.01 and Knudsen numbers from 0.01 to 1.0. A typical six-dimensional computational grid consists of 60,000 points for the three-dimensional physical space and 1,000 points for the three-dimensional velocity space, in total sixty million points.

Figure 17 shows the pressure contours on the micro sphere surface at a free stream Mach number of 0.01 and a Knudsen number of 0.1. Though the computation is carried out using an unstructured grid for the whole

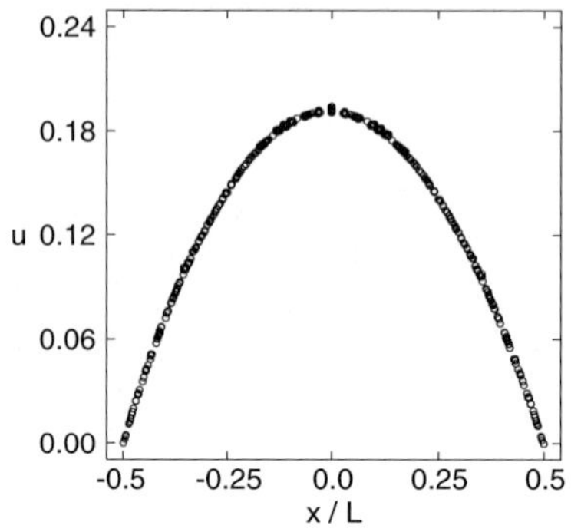

Fig. 19. Normalized x-component Slip velocity distribution as a function x on a micro sphere

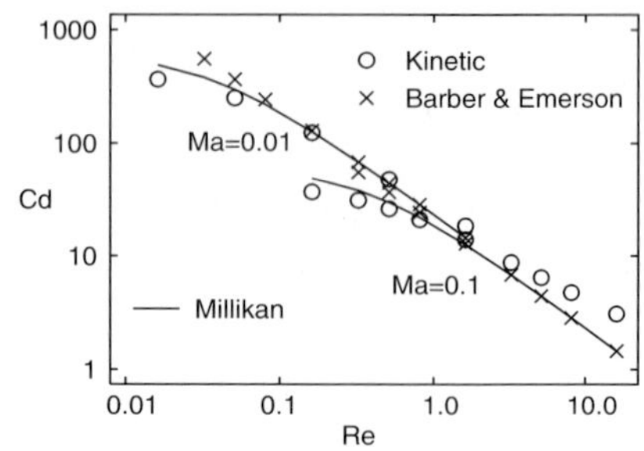

Fig. 20. Drag coefficient as a function of Reynolds number

three-dimensional computational domain without any symmetric condition, axisymmetric pressure distributions expected are obtained. Surface pressure and x-component slip velocity are plotted as a function of x-coordinate in Figs. 18 and 19, respectively. All surface grid point data are plotted in these figures. The pressure monotonously decreases from the front stagnation to rear stagnation. The slip velocity, on the other hand, has a peak at $x = 0$.

The drag coefficients of micro sphere are plotted as a function of the Reynolds number in Fig. 20. Data of Millikan's empirical formula [14]:

$$Cd = \frac{1}{1 + 2Kn\left[A + Bexp(-0.5c/Kn)\right]} \frac{24}{Re} \tag{39}$$

where

$$A = 0.864, B = 0.290, C = 1.25$$

and an analytic solution of Barber and Emerson [15]:

$$Cd = \frac{1 + 4Kn}{1 + 6Kn} \frac{24}{Re} \tag{40}$$

are plotted in the figure for comparison. Present results generally reproduce the Millikan's prediction.

6 Kinetic/Continuum Hybrid Approach

Since the same numerical methods, the gridless method and LU-SGS method, are used for both the Boltzmann equation and the Navier-Stokes equations, construction of the kinetic/continuum hybrid scheme is quite simple. At each computational point either the kinetic approach or the continuum approach is selected. At the interface the Chapman-Enskog distribution functions of the kinetic model equation are passed from the continuum side to the kinetic

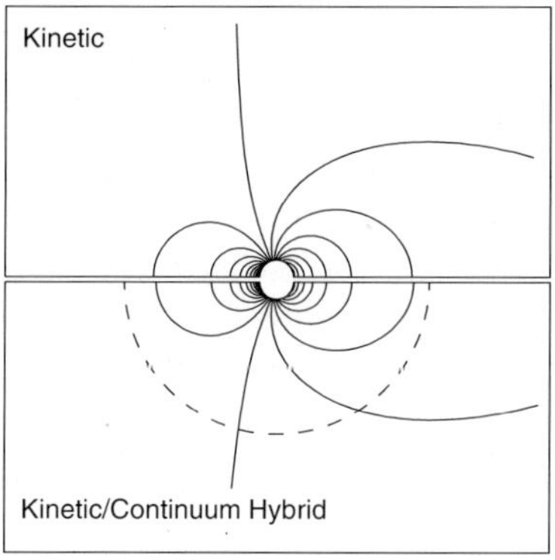

Fig. 21. Pressure contour comparison at $M_\infty = 0.1$ and $kn = 0.1$

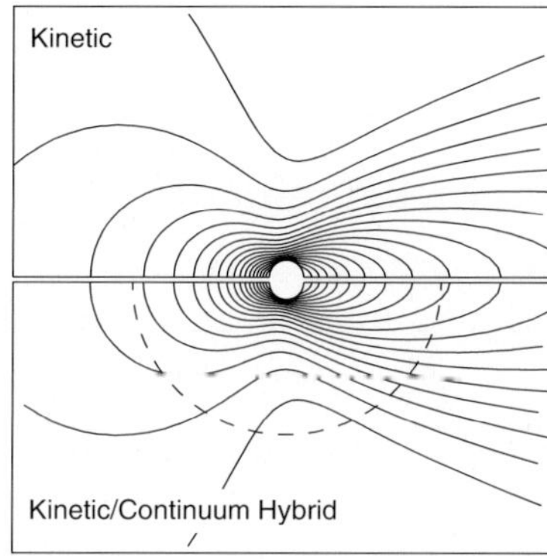

Fig. 22. Comparison of x-component velocity at $M_\infty = 0.1$ and $kn = 0.1$

side, while macroscopic flow quantities are simply passed from the kinetic side to the continuum side. The Chapman-Enskog distribution functions of the kinetic model equation is given by

$$f = f_e \left\{ 1 - Kn \frac{5\sqrt{\pi}}{8T} \left[\frac{1}{Pr} \left(\frac{\boldsymbol{C}^2}{T} - \frac{5}{2} \right) \boldsymbol{C} \cdot \boldsymbol{q} + 2\boldsymbol{C}^0 \boldsymbol{C} : \boldsymbol{\tau} \right] \right\} \tag{41}$$

Comparisons of the density and x-component velocity contours obtained with the kinetic model Boltzmann equation method and kinetic/continuum hybrid method at a free stream Mach number of 0.1 and a Knudsen number of 0.1 are made in Figs. 21 and 22, respectively. In the hybrid method, the solution is computed with the kinetic model Boltzmann equation method on the inner field of broken line circle and with the Navier-Stokes equations on the outer field. The comparisons between the two results are quite good in the whole flow field. Using the hybrid method, the number of time steps for the steady state solution obtained as well as the operational counts a step can be reduced.

7 Conclusions

Computational method based on the kinetic model Boltzmann equation has been developed for micro scale low speed flows. The results obtained with the method are well compared with those of the direct simulation Monte Carlo method and experiments for supersonic flows. Reliable results are also obtained with the method for low speed flows over a circular cylinder and a sphere, while it is difficult to obtain the results with the direct simulation Monte Carlo method. Results of the compressible Navier-Stokes equations with the first order slip boundary are also obtained for comparison. The results generally agree well with those of the kinetic model equation if the Knudsen number is less than 0.1. The kinetic/continuum hybrid method has also been developed. The hybrid method must be the promising tool for analyzing whole flow regimes from free molecule to continuum flows.

Acknowledgements

This study is supported in part by a Grant-in Aid for Scientific Research (15560140) from Japan Society for the Promotion of Science.

References

1. Bird, G.A., Molecular Gas Dynamics, Oxford University Press (1976).
2. Oran, E.S., Oh, C.K., and Cybyk, B.Z., Direct Simulation Monte Carlo: Recent Advances and Applications, Annual Review of Fluid Mechanics, 30 (1998), 403-442.
3. Fan, J., Boyd, I.D., and Cai, C.P., Computation of Rarefied Gas Flows Around a NACA0012 Airfoil, AIAA Journal, 39-4 (2001), 618-625.
4. Morinishi, K. and Oguchi, H., A Computational Method and Its Application to Analyses of Rarefied Gas Flows, Proc. 14th International Symposium on Rarefied Gas Dynamics, University of Tokyo Press, 1 (1984), 149-158.
5. Oguchi, H., Morinishi, K., and Satofuka, N., Time-Dependent Approach to Kinetic Analyses of Two-dimensional Rarefied Gas Flows, Proc. 13th International Symposium on Rarefied Gas Dynamics, Plenum Publishing Corporation, 1 (1985), 293-302.
6. Satofuka, N., Morinishi, K., and Oishi, T., Numerical Solution of the Kinetic Model Equations for Hypersonic Flows, Computational Mechanics, 11 (1993), 452-464.
7. Morinishi, K. and Satofuka, N., Parallel Implementation of the Boltzmann Equation Solvers Using PVM, Parallel Computational Fluid Dynamics, Elsevier, (1995) 339-346.
8. Bhatnager, P.L., Gross, E. P., and Krook, M., A Model Collision Processes in Gases. I. Small Amplitude Processes in Charged and Neutral One-Component System, Physical Review, 94 (1954), 511-525.
9. Morinishi, K., A Gridless Type Solver - Generalized Finite Difference Method -, Computational Fluid Dynamics for the 21st Century, Notes on Numerical Fluid Mechanics, Springer, 78 (2001), 43-58.
10. Morinishi, K., Parallel Computation of Gridless Type Solver for Unsteady Flow Problems, Parallel Computational Fluid Dynamics, Elsevier, (2002) 275-284.
11. Morinishi, K., Parallel Computation of Higher Order Gridless Type Solver, Parallel Computational Fluid Dynamics, Elsevier, (2003) 427-434.
12. Jameson, A. and Yoon, S., Lower-Upper Implicit Schemes with Multiple Grids for the Euler Equations, AIAA Journal, 25 (1987), 929-935.
13. Maslach, G.J. and Schaaf, S.A., Cylinder Drag in the Transition from Continuum to Free-Molecule Flow, Physics of Fluids, 6 (1963), 315-321.
14. Millikan, R.A., The General Law of Fall of a Small Spherical Body Through a Gas, and Its Bearing upon the Nature of Molecular Reflection from Surfaces, Physical Review, 22 (1923), 1-23.
15. Barber, R.W. and Emerson, D.R., Analytical Solution of Low Reynolds Number Slip FLow Past a Sphere, CLRC Technical Report, DL-TR-00-001 (2000).

Part II

Acoustics

A Hybrid FE/Spectral Analysis of Turbofan Aeroacoustics

M. C. Duta, M. B. Giles, and A. Laird

University of Oxford, Computing Laboratory, Oxford OX1 3QD, U.K.
Mike.Giles@comlab.ox.ac.uk

1 Introduction

This paper introduces a novel method for analysing the tonal noise radiated from non-axisymmetric turbofan inlets. This method combines a standard finite element (FE) discretisation of the acoustic field in the axial and radial direction with a Fourier spectral representation in the circumferential direction. Because relatively few Fourier modes need to be retained for an accurate field representation, the method involves many fewer discrete unknowns than a conventional 3D finite element analysis.

The noise source due to the fan blades is characterised by a single circumferential mode number, which in the case of an axisymmetric inlet would produce a radiated acoustic field with the same circumferential mode number. In practice, real engine inlets are not axisymmetric, so the different circumferential modes of the acoustic field are coupled and all are excited by the noise source. However, the level of asymmetry is quite moderate and this coupling is relatively weak. This feature is used to construct a very efficient preconditioner for the iterative solution of the aeroacoustic problem.

2 Aeroacoustic Equations

Given the standard assumption of irrotational flow with uniform entropy, the velocity field is represented as the gradient of a potential function ϕ and the density ρ and speed of sound c are given by

$$(\rho\,/\,\rho_\infty)^{\gamma-1} = \;(c\,/\,c_\infty)^2 = \;1-(\gamma-1)\;(q-q_\infty)\,/\,c_\infty^2,$$

where $q=\frac{1}{2}|\nabla\phi|^2$ and $\rho_\infty, c_\infty, q_\infty$ are the freestream values. Integrating by parts the steady mass equation gives the weak form of the steady equation,

$$\int_V \rho\,\nabla\phi\cdot\nabla w\;\boldsymbol{dV} - \int_{\partial V} \beta\;w\;\boldsymbol{dS} = 0, \qquad (1)$$

for all smooth test functions w which are zero on the far-field boundary where Dirichlet boundary conditions are specified for ϕ. The other two boundaries have the Neumann boundary condition $\partial\phi/\partial n = \beta$, with $\beta = 0$ on the inlet surface, and a prescribed massflow $\beta \neq 0$ on the fan face – see Fig. 1(a).

The propagation of acoustic waves of a single frequency ω is modelled as the real part of a harmonic perturbation $\widehat{\phi}\,\exp(\mathrm{i}\omega t)$ superimposed on the steady mean flow given by (1). The weak form of the acoustic equation comes from the linearisation of the unsteady mass conservation equation,

$$\int_V \rho\,\nabla\widehat{\phi}\cdot\nabla w - \frac{\rho}{c^2}\left(\nabla\phi\cdot\nabla\widehat{\phi}+\mathrm{i}\,\omega\,\widehat{\phi}\right)(\nabla\phi\cdot\nabla w - \mathrm{i}\,\omega\,w)\;\boldsymbol{dV} - \int_{\partial V}\hat{\beta}\;w\;\boldsymbol{dS} = 0, \tag{2}$$

where

$$\hat{\beta} = \rho\,\frac{\partial\widehat{\phi}}{\partial n} - \frac{\rho}{c^2}\left(\nabla\phi\cdot\nabla\widehat{\phi}+\mathrm{i}\,\omega\widehat{\phi}\right)\frac{\partial\phi}{\partial n}.$$

All the boundary conditions of practical interest in engine inlet aeroacoustics are represented by $\hat{\beta}$. At the fan boundary, the field $\widehat{\phi}$ is decomposed into a sum of incident and radiated eigenmodes [1, 7]. $\hat{\beta}$ is in that case a function of the incident mode prescribed to model the presence of the downstream fan as the source of noise. At the far-field, ray theory is used to determine the angle at which the acoustic waves cross the boundary and establish an expression for $\hat{\beta}$ which minimises the reflection of acoustic waves back into the computational domain. Finally, $\hat{\beta}=0$ at the solid inlet surface but if acoustic liners are present then additional modelling yields a modified boundary integral [3].

3 Axisymmetric Analysis

The standard approach for axisymmetric inlets is based on cylindrical coordinates (x, r). Using quadrilateral elements with bi-quadratic shape functions $N_n(\xi,\eta)$, the iso-parametric finite element representation for the coordinates and the steady potential flow field is

$$\begin{pmatrix} x(\xi,\eta) \\ r(\xi,\eta) \\ \phi(\xi,\eta) \end{pmatrix} = \sum_n N_n(\xi,\eta) \begin{pmatrix} x_n \\ r_n \\ \phi_n \end{pmatrix}.$$

Given this representation, a Galerkin discretisation of equation (1) yields a system of discrete nonlinear steady flow equations $\mathcal{R}(\partial)=0$, where ∂ is the vector of unknown potentials at the nodes of the computational mesh. These nonlinear equations can be solved very efficiently using Newton iteration, with the linear system of equations at each Newton step being solved by direct Gaussian elimination.

The acoustic solution has the circumferential variation $\exp(\mathrm{i}\kappa\theta)$, where κ is the circumferential mode number of the specified incoming mode due to the rotating fan. Using the finite element approximation

$$\widehat{\phi}(\xi,\eta,\theta) = \sum_n \exp\left(\mathrm{i}\kappa\theta\right) N_n(\xi,\eta)\,\widehat{\phi}_n,$$

equation (2) yields a linear system of equations $\widehat{\mathbf{L}}\,\widehat{\mathbf{\Phi}} = \widehat{\mathbf{f}}_0$ which can again be efficiently solved by direct means.

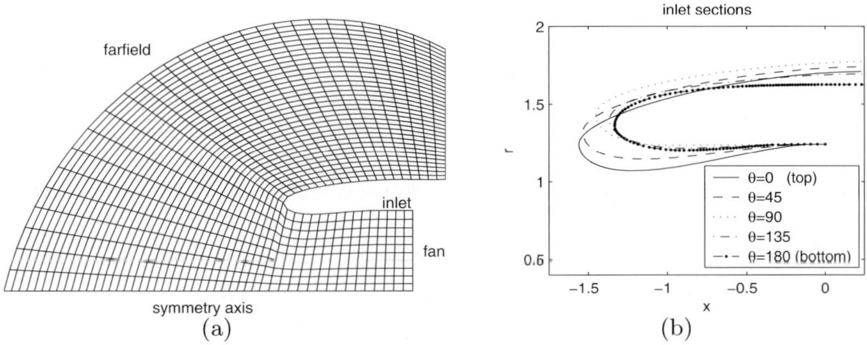

Fig. 1. (a) A coarse quadrilateral mesh around an axisymmetric inlet; a much finer grid is used for acoustic calculations. (b) An asymmetric inlet geometry described by a set of axial sections at different circumferential angles θ.

4 Non-Axisymmetric Analysis

The key to the non-axisymmetric discretisation is a spectral Fourier representation in the circumferential direction [9],

$$\begin{pmatrix} x(\xi,\eta,\theta) \\ r(\xi,\eta,\theta) \\ \phi(\xi,\eta,\theta) \end{pmatrix} = \sum_m \sum_n \exp\left(im\theta\right) N_n(\xi,\eta) \begin{pmatrix} x_{mn} \\ r_{mn} \\ \phi_{mn} \end{pmatrix}.$$

This uses the same shape functions as in the axisymmetric case to interpolate in the (ξ, η) coordinates, but there is an additional summation over the different circumferential modes. The modal coefficients are obtained for the nodes on the inlet geometry by selecting corresponding points on a number of axial sections of the inlet geometry, as shown in Fig. 1(b), for example by choosing the same percentage arc length around the section, and using a Fast Fourier transform (FFT) to perform a Fourier decomposition. The axisymmetric components x_{0n}, r_{0n} of the inlet geometry are used to construct the axisymmetric grid shown in Fig. 1(a) and the non-axisymmetric components $x_{mn}, r_{mn}, m{\neq}0$ are interpolated onto the interior grid nodes.

The unknowns are now the nodal values of each of the circumferential modes (including the axisymmetric mode $m = 0$) of the steady potential. Only a limited number of modes $-M{\leq}m{\leq}M$ need to be retained in the solution; in practice, this number is very small relative to the number of nodes in the circumferential direction which would be required by an accurate standard 3D FE solution. With the above spectral representation, the Galerkin discretisation of the steady equation (1) gives a discrete system of nonlinear equations of the form $\mathcal{R}(\partial) = 0$, with the steady potential values grouped by circumferential mode within the vector $\partial = (\partial_{-M}, \ldots, \partial_0, \ldots, \partial_M)$.

There are now two important questions, how to efficiently evaluate the residual $\mathcal{R}(\partial)$, and how to solve the equations to obtain the steady solution.

The finite element construction of $\mathcal{R}(\partial)$ requires for each element the evaluation of a 3D integral over (ξ, η, θ). The integration over (ξ, η) is performed using Gauss quadrature, while the integration over θ follows a pseudo-spectral approach [9]. This starts with the circumferential modal values $\widehat{\phi}_{mn}$, and performs an FFT to obtain the values and circumferential derivatives at a number of equally spaced points circumferentially. These are used to form residuals at the same set of circumferential points, which are then combined through an inverse FFT to obtain the modal residuals. It is important that enough circumferential points are used to avoid aliasing errors.

The equations are again solved by Newton iteration, but it is no longer efficient to use direct solution methods to solve the resulting linear Newton update equations, since the solution cost is approximately proportional to M^3. Instead, since the Jacobian matrix $\partial\mathcal{R}/\partial\partial$ is symmetric and positive definite, the linear equations are solved iteratively using the Conjugate Gradient (CG) method [2, 5, 8]. The iterative convergence rate is greatly improved through the use of a preconditioner. The preconditioner is based on the observation that if the geometry is axisymmetric, then the Jacobian matrix $\partial\mathcal{R}/\partial\partial$ becomes block-diagonal, with each of the circumferential modes becoming decoupled. It is this block-diagonal matrix, based on the axisymmetric component of the geometry and the steady flow field, and constructed in the standard way using Gauss quadrature, which is therefore used as the preconditioner. Inverting the preconditioner as part of the preconditioned CG iteration requires the solution of a separate 2D system of equations for each circumferential mode, which is done efficiently by direct solution.

The acoustic field is represented by

$$\widehat{\phi}(\xi, \eta, \theta) = \sum_m \sum_n \exp\left(\mathrm{i}(\kappa + m)\theta\right) N_n(\xi, \eta)\, \widehat{\phi}_{mn}.$$

The Galerkin discretisation of equation (2) yields the discrete acoustic equations $\widehat{\mathbf{L}}\,\widehat{\mathbf{\Phi}} = \widehat{\mathbf{f}}$, with $\widehat{\mathbf{\Phi}} = (\widehat{\mathbf{\Phi}}_{-M}, \ldots, \widehat{\mathbf{\Phi}}_0, \ldots, \widehat{\mathbf{\Phi}}_M)$ comprising the complex amplitudes grouped by circumferential mode number, coupled together by the matrix $\widehat{\mathbf{L}}$. The forcing term is $\widehat{\mathbf{f}} = (0, \ldots, \widehat{\mathbf{f}}_0, \ldots, 0)$ and reflects the fact that the forcing duct mode $(m{=}0)$ is prescribed at the fan face which is necessarily axisymmetric. The Quasi-Minimal Residual (QMR) [2, 4] method is used to solve the equations iteratively. Although not guaranteed to converge, it has been found to behave very well on this acoustic problem. As with the solution of the linear Newton equations for the steady problem, the spectral acoustic problem is preconditioned using the block-diagonal matrix given by the axisymmetric components of the geometry and the steady flow field. This effectively requires the direct solution of a separate 2D axisymmetric equation for each of the circumferential modes for each step of the preconditioned QMR iteration. However, the CPU cost can be reduced by performing an LU factorisation of each of the axisymmetric matrices, so that then each QMR iteration requires only a back-solve for each circumferential mode.

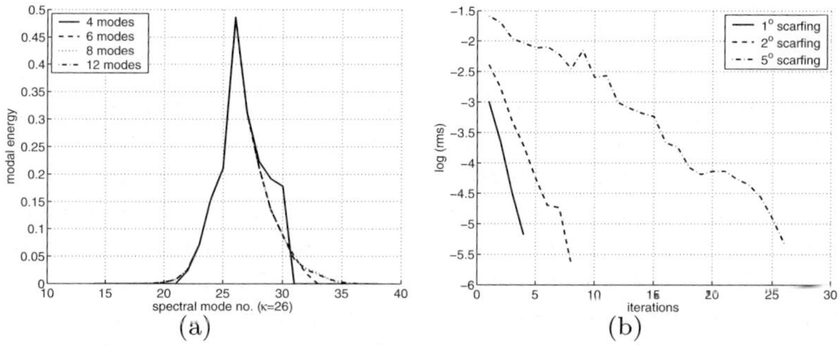

Fig. 2. (a) The variation of the modal "energy" distribution with the number of circumferential modes used to represent the solution. (b) The QMR iteration convergence histories for the inlet with 1°, 2° and 5° scarfing.

5 Example

The numerical results are for the engine inlet shown in Fig. 1, which has a typical scarfing angle of 5° (the angle with which the plane of the inlet front is tilted with respect to the engine axis) and was defined by a series of axial sections equally spaced circumferentially at 22.5° increments.

The Mach number of the steady flow is specified to be 0.3 in freestream and 0.4 at the fan face. The source of acoustic excitation is the rotation of the shocks attached to the fan blades and considering 26 fan blades with a tip Mach number 1.2, the circumferential mode number is $\kappa = 26$ and the reduced frequency of the first blade passing frequency (based on engine radius and freestream speed of sound) is $\omega R/c_\infty = 30$.

Using bi-quadratic elements and a minimum of 8 nodes per wavelength, the axisymmetric mesh has 14,000 nodes. To determine the number of circumferential modes required to accurately represent the acoustic field, a series of calculations were performed with $M = 4$, 6, 8 and 12 modes. The distribution of "energy" (the RMS of the solution vector in each mode) across modes was computed, Fig. 2(a). It was thus found that $M = 8$ is sufficient to represent the circumferential variation. This corresponds to 0.25 million unknowns and the advantage of the spectral discretisation becomes clear as the equivalent acoustic calculation using a standard 3D FE formulation with at least 8 nodes per wavelength in the circumferential direction would require approximately 3 million unknowns.

Finally, keeping the axisymmetric mean of the inlet geometry and scaling the asymmetry modes, configurations with scarfing angles of 1° (almost axisymmetric), 2° and 5° (original geometry) were obtained. Fig. 2(b) depicts the QMR convergence histories for the acoustic problems in these three cases, demonstrating clearly how the convergence rate is fastest when the preconditioner is most effective, *i.e.* when the degree of asymmetry is smallest.

6 Conclusions

This paper has introduced a new method for the analysis of the tonal noise radiated from non-axisymmetric turbofan inlets. It combines a standard FE discretisation of the acoustic field in the axial and radial coordinates with a Fourier spectral representation in the circumferential direction. As a relatively small number of Fourier modes are enough for an accurate field representation, the method is far less costly than the conventional 3D FE approach. The most novel feature of the work is the iterative solution technique using a preconditioner based on an axisymmetric geometry and steady flow field. A numerical example illustrates how few Fourier modes need to be retained, and demonstrates the effectiveness of the preconditioner.

Acknowledgements

This research was supported by EPSRC through the GEODISE e-Science project (*www.geodise.org*). We are very grateful to R. J. Astley, A. J. Keane, L. Lafronza, W. Song and R. Sugimoto of Southampton University and A. J. Kempton of Rolls–Royce plc. for their assistance with various aspects of this work.

References

1. R. J. Astley. Wave envelope and infinite elements for acoustical radiation. *Int. J. Num. Methods Fluids* 3:507-526, 1983.
2. R. Barrett, M. Berry, T. F. Chan *et al.*. *Templates for the Solution of Linear Systems: Building Blocks for Iterative Methods.* SIAM, 1994.
3. W. Eversman. The boundary condition at an impedance wall in a non-uniform duct with potential mean flow. *J. Sound and Vibration*, 246(1):63–69, 2001.
4. R. W. Freund and N. M. Nachtigal. QMR–a quasi-minimal residual method for non-Hermitian linear systems. *Numerische Mathematik*, 60(3):315–339, 1991.
5. G. H. Golub and C. F. van Loan. *Matrix Computations (third edition).* Johns Hopkins, 1996.
6. A. Laird. *A Hybrid Spectral Discretisation and Iterative Solution Methods for Acoustic Models in Potential Flow.* D.Phil Thesis, University of Oxford, 2004.
7. A. V. Parrett and W. Eversman. Wave envelope and finite element approximations for turbofan noise radiation in flight. *AIAA J.*, 24(5):753–760, 1986.
8. L. N. Trefethen and D. Bau III. *Numerical Linear Algebra.* SIAM, 1997.
9. L. N. Trefethen. *Spectral methods in MATLAB.* SIAM, 2000.

Numerical Simulation of the Oscillations in a Mixer - An Internal Aeroacoustic Feedback System

Philip C. E. Jorgenson[1] and Ching Y. Loh[2]

[1] NASA Glenn Research Center at Lewis Field, Cleveland, OH 44135, USA
[2] Taitech Inc., Cleveland, OH 44135, USA

Summary. The space-time conservation element and solution element method is employed to numerically study the acoustic feedback system in a high temperature, high speed wind tunnel mixer. The computation captures the self-sustained feedback loop between reflecting Mach waves and the shear layer. This feedback loop results in violent instabilities that are suspected of causing damage to some tunnel components. The computed frequency is in good agreement with the available experimental data. The physical phenomena are explained based on the numerical results.

1 Introduction

In modern hypersonic test facilities, mixers [1] can be used to allow a wind tunnel to operate at different Mach numbers thus making the test facility more versatile. An example of such a tunnel is the NASA Langley eight foot high temperature tunnel. The original axisymmetric nozzle was designed to provide a Mach 7 flow condition at the test section with a diameter of eight feet. A tunnel modification shortens the nozzle, adds a mixer, and an additional nozzle to provide the Mach 4 flow condition at the second nozzle exit where it enters the test section. The mixer is used to combine the high pressure, and high temperature jet core with the ambient air to generate an appropriate flight environment for wind tunnel testing. The mixer must be carefully designed to avoid any unwanted performance degradation caused by unsteady flow phenomena such as high amplitude oscillation. Such oscillations and consequential waves could potentially damage or destroy the tunnel as already observed in the NASA Langely mixer. Figure 1 illustrates the streamwise cross section of the wind tunnel upstream of the experimental test section where the flow exhibits such unwanted unsteady characteristics. This section of the wind tunnel consists of two converging-diverging (C-D) nozzles serially connected by a large mixing chamber. The inlet (upstream condition for the first C-D nozzle) is supplied by a high temperature and high pressure combustor. In the diverging section of the first C-D nozzle, the flow expands and then separates at the expansion lip as it enters the mixing chamber. Vortices or shocklets are shed and grow in strength along the shear layer. Consequently, strong Mach radiation waves are generated in a feedback

Fig. 1. Computational domain and instantaneous isomach contours, Mach number range 0.0 to 5.5.

cycle in which the reflected Mach waves from the mixing chamber wall propagate upstream around the jet core to the exit of the first C-D nozzle to excite new vortices or shocklets. The phenomenon is complicated and interesting, and to our best knowledge, has not yet been studied numerically.

Numerical investigation of such a problem with mixed aeroacoustic waves and shocks poses stringent requirements that the scheme must be able to: (i) resolve acoustic waves with low dispersion and dissipation, (ii) capture shock waves, and (iii) have an effective non-reflecting boundary condition. The space-time conservation element and solution element method (CE/SE), is a numerical method that meets the above requirements and is adopted in this paper.

Background information on the governing equations and computational grid are provided first followed by the problem description, results, and conclusions in Sections 2 through 5 respectively.

2 Governing Equations, Numerical Scheme, and Grid

The unsteady axisymmetric Navier-Stokes equations are used to model the viscous problem in this study. As explained in Section 1, the nonlinear aeroacoustic feedback loop lies primarily in the streamwise direction. The two-dimensional axisymmetric Navier-Stokes equations are chosen as the governing equations for computational efficiency. Therefore, some possible three-dimensional flow phenomena of secondary importance such as the helical mode, etc. may be excluded. In nondimensionalized differential form the conservation laws are written as:

$$\mathbf{U_t} + \mathbf{F_x} + \mathbf{G_y} = \mathbf{Q}, \tag{1}$$

where x, y, and t are the streamwise and radial coordinates and time, respectively. The conservative flow variable vector is $\mathbf{U}$, and the flux vectors are $\mathbf{F}$ in the streamwise direction and $\mathbf{G}$ in radial direction. The flux vectors are a composite of the inviscid and viscous fluxes. The source term, $\mathbf{Q}$, appearing on the right hand side of the equation, Eq. (1), is a result of the axisymmetric formulation [3].

Considering (x, y, t) as coordinates of a three-dimensional Euclidean space, E_3, the Gauss divergence theorem can be used to give the following equivalent integral form of the conservation laws, Eq. (1):

$$\oint_{S(V)} \mathbf{H_m} \cdot \mathbf{dS} = \int_{\mathbf{V}} \mathbf{Q_m dV}, \qquad \mathbf{m} = \mathbf{1}, \mathbf{2}, \mathbf{3}, \mathbf{4}, \tag{2}$$

where $S(V)$ denotes the surface of the control volume V in E_3 and $\mathbf{H_m} = (\mathbf{F_m}, \mathbf{G_m}, \mathbf{U_m})$ represents the total flux leaving the boundary of space-time region. The integral form of the conservation laws allows the capture of shocks and other flow discontinuities naturally. This integral form of the equations is solved numerically by a recently developed finite volume method, the space-time conservation element and solution element (CE/SE) method. Details of the CE/SE method can be found in [4, 5]

The CE/SE method is constructed to take advantage of unstructured grids for complex geometries. In this case the geometry is relatively simple, so a structured (quadrilateral) element based grid is first generated. The cells are then subdivided into four separate triangular elements and converted to an unstructured format compatible with the flow code. The computation is axisymmetric, so only the upper half of the domain needs to be considered.

For aeroacoustic problems, large numbers of cells are typically needed to resolve the unsteady waves. In this case a subdivided $50X655$ point mesh results in a 128184 triangular cell grid. The grids are made more manageable through the use of parallel computer architecture. The unstructured mesh is decomposed into eight subdomains to take advantage of computers with multiple processors. A message passing library is used for point-to-point communication between processors.

3 Problem Description

The physical domain of interest for the problem is shown in Fig. 1. The inlet conditions for this problem are specified from the exit of the combustor. The high pressure high temperature combustor flow enters the nozzle from the left, chokes in the nozzle throat, expands, and enters the mixer. The flow then passes through the second nozzle and expands into the downstream test section.

We are interested in predicting unsteady flow field upstream of the test section. The diameter, D, at the throat of the upstream nozzle is chosen as the length scale. The density, speed of sound, and temperature, at ambient conditions are taken as scales for the dependent variables.

Initially, the flow in the computation domain is set to ambient flow conditions, *i.e.*, (using nondimensional variables) $\rho_a = 1, \quad p_a = \frac{1}{\gamma}, \quad u_a = 0, \quad v_a = 0$.

At the inlet boundary the conservative flow variables and their spatial derivatives are specified to be those of the combustor exit flow conditions: $\rho_0 = 25.099, \quad p_0 = 124.294, \quad u_0 = 0, \quad v_0 = 0$, with all the derivatives set to zero. The nondimensional flow variables at the exit, downstream of the second C-D nozzle, are specified as ambient.

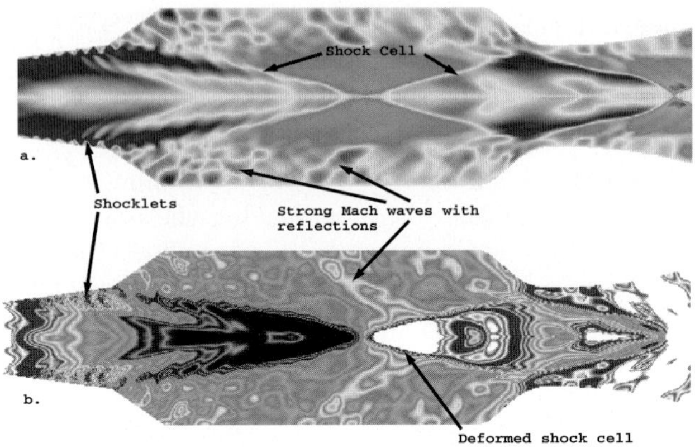

Fig. 2. Snapshot of (a.) radial velocity and (b.) isobar contours showing deformed shock cells, Mach waves, and shocklets/vortices, plotting contours -1.70 to 1.45 and 0.1 to 0.8 respectively.

At the symmetry axis, *i.e.*, $y = 0$, a simple reflective boundary condition is applied. The no-slip boundary condition is applied on the nozzle walls.

4 Results

Figure 1 is a snapshot of the computational domain illustrating the Mach number contours in the domain. The flow enters the first C-D nozzle at a very low Mach number ($M = 0.005$). The flow chokes in the nozzle throat and expands in the diverging section to about $M = 5$ in the core. At the sudden expansion lip, entering the mixing section, the flow separates and shock cells form. The core does not fill the mixer but passes through and enters the second C-D nozzle. The resulting separated annular region surrounding the core remains subsonic allowing waves to propagate upstream. The Mach radiation waves Fig. 2 are formed when the local phase speed of the outer skirt of the unstable shear layer reaches supersonic. This Mach wave strikes the converging section of the second C-D nozzle and the reflected wave propagates upstream. This wave reaches the sudden expansion lip in the mixer section where the receptivity is high triggering another vortex street and the feedback loop is thus completed. The presence of screech in this calculation is indeterminate.

A sequence of snapshots is illustrated in Fig. 3. Here the shock cell deformation is more obvious. Also the effect of the wave propagation in the subsonic region can be observed. The streamwise propagating Mach waves show the interference from the upstream propagating reflected waves. Shocklets/vortices can be seen along the shear layer causing a rippling effect apparent in the shock cell. These vortices or shocklets are shed and grow in

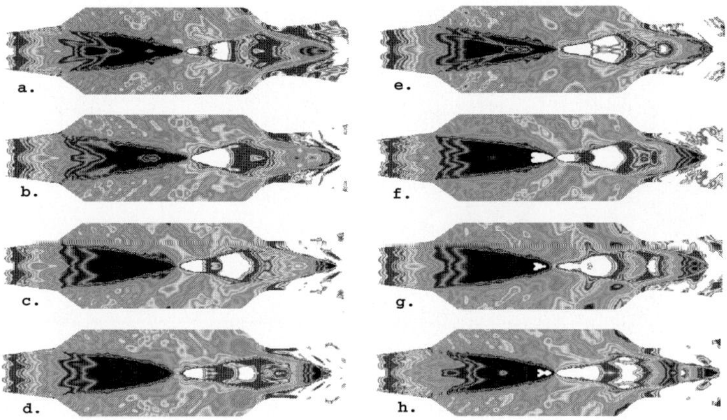

Fig. 3. Sequence of snapshots with isobars contours a-h, pressure contour range 0.1 to 0.8.

strength, propagating downstream to reflect from the converging section of the second C-D nozzle.

A time history of pressure data is recorded, Fig. 4, at several specified points in the computational domain. This data is then post processed by Fast Fourier Transform techniques to obtain spectral information. A large number of times steps are required to achieve appropriate accuracy in the Fourier analysis of the time series data. Small time steps are required initially to keep the code stable. Over three million time steps are needed to remove the initial transients from the computational domain. Another 1.3 million time steps are required to record the time series data for spectral analysis. The computation and spectral analysis are continued until a pervasive fundamental frequency appears. Over seven million time steps are computed for this test case.

Data was recorded at several points in the flow field. The most interesting spectrum was obtained at the point $x/D = 30.0, y/D = 2.5$ which is in the subsonic region of the mixer near the second C-D nozzle. A plot of the analysis is illustrated in Fig. 5. Two very strong peaks at $730Hz$ and $1440Hz$ are indicated. The high frequency is obviously the harmonic of the fundamental frequency due to strong nonlinearity. The reported experimental data [2] for this configuration gave peak values at $800Hz$ and $1400Hz$. The strength of these waves ($\sim 160dB$) is suspected to be strong enough to loosen retaining bolts and destroy the liner of the mixer, as observed in the experiment.

5 Concluding Remarks

In this paper, the unstructured space-time CE/SE Navier-Stokes method is applied to a modified wind tunnel mixer to understand the physics of the acoustic feedback system. The results are analyzed for the spectral content.

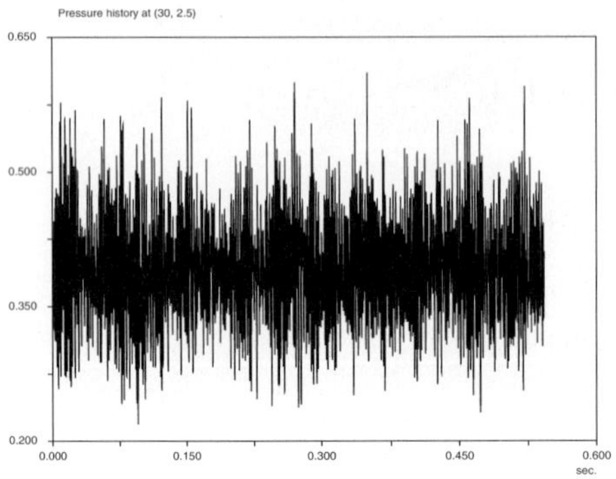

Fig. 4. Time history of pressure at the point $x/D = 30.0, y/D = 2.5$.

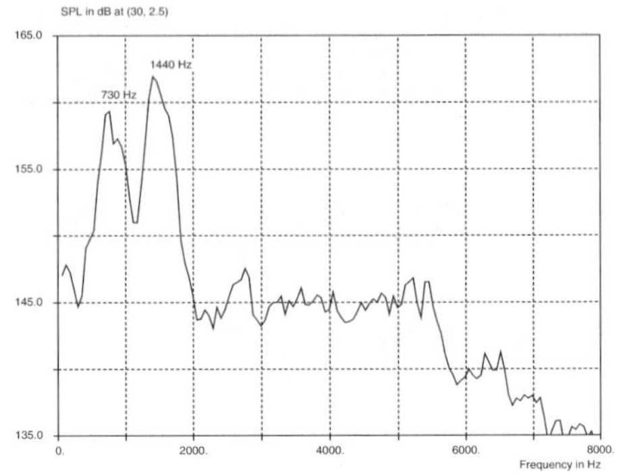

Fig. 5. Spectral analysis at the point $x/D = 30.0, y/D = 2.5$.

The computed frequency corresponds well with the experimental findings [2] at Mach 4. For further work, it may be necessary to model more of the tunnel operating conditions at a higher fidelity and dimension to truly understand the flow physics of the wave structure in the tunnel in conjunction with physical experiments.

A computational tool that can analyze wind tunnels over a wide Mach number range capable of accurately capturing the unsteady wave propagation and interactions could circumvent costly structural damage.

Acknowledgements

This work was supported by the Quiet Aircraft Technology Program at the NASA John H. Glenn Research Center, Lewis Field.

References

1. J.S.Hodge and S.F.Harvin, 'Test Capabilities and Recent Experiences in the NASA Langley 8-Foot High Temperature Tunnel', AIAA Paper 2000-2646.
2. R. L. Puster, Personal communication.
3. C. Y. Loh, L. S. Hultgren, and P. C. E. Jorgenson, "Near Field Screech Noise Computation for an Underexpanded Supersonic Jet by the CE/SE Method," AIAA Paper 2001-2252 (2001).
4. S.-C. Chang, X.-Y. Wang, and C.-Y. Chow, "The Space-Time Conservation Element and Solution Element Method—A New High Resolution and Genuinely Multidimensional Paradigm for Solving Conservation Laws," J. Comp. Phys. vol. 159, pp. 89-136 (1999).
5. X.-Y. Wang, and S.-C. Chang, " A 2-D Non-splitting Unstructured Triangular Mesh Euler Solver Based on the Space-Time Conservation Element and Solution Element Method," C.F.D. J. vol. 8, pp309-325 (1999).

CFD Simulations of Acoustic Wave Phenomena in Combustion Chambers

Venkateswaran Sankaran, Guoping Xia, Matthew Ellis, and Charles Merkle

School of Aeronautics and Astronautics, Chaffee Hall, 500 Allison Road, Purdue University, West Lafayette, IN 47907, venke@purdue.edu

1 Introduction

Computational fluid dynamics (CFD) simulations of acoustic instabilities in rocket chambers are complicated by the presence of significant low Mach number regions in the flow. These may arise for a variety of reasons including the presence of recirculation regions, due to one or more of the propellants being in the liquid phase and/or because of mass-flow limitations in experimental test chambers. Moreover, the high frequency of these instabilities leads to further issues regarding the proper resolution of these waves in the chamber. The present studies are part of an effort towards developing an accurate and efficient CFD capability using exact analytical solutions and companion experimental studies as a guide in developing the underlying algorithms.

Low Mach number computations are typically carried out using a preconditioned time-marching approach, wherein the pressure updates are scaled by the Mach number squared (M^2) to eliminate the stiffness between the fluid and acoustic speeds. For steady-state computations such a formulation is known to insure both accuracy of the discrete formulation and efficiency of the linear solution [1]. The preconditioning methodology can be extended to time-accurate flows by invoking a dual-time strategy. Here, the physical time advances the equations in a time-accurate manner, while the pseudo-time provides the framework for non-linear iterations at each physical time-step. For flows involving acoustic time-scales, asymptotic analysis shows that the proper pressure scaling is the Mach number (M) and not M^2 as in the steady-state case. To account for such scales, the present authors have recently devised a generalized formulation that extends uniformly from steady to unsteady problems for all speeds, Reynolds numbers and time-scales [2]. The procedure involves using an AUSM-like flux formulation with properly defined pressure-velocity coupling terms in the continuity equations that insures accuracy under all limits. Further, proper scaling of the preconditioning matrix is accomplished through a single-parameter generalization that insures uniform convergence efficiency as well.

In this paper, we assess the capabilities of the modified model for the simulation of acoustic wave phenoemna in rocket combustion chambers. Typical chamber Mach numbers of interest range from 0.01 to 0.5 and the acoustic frequencies are in the range of several thousand Hz. To aid in the numerical assessments, we obtain linearized acoustic solutions for simple geometries and verify the numerical predictions against them. Our results show that the numerical method is capable of obtaining accurate and efficient solutions at all speeds and time-scales of interest.

2 Numerical Formulation

The governing equations for time-accurate Euler equations may be written in the following dual-time form:

$$\Gamma_p \frac{\partial Q_p}{\partial \tau} + \frac{\partial Q}{\partial t} + \frac{\partial E_i}{\partial x_i} = 0 \tag{1}$$

where the first time-derivative is introduced for purely numerical reasons, while the seond time-derivative represents the physical time-derivatives. The flux vectors, Q_p, Q and E_i are defined in the usual manner:

$$Q = \begin{pmatrix} \rho \\ \rho u_j \\ e \end{pmatrix} \qquad Q_p = \begin{pmatrix} p \\ u_j \\ T \end{pmatrix} \qquad E_i = \begin{pmatrix} \rho u_i \\ \rho u_i u_j + p\delta_{ij} \\ (e+p)u_i \end{pmatrix}$$

The preconditioning matrix that is chosen to properly condition the time-marching system is given by:

$$\Gamma_p = \begin{pmatrix} \rho_p' & 0 & \rho_T \\ u_j \rho_p' & \rho & u_j \rho_T \\ h_0 \rho_p' - (1 - \rho h_p) & \rho u_i & h_0 \rho_T + \rho h_T \end{pmatrix}$$

where the definition of the scaled property is given as:

$$\frac{1}{\rho_p'} = Min(Max(V^2, V^2 Str^2), c^2)$$

where V is some appropriate local convective velocity scale and Str is the local Strouhal number. For low Mach number acoustic problems, $Str \to 1/M$ and, therefore, $\rho_p' \to 1/c^2$. In other words, for low Mach number acoustics, the preconditioning formulation essentially reverts to the physical time-derivatives. Naturally, such a scaling has consequences on the accuracy of the traditional upwind flux-difference formulation. We therefore consider a modified discrete formulation as proposed in Ref. [2]. The discretized equations may be written in the following form:

$$\Gamma_1 \frac{Q_p^{m+1} - Q_p^m}{\Delta\tau} + \frac{Q^{n+1} - Q^n}{\Delta t} + \frac{1}{2}\sum_{faces} [(E_L + E_R)] - D_{LR}] \cdot \mathbf{n} = 0 \tag{2}$$

with the matrix dissipation term being defined as:

$$D_{LR} = \frac{\Delta x_j}{2} \Gamma_1 |\Lambda| (Q_R - Q_L) \tag{3}$$

where $\Lambda = diag(\sigma[\Gamma_1^{-1} A], u_i, u)$. In other words, the spectral radius of the preconditioned Jacobian is used only in the dissipation terms of the continuity equation, while the standard convective velocity scale is used in the momentum and energy equations. In this manner, the proper low Mach scaling is introduced into the momentum equations, while the continuity equation is scaled for resolving the acoustic disturbances. Further, we point out that the scheme can be extended to higher-order formulations in the usual manner by introducing higher-order derivatives.

3 Analytical Solution

As a validation test case for the resolution of low Mach number acoustic wave phenomena, we consider the asymptotic solution of waves in a chamber that is forced by a harmonic pressure oscillation either at the inlet or exit section of the chamber. The oscillations are introduced by varying the pressure as follows:

$$p(t) = \overline{p}(1 + \epsilon e^{i\omega t}) \tag{4}$$

where ϵ is chosen to be sufficiently small so that the oscillations are linear in character.

For the linearized, constant mean flow problem, analytical solutions may be readily obtained [3]. The analytical solution is used to map the topography of the stability plane and discern the characteristics of the various mode shapes. In addition, these analytical solutions serve as the first validation step for the computational simulations. It is, however, necessary that the analytical solution be formulated for boundary conditions that are identical to those that will be used in the numerical simulations (to ensure that the boundary conditions are validated along with the global formulation) and that it be identical to the numerical solution in other aspects as well. For our formulation, we have specified boundary condition forcing at the upstream end or the downstream end. Boundary condition pairs at the upstream end are either the stagnation temperature and stagnation pressure or the mass flow and stagnation pressure. In both cases, the v-velocity component is set to zero. At the downstream boundary, the static pressure is specified. Boundary conditions at either end can be specified as constant or an arbitrary function of cross-stream distance and time (expressed as a Fourier series) to see the response to forcing. Results given below refer to the downstream forcing instance.

Numerical solutions are obtained with an in-house CFD code that uses the preconditioned dual time scheme described earlier for time-accurate results. The present solutions are formally second-order accurate in time and space.

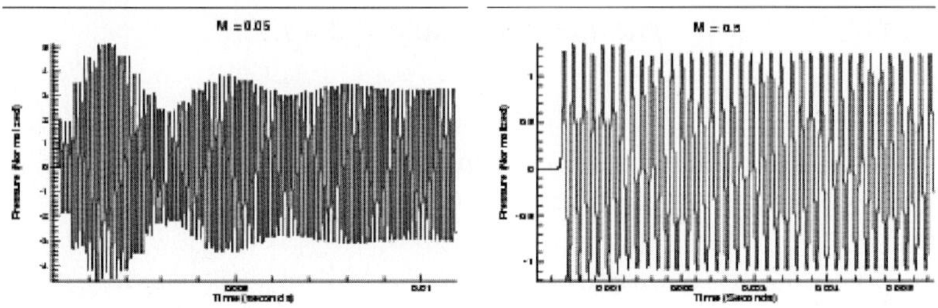

Fig. 1. Pressure response in head-end of chamber for pressure forcing at the downstream end.

4 Results

Representative CFD validation solutions are obtained for acoustic oscillations in a two-dimensional chamber. Figure 1 shows the pressure response at the chamber head-end when pressure forcing is applied at the exit section. Results are shown for two different chamber mean Mach numbers—0.05 and 0.5. The time-history shows that there is a certain amount of relaxation time for the pressure response to achieve a harmonic asymptotic solution. In particular, this relaxation time is significantly longer for the lower Mach number case. This is due to the presence of the slower moving particle waves, which take much more time to traverse the chamber than the faster moving acoustic waves. At low Mach numbers, this time-scale difference becomes more pronounced, thereby leading to longer relaxation times.

A key parameter in unsteady flow solutions is the degree of grid resolution. The representative parameter is the points-per-wavelength (ppw), but since the wavelength in these computations gets consistently shorter as the frequency is increased, we present a grid refinement study in terms of the number of grids along the length of the combustor. The results are shown in Fig. 2 for grids of 20, 40 and 80 points in the longitudinal direction. The 20 grid-point results correspond to approximately 40 points per wavelength at the fundamental frequency of the chamber and to correspondingly fewer ppw as the frequency is increased. The comparison of accuracy with frequency clearly shows these trends. Numerical results are compared with the exact solution for a frequency range of 2500-8000 Hz. For the 20 grid point case, the accuracy of the numerical solution remains good for the first mode, but as the second mode is approached, the errors begin to grow and the predictions become increasingly poorer. In addition, the amplitude of the second mode is considerably under-predicted on this coarse grid. For the 40 and 80 grid-point cases, the numerical results remain very similar to each other and show good agreement with the exact solution to the highest frequencies shown. There is noticeable dispersion error at the highest frequencies on all grids, although the finest grid is the best and the coarsest is worst. In all cases, the temporal resolution was 100 points per period.

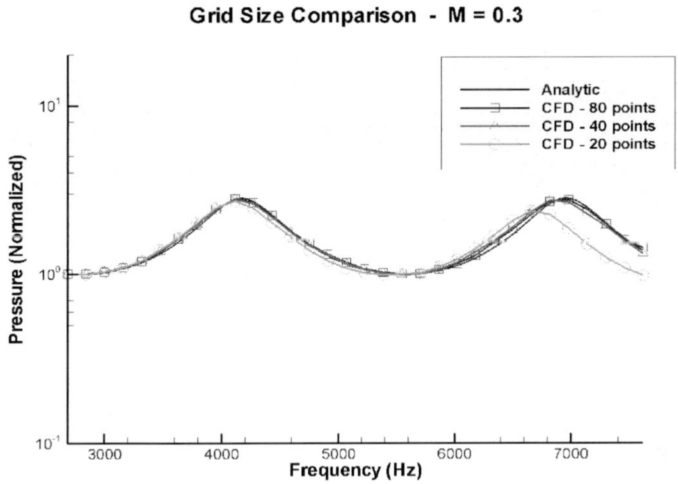

Fig. 2. Comparison of numerical and analytical pressure amplitudes at the head-end as a function of the forcing frequency for different grid resolutions.

The effect of Mach number on both the amplitude and resonant frequency of the disturbance is readily seen from the four plots in Fig. 3 which show the normalized pressure at the head end resulting from forcing at the downstream end for four Mach numbers. The computational results are again obtained from a sequence of numerical solutions corresponding to different discrete frequencies. In terms of the comparison between numerical and analytical results, we note that the numerical comparison for the lowest Mach number case in Fig. 3 exhibits larger errors for the higher frequencies. These errors are a direct result of the slow convergence to a stationary iteration demonstrated above and inspection of these simulations show they have not quite reached stationarity. Efforts to carry out the computations to longer times indeed confirm that very good accuracy is obtained at the low Mach number instances as well.

In addition to demonstrating the accuracy of the numerical results, the results in Fig. 3 also demonstrate the effect of mean flow Mach number on the pressure response. Both the pressure scale and the frequency scale are identical in all four plots so the relative magnitudes and frequencies can be obtained by visually comparing the heights of the frequency response. The pressure amplitude decreases continuously from M = 0.05 to M = 0.5 with a total reduction in amplitude of nearly an order of magnitude. We also note that there is a considerable downward shift in the frequency of both the first and second modes as the Mach number is increased. At a mean flow Mach number of 0.05, the first longitudinal mode is very close to the classical result for acoustic modes in a quiescent medium. As the Mach number is increased to 0.5, this first mode frequency is decreased by nearly 50 %.

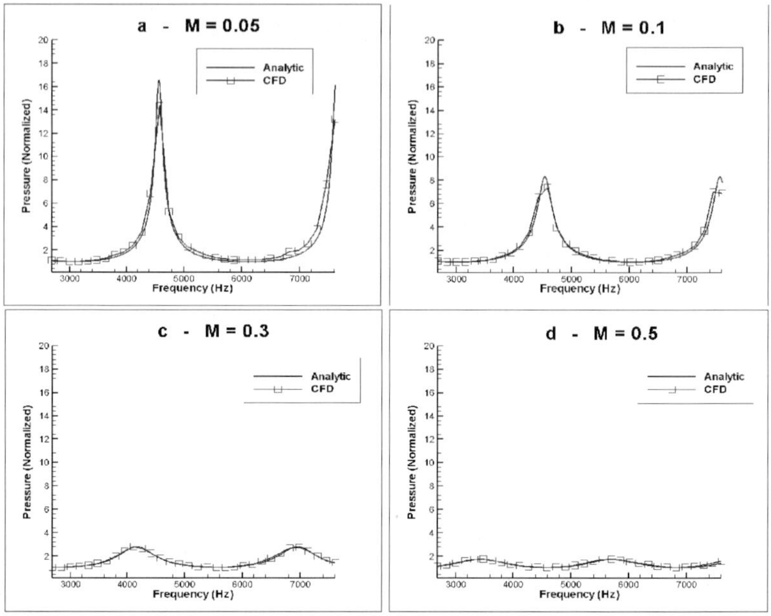

Fig. 3. Comparison of numerical and analytical pressure amplitudes at the head-end for different chamber Mach numbers.

Acknowledgements

Part of this research was funded by a grant from NASA Ames Research Center (Grant No. NCC2-5493).

References

1. V. Sankaran and C. L. Merkle, "Efficiency and Accuracy Issues in Navier-Stokes Computations," AIAA 2000-2251, Fluids 2000 Conference, Denver, CO, June, 2000.
2. V. Sankaran and C. L. Merkle, "Artificial Dissipation Control for Unsteady Computations," AIAA 2003-3695, 16th AIAA Computational Fluid Dynamics Conference, Orlando, FL, June 23-26, 2003.
3. V. Sankaran, J. M. Grenda and C. L. Merkle, "Computational Fluid Dynamic Analysis of Liquid Rocket Combustion Instability," AIAA Paper 91-1609, June, 1991.

Computation on Wake/Stator Interaction in a 2D Cascade

X.Y. Wang[1], A. Himansu[2], S.C. Chang[3], and P. Jorgenson[4]

[1] NASA Glenn Research Center, Cleveland, OH 44135,
xiao-yen.wang@grc.nasa.gov
[2] Taitech Inc., NASA Glenn Research Center, Cleveland, OH 44135,
ananda.himansu@grc.nasa.gov
[3] NASA Glenn Research Center, Cleveland, OH 44135,
sin-chung.chang@grc.nasa.gov
[4] NASA Glenn Research Center, Cleveland, OH 44135,
jorgenson@grc.nasa.gov

Summary. Numerical simulations of rotor-stator interaction using the space-time conservation element and solution element (CE/SE) method are presented. The wake generated by the rotor is specified through the unsteady inlet boundary condition of the stator, over which the time-marching is performed. The Giles approach is incorporated with the CE/SE method such that numerical simulations involving rotor and stator rows with unequal blade counts can be carried out using a single or a few flow passages instead of using the full set of passages. The benchmark problems 3.2 of the 2nd and 4th CAA Workshops are used to validate the Giles approach with the CE/SE method. For Prob. 3.2 of the 2nd CAA Workshop, both numerical results obtained by using a single passage and the full set of passages are presented and compared with the analytical solution, while for the problem of the 4th CAA Workshop, the numerical solution obtained by using two passages is presented.

1 Introduction

The noise generated by rotor-stator interaction is one of the major turbomachinery noise sources. The vortices generated by the rotor blades swept into the next stator row and generated acoustic waves. Numerical simulation of the unsteady wake/stator interaction has been performed by many researchers [1–5]. One major difficulty is related to the unequal blade counts of rotor and stator. Numerical approaches to overcome this difficulty were developed by Erdos[2], Hodson[3], and Giles[4]. In this work, Giles approach is incorporated with the CE/SE method and validated by solving two benchmark problems regarding wake/stator interaction. The CE/SE method is a second-order accurate finite volume method with low numerical dissipation and dispersion errors. Details of the numerical algorithm based on the CE/SE method are referred to [6-7].

In the following, the Giles approach is described briefly, which is followed by the numerical results of test problems.

2 Giles Approach for Non-equal Pitch Treatment

In Giles approach [4], the non-equal pitch is handled by using computational planes that are inclined in time. Its mathematical implementation is described as follows. Consider the 2D nonlinear Euler equations described as follows:

$$\frac{\partial u_m}{\partial t} + \frac{\partial f_m}{\partial x} + \frac{\partial g_m}{\partial y} = 0, \quad m = 1, 2, 3, 4 \tag{1}$$

Here

$$u_1 = \rho, \quad u_2 = \rho u, \quad u_3 = \rho v, \quad u_4 = E_t \tag{2}$$

$$f_1 = \rho u,\ f_2 = \rho u^2 + p,\ f_3 = \rho uv,\ f_4 = (E_t + p)u \tag{3}$$

$$g_1 = \rho v,\ g_2 = \rho uv,\ g_3 = \rho v^2 + p,\ g_4 = (E_t + p)v \tag{4}$$

with $E_t = p/(\gamma - 1) + \rho(u^2 + v^2)/2$. By introducing the coordinate transformation:

$$x' = x, \quad y' = y, \quad t' = t - \lambda y \tag{5}$$

where $\lambda = \frac{P_s - P_r}{V P_s}$ with V being the rotor wheel speed, P_s, P_r being the pitches of stator and rotor. The 2D Euler equations become

$$\frac{\partial (u_m - \lambda g_m)}{\partial t'} + \frac{\partial f_m}{\partial x'} + \frac{\partial g_m}{\partial y'} = 0 \tag{6}$$

Thus the conservation state variables have changed from u_m to $u_m - \lambda g_m$. The numerical scheme for Eq. (6) can be constructed based on the CE/SE method. The time-marching is performed in the x'-y'-t' coordinate. Note that at a constant t', the physical time t is not necessarily the same for different y locations. The periodic boundary condition:

$$u_m(x', y', t') = u_m(x', y' + P_s, t') \tag{7}$$

is imposed at the upper and lower boundaries in the y direction of the computational domain.

3 Numerical Results

3.1 Numerical Results of Problem 3.2 of the 2nd CAA Workshop

The problem 3.2 of the 2nd CAA Workshop concerning vortical waves interaction with a flat-plate 2D cascade is used here to validate the Giles approach with the CE/SE method. Details of this benchmark problem are referred to [8]. Numerical results obtained by using 4 passages and single passage are presented in Figs. 1–4. When 4 passages are used, the original Euler equations are solved, while the Giles approach is used when the single passage is used in the computation. In this problem, $P_s = 1.0, P_r = 0.8, V = 1.0$, then $\lambda = 0.2$. In Fig. 1, the acoustic pressure contour at one time instance is plotted in the 4 passages to show the wave pattern. In Fig. 2, the acoustic pressure on the upper surface of the flat-plate blade at $y = 0$ is plotted for both solutions obtained by using 4 passages and single passage, showing a good agreement except a slight difference near the leading edge. The RMS pressure at the inlet and outlet planes obtained using 4 passages is plotted in Fig. 3 with the analytical solution, while the corresponding solution obtained using single passage is plotted in Fig. 4, showing slightly better agreement with the analytical solution because of less numerical reflections from the open inlet and outlet boundaries.

3.2 Numerical Results of Problem 3.2 of the 4th CAA Workshop

The problem 3.2 of the 4th CAA Workshop describes wake/stator interaction in a 2D cascade. Details of this benchmark problem are referred to [9]. In this problem, $P_r = 18/11$, $P_s = 2/3$, and $V = 27/44$. Two passages of the stator vane are used in the computation, thus $\lambda = -10/27$. The steady-state pressure contour is plotted in Fig. 5. The mean pressure distribution on the blade surface is compared with the Turbo code solution in Fig. 6, showing a good agreement. The sound pressure level at different rotor blade passing frequency (BPF) ($n = 1, 2, 3$) is obtained by performing a Fast Fourier Transform (FFT) analysis based on the time-domain solutions. And a joint FFT analysis in time and spatial y- direction is performed to investigate the sound pressure level as a function of rotor harmonic n, spatial mode order m, and axial location x [5]. As shown in Fig. 7, the acoustic response exists only at the excitation frequencies ($n = 1, 2, 3$). On the blade surface, the acoustic wave at $n = 1$ is dominant, while at the inlet and outlet planes, the sound pressure level at $n = 2$ becomes the largest. In Fig. 8, the distribution of sound pressure level at different spatial modes along the x- direction is plotted for $n = 1, 2, 3$, respectively. It shows that the spatial modes $m = -32$ and 22 at $n = 1$ exponentially decay, and the spatial modes $m = 10$ at $n = 2$, $m = -42$ and 12 at $n = 3$ propagate both upstream and downstream. Some oscillations are observed in Fig. 8(c), which needs to be investigated further.

4 Conclusions

The problems 3.2 of the 2nd and 4th CAA Workshop have been solved using the space-time (CE/SE) method with Giles approach. Numerical results of both problems have been presented. For the first problem, the CE/SE solutions are compared with analytical solution. A fairly good agreement is achieved. For the second problem, no direct comparison between CE/SE and other solutions is made. The CE/SE solutions are similar to the results presented in [5] and agree with the prediction based on the linearized theory [10].

Acknowledgement

This work was supported by Engine System Noise Reduction Project which is part of Quiet Aircraft Technology Program at NASA Glenn Research Center.

References

1. M.M. Rai, "Navier-Stokes Simulations of Rotor-Stator Interaction using Patched and Overlaid Grids,", AIAA Paper 85-1519, 1985.
2. J.I. Erdos, E. Alaner, and W. McNally, "Numerical Simulations of Periodic Transonic Flow through a Fan Stage,", *AIAA Journal*, **Vol. 15**, pp. 1559–1568, Nov. 1977.
3. H.P. Hodson, "An Inviscid Blade-to-Blade Prediction of a Wake-generated Unsteady Flow," ASME Paper 84GT-43, 1984.
4. M. B. Giles, "Generalized Conservation Cells for Finite Volume Calculations," AIAA-87-1118, July, 1987.
5. S. Sawyer, M. Nallasamy, R. Hixon, R.W. Dyson, L.D. Koch, "Computational Aeroacoustic Prediction of Discrete-Frequency Noise Generated by a Rotor-Stator Interaction," AIAA-2003-3268, May 12–14, 2003.
6. S.C. Chang, X.Y. Wang, and C.Y. Chow, "The Space-Time Conservation Element and Solution Element Method – A New High-Resolution and Genuinely Multidimensional Paradigm for Solving Conservation Laws," *J. Comput. Phys.*, **156**, pp. 89–136, (1999).
7. X.Y. Wang and S.C. Chang, "A 2D Non-splitting Unstructured-triangular-mesh Euler Solver based on the Method of Space-Time Conservation Element and Solution Element," Vol. 8, No. 2, pp. 326-340, 1999, *Computational Fluid Dynamics JOURNAL.*
8. X.Y. Wang, C.Y. Chow and S.C. Chang, "Numerical Simulation of Gust Generated Aeroacoustics in a Cascade Using the Space-Time Conservation Element and Solution Element Method," AIAA Paper 98-0178, January 12-15, 1998, Reno, NV.
9. X.Y. Wang, A. Himansu, and P. Jorgenson, "Gust-driven Cascade Acoustics via a CE/SE Numerical Scheme," HT-FED2004-56914, in *Proceedings of 2004 ASME Heat Transfer/Fluids Engineering Summer Conference*, July 11–15, 2004.
10. S.N. Smith, "Discrete Frequency Sound Generation in Axial Flow Turbomachines," Aeronautical Research Council Reports and Memorandum, No. 3709, 1973.

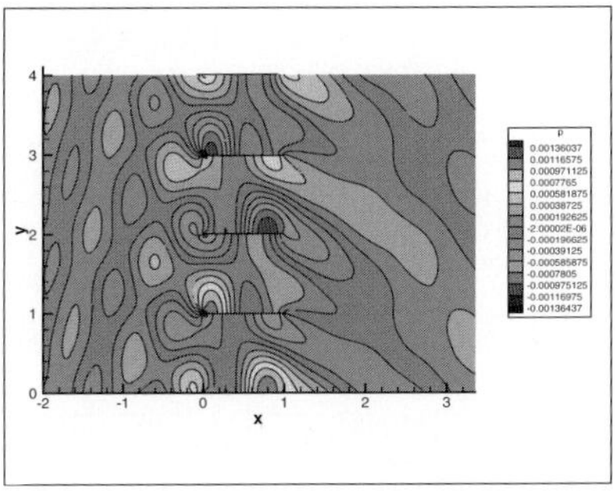

Fig. 1. Acoustic pressure contour at one time instance.

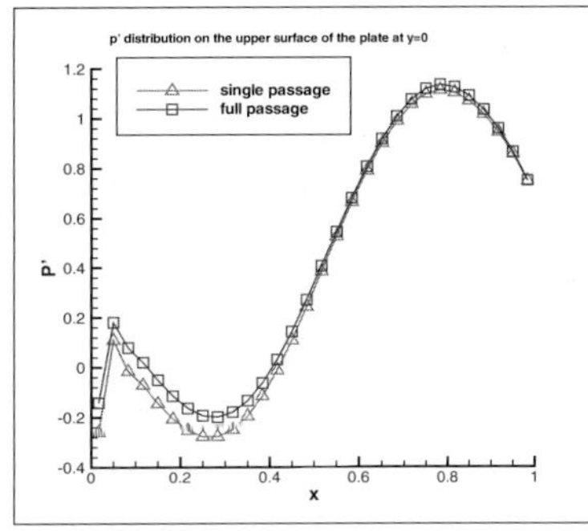

Fig. 2. Acoustic pressure distribution on the upper surface of the flat plate located at $y = 0$.

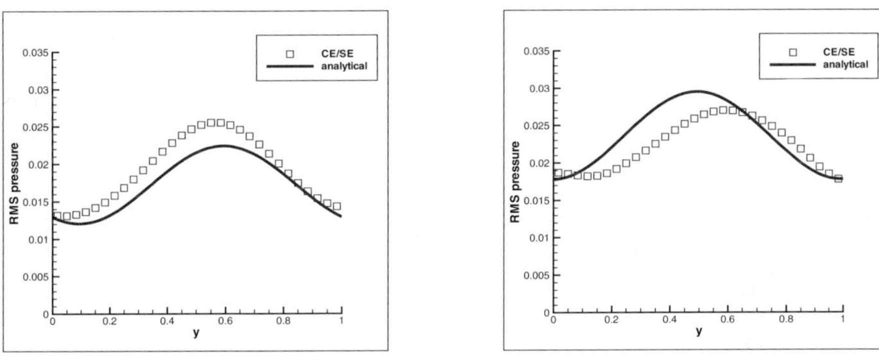

(a) inlet (b) outlet

Fig. 3. RMS pressure at the inlet and outlet planes obtained using 4 passages.

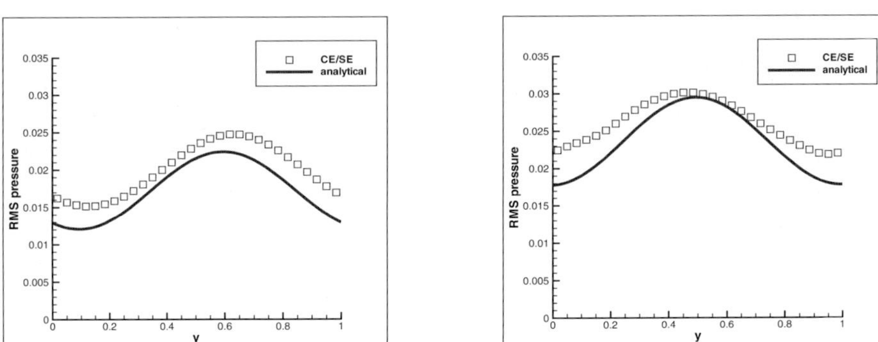

(a) inlet (b) outlet

Fig. 4. RMS pressure at the inlet and outlet planes obtained using single passage.

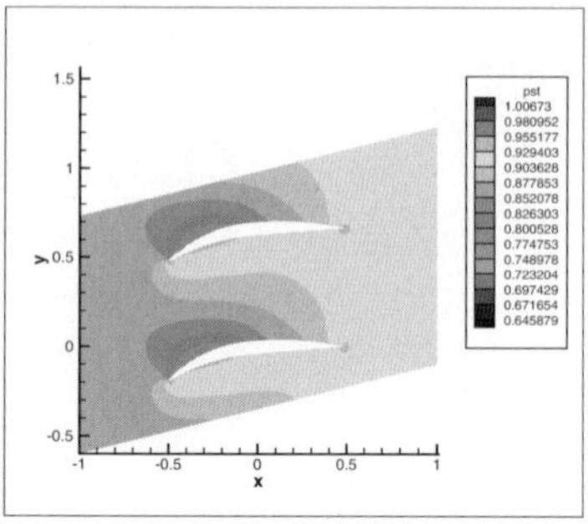

Fig. 5. Numerical solution of the mean pressure contours.

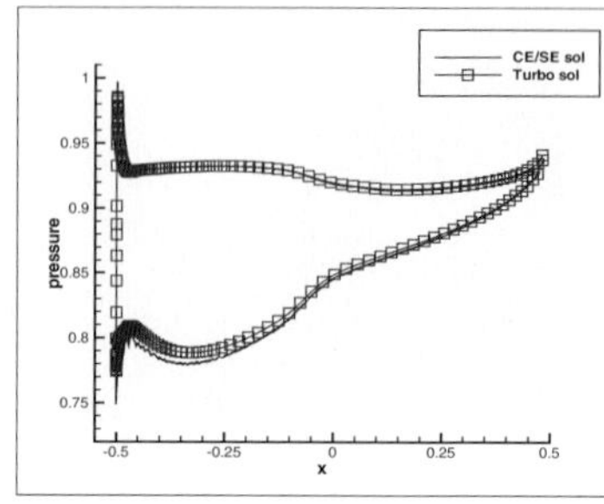

Fig. 6. Mean pressure on the blade surface compared with Turbo solution.

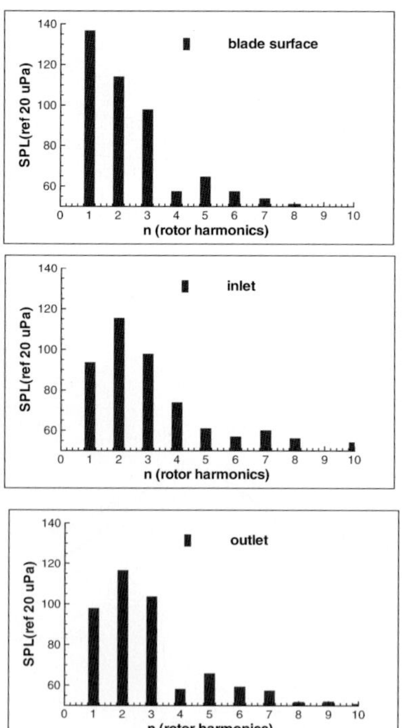

Fig. 7. Sound pressure level distribution on the blade surface at $n = 1, 2, 3$.

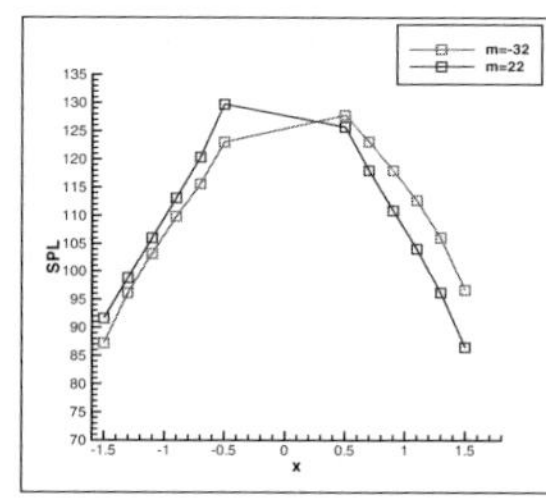

(a) n=1

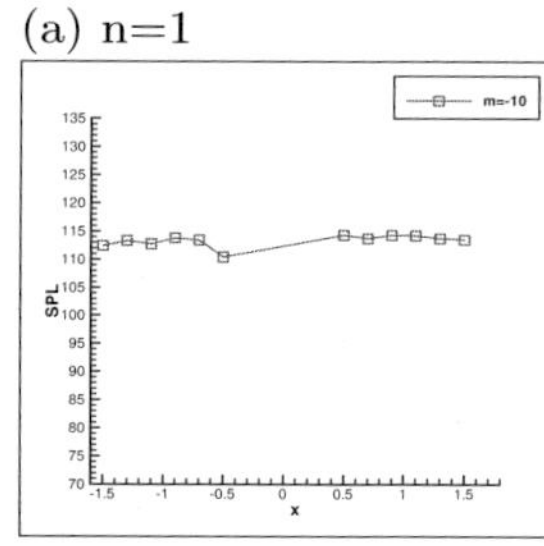

(b) n=2

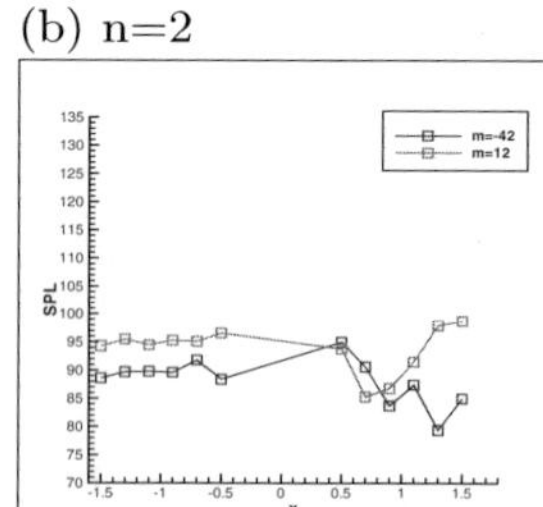

(c) n=3

Fig. 8. Sound pressure level distribution along the cascade for various spatial modes (m) at $n = 1, 2, 3$.

Part III

Adaptive Meshing

An Error Indicator for Semidiscrete Schemes

Daniele Marobin[1] and Gabriella Puppo[2]

[1] Aerospace Department, Politecnico di Torino, Corso Duca degli Abruzzi 24, 10129 Torino, Italy. daniele.marobin@studenti.polito.it
[2] Mathematics Department, Politecnico di Torino, Corso Duca degli Abruzzi 24, 10129 Torino, Italy gabriella.puppo@polito.it

1 Motivation

Much work has been concentrated on the development of Adaptive Mesh Refinement (AMR) codes for gas-dynamics, see for instance [1] for one of the most classic examples and [2] for a more recent result. The idea is to concentrate degrees of freedom where the solution is rich in structure and thus needs a high resolution.

The effectiveness of the AMR strategy relies heavily on the quality of the indicator that drives the process of mesh refinement and coarsening. In this work, an indicator based on the local production of entropy is proposed, extending previous work [5] on one dimensional schemes, based on staggered cells.

In the following, we start with a brief description of the entropy indicator and we conclude with a few numerical examples from gas dynamics, comparing AMR versus fixed grid computations.

2 The Entropy Indicator

We consider the hyperbolic system of conservation laws:

$$u_t + f_x(u) + g_y(u) = 0 \tag{1}$$

First, we write the finite volume formulation for Eq. (1). For simplicity, we start with a uniform grid, composed of square cells of the form $Q_{j,k} = [x_j - h/2, x_j + h/2] \times [y_k - h/2, y_k + h/2]$. Integrating Eq. (1) over the cells $Q_{j,k}$, we find the semidiscrete formulation, which gives the evolution of the cell averages:

$$\begin{aligned} \frac{d}{dt}\overline{u}_{j,k} = -\frac{1}{h^2}\Big\{ & \int_{-h/2}^{h/2} \left[f(u(x_j + h/2, y_k + s, t) - f(u(x_j - h/2, y_k + s, t)\right] \, ds \\ & + \int_{-h/2}^{h/2} \left[g(u(x_j + s, y_k + h/2, t) - g(u(x_j + s, y_k - h/2, t)\right] \, ds \Big\} . \end{aligned} \tag{2}$$

This system of equations is discretized with a second order semidiscrete scheme. Space discretization is obtained reconstructing point values from cell averages through a piecewise linear function. The approximate slopes, in the numerical examples, are computed with the MinMod limiter. The fluxes appearing in Eq. (2) are approximated with the Lax-Friedrichs numerical flux, while the integrals on the edges of each cell are evaluated with the mid-point rule. Time integration is performed with a TVD second order Runge-Kutta scheme. See [3] for further details.

As error indicator to drive the choice of the adaptive grid, the entropy indicator proposed in [5] for central schemes on staggered grids was adapted to the 2D setting.

To construct the entropy indicator, we associate to Eq. (1) the entropy inequality:

$$\eta_t(u) + \psi_x(u) + \phi_y(u) \le 0, \tag{3}$$

where η is a convex function and ψ and ϕ are entropy fluxes defined by the compatibility conditions:

$$\nabla_u^T \eta \; Jf = \nabla_u^T \psi \qquad \nabla_u^T \eta \; Jg = \nabla_u^T \phi,$$

where Jf and Jg are the Jacobians of f and g respectively. For gas dynamics, in the following examples, η was chosen as:

$$\eta = -\rho \log\left(\frac{e}{\rho^{\gamma-1}}\right); \quad \psi = v_x \eta; \quad \phi = v_y \eta,$$

where ρ is the density, e is the internal energy, γ is the ratio of specific heats, and v_x and v_y are the x and y components of the gas velocity.

Next, Eq. (3) is written in finite volume formulation, integrating on the control volumes $V_{j,k}^n = Q_{j,k} \times [t^n, t^{n+1}]$:

$$\begin{aligned} &\frac{1}{h^2} \int_{Q_{j,k}} \left[\eta(u(x,y,t^{n+1})) - \eta(u(x,y,t^n))\right] \, dx\, dy \\ &\quad + \tfrac{1}{h^2} \int_{t^n}^{t^{n+1}} \int_{y_{k-1/2}}^{y_{k+1/2}} \left[\psi(u(x_j + \tfrac{h}{2}, y, t)) - \psi(u(x_j - \tfrac{h}{2}, y, t))\right] \, dy\; dt \\ &\quad + \tfrac{1}{h^2} \int_{t^n}^{t^{n+1}} \int_{x_{j-1/2}}^{x_{j+1/2}} \left[\phi(u(x, y_k + \tfrac{h}{2}, t)) - \psi(u(x, y_k - \tfrac{h}{2}, t))\right] \, dx\; dt \le 0. \end{aligned} \tag{4}$$

It is well known that, if the solution is smooth, the LHS of Eq. (4) is identically zero, while, if u has a shock, the LHS of Eq. (4) is strictly negative, and its sign identifies the unique physically relevant weak solution of Eq. (1).

We define the numerical entropy production in the $V_{j,k}^n$ control volume as the quantity:

$$\begin{aligned} S_{j,k}^n = \tfrac{1}{\Delta t\, h^2} \Big\{ &\textstyle\int_{Q_{j,k}} \left[\eta\left(u^{n+1}(x,y)\right) - \eta\left(u^n(x,y)\right)\right] \, dx\; dy \\ &+ h \textstyle\int_{t^n}^{t^{n+1}} \left[\Psi_{j+1/2,k}(t) - \Psi_{j-1/2,k}(t) + \Phi_{j,k+1/2}(t) - \Phi_{j,k-1/2}(t)\right] \, dt \Big\}. \end{aligned} \tag{5}$$

Here Ψ and Φ are smooth numerical entropy fluxes, consistent with the physical entropy fluxes. In the present case, we pick:

$$\begin{aligned}\Psi_{j+1/2,k}(t) &= \Psi\left(u(x_j+\tfrac{h}{2}^+, y_k, t), u(x_j+\tfrac{h}{2}^-, y_k, t)\right)\\ \Psi(a,b) &= \tfrac{1}{2}\left(\psi(a)+\psi(b)\right),\end{aligned}$$

with similar definitions for the other numerical entropy fluxes appearing in Eq. (5). The functions $u^n(x,y)$ and $u^{n+1}(x,y)$ are the reconstructed data at time t^n and t^{n+1} obtained from the cell averages $\{\overline{u}^n_{j,k}\}$ and $\{\overline{u}^{n+1}_{j,k}\}$ respectively. The space integrals for η in Eq. (5) are approximated with the midpoint rule:

$$\int_{Q_{j,k}} \eta\left(u^n(x,y)\right)\, dx\, dy \simeq h^2\eta\left(u^n(x_j,y_k)\right) \simeq h^2\eta(\overline{u}^n_{j,k}),$$

while the time integrals of the fluxes are computed with the same quadrature rule induced by the Runge-Kutta scheme.

The quantity $S^n_{j,k}$ defined in Eq. (5) is the error/smoothness indicator proposed in this work for two-dimensional systems of conservation laws. In [5], it is proven that the entropy indicator $S^n_{j,k}$ satisfies the following estimates:

$$\begin{aligned} S^n_{j,k} &= O(h)^p \quad \text{on smooth solutions}\\ S^n_{j,k} &\leq \tfrac{1}{h} \qquad\quad \text{on discontinuous solutions}\end{aligned} \tag{6}$$

where p is the order of accuracy of the underlying numerical scheme, in the examples shown in this work therefore $p = 2$. Thus $S^n_{j,k}$ is of the same order of magnitude of the local residual for smooth flows, and it decays as h^p when the grid is fine enough to resolve the smoothness of the solution. On the other hand, $S^n_{j,k}$ is very large on discontinuous solutions.

This behaviour makes $S^n_{j,k}$ an effective a posteriori error/regularity indicator for grid adaptivity.

Remark 1. Note that the evaluation of $S^n_{j,k}$ involves quantities that must be computed to update the numerical solution itself: reconstructed values of u^n and u^{n+1}, intermediate values in time of the solution, computed within the Runge-Kutta procedure. Thus the computation of $S^n_{j,k}$ does not involve a heavy computational overhead.

Remark 2. The construction of the quantity $S^n_{j,k}$ applies, in principle also to higher order schemes, see [5]. Only, the integrals in Eq. (5) must be evaluated with higher order quadrature rules.

3 Numerical Results

We illustrate the technique proposed in this work on a few examples drawn from 2D gas dynamics. The solutions appearing in Figs. 1, 2 and 3 are the result of the evolution of 2D Riemann problems. We consider, respectively, the initial conditions labelled as configuration 4, 5 and 16 in [4]. Configuration 4 contains the interaction of four shock waves, configuration 5 results from the evolution of four contact discontinuities, while configuration 16 is the solution of a two-dimensional Riemann problem containing two contact discontinuities, a rarefaction wave and a shock.

The figures show a contour plot of the density component of the solution, together with the mesh chosen by the adaptive algorithm. It is clear that in all cases the indicator identifies the cells crossed by a discontinuity. Moreover, Fig. 3 shows that the indicator is able to distinguish between regions containing sharp gradients from wave fronts: here in fact refinement occurs on the shock, the two contacts and the corners of the rarefaction wave, while no refinement is needed in the body of the rarefaction wave, where the solution is smooth.

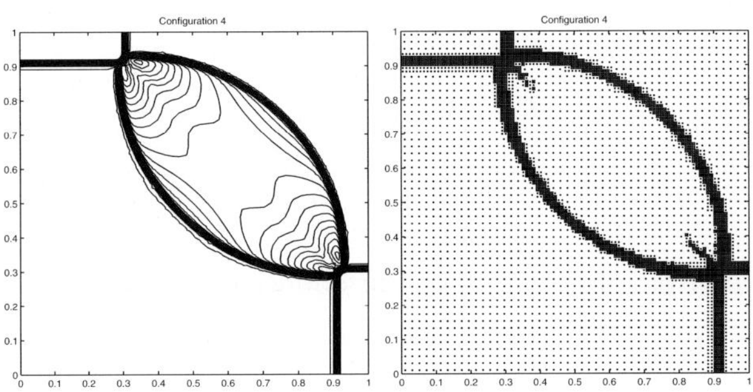

Fig. 1. Configuration 4: interaction of 4 shocks

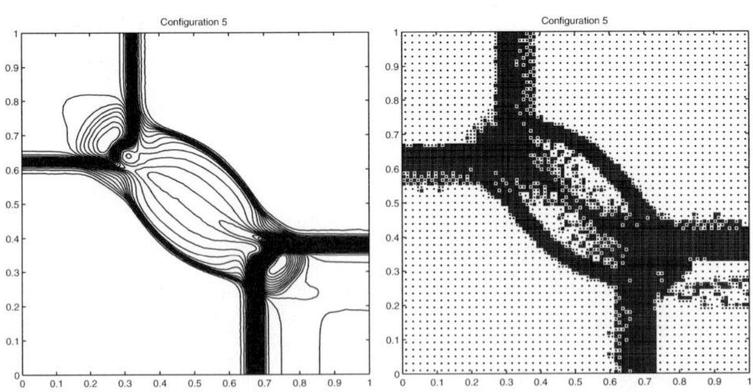

Fig. 2. Configuration 5: interaction of 4 contacts

The computational domain is the square $[0,1] \times [0,1]$. The adaptive grid has three levels of refinement: the coarsest grid contains square cells with side $h_0 = 1/53$, while the finest cells have $h_2 = 1/212$. In the results shown

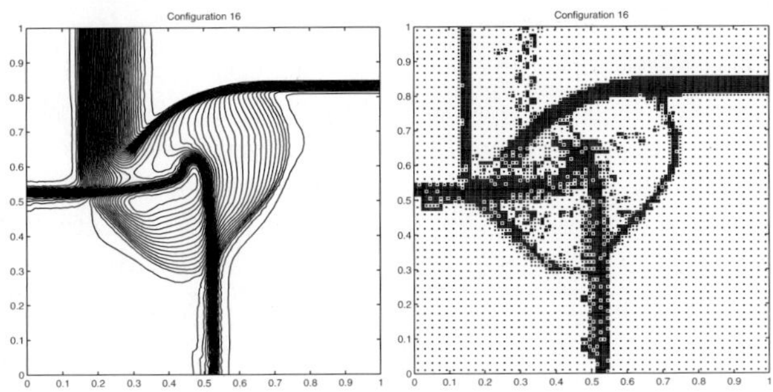

Fig. 3. Configuration 16: interaction of a rarefaction, two contacts and a shock (counterclockwise)

here there is no adaptivity in time, i.e. the time step is the same for all cells. For stability, this restriction forces to use the time step dictated by the CFL condition on the finest grid. In semidiscrete schemes, artificial diffusion does not depend on the local mesh ratio $\lambda = \Delta t/h$, see [3]. Thus we do not expect a deterioration in resolution caused by the fact that the solution on the coarse cells marches at a time step well below the CFL limit.

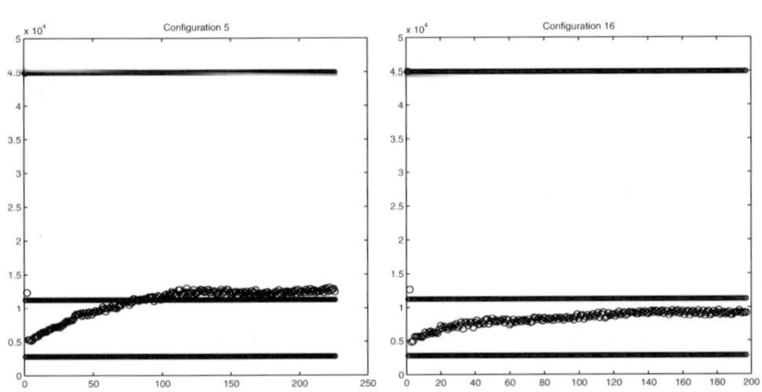

Fig. 4. Number of cells used for Configuration 5 (left) and 16 (right)

Fig. 4 shows the behaviour of the number of cells present at each time step, as a function of time for Configurations 5 and 16, respectively. The horizontal lines represent the number of cells that would be needed if the whole computation were carried out using only cells with size $h_0 = 1/53$, $h_1 = 1/106$ and $h_2 = 1/212$ (from bottom to top).

Finally, Fig. 5 contains as a comparison the results obtained for Configuration 16, using a uniform grid containing, respectively, 53, 106 and 212 points per side. Apparently, the results with the adaptive grid have the same resolution than the numerical solution computed on the finest fixed grid: see expecially the structure formed by the two contacts at the center of the fig-

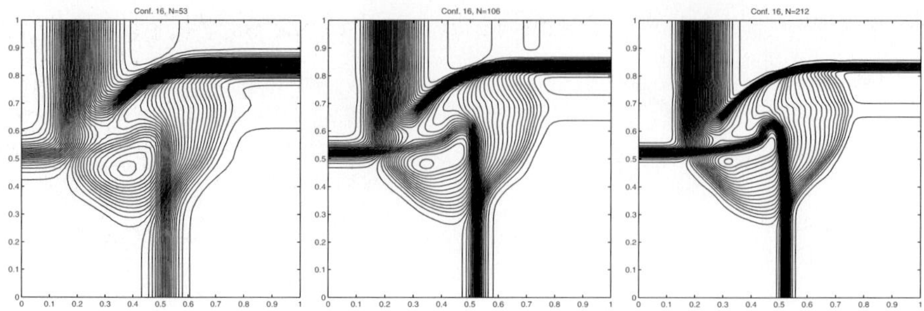

Fig. 5. Configuration 16: fixed mesh computations with $N = 50$, $N = 100$ and $N = 200$ cells per side.

ure. Thus it appears that no important details are lost, even though most of the computation of the adaptive code was carried out with the coarse grid.

We are currently extending this work to consider also adaptive time stepping, in order to have the same CFL ratio in all cells. However the results shown here already illustrate the effectiveness of the entropy error indicator in constructing an adaptive grid which follows the evolution in time of the solution.

References

1. Berger M.J., Colella P., *Local adaptive mesh refinement for shock hydrodynamics*, J. Comp. Phys., **82**, (1989), 67-84.
2. Karni S., Kurganov A., Petrova G., *A Smoothness Indicator for Adaptive Algorithms for Hyperbolic Systems*, JCP, **178**, (2002), 323-341.
3. Kurganov A., Tadmor E., *New High Resolution central Schemes for Non-Linear Conservation Laws and Convection-Diffusion equations*, J. Comp. Phys. **160** (2000), 241-282.
4. Lax P.D., Liu X.D., *Solution of two-dimensional Riemann problems of gas dynamics by positive schemes*, SIAM J. Sci. Comp. **19**, 1998, N° 2, 319-340.
5. Puppo G., *Numerical Entropy Production for Central Schemes*, SIAM J. Sci. Comp. **25**, 2003, N° 4, 1382-1415.
6. Shu C. W., *Essentially Non-Oscillatory and Weighted Essentially Non-Oscillatory Schemes for Hyperbolic Conservation Laws* in Advanced Numerical Approximation of Nonlinear Hyperbolic Equations, Lecture Notes in Mathematics (A. Quarteroni editor), Springer, Berlin, 1998.

A Mesh Adjustment Scheme for Embedded Boundaries

J. S. Sachdev and C. P. T. Groth

University of Toronto Institute for Aerospace Studies
Toronto, ON, Canada, M3H 5T6
j.sachdev@utoronto.ca, groth@utias.utoronto.ca

1 Introduction

In this paper, a scheme is described in which a body-fitted multi-block mesh is locally adjusted to arbitrarily embedded boundaries that are not necessarily aligned with the mesh. Not only does this scheme allow for rapid and robust mesh generation involving complex embedded boundaries, it also enables the solution of unsteady flow problems involving bodies and interfaces moving relative to the flow domain. In addition, the application of block-based adaptive mesh refinement (AMR) can provide a more detailed representation of the embedded boundary as well as accurately resolve multiple flow scales. The proposed mesh adjustment algorithm for stationary or moving embedded boundaries has similarities with the Cartesian cut-cell methods developed by De Zeeuw and Powell [1], Bayyuk *et al.* [2], and Murman *et al.* [3]. Alternative approaches include overset mesh (e.g., [4]), arbitrary Lagrangian-Eulerian (ALE) methods [5], the interface tracking methods of Glimm *et al.* [6], and immersed boundary methods [7]. Though the discussion of the scheme presented here is limited to the solution of the Euler equations governing compressible in two space dimensions, it is fully extendable to the three dimensional case and other systems of equations.

2 Mesh Adjustment Algorithm

A description of the mesh adjustment scheme is presented in this section. The motivation for developing a mesh adjustment scheme for arbitrarily embedded boundaries stemmed from the desire to numerically predict flow involving complex bodies which can be moving relative to the computational domain. Design criteria for the algorithm include the use of block-based adaptive mesh refinement on body-fitted multi-block quadrilateral mesh and that the scheme only produces cells of the same order of magnitude. An example of the mesh adjustment algorithm is given in Figures 1(a)–1(d). Here, the initial configuration shows a curved boundary embedded through the computational domain. The mesh adjustment is comprised of two main stages. The primary adjustment involves merely relocating the mesh nodes that are closest to the intersection point between the spline defining the embedded boundary and the mesh lines. The resulting adjusted grid is shown in Figure 1(b).

A secondary adjustment is required to account for computational cells that are bisected diagonally by the embedded boundary. This is accomplished by choosing the closest not-aligned node to the embedded boundary and relocating it to that boundary point, as shown in Figure 1(c). The solid lines in the adjusted grid denote the "active" computational mesh and the dashed lines correspond to the portion of the mesh that is not needed for the computation and are deemed "in-active." Note that the in-active nodes are retained to maintain the block-based data structure and may be reactivated in computations involving moving boundaries. Mesh adjustment for moving embedded boundaries is accomplished by returning the grid to the original unadjusted form and then readjusting the mesh to the new location.

The preceding mesh adjustment scheme will generate a piece-wise linear representation of the embedded boundary while maintaining the (i, j) data structure of the original body-fitted mesh, as shown by Figure 1(d). The resulting mesh allows for accurate calculation of cell areas and straight-forward application of boundary conditions. It can be seen from the adjusted mesh in Figure 1(c) that some triangular cells can be generated from this scheme. These cells are treated as quadrilateral cells with two coincident nodes. An advantage of this mesh adjustment algorithm is that very small cut-cells are not introduced and cell merging techniques are not required. Since only local alterations are made to the mesh, the need for inter-solution-block communication is not required and is, therefore, transparent to the parallel block-based adaptive mesh refinement scheme.

3 Finite Volume Scheme

The embedded mesh algorithm is combined with a higher-order finite-volume scheme to solve the Euler equations for inviscid flow on the adjusted body-fitted multi-block quadrilateral mesh. An upwind-biased finite-volume scheme with the use of linear limited solution reconstruction and Riemann solver based flux functions are used to determine the inviscid fluxes at cell interfaces. The finite-volume procedure applied to cell (i, j) results in the following semi-discrete form of the equations,

$$\left.\frac{d\mathbf{U}}{d\tau}\right|_{ij} + \left.\frac{d\mathbf{U}}{dt}\right|_{ij} + \left[\frac{\mathbf{U}}{A}\frac{dA}{dt}\right]_{ij} + \frac{1}{A_{ij}}\sum_{k}\left(\boldsymbol{F}_{ij_k} - \boldsymbol{w}_k\mathbf{U}_{ij_k}\right)\cdot\hat{\mathbf{n}}_{ij_k}\,\Delta\ell_k = 0, \quad (1)$$

where A_{ij} is the time-dependent area of cell (i, j), $\Delta\ell$ is the length of the cell face k, and $\hat{\mathbf{n}}$ is the unit vector normal to the cell face k. The conserved solution state vector is given by $\mathbf{U} = [\rho, \rho u, \rho v, E]$ and $\boldsymbol{F}$ is the inviscid flux dyad. To maintain conservation in moving embedded boundary problems, the flux computed at cell interfaces must include the velocity of the interface, $\boldsymbol{w}_k$. In addition, an approximation of the rate-of-change of cell area is required and can be determined from the *Geometric Conservation Law* [8] given in semi-discrete form for cell (i, j) by $\mathrm{d}A/\mathrm{d}t|_{ij} = \sum_k \boldsymbol{w}_k \cdot \hat{\mathbf{n}}_{ij_k}\,\Delta\ell_k$.

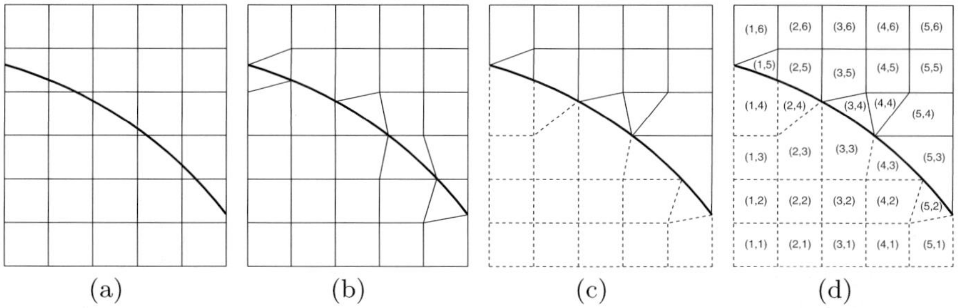

Fig. 1. Mesh adjustment algorithm: (a) Initial mesh and embedded boundary (thick line), (b) result of primary adjustment, (c) result of secondary adjustment (dashed lines indicate inactive cells), and (d) example of (i, j)-indexing on an adjusted mesh.

The inviscid numerical fluxes at the faces of each cell are determined from the solution of a Riemann problem. Given the left and right solution state vectors, $\mathbf{U}_l$ and $\mathbf{U}_r$, and embedded boundary velocity $\boldsymbol{w}$ at the cell interfaces, the numerical flux is given by $(\boldsymbol{F} - \boldsymbol{w}\mathbf{U}) \cdot \hat{\mathbf{n}} = \mathcal{F}(\mathbf{U}_l, \mathbf{U}_r, \boldsymbol{w}, \hat{\mathbf{n}})$ where $\mathcal{F}$ is evaluated by solving a Riemann problem in a direction defined by the normal to the face, $\hat{\mathbf{n}}$ in the reference frame of the cell interface. The left and right solution states are determined using limited piece-wise linear solution reconstruction [9]. Both Roe's approximate flux function [10] and the efficient exact Riemann solver of Gottlieb and Groth [11] can be used to determine the numerical flux at cell interfaces.

A pseudo-time derivative, τ, is included in equation (1) to facilitate solution of steady-state problems or dual time-stepping for unsteady problems. For steady-state calculations, the derivative of the conserved state variables with respect to the physical time, t, is ignored and the steady-state solution is achieved ($\tau \to \infty$) through the use of a parallel multigrid method with an explicit optimally-smoothing multi-stage temporal discretization scheme as the smoother [12]. Both V and W multigrid cycles have been implemented and a full-multigrid cycle can be used for start-up purposes. The coarse grids required for the parallel multigrid method are generated by the standard approach on the unadjusted initial mesh. The mesh on each level is then adjusted to the embedded boundary using the mesh adjustment algorithm. This method of coarse grid generation is straight-forward, however, the standard restriction and prolongation operators near the embedded boundary are no longer valid since the $(2i, 2j)$-indexing on the fine mesh may not correspond to the (i, j)-indexing on the coarser mesh as indicated in Figure 2. The restriction operator used here for cells near the embedded boundary requires the calculation of the area of intersection between the coarse grid and fine grid cells allowing for an accurate area-weighted average, $U_{i,j}A_{i,j} = \sum_m \sum_n U_{2i+m,2j+n}(A_{2i+m,2j+n} \cap A_{i,j})$, where $A_{i,j}$ and $U_{i,j}$ are the area and solution state of the coarse grid cell and $A_{2i+m,2j+n}$ and $U_{2i+m,2j+n}$ are the area and solution state of the contributing fine grid cells. The area of intersection between the fine and coarse grid cells, $A_{2i+m,2j+n} \cap A_{i,j}$, is determined using a polygon clipping algorithm [13]. The search space for

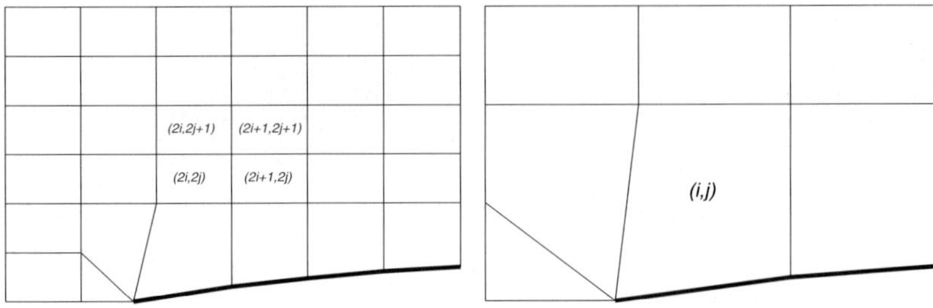

Fig. 2. The fine and coarse grids for a two-stage multigrid method are shown in the left and right panels, respectively. The embedded boundary is indicated by the thick solid line and the corresponding indexing of the fine and coarse grids are indicated.

contributing fine grid cells can be limited to $(m, n) \in [-2, 3]$. The prolongation operator is reduced to injection for cells near the embedded boundary. Bilinear interpolation is used elsewhere.

Unsteady calculations are performed either through the use of the full dual time-stepping approach or through explicit time-marching techniques (the pseudo-time, τ, is ignored). In the dual time-stepping approach, the pseudo-time residual is converged to steady-state using the parallel multigrid method described above. The physical time derivative can be formed using standard explicit or implicit time discretizations. Note that in both approaches, the physical time-step is restricted by the motion of the embedded boundary.

4 Parallel Adaptive Mesh Refinement

The mesh adjustment scheme has been developed for use in combination with block-based AMR [14, 15]. Mesh refinement techniques which automatically adapt the computational grid to the solution of the governing equations can be very effective in treating problems with disparate length scales. In addition, mesh refinement will allow for a more accurate representation of the embedded boundaries. A flexible block-based hierarchical data structure has been developed and is used in conjunction with the finite-volume scheme to facilitate automatic solution-directed block-based mesh adaptation on body-fitted multi-block quadrilateral mesh according to physics-based refinement criteria. This block-based data structure also lends itself naturally to domain decomposition and thereby enables efficient and scalable implementations of the algorithm on distributed-memory multi-processor architectures.

5 Numerical Examples and Validation

5.1 Accuracy Assessment: Ringleb's Flow

Ringleb's flow is a hodograph solution to the Euler equations that is widely used to demonstrate the accuracy of the spatial discretization scheme (e.g.,

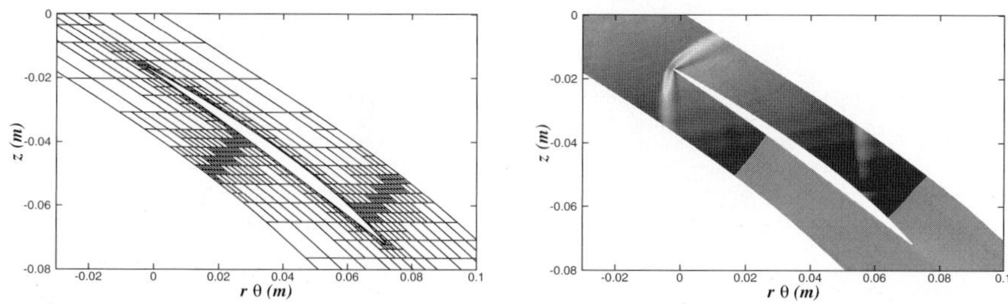

(a) (b)

Fig. 3. NASA rotor 67 blade at 70% span: (a) Final mesh (705 blocks, 360,960 cells) and (b) computed steady-state Mach contours.

[16]). The spatial accuracy of the finite-volume scheme was assessed by comparing the computed solution on a series of four uniformly refined meshes to the analytic solution. The L_1- and L_2-norms of the difference in the solution densities were used as the measure of solution accuracy. The error-norms were computed for a body-fitted mesh and for a Cartesian mesh with an embedded boundary. The slopes of the L_1- and L_2-norms (plot not shown) are 2.11 and 1.94 for the body-fitted mesh, respectively. For the Cartesian mesh, the slopes of the L_1- and L_2-norms are 1.93 and 1.81, respectively. However, considering only the two finest meshes, the L_1- and the and L_2-norms are 2.03 and 1.89, respectively. This indicates that the scheme is indeed second order accurate, even with embedded boundary treatment.

5.2 NASA Rotor 67

To demonstrate the capability of the mesh adjustment scheme on non-Cartesian mesh, the inviscid flow around an embedded NASA rotor 67 blade at 70% span was computed where the underlying mesh is body-fitted to the upper and lower periodic boundaries. A complete description of the geometry and experimental results for this case were reported by Strazisar *et al.* [17]. The steady-state solution was computed using the parallel multigrid method described above. A three stage V-cycle was used with full multigrid start-up. The steady-state adapted mesh and Mach contours are shown in Figures 3(a) and 3(b). Five levels of AMR were performed, followed by full multigrid cycles. It can be seen that the AMR scheme has clustered solution blocks at the shocks on the upper and lower surfaces for the rotor blade. The flow solution agrees reasonably well with the experimental results.

5.3 Translating Ellipse

The motion of an ellipse translating at Mach 1.5 in a channel is used as an example to illustrate the effectiveness of the proposed algorithm for moving boundary flow computations. The pressure contours are shown in Figures 4(a) and 4(b) at two different times during the computation. It is evident from the figures that the AMR scheme has successfully detected the bow

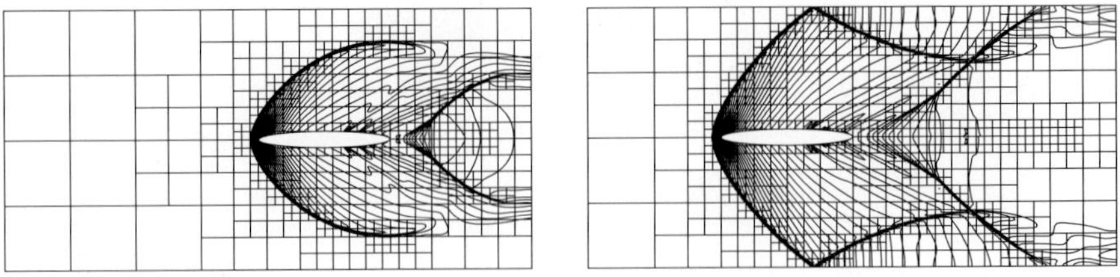

(a) (b)

Fig. 4. An ellipse translating at Mach 1.5 at (a) $t = 0.735$ms (574 blocks, 40,600 cells), and (b) $t = 1.470$ms (838 blocks, 83,800 cells). Pressure contours and solution block boundaries are shown.

shock which forms at the leading edge of the ellipse and the reflected and wake shock structures that form at solid boundaries and behind the ellipse.

6 Concluding Remarks

A mesh adjustment scheme has been described in which a body-fitted multi-block mesh is locally adjusted to embedded boundaries that are not aligned with the mesh. This scheme allows for quick and robust mesh generation involving complex embedded boundaries. The viability of this scheme has been demonstrated for stationary and moving embedded boundaries involving inviscid flow. The application of block-based AMR allows for a more detailed representation of the embedded boundary and accurate resolution of flows having multiple scales. An accuracy assessment of viscous discretization operators on the adjusted mesh is currently under investigation.

References

1. D. De Zeeuw and K. G. Powell. *J. Comput. Phys.*, 104:56–68, 1993.
2. S. A. Bayyuk, K. G. Powell, and B. van Leer. AIAA Paper 1993–3391.
3. S. M. Murman, M. J. Aftosmis, and M. J. Berger. AIAA Paper 2003–1119.
4. W. D. Henshaw. *J. Comput. Phys.*, 113:13–25, 1994.
5. C. Hirt, A. Amsden, and J. Cook. *J. Comput. Phys.*, 14(3):227–253, 1974.
6. J. Glimm, M. J. Graham, J. Grove, X. L. Li, T. M. Smith, D. Tan, F. Tangerman, and Q. Zhang. *Computers Math. Applic.*, 35(7):1–11, 1998.
7. C. S. Peskin. *Acta Numer.*, pages 1–39, 2002.
8. P. D. Thomas and C. K. Lombard. *AIAA J.*, 17(10):1030–1037, 1979.
9. T. J. Barth. AIAA Paper 1993-0668.
10. P. L. Roe. *J. Comput. Phys.*, 43:357–372, 1981.
11. J. J. Gottlieb and C. P. T. Groth. *J. Comput. Phys.*, 78:437–458, 1988.
12. B. van Leer, C. H. Tai, and K. G. Powell. AIAA Paper 1989-1933-CP.
13. I. E. Sutherland and G. W. Hodgman. *Commun. ACM*, 17(1):32–42, 1974.
14. J. S. Sachdev, C. P. T. Groth, and J. J. Gottlieb. AIAA Paper 2003–4106.
15. C. P. T. Groth. ICCFD3 Proceedings, Toronto. Springer-Verlag, 2004.
16. W. J. Coirier and K. G. Powell. *J. Comput. Phys.*, 117:121–131, 1995.
17. A. Strazisar, J. Wood, M. Hathaway, and K. Suder. NASA TP–2879, 1989.

Numerical Simulations of Flows Past 2D Complex Shapes Using Building Cube Method

Siro Kitamura[1], Yoshinori Inoue[2], and Takehisa Iwai[3]

[1] Tokyo Medical and Dental University, Graduate school, Vascular Surgery `kitamura@icfd.co.jp`
[2] Tokyo Medical and Dental University, Vascular Surgery `yoshi.inoue.srg1@tmd.ac.jp`
[3] Tokyo Medical and Dental University, Vascular Surgery `iwai.srg1@tmd.ac.jp`

1 Introduction

Many adaptive mesh generation systems have been developed so far. In generally, exactly body-fitted meshes can be generated using a curvilinear coordinate system. But it is difficult to generate automatically and quickly. It needs many experience to make a sufficient mesh. It can be also generated using unstructured mesh system (like tetra mesh). But higher order schemes may not be applied.

Building Cube Method (BCM), which was proposed by Nakahashi [1], [2] in recent years, is one of adaptive mesh generation methods. This has the advantages of quick and easy mesh generation around complex objects. In addition, no ability to make meshs is needed.

In this method, a calculation region is separated into many small regions of which size are difference, but similar each other. They are called as "cube". Number of grid in a cube is same (for example, 33x33, 65x65 in 2D). It is easy to apply higher order schemes because it is an orthogonal uniform mesh in each cube although a whole mesh is unstructured. And this is a simple domain decomposition method so that it is suitable to the parallel computation mechanism [3].

On the other hand, Kuwahara[4] has been developed the scheme of which the discretization method of the convective term is the third-order upwind scheme on orthogonal uniform mesh. With this scheme, we don't need so many grid points and iteration time in temporal direction to get the reasonable solutions[4]. And it is very easy to reconstruct the shape of objects in uniform mesh. Kuwahara has calculated the flow around a protoceratopus that is one of complex objects[4]. However, of course, it is difficult to distribute grid points around objects or considerable regions effecitively because of the uniformity of mesh. For example, it cannot be resolve the distribution of velocity or pressure in a boundary layer.

Kuwahara has simulated many flow fields by the same method on curviliner coordinate system[5],[6]. But, the shapes of calculated objects, for example airfoils of NACA series or 2D cylinder, are no complex. These shapes are made functionally.

In this study, the governing equations are Naver-Stokes equation system. The third-order upwind scheme is employed as the discretization method of the convective term. The flow field around objects are solved on the mesh generated by Building-Cube method. It will improve the fault of Kuwahara's scheme on orthogonal uniform mesh that is no efficient disposition of gird points.

In this presentation, some calculation results of flow fields around 2D complex objects are shown. It can be found that the effectivity and advantages of Kuwahara's scheme for complex shapes with Building-Cube method.

2 Mesh Generation Method

The details of BCM are mentioned in Nakahashi[2]. Figure 1 shows the simple example of a mesh generated by BCM. A calculation region is filled with veriously scaled cubes which are similar each other. The scale factor is restricted in 2^n or $(1/2)^n$. And that with neighbor cubes must be 1.0, 2.0 or 1/2. Generally, cubes near objects or considerable regions are divided, becomes small, and mesh in the cubes is fine, while these far from objects are large, and mesh is coarse.

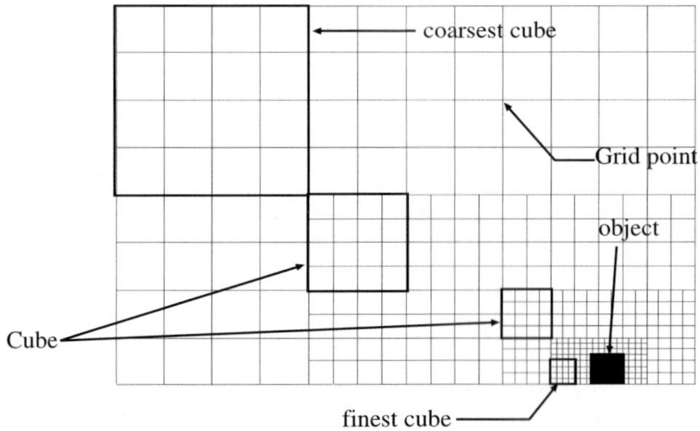

Fig. 1. Example of a mesh generated by BCM.

Automatic mesh generation can be made easily. Fig.2 shows the dividing process of cubes. A cube is divided into the four small cubs in 2D. These grid spacing are half of that in the previous cube. Figure 3 shows the model of a cube in 2D, which has the buffer and sweeping region that is surrounded by buffer region. A flow is simulated in the sweeping region. Boundary conditions for a cube are imposed in the buffer region. Values on a buffer region are substituted after each iteration in time from neighbor cubes.

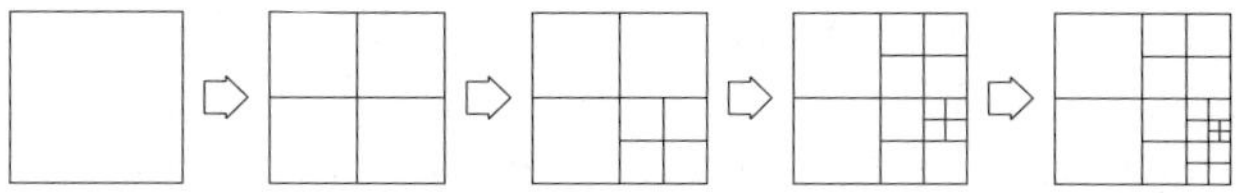

Fig. 2. Dividing process of cubes.

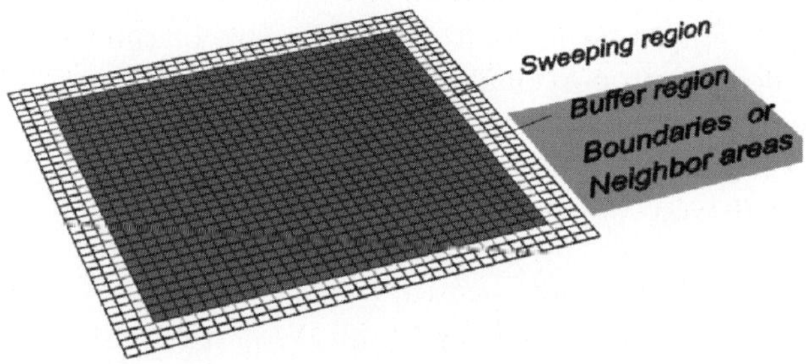

Fig. 3. Structure of 2D cubes which includes the buffer and sweeping region. The flow is solved in the sweeping region.

3 Governing Equations

In this presentation, incompressible 2D Navier-Stokes equations are solved as the governing equations, which are;

$$\mathrm{div}\mathbf{u} = 0 \tag{1}$$

$$\frac{\partial \mathbf{u}}{\partial t} = -\mathbf{u} \cdot \mathrm{grad}\mathbf{u} - \mathrm{grad}p + \frac{1}{\mathrm{Re}}\Delta\mathbf{u} \tag{2}$$

$$\Delta p = -\mathrm{div}(\mathbf{u} \cdot \mathrm{grad}\mathbf{u}) + \frac{D^n}{\delta t} \tag{3}$$

where $\mathbf{u}$ is velocity vector, p is pressure, D^n is divergence of $\mathbf{u}$ at time n. δt is time step. Re is Reynolds number. In these equations, Eq.(1) is continuity equation, Eq.(2) Navier-Stokes equation, Eq.(3) pressure Poisson equation.

4 Calculation method

The convective term is discretized by the third-order upwind scheme, diffusive term by the second-order central scheme which are;

$$\begin{aligned} u\frac{\partial u}{\partial x} =& u_i(-u_{i+2} + 8(u_{i+1} - u_{i-1}) + u_{i-2})/(12\delta x) \\ &+ |u_i|(u_{i+2} - 4u_{i+1} + 6u_i - 4u_{i-1} + u_{i-2})/(4\delta x) \end{aligned} \tag{4}$$

$$\frac{\partial^2 u}{\partial x^2} = (u_{i+1} - 2u_i + u_{i-1})/(\delta x^2) \quad (5)$$

where u_i is velocity in x direction on the ith grid point. Those in y direction are treated in the same method. The detailed property of the third-order upwind scheme for convective term in Navier-Stokes equation has been discussed in Kuwahara[4]. Velocity and pressure field are coupled by smac method.

The following method is applied to the time integration. The explicit Euler scheme is employed in each cube. The size of time steps on cubs are difference, which are proportional to the size of the grid spacing on respective cubes. In the following equations, l denotes the dividing level of cubes. A cube of l is divided into four cubes of $l+1$ in 2D. The time step is expressed as follows;

$$\delta t_l = 2^n \delta t_m, \qquad n = m - l \quad m = \cdots, l-1, l, l+1, \cdots$$

where δt_l is the time step on a cube of which the dividing level is l, δt_m is a time step on any cube. Fig.4 shows the time-integration method. In this figure, it is shown that the relations among three different cubes that are divided successively (from right to left). Two time integration processes on the smallest cube correspond to one process in the middle cube. And two processes in the middle correspond to unity on the biggest cube. At the time 1 in the figure, the evaluations of the terms in spatial direction and boundary conditions are done on only the smallest cube. The delivery of boundary conditions among only the adjoining smallest cubes (only unity dividing level) are carried out. The delivery between the different levels doesn't made. At the time 2, that is calculated on both the smallest and middle. And, the delivery among adjoining cubes, whcih are smallest and middle, are made. All cubes are evaluated at the time 3. On this time, the values on boundaries of all adjoining cubes are delivered. It is can be employed that the cubes on which the time integration is carried out by the following method. On a cube

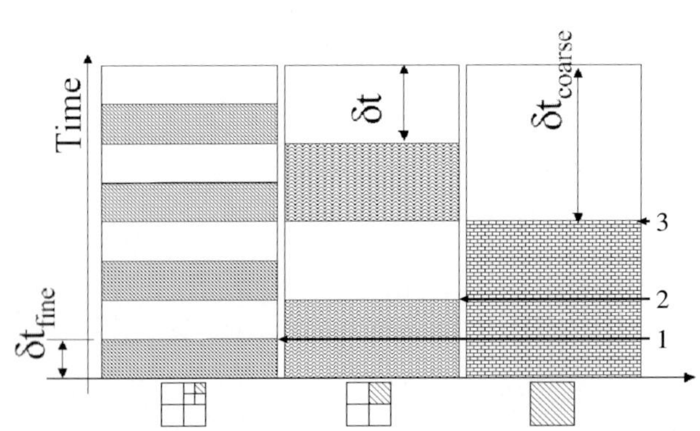

Fig. 4. Time integration. It shows the relations among three different cubes which are represented in the bottom of figure.

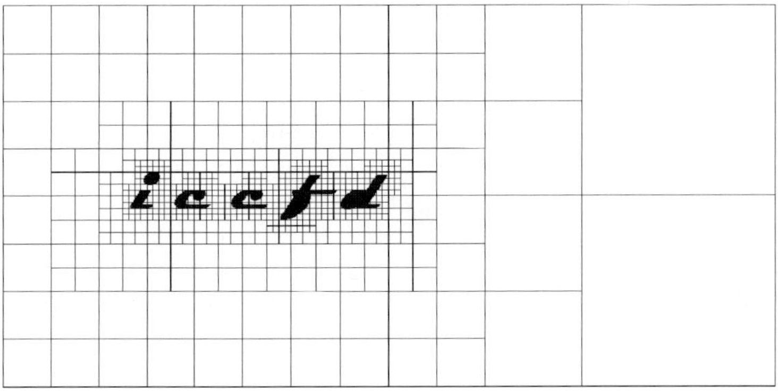

Fig. 5. (a) Distribution of cubes around "iccfd" that includes 626 cubes.

of which dividing level is l, the following equation represents a point in time, that is expressed by δt of any cube, on which spatial terms are evaluated;

$$2^{m-l} \cdot p \cdot \delta t_m \qquad m = \cdots, l-1, l, l+1, \cdots \tag{6}$$

where p is positive real number that divides by 0.5. The time integration is made on cubes where the following condition is valid.

$$p \cdot 2^{m-l} - \lfloor p \cdot 2^{m-l} \rfloor = 0 \tag{7}$$

where $\lfloor x \rfloor$ is the floor function, which returns the maximum integer that is less than or equal to x.

5 Calculation Results

Figure 5 shows the distributions of cubes around "iccfd" that may be one of complex objects. Number of grid points in a cube may be determined after dividing the calculation region into cubes. 17×17 grid points are included in each sweeping region of cube. Fig.6 shows the pressure contours around iccfd.

6 Conclusions

In this paper, it is proposed that the scheme which employs the third-order upwind scheme as the distretization method of the convective term and Buliding-Cube method as the adaptive mesh generation system. And the time integration method on BCM is also suggested.

Fig. 6. Pressure distribution around iccfd. Re=10000, 626 cubes and 10349 grid points exist in the calculation region.

References

1. Nakahashi, K, Building-Cube Method for Flow Problems with Broadband Characteristic Length, ICCFD2, 2002.
2. Nakahashi, K and Kim, L, Building-Cube Method for Large-scale, High Resolution Flow Computations, AIAA paper, 2004-0434, 2004.
3. Nakahashi, K, Kim, L.S., Fujita, T, Building-Cube Method for Large Scale Flow Computations on Parallel Computers, Parallel CFD, 2002.
4. Kuwahara, K., Unsteady Flow Simulation and Its Visualization, AIAA paper, 99-3405, 1999.
5. Kuwahara, K and Komurasaki, S, Semi-Direct Simulation of a Flow aroud a Subsonic Airfoil, AIAA paper, 2000-2656, 2000.
6. Kuwahara, K and Komurasaki, S, Direct Simulation of a Flow around a Subsonic Airfoil, AIAA paper, 2001-2545, 2001.

High-Density Mesh Flow Computations by Building-Cube Method

Kazuhiro Nakahashi[1] and LaeSung Kim[2]

[1] Department of Aerospace Engineering, Tohoku University, Sendai 980-8579, JAPAN; naka@ad.mech.tohoku.ac.jp
[2] Department of Aerospace Engineering, Tohoku University, Sendai 980-8579, JAPAN; lskim12@ad.mech.tohoku.ac.jp

1 Introduction

CFD has become an indispensable tool for designing an airplane. However, wind tunnel experiments are still required to confirm the computed results. For flows at off-design conditions, the experiment is the central player and the CFD plays a subordinate part. This situation is mainly because of the lack of the reliability in the current CFD due to the dependencies of computed results on the grid and on the turbulence models. To remove the grid dependency, the most straightforward way is to use a highly dense and regular grid. To remove the dependency on physical models, we have to move from the Reynolds-averaged Navier-Stokes equations to LES and DNS. This also means to use a high-density mesh and a spatially higher-order numerical scheme.

Meanwhile, the advancement of the computer performance is expected to keep the current pace for a while. The allowable mesh size will become an order of one giga points (1×10^9) soon, and will be one tera points in near future. To meet the near-future expectation of the advanced computers, however, a simple extension of the conventional numerical scheme will be stuck soon. CFD using structured meshes has a difficulty in the mesh generation for 3D complex geometries. CFD using unstructured meshes can now treat really complex geometry [1], but has a difficulty in constructing a higher-order numerical scheme in space. Both approaches will also have a difficulty in the post processing of huge data produced by the large-scale computations. For the next-generation CFD applicable for large-scale LES/DNS, the simplicity of the numerical algorithm in all aspect of the computations, from the mesh generation to the post processing, is most important.

The final goal of the present study is to construct a next-generation CFD that does not need to worry about the grid and physical model dependencies. To achieve this goal, a new approach named 'Building-Cube Method (BCM)', whose concept was proposed by the present author [2], [3], is developed in this paper.

2 Building-Cube Method

2.1 Approach

The Building-Cube Method basically employs a uniformly-spaced Cartesian mesh because of the simplicity in the mesh generation, the solution algorithm, and the post processing. The most critical issue of the conventional Cartesian mesh, however, is how to fit the grid spacing to the local characteristic flow length without introducing the algorithm complexity.

Here, a flow field is divided into a number of regular hexahedron subdomains, named 'Cube' as shown in Fig.1. The geometrical size of each cube is determined by adapting to the geometry and the flow features. In each cube, a uniformly-spaced Cartesian mesh is used. All cubes have the same mesh density so that the local computational resolution is determined by the cube size. The same mesh density among all cubes also simplifies the parallelization.

Wall boundary is defined by a staircase representation in order to keep the simplicity of the algorithm and to minimize the memory requirement per node. To keep the geometrical accuracy by the staircase representation, very fine mesh that resolves viscous sublayers in the boundary layers is used.

With this approach, it is capable (1) to treat complex geometries, (2) to use locally dense grid fitted to the local characteristic flow length, (3) to implement a higher-order scheme, (4) to use parallel computers, (5) to treat extremely huge data in the post-processing.

2.2 BCM-Mesh Generation

The BCM-mesh generation has two steps, the cube-frame generation (Fig.1(a)), and the Cartesian mesh generation in cubes near the body (Fig.1(b)).

The cube-frame is generated by the geometry-adaptive refinement method [4], which is similar to the quadtree/octtree method. Cartesian-mesh generation in the second step is required for information about the wall boundary location. Here, cells inside of the body are marked 0, and those outside are marked 1. The body boundary is defined by the cell boundaries between adjacent cells having different marks. This in-out information of cells is compactly kept by the runlength data structure [3]. With this data compression, the file size of a mesh about 3×10^7 cells (500 cubes with 256x256 cells in each cube), for example, is only 87 Kbytes in ASCII text format.

2.3 Numerical Method

The compressible Navier-Stokes equations in a non-dimensional form are solved by a cell-centered, finite-difference scheme on an equally-spaced Cartesian mesh. The inviscid numerical flux vector is computed using an approximate Riemann solver of Harten-Lax-van Leer- Einfeldt-Wada (HLLEW) [5].

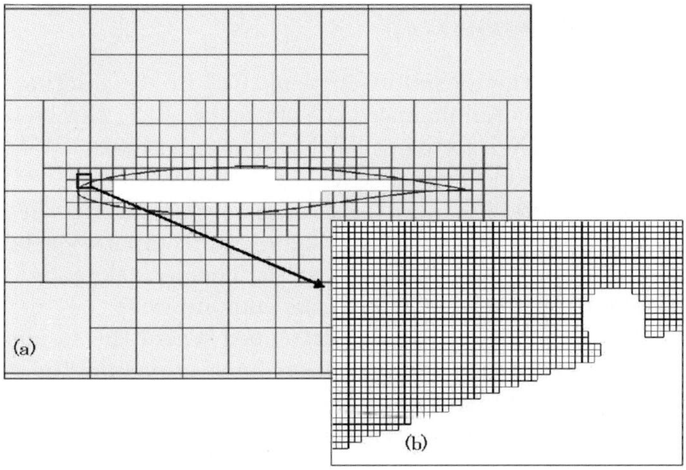

Fig. 1. BCM mesh of RAE2822 airfoil with transition-trip wire; (a) cube-frame, (b) Cartesian mesh near the transition trip.

The primitive variables on the cell interface are evaluated by the third-order MUSCL scheme with the differential limiter or the fourth-order compact MUSCL scheme [6].

The LU-SGS implicit method with the sub-iteration scheme for time accuracy [7] is used for the time integration. The sub-iteration is also utilized to update the values on the cube boundaries.

2.4 Boundary conditions

At wall boundaries, the density and pressure at the ghost cell, which locates inside of the body and next to the flow-cells, are given by the following manner.

$$\rho_{ghost} = \sum_{i(\text{adjacent cells})} (\rho_i \cdot flag_i) / \sum_{i(\text{adjacent cells})} flag_i \quad (1)$$

Here the value of flag is 0 for a cell in the body and 1 in the flow field. The velocity components at ghost cells for the solid wall are set to be zero.

At cube boundaries, three ghost cells are added beyond the cube boundary and used for the information exchanges between adjacent cubes. If the sizes of two adjacent cubes are same, the six exact overlapping of cells at boundary ensure the accurate transfer of the information. The information transfer from smaller cube to larger one is performed by assigning the average values of smaller cube near the boundary. The transfer from the larger cube to smaller one is by just taking the value in the larger cells.

2.5 Efficiency improvements

The parallel computation is simply implemented by distributing the cubes to the CPUs in a sequential manner. The efficiency of the parallelization was checked on the NEC SX-7 vector-parallel computer with 32 CPUs and shown to be very close to the ideal one.

In order to accelerate the convergence, a mesh sequence that is to use the pixel skipped mesh at the early stage of the computation is employed utilizing the advantage of the Cartesian mesh. The pixel skipping is also very useful for the post-processing of large-scale computations.

An active-cube selection was also developed where the residual at each cube was monitored at the end of every time integration, and the small residual cubes, named non-active cubes, were skipped in the next time integration. This is especially effective for supersonic flows.

3 Numerical Results

3.1 RAE2822 airfoil

The method was applied to RAE2822 airfoil at $Re = 6.56 \times 10^6$, $M_\infty = 0.73$, $\alpha = 2.79°$. The cube-frame is shown in Fig.1(a) and the Caltesian mesh near the wall boundary is shown in Fig.1(b). The total number of cubes is 486 with 256×256 cells in each cube. The minimum spacing near the airfoil surface is 1.14×10^{-4}.

In the experiment [8], the boundary layer transition trip of diameter 1.25×10^{-3} to the chord length was attached at 3% chord length on the upper

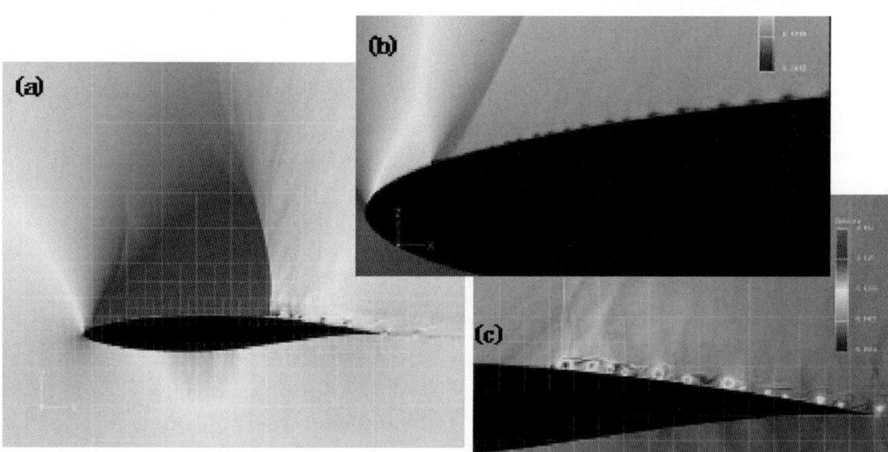

Fig. 2. Computed flows of RAE2822 airfoil with trip wire ; (a) Mach contours, (2) density contours near trip wire, (c) density contours near trailing edge.

surface of the airfoil. As shown in Fig.1(b), the trip wire is described by about 10 cells of the present mesh in x and y coordinates directions.

The flow was computed by the third order spatial accuracy. The required memory of this computation with about 30×10^6 mesh points is about 2 GBytes.

The computed result with this fine mesh shows the detailed flow features (Fig.2). In Fig.2(b), a thin boundary layer in the leading edge region is thickened in the upstream of the trip wire. In the downstream of the trip wire, very small vortices are shed periodically. The flow field at the root of the shock wave becomes more complex due to the interaction between the oncoming vortices and the shock wave (Fig.2(c)). Although the turbulence at high-Reynolds number always has three-dimensionality, here two-dimensional computation shows the essential features of the flow field.

3.2 Four-element airfoil

NASA supercritical four-element airfoil [9] was computed by the present method at $Re = 2.83 \times 10^6$, $M_\infty = 0.201$, $\alpha = 8.16°$. The number of cubes is 481 with Cartesian mesh of 256x256 in each cube. The minimum spacing near the wall is 1.4×10^{-4}.

The flow was computed by the fourth-order spatial accuracy and second-order time accuracy. The time-averaged pressure contoures is shown in Fig.3(a) and a comparison of the pressure coefficients with the experiment is shown in Fig.3(b). The computation did not use any turbulence models but the result showed good agreement with the experiment. This indicates that, with the enough resolution, the computation without physical models can accurately simulate the flows.

4 Conclusion

In this paper, an approach named Building-Cube Method, aimed for large-scale computation on near-future advanced parallel computers, was applied to flow computations of two airfoils; RAE2822 airfoil at transonic speed and four-element airfoil at low Mach number. The method is based on the Cartesian mesh, and the local grid density is adapted to the flow characteristic length by changing the cube size. Equal spacing and equal number of Cartesian grid in each cube make it easy to parallelize the flow solver and to handle huge data output. The computed results showed detailed flow features near the airfoil surfaces owing to the high-density and isotropic mesh. The pressure coefficient distributions of the time-averaged result of the four-element airfoil showed good agreement with the experiment data. It is interesting that, with the high-density and isotropic mesh, the computations without turbulence models show reasonable results.

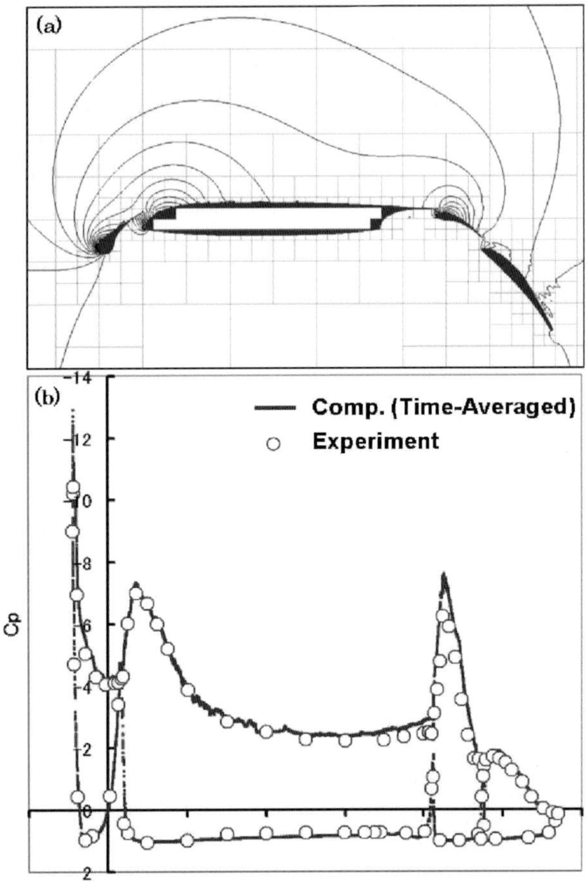

Fig. 3. Computed result of 4 element airfoil; (a) time-averaged pressure contours, (b) Cp comparison with experiment.

References

1. K. Nakahashi, Y. Ito, F. Togashi: Int. J. Numer. Meth. Fluids, **43**, pp.769-783, 2003.
2. K. Nakahashi: *Computational Fluid Dynamics 2002*, Eds. S. Armfield, R. Morgan, K. Srinivas, (Springer, 2003) pp.77–81.
3. K. Nakahashi, LaeSung Kim: *AIAA Paper 2004-0434*, January 2004.
4. K. Nakahashi, K. Egami: Computers and Fluids, **19**, 3/4, 1991, pp.273-286.
5. S. Obayashi, G.. P. Guruswamy: AIAA J. **33**, 6, 1995, pp.1134-1141.
6. S. Yamamoto, H. Daiguji: Computer and Fluids, **22**, 2/3, 1993, pp.259-270.
7. K. Matsuno: Computer and Fluids, **22**, 2/3, 1993, pp.311-322.
8. P. H. Cook, M. A. McDonald, M. C. P. Firmin: AGARD-AR-138, 1979.
9. E. Omar, T. Zierten, M. Habn, E. Szpiro, and A. Mabal: NASA CR-2215, 1979.

Part IV

Algorithms

A Matrix-Free Implicit Method for Flows at All Speeds

Alberto Beccantini[1,2], Christophe Corre[2] and Thibaud Kloczko[1,2]

[1] CEA Saclay / DM2S / LTMF, 91 191 Gif-sur-Yvette Cedex, FRANCE
[2] Laboratoire SINUMEF, ENSAM, 151 Bd de l'Hôpital 75013 Paris, FRANCE.
beccantini@cea.fr, corre@paris.ensam.fr, kloczko@cea.fr

Summary. Matrix-free implicit treatments are now commonly used for computing compressible flow problems : a reduced cost per iteration and low memory requirements are their most attractive features. This paper explains how it is possible to preserve these features for all-speed flows, in spite of the use of a low Mach preconditioning matrix. The proposed approach exploits a particular property of a widely used low-Mach preconditioner proposed by Turkel. Its efficiency is demonstrated on some steady and unsteady applications.

1 Introduction

The computation of steady flows can be considered efficient not only if a steady state is reached for a reduced CPU time but also if a low memory storage is used in this process; besides, for complex industrial applications involving grids with very large numbers of points this latter requirement may become especially critical. The same comment holds of course for unsteady flows, now classically computed by a dual time-stepping approach in which a steady-state with respect to the dual time must be reached at each physical time-step. The strong need for low-storage efficient implicit treatments has led to the development of so-called matrix-free (MF) methods [1], [2]. In the particular applications the CEA is interested in, namely the simulation of buoyant multicomponent reactive flow in a nuclear reactor containment, the methods should be versatile enough to deal with flow regimes ranging from nearly incompressible to highly compressible [3]. It has been known for more than a decade now that a proper preconditioning, so-called low-Mach preconditioning, of the equations governing compressible flows enables the application of schemes initially developped for compressible flows simulation to the incompressible regime (see [4] for a review). The introduction of such a preconditioning matrix in an implicit scheme is not really an issue if a standard block factorization or relaxation is used to solve the linear system associated with the implicit stage (see [5], [6] for instance). However, one may wonder whether it is possible to preserve a *matrix-free* implicit treatment for flows at all speeds when the implicit stage involves a *preconditioning matrix*; the present paper aims at answering this question.

2 A Not-so-Cheap Implicit Treatment for All-Speed Flows

2.1 General principles

A time-accurate solution of the Navier-Stokes equations is computed for flows at all speeds by solving :

$$P_c^{-1}\frac{\partial w}{\partial \tau}+\frac{\partial w}{\partial t}+\sum_{p=1}^{d}\frac{\partial f_p^E}{\partial x_p}=\sum_{p=1}^{d}\frac{\partial f_p^V}{\partial x_p}, \tag{1}$$

where τ is a pseudo or dual-time, t is the physical time, f_p^E and f_p^V are respectively the convective and viscous fluxes in the pth space-direction (d is the dimension of the problem) and P_c is a preconditioning matrix which takes, in the present work, the form proposed in [7], which will be detailed in the next section. System (1) is solved by the following scheme :

$$P_c^{-1}\frac{\Delta w_j^{n,m}}{\Delta\tau}+\frac{\frac{3}{2}(w_j^{n,m}-w_j^n)-\frac{1}{2}\Delta w_j^{n-1}}{\Delta t}+\sum_{p=1}^{d}(\frac{\delta_p h_p}{\delta x_p})_j^{n,m}=0, \tag{2}$$

where m is the pseudo-iteration (on dual-time) counter, n is the time step counter, $\Delta w^{n,m}=w^{n,m+1}-w^{n,m}$, $\Delta w^{n-1}=w^n-w^{n-1}$, $j=(j_1,j_2,\dots,j_d)$ is a multi-integer associated with a point $x_j=(j_1\delta x_1,\dots,j_d\delta x_d)$ of a regular Cartesian grid; $(\delta_p h_p)$ is the difference operator on a grid cell δ_p in the pth space-direction applied to the numerical flux h_p approximating the physical flux $(f_p^E-f_p^V)$: $(\delta_p h_p)_j=(h_p)_{j+\frac{e_p}{2}}-(h_p)_{j-\frac{e_p}{2}}$, where e_p is a multi-integer with components e_{pq} equal to 0 if $q\neq p$ and to 1 if $q=p$. Viscous fluxes are classically approximated at second-order using simply centred formulae while the inviscid fluxes are computed using an upwind numerical flux (e.g. Roe, Rusanov or AUSM+) h_p^E, extended to high-order using a variable reconstruction approach (MUSCL) and with a numerical dissipation taking into account the preconditioning of (1). It is easy to check that, at steady-state on τ, scheme (2) yields a second-order time-accurate solution of the preconditioned Navier-Stokes equations (1). In order to speed up the convergence of the scheme to a pseudo-steady state, (2) is made implicit with respect to τ, that is $h_p^{n,m}$ is replaced by $h_p^{n,m+1}$, leading to :

$$P_c^{-1}\frac{\Delta w_j^{n,m}}{\Delta\tau}+\frac{3}{2}\frac{\Delta w_j^{n,m}}{\Delta t}+\sum_{p=1}^{d}(\frac{\delta_p(\Delta h_p)}{\delta x_p})_j^{n,m}=\Delta w_j^{exp}, \tag{3}$$

with an explicit stage given by :

$$\Delta w_j^{exp}=-\sum_{p=1}^{d}(\frac{\delta_p h_p}{\delta x_p})_j^{n,m}-\frac{\frac{3}{2}(w_j^{n,m}-w_j^n)-\frac{1}{2}\Delta w_j^{n-1}}{\Delta t}. \tag{4}$$

The so-called matrix-free approach used in [1], [2] may be interpreted as using in the LHS of (3) an inviscid numerical flux which is not consistent, in the general case, with the one used in the explicit stage but which makes the implicit stage particularly simple to deal with, viz the Rusanov flux, whose first-order formula reads as follows for the preconditioned system (1) :

$$h_p^E = \mu_p f_p^E - \frac{1}{2} P_c^{-1} \rho(P_c A_p^E) \delta_p w, \tag{5}$$

where μ_p is the average operator over a grid cell in the pth space-direction ($(\mu_p v)_j = \frac{1}{2}(v_{j-\frac{e_p}{2}} + v_{j+\frac{e_p}{2}})$) and $\rho(P_c A_p^E) = \rho(\tilde{A}_p^E) = \tilde{\rho}_p^E$ denotes the spectral radius of the preconditioned Jacobian matrix $\tilde{A}_p^E = P_c \cdot (df_p^E/dw)$. Inserting (5) into (3) and introducing the spectral radius ρ_p^V of the viscous Jacobian matrix $(\partial f_p^V)/(\partial w_{x_p})$ yields eventually the following simple implicit scheme :

$$D \Delta w_j^{n,m} + \sum_{p=1}^{d} [\sigma_p (\mu_p A_p^E \delta_p \Delta w^{n,m})_j - C_p^- \Delta w_{j-e_p}^{n,m} - C_p^+ \Delta w_{j+e_p}^{n,m}] = \Delta w_j^{exp} \tag{6}$$

where $\sigma_p = \Delta\tau/\delta x_p$, $\lambda = \Delta\tau/\Delta t$ and $\dot{\tilde{\rho}}_p^E = \sigma_p \rho_p^E$, $\dot{\rho}_p^V = \sigma_p \rho_p^V / \delta x_p$ and the coefficients appearing in the implicit stage are defined by :

$$C_p^{\pm} = (\frac{\dot{\tilde{\rho}}_p^E}{2} P_c^{-1} + \dot{\rho}_p^V Id)_{j \pm \frac{e_p}{2}}, \; D = (P_c^{-1})_j + \tfrac{3}{2}\lambda Id + \textstyle\sum_{p=1}^{d} [C_p^+ + C_p^-] \tag{7}$$

2.2 Impact of preconditioning on the implicit treatment

In order to produce a matrix-free method, products $A_p^E \delta_p \Delta w^{n,m}$ appearing on the LHS of (6) are replaced by $\delta_p \Delta (f_p^E)^{n,m}$ and all the non-diagonal terms are relaxed to yield :

$$D \Delta w_j^{(l+1)} = \Delta w_j^{exp} - \sum_{p=1}^{d} [\sigma_p (\delta_p \mu_p \Delta (f_p^E)^{(l)})_j - C_p^- \Delta w_{j-e_p}^{(l)} - C_p^+ \Delta w_{j+e_p}^{(l)}] \tag{8}$$

Now, for standard compressible flows, in which no preconditioning is required, $P = Id$ so that D reduces to $D = (a+b)Id$ with :

$$a = 1 + \sum_{p=1}^{d} [\frac{1}{2} (\dot{\tilde{\rho}}_p^E)_{j-\frac{e_p}{2}} + \frac{1}{2} (\dot{\tilde{\rho}}_p^E)_{j+\frac{e_p}{2}}] \;\; , \;\; b = \frac{3}{2}\lambda + \sum_{p=1}^{d} [(\dot{\rho}_p^V)_{j-\frac{e_p}{2}} + (\dot{\rho}_p^V)_{j+\frac{e_p}{2}}],$$

(and $\tilde{\rho}_p^E = \rho_p^E$) while $C_p^{\pm}$ are scalar coefficients depending on the Jacobian spectral radii, so that the implicit scheme (8) is indeed matrix-free. However, when the preconditioning [7] is applied, P_c becomes of the form :

$$P_c = Id + (\beta^2 - 1) Q_c \;\text{ with }\; Q_c = \frac{\gamma - 1}{c^2} \begin{pmatrix} q^2 = \frac{1}{2}(u^2+v^2) & -u & -v & 1 \\ uq^2 & -u^2 & -uv & u \\ vq^2 & -uv & -v^2 & v \\ Hq^2 & -uH & -vH & H \end{pmatrix} \tag{9}$$

where the preconditioning parameter β is a function of the local Mach number, c is the speed of sound, u and v the velocity components and H the total enthalpy. Thus, the diagonal coefficient D given by (7) is now a full matrix, to be inverted, and $C^{\pm}$ are also block coefficients, so that storage requirement and operations count of the implicit treatment increase for low Mach number flows. The next section explains how it is possible to take advantage of the properties of P_c to keep the preconditioned approach as simple as its compressible flow version.

3 Preserving the MF Property with Preconditioning

3.1 Detail of the preconditioning matrix

If, in a first step, D is computed with a single evaluation of matrix P_c^{-1} at point j, it takes the simple form $D = a(P_c^{-1})_j + bId$. Therefore, keeping the implicit treatment (8) simple comes to invert $aP_c^{-1} + b$ in the most efficient manner. The preconditioner P_c given in (9) is associated with the Euler equations expressed in terms of conservative variables (see [9]) while it is computed in [8] a set of primitive variables. Originally, this preconditioner has been proposed by Turkel [7] starting from the Euler equations expressed in entropic variables. Using this last set of variables yields a very simple expression for the preconditioning matrix : $P_e = Id + (\beta^2 - 1)Q_e$ where $Q_e = Diag(1, 0, 0, 0)$; clearly the matrix Q_e is idempotent, *i.e.* such that $Q_e^2 = Q_e$. Naturally, since the matrices P_c and P_e are related through the relationship $P_c = (\frac{\partial w}{\partial V}) \cdot P_e \cdot (\frac{\partial V}{\partial w})$, where V denotes the vector of entropic variables, it follows immediately that Q_c is also idempotent. Exploiting this property allows to compute explicitely the inverse of D and that of P_c :

$$D^{-1} = \frac{1}{a+b}[Id + \frac{a(\beta^2-1)}{a+b\beta^2}Q_c] \ , \ P_c^{-1} = Id + (\frac{1}{\beta^2} - 1)Q_c. \tag{10}$$

On the other hand, it is also possible to rewrite (8) in the form $D\Delta w_j^{(l+1)} \approx (\Delta w_1^{(l)})_j + (P_c^{-1})_j(\Delta w_2^{(l)})_j$ with the following definitions for the increments $\Delta w_1^{(l)}$ and $\Delta w_2^{(l)}$:

$$\begin{cases} (\Delta w_1^{(l)})_j = \Delta w_j^{exp} \\ \qquad - \sum_{p=1}^{d} [\sigma_p(\delta_p \mu_p \Delta (f_p^E)^{(l)})_j - (\dot{\rho}_p^V)_{j-\frac{e_p}{2}} \Delta w_{j-e_p}^{(l)} - (\dot{\rho}_p^V)_{j+\frac{e_p}{2}} \Delta w_{j+e_p}^{(l)}] \\ (\Delta w_2^{(l)})_j = \sum_{p=1}^{d} [(\frac{\dot{\rho}_p^E}{2})_{j-\frac{e_p}{2}} \Delta w_{j-e_p}^{(l)} + (\frac{\dot{\rho}_p^E}{2})_{j+\frac{e_p}{2}} \Delta w_{j+e_p}^{(l)} \end{cases}$$

3.2 A truly MF treatment for flows at all speeds

Using expressions (10) to evaluate $\Delta w_j^{(l+1)}$ yields the following implicit treatment :

$$\Delta w_j^{(l+1)} = \frac{1}{a+b}[\Delta w_1^{(l)} + \Delta w_2^{(l)}]_j + \frac{\beta_j^2 - 1}{(a+b)(a+b\beta_j^2)}(Q_c)_j[a\Delta w_1^{(l)} + b\Delta w_2^{(l)}]_j \tag{11}$$

which allows to decouple nicely the standard matrix-free implicit treatment used for compressible flow from the added treatment specific to low-Mach number flows ($\beta \neq 1$). Moreover, treatment (11) becomes wholly matrix-free if the specific expression of matrix Q_c given in (9) is used to compute a matrix-vector such as $Q_c \cdot X$ under the form :

$$Q_c.X = \frac{\gamma - 1}{c^2}(q^2 X^{(1)} - uX^{(2)} - vX^{(3)} + X^{(4)}) \cdot \left(1\ u\ v\ H\right)^T$$

with $X^{(m)}$ the components of vector X, as previously pointed out in [4]. Eventually, $\Delta w_j^{(l+1)}$ may be computed at a low cost and with a reduced memory storage using the following expression :

$$\Delta w_j^{(l+1)} = \frac{1}{a+b}[\Delta w_1^{(l)} + \Delta w_2^{(l)}]_j + \Delta\phi_j^{(l)} \left(1\ u\ v\ H\right)_j^T \tag{12}$$

where $\chi_j = \frac{\gamma-1}{c^2} \frac{\beta_j^2-1}{(a+b)(a+b\beta_j^2)}$ and

$$\begin{aligned}\Delta\phi_j^{(l)} &= \chi_j[q^2(a\Delta w_1^{(1)} + b\Delta w_2^{(1)}) - u(a\Delta w_1^{(2)} + b\Delta w_2^{(2)}) \\ &\quad -v(a\Delta w_1^{(3)} + b\Delta w_2^{(3)}) + (a\Delta w_1^{(4)} + b\Delta w_2^{(4)})]^{(l)}.\end{aligned}$$

4 Example

The MF treatment defined by (12) has been successively applied with Roe, Rusanov and AUSM+ flux formula for h^E and compared with a block alternate-line Gauss-Seidel (ALGS) relaxation of (3), in which the numerical flux appearing in the LHS was consistent with the one used in the explicit stage but of order 1 only, as is usually done to retain simple tridiagonal to solve in each space direction. The higher intrinsic efficiency of the block relaxed treatment is clearly observed on Fig. 1 where a significant cutline of $\rho(G(\xi, \eta))$, spectral radius of the amplification matrix associated with both treatments, is plotted in the case of the preconditioned linearized Euler equations with $M = 1 \times 10^{-4}$ (and $\beta = M$). The loss of intrinsic efficiency for the MF treatment comes naturally from the use of the spectral radii $\tilde{\rho}_p^E$ in the implicit stage instead of the true preconditioned dissipation matrices associated with each explicit stage. The better damping of the high-frequency errors by the block treatment leads to a faster convergence to steady-state when both preconditioned block and MF treatments are applied to the computation of the inviscid flow at $M = 1 \times 10^{-4}$ in a channel with a bump : typically, a single sub-iteration of the block relaxed treatment at each time-step allows to converge to steady-state using roughly 1.75 times less iterations than 15

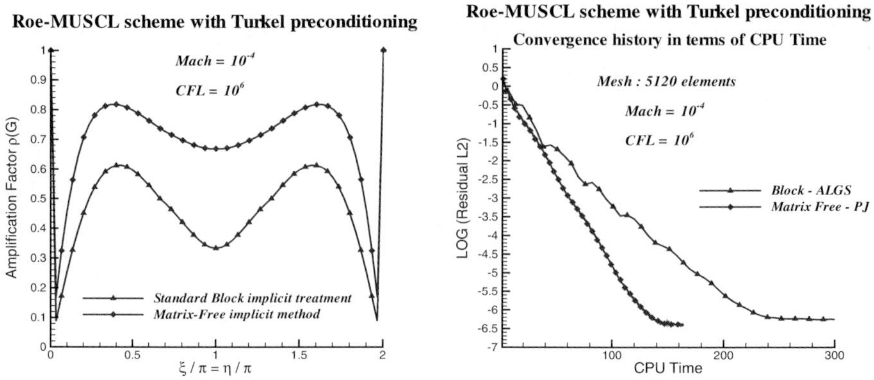

Fig. 1. Left : comparison of amplification factor for block and MF treatments applied to the preconditioned linearized Euler equations; cutline is along $\xi = \eta$ in the wavenumbers plane. Right : convergence history with block and MF treatments for the inviscid flow at $M = 0.1$ in a channel with a bump.

sub-iterations of the MF-PJ approach. However, this latter approach offering a cost per point per iteration which is almost 3 times less than the cost of the block approach, the MF treatment is eventually more cost-efficient, as seen on Fig. 1. Moreover, the global memory storage induced by the MF treatment coupled for instance with a Roe-MUSCL explicit stage is about half of that required by the block treatment. The same conclusions could also be drawn for viscous steady and unsteady flow problems. The cost of the MF treatment can be further reduced by using a set of primitive variables, rather than w, which allows to minimize the operation count associated with (12). Note that the proposed approach remains valid for other preconditioning matrices than the specific one used in this study, as long as the idempotence property is satisfied : this is in particular the case for the preconditioner proposed in [10].

References

1. H. Luo, J.D. Baum, R. Löhner, *Comp. & Fuids*, **30** (2001), 137-159.
2. Y. Zhao, *Comp. & Fluids*, **33** (2004), 119-136.
3. A. Beccantini, F. Dabbene, S. Kudriakov et al., Int. Conf. on Supercomputing in Nuclear Application, Paris, France, Sept. 22-24, 2003.
4. E. Turkel, *Annu. Rev. Mech.* **31** (1999), 385-416.
5. W.R. Briley, L.K. Taylor, D.L. Whitfield, *J. Comput. Phys.* **184**(2003), 79-105.
6. S.A. Pandya, S. Venkateswaran, T.H. Pulliam, AIAA Paper 2003-0072, Jan. 2003.
7. E. Turkel, *J. Comput. Phys.* **72**, 277-298 (1987).
8. J.M. Weiss, W.A. Smith, *AIAA J.*, **33**, n° 11 (1995).
9. C. Viozat, INRIA Report n° 3084 (1997).
10. Y.H. Choi, C.L. Merkle, *J. Comput. Phys.* **105**, 207-223 (1993).

An Efficient and Accurate Pressure-Correction Method for All Mach Numbers

Krista Nerinckx, Jan Vierendeels, and Erik Dick

Ghent University, Sint-Pietersnieuwstraat 41, B-9000 Ghent, Belgium
Krista.Nerinckx@UGent.be, Jan.Vierendeels@UGent.be, Erik.Dick@UGent.be

1 Introduction

Mach-uniform algorithms are an indispensible tool in numerous flow situations. With preconditioning, the originally high speed density-based algorithms were extended towards the low Mach number regime. From the other side, several attempts have been made to develop compressible pressure-correction methods, but they are not satisfying: either they are not Mach-uniform, or they are not applicable in general flow situations. In this paper, we aim to construct a collocated pressure-correction method that does have these features. Analysis of the Euler and Navier-Stokes equations, valid for a general fluid, reveals how to reach Mach-uniformity. For a general case, this leads to an algorithm that finds its place in between the fully coupled and the fully segregated approach: the *coupled pressure and temperature correction algorithm.*

2 Mach-uniform accuracy

The governing Navier-Stokes equations are nondimensionalized by choosing reference quantities for pressure, temperature and length scale. A collocated vertex-centered finite volume method and an first order time integration scheme are applied to discretize them. The conductive heat flux and the viscous terms are discretized centrally. The spatial flux definitions are AUSM+ [1], where the transported quantities are first order upwinded. At high Mach numbers, the AUSM-flux performs very well. At low Mach numbers special measures have to be taken with respect to the scaling and decoupling problem [2]. Doing so, it enables us to reach Mach-uniform accuracy.

3 Mach-uniform efficiency

Mach-uniform efficiency implies a good convergence rate, whatever the Mach number is. Especially the low speed limit is critical, due to severe time step restrictions imposed by the stability of the scheme. They can be of an acoustic as well as of a diffusive nature.

The first of them is expressed by the acoustic CFL-number $CFL_{u+c} = (u+c)\Delta t/\Delta x$. To remedy the low Mach stiffness problem, the acoustic CFL-limit has to be removed. This can be done by treating the terms that carry acoustic information in an implicit way. Considering the conservative Euler equations, the question raises how the acoustic terms can be identified. This becomes clear by transforming the equations into a quasi-linear set,

$$p_t + up_x + \underline{\rho c^2 u_x} = 0, \quad (1)$$

$$u_t + uu_x + \underline{p_x/\rho} = 0, \quad (2)$$

$$s_t + us_x = 0. \quad (3)$$

The underlined terms represent the acoustic part of the system. Using the equations of state $\rho = \rho(p,T)$ and $\rho e = \rho e(p,T)$, and a general definition for the speed of sound c, the pressure equation (1) can be written as

$$\left[\rho_p(\rho e)_T - (\rho e)_p \rho_T\right](p_t + up_x) + (\rho e)_T \underline{\rho u_x} - \rho_T \underline{(\rho e + p)u_x} = 0. \quad (4)$$

Note that the first underlined term finds its origin in the continuity equation, while the second one comes from the energy equation. Returning to the conservative Euler equations gives

$$\rho_t + \underline{(\rho u)_x} = 0, \quad (5)$$

$$(\rho u)_t + (\rho uu)_x = -\underline{p_x}, \quad (6)$$

$$(\rho E)_t + (\rho Hu)_x = 0, \quad (7)$$

with $\rho Hu = \underline{(\rho e + p)u} + \rho u^2 u/2$. The underlined terms will be treated implicitly in order to remove the acoustic CFL-limit. In the static enthalpy flux, only the velocity u has to be treated implicitly, since only this variable appears under a derivative in the acoustic term.

At this point, it is very instructive to consider some special cases. If the density ρ can be considered as constant, the second acoustic contribution in (4) disappears. It means that for this case, the energy equation contains no acoustic information. Furthermore, the pressure equation (4) originates from the continuity equation alone, and reduces to the incompressible constraint $u_x = 0$. On the other hand, for a perfect gas, ρe only depends on the pressure. The first acoustic part in (4) therefore becomes zero, so that the continuity equation doesn't contain acoustic information. The pressure equation (4) is derived from the energy equation alone. We stress, however, that this doesn't hold anymore when conductive temperature terms occur in the energy equation.

The diffusive time step limit is expressed by the Von Neumann number $Ne = \alpha\Delta t/\Delta x^2$, with α the thermal diffusivity or the kinematic viscosity. To remove it, also the diffusive conductive and viscous terms have to be treated implicitly.

4 Implementation

The first step is a predictor step for density and momentum, from the continuity and momentum equation respectively. The predicted values ρ^*, $(\rho u)^*$, together with the old pressure values, determine an intermediate state $*$. The predictor step takes place under a frozen pressure and in essence determines a velocity u^*. It therefore can be considered as a convective step.

Secondly, corrections with respect to the predictor values are defined. They are introduced into the discretized equations, namely in the terms at the new time level. Amongst them are the implicit diffusive and acoustic terms identified in paragraph 3. The corrector step can therefore be considered as an acoustic/thermodynamic step.

In the continuity equation, the density is expanded as $\rho^{n+1} = \rho^* + \rho^*_p p' + \rho^*_T T'$. The mass flux, which is an acoustic term, is corrected as $(\rho u)^{n+1} = (\rho u)^* + (\rho u)'$, and the momentum correction is related to pressure corrections through the momentum equation. The continuity equation thus becomes an equation for both pressure and temperature corrections.

In the energy equation, the total energy is expanded as $(\rho E)^{n+1} = (\rho E)^* + (\rho e)^*_p p' + (\rho e)^*_T T'$. According to the identification of the acoustic terms, the enthalpy flux is corrected as $\rho H u = (\rho e + p)^* \left[(\rho u)^* + (\rho u)'\right] / \rho^* + \left(\rho u^2\right)^* u^*/2$, and the momentum correction is replaced. In the heat flux, temperature corrections are introduced to avoid a diffusive time step limit, $q^{n+1} = \kappa(T^* + T')_x$. Introducing these terms into the energy equation, results in a second equation containing both pressure and temperature corrections. The two correction equations, originating from the continuity and energy equation, are solved in a coupled way. We therefore refer to this method as *coupled pressure and temperature correction algorithm.* After updating the pressure and the temperature, momentum is updated through the momentum equation. Density is updated through the equation of state.

For a perfect gas, in absence of heat transfer, all temperature corrections disappear from the energy equation. Thus, no coupled solution with the continuity equation is needed anymore: after the predictor step, pressure corrections can be determined from a pressure-correction equation based on the energy equation. However, when heat conduction is present, a coupled solution is needed anyhow, due to the temperature corrections in the heat flux. If the energy equation is used as a pressure-correction equation, with a heat flux calculated with known temperature values T^*, the explicit treament of the diffusive temperature terms causes a diffusive time step limit.

5 Results

5.1 Adiabatic flow

First the Euler equations are considered, i.e. heat conduction and viscosity are neglected. As a test case we take a one-dimensional nozzle flow of a perfect

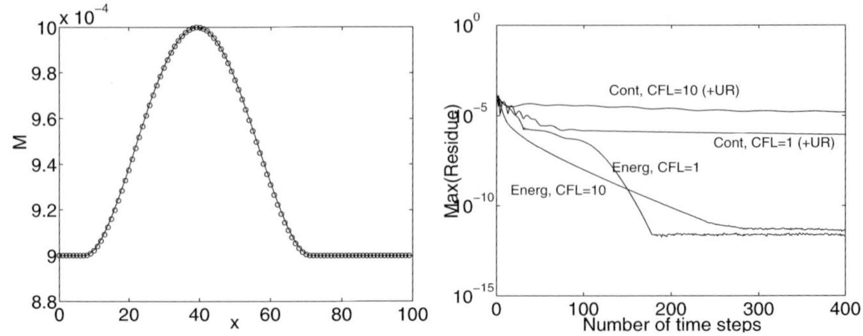

Fig. 1. Subsonic nozzle flow, $M_t = 0.001$. (a) Mach number distribution. (b) Convergence plot, $CFL_u = 1$ and 10. *Energ*: Mach-uniform algorithm with pressure-correction equation based on the energy equation. *Cont*: equivalent algorithm based on the continuity equation (computation with underrelaxation (UR))

gas. The fully segregated algorithm with a pressure-correction equation based on the energy equation was used.

As a low speed test, we consider a subsonic nozzle flow with a throat Mach number M_t of 10^{-3}. The time step is calculated from a chosen convective CFL-number, $\Delta t/\Delta x = CFL_u/max(u)$. We remark that this corresponds with an acoustic CFL_{u+c} number, that is about 1000 times higher. The CFL_u number could be taken arbitrarily high: there is no stability limit. Figure 1 shows the Mach number distribution, and the convergence plot for both the Mach-uniform algorithm and an equivalent algorithm where the pressure-correction equation was derived from the continuity equation.

The continuity-based algorithm has a very bad convergence rate at this low Mach number. Furthermore, the computations could only be made stable under a severe underrelaxation. It clearly suffers from the stiffness problem, which results in a convergence breakdown. The Mach-uniform algorithm, on the other hand, performs very well for this low speed flow, with regard to accuracy as well as efficiency.

Also for the high-speed case of a transonic nozzle, there is no CFL-limit. Figure 2 shows the Mach number distribution and the convergence plot. Again, a good convergence rate and accuracy are obtained. Clearly, the algorithm shows Mach-uniform efficiency and accuracy.

5.2 Non-adiabatic flow

First, we consider again the test cases of a one-dimensional nozzle flow, but now heat conduction is taken into account. Just like for the adiabatic computations, no acoustic time step limit occurs. All the simulations are done at a convective CFL_u number of 1. If also a diffusive time step limit is to be avoided, the coupled solution of energy and continuity equation for both pressure and temperature corrections is needed. Table 1 shows the stability

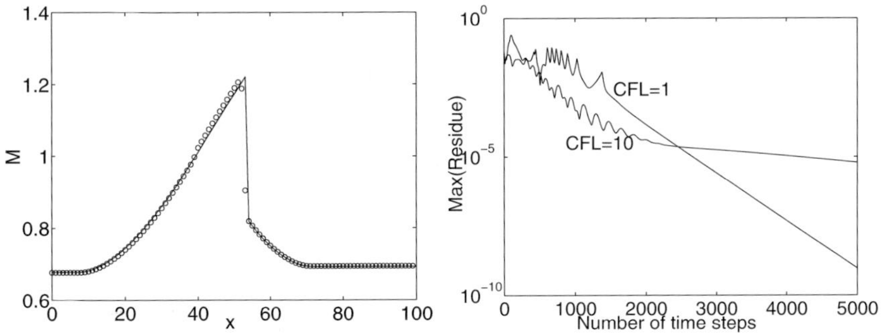

Fig. 2. Transonic nozzle. (a) Mach number distribution. (b) Convergence plot, $CFL_u = 1$ and 10.

results for different nozzle flows. The columns indicated with *exp* refer to an explicit calculation of the heat flux, which is added as extra term in the RHS of the pressure-correction equation based on the energy equation. *Coup* refers to the coupled pressure and temperature correction method. Different values for the nondimensional conduction coefficient κ were tested. The results show clearly that the method with explicit heat flux becomes unstable as soon as the Von Neumann number becomes higher than order unity. The coupled method, however, stays stable no matter how high κ is taken.

Table 1. *Exp*: pressure-correction equation based on the energy equation, heat flux calculated with T^*. *Coup*: coupled pressure and temperature correction method.

		stable?				*stable?*	
κ	Ne	*exp*	*coup*	κ	Ne	*exp*	*coup*
10^{-5}	$2.42\text{x}10^{-4}$	yes	yes	10^{-5}	$2.42\text{x}10^{-6}$	yes	yes
0.01	0.242	yes	yes	1	0.242	yes	yes
0.1	2.42	no	yes	10	2.42	no	yes
1	24.2	no	yes	100	24.2	no	yes
Subsonic $M_t = 0.01$				**Transonic**			

As a second test case for non-adiabatic flow, the two-dimensional thermal driven cavity problem is considered [3], with $Ra = 10^3$ and $\epsilon = 0.6$. This is a very challenging problem for our algorithm. Indeed, due to the very low Mach numbers, no acoustic CFL-limit is allowed. Furthermore, in wall vicinity, the convective speeds are very small and the conduction becomes the dominating phenomenon. Therefore, also a diffusive Von Neumann-limit is to be avoided, so that a coupled solution of the continuity and the energy equation is needed. Streamline patterns, temperature contours and a convergence plot are shown in figure 3. From the used values for α, Δt and grid dimensions, we obtain for the maximum CFL_{u+c} and Neumann number $O(10^7)$ and $O(10^4)$ respec-

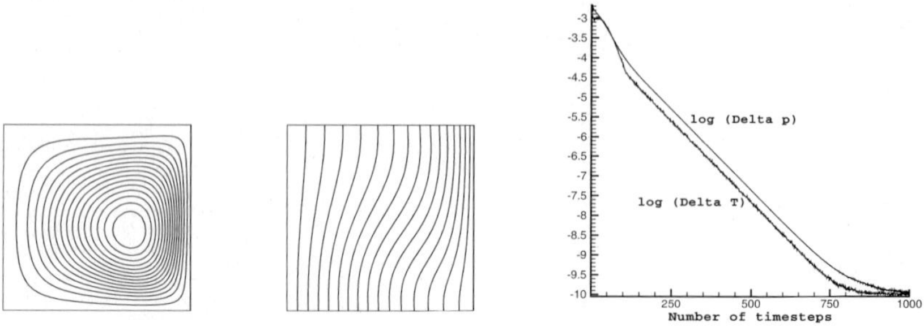

Fig. 3. (a)Streamline patterns. (b)Temperature contours. (c)Convergence plot.

tively. Clearly, the acoustic and diffusive time step limits have been removed.

6 Conclusion

We have presented a new type of algorithm: the coupled pressure and temperature correction algorithm. The essential idea is the separation of the convective phenomenon on the one side, and the acoustic/thermodynamic phenomenon on the other side. Based on a theoretical analysis, the algorithm was constructed so that Mach-uniform accuracy and efficiency are obtained, which was confirmed by the test results. Especially the removal of the acoustic and diffusive time step limits is an important feature. When the special case of a perfect gas without heat transfer is considered, the algorithm reduces to a fully segregated method with a pressure-correction equation based on the energy equation.

Acknowledgements

Research funded with a fellowship granted by the Flemish Institute for the Promotion of Scientific and Technological Research in the Industry (IWT).

References

1. M.-S. Liou. Mass flux schemes and connection to shock instability. *Journal of Computational Physics*, **160**, pp. 623-648, 2000.
2. J.R. Edwards, M.-S. Liou. Low-diffusion flux-splitting methods for flows at all speeds. *AIAA Journal*, **36**, pp. 1610-1617, 1998.
3. J. Vierendeels, B. Merci and E. Dick. Benchmark solutions for the natural convective heat transfer problem in a square cavity with large horizontal temperature differences. *International Journal of Numerical Methods for Heat and Fluid Flow*, **13**(8), pp. 1057-1078, 2003.

The Analysis of Electromagnetic Waves Using CIP Scheme with Soroban Grid

Yoichi Ogata[1], Takashi Yabe[1], Kenji Takizawa[2], and Tomomasa Ohkubo[2]

[1] Department of Sciences and Enginnering, Tokyo Institute of Technology, yogata@mech.titech.ac.jp,yabe@mech.titech.ac.jp
[2] Department of Energy Sciences, Tokyo Institute of Technology, ktaki@es.titech.ac.jp,tohkubo@es.titech.ac.jp

Summary. In the regime of finite-difference solutions for Maxwell's equations, all kinds of methods like FDTD (Finite-Difference Time-Domain) method, LBS (Linear Bicharacteristics Scheme) and so on, have been standard. However, in order to deal with large scale and complicated structure in future, higher order scheme and adaptive mesh that does not lose accuracy are required. In this paper, we propose the new method that uses the CIP (constrained interpolation profile/cubic interpolated propagation) method in combination with the numerical method of characteristics (hereafter, CIP-MOC) and the Soroban grid.

1 The Numerical Scheme for Maxwell's Equations

1.1 The Fundamental Concept of CIP Method

The CIP method[1][2] is able to solve hyperbolic equations like Eq.(1) with the third-order accuracy in time and space,

$$L(f) \equiv \frac{\partial f}{\partial t} + u\frac{\partial f}{\partial x} = 0 \tag{1}$$

and the CIP uses the spatial derivative of f. When u is constant, the differentiated Eq.(1) with spatial variable x will be

$$\partial_x L(f) \equiv \frac{\partial g}{\partial t} + u\frac{\partial g}{\partial x} = 0 \tag{2}$$

where g is $\partial_x f$. If two values of f and g are given at two grid points, the profile between these points can be interpolated by cubic polynomial $F_i(x) = ax^3 + bx^2 + g_i^n x + f_i^n$. Thus, the profile at next $n+1$-step can be obtained according to Eqs.(1) and (2), that is, by shifting $f_i^{n+1} = F_i(x_i - u\Delta t)\,, g_i^{n+1} = dF_i(x_i - u\Delta t)/dx$.

$$f_i^{n+1} = a_i\xi^3 + b_i\xi^2 + g_i^n\xi + f_i^n \tag{3}$$

$$g_i^{n+1} = 3a_i\xi^2 + 2b_i\xi + g_i^n \tag{4}$$

where $\xi = -u_i\Delta t$. The coefficients a_i and b_i are easily obtaind as

$$a_i = \frac{g_i + g_{iup}}{D^2} + \frac{2\left(f_i - f_{iup}\right)}{D^3} \tag{5}$$

$$b_i = \frac{3\left(f_{iup} - f_i\right)}{D^2} + \frac{2g_i + g_{iup}}{D} \tag{6}$$

where $D = -\Delta x \cdot sgn\left(u_i\right)$,$iup = i - sgn\left(u_i\right)$,$sgn\left(u_i\right) = 1\left(u_i \geq 0\right), -1\left(u_i < 0\right)$.

This 1D-CIP will be directly applied to the multi-dimensional Soroban grid using directional splitting.

1.2 Method of Characteristics with CIP Method

Maxwell's equations in an isotropic medium are

$$\frac{\partial E_y}{\partial t} + \frac{1}{\epsilon}\frac{\partial H_z}{\partial x} = 0 \tag{7}$$

$$\frac{\partial H_z}{\partial t} - \frac{1}{\mu}\frac{\partial E_y}{\partial x} = 0 \tag{8}$$

in one-dimensional TE case(taking $\partial/\partial y = \partial/\partial z = 0$) and $\sigma = 0$(no-conductivity). ϵ and μ are the permittivity and permeability. Since Maxwell's equations (7) and (8) are also hyperbolic equations, the CIP method can be directly applied to them using characteristics theory. They can be changed to the following characteristic equations[3].

$$C^{\pm} : \frac{D^{\pm} E_y}{\partial t} \pm \sqrt{\frac{\epsilon}{\mu}} \frac{D^{\pm} H_z}{\partial t} = 0 \tag{9}$$

$$\frac{D^{\pm}}{\partial t} \equiv \frac{\partial}{\partial t} \pm c^{\pm} \frac{\partial}{\partial x}$$

where $c^{\pm} = 1/\sqrt{\epsilon^{\pm}\mu^{\pm}}$ is the speed of light of each characteristic $C^{\pm}$.

Although we have already applied CIP-MOC to the shallow water equations in which similar equations to Eq.(9) can be derived[4], we can apply the same procedure to Maxwell's equations.

When we analyze Maxwell's equations, there are some dielectrics in simulation region in general. In the case, each parameter like ϵ, μ and the speed of light c is different at the material boundary. The conventional FDTD algorithm[5] has a problem in treatment of material discontinuities. However, the CIP-MOC scheme is able to implement the dielectric material interface naturally because the present scheme uses co-located E_y and H_z that are used for physical boundary conditions.

Equation (9) means that time evolution of (E_y, H_z) at grid point i is determined from two characteristics along $C^{\pm}$ in terms of Riemann invariants(Fig.1). Therefore, the relations between $(E_y, H_z)^{n+1}$ and $(E_y, H_z)^{\pm}$ can be derived like

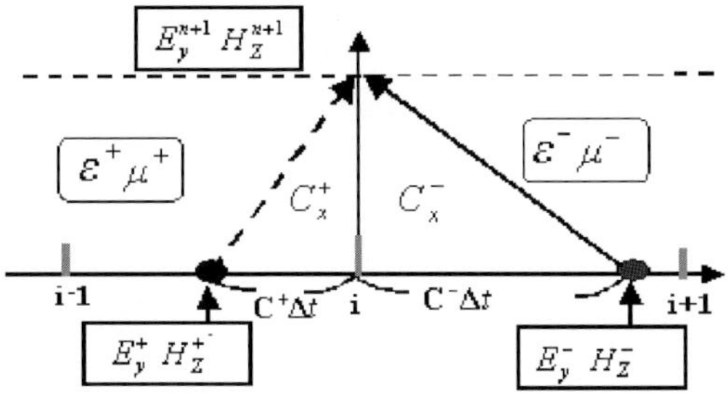

Fig. 1. Space-time diagram showing the characteristic curves and trajectories for one-dimensional Maxwell's Equations.

$$E_y^{n+1} - E_y^{\pm} \pm \sqrt{\frac{\epsilon^{\pm}}{\mu^{\pm}}} \left(H_z^{n+1} - H_z^{\pm}\right) = 0 \tag{10}$$

and then, the values at $n+1$ step can be easily obtained by Eq.(10).

Since the upstream departure points ($E_y^{\pm}$, $H_z^{\pm}$) are not necessarily located at the grid point, the CIP method can be used to get cubic-polynomial approximation inside the grid cell, that is, when it comes to values,

$$E_y^{\pm} = E_{yCIP}\left(x_i \pm c_i^{\pm} \Delta t\right), \quad H_z^{\pm} = H_{zCIP}\left(x_i \pm c_i^{\pm} \Delta t\right) \tag{11}$$

where $f\left(= E_y, H_z\right)_{CIP}(x)$ is Eqs.(3)-(6). In addition, the characteristics of spatial derivatives ($\partial_x E_y^{\pm}$, $\partial_x H_z^{\pm}$) used in the CIP method can also be solved by differentiating Eq.(9) with spatial variable x.

This method based on the CIP method, which is Semi-Lagrange scheme, is also third-order accurate while FDTD and quadratic interpolation[3] are second-order accurate. Furthermore, it has already been proved that phase error and damping rate of the CIP even at $k\Delta x = \pi$ are extremely low, which means that the CIP-MOC is suitable for analysis of electromagnetic wave that could generally include various wavelength. The present scheme can be easily extended to multi-dimensional case using directional splitting.

2 Soroban Grid

When there are dielectrics, grid size should be determined by minimum wavelength that will be shorter in dielectrics Furthermore, if some objects have complicated structures, it will be better to adjust grid points to structures in terms of material physical boundary conditions, large variation near the boundary and so on. Therefore, if we use uniform mesh in such cases, grid number will be very large. That is why adaptive grid and local mesh refinement are thought to be very promising for efficient simulations.

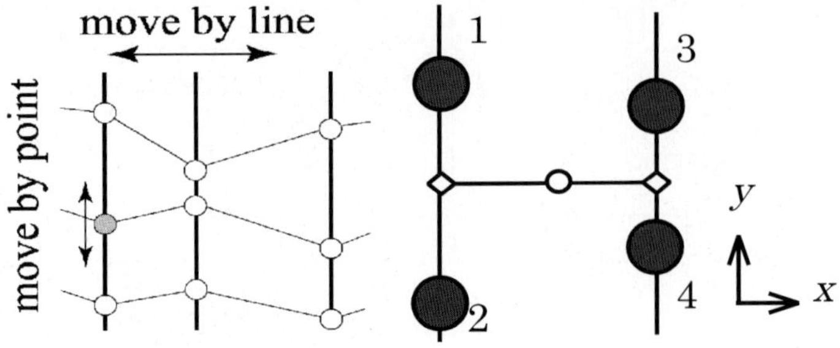

Fig. 2. Left: The straight lines can move in horizontal direction and grid points move along the straight line. Right: Interpolation in the Soroban grid.

However, they usually require coordinate-transformation with solving partial differential equations, which will make accuracy reduced. In order to resolve this problem, the Soroban grid[6] was proposed by Yabe et.al. Figure 2 shows the principle of the Soroban grid. The straight line can move in horizontal direction and grid points move along the straight line like abacus - Soroban in Japanese. As for interpolation in the Soroban grid, when we try to obtain the value at white circle, first, we use 1D(y-direction) CIP between grid 1 and 2, 3 and 4, respectively, then, use 1D(x-direction) CIP between two white diamonds. This method enables the present scheme to keep the third-order even in non-uniform mesh[6].

For the simplest choice of the monitoring function to the variation, the following quantity is used:

$$M\left(x,t\right) \equiv \sqrt{1+\alpha\left(\frac{\partial f}{\partial x}\right)}+\beta\left(\frac{\partial^2 f}{\partial x^2}\right) \tag{12}$$

Therefore monitoring function M becomes large for larger gradient region. Since the Soroban grid is straight in one dimension, it is much easier to generate the adaptive grid points along the line.

In the two-dimensional mesh shown in Fig.2, mesh moving is performed as follows.

1. Calculate $M\left(y,t\right)$ along each line.
2. Generate the points along each line.
3. Calculate the average $M\left(x,t\right)$ from all the points along each line.
4. Move the lines.

Three dimension is also straightforward, that is, the average $M\left(z,t\right)$ in xy-plane is simply added. Therefore, the Soroban grid is suitable for the CIP method, that is, it is almost impossible to apply the other schemes that use Although all kinds of f in Eq.(12) can be taken, energy density $\left(=\epsilon\mathbf{E}^2+\mu\mathbf{H}^2\right)$ will be one of the best choices as f in Maxwell's equations.

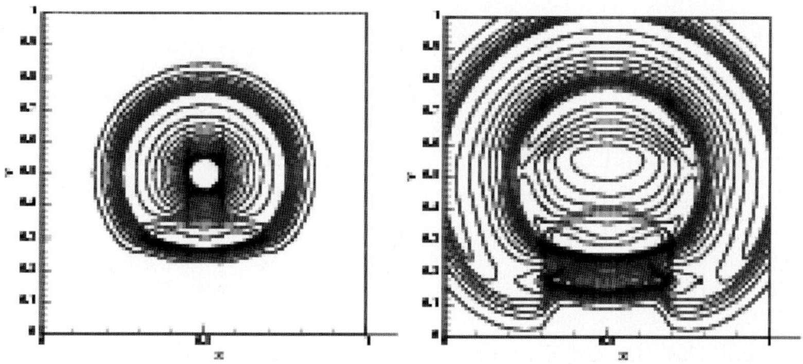

Fig. 3. The contours of H_z. Left : $t = 0.125$, Right : $t = 0.8$

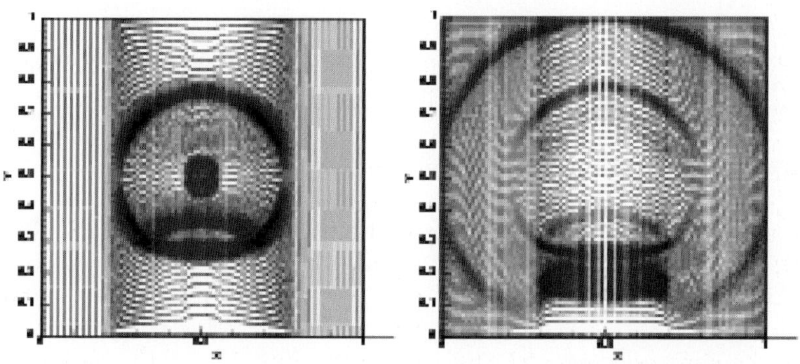

Fig. 4. The Soroban grid. Left : $t = 0.125$, Right : $t = 0.8$

The Soroban grid is applied to Maxwell's equations in two-dimensional case of TE wave mode. The initial condition is

$$H_z\,(x, y, y = 0) = \exp\{-\left(\frac{r^2 - r_0^2}{\sigma^2}\right)\} \quad r \geq r_0 \tag{13}$$
$$= 1.0 \quad r < r_0$$

where $r_0 = 0.1, \sigma = 0.1, r^2 = (x - 0.5)^2 + (y - 0.5)^2$, $E_x = E_y = 0.0$. A dielectric is set in $[0.3 < x < 0.7]$ and $[0 < y < 0.3]$. (ϵ, μ) is $(1.0, 4.0)$ inside wall, $(1.0, 1.0)$ outside wall, respectively. The total number of grid points is 10000. The mesh size is initially $\Delta x = \Delta y = 0.01$.

Figures 3 and 4 show the time evolution of H_z and the Soroban grid. We can see that grid points concentrate near the wall surface where wavelength is short and electromagnetic wave fronts where spatial gradient is large. Meanwhile, there are few grid points in the wall where electromagnetic waves have not come yet. Furthermore, note that wave can pass through the boundary without any absorbing boundary condition like FDTD because MOC is used in our method.

3 Conclusion

We bring forward the new Maxwell's equations solver using the CIP in combination with the Soroban grid and MOC. The CIP and the Soroban grid make it easy to build higher order adaptive mesh Maxwell's solver without subgrid or coordinate-transformation that will reduce accuracy.

In the other paper[6], we have already proposed new Soroban grid that incorporates the local mesh refinement so that we can more effectively put the grid points. These applications will be shown in the future paper.

References

1. T.Yabe, T.Aoki: Comput.Phys.Commun. **661**, 219 (1991)
2. T.Yabe, T.Ishikawa, P.Wang, T.Aoki, Y.Kadota, F.Ikeda: Comput.Phys.Commun. **66**, 233 (1991)
3. J.H.Beggs, D.L.Marcum, S.L.Chan: Applied.Comput.Electro.Soc.Journal. **14**, no.2, 25 (1999)
4. Y.Ogata, T.Yabe: Int.J.Comput.Eng.Sci. to be published(2004).
5. K.S.Yee: IEEE Trans.Antennas Propagation. **141**, 302 (1966)
6. T.Yabe, H.Mizoe, K.Takizawa, H.Moriki, Hyo-Num Im, Y.Ogata: J.Comput.Phys. **194**, 57 (2004)

Multigrid Third-Order Least-Squares Solution of Cauchy-Riemann Equations on Unstructured Triangular Grids

Hiroaki Nishikawa

Department of Aerospace Engineering,
University of Michigan, Ann Arbor, MI 48109, USA

1 Introduction

In this paper, we present a multigrid third-order accurate Cauchy-Riemann solver on triangular grids. This work is motivated by the wish to develop a highly accurate and efficient algorithm for solving the elliptic (Cauchy-Riemann type) subsystem of the two-dimensional Euler equaitons. The decomposition approach, where a partial differential equation is split into elliptic and hyperbolic equations, is an attractive strategy as we can apply a technique specially tailored to the physics of each subproblem. Such a strategy has been shown to yield highly accurate numerical solutions over the conventional finite-volume scheme [12, 4, 5, 6]. It has also been shown that such decomposition can lead to optimal multigrid convergence of the Euler equations on structured grids [10], employing suitable grid hierarchy for each part: full-coarsening for elliptic part and semi-coarsening for hyperbolic part. This work contributes toward the same optimal convergence of the Euler computation on unstructured grids.

The least-squares method has been successfully used to solve the elliptic part of the Euler equations [9], demonstrating its superior properties over the Lax-Wendroff type solver which is known in particular to break the symmetry of the solution and not to attain a typical multigrid convergence rate[12, 13]. The advantage of the least-squares method is not only that it provides a symmetric stencil, but also that the spurious entropy generation can be suppressed, in a natural way, in the form of a constrained minimization[12]. Also an adaptive grid technique can be devised, again in the framework of minimization[11, 8]. Moreover it can be upgraded to higher-order by a simple modification; the third-order scheme has already been developed in [9] following the work by Caraeni and Fuchs [3].

Similar work has been reported by Borzi *et. al.* [7]. Our method has three distinctive features. First, our method is third-order accurate while their method is second-order. Second, our least-squares formulation is directly applicable to unstructured grids while such extension is not clear for their method. Third, we restrict the nodal residual to define the coarse grid problems while they restrict the cell-residual.

2 Least-Squares Scheme

Consider the Cauchy-Riemann equations

$$\partial_x u + \partial_y v = 0, \quad \partial_x v - \partial_y u = 0 \tag{1}$$

where (u, v) is taken to be two velocity components. On the boundary, zero normal component ($u_n = 0$) is enforced. To discretize the system, we triangulate the domain of interest into a set of triangles $\{T\}$. Storing the solutions at the vertices, we define the residual on a triangular element T as integrals of (1). Assuming piecewise linear variation of u and v within the element, we obtain the 2nd-order version of the residuals.

$$\Theta_T = \frac{1}{2} \sum_{i \in \{j_T\}} (u_i \Delta y_i + v_i \Delta x_i), \quad \Omega_T = \frac{1}{2} \sum_{i \in \{j_T\}} (-v_i \Delta y_i + u_i \Delta x_i) \tag{2}$$

where $\{j_T\}$ is the set of vertices forming the triangle T and $\Delta\{\}_i$ denotes a difference taken counterclockwise along the side opposite to j. On triangular grids, the number of triangles is typically twice the number of vertices. This results a highly overdetermined problem, and therefore residuals cannot be driven to zero everywhere. We then seek to find a solution that minimizes the residuals in a least-squares norm in the form

$$\mathcal{F} = \sum_{T \in \{T\}} F_T = \frac{1}{2} \sum_{T \in \{T\}} \left(\Theta_T^2 + \Omega_T^2\right). \tag{3}$$

We remark that by weighting F_T with $1/S_T$, where S_T is the area of the triangle T, we obtain the standard Galerkin discretization of the associated Laplace equations for u and v. This discretization however loses its accuracy, yielding even less than first-order accuracy. Although equivalent on uniform grids, the norm without such a weight as in (3) cures the problem, and retain the formal accuracy of the method. See [8] for details.

To minimize the norm, we employ the steepest descent method,

$$u_j^{n+1} = u_j^n - c_j \frac{\partial \mathcal{F}}{\partial u_j} = u_j^n - \frac{c_j}{2} \sum_{T \in \{T_j\}} (\Delta y_T \Theta_T - \Delta x_T \Omega_T), \tag{4}$$

similarly for v, where $\{T_j\}$ is the set of triangles that share the vertex j, c_j is a constant and $\Delta\{\}_T$ denotes a difference taken counterclockwise along the side opposite to j on the triangle T. This method is however very slow to converge. To improve on it, we use Newton's method, but with only diagonal elements of the Hessian matrix, which defines c_j as

$$c_j = \omega / \frac{\partial^2 \mathcal{F}}{\partial u_j^2} = \omega / \frac{\partial^2 \mathcal{F}}{\partial v_j^2} = \omega / L_j^2, \; L_j^2 = \frac{1}{4} \sum_{T \in \{T_j\}} \left\{ (\Delta x_T)^2 + (\Delta y_T)^2 \right\} \tag{5}$$

where ω is the relaxation parameter. On the boundary where the tangency condition must be imposed, we simply ignore the updates in the normal component of the solution.

To extend the method to third-order, we first recover the solution gradients at each vertex by the Green-Gauss formula, interpolate the solutions at the midpoint of each edge (Hermite interpolation), and then evaluate the residuals to the fourth-order accuracy. The results are

$$\Theta_T^H = \Theta_T - \frac{1}{12}\sum_{edges}(\Delta p \Delta y - \Delta q \Delta x)\,, \quad \Omega_T^H = \Omega_T - \frac{1}{12}\sum_{edges}(\Delta q \Delta y + \Delta p \Delta x) \quad (6)$$

where Δ on the right hand side denotes the difference taken clockwise along the edge, and p_i and q_i are the gradients evaluated at vertex i projected onto the edge. These residuals are however third-order accurate in reality because the gradient estimate is at best second-order. As clearly seen, the second terms on the right can be considered as a third-order correction to the second-order residual. With this correction taken as a numerical correction, the solution procedure remains identical to the second-order method except that we replace the second-order residuals (Θ_T, Ω_T) by the third-order counterparts (Θ_T^H, Ω_T^H) in the update formulas (4). See [8] for more details.

3 Multigrid Algorithm

We now describe the multigrid algorithm that accelerates the convergence of the method to reach the ultimate O(N) convergence where N is the number of unknowns. Coarse grids are generated by removing every other point on a fine grid, resulting vertex-nested structure. Coarse grid problems are then independently discretized on their grids. It is crucial at this point that the source term in the coarse grid problem (i.e. the restricted nodal residual) must be properly scaled [1]. Defining the nodal residual on a fine grid $\{T^h\}$ to be restricted onto a coarse grid $\{T^{2h}\}$ by

$$R_j^h = \frac{\sum_{T \in \{T_j^h\}} \left(\Delta y_T \Theta_T^H - \Delta x_T \Omega_T^H \right)}{2\,(L_j^h)^4} + R^{h/2}|_j^h \quad (7)$$

where $R^{h/2}|_j^h$ is the nodal residual restricted onto the node j on $\{T^h\}$ from the next finer grid (which is zero if $\{T^h\}$ is the finest grid), we have the update formula for the solution on the coarse grid,

$$\left(u_j^{2h}\right)^{n+1} = \left(u_j^{2h}\right)^n - \omega\,(L_j^{2h})^2\left[R_j^{2h} + \alpha\, R^h|_j^{2h}\right], \quad (8)$$

similarly for the other variable. Note that the coarse grid problems can be badly scaled if the coarsening ratio is locally too large. We may modify the coarse grids, but instead we modify the algorithm by introducing a parameter α: set $\alpha = 1$ if $(L_j^{2h}/L_j^h)^2 < 4.0$, otherwise $\alpha = 4.0 \times (L_j^h/L_j^{2h})^2$.

As for intergrid transfer operators, following [2], we construct node-based intergrid transfer operators as follows. Prolongation is based on injection and linear interpolation,

$$u^{2h}|_j^h = \begin{cases} u_j^{2h} & \text{if } j \text{ is a nested vertex} \\ \sum_{i \in \{j_T\}} u_i^{2h} w_i^T & \text{otherwise} \end{cases} \tag{9}$$

where $u^{2h}|_j^h$ denotes the value interpolated from a coarse grid $\{T^{2h}\}$ to a fine grid $\{T^h\}$, T is a coarse-grid triangle enclosing a fine-grid node j and $\{w^T\}$ is a set of weights for the linear interpolation over that triangle. Restriction is taken as the transpose of the prolongation

$$R^h|_j^{2h} = \frac{R_j^h + \sum_{T \in \{T_j^{2h}\}} \sum_{\{i_T\}} R_i^h w_j^T}{1 + \sum_{T \in \{T_j^{2h}\}} \sum_{\{i_T\}} w_j^T} \tag{10}$$

where $\{i_T\}$ is a set of fine-grid vertices in a coarse-grid triangle T. These operators are accurate enough to satisfy the well-known necessary condition for the multigrid efficiency [1] for the Cauchy-Riemann equations.

4 Results

We use W-cycle with 2 relaxation sweeps on each grid level after both restriction and prolongation. Relaxation scheme is SOR with $\omega = 1.6$. The method is taken to be converged when the L_2 errors of the solutions do not change more than 1% per cycle. CPU time was measured on a PC with Pentium III 800MHz processor (3.06 GHz for unstructured grid cases).

The algorithm was first tested for flows around a cylinder (with circulation 2π) and a Joukowsky airfoil at an angle of attack $10°$. In both cases, the grids are O-type 40x20, 80x40, and 160x80 grids with unit chord length, extending to 10-chord length to the outer boundary (Figures 1 and 4). On the outer boundary the exact solution is given, while on the inner boundary we simply ignore the solution updates in the normal direction. Figures 2 and 5 show the error convergence. 3rd-order accuracy is observed for both cases, where the solid line has the slope of 3. The method converged in about 12 W-cycles. The average convergence rates are 0.35 for the cylinder case and 0.6 for the airfoil case, which are very small compared with that of the single grid case for which the rate is about 0.996 for both cases. Figures 3 and 6 shows that the method achieves O(N) convergence in both cases as the data fit the line of slope 1, in which sense the method is optimal.

Another test case is the same flow around the same airfoil on unstructured grids. Grids were generated by "Delaundo" developed by Muller[13]: 3848,

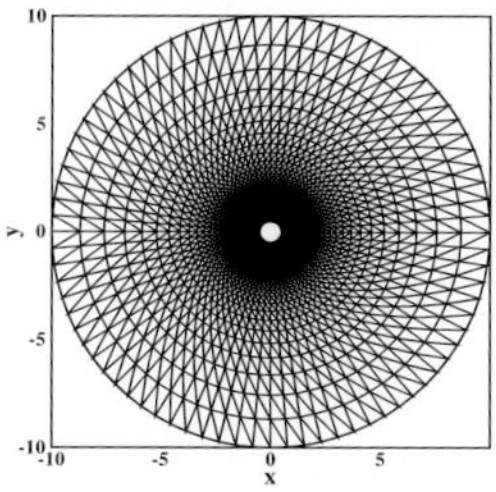

Fig. 1. 80x40 O-grid around a cylinder.

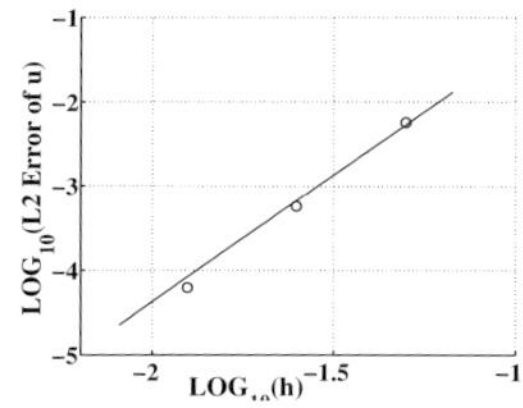

Fig. 2. Error Convergence: Cylinder Case

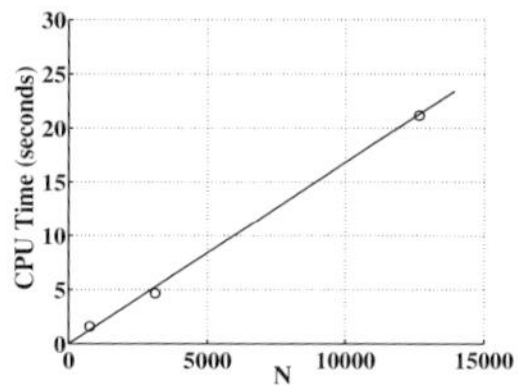

Fig. 3. CPU Time vs. Grid Size: Cylinder Case.

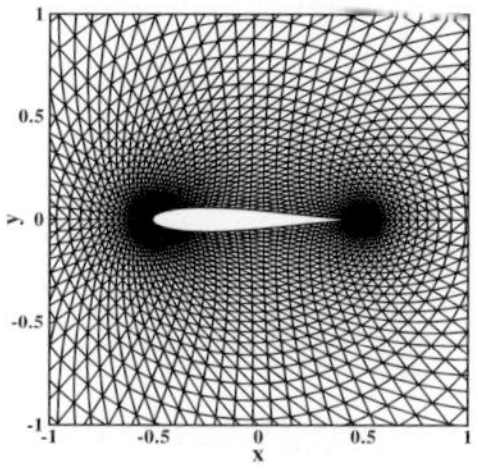

Fig. 4. 80x40 O-grid around an airfoil.

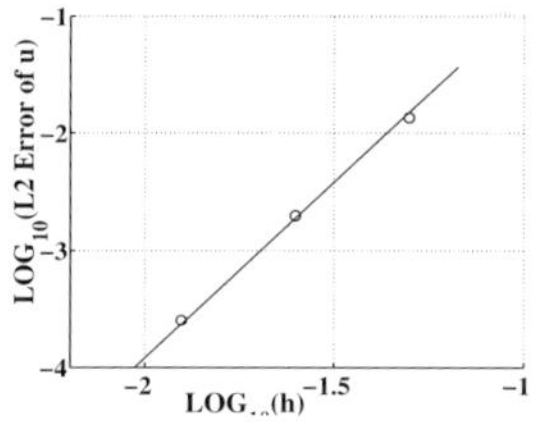

Fig. 5. Error Convergence: Airfoil Case

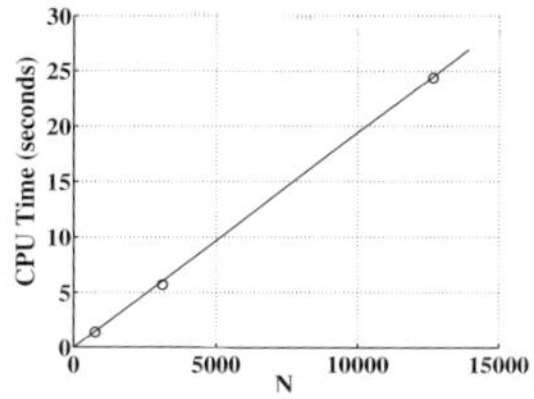

Fig. 6. CPU Time vs. Grid Size: Airfoil Case.

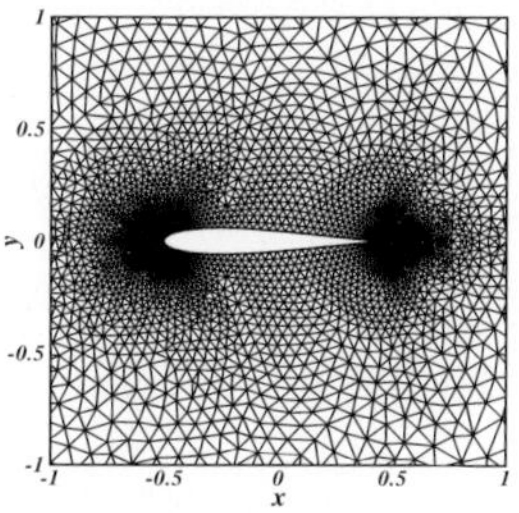

Fig. 7. 80 nodes around an airfoil.

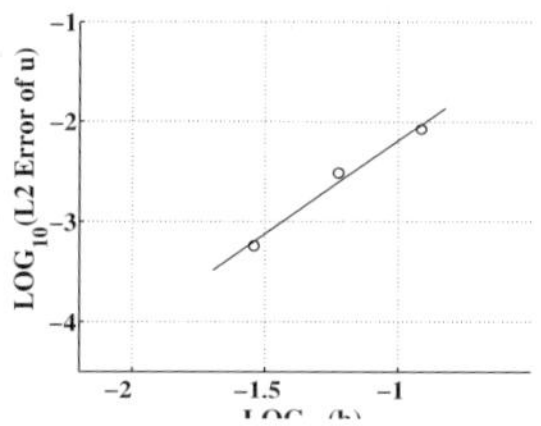

Fig. 8. Error Convergence: Airfoil Case

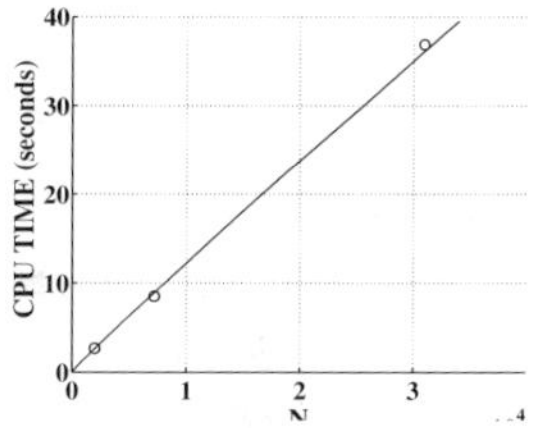

Fig. 9. CPU Time vs. Grid Size: Airfoil Case.

14110, 61468 triangles with 80, 160, 320 nodes to represent the airfoil and 3, 4, 5 multigrid levels respectively. The method converged in about 15 W-cycles. The average convergence rate is 0.5. Again, O(N) convergence is observed (Figure 9), but 3rd-order accuracy is lost (Figure 8; Slope is 1.8) possibly because the gradient is not evaluated accurately on unstructured grids.

C_p distributions around the bodies on the 80x40 grids are plotted in Figures 10-12 show excellent agreement between the exact (solid curve) and computed values (circles). The actual solution errors are below 1%. In particular, this shows that the method preserves the circulation which is very

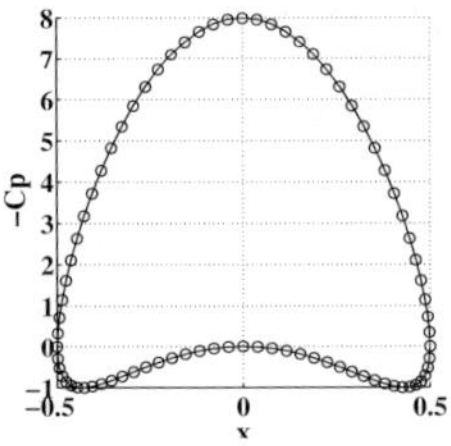

Fig. 10. C_p: Cylinder: 80x40 grid.

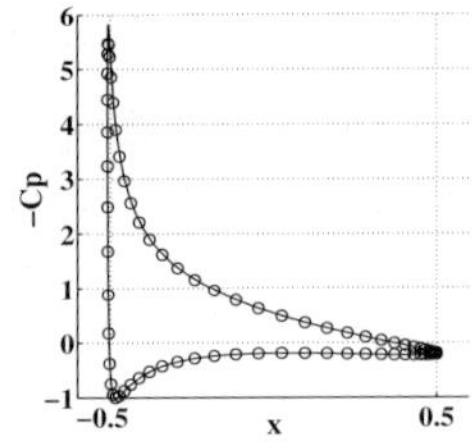

Fig. 11. C_p: Airfoil: 80x40 grid.

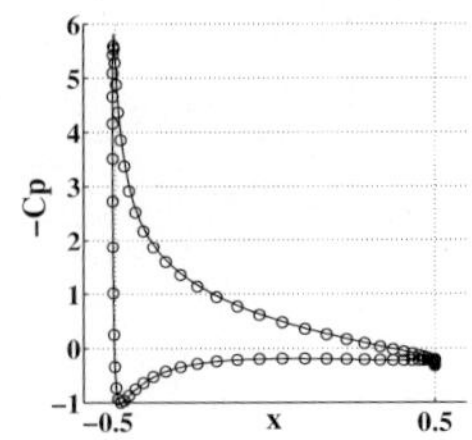

Fig. 12. C_p: Airfoil: 80 nodes on the airfoil.

important for accurate estimate of the lift force but is very difficult to achieve in the 2nd-order version [8]. We remark also that the solution for unstructured grid case is not as clean as the one for O-grid near the trailing edge. This can be improved by using least-squares quadratic gradient recovery.

5 Concluding Remarks

The method can be used for solving the Laplace (or Poisson) equation by ignoring one of the variables. This can serve also as a high-order compact discretization of the viscous operator of the Navier-Stokes equations [14].

Future work includes the application of the method for the elliptic part of the two-dimensional Euler equations. Combining the method with a hyperbolic solver with semicoarsening multigrid method, we expect to obtain a highly accurate and optimaly efficient 2D Euler solver.

References

1. P. Wesseling, An Introduction to Multigrid Methods, John Wiley & Sons, 1992.
2. D. Mavriplis and A. Jameson,AIAA Paper 0353, 1987.
3. D. Caraeni, L. Fuchs. *Proceedings of the 4th World CSCC Conference*, Vouliagmeni, Greece, July 10-15 2000.
4. L. M. Mesaros, P. L. Roe, *AIAA paper 95-1699*, 1995.
5. S. Ta'asan, *14th International Conference on Numerical Methods for Fluid Dynamics*, Bangalore India, 1994.
6. M. Hafez and W. H. Guo, Int. J. Numer. Meth. Fluids 1998; **27**: pp. 127-138.
7. A. Borzi, K. W. Morton, E. Suli, and M. Vanmaele. *SIAM J. Sci. Comp.* 1997, **18**(2):441-459.
8. H. Nishikawa, *Ph.D. Thesis*, Aerospace Engineering, University of Michigan, 2001.
9. H. Nishikawa, M. Rad, and P. L. Roe, AIAA Paper 2001-2595, 2003.
10. H. Nishikawa and B. van Leer, *J. Comput. Phys.* 2003, **190**: 52-63.
11. P. L. Roe and H. Nishikawa,Int. J. Numer. Meth. Fluids 2002; **40**: 121-136.
12. M. Rad, *Ph.D. Thesis*, Aerospace Engineering, University of Michigan, 2001.
13. J. Muller, *Ph.D. Thesis*, Aerospace Engineering, University of Michigan, 1996.
14. H. Nishikawa and P. L. Roe, ICCFD Proceedings, Toronto. Springer-Verlag, 2004.

Solution of the Fluid Dynamical Formulation of Nonlinear Schrödinger Equation with Radial Basis Function Interpolation

T. Y. Hsieh[1], J. C. Huang[2], and J. Y. Yang[1]

[1] Institute of Applied Mechanics, National Taiwan University, 106 Taipei, TAIWAN, Corresponding author:yangjy@spring.iam.ntu.edu.tw
[2] Chung Shan Institute of Science and Technology, 325 Lung-Tan, TAIWAN

1 Introduction

Recently, interest in de Broglie-Bohm formulation of quantum mechanics has increased dramatically within various fields [5]. With the wave function written in Madelung's form, the de Broglie-Bohm theory possesses an intuitive physical representation as quantum fluid dynamics (QFD), reminiscent of classical fluid dynamics [4]. Besides its conceptual importance, the potential numerical advantage of the QFD formulation over working with the Schrödinger equation can be exploited to provide an alternative solution method. Radial basis function (RBF) interpolation has been widely used as a spatial approximation scheme for PDEs [1]. The RBF is a very high order scheme, one can obtain excellent results using a coarser distribution of data points than with a finite difference (FD) approximation. It has been demonstrated multiquadrics (MQ) is an excellent spatial approximation scheme for nonlinear system of hyperbolic PDEs. Applications of RBF to QFD based on linear Schrödinger equation have been recently presented [2]. In this work, we consider the RBF scheme for QFD based on Gross-Pitaevskii(GP) equation which is a nonlinear Schrödinger equation (NLSE). After substituting the polar form of the wave function into the GP equation, one obtains a set of governing equations reminiscent of compressible gas dynamics.

2 Fluid Dynamical Formulation of Nonlinear Schrödinger Equation

We consider the Gross-Pitaevskii (GP) equation:

$$i\hbar\frac{\partial\psi}{\partial t} = -\frac{\hbar^2}{2m}\Delta\psi + V(\boldsymbol{x})\psi + g|\psi|^2\psi \tag{1}$$

where g reflects whether the interaction is repulsive ($g > 0$) or attractive ($g < 0$).

Starting with the GP equation for a dilute gas, we write the wavefunction $\psi(\boldsymbol{x},t) = \sqrt{\rho(\boldsymbol{x},t)}e^{iS(\boldsymbol{x},t)/\hbar}$ in terms of the phase and amplitude. This transformation is usually called Madelung's transformation and was originally introduced in the context of the linear Schrodinger equation for quantum mechanics. After substitution in (1) and separation of the real and imaginary parts of the equation, one obtains Quantum Fluid Dynamical Equations

$$\rho_t + \nabla \cdot (\frac{\rho \nabla S}{m}) = 0 \tag{2}$$

$$S_t + \frac{1}{2m}|\nabla S|^2 + g\rho + V - \frac{\hbar^2}{2\sqrt{\rho}}\Delta\sqrt{\rho} = 0$$

Here ρ is the probability density and S is the hydrodynamic potential. The last term at the 2nd equation of (2) corresponds to the "quantum potential ".

Madelung's transformation is, however, singular when ψ vanishes. Since this field is complex, these " topological defects" are generically located on points in two space dimensions and on lines in three dimensions. In the context of superfluidity, they are known as " quantum vortices ". The circulation of the velocity $\boldsymbol{v} = \nabla S/m$ around each of them is $\pm 2\pi$.

The advantage of the ρ-based QFD formulation is the assumption that the density ρ and phase S are more slowly varying functions than the generally oscillatory real and imaginary parts of the corresponding complex-valued wavefunction. The discussion above further suggests that the density ρ can be replaced by $ln\rho$ without altering the nature of the QFD equations. This replacement has several suggestive numerical advantages. First $ln\rho$ should be a more slowly varying function than its argument ρ, and working with the former should make the QFD method numerically even more expedient. Second, the practical dynamical range of $ln\rho$ should be much smaller than that of ρ. Third the structure of $ln\rho$ is in general simpler than that of ρ itself. So, with the wavefunction written as $\psi(\boldsymbol{x},t) = exp\{R(\boldsymbol{x},t) + iS(\boldsymbol{x},t)/\hbar\}$ and following the same procedure leading to an equivalent QFD formulation:

$$R_t + \frac{1}{m}\{\nabla R \cdot \nabla S + \frac{\hbar}{2}\nabla^2 S\} = 0 \tag{3}$$

$$S_t + \frac{(\nabla S)^2}{2m} + V + V_q + gexp(\frac{2R}{\hbar}) = 0$$

where $V_q = -\{(\nabla R)^2 + \hbar\nabla^2 R\}/2m$ is the quantum potential.

3 Numerical Scheme

We apply radial basis function (RBF) interpolation in a method-of-line (MOL) approach for solving QFD equations.

3.1 Spatial derivatives

Consider an arbitrary set of N distinct scattered points $\boldsymbol{x}_1, \boldsymbol{x}_2, \cdots, \boldsymbol{x}_N$, and the corresponding data values $v_1, v_2), \cdots, v_N)$. The RBF interpolation can approximate the data values as a function of space

$$f(\boldsymbol{x}_j) = \sum_{i=1}^{N} \alpha_i \phi(||\boldsymbol{x} - \boldsymbol{x}_i||) = \sum_{i=1}^{N} \alpha_i \phi(\boldsymbol{x}) \tag{4}$$

where ϕ_i is basis function centered at x_i and the real coefficients $\alpha_1, \alpha_2, \cdots, \alpha_N$ can be solved from the linear system:

$$y_j = f(\boldsymbol{x}_j) = \sum_{i=1}^{N} c_i \phi(||\boldsymbol{x}_j - \boldsymbol{x}_i||) = \sum_{i=1}^{N} c_i \phi_i(\boldsymbol{x}) \tag{5}$$

The gradient and Laplacian are calculated by directly applying them to the interpolated function (5).

$$\nabla f(\boldsymbol{x}_j) = \sum_{i=1}^{N} \alpha_i \nabla \phi(\boldsymbol{x}) \qquad \nabla^2 f(\boldsymbol{x}_j) = \sum_{i=1}^{N} \alpha_i \nabla^2 \phi(\boldsymbol{x}) \tag{6}$$

In this paper, multiquadrics (MQ) are employed for spatially approximating the QFD variables R and S,

$$\phi_i(\boldsymbol{x}) = \sqrt{(\boldsymbol{x} - \boldsymbol{x}_i)^2 + c_i^2} \tag{7}$$

where c_i are called shape parameters whose magnitudes are a key factor for obtaining accurate solutions.

3.2 Time Integration

The time integration is done using a 4th order Runge-Kutta method. In this paper, a system of ODEs to be solved is:

$$\frac{du_l}{dt} = f_l(t, u_1, u_2, \cdots, u_k) \quad l = 1, 2, \cdots, k \tag{8}$$

The implementation of the 4th order Runge-Kutta method is as follows:

$$\begin{aligned}
u_l^{n+1} &= u_l^n + \frac{\Delta t}{6}(a_l^n + 2b_l^n + 2c_l^n + d_l^n) \\
a_l^n &= f_l(t^n, u_1^n, u_2^n, \cdots, u_k^n) \\
b_l^n &= f_l(t^n + \frac{1}{2}\Delta t, u_1^n + \frac{1}{2}a_1^n \Delta t, u_2^n + \frac{1}{2}a_2^n \Delta t, \cdots, u_k^n + \frac{1}{2}a_k^n \Delta t) \\
c_l^n &= f_l(t^n + \frac{1}{2}\Delta t, u_1^n + \frac{1}{2}b_1^n \Delta t, u_2^n + \frac{1}{2}b_2^n \Delta t, \cdots, u_k^n + \frac{1}{2}b_k^n \Delta t) \\
d_l^n &= f_l(t^n + \Delta t, u_1^n + c_1^n \Delta t, u_2^n + c_2^n \Delta t, \cdots, u_k^n + c_k^n \Delta t)
\end{aligned} \tag{9}$$

4 Numerical Simulations and Discussions

The first case is the focusing ($g = -2$) 1D GP equation with the initial conditions of the form $\psi(x,0) = a(1 + \epsilon cos\mu x)$ where $a = 0.5$, $\epsilon = 0.1$, $\mu = 2\pi/L$, mass $m = 1$, and $L = 4\sqrt{2}\pi$[3]. This is a two mode case. We used $N = 32$ points in a period. The results are shown in Fig. 1 for probability density and velocity. Our results compare well with the multi-symplectic solution [3].

The second case is the defocusing ($g = +2$) 1D GP equation with the same initial conditions as in case 1. We used $N = 32$ points in one period. The results for density and velocity are shown in Fig. 2. This is a stable evolution solution.

The third case is the standard 1D Gaussian wave packet propagation in free space with mass $m = 20$ and momentum $p = 10$ (moving grids were used). This case is based on the linear Schrödinger equation and our results shown in Fig. 3 are in good agreement with [1].

The fourth case is 1D GP equation with an external standing light wave potential [13], $i\psi_t = -\frac{1}{2}\psi_{xx} + V\psi + |\psi|^2\psi$, external potential, $V(x) = -sin^2(kx)$, with initial conditions $\psi(x,0) = \sqrt{1 + sin^2(x)}e^{i\theta(x)}$, $tan[\theta(x)] = \sqrt{2}tan(x)$, the computational domain is $x \in (-4\pi, 4\pi)$, and in this case we use domain decomposition to separate the total domain into 8 sub-domains, each sub-domain contains 20 equally spaced grids.

The fifth case is 2D GP equation with an external standing light wave potential [14], $i\psi_t = -\frac{1}{2}\psi_{xx} - \frac{1}{2}\psi_{yy} + V\psi + |\psi|^2\psi$, $V(x,y) = -(sin^2(x) + 1)(sin^2(y)+1)$, with initial conditions $\psi(x,y,0) = \sqrt{(sin^2(x) + 1)(sin^2(y) + 1)}$ $e^{i\theta_1(x)+i\theta_2(y)}$, $tan[\theta_1(x)] = \sqrt{2}tan(x)$, $tan[\theta_2(y)] = \sqrt{2}tan(y)$, the computational domain is $x \in (-2\pi, 2\pi)$, $y \in (-2\pi, 2\pi)$, and in this case we use domain decomposition to separate the total domain into 64 sub-domains, each sub-domain contains 16 equally spaced grids.

The radial basis function interpolation developed for the quantum trajectory method has been extended in this work to solve the quantum hydrodynamics based on the Gross-Pitaevskii equation for Bose-Einstein condensation. The spatial approximation is done using radial basis functions together with their derivatives and a 4th order Runge-Kutta time integrating scheme

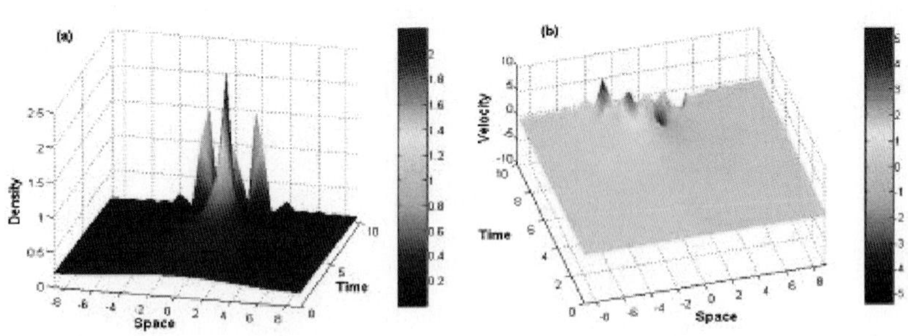

Fig. 1. Solution of case 1:, (a)$\rho = |\psi|^2 = e^{2R}$,(b) $u = \nabla S$

was adopted to yield the present solution method. Numerical experiments with several typical problems including Gaussian wavepacket propagation , focusing and defocusing without external potential have been successfully carried out and our results compare well with other solution methods. Solutions to the Gross-Pitaevskii equation with an external standing light wave potential are also presented using domain decomposition. Potential application of the present method to airfoil flows in perfect superfluid wind tunnel will be the main goal of this research.

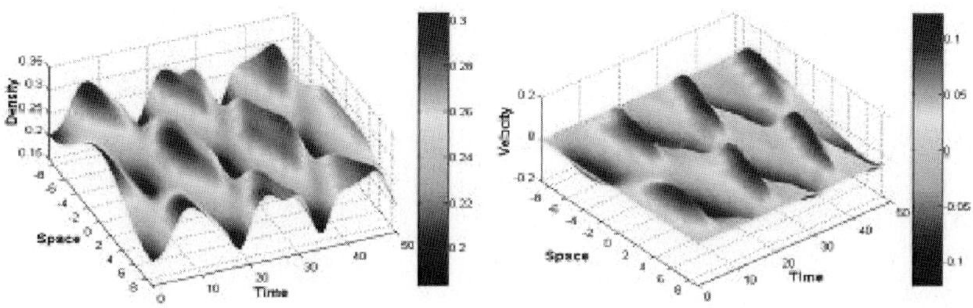

Fig. 2. Solution of case 2:, (a)$\rho = |\psi|^2 = e^{2R}$,(b) $u = \nabla S$

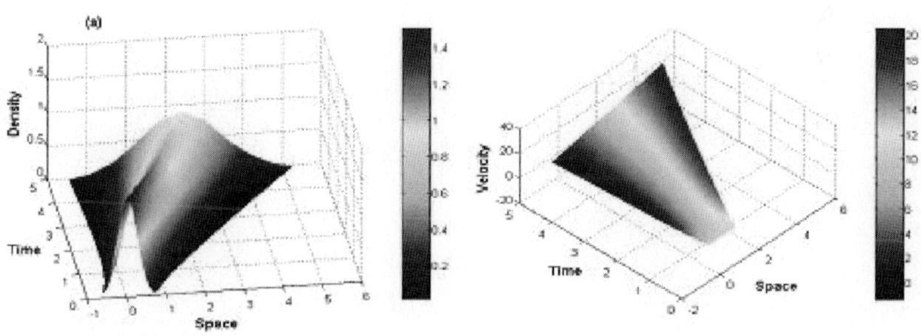

Fig. 3. Solution of case 3:, (a)$\rho = |\psi|^2 = e^{2R}$,(b) $u = \nabla S$

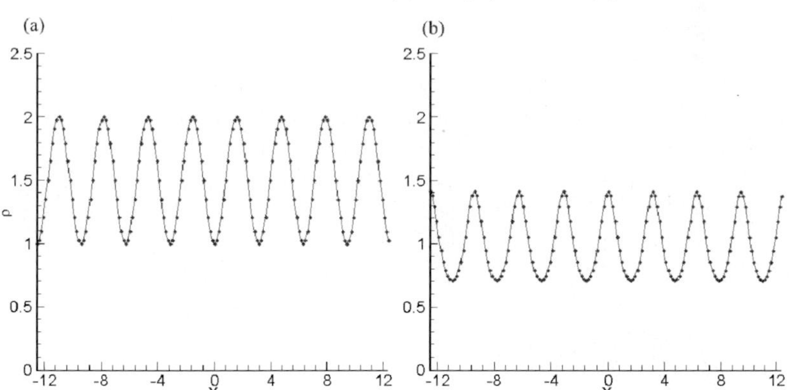

Fig. 4. Solution of case 4 at t=30:, (a)$\rho = |\psi|^2 = e^{2R}$,(b) $u = \nabla S$

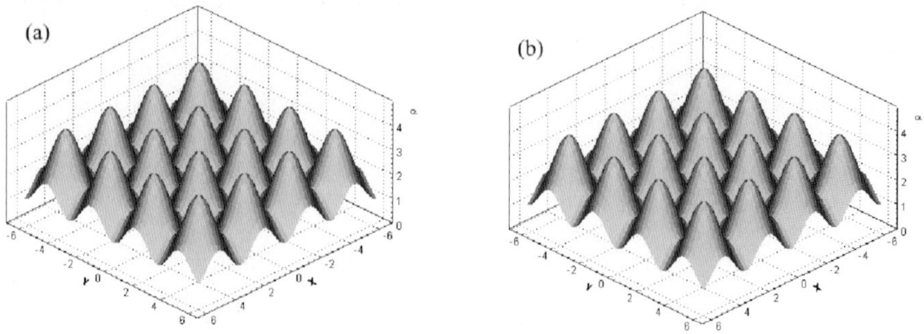

Fig. 5. Solution of case 5:, $\rho = |\psi|^2 = e^{2R}$ at (a)$t = 0$,(b)$t = 40$

Acknowledgements

This work was supported by the National Science Council, Taiwan through grant NSC-92-2212E002-094.

References

1. X.-G. Hu, T.-S. Ho and H. Rabitz, Computers Math. Appli. 43, 525-537, 2002.
2. E.J. Kansa, Computers Math. Appli. 19 , 127-145, 1990.
3. A.L. Islas, D. A. Karpeev and C.M. Schober, J. Comp. Phys., 173, 116-148, 2001.
4. C. Coste, Eur. Phys. J. B1, 245, 1998.
5. R. E. Wyatt and E.R. Bittner, Computing in Sci. & Engi., 22, 2003.
6. P.R. Holland, Cambridge University Press, New York, 1993.
7. C. J. Trahan and R. E. Wyatt, J. Comput. Phys., 185, 27-49, 2003.
8. W. Bao, S. Jin and P.A. Markowich, SIAM J. Sci. Comput., 2003.
9. P.P. Craig and J.R. Pellam, Phys. Rev., 108, No. 5, pp. 1109-1112, 1957.
10. L.D. Landau and E.M. Lifshitz, 2nd Ed., Pergamon Press, 1987.
11. P.H. Roberts and R.J. Donnelly, Annu. R. Fluid Mech., 6, 179, 1974.
12. J.Y. Yang, J.C. Huang and T.Y. Hsieh, High resolution schemes for quantum hydrodynamics based on nonlinear Schrödinger equation, accepted for oral presentation at the ICCFD3, Toronto, Canada, July 12-16, 2004.
13. B. Deconinck and J.N. Kutz, Phys. Lett. A 319, 97, 2003.
14. B. Deconinck, B. A. Frigyik, J.N. Kutz, Phys. Lett. A 283, 177, 2001.

High Resolution Schemes for Quantum Hydrodynamics Based on Nonlinear Schrödinger Equation

J. Y. Yang[1], J. C. Huang[2], and T. Y. Hsieh[1]

[1] Institute of Applied Mechanics, National Taiwan University, 106 Taipei, TAIWAN, Corresponding author:yangjy@spring.iam.ntu.edu.tw
[2] Chung Shan Institute of Science and Technology, 325 Lung-Tan, TAIWAN

1 Introduction

Recently, interest in the de Broglie-Bohm formulation of quantum mechanics has increased dramatically within various fields [1]. In the de Broglie-Bohm approach, the complex wave function is expressed in polar form and substituted into the time-dependent linear or nonlinear Schrödinger equation results a set of hydrodynamic-like equations which describe the flow of the probability. This set of equations are very similar to those of classical hydrodynamics except that an additional quantum potential term is present. The quantum potential and its associated quantum force give rise to all quantum effects such as tunneling and interference [2]. Despite its conceptually attractive features, there are major computational problems inherent to the de Broglie-Bohm approach when a direct numerical solution of the quantum hydrodynamics is attempted. In particular, the quantum potential possess some unique features, e.g., it does not respond to the intensity of the wave but rather depends upon its form, i.e., the shape and curvature of the wave function. It is also inversely proportional to the magnitude of the wave function and can often become singular whenever the magnitude becomes too small. This singularity also often causes the propagation in time to become unstable. The quantum potential and force also become large near the nodes and eventually errors accumulate and the propagation become unstable. Singularities in quantum potential gives rise to kinks in the velocity field which are similar to those of shock fronts in classical gas dynamics [2,3]. A desirable numerical method for solving the quantum hydrodynamics based on Schrödinger equation not only requires high accuracy but also the ability to handle discontinuities caused by the singularity of quantum potential. Very recently, many solution methods have been proposed for solving quantum fluid dynamics, including the quantum trajectory method for wavepacket dynamics, Lagragian method using multivariate radial basis function interpolation, and arbitrary Lagrangian-Eulerian method with moving least squares approach [4-8]. These methods are mostly applied to linear Schrödinger equation. Recently, there also has been much interest in the study of superfluid properties based on the nonlinear Schrödinger equation, such as the excitation spectrum, sound scattering by quantum vortices and vortex nucleation [3]. One of the important classes of nonlinear Schrödinger equation is the Gross-Pitaevskii equation, which describes the zero-temperature properties of the nonuniform Bose gas when the scattering length is much less than the mean interparticle spacing. This GP model assume weakly interacting Bose gas and ensures that almost all helium atoms are Bose-condensed in the state of lowest energy. After substituting the polar form of the wave function into the Gross-Pitaevskii equation, one can obtain a set of governing equations in the form of hyperbolic system of conservation laws with source terms which is reminiscent of the classical compressible gas dynamics. The source term is very stiff since it contains the quantum potential which gives all the quantum effects. The objective of this work is to devise a class of high resolution schemes for the numerical solution of the quantum hydrodynamics based on the Gross-Pitaevskii equation. Our method relies upon the conservation law form of the formulation. The time-dependent Gross-Pitaevskii equation differs from the time-dependent linear Schrödinger equation in that an additional nonlinear term exists in the GP equation which resulting the pressure term and makes the the definition of speed of sound available. The characteristic properties of the system are exploited to implement the numerical method which are quite different to the above mentioned quantum trajectory method and the multivariate radial basis function interpolation method.

2 Quantum Fluid Dynamic Equations & Its Characteristic Properties

The Gross-Pitaevskii equation for a dilute and cold gas is

$$i\hbar\frac{\partial\psi(\boldsymbol{x},t)}{\partial t} = -\frac{\hbar^2}{2m}\nabla^2\psi(\boldsymbol{x},t) + V(\boldsymbol{x})\psi(\boldsymbol{x},t) + g|\psi(\boldsymbol{x},t)|^2\psi(\boldsymbol{x},t) \tag{1}$$

where $g = 4\pi\hbar^2 a/m$ and $\psi = \psi(\boldsymbol{x},t)$ is wave function, $\boldsymbol{x} = (x,y,z)$ is the position vector, m is atomic mass,$\hbar$ is the Planck constant and a is scattering length. Assume the polar form of the wave function $\psi(\boldsymbol{x},t) = \sqrt{\rho(\boldsymbol{x},t)}e^{iS(\boldsymbol{x},t)/\hbar}$ and substitute it into the GP equation (1) and the imaginary and real parts are:

$$\frac{\partial\rho}{\partial t} + \nabla\cdot(\rho\nabla S) = 0; \quad \frac{\partial S}{\partial t} = -\frac{1}{2m}(\nabla S)^2 + \frac{\hbar^2}{2m}\frac{\nabla^2\sqrt{\rho}}{\sqrt{\rho}} - V - \rho g \tag{2}$$

Let $\nabla S = \boldsymbol{u}$ be the fluid velocity and taking gradient to the real part, one can obtain the quantum fluid dynamics equations in conservation law form with source terms:

$$\frac{\partial Q}{\partial t} + \frac{\partial F}{\partial x} = S \tag{3}$$

where $Q = (\rho, \rho u)^T$ is the state vector and $F = (\rho u, \rho u^2 + p)^T$ is the flux vector and $S = (0, \rho\frac{\partial}{\partial x}(V + V_q))^T$ is a stiff source term consisted of the quantum potential. It is found that the set of equation on the left hand side is hyperbolic and its characteristic structure can be found and the high resolution methods developed for solving hyperbolic system of conservation laws can be adopted to solve them. The flux Jacobian of (3) is :

$$A(Q) = \frac{\partial F}{\partial Q} = \begin{bmatrix} 0 & 1 \\ c^2 - u^2 & 2u \end{bmatrix} \tag{4}$$

Where the equivlent speed of sonic c is defined as $c^2 = \frac{\partial p}{\partial \rho} = \frac{g}{m}\rho$ The eigenvalues of flux jacobian matrix $A(Q)$ are $\lambda_1 = u - c$ and $\lambda_2 = u + c$ The matrices of right eigenvectors and its inverse:

$$R = \begin{bmatrix} 1 & 1 \\ u - c & u + c \end{bmatrix} \qquad R^{-1} = \begin{bmatrix} \frac{1}{2c}(u+c) & \frac{-1}{2c} \\ \frac{-1}{2c}(u-c) & \frac{1}{2c} \end{bmatrix} \tag{5}$$

Formulations for multiple-space dimensions can be similarly expressed.

3 High Resolution Schemes

Here, we adopt the method presented in [9] to solve the quantum fluid dynamics equations. Other class of methods such as weighted ENO methods [10,11] can also be employed. The Eq. (3) is discretized and expressed in the form of a conservative scheme :

$$Q_j^{n+1} = Q_j^n - \lambda(F^N_{j+\frac{1}{2}} - F^N_{j-\frac{1}{2}}), \quad F^N_{j+\frac{1}{2}} = \frac{1}{2}[F_j^n + F_{j+1}^n + R_{j+\frac{1}{2}}\Phi_{j+\frac{1}{2}}] \tag{6}$$

Here $\phi^l_{j+\frac{1}{2}}$ are the elements of vector $\Phi_{j+\frac{1}{2}}$ and are defined by:

$$\begin{aligned}\phi^l_{j+\frac{1}{2}} &= \sigma(a^l_{j+\frac{1}{2}})(g^l_j + g^l_{j+1}) + \xi\overline{\sigma}(a^l_{j+\frac{1}{2}})(d^l_j + d^l_{j+1}) \\ &\quad - \psi(a^l_{j+\frac{1}{2}} + \gamma^l_{j+\frac{1}{2}} + \xi\delta^l_{j+\frac{1}{2}})\alpha^l_{j+\frac{1}{2}}, \qquad l = 1,\ 2,\ 3\end{aligned} \tag{7}$$

where $\alpha_{j+\frac{1}{2}} = R^{-1}_{j+\frac{1}{2}}(Q_{j+1} - Q_j)$ is the characteristic variable difference. The above formula covers TVD, ENO2 and ENO3 methods, see [9] for details.

4 Numerical Experiments and Discussion

Some numerical experiments are tested to illustrate the method. The first example is a model for quasi-1D confinement in a standing light wave[12]. We use a sinusoidal potential $V(x) = -V_0 sin^2(x)$ where $V(x) = 0.5$. In this case, $V(x)$ is exactly a standing light wave. The initial conditions are

$$\rho(x,0) = V_0 sin^2(x) + B \qquad u(x,0) = \frac{\pm\sqrt{1+\frac{V_0}{B}}}{1+\frac{V_0}{B} sin^2 x} \tag{8}$$

where the parameter B is set as 1.0. The exact solutions are same as the initial values. This example was computed using 401 grid points, with computational domain: $-4\pi \le x \le 4\pi$, and periodic boundary condtions were employed at both ends. The results were output after 12000 iterations at time 30. We compare the results of 1st order Roe's scheme (Roe1), 2nd order ENO scheme (ENO2) and WENO scheme with the exact solutions. The results are shown in Figure 1 and 2. It is noted that long time integration is possible with the present characteristics-based high resolution method and the higher order accuracy of results can be obtained. We also note that the treatment of source term due to quantum potential is crucial to the success of computation.

The second testing case we considered was the one presented in [13]. We used atomic units, and atomic mass m = 1a.u., $g = 1/2$, and let potential $V(x) = 0$. The exact solutions are given by

$$\rho(x,t) = 4 - 2sech^2(x-t) \qquad u(x,t) = -[sinh^2(x-t) + cosh^2(x-t)]^{-1} \tag{9}$$

The initial conditions are obtained from above equations by letting $t = 0$. We employed 901 grid points with domain $-30 \le x \le 30$, and free boundary condition at two ends. The results were output after 6000 time iterations as time marched to 20. The results are depicted in Figure 3. We compare the results of 1st order Roe's scheme (Roe1) and 2nd order WENO scheme (WENO2) with the exact solutions. Again, it can be easily seen the effect of higher order method and the proper working of the present methods. Our solution compares very well with that in [13].

The third test case is a quantum hydrodynamic shock tube flow initially with $\rho = 4 (x \le 0.)$, $\rho = 1 (x > 0)$ and $\nabla S = 0$ on both sides. The computational domain is $-0.5 \le x \le 0.5$ with 101 grid points and output at time $t = 0.1$. The computational results using first order upwind, second-order ENO scheme, second and third order WENO schemes are shown in Figure

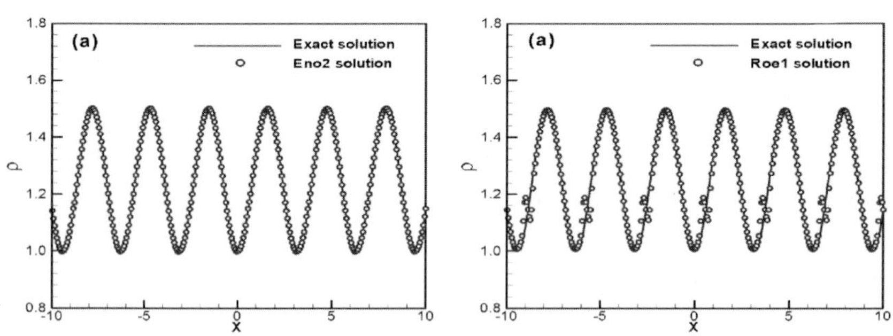

Fig. 1. Solution of test case 1: $\rho = |\psi|^2$. (a)Roe1, (b)ENO2.

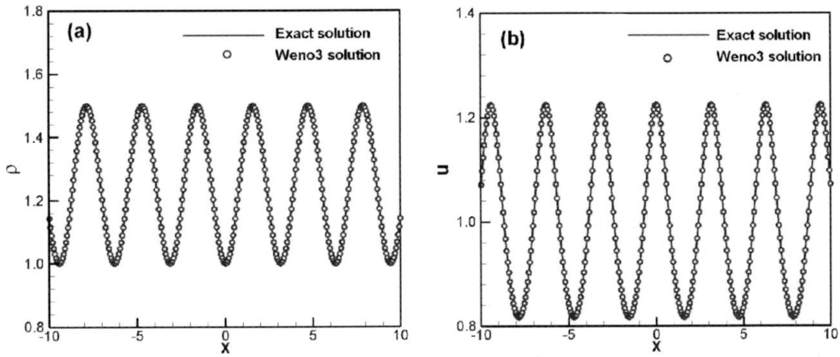

Fig. 2. Solution of test case 1 by 3rd order WENO scheme. (a)$\rho = |\psi|^2$, (b) $u = \nabla S$

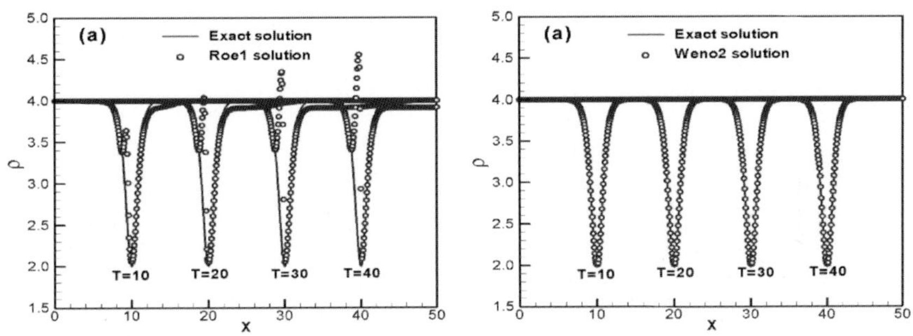

Fig. 3. Solution of test case 2: $\rho = |\psi|^2$. (a)Roe1, (b)WENO2.

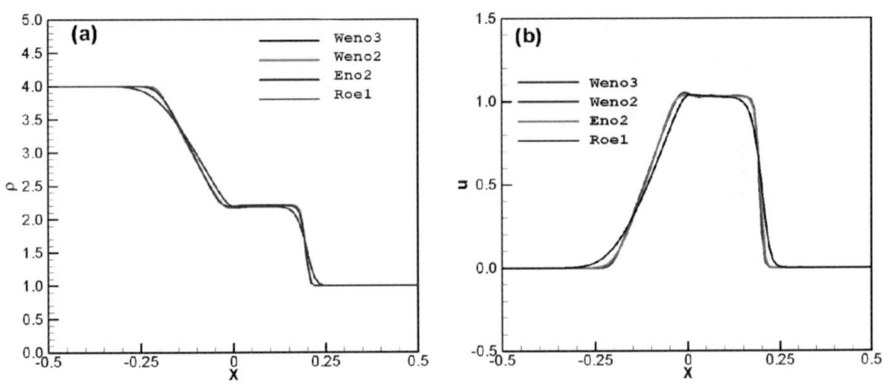

Fig. 4. A comparison of solutions for quantum shock tube flow. (a)$\rho = |\psi|^2$,(b) $u = \nabla S$

4. The high order methods give more crisp shock profile compared with the first order upwind method.

Based on the present although preliminary, however, encouranging results for the quantum fluid dynamics, it is fair to conclude that the present method may provide an robust solution method to the nonlinear Schrödinger equations. The real challenge will be in the multi-dimensional superfluid flow past a finite obstacle, such as cylinder or airfoil with boundary conditions [14]. Progress towards this goal is currently underway and will be reported in a sequel.

Acknowledgements

This research was supported by the National Science Council, Taiwan, through grant NSC-92-2212E002-094.

References

1. P. R. Holland, The Quantum Theory of Motion (Cambridge University Press, New York, 1993)
2. B. K. Kendrick, J. Chem. Phys. Vol. 119, 5805-5817, 2003
3. A. L. Islas, D.A. Karpeev and C.M. Schober, J. Comput. Phys., 173, 116-148 (2001)
4. X.G. Hu, T.S. Ho and H. Rabitz, Computers Math. Appli. 43, 525-537 (2002)
5. E.J. Kansa, Computers Math. Appli., 19, 127-145 (1990).
6. R. E. Wyatt and E.R. Bittner, Computing in Sci & Engi., p.22, May (2003).
7. C.J. Trahan and R.E. Wyatt, J. Comput. Phys., 185, pp. 27-49 (2003).
8. W. Bao, S. Jin and P.A. Markowich, SIAM J. Sci. Comput., (2003).
9. J.Y. Yang and C.A. Hsu, High resolution nonsocillatory schemes for unsteady compressible flows, AIAA J. Vol.30, pp.1570-1575 (1992)
10. G.S. Jiang and C.W. Shu, J. Comput. Phys., 126, 202-228, (1996)
11. J.Y. Yang, Y.C. Perng and R.H. Yen, AIAA J., Vol.39, 2082-2090 (2001)
12. J. C. Bronski, L. D. Carr, B. Deconinck, and J. N. Kutz, Physical Review Letters, Vol.86, No.8 Feb. (2001)
13. L. D. Carr, C. W. Clark, and W. P. Reinhardt, , Physical Review A, Vol.62, 063610 (2000)
14. P.H. Roberts and R.J. Donnelly, Annu. Rev. Fluid Mech., Vol. 6 (1974)

Multigrid Acceleration for Transonic Aerodynamic Flow Simulations Based on a Hierarchical Formulation

Mohamed Hafez and Essam Wahba

Department of Mech. and Aero. Eng., University of California, Davis CA 95616, USA, mhafez@ucdavis.edu

Summary. In [1], the authors used a hierarchical formulation based on a Helmholtz velocity decomposition to simulate transonic flows over airfoils. The potential flow formulation is augmented with entropy and vorticity corrections and the numerical results are compared to standard Euler and Navier-Stokes calculations. For many aerodynamic applications, the corrections are limited to relatively small regions; the flow in the near and far fields is irrotational and isentropic. The entropy and vorticity corrections are governed by convection/diffusion equations while the non-homogeneous potential equation is of mixed type; elliptic in the subsonic domain and hyperbolic in the supersonic one. The forcing function represents the necessary correction for mass conservation. Upwind schemes are used for the convection terms of the scalar equations of the corrections and for the potential equation in the supersonic region. The formulation can be viewed as an implementation of a viscous/inviscid interaction procedure which is equivalent to Navier-Stokes equations in the inner field.

In this paper, convergence acceleration techniques are applied for such a formulation using single and multiple grids. For a single grid, optimal relaxation parameters for subsonic potential flow regions and local artificial time steps for the scalar correction equations have been used. A full multigrid technique is implemented to the augmented potential equation. Only three grids: coarse, intermediate and fine meshes are used. Results for both inviscid and viscous transonic flows are presented. It is noticed that on a single grid, the potential and the viscous flow calculations, based on the present formulation, have comparable convergence histories. The limited applications of multigrid result in an order of magnitude saving of the work units for both calculations. More savings should be achievable with more sophisticated multigrid procedures.

1 Introduction

In the last three decades, multigrid convergence acceleration techniques have been used in computational fluid dynamics. It has been proven to be a very effective tool for elliptic problems. Applications to flow simulations, governed by more complicated equations have been advocated by A. Brandt. His early ideas were summarized in [2]. Since then, progess has been made by many researchers. Recently, textbook multigrid efficiency has been obtained by Thomas et al [3]. Very impressive results for two dimensional transonic inviscid flows are reported in [4] by Caughey and Jameson. They used LU factorization with local iterations, rather than Runge Kutta schemes adopted earlier in [5], as a smoothing operator. Further investigations are, however, needed for viscous flow problems.

In the present paper, convergence acceleration techniques are applied for transonic aerodynamic steady flow simulations based on the hierarchical formulation of [1]. The full multigrid procedure is applied to accelerate the convergence of iterative line relaxation methods to a steady state solution of a potential equation (with nonhomogeneous terms). Only three grids (course, intermediate and fine) are used. The calculations start with uniform flow as an initial guess for the coarse mesh. Mesh refinement is adopted to obtain a good initial guess for the fine mesh calculations. Multigrid is then applied with a V-cycle and standard restriction and prolongation schemes. The non-homogeneous terms are updated only on the fine mesh through solving scalar equations for the corrections to the potential field. The density is calculated in terms of the speed, augmented with entropy.

Numerical results for inviscid and viscous flows with single and multiple grids are presented. The results of standard potential flow calculations (i.e. without corrections) are comparable to the early work of Jameson where he introduced the MAD scheme, see [6]. On the other hand, for viscous flow problems, the efficiency of the present results is comparable to that of [7].

The details of the hierarchical formulation can be found in [1]. In the following, the numerical methods are given and then some numerical results are discussed.

2 Numerical Methods

In this section, the grid, the scheme and the solver are briefly outlined. To discretize the domain, an algebraic grid generation is adopted. A C-grid is used as in [1]. Finite volumes are chosen to discretize the equations, however, finite elements can be used as well. Artificial time dependent terms are added to the momentum equations to help the overall convergence. Artificial viscosity is needed for numerical stability for the potential equation in the supersonic region and for the convection terms in the momentum equations. Deferred correction approach is implemented where the original equations are written in conservation forms.

The multigrid acceleration techniques are used only for the augmented potential equation. The momentum equations are solved only on the fine mesh. This simple strategy is a very convenient way to avoid many problems associated with high speed flows. The restriction of multigrid application to the potential equation may not, however, provide the optimal convergence rate and other strategies dealing with the convection/diffusion equations will be studied elsewhere.

The present formulation allows easily the identification of the acoustic mode represented by the potential equation, while the entropy and vorticity modes are governed by scalar convection/diffusion equations. For many transonic aerodynamic problems, most of the near and far flow fields are subsonic, hence an effective treatment of the elliptic potential equation is very attractive. On the other hand, for standard Euler and Navier-Stokes equations, the identification of the modes are not obvious nor simple in general. Many attempts to exploit the structure of certain schemes to achieve splitting of the modes on the discrete level are encouraging but usually there are restrictions on the grid and calculations are more complicated.

A type dependent line relaxation procedure with optimal parameters is used for single grid calculations for comparison. An order of magnitude reduction in work units is possible with the present multigrid acceleration. Full weighting for restriction and bilinear interpolation for prolongation are implemented in a straightforward manner.

3 Numerical Results

Several cases are calculated for transonic flows over NACA0012 airfoil at different angles of attack, Mach and Reynolds numbers. These cases include:

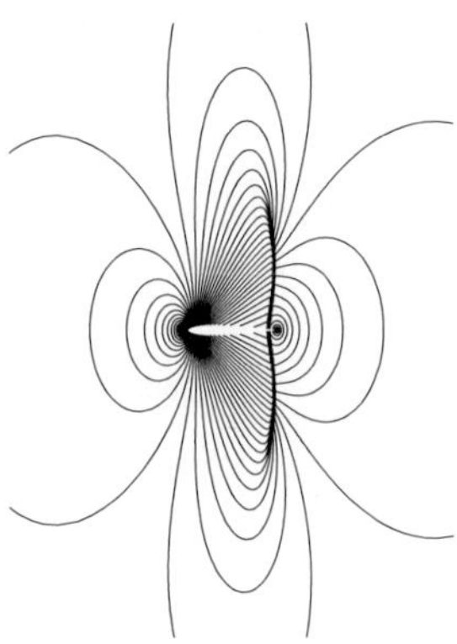

Fig. 1. Mach contours ($M_\infty = 0.86$, Potential Flow, $\alpha = 0^o$)

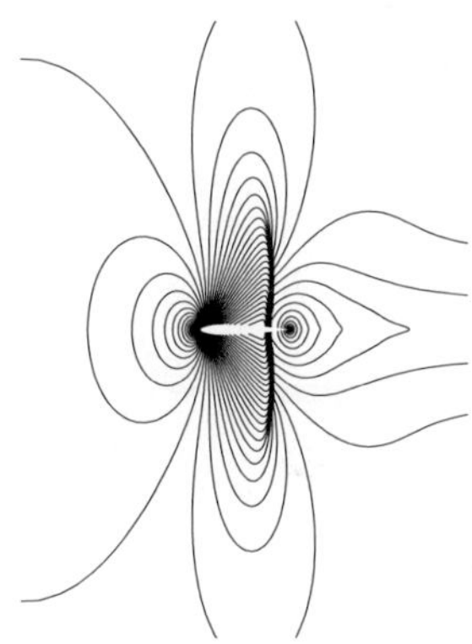

Fig. 2. Mach contours ($M_\infty = 0.86$, Rotational Flow, $\alpha = 0^o$)

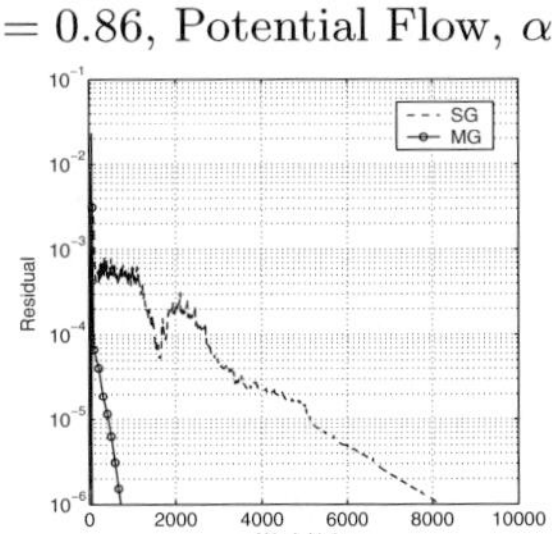

Fig. 3. Convergence History ($M_\infty = 0.86$, Potential Flow, $\alpha = 0^o$)

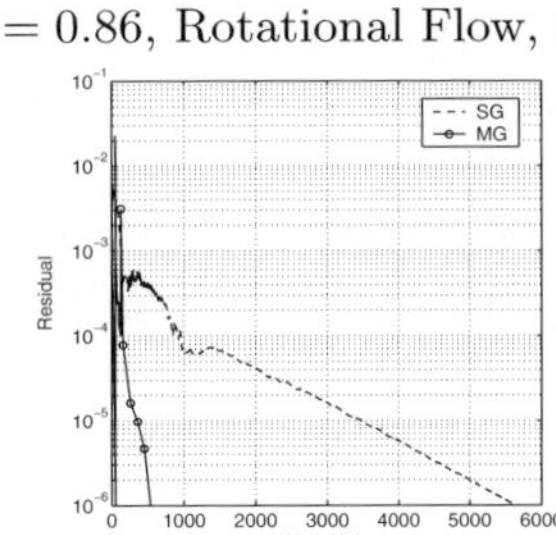

Fig. 4. Convergence History ($M_\infty = 0.86$, Rotational Flow, $\alpha = 0^o$)

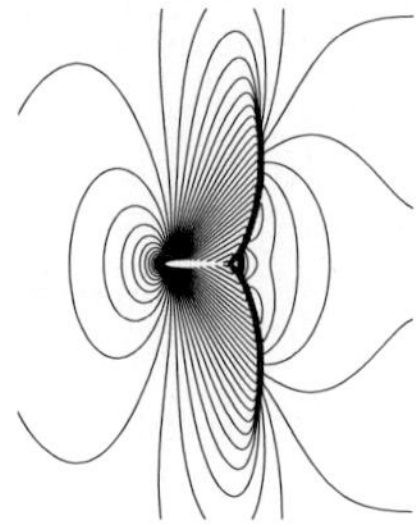

Fig. 5. Mach contours
($M_\infty = 0.9$, Potential Flow, $\alpha = 0^o$)

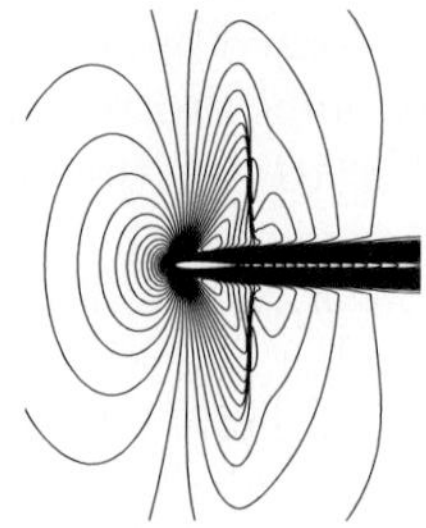

Fig. 6. Mach contours
($M_\infty = 0.9$, $Re = 500$, $\alpha = 0^o$)

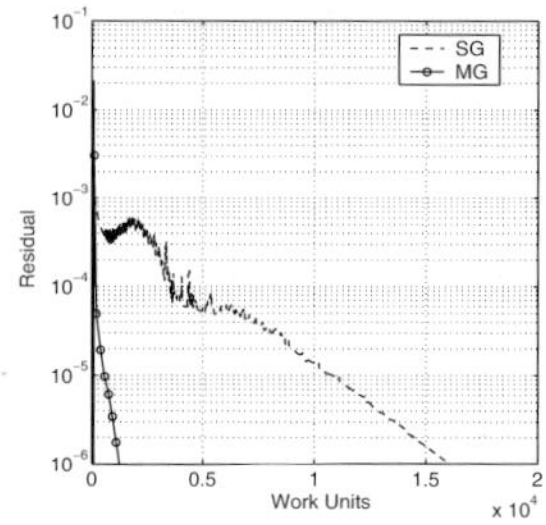

Fig. 7. Convergence History
($M_\infty = 0.9$, Potential Flow, $\alpha = 0^o$)

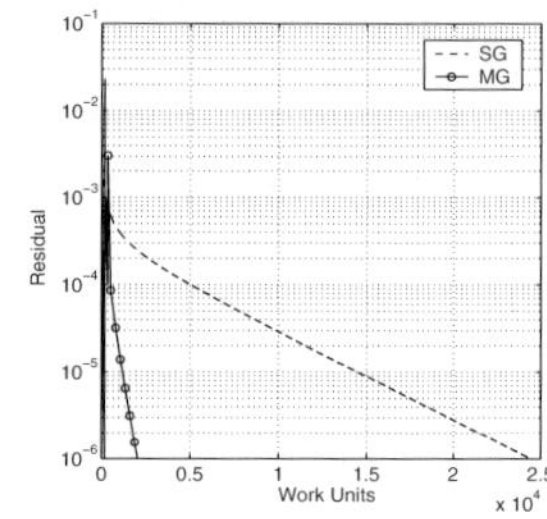

Fig. 8. Convergence History
($M_\infty = 0.9$, $Re = 500$, $\alpha = 0^o$)

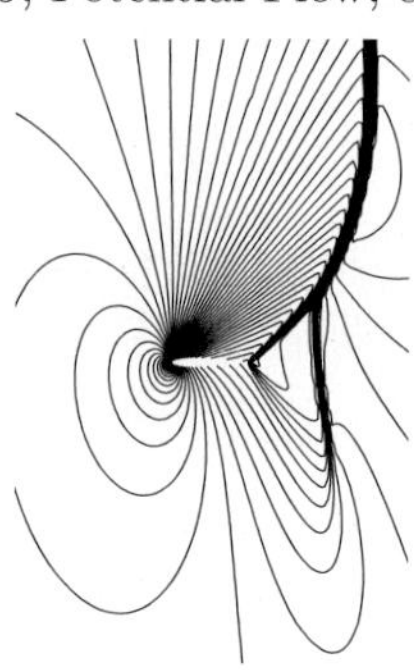

Fig. 9. Mach contours
($M_\infty = 0.9$, Potential Flow, $\alpha = 10^o$)

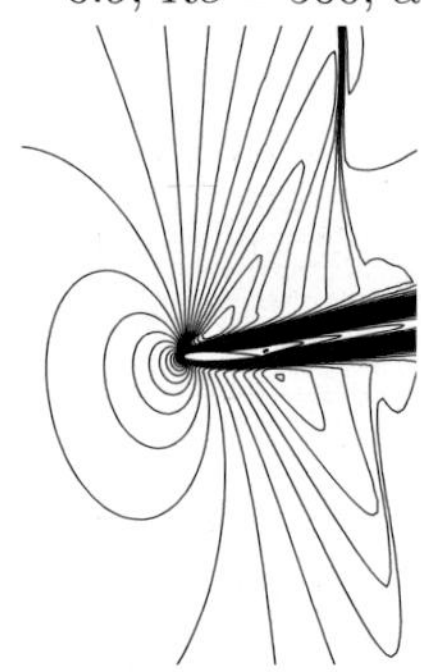

Fig. 10. Mach contours
($M_\infty = 0.9$, $Re = 500$, $\alpha = 10^o$)

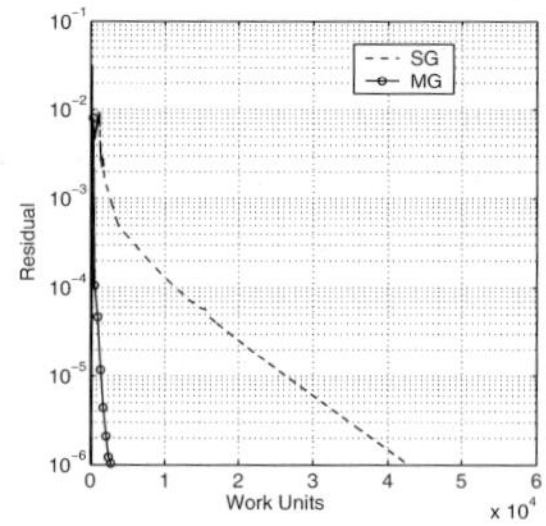

Fig. 11. Convergence History
($M_\infty = 0.9$, Potential Flow, $\alpha = 10^o$)

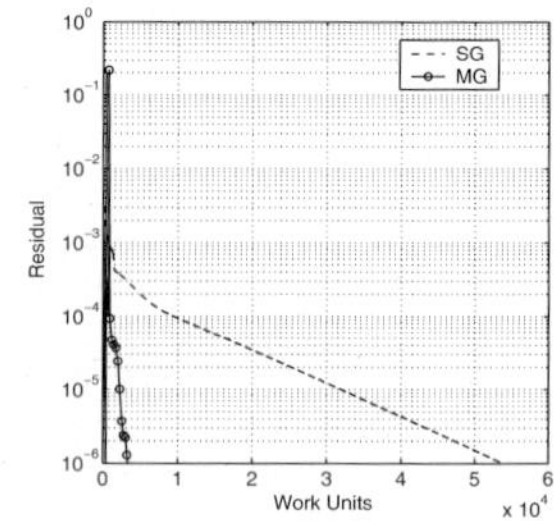

Fig. 12. Convergence History
($M_\infty = 0.9$, $Re = 500$, $\alpha = 10^o$)

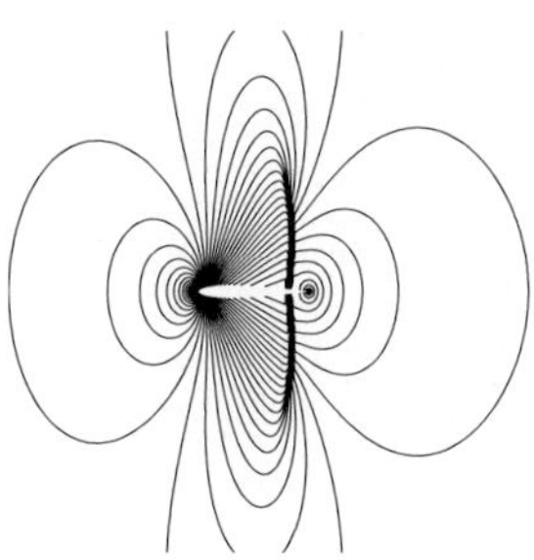

Fig. 13. Mach contours
($M_\infty = 0.85$, Potential Flow, $\alpha = 0^o$)

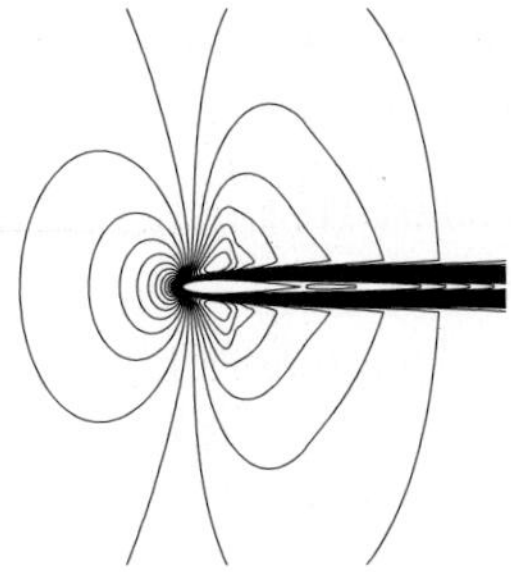

Fig. 14. Mach contours
($M_\infty = 0.85$, $Re = 2000$, $\alpha = 0^o$)

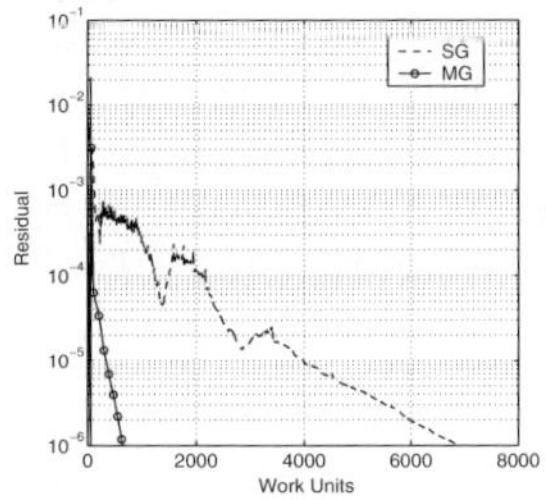

Fig. 15. Convergence History
($M_\infty = 0.85$, Potential Flow, $\alpha = 0^o$)

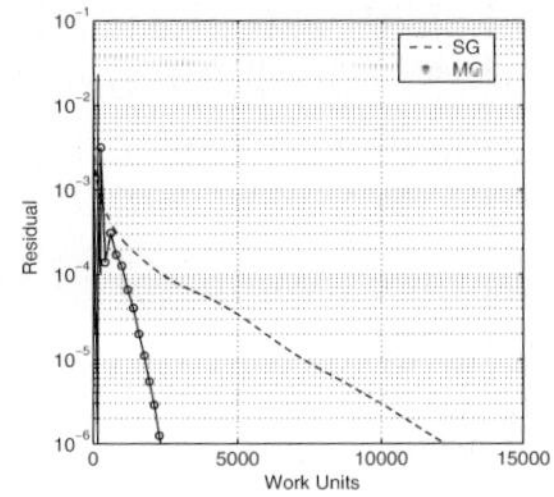

Fig. 16. Convergence History
($M_\infty = 0.85$, $Re = 2000$, $\alpha = 0^o$)

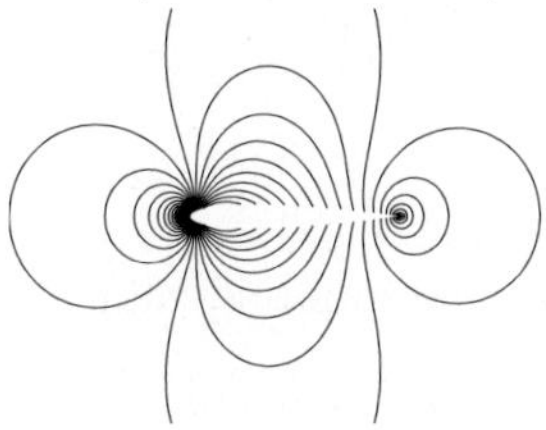

Fig. 17. Mach contours
($M_\infty = 0.5$, Potential Flow, $\alpha = 0^o$)

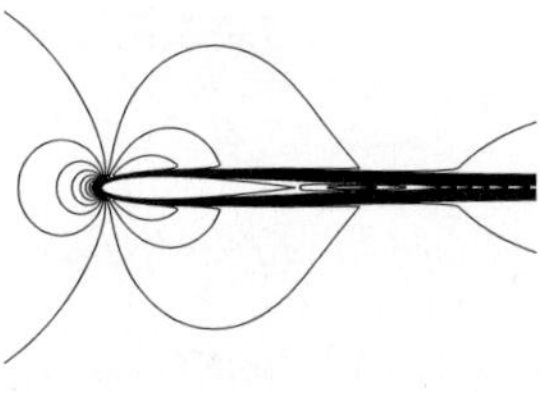

Fig. 18. Mach contours
($M_\infty = 0.5$, $Re = 10000$, $\alpha = 0^o$)

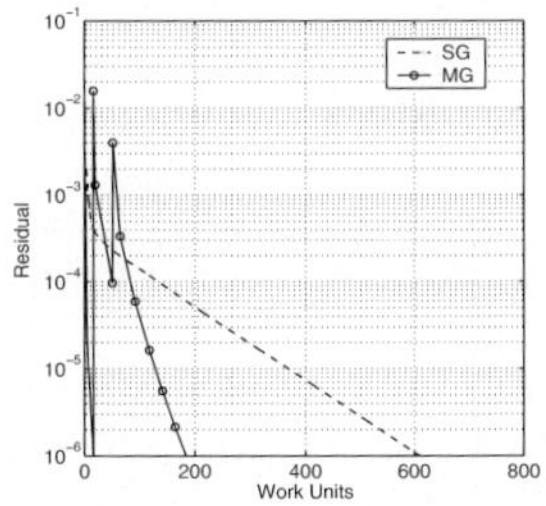

Fig. 19. Convergence History
($M_\infty = 0.5$, Potential Flow, $\alpha = 0^o$)

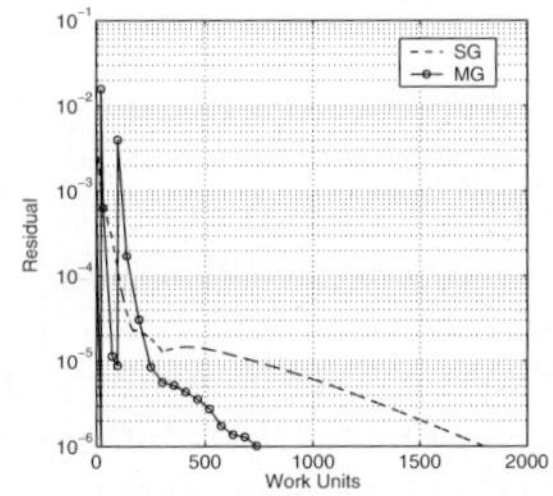

Fig. 20. Convergence History
($M_\infty = 0.5$, $Re = 10000$, $\alpha = 0^o$)

potential flows, inviscid rotational flows and viscous flows. The residual histories for single and multiple grids are compared for each case.

3.1 Inviscid Rotational Flows

Figures 1 and 2 show the Mach contours for potential and inviscid rotational flows at $M_\infty = 0.86$ and $\alpha = 0^o$. The corresponding convergence histories are plotted in Figs. 3 and 4. Notice that the convergence rates of the potential and rotational flows are very similar for single and multiple grids.

3.2 Viscous Flows

The results for $M_\infty = 0.9$ at $\alpha = 0^o$ and $\alpha = 10^o$ for potential and viscous flows ($Re = 500$) are plotted in Figs. 5-12.

The results for $M_\infty = 0.85$ at $\alpha = 0^o$ for potential and viscous flows ($Re = 2000$) are plotted in Figs. 13-16. The work unit for viscous flow simulations includes the residual evaluations of the augmented potential equation and the momentum equations on the fine mesh. The latter are updated only in the viscous layer.

Higher Reynolds number flows ($Re = 10000$) are calculated for $M_\infty = 0.5$ at $\alpha = 0^o$, see Figs. 17-20. For these cases, a highly stretched grid is used without noticable degeneration of multigrid performance. The above numerical results are in good agreement with those available in the literature for the same test cases, see [8], [9].

References

1. M. Hafez and E. Wahba: *Numerical Simulations of Transonic Aerodynamic Flows based on a hierarchical formulation*, To appear in Int. J. Numer. Methods Fluids
2. A. Brandt: *Multigrid Techniques: 1984 Guide with Applications to Fluid Dynamics*, GMD-Studie, vol. 85. GMD-FIT, 1985
3. J. L. Thomas, B. Diskin and A. Brandt: *Textbook Multigrid Efficiency for Fluid Simulations*, Annu. Rev. Fluid Mech. Vol. 35, pp 317-340, 2003
4. D. A. Caughey and A. Jameson: *Fast Precondintioned Multigrid Solution of the Euler and Navier-Stokes equations for steady compressible flows*, Int. J. Numer. Meth. Fluids vol. 43, pp 537-553, 2003
5. A. Jameson : *Solution of the Euler equations for two-dimensional, transonic flow by a multigrid method*, Applied Math. and Computation, vol. 13, pp 327-356, 1983
6. A. Jameson : *Accelerations of Transonic Potential Flow Calculations on Arbitrary Meshes by the Multiple Grid Method*, AIAA Paper 79-1458, 1979
7. D. Drikakis, O. Iliev and D. Vassileva: *On Multigrid Methods for the Compressible Navier-Stokes Equations*, ICLSSC 2001, Springer-Verlag, pp 344-352, 2001
8. H. Hollanders, A. Lerat, R. Peyert: *3-D Calculations of Transonic Viscous Flows by an Implicit Method*, AIAA Paper 83-1953, 1983
9. M. M. Hafez and W. H. Guo: *Simulation of Steady Compressible Flows Based on Cauchy/Riemann Equations and Crocco's Relation, Part II: Viscous flows*, Int. J. Numer. Meth. Fluids, Vol. 31, pp 325-343, 1999

A New Accurate and Stable Least-Square Method to Compute the Gradient on Non-Orthogonal Meshes

Céline Béchaud, Khalid Yaqobi, Frédéric Archambeau, and Namane Méchitoua

EDF R&D, Département Mécanique des Fluides et Transferts Thermiques, 6 quai Watier -BP 49- 78401 Chatou cedex, FRANCE, celine.bechaud@edf.fr

1 Introduction

Code_Saturne is used by EDF in various industrial fields such as process engineering, aeraulics, coal combustion and nuclear applications [1]. It is a general finite volume CFD code developed by EDF under quality assurance for 2D and 3D simulations, laminar and turbulent flows, conjugate heat transfer (coupling with SYRTHES), including combustion modeling and a Lagrangian module. The solver relies on a finite volume method on arbitrary meshes (hybrid, with hanging nodes, any type of element). All variables are located at the cell centers. The solver is time marching, with a predictor-corrector scheme for Navier-Stokes equations. Standard RANSE models are included (k-epsilon, RSM) as well as a Large Eddy Simulation.

To solve the equations for convection-diffusion, the cell gradient of different variables, such as the pressure, the velocity or the scalars, has to be determined. Two discrete methods are available in the code for this calculation:

- an iterative method, called the midpoint rule integral approximation with deferred correction, MRDC [2]. This method is accurate but relatively time consuming.
- the least-square method [3], LS, which is faster but appeared less robust for distorted meshes.

The present paper describes a new approach implemented in *Code_Saturne* to compute cell gradients. This method is a variation on the least square method, denoted LSMod, characterized by a careful selection, within the whole neighborhood of a cell, of the cells that are best located to allow for better gradient calculation.

2 Numerical Methods to Compute the Cell Gradient

The aim of this study is to compute the cell gradient of a variable P. The notations used in the following descriptions are summarized in Figure 1. The first method presented is the midpoint rule integral approximation, MRDC, which gives:

Fig. 1. Geometrical points for the internal faces (left) and boundary faces (right).

$$\int_{\Omega_i} \underline{\text{grad}}\, P\, d\Omega = |\Omega_i|(gradP)_I = \sum_{j \in Neighbor(i)} P_{ij}.\underline{S}_{ij} + \sum_{k \in \gamma_b(i)} P_{b_{ik}}.\underline{S}_{b_{ik}} \quad (1)$$

where $Neighbor(i)$ is the set of cell numbers of the neighbors of the cell I and $\gamma_b(i)$ is the set of boundary faces of the cell I pertaining to the boundary of the domain.

We use Taylor development to express the value of the variable P at the faces ij and b_{ik} and we assume a first order approximation for the gradient as follows:

$$\begin{aligned} P_{ij} = P_{O_{ij}} + \underline{O_{ij}F_{ij}}.(gradP)_{O_{ij}} \text{ with } P_{O_{ij}} &= \alpha_{ij}P_I + (1-\alpha_{ij})P_J \\ (gradP)_{O_{ij}} &= \tfrac{1}{2}[(gradP)_I + (gradP)_J] \\ P_{b_{ik}} = a + b * P_{I'} \text{ with } P_{I'} &= P_I + \underline{II'}.(\underline{\text{grad}}\ P)_I \end{aligned} \quad (2)$$

where a and b are boundary condition coefficients defined by the users. When the mesh is orthogonal, we have $O_{ij} = F_{ij}$ and $I = I'$, then, the gradient correction for P_{ij} and $P_{I'}$ vanishes. In the more general case, when the mesh is not orthogonal, this formulation couples gradient of the cell I to the gradient of the neighbor cells J. To make the system easier to solve, an iterative procedure is adopted. The gradient in cell I, $\underline{G}_{c,i}$, is treated implicitly while the gradient in the neighboring cells, $\underline{G}_{c,j}$, is handled with a deferred correction technique. The solution $\underline{G}_{c,i}$ is assumed to be the limit of a series $\underline{G}^k_{c,i}$ (k tending to infinity) with each $\underline{G}^{k+1}_{c,i}$ solution of such a linear system:

$$A_i^{k+1}\underline{G}^{k+1}_{c,i} = B_i^k \underline{G}^k_{c,j} + C_i^k \quad (3)$$

This iterative procedure is time consuming.

Then, the least square method (LS) is detailed with the new developments which take into account the extended neighborhood (LSMod). We remind the reader that the cell center method used in *Code_Saturne* allows us to compute easily the face gradient. Then we look forward to use this value to determine the gradient at the cell center. The rationale of this approach is to find a vector $\underline{G}_{c,i}$ such as for each internal face ij and boundary face ik, we have the following equalities:

$$\begin{cases} \underline{G}_{c,i} \cdot \underline{d}_{ij} = \underline{G}_{f,ij} \cdot \underline{d}_{ij} = \frac{P_J - P_I}{||\underline{IJ}||} \\ \underline{G}_{c,i} \cdot \underline{d}_{b,ik} = \underline{G}_{f,b,ik} \cdot \underline{d}_{b,ik} \frac{P_{b,ik} - P_{I'}}{||\underline{I'F}||} \end{cases} \quad (4)$$

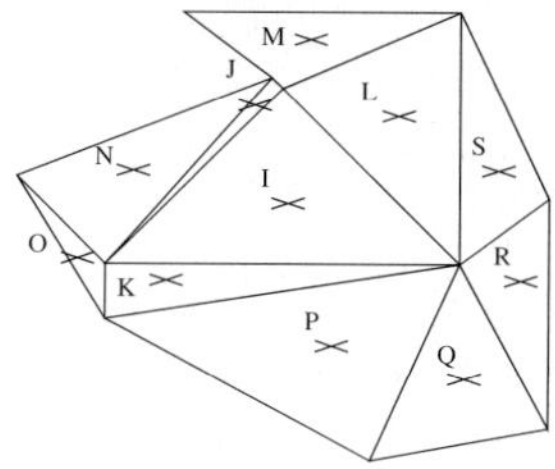

Fig. 2. Non-orthogonal mesh.

where $\underline{G}_{f,ij} \cdot \underline{d}_{ij}$ and $\underline{G}_{fb,ik} \cdot \underline{d}_{b,ik}$ are estimates of the gradient projected on the directions $\underline{d}_{ij}$ and $\underline{d}_{b,ik}$ respectively. As it is not always possible to obtain equality, we minimize the following function $\mathcal{F}_i$:

$$\mathcal{F}_i(\underline{G}_{c,i}, \underline{G}_{c,i}) = \frac{1}{2} \sum_{j \in Neighbor(i)} \left[\underline{G}_{c,i} \cdot \underline{d}_{ij} - \underline{G}_{f,ij} \cdot \underline{d}_{ij}\right]^2 + \frac{1}{2} \sum_{k \in \gamma_b(i)} \left[\underline{G}_{c,i} \cdot \underline{d}_{b,ik} - \underline{G}_{f,b,ik} \cdot \underline{d}_{b,ik}\right]^2 \tag{5}$$

This is possible by driving to zero the derivatives of $\mathcal{F}_i(\underline{G}_{c,i}, \underline{G}_{c,i})$ according to each component $(G_{c,i,x}, G_{c,i,y}, G_{c,i,z})$ of the unknown vector $\underline{G}_{c,i}$. Using (4) leads to a $3 * 3$ system which is solved by a direct method. This method is efficient only when $\underline{G}_{f,ij} \cdot \underline{d}_{ij}$ and $\underline{G}_{f,b,ik} \cdot \underline{d}_{b,ik}$ are good approximations of the face normal gradients.

The aim of the new method presented in this paper is to take into account an extended neighborhood which corrects the errors induced by the arbitrary mesh. Indeed, when strong non-orthogonality appears in the mesh (Figure 2), the face gradient projection $\underline{G}_{f,ij} \cdot \underline{d}_{ij}$ is not the most appropriate to predict the cell gradient. On Figure 2, the gradient $\underline{G}_{f,in} \cdot \underline{d}_{in}$ is a better approximation of the gradient normal to the face IJ than $\underline{G}_{f,ij} \cdot \underline{d}_{ij}$. For this reason, when strong non-orthogonality appears at the face IJ we add to the fonction $\mathcal{F}_i$ the gradients based on the extended neighborhood. Now, we should define the best way to choose the extended neighborhood.

In a first phase, all the cells sharing at least one vertex with cell I were considered. This solution turned out to be memory and time consuming. Hence, in a second phase, the angle between the vectors $\underline{IJ}$ and the normal face $\underline{n}_{ij}$ for each face ij has been used as a criterion. If this angle is large, all the cells sharing at least one of the face vertices are included in the extended neighborhood, otherwise only the cell J is included. We chose this criterion because the main difficulty occurring in our meshes comes from the non-orthogonality and not from large discrepancies in cell sizes.

These new definitions of the neighborhood of the mesh cells still yield to a $3 * 3$ system and improve the accuracy and stability of this method as it is shown in the next section.

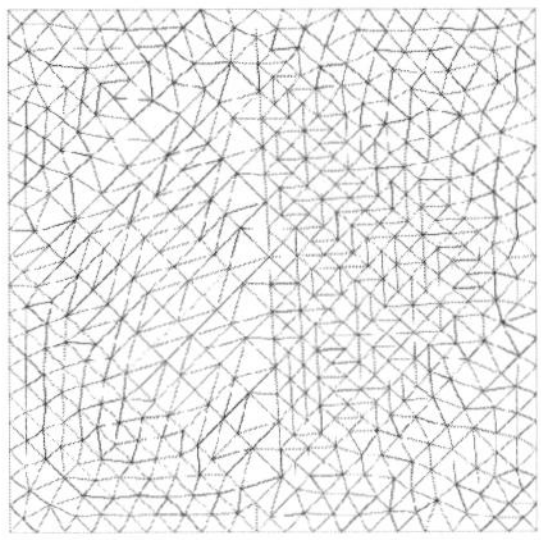

Fig. 3. Coarser mesh for the analytical test case.

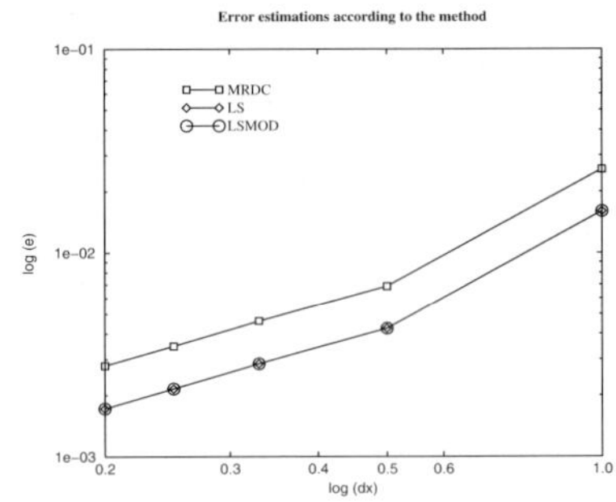

Fig. 4. Estimation of the errors according to the method.

3 Results

The calculations have been carried out using a pentium IV xeon computer of 1,7 Ghz for the analytical case and a Fujitsu VPP 5000 vector computer for the semi-industrial case. The previous notations, MRDC, LS and LSMod, are used on the following figures.

The analytical case consists in calculating the gradient of a scalar function. The function studied is defined as follows:

$$f(x, y, z) = x^2 + 2y^2 + 3 * (0.025)^2 \qquad (6)$$

The 2D grid mesh used for this test case is non-orthogonal and the origin is located on the bottom left corner (*cf* Figure 3). The accuracy of the different methods is compared in Figure 4. Indeed, different computations have been done refining the initial mesh by 2, 3, 4 and 5 and for each simulation, the L norm of the error has been calculated. All three curves exhibit the same slope, characteristic of a first order method. For this mesh, the least square methods present results which are a little more accurate than those of the MRDC method. These curves also show that, here, the extension of the neighborhood for the LS method (LSMod) does not affect the accuracy.

The industrial test case is the injection of a turbulent gas mixture ($He + CO_2$) in a pipe traversed by a main flow of CO_2 (*cf* Figure 5). This case is interesting as we can confront the calculated results with the experimental ones [4]. Besides, this test case is more complex in terms of physics and geometry than the previous analytical one. All cell gradients necessary throughout the whole calculation (velocity, scalars, pressure...) are calculated using the same method (MRDC, LS or LSMod).

The mesh (*cf* Figure 6) has been generated on purpose without trying to obtain a good quality of the cells. Thus 11% of the faces are non-orthogonal with an angle greater than 45. These non-orthogonalities induce important numerical difficulties; for instance, the calculation carried out with the LS method for computing all cell gradients required throughout the simulation does not converge. This explains the lack of data on the figures for this

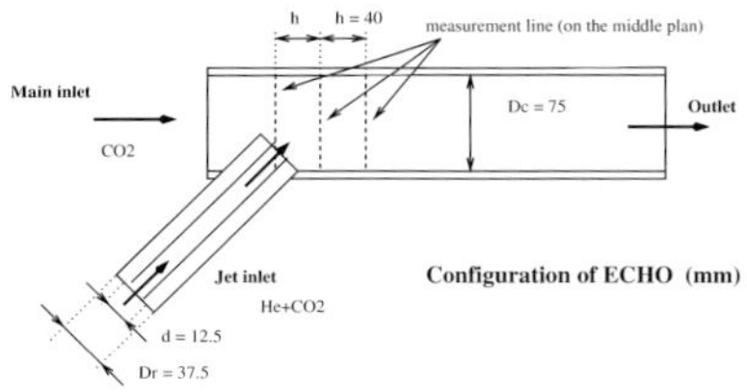

Fig. 5. Configuration of the calculus

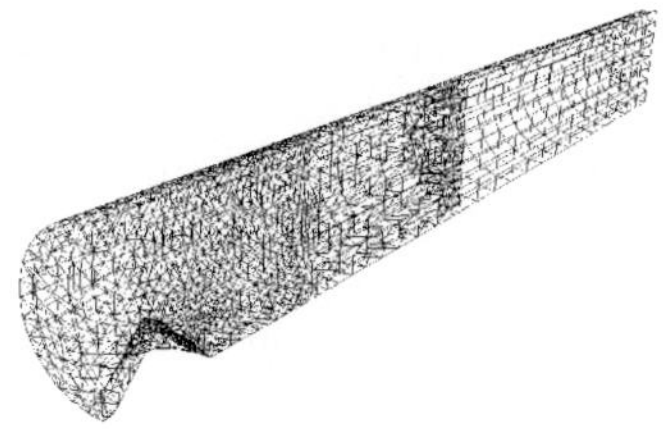

Fig. 6. Mesh of the computed domain

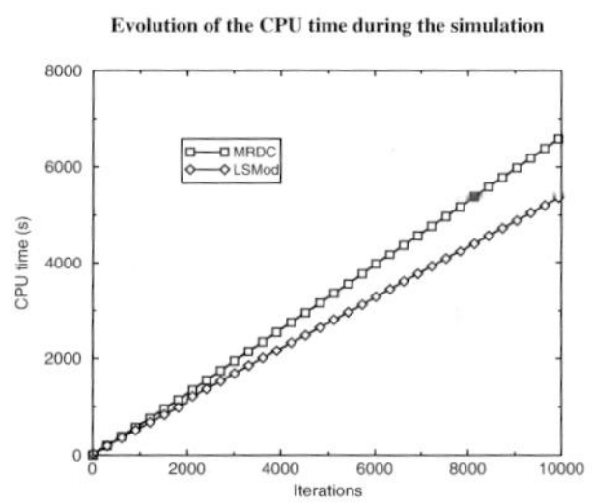

Fig. 7. Evolution of the computation time

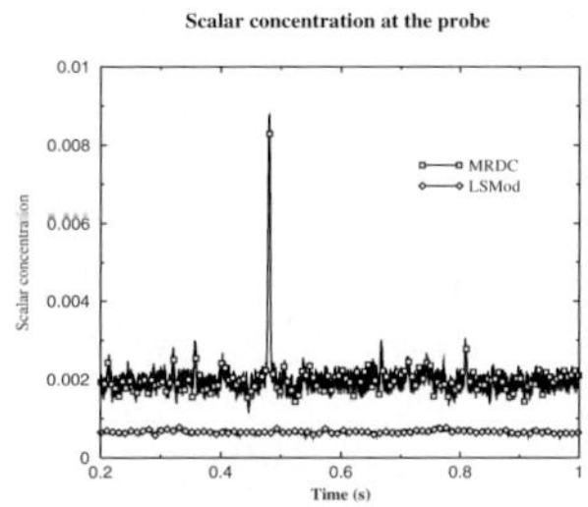

Fig. 8. Time evolution of the scalar concentration at the probe located just after the jet $X(0.22; 0.014; 0.002)$

method. On the contrary, the LSMod method converges which means that the extended neighborhood provides stability.

In Figure 7, the evolution of the CPU time during the simulation is drawn. These curves shows that the LSMod method provides a reduction of 10% on the total CPU time after 2000 iterations where the steady state is not yet established and 20% when the steady state is established. Besides, the convergence of the variables is obtained faster with the LSMod method than with the MRDC method as it is shown for the scalar concentration field in Figure 8. (On the curves associated with the MRDC method, strong instabilities are noted, that do not appear with the LSMod method.)

The computed values of the concentration field and the velocity field are then confronted to experiment (Figures 9). These curves show that computed values with *Code_Saturne* are in good agreement with the experimental data. The LSMod method seems to better predict the maxima than the MRDC one. Nevertheless, this phenomenon could be an artefact of the result visualization since the value of these fields varies significantly when the probe is displaced in the neighboring cell.

4 Conclusion

In this paper, we have described an innovative way to compute the cell gradient which gives better results in accuracy, stability and is less time consum-

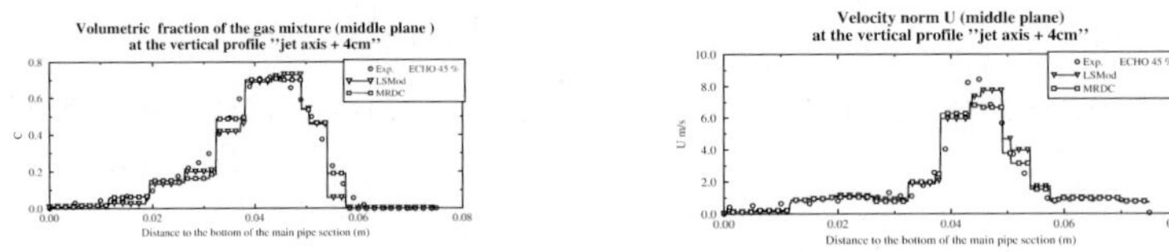

Fig. 9. Volumetric fraction of the gas mixture (left) and velocity norm (right) at profile "jet axis + 4cm"

ing for distorted meshes. This method, denoted the LSMod, is a variation on the least-square method with an extended neighborhood to deal with non-orthogonal meshes. After describing this method, three test cases have been presented. The first one is an analytical case for which the order in space of the LSMod method is the same as the order of the MRDC and LS methods. Nevertheless, the LS and LSMod methods seem slightly more accurate.

The two other cases are semi-industrial and have been chosen because of the numerical difficulties associated with grids of poor quality. Both cases indicate that the LSMod method provides stability, is less time consuming than the MRDC method and at least as satisfactory with respect to the agreement of the numerical results to the experimental data. It has also been demonstrated that the LSMod is better than the standard LS method.

These results lead us to the conclusion that the LSMod method is more efficient than the two methods currently available in *Code_Saturne* for non-orthogonal meshes. For this reason, it will be integrated in the next version of the code. Nevertheless, a more specific study on the way to choose the extended neighborhood could be interesting to try to improve the results. It could be also interesting to try the new efficient method using a sort of equivalence of a finite element method and a finite volume method [5].

References

1. F. Archambeau, N. Méchitoua and M. Sakiz : "*Code_Saturne*: a Finite Volume Method for the Computation of Turbulent Incompressible Flows - Industrial Applications", to be published in Int. J. of Finite Volume.
2. J.H. Ferziger and M. Perić : *Computational Methods for Fluid Dynamics*, Second Edition, Springer 1999.
3. S. Musaferija, D. Gosman : "Finite-Volume CFD Procedure and Adaptive Error Control Strategy for Grids of Arbitrary Topology", J. Comp. Phys., pp. 766-787, Vol. 138, 1997.
4. P.L. Viollet, D. Flamain, F. Lana, J.C. Olivier : "Etude du mélange d'un jet émis à 45 degrés dans un écoulement gazeux plus dense (Expérience ECHO)", Internal Report EDF HE-44/87.22, 1987, in French.
5. P. Omnes and K. Domelevo : "Equivalence of a finite element method and a finite volume method for the Laplace equation on almost almost arbitrary two-dimensional grids, to be published.

Robustness of a Characteristic Finite Element Scheme of Second Order in Time Increment

Masahisa Tabata[1] and Shoichi Fujima[2]

[1] Department of Mathematical Sciences, Kyushu University Hakozaki, Fukuoka, 812-8581 Japan. tabata@math.kyushu-u.ac.jp
[2] Department of Mathematical Sciences, Ibaraki University Bunkyo, Mito, 310-8512 Japan. fujima@mx.ibaraki.ac.jp

1 Introduction

In constructing computational scheemes for flow problems, a crucial point is how to approximate the convection terms. Mainly there are two ways of the approximation, one is of upwind type and the other is of characteristic type [1]. In this paper the latter is treated.

We consider the characteristic finite element method, which is a combination of characteristic method and the finite element method. As a solver of flow problems this method has such a nice property that the obtained linear system is symmetric. We can, therefore, solve large-scale problems using efficient iterative solvers, e.g., preconditioned CG method. Every difficulty of the scheme is concentrated in the integration of composite functions derived from the convection terms. Since those functions are not smooth in each element, numeirical integration does not always work well.

We investigate the effect of errors caused by numerical integration in the convection-diffusion equation. Theoretically characteristic finite element schemes are proved to be unconditionally stable and the finite element solutions converge to the exact one when the discretization paremeters tend to zero. Hence we are sure that the equation is solved correctly by those shcemes if the integrations of the composite functions are done exactly or alomost exactly. In real computation, however, we often encounter that the characteristic finite element scheme of first order in time increment, which is the most familiar in those methods, is unstable. This shows that the scheme is not robust with respect to errors introduced from numerical integration of the composite functions. Recently we have developed a characteristic finite element scheme of second order in time increment [2]. We show that this scheme is not only stable theoretically but also robust in practical computation.

2 Characteristic Finite Element Schemes with First and Second Orders in Time Increment

Let us consider a convection-diffusion problem in a bounded domain Ω in $\mathbf{R}^d$ $(d = 2, 3)$ and a time interval $[0, T]$,

$$\frac{\partial \phi}{\partial t} + (u \cdot \nabla)\phi - \nu \Delta \phi = f, \qquad (x,t) \in \Omega \times (0,T)$$
$$\phi = 0, \qquad (x,t) \in \partial\Omega \times (0,T)$$
$$\phi = \phi^0, \qquad x \in \Omega,\ t = 0$$

where ϕ is an unknown function, u is a given velocity, f is a given source, ν is a viscosity, and ϕ^0 is an initial function. Let V_h be the P_1 finite element space whose functions vanish on the boundary. Let Δt be a time increment and $N_T = \lfloor T/\Delta t \rfloor$. The characteristic finite element method of first order in time increment is to find $\phi_h^n \in V_h$, $n = 1, ..., N_T$, such that

$$\left(\frac{\phi_h^{n+1} - \phi_h^n \circ X_1^n}{\Delta t}, \psi_h\right) + \nu(\nabla \phi_h^{n+1}, \nabla \psi_h) = (\Pi_h f^{n+1}, \psi_h), \quad \forall \psi_h \in V_h \tag{1}$$
$$\phi_h^0 = \Pi_h \phi^0 \tag{2}$$

where $X_1^n(x) \equiv x - u^{n+1}(x)\Delta t$ and Π_h is the interpolation operator. Scheme (1) and (2) is proved to be unconditionally stable and convergent in $O(\Delta t + h)$, where h is the representative element size. See [1]. In real computations, however, it often become unstable when numerical integration is used for the integration of the composite function term $\phi_h^n \circ X_1^n$.

In a recent paper [2] we have developed a characteristic finite element scheme of second order in time increment by using the second-order Runge-Kutta method

$$X_2^n(x) \equiv x - u^{n+1/2}(x - u^n x \Delta t/2)\Delta t$$

in place of X_1^n. We have also pointed out that a correction term is necessary for the real second order scheme in Δt. The scheme is

$$\left(\frac{\phi_h^{n+1} - \phi_h^n \circ X_2^n}{\Delta t}, \psi_h\right) + \frac{\nu}{2}(\nabla \phi_h^{n+1} + \nabla \phi_h^n \circ X_1^n, \nabla \psi_h)$$
$$+ \frac{\nu \Delta t}{2}(J^n \nabla \phi_h^n \circ X_1^n, \nabla \psi_h)$$
$$= \frac{1}{2}\left(\Pi_h f^{n+1} + \Pi_h f^n \circ X_1^n, \psi_h\right), \qquad \forall \psi_h \in V_h \tag{3}$$

where $J \equiv \partial u/\partial x$ is the Jacobian matrix. Scheme (3) and (2) is proved to be unconditionally stable and the error estimate

$$\max\{||\phi_h^n - \phi^n||_{L^2(\Omega)};\ n = 0, ..., N_T\} \leq c(\Delta t^2 + h) \tag{4}$$

can be obtained, where ϕ is the exact solution and ϕ_h is the finite element solution found in the P_1 space. When the P_k finite element space is used, the right-hand side of (4) is replaced by $c(\Delta^2 + h^k)$.

3 Effect of Numerical Integrations

In order to get $O(h)$ approximate solutions we take $\Delta t = O(h)$ in (1) and $\Delta t = O(\sqrt{h})$ in (3). For both schemes we use numerical integration for the composite function terms. We observe effect of numerical integrations in a model problem of the rotating Gaussian hill. Let Ω be a square $[-1,1] \times [-1,1]$, $T = 2\pi$, $\nu = 5 \times 10^{-4}$, $u = (-x_2, x_1)$, $f = 0$, and

$$\phi^0(x) = \exp\left(-\frac{|x - x_c|^2}{\sigma}\right), \qquad x_c = (0.25, 0), \qquad \sigma = 0.01.$$

Let us divide each side into N segments and get N^2 small squares. Connecting diagonals, we get $2N^2$ triangles. The maximum side length h is equal to $2\sqrt{2}/N$. On this mesh we use the P_1 finite element. Δt is taken as mentioned above. We use the following numerical integration, which is denoted by $I_h(k)$, $k = 1, 2, 3$, to the conposite convection terms. We divide each element into congruent k^2 triangles and on the small triangles we use the numerical integration of the first order. For example, the numerical integration $I_h(1; f)$ of a function f on element K is

$$I_h(1; f) = \frac{\text{meas}(K)}{3}\{f(A_1) + f(A_2) + f(A_3)\}$$

where A_i, $i = 1, 2, 3$, are vertices of K. Taking $N = 64, 96, 128, 192$, we calculate the relative errors

$$\frac{||\phi_h - \Pi_h \phi||_{\ell^\infty(L^2)}}{||\Pi_h \phi||_{\ell^\infty(L^2)}}$$

where

$$||\phi_h||_{\ell^\infty(L^2)} = \max\left\{||\phi_h^n||_{L^2(\Omega)}; n = 0, \cdots, N_T\right\}.$$

Table 1 shows the relative errors, where $F(k)$ and $S(k)$ means that the numerical integration $I(k)$ is employed for the first and the second order characteristic finite element method, respectively. The symbol X shows no result is obtained by the instability.

Table 1. Relative errors

N	F(1)	F(2)	F(3)	S(1)	S(2)	S(3)
64	X	7.74e-01	2.40e-01	2.82e+00	6.91e-02	5.44e-02
96	X	2.00e-01	1.60e-01	1.36e-01	4.28e-02	3.59e-02
128	X	1.45e-01	1.19e-02	8.39e-02	3.07e-02	2.69e-02
192	X	9.36e-02	7.94e-02	4.44e-02	1.94e-02	1.77e-02

Fig. 1 depicts the results of Table 1 in the logarithmic scale, where the dotted line shows the theoretical convergence order h. In both cases $O(h)$ convergence is obtained when the solutions are calculated without overflow. These results shows that the second order scheme is more robust than the

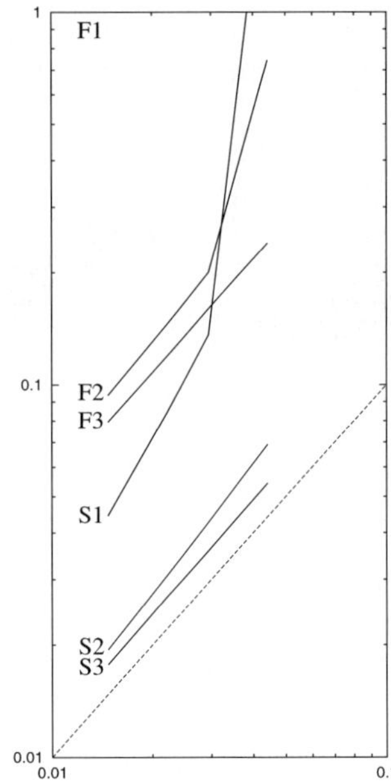

Fig. 1. Relative errors vs. element size h

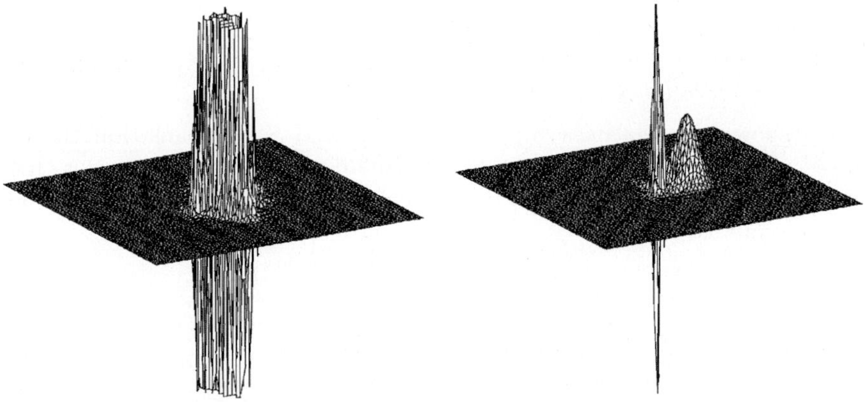

Fig. 2. Stereographs of F1 and F2(2π)

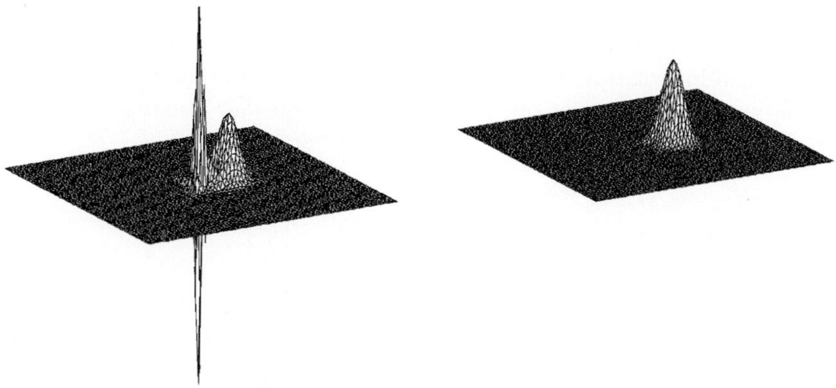

Fig. 3. Stereographs of S1(2π) and S2(2π)

first order scheme with respect to the numerical integration error and produce better numerical results.

In Fig. 2 the stereograph of F1 before the overflow and that of S2 at $t = 2\pi$ are shown, and in Fig. 3 those of S1 and S2 at $t = 2\pi$. Those of the other convergent cases are almost same as that of S2.

We can explain the numerical results as follows. To obtain the L^2-stability we substitute ϕ_h^{n+1} into ψ_h in (3). We denote by $(\cdot,\cdot)_h$ the value obtained by numerical integration of $(\cdot,\cdot)$. We have

$$|\frac{(\phi_h^n \circ X_2^n, \phi_h^{n+1}) - (\phi_h^n \circ X_2^n, \phi_h^{n+1})_h}{\Delta t}| \le c\frac{h}{\Delta t}||\nabla\phi_h^n||_{L^2(\Omega)}||\phi_h^{n+1}||_{L^2(\Omega)}. \quad (5)$$

In the second order scheme the right-hand side is evaluated by

$$c\Delta t||\nabla\phi_h^n||_{L^2(\Omega)}||\phi_h^{n+1}||_{L^2(\Omega)}$$

by virtue of $\Delta t = O(\sqrt{h})$. In the first order scheme, however, $\Delta t = O(h)$ and such an estimate does not hold. A simple estimate of the right-hand side of (5) is

$$c||\nabla\phi_h^n||_{L^2(\Omega)}||\phi_h^{n+1}||_{L^2(\Omega)}$$

which implies that the instability may occur as the time-step number increases.

4 Concluding Remarks

In using the characteristic finite element method we have to pay much attention to the errors caused by numerical integration. Compared to the first order scheme, the second order scheme is robust with respect to the numerical integration error, which has been recognized by numerical results. We have also given a rough explanation for the robustness. Under a mild condition on the numerical integration we can show that the second order scheme is stable with the numerical integration. A detail discussion and a complete proof of the stability will be presented in the forthcoming paper.

Acknowledgments

The first author wishes to express his thanks to support from the Japan Society for the Promotion of Science under Grant-in-Aid for Scientific Research (S), No. 16104001 and from the Ministry of Education, Culture, Sports, Science and Technology of Japan under Kyushu University 21st Century COE Program, Development of Dynamic Mathematics with High Functionality. The second author was supported by the Japan Society for the Promotion of Science under Grant-in-Aid for Scientific Research (C), No. 16540093.

References

1. O. Pironneau, Finite Element Methods for Fluids, John Wiley & Sons, 1989.
2. H. Rui, and M. Tabata, A second order characteristic finite element scheme for convection-diffusion problems, Numer. Math., 92(2002), 161–177.

Gas-Kinetic BGK Scheme for Hypersonic Viscous Flow

Kun Xu[1] and Meiliang Mao[2]

[1] The Hong Kong University of Science & Technology, Hong Kong
[2] China Aerodynamics Research and Development Center, China

Summary. This paper is about the application of a gas-kinetic scheme based on the Bhatnagar-Gross-Krook (BGK) model for the Navier-Stokes equations in the study of hypersonic viscous flow. In the laminar hypersonic viscous flow computations, complicated flow phenomena, i.e., shock boundary layer interaction, flow separation, and viscous/inviscid interaction, will be encountered. The cases studied include the type IV shock-shock interaction around a circular cylinder and hypersonic flow passing through a double-cone geometry.

1 Introduction

An important issue in the design and development of aerospace vehicles is the effect of various types of flow phenomena on aerodynamic performance and aerothermal loads. Evaluation of aerodynamic heating during reentry flight is one of the key issues. To the current stage, the computation of the interaction between a shock wave and a separated region in the hypersonic flow is still a very challenging problem in computational fluid dynamics. In terms of heat transfer, significant differences between the computational results and experiments indicate that further investigation and design of more accurate viscous flow solvers are needed to understand these phenomena. Over the past 15 years there have been concerted efforts both in Europe and America to validate the Navier-Stokes and DSMC-based methods in the description of complex hypersonic flows. Extensive codes validation have been conducted. A number of experiments specifically designed with simple model configuration in the laminar hypervelocity flow have been constructed. In this paper, we are going to first introduce a multidimensional gas-kinetic BGK scheme, where the gradients of flow variables in both normal and tangential directions around a cell interface are accounted in the flux evaluation. Then, the scheme is applied to the study of laminar hypersonic viscous flows.

2 BGK Flow Solver

The BGK model can be written as [1]

$$f_t + uf_x + vf_y = \frac{g-f}{\tau}, \tag{1}$$

where f is the gas distribution function and g is the equilibrium state approached by f. Both f and g are functions of space (x, y), time t, particle velocities (u, v), and internal variable ξ. The particle collision time τ is related to the viscosity and heat conduction coefficients. The equilibrium state is a Maxwellian distribution,

$$g = \rho(\frac{\lambda}{\pi})^{\frac{K+2}{2}} e^{-\lambda((u-U)^2+(v-V)^2+\xi^2)},$$

where ρ is the density, U and V are the macroscopic velocities in the x and y directions, and λ is related to the gas temperature $m/2kT$. For a 2D flow, the particle motion in the z direction is included into the internal variable ξ, and the total number of degrees of freedom K in ξ is equal to $(5-3\gamma)/(\gamma-1)+1$. In the equilibrium state, ξ^2 is equal to $\xi^2 = \xi_1^2 + \xi_2^2 + ... + \xi_K^2$. The relation between mass ρ, momentum $(n = \rho U, m = \rho V)$, and energy E densities with the distribution function f is

$$w = \begin{pmatrix} \rho \\ n \\ m \\ E \end{pmatrix} = \int \psi_\alpha f d\Xi, \qquad \alpha = 1, 2, 3, 4, \tag{2}$$

where ψ_α is the component of the vector of moments

$$\boldsymbol{\psi} = (\psi_1, \psi_2, \psi_3, \psi_4)^T = (1, u, v, \frac{1}{2}(u^2 + v^2 + \xi^2))^T,$$

and $d\Xi = dudvd\xi$ is the volume element in the phase space with $d\xi = d\xi_1 d\xi_2 ... d\xi_K$.

The general solution f of the BGK model at a cell interface $(x_{i+1/2}, y_j)$ and time t is

$$f(x_{i+1/2}, y_j, t, u, v, \xi) = \frac{1}{\tau} \int_0^t g(x', y', t', u, v, \xi) e^{-(t-t')/\tau} dt' + e^{-t/\tau} f_0(x_{i+1/2} - ut, y_j - vt), \tag{3}$$

where $x' = x_{i+1/2} - u(t-t'), y' = y_j - v(t-t')$ are the trajectory of a particle motion and f_0 is the initial gas distribution function f at the beginning of each time step $(t = 0)$. Two unknowns g and f_0 must be specified in Eq.(3) in order to obtain the solution f. In order to simplify the notation, $(x_{i+1/2} = 0, y_j = 0)$ will be used in the following text.

Different from the directional splitting BGK-NS code [8], the multidimensional gas-kinetic scheme has the initial gas distribution function f_0,

$$f_0 = \begin{cases} g^l \left(1 + a^l x + b^l y - \tau(a^l u + b^l v + A^l)\right), & x \leq 0 \\ g^r \left(1 + a^r x + b^r y - \tau(a^r u + b^r v + A^r)\right), & x \geq 0 \end{cases} \tag{4}$$

where b^l and b^r are related to the gradients in the tangential direction, and terms proportional to τ represent the non-equilibrium parts in the Chapman-Enskog expansion of the BGK model. Note that the non-equilibrium parts have no direct contribution to the conservative flow variables, i.e.,

$$\begin{aligned} \int (a^l u + b^l v + A^l)\psi_\alpha g^l d\Xi &= 0, \\ \int (a^r u + b^r v + A^r)\psi_\alpha g^l d\Xi &= 0. \end{aligned} \tag{5}$$

In the above f_0, g^l, g^r, a^l, and a^r have the same definition as the corresponding parameters in the BGK-NS method [8].

After having f_0, the equilibrium state g around $(x = 0, y = 0, t = 0)$ is constructed as

$$g = g_0 \left(1 + (1 - \mathrm{H}[x])\bar{a}^l x + \mathrm{H}[x]\bar{a}^r x + \bar{b}y + \bar{A}t\right), \tag{6}$$

where $\bar{b}$ is the additional term related to the flow variations in the tangential direction, and $\mathrm{H}[x]$ is the Heaviside function. Here g_0 is a local Maxwellian distribution function located at $(x = 0, y = 0)$. In both f_0 and g, a^l, A^l, a^r , A^r, $\bar{a}^l$, $\bar{a}^r$, $\bar{b}$, and $\bar{A}$ are related to the derivatives of a Maxwellian in space and time.

After substituting Eq.(4) and Eq.(6) into Eq.(3), the gas distribution function f at a cell interface can be expressed as

$$\begin{aligned} f(x_{i+1/2}, y_j, t\ , u, v, \xi) = & (1 - e^{-t/\tau})g_0 \\ & + \left(\tau(-1 + e^{-t/\tau}) + te^{-t/\tau}\right)\left(\bar{a}^l \mathrm{H}[u] + \bar{a}^r(1 - \mathrm{H}[u]) + \bar{b}v\right) ug_0 \\ & + \tau(t/\tau - 1 + e^{-t/\tau})\bar{A}g_0 \\ & + e^{-t/\tau}\left((1 - (t + \tau)(ua^l + vb^l))\mathrm{H}[u]g^l\right. \\ & \left. + (1 - (t + \tau)(ua^r + vb^r))(1 - \mathrm{H}[u])g^r\right) \\ & + e^{-t/\tau}\left(-\tau A^l \mathrm{H}[u]g^l - \tau A^r(1 - \mathrm{H}[u])g^r\right), \end{aligned} \tag{7}$$

where

$$\int_0^{\Delta t} \int (g - f)\psi_\alpha dt d\Xi = 0,$$

is used to evaluate $\bar{A}$ uniquely in the above equation. The above gas-kinetic method is similar to that in [8], except additional terms b^l, b^r, and $\bar{b}$ terms related to the flow variations in the tangential direction are included in the current method. Based on the above gas distribution function, the numerical fluxes across a cell interface can be evaluated. In order to study the flow with any realistic Prandtl number, a Prandtl number fix based on the modification of energy flux can be used [8]. In all simulations for the air flow in the next section, the Prandtl number is equal to 0.72. The collision time τ is determined by $\tau = \mu/p$, and the Sutherland's law is used for the viscosity coefficient,

$$\mu = \mu_\infty (\frac{T}{T_\infty})^{3/2} \frac{T_\infty + S}{T + S},$$

where $T_\infty = 285K$ and $S = 110.4K$.

3 Numerical Experiments

Case(1) Type IV shock-shock interaction around a circular cylinder

Shock/shock interaction was classified by Edney [2] into six patterns, which depend on the impinging position and angle. It results in a very complex flow field with high pressure and heat flux peak in localized region. Of special interest in this investigation is the type IV interaction, which occurs when the incident shock impinges on the bow shock in a region where the flow behind it is subsonic. The type IV interaction is the most severe case to form the hot spot on the surface of the cylinder due to the supersonic jet hitting on the wall.

In the current calculation, before imposing the incident oblique shock, the initial condition for the inflow is

$$M_\infty = 8.03, T_\infty = 122.11K, T_w = 294.44K, R_e = 1.94 \times 10^5.$$

Based on the experimental measurement and the numerical calculations, Thareja et. al. [5] summarized that the position of incident impinging shock on the cylinder can be approximated by the curve $y = 0.3271x + 0.4147$ for the experiment (Run 21) [6, 7], where $(x = 0, y = 0)$ is the center of the cylinder. Our simulations are based on 181×300 mesh points with a cell Reynolds number 4.6. A steady state solution is obtained from the explicit scheme after a long time integration with the iterative steps on the order of 10^7, where the residual is decreased to the order of 10^{-7} and the flow structure keeps the same form. The measured as well as computational pressure and heat flux along the cylindrical surface are shown in Fig.(1), where the symbols are the experimental data [6, 7] and solid lines are computational results.

Case(2) Axis-symmetric double cone geometry with flow separation

The double-cone configuration has a first cone half-angle 25^0 and the second cone angle 55^0. Under the experimental condition (Run 28) [3], the incident flow has

$$\rho_\infty = 0.6545 \times 10^{-3} kg/m^3, U_\infty = 2664.0m/s, T_\infty = 185.56K, T_{wall} = 293.33K.$$

The corresponding Mach and Reynolds number are $M_\infty = 9.59, R_e = 13090$. Under this flow condition, the first cone produces an attached shock wave, and the second cone with larger angle produces a detached bow shock. These two shocks interact to form a transmitted shock that strikes the second cone surface near the cone-cone juncture. The adverse pressure gradient due to the

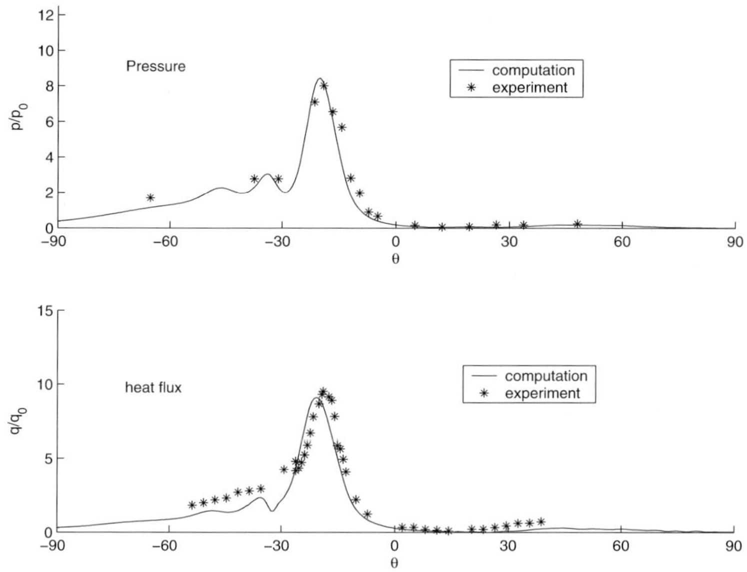

Fig. 1. Pressure (upper) and heat flux (lower) along the cylindrical surface, where the symbols are the experimental data [6] and the solid lines are the computational results.

cone juncture and the transmitted shock generates a large region of separated flow that produces its own separation shock. This shock interacts with the attached shock from the first cone, altering the interaction with the detached shock from the second cone. This in turn effects the size of the separation region. The shock interaction produces very high surface pressure and heat transfer rates where the transmitted shock impinges on the second cone.

Along the cone surfaces, the measured and computational pressure and heat flux with 500×200 mesh points are presented in Fig.(2), where the symbols are the experimental results [3]. As shown in these figures, the size of the separation region and the heat fluxes on the cone surfaces match the experimental data excellently. Especially along the surface of the 2nd-cone, the complicated flow structures are captured. As analyzed in [4], the deviation in the heat flux before the separation shock along the first cone surface is due to the non-equilibrium nature in the incoming expansion gas in the experiment device. In our simulation, the ideal equilibrium incoming gas is assumed.

4 Conclusion

In this paper, we used a gas-kinetic BGK flow solver for the compressible Navier-Stokes equations in the study of laminar hypersonic viscous flows. The cases studied include the Run 28 of the double cone geometries and the shock shock interactions. The complicated flow patterns, such as shock-shock,

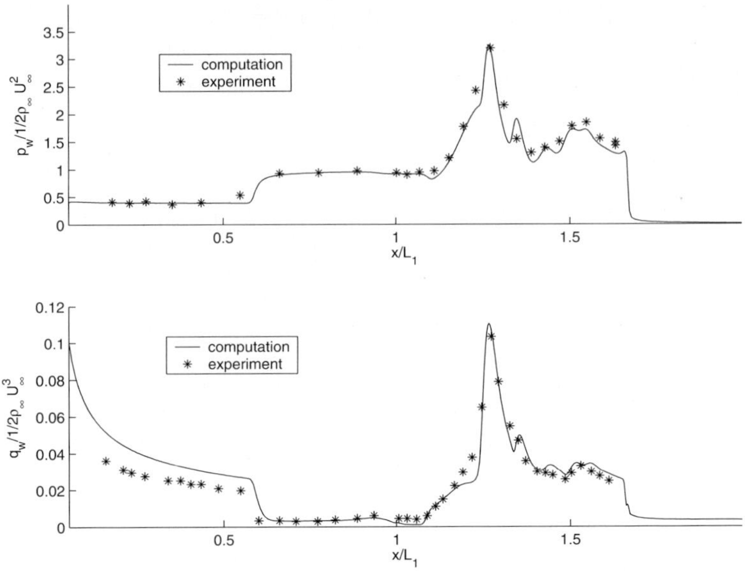

Fig. 2. Pressure (upper) and heat flux (lower) along cone surface, where the symbols are the experimental data [3] and the solid lines are the computational results.

shock boundary layer, and inviscid-viscous interactions, are well captured. For example, a steady state flow structure in type IV shock-shock interaction around a circular cylinder is obtained. The good agreement between the simulation results and the experimental measurements, especially the heat flux around body surfaces, shows clearly that the gas-kinetic scheme is a reliable viscous flow solvers, even though it does not solve the compressible Navier-Stokes equations directly.

Acknowledgements

This work was supported by Hong Kong Research Grant Council (RGC).

References

1. P.L. Bhatnagar, E.P. Gross, and M. Krook, Phys. Rev., **94** (1954), pp. 511-525.
2. B. Edney, FFA Report 115, Aero. Res. Institute of Sweden (1986).
3. M.S. Holden and T.P. Wadhams, AIAA 2003-3641 (2003).
4. I. Nompelis, G.V. Candler, and M.S. Holden, AIAA J. **41**, 2162-2169 (2003).
5. R.R. Thareja, et. al., Int. J. Numer. Methods Fluids **9** (1989), pp. 405-425.
6. A.R. Wieting, NASA TM-100484 (1987).
7. A.R. Wieting and M.S. Holden, AIAA Journal **27** (1989), pp. 1557-1565.
8. K. Xu, J. Comput. Phys. **171** (2001), pp. 289-335.

A Comparison of Space-Time Variational-Multiscale Discretizations

S. J. Hulshoff, E. J. Munts, and R. de Borst

Faculty of Aerospace Engineering, Delft University of Technology, Kluyverweg 1, 2629 HS Delft, The Netherlands, S.J.Hulshoff@LR.TUDelft.NL

Summary. Three variational-multiscale space-time finite-element discretizations are compared. The first is based on the modal spectral-element method, the second on the partition-of-unity concept, and the last on the discontinuous-Galerkin method. Their performance is evaluated using one-dimensional viscous Burgers computations, which allows subgrid-scale modelling parameters to be clearly identified. The effects of mesh refinement, number of scales and scale partitioning on solution accuracy are illustrated and discussed.

1 Introduction

There is substantial motivation for the development of accurate large-eddy simulation (LES) techniques applicable to complex flow problems. A promising approach is based on the unification of the variational multiscale (VMS) method [1] with finite-element discretizations. The advantages of the VMS method over traditional techniques include its consistency with the governing equations at large scales, and its consistent treatment of scale separation near boundaries. Within a finite-element formulation [2] it can also be applied to complex mesh geometries, and be naturally incorporated into *h-p* adaptive solution procedures.

Up to now, the VMS technique has been applied using separate space and time discretizations. In this paper, we consider three alternative space-time VMS discretizations. Space-time discretizations are advantageous in that they provide natural and accurate treatment of moving boundaries. If the solution is assumed to be time-discontinuous, they also allow arbitrary re-meshing of the spatial domain from one time step to the next.

The first discretization considered is based on the modal spectral-element method (MSE) [3], the second on the partition-of-unity method (PUM) [4], and the last on the space-time discontinuous Galerkin method (STDG) [5, 2]. The performance of the discretizations is compared using "large-eddy" simulations of the one-dimensional Burgers equation. This allows well-resolved direct numerical simulations to be used to clearly identify subgrid-scale modelling parameters, simplifying the interpretation of results.

2 Space-Time VMS Discretizations

For a space-continuous, time-discontinuous Galerkin discretization on a space-periodic domain, the viscous Burgers equation may be represented as:

$$B(w,u) = (w, u_t + uu_x - \nu u_{xx} - f)_Q = 0 \quad (1)$$

$$= -(w_t, u)_Q - (w_x, \frac{u^2}{2} - \nu u_x)_Q - (w, f)_Q$$
$$+(w, u^-)_{\Omega^{n+1}} - (w, u^+)_{\Omega^n} + (w, (u^+ - u^-))_{\Omega^n} \quad (2)$$

Here $(\, , \,)_Q$ represents the L_2 inner product on the space-time domain, while $(\, , \,)_{\Omega^n}$ and $(\, , \,)_{\Omega^{n+1}}$ are the L_2 inner products on the boundary of the domain at times $t = t^n$ and $t = t^{n+1}$. The solution is advanced in time by solving a sequence slabs, with the Ω^{n+1} boundary of each slab providing the initial condition for the following slabs. The initial condition is imposed weakly, using the last term in (2).

The discretization is implemented as a VMS method by employing a hierarchical basis, and interpreting its components in terms of large and small scales. The discrete solution, u, and the test function w are assumed to lie in the function space $\mathcal{V}$. Following [1, 2], $\mathcal{V}$ is partitioned so that $\mathcal{V} = \bar{\mathcal{V}} + \mathcal{V}' + \hat{\mathcal{V}}$, $u = \bar{u} + u' + \hat{u}$, and $w = \bar{w} + w' + \hat{w}$, where $\bar{}$, $'$ and $\hat{}$, denote the large, small, and unresolved scales. From the argument of scale separation, it is assumed that the unresolved scales have negligible influence on the large scales. In contrast, the effects of the unresolved scales on the small resolved scales are represented by an additional physical SGS model, M. The large- and small-scale equations are then:

$$B(\bar{w}, \bar{u}) = -B(\bar{w}, u') + (\bar{w}_x, (\bar{u}u'))_Q \quad (3)$$
$$B(w', u') = -B(w', \bar{u}) + (w'_x, (\bar{u}u'))_Q + M \quad (4)$$

In the current study a Smagorinsky-like physical SGS model is employed, $M = -(w'_x, \nu_t u'_x)_Q$, with ν_t defined to reproduce the dissipation of the unresolved scales.

2.1 MSE Discretization

For the MSE discretization, the hierarchical basis consists of the standard continuous bilinear functions supplemented with r Legendre polynomials within each element. These polynomials are defined to be zero at the nodes (see [3]). The solution interpolation within an element is then:

$$u(x,t) = \sum_{i=1}^{n} N_i(x,t) a_i + \sum_{j=1}^{r} \beta_j(x,t) b_j \quad (5)$$

where n is the set of all nodes in the domain, $N_i(x,t)$ are the bilinear “tent” functions, and β_j are products of Legendre polynomials in space with a linear basis in time. In multiple dimensions, the edge traces of $\beta_j(x,t)$ are constrained between elements to ensure continuity.

2.2 PUM Discretization

The partition-of-unity-method [4] is similar to the MSE method in that it allows for local changes in solution interpolation. In this case, however, the enrichments are applied node-wise rather than element-wise. We consider a PUM-VMS discretization where the hierarchical basis is established by enriching each node with Legendre polynomials multiplied by the local "tent" function. The resulting solution interpolation is then:

$$u(x,t) = \sum_{i=1}^{n} N_i(x,t)a_i + \sum_{i=1}^{n}\sum_{j=1}^{r} N_i(x,t)\gamma_{ij}(x,t)b_{ij} \tag{6}$$

$\gamma_{ij}(x)$ are the r Legendre basis functions centred at node i.

2.3 STDG Discretization

For space-discontinuous, time-discontinuous Galerkin discretizations on a space-periodic domain, $B(w,u)$ is instead defined by:

$$\begin{aligned} B(w,u) = \textstyle\sum_{Q_e} -(w_t,u)_{Q_e} - (w_x, \frac{u^2}{2} - \nu u_x)_{Q_e} + (w, \left(\hat{F}_i - \hat{F}_v\right) \cdot n)_{P_e} \\ +(\frac{\nu w_x}{2}, [u])_{P_e} + (w,u^-)_{\Omega_e^{n+1}} - (w,u^-)_{\Omega_e^n} - (w,f)_{Q_e} \end{aligned} \tag{7}$$

Here the operator is applied to the space-time domain of each element Q_e. The element spatial boundaries are denoted by P_e, n is the boundary normal vector, and $[u]$ is the jump in u across an intra-element boundary. $\hat{F}_i$ and $\hat{F}_v$ are the inviscid and viscous Riemann fluxes:

$$F_i = \frac{u_L^2}{4 \mid u \mid}(\mid u \mid + u) + \frac{u_R^2}{4 \mid u \mid}(\mid u \mid - u) \ ; \qquad F_v = \{\nu u_x\} - \frac{\eta}{h}[u] \tag{8}$$

Where { } denotes the average of the left and right states, and h is the length scale of the element. Equations (7) and (8) correspond to the interior-penalty method [5], which provides a stable discretization for the viscous terms when $\eta > 0$. The solution within the element is represented by a constant, augmented by a linear basis in time and Legendre polynomials in space.

3 DNS results

Although the Burgers cascade relies on very different mechanisms than those present in the Navier-Stokes cascade, its characteristic transfer of energy towards dissipation-dominated small scales makes it sufficiently analogous for use in the testing of multiscale methods. We consider a model problem defined by $\nu = \frac{2\pi}{1000}$ and $f(x,t) = A\ sin(x - Ut)$, where $U = 1$, $A = 0.1$. This produces a moving wave with the fixed geometry shown in figure 1.

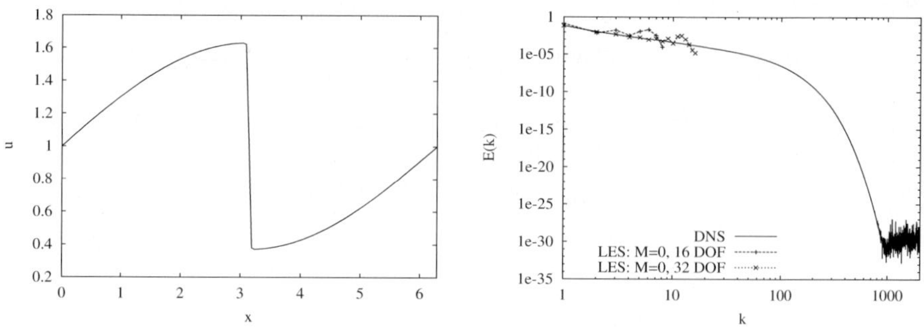

Fig. 1. DNS solution (left) and DNS and LES energy spectra (right).

DNS computations were performed on a sequence of meshes to ensure grid independence. The finest of these used 8192 bilinear space-continuous elements on a domain of length 2π, with a constant time step of $\frac{\pi}{8192}$. The computations were started with a uniform flow of $U = 1$, and advanced to $t = 8\pi$. The solution remains essentially steady after $t = 6\pi$. The fine-mesh energy spectrum, along with the spectra of two LES discretizations with $M = 0$, is shown in figure 1.

As the shape of the profile remains constant, the parameters of the SGS model can be accurately estimated. In particular, given the wave-number range to which the physical SGS model will be applied, the final DNS solution can be used to estimate ν_t with

$$\nu_t = \nu \; \frac{\int_{k_c}^{\infty} k^2 \; E(k) \; dk}{\int_{k_m}^{k_c} k^2 \; E(k) \; dk} \tag{9}$$

where $E = \frac{u^2}{2}$ is the kinetic energy, k_m is the starting wavenumber of the physical SGS model range, k_c is the cutoff wavenumber, and the spectral definition of the energy dissipation rate has been used.

4 Time-Resolved LES Results

The goal of LES is to accurately predict large-scale functionals while computing with a minimum of resolved wave numbers. The computations performed here consider wave cut-off numbers from 8 to 64, well below the 4096 cut-off number of the DNS. As a large-scale functional, we use the integral of kinetic energy over the spatial domain, time-averaged from $t = 6\pi$ to 8π. The balance of energy between body forces and viscous dissipation which characterises the steady solution makes this quantity sensitive to the model of unresolved-scale energy dissipation. Figure 2 shows the convergence of the error in averaged energy versus the inverse of the number of degrees of freedom in space computed using the MSE discretization. Here q is the number of scales per element, and N is the number of elements. The solid line in both

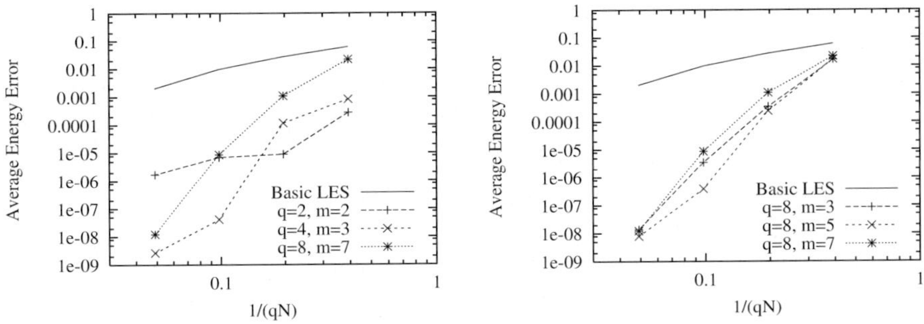

Fig. 2. Average energy error for MSE-VMS

plots corresponds to the basic LES result, for which the model, M, is applied to all scales. Although ν_t is accurately estimated, the large scales are strongly affected by applying a viscous term to them directly. For the test case, the basic LES results are indistinguishable when computed with either 1, 2 or 4 scales per element.

The remaining lines in figure 2 show VMS-LES results with the model applied only to scales m and higher. When large scales are released from model, accuracy improves dramatically. This occurs in discretizations with as little as two scales, although adding more scales increases the convergence rate. The latter effect is offset by a reduction in coarse-mesh accuracy due to the incompatibility of the higher-order bases with the "roughness" of the LES problem.

PUM displays similar trends to MSE (figure 3), but requires more DOFs to achieve equivalent accuracy. A possible cause is the reduced scale separation inherent in the basis obtained by multipling enrichments with the local "tent" function. The latter also leads to poor conditioning of the stiffness matrix.

The STDG results are shown in figure 3 for $\eta = 64$. STDG is superior to PUM, although not as accurate per DOF as the MSE discretization. The expense of STDG is higher than that of MSE due to the additional evaluation

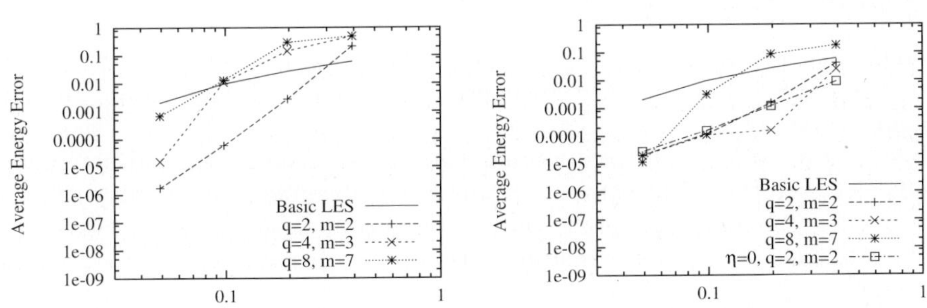

Fig. 3. Average energy error for PUM-VMS (left) and STDG-VMS (right)

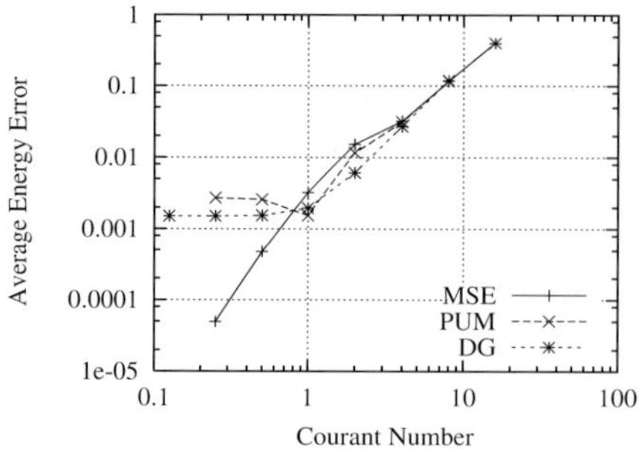

Fig. 4. Average energy error vs Courant number for q=2, m=2 (left), and q=4, m=3 (right)

of boundary flux integrals. The reduced accuracy of STDG is partially due to the contribution of the standard Riemann fluxes to the implicit SGS model. When a single scale is used, for example, (8) leads to a highly dissipative discretization. Similarly, varying η also changes the implicit SGS model, as can be seen by comparing the second and last curve in the right plot of figure 3.

5 LES Results for Large Time Steps

For the linear-in-time basis used, the discontinuous space-time discretization should deliver third-order time accuracy. Whether this is achieved or not for LES computations depends on the relative accuracy of the spatial discretization. For the present study, the basic LES method did not approach the asymptotic range of the time discretization. In contrast, the more accurate of the VMS-LES results were able to achieve third-order time accuracy (figure 4). Since LES by definition introduces significant spatial discretization errors, however, there remains a limit to the benefits of refining in time alone.

6 Conclusions

Of the discretizations compared, the MSE method appears to be best in terms of both speed and accuracy. In contrast, the PUM discretization was not competitive, being both less accurate than the other methods and more expensive due to its poor conditioning. The STDG discretization, although less efficient then the MSE method, retains some advantages. Its highly local nature offers maximum suitability for adaptation and parallelization. However, the influence of the flux and stabilization terms on its implicit subgrid-scale model implies that a integrated design is required in order to fully exploit this method for LES applications.

References

1. T. J. R. Hughes, L. Mazzei, K. E. Jansen (2000) Large Eddy Simulation and the Variational Multiscale Method. Comput.Visual Sci. 3, 47–59.
2. S. Scott Collis (2002) The DG/VMS Method for Unified Turbulence Simulation. AIAA Paper 2002-3124.
3. G. E. Karniadakis (1999) Spectral/hp Element Methods for CFD. Oxford University Press.
4. I. Babuška, J. M. Melenk (1997) The Partition of Unity Method. Int. J. Num. Meth. Eng., 40, 727–758.
5. Arnold, D. N., Brezzi, F., Cockburn, B, and Marini, L. D. (2002) Unified Analysis of Discontinuous Galerkin Methods for Elliptic Problems. SIAM J. Numer. Anal., Vol 39, No 5, pp 1749-1779.

Part V

Algorithms for Unsteady Flows

A Third-Order-Accurate Multidimensional Residual-Distribution Scheme for Unsteady Problems

P. De Palma, G. Pascazio, G. Rossiello, and M. Napolitano

DIMeG, Sez. Macchine ed Energetica, & CEMeC,
Politecnico di Bari, Via Re David 200, 70125 Bari, Italy

1 Introduction

In the last years, residual distribution Fluctuation Splitting (FS) schemes using a compact stencil have been developed and applied successfully to compute steady flows with shocks [1]. Recently, these methods have been extended to compute unsteady flows, using explicit as well as implicit integration schemes, achieving second-order accuracy in both space and time [2, 3]. In particular, the authors have proposed a second-order-accurate FS scheme based on a dual-time stepping procedure and a linear reconstruction for the unsteady residual [4], which enjoys a smaller phase error with respect to current state-of-the-art FS schemes based on the Lax–Wendroff approach [2]. Here, such a methodology is improved to obtain third-order-accurate FS schemes for scalar-advection problems and inviscid flows in two dimensions.

2 Scalar advection

Consider a 2D advection equation for the scalar variable u, $u_t + \nabla \cdot \mathbf{F} = 0$, and a discretization of the computational domain by triangular elements. Explicit Fluctuation Splitting (FS) schemes are obtained by three main steps: i) evaluation of the residual , ϕ_T, over each cell; ii) distribution of the residual among the nodes j of each triangle; iii) update of the solution at each node i of the computational domain by summing up all nodal contributions from the triangles $\triangle_i$ sharing node i:

$$u_i^{k+1} = u_i^k - \frac{\Delta\tau}{S_i} \sum_{T \in \cup \Delta_i} (\beta_{T,i}\phi_T)^k, \tag{1}$$

where $\Delta\tau$ is a pseudo-time step and S_i is the dual-cell area. The first two steps of an FS scheme, namely, the residual evaluation and the distribution, are crucial for its accuracy. In particular, second-order accuracy is achieved if the residual is evaluated assuming linear variation of the unknown over each cell and the distribution coefficients are bounded. In this case a Linearity Preserving ($\mathcal{LP}$) scheme is obtained, namely, a scheme which preserves an initial exact linear solution and is second-order-accurate in space for homogeneous advection equations [5]. The general FS procedure described above can be also employed to design higher-order-accurate space-discretization schemes, using a cell-residual evaluation based on a higher-order polynomial reconstruction and a distribution step with bounded coefficients. The following conditions are required by an FS scheme to achieve third-order accuracy at steady state for smooth solutions [6] on a grid with spacing h:

R1: the fluctuation over each triangle T needs to be computed such that

$$\phi_T = \int_T \nabla \cdot \mathbf{F}(u)\, d\Omega = -\oint_{\partial T} \mathbf{F}^h \cdot \mathbf{n}\, d\ell = 0 + \mathcal{O}\left(h^4\right),$$

$\mathbf{n}$ being the inward unit vector normal to the edges of T.

R2: the distribution coefficients, β_i^T, must be bounded, so that, because of R1, $\phi_{T,i} = \beta_{T,i}\,\phi_T = 0 + \mathcal{O}\left(h^4\right)$.

Condition R2 is satisfied using *classical* FS schemes with bounded coefficients, such as the LDA, LW, and UCV schemes [5]. Condition R1 is fulfilled via a quadratic polynomial reconstruction of u and a contour integration using only cell-vertex values, the mid-point values on each edge being evaluated by a reconstruction involving the neighbouring cells [7]. To this purpose, the gradient of the solution at the vertices is employed:

$$\boldsymbol{\nabla} u_i = \frac{1}{\sum_{T\in\cup\Delta_i} S_T^{-1}} \sum_{T\in\cup\Delta_i} S_T^{-1} \boldsymbol{\nabla} u_T, \quad \text{with} \quad \boldsymbol{\nabla} \mathrm{u_T} = \frac{1}{2\,\mathrm{S_T}} \sum_{\mathrm{j}=1}^{3} \mathrm{u_j} \boldsymbol{n}_\mathrm{j} \ell_\mathrm{j}, \tag{2}$$

where the gradient $\boldsymbol{\nabla} u_T$ is constant over each triangle, $\boldsymbol{n}_j$ and ℓ_j being the inward unit vector normal to the edge opposing node j and its length, respectively. The parabolic function along the edge can thus be defined, allowing to evaluate the sought mid-point value as:

$$u_{\mathrm{mid},j} = \frac{u_m + u_p}{2} + \frac{\boldsymbol{\nabla} u_m - \boldsymbol{\nabla} u_p}{8} \cdot (\boldsymbol{x}_p - \boldsymbol{x}_m), \tag{3}$$

where $m = (j-1) \bmod 3$ and $p = (j+1) \bmod 3$. The same formula was derived in [7] using a least-square-fitting technique.

The fluctuation is computed using the Simpson formula on each edge; for the case of constant advection velocity, one has:

$$\phi_T = -\sum_{j=1}^{3} \frac{\boldsymbol{\lambda} \cdot \boldsymbol{n}_j\, \ell_j}{6} \left[u_m + 4\, u_{\mathrm{mid},j} + u_p\right]. \tag{4}$$

In the present study, all computations have been performed using uniform Cartesian grids, the quadrilateral cells being divided into triangles by alternating diagonals. Figure 1 provides a mesh refinement study for the steady linear advection of a sine-shaped function with periodic boundary conditions in the $[0,1] \times [0,2]$ domain, using a sequence of five grids starting from $\Delta x = \Delta y = 1/10$. For this case (linear advection and uniform Cartesian grid), the fluctuation is seen to be $\mathcal{O}\left(h^5\right)$ instead of $\mathcal{O}\left(h^4\right)$, due to a well known cancellation property of the Simpson formula. Writing $\phi_T = \phi_{T,\ell} + \phi_{T,q}$, where $\phi_{T,\ell}$ is the fluctuation computed by a bilinear reconstruction through the nodes of T and $\phi_{T,q}$ is an higher-order correction, one has that $\phi_{T,\ell}$ and $\phi_{T,q}$ are $\mathcal{O}\left(h^3\right)$, as shown in figure 1.

For the case of unsteady problems, third-order accuracy in time is achieved by employing an implicit scheme based on a dual-time-stepping technique [8]:

$$u_i^{n+1,k+1} = u_i^{n+1,k} - \frac{\Delta\tau}{S_i} \sum_{T \in \cup \Delta_i} \left(\alpha_{T,i}\phi_T^t + \beta_{T,i}\phi_T\right)^{n+1,k}. \tag{5}$$

In equation (5): the unsteady residual, $\phi_T^t = \int u_t d\Omega$, is evaluated using a quadratic variation of the time derivatives over the cell as:

$$\phi_T^t = \frac{S_T}{3} \sum_{j=1}^{3} \left(\frac{\partial u}{\partial t}\right)_{\mathrm{mid},j}, \tag{6}$$

where the time derivative is computed using a third-order-accurate backward four-level scheme:

$$\left(\frac{\partial u}{\partial t}\right)_i = \frac{11u_i^{n+1} - 18u_i^n + 9u_i^{n-1} - 2u_i^{n-2}}{6\Delta t}; \tag{7}$$

superscripts n and k indicate the physical and fictitious time levels, respectively. At each new time level $n+1$, equation (5) is iterated untill the solutions at levels $k+1$ and k coincide within a prescribed tolerance. The two conditions for third-order accuracy, R1 and R2, can be generalized to the unsteady case by requiring that $\phi_T^t + \phi_T = 0 + \mathcal{O}\left(h^4\right)$, the distribution coefficients being bounded. Using the simplest choice of distributing the entire residual by means of the same scheme, namely, taking $\alpha_{T,i} = \beta_{T,i}$, stability problems are experienced due to insufficient dissipation. Therefore, the following distribution procedure is proposed and tested. The fluctuation is written as:

$$\phi_T^t + \phi_T = \phi_{T,\ell}^t + \phi_{T,q}^t + \phi_{T,\ell} + \phi_{T,q}, \tag{8}$$

where $\phi_{T,\ell}^t$ is the unsteady residual evaluated using a linear variation of the time derivatives over the cell and $\phi_{T,q}^t = \phi_T^t - \phi_{T,\ell}^t$. It can be shown that $(\phi_{T,\ell}^t + \phi_{T,\ell})$ is $0 + \mathcal{O}\left(h^3\right)$, whereas $\phi_{T,q}^t$ is $0 + \mathcal{O}\left(h^4\right)$. Defining the signal as:

$$\left(\phi_T^t + \phi_T\right)_i = \beta_{T,i}\left(\phi_{T,\ell}^t + \phi_{T,q}^t + \phi_{T,\ell}\right) + \frac{1}{3}\phi_{T,q}, \tag{9}$$

provides the necessary dissipation to render the scheme stable and does not alter its order of accuracy since the second requirement is still fulfilled. In this work, the distribution coefficients, $\beta_{T,i}$, of the UCV have been used.

Figure 2 provides a mesh refinement study on a uniform Cartesian grid for the unsteady linear advection of a double-sine-shaped function [2]. Computations are performed in the domain $[0,1]^2$ using a sequence of six grids, starting from $\Delta x = \Delta y = 1/8$ and $\Delta t = 0.0125$, and halving both the space and time steps. The global fluctuation, $\phi_T^t + \phi_T$, is seen to vanish as h^4.

Then, the circular advection of a smooth hump in the square domain $[-1,1]^2$ has been considered. The hump follows a circular path and returns to its initial position at $t = 1$. A uniform Cartesian grid has been employed with $\Delta x = \Delta y = 1/32$ and $\Delta t = 0.0025$. Figure 3 provides the vertical cuts through the exact centre of the hump, $(-0.5, 0)$, at $t = 1$, obtained using the present scheme, as well as the second-order-accurate FS scheme of [4] and the classical Lax–Wendroff one; the exact solution is also given for comparison. Both FS schemes are markedly less dispersive than the LW one and the proposed third-order-accurate scheme is seen to enjoy a negligible dissipation while remaining nearly monotone.

3 Euler equations

The proposed implicit third-order FS scheme is extended to the Euler equations by defining the steady and unsteady residual vectors as follows:

$$\Phi_T = \int_T \nabla \cdot \mathcal{F}(U)\, d\Omega = -\oint_{\partial T} \mathcal{F} \cdot \mathbf{n}\, d\ell, \qquad \Phi_T^t = \int_T \frac{\partial U}{\partial t}\, d\Omega, \tag{10}$$

where U is the vector of the conservative variables and $\mathcal{F}$ is the flux vector. The steady residual, Φ_T, is computed by a contour integral by assuming a quadratic reconstruction of the parameter vector, $z = \sqrt{\rho}\,(1, u, v, H)$, and using the Simpson formula:

$$\Phi_T = -\sum_{j=1}^{3} \frac{1}{6} \left[\mathcal{F}(z_m) + 4\,\mathcal{F}(z_{\mathrm{mid},j}) + \mathcal{F}(z_p)\right] \cdot \mathbf{n}\, \ell_j. \tag{11}$$

Third-order accuracy in time is achieved again by employing a dual-time-stepping technique:

$$U_i^{n+1,k+1} = U_i^{n+1,k} - \frac{\Delta\tau}{S_i} \sum_{T \in \cup \Delta_i} \left(A_{T,i}\Phi_T^t + B_{T,i}\Phi_T\right)^{n+1,k}; \tag{12}$$

each component of the unsteady residual, Φ_T^t, is evaluated employing equation (6), the time derivatives being computed using a third-order-accurate backward four-level scheme, as in equation (7). As in the case of scalar advection equation, choosing the same bounded matrix distribution coefficients (see [5] for details) for the the steady and unsteady residuals, $A_{T,i} = B_{T,i}$, leads to stability problems due to insufficient dissipation. Therefore, the above distribution procedure is employed, applying equation (9) in vector form.

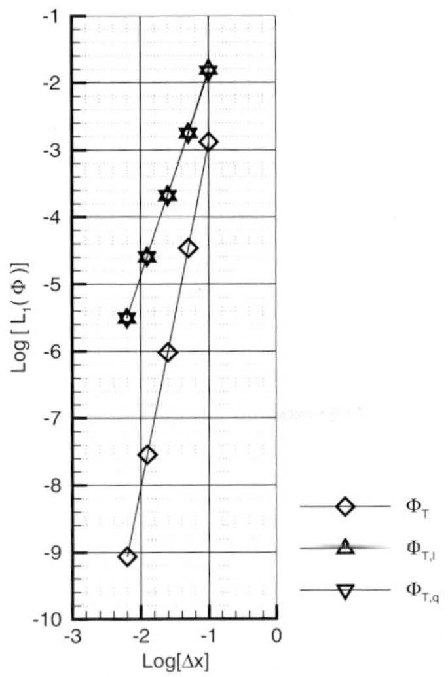

Fig. 1. Mesh refinement study for the steady linear advection of a sine-shaped function.

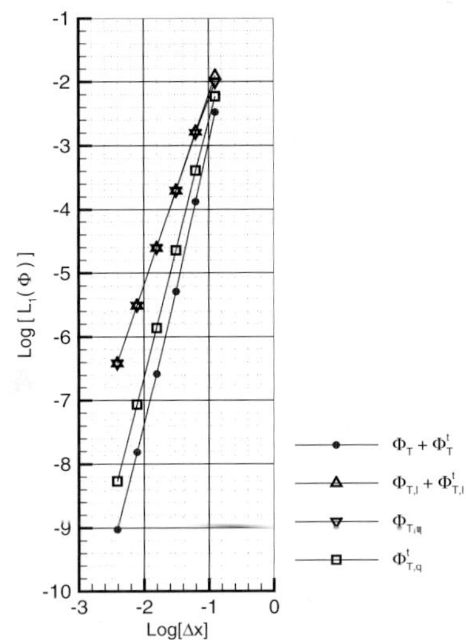

Fig. 2. Mesh refinement study for the unsteady linear advection of a double-sine-shaped function.

The accuracy of the proposed third-order schemes has been verified by computing the advection of a two-dimensional vortex superposed to a uniform flow with $(\rho, u, v, p) = (1, \sqrt{\gamma}, 0, 1)$, with $\gamma = 1.4$. The vortex is given in polar coordinates (r, θ) as:

$$u^* = \epsilon\, r\, e^{\alpha(1-\tau^2)} \sin\theta, \quad v^* = -\epsilon\, r\, e^{\alpha(1-\tau^2)} \cos\theta, \quad T^* = -\frac{(\gamma-1)\epsilon^2}{4\,\alpha} e^{2\,\alpha(1-\tau^2)}, \tag{13}$$

where $\tau = r/0.05$, $\epsilon = 0.3$, $\alpha = 0.204$, and θ is measured with respect to the horizontal direction. Computations are performed in the $[0, 2] \times [0, 1]$ domain up to $t = 0.2$, on a sequence of five grids, starting from $\Delta x = \Delta y = 1/20$ and $\Delta t = 0.0125$. Figure 4 provides the L_1 norms of the density errors for the explicit Lax–Wendroff scheme, the second-order-accurate FS scheme of [4], and the proposed scheme. The present scheme is shown to be third-order accurate and provides markedly lower errors.

Finally, the interaction between a stationary shock, with upstream conditions $(\rho, u, v, p) = (1, 1.1\sqrt{\gamma}, 0, 1)(M_1 = 1.1)$, and the vortex described in equation (13) has been computed in the $[0, 2] \times [0, 1]$ domain using $\Delta x = \Delta y = 1/100$ and $\Delta t = 0.0025$. The limiting procedure proposed in [4] has been employed in order to dump spurious oscillations. Figures 5 and 6 provide the pressure contours at times $t = 0.2$ and $t = 0.4$, respectively. Due to its low dissipative error, the proposed scheme is able to provide a sharp shock and a well preserved vortex after the interaction.

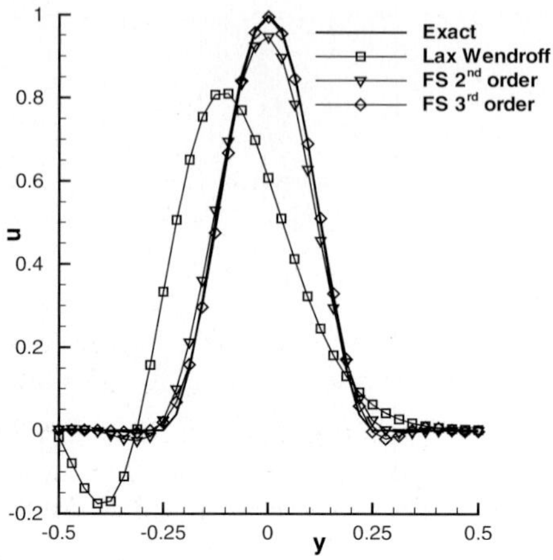

Fig. 3. Rotating hump test case.

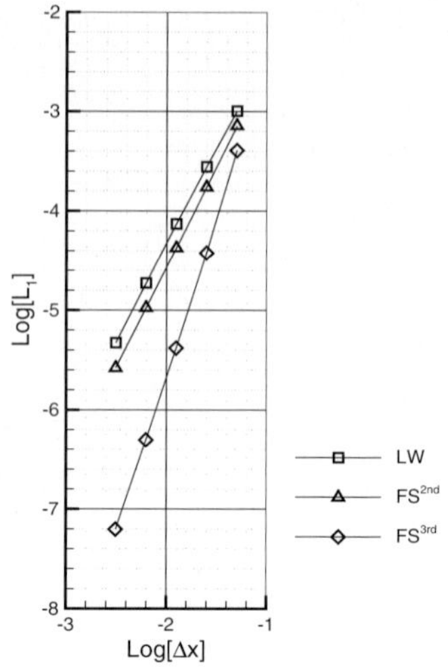

Fig. 4. Mesh refinement study for the unsteady Euler equations.

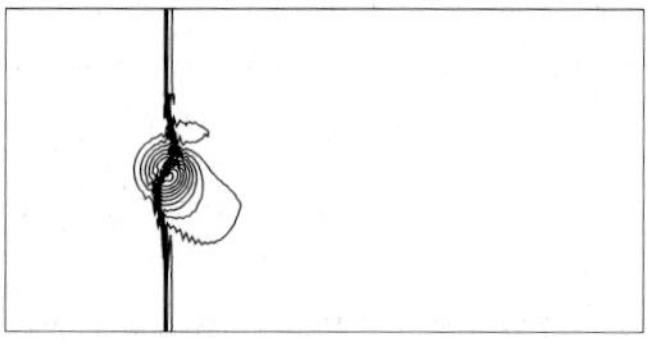

Fig. 5. Vortex-shock interaction: pressure contours at $t = 0.2$ using the FS-UCV-N scheme ($\Delta p = 0.02$)

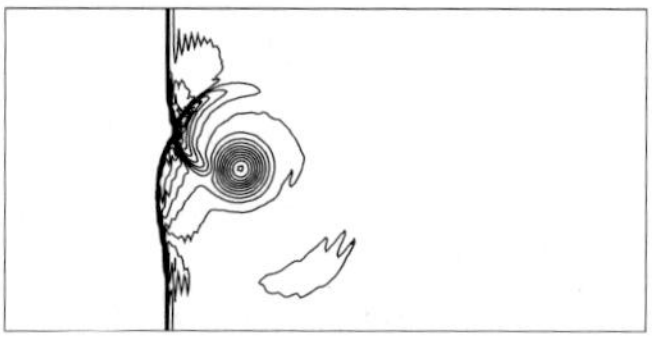

Fig. 6. Vortex-shock interaction: pressure contours at $t = 0.4$ using the FS-UCV-N scheme ($\Delta p = 0.02$)

References

1. P. De Palma, G. Pascazio, and M. Napolitano: 'A hybrid fluctuation splitting scheme for two-dimensional compressible steady flows'. In: *Innovative methods for numerical solution of partial differential equations.* ed. by M. M. Hafez and J. J. Chattot (World Scientific Publishing Co. Inc., New Jersey 2002) pp. 305-333.
2. M. E. Hubbard and P. L. Roe: Int. J. Numer. Methods Fluids **33**, 711 (2000).
3. R. Abgrall, M. Mezine: J. Comp. Phys. **188**, 16 (2003).
4. P. De Palma, G. Pascazio, G. Rossiello, and M. Napolitano: 'A second-order-accurate monotone implicit fluctuation splitting scheme for unsteady problems'. In *Computational Fluid Dynamics 2002*, ed by S. Armfield, P. Morgan, and K. Srinivas (Springer, New York 2003) pp. 731-736.
5. H. Paillère: Multidimensional upwind residual distribution schemes for the Euler and Navier-Stokes equations on unstructured grids. Ph.D Thesis, Université Libre de Bruxelles, Belgium (1995).
6. R. Abgrall and P.L. Roe: J. Sci. Comput. **19**, 3 (2003).
7. D. Caraeni and L. Fuchs: Theoret. Comput. Fluid Dynamics **15**, 373 (2002).
8. A. Jameson: 'Time dependent calculations using multigrid with applications to unsteady flows past airfoils and wings'. *AIAA Paper 91-1596*, 1991.

Unsteady Simulations for Flutter Prediction

Julien Delbove

Aerodynamic Design & Data Domain,
AIRBUS France, 316 route de Bayonne, 31060 Toulouse Cedex, France
julien.delbove@airbus.com

1 Introduction

Accurately predicting flutter phenomenon on full aircraft configurations still remains a challenging task. Within the Airbus industrial framework, the flutter predictions rely on the P-K method [3]. The Finite Element Method applied to the structure dynamic equations leads to the second order differential system $M\ddot{q}(t) + Kq(t) = F(t)$ where M is the mass matrix, K is the stiffness matrix, F represents the aerodynamic loads and q the generalized displacements. The unsteady aerodynamic loads for each of the eigenmodes are required as input of the P-K method. Until now, the unsteady aerodynamic loads were predicted by doublet lattice method [11], linearized solver [10] or strong coupling Euler/boundary layer computation. With the increasing flight domain, flutter predictions are required until Mach number equal to 0.96. As the classical methods are no more valid for this range of Mach number, a Navier-Stokes unsteady solver with mesh deformation algorithm has to be used. In order to get afordable turn around time, numerical methods have to be enhanced and combined. After the presentation of the numerical methods required for the unsteady solver, and of some of the validation cases performed with it, flutter applications will be presented.

2 Numerical Methods

2.1 U-RANS Solvers

The present flow solver, elsA [1] co-developed by ONERA and CERFACS is a classical Multi-Block Structured Navier-Stokes solver developed with an object oriented approach. The equations in their Arbitrary Lagrangian-Eulerian formulation (ALE) [2] are solved using a classical JST [5] scheme. Time integration is carried out with a Dual Time Stepping [6] procedure consisting of a second order implicit backward differentiation combined with a Multigrid algorithm and a point-implicit LU-SGS smoother for convergence acceleration purpose. The representation of the turbulence is achieved via the Spalart-Allmaras model [12].

2.2 Grid Deformation Algorithm

In the case of direct unsteady aeroelastic simulations, the interface between the fluid and the structure domains is deformed at each physical time step. It requires the search of a deformation algorithm as efficient as possible. For this purpose, an analytical deformation method ([8, 7]) has been mixed with a transfinite interpolation [13] algorithm. This hybrid approach provides a good compromise between computational effort and grid quality since the good behavior properties of both methods are kept: i.e. the deformation algorithm is parallel, the computational cost is relatively low and the mesh quality is preserved.

2.3 P-K Method

In 1965, Irwin and Guyett [3] presented a method for the flutter analysis which has been widely adopted by aeroelasticians as the primary tool for finding flutter solutions. This method is named the P-K method. By transforming the general dynamic structure problem $M\ddot{q} + Kq = F(t)$ in the Laplace domain, the flutter prediction is reduced to an eigenvalue problem. M, K, $F(t)$ and $q(t)$ are respectively the mass matrix, the stiffness matrix, the aerodynamic loads and the degrees of freedom vector. The Laplace transformation of the aerodynamic loads $F(t)$ can be approximated by the evolution of their Fourier transform in function of the frequency for all of the structure eigenmodes. This is the reason why an unsteady aerodynamic prediction tool is required.

3 Numerical Simulation

3.1 Unsteady Simulations Validation

Validation of the unsteady capability of the elsA software has been performed on a large number of test cases. Two test cases are presented here: the NACA64A010 airfoil and the LANN wing [14]. These test cases consist in pitching movement of the configuration. The angle of attack is described by the function of time $\alpha(t) = \alpha_0 + \alpha_m \sin(\omega t)$. The results obtained with the elsA software have been compared with experimental data.

The NACA64 airfoil is a two dimensional test case. Two unsteady simulations have been performed. The first one is named the CT6; the parameters of the simulation are $M = 0.789$, $\alpha_0 = -0.21^o$, $\alpha_m = 1.01^o$, $k = 0.202$ and $R_e = 13\,10^6$, $k = \omega c/V$ where ω is oscillation frequency, c the airfoil chord and V the velocity at infinity. The second one is the shock stall case, $M = 0.789$, $\alpha_0 = 4.0^o$, $\alpha_m = 1.01^o$, $k = 0.204$ and $R_e = 12\,10^6$.

As a temporal periodic signal can be transformed into its Fourier series, $C_p(t)$ can be expressed $C_p(t) = \sum_{k=-N/2}^{N/2} \widetilde{C_p}(k) e^{-i\frac{2\pi k t}{T}}$. The comparison with

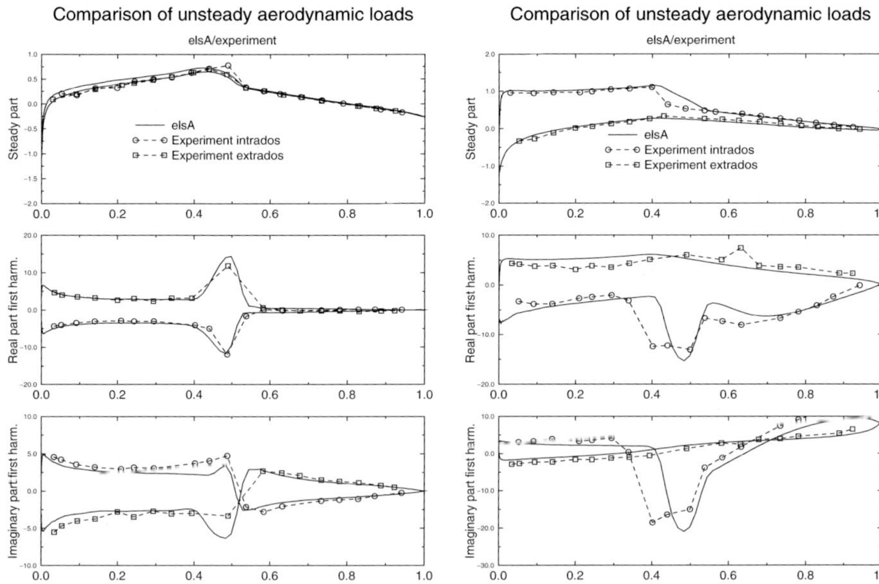

(a) CT6, M=0.789, $\alpha_0 = -0.21^o$ (b) Shock stall, M=0.789, $\alpha_0 = 4.0^o$

Fig. 1. NACA64A010, comparison of average and first harmonic of C_p

the experimental data consist in the comparison of the average part of the signal $\widetilde{C_p}(0)$ and of its first harmonic $\widetilde{C_p}(1)$. The results depicted on figures 1(a) and 1(b) compare these values for the CT6 and the shock stall case. Conserning the CT6, the good agreement between the simulation and the experimental data proves that the unsteady C_p is well predicted for this attached flow. The top graph of figure 1(b) shows some slight differences of the shock positions for the shock stall case. They are due to the massively separated boundary layer which is not well predicted by the Spalart-Allmaras turbulence model. As a consequence, the comparison of the first harmonic presents differences concerning the peak positions.

Conserning the LANN configuration, the test case presented is the CT9 one, $\alpha_0 = 2.6^o$, $\alpha_m = 0.25$, $k = \omega c/V = 0.103$ with c the root chord and V the velocity at infinity, the Mach number is equal to 0.822 and the Reynolds number is $5.32\,10^6$. The figure 2 depictes the comparison of the real and imaginary part of the first C_p harmonic for several sections in span. From 30% of the span to the wing tip the boundary layer is separated. This test case is well known to be a critical one because of the strong nonlinearities of the unsteady flow. Moreover the comparison to the experimental data must be carefully done because of the low number of pressure transducers. For all the sections and for both real and imaginary parts, the results obtained give the same qualitative behavior. Some slight differences can be observed. The shock positions and strenghs, indicated by the peak positions and highs of the first harmonic curves, are not exactly predicted because of the turbulence

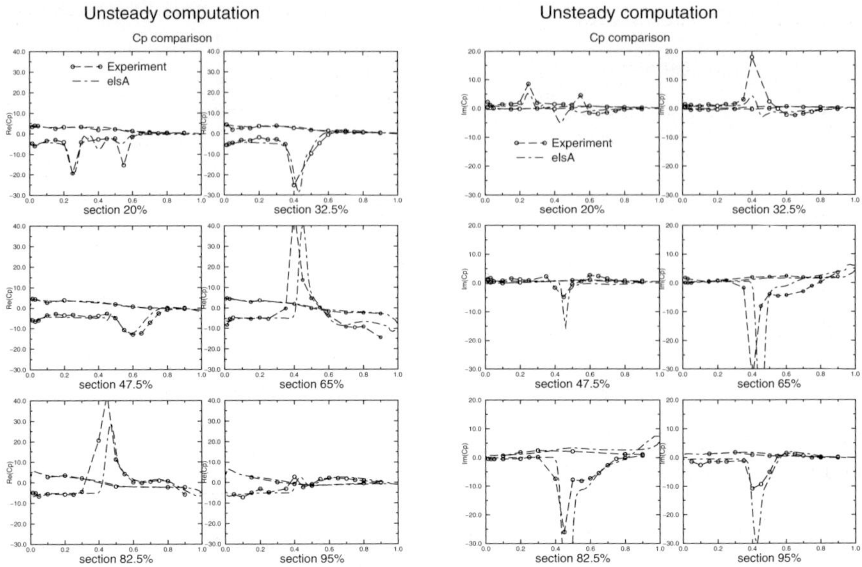

(a) Real part of C_p first harmonic

(b) Imaginary part of C_p first harmonic

Fig. 2. LANN CT9, M=0.822, $\alpha_0 = 2.6^o$, $\alpha_m = 0.25^o$

model. They are due to the difficulty to completly and accurately represent the complex physics with a RANS Spalart-Allmaras simulation. Nonetheless most of the physics is captured and the sign of the phase difference, indicated by the imaginary part of first harmonic, is correctly predicted.

Within an industrial context, in addition to the accurate results of the unsteady simulations the turm around time of simulations must be acceptable. The table 1 indicates the computational time required for the previous test cases. These performances are compatible with the limit of a complete flutter computation performed during a week-end.

3.2 Two Dimensional Flutter Prediction

The studied configuration is a symmetric airfoil NACA64A010 representing a three dimensional wing section close to the wing tip. The aeroelastic system has two degrees of freedom, one in bending and the other in torsion. Isogai [4] proposed this configuration for the flutter studies of swept wings. The

Table 1. Computation performance on fujitsu VPP700

Test case	Node Nb	Nb of proc	Elapse time
NACA64	81 092	1	3h9min
LANN	2 286 240	4	14h8min

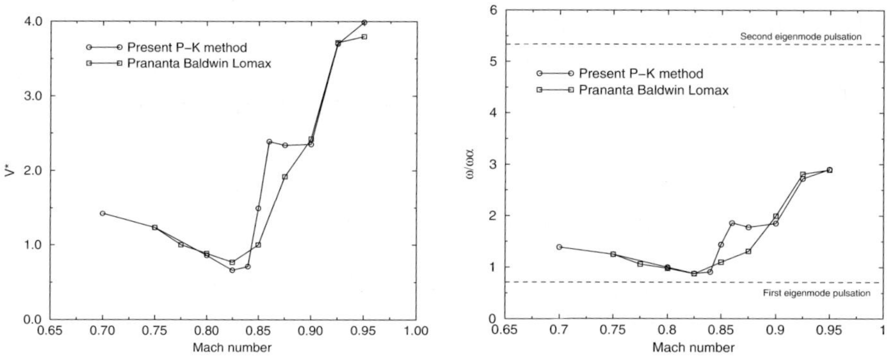

(a) $M - V^*$ plan (b) $M - \omega$ plan

Fig. 3. Two dimensional flutter prediction, comparison of P-K method to Prananta Baldwin Lomax prediction [9]

evolution of the unsteady aerodynamic loads with the oscillation frequency is required as an imput of the P-K method. For each structure mode, the evolution of the aerodynamic loads with the frequency is obtained with an interpolation of three unsteady simulations for different frequencies. For a given Mach number, the P-K method gives the evolution of the damping coefficient γ with V^*, the nondimensional velocity. The flutter boundary is the curve representing the points for which the damping is equal to zero. The perturbations are damped if $\gamma < 0$ and amplified if $\gamma > 0$. The figure 3 compares the flutter boundary obtained during the present work and the one obtained by Pranauta [9]. Figure 3(a) depictes in the plan $M - V^*$, the points for which $\gamma = 0$. The figure 3(b) represents the frequency of the neutral solution as a function of the Mach number. There is a global good agreement of the flutter boundaries except for the transonic dip Mach numbers. Pranauta used a thin-layer Navier-Stokes method whereas full Navier-Stokes simulations were performed in this work. The transonic dip prediction is highly dependent on the boundary layer separation simulation. As more physics is captured by the Navier-Stokes solver, more accurate predictions are expected but it has to be demonstrated by validation versus experimental results.

3.3 Complex Configuration Flutter Prediction

Exactly the same procedure has already been applied at Airbus for the flutter prediction of complex configurations. The finite element model supplies the structure eigenmodes. The current process involves a combinaison of elsA unsteady CFD inputs on the wing and cheaper methods inputs on the rear part of the aircraft. The results are finally used by the P-K method for the flutter prediction.

4 Conclusion

An implicit solver enhanced for the unsteady simulations on deformable grid has been presented. On two dimensional test cases the direct aeroelastic simulations could provide an accurate prediction of the flutter boundary. However in an industrial context, direct aeroelastic simulations on complex three dimensional configurations are too expensive. The P-K method used in association with the aerodynamic loads predicted by the unsteady Navier-Stokes solver elsA is very efficient, even for high Mach number. Futter predictions are accurately performed for the overall flight domain Mach number range. Thanks to the decrease of the unsteady simulation computational time, such applications are becoming compatible with industrial constraints.

References

1. L. Cambier, M. Gazaix: *elsA : An Efficient Object-Oriented Solution to CFD Complexity*, (AIAA paper 2002-0108, 40th AIAA Aerospace Sciences Meeting and Exhibit, Reno, Nevada 2002)
2. J. Donea, S. Giuliani and J-P Halleux: An Arbitrary Lagrangian-Eulerian Finite Element Method For transient Dynamic Fluid-Structure Interactions, *Comp. meth. in Applied Mechanics and Engeering*, vol. 33, (1982) 689–723
3. C. A. Irwin and P. R. Guyett: *The Subcritical Response and Flutter of a Swept Wing Model*, (Royal Aircraft Establishment, August 1965)
4. K. Isogai: *On the Transonic-Dip Mechanism of Flutter of a Sweptback Wing*, *AIAA journal*, Vol. 17, No. 7, July 1979)
5. A. Jameson, W. Schmidt and E. Turkel: *Numerical Solutions of the Euler Equations by Finite Volume Methods Using Runge-Kutta Time Stepping*, (AIAA Paper 81-1259)
6. A. Jameson: *Time Dependant Calculations Using Multigrid, with Applications to Unsteady Flows Past Airfoils and Wings*, (AIAA paper 91-1596, 1991)
7. M. Meaux: Amélioration d'une chaine d'optimisation numérique de formes aérodynamiques en vue du traitement de configurations complexes en écoulements visqueux. MA Thesis, ENSICA, Toulouse (2001)
8. V. Pavan: Le problème de remaillage en optimisation de formes aérodynamiques. MA Thesis, INPL (2000)
9. B. Prananta and M. Hounjet and R. Zwaan: *Two-Dimensional Transonic Aeroelastic Analysis Using Thin-Layer Navier-Stokes Method*, (Journal of Fluids and Structure 12(6), 655-676, 1998)
10. G. D. Mortchéléwicz: *Résolution des équations de Navier-Stokes moyennées linéarisées*, (Rapport Technique RT16/05284 DDSS 2002)
11. W. P. Rodden and J. P. Giesing and T. P. Kalman: *New Developments and Applications of the Subsonic Doublet-Lattice Method for Nonplanar Configurations*, (Douglas Aircraft Company, Tecnical report)
12. P. R. Spalart and S. R. Allmaras: *A One-Equation Turbulence Model for Aerodynamic Flows*, (AIAA Paper 92-0439, 1992)
13. A.S.F. Wong, H.M. Tsai, J. Cai, Y. Zhu, F. Liu: *Unsteady Flow Calculations with a Multi-Block Moving Mesh Algorithm*, (AIAA Paper 2000-1002)
14. R. Zwaan : Verification of Calculation Methods for unsteady Airloads in the Prediction of transonic Flutter, *Journal of Aircraft*, vol 22 (10), 1985, 833–839.

Time-Accurate Navier-Stokes Calculations with Approximately Factored Implicit Schemes

Richard P. Dwight

German Aerospace Center (DLR), Braunschweig, Germany.
richard.dwight@dlr.de

Summary. The performance of the dual time method is improved by means of an inner iteration based on an LU-SGS driven FAS multigrid method, and by using the same LU-SGS method in a time-accurate mode to provide a starting solution for each dual time step. This hybrid scheme is applied to an aerofoil undergoing dynamic stall, whereby considerable performance improvements are demonstrated over the original dual time method.

1 Introduction

One of the most computationally demanding tasks in modern CFD is the simulation of unsteady flows. The lack of efficient algorithms for non-stationary problems either limits the accuracy to which cases may be solved, or creates the demand for more computing resources than the engineer is able to provide.

Typically modern flow solvers tackle unsteady problems by employing a fully implicit time discretization, in which the non-linear system is solved by an inner iteration, the *dual time* method [4]. The inner iteration strongly resembles the iteration for convergence to steady state, so that all techniques developed for efficient steady state convergence (most of which sacrifice time accuracy) may be applied. However even with the application of local time stepping, FAS multigrid, residual smoothing and an explicit Runge-Kutta scheme, typically 50-100 inner iterations are necessary for convergence. In the case that particularly sensitive integrated quantities are of interest, for example the rolling moment of a delta wing, even more iterations are required.

The alternative approach, of linearizing the implicit time discretization and solving the resulting linear system at each time step, has the serious disadvantage that the linear system must be constructed and solved exactly if time accuracy is to be preserved. Most practical implicit solvers on the other hand make use of an approximate system matrix and an inexact linear solver to reduce memory requirements and remain competitive with the efficiency of explicit solvers [1]. Even if everything is formulated exactly, the linearization limits the time accuracy to at most second order.

In this paper a combination of a dual-time method with a first order time accurate LU-SGS method [7] is presented. The hybrid scheme resembles the proposed scheme of Hsu et al. [3] in that an initial approximate implicit step provides a starting solution for a dual time inner iteration. The convergence behavior of the inner iteration is improved by applying LU-SGS as a multigrid smoother. The use of LU-SGS ensures that the memory requirements of the implicit solver remain low. A performance improvement over the dual time scheme when used in isolation is demonstrated for a two-dimensional dynamic stall test case, for an implementation in the DLR unstructured solver *TAU* [2].

2 Theory

The governing equations are the Reynolds Averaged Navier-Stokes (RANS) equations closed by a one- or two-equation turbulence model. The computational domain is discretized by an unstructured grid with a cell-vertex metric, and the finite volume method is applied. It is assumed for simplicity that the grid undergoes only rigid body motion.

2.1 Temporal Discretization

Consider the semi-discretization of the governing equations

$$\frac{dW_i}{dt} + R_i(W) = 0, \tag{1}$$

where the subscript i indicates the index of a point in the grid, and the residual R contains contributions from both fluid fluxes and whirl fluxes due to the motion of the grid. Discretize (1) in time by means of the generalized trapezoidal scheme

$$\Omega_i \frac{W_i^{(n+1)} - W_i^{(n)}}{\Delta t_i} = -\beta R_i(W^{(n+1)}) - (1-\beta) R_i(W^{(n)}). \tag{2}$$

which is first order accurate in time for $\beta = 1$ and second order accurate for $\beta = \frac{1}{2}$. For both values of β the scheme is A-stable. It has the advantage over backward difference formulae (BDF) that the solution at time levels $n-1$ and lower are not needed.

2.2 Linearization of the Implicit Scheme

To reduce the non-linear algebraic system of (2) to a linear system it is necessary to linearize $R_i(W^{(n+1)})$ for each grid point about the known solution $W^{(n)}$. This follows as

$$R_i(W^{(n+1)}) = R_i(W^{(n)}) + \frac{\partial R_i(W^{(n)})}{\partial t}\Delta t + \mathcal{O}(\Delta t^2), \tag{3}$$

$$= R_i(W^{(n)}) + \sum_{j \in \mathcal{N}(i)} \frac{\partial R_i(W^{(n)})}{\partial W_j}\frac{\partial W_j}{\partial t}\Delta t + \mathcal{O}(\Delta t^2), \tag{4}$$

where $\mathcal{N}(i)$ is the set of grid points in the stencil of R_i. Substituting (4) into (2), and applying

$$\frac{\partial W_j}{\partial t}\Delta t = \Delta W_j + \mathcal{O}(\Delta t^2), \tag{5}$$

results in the linear algebraic system

$$\left\{\frac{\Omega_i}{\Delta t_i}\delta_{ij} + \beta\frac{\partial R_i(W^{(n)})}{\partial W_j}\right\} \cdot \Delta W_j^{(n)} = -R_i(W^{(n)}), \tag{6}$$

where the summation convention is used on j. Note that the linear system takes the same form for first and second order time accuracy, allowing a general implementation of linear solver and left-hand side evaluation. This expression retains the second order time accuracy of (2) for $\beta = \frac{1}{2}$ provided that the flux Jacobian of the discrete equations $\partial R/\partial W$ is evaluated exactly. Otherwise the method is first order accurate, with leading order error controlled by the approximation of the flux Jacobian.

2.3 Solution of the Linear System by LU-SGS

The LU-SGS method is applied to equation (6) to obtain an approximation for ΔW. Rather than solving the original system, which may be written $A \cdot x = b$, an *approximately factorized* system, in which the system matrix is replaced by

$$A' = (L + D)D^{-1}(U + D) \approx (L + D)D^{-1}(U + D) - LD^{-1}U = A \tag{7}$$

where L, D and U are the block lower triangular, diagonal and upper triangular parts of A, is solved. The solution operation may be performed using two Gauss-Seidel sweeps, equivalent to two triangular system solves, one lower and one upper. The error introduced due to the approximate factorization is $LD^{-1}U\Delta W \approx \mathcal{O}(\Delta t^2)$.

A second approximation is to use a flux Jacobian based on first order fluxes in (6). These two approximations in combination have the important consequence of allowing the Gauss-Seidel sweeps to be performed without explicit storage of the system matrix [6], reducing the memory requirements to that of an explicit scheme. Furthermore a complete step may be performed in a time equivalent to that of a 3-stage Runge-Kutta step.

At this point there are four sources of error in the time discretization:

- the discretization error of the trapezoidal scheme,
- the truncation error of the linearization of $R(W^{(n+1)})$,
- the factorization error of the LU-SGS scheme $LD^{-1}U\Delta W$,
- the error due to the first order approximation of the flux Jacobian.

All these errors, with the exception of the last, are nominally of size $\mathcal{O}(\Delta t^2)$, but the cost of increasing the accuracy of the last to second order is high in terms of memory and storage on unstructured grids. This approach leads to the high-power implicit methods discussed in Sect. 1.

2.4 Dual Time Stepping

Dual time is an alternative means of time integration. The time derivative of (1) is discretized with a BDF for example, and the resulting algebraic system is solved by iterating the equation

$$\frac{\partial \bar{W}}{\partial \tau} = -\left\{ \frac{\Omega(3\bar{W} - 4W^{(n)} + W^{(n-1)})}{2\Delta t} + R(\bar{W}) \right\} = -\bar{R}(\bar{W}) \qquad (8)$$

to a steady state in τ. Here τ is an artificially introduced pseudo time, $\bar{W}$ is the iterate, and the iterations are henceforth known as *inner iterations*. At a steady state $\partial \bar{W}/\partial \tau = 0$ and hence $\bar{W} = W^{(n+1)}$. This system strongly resembles the original problem (1) for the steady state case, and so methods developed for the solution of that problem may be applied here also. Implicit methods are particularly suited to inner iterations as the extra term in the modified residual $\bar{R}$ increases the diagonal dominance of the flux Jacobian, improving the stability of e.g. LU-SGS.

2.5 Hybrid Time Stepping

At each step, dual time requires an initial estimate of the solution for that step, $\bar{W}^0$. This is typically provided by polynomial extrapolation of the solution at previous time levels to the new time level, $\bar{W}^0 = W^{(n)} + \left(W^{(n)} - W^{(n-1)}\right)$, for linear extrapolation.

An alternative means of start-up might be to apply the LU-SGS method of Sect. 2.3 once in real time. This method has been demonstrated to be nominally first order in time, which is typically insufficient for direct use, but which may lead to a more accurate $\bar{W}^0$. LU-SGS has the ability to perform large steps stably, so the time step size will not be limited by the start-up step, and as seen in Sect. 2.2, the algorithm can be run in time accurate mode without modification, allowing very straightforward implementation. The nature of dual time means that the hybrid scheme will have the higher order time properties of the original BDF, provided inner iterations converge.

In what follows, *dual time* refers to (8) with linear extrapolation start-up, solved by explicit Runge-Kutta smoothed FAS multigrid iterations. *Hybrid time* refers to (8) with start-up by one or more iterations of LU-SGS with $\beta = \frac{1}{2}$, solved by LU-SGS smoothed multigrid with $\beta = 1$.

3 Results

The algorithms of the preceeding sections have been implemented in the unstructured grid, compressible RANS solver *TAU* [2], which is developed at the German Aerospace Center (DLR). A dual time algorithm and general rigid body motion were already available in this code, and these enabled the rapid implementation and testing of the hybrid method.

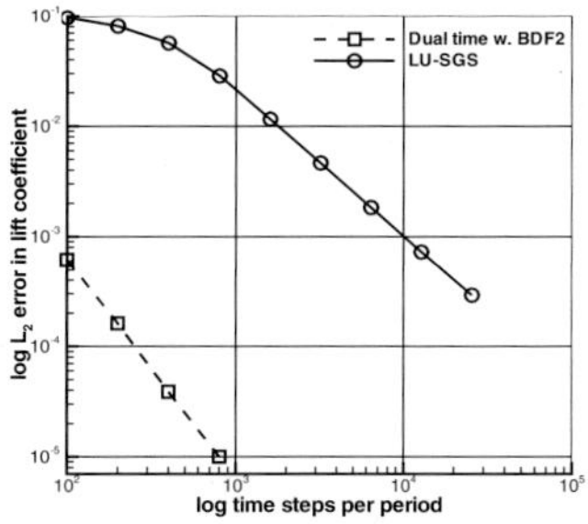

Fig. 1. Comparison of the time accuracy of second order dual time with LU-SGS.

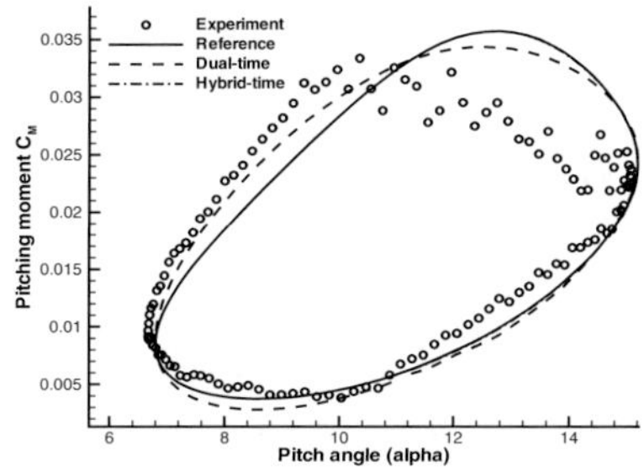

Fig. 2. Pitching moment of the NACA0015 test case. The test schemes use 30 inner iterations.

The first test case is designed to determine the time accuracy of the LU-SGS scheme alone. This will help decide at what range of Δt the hybrid method is likely to be effective. It is essential that this test be viscous, as the approximation of the viscous terms on the LHS of (6) will affect the time accuracy, as will the stretched boundary-layer cells of a viscous grid which will give locally very small Δx.

A harmonically oscillating RAE2822 aerofoil is chosen, with pitch angle $\alpha = 3^\circ \pm 2.51^\circ$, at a reduced frequency of $k = \omega l_k / V_\infty = 0.163$. Under these conditions the flow remains fully attached at all stages of the motion. A third order accurate dual time method, with 200 inner iterations per time step and 1024 time steps per period (TSPP), was used to obtain a reference solution. Unsteady solutions obtained with LU-SGS alone, and second order accurate dual time alone, for varying Δt were compared with the reference solution using the L_2 norm of the difference in calculated lift coefficients over one period. The results are plotted in Fig. 1.

Both the second order behavior of dual time and the first order behavior of LU-SGS are readily apparent, confirming the theory of Sec. 2.3. However the breakdown of the time accurate behavior of LU-SGS occurs at relatively large values of TSPP (about 300), which is a result of the Δx dependence of the factorization error in very thin viscous wall cells. Further note that, in terms of absolute error, LU-SGS reaches the accuracy of 100 TSPP dual time with 10,000 TSPP. Since one dual time inner iteration costs as much as one LU-SGS step, the two methods have roughly equal efficiency (assuming 100 dual time inner iterations per step) for this level of accuracy.

The second test case was chosen to be a demanding problem of practical interest where polynomial extrapolation performs poorly. Dynamic stall refers to unsteady flow separation occurring on aerofoils executing unsteady motion, and is of great importance in rotorcraft aerodynamics where it can have a significant effect on the flight envelope of helicopters. The case considered here is that of a 2d NACA0015 aerofoil pitching harmonically with an angle

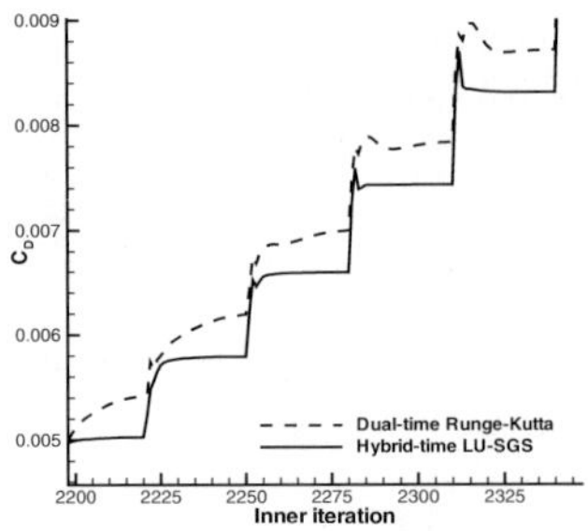

Fig. 3. Convergence for dual- and hybrid-time for 30 inner iterations.

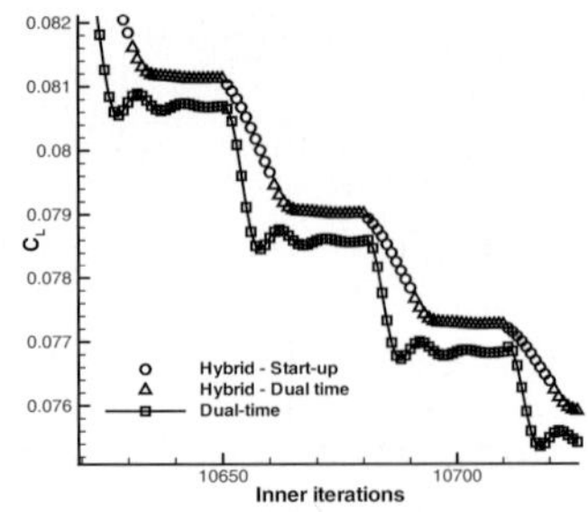

Fig. 4. Demonstration of the combination of LU-SGS with dual-time.

of $\alpha = 11^\circ \pm 4.2^\circ$ and a reduced frequency of $k = 0.2$, being one of the cases investigated experimentally by Piziali [5].

Calculations were performed with the dual- and hybrid-time schemes with only 30 inner iterations, the results can be seen in Fig. 2, compared with a reference calculation using 200 inner iteration dual time, and experimental results. While the dual time scheme deviates from the reference result significantly, the hybrid scheme shows no such difference. The cause can be seen in the convergence of the inner iterations for the two schemes, shown in Fig. 3. Integrated values for dual time do not converge within the 30 iteration limit. Corresponding results have been obtained in three dimensions.

One question raised by hybrid time is, how many start-up LU-SGS steps should be taken for optimal performance? Shown in Fig. 4 is a hybrid scheme with 10 start-up steps each taking one-tenth the time step of the outer scheme, followed by 20 continuation steps. Measurements of the time accuracy of the hybrid scheme for varying numbers of start-up steps produce the rule of thumb that about 10% of the total steps should be start-up steps.

4 Conclusion

An original time stepping method has been introduced based on a combination of a nominally first order time accurate formulation of the LU-SGS implicit scheme, and a dual time method driven by LU-SGS inner iterations. This hybrid scheme has been implemented in an unstructured grid RANS code and applied to dynamic stall on a two-dimensional aerofoil. It has been found to offer moderate efficiency improvements over the dual time scheme used in isolation.

To improve the efficiency of the hybrid scheme further it will be necessary to improve the time accuracy of the LU-SGS start-up step. This may be accomplished with a better flux Jacobian approximation.

References

1. T. Chisholm and D.W. Zingg. A fully coupled Newton-Krylov solver for turbulent aerodynamic flows. *ICAS 2002 Conference, Toroto*, (Paper 333), 2002.
2. T. Gerhold, M. Galle, O. Friedrich, and J. Evans. Calculation of complex 3D configurations employing the DLR TAU-Code. *AIAA Paper*, (AIAA-97-0167), 1997.
3. J.M. Hsu and A. Jameson. An implicit-explicit hybrid scheme for calculating complex unsteady flows. *Proceedings of 40th AIAA CFD Conference, Reno*, (AIAA-2002-0714), 2002.
4. A. Jameson. Time dependant calculations using multigrid with applications to unsteady flows past airfoils and wings. *AIAA Paper*, (AIAA-91-1596), 1991.
5. R.A. Piziali. An experimental investigation of 2D and 3D oscillating wing aerodynamics for a range of angle of attack including stall. *NASA Technical Memorandum*, (4632), 1993.
6. D. Sharov and K. Nakahashi. Reordering of hybrid unstructured grids for Lower-Upper Symmetric Gauss-Seidel computations. *AIAA Journal*, 36(3):484–486, 1997.
7. S. Yoon and A. Jameson. An LU-SSOR scheme for the Euler and Navier-Stokes equations. *AIAA Journal*, 26:1025–1026, 1988.

Conservative Residual Distribution Schemes for General Unsteady Systems of Conservation Laws

Mario Ricchiuto[1], Árpád Csík[1,2], and Herman Deconinck[1]

[1] von Karman Institute for Fluid Dynamics, Belgium
[2] Katholieke Universiteit Leuven, Belgium
ricchiut@vki.ac.be, arpi@vki.ac.be, deconinck@vki.ac.be

1 Introduction

We present the construction of non-linear monotone schemes for the approximation of weak solutions of time-dependent non-linear systems of conservation laws lacking a Roe linearization [1]. The schemes we consider are of the so-called *Residual Distribution* ($\mathcal{RD}$) or *Fluctuation Splitting* ($\mathcal{FS}$) class [2]. Due to their very compact stencil (only nearest neighbors) and to inherent monotonicity properties (no *tuning* of the numerical dissipation), $\mathcal{RD}$ schemes represent an appealing alternative to both finite volume and standard finite element methods [2, 3]. In this work, we make use of the space-time formulation of residual distribution [4, 5]. In particular, consider the scalar problem

$$\frac{\partial u}{\partial t} + \boldsymbol{\lambda} \cdot \nabla u = 0 \quad \text{on} \quad \Omega \times [t_0, t_f] \subset \mathbb{R}^2 \times \mathbb{R}^+ , \quad \boldsymbol{\lambda} = \text{const} . \tag{1}$$

Let $\mathcal{T}_h$ be an unstructured triangulation of Ω and $\{t^1, \ldots, t^M\}$ a sequence of M discrete time levels. Given the nodal values at time t^n, $\{u_i^n\}_{i\in\mathcal{T}_h}$, the unknowns $\{u_i^{n+1}\}_{i\in\mathcal{T}_h}$ are computed by the space-time residual distribution method as the solution of the algebraic system

$$\sum_{T\in\mathcal{D}_i} \phi_i = 0 \quad \forall i \in \mathcal{T}_h \tag{2}$$

System (2) is assembled as follows. First a local *space-time residual* is computed in every every triangle $T \in \mathcal{T}_h$:

$$\phi^h = \int_{t^n}^{t^{n+1}} \int_T \left(\frac{\partial u^h}{\partial t} + \boldsymbol{\lambda} \cdot \nabla u^h \right) d\Omega \, dt , \tag{3}$$

with u^h a *continuous* numerical approximation of the unknown u. Assuming u^h to be piecewise continuous in space and linear in time one has

$$\phi^h = \sum_{j\in T} \left\{ \frac{|T|}{3} \left(u_j^{n+1} - u_j^n \right) + \frac{\Delta t}{2} k_j u_j^n + \frac{\Delta t}{2} k_j u_j^{n+1} \right\} \tag{4}$$

where the *inflow parameter* k_j is given by

$$k_j = \frac{\boldsymbol{\lambda} \cdot \boldsymbol{n}_j}{2} \tag{5}$$

being $\boldsymbol{n}_j$ is the scaled inward pointing vector normal to the edge of T in front of node j. The inflow parameters (5) allow to easily detect upstream ($k_j < 0$) and downstream nodes ($k_j > 0$). The cell-residual ϕ^h is then distributed to each node of T. The fraction of ϕ^h distributed to a node i (*local nodal residual*) is denoted by ϕ_i. We introduce the *distribution coefficients* $\beta_i = \phi_i/\phi^h$.

The properties of the discretization are determined by the definition of the ϕ_i. For the scope of this paper, we are interested in the following properties:

Consistency: $\sum_{j\in T} \phi_j = \phi^h$ or equivalently $\sum_{j\in T} \beta_j = 1$

Linearity Preservation: a scheme is said to be *linearity preserving* if the β_i coefficients are bounded. In [5, 8] is shown that linearity preserving schemes are second order accurate.

Positivity: rewriting (2) as $A\,U^{n+1} = B\,U^n$, a scheme is *positive* if A is an invertible $\mathcal{M}$-matrix and if B is positive ($B_{ij} \geq 0\;\forall\, i, j$) [5]. Positive schemes exhibit a discrete maximum principle and are essential for a monotone approximation of discontinuous solutions.

It is useful to introduce here the positive linear schemes used in this work. They are space-time extensions of the optimal positive N-scheme [2]. In particular, we will refer to the N1-scheme as to the space-time linear scheme defined by the local nodal residuals [4]:

$$\phi_i^{\mathrm{N1}} = \overline{k}_i^+ (u_i^{n+1} - \overline{u}_{in}); \tag{6}$$

Apart from the notation, it is important to note that the *inflow state* $\overline{u}_{in}$ represents the linearly interpolated value of u^h in the most upstream point of the space-time prism $T \times [t^n, t^{n+1}]$, *i.e.* the most upstream point with respect to the space-time characteristic line crossing the prism (left on figure 1). The parameters $\overline{k}_j$ are space-time equivalents of the inflow parameters (5). Scheme (6) represents a truly space-time generalization of the linear N-scheme. It is consistent and positive provided that the residual can be expressed as in (4) and under a time-step constraint [4]. A different scheme has been introduced in [5], defined by the local nodal residual

$$\phi_i^{\mathrm{N2}} = \frac{|T|}{3}(u_i^{n+1} - u_i^n) + \frac{\Delta t}{2} k_i^+ (u_i^n - u_{in}^n) + \frac{\Delta t}{2} k_i^+ (u_i^{n+1} - u_{in}^{n+1}) \tag{7}$$

In this case, the state u_{in} represents the linearly interpolated value of u in the most upstream point in triangle T, *i.e.* the most upstream point on the streamline crossing the element (right on figure 1). Scheme (7) corresponds to the standard N-scheme with Crank-Nicholson time integration. It is consistent and positive provided that the residual can be expressed as in (4) and under a time-step constraint [5]. We will refer to scheme (7) as to the N2-scheme. Note that for both the N1 and the N2 scheme consistency is achieved provided that the residual can be expressed as in (4).

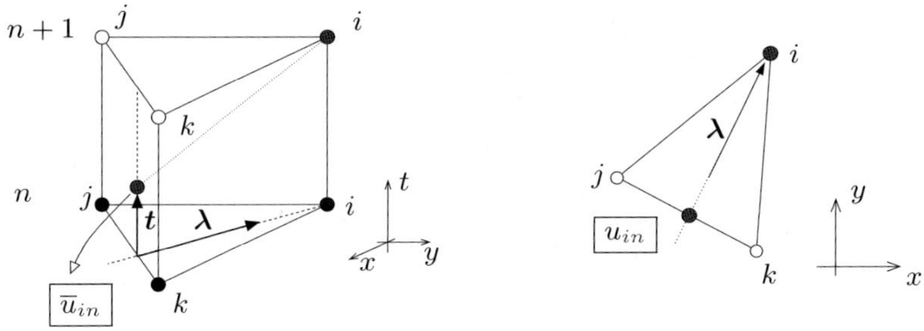

Fig. 1. Space-time inflow state $\overline{u}_{in}$ (left) and space inflow state u_{in} (right)

2 Limited Residual Distribution Schemes

An analog of Godunov's theorem for residual distribution [8, 9] states that linear schemes cannot be at the same time linearity preserving and positive. As a consequence, non-linear schemes must be considered if one is interested in a monotone and sharp approximation of discontinuous solution. Here we follow [5, 8, 10]: given a linear first order positive scheme with local residuals $\phi_i^{\mathcal{P}}$ and distribution coefficients $\beta_i^{\mathcal{P}} = \phi_i^{\mathcal{P}}/\phi^h$, one can introduce the class of nonlinear schemes whose distribution coefficients β_i respect the conditions:

$$\begin{cases} \beta_i \beta_i^{\mathcal{P}} \geq 0 & \text{(for positivity)} & (8.a) \\ |\beta_i| < C < \infty & \text{(for linearity preservation)} & (8.b) \\ \sum\limits_{j \in T} \beta_j = 1 & \text{(for consistency)} & (8.c) \end{cases} \tag{8}$$

Mappings satisfying (8) are presented in [5, 8, 10]. Here we give a condition on the $\phi_i^{\mathcal{P}}$'s guaranteeing the well-posedness of the procedure. The consistency constraint (8.c) requires the existence of at least one positive β_i, hence, due to (8.a), at least one positive $\beta_i^{\mathcal{P}}$ must exist. If the linear scheme is consistent, that is if $\sum_{j \in T} \beta_j^{\mathcal{P}} = 1$, this condition will be met. However, if $\sum_{j\in T} \phi_j^{\mathcal{P}} = \phi^1 \neq \phi^h$, one could run into the unfortunate case $\phi^1 \phi^h \leq 0$ and

$$\sum_{j \in T} \beta_j^{\mathcal{P}} = \frac{\sum\limits_{j \in T} \phi_j^{\mathcal{P}}}{\phi^h} = \frac{\phi^1}{\phi^h} = \Gamma \leq 0\,, \quad \text{sign}(\beta_j^{\mathcal{P}}) = -1 \quad \forall j \in T \tag{9}$$

In this case, (8) cannot be satisfied, since either the first or the last relation would have to be violated compromising the positivity or the consistency of the nonlinear scheme. So we have the general compatibility requirement

Proposition 1. *$\forall$ mappings $\{\beta_j^{\mathcal{P}}\} \to \{\beta_j\}$ respecting (8), a sufficient condition for the existence of the scheme defined by $\phi_i = \beta_i \phi^h$, is $\sum_j \phi_j^{\mathcal{P}} = \phi^h$.*

Using the argument that the only role of the positive linear scheme is to give the correct sign of the β_is, nonlinear schemes based on inconsistent linear ones have been proposed in [8, 10, 11]. Even though the compatibility requirement is not a necessary condition, and one could find an inconsistent positive linear scheme for which (9) are never met, in [8, 10, 11] fixes compromising positivity are introduced to retain consistency.

3 Conservative Schemes for Multidimensional Systems

Consider the hyperbolic system of conservation laws

$$\frac{\partial \boldsymbol{u}}{\partial t} + \nabla \cdot \mathcal{F}(\boldsymbol{u}) = 0 \quad \text{on} \quad \Omega \times [t_0, t_f] \subset \mathbb{R}^2 \times \mathbb{R}^+ \; ; \quad \mathcal{F} = (F, G) \tag{10}$$

Writing (10) in the quasilinear form

$$\frac{\partial \boldsymbol{u}}{\partial t} + A_x \frac{\partial \boldsymbol{u}}{\partial x} + A_y \frac{\partial \boldsymbol{u}}{\partial y} = 0 \; ; \quad A_x = \frac{\partial F}{\partial \boldsymbol{u}} \, , A_y = \frac{\partial G}{\partial \boldsymbol{u}} \tag{11}$$

its hyperbolicity guarantees that $\forall \boldsymbol{\xi} \in \mathbb{R}^2$ the matrix $K_{\boldsymbol{\xi}} = A_x \xi_x + A_y \xi_y$ has real eigenvalues and linearly independent eigenvectors. The *matrix* variant of space-time $\mathcal{RD}$ approximates solutions of linear hyperbolic systems [4, 5] as described in sections 1 and 2 except that the k_j parameters (5) become matrices defined as $K_j = K_{\boldsymbol{n}_j}/2$, the distribution coefficients become distribution matrices and the scalar u^h is replaced by a discrete unknown vector U^h. The N1 and N2 schemes are defined as in (6) and (7) with the proper change of notation. Non-linear schemes are built through a wave decomposition procedure [5, 8]. The case of a non-linear system is more complex. Non-linear systems can evolve discontinuities and it is essential that across these discontinuities the discretization consistently approximates the integral weak form of (10). In the case of the Euler equation for a perfect gas, the existence of a conservative Roe linearization [1, 7] implies an equivalence between the integral and the quasi-linear form of the equations, so that conservation is guaranteed by evaluating the K_j matrices in the Roe averaged state [7]. This allows to compute the element residual using a formula analog to (4) and to use the N1 and N2 schemes. Due to the conservative linearization, these are conservative and consistent, hence the limiting procedure can be applied and is well-posed. In absence of a conservative linearization, the N1-scheme and N2-scheme cannot be conservative since the residual cannot be expressed as in (4). In this case, we propose to compute ϕ^h as

$$\phi^h = \sum_{j \in T} \frac{|T|}{3} \left(\boldsymbol{u}_j^{n+1} - \boldsymbol{u}_j^n \right) + \frac{\Delta t}{2} \oint_{\partial T} \mathcal{F}^n \cdot \hat{n} \, \partial T + \frac{\Delta t}{2} \oint_{\partial T} \mathcal{F}^{n+1} \cdot \hat{n} \, \partial T \, , \tag{12}$$

approximating the boundary integrals with Gauss integration. The problem is now to define consistent monotone linear schemes. Here we follow [6]. Consider the case of the N1-scheme. For a non-linear system, the local residuals obtained with the matrix version of (6) would not yield a consistent scheme. Nevertheless, it is easy to see that the two conditions

$$\phi_i^{\mathrm{N1}} = \overline{K}_i^+ \left(\boldsymbol{u}_i - \overline{\boldsymbol{u}}_{in} \right) ; \quad \sum_{j \in T} \phi_j^{\mathrm{N1}} = \phi^h ,$$

with ϕ^h given by (12), uniquely define $\overline{U}_{in}$. In particular, a unique conservative matrix variant of the N1-scheme (6) is obtained defining $\overline{\boldsymbol{u}}_{in}$ as [4, 6]

$$\overline{\boldsymbol{u}}_{in} = \Big(\sum_{j\in T} \overline{K}_j^+\Big)^{-1} \Big(\sum_{j\in T} \overline{K}_j^+ \boldsymbol{u}_j^{n+1} - \phi^h\Big), \quad \text{with } \phi^h \text{ given by (12)}$$

Proceeding in a similar way, a unique conservative matrix variant of the N2-scheme (7) is obtained, by defining the inflow states as [5, 6, 12]

$$\boldsymbol{u}_{in}^k = \Big(\sum_{j\in T} K_j^+\Big)^{-1} \Big(\sum_{j\in T} K_j^+ \boldsymbol{u}_j^k - \phi^k\Big); \quad \phi^k = \oint_{\partial T} \mathcal{F}^k \,\partial T; \quad k = n, n+1$$

The schemes obtained in this way are indeed conservative but their monotonicity is to be verified numerically [6]. Starting from the conservative variants of the linear schemes we construct limited schemes, which we will refer to as limited N1-scheme and limited N2-scheme. The well-posedness of the limiting is guaranteed by the conservative formulation of the linear schemes.

4 Results

We apply the schemes to the hyperbolic two-phase flow model defined by

$$\boldsymbol{u} = \begin{pmatrix} \alpha_g \rho_g \\ \alpha_l \rho_l \\ \rho u \\ \rho v \end{pmatrix}, \; \mathcal{F} = \begin{pmatrix} \alpha_g \rho_g u & \alpha_g \rho_g v \\ \alpha_l \rho_l u & \alpha_l \rho_l v \\ \rho u^2 + p & \rho u v \\ \rho u v & \rho v^2 + p \end{pmatrix}; \; \begin{cases} \rho = \alpha_g \rho_g + \alpha_l \rho_l \\ \alpha_g + \alpha_l = 1 \\ p = \Gamma_g \left(\dfrac{\rho_g}{\rho_{g0}}\right)^{\gamma_g} \\ p = \Gamma_l \left[\left(\dfrac{\rho_l}{\rho_{l0}}\right)^{\gamma_l} - 1\right] + p_{l0} \end{cases}$$

with α_g and α_l gas and liquid *volume fractions*, ρ_g and ρ_l gas and liquid densities, p the pressure, $\boldsymbol{u} = (u, v)$ the flow speed and ρ the mixture density. The constants in the equations of state are taken as in [13]. No conservative linearization is available for this model but, being in strong conservative form, one can compute exact Rankine-Hugoniot relations. On the first row in figure 2 we show the computation of a planar shock moving in a mixture with $\alpha_g = 0.5$ performed on a 2D mesh ($h \approx 1/100$) with periodic boundary conditions in the y direction. The shock speed u_s is defined by the Mach number $M_s = u_s/\sqrt{p_R/\rho_R}$. The results in the picture show the mixture density distribution in the middle of the 2D domain for $M_s = 10$. The shock is correctly captured by all schemes, confirming their conservative character. Very sharp shock capturing is obtained with the non-linear schemes. On the last row of figure 2 we present the computation of the interaction of a $M_s = 3$ shock with a circular discontinuity in the volume fraction. The solutions are obtained with the non-linear schemes (Top: limited N1. Bottom: limited N2) on a 2D mesh ($h \approx 1/200$). Both schemes give a very good resolution of the interaction between the shock and the *bubble* comparing to similar results available in literature [14, 15, 16, 17].

5 Concluding Remarks

We propose conservative, non-linear and linearity preserving space-time $\mathcal{RD}$ schemes for the discretization of time-dependent systems lacking a conser-

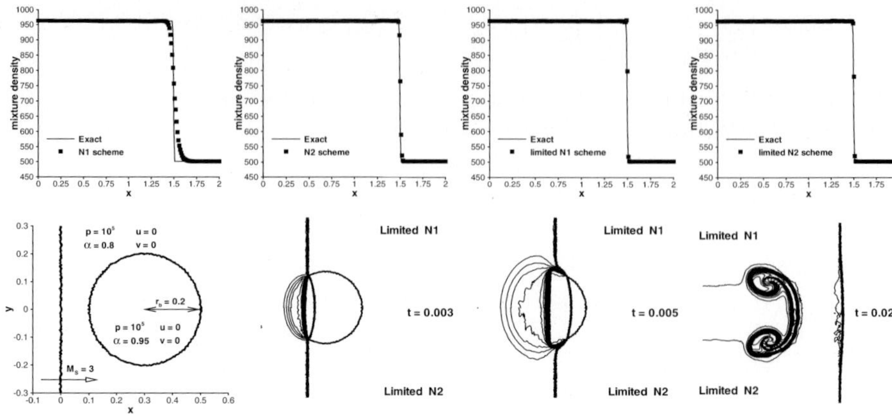

Fig. 2. First row: $M_s = 10$ moving shock. Second row: Shock bubble interaction.

vative linearization [1, 7]. Making use of the technique proposed in [6] we obtained conservative variants of the linear first order schemes of [4] and [5] to be used as a basis for the construction of non-linear schemes through the limiting procedure of [5, 10, 8]. We have shown that the well-posedness of this procedure is guaranteed by the conservative formulation of the linear schemes. Results involving the solution of a hyperbolic two-phase flow model have shown promising features of the schemes proposed: discrete conservation, generality, monotone and sharp capturing of discontinuities.

References

1. P.L. Roe, J. of Comput. Phys. **43** (1981)
2. H. Deconinck *et al.*: AIAA paper 2000-2328 (2000)
3. E. van der Weide *et al.*: Computational Mechanics **23** (1999)
4. Á. Csík *et al.*: In: *VKI LS 2003-05*, von Karman Institute (2003)
5. R. Abgrall and M. Mezine: J. of Comput. Phys. **188** (2003)
6. Á. Csík *et al.*: J. of Comput. Phys. **179** (2002)
7. H. Deconinck *et al.*: Computers and Fluids **22** (1993)
8. R. Abgrall and M. Mezine. In: *VKI LS 2003-05*, von Karman Institute (2003) http://www.ufr-mi.u-bordeaux.fr/ ~abgrall/LS/ls1.ps.gz
9. S.K. Godunov: Mat. Sb. **47** (1959)
10. R. Abgrall and P.L.Roe, J. of Sc. Comput. **19** (2002)
11. M. Mezine *et al.*: In: *VKI LS 2003-05*, von Karman Institute (2003) http://www.ufr-mi.u-bordeaux.fr/ ~abgrall/LS/ls2.ps.gz
12. T. Quintino *et al.*, In *ICCFD 2002*, Springer Verlag (2002)
13. H. Paillere *et al.*: Computers and Fluids **32** (2003)
14. H. Holden *et al.*: J. of Comput. Phys. **150** (1999)
15. J.O. Langseth and R.J. Leveque: J. of Comput. Phys. **165** (2000)
16. R.K.S. Hankin: J. of Comput. Phys. **172** (2001)
17. R. Abgrall *et al.*: Computers and Fluids **32** (2003)

Part VI

Applications

Transonic Flows of BZT Fluids Through Turbine Cascades

P. Cinnella, P.M. Congedo, and D. Laforgia

Università di Lecce, Dipartimento d'Ingegneria dell'Innovazione via Monteroni, 73100, Lecce, Italy, paola.cinnella@unile.it

1 Introduction

Bethe–Zel'dovich–Thompson (BZT) fluids are heavy polyatomic compounds which exhibit, in the vapor phase, a region of negative values of the Fundamental Derivative of Gasdynamics [1]:

$$\Gamma := 1 + \frac{\rho}{a}\left(\frac{\partial a}{\partial \rho}\right)_s \tag{1}$$

In the above, ρ is the fluid density, a the sound speed, and s the entropy. The thermodynamic region characterized by negative values of Γ is referred to as the *inversion zone*, and is included between the upper saturation curve and the $\Gamma = 0$ contour (called the *transition line*). The Fundamental Derivative Γ represents a measure of the rate of change of the sound speed with density in isentropic perturbations. If $\Gamma < 1$, the flow exhibits an uncommon sound speed variation in isentropic perturbations: a grows during expansions and decreases in compressions, the opposite of what happens in "common" fluids. For perfect gases, $\Gamma = \frac{\gamma+1}{2}$, with the specific heats ratio $\gamma > 1$ for thermodynamic stability reasons; therefore, $\Gamma > 1$ as well. Examples of BZT fluids are given by heavy hydrocarbons (decane or higher), commercially available heat transfer fluids such as FC-71 ($C_{18}F_{39}N$), FC-72 (C_6F_{14}), PP9 ($C_{11}F_{20}$), PP10 ($C_{13}F_{22}$), and some methylsiloxanes. The dynamics of Bethe-Zel'dovich-Thompson fluids has been studied extensively in recent years (e.g. [2] and references cited therein). The most impressive phenomenon is the disintegration of compression shocks in flows with $\Gamma < 0$. The entropy change through a weak shock can be written [3]:

$$\Delta s = -\frac{a^2 \Gamma}{v^3}\frac{(\Delta v)^3}{6T} + O\left((\Delta v)^4\right), \tag{2}$$

where Δ represents a change in a given fluid property through the shock and T is the absolute temperature. If $\Gamma < 0$, the second law of thermodynamics requires that compression shocks (characterized by a negative change in the specific volume) cannot form, and necessarily split into fans if inserted within the flow. Conversely, expansion shocks are physically admissible. Another important property is that flow discontinuities have limited strength for thermodynamic conditions where $\Gamma \simeq 0$. Cramer and Kluwick [3] showed that $\Gamma = O(\Delta v)$ for small volume changes in the vicinity of the transition line: this implies, because of Eq. (2), that shock waves having jump conditions close to the $\Gamma = 0$ contour are one order of magnitude weaker than normal.

The nonclassical phenomena typical of BZT fluids have several practical outcomes; prominent among them is an active research effort to reduce losses due to wave drag and shock/boundary layer interactions in turbomachines and nozzles [4, 5, 6], with particular application to Organic Rankine Cycles (ORCs). One of the major loss mechanisms in transonic and supersonic turbomachinery is related to the generation of shock waves. The use of a BZT fluid could avoid shock formation and, ideally, allow isentropic turbine expansion. However, simply utilizing a BZT working fluid is not sufficient to maximize the reduction in losses: it is also necessary to operate the turbine cascade at a pressure and temperature near the inversion zone. But, how near? The inversion zone has a quite limited extent. Therefore, a reduction in the temperature jump between the heater and condenser stages is generally needed in order to operate the turbine cascade in the BZT regime. However, if the temperature jump is taken too small, the overall cycle efficiency and global power output tend to decrease. The optimum choice depends basically on fluid properties (extent of the inversion zone), and blade geometry.

In the present study, a parametric investigation of transonic BZT flows through linear transonic turbine cascades is presented. Specifically, the aim is to understand how BZT effects influence the performance of modern turbomachinery bladings. Two linear turbine cascades are considered: a low-pressure stator blade cascade, and a rotor blade cascade. In both cases, the free-stream thermodynamic state is changed, starting from operation points located in the neighborhood of the transition line and adjacent to the vapor saturation curve, and progressively increasing inlet pressure and temperature until BZT effects disappear from the flow. The corresponding system efficiency is then evaluated. Numerical results for BZT fluids are compared with those obtained for a diatomic perfect gas and for water vapor.

2 Governing Equations and Numerical Method

As BZT effects influence essentially the inviscid behavior of the flow, the two-dimensional compressible Euler equations for single-phase, non-reacting flows are considered. Two equations of state are taken into account, in addition to the perfect gas law. Saturated water vapor is modeled through the van de Waals equation of state for polytropic gases:

$$p = \frac{(\gamma - 1)(e + \alpha/v)}{v - \beta} - \frac{\alpha}{v^2} \tag{3}$$

(with e the specific internal energy, v the specific volume, and α and β gas-dependent constants). For BZT fluids, the more realistic Martin-Hou equation of state [7] is used, involving five virial expansion terms. The equation relates the fluid pressure to the absolute temperature T and specific volume through the relation:

$$p = \frac{RT}{v-b} + \sum_{i=2}^{5} \frac{f_i(T)}{(v-b)^i}, \tag{4}$$

where the functions $f_i(T)$ are of the form $f_i(T) = A_i + B_i\, T + C_i\, e^{-kT/T_c}$, with T_c the critical temperature and $k = 5.475$. The gas-dependent coefficients A_i, B_i, and C_i can be expressed in terms of the critical temperature and pressure, the critical compressibility factor, the Boyle temperature, and one point on the vapour pressure curve. A compatible caloric equation of state is constructed by assuming a power-law variation of the specific heat in the low-pressure limit.

The governing equations are solved numerically using a finite volume third-order accurate centred scheme [8, 9]. The scheme uses a fourth-order accurate centred approximation of the fluxes, with addition of a scalar numerical dissipation term [10]. Such a choice greatly simplifies the implementation with complex equations of state and reduces computational costs. The resulting scheme is globally third-order accurate in smooth flow regions. The solution is advanced in time using a four-stage Runge-Kutta scheme. Local time-stepping, implicit residual smoothing and multigrid are used to efficiently drive the solution to the steady state. For internal flows, non-reflecting boundary conditions using the method of characteristic are applied at the inlet and outlet boundaries; a slip condition is imposed at the wall, which uses multidimensional linear extrapolation from interior points for the pressure; periodicity conditions are prescribed at the inter-blade-passage boundaries.

3 Results and Discussion

Transonic BZT flows at different operating conditions have been studied for two linear turbine cascades. The first one represents a low pressure stator turbine profile (11th Standard Aeroelastic Configuration, SC11); the second-one is the von-Karman Institute LS-59 rotor blade cascade (VKI LS-59). Geometry and experimental data (for air flow) are available in Refs [11, 12]. The first cascade is caracterized by an inlet angle of 15.2^0, and an inlet-to-outlet pressure ratio equal to 1.64. For air flow, this corresponds to an exit isentropic Mach number of 0.9. The second cascade has inlet angle equal 30^0, and pressure ratio 1.82; the corresponding exit Mach number (for air) is about 1. All computations, concerning both perfect and real gas flows, have been performed at fixed inlet angle and pressure ratio. The computations are performed using C-grids of increasing density, composed by 71x8, 142x16, and 284x32 cells for the first cascade and by 86x8, 192x16 and 384x32 cells for the second one. The computed convergence order of the numerical solver, based on cascade efficiency, is in the range 2.2–2.5 for all the computed cases; grid convergence index (GCI) on the medium grid is less than 1%. Firstly, perfect diatomic gas flows have been computed. The perfect gas flow is characterized by a weak shock wave located at about 75% the chord on the suction side for th first cascade, and by a weak shock at mid-chord and a trailing edge shock for the second one (see Fig. 2a and 4a). The computed efficiencies (real-to-ideal static enthalpy drop, η) are about 98.0% for the first cascade and 93.8% for the second one. Water vapor flows, with reduced inlet thermodynamic conditions $p_1/p_c = 0.9$, $\rho_1/\rho_c = 0.6$, $\Gamma_1 = 1.26$, are characterized by lower efficiencies: η equals 89.3% for cascade SC11, and 87.7% for LS-59. In both cases, efficiency losses are related to the formation of strong shocks (upstream Mach number about 1.8) close to the trailing edge (Figs 2b and 4b). Then, computations are performed using the BZT fluorocarbon PP10 (pf-perhydrofluorene) as the working fluid. Several computations, corresponding to different thermodynamic inlet conditions are carried out. Figure 1 represents cascade efficiencies for different values of the inlet Fundamental Derivative. This one is changed at constant inlet entropy, and three different entropy values are considered. The first value corresponds to an isentropic line crossing the inversion zone, the second to a line almost tangent to the transition line, and the last one to a line outside the inversion zone, but within the $\Gamma = 1$ contour. For both cascades, the efficiency tends initially to increase with Γ_1, reaching an optimum. This roughly corresponds to the lower mean Γ value at the blade wall. In such conditions, average Γ is less than 1 at the wall, and the sound speed increases with dropping pressure. This limits the growth of the Mach number during the flow expansion and reduces shock strength. For higher values of Γ_1, expansion shock waves appear for the two lower entropy values, and efficiency decreases. This does not happens for the higher entropy value, the flow evolving entirely outside the inversion zone. On the other hand, for such conditions, the average Fundamental Derivative at the wall is below 1, which contributes to weaken or even remove shock waves. For the VKI cascade, characterized by a higher pressure drop, peak efficiency is attained for higher values of Γ_1 as compared with cascade SC11. In fact, Γ quickly drops outside the inversion zone during the flow expansion, and it is necessary to start from higher inlet Γ values in order to have Γ in the right range close to the shock region. Figures 2c and 4c present pressure contours for PP10 corresponding to maximum cascade efficiency. Mach number, pressure, and Γ distributions at the wall for the three working fluids and the two cascades under investigation are depicted in Figs 3 and 5.

4 Final Remarks

The present work has provided a study of transonic flows of Bethe–Zel'dovich–Thompson (BZT) gases through modern turbine cascades. The effectiveness of the use of such fluids for improving cascade efficiency has been demonstrated. Present results show that the choice of turbine operation conditions within or very close the BZT region is not mandatory in order to achieve high turbine efficiency: very good performances can be also obtained by just operating the turbine inside the dense region of the Amagat plane, where the Fundamental Derivative is less than 1, and the flow exhibits reversed sound speed variation with density. The last result is particularly important in view of the development of Dense-gas Organic Rankine Cycles of practical interest.

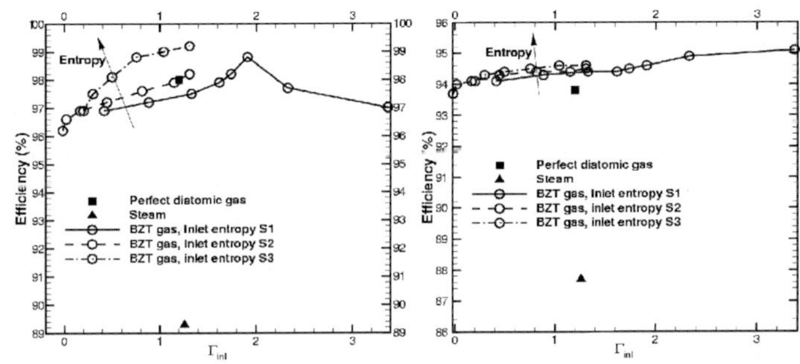

Fig. 1. Cascade efficiency versus inlet Fundamental Derivative. Left: SC11 cascade; right: VKI LS-59 cascade.

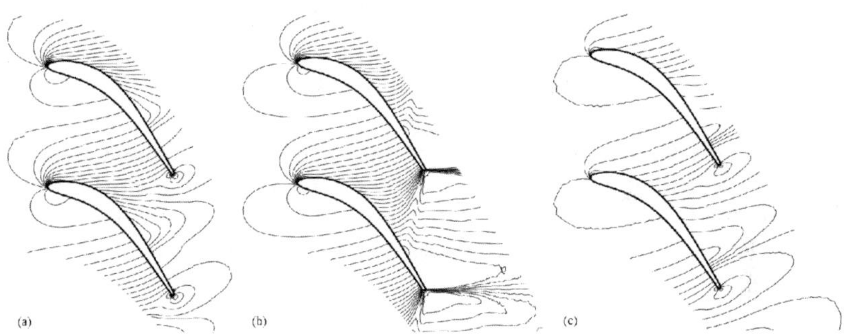

Fig. 2. Isobares for a perfect gas (a), steam (b), and PP10 (c). SC11 cascade.

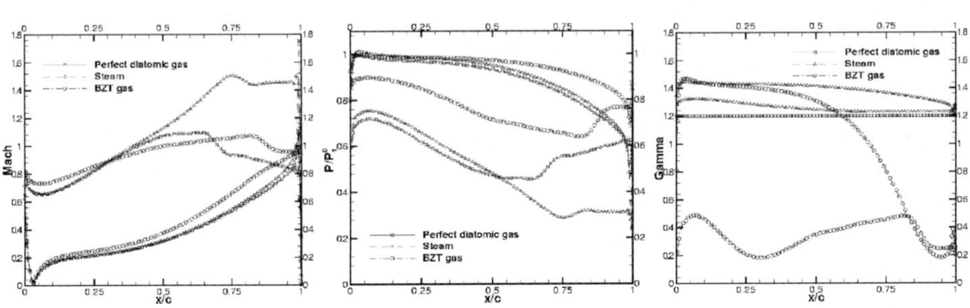

Fig. 3. Mach, pressure, and Γ distributions at the wall. SC11 cascade.

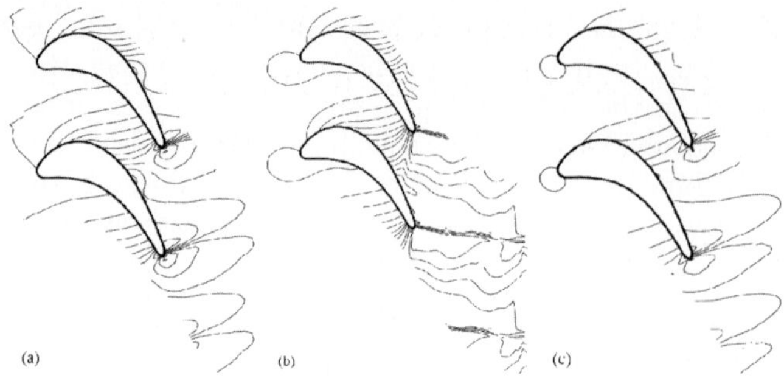

Fig. 4. Isobares for a perfect gas (a), steam (b), and PP10 (c). VKI LS-59 cascade.

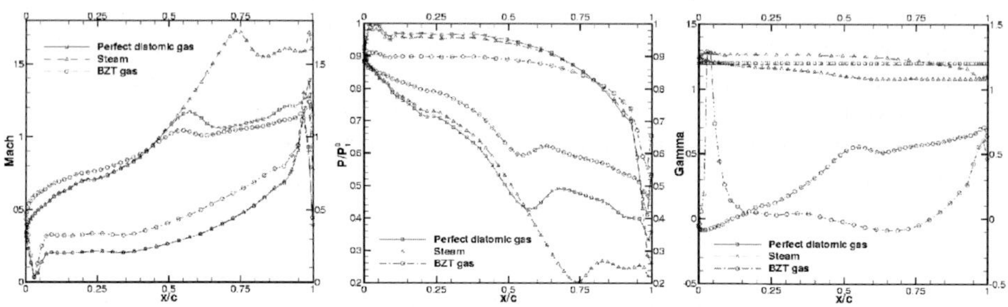

Fig. 5. Mach, pressure, and Γ distributions at the wall. VKI LS-59 cascade.

References

1. P.A. Thompson: Phys. Fluids **14**, 1843 (1971)
2. M.S. Cramer: Nonclassical dynamics of classical gases. In: *Nonclassical waves in real fluids*, ed by A. Kluwick (Springer-Verlag, Berlin Heidelberg New-York 1991) pp 91–145
3. M.S. Cramer, A. Kluwick: J. Fluid Mech. *142*, 9 (1984)
4. M.S. Cramer, L.M. Best: J. Fluid Mech. *240*, 197 (1992)
5. J.F. Monaco, M.S. Cramer, L.T. Watson: J. Fluid Mech. *330*, 31 (1997)
6. B.P. Brown, B.M. Argrow: J. Prop. Power *16*, 1118 (2000)
7. J.J. Martin, Y.C. Hou: A.I.Ch.E. J., *1*, 142 (1955)
8. Y. Huang, P. Cinnella, A. Lerat: A third-order accurate centered scheme for turbulent compressible flow calculations in aerodynamics. In: *Numerical Methods for Fluid Dynamics*, vol VI, ed by M.J. Baines (Will Print, Oxford, UK 1998) pp 355–361
9. P. Cinnella, P.M. Congedo: A numerical solver for dense gas flows. AIAA Paper 2004-2137
10. A. Jameson, W. Schmidt, E. Turkel: Solutions of the Euler equations by Finite Volume methods using Runge–Kutta time stepping schemes. AIAA Paper 81-1259
11. T.H. Fransson *et al.*: ASME J. Turbomach. *121*: 717 (1999)
12. R. Kiock *et al.*: ASME J. Eng. Gas Turb. Power *108*: 810 (1996)

CFD Simulation of the Space Shuttle Launch Vehicle with Booster Separation Motor and Reaction Control System Plumes

L. M. Gea[1] and D Vicker[2]

[1] The Boeing Company, 5301 Bolsa Ave. MC H013-B308, Huntington Beach, CA, lie-mine.gea@boeing.com
[2] NASA Johnson Space Center, EG3-Applied Aeroscience, 2101 NASA Parkway, Houston, TX, darby.vicker@nasa.gov

Summary. The primary objective of this paper is to demonstrate the capability of computational fluid dynamics (CFD) to simulate a very complicated flow field encountered during the space shuttle ascent. The flow field features nozzle plumes from booster separation motor (BSM) and reaction control system (RCS) jets with a supersonic incoming cross flow at speed of Mach 4. The overset Navier-Stokes code OVERFLOW [1], was used to simulate the flow field surrounding the entire space shuttle launch vehicle (SSLV) with high geometric fidelity. The variable gamma option was chosen due to the high temperature nature of nozzle flows and different plume species. CFD predicted Mach contours are in good agreement with the schlieren photos from wind tunnel test. Flow fields are discussed in detail and the results are used to support the debris analysis for the space shuttle Return To Flight (RTF) task.

1 Introduction

Following the Columbia disaster, an investigation identified a debris impact on the wing leading edge during ascent as the major cause [2]. In order to predict the potential debris path, a state-of-the-art debris analysis process was developed by NASA for debris analysis across a wide range of flight conditions. CFD is used to provide the flow fields at various flight conditions with a high level of the geometric fidelity. Solid Rocket Booster (SRB) separation is initiated when the vehicle Mach number reaches about 4.0. The BSMs are fired to push the SRBs away from the orbiter. In addition, the forward RCS jets are also fired to prevent BSM plume by-products from hitting the orbiter windshield. The interaction of these plumes with the supersonic cross flow presents a very complicated aerodynamic flow field and posts a great challenge for CFD simulation.

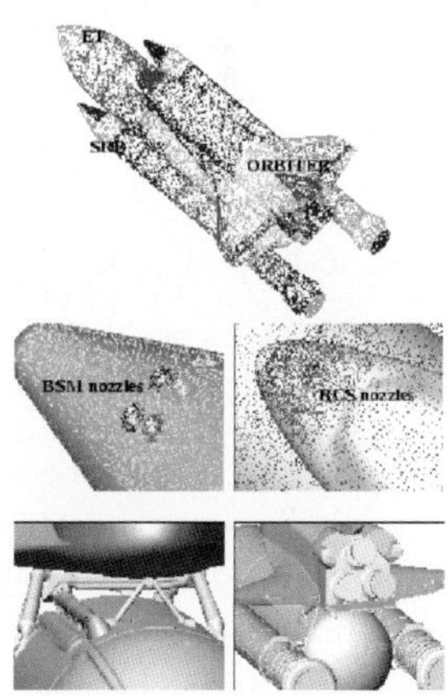

Fig. 1. SSLV CFD geometry and surface grids.

NASA's Johnson Space Center (JSC) and Ames Research Center have developed a grid system of the SSLV with high geometric fidelity (see Fig. 1) for extensive CFD simulations [3-6]. However, these simulations are in general limited to the conditions prior to SRB separation, therefore, no BSM or RCS plumes were considered. In the current work, grid systems for the BSM and RCS nozzle geometries are generated and corresponding plume grids are also introduced to capture the flow features. OVERFLOW, a NASA developed Navier-Stokes code which uses the Chimera overset grid scheme [7], was used to compute the steady state solutions at five different time stages after SRB separation. One important aspect of the current work is to include gamma variation due to the different plume gases in the flow field and the high temperature in these plumes. The species continuity equations are solved to track the mass fractions of the different gases. The resulting gamma of the gas mixture can be computed by the mass-average of the individual gases. CFD predicted Mach contour distributions with and without the BSM plumes are in good agreement with schlieren photos from wind tunnel tests. Mach number and pressure contours for T=0.0 sec (beginning of SRB separation) are analyzed to assess some unique flow features. Mach contours for the SSLV configuration during five time stages of SRB separation are presented to show the plume's effect on the overall flow field.

2 Numerical Approach

Due to the high temperature nature of nozzle flows and the different species of the BSM and RCS plumes, a variable gamma (γ) option [5] implemented in OVERFLOW code was adopted. The species continuity equation, without diffusion or source terms, is solved in strong conservative-law form as:

$$\frac{\partial \rho}{\partial t} + \frac{\partial \rho c U}{\partial \xi} + \frac{\partial \rho c V}{\partial \eta} + \frac{\partial \rho c W}{\partial \zeta} = 0\,, \tag{1}$$

where c is the species mass fraction and U, V, and W are the contravariant velocities in the ξ, η, and ζ directions, respectively. The variation of γ with the temperature is computed using polynomial curve fits of the normalized specific heat, C_p/R versus temperature:

$$\frac{C_p}{R} = a_0 + a_1 T + a_2 T^2 + a_3 T^3 + a_4 T^4\,, \tag{2}$$

where R is the gas constant and T is the local dimensional temperature. The coefficients, a_0 to a_4, of the polynomial equation are calculated based on the molecular weights and the mass fraction of different chemical species within the plume. The resulting γ for the gas mixture can be computed from mass-averaged specific heats

$$C_p^* = \sum_i^{n_{\text{gas}}} c_i C_{p_i}\,, \quad C_v^* = \sum_i^{n_{\text{gas}}} c_i C_{v_i}\,, \quad \gamma^* = \frac{C_p^*}{C_v^*} \tag{3}$$

where C_p and C_v are the constant pressure specific heat and constant volume specific heat coefficients. The detailed implementation and formulations for this approach can be found in [1].

The implementation of the nozzle plumes was identical for both BSM and RCS nozzles. NASA's Marshall Space Flight Center (MSFC) performed a simulation of the nozzle flow field with the RAMP code [8], which included chemistry characteristics. A computational plane was constructed a short distance upstream of the nozzle exit plane, which has a double fringe and is point-matched with the nozzle grid. An OVERFLOW boundary condition was calculated from the RAMP solution and imposed on this plane. Each of the BSM thrust levels required a separate RAMP solution. Steady state CFD solutions for five different time stages after SRB separation were carried out, the respective locations of the SRB (Fig. 2) were determined from Monte Carlo analysis. The corresponding BSM and RCS nozzle thrust levels are given in Table 1. The Spalart-Allmaras one equation turbulence model [9] was used for all the simulations.

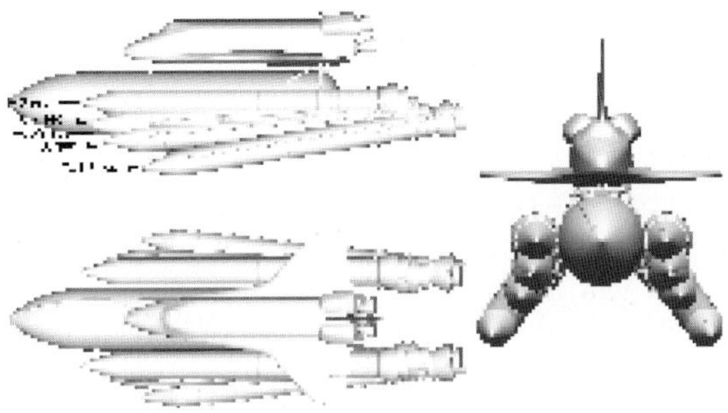

Fig. 2. SRB locations at five different time stages after separation.

3 Results And Discussion

Earlier simulations conducted by the second author showed excellent agreement between CFD and wind tunnel test pressure data from IA-613 [10]. Figure 3 highlights some of the comparisons. To further validate the plume features of the CFD results, two additional BSM/RCS CFD solutions were generated to compare with the existing wind tunnel photos from IA-193 [11].

Table 1. BSM and RCS thrust levels.

Free Stream Mach Number	Time After SRB Seperation	BSM Thrust Level	Forward RCS Thrusters
4.0	0.0 sec	100%	on
4.0	0.7 sec	100%	on
4.0	0.75 sec	52%	on
4.0	1.0 sec	13%	on
4.0	1.5 sec	3%	off

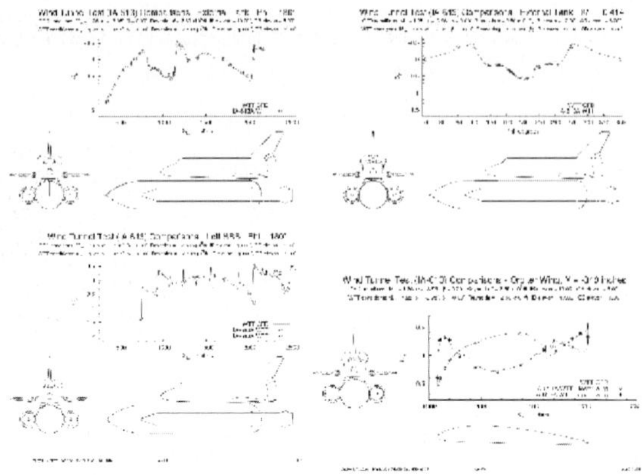

Fig. 3. CFD predicted C_p vs wind tunnel test data.

For the first case (BSM and RCS off), the CFD predicted shapes of the oblique shocks originating from the nose of the external tank (ET) and SRB agree very well with the test photo. Refer to Fig. 4(a). For the second case (BSM on, RCS off), the oblique shock on the orbiter side of the ET is pushed out by the BSM plume to form a normal shock on the nose cone region. See Fig. 4(b). The low Mach number area downstream of the normal shock correlates well with the disturbance observed from the schlieren photo. On the bottom side of the ET, the oblique shock from the SRB nose is pushed upstream by the BSM plume to merge with the oblique shock from the ET nose to form a lambda type of shock structure. This phenomenon can be observed in both CFD result and the schlieren photo.

Following satisfactory comparisons with wind tunnel test data, CFD solutions for the five conditions listed in table 1 were carried out. Mach number and pressure contours at two constant Y cuts (Y=0 and -200) for a representative case (T=0.0 sec, SRB at mated position with BSM at 100% thrust and RCS on) are shown in Fig. 5(a). Driven by the combined effect of the su-

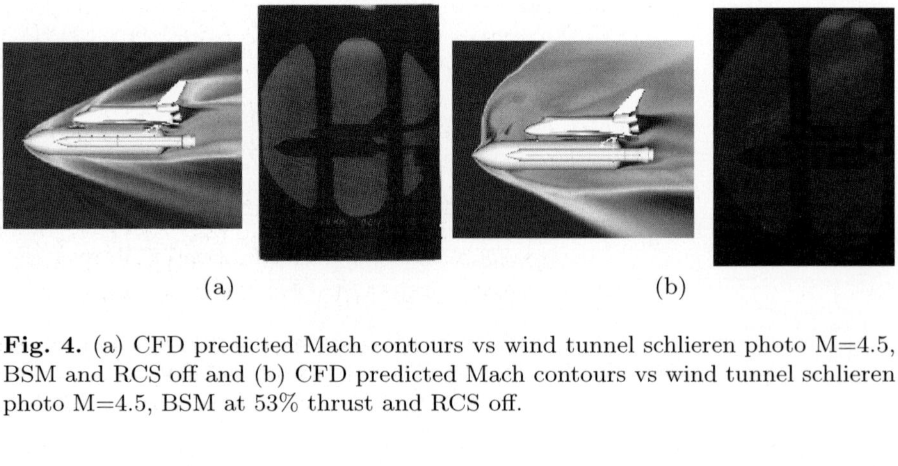

(a) (b)

Fig. 4. (a) CFD predicted Mach contours vs wind tunnel schlieren photo M=4.5, BSM and RCS off and (b) CFD predicted Mach contours vs wind tunnel schlieren photo M=4.5, BSM at 53% thrust and RCS off.

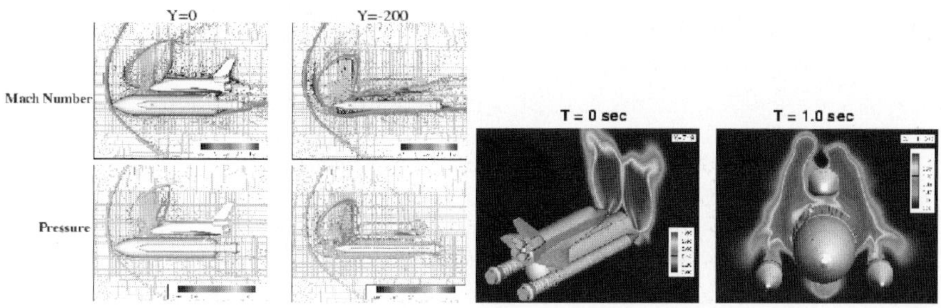

(a) (b)

Fig. 5. (a) Mach number and pressure contours at constant Y planes for T=0.0 sec case M=4.0, BSM at 100% thrust and RCS on and (b) CFD predicted mass fraction for BSM gas at T=0.0 and 1.0 sec M=4.0, BSM and RCS on.

personic inflow and the BSM and RCS plumes, a bow shock stands ahead of the ET nose. Compressed by the bow shock and the plumes, the area downstream of the bow shock has low Mach number and high pressure. Further downstream, due to the plume expansion into the supersonic flow, a high Mach number region associated with low pressure (and static temperature) roughly represents the extent of the plumes. At the forward edge of this region the flow is moving upstream and compressed into a subsonic flow to form a strong shock with a very high pressure gradient. This unique feature of two shocks forming in the opposite direction highlights the complexity of the flow field. Downstream of the BSM and RCS plumes, a low Mach number region indicates flow separation on top of the orbiter and the leeside of the SRB. Figure 6 compares the Mach number contours from front, side, and top views at different time stages after the SRB separation. In general, the BSM thrust level dominates the shape of the bow shock in front of the ET. The RCS plumes are much weaker than the BSM plumes and have very little impact to the overall shock structure. The bow shock moves downstream toward an oblique shock as the SRB moves away and the BSM thrust diminishes. Finally, the mass fractions in an X-constant cut for the T=0.0 and 1.0 sec cases are shown in Fig. 5(b) to illustrate the extent of the BSM plumes.

4 Conclusion

The current analysis has demonstrated the CFD capability to model the flow fields for the SSLV configuration with both BSM and RCS nozzle plumes. Validation results show promising correlations between the CFD and wind tunnel results. On the practical side, the effects of these plumes during SRB separation are included in the current CFD simulation, and the results have been provided for the current debris analysis supporting the shuttle return-to-flight effort.

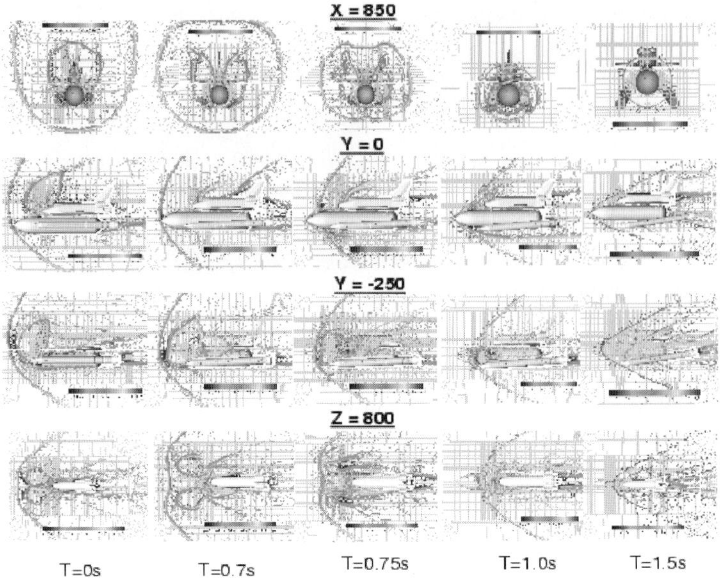

Fig. 6. CFD predicted Mach contours for different time stages after SRB separation M=4.0, BSM and RCS on.

References

1. Buning, P.G., et al., OVERFLOW Use's Manual, Version 1.8j, February 1999.
2. Columbia Accident Investigation Board Report, August 2003.
3. Buning, P.G., et al.., Flow field Simulation of the Space Shuttle Vehicle in Ascent, Proc. Fourth International Conference on Supercomputing, April 1989.
4. Buning, P.G., Chiu, I.T., Obayashi, S., Rizk, Y.M., and Steger, J.L., Numerical Simulation of the Integrated Space Shuttle Vehicle in Ascent, AIAA-88-4359-CP, August 1988.
5. Slotnick, J.P., Kandula, M., Buning, P.G.,and Martin, F.W., Numerical Simulation of the Space Shuttle Vehicle Flow Field with Real Gas Solid Rocket Motor Plume Effects, AIAA-93-0521, January 1993.
6. Gomez, R.J. and Ma, E.C., Validation of a Large Scale Chimera Grid System for Space Shuttle Launch Vehicle, AIAA-94-1859, August 1994.
7. Suhs, N.E., Dietz, W.E., Nash, S.M., Onufer, J., Baker, D., and Rogers, S.E., PEGAUS User's Guide Version 5.1c, July 2000.
8. Smith, S.D., High Altitude Chemical Reaction Gas Particle Mixture Code User's Manual, NAS9-16256, 1984.
9. Spalart, P.R., and Allmaras, S.R., A One-Equation Turbulence Model for Aerodynamic Flows, AIAA 92-0439, AIAA paper 92-0439, January 1992.
10. Crosby, W. A. and Lanham, D. D., Integrated aerodynamic tests of the space shuttle vehicle during solid rocket booster separation at Mach 4.5 (IA193), NASA Center for Aerospace Information, AEDC-TSR-82-V15, June 1982.
11. Marroquin, J. and Lemoine, P., Results of wind tunnel tests of an ASRM configured 0.03 scale Space Shuttle integrated vehicle model (47-OTS) in the AEDC 16-foot transonic wind tunnel, NASA Center for Aerospace Information, NASA-CR-185697, October 1992.

An Efficient Numerical Method for 3D Viscous Ship Hydrodynamics with Free-Surface Gravity Waves

Mervyn Lewis and Barry Koren

CWI, P.O. Box 94079, 1090 GB Amsterdam, The Netherlands
[Mervyn.Lewis,Barry.Koren]@cwi.nl

Summary. A new numerical method for water flows with free-surface gravity waves is investigated. The method is first analyzed with respect to the existence of steady free surface waves, and the dispersion properties of these waves. Next, the method is used to compute the free water surface generated by a standard ship hull.

1 Introduction

Most of the current numerical methods for solving steady, viscous free-surface flows in ship hydrodynamics employ a time-dependent formulation and integrate until a steady state is reached. This approach typically displays the defect of high computational costs due to persistent transient behaviour of the long gravity waves. The attenuation of these waves behaves like $\mathcal{O}(t^{(1-d)/2})$ in $\mathbb{R}^d$. To reduce the computational effort of solving free-surface Navier-Stokes flow, an efficient iterative method, employing a stationary formulation of the problem, has been introduced for 2D in [1] and has been extended to 3D in [3]. The method relies on a novel formulation of the free-surface flow problem, involving the so-called quasi free-surface condition (QFSC):

$$\boldsymbol{u} \cdot \nabla \varphi - \mathrm{Fr}^{-2} \boldsymbol{u} \cdot \boldsymbol{e}_z = 0, \tag{1a}$$

where $\boldsymbol{u}$ is the velocity vector, φ the hydrodynamic pressure:

$$\varphi = p + \mathrm{Fr}^{-2} z, \tag{1b}$$

Fr the Froude number and $\boldsymbol{e}_z$ the unit vector in vertical direction, positive upward. Note that the boundary condition (1a), see [1] for a derivation, is nonlinear. So far, this method has not yet been applied to a real 3D ship-hydrodynamics problem. In the present paper we will do so. Moreover, a Fourier analysis will be made of a semi-discrete generic model problem representing our discrete, 3D Navier-Stokes flow problem. In the Navier-Stokes equations considered, streamwise diffusion, i.e., diffusion in x-direction, is neglected. This simplification turns out to have far-reaching consequences for the well-posedness of the continuous and semi-discrete initial boundary-value problem.

2 Fourier Analysis of Semi-Discrete Problem

As mentioned above, the aim is to compute the complete stationary flow field, i.e., the shape of the free surface and the underlying viscous flow field, generated by a ship. We start by analyzing the wave pattern by performing a Fourier analysis for a 2D, uniform horizontal flow in infinitely deep water. The analysis uses classical perturbation methods, linearization and Fourier techniques. We assume perturbation expansions in powers of ϵ, $\epsilon \ll 1$, of the form

$$\boldsymbol{q}(\boldsymbol{x}, t; \epsilon) = \boldsymbol{q}_0 + \epsilon \boldsymbol{q}_1(\boldsymbol{x}, t) + \mathcal{O}(\epsilon^2), \tag{2a}$$

$$\eta(x, t; \epsilon) = \eta_0 + \epsilon \eta_1(x, t) + \mathcal{O}(\epsilon^2), \tag{2b}$$

where the solution vector is represented by $\boldsymbol{q} = (\boldsymbol{u}(\boldsymbol{x}, t), \varphi(\boldsymbol{x}, t))^T$ and where $\eta = \eta(x, t; \epsilon)$ denotes the water height. The generating solution, denoted by $\boldsymbol{q}_0 = (\boldsymbol{U}, \Phi)^T$ and η_0, consists of the unperturbed flow velocity $\boldsymbol{U} = (U, 0)^T (U = \text{constant})$, the unperturbed hydrodynamic pressure $\Phi = 0$ and the unperturbed water height $\eta_0 = 0$. Our spatial domain Ω is defined as

$$\Omega = \{\boldsymbol{x} \in \mathbb{R}^2 : x \in (-\pi, \pi), z \in (\eta, -\infty)\}. \tag{3}$$

Since we are considering the flow in infinitely deep water, in z-direction the domain Ω is only bounded by the free surface. We assume that

$$\lim_{z \downarrow -\infty} \boldsymbol{q}_1(\boldsymbol{x}, t) = \boldsymbol{0}. \tag{4a}$$

Further, we take

$$\boldsymbol{q}(-\pi, z) = \boldsymbol{q}(\pi, z), \tag{4b}$$

i.e., periodicity of the solution in x-direction. To study the combined effect of the different truncation errors on the numerical approximation of the free-surface flow, we consider the system of modified equations resulting from a semi-discretization of the reduced Navier-Stokes equations in $\mathbb{R}^2$, on a uniform Cartesian grid with mesh width h. This system reads, neglecting the $\mathcal{O}(h^4)$ terms,

$$u_t + U u_x + \varphi_x - \mathrm{Re}^{-1} u_{zz} = U \frac{h^2}{3} u_{xxx} + \frac{h^3}{12} \varphi_{xxxx} + \mathrm{Re}^{-1} \frac{h^2}{12} u_{zzzz}, \tag{5a}$$

$$w_t + U w_x + \varphi_z - \mathrm{Re}^{-1} w_{zz} = U \frac{h^2}{3} w_{xxx} + \frac{h^3}{12} \varphi_{zzzz} + \mathrm{Re}^{-1} \frac{h^2}{12} w_{zzzz}, \tag{5b}$$

$$u_x + w_z = \frac{h^3}{3} u_{xxx} + \frac{h^3}{12} w_{zzzz}. \tag{5c}$$

2.1 Second-Order Accurate Upwind Discretization of QFSC

On the free-surface boundary $z = \eta_0$, the modified quasi-free surface condition resulting from an $\mathcal{O}(h^2)$ upwind-biased spatial discretization is imposed, i.e.,

$$\varphi_t + U\varphi_x - \mathrm{Fr}^{-2} w = U\frac{h^2}{3}\varphi_{xxx} + \mathcal{O}(h^3) \tag{6}$$

is imposed. To construct a Fourier representation of the solution $\boldsymbol{q}_1(\boldsymbol{x}, t)$, consider the following isolated mode:

$$\boldsymbol{q}_1(\boldsymbol{x}, t) = \hat{\boldsymbol{q}}(k, s, \omega) e^{ikx + sz - i\omega t}, \tag{7}$$

where $k, s \in \mathbb{R}$ are the wave numbers in x- and z-direction, respectively, and where $\omega \in \mathbb{R}$ is the radial frequency. Substitution of (7) into (5), neglecting the higher than $\mathcal{O}(h^2)$ terms, yields

$$\hat{\boldsymbol{L}}_h(k, s, \omega)\hat{\boldsymbol{q}} e^{ikx+sz-i\omega t} = \boldsymbol{0}, \tag{8a}$$

$$\hat{\boldsymbol{L}}_h(k, s, \omega) = \begin{pmatrix} \hat{H}_h(k, s, \omega) & 0 & \hat{G}_{x,h} \\ 0 & \hat{H}_h(k, s, \omega) & \hat{G}_{z,h} \\ \hat{D}_{x,h} & \hat{D}_{z,h} & 0 \end{pmatrix}, \tag{8b}$$

with

$$\hat{H}_h(k, s, \omega) = -i\omega + Uik\left(1 + \frac{h^2k^2}{3}\right) - \mathrm{Re}^{-1}s^2\left(1 + \frac{h^2s^2}{12}\right) \tag{8c}$$

the semi-discrete convection-diffusion operator, and with

$$\begin{pmatrix} \hat{G}_{x,h} \\ \hat{G}_{z,h} \end{pmatrix} = \begin{pmatrix} ik \\ s \end{pmatrix}, \quad \begin{pmatrix} \hat{D}_{x,h} \\ \hat{D}_{z,h} \end{pmatrix} = \begin{pmatrix} ik\left(1 + \frac{h^2k^2}{3}\right) \\ s \end{pmatrix} \tag{8d}$$

the discrete pressure gradient operator and divergence operator, respectively. A nontrivial solution of (8a) exists only if

$$\det(\hat{\boldsymbol{L}}_h) = 0, \tag{9}$$

which results in

$$s = k\left(1 + \frac{h^2k^2}{6}\right). \tag{10}$$

Here we have used the fact that for $z \downarrow -\infty$ the solution perturbations are zero. So, the corresponding solution of the system of modified equations is

$$\boldsymbol{q}_1(\boldsymbol{x}, t) = \begin{pmatrix} \hat{G}_{x,h} \\ \hat{G}_{z,h} \\ -\hat{H}_h(k, s(k), \omega) \end{pmatrix} e^{ikx + k\left(1+\frac{h^2k^2}{6}\right)z - i\omega t}. \tag{11}$$

The dispersion relation for the modified problem follows by substitution of this solution into (6) and solving for ω. The roots $\omega_{1,2}$ can be calculated explicitly, giving

$$\omega_{1,2} = U\tilde{k} \pm \frac{1}{2}\sqrt{4\mathrm{Fr}^{-2}s - \mathrm{Re}^{-2}\tilde{s}^4} + i\frac{1}{2}\mathrm{Re}^{-1}\tilde{s}^2, \tag{12a}$$

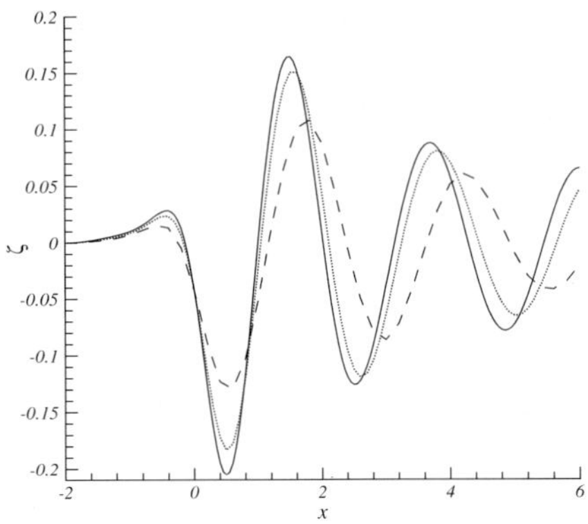

Fig. 1. Wave elevation in some plane y = constant for a 3D test case from Chapter 2 of [2], with $U = 1$ and $\mathrm{Fr} = 0.6$.

$$\tilde{k} \equiv k\left(1 + \frac{h^2k^2}{3}\right), \quad \tilde{s}^2 \equiv s^2\left(1 + \frac{h^2s^2}{12}\right). \tag{12b}$$

Note that the presence in (12a) of the positive imaginary term $i\mathrm{Re}^{-1}\tilde{s}^2$ gives rise to exponentially growing solutions. Hence, here steady free-surface waves cannot exist for finite Reynolds number. It can be shown that this is *not* due to the discretization but due to the continuous formulation of the free-surface flow problem. This suggests that this formulation of the free-surface flow problem is formally ill-posed. However, the measure of ill-posedness is small for large Re.

In the inviscid limit a stationary wave is found. For $\mathrm{Re} \uparrow \infty$ and $h \downarrow 0$, $\omega_{1,2} = 0$ in (12a) yields

$$U^2k\left(1 + \frac{h^2k^2}{2}\right) = \mathrm{Fr}^{-2}. \tag{13}$$

This relation implies that for fixed U and Fr, k decreases with increasing h. I.e., a coarser mesh results in a larger length of the free-surface wave. Solving the previous expression for k, assuming a solution of the form $k = (U\mathrm{Fr})^{-2}(1+\epsilon)$, $\epsilon \ll 1$, yields, neglecting higher-order terms of ϵ,

$$k = \frac{1}{(U\mathrm{Fr})^2}\left(1 - \frac{\frac{1}{2}h^2}{(U\mathrm{Fr})^4}\right). \tag{14}$$

We investigate the correctness of (14) for the numerical results depicted in Figure 1. Here we have applied three grids with in x-direction the mesh sizes $\frac{h}{2}$, h, and $2h$. For $U = 1$, $\mathrm{Fr} = 0.6$ and through the wavelength relation $k = \frac{2\pi}{\lambda}$, according to (14) it should then hold: $\lambda_{\frac{h}{2}} = 2.27$, $\lambda_h = 2.31$, $\lambda_{2h} = 2.48$, with $\lambda_{h=0} = 2.26$. Detailed observation of the results shown in Figure 1 shows that the above, analytically found wavelengths are quite accurate, thus proving that the Fourier analysis is not only of *qualitative* value but even of *quantitative* value.

2.2 First-Order Accurate Upwind Discretization of QFSC

In this section we analyze the effect of an $\mathcal{O}(h)$-upwind discretization of the quasi free-surface condition. The modified equation belonging to the $\mathcal{O}(h)$-upwind discretization is

$$\varphi_t + U\varphi_x - \mathrm{Fr}^{-2}w = U\frac{h}{2}\varphi_{xx} + \mathcal{O}(h^2). \qquad (15)$$

Applying the Fourier transform to the previous expression and substitution of the solution (11) into the Fourier-transformed quasi free-surface condition (15) yields again a dispersion relation. The solution of this quadratic equation for ω, neglecting the $\mathcal{O}(h^2)$ term, results in the following roots

$$\omega_{1,2} = \frac{2Uk + i(\mathrm{Re}^{-1}s^2 - \frac{h}{2}Uk^2) \pm \sqrt{4\mathrm{Fr}^{-2}s - (\mathrm{Re}^{-1}s^2 + \frac{h}{2}Uk^2)^2}}{2}. \qquad (16)$$

Note that now, as real part of the solution, we have (at least) $e^{\frac{1}{2}(\mathrm{Re}^{-1} - \frac{h}{2}U)k^2 t}$, where we have again neglected the $\mathcal{O}(h^2)$ terms. Whereas for both the exact equations and the $\mathcal{O}(h^2)$ modified equations, the time behaviour is always exponentially growing, with the $\mathcal{O}(h)$ upwind discretization of the QFSC, it can be exponentially decreasing, i.e., it can lead to a steady wave. From (16), it appears that a steady wave is obtained if

$$U\mathrm{Re}h > 2. \qquad (17)$$

Considering the case $U = 1$, $\ell = 1$, where ℓ is the reference length, this means that for mesh Reynolds numbers $\mathrm{Re}h > 2$ we may obtain steady waves due to the numerical diffusion introduced by the $\mathcal{O}(h)$ upwind discretization of the pressure derivative in the QFSC. In our practical computations, with $h \div \mathrm{Re}^{-\frac{1}{2}}$, this condition is easily satisfied.

3 Numerical Results

The free-surface algorithm is now applied to a standard test case originating from ship hydrodynamics: the computation of the complete flow field generated by a Series 60 hull at $\mathrm{Fr} = 0.316$ and $\mathrm{Re} = 10^6$.

The computational domain for this test case contains $353 \times 69 \times 45$ grid points and for the initial Γ_{FS} we take $z = 0$. Here, in the solution process, (1a) is treated using Newton's method and $\nabla\varphi$ is discretized employing the $\mathcal{O}(h)$-upwind scheme. The solution method corresponds to the modified algorithm as described in Chapter 2 of [2]. Two types of solutions are computed: (*i*) solutions belonging to a linearization of the free-surface flow, the so-called 'uniform-flow' linearization (described in Chapter 4 of [2]) and (*ii*) solutions of the fully nonlinear free-surface flow problem. A comparison of the computed results with the experimental data, obtained from [4], is shown in Figure 2.

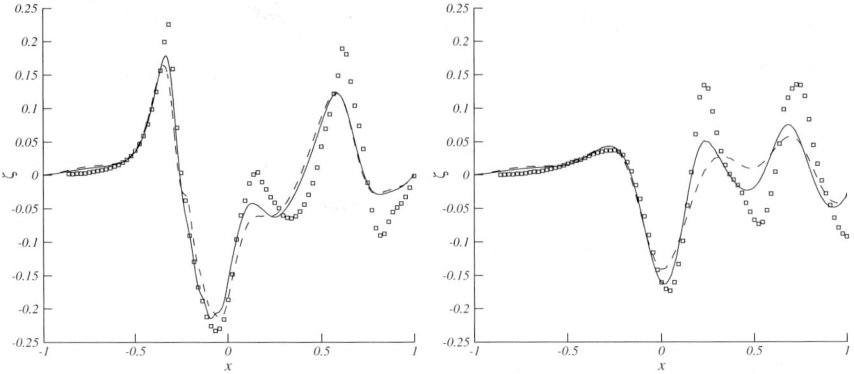

Fig. 2. Comparison of the computed longitudinal wave cuts, obtained from the nonlinear method (solid), the 'uniform-flow' linearization method (dashed), with experimental results (markers). **Left:** $y = 0.0755\ell$, **right:** $y = 0.2067\ell$, $y = 0$ corresponds to the plane of symmetry.

4 Conclusions

A Fourier-type analysis shows that for finite Reynolds number, steady free-surface waves cannot exist for our reduced Navier-Stokes equations, in combination with the quasi free-surface condition, neither in the continuous case nor in the fully second-order accurate discrete case. In agreement with our computational findings, the Fourier analysis also shows that for the reduced Navier-Stokes equations steady waves *can* exist when discretizing the quasi free-surface condition at first-order accuracy, under the constraint that the mesh Reynolds number does not exceed 2.

We have applied our iterative free-surface method to a real 3D ship hydrodynamics problem. The computed results reveal that the new formulation of the free-surface flow problem yields the correct wave physics and offers a fast convergence towards the nonlinear solution.

Acknowledgement: This work was supported by the Dutch Technology Foundation STW under grant CWI.4883.

References

1. E.H. van Brummelen, H.C. Raven, B. Koren: J. Comput. Phys **174**, 120 (2001)
2. M.R. Lewis: Numerical methods for water flows with free-surface gravity waves. PhD Thesis, Delft University of Technology, Delft (2004)
3. M.R. Lewis, B. Koren, H.C. Raven: Computation of 3D steady Navier-Stokes flow with free-surface gravity waves. In: Proceedings of *Computational Fluid Dynamics 2002*, ed by S. Armfield, P. Morgan and K. Srinivas (Springer, Berlin 2003) pp 100–105
4. Y. Toda, F. Stern, J. Longo: Mean-flow measurements in the boundary layer and wake field of a Series 60 $c_b = .6$ ship model for Froude numbers .16 and .316. Report 352, Iowa Institute of Hydraulic Research, Iowa City (1991)

Numerical Approach to the Analysis of Internal Pressure of M-V Rocket Fairing

Ayako Yamamoto[1], Keiichiro Fujimoto[2], Kozo Fujii[3] and Nobuyuki Tsuboi[4]

[1] Graduate student-Ochanomizu University, 2-1-1 Ootsuka Bunkyo-ku, Tokyo, 112-8610, JAPAN aya@ns.is.ocha.ac.jp
[2] Graduate student-University of Tokyo, 3-1-1 Yoshinodai, Sagamihara, Kanagawa 229-8510, JAPAN fujimoto@flab.eng.isas.jaxa.jp
[3] Institute of Space and Astronautical Science, Japan Aerospace Exploration Agency, 3-1-1 Yoshinodai, Sagamihara, Kanagawa 229-8510, JAPAN fujii@flab.eng.isas.jaxa.jp
[4] Institute of Space and Astronautical Science, Japan Aerospace Exploration Agency, 3-1-1 Yoshinodai, Sagamihara, Kanagawa 229-8510, JAPAN tsuboi@flab.eng.isas.jaxa.jp

Abstract. Navier-Stokes simulations are performed for time variation of the pressure inside the nose-fairing of M-V rocket. As the altitude increases during the launch of the rocket, flight Mach number and ambient conditions change with time. The numerical method to treat such an unsteady phenomenon is described in detail. The external pressure around the venting hole rapidly drops due to the passage of the local shock-wave over the rocket surface at transonic speed. The pressure inside the nose-fairing changes, because the external pressure is weakened through the venting hole. The effect of the time-accuracy is investigated.

1 Introduction

Controlling the pressure inside the nose-fairing is one of the key issues for the launching, since the time variation inside the nose-fairing influences the structural design of satellites. Venting holes are installed on the nose-fairing to equalize the pressure between the outside and inside of the nose-fairing. In the past, many investigations have carried out research for the time variation of the pressure over the nose-fairing surface caused by the local shock waves and shock-induced-separation[1] at transonic speeds. It is empirically known that the external pressure around the venting hole rapidly drops at the transonic condition due to the passage of the local shock-wave over the rocket surface. The pressure inside the nose-fairing changes in response to the external pressure as it propagates through a venting hole. However, there is little data about the characteristics of the pressure variation inside the nose-fairing.

In this study, the variation of the pressure inside the nose-fairing is numerically investigated. In the practical launch of the M-V rocket, flight velocity of the rocket increases as time passes, but it is difficult to consider the change of the flight velocity in the experiment. In addition, the ambient pressure and density also change. Such a situation requires special treatment of the boundary condition for the flow simulation. The numerical method to treat such unsteady effects is presented in detail, and the effect of the time accuracy is also investigated.

2 Numerical Method

To consider unsteady boundary conditions, variables used in the present calculations are normalized by reference values at a certain iteration step ($\circ^{ref}$). Length, ρ, p and velocity components u,v,w are normalized by the length of the M-V rocket L^{ref}, ρ_∞^{ref}, p_∞^{ref} and c_∞^{ref} respectively. ρ_∞^{ref}, p_∞^{ref} and c_∞^{ref} are fixed in the calculation.

$$c^* = \frac{c}{c_\infty^{ref}} \tag{1}$$

$$\rho^* = \frac{\rho}{\rho_\infty^{ref}} \tag{2}$$

$$p^* = \frac{p}{\gamma p_\infty^{ref}} \tag{3}$$

$$(u^*, v^*, w^*) = \frac{(u, v, w)}{c_\infty^{ref}} \tag{4}$$

$$dt^* = \frac{dt}{L^{ref}/C_\infty} \tag{5}$$

The three-dimensional compressible Navier-Stokes equations in the generalized coordinate system can be written in the conservation-law form:

$$\frac{\partial \hat{Q}}{\partial t} + \frac{\partial \hat{E}_{eul}}{\partial \xi} + \frac{\partial \hat{F}_{eul}}{\partial \eta} + \frac{\partial \hat{G}_{eul}}{\partial \zeta} = \frac{1}{Re_{L,c}^{ref}} \frac{\partial \hat{S}_{vis}}{\partial \zeta} \tag{6}$$

Here,

$$\mathrm{Re}_{L,c}^{ref} = \frac{c_\infty^{ref} L}{\nu_\infty^{ref}} = \frac{V_\infty^{ref} L}{\nu_\infty^{ref}} \frac{c_\infty^{ref}}{V_\infty^{ref}} = \frac{\mathrm{Re}_L^{ref}}{M_\infty^{ref}} \tag{7}$$

Numerical fluxes for the convective terms are evaluated by the SHUS (Simple High-resolution Upwind Scheme) [2], extended to high-order space accuracy by the 3rd-order upwind-biased MUSCL interpolation[3] based on the primitive variables. The viscous terms are evaluated by the 2nd-order central differences. The flow-field is considered to be fully turbulent, and the Baldwin-Lomax's algebraic turbulence model[4] is applied. The LU-SGS factorized implicit algorithm[5] is used for the time integration with inner iterations.

Non-dimensional time step is set $dt^* = 0.003$, and a CFL number at outer boundary of the computational domain is less than one.

2.1 Computational Grids

Computational grids are generated for an external area over the rocket, inside of the nose-fairing and three venting holes. All geometries are normalized based on the length of the M-V rocket (29.214 m) . The flow-fields are assumed to be symmetric about the pitch-plane (y=0) . Therefore, the computational domain covers only half of the body. The number of grid points are $(141(\xi) \times 83(\eta) \times 71(\zeta))$ for the external area, $(61 \times 83 \times 41)$ for the inside of nose-fairing. The geometry inside the nose-fairing is simplified as a cylinder, its volume is set as the net volume, which is the same as the volume of the real model of M-V rocket with the pay-load volume subtracted. Four venting holes are installed at 8100 mm aft from the nose-edge (STA8100) . The diameter of the nose-fairing is 2500 mm and the diameter of the venting hole is 100 mm. In the present study, the roll angle is considered as two venting holes are located in the symmetric plane. Therefore, the computational domain includes three venting holes. The number of the grid points for lee-side (VH1) and wind-side (VH3) venting holes are $(36 \times 12 \times 21)$, and $(36 \times 21 \times 21)$ for middle venting hole (VH2) . An unified zonal method based on the fortified solution algorithm interface scheme[6] is used, which is an interface scheme to exchange the computed physical data between zonal grids.

2.2 Boundary Conditions

For the computations to simulate the time variations of the internal pressure, flight Mach number of the M-V rocket is changed from 0.7 to 1.2 with time based on the actual flight data of the M-V rocket. The pitch angle α is 7°. External flow-field over the M-V rocket is computed at $M_\infty = 0.7$ in advance, and used as the initial state of the flow-field. The initial state of density and pressure inside the nose-fairing and venting holes are same as ambient conditions; $\rho = \rho_\infty, p = p_\infty$, and the velocity is same as the flight velocity.

The body surface has non-zero velocity components in the calculation to prevent the non-physical compression wave from the computational outer boundary. The ambient pressure p_∞ and density ρ_∞ change at the outer boundary of the computational domain ($L = L_{max}$) based on the International Standard Atmosphere (ISA) for dry air, and the velocity changes at the outer boundary of the computational domain based on the M-V flight trajectory data as shown in Fig. 1. Thus, the gradient of the ambient pressure and density relative to the altitude is small. Therefore, the gravitational force and the pressure difference over the outer boundary are not considered in the computations.

3 Results and Discussions

3.1 Fixed Mach number calculation

At $M_\infty < 0.70$, there is no local shock-wave and no flow separation. At the transonic flow condition, a local shock-wave appears over the body surface and there is shock-induced-separation in the lee-side of the nose-fairing. The computed result for $M_\infty = 0.95$, $\alpha = 7°$ is shown in Fig. 2. The pressure gradient on the surface is large due to the local shock-wave. As a result of this large pressure gradient region moving downstream, a strong time variation of the external pressure around the venting hole occurs. The shock-wave moves backward as the Mach number (speed of the rocket) increases. After the Mach number reaches $M_\infty = 1.10 \sim 1.20$, the local shock-wave disappears over the body surface and a detached shock-wave appears in front of the rocket.

3.2 The effect of time accuracy

The effect of time accuracy on the history of the pressure outside the leeward venting hole (VH1: Fig 3) is shown in Fig. 4. Four computations: without the inner-iteration No-I, with the 3-time inner-iteration IN03, 5-time inner-iteration IN05 and 10-time inner-iteration IN10, are compared. Little effect on time accuracy at middle venting hole (VH2: Fig 3) and windward venting hole (VH3: Fig 3) is observed. There is, however, much difference in the pressure history of the external pressure at VH1. The flow structures of the shock-induced-separation and flow inside the venting hole are different. There is the adverse flow in VH1 for IN03 and IN05 as shown in Fig. 5. But there is not for IN10. Moreover, the shock-induced-separation decreases. The implementation of about 10 times inner iteration is required for more accurate time responses. These results indicate the importance of the time accuracy improvement by the inner iteration method.

3.3 Analysis of the pressure variation inside of the nose-fairing

The number of the inner iterations is set to be ten times. Pressure variations outside the venting holes (VH1, VH2 and VH3: Fig. 3) and 35 mm inside of the nose-fairing near the venting hole(Station A1-1, A1-7 and A2-1 :Fig. 3) are shown in Fig. 6. In all stations outside of the venting hole, external pressure drops roughly 3 kPa in about one second ($dp/dt = 3.58kPa/sec$) under the transonic condition ($0.95 < M_\infty < 1.0$) at STA A1-1. This is due to the passage of the local shock-wave over the venting hole as described in the previous section. On the other hand, the pressure inside of the nose-fairing gradually changes in all stations as shown in Fig. 6. These results indicate that the effect of the external pressure change is weakened through the venting hole and pressure inside of the nose-fairing gradually changes, even at the station very close to the venting hole (35 mm inside).

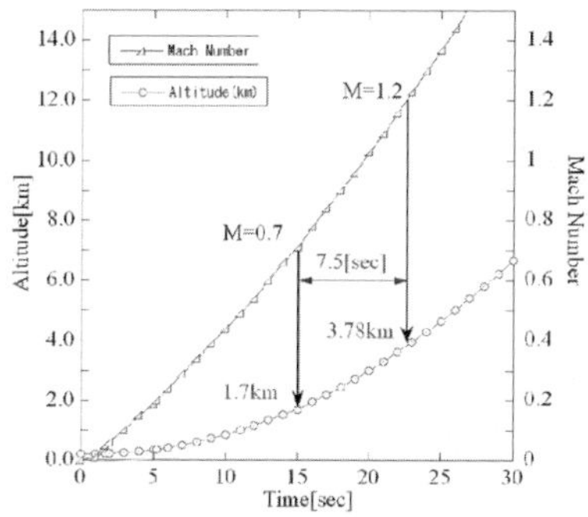

Fig. 1. Flight trajectory of the M-V rocket.

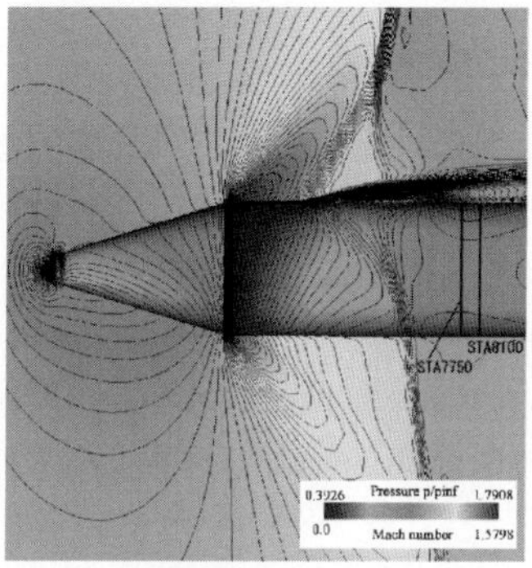

Fig. 2. Pressure distributions over the surface and local Mach number distributions in the symmetric plane at $M_\infty = 0.95$, $\alpha = 7°$.

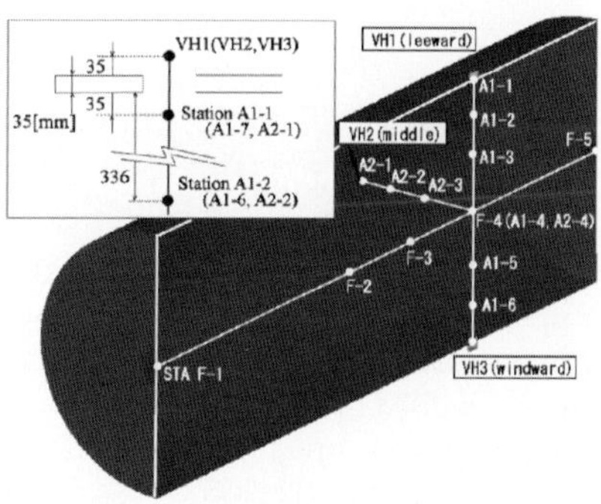

Fig. 3. Location of the reference station in the nose-fairing.

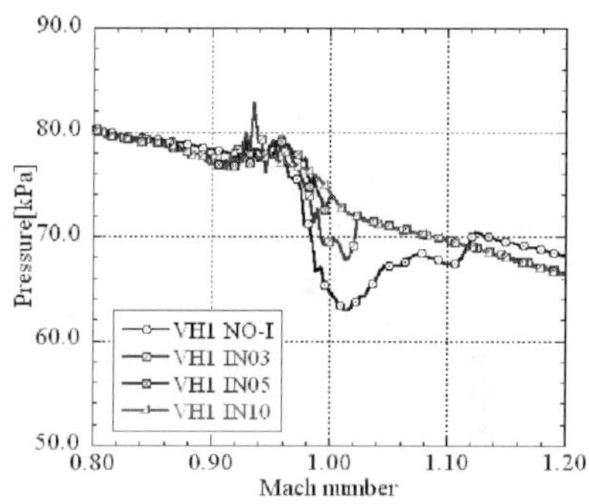

Fig. 4. The effect of the time accuracy.

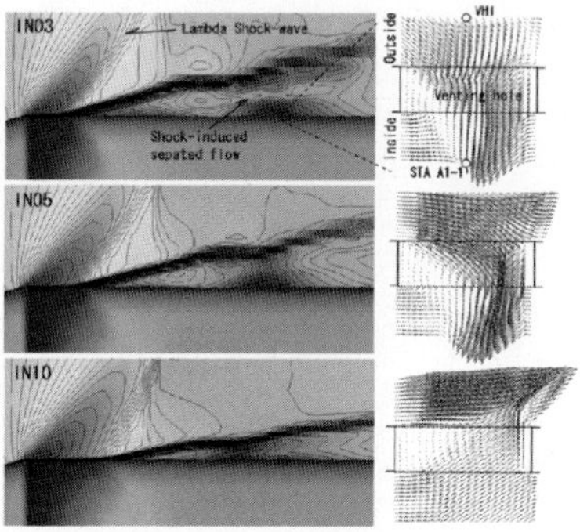

Fig. 5. Flow structure at VH1.

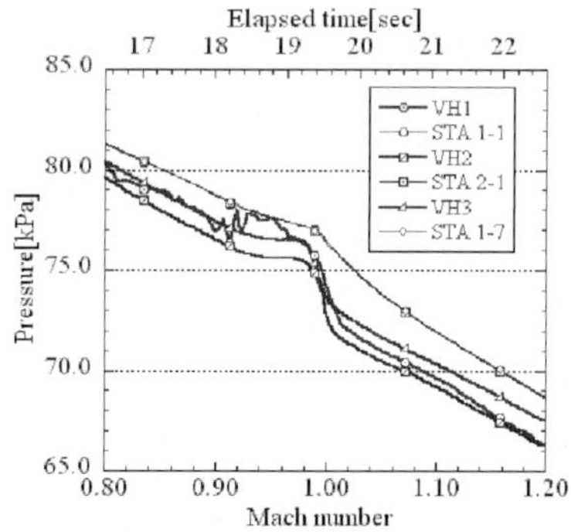

Fig. 6. Pressure histories at VH1-VH3, Station A1-1, A2-1 and A1-7.

4 Conclusion

Navier-Stokes simulations were performed and the investigation of the time variation of the pressure inside the nose-fairing of the M-V rocket was investigated. The external pressure around the venting hole dropped rapidly at transonic condition due to the passage of the local shock-wave over the rocket surface. The pressure inside of the nose-fairing changed in response to the external pressure change. The external pressure dropped due to the passage of the local shock-wave, and the pressure inside of the nose-fairing decreased. Since the effect of the external pressure is weakened through the venting hole,the internal pressure changed gradually compared with the external pressure change even at the station very close to venting hole.

Time accuracy of the present flow simulations was discussed. With the number of inner-iterations set to 3 and 5, there were the adverse flows, but not at 10. When the number of the inner-iteration was different, flow structures and shock-induced-separation differed greatly.

References

1. Mizuguchi, H., Aihara, Y., Ebihara, M., Suzuki, K., Kokuni, Y., Koike, Y., and Hosoe, N.: *19th Fluid Dynamics Conference*,(1987) pp 244–247.
2. Shima E. and Jounouchi T.: *Proceedings of the 14th NAL Symposium on Aircraft Computational Aerodynamics*(1997) pp 7–12.
3. van Leer, B.: *Journal of Computational Physics*,(1997) **23** 276–299 .
4. Baldwin, B. S., and Lomax, H.: AIAA Paper 78–257,(1978).
5. Yoon, S., and Jameson, A.: *AIAA Journal*, **26**, (1988), pp. 1025–1026.
6. Fujii, K.: *Journal of Computational Physics*, **118**, (1995) 92–108.

Automated Euler and Navier-Stokes Database Generation for a Glide-Back Booster

Neal M. Chaderjan[1], Stuart E. Rogers[2], Mike J. Aftosmis[3], Shishir A. Pandya[4], Jasim U. Ahmad[5], and Edward Tejnil[6]

[1] NASA Ames Research Center,Moffett Field,CA, Neal.M.Chaderjian@nasa.gov
[2] NASA Ames Research Center,Moffett Field,CA, rogers@nas.nasa.gov
[3] NASA Ames Research Center,Moffett Field,CA, Michael.J.Aftosmis@nasa.gov
[4] NASA Ames Research Center,Moffett Field,CA, Shishir.A.Pandya@nasa.gov
[5] NASA Ames Research Center,Moffett Field,CA, ahmad@nas.nasa.gov
[6] NASA Ames Research Center Moffett Field,CA, tejnil@nas.nasa.gov

Summary. AeroDB is a database generation framework that is used to compute thousands of Euler and Navier-Stokes solutions for a 2nd generation glide-back booster in one week using the NASA Information Power Grid. Process automation and web-based access is used to greatly simplify and reduce the user workload. The solutions are validated with experimental data, and stability derivatives are computed using a monotone cubic-spline procedure. Flow visualization and three-dimensional surface plots are used to interpret and characterize the nature of computed flow fields.

1 Introduction

The past two decades have seen a sustained increase in the use of high fidelity Computational Fluid Dynamics (CFD) in basic research, aircraft design, and the solution of post-design issues. As the fidelity of a CFD method increases, the number of cases that can be readily and affordably computed greatly diminishes. However, computer speeds now exceed 2 GHz, and hundreds of processors are currently available and more affordable. Moreover, advances in parallel CFD algorithms scale more readily with large numbers of processors. All of these factors make it feasible to compute thousands of high fidelity cases. However, there still remains the overwhelming task of monitoring the solution process.

Automation can reduce the tedious and error prone nature of the process, which can easily overwhelm a team of engineers. ILab [1] is an example of a general purpose parameter analysis tool, but it's generality requires a significant amount of user input. Chaderjian et al. [2] describe a solution automation approach in which PERL scripts were used to generate a database of time-dependent, Reynolds-averaged, Navier-stokes (RANS) solutions for a Harrier in ground effect. These scripts greatly reduced the user workload, but used only one CFD code and one geographical site.

The objective of this paper is to present a database generation framework that improves on the previous examples, and demonstrates its capabilities by computing at least 100 RANS solutions and 1000 Euler solutions in one week for a 2nd generation Liquid Glide-Back Booster (LGBB) design. The solution method is described in Section 2, results of the database generation are discussed in Section 3, and concluding remarks are made in Section 4.

2 Solution Procedure

Grid computing [3] is based on the concept that one can gain significant increases in computational throughput by accessing any number of computers at remote sites. One example is NASA's Information Power Grid (IPG), which consists of distributed heterogeneous computer systems at different NASA and supercomputer centers in the United States. AeroDB is a database generation framework presented in this paper that utilizes NASA's IPG and the Globus toolkit [4], which provides common secure services for user authentication over an open network.

A flowchart of AeroDB is shown in Fig.1 and consists of a system of PERL modules, MySQL database, and web portal. The web portal is used to submit a run matrix to AeroDB anywhere there is internet access. It is also used to select the flow solver, number of CPUs per case, etc. A Job Launcher (JL) script is continually running on a front-end machine in the background, and checks the database for cases to run. The JL uses the Globus toolkit for remote site authentication, and also utilizes a Resource Discovery Broker to decide where to submit the jobs. The present application uses 13 computers at 4 different geographical sites across the United States.

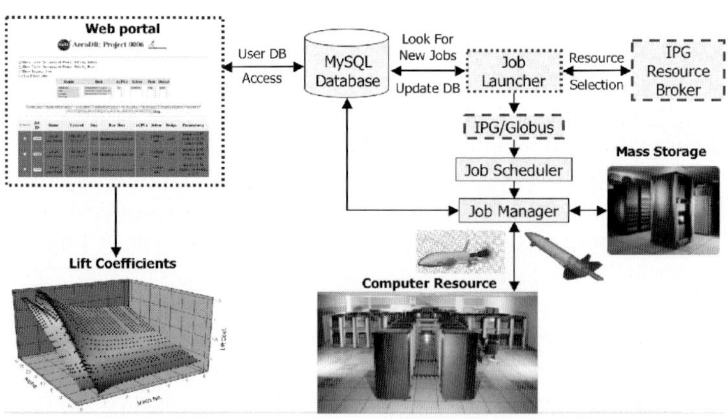

Fig. 1. AeroDB flowchart.

The JL submits the job to the Job Scheduler (JS), which starts the Job Manager (JM) at the remote site. The JM gets job attributes from the MySQL database. Once a job begins to run, a Run Manager (RM) in the

CFD code monitors flow solver convergence and the time remaining in the execution queue. If the RM determines that the solution is converged or the queue time is depleted, it sends a signal to the JM to stop the run, and if needed, resubmit the job for continued execution. The JM also transfers updated solution files to the mass storage system and MySQL database. AeroDB can be run on non-grid enabled systems using ssh for authentication. In such cases, the Resource Discovery Broker is not available.

The current application solves the steady Euler equations with the Cart3D Euler code [5] using unstructured Cartesian grids. The steady RANS equations are solved with the Overflow code,[6] which uses overset structured grids to model complex geometries. Other CFD codes are also available in AeroDB. The RM is installed in each CFD code.

3 Numerical Results

AeroDB is used to generate a database of Euler and Navier-Stokes solutions for the LGBB geometry, (see Fig. 3). Inviscid Cart3D computations were carried out using 38 Mach numbers ($0.2 \leq M_\infty \leq 6.0$), five sideslip angles ($0 \leq \beta \leq 4$ deg), and angles of attack $-5 \leq \alpha \leq 30$ deg. Navier-Stokes Overflow computations were also carried out using 14 Mach numbers ($0.2 \leq M_\infty \leq 3.0$), and five sideslip angles ($0 \leq \beta \leq 4$ deg). The angles of attack for viscous cases were $0 \leq \alpha \leq 20$ deg for subsonic flows, and $0 \leq \alpha \leq 30$ deg for supersonic flows. The goal of computing at least 100 Overflow cases and 1000 Cart3D cases in one week was fully met within a 72 hour time period using AeroDB. At the end of seven days, 211 Overflow cases and 2863 Cart3D cases were completed. The current LGBB CFD database consists of 3666 cases, (499 Overflow solutions and 3167 Cart3D solutions). Overflow runs typically used 32 CPUs and required 100-200 cpu-hr/case on an SGI Origin 3000. Cart3D typically required 2-10 cpu-hr/case.

The CFD database is validated through a grid refinement study and comparison of the computed results with wind tunnel data. The Overflow RANS solutions are computed at flight Reynolds number (Re) conditions using the Spalart-Allmaras turbulence model.[7] The Reynolds number is chosen according to a flight trajectory scenario. The Overflow grid consists of 8.5 million grid points, and the Cart3D grid consists of 1.4 million cells, which provides good grid support for the present parameter study, (see Ref. [8]).

Figure 2 compares the computed lift coefficient (C_L), drag coefficient (C_D), and pitching moment coefficient (C_m) with transonic and supersonic wind-tunnel data. The Overflow lift, drag and pitching moment coefficients compare well with the experiment. The Cart3D lift and drag coefficients compare equally well with the experiment, but under predict the transonic pitching moment coefficient somewhat. Transonic shock positions can be sensitive to viscous effects.

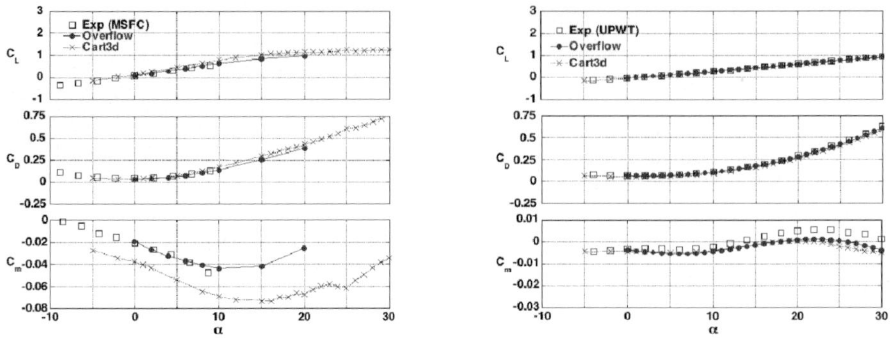

(a) $M_\infty = 0.9, Re = 60 \times 10^6$.

(b) $M_\infty = 3.0, Re = 32 \times 10^6$.

Fig. 2. Comparison of computed C_L, C_D, and C_m with wind-tunnel data.

Visualization of a viscous supersonic case is shown in Fig. 3. Figure 3(a) shows Overflow pressure coefficient (C_p) contours at the symmetry plane, on the LGBB surface, and two cutting planes through the canard and wing. Shocks occur near the vehicle nose, leading and trailing edges of the canard and wing, and along the vertical tail. The Cp contours at the tail end of the fuselage also indicate separated flow. Figure 3(b) shows the viscous surface flow topology (white), and off-surface vortical flows highlighted by helicity-density contours (yellow). It is remarkable how complex the surface-flow topology is. However, due to domain-of-influence (DOI) effects, these separated regions are steady and confined very close to the body. This helps explain why there are very little viscous effects shown in the supersonic Euler and Navier-Stokes C_L and C_D, but more significant effects for C_m, (see Fig.2).

Figure 4 shows the variation of Overflow C_L and C_m with Mach number and angle of attack. Symbols indicate computed cases and lines represent values obtained by using a monotone cubic-spline interpolation procedure.[8] The lift coefficient shows a trend of increasing lift with angle of attack, and a compressibility rise/fall near $M = 1$, as expected. The pitching moment coefficient shows a relatively flat behavior in the supersonic region due to shock positions being fixed near wing/canard trailing edges, and a dramatic valley in the transonic region. Here, the shock position plays an important role in determining the shape of this valley.

Figure 5 shows the variation of longitudinal stability derivatives, C_{L_α} and C_{m_α}, with Mach number and angle of attack. These surfaces were generated by numerically differentiating the data with the monotone procedure. A monotone procedure helps control spurious oscillations, and provides reasonable slope information.

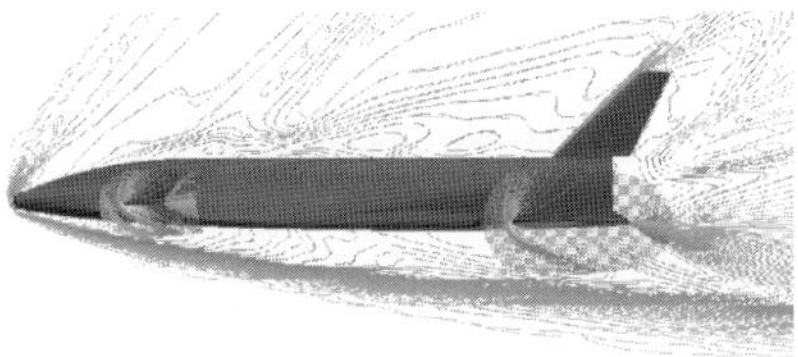

(a) C_p contours, (blue $\rightarrow$ low, green $\rightarrow$ high).

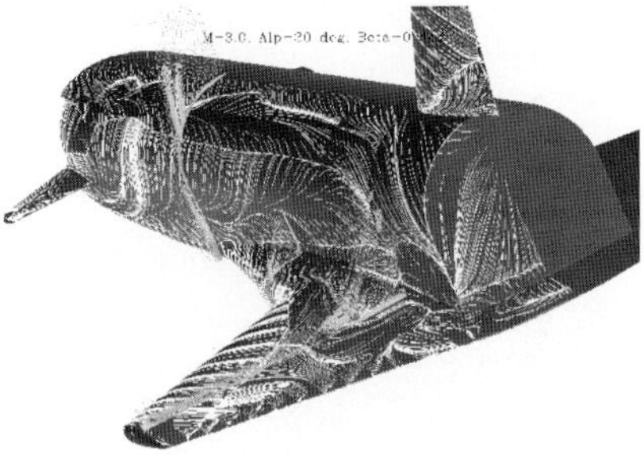

(b) Surface flow (white), helicity-density (yellow).

Fig. 3. Viscous flow visualization. $M_\infty = 3.0, \alpha = 30°, Re = 32 \times 10^6$.

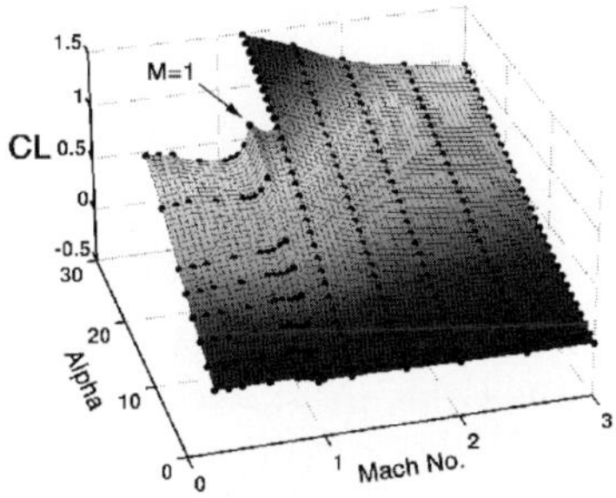

(a) Overflow C_L.

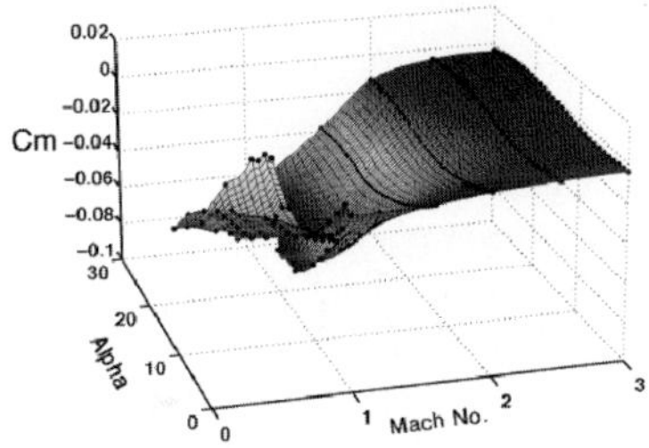

(b) Overflow C_m.

Fig. 4. Viscous force/moment database.

Inviscid Cart3D results show similar trends as the viscous Overflow cases, (see Chaderjian et al.[8]).

4 Conclusions

The ability to automate and manage the process for generating thousands of Euler and Navier-Stokes CFD solutions has been demonstrated using the AeroDB database generation framework. The primary goal of computing at least 100 Navier-Stokes solutions and 1000 Euler solutions for a liquid glide-back booster vehicle in one week was fully met in 72 hours, using 13 computers at 4 different geographical sites across the United States (the NASA IPG). After seven days, 211 viscous cases and 2863 inviscid cases were completed, and the current database consists of 3666 cases, (499 viscous and 3167 inviscid). The results compared well with experimental data and showed ex-

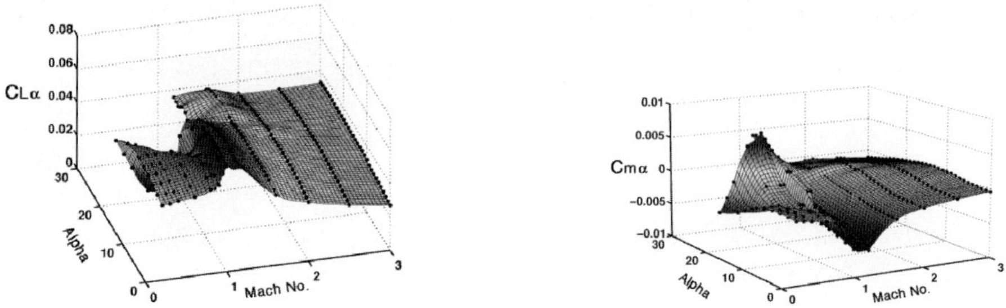

(a) Overflow C_{L_α}. (b) Overflow C_{m_α}.

Fig. 5. Viscous longitudinal stability derivatives.

pected trends. The ability to compute static stability derivatives from the CFD database was demonstrated using a monotone cubic-spline procedure.

References

1. Yarrow, M., McCann, K. M., DeVivo, A., and Tejnil, E., "Production-Level Distributed Parametric Study Capabilities for the Grid," NAS Technical Report NAS-01-009, NASA Ames Research Center, 2001.
2. Chaderjian, N. M., Pandya, S. A., Ahmad, J. U., and Murman, S. M., "Progress Toward Generation of a Navier-Stokes Database for a Harrier in Ground Effect," AIAA Paper 2002-5966, November 2002.
3. Foster, I., "The Grid: A New Infrastructure for 21st Century Science," Physics Today, Vol. 55, No. 2, pp. 42-47, 2002.
4. Foster, I., and Kesselman, C., "Globus: A Metacomuting Infrastructure Toolkit," Int. J. Supercomputer Applications, Vol. 11, No. 2, pp. 115-128, 1997.
5. Aftosmis, M. J., Berger, M. J., and Adomavicius, G., "A Parallel Multilevel Method for Adaptively Refined Cartesian Grids with Embedded Boundaries," AIAA Paper 2000-0808, January 2000.
6. Jespersen, D. C., Pulliam, T. H., and Buning, P. G., "Recent Enhancements to Overflow," AIAA Paper 97-0644, January 1997.
7. Spalart, P. R., and Allmaras, S. R., "A One-Equation Turbulence Model for Aerodynamic Flows," AIAA Paper 92-0439, January 1992.
8. Chaderjian, N. M., Rogers, S. E., Aftosmis, M. J., Pandya, S. A., Ahmad, J. U., and Tejnil, E., "Automated CFD Database Generation for a 2nd Generation Glide-Back Booster," AIAA Paper 2003-3788, June 2003.

Numerical Simulation of Radiative Heating for Atmospheric Reentry in Martian Atmosphere

O. Rouzaud[1], J. Hylkema[1], L. Tessé[2], and F. Longueteau[1]

[1] ONERA, DMAE, BP4025, 31055 TOULOUSE, FRANCE, rouzaud@onecert.fr
[2] ONERA, DEFA, BP 92, 92322 CHÂTILLON, FRANCE, tesse@onera.fr

1 Introduction

The future space exploration missions need new and thorough studies on the radiative heating of space vehicles during atmospheric re-entry. For instance, preliminary computations in the Mars Premier project, tended to prove that the radiative heating emanating from the wake region of the orbiter could overheat the payload [1].

CNES (Centre National d'Etudes Spatiales) decided to launch activities on the gas radiation of high-temperature gas involving ONERA, CNRS (Centre National de la Recherche Scientifique) laboratories and the RTech company. One of the actions was to develop a computational chain for the numerical simulation of radiative heating. The present chain consists in a Navier-Stokes solver, a Radiative Heat Transfer (RHT) solver and a software called PHARAON. The platform has to be regarded as a tool managing radiation calculations between different kind of solvers and is now operable for martian applications.

The first part of the paper describes each component of the computational chain. In the next part, we present first step results obtained on the generic TC3 testcase [2] defined within the framework of the European Working Group on gas radiation conducted by ESA and CNES.

2 Computational Chain

2.1 Navier-Stokes Solver

The family of CELHyO codes has been developed in ONERA for the computation of hypersonic high-enthalpy flows. The codes solve the Euler or the Navier-Stokes equations using a finite-volume method based on structured meshes. The discretization is cell-centered. Various geometrical configurations are treated from the 1D stagnation streamline to 3D flows. The codes deal with gas mixtures representative of air or martian atmospheres. The flow is assumed to be in chemical nonequilibrium while thermal equilibrium or thermal nonequilibrium are possible.

Assuming the thermal equilibrium of the flow, the governing equations for a mixture of ns species reduce to the conservation of mass for each species, the conservation of momentum $\rho\mathbf{V}$ and the conservation of the total energy $\rho\mathsf{E}$ of the mixture, written under the following form

$$\frac{\partial \rho_\alpha}{\partial t} + \nabla.(\rho_\alpha \mathbf{V}) = \nabla.(\rho \mathsf{D}_\alpha \nabla \mathsf{Y}_\alpha) + \dot{w}_\alpha$$

$$\frac{\partial \rho \mathbf{V}}{\partial t} + \nabla.(\rho \mathbf{V} \otimes \mathbf{V}) = -\nabla \mathsf{p} + \nabla.\sigma$$

$$\frac{\partial \rho \mathsf{E}}{\partial t} + \nabla.(\rho \mathsf{H} \mathbf{V}) = \nabla.(\lambda \nabla \mathsf{T}) + \nabla.(\mathbf{V}.\sigma) + \nabla.\left(\rho \sum_{\alpha=1}^{ns} \mathsf{h}_\alpha \mathsf{D}_\alpha \nabla \mathsf{Y}_\alpha\right)$$

The RHS term of the continuity equations contains two source terms representing respectively the mass diffusion of species α and the mass production rate of species α due to chemical reactions. In the diffusion term, the effects of the thermal and pressure gradients are ignored. The term D_α represents the diffusion coefficient. The value Y_α represents the mass fraction ρ_α/ρ of the species α.

In the RHS term of the momentum, appears the pressure p, its expression is given by the Dalton's law. The RHS also contains the viscous stress tensor which depends on the strain tensor and on the mixture viscosity μ.

Finally, the RHS of the total energy equation contains three terms representing the effect of the thermal conduction, the work done by shear forces and the diffusion of enthalpy due to mass fraction gradients. The additional variables are the total enthalpy of the mixture H, the enthalpy h_α of the species, the temperature T and the thermal conductivity λ. Note that an additional term should be present in this equation if radiative effects had to be taken into account.

Concerning the physico-chemical modelling, the CELHyO codes offer several possibilities but we will restrict the presentation to the choices used in the numerical simulation of the testcase. For the transport coefficients, their evaluations are rather classical. The mixture viscosity μ is given by the Armaly-Sutton formula [3] while the viscosity of each species depends on the Blöttner [4] relation. Determination of the thermal conductivity relies on the assumption of a fixed Prandtl number Pr equal to 0.66, $\lambda = \mu C_p/Pr$. Lastly, we prescribe that the diffusion coefficient D_α is not dependent on the species and assume a fixed Lewis number $Le = 1$ leading to a general diffusion coefficient $D = \lambda Le/\rho C_p$. In the last two formulae, the specific heat coefficient C_p represents the coefficient value for the mixture. Concerning the chemical model, the Park chemistry [5] is precribed with 5 species and 18 reactions [2]. The five species are CO_2,CO,O_2,C,O.

From a numerical point of view, aerothermodynamic computations have been performed using the HUS scheme [6] with a combination of van Leer and Osher methods. The minmod limiter ensures a second-order accuracy in space. To improve the robustness and the convergence rate, we employ an Euler implicit method. Using the van Leer scheme provides the implicit system. Its solution is based on a Krylov-GMRES method.

The reader, interested in further details, will refer to the works of Flament [7] and of William [8] on CelHyo2D.

2.2 Radiative Heat Transfer Solver

The RHT solver ASTRE is a Monte-Carlo solver enabling the treatment of 3D geometries on structured or unstructured meshes. The code has been primarily developed by ONERA with the collaboration of EM2C laboratory to study the radiative transfer in turbulent flames in the presence of soot particles [9]. Its capabilities have now been extended to reentry problems in the martian atmosphere due to the implementation of a Statistical Narrow-Band (SNB) model proposed by EM2C for CO and CO_2 molecules [10].

One peculiarity of the code is to provide the results of three different formulations of the Monte-Carlo method [11]. Two of them, the Emission-based Reciprocal Method and the Absorption-based Reciprocal Method, relies on the exchange formulation of radiative transfer which fulfills systematically the reciprocity principle. The third one is a conventional Forward Method (FM). In our case of application, the last method is the most appropriate and we will only present results using the FM method.

Concerning the algorithm, the first step consists in evaluating the emitted power P_e in each cell or surface of the computational mesh. Afterwards, the number N_i of optical paths originating from a cell i is calculated. The spatial distribution of the optical paths can be uniform or non-uniform, the latter one relying on the previously calculated emitted power. In a third time, the code performs a global loop on all cells and all optical paths. The wavenumber, the source point and the direction of each optical path are stochastically generated. Besides, the absorption is treated as a continuous phenomenon in a deterministic manner. Once the global loop is over, the code yields the radiative heat fluxes on the boundary surfaces, the radiative power in the flowfield and the associated standard deviations.

Note that computations with the three methods are simultaneous. Furthermore, it is not more expensive, in terms of computational time to obtain the results from all the methods than from one of them. In fact, most of the computational time is due to the generation of the optical paths and to the initialization steps before the global loop.

To conclude this part, the SNB model [10] is shortly described. The model takes into account the radiative properties of the CO and CO_2 molecules in the infrared part of the spectrum. The parameters of the model have been generated for the following temperature and pressure ranges

$$1\,\text{Pa} \leq p \leq 500\,\text{Pa} \qquad 1000\,\text{K} \leq T \leq 4000\,\text{K}$$

These intervals are representative of the values expected in the wake region of the vehicle.

2.3 PHARAON Platform

The PHARAON platform is a C++ software developed by ONERA and RTech. Its present functionnalities include the standard CGNS format for the IO data and the development of interpolation methods to treat different computational meshes, if required. Structured or unstructured meshes

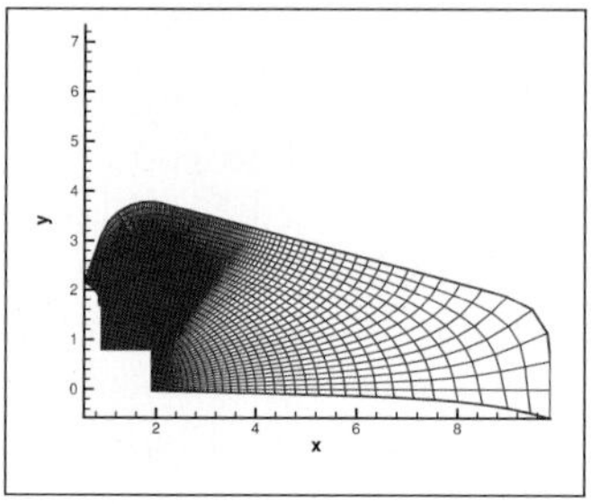

Fig. 1. Computational grid

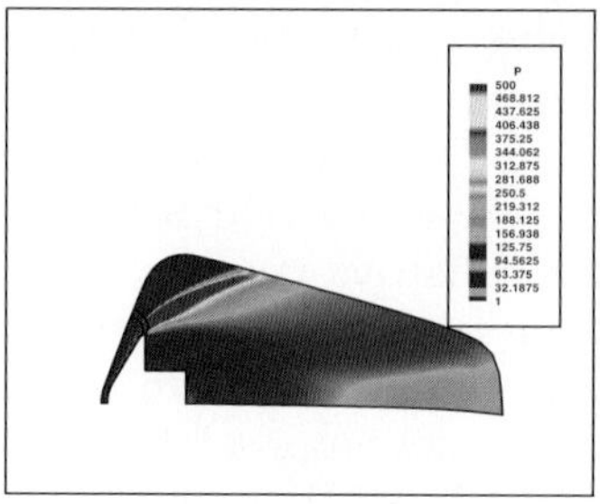

Fig. 2. Pressure contours

are treated. Due to its original design, use of different Navier-Stokes solvers or RHT solvers should not create any problem, facilitating cross-checkings between different groups of users.

Further extensions could conduct to the creation of a general database gathering the most relevant results (calculation history, convergence, data storage), to the management of calculations on remote computers and to the inclusion of post-processing tools.

3 Numerical Results

3.1 Flowfield Conditions

We consider the testcase TC3-1 [2] of the ESA-CNES Gas Radiation of High Temperature Gases Workshop held at Lisbon in 2003. The trajectory point is chosen at t=115 s with the following associated conditions

$$Y_{CO_2\infty} = 1$$

$$\rho_\infty = 2.933\text{x}10^{-4}\text{kg/m}^3 \quad V_\infty = 5223\text{m/s} \quad T_\infty = 140\text{K} \quad P_\infty = 7.87\text{Pa}$$

The wall is assumed to be non-catalytic and isothermal. The shield temperature is equal to 1500K while, along the rear part of the vehicle, the temperature is only 500K. Note that, for the RHT calculations, the wall is a black body and we have assumed a temperature of 1K or of 500K. The first value allows to determine the incident radiative flux on the wall while the second one yields the global balance. Finally, the flow is under chemical nonequilibrium and thermal equilibrium.

3.2 Aerothermodynamic Simulation

The numerical conditions are summarized in the previous section. 2D axisymmetric computations have been performed on a monoblock grid 1 with ONERA CELHyO2D solver [12]. The grid contains approximately 19200 points (figure 1). The height of the first cell is close to 10^{-6}m.

Figure2 presents the pressure contours around the vehicle. The main features of the flow are a strong shock wave in front the shield, a quite complex flow in the rear part of the vehicle involving recirculating vortices and a shear layer. The closure of the shear layer on the axis is located near $x = 6$ m.

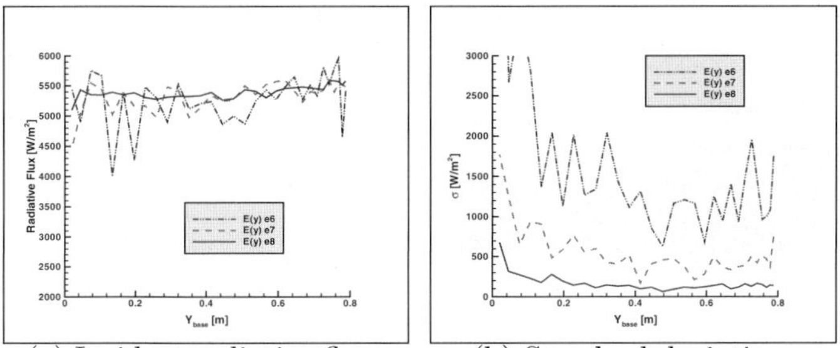

(a) Incident radiative flux (b) Standard deviation

Fig. 3. Wall values - Influence of the paths number

3.3 Radiative Heat Transfer Simulation

Firstly, we present results with a wall temperature $T_{wall} = 1\,\mathrm{K}$. Since the wall is considered as a black body, the radiative flux at the wall corresponds to its incident value. Fig. 3 describes the influence of the number of optical paths on the heat flux values and the associated standard deviations along the cylinder base using the FM method. The number of paths varies from 10^6 to 10^8. Increasing the number of paths allow to reduce both the heat flux variations and the standard deviations. The most important standard deviations appear near both ends of the base since the grid is refined in these regions. Thus, the surface cells receive less optical paths than cells located in the middle of the base. We also present the radiative power for 10^8 optical paths in Fig. 4. The most emissive regions of the flow are located in the shock region and in the vicinity of the wake closure. Finally, Fig. 5 compares the distribution of convective and radiative heat fluxes along the rear part of the vehicle, the value s representing the curvilinear abscissa from the shield shoulder to the symmetry axis. Results demonstrate that the levels of both are comparable, meaning that the radiative contribution can have a major impact on the thermal protection system. It also confirms a previous numerical study [1] on the martian orbiter.

4 Conclusion

The paper presents the ONERA computational chain PHARAON for the numerical simulation of radiative heating. The chain is operational for martian reentry applications. First test results on a generic vehicle demonstrate the importance of the radiative heat flux on the rear part of the vehicle, due to the infrared radiation of CO and CO_2 molecules.

Further extension of the computational chain PHARAON is now in progress, in particular for Earth reentries.

Acknowledgments

The authors wish to thank CNES for its support, EM2C laboratory and RTech society for their technical and scientific contribution to the computational chain PHARAON.

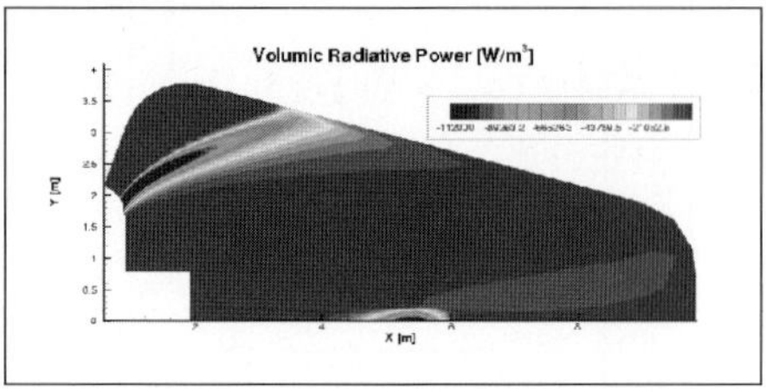

Fig. 4. FM method - ONERA grid

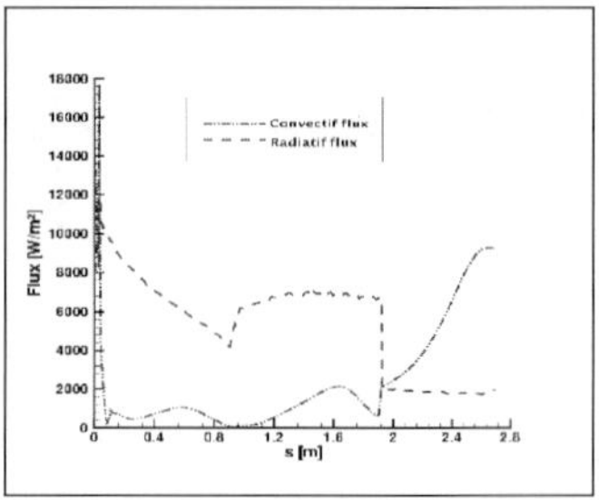

Fig. 5. Heat flux comparison

References

1. V. Gromov and S. Surzhikov: Convective and Radiative Heating of a Martian Space Vehicle Base Surface. In: *Proceedings of the AAAF conference*, Arcachon (2002)
2. J.-M. Charbonnier, J. Couzi, W. Dieudonné, J.-L. Vérant: Definition of an Axially Symmetric Testcase for High Temperature Gas Radiation Prediction in Mars Atmosphere Entry. Rapport CNES NG104-07-TF-001-CNES, 2003.
3. B.F. Armaly, K. Sutton: Viscosity of Multicomponent Partially Ionized Gaz Mixtures. AIAA Paper 80-1495 (1980)
4. F.G. Blöttner, M. Johnson, M. Ellis: Chemically Reacting Viscous Flow Program for Multicomponent Gas Mixtures. Sandia Laboratories, Technical Report Sc-RR-70-754 (1971)
5. C. Park, J.T. Howe, R.L. Jaffe, G.V. Candler: J. Thermophysics and Heat Transfer **8**, 33 (1994)
6. F.Coquel, V. Joly, C. Marmignon: Méthodes de décentrement hybride pour la simulation d'écoulements en déséquilibre thermique et chimique. AGARD Conference, Progress and Challenges in CFD Sevillia (1995)
7. C. Flament: Ecoulements de fluide visqueux en déséquilibre chimique et vibrationnel: modélisation, applications internes et externes. PhD Thesis, Université Paris (1991)
8. J. William: Etude des processus physico-chimiques dans les écoulements détendus à haute enthalpie: application à la soufflerie F4. PhD Thesis, Université de Provence (1999)
9. L. Tessé: Modélisation des transferts radiatifs dans les flammes turbulentes par une méthode de Monte-Carlo. PhD Thesis, Ecole Centrale de Paris (2001)
10. Ph. Rivière, A. Soufiani, M.-Y. Perrin: Line-by-Line and Statistical Narrow-Band Calculations of Radiative Transfer in some Atmospheric Entry Problems. In: *Proceedings of the ESA-CNES Wopkshop for Gas Radiation*, ESA-SP 533, pp 189–196 (2003)
11. L. Tessé, F. Dupoirieux, B. Zamuner, J. Taine: Int. J. Heat and Mass Transfer **45**, pp 2797-2814 (2002)
12. O. Rouzaud, J. Hylkema, J.-L. Vérant, L. Tessé: Development of the PHARAON planform and ONERA numerical solvers for Gas Radiation. In: *Proceedings of the ESA-CNES Wopkshop for Gas Radiation*, ESA-SP 533, pp 181–188 (2003)

An Implicit Preconditioned JFNK Method for Fully Coupled Radiating Flows. Application to Superorbital Re-Entry Simulations

R. Turpault

MAB, University Bordeaux 1, Cours de la Liberation, 33405 Talence, France,
Rodolphe.Turpault@math.u-bordeaux1.fr

Summary. The effects of radiation sometimes have a great impact on the aerodynamics flows. This happens in several applications such as in superorbital atmospheric entry. In this case, it is important to have a model that fully couples hydrodynamics and radiation in order to have a good behaviour of the solution. However, coupling with the full radiative transfer equation is usually very expensive hence it is not reasonable for multidimensional unsteady computations. In this paper, we briefly introduce the radiation model we use to reduce the computation cost and then discuss the way we make the full coupling between radiation and hydrodynamics. The example of a venusian early entry is given which shows the interest of the method.

1 Introduction to the $M1$-Multigroup Model

As we are interested in computing flows where radiation plays a crucial role, it is important to modelize it with a model that is both accurate enough and as cheap as possible in terms of computational cost. Solving the radiative transfer equation (RTE) would be too expensive for our purpose, we then consider a moments model. The system that has to be solved is the following:

$$\partial_t E_q + \nabla_x . F_q = c[\sigma_q^e a\theta_q^4(T) - \sigma_q^a E_q] \;\; 1 \leq q \leq Q \tag{1}$$

$$\frac{1}{c}\partial_t F_q + c\nabla_x . P_q = (-\sigma_q^f - \sigma_q^d(1 - \breve{g}_q))F_q \;\; 1 \leq q \leq Q \tag{2}$$

where E, F and P are the first three moments of the radiative intensity respectively the radiative energy, the radiative flux vector and the radiative pressure tensor. One can notice that this model is multigroup: instead of integrating the RTE over all frequencies, we rather integrate over intervals of frequency called groups. This degree of freedom is necessary to minimize the error due to frequency approximation since several parameters, such as the opacities σ are highly frequency-dependant. The system (1)-(2) is not closed. To do so, we have to make an asumption that allows to consider the radiative pressure as a function of the radiative energy and flux. We then use a minimum entropy principle a la Levermore [4] ie choose an underlying radiative intensity that minimizes the radiative entropy H:

$$H(\mathcal{I}) = \min_I \{H(I) = \sum_q < \check{h}(I) >_q /\forall q, < I >_q = E_q \; and \; c < \Omega I >_q = F_q\} \tag{3}$$

This leads to the following form of the underlying radiative intensity:

$$\mathcal{I} = \mathcal{I}(\Omega, \nu) = \sum_q 1_{[\nu_{q-\frac{1}{2}};\nu_{q+\frac{1}{2}}[} \frac{2h\nu^3}{c^2} \left[exp(\frac{h\nu}{k}\boldsymbol{m}.\boldsymbol{\alpha}_q) - 1\right]^{-1}, \tag{4}$$

where $\boldsymbol{\alpha}$ are Lagrange's multipliers of the minimization problem. The radiative pressure is then chosen as the third moment of this function and can be written as $P_q = D_q E_q$ where:

$$D_q = \frac{1-\chi_q}{2} I_d + \frac{3\chi_q - 1}{2} \check{f}_q \otimes \check{f}_q \tag{5}$$

$\check{f}_q = \dfrac{F_q}{\|F_q\|}$ and χ_q is the eigenvalue of D_q relative to $\check{f}_q$.

The model we have obtained, called $M1$-multigroup (see [1] and references therein) proves to be satisfying in terms of properties. First, it allows to keep the fundamental physical properties such as energy conservation and entropy dissipation. It is also an hyperbolic system with bounded eigenvalues ($< c$) that naturally has the very important property of flux limitation: physically, the norm of the radiative flux is bounded by c times the radiative energy. Since the $M1$ model has this property, it always predicts physically admissible results, which is not always the case of other models. Finally, it is possible to make a consistant choice of mean opacities by using the underlying radiative intensity. Since this choice is far from being trivial and has huge consequencies over the results, this model proves to be very efficient.

2 Solving the System

We then consider the equations of coupled hydrodynamics and radiation. If we use the Navier-Stokes system for the hydrodynamics part, it can be written as:

$$\partial_t \mathcal{U} + \partial_x \mathcal{F}(\mathcal{U}) + \partial_y \mathcal{G}(\mathcal{U}) = \mathcal{S}(\mathcal{U}) \tag{6}$$

where:

$$\mathcal{U} = \begin{pmatrix} (\rho_i)_{i=1..n_s} \\ \rho u \\ \rho v \\ \mathcal{E} \\ E_R \\ F_R^x \\ F_R^y \end{pmatrix}$$

$$\mathcal{F}(\mathcal{U}) = \begin{pmatrix} (\rho_i u - \rho \mathcal{D} \partial_x c_i)_{i=1..n_s} \\ \rho u^2 + p - \mathcal{T}_{xx} \\ \rho uv - \mathcal{T}_{xy} \\ \rho u H - \mathcal{T}_{xx} u - \mathcal{T}_{xy} v + q_x \\ F_R^x \\ c^2 D_R^{xx} E_R \\ c^2 D_R^{yx} E_R \end{pmatrix}, \mathcal{G}(\mathcal{U}) = \begin{pmatrix} (\rho_i v - \rho \mathcal{D} \partial_y c_i)_{i=1..n_s} \\ \rho uv - \mathcal{T}_{yx} \\ \rho v^2 + p - \mathcal{T}_{yy} \\ \rho v H - \mathcal{T}_{yx} u - \mathcal{T}_{yy} v + q_y \\ F_R^y \\ c^2 D_R^{xy} E_R \\ c^2 D_R^{yy} E_R \end{pmatrix}$$

$$\mathcal{S}(\mathcal{U}) = \begin{pmatrix} (\dot{\omega}_i)_{i=1..n_s} \\ -\rho u g_x \\ -\rho v g_y \\ -\rho V.g - c(\sigma^e a T^4 - \sigma^a E_R) \\ (\sigma^e a T^4 - \sigma^a E_R) \\ -(\sigma^f + (1-\breve{g}_\nu)\sigma^d) F_R^x \\ -(\sigma^f + (1-\breve{g}_\nu)\sigma^d) F_R^y \end{pmatrix}$$

This system is clearly divided into two sub-systems: the hydrodynamics part is given by multi-species Navier-Stokes equations for the sake of simplicity our code can handle several other situations such as basic turbulence, and the radiation part is the multigroup-$M1$ model with q_{max} frequency groups.

2.1 Implicit Time Scheme

We use a structured multibloc structure where the unknown vector is defined by:

$$\mathcal{U}(i + (l-1)n_{fr} + (m-1)l_{max}n_{fr})_{i=1..n_{fr}, l=1..l_{max}, m=1..m_{max}} \quad (7)$$

With this spatial discretization, the spatial scheme has the following form:

$$\partial_t \mathcal{U}_{l,m}(t) + \frac{\mathcal{F}_{l+\frac{1}{2},m}(t) - \mathcal{F}_{l-\frac{1}{2},m}(t)}{\Delta x_{l,m}} + \frac{\mathcal{G}_{l,m+\frac{1}{2}}(t) - \mathcal{G}_{l,m-\frac{1}{2}}(t)}{\Delta y_{l,m}} = S_{l,m}(t) \quad (8)$$

which may be written as:

$$\partial_t \mathcal{U}_{l,m}(t) = \Theta_{l,m}(\mathcal{U}(t)) \quad (9)$$

For time discretization, we use the following schemes: Since the characteristic

Table 1. Time discretization

Explicit Euler	$\mathcal{U}_{l,m}^{n+1} - \mathcal{U}_{l,m}^{n} - \Delta t \Theta(\mathcal{U}^n) = 0$
Implicit Euler	$\mathcal{U}_{l,m}^{n+1} - \mathcal{U}_{l,m}^{n} - \Delta t \Theta(\mathcal{U}^{n+1}) = 0$
Gear	$3\mathcal{U}_{l,m}^{n+1} - 4\mathcal{U}_{l,m}^{n} + \mathcal{U}_{l,m}^{n-1} - 2\Delta t \Theta(\mathcal{U}^{n+1}) = 0$

speeds of radiation and hydrodynamics are very different (speed of the light vs Mach number), it is necessary to use an implicit scheme to avoid too restrictive a CFL condition which would lead to spend most of the time computing radiation. Hence the code relies on the implicit Euler or Gear schemes depending on the time order required.

2.2 Computation

All the previous schemes have the shape $\Phi(\mathcal{U}^{n+1}) = 0$. To know $\mathcal{U}^{n+1}$, one has to solve this equation. To do so, we use a Newton method. Then, if we set $A(\mathcal{U}) = \Phi'(\mathcal{U})$, we have to iterate to obtain the solution of the linear system:

$$A(\mathcal{U}^k)(\mathcal{U}^{k+1} - \mathcal{U}^k) = -\Phi(\mathcal{U}^k) \quad (10)$$

The solution of these systems are made thanks to a GMRes(Generalized Minimal Residual)-type Krylov method. In a classical approach the jacobian matrix A, which is usually quite huge, has to be built and stocked. Here we use a Jacobian-Free method which does not require the knowledge of the Jacobian matrix. Finally, as we often wish to use large time steps which destroy the conditionning, we develop preconditioners to speed up the convergence of the GMRes algorithm.

As we mentionned, it is not necessary to stock the jacobian matrix. The Az products are computed as follows:

$$Az = A(\mathcal{U}^k)z = \lim_{\varepsilon \to 0} \frac{\Phi(\mathcal{U}^k + \varepsilon z) - \Phi(\mathcal{U}^k)}{\varepsilon} \tag{11}$$

For real simulations, ε cannot be fixed as one scalar. In fact, the differences of scales between radiation and hydrodynamics do not allow to find a satisfying value of ε. At least 2 different ε are indeed used: one to get the derivatives of the hydrodynamics' values and the other for the radiative ones.

2.3 Preconditioning

Several preconditioners were developed in order to speed up the convergence of the GMRes method. We can use them in any order by defining a preconditioning matrix as the product of any of these preconditioners.

Diagonal preconditioning

The first problem induced by the method lies in the sometimes huge scale differences that exist between the variables taken into account. The most usual are $F_R \simeq cE_R$ and $\mathcal{E} >> \rho$. This numerical stiffness may induce an error that is far from being neglectable on the smallest variables. However, it can easily be overcome by introducing a very simple diagonal preconditioning. The preconditioning matrix is defined by:

$$M^d = diag\{(u_\eta^{ref})\} \tag{12}$$

where u_η^{ref} are reference values. For example, we can choose: $u_\eta^{ref} = \max_{l,m} u^0_{\eta,l,m_i}$. This choice is almost everytime good enough to have values of $\mathcal{U}$ that are of the order of 1.

Bloc-diagonal (physical) preconditioning

Apart from the scale differences on the values, there are also scale differences on the evolution speed of the phenomena. Most of the times, the radiative variables evolve much faster than the hydrodynamic ones. These scale differences do not lead to badly converged solutions but rather hugely slow down the convergence of the algorithm. To overcome this problem, we introduce a

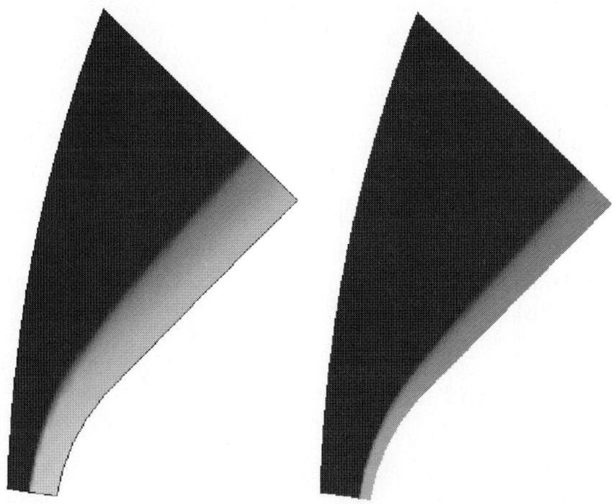

Fig. 1. Temperatures (left) coupled, (right) uncoupled computations

bloc-diagonal (physical) preconditioning that approximates the jacobian matrix inside each cell. In the case of one chemical species in 2D, each bloc has the following form:

$$(M^{\phi})_{l,m} = \begin{pmatrix} m^{\phi}_{1,1} & m^{\phi}_{1,2} & m^{\phi}_{1,3} & m^{\phi}_{1,4} & 0 & 0 & 0 \\ m^{\phi}_{2,1} & m^{\phi}_{2,2} & m^{\phi}_{2,3} & m^{\phi}_{2,4} & 0 & 0 & 0 \\ m^{\phi}_{3,1} & m^{\phi}_{3,2} & m^{\phi}_{3,3} & m^{\phi}_{3,4} & 0 & 0 & 0 \\ m^{\phi}_{4,1} & m^{\phi}_{4,2} & m^{\phi}_{4,3} & m^{\phi}_{4,4} & 0 & 0 & 0 \\ 0 & 0 & 0 & 0 & m^{\phi}_{5,5} & m^{\phi}_{5,6} & m^{\phi}_{5,7} \\ 0 & 0 & 0 & 0 & m^{\phi}_{6,5} & m^{\phi}_{6,6} & m^{\phi}_{6,7} \\ 0 & 0 & 0 & 0 & m^{\phi}_{7,5} & m^{\phi}_{7,6} & m^{\phi}_{7,7} \end{pmatrix} \tag{13}$$

where :

$$m^{\phi}_{i_1,i_2} = \frac{\Phi(u_{i_1} + \varepsilon \delta u_{i_2}) - \Phi(u_{i_1})}{\varepsilon_{i_2}} \tag{14}$$

Experience shows that defining this preconditioning matrix only once every time step is the best solution. In fact, computing this matrix inside the Newton loop would increase too much the calculation cost. The time saved by using a preconditioning technique would vanish compared to the cost of these updates. This bloc-diagonal preconditioning proves to be very efficient in several applications.

3 Application

This application is an example of the importance of radiation in some problems. It re-creates with simplifications, the early entry of the Pioneer probe in the Venusian atmosphere. As we mentionned earlier, Venus' atmosphere is dense so that the opacities become important. Here, we consider a simplified atmosphere composed only of carbon dioxyde (neglecting nitrogen). The

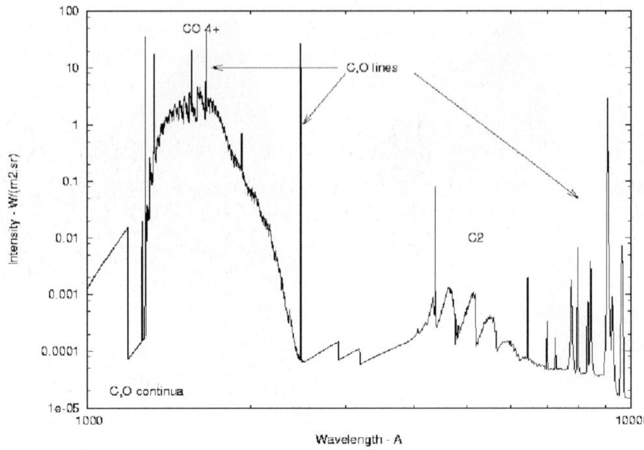

Fig. 2. Radiation spectrum reaching the wall on the axis

chemical species taken into account are CO_2, CO, C_2, O_2, C and O. We arbitrarily chose to take 4 frequency groups. The opacities where computed from PARADE [7] database for monoatomic species and a CEA internal database for others [3]. This early entry occurs above 80 km from the surface at a speed of a little bit more than 11 $km.s^{-1}$. The temperature is 142 K and the pressure is roughly 300 Pa. First, we introduce the material temperatures predicted by the full system (fully coupled result) and by the hydrodynamics alone (uncoupled result). These results are shown in Figure 1.

At the first glance, one can see that the flow is greatly modified by radiation effects. The position of the shock is closer from the body (3.7 cm instead of 11 cm in the uncoupled case). The temperature profiles are also very different from each other. The maximal temperatures are less than 10500 K in the coupled case compared to nearly 12500 K in the uncoupled case. A lot of enery is then dissipated due to the radiation as figure shows: the anisotropic factor is quite high in the cold region, revealing a loss of energy from the hot zone. Since in the radiative equilibrium the energy is proportionnal to T^4, we see that approximately one half of the energy is lost from the uncoupled to the coupled case.

Once we get the flow, it is easy to do a postcalculation using models with an high level of precision if more details are needed. For instance, it is possible to use narrow band or even line-by-line models for radiation. As an example, we have constructed the radiation spectrum reaching the wall on the axis. The spectrum is shown in Figure 2. Even if it only represent a simplificated phenomenon -a real simulation would involve turbulence and a more detailed chemistry-, we can conclude from this test-case on the importance of being able to adequatly take into account the effects of radition in some cases.

Another interesting point to look at is the CPU time required to carry on the computations. Where a simple hydrodynamics simulation with chemical equilibrium took 37 min, the time required to run a multi-species hydrodynamics simulation was roughly 253 min. The fully coupled multi-species simulation ran during 271 min. All these computations were carried on a axisymetric 120×40 grid on one 375 Mhz processor of an IBM SP2 computer. The last two computations provided the results showed above in this paragraph. It is to notice that the cost of radiation is very low compared to the cost of doing a multi-species computation. Of course, this cost depends on the number of groups chosen for the simulation, but the experience shows that only a few number of them is necessary to get a good approximation on the flux in this kind of test-case [1]. In fact, the bloc-diagonal preconditioner proves to be very useful in this kind of problems, allowing to consider huge times steps (up to 10^{12} times the explicit time-step) and then converge more quickly.

References

1. Charrier P, Dubroca B, Duffa G and Turpault R, *Schéma implicite pour un modèle d'hydrodynamique radiative multigroupe 2D.*, rapport LRC-03.16 (available on www.math.u-bordeaux.fr/LRC-CEA/) (2003).
2. Charrier P, Dubroca B, Duffa G and Turpault R, *Multigroup model for radiating flows during atmospheric hypersonic re-entry*, proceedings of International Workshop on Radiation of High Temperature Gases in Atmospheric Entry, Lisbon, Portugal :103-110 (2003).
3. Chevalier JM (1995), *Calculs 'ab-initio' des spectres d'absorption des modécules diatomiques: application aux spectres IR, visible et UV de CN, CH, CO, NO, OH,* N_2, N_2^+ *et* O_2, personnal communication.
4. Levermore, D (1996), *Moment Closure Hierarchies for Kinetic Theories*, J. Stat. Phys., vol. 83 pp 1021-1065.
5. Nicolet, WE et al. (1975), *Analytical and design study for a high-pressure high-enthalpy constricted arc heater*, Aerotherm division, Acurex Corp.
6. Sakai T, Tsuru T and Sawada K (2001), *Computation of Hypersonic Radiating Flowfield over a Blunt Body*, Journal of Thermophysics and Heat Transfer Vol 15 no1.
7. Smith AJ, Gogel T, Vandevelde P and Marraffa L, *Plasma Radiation Database PARADE Final Report*, ESTEC contract 11148/94/NL/FG (1996).

Two- and Three-Dimensional Flow Optimization in Chemical Engineering

Günter Bärwolff

TU Berlin, Institut of Mathematics, D-10623 Berlin, Str. d. 17. Juni 135, Germany

1 Introduction

In several mathematical models of engineeering processes the Boussinesq equation system occurs. Especially for melting processes the heating by the realisation of certain boundary conditions is responsible for the resulting flow and temperature field. Beside the thermal boundary conditions non homogeneous velocity boundary conditions influence also the melt flow. Thus the control of certain boundary conditions is a possibility to optimize the process with the aim of reaching flow and temperature fields which guarantee desirable results of the melting process.

We will discuss this optimization with partial differential equations for the problem of crystal melts with the aim of having single crystals without defects in the result of the melting process.

This leads to tracking type optimization problems with a functional like

$$J(\boldsymbol{u}, \theta_c) = \frac{1}{2}\int_0^T \int_\Omega |\boldsymbol{u} - \overline{\boldsymbol{u}}|^2 \, d\Omega dt + \frac{1}{2}\int_0^T \int_{\Gamma_c} (\theta_c^2 + \theta_{c_t}^2) \, d\Omega dt \, . \qquad (1)$$

$\boldsymbol{u}$ is the velocity vector field in the melt and $\overline{\boldsymbol{u}}$ is the state, which we want to have, θ_c is the control temperature on the control boundary Γ_c.

2 Mathematical Model

The crystal melt is described by the Navier-Stokes equation for an incompressible fluid using the Boussinesq approximation coupled with the convective heat conduction equation and the diffusion equation. Heat conductivity and viscosity depend on the temperature. Because of the axisymmetric situation of the melting zone we write down the equations in cylindical coordinates. Thus we have a Boussinesq equation system in cylindrical coordinates for the velocity $\boldsymbol{u} = (u, v, w)$, the pressure p and the temperature θ.

For the velocity no slip boundary conditions are used. But in the case of rotating crystals and crucibles we have in the circumferential direction inhomogeneous velocity boundary conditions given by rotation numbers. At the interfaces between the solid material and the fluid crystal melt we have for the temperature inhomogeneous Dirichlet data, i.e. the melting point temperature. On the heated coat (control boundary) of the ampulla the experimentators gave us measured temperatures and we will look further for optimal profiles during the optimization. The initial state was assumed as the neutral position of the crystal melt ($\boldsymbol{v} = \mathbf{0}$) and a temperature field, which solves the non convective heat conduction equation with the given temperature boundary conditions. The material properties and the dimensionless parameters for the investigated crystal close the initial boundary value problem for the description of the melt flow.

3 Optimization

For the calculus of optimization and the derivation of an optimization system we start with the Boussinesq equation system for the velocity vector $\boldsymbol{u}$ and the temperature θ in the space time cylinder $\Omega_T = \Omega \times (0,T)$. For the boundary conditions we have

$$\boldsymbol{u} = \boldsymbol{u}_g \quad \text{on } \Gamma \times (0,T), \quad \theta = \theta_c \quad \text{on } \Gamma_{cT}, \quad \text{and} \quad \theta = 0 \quad \text{on } \Gamma_d \times (0,T) \tag{2}$$

where Γ is the boundary of the spatial region $\Omega \subset \mathrm{I\!R}^3$, on which the problem lives, and Γ_c is the control boundary, Γ_d is the Dirichlet part of the boundary and $\Gamma_{cT} = \Gamma_c \times (0,T)$. That means in this paper our interest is focussed on the temperature boundary control. For $t = 0$ we have the initial condition $\boldsymbol{u} = \mathbf{0}$ and a temperature field, which solves the non convective heat conduction equation with the given temperature boundary conditions $\theta = \theta_0$ on Ω.

The use of formal Lagrange parameters technique with respect to the functional of type (1) means the consideration of the Langrange functional

$$L(\boldsymbol{u}, p, \theta, \theta_c, \boldsymbol{\mu}, \xi, \kappa, \chi) = J(\boldsymbol{u}, \theta_c) + < \boldsymbol{\mu}, mo >_{\Omega_T}$$
$$- < \xi, div\, \boldsymbol{u} >_{\Omega_T} + < \kappa, en >_{\Omega_T} + < \chi, \theta - \theta_c >_{\Gamma_{cT}} .$$

mo and en stand for the left sides of the impulse and energy balance. The Lagrange functional means the objective of the optimization

$$J(\boldsymbol{u}, \theta_c) = \frac{1}{2} \int_0^T \int_\Omega |\boldsymbol{u} - \overline{\boldsymbol{u}}|^2 \, d\Omega dt + \frac{1}{2} \int_0^T \int_{\Gamma_c} (\theta_c^2 + \theta_{c_t}^2) \, d\Omega dt ,$$

the restrictions i.e. the weak formulation of the occuring partial differential equations

$$< \boldsymbol{\mu}, mo >_{\Omega_T} = \int_{\Omega_T} [\boldsymbol{u}_t + (\boldsymbol{u} \cdot \nabla)\boldsymbol{u} - \Delta \boldsymbol{u} + \nabla p - \rho(\theta)\boldsymbol{g}] \cdot \boldsymbol{\mu} \, d\Omega \, dt ,$$

$$< \xi, div\, \boldsymbol{u} >_{\Omega_T} = \int_{\Omega_T} \xi \, div\, \boldsymbol{u} \, d\Omega dt ,$$

$$< \kappa, en >_{\Omega_T} = \int_{\Omega_T} [\theta_t + \boldsymbol{u} \cdot \nabla\theta - \frac{1}{Pr} \Delta\theta] \, \kappa \, d\Omega dt ,$$

and the condition for the boundary control

$$< \chi, \theta - \theta_c >_{\Gamma_{cT}} = \int_{\Gamma_{cT}} [\theta - \theta_c] \chi \, d\Gamma dt \ ,$$

using the Lagrange parameters $\boldsymbol{\mu}$, ξ, κ and χ.

We will not discuss the functional analytical aspects of the used Lagrange method, i.e. function spaces, smoothness properties etc. A very good overview over the functional analytical background and the fundation of the optimization of Navier-Stokes problems is developed in [2].

To find candidates $\boldsymbol{u}(\theta_c)$ and θ_c, which minimize the functional (1) we have to analyze the necessary conditions

$$\begin{aligned} L_{\boldsymbol{u}}\tilde{u} &= J_{\boldsymbol{u}}\tilde{u} + < \boldsymbol{\mu}, mo_{\boldsymbol{u}} >_{\Omega_T} - < \xi, div\tilde{u} >_{\Omega_T} + < \kappa, en_{\boldsymbol{u}} >_{\Omega_T} = 0, \\ L_p\tilde{p} &= < \nabla\tilde{p}, \boldsymbol{\mu} >_{\Omega_T} = 0, \\ L_\theta\tilde{\theta} &= < -\rho_\theta \boldsymbol{g}\tilde{\theta}, \boldsymbol{\mu} >_{\Omega_T} + < \kappa, en_\theta >_{\Omega_T} + < \chi, \tilde{\theta} >_{\Gamma_{cT}} = 0, \\ L_{\theta_c}\tilde{\theta}_c &= J_{\theta_c}\tilde{\theta}_c + < -\chi, \tilde{\theta}_c >_{\Gamma_{cT}} = 0 \ , \end{aligned}$$

i.e. the Frechet derivative of the Lagrange functional equal to zero. The evaluation of the necessary optimality condition gives the equation for the adjoint velocity

$$-\boldsymbol{\mu}_t - \Delta\boldsymbol{\mu} + (\nabla\boldsymbol{u})^t\boldsymbol{\mu} - (\boldsymbol{u}\cdot\nabla)\boldsymbol{\mu} + \nabla\xi = -(\boldsymbol{u} - \overline{\boldsymbol{u}}) - \kappa\nabla\theta \quad \text{in} \quad \Omega_T \ , \quad (3)$$

with the boundary condition and the final condition

$$\boldsymbol{\mu} = \boldsymbol{0} \quad \text{on} \quad \Gamma \times (0,T), \quad \text{and} \quad \boldsymbol{\mu}(T) = \boldsymbol{0} \quad \text{in} \quad \Omega \ , \quad (4)$$

and the equation

$$-div \ \boldsymbol{\mu} = 0 \quad \text{in} \quad \Omega_T \ . \quad (5)$$

For the adjoint temperature κ we get

$$-\kappa_t - \frac{1}{Pr}\Delta\kappa - (\boldsymbol{u}\cdot\nabla)\kappa = -\rho_\theta \boldsymbol{g}\cdot\boldsymbol{\mu} \quad \text{in} \quad \Omega_T \ , \quad (6)$$

with the boundary condition and the final condition

$$\kappa = 0 \quad \text{on} \quad \Gamma \times (0,T) \ , \quad \text{and} \quad \kappa(T) = 0 \quad \text{in} \quad \Omega \ , \quad (7)$$

and with the choice of the free χ as $\chi = \frac{1}{Pr}\frac{\partial\kappa}{\partial\mathbf{n}}$ on Γ_{cT} we get for the control the equation

$$-\theta_{c_{tt}} + \theta_c = \frac{1}{Pr}\frac{\partial\kappa}{\partial\mathbf{n}} \quad \text{on} \quad \Gamma_{cT} \ , \quad (8)$$

with the time boundary conditions

$$\theta_{c_t}(0) = \theta_{c_t}(T) = 0 \ . \quad (9)$$

The optimality system consists of the forward model with the Boussinesq equation system and the boundary conditions (2) and the given initial state for the velocity field $\boldsymbol{u}$, the pressure p and the temperature θ, and the adjoint model with the equations (3),(5),(6),(8), and the conditions (4),(7),(9) for the adjoint variables $\boldsymbol{\mu}$, ξ, κ and the control θ_c.

4 To the Numerical Solution Method

The optimization system is now under consideration for a numerical solution. The Boussinesq equation system is solved with a finite volume method [1]. In the present case of axisymmetric conditions we can transform the adjoint equations into a cylindrical coordinate system. Using the adjoint divergence condition $div\,\boldsymbol{\mu} = 0$ we can write the adjoint equations in the following quasi conservative form. For adjoint velocity $\boldsymbol{\mu} = (\mu, \nu, \omega)$ in the cylindrical coordinate system with the radial component μ, the azimuthal component ν and the z-component ω we get from (3) for example for μ

$$\begin{aligned} -\mu_t - ((r\mu)_r/r)_r - \mu_{\varphi\varphi}/r^2 + 2\mu_\varphi/r^2 - \mu_{zz} + \mu u_r + \nu v_r + \omega w_r \\ -(ru\mu)_r/r - (v\mu)_\varphi/r - (w\mu)_z + v\nu/r + \xi_r = -(u - \overline{u}) - \kappa\theta_r \,. \end{aligned} \tag{10}$$

¿From equation (6) we get for the adjoint temperature κ

$$-\kappa_t - \frac{1}{Pr}[(r\kappa_r)_r/r + \kappa_{\varphi\varphi}/r^2 + \kappa_{zz}] - (ru\kappa)_r/r - (v\kappa)_\varphi/r - (w\kappa)_z = -\rho_\theta g w \,. \tag{11}$$

Equation (11) is a convective heat conduction equation and the discretization can be done like those in [1]. In the equation (10) and the equations for ν, ω the terms $(\nabla\boldsymbol{u})^t\boldsymbol{\mu}$ and $\kappa\nabla\theta$ are not known from the classical Navier-Stokes equations and they will by discretized in a canonical way for staggered grids. The solution of the discretized Boussinesq equation system and the equations for ν, ω and ξ (3) is difficult and expensive, because of the opposite time direction of the forward system and the adjoint system. That means we know the forward solution $\boldsymbol{u}, \theta$ on the whole time interval $[0, T]$ to get the adjoint solution $\boldsymbol{\mu}, \kappa$ and vice versa. If we have discretized the time interval $[0, T]$ by Z timesteps and the dimensions of the spatial discretizations are N, M and P a direct solution of the whole system means the solution of an algebraic equation system with $2Z \times N \times M \times P \times 10$ equations. Iterative methods of the form

i) choose a suitable start value of $\boldsymbol{u}, \theta$,
ii) solve the adjoint problem and get $[\boldsymbol{\mu}, \kappa, \theta_c](\boldsymbol{u}, \theta)$
iii) solve the forward problem and get $[\boldsymbol{u}, \theta](\theta_c)$
iv) until convergence, go to ii),

are realized and we will now discuss the results of the proposed method.

5 Results of the Optimzation

In the table 1 the used geometrical and material parameters for the crystal $(Bi_{0.25}Sb_{0.75})_2Te_2$, a composition of bismuth point fifty antimony one point fifty telurium two, are summarized ([1]). We consider the temperature boundary control of the the zone melting method. The melting zone (integration region Ω) is shown in figure 1. For the temperature we have the boundary

Table 1. Parameters of $(Bi_{0.25}Sb_{0.75})_2Te_2$-melt and of the melt geometry

parameter	symbol	value
radius of the ampulla	R	0.004 m
height of the ampulla	H	0.016 m
melting point temperature	θ_s	613 K
thermal diffusivity	a	0.44000e-05 $\frac{m^2}{s}$
kinematic viscosity	ν	0.36310e-06 $\frac{m^2}{s}$
thermal expansion coefficient	β	0.96000e-04 K^{-1}

conditions

$$\theta = \theta_c \quad \text{for } r = R, 0 \leq z \leq H, \varphi \in (0, 2\pi), \text{ (control boundary } \Gamma_c)$$
$$\theta = \theta_s, \quad \text{for } 0 \leq r \leq R, z = H, \quad \theta = \theta_s, \quad \text{for } 0 \leq r \leq R, z = 0.$$

For $t = 0$ we start with a given temperature profile $\theta_c = \theta_{c0}$ on Γ_c and with $\theta_s = 613\,K$, $\delta\theta = 25\,K$ for θ_{c0} we have $\theta_{c0}(z) = \theta_s + 4\frac{z}{H}(1-\frac{z}{H})\delta\theta$. The velocity field $\overline{\mathbf{u}}$, which we want to reach is a typical twodimensional toroidal flow and We consider a time interval $[0, T] = [0, 4\,\text{seconds}]$ with $Z = 60$ time steps of 0.0661 seconds. For the spatial discretization we use 20×25 finite volumes. The figures 2 and 3 show the result of the optimization. As a threedimensional testproblem we consider the above discussed zone melting configuration. The aim of the optimization is to reach a velocity field $\overline{\mathbf{u}} = \mathbf{0}$. This an artificial but a good test case. The size of the spatial discretization of $[0, 2\pi]x[0, R]x[0, H]$ is $20 \times 20 \times 30$. We consider the time interval $[0, T] = [0, 4\,\text{seconds}]$ with $Z = 60$ time steps of 0.0661 seconds. The figure 4 shows the control temperature $\theta_c(\varphi, z)$ at the time $t = T$ on Γ_c and the development of the functional values during The figures 4 shows the twodimensionality of the control temperature θ_c, which does not depend on the angle φ.

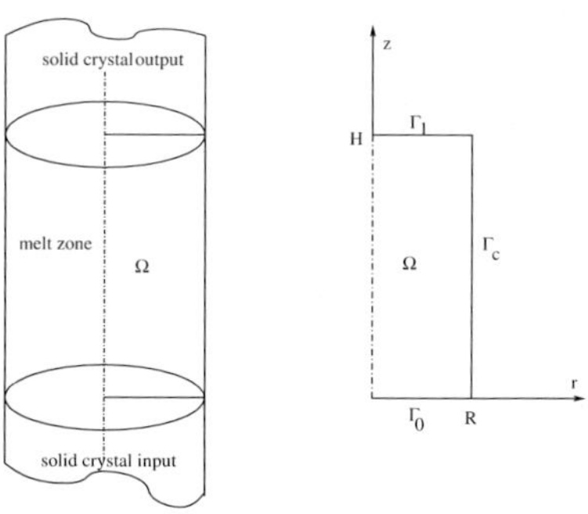

Fig. 1. Physical domain for the zone melting growth

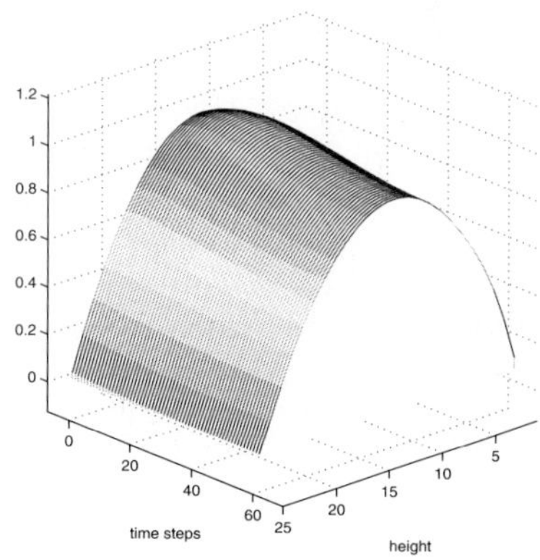

Fig. 2. control on the boundary time cylinder

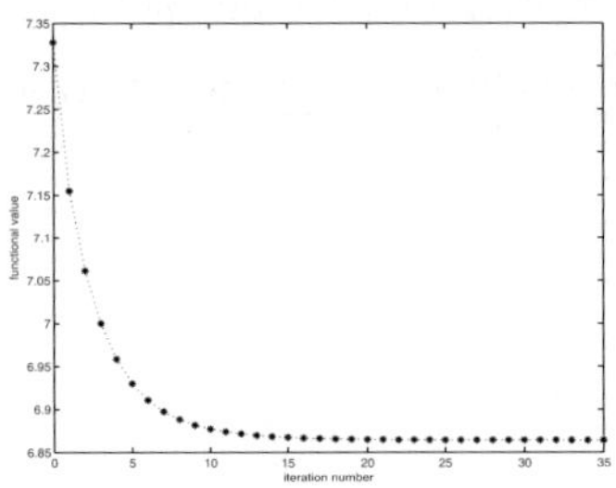

Fig. 3. decreasing functional values during the fixpoint iteration

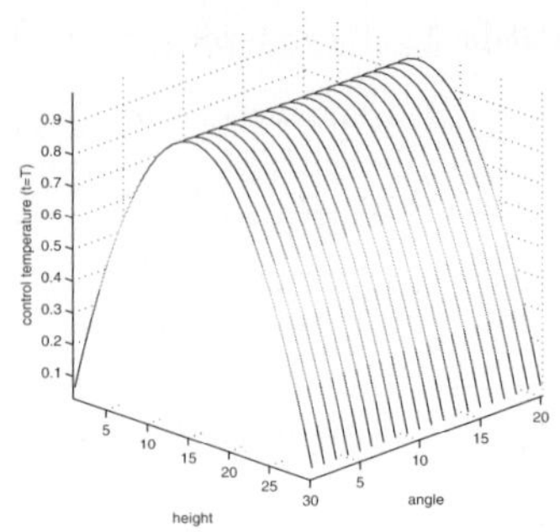

Fig. 4. optimal control in the 3d case

6 Conclusion

With the Lagrange parameter technique it's possible to derive an optimization system for a given functional, which solution gives an optimal control. The numerical examples of the complete time-depend 2.5d optimization system show the possibility of the practical optimization of a thermal coupled flow problem in the crystal growth field. The results show the possibility of boundary control especially in the case of the zone melting technique. Based on the results the proposed strategies it is now possible to do a fully 3d optimization. It is necessary to continue numerical experiments to investigate if the optimization during a boundary control only will be successful technology. There are some experiences with other optimization problems which show the efficiency of volume control, if there is a possibility of the production of volume forces (for example by a magnetic field).

Acknowledgements

The presented techniques and the resulting optimization systems are based on the results of a close cooperation with PROF. DR. MICHEAL HINZE, Dresden/Berlin and I have to thank him very much.

References

1. BÄRWOLFF, G., KÖNIG, F. AND G. SEIFERT: Thermal buoyancy convection in vertical zone melting configurations, ZAMM 77 (1997) 10
2. HINZE, M.: Optimal and instantaneous control of the instationary Navier-Stokes equations, habilitation thesis, Berlin, August 2000 (available on the webpage `http:\\www.math.tu-dresden.de\~hinze`)
3. HINZE, M.: Optimization of the Navier-Stokes equation, Adjoints workshop, Decin/Czech Republic, September 2001
4. CONSTANTIN, P. AND C. FOIAS: Navier-Stokes Equations, The University of Chicago Press, 1988

Numerical Simulation of the Shock Wave / Boundary Layer Interaction in a Shock Tube by Using a High Resolution Monotonicity-Preserving Scheme

V. Daru[1,2] and C. Tenaud[2]

[1] LIMSI - UPR CNRS 3251, BP.133, 91403 ORSAY Cedex, FRANCE
[2] Laboratoire SINUMEF, ENSAM, 151 Bd de l'Hôpital 75013 Paris, FRANCE.
Virginie.Daru@paris.ensam.fr, Christian.Tenaud@limsi.fr

Summary. This work concerns the reflected shock wave/boundary layer interaction in a shock tube. The numerical simulation of this problem necessitates numerical schemes which are robust and very accurate. Thanks to the use of a one step high order accurate scheme that we recently developed [3], we have obtained converged results for several values of the Reynolds number. We then analyze in details the interaction mechanisms and the flow dynamics.

1 Introduction

The shock tube problem may be found in various real applications such as, for instance, in high speed aerodynamic flow facilities either to study the chemical relaxation process in high temperature gas mixture or to generate high enthalpy reservoir conditions. In such a configuration, one wishes ideally to get a uniform state downstream of the reflected shock wave near the end wall. In fact, the reflected shock wave interacts with the incident boundary layer, and this process modifies the flow properties in the end wall region. When the stagnation pressure within the boundary layer is lower than the one downstream of the normal reflected shock wave, the boundary layer separates and a lambda-like shock wave pattern occurs. The numerical simulation of this problem necessitates numerical schemes which are both robust and very accurate, meaning those which can represent with high accuracy both the smooth regions and capture discontinuities with the robustness that is common to Godunov-type methods. In a previous study [1], we exhibited the convergence problems encountered by using classical shock capturing TVD approaches on such a test-case. We recently developed an original high order coupled time-space scheme (named OSMP7) which guaranties Monotonicity Preserving properties [2, 3]. The OSMP7 scheme gives very accurate results on numerous classical test cases . Compared to classical WENO schemes, the OSMP7 scheme better captures the discontinuities while showing a comparable accuracy in the regular regions, and gives a sixth fold lower CPU cost than the WENO schemes.

In this work, the OSMP7 scheme was applied to the shock tube problem. Converged results were obtained and are presented for several Reynolds num-

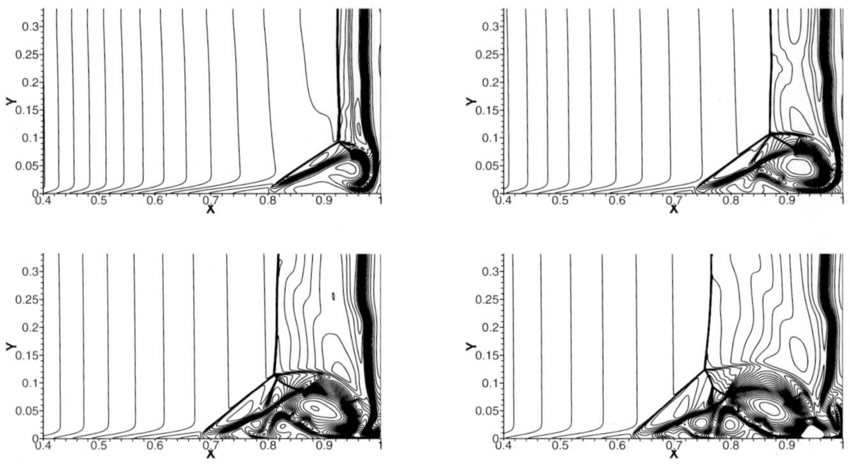

Fig. 1. Density contours for $Re = 200$ at several dimensionless time (from left to right and top to bottom): $t = 0,4$, $t = 0,5$, $t = 0,6$, $t = 0,7$.

bers, with a grid independency assessment, and compared with the results published by Sjögreen et al. [5]. The present results allow us to analyze the dynamics of the shock wave / boundary layer interaction at the end wall of the shock tube.

2 Numerical Procedure

We solve the non-dimensionalized Navier Stokes equations in cartesian coordinates. A perfect gas law is used, assuming a constant specific heat ratio $\gamma = 1.4$. The discretization of the Euler fluxes is performed by using the OSMP7 scheme. This scheme was developed following a coupled time and space (Lax-Wendroff type) approach. It is seventh order accurate in time and space, at least in the scalar case. For shock capturing purpose, Monotonicity-Preserving (MP) constraints, derived from [6], are incorporated to the scheme. The MP constraints were demonstrated to be equivalent to the TVD conditions [3] for discontinuities, but retains the original high order scheme in the vicinity of extrema at the difference of classical TVD schemes. In this way, the OSMP7 scheme allows both to capture sharp discontinuities without producing spurious oscillations and to preserve high order of accuracy in the smooth regions. The viscous fluxes are discretized through the use of a classical second-order centered scheme. In the multidimensional case, a Strang splitting is used. For more details about the numerical procedure, see references [2, 3].

3 Analysis of the Flow Dynamics in the Interaction Zone

We consider a unit side length square shock tube with insulated walls. The diaphragm is initially located in the middle of the tube ($x = 0.5$). The initial

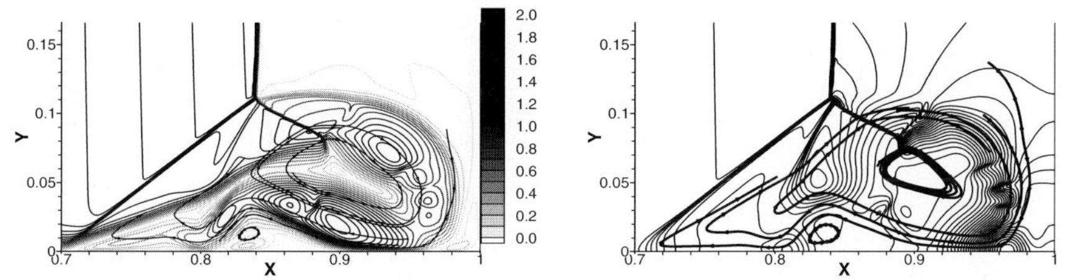

Fig. 2. Mach number (left) and static pressure (right) contours for a Reynolds number $Re = 500$ at a dimensionless time $t = 0,55$.

state, in terms of dimensionless quantities, is on the left of the diaphragm: $\rho_L = 120$, $p_L = \rho_L/\gamma$, $u_L = v_L = 0$, and on the right: $\rho_R = 1.2$, $p_R = \rho_R/\gamma$, $u_R = v_R = 0$. At the initial time, the diaphragm is broken. A shock wave, followed by a contact discontinuity, moves to the right (the shock Mach number is equal to 2.37). The incident shock wave is weak, and reflects at the right end wall approximately at time $t = 0.2$. After this reflection, it interacts with the contact discontinuity. Then, complex interactions occur. The contact discontinuity stays stationary, close to the right end wall. Afterwards, the reflected shock wave begins to interact with the rarefaction wave at time $t = 0.4$, and the rarefaction wave reflects on the left end wall at time $t = 0.5$.

In the viscous case, a boundary layer is created at the bottom wall. After its reflection on the right end wall, the shock wave interacts with this boundary layer. As the stagnation pressure in the boundary layer is lower than the one within the outflow region, a dead-water region (named "bubble") takes place over a large extent within the boundary layer, resulting in a major modification of the flow pattern and the formation of a " *lambda-shape like shock pattern*". The triple point generates a slip line which rolls up in the bottom right corner.

The development of the interaction is reported in Fig. 1, showing the time evolution of the density contours in the interaction zone for a Reynolds number $Re = 200$. From $t = 0,4$, the main stream, flowing in between the lambda-shock and the bubble, wraps around the dead-water region by forming a high-speed jet which impacts the bottom wall (Fig. 1 top-left). At $t = 0.5$, the jet, after reflection on the solid wall, impacts the supersonic shear layer (slip line) coming from the separation point (Fig. 1 bottom-left) and reflects down toward the bottom wall. As time goes on, successive alternate reflections between the bottom wall and the slip line are carried on (Fig. 1). This jet flow is very energetic since it can reach supersonic speeds at several places as we can see on the Mach number contours (Fig. 2-left). Streamlines are superimposed to show the path of the jet. Each impact of the jet, either on the solid wall or on the slip line, corresponds to a relatively high pressure region (Fig. 2-right) and in between, relatively low pressure regions are created. Hence, the flow wraps around these relatively low pressure zones to create counter-rotative secondary vortices.

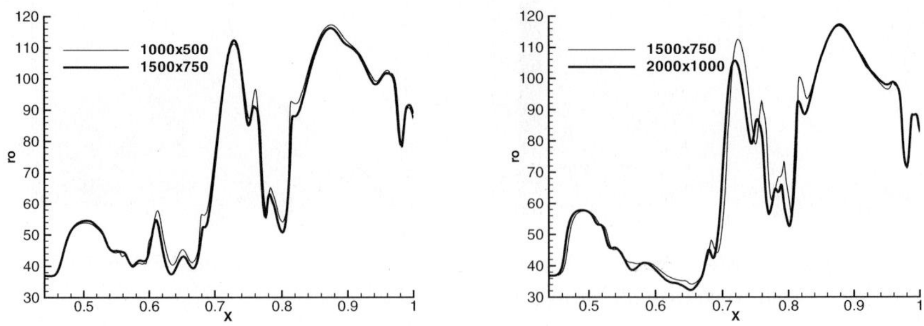

Fig. 3. Density distributions along the bottom wall of the shock tube, obtained on several fine grids for $Re = 500$ (left) and $Re = 750$ (right).

The slip line is deformed due to jet impacts. The flow over the slip line being supersonic, this induces shocklets which interact with the main shock pattern to produce small Mach stems, as we can see just under the triple point (Fig. 2).

4 Influence of the Reynolds Number

4.1 Grid Convergence Study

The simulations have been carried out for four Reynolds numbers: $Re = 200$, 500, 750 and 1000. Several uniform meshes with increasing resolutions have been used . Grid convergence is demonstrated for instance on the density distribution along the bottom wall for two Reynolds numbers: $Re = 500$ and $Re = 750$ (Fig. 3). While the results do not fit exactly at the highest resolutions, the differences remain small and we can consider that the results are converged when the grid resolutions are: 1000x500 for $Re = 200$, 1500x750 for $Re = 500$ and 2000x1000 for $Re = 750$. For $Re = 1000$, the discrepancies between the results obtained on the 2000x1000 and 3000x1500 grids are still significative and we cannot conclude that convergence is completely reached. Calculations on finer grids could not be carried out because of the prohibited CPU time required. For comparison, we also performed calculations using a more classical scheme, based on a third order Runge-Kutta time integration and a 6th order WENO (r=5) scheme [4] as spatial discretization. Using this approach, results have been obtained very close to the OSMP7 ones. Nevertheless, the grid convergence cannot be reach in several grid resolutions using the WENO scheme because of the CPU time required which is at least six times greater than for the OSMP7 scheme. Moreover, the OSMP7 scheme exhibits a more rapid grid convergence than the WENO scheme.

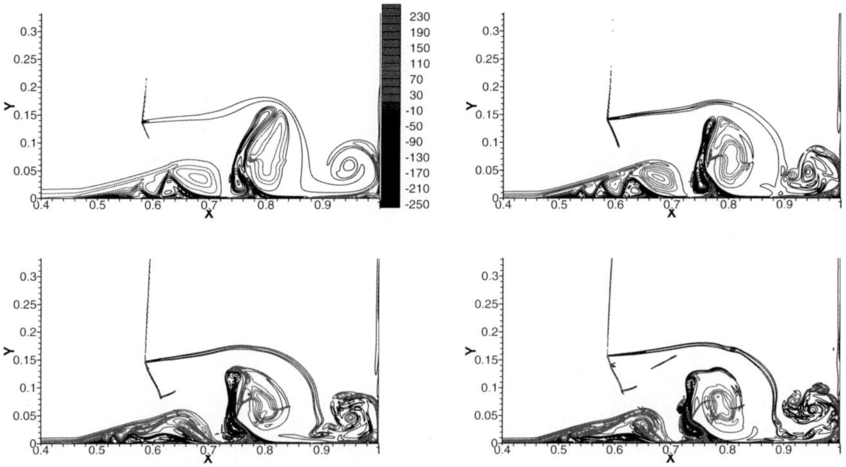

Fig. 4. Vorticity contours at a dimensionless time $t = 1$ for several values of Re; from left to right and top to bottom and: $Re = 200$; $Re = 500$; $Re = 750$; $Re = 1000$.

4.2 Influence on the Flow Dynamics

When the Reynolds number increases, the velocity of the jet also increases. Thus during a fixed period an increased number of alternate reflections between the solid wall and the slip line can take place. Consequently, the number of counter rotative vortices produced in this interaction region might increase too. For the lowest Reynolds numbers studied $Re = 200$ and $Re = 500$, this dynamical process has been recorded and is visible in Fig. 4 on the vorticity contours. Nevertheless, this process does not persist from $Re = 750$ (Fig. 4) since the regular arrangement of the largest eddy structures seems to disappear. During the set up of the large scale eddy structures (Fig. 5, in grey), eddy structures with positive vorticity (in black in Fig. 5) are created close to the bottom wall, with an enstrophy increasing with the Reynolds number. In a similar manner as when a vortex interacts with a boundary layer, when the enstrophy of these eddy structures is high enough, they are repulsed out of the boundary layer. Hence, the structures are stretched and distorted as an intense vortex filament. Due to the shearing effect of the outflow, this filament is elongated up to the bursting of the filament. When this filament is broken up, the enstrophy intensity decreases close to the wall which results in the retraction of the structure with a "spring back effect". The two consecutive structures with a negative vorticity (in grey) then merge. This dynamical process described herein is carried out with strengthening, up to a pair of counter-rotative vortices could be burst out of the boundary layer. Let us notice that the merging process disappears if the grid resolution is not sufficient, which emphasizes the importance of grid convergence.

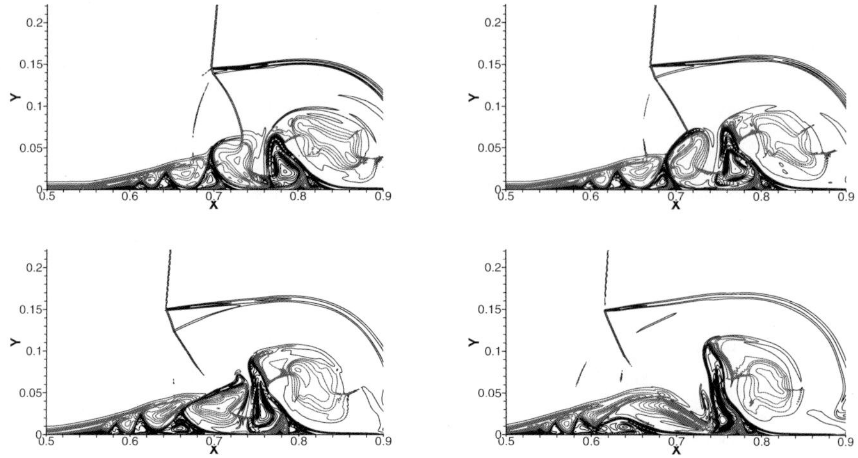

Fig. 5. Vorticity contours obtained for $Re = 750$ at several dimensionless time; from top to bottom and left to right: $t = 0,80$; $t = 0,85$; $t = 0,90$; $t = 0,95$.

5 Conclusion

The present study shows that the calculation of such a complex shock tube problem necessitates numerical schemes which are both robust and very accurate. The OSMP7 scheme allowed us to reach converged solutions for several Reynolds numbers, which was not possible using WENO schemes because of the high CPU cost. The present shock tube problem can be considered as a good test-case for checking high-resolution schemes. The solutions we obtained in this study with the OSMP7 scheme could constitute reference solutions.

References

1. V. Daru & C. Tenaud 2001 'Evaluation of TVD high resolution schemes for unsteady viscous shocked flows', *Computers and Fluids*, **30**, 89–113.
2. V. Daru & C. Tenaud 2002 'High resolution monotonicity-preserving schemes for unsteady compressible flows', *International Conference on Conputational Fluid Dynamics*, July 2002, Sydney, Australia.
3. V. Daru & C. Tenaud 2004 'High order one-step monotonicity-preserving schemes for unsteady compressible flow calculations', *Journal of Computational Physics*, **193**, 563–594.
4. G.S. Jiang & C.W. Shu 1989 'Efficient implementation of weighted ENO schemes', *Journal of Computational Physics*, **83**, 32–79.
5. B. Sjogreen, H.C. Yee: 'Grid convergence of high order methods for multiscale complex unsteady viscous compressible flows', *Journal of Computational Physics*, **185**, 1–26 (2003).
6. A. Suresh & H.T. Huynh 1997 'Accurate Monotonicity-Preserving Schemes with Runge-Kutta Time Stepping', *J. Computational Physics*, **136**, 83–99.

Accurate Flow Prediction for Store Separation from Internal Bay

M. Mani, A. Cary, and W. W. Bower

The Boeing Company, P.O. Box 516 MC S111-1240, St. Louis, Missouri,
mori.mani@boeing.com

1 Introduction

Accurate prediction of store separation from internal aircraft bays is of great importance in commercial and military aircraft design. The flow-field inside and over the bay is very complex, especially for supersonic flows involving shock-boundary layer interaction, compressible shear layer, and strong acoustics. Understanding the physics of the flow in and surrounding the bay is of great importance for safe release of stores and controlling the acoustic environment.

The supersonic flow-field within and surrounding the bay is unsteady, and accurate prediction of the shear layer requires dynamic simulations with advanced numerical and physical models. The objective of this paper is to demonstrate accurate prediction of the effect of the flow and a control method by employing a turbulence model that is not cost-prohibitive and is accurate enough to simulate such complex flow phenomena. The turbulence model employed for this study is a two-equation hybrid Reynolds-Averaged Navier-Stokes (RANS)[1] and Large Eddy Simulation (LES)[2] formulation known as the Large Eddy Stress balanced (LESb) model[3].

2 Turbulence Model

The two-equation RANS models use k to represent all of the unsteady fluctuations, while the LES models use k to represent only the spatial average of the fluctuations within a filter width. The length scale l_B in the balanced LES model (LESb) is defined as:

$$l_B = \min(l_\epsilon, C_B \Delta) \tag{1}$$

and the turbulent viscosity is represented as:

$$\mu_t = \bar{\rho} C_\mu l_B \sqrt{k} \tag{2}$$

For unresolved length scales ($\Delta \gg l_\epsilon$) the model reverts to the standard two-equation model. However, for resolved length scales ($\Delta \ll l_\epsilon$), the model reverts to the LES model, and as the resolved length scale goes to zero (full resolution of all pertinent scales) the model approaches a Direct Numerical Simulation (DNS) of turbulence.

Implementation of the LESb model in an existing RANS code with a two-equation model (and the ability to run time-accurate simulations) is straightforward. The existing model solves for ω, which implies an effective length scale of the turbulence and enters the k equation as a dissipation rate of the turbulent kinetic energy (TKE). To implement the model, one should compare the computed effective length scale with the grid scale, and limit the dissipation term accordingly. Then one can implement the filter as follows:

$$\omega_B = \max\left(\omega, \frac{\sqrt{k}}{C_\mu C_B \Delta}\right) \tag{3}$$

The effect of the limiter is to increase the dissipation of turbulent kinetic energy. Thus the TKE is reduced from that predicted by the traditional turbulence model such that length scales that are resolved do not contribute to the Reynolds stress terms.

For isotropic grids, Δ is defined as

$$\Delta = (\text{Volume})^{1/3} = (1/J)^{1/3} \tag{4}$$

where J is the Jacobian of the grid transformation. For stretched grids, the model assumes that the smallest resolved eddies should be roughly isotropic and so must be resolved in all three coordinate directions, and in time. Thus the filter is defined as:

$$\Delta = \max(dx, dy, dz, u^*\, dt, \sqrt{k}\, dt) \tag{5}$$

The limiting length scales based on time-step represent the scales based on convection velocity and SGS turbulence respectively. These scales are included to ensure that there is sufficient time resolution to resolve the captured physical phenomena. That is, if the time steps are too large, the unsteady phenomena cannot be resolved, and the RANS model should be used.

Near the wall in typical boundary layer grids, dy will be small to resolve the mean shear, but dx is typically very large. Thus the LESb model should revert to the underlying two-equation model near walls. However, as the boundary layer is traversed into grids that are more isotropic and resolve the large-scale structures, the model should smoothly transition to an LES model.

3 Grid and Solutions

To understand the impact of the shear layer from an internal bay, we will begin our study with a supersonic flow over an empty cavity; then we will investigate how to actively control the shear layer and suppress resonance in the bay by blowing with a sonic slot-jet upstream of the bay. A similar study was carried out experimentally by Zhuang, et.al[4] and a numerical investigation of approximately similar geometry using LES turbulence model was carried out by Rizzetta, et.al[5].

A 3D grid was generated for a cavity (L=12, D=2.4, W=2.1 centimeters) with an aspect ratio of L/D=5.0 The geometry contains the upstream plane to capture the boundary layer accurately and side panels to include the boundary layer interaction with the shear layer. The computational grid contains 2.7 million points. To avoid a small time-step in advancing the governing equation in time, the grid spacing at the wall is set to $y^+=20$. The wall function[6] is invoked point by point when the y^+ is greater than 15. The free-stream flow conditions are Mach=2.0, P_0 = 217 kPa, and T_0 = 336 K, Re/cm= 2.33_10^5. The unsteady cavity solutions were obtained with symmetric and full configurations with and without control to investigate the asymmetric effect of the turbulence model. The plenum total pressure and temperature for the slot-jet are 1496 kPa and 345.3 K.

The solutions were obtained using the BCFD code[7]. The BCFD code is an implicit general- purpose Euler and Navier-Stokes solver with hybrid unstructured and structured / unstructured grids. Numerous special algorithms from first-order to fifth-order accurate are available for structured grids applications. The Reynolds average Navier-Stokes equations with the hybrid turbulence model of Reference [3] are used to advance the governing equations in time. The Roe scheme with total variation diminishing, which is second-order accurate in the physical domain, was employed in this study. Since the computational domain was split into 26 zones, a global Newton with five sub-iterations was employed to unify the zonal solution in time. A Δt of 1.0 microsecond was employed to advance the solution in time, which will provide one free-stream convective time scale in 200 time-steps.

4 Results

Two unsteady solutions with the hybrid turbulence model were obtained. The time-mean of pressure contours at the center plane of the cavity is shown in Figure 1. The pressure in and over the cavity clearly has been suppressed by the slot-jet. There is a region of high pressure at the aft wall followed by a low pressure region in the uncontrolled case. The pressure in the same region for the controlled case is lower. The pressure along the cavity wall at the centerline is plotted for both cases in Figure 2. The pressure is lower at all locations for the controlled case. However, the lowest pressure is at x=12.0 inches. There is a separation region upstream of the slot-jet which causes a secondary shock that combines with the slot-jet shock and forms a lambda shock. A thickening of the boundary layer can be observed upstream of the lambda shock as a result of this separation.

Instantaneous snapshots of the density gradient without and with control at $t = t_1$ are shown in Figure 3. The shock structure due to mass injection substantially changes the flow structure in the shear layer and cavity. The shear layer is lifted off the bay, and as a result there is reduced interaction between the shear layer and the aft end of the bay. This would substantially diminish the sonic fatigue by damping the feedback loop generated by the shear layer interacting with the aft face of the cavity. In the controlled case, the cavity leading edge shock is a pulsating shock. This shock is formed due to the entrainment of the flow across the leading edge of the bay. The iso-surfaces of vorticity magnitudes are shown with the density gradient in Figure 4 with and without mass injection at time t_1. The iso-surface for the uncontrolled flow shows substantially more three-dimensionality in the shear layer than the controlled flow shear layer. The interaction of the cavity side on the shear layer flow and the extent that the free shear layer extends beyond the edge into the bay can be seen. In the downstream of the cavity, flow is being sucked into the cavity from the side wall, while in the upstream region, flow is spilling out of the cavity over the side wall, creating a weak vortex.

A snapshot of the Mach number for the two cases is shown in Figure 5. The dynamic of the cavity flow is formed by the large structures formed in the upstream section of the cavity that convects downstream as he unsteady shear layer oscillates over the cavity.

The pressure spectra at three different locations at mid-span on the upstream, center and aft cavity walls are shown in Figure 6. The coordinate

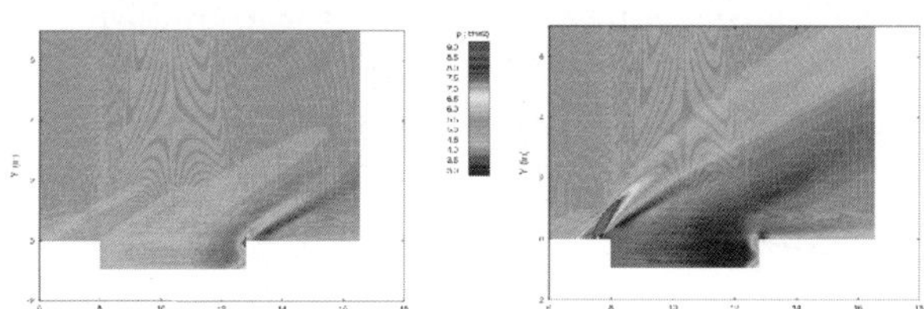

Fig. 1. Pressure contours with and without control at mid-span.

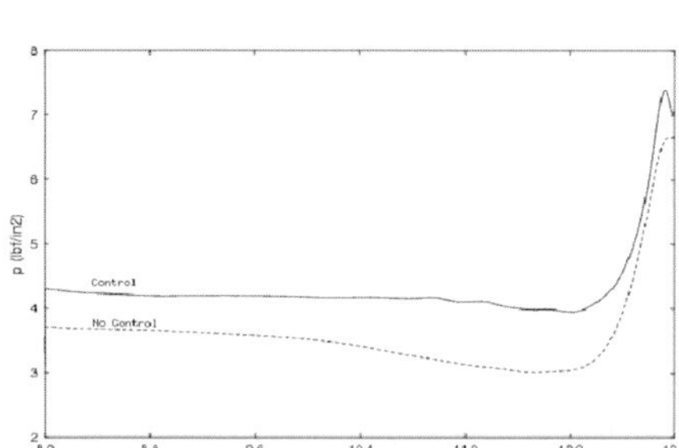

Fig. 2. Average pressure along the cavity floor.

Fig. 3. Instantaneous density gradient without and with slot-jet at time = t_1.

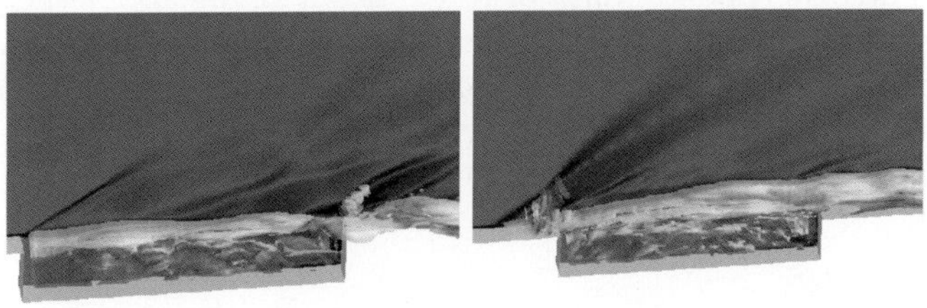

Fig. 4. Instantaneous density gradient with iso-surfaces of vorticity magnitude at time t_2 without and with control.

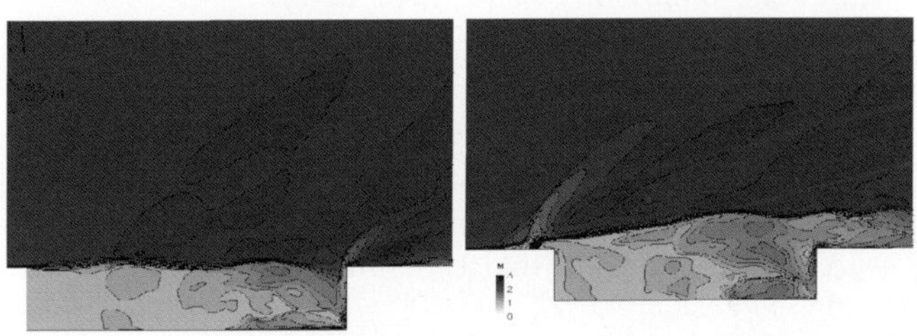

Fig. 5. Instantaneous Mach contours without and with AFC at time t_2

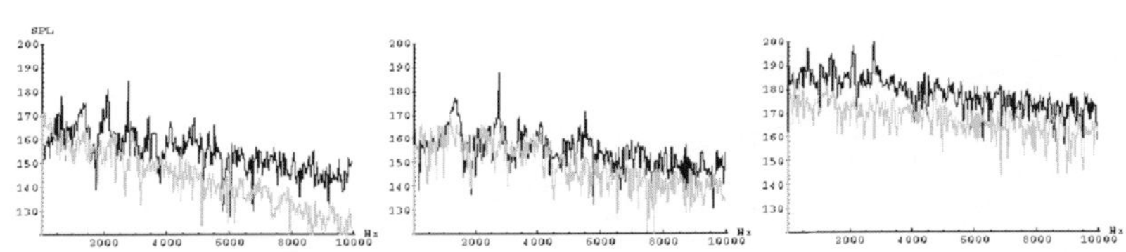

Fig. 6. Pressure spectra at mid-span on the upstream, bottom center, and aft cavity wall.

locations are x=20.3, 25.4, 32.5; y=-0.3, -2.4, -0.3; z=0.0 centimeters accordingly. The primary cavity tones are seen in the 1000 to 3000 Hz frequencies, with primary peak at 3000 Hz. There are no significant differences in the magnitude of the frequencies in the upstream and cavity floor. However, the magnitude of the SPL is greater by about 20 dB on the downstream wall just below the cavity lip. The comparison of the SPL between no flow control and flow-control shows that the primary peaks have been eliminated and the broadband spectrum has been reduced by about 10 to 20 dB when the control is on. The experimental data of Reference 2 shows about the same order of magnitude reduction in the SPL as we have observed in the CFD study. However, in their cases the pressure peaks are about 20 dB below what CFD has predicted.

5 Conclusions

The hybrid LESb turbulence model is a cost-efficient approach to predicting complex high shear flow-field and massively separated flows. The results indicate that the slot-jet has significant impact (20 dB) in suppressing the acoustic load and the behavior of the shear layer.

References

1. F. R. Menter, "Zonal Two Equation k-ω Turbulence Models for Aerodynamic Flows," AIAA 93–2906, 24th Fluid Dynamics Conference, July 6–9, 1993, Orlando Florida
2. J. Smagorinsky, "General Circulation Experiments With the Primitive Equations," Mon. Weather Rev., **91**, 99–164.
3. R. H. Bush and M. Mani, "A Two-equation LES/RANS Hybrid Turbulence Model for High Sub-grid Shear Model," AIAA 2001–2561.
4. N. Zhuang, F. S. Alvi, M. B. Alkislar, and C. Shih, "Aeroacoustic Properties of Supersonic Cavity Flows and Their Control," AIAA 2003–3101, 9th AIAA/CEAS Aeroacoustics Conference, May 2003.
5. D. P. Rizzetta and M. R. Visbal, "Large-Eddy Simulation of Supersonic Cavity Flowfields Including Flow Control," AIAA Journal, Vol. 41, No. 8, August 2003.
6. M. Mani and D. Ota, "A Compressible Wall Function for Steady and Unsteady Flow Applications," AIAA-99-3216. Norfolk, VA. 1999.
7. M. Mani, A. W. Cary, and S. V. Ramakrishnan, "A Structured and Hybrid-unstructured Grid Euler and Navier-Stokes Solver for General Geometry," AIAA 2004–524.

A Coupled Navier-Stokes/Vortex-Panel Solver for the Numerical Analysis of Wind Turbines

Sven Schmitz[1] and Jean-Jacques Chattot[2]

[1] University of California, Davis shschmitz@ucdavis.edu
[2] University of California, Davis jjchattot@ucdavis.edu

1 Introduction

Navier-Stokes codes require artificial viscosity in order to be numerically stable. Artificial viscosity is responsible for rapid numerical dissipation of vortical structures such as tip vortices and vortex sheets behind wind turbine blades. However, this dissipation is not physical and does not simulate a high Reynolds number flow, as Figure 1 reveals. Results obtained for lift and induced drag are highly sensitive to the capability of the simulation method for maintaining the vortex sheet in the far wake of the wind turbine. High order discretization schemes combined with grid refinement or vorticity confinement due to Steinhoff [1] are two approaches for reducing artificial dissipation, yet numerically very expensive. For incompressible high Reynolds number flows, the convection of shed vorticity to the far-field is best carried out by a vortex method that is dissipation free. However, a vortex method cannot predict complex 3D viscous flow phenomena in the immediate vicinity of a wind turbine blade. A novel approach has been developed and validated, in which the flow domain is split into near-field and far-field respectively. The near-field is predicted by a commercially available Navier-Stokes solver, CFX, and coupled with an in-house developed incompressible Vortex Panel Method to account for the far-field.

2 Numerical Methods

2.1 Coupling Methodology

Starting with an undisturbed flow, the Navier-Stokes code solves for the flow field around the wind turbine blade. The spanwise distribution of circulation Γ_j along the blade is determined by performing a velocity contour integral

$$\Gamma_j = \oint_{L_j} \boldsymbol{v}\, \boldsymbol{ds} = \int\int_{A_j} \boldsymbol{\omega}\, \boldsymbol{dA} \qquad (1)$$

where L_j is a closed contour around a radial section of blade and passes through the trailing edge. For Euler flow, L_j can be chosen as the sectional profile contour itself or any other closed contour around the wind turbine blade. For high Reynolds number viscous flow, L_j has to be sufficiently large to include all sources of vorticity, namely the boundary layer. This is stated by Stokes' theorem in equation (1). The distribution of circulation Γ_j is then used to create a helicoidal vortex structure behind the wind turbine blade. The vortex structure consists of helicoidal vortex filaments as well as a bound vortex of variable strength, see section 2.2. A discrete form of the Biot-Savart law

$$\boldsymbol{v} = \int \frac{\Gamma}{4\pi} \cdot \frac{\boldsymbol{dl} \times \boldsymbol{r}}{r^3} \tag{2}$$

is used to calculate induced velocities on the boundary of the Navier-Stokes zone. Induced velocities are superimposed to the undisturbed flow consisting of the incoming flow speed and blade rotation entrainment speed. Next, the Navier-Stokes zone is solved again with an updated set of boundary conditions. Determining Γ_j serves as input for another coupling step. The coupling procedure is converged when the maximum change in Γ_j is less than 10^{-5}. Convergence is achieved in 3-5 coupling steps where coupling is performed after 10-20 time steps of the Navier-Stokes solver.
The Coupled Solver is parallelized on a cluster using MeTiS [2] as a partitioning algorithm. Higher than 90 percent efficiencies are obtained on 4 processors.

2.2 Vortex Model

The vortex model is described in a rotating frame of reference, with the y-axis being the quarter chord line of the blade, and the wind blowing in positive x-direction. The influence of the vortex sheets and bound vortices of the two blades is determined at an arbitrary point on the outer boundary of the Navier-Stokes zone $\boldsymbol{c} = (x_c, y_c, z_c)^T$. Length scales are non-dimensionalized by the blade radius R, velocity scales by the wind speed U_∞.

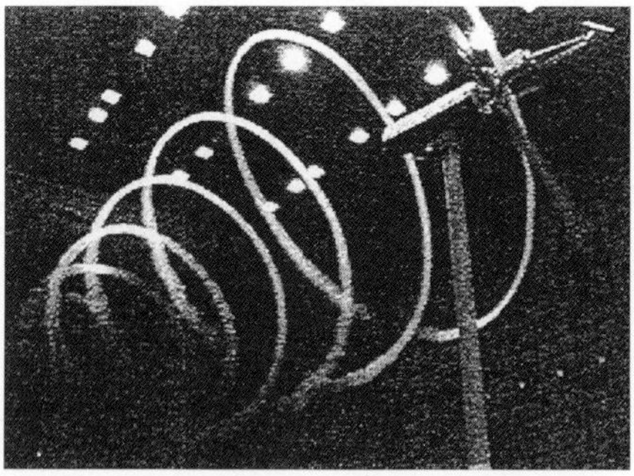

Fig. 1. NREL Phase VI Rotor [3]

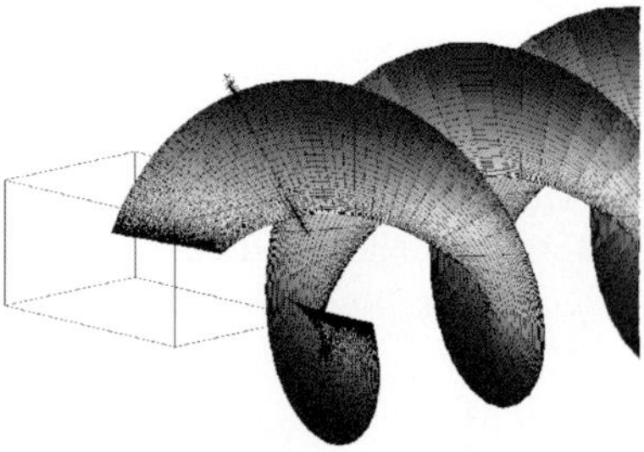

Fig. 2. Helicoidal Vortex Sheet

Helicoidal Vortex Sheet

Each vortex sheet consists of jx helicoidal vortex filaments originating from the blade's trailing edge and extending to infinity with the vortex strength being constant along each filament. Denoting by $\boldsymbol{\Delta l_{i,j}} = (\Delta x_{i,j}, \Delta y_{i,j}, \Delta z_{i,j})^T$ a discrete piece of vortex filament and by $\boldsymbol{r} = (x_c - x_{i,j}, y_c - y_{i,j}, z_c - z_{i,j})^T$ the distance of a piece of vortex filament to an arbitrary point on the boundary of the Navier-Stokes zone, a discrete form of equation (2) results in influence coefficients for the cartesian velocity components:

$$
\begin{aligned}
a_j^{helix} &= \sum_{i=2}^{ix} \left. \frac{(z_c - z_{i,j})\,\Delta y_{i,j} - (y_c - y_{i,j})\,\Delta z_{i,j}}{[(x_c - x_{i,j})^2 + (y_c - y_{i,j})^2 + (z_c - z_{i,j})^2]^{3/2}} \right|_{Blade\,I,II} \\
b_j^{helix} &= \sum_{i=2}^{ix} \left. \frac{(x_c - x_{i,j})\,\Delta z_{i,j} - (z_c - z_{i,j})\,\Delta x_{i,j}}{[(x_c - x_{i,j})^2 + (y_c - y_{i,j})^2 + (z_c - z_{i,j})^2]^{3/2}} \right|_{Blade\,I,II} \\
c_j^{helix} &= \sum_{i=2}^{ix} \left. \frac{(y_c - y_{i,j})\,\Delta x_{i,j} - (x_c - x_{i,j})\,\Delta y_{i,j}}{[(x_c - x_{i,j})^2 + (y_c - y_{i,j})^2 + (z_c - z_{i,j})^2]^{3/2}} \right|_{Blade\,I,II}
\end{aligned}
\tag{3}
$$

The location of the Trefftz plane is $x_{T,j} = x_{ix,j}$. Asymptotic remainder terms, taking into account two rotor blades, are found to be:

$$
\begin{aligned}
a_{j,rem}^{helix} &= \frac{r_j^2}{adv_T\, x_{T,j}^2} \qquad\qquad b_{j,rem}^{helix} = -\frac{z_c - 2\,r_j\,\sin\frac{x_{T,j}}{adv_T}}{x_{T,j}^2} \\
c_{j,rem}^{helix} &= \frac{y_c - 2\,r_j\,\cos\frac{x_{T,j}}{adv_T}}{x_{T,j}^2}
\end{aligned}
\tag{4}
$$

The error in the remainder terms is of the order $1/x_{T,j}^3$. In equations (4), $adv_T = (U_\infty + u_T)/(\Omega\, R)$ is the advance ratio in the Trefftz plane and $r_j = y_{j-\frac{1}{2}}$ denotes the spanwise origin of the vortex filament. It should be stated that the vortex sheets shown in Figure 2 account for contraction of the vortex sheet with $u_B = 0.5\, u_T$, $u_B < 0$ being the average induced velocity in x-direction at the trailing edge obtained from the rotor power using rotor disc theory. The Trefftz plane $x_{T,j}$ is typically set at approximately 20 blade radii downstream of the rotor disc. The remainder terms in equation (4) are then 3 orders of magnitude smaller than the influence coeffcients in equation (3).

Bound Vortex

The bound vortex consists of jx vortex filaments of finite length $y_j - y_{j-1}$ located on the y-axis which coincides with the quarter chord of the blade. Each bound vortex filament is divided into kx pieces allowing for accurate calculation of induced velocities. Thus, the influence coefficients become:

$$
\begin{aligned}
a_j^{bound} &= \sum_{k=1}^{kx} \left. \frac{(z_c - z_{k,j})\,\Delta y_{k,j}}{[x_c^2 + (y_c - y_{k,j})^2 + z_c^2]^{3/2}} \right|_{Blade\,I,II} \\
c_j^{bound} &= \sum_{k=1}^{kx} \left. -\frac{(x_c - x_{k,j})\,\Delta y_{k,j}}{[x_c^2 + (y_c - y_{k,j})^2 + z_c^2]^{3/2}} \right|_{Blade\,I,II}
\end{aligned}
\tag{5}
$$

with $b_j^{bound} = 0$, as the bound vortex lies on the y-axis. The total induced velocities due to the helicoidal vortex sheets as well as bound vortices of two blades are found to be:

$$
\begin{aligned}
u_c &= \frac{1}{4\pi} \sum_{j=1}^{jx} (\Gamma_{j-1} - \Gamma_j) \cdot (a_j^{helix} + a_{j,rem}^{helix}) + \frac{\Gamma_{j-1} + \Gamma_j}{2}\, a_j^{bound} \\
v_c &= \frac{1}{4\pi} \sum_{j=1}^{jx} (\Gamma_{j-1} - \Gamma_j) \cdot (b_j^{helix} + b_{j,rem}^{helix}) \\
w_c &= \frac{1}{4\pi} \sum_{j=1}^{jx} (\Gamma_{j-1} - \Gamma_j) \cdot (c_j^{helix} + c_{j,rem}^{helix}) + \frac{\Gamma_{j-1} + \Gamma_j}{2}\, c_j^{bound}
\end{aligned}
\tag{6}
$$

3 Results

The Parallelized Coupled Solver (PCS) is applied to an optimal wind turbine configuration obtained by an incompressible vortex line model (VLM), [4] and [5]. The two-bladed rotor has a radius of 5m equipped with the S809 profile and performs 72 revolutions per minute. The root section starts at 25 percent of the blade radius. Taper varies from 10 to 13 percent of the blade radius, twist between 9-39 degrees. The wind speed is 10m/s. The Navier-Stokes solver, CFX, solves the 3D incompressible Navier-Stokes equations using a second-order accurate scheme. Fully turbulent flow is assumed using the two-equation k-epsilon turbulence model. Near-wall treatment is handled

Table 1. Comparison of VLM with PCS

		Inviscid		Viscous	
		VLM	PCS	VLM	PCS
Thrust	[N]	509.62	508.31	472.41	458.60
Tangential Force	[N]	-183.63	-179.89	-163.26	-150.80
Bending Moment	[Nm]	1803.1	1814.8	1670.2	1636.4
Torque	[Nm]	-588.82	-583.80	-519.58	-485.50
Power	[kW]	8.879	8.804	7.835	7.321

by scalable wall functions. The Navier-Stokes zone consists of a hybrid mesh with prisms and hexahedrons, the prisms being extruded from the blade tips and the hexahedrons forming a C-type mesh around the blade, extruded tips, and a wake plane, see Figure 3. In order to capture the vortex sheet within the Navier-Stokes zone, the wake plane coincides with the first approximately 60 degrees of a vortex helix sheet such as in Figure 2, whose geometry was obtained by a power estimate prior the PCS run. The surface mesh on the blade has 60 points in the chordwise and 40 points in the spanwise directions respectively. The size of the Navier-Stokes zone varies from 1 to 3 reference chord lengths around the blade and 3 to 6 reference chord lengths in the wake. The hybrid mesh contains a total of approximately 800,000 nodes. Results are compared between the two methods for inviscid as well as viscous flow simulation in Table 1.

4 Discussion

The two results show excellent agreement for the inviscid case. The difference in rotor power is less than 1 percent. Figure 4 shows the spanwise distribution of moments in the normal and tangential direction. The integral under the torque curves corresponds to half of the power of the two-bladed wind turbine. For viscous flow, PCS predicts 6.5 percent less power than the vortex model including viscous correction due to three-dimensional effects not accounted for in VLM. The spanwise distributions of circulation and moments are presented in Figures 5 and 6. Strong convergence of the coupling is revealed by Figure 7. The coupling procedure hardly increases the computational cost. Performing a PCS run with the vortex sheet ending at the Navier-Stokes zone, the load distribution changes according to Figure 8 with a change in rotor power of 12 percent, revealing the importance of maintaining the vortex sheet far into the wake. The main advantages of the PCS solver are the small size of the Navier-Stokes domain, the efficient coupling, the parallelization, and a better modelization of the high Reynolds number flow behind the wind turbine. Up to this stage, only fully attached flow was

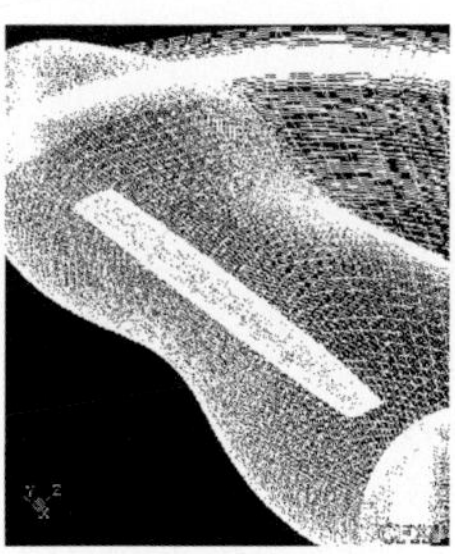

Fig. 3. Navier-Stokes Domain

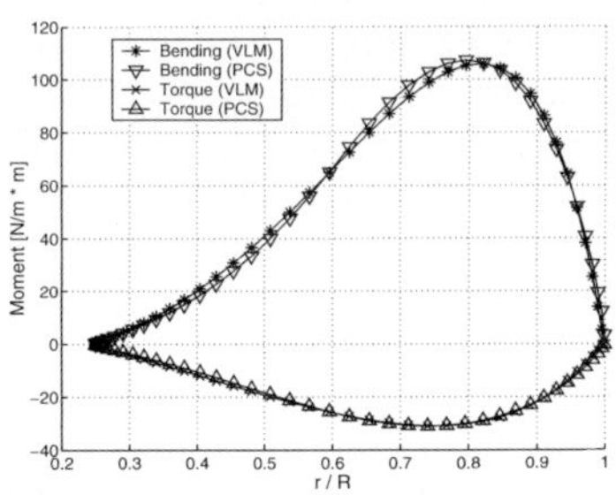

Fig. 4. Spanwise Moments (Inviscid)

considered in order to validate the coupling methodology. In the future, the PCS solver will be applied to off-design conditions of wind turbines such as the NREL PhaseVI Rotor [3].

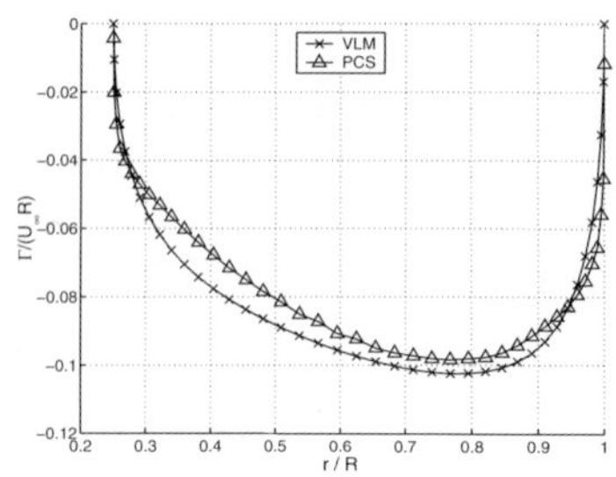

Fig. 5. Spanwise Circulation (Viscous)

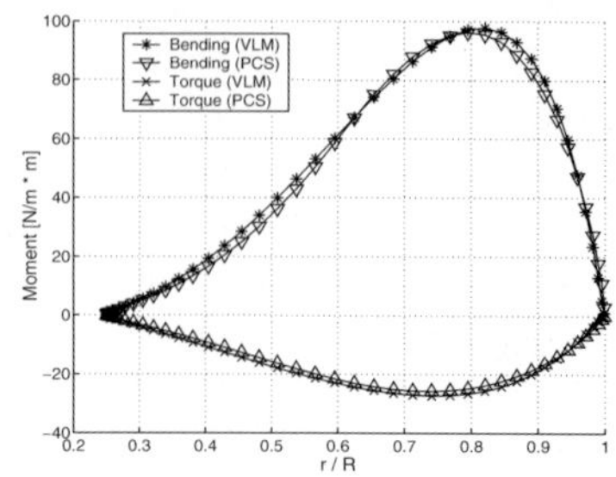

Fig. 6. Spanwise Moments (Viscous)

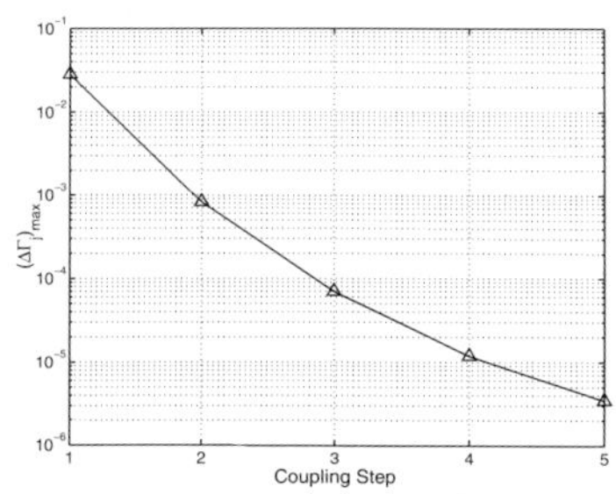

Fig. 7. Convergence of PCS (Viscous)

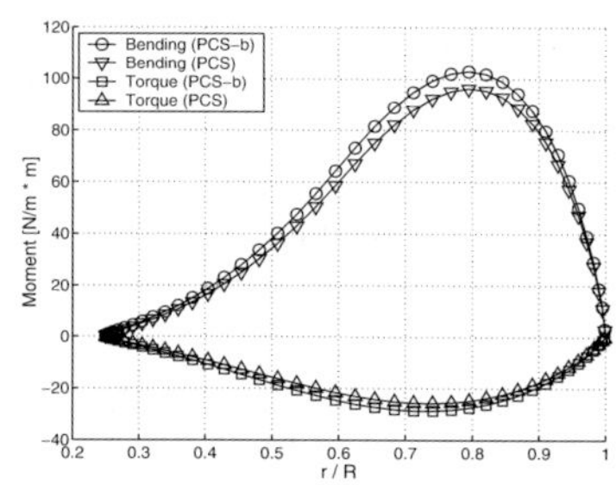

Fig. 8. Spanwise Moments 2 (Viscous)

References

1. Steinhoff, J., Underhill, D., Modification of the Euler Equations for Vorticity Confinement: Application to the computation of interacting vortex rings, *Physics of Fluids*, Vol. 6, 1994, pp. 2738-2744
2. Karypis, G., Kumar, V., Parallel Multilevel k-way Partitioning Scheme for Irregular Graphs, University of Minnesota, Department of Computer Science, Technical Report 96-036
3. Fingersh, L. J., Simms, D., Hand, M., Jager, D., Cotrell, J., Robinson, M., Schreck, S., Larwood, S., Wind Tunnel Testing of NREL's Unsteady Aerodynamics Experiment, *AIAA-2001-0035*, pp.194-200
4. Chattot, J. J., Design and Analysis of Wind Turbines Using Helicoidal Vortex Model, *Journal of Computational Fluid Dynamics*, April 2002
5. Chattot, J. J., Optimization of Wind Turbines Using Helicoidal Vortex Model, *Journal of Solar Energy Engineering*, Vol. 125, November 2003

Matematical Modeling of Supersonic Turbulent Flows in Inlets with Rotating Cowl

Bedarev I.A.[1], Fedorova N.N.[1], Goldfeld M.A.[1], and Falempin F.[2]

[1] Institute of Theoretical and Applied Mechanics, Novosibirsk, 630090, Russia
nfed@itam.nsc.ru
[2] MBDA, Chatillon, France

1 Introduction

A lot of key issues have to be addressed for developing the dual mode ramjet technology. One of them consists in designing an inlet able to operate in a very large Mach number range with enough pressure recovery while providing the right air mass-flow to the combustion chamber. The use of a sophisticated variable geometry simplifies the aerodynamic design of the inlet. But, in the same time, the mechanical feasibility of the inlet is more and more questionable considering the very severe environment the inlet will encounter during the flight. At the contrary, a fixed geometry inlet will be much more feasible from the mechanical point of view but it will lead to a very difficult trade–off for the choice of the design Mach number. The present study is related to an intermediate solution, which uses a simple geometry variation (single rotation of the cowl around an axis placed upstream of the minimum section) to adapt the contraction ratio to the flight conditions in the Mach number range from 2 to 8.

The computer design tools gave a preliminary design of inlet, which predicted performances fulfill engine requirements. Automatic designing performed in the Center for Computation of Rutgers University (USA) in cooperation with MBDA France and United Technologies Research Center (USA) permits one to get the hypersonic aircraft inlets shapes which must guarantee inlet start and effective operation within the wide range of Mach numbers [1].

The model design is made in such a manner that the motion of regulated surface may be carried out during wind tunnel operation in order to ensure the inlet start in the wind tunnel. Then, the definition of a first test program led to chose the main options for the inlet model conception. The results of experimental investigations for different Mach numbers [2], however, have shown a significant discrepancy between the engine characteristics and anticipated values, in particular, in total pressure losses.

2 Channel Configurations and Experimental Conditions

The experimental model presents a channel with flat walls and 3D forebody. The quasi–two–dimension configuration of the forebody allows one to simulate a boundary layer developed on the aircraft nose part and guarantees preliminary flow deceleration in two shock waves. The variation of inlet geometry was performed by rotating cowl [2]. Several model configurations with different cowl inclination angle have been considered (Fig. 1) for the corresponding Mach numbers to provide the inlet "start". Numbers 1 to 7 in Fig.1 denote the cowl positions relevant to the angles $\theta = 12^o, 10^o, 5.5^o, 2.5^o, 2^o, 0^o$ and -1° that were chosen as basic configuration for the flows studied.

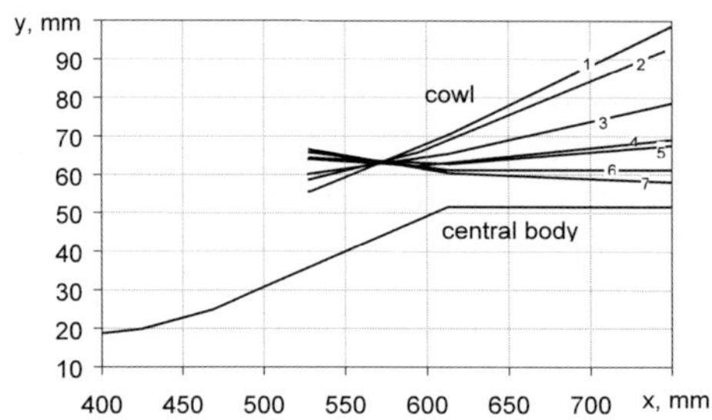

Fig. 1. Channel configuration

The experimental investigations of the model have been carried out in the ITAM blow down wind tunnel T–313 at Mach numbers from 2 to 6 and in the hot–shot wind tunnel IT–302 at Mach numbers from 6 to 8. The conditions of the experiments simulated in the present paper are given in Table 1. The conditions in the hot–shot facility were changing during the run. The flow Mach number was constant but the Reynolds number can vary by a factor of 3 to 5, as well as the temperature of the walls. In the computations, the fixed conditions were used that correspond to some median run moment.

3 Mathematical Model and Calculation Method

The calculations have been performed on the basis of the full Favre–averaged Navier–Stokes equations and Wilcox two equation turbulence model. The original numerical algorithm [3] used earlier for the 2–D turbulent supersonic flow computations was extended to a new class of internal flows. The details of the computation algorithm could also be found in [4]. In the input section, the incoming flow conditions and the profiles of all gas dynamic and turbulent parameters obtained from the computation of the turbulent boundary

Table 1. Conditions of the experiments

M	P0, bar	T0, K	$Re_1, 10^6/m$
2	2	280	28
3	4.2	280	36
4	10.5	270	54
5	8	373	17
6	8.8	540	9.4
6	$20 \div 45$	$1500 \div 1780$	$3 \div 6$
7	$50 \div 140$	$1300 \div 1700$	$6 \div 13$
8	$100 \div 215$	$1600 \div 2000$	$6.5 \div 12$

layer on the plate have been assigned. The coincidence with experimental boundary layer integral parameters and skin friction coefficient has been provided in computations if such data were available from the experiments. The no–slip conditions for velocities have been specified on inlet walls. Adiabatic condition for temperature was used in the computation for $M = 2 \div 6$ corresponding T–313 facility and constant temperature ("cold wall") for $M = 6 \div 8$ computations in condition of IT–302.

In inlets, many types of Shock Wave /Boundary Layer Interactions (SWBLI) take place, making the flow field picture rather complex. The verification of the turbulence model and computation methods have been carried out for simple SWBLI configurations such as forward– and backward– facing steps [3],[4], impinging shock wave [5] and single ramp [6] etc.

4 Inlet flow simulations

The computation were performed for the inlet configurations shown in Fig. 1 under the condition of experiments described in Tables 1. Computations for Mach 2 to 6 carried out under the conditions of T–313 facility indicated a big lengthwise pressure non uniformity and high pressure loads that increase together with Mach number (Fig. 2a). For $M = 4$ a saw–toothed structure appears in the pressure distribution related to the alternation of shock waves and rarefaction waves inside the channel. The increment of shock wave intensity with Mach number rise results in the formation of an separation region on the channel bottom wall that could be seen in the Fig. 2b where the skin friction distributions along the central body are presented. The separation region covers the channel section and causes the significant pressure rise. The static pressure distributions on the model bottom and top walls for $M = 5$ given in Fig. 3 has demonstrated the good agreement between computed and experimental data.

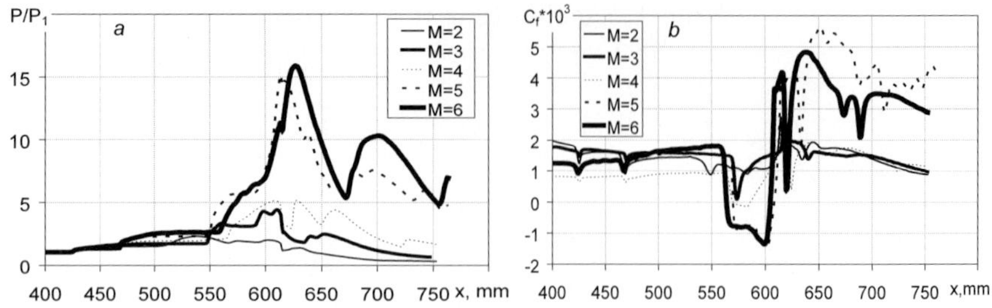

Fig. 2. The computed pressure (a) and skin friction (b) distributions along the centerbody for varios Mach numbers

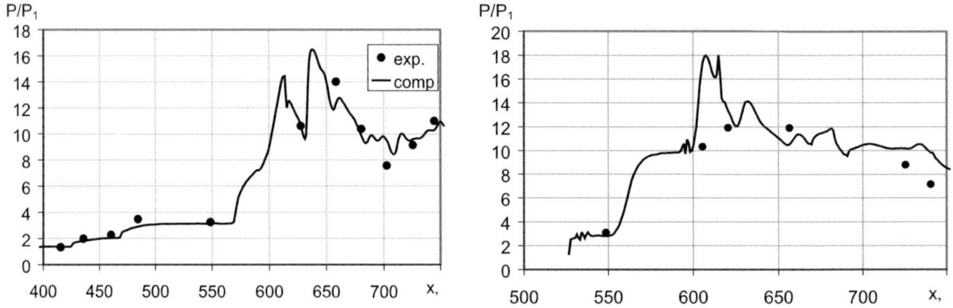

Fig. 3. The experimental and computed pressure distributions along the centerbody (a) and the cowl (b) for $M = 5$

The flow in inlet has a very complex structure because of multiple shock wave and expansion fan / boundary layer interactions which is difficult to analyze on the basis of only measurements and even visualization. And here mathematical modeling gives the instrumentation that helps one to construct the flowfield picture.The gas–dynamic patterns of the flows constructed on the basis on a joint analysis of the computation and the visualization and measurements gives a deep understanding of the processes in inlet and helps one to design the efficient devices. The computations have shown that sudden flow regime modification takes place when Mach number decreasing from $M = 8$ to $M = 6$ that leads to a sharp pressure rise inside inlets. The process may result in an inlet "unstart". The Mach number range which is the most dangerous for "unstart" is from 6 to 7.

In Fig. 4, a, b, the computed pressure distributions along the center body and the cowl are presented for $M = 7$ comparing to the experimental data. The figure shows that a very high static pressure level is observed in the experiments predicted rather well by computations. Figure 5 presents the computed static contours for this case. Here 1 and 2 denote the shocks from the external compression ramps and 3 is the shock from the cowl edge. These three shocks stick together forming the single compression wave system that further downstream travels along the channels refracting from its upper and lower walls. When the Mach number decreasing, the point where this compression wave system first comes to the center body, moves upstream. The change of the flow regime is observed when this point reaches the throat (4), and the angle between the shock and the surface changes suddenly. Downstream the throat, the shock reflects from the surface regularly, but before the throat we have the irregular reflection with a Mach stem. In this situation, the pres-sure gradient is much stronger, that results in a flow separation. The separation zone denoted by 5 in the picture, cover the channel that leads to a pressure rise inside the inlet.

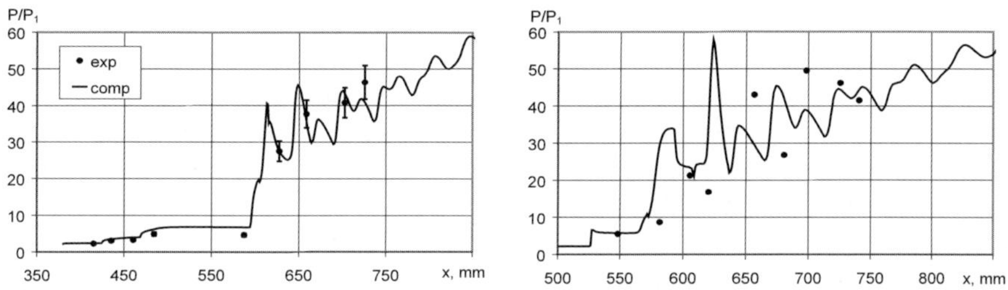

Fig. 4. The experimental and computed pressure distributions along the center-body (a) and the cowl (b) for $M = 7$

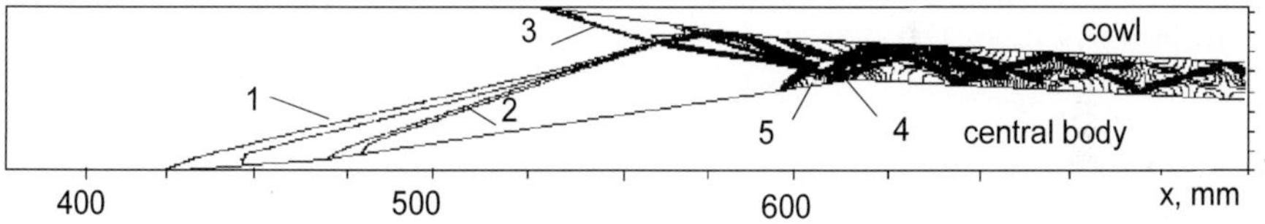

Fig. 5. The computed pressure contours for $M = 7$

The experimental and computed pressure recovery coefficient for inlets of various geometry are shown together with USA standard curve MIL–500 in Fig. 6. This data demonstrates that the experiments underpredict values whereas the computations get the values over the standard curve. The double and triple points at the same Mach number show the results for different experimental runs and inlet configuration studied.

5 Conclusions

Thus, the methods of physical experiment and computational modeling have been used in this paper to study the properties of complex flows in the inlets

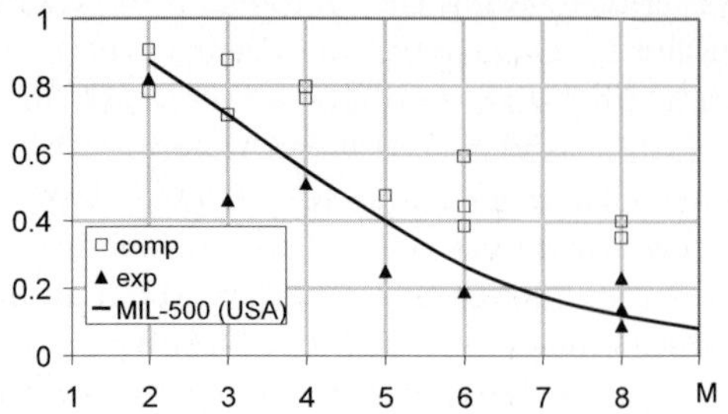

Fig. 6. The experimental and computed pressure recovery coefficient

with rotating cowl in a wide range of Mach number. The computations were performed for the Mach number 2 to 8 and the results were compared to experiments carried out in the blow down and hot shot wind tunnels. The experimental findings present a basis for the mathematical model and calculation algorithm verification. At the same time, the parametric computations performed within the wide range of flow and geometric parameters help one to carry out the experiment and provide a basis for the choice of optimum configurations and explanation of flow features.

Acknowledgements

These investigations has been supported by MBDA (France) and European Commission within the frame of ISTC project # 887.

References

1. Bourdeau C., Blaize M. and Knight D.: *AIAA Paper 99–0611.*
2. Falempin F., Goldfeld M., Nestoulia R., Starov A.: *AIAA Paper No 2002-5232.*
3. Borisov A.V., Fedorova N.N.:*Thermophisics and Aeromechanics*, **4**, 1, pp. 69–83 (1996).
4. Bedarev I.A., Borisov A.V., Fedorova N.N.: *Comp. Fluid Dyn. Journ.*, **10**, Special Number, pp. 194 – 202 (2001).
5. Fedorova N.N., Fedorchenko I.A., Shuelein E.: *Comp. Fluid Dyn. Journ.* **10**, 3, pp. 376–381 (2001).
6. Bedarev I.A., Fedorova N.N.: *First Int. Conf. CFD* (Kyoto, Japan, 2000) pp. 343–344.

Computational Fluid Dynamics of Crossflow Filtration in Suspension-Feeding Fishes

A. Y. Cheer[1], S. Cheung[2], and S. L. Sanderson[3]

[1] Dept. of Math, UC, Davis, CA. cheer@math.ucdavis.edu
[2] NASA Ames Research Center, Moffett Field, CA. cheung@nas.nasa.gov
[3] Dept. of Biology, College of William and Mary, Williamsburg, VA. slsand@wm.edu

Summary. Suspension-feeding fishes such as herring and anchovies engulf particle-concentrated water through their mouths and release the water through the posterior oral cavity. Food particles are separated from the water at the gill rakers that act like modern crossflow filters. This paper uses Computational Fluid Dynamic (CFD) techniques to study the feeding mechanism of these suspension feeders. By understanding how food is separated from the water, we can elucidate why fish gill rakers do not get clogged with particles in the same manner that industrial crossflow filters eventually become fouled.

Key words: Crossflow filtration, Suspension feeding

1 Introduction

Suspension-feeding fishes filter small particles from enormous volumes of water that enter their mouth. These fishes, belonging to 21 families in 12 orders, are critical components of many ecological communities [1] and comprise approximately 25% of the annual world fish catch [2]. As the details of oral cavity structure are very complex, we focus on the morphological features that are generally considered to serve the most critical functions during suspension feeding [3]. During suspension feeding, water that enters the mouth travels posteriorly to exit the oral cavity via branchial slits between the gill arches on both sides of the head. The gill rakers are finger-like projections of the gill arch on the opposite side from the gill filaments. Figure 1 shows the arrangement of the gill rakers and gill arch of a bony fish [4].

In this paper, CFD techniques are used to study the feeding mechanism of suspension-feeding fishes [5-6]. Gill rakers have been postulated to function as a 'dead-end' filter, collecting small food items by sieving [7] or by hydrosol filtration, in which particles smaller than the size of the filter pores stick to the filter's elements [8]. A potential problem with dead-end or hydrosol filtration, however, is that the filter gets clogged as tiny food particles become trapped on the gill rakers. Moreover, Sanderson, et al. [9] observed that more than 95% of food particles do not come into contact with the gill rakers but instead remain suspended in fluid flows parallel to the filter surface as water leaves between the gill rakers. Particles become more concentrated as they travel towards the esophagus, and the fish swallows a concentrated slurry of food particles. Thus, the gill rakers act like a modern crossflow filtration [10] mechanism similar to those in our water treatment plants. However, unlike most industrial crossflow filtration, particles do not accumulate on the gill rakers. Sanderson *et al.* [9] estimated the value for one theoretical transport mechanism— radial inertial migration — and found that it is inadequate, by an order of magnitude, to explain particle transport in fishes that feed by crossflow filtration. The underlying physical mechanisms are not yet fully understood. Our goal is to investigate the underlying mechanism of crossflow filtration in suspension feeders.

In this study, a 3-D Navier-Stokes flow solver is used to calculate the flow field around the fish rakers and small spherical food particles are released into the flow field so that we can track the particle trajectories. This enables us to compare the effects of different flow conditions (speed and incident angles) on particles of different size and mass inside the oral cavity of the fish. For these calculations, we assume Newtonian flow since the particles we are simulating are small enough that they do not influence the flow field.

2 Overflow

In this study, the OVERFLOW code developed at NASA is used to solve the Navier-Stokes (N-S) equations for structured Chimera overset grids (see Figure 2). This code is an outgrowth of the previous computer codes named ARC3D [11] and F3D [12] which have various options of numerical algorithms. Local time step scaling, grid sequencing and multi-grid are also implemented for convergence acceleration. Our calculations require the Low-Mach number preconditioning technique where the resulting matrix equation is solved using the Pulliam-Chaussee diagonalized (scalar pentadiagonal) scheme [13] and the Roe upwind scheme [14].

3 Particle Traces Algorithm

The following algorithm is used to trace the trajectory of food particles that enter the oral cavity of the fish. The velocity field $v(\mathbf{x})$ is obtained from solving the N-S equations and the initial position of a particle $\mathbf{x}(0)$ is prescribed. The particle path $\mathbf{x}(t)$ is then obtained by first solving the following equation $\frac{d\mathbf{x}}{dt} = v(\mathbf{x})$. Discretizing this equation using the second-order Runge-Kutta integration scheme with adaptive step sizing [15], we get

$$\begin{aligned} \mathbf{x}^* &= \mathbf{x}^k + dt v(\mathbf{x}^k) \\ \mathbf{x}^{k+1} &= \mathbf{x}^k + dt \tfrac{1}{2}(v(\mathbf{x}^k) + v(\mathbf{x}^*)) \end{aligned} \quad (1)$$

The value of dt is given by $dt = c/\max(V_x, V_y, V_z)$; Vx, Vy, and Vz are the three components of the velocity v; and $0<c<1$. The constant c is used for adaptive step sizing and controls the number of steps that x will advance in each grid cell. Adaptive step sizing is required when the velocity varies rapidly in some regions of the grid. The algorithm described above assumes that the flow is steady (time-independent).

When the particle has finite mass, additional formulas are needed due to Newton's Second Law. Let m be the mass of the particle and let $\mathbf{F}(\mathbf{v})$ be the force applied to the particle. Newton's Second Law states that $\frac{d\mathbf{v}}{dt} = \frac{1}{m}\mathbf{F}(\mathbf{v})$. Again using the second-order Runge-Kutta integration scheme, we have

$$\begin{aligned} \mathbf{v}^* &= \mathbf{v}^{k-1} + dt\mathbf{F}(\mathbf{v}^{k-1})/m \\ \mathbf{v}^k &= \mathbf{v}^{k-1} + dt\tfrac{1}{2}(\mathbf{F}(\mathbf{v}^{k-1}) + \mathbf{F}(\mathbf{v}^*))/m \end{aligned} \tag{2}$$

Thus for particle with finite mass, the trajectory (path) is obtained by combining (1) and (2). The values of $v(\mathbf{x}^k)$and $v(\mathbf{x}^*)$ in (1) are replaced by $\mathbf{v}^k$and $\mathbf{v}^*$in (2) respectively.

The forces applied to a moving particle of velocity v at a given instant are the sum of the buoyant force, gravitational force, and drag. Thus, the force vector $\mathbf{F}$ is expressed by

$$\mathbf{F}(\mathbf{v}) = \underbrace{-V\rho_w g\hat{\mathbf{z}}}_{buoyant\ force} + \underbrace{V\rho_p g\hat{\mathbf{z}}}_{gravitation} \underbrace{+C_D\tfrac{1}{2}\rho_w S\left|w\right|^2\hat{\mathbf{w}}}_{drag\ force} \tag{3}$$

where V is the volume of the particle, $g\hat{\mathbf{z}}$ is the gravitation acting in the $\hat{\mathbf{z}}$ direction, ρ_w is the water density, and ρ_p is the particle density. The drag force is a function of water density (ρ_w), particle wetted area (S), drag coefficient (C_D) of the particle in water, and the traveling velocity (w, is the fluid velocity u minus particle velocity v; $\hat{\mathbf{w}}$ is the unit vector). The mass (m) of the particle is $V\rho_p$, thus

$$\frac{\mathbf{F}(\mathbf{v})}{m} = (1 - \frac{\rho_w}{\rho_p})g\hat{z} + \frac{C_D\frac{1}{2}\rho_w S\left|w\right|^2\hat{\mathbf{w}}}{m} \tag{4}$$

Note that if the particle density is the same as the water density, the buoyancy and the gravitation terms cancel and the first term in (4) is zero.

In order to calculate the particle trace under the influence of the flow pattern $v(t)$, we need to calculate the drag coefficient (C_D). To do this, consider a spherical particle dropped into a tank of water. It will reach a constant velocity, called terminal velocity (v_t). At terminal velocity, the submerged weight of the particle is the same as the drag (D) on the particle. The submerged weight is no more than the difference of buoyancy and gravity; that is the first two terms in (3). Therefore, from (3), $V\rho_p g - V\rho_w g = D$.

Since the food particles are very small (~100 microns) and the Reynolds number is very low (<0.2), we can use the well-known Stokes relation for drag [16], $D = 6\pi\mu r v_t$. Here r is the radius of the sphere and μ is the absolute viscosity of the fluid. Therefore,

$$C_D = \frac{D}{\frac{1}{2}\rho_w S v_t^2} = \frac{V\rho_p g - V\rho_w g}{\frac{1}{2}\rho_w S v_t^2} = \frac{(6\pi\mu r)^2}{\frac{1}{2}\rho_w S V g(\rho_p - \rho_w)} \tag{5}$$

4 The Simulation

To simulate flow through the rakers of a generalized suspension-feeding fish, the velocity field is calculated through an infinite array using periodic boundary conditions on a finite array of elliptical cylinders (Fig. 2). Each elliptical cylinder has dimensions 375 microns in major axis diameter, 250 microns in minor axis diameter. The gap between two cylinders is 250 microns. The suspended food particles are represented in our calculations by spheres with radius 82.5 microns ($8.25 \text{x} 10^{-5}$ m) and density (ρ_p) is 1.05gm/cc (1050kg/m3). The water density (ρ_w) and viscosity (m) are 1000kg/m3 and 0.001kg/ms respectively. The gravitation constant (g) is 9.81, the volume $(4/3)\pi r^3$ and the wetted area of the particle is πr^2. With this information, the drag coefficient of the particles can be obtained from Eqn. 5.

5 Computation Results

Figures 3a and b show a 2-D view of a 3-D calculation of flow at 0.6m/s with incident angle 75 and 85 degrees, respectively for an infinite array. The upper figures show the velocity contours and the lower figures show the streamlines. The vortex between the rakers in Fig. 3b suggests that it may be an impediment to particle movement between the rakers in contrast to Fig. 3a where the center of the large vortex is not in the gap.

The differences in vortex structure at the two different incident angles are consistent because complex vortex structures are always created in the aft area of an obstacle and the circled areas in Fig. 3 in front of the leading edge of the ellipse indicate the highest velocity region.

When a particle with density ρ_p =1.05gm/cc is introduced into the flow, its trajectory deviates from the streamlines as shown in Fig. 4 and it tends to keep a larger distance from the leading edge of the raker. This means that the particle tends to travel with the high velocity flow rather than escaping through the gap between rakers. Examination of the velocity vector shows a very slow recirculating flow inside the gap between the rakers. In contrast, the flow in front of the rakers accelerates to 0.682 m/s which is higher than the free-stream velocity of 0.6 m/s at the leading edge of the rakers. Since the force applied to a particle is tangential to the velocity vector and since there is a low circular velocity field in the gap parallel to the high velocity flow outside, food particles will not cross high flow regions to escape between the gaps. Thus, although the gap is large (250 μm) compared to the particle size, the effective gap is small, only about 50 μm, indicating that particles larger than 1/5 of the gap will not tend to escape.

Following the trajectory of two particles starting at the same position, one massless particle that follows the streamlines and one with density ρ_p =1.05gm/cc that deviates from the streamlines, we note that food particles with drag will travel further downstream (Fig. 5). This is consistent with

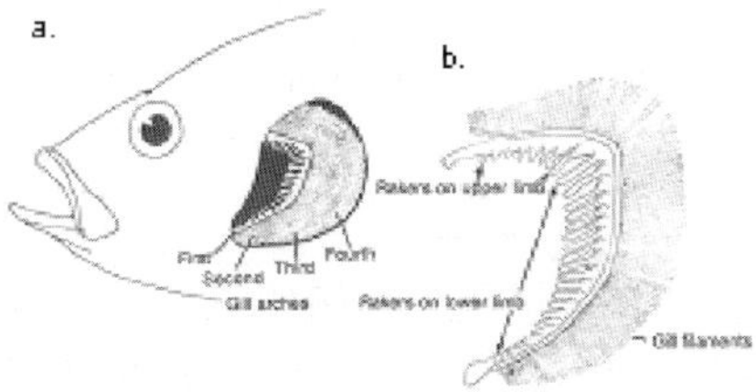

Fig. 1. Structure of gill rakers and gill arches: (a) the position of gill arches inside the gill cavity; (b) a side view of the first gill arch. (Miler and Lea 1972)

Fig. 2. The Computational Grids

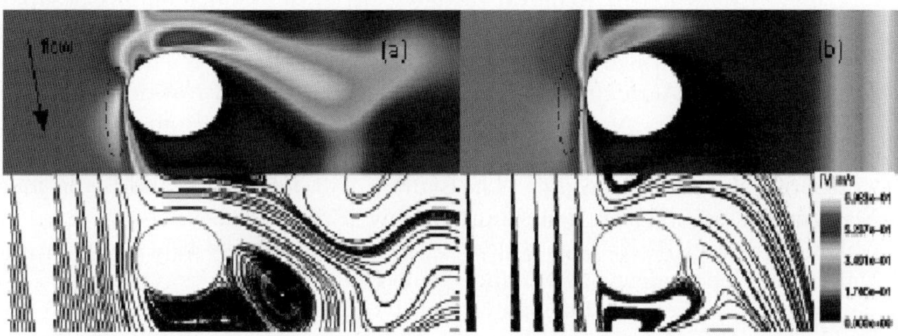

Fig. 3. a, b Steady flow solutions with incident angle of 75 and 85 degrees

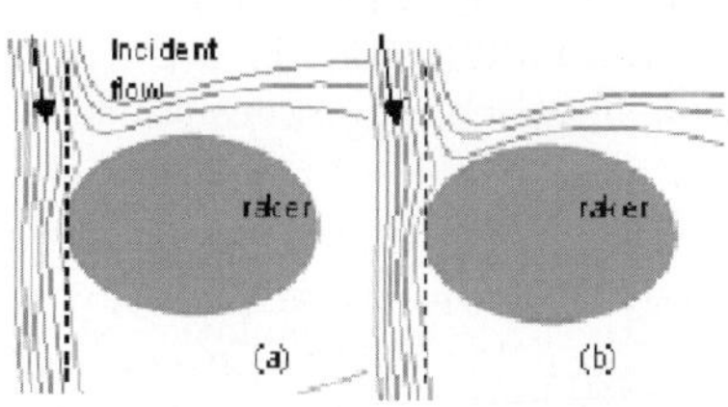

Fig. 4. A side-by-side comparison of particle traces of massless particles (a) and finite-weight particles (b). The incident angle is 85 degrees

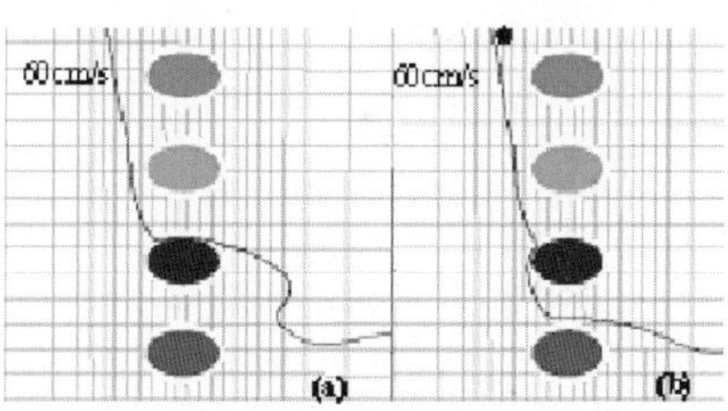

Fig. 5. Flow with speed 60cm/s. (a) A massless particle is released at a point upstream; it passes betweeen 2nd and 3rd rakers. (b) A particle with drag is released at the same physical point; it travels further downstream before exiting between 3rd and 4th rakers

the hypothesis that the gill rakers act like a crossflow filter rather than a dead-end filter and with the observations of Sanderson et al. [9].

Acknowledgements

This study is supported in part by NSF grant DMS-0079760, DOE: DE-FG03-99ER25375 and UCD Fellowship to A.Y.C. and by NSF grant IBN-0131293 to S.L.S.

References

1. Gottlieb, S.J., Nutrient Removal by age-O Atlantic Menhaden in Chesapeake Bay and Implications for Seasonal Management of the Fisheries. Ecol. Model **112**, pp. 111-130. 1998.
2. FAO (Food and Agriculture Organization of the United Nations) Year Book of Fishery Statistics, Vol. **86**, Capture Production, 1998. Rome:FAO. 2000.
3. Sanderson, S.L. and Wassersug, R. Convergent and Alternative designs for Vertebrate Suspension Feeding. The Skull, Vol. 3, Functional and Evolutionary Mechanisms, pp. 37-112. The Chicago Univ. Press, 1993.
4. Miller D. J. and R. N. Lea. Guide to the coastal marine fishes of California. California Department of Fish and Game Bulletin. 1972.
5. Gerking, S. D., Feeding Ecology of Fish (Academic, San Diego, 1994).
6. Sanderson, S. L., Cech, J. Jr., and Patterson, M. R., Fluid dynamics in suspension-feeding blackfish. Science 251, 1346-1348 (1991).
7. Hoogenboezem, W., Lammens, E. H. R. R., MacGillavry, P. J. & Sibbing, F. A. Prey Retention and Sieve Adjustment in Filter-Feeding Bream (Abramis brama) (Cyprinidae). *Can. J. Fish. Aquat. Sci.* **50,** 465–471 (1993).
8. Sanderson, S. L. & Wassersug, R. *Sci. Am.* **262,** 96–101 (1990).
9. Sanderson, S. L., Cheer, A., et al., Crossflow filtration in suspension-feeding fishes. Nature 412, 439-441 (2001).
10. Scott, K. & Hughes, R. (eds), *Industrial Membrane Separation Technology* (Blackie, London, 1996).
11. Pulliam, T. H. and Steger, J. L., Implicit Finite Difference Simulation of 3-D Compressible Flow. AIAA J., Vol 18, No. 2, 1980 pp.159-167.
12. Steger, J. L., Ying, S. X., and Schiff, L. B., A partially Flux Split Algorithm for Numerical Simulation of Compressible Inviscid and Viscous Flow. Proceedings of Workshop on Computational Fluid Dynamics, UC, Davis, CA 1986.
13. Pulliam, T.H. and Chaussee, D. S., A Diagonal Form of an Implicit Approximate Factorization Algorithm. J. Comp Physics Vol. 39, 1981, pp. 347-363.
14. Roe, P. L., *Approximate Riemann solvers, parameter vectors and difference schemes,* J. of Comp. Phys., Vol. 43, 1981, pp. 357-372.
15. Lane, D., UFAT - a Particle Tracer for Time-Dependent Flow Fields, in: D. Bergeron and A. Kaufman, eds., *Proceedings of Visualization '94,* Washington, D.C., October 1994.
16. Batchelor, G.K., An Introduction to Fluid Dynamics (Cambridge University Press, 1983).

Numerical Simulation of R-M Instability

Fu Dexun, Ma Yanwan, and Tian Baolin

LNM Institute of Mechanics, Chinese Academy of Sciences, Beijing, China

Summary. In order to capture shock waves and contact discontinuities in the field and easy to program with parallel computation a new algorithm is developed to solve the N-S equations for simulation of R-M instability problems. The method with group velocity control is used to suppress numerical oscillations, and an adaptive non-uniform mesh is used to get fine resolution. Numerical results for cylindrical shock-cylindrical interface interaction with a shock Mach number Ms=1.2 and Atwood number A=0.818, 0.961, 0.980 (the interior density of the interface/ outer density $\rho_1/\rho_2 = 10, 50, 100$, respectively), and for the planar shock-spherical interface interaction with Ms=1.2 and $\rho_1/\rho_2 = 14.28$are presented. The effect of Atwood number and multi-mode initial perturbation on the R-M instability are studied. Multi-collisions of the reflected shock with the interface is a main reason of nonlinear development of the interface instability and formation of the spike-bubble structures In simulation with double mode perturbation vortex merging and second instability are found. After second instability the small vortex structures near the interface produced. It is important factor for turbulent mixing.

1 Introduction

Consider a material interface between two different media. As an incident shock interacts with the material interface the interface becomes unstable due to shock acceleration. The small disturbances at the interface start to grow. This kind of instability is called as a Richtmyer-Meshkov instability(R-M instability). The R-M instability problems can be met in many important practical applications, for example, the propagation of sound boom in turbulent atmosphere, the Inertial Confinement Fusion(ICF), and the explosion of the supernova. The R-M instability is also a model problem for studying the physical mechanism of fluid motion from instability to turbulence. Many works have been done for studying R-M instability, but mainly for the case of interaction between plane shock and plane interface with perturbation, interaction between plane shock and cylindrical interface with single mode perturbation. In order to get numerical results of the R-M instability with high density ratio a new algorithm is developed in this paper, and it is used to simulate cylindrical shock-cylindrical material interface interaction and the planar shock-spherical interface interaction. The shock is going from the heavy gas through the interface to the light gas. The flow structures of shock refraction, reflection, interaction of the reflected shock with the material interface, and effect of Atwood number and the initial perturbation modes on R-M instability are investigated.

2 Numerical Method

A sixth order traditional difference approximation with group velocity control(GVC) is used to simulate the physical problem. Consider the convection term for x-direction in the N-S equations $\partial f/\partial x$ discretized with $F_j/\triangle$ for positive characteristics:

$$F_j^{(6,+)} = H_{j+1/2}^{(6,+)} - H_{j-1/2}^{(6,+)} \quad (1)$$

$$H_{j+1/2}^{(6,+)} = \frac{1}{2}[1 + SS(u_{j+1/2})]\overline{h}_{j+1/2}^{(6,+)} + \frac{1}{2}[1 - SS(u_{j+1/2})]h_{j+1/2}^{(6,0)} \quad (2)$$

$$\overline{h}_{j+1/2}^{(6,+)} = (1+\sigma)h_{j+1/2}^{(7,+)} - \sigma h_{j+1/2}^{(6,0)} \quad (3)$$

$$h_{j+1/2}^{(6,0)} = [f_{j+3} - f_{j-3} - 9(f_{j+2} - f_{j-2}) + 45(f_{j+1} - f_{j-1})]/60$$

$$h_{j+1/2}^{(7,+)} = h_{j+1/2}^{(6,0)} - \delta_x^6 f_j/140$$

$$\delta_x^6 f_j = f_{j+3} + f_{j-3} - 6(f_{j+2} + f_{j-2}) + 15(f_{j+1} - f_{j-1} - 20f_j$$

The upper index (k,i) shows that the scheme is kth order accurate, and it is symmetrical for case of i=0 and upwind biased for case i=+. The scheme with $h_{j+1/2}^{(7,+)}$is SLW, and a parameter $\sigma > 0$ is introduced to make the scheme MXD. The function $SS(u)$ is used to control the group velocity[1] .

3 Numerical results of R-M instability

3.1 Interaction of Cylindrical Shock with Cylindrical Interface

Consider R-M instability produced by interaction of a cylindrical shock with cylindrical material interface. The shock is going inward from the heavy gas with density ρ_1 into the light gas with density ρ_2 . This problem is solved by using the two dimensional compressible N-S equations in the cylindrical coordinate discretized with above presented method. The interface is tracked with time according to the maximal gradient of scale function g: $[gradg]_{max}$.

General characteristics of R-M instability

When the interface is perturbed with small amplitude the interface becomes unstable after collision of the incident shock with the interface. Now we are going to discuss the general characteristics of R-M instability with interaction between a cylindrical shock and a cylindrical material interface based on initial single mode perturbation. Consider a case with Ms=1.2 ,Re=50000,$\rho_1/\rho_1 = 10$. The initial cylindrical surface with perturbation is defined as $r_0 = 1 + \alpha\cos(n\varphi)$,where $\alpha = 0.033$, $n = 12$, and the length scale is normalized with the radius of initial interface without perturbation. Fig.1 shows the density contours at different times. We see that at t=0.10 the

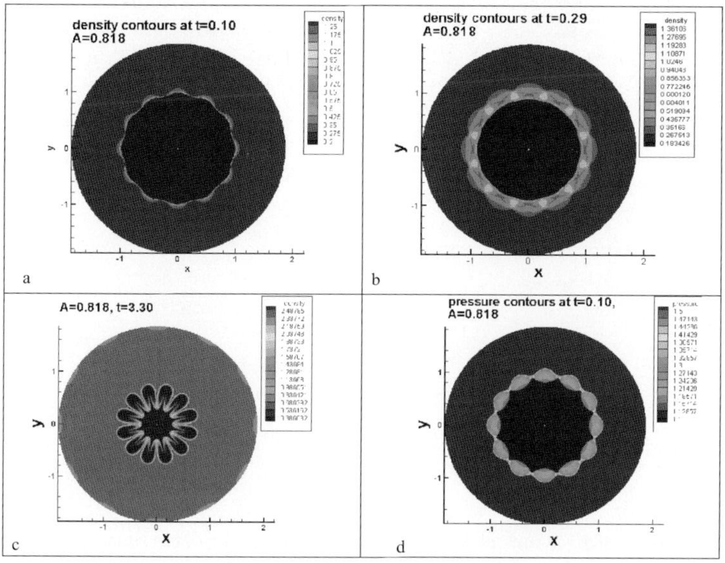

Fig. 1. Density and Pressure contours: a. Density contours at t=0.10; b. Density contours at t=.0.29 c. Density contours at t=3.30; d. Pressure contours at t=0.10.

discontinuity bifurcates into reflected rarefaction wave and refracted shock after collision with interface. The discontinuity bifurcation can be seen clearly from Fig.1d for pressure contours at t=0.10. The perturbation starts to grow linearly, and the interface starts to deform. Existence of perturbation on the interface leads to non- parallelism of the pressure gradient $[gradp]$ with the density gradient $[gradp]$ which produces the vorticity near the interface. Further development nonlinear growth of perturbation leads to change the phase of the perturbed surface(see Fig.1b). The transmitted shock reaches the center and reflects. Collision of this reflected shock with the interface increases nonlinear development of the perturbation. After multiple collision of the reflected shock with interface the spike-bubble structures are formed(Fig.1c). The spike is a portion of heavy gas penetrating into the light gas, and the bubble is a light gas penetrating into the heavy gas. fig1

Effect of Atwood number on R-M instability

The cases Ms=1.2Re=50000 with different Atwood number ($A = [\rho_1 - \rho_2]/[\rho_1 + \rho_2]$) are computed in order to study the effect of density ratio on R-M instability. The initial cylindrical surface with perturbation is defined as above. The variation of pressure with the time at the cylinder center for different Atwood number is given in Fig.2a-c. From it we can see that when the transmitted shock approaches the center the center pressure increases, and then reflected shock is formed. This reflected shock is going outward,

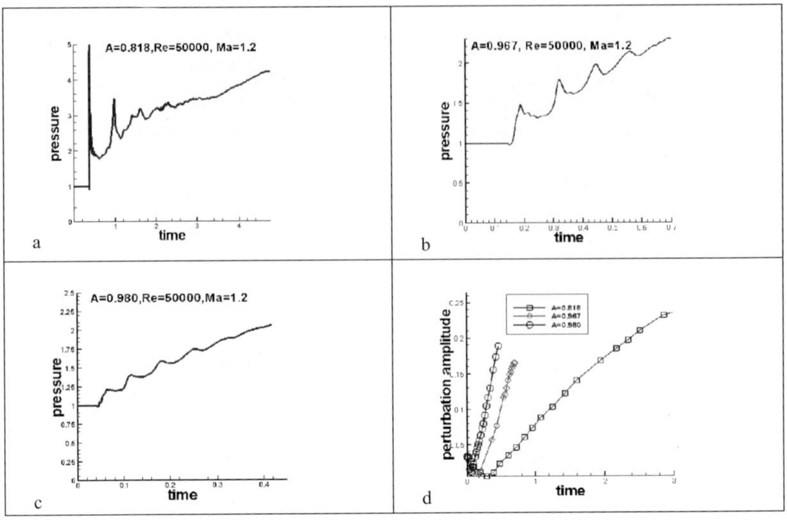

Fig. 2. Variation of pressure at the cylinder center and perturbation amplitude with different Atwood numbers (A: a. A=0.818;b. A=0.967; c. A=0.980; d: perturbation amplitude).

and the gas in central region starts to rarefy with pressure decreasing. This outward going shock in light media will collide the interface behind which the gas is heavy. After interaction an inward going reflected shock is formed. The inward going shock will reach the center, then it reflects, and it collides with the interface again.From Fig.2a we can see that there are four times of this kind of collisions for the case A=0.818. After the collisions the pressure at the center increases fast. For the case of A=0.967 and 0.980 there are five six times of this kind of collisions respectively, but the pressure increases much less than the case of A=0.818 (see Fig. 2a-c). It means that pressure increase decays with Atwood number increasing. In Fig.2d the variation of perturbation amplitude with the time for different Atwood number is given. It can be seen that at first stage before the phase change of the interface the perturbation amplitude decreases, after phase changing it is increases. The perturbation amplitude with large Atwood number grows much faster because of faster sound speed in the light gas.

Effect of initial perturbation on R-M instability

The cases Ms=1.2, Re=50000, ρ_1/ρ_2=10 with both single and double initial perturbation modes at interface are computed. The initial perturbed interface for single mode is as above. The double mode perturbed interface is given as $r_0 = 1 + a[cos(n_1\varphi) + cos(n_2\varphi)]$ where $n_1 = 12$,$n_2 = 60$, $a = 0.033$. The vorticity contours with two different initial perturbation mode are given in Fig.3. Comparing the results with different initial perturbation modes we

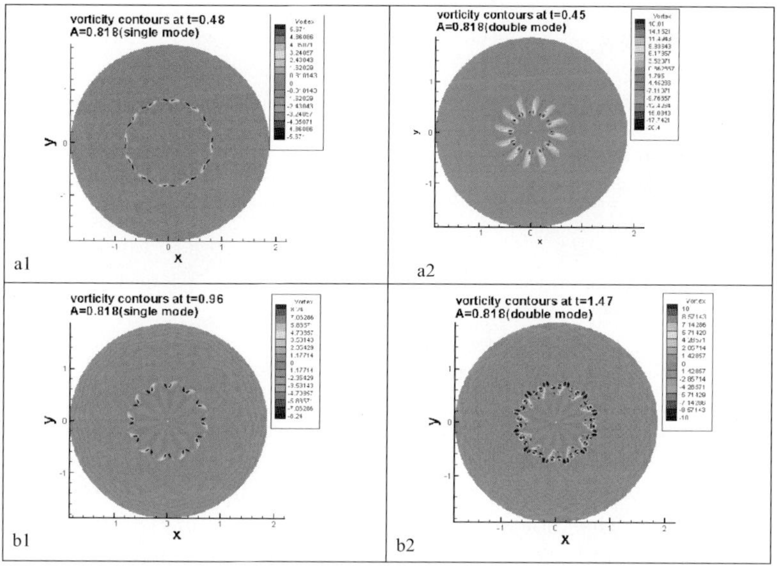

a. Single mode b. Double mode

Fig. 3. Vorticity contours with two initial mode perturbations at different time.

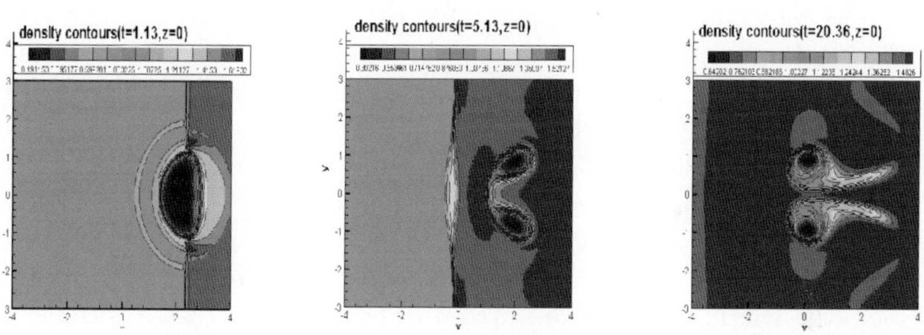

Fig. 4. Density contours on section z=0 for different time (t=1.33, 5.13, 20.36).

see that at first stage the initial subharmonic is appended on the basic mode perturbation, and vorticity is much increased . With nonlinear growth of perturbation secondary instability occurs, and vortex merging can be seen in Fig. 3a2. With further development of perturbation in Fig.3b2 we can see that small vortex structures near the interface are formed. This is important factor for turbulence development.

3.2 Interaction of Planar Shock with Spherical Interface

The numerical method presented above is used to solve the 3-D compressible N-S equations in the Cartesian coordinate for simulating interaction of a planar shock with spherical interface. The Mach number is and the density

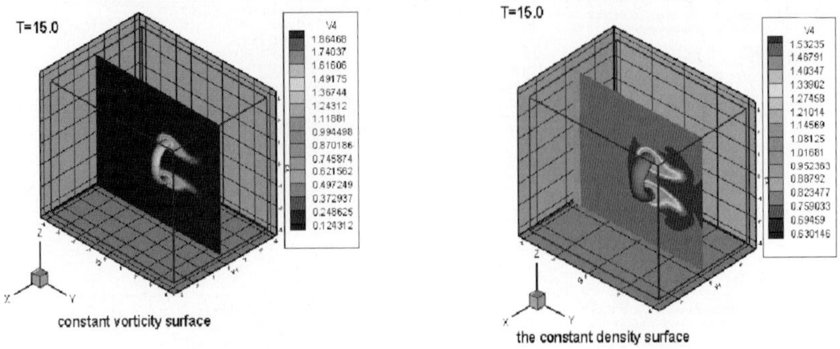

Fig. 5. Three Dimension numerical results.

ration is . Figure 4 shows density contours on section z=0 for different times (t=1.33, 5.13, 20.36). From Fig.4a it can be seen that as the shock sweeps over the interface the shape of the interface at the upstream face changes due to compression by the shock and the interfacial instability. From here also we can see that first and second transmitted waves at the left are followed by the interface and the reflect waves are going to the right. At the same time due to the shock going from the heavy gas to the light gas, the gas inside the interface has a higher wave speed, and the effect of the interfacial instability, the incident shock is deformed. This phenomenon can be seen from the pressure contours (there is not presented). From Fig.4b it can be seen that after interaction due to interface instability the heavy air jet with conical shear layer is formed, then this jet impinges on the downstream interface and pierces it (see Fig. 4c). During the time of interaction the vorticity is produced due to non-parallelism of the gradients for pressure and density. Because the shock is incident from the right to the left and the light gas is relatively easier to accelerate, clockwise vorticity is produced at the top and anticlockwise vorticity is at the bottom of the interface. With further development of flow structure a vortex ring is formed. Fig.5 shows the constant vorticity surface and vorticity contours on section z=0 at t=15.0. This vortex ring for the sphere-interface is more distinct than the case for the cylinder interface. From numerical results we can also see that most of the vorticity in the flow is concentrated in the vortex ring and along the conical shear layer at the boundary of the heavy air jet. Obtained numerical results agree well with experiment results [2].

Acknowledgements

This work is supported by NSFC and SSFMSBRP (G1999032805).

References

1. Fu DX, Ma YW. J. Compt. Phys. 134, 1997,1-15.
2. Haas JF, Sturteavant B. J. Fluid Mech. 181,1987, 41-76

Thrust and Efficiency of Propulsion by Oscillating Foils

J. Young[1], J.C.S. Lai[1], M.Kaya[2], and I.H. Tuncer[2]

[1] School of Aerospace, Civil and Mechanical Engineering, UNSW@ADFA, Australian Defence Force Academy, Canberra, ACT, 2600, Australia, j.young@adfa.edu.au, j.lai@adfa.edu.au

[2] Department of Aerospace Engineering, Middle East Technical University, 06531, Ankara, Turkey, mkaya@ae.metu.edu.tr, tuncer@ae.metu.edu.tr

1 Introduction

Oscillating foils are being considered for propulsion of submersible vehicles and micro air vehicles (MAVs), at low Reynolds numbers where conventional propulsion techniques become inefficient. Linearized theories and potential flow methods provide good estimates of thrust and propulsive efficiencies at higher Reynolds numbers, but cannot account for the leading-edge flow separation and vortex formation that can occur in viscous flows. A 2D unsteady Navier Stokes solver was used to explore the effect of flow separation on the thrust and efficiency of an airfoil undergoing various combinations of sinusoidal pitching and plunging motions.

The Navier Stokes solver has been validated against a number of flow visualizations for pure plunging motion, and quantitative experimental and computational studies of pure pitching motion, with good results [1]. Here combined pitching and plunging motions are computed, with thrust and efficiency results compared to small-amplitude potential theory (Garrick [2]), large-amplitude unsteady panel method (UPM) potential flow simulations [1, 3], and experimental results of Anderson et al [4].

2 Numerical Method

The unsteady flow field around the oscillating airfoil is calculated using a 2nd-order accurate unsteady 2D compressible Navier Stokes solver. Motion of the airfoil is introduced by a combination of rigid-body motion and deformation of the grid, resulting in time-dependent grid metrics and Jacobian. Details of the method may be found in Young and Lai [1, 3], and Tuncer and Platzer [5]. Versions of the code have been parallelized with OpenMP for use on an Alpha cluster, and Parallel Virtual Machine (PVM) for use on a cluster of Pentium-based workstations running Linux.

The motion of the airfoil is a combination of plunging, $y = a\sin(\omega t)$ and pitching, $\theta = \theta_0 \sin(\omega t + \phi)$. Parameters of interest are the reduced frequency

$k = \omega c/2U_\infty$, non-dimensionalized plunge amplitude $h = a/c$, pitch amplitude θ_0, pivot point of the pitching motion, and the phase angle ϕ (c is the airfoil chord and U_∞ is the free-stream velocity). The Strouhal number is calculated based on the frequency of motion and the total (peak to peak) excursion A_{TE} of the trailing edge, $St = \omega A_{TE}/2\pi U_\infty = kA_{TE}/\pi c$. Outputs include the time-averaged thrust coefficient $\overline{C_T} = -1/T \int_t^{t+T} C_D(t)dt$, input power coefficient $\overline{C_{P_{in}}} = -1/T \int_t^{t+T} [C_L(t)\dot{y}(t)/c + C_M(t)\dot{\theta}(t)]dt$, and propulsive efficiency, $\eta = \overline{C_T}U_\infty/\overline{C_{P_{in}}}$. C_D, C_L and C_M are the drag, lift and moment coefficients calculated by integrating viscous and pressure forces around the airfoil surface.

Optimization of time-averaged thrust and propulsive efficiency is performed using a gradient-descent based method. The objective function is a linear combination of the thrust coefficient and the efficiency, and the optimization variables are h, θ_0, k, and ϕ. Evaluation of the gradient vector involves unsteady flow computation over a number of motion cycles until periodic behavior is established. Each of the gradient vector components is calculated in parallel. Calculations were typically performed on a 541×61 grid, with 189 points on the airfoil, a first grid point at 8.75×10^{-6} chords from the airfoil surface, and a boundary 20 chords from the airfoil in all directions.

3 Comparison of Numerical and Experimental Results

Figs. 1, 2, and 3 show the time-averaged thrust coefficient and the propulsive efficiency vs Strouhal number for three different motions. The parameter $\alpha_0 = -\tan^{-1}(2kh) + \theta_0$ is kept constant at 15° for all cases. Reynolds number $Re = 40,000$, and the airfoil is pitching about a point 1/3 chords aft of the leading edge. These conditions were chosen to match those of the experiment performed by Anderson et al [4]. Navier Stokes calculations were performed using fully laminar flow, fully turbulent flow with the Baldwin-Lomax turbulence model, and a transition model using the Spalart-Allmaras turbulence model. Negligible differences in thrust and efficiency were found between the Baldin-Lomax and Spalart-Allmaras models. Fig. 1, for which leading edge separation is minimal across the range of Strouhal number, shows relatively good agreement with the measurements of Anderson et al [4], and little difference between laminar and turbulent flows. Figs. 2 and 3, for which large leading edge separation is apparent at some Strouhal numbers, show lower efficiencies than measured by Anderson et al, and indicate a high degree of sensitivity to the precise details of leading edge vortex formation, dependent on whether the flow is assumed laminar or turbulent. In all three cases the level of agreement with experiment is good for the thrust coefficient, and moderate for the propulsive efficiency although the trends are well predicted.

It should be noted that small discrepancies in time-varying quantities may become relatively larger errors in the time-averaged quantities.

4 Optimization of Thrust and Efficiency

Figs. 1, 2 and 3 taken together show that the propulsive efficiency of an oscillating airfoil is sensitively dependent on the phase angle between pitching and plunging motion. This is in agreement with the findings reported by Isogai et al [6] for $Re = 10^5$ and Tuncer and Kaya [7], suggesting that high efficiencies and thrusts may be simultaneously achieved by optimizing interaction of the leading edge vortex with the foil motion.

The optimization procedure was checked by allowing the frequency and pitch amplitude to vary while holding other parameters fixed, and optimizing against efficiency. The results are shown in Fig. 2. In both laminar and turbulent cases the optimized solution converges to a point close to the peak efficiency found by a parametric search. In Table 1 the plunge amplitude is held constant at $h = 0.75$, $\alpha_0 = 15°$, and the oscillation frequency and pitch amplitude are optimized. The procedure is repeated both with phase angle held constant, and optimized. Laminar and turbulent flows are considered. The table shows that the optimum phase angle between pitch and plunge motion is somewhere between $\phi = 75°$ to $85°$, again in good agreement with Anderson et al [4] and Isogai et al [6].

The effect of leading edge separation on thrust and efficiency may be seen in Figs. 4 and 5. Here numerical particle traces are shown for four points in the motion cycle for turbulent flow at two different Strouhal numbers. Fig. 4 corresponds to the highest efficiency point of Fig. 3 ($St = 0.184$, $\overline{C_T} = 0.24$, $\eta = 0.72$), and Fig. 5 is a high thrust, low efficiency case ($St = 0.70$, $\overline{C_T} = 1.22$, $\eta = 0.44$). Optimization of the flapping motion for highest efficiency appears to minimize leading edge separation. Optimization for high thrust promotes leading edge separation on the airfoil surface which is tilted upstream, so the low pressure of the separated vortex results in thrust rather than drag. The low thrust, low efficiency cases in Figs. 2 and 3 correspond to leading edge separation on the airfoil surface that is is tilted downstream.

Table 1. Optimization results, $Re = 40,000$, $\alpha_0 = 15°$, pivot at 1/3 chord, $h = 0.75$.

Case	St	k	θ_0	ϕ	η
Turbulent, ϕ fixed	0.244	0.97	21.1°	75.0°	0.722
Turbulent, ϕ fixed	0.198	0.80	16.1°	90.0°	0.705
Turbulent, ϕ optimized	0.252	1.00	21.9°	78.6°	0.726
Laminar, ϕ fixed	0.280	1.10	24.5°	75.0°	0.598
Laminar, ϕ fixed	0.287	1.12	25.0°	90.0°	0.636
Laminar, ϕ optimized	0.258	1.02	22.5°	83.3°	0.643

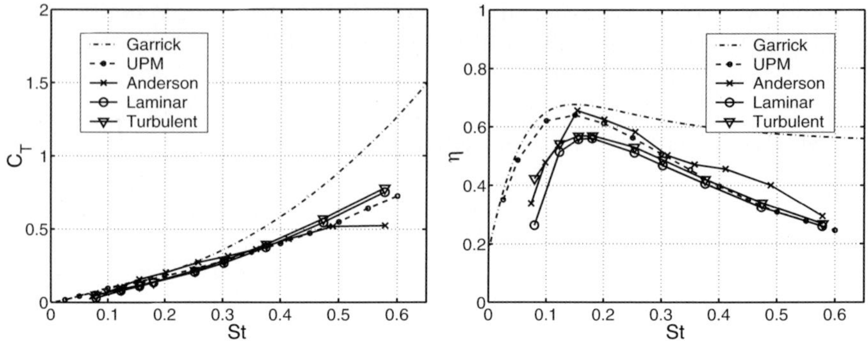

Fig. 1. Mean thrust coefficient and propulsive efficiency, $Re = 40,000$, $\alpha_0 = 15^\circ$, pivot at 1/3 chord, $h = 0.25$, $\phi = 90^\circ$.

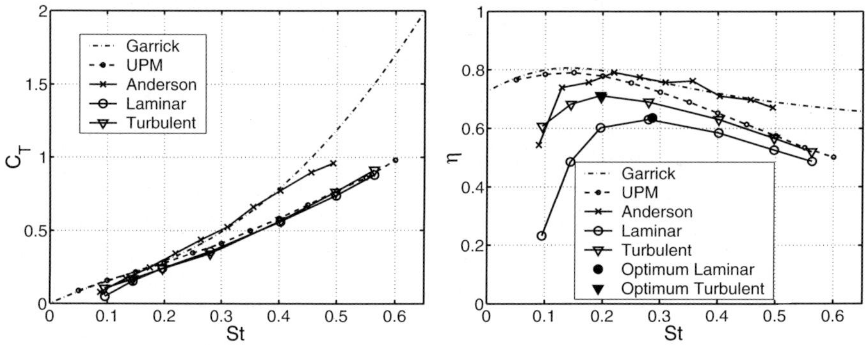

Fig. 2. Mean thrust coefficient and propulsive efficiency, $Re = 40,000$, $\alpha_0 = 15^\circ$, pivot at 1/3 chord, $h = 0.75$, $\phi = 90^\circ$.

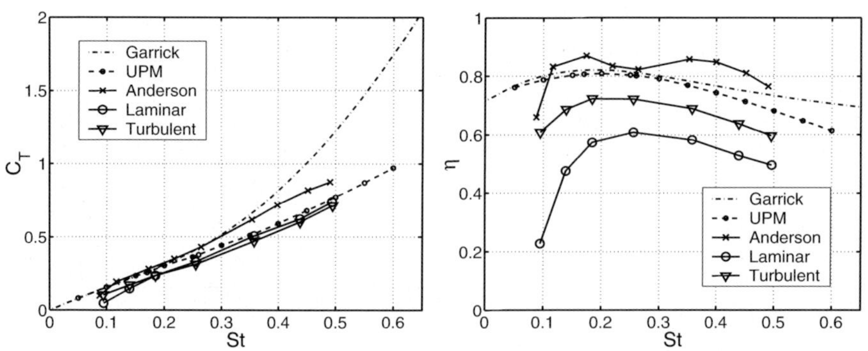

Fig. 3. Mean thrust coefficient and propulsive efficiency, $Re = 40,000$, $\alpha_0 = 15^\circ$, pivot at 1/3 chord, $h = 0.75$, $\phi = 75^\circ$.

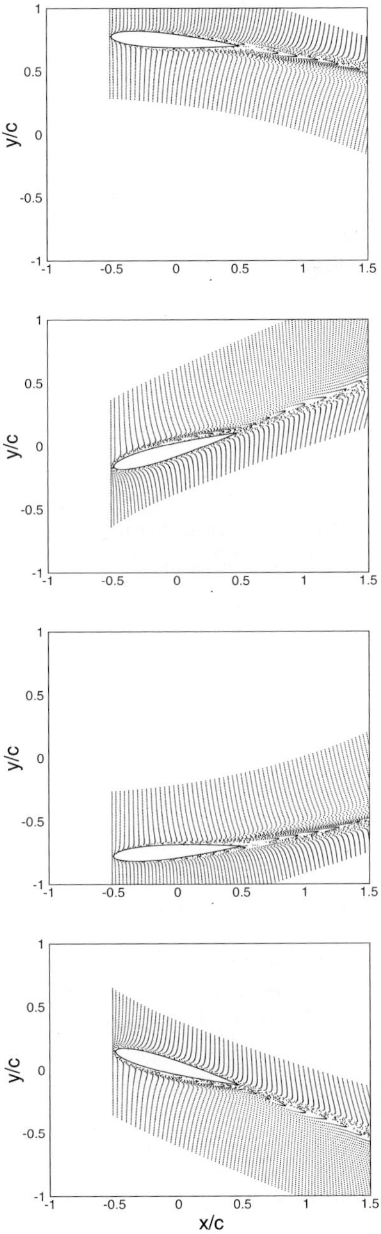

Fig. 4. Particle traces, $\alpha_0 = 15°$, $h = 0.75$, $\phi = 75°$, $St = 0.184$

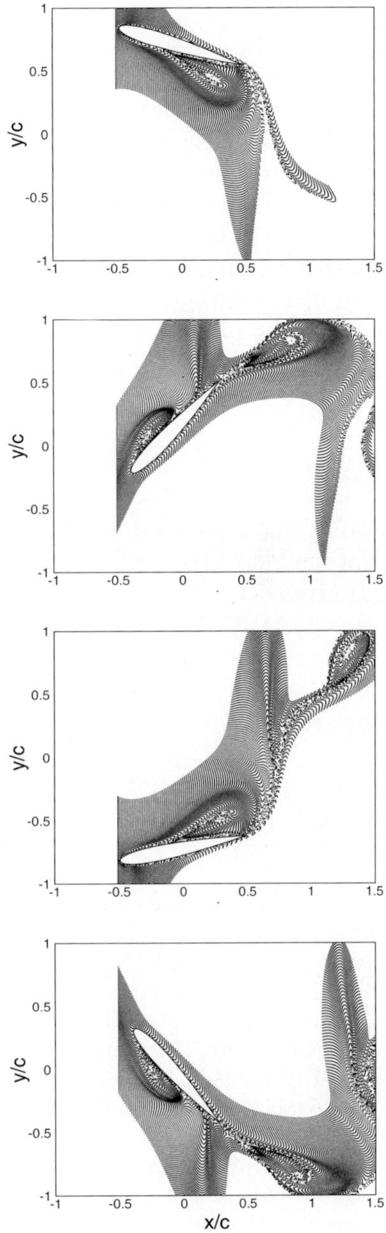

Fig. 5. Particle traces, $\alpha_0 = 15°$, $h = 0.75$, $\phi = 75°$, $St = 0.70$

5 Conclusion

Simulations of the flow over an oscillating airfoil show good agreement with the experimental results of Anderson et al [4], for a range of different flapping parameters. The propulsive efficiency peaks at a Strouhal number between 0.15-0.3, as found by Anderson et al and other researchers. Gradient-descent based optimization of the phase angle between the xpitching and plunging components of the motion shows a phase angle between $\phi = 75°$ and $85°$ for best propulsive efficiency, in agreement with Anderson et al [4] and Isogai et al [6]. Highest efficiency propulsion occurs when leading edge separation is avoided or minimized, however thrust is low in this case. High thrust corresponds to long periods of high effective angle of attack, and large separated leading edge vortices.

Acknowledgements

This work was supported by an award under the Merit Allocation Scheme on the National Facility of the Australian Partnership for Advanced Computing. I.H. Tuncer acknowledges the support of a visiting fellowship from the Rector of UNSW@ADFA for part of this work.

References

1. J. Young, J.C.S. Lai: Oscillation Frequency and Amplitude Effects on the Wake of a Plunging Airfoil. AIAA Journal (2004) (in press)
2. I.E. Garrick: Propulsion of a Flapping and Oscillating Airfoil. NACA Report 567 (1937)
3. J. Young, J.C.S. Lai: Thrust and Lift Dependence of a Plunging Airfoil on Oscillation Frequency and Amplitude. In: *Computational Fluid Dynamics 2002* ed by S. Armfield, P. Morgan, K. Srinivas (Springer, Berlin Heidelberg New York 2002) pp 273–278
4. J.M. Anderson, K. Streitlien, D.S. Barrett, M.S. Triantafyllou: Journal of Fluid Mechanics **360** (1998) pp 41–72
5. I.H. Tuncer, M.F. Platzer: Journal of Aircraft **37**, 3 (2000) pp 514–520
6. K. Isogai, Y. Shinmoto, Y. Watanabe: AIAA Journal **37**, 10 (1999) pp 1145–1151
7. I.H. Tuncer, M. Kaya: Optimization of Flapping Airfoils For Maximum Thrust and Propulsive Efficiency. In: *3rd International Conference on Advanced Engineering Design, Prague, Czech Republic, June 1-5 2003*

Part VII

Biological Flows

Towards Numerical Simulation of Blood Flow in Small Vessels

Zinedine Khatir and Adélia Sequeira

Centro de Matemática e Aplicações, Departamento de Matemática, Instituto Superior Técnico, Av. Rovisco Pais, 1, 1049-001 Lisboa, Portugal. Email: zkhatir@math.ist.utl.pt / zkhatir@hotmail.com

Summary. A methodology for the numerical simulation of a non-Newtonian fluid model is presented. A particle method is combined with a mixed finite element method to study blood flow in small vessels. This is simulated using a Lagrangian description of the viscoelastic stress contribution, whereas momentum and continuity equations are solved in an Eulerian coordinates system of reference with the viscoelastic stress treated as a fixed body force. Numerical results are presented to assess the validity of the proposed approach.

1 Introduction

Biological and industrial flows in complex geometries are quite common and have many applications, which leads to the investigation of both Newtonian and non-Newtonian fluids. Such endeavor arises in biomechanical engineering when studying blood flow in arteries and in smaller vessels, or in the area of chemical and process engineering when considering flows in porous media.

An important class of non-Newtonian models are shear-dependent viscosity fluids whose viscosity depends on the modulus of the symmetric velocity gradient. If the viscosity function is increasing with the shear rate, the corresponding fluids are called shear-thickening, while fluids where the viscosity decreases for increasing shear rate are named shear-thinning fluids. The latter has wide engineering applications that ranges from chemical engineering, geology or glaciology to blood rheology (see e.g. [1]).

Numerical simulation is certainly considered as an important tool for prediction of non-Newtonian phenomena. In the last two decades, intensive research has been performed in this area, mainly for differential and rate-type models, using finite element or spectral methods for steady flows, finite differences in time and finite element or finite volume approximations in space for unsteady flows (see *e.g.* [2], the monograph [3]).

Therefore, as mentioned for example in [4], investigations of the behaviour of viscoelastic fluids in transient flows are very important in studying the rheological properties of viscoelastic materials and their processing, as well as evaluating the predictive capability of the constitutive equations used to model the materials. In this paper, a numerical procedure to investigate non-Newtonian fluids is presented. Some results embedding the new scheme for a simplified model problem applied to blood flow rheology are discussed and are in agreement with previous studies.

2 Mathematical Formulation

While the Newtonian assumption for blood is acceptable in modelling flow in large arteries, the viscoelasticity and shear-thinning behaviour cannot be neglected when describing flow in small vessels or at low shear rates (less than $100s^{-1}$).

A possible approach is to use the generalized Oldroyd-B model with shear-dependent viscosity (as described in [5]) where the extra-stress tensor $\mathbf{S}$ is related to the kinematic variables :

$$\mathbf{S} + \lambda_1 \overset{\triangledown}{\mathbf{S}} = 2\mu(|D(\mathbf{u})|^2)D(\mathbf{u}) + 2\lambda_2 \overset{\triangledown}{D}(\mathbf{u}) \, .$$

Here $\overset{\triangledown}{\mathbf{S}}$ denotes the upper convected derivative: $\overset{\triangledown}{\mathbf{S}} = \frac{\partial \mathbf{S}}{\partial t} + \mathbf{u} \cdot \nabla \mathbf{S} - \mathbf{S}\,(\nabla \mathbf{u})^t - \nabla \mathbf{u}\, \mathbf{S}$, $\mathbf{u}$ is the velocity, $D(\mathbf{u}) = \frac{1}{2}[\nabla \mathbf{u} + (\nabla \mathbf{u})^t]$ the symmetric part of the velocity gradient tensor, $\lambda_1 > 0$, $\lambda_2 > 0$ are viscoleastic constants and $\mu : \mathbb{R}^+ \to \mathbb{R}^+$, the viscosity function describing the shear-thinnig behaviour such that $\mu(|D(\mathbf{u})|^2) = \nu + \nu_0(1 + |D(\mathbf{u})|^2)^q$ [1] where $q < 0$ and $\mu_0 = (\nu + \nu_0) > 0$.

$\mathbf{S}$ can be decomposed into the sum of its Newtonian part $\boldsymbol{\tau}_n$ and its viscoelastic part $\boldsymbol{\tau}_v$ such that:

$$\mathbf{S} = \boldsymbol{\tau}_n + \boldsymbol{\tau}_v = 2\frac{\lambda_2}{\lambda_1} D(\mathbf{u}) + \boldsymbol{\tau}_v$$

$$\boldsymbol{\tau}_v + \lambda_1 \overset{\triangledown}{\boldsymbol{\tau}_v} = 2\left(\mu(|D(\mathbf{u})|^2) - \frac{\lambda_2}{\lambda_1}\right) D(\mathbf{u})$$

For $\Omega \subset \mathbb{R}^n$ ($n = 2, 3$), by applying the momentum and mass conservation laws and introducing the non-dimensional variables: $x = \frac{\tilde{x}}{L}$, $\mathbf{u} = \frac{\tilde{\mathbf{u}}}{V}$, $p = \frac{\tilde{p}L}{\mu_0 V}$, $\lambda_1 = \frac{\tilde{\lambda_1}}{t}$, $\lambda_2 = \frac{\tilde{\lambda_2}}{t}$ (where tildes are attached to dimension variables), the dimensionless form of the governing equations to be solved are given by:

$$\begin{aligned} &Re\left(\frac{\partial \mathbf{u}}{\partial t} + \mathbf{u} \cdot \nabla \mathbf{u}\right) + \nabla p - (1-w)\Delta \mathbf{u} = f + \nabla \cdot \boldsymbol{\tau}_v \quad \text{in } \Omega \\ &\nabla \cdot \mathbf{u} = 0 \quad \text{in } \Omega \\ &We\, \mathcal{N}(\mathbf{u}, \boldsymbol{\tau}_v) + \boldsymbol{\tau}_v = 2\left(\frac{\mu(|D(\mathbf{u})|^2)}{\mu_0} - (1-w)\right) D(\mathbf{u}) \quad \text{in } \Omega \end{aligned} \qquad (1)$$

Here $\mathcal{N}(\mathbf{u}, \boldsymbol{\tau}_v) = \left[\frac{\partial \boldsymbol{\tau}_v}{\partial t} + \mathbf{u} \cdot \nabla \boldsymbol{\tau}_v - \boldsymbol{\tau}_v\,(\nabla \mathbf{u})^t - \nabla \mathbf{u}\, \boldsymbol{\tau}_v\right]$, $We = \frac{\lambda_1 V}{L}$ is the Weissenberg number, $Re = \frac{\rho V L}{\mu_0}$ the Reynolds number and $(1-w) = \frac{\lambda_2}{\lambda_1 \mu_0}$, with ρ the constant density of the fluid and f the external forces.

The system is completed with boundary and initial conditions:

$$\begin{aligned} &\mathbf{u}|_\Gamma = \mathbf{g}(t) \text{ (with } \mathbf{g}(t) \cdot \hat{\mathbf{n}} = 0 \text{ on } \Gamma), \qquad && \boldsymbol{\tau}_v|_{\Gamma^{in}} = \boldsymbol{\varphi}(t), \\ &\mathbf{u}(\mathbf{x}, 0) = \mathbf{u}_0(\mathbf{x}), && \boldsymbol{\tau}_v(\mathbf{x}, 0) = \boldsymbol{\tau}_{v_0}(\mathbf{x}) \end{aligned} \qquad (2)$$

[1] "power-law" model (of Carreau-Yasuda type)

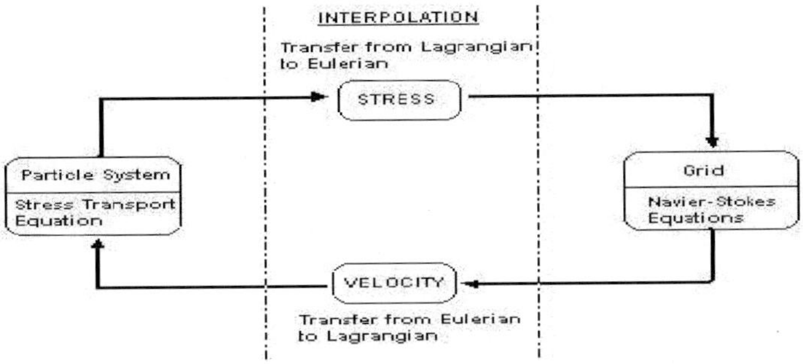

Fig. 1. Hybrid scheme.

where Γ is the boundary of the flow domain and $\Gamma_{in} = \{\mathbf{x} \in \Gamma,\ (\mathbf{u}\cdot\hat{\mathbf{n}})(\mathbf{x}) < 0\}$ with $\hat{\mathbf{n}}$ the outward unit normal vector to Ω on the boundary.

3 Numerical Method

The idea is solve equation $(1)_3$ for the unknown stress $\boldsymbol{\tau}_v$ in a Lagrangian frame of representation of the flow field, whereas the first two equations of (1) are to be solved in an Eulerian coordinates system of reference to obtain the velocity $\mathbf{u}$ and pressure p ([6]). The hybrid scheme is summarized in Fig. 1. The equations for conservation of momentum and mass are solved by means of a mixed FEM (i.e P2-P1 elements for velocity and pressure evaluation), whereas a particle method (in Lagrangian coordinates) is used for the stress transport equation, namely

$$\frac{D\hat{\boldsymbol{\tau}}}{Dt} = \frac{2}{We}\left(\frac{\mu(|D(\hat{\mathbf{u}})|^2)}{\mu_0} - (1-w)\right)D(\hat{\mathbf{u}}) - \frac{\hat{\boldsymbol{\tau}}}{We} + \hat{\boldsymbol{\tau}}\,(\nabla\mathbf{u})^t + \nabla\mathbf{u}\,\hat{\boldsymbol{\tau}} \equiv |!F(D(\hat{\mathbf{u}}),\hat{\boldsymbol{\tau}}) \quad (3)$$

With $\{\mathbf{x}_i(t)\ : 1 \leq i \leq N\}$ denoting the positions of a set of particles at time t, then for a particle i, its stress and velocity are given respectively by

$$\hat{\boldsymbol{\tau}}(\mathbf{x}_i(t)) = \int_\Omega \hat{\boldsymbol{\tau}}(\mathbf{x}')W(\mathbf{x}_i(t)-\mathbf{x}',\epsilon)dx' \ \text{ and } \ \hat{\mathbf{u}}(\mathbf{x}_i(t)) = \int_\Omega \hat{\mathbf{u}}(\mathbf{x}')W(\mathbf{x}_i(t)-\mathbf{x}',\epsilon)dx'$$

where the kernel function $W(\mathbf{x},\epsilon)$ ([7]) is such that

$$\int_{\mathbb{R}^n} W(\mathbf{x}-\mathbf{x}',\epsilon)dx' = 1 \ \text{ and } \ \lim_{\epsilon\to 0} W(\mathbf{x}-\mathbf{x}',\epsilon) = \delta(\mathbf{x}-\mathbf{x}')$$

Therefore, the approximate integrals for a particle i are:

$$\hat{\boldsymbol{\tau}}(\mathbf{x}_i(t)) = \sum_j \hat{\boldsymbol{\tau}}(\mathbf{x}_j(t))W(\mathbf{x}_i(t)-\mathbf{x}_j(t),\epsilon)\,V_j$$

and

$$\hat{\mathbf{u}}(\mathbf{x}_i(t)) = \sum_j \hat{\mathbf{u}}(\mathbf{x}_j(t))W(\mathbf{x}_i(t)-\mathbf{x}_j(t),\epsilon)\,V_j$$

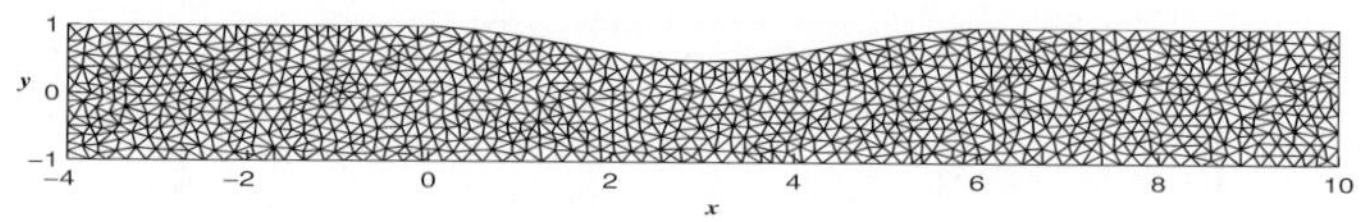

Fig. 2. Geometry of the stenosis.

with V_j the volume of particle j. Accordingly, discretising in time we obtain

$$\begin{cases} \frac{\mathbf{x}_i(t^{n+1})-\mathbf{x}_i(t^n)}{t^{n+1}-t^n} = \mathbf{u}_i(t^n) \\ \mathbf{x}_i(t^n) = \mathbf{x}_i^n \end{cases}$$

$$\begin{cases} \frac{\hat{\boldsymbol{\tau}}(\mathbf{x}_i(t^{n+1}))-\hat{\boldsymbol{\tau}}(\mathbf{x}_i(t^n))}{t^{n+1}-t^n} = F(D(\hat{\mathbf{u}}), \hat{\boldsymbol{\tau}})(\mathbf{x}_i(t^n)) - \frac{\hat{\boldsymbol{\tau}}(\mathbf{x}_i(t^n))}{We} \\ \hat{\boldsymbol{\tau}}(\mathbf{x}_i(t^n)) = \hat{\boldsymbol{\tau}}_i^n \end{cases}$$

Once the stress is computed on the particles, it is then transferred onto the mesh nodes by means of an interpolation procedure ([7], [8], [9]).

4 Numerical Results and Discussion

Let us consider incompressible, isothermal and inertialess Oldroyd-B viscoelastic fluid with constant viscosity with no body forces in a stenosis as shown in Fig. 2. For our numerical simulation the FE mesh is made of 1997 elements with 1082 P1-nodes and 1271 P2-nodes. There are around 13 particles per element. We impose a Poiseuille inflow velocity field given by $\mathbf{u} = (u_x, u_y)$ with $u_x = \frac{3}{32}(1 - y^2)$ and $u_y = 0$. The time-step has been selected as small as $dt = 10^{-6}$ in order to guarantee the convective and diffusive stability condition. The parameters in Eq. 1 are fixed as: $We = 0.25$, $w = 0.1$.

In Fig. 3, velocity vector plots are shown and are of some interest since they provide a detailed description on the flow field. The profile of the velocity is clearly emphasized together with vortex structures and the formation of secondary flow. Regions of reversal flow are evident along the domain in front and behind the stenosis. Fig. 4 is a close-up of some of thoses regions. The results compare well with previous studies of this kind (see [1] and [10]). The isosurface of the u_y-velocity component together with the velocity field and the isosurface of the first normal normal stress difference $\mathrm{N}_1 = \tau_{xx} - \tau_{yy}$ are plotted in Fig. 5 for a region behind the stenosis. The recirculation zone is evidenced by the negative values of u_y-velocity component and the shear-stress in Fig. 5a and 5c, respectively.

The present study has to be interpreted as a preliminary model study. The results, however, indicate the importance of the blood flow rheology and are in qualitative agreement with those of other models existing in the literature (see e.g. [1], [10], [11]).

5 Conclusion

A hybrid Lagrangian particle in cell method for the numerical simulation of a non-Newtonian model has been presented. The preliminary investigation

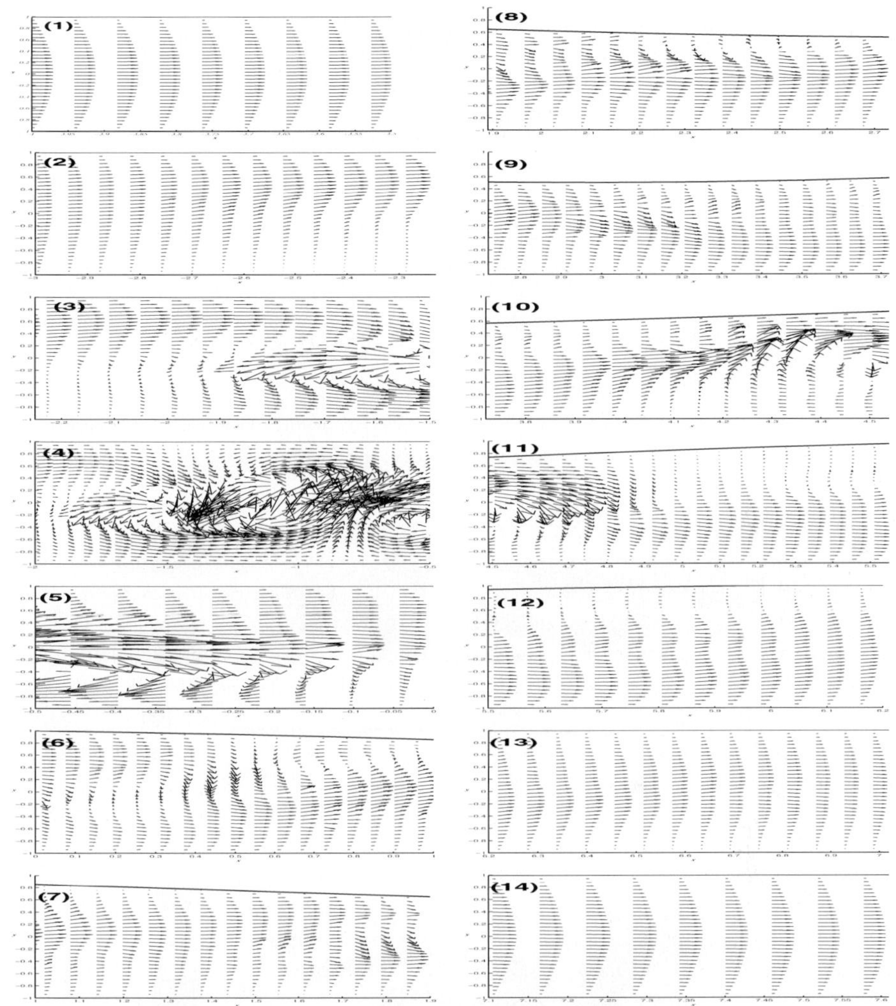

Fig. 3. Velocity vector plot inside the computational domain embedding the stenosis. The profile of the velocity is clearly emphasized together with vortex structures and secondary flow along the vessel.

and computational results of blood flow in a stenosis are very promising and compare wellwith previous studies. Further research towards a better understanding is needed and is still ongoing to improve the novel numerical scheme to study blood flow rheology in small vessels.

Acknowledgements

This work is done within the Framework of the European Network HAEMODEL - *Mathematical Modelling of the Cardiovascular System* through contract:

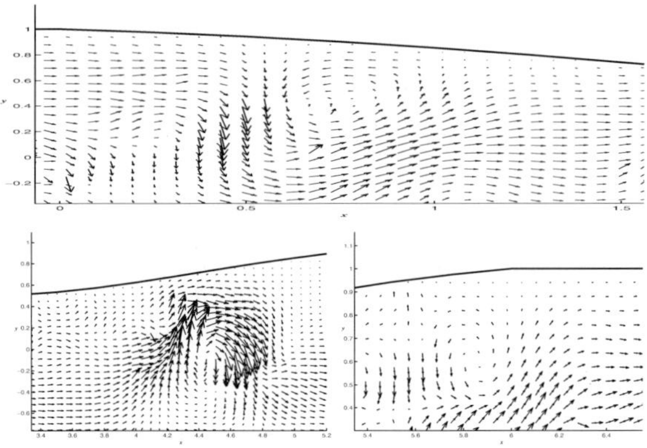

Fig. 4. Velocity vector field: emphasis of vortex structures being generated as shown in Fig. 3 due to the formation of secondary flow in front of and behind the stenosis.

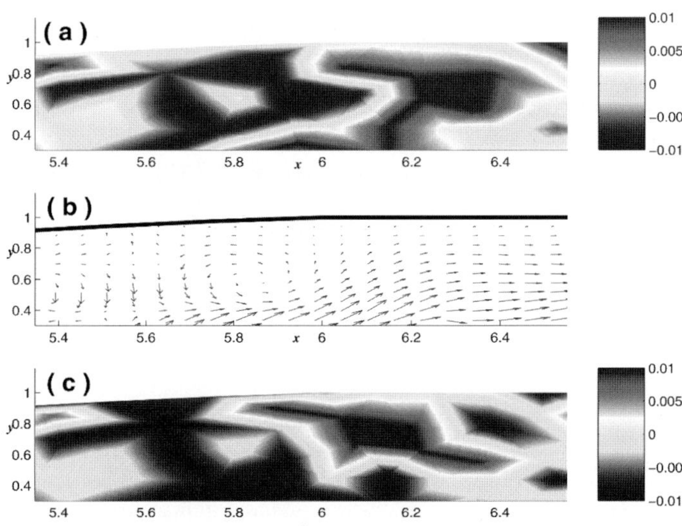

Fig. 5. (a) Isosurface of the u_y-velocity component; (b) Velocity vector field; and (c) Isosurface of the first normal normal stress difference $\mathrm{N}_1 = \tau_{xx} - \tau_{yy}$.

HPRN-CT-2002-00270 and project POCTI/MAT/41898/2001. The support of Center for Mathematics and its Applications (CEMAT) through FCT's Funding Program is highly appreciated.

References

1. J. Hron, J. Málek and S. Turek, A numerical investigation of flow of shear-thinning fluids with applications to blood rheology, *Int. J. Num. Meth. in Fluids*, Vol. 32, No. 7, 2000, 863–879.
2. R. Keunings, A survey of computational rheology, in: Proceedings of the XIIIth International Congress on Rheology (D.M. Binding *et al.* ed.), British Soc. Rheol., 1, 2000, 7–14.
3. R. G. Owens, T. N. Phillips, *Computational Rheology*, Imperial College Press/World Scientific, London, UK, 2002.
4. R. I. Tanner and S-C Xue, Computing transient flows with high elasticity, *Korea-Australia Rheology Journal*, Vol.14, No. 4, December 2002, 143–159.
5. N. Arada and A. Sequeira, Strong steady solution for a generalized Oldroyd-B model with shear-dependent viscosity in a bounded domain,*M3AS*, Vol. 13, No. 9, Sept. 2003, 1303–1333.
6. Z. Khatir, A Lagrangian-Eulerian method for the simulation of blood flow in small vessels, Oral communication at *European RTN Haemodel Workshop*, Graz, Austria, 13-15 April 2004.
7. A. K. Chaniotis, D. Poulikakos and P. Koumoutsakos, Remeshed Smoothed Particle Hydrodynamics for the simulation of viscous and heat conductiong flows, *J. Comp. Physics*, Vol. 182, 2002, 67–90.
8. G. H. Cottet and P. Koumoutsakos, *Vortex Methods*, Cambridge University Press, Cambridge, 2000.
9. M. L. Ould-Salihi, G. H. Cottet and M. El Hamraoui, Blending finite-difference and vortex methods for incompresible flow computations, *SIAM J. Sci. Comput.*, Vol. 22, No. 5, 2000, 1655–1674.
10. G. Pontrelli, Blood flow through an axisymmetric stenosis, *Proc. Instn. Mech. Engrs., Part H J. Eng. in Medicine*, Vol. 215, No. 1, 2001, 1-10.
11. J. R. Buchanan Jr, C. Kleinstreuer and J. K. Comer, Rheological effects on pulsatile hemodynamics in a stenosed tube, *Computers & Fluids*, 29, 2000, 695–724.

Computational Fluid Dynamics and Wall Mechanics of Pre- and Post-Operative Abdominal Aortic Aneurysms

Christine M. Scotti, Ender A. Finol, and Cristina H. Amon

Institute for Complex Engineered Systems, Biomedical Engineering Mechanical Engineering, Carnegie Mellon University, Pittsburgh, PA 15213, USA,
cscotti@andrew.cmu.edu, finol@andrew.cmu.edu, camon@andrew.cmu.edu

1 Introduction

An abdominal aortic aneurysm (AAA) is an enlarged region of the aorta, typically occurring between the renal arteries and the iliac bifurcation. Due to its largely asymptomatic nature, AAAs are commonly left untreated and therefore continue to expand until rupture, an event associated with a 90% mortality rate that represents the 13th most common cause of death in the Unites States [1]. However, with early detection these aneurysms can be treated by surgical repair involving either an open or endovascular technique. Endovascular aneurysm repair (EVAR) is a clinically advantageous technique whereby a graft is expanded and attached to the arterial wall, providing a conduit for blood flow through the abdominal aorta that protects the aneurysm wall from hemodynamic stresses and forces. However this method of repair has an increased rate of secondary intervention due to the development of endoleaks and graft migration; two events which are characteristic of EVG failure and a consequence of the biomechanical environment within the AAA and EVG.

Recent publications report on the characterization of the fluid dynamics of hypothetical AAAs [2] and endovascular graft (EVG) models [3], as well as patient-specific aneurysm models with a cylindrical inlet and outlet [4, 5, 6]. These studies have indicated increasing AAA asymmetry and patient-specific tortuosity significantly alter the fluid dynamics and resulting shear stress experienced along the lumen, as well as the displacement force exerted on the EVG. Patient-specific solid stress analyses have been presented for different pre-operative conditions [4, 7], studying in detail the impact of aneurysm geometry on peak wall stresses. In this work we present a combination of these approaches, utilizing accurate patient-specific models to quantify the changes in pulsatile fluid dynamics and static wall mechanics between the pre- and post-operative scenarios while identifying regions where failure in both may occur.

2 Methods

Computed tomography (CT) scans of an AAA patient who underwent EVAR were obtained from the University of Pittsburgh Medical Center. These scans were completed at 3 months prior to and 30 days after the procedure. The images were processed and converted into a point cloud representation of the aneurysm utilizing edge-detection software and a series of algorithms as outlined by DiMartino et al [8]. From this point cloud, smoothed surfaces were generated and the lumen of the aneurysm or graft of the patient created. The defined thickness of the arterial wall could not be detected from the CT scans, so a uniform value of 1.5 mm was assumed for both the pre- and post-operative solid models.

The smoothed surfaces were then meshed with linear tetrahedral elements using the commercial software ADINA (v.8.0, ADINA R&D, Inc., Cambridge, MA) and subjected to finite element analysis. For the fluid flow simulations we assumed pulsatile, incompressible, homogeneous, Newtonian flow within a rigid wall geometry. The properties of blood were obtained from previous studies [2]. The simplified Navier-Stokes equations are as follows:

$$\rho \nabla \cdot \boldsymbol{v} = 0 \tag{1}$$

$$\rho \frac{\partial \boldsymbol{v}}{\partial t} + \rho \boldsymbol{v} \cdot \nabla \boldsymbol{v} - \nabla \cdot \tau = 0 \tag{2}$$

with a stress tensor that reduces to $\tau = -p\delta_{ij} + 2\mu\varepsilon_{ij}$, where v is the velocity, ε is the strain rate, and τ is the fluid stress tensor. These equations were then discretized and solved implementing boundary conditions of a stress free outflow, no-slip conditions at the walls, and uniform inlet velocity profile. The applied velocity waveform is shown in Figure 1, which depicts the in vivo ultrasound measurement of blood flow through the infrarenal abdominal aorta of a different patient under resting conditions. The time-average Reynolds (Re_m) and Womersley (α) numbers for the pre- and post-operative models are shown in Table 1.

The solid stress analyses were completed assuming a peak sytolic pressure of 116.8 mmHg ($1.6 \cdot 10^5$ dynes/cm^2) applied uniformly along the lumen of

Table 1. Blood flow parameters

	d(cm)	Re_m	Re_{peak}	α
Pre-op	2.54	474	3164	15.8
Post-op, lumen	2.98	554	3713	13.8
Post-op, graft	2.22	414	2766	18.5

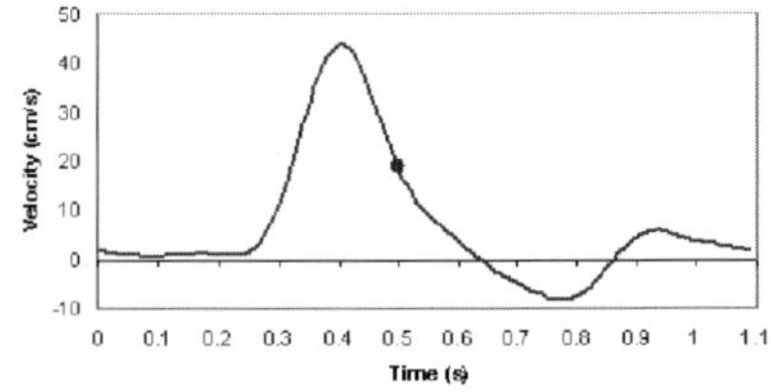

Fig. 1. Velocity waveform for average resting conditions

the AAA and EVG in the normal direction. For average resting conditions, the flow is at a mid-systolic stage when peak systolic pressure is achieved, as indicated in Figure 1.

Incompressible, linearly elastic properties were used to model the suprarenal neck and iliac arteries with a bulk Young's modulus of $0.7 \cdot 10^7$ dyn/cm^2 while hyperelastic material properties were specified for the aneurysm wall and intraluminal thrombus ($E_{bulk} = 2.7 \cdot 10^7$ dyn/cm^2 and $E_{bulk} = 1.1 \cdot 10^6$ dyn/cm^2, respectively). A formulation similar to the Mooney-Rivlin equation was used to model these properties of the AAA wall and thrombus in accordance with previously published data [4, 7].

We used equivalent PTFE/polyester graft material properties [9] to simulate the implant of the now commercially-unavailable Ancure® device (Endovascular Solutions, Guidant Corporation, Santa Clara, CA). The inlet and outlet sections of the computational domain were fixed in order to simulate the tethering effects of the suprarenal and iliac arteries. Tied contact was applied between the graft and arterial wall along a 3 mm segment at the proximal end and 5 mm segment of the distal ends of the EVG in accordance with the specifications for attachment of an Ancure graft. Additional contact surfaces were specified for the other entities, including the EVG and thrombus. The resulting simplified static equilibrium equation was discretized and solved using the direct sparse solver of the ADINA software.

3 Results and Discussion

Table 2 summarizes the clinically measurable changes that occurred between the pre- and post-operative aneurysms. Proximal neck diameter was measured as the dimension of the pre- and post-operative aneurysm at the graft attachment site. This EVG utilizes an active attachment mechanism such that a system of hooks secure the graft to the arterial wall, thereby constricting the aneurysm neck slightly, resulting in a decrease of the measured diameter. A qualitative inspection of the aneurysm shape showed a change along the anterior wall of the aneurysm, yielding the reduced aneurysm volume. Figures 4a and 4b adequately demonstrate the morphological changes between the pre- and post-operative aneurysm and the noticeable difference at the proximal neck and graft attachment site, indicated by the arrows.

Table 2. Quantitative measurements of pre- and post-operative AAA

	Pre-operative	Post-operative	% Change
Proximal Neck Diameter (cm)	2.64	2.18	17.4
Aneurysm Volume (cm^3)	102.94	88.85	13.7
Maximum Diameter (cm)	5.24	4.90	6.5

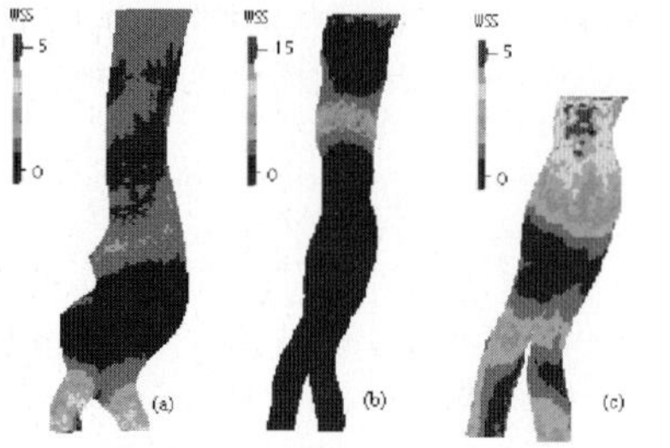

Fig. 2. Wall shear stress (Pa) for the (a) pre-operative lumen, (b) post-operative lumen, and (c) EVG model

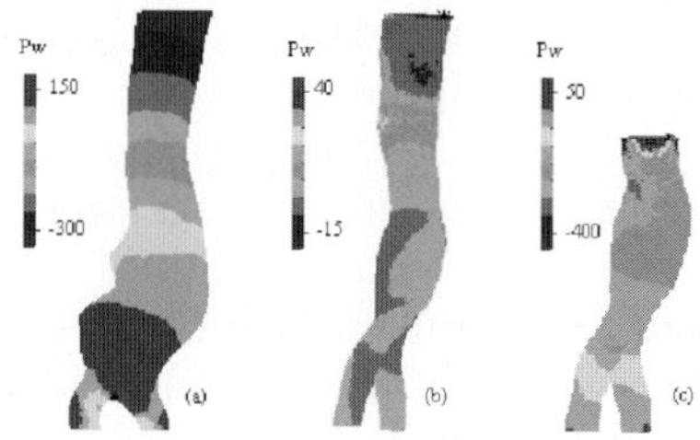

Fig. 3. Fluid wall pressure (Pa) for the (a) pre-operative lumen, (b) post-operative lumen, and (c) EVG model

The results of the fluid dynamics analyses are shown in Figures 2 and 3 for the pre- and post-operative lumen as well as the isolated endovascular graft at peak systolic pressure ($t = 0.5\,\mathrm{s}$ in the velocity waveform). The wall shear stress (WSS) distributions shown in Figure 2 indicate elevated values along the aortic neck and iliac bifurcation due to the regions of recirculating blood flow and altered, disorganized dynamics created by the presence of intraluminal thrombus (ILT) and onset of the diastolic phase of the cardiac cycle.

The angulation of the aortic neck and the graft attachment site of the post-operative aneurysm result in skewed flow patterns and a new location of the maximum WSS. The sharp increase and localization of the maximum WSS values between the pre- and post-operative lumen are a direct consequence of the changes in geometry, a reduced lumen of the graft, and a more organized flow pattern in the post-operative model. Only minor recirculation areas are generated along the posterior wall of the graft while the anterior wall experiences a more concentrated wall shear stress, as shown in Figure 2c.

The overall deceleration of fluid flow caused by the initiation of diastolic conditions produces the pressure distributions shown in Figure 3. Within the pre-operative model this effect is compounded by the flow reversal which results from the localized thrombus present in the aneurysmal sac. The uneven distribution present along the iliac arteries reflects the influence of the bifurcation as well as the recirculating blood flow found along the distal end of the thrombus region. The post-operative pressure distribution is more uniform, the result of the restoration of a normal pathway of forward-moving blood through the aneurysm., while within the EVG the pressure is well distributed, with a small concentration towards the anterior wall of the graft attachment. Comparing the pressure and WSS results, the pressure-dominated characteristic of the flow is confirmed. Utilizing the pressure distribution and a control volume analysis [3], the displacement force as calculated over the EVG was

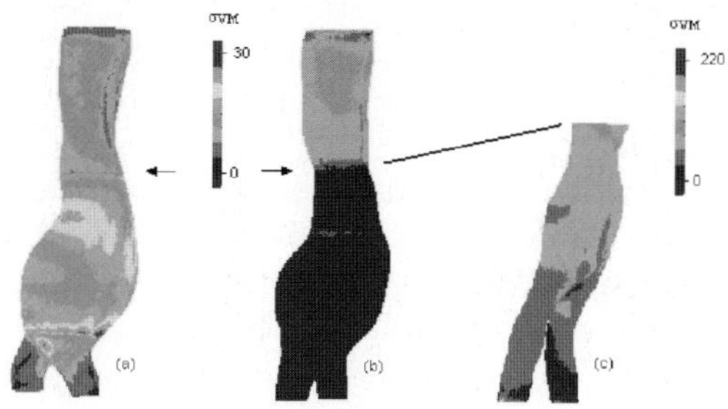

Fig. 4. Von Mises stress (N/cm^2) distributions for the (a) pre-operative, (b) post-operative with EVG, and (c) EVG model

determined to be 3.4 N under these conditions, a value that may be sufficient to cause graft migration.

Figure 4 illustrates the sharp reduction in peak Von Mises stress experienced by the aneurysm after surgery. The thrombus shields the aneurysm wall in the pre-operative aneurysm, indicated by the area of reduced stress in the aneurysmal sac of Figure 4a, while stress is elevated where the geometry and material properties shift. The maximum Von Mises stress under peak systolic conditions with the thrombus included is 94 N/cm^2; when the computed stresses are averaged over the entire model to accommodate the variation of material properties and any resulting incompatability this maximum is reduced to 68 N/cm^2.

Figure 4b shows the post-operative Von Mises stress distribution, illustrating the effectiveness of the graft at shielding the aneurysm wall. The stresses on the graft wall are shown in Figure 4c, with a maximum value of 220 N/cm^2, about three times the stress shown in the pre-operative case. While the stress appears to be evenly distributed along the body of the EVG, a localized Von Mises stress is found near the proximal attachment site, in part due to reaction forces generated at the contact region between the graft and the inner AAA wall. However, the maximum Von Mises stress occurs at the bifurcation point which may increase its susceptibility to biomechanical failure.

4 Conclusions

In the present work we numerically evaluated the biomechanical environment of pre- and post-operative aneurysms and examined the cumulative ffect of factors such as the presence of localized thrombus, non-uniform material properties, and patient-specific geometry. Flow disturbance was intensified through the pre-operative aneurysm sac with the onset of diastolic conditions and the obstruction of the localized thrombus. This led to increased flow-induced forces, particularly in the normal direction to the wall. Similarly, the thrombus localized the Von Mises stress where changes in geometry and material property occur, while simultaneously shielding the AAA wall from elevated mechanical forces.

Under post-operative conditions, the EVG restores normal blood flow through the arterial lumen. A region of increased wall shear stress gradient exists along the posterior neck of the graft attachment site due to angulation of the aneurysm neck. At peak systolic pressure, the flow-induced displacement force on the graft may be sufficient to cause graft migration. The implantation of the EVG reduces the stresses exerted on the aneurysm wall, decreasing the Von Mises stress within the aneurysm by 83% in comparison with the pre-operative conditions. The maximum Von Mises stress at the post-operative aneurysm wall is 96% lower than that experienced by the endovascular graft. Stress remained uniformly distributed on most of the graft, but localized at the proximal attachment site due to contact forces between the graft and the arterial wall.

Acknowledgements

Funding for this work was provided in part by a grant from the Pennsylvania Infrastructure Technology Alliance (PITA) of the Commonwealth of Pennsylvania's Department of Community and Economic Development, and by a computing grant from the Pittsburgh Supercomputing Center. The authors gratefully acknowledge Dr. Elena S. DiMartino at the University of Pittsburgh for providing the patient's CT images and assisting in the reconstruction of the geometries.

References

1. Crawford C.M. et al. *Journal of Manipulative and Physiological Therapeutics*, 26(3):184–195, 2003.
2. Finol E.A. et al. *ASME Journal of Biomechanical Engineering*, 125(2):207–217, 2003.
3. Finol E.A. et al. In *Simulations in Biomedicine V: Advances in Computational Bioengineering*, pages 231–241. WIT Press, 2003.
4. Raghavan M.L. et al. *Journal of Vascular Surgery*, 31(4):760–769, 2000.
5. Finol E.A. et al. In *Computer Methods in Biomechanics and Biomedical Engineering - 4*. Gordon and Breach Science Publishers, cd-rom edition, 2002.
6. Finol E.A. et al. In *Proceedings of the 28th Annual Northeast Bioengineering Conference*, pages 191–192.
7. DiMartino E.S. et al. *Annals of Biomedical Engineering*, 31:804–809, 2003.
8. DiMartino E.S. et al. In *2004 Advances in Bioengineering*. ASME BED-Vol. 57, IMECE2004-61556, 2004.
9. Suzuki K. et al. *Cardiovascular and Interventional Radiology*, 24:94–98, 2001.

Geometrical Considerations in Patient Specific Models of a Human Aorta with Stenosis and Aneurysm

Johan Svensson[1], Roland Gårdhagen[1], and Matts Karlsson[2]

[1] Department of Mechanical Engineering, Linköping University, SE-581 83 Linköping SWEDEN, johsv@ikp.liu.se, rolga@ikp.liu.se
[2] Department of Biomedical Engineering, Linköping University, SE-581 85 Linköping SWEDEN, matka@imt.liu.se

Summary. The most important artery in the human body is the aorta that supplies the rest of the body with blood. Lesions in the aorta can cause serious complications, which can even lead to death. How the flow is behaving in lesions, what causes the problems and which lesions are dangerous are highly interesting to determine. Laminar, stationary CFD calculations are performed on two geometrically different models of the human aorta created from the same set of patient MRI (Magnetic Resonance Imaging) data. Differences in the CFD results due to different geometries are evaluated. Overview results e.g. pressure variations throughout the artery are not dependent on an exact description of the geometry. If absolute and local values e.g. wall shear stress are sought more robust geometry creation procedure is needed in order to get more reliable results.

Key words: Human Aorta, Influence of Geometry, Stenosis, Aneurysm

1 Introduction

Understanding the blood flow and its distribution of pressure and wall shear stress (WSS) at the wall is important for diagnosis as well as intervention planning. WSS is believed to be important in atherosclerosis and plaque forming. Extensive research in the area has resulted in the current consensus concerning atherogenesis and its association with low mean WSS, low peak WSS, high oscillation and large temporal or spatial gradients of WSS [1, 2, 3, 4]. Furthermore, local geometry is known to be a major determinant of WSS within the individual arteries [5]. Here we analyze the aorta from the middle of the arch to the mid thoracic region, Figure 1. It is well known that lesions in the aorta can result in very serious complications as the aorta supplies the whole body with blood. This study focuses on an aortic

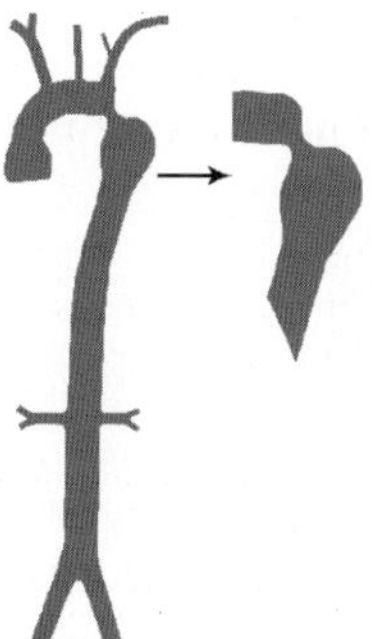

Fig. 1. Part of the aorta used for the models.

Fig. 2. Workflow overview for performing a CFD analysis on the geometry of human aorta. *From left:* **(a-e)** MRI image, Point cloud, 3D geometry, Mesh and CFD result.

coarctation, a stenosis (narrowing), followed by an aneurysm (widening).

2 Method

To perform CFD analyses of the blood flow a geometrical model of the artery is needed. The geometrical model here is created from MRI (Magnetic Resonance Imaging) images that are used to obtain information about the arterial wall location. There are uncertainties in the location of the arterial wall as well as in the 3D volume creation. Thus, if the same workflow for geometry creation is performed twice (Figure 2 a-c) the geometries created (denoted A and B) will be quite different. This study aims at analyzing the influence that these differences will have on derived parameters such as pressure and WSS. This knowledge is of crucial importance to determine which data can be trusted and which methods in the workflow that need improvements.

The location of the arterial wall is determined from the MRI images by semiautomatic segmentation. A number of points (maximum 40) are placed on the arterial wall in each image, Figure 3 left. Cartesian coordinates for each point are stored. The set of points collected is called point cloud, Figure 3 right.

The point clouds are processed in a CAD software (Catia) in order to get a 3D volume model of the aorta. Splines are constructed from a number of the points and these lines are used to create a surface approximating the arterial wall. Locating the arterial wall and 3D volume creation are in the following called segmentation (which here contains some manual work). Segmentation is performed by two different individuals having no mutual definition of wall location in the MRI images (worst case scenario). Segmentation is performed on the same MRI data set from one patient for comparison.

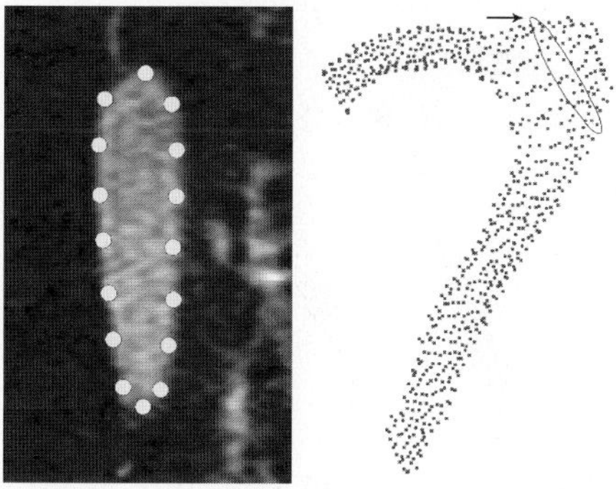

Fig. 3. *Left:* Delineation points in an MRI image. *Right:* Points cloud with delineated points marked (arrow).

Unstucturated meshes are created of the 3D geometries of the aorta, using mainly tetrahedral volumes. The mesh operations are performed with the software Gambit. A spacing of $1\,mm$ is used for both meshes.

The wall is considered rigid, flow is laminar and the blood is incompressible. The Navier-Stokes equations are discretized and solved with a finite volume method in the CFD solver Fluent. Body forces such as gravity are neglected. The Navier-Stokes equations read:

$$\nabla \cdot \mathbf{u} = 0 \tag{1}$$

$$\rho(\mathbf{u} \cdot \nabla)\mathbf{u} = -\nabla p + \nabla \cdot \boldsymbol{\tau} \tag{2}$$

where $\mathbf{u}$ is the velocity vector, p is the pressure and $\boldsymbol{\tau}$ is the shear stress tensor. Density is set to $\rho = 1060\,kg/m^3$ and the value used for dynamic viscosity is $\mu = 0.004\,Ns/m^2$. Due to the assumtion of laminar flow and the fact that the flow is studied in a large artery the blood can be considered as Newtonian [6], which implies that $\nabla \cdot \boldsymbol{\tau} = \mu\nabla^2\mathbf{u}$. The inlet boundary condition is a specified velocity, where the velocity profile being a fully developed laminar profile. This profile is used in order to get similar boundary conditions for the two models. The profile is constructed from an ellipse shape to fit the inlet geometry. The outlet is treated with a conserved mass flow boundary condition. The arterial wall is considered rigid with a no-slip boundary condition.

Parameters for the comparative analysis of the CFD results from the models are static pressure and WSS. These parameters are physiologically important and describe the load that acts on the arterial wall [7] and influence development of lesions [1]. To make comparisons between geometrical different models a Reynolds number in the stenosis area (smallest cross sectional area in the model) is defined as coarctation Reynolds number (Re_{coa}). The characteristic length is determined with the assumption that the cross sectional area at the stenosis is circular.

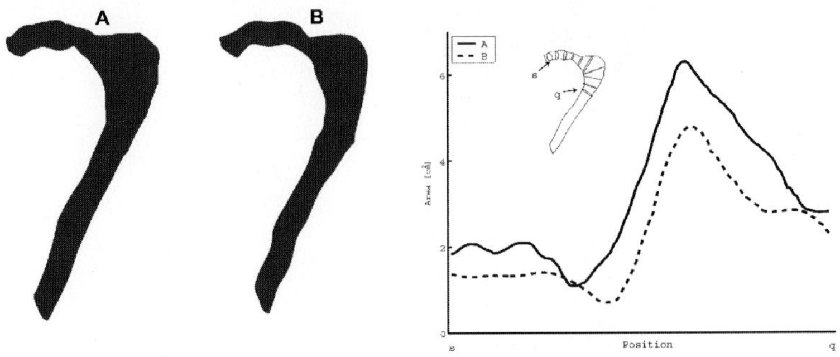

Fig. 4. *Left:* Three dimensional geometry of the two models. *Right:* Cross sectional area between point **s** and **q**.

The MRI images used are obtained with a GE Signa Horizon MRI scanner at the University Hospital, Linköping University. The patient is a juvenile with a coarctation and an aneurysm. Slice thickness in the scans is $2.6\,mm$ and the distance between center of slices is $3.9\,mm$, 35 slices were used to create the models. The resolution in the plane is 512×512 pixels and each pixel has a size of $0.56 \times 0.56\,mm$.

3 Results

The models contain relatively different amount of cells (calculation volumes). The number of cells in the models were approximately 388000 for the A-model and 287000 for the B-model. Total volume in the models also differs: $70.8\,cm^3$ for A and $52.0\,cm^3$ for B. This indicates that the models are significantly different geometrically, which also can be seen in Figure 4. The volume differences depend mainly on differences in cross sectional area in the aortic models, Figure 4 right, also showing that the coarctation location differs slightly between the models.

Calculations are made using six different flows for the A model and five for B model. The different number of flow simulations in the models are made in order to get a similar range of Re_{coa}. Figure 5 left shows that the static pressure (mean value of five streamlines in the main flow) has a similar behavior for the two models. Two parameters are introduced in Fig. 5 left, pressure drop (p_{drop}) that is the difference in pressure between the inlet pressure and the lowest pressure that occurs near the coarctation and a pressure recovery (p_{recov}) that is the difference in pressure between the coarctation minimum pressure and maximum pressure that occurs in the aneurysm region. These two parameters p_{drop} and p_{recov} are plotted against Re_{coa}, Fig. 5 right: there are significant differences in pressure drop and pressure recovery for model A and B and the differences become larger with increasing Re_{coa}. To investigate

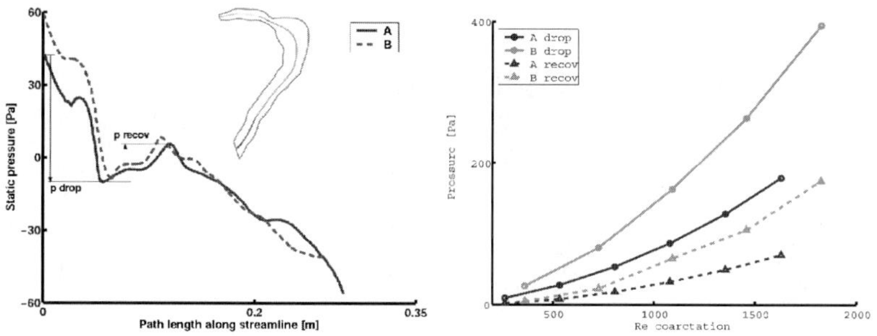

Fig. 5. *Left:* Mean pressure from five streamlines for model A and B. Definition of the measurements pressure drop and pressure recovery. *Right:* Varations of pressure drop and recovery as function of Re_{coa}.

how WSS varies with choice of geometrical model and Re_{coa} the maximum WSS in a region around the coarctation and minimum WSS at a region in the aneurysm (section with the largest cross sectional area) are plotted, Fig. 6. The two curves for maximum WSS are very similar for models A and B. Minimum WSS in the aneurysm are also quite similar but there are a rapid change in the parameter at high Re_{coa}, which depend on changing flow field in the aneurysm.

Distribution of WSS magnitude is presented as color coded maps of the arterial walls in Fig. 7 (here presented in black and white). Distribution of WSS in model A is more scattered then in model B indicating that the WSS patterns are different.

4 Discussion

The geometrical difference can be considered as a worst case due to the fact that the persons performing the segmentation did not have any mutual definition of wall location in the MRI images. This resulted in two significantly different geometries. For the parameters used for the comparison it seems

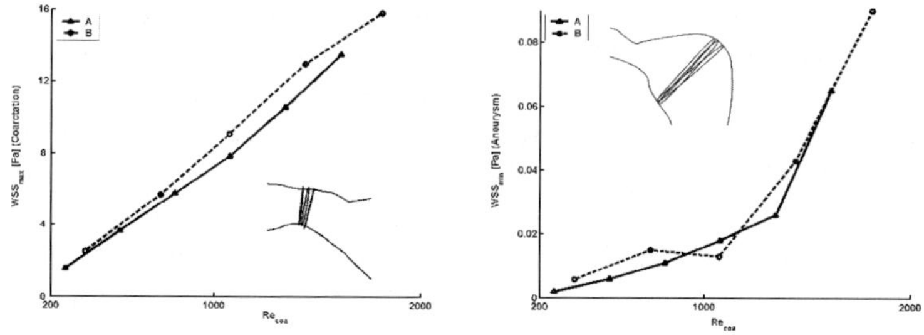

Fig. 6. *Left:* Varations of maximum WSS at the coarctation site. *Right:* Variations of minimum WSS in the defined aneurysm area.

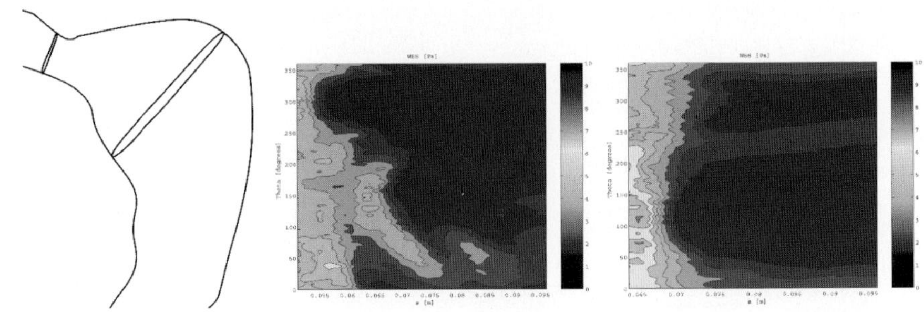

Fig. 7. Distribution of WSS as maps from coarctation (A_{min}) to aneurysm area (A_{max}).

that the overall pressure behavior is quite similar and does not depend on an exact geometrical description. For absolute values and for parameters such as the p_{drop} and p_{recov} differences are clearly seen between the models. Thus, if analyses of pressure recovery after a stenosis and pressure drop are of interest, an accurate geometrical description is needed. The maximum value of WSS in the coarctation area is quite similar for both the models when compared with Re_{coa}. The minimum WSS in the aneurysm region are similar for both the models, but a more complex behavior due to changes in the flow distribution. The WSS distribution seems to depend very much on the local geometry that are different between the models, indicating the necessity of an accurate segmentation method.

References

1. Henry M. Honda, Tzung Hsiai, Charles M. Wortham, Mingdar Chen, Hank Lin, Mohamad Navab, and Linda L. Demer. A complex flow pattern of low shear stress and flow revarsal promotes monocyte binding to endothelial cells. *Atherosclerosis*, (158):385–390, 2001.
2. Arnold P.G. Hoeks Robert S. Reneman Lilian Kornet, Jacques Lambregts. Differences in near-wall shear rate in the carotid artery within subjects are associated with different intima-media thicknesses. *Arterioscler Thromb Vasc Biol.*, 18:1877–1884, 1998.
3. Sumpio B. Frangos SG, Gahtan V. Localization of atherosclerosis: role of hemodynamics. *Archives of Surgery*, 134:1142–1149, 1999.
4. Brands PJ Reneman RS. Hoeks AP, Samijo SK. Noninvasive determination of shear-rate distribution across the arterial lumen. *Hypertension*, 26:26–33, 1995.
5. Mark FF Bargeron CB Hutchins GM Friedman MH, Deters OJ. Arterial geometry affects hemodynamics. a potential risk factor for athersoclerosis. *Atherosclerosis*, 46:225–231, 1983.
6. R.L. Whitmore. *Rheology of the Circulation.* Pergamon Press, first edition, 1968.
7. Jay D. Humprey. *Cardiovascular Solid Mechanics Cells, Tissues, and Organs.* Springer Verlag, 2002. Figure 8.20 p.408.

Part VIII

Flow Control

Numerical Simulation and Control of Bluff-Body Flows Using the Penalization Method

Charles-Henri Bruneau, Iraj Mortazavi, and Gwendal Wilczyk

Mathématiques Appliquées de Bordeaux, Université Bordeaux 1
351 cours de la Libération, F-33405 Talence, France,
bruneau@math.u-bordeaux1.fr

Summary. A penalization method is used on Cartesian meshes to take into account bluff-bodies immersed in an incompressible flow. Realistic numerical simulations of 2D flows are performed using in particular accurate schemes and relevant boundary conditions on artificial frontiers. Then the penalization method is extended to three media namely the solid obstacles, a porous medium and the fluid. Therefore a porous layer is added to some parts of the obstacles to modify the shear layer effects and control the drag and lift forces as well as the vorticity. Finally an active control strategy is also presented.

1 Introduction

Among the fictitious domain methods the penalization method is a very efficient and easy to handle one. It consists only in adding in the equation of conservation of momentum a mass term to represent the obstacles. Then the new model is solved directly in both the solid domain and the fluid domain without addition of any boundary condition or second member. The result is that Darcy equations are automatically solved in the solid domain and Navier-Stokes equations are solved in the fluid domain [1]. As a consequence the pressure is a continuous field computed on the whole domain. Besides the velocity field computed inside the obstacles is used to get the drag and lift forces exerted on the bodies.

The penalization method can be extended to simulate also a porous medium that will be used here to separate the bodies and the fluid in order to reduce the shear forces. This extension of the method is referred as the double penalization method, it was introduced and numerically used in [3, 4] and mathematically studied in [6]. Actually the simple or double penalization method is a very convenient way to simulate complex flows at high Reynolds numbers around obstacles with Cartesian meshes. Besides passive and active control strategies are easy to handle with this method. In the first part a brief description of the simulation method including performant non-reflecting boundary conditions is presented. Complex flows around the Ahmed body (simplified car bench) moving on the top of a road are simulated at Reynolds number $Re = 30000$. Then, a passive control technique using porous interfaces on bluff-bodies is introduced and applied to several geometries. Finally, an active control of the flow around the Ahmed body is performed.

2 Immersing Obstacles by the Penalization Method

We present in this section the penalization method that consists in adding a term U/K in the adimensional incompressible Navier-Stokes equations and to solve them in a large domain Ω including the fluid, the obstacles and eventually the porous layers. To derive the model we assume that there is a porous medium saturated by a Boussinesq fluid. Starting from Brinkman equation

$$\nabla p = -\frac{\mu}{k}\Phi U + \tilde{\mu}\Phi\Delta U$$

we get with the Dupuit-Forchheiner relationship and after nondimensionalisation of the Brinkman-Navier-Stokes or penalized Navier-Stokes equations (see for more details [9] and [4])

$$\begin{aligned} \partial_t U + (U\cdot\nabla)U - \frac{1}{Re}\Delta U + \frac{U}{K} + \nabla p &= 0 \; in \; \Omega_T = \Omega\times(0,T) \\ divU &= 0 \; in \; \Omega_T \end{aligned}$$

where $U = (u, v)$ is the velocity, p the pressure, μ the viscosity of the fluid, $\tilde{\mu}$ the Brinkman effective viscosity, Φ the porosity of the medium and Re the adimensional Reynolds number. The penalization is achieved by the mass term where $K = \frac{\rho k \Phi \overline{U}}{\mu H}$ is the adimensional coefficient of permeability of the medium with ρ the density of the fluid, k the intrinsic permeability and $\overline{U}$ the mean velocity. To simulate the flow, the only thing to do is to specify K in each medium, the fluid being considered as a porous medium with a very high permeability $K = 10^{16}$, the obstacles being considered as porous media with a very low permeability $K = 10^{-8}$ and the porous medium being represented by an intermediate value $K = 10^{-1}$ [4].

To get a well-posed problem we have to impose boundary conditions at the artificial limits of the domain. On upstream boundary a constant flow $U = (1, 0)$ is imposed while for other boundaries an efficient artificial boundary condition is implemented [2]. This boundary condition reads

$$\sigma(U,p)\cdot n + \frac{1}{2}(U\cdot n)^-(U - U^{ref}) = \sigma(U^{ref}, p^{ref})\cdot n$$

where $\sigma(U,p) = \frac{1}{2\,Re}(\nabla U + \nabla U^t) - pI$ is the stress tensor, n is the unit normal vector pointing outside of the domain, $a^- = -min(a, 0)$ and (U^{ref}, p^{ref}) is the reference flow which is taken equal to the flow on the previous cell.

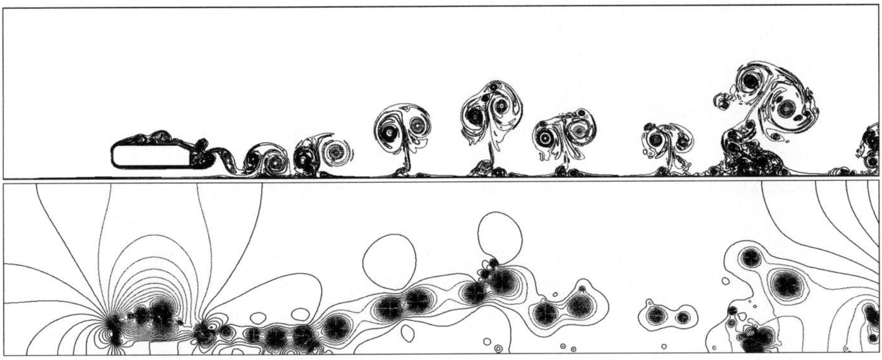

Fig. 1. Vorticity (top) and pressure (bottom) fields around Ahmed body moving over a road at $Re = 30000$

This boundary condition enables the vortices to convey properly outside the computational domain without any reflection on the artificial frontiers.

The discretization is achieved by a second-order Gear scheme in time with explicit treatment of the convection term and by finite differences in space on a uniform mesh. The linear terms are discretized by centered second-order finite differences whereas an upwind third-order Murman-like scheme is used for the convection terms [5]. The resolution is achieved by a multigrid method on a sequence of 7 or 8 grids. This whole method is efficient, accurate and stable enough for 2D direct numerical simulations of complex flows.

As an illustration, Figure 1 shows the vorticity field of the flow around and in the wake of Ahmed body moving on top of a static road at Reynolds number $Re = 30000$. The vortex shedding and the interaction of the vortices with the road generate a complex dynamics of vortex wall interaction. The vortex dipoles are regularly generated on the top of the road due to the Couette type behaviour of the body-road interface. They progressively detach in the wake of the obstacle and move away from the road, creating high vorticity shear areas that are clearly seen on the pressure field.

3 Passive and Active Control of Bluff-Body Flows

In this section we use both passive and active control methods to reduce the oscillations induced by the non-symmetric and unsteady shedding of vortices, high pressure drag and wake buffeting of vortical structures. To study the control effects on the flow we analyze the vorticity fields as well as the global quantities C_d (drag coefficient), CL_{rms} (root mean square of the lift coefficient) and Z (enstrophy). The first case is a square cylinder in an open flow at $Re = 300$. It is chosen in order to point out the regularization effects of the passive control technique. Two porous layers are introduced on the top and on the bottom sides. Figure 2 shows how the wake is regularized since a

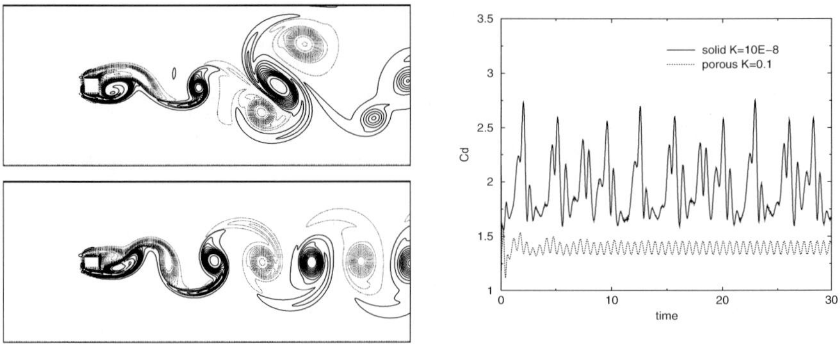

Fig. 2. Vorticity field around a square cylinder for solid (top left) and porous (bottom left) layers at the same time for $Re = 300$ and corresponding drag histories (right)

Karman street is recovered and the drag history becomes periodic with a low amplitude. Another consequence is the drastic reduction of the mean drag coefficient.

The same method is applied to a more realistic bluff-body at higher Reynolds number $Re = 30000$ in an open flow (Figure 3). On such a stretched body (case 1) the boundary layer is developed and we can quantify the reduction of the shear forces. First, the porous layer is only implemented on the upper side of the obstacle (non-symmetric case 2). Then, a second control is achieved introducing a porous layer on both sides (symmetric case 3). Figure 3 shows the modification of the vorticity field and the laminar flow inside the porous layer that changes the shear effects. In Table 1 the reduction of global quantities is summarized, it appears that substantial gains up to more than 30% are obtained in particular for the drag coefficient.

Then a test of passive control is achieved when the same body is located at a given distance of a road. We can see on Figure 4 that the laminar flow in the porous layer changes the roll-up at the rear of the body as it induces a small blowing jet. The reduction of the mean drag coefficient and the enstrophy is significant as it reaches 29% and 27% respectively (see Table 2). But if

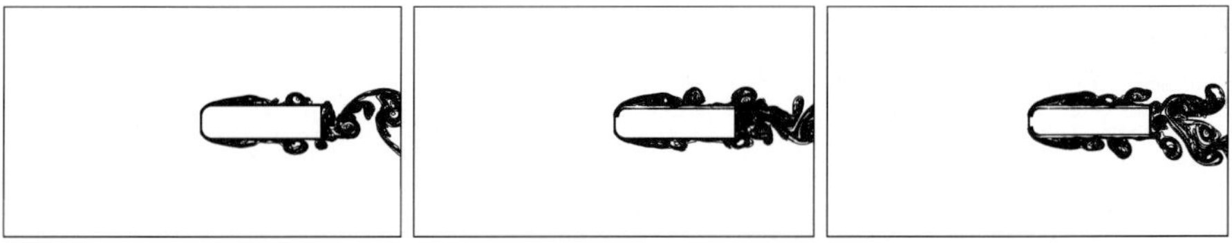

Fig. 3. Vorticity field around a stretched body (left) with a porous layer on the top side (middle) and with two symmetric porous layers on the top and the bottom sides (right)

Table 1. Mean values for the stretched body flow at $Re = 30000$. Case 1 without passive control, case 2 with a porous layer on the top side and case 3 with two symmetric porous layers

Case	CL_{rms}	C_d	Enstrophy
1	0.196	0.45	7913
2	0.182	0.34	7846
3	0.164	0.30	7881

instead of the stretched body we take Ahmed body with a porous layer on the roof, the resulting jet is located on the top of the rear-window. In that case the effect is negative as the mean drag increases because of the jet which yields a larger detachment of the flow.

To control the flow over Ahmed body, we perform an active control by means of blowing jets. The best part of the body to get a significant action is to blow on the boot as it is shown by the passive control above and previous studies [7, 8]. We present here two results with blowing on the top and in the middle of the boot as it is drawn on Figure 5. The drag histories show that for both cases a drag reduction occurs. But the position of the jet is very

Fig. 4. Vorticity field around a stretched body moving over a road (left) and with a porous layer on the top side (right)

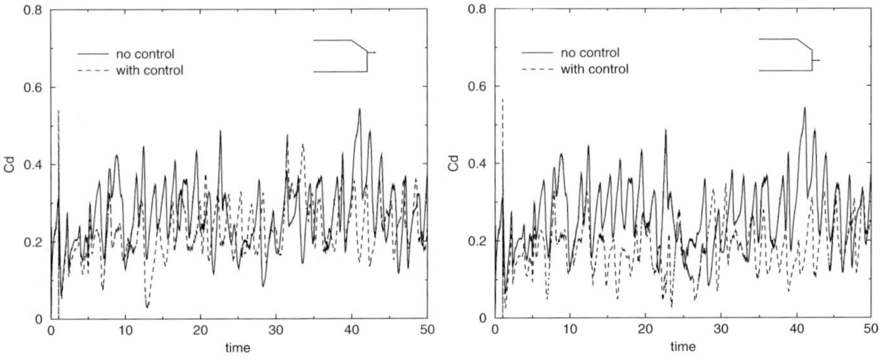

Fig. 5. Drag history for Ahmed body over a road with control by a blowing jet on the top of the rear (left) and in the middle of the rear (right)

Table 2. Mean values for the stretched body flow over a road at $Re = 30000$. Case 1 without passive control and case 2 with a porous layer on the top side

Case	CL_{rms}	C_d	Enstrophy
1	0.349	0.52	823
2	0.320	0.37	614

important as the reduction of the mean value is about 14% in the first case and about 35% in the second one. Indeed when blowing in the middle, there is a stronger action on the under-pressure area located behind the boot that is responsible of the main part of the drag.

4 Conclusion

The penalization method is an easy way either to simulate flows around obstacles without body fitting or to implement passive or active control procedures. A realistic simulation of the flow around Ahmed body is efficiently achieved on a Cartesian grid in two dimensions. Then the method is used to add a porous interface between the flow and the obstacle in order to change the shear forces, and thus to passively control the flow around bluff-bodies. Depending on the geometry and the Reynolds number a reduction up to 30% of the global quantities, in particular the drag coefficient, is obtained. Similar performances using active control by a blowing jet more convenient for Ahmed body are also observed.

Acknowledgment

This work is partially supported under grants by RENAULT SA. The authors wish to thank P. Gilliéron for his interest in this work and fruitful discussions.

References

1. P. Angot, C. H. Bruneau and P. Fabrie: *Numer. Math.*, **81** (1999).
2. C. H. Bruneau and P. Fabrie: *Int. J. Num. Meth. Fluids*, **19** (1994).
3. C. H. Bruneau and I. Mortazavi: *C. R. Acad. Sci.*, **329** (2001).
4. C. H. Bruneau and I. Mortazavi: *Int. J. Num. Meth. Fluids*, (to appear 1994).
5. C. H. Bruneau and M. Saad: submitted.
6. G. Carbeau: submitted.
7. P. Gilliéron and F. Chometon, *ESIAM Proc.*, **7** (1999).
8. P. Gilliéron: *J. Mech. Eng.*, **53** (2003).
9. D. A. Nield and A. Bejan, *Convection in Porous Media* (Springer, 1999).

Application of Genetic Algorithm to Two-jet Control System On NACA 0012 Airfoil

L. Huang[1], R.P. LeBeau[2], and T. Hauser[3]

[1] University of Kentucky, 151 RGAN Building, Lexington, KY, 40506, liang.huang@gmail.com
[2] University of Kentucky, 151 RGAN Bldg, Lexington, KY 40506, rplebeau@engr.uky.edu
[3] Utah State University, Logan, UT 84322-4130, thomas.hauser@usu.edu

1 Introduction

Computational Fluid Dynamics (CFD) in conjunction with a searching algorithm like a genetic algorithm (GA) potentially offers an efficient and robust optimization method and is a promising solution for current flow control designs. However, the traditional binary GAs and its operators need to be transformed or re-defined to meet the requirements of real world engineering problems. Three critical challenges are the length of binary substring encoding for high-precision, many parameter problems, the limited fine-grain searching ability of the basic GA approach, and the danger of finding local, not global, optima. Generally speaking, the methods dealing with the second challenge and those dealing with the third challenge will counter each other at some level. For example, dealing with the second challenge will require maintaining higher diversity among genomes during at least part of the GA evolution, but this is likely to slow down the convergence rate. Therefore, a key goal of this project is to improve the robustness of the search without significantly compromising the rate of convergence.

2 EARND Genetic Algorithm with Diversity Control

The current approach is to combine different existing GA techniques into a real-encoded 'Explicit Adaptive Range Normal Distribution' (EARND) genetic algorithm with an additional diversity control technique during the first 20% of the evolution to solve local convergence problems. First, a traditional binary-encoded GA is replaced by a real-encoded algorithm (Fig. 1) in which the corresponding design variables are encoded into a vector of real numbers that is conceptually closest to the real design space. Second, to improve convergence, an additional normal distribution scheme [1] is added into the basic GA to monitor the global optimization and balance the workload between the global and local searching process. Moreover, design parameter boundaries are explicitly updated to eliminate unnecessary evaluations (computations) in unpromising areas. Finally, diversity control is added to the selection function during the initial 20% of the evolution to reduce early local convergence.

The procedure for the normal distribution function is that at selected generations (every *NUpdate* generations), the next generation will be born according to a normal distribution rule based on the statistics of the set of the previously-generated best individuals. The set size is equal to the number of individuals in each generation times the number of generations between the normal distribution generations plus one (*NUpdate* × *NPopSize*). As an example, consider the creation of an individual consisting of *NVariable* variables using the normal distribution approach. For the jth variable of this new i^{th} individual, first randomly generate a value r_j^i:

$$r_j^i \in [0,1],\ 1 \leq i \leq NPopSize,\ 1 \leq j \leq NVariable\ . \tag{1}$$

Tnen find the corresponding value p_j^i from:

$$r_j^i = \int_{-\inf}^{p_j^i} N(0,1)(z)dz\ , \tag{2}$$

where $N(z)$ is the normal distribution. With the mean value μ_j and the standard deviation σ_j of the jth variable calculated from the previous best *NUpdate* × *NPopSize* individuals calculate the value of the jth variable of the new i^{th} individual, x_j^i, as following:

$$x_j^i = \mu_j + \sigma_j p_j^i\ . \tag{3}$$

In general, *NUpdate* is about one-tenth of the total number of generations.

The design parameter boundary updates occur at each *NUpdate* generation, but only in the second half of the evolution. The boundaries of each variable's design space is explicitly updated according to the statistics of the previous best *NUpdate* × *NPopSize* individuals. The new range, x_j^{Upper} and x_j^{Lower}, for the jth variable is defined by:

$$x_j^{Upper} = \min(x_j^{Upper_{old}}, \mu_j + \kappa\sigma_j),\ x_j^{Lower} = \max(x_j^{Lower_{old}}, \mu_j - \kappa\sigma_j), \tag{4}$$

with the previous boundary values designated by the old subscript. These new bounds are used until the next *NUpdate* generation is reached, at which point this process is repeated. In the current paper, $\kappa = 5.0$ which yields a conservative but robust searching process.

The diversity control is added to the selection function during the initial 20% of the evolution by adding a denominator d to the calculation of the scale fitness (5). The value of d is proportional to the number of similar individuals in that generation. This penalizes an given individual's scale fitness if multiple individuals fall in the same section of parameter space. The scaled fitness for an individual, an important parameter to create the next generation, is given by:

$$fit_{scale} = \frac{fit - fit_{min} + \gamma}{fit_{max} - fit_{min} + \gamma} \frac{1}{d} . \tag{5}$$

where γ is changing from $0.5(fit_{max} - fit_{min})$ to $0.25(fit_{max} - fit_{min})$ during the evolution process.

3 Two Suction Jets Control System Optimization

The results presented in this paper address flow-control over a NACA 0012 airfoil at a Reynolds number of 500,000 and an angle of attack of 18 deg using a two jet system. The computational approach is consistent with previous publications by the authors [2], and as such will only be described in brief. Likewise, a short summary of a study on diversity control on the jet optimization problem will be presented to demonstrate its effects before addressing the focus study of this paper, optimization of flow control system consisting of two suction jets.

3.1 Computational Details

The current grid setup is the same as the one used in the flow control study presented in reference [2]. The basic grid, without the jets implemented on the airfoil, consists of 15, two-dimensional blocks (Fig. 2). The dimensionless outer boundary of the computational area is chosen as $A_H \times A_W = 12c \times 12c$ in terms of chord length, c. The CFD code used is called GHOST, an in-house solver developed by P.G. Huang at the University of Kentucky. This code has been used in previous GA-CFD studies [3], as well as in numerous other research projects [4] [5]. For the GA-CFD simulations Menter's SST two-equation turbulence model is used due to its good predictive capabilities in separated flows.

3.2 Effect of Diversity Control

We investigated a single suction jet and a single blowing jet control systems on a NACA 0012 airfoil previously [2]. The five control parameters selected in the single jet research were suction location (L_{jS}), suction angle (θ_S), blowing location (L_{jB}), blowing angle (θ_B) and blowing amplitude (A_B), with a fixed suction velocity amplitude (A_S) of 3% of the free stream velocity. The suction amplitude was fixed so that the effects of suction did not completely dominate those of blowing. The EARND genetic algorithm without and with diversity control were applied to these systems to search for optimal lift/drag characteristics. From the converged history and final optimization results in Fig. 3, it can be seen that EARND genetic algorithm with diversity control has the best performance. The current optimization of a two suction jet system uses the same EARND genetic algorithm with diversity control.

3.3 The Two Suction Jet Control System GA Optimization

To stay consistent with the previous optimization, the current study focuses on a NACA 0012 airfoil with two fixed amplitude suction jets ($A_S = 0.03$). The locations of the suction jets, L_{j1} and L_{j2}, initially range from 5% to 80% of chord length and the angles, θ_1 and θ_2, range from -90 deg to 0 deg. The individuals of the initial generation are randomly generated. The evolution develops over 100 generations. Each generation consists of 16 individuals ($NPopSize = 16$) described by 4 variables ($NVariable = 4$) each. The other parameters of the GA are set as following: update frequency $NUpdate = 8$, crossover rate $P_c = 0.2$, mutation rate $P_m = 0.1$, and scale factor $s_j = 1.0$. The goal is to maximize the system aggregate fitness function

$$(Fit_A)_{\max} = a\frac{C_l}{C_{lB}} + b\frac{C_{dB}}{C_d}, \; a = 1, \; b = 1 \,, \tag{6}$$

where the coefficients of lift and drag (C_l, C_d) are compared to the baseline case with no jets (C_{lB}, C_{dB}).

Starting with a 0th generation of 16 random individuals, the GA is allowed to evolve for 100 generations. The best fitness value in the initial generation computed from eq. (6) is 2.17. At the end, the final generation has an optimum value of over 2.23. During the evolution, every 8 ($NUpdate$) generations the new generation will be generated according to the statistics of the best 128 ($NUpdate \times NPopSize$) individuals; their means (μ) and deviations (σ) are plotted in Fig. 4. It can be seen that by the 48th generation, the parameters are largely converged, although further evolution does occur in the second half.

4 Results and Conclusions

In Fig. 4, the values of four control parameters of the 100 most fit individuals within the 100 generations are plotted in sequence according to their fitness ranking. The best overall solution is $L_{j1} = 0.10183$, $\theta_1 = -89.917$, $L_{j2} = 0.14001$, $\theta_2 = -86.052$, with a fitness of 2.2379. Given that the jet width is $0.025c$, the jets notably do not overlap. The reduction of the separation bubble due to this suction is clearly seen in Fig. 5. The location of the jets near the leading edge with near-perpendicular suction is consistent with previous results from single jet [2] and one suction/one blowing jet studies [3].

The two main challenges faced by the GA have been meet by the EARND genetic algorithm with diversity control The GA-CFD combination is capable of determining an optimal configuration for a two jet system. With further validation against experiments and the increasing power of computers, this approach will be useful for the arrangement of jet arrays on an airfoil as well as similar parametric CFD challenges.

Support for this research was provided by Kentucky NASA EPSCoR, grant WKU 516140-02-06.

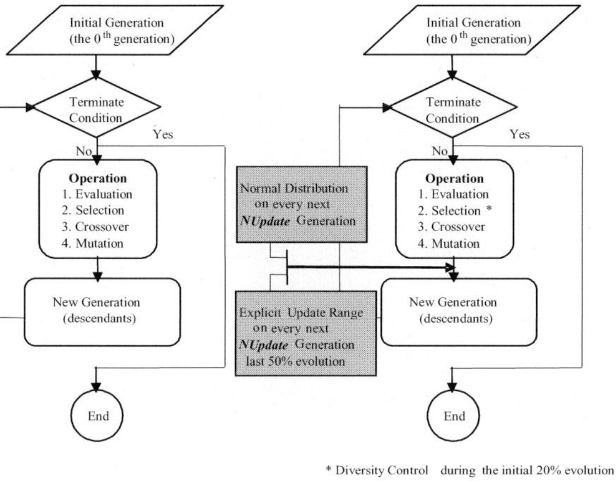

Fig. 1. Flow chart of the base (left) and the EARND (right) algorithm

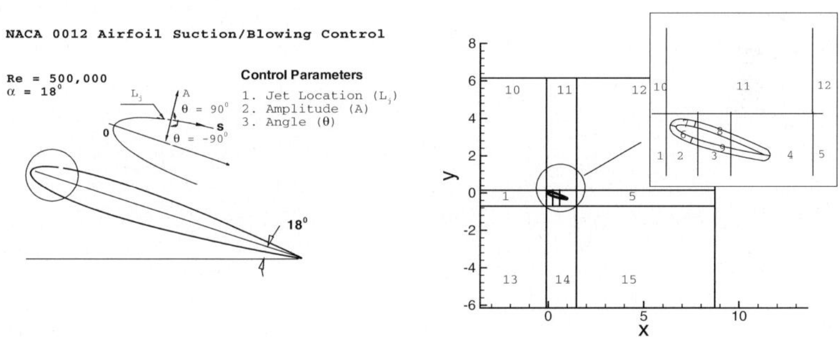

Fig. 2. Multi-block grid and airfoil parameters

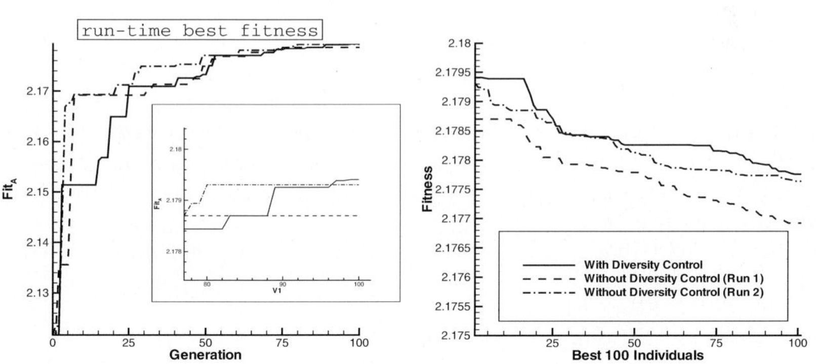

Fig. 3. Fitness over time (left) and overall (right) of the one suction and one blowing jet study, illustrating the effect of diversity control

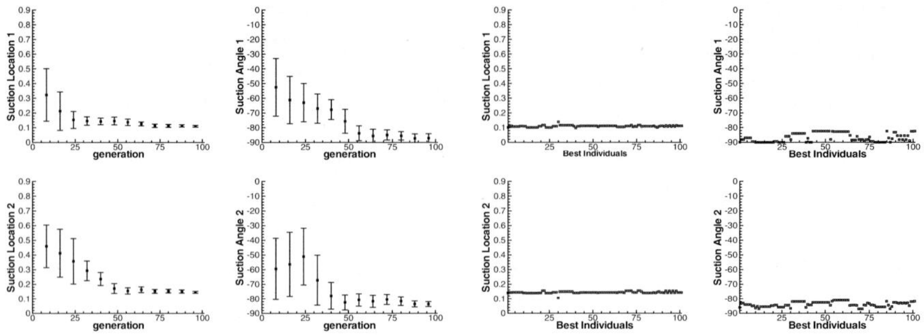

Fig. 4. The mean and deviation at every 8th (*NUpdate*) generation (left) and the hundred most fit individuals from the complete two suction jet evolution

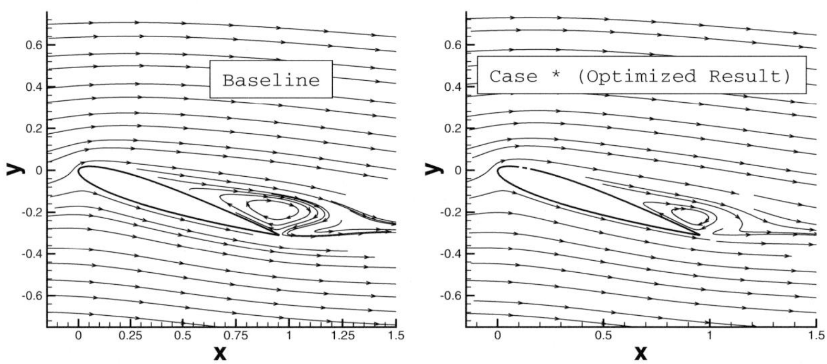

Fig. 5. Streamline plot comparing the no suction case (left) to the optimal two-suction jet solution from the GA (right), showing the reduction of the separation region

References

1. A. Oyama, S. Obayashi, K. Nakahashi: JSME Int. J., Series A **43**, 124 (2000)
2. L. Huang, P.G. Huang, R.P. LeBeau, T. Hauser: Numerical Study of Blowing and Suction Control Mechanism on NACA0012 Airfoil, accepted to AIAA J. Aircraft, assigned to **41** (2004)
3. L. Huang, P.G. Huang, R.P. LeBeau, T. Hauser: AIAA 2004-0225 (2004).
4. Y.B. Suzen, P.G. Huang, R.J. Volino, T.C. Corke, F.O. Thomas, J. Huang, J.P. Lake, P.I. King: AIAA 2003-3591 (2003)
5. Y.B. Suzen, P.G. Huang: J. Turbomachinery **125** 455 (2003)

Reynolds-Averaged Navier-Stokes Computations of a Synthetic Jet in a Turbulent Boundary Layer

Christopher L. Rumsey

NASA Langley Research Center, Hampton, VA 23681, c.l.rumsey@nasa.gov

Summary. An unsteady Reynolds-averaged Navier-Stokes (URANS) computation is performed for the flow of a synthetic jet issuing into a turbulent boundary layer through a circular orifice. Several numerical parameters are investigated, and the effects of three different turbulence models are explored. Results are compared to experiment.

1 Introduction

A CFD validation workshop related to synthetic jets (CFDVAL2004) [1] was held in March 2004 in Williamsburg, VA. The purpose of the workshop was to assess the current capabilities of different classes of turbulent flow solution methodologies to predict flow fields induced by synthetic jets and separation control geometries. Case 2 of the workshop was a synthetic jet issuing into a turbulent boundary layer through a circular orifice of diameter 6.35 mm in the floor [2]. The flow was characterized by a forcing frequency of 150 Hz, with a maximum discharge vertical velocity of approximately $1.3U_\infty$. The approach boundary layer thickness was approximately 21 mm, the freestream Mach number was $M = 0.1$, and the Reynolds number was approximately $\text{Re} = 14160$ per orifice diameter.

The problem of synthetic jet flow in a turbulent boundary layer has been studied both experimentally [3, 4, 5, 6, 7, 8] and computationally [9, 10, 11, 12, 13]. In general, this type of flow control may be useful for reducing separation or for achieving virtual aeroshaping. This paper reports the results obtained with the code CFL3D [14]. Several other participants also computed this test case for the workshop; their results are summarized in Rumsey et al.[15], as well as on the CFDVAL2004 website.

2 Numerical Method and Implementation Details

CFL3D is a finite volume code for solving the compressible Navier-Stokes equations. In time-accurate mode, CFL3D uses pseudo-time stepping with multigrid and achieves second order temporal accuracy. It employs third-order upwind-biased spatial differencing on the convective and pressure terms, and second-order differencing on the viscous terms; it is globally second-order spatially accurate. The flux difference-splitting (FDS) method of Roe is employed to obtain fluxes at the cell faces. Turbulence models are implemented uncoupled from the mean-flow equations. In this study, three different turbulence models were utilized: the one-equation Spalart-Allmaras model (SA) [16], the two-equation shear-stress transport model of Menter (SST) [17], and the explicit algebraic stress model (EASM) in k-ω form [18]. The first two models are both linear eddy-viscosity models that make use of the Boussinesq eddy-viscosity hypothesis. The EASM is a nonlinear model. The equations describing these models are not given here; they can be found in the references cited above.

The fine grid used for this case was a structured grid made up of 7 zones connected in a 1-to-1 fashion, and approximately 4.09 million grid points. A medium grid was made from the fine grid by extracting every-other point in each coordinate direction. The SA model was solved on both the fine and medium grids, whereas the other models were only solved on the medium grid. With pseudo-time stepping, subiterations are used to reduce the linearization and factorization errors, and advance the solution in pseudo-time to the next physical time. Details are given in Krist et al.[14]. In the current study, several different physical time steps and number of subiterations were employed. These are listed in Table 1.

Table 1. Summary of nomenclature associated with computations performed

Time advancement method	SA (fine)	SA (med)	SST (med)	EASM (med)
720 steps/cycle, 5 subits	SA-720-5f	SA-720-5m	SST-720-5m	
720 steps/cycle, 10 subits		SA-720-10m	SST-720-10m	
720 steps/cycle, 20 subits		SA-720-20m		
1440 steps/cycle, 10 subits			SST-1440-10m	EASM-1440-10m

For the SST model, an instability was noted in the solution when 5 subiterations were employed with 720 steps per cycle. This instability showed up as a series of un-physical oscillations in the flowfield variables at certain points in space during part of the cycle. By increasing the number of subiterations and halving the time step, the magnitude of the instability was dramatically reduced. No oscillations were evident for the SA or EASM models, although the EASM model would not run with 720 steps per cycle.

The boundary conditions were as follows. Upstream, the streamwise velocity component was specified to approximately match the experimental boundary layer thickness, and the turbulent shear stress was specified to approximate a fully-developed turbulent boundary layer. At the tunnel floor as well as on the inner lip of the orifice and upper wall of the cavity, solid wall adiabatic boundary conditions were applied. The side walls of the cavity were modeled as slip walls, and the bottom surfaces of the cavity employed a time-dependent boundary condition in which the moving diaphragm was modeled on the stationary grid through an imposed a vertical velocity: $w = [(\rho w)_{\max}/\rho] cos(2\pi Ft)$, where F is the frequency and t is the time. The $(\rho w)_{\max}$ was adjusted in order to achieve an approximate match of the vertical velocity component at the outflow of the orifice with the experiment. The resulting vertical component of velocity at the orifice exit as well as at a point 1D downstream can be seen in Fig. 1. The computations failed to capture the secondary peak in w seen at the orifice in the experiment during the downstroke. It should be noted that the experimental data exhibited significant spanwise velocities (order U_∞) during part of the cycle at this location, whose cause has not been accounted for. The current CFD method made no attempt to duplicate this v-velocity component. Also shown in this figure are the effects of fine vs. medium grids, the effects of different time steps and numbers of subiterations, and the effects of different turbulence models. These variations had almost no influence over the orifice, and only a relatively small influence 1D downstream.

3 Results

Representative results for this case are given here. Fig. 2 shows long-time-averaged u-velocity profiles at two locations downstream of the orifice. The

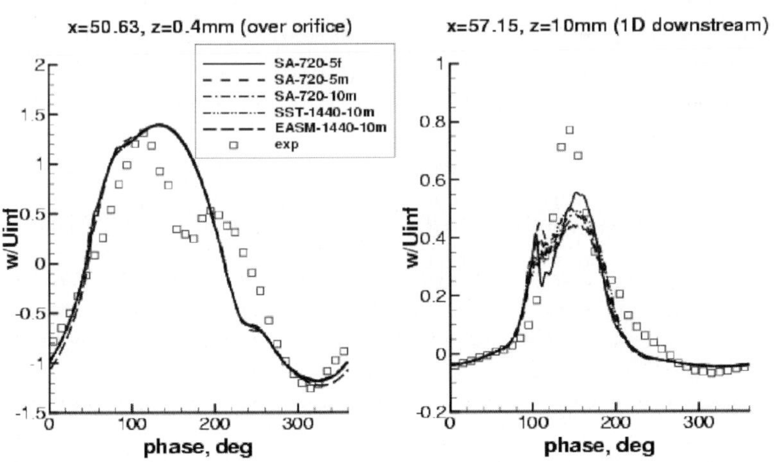

Fig. 1. Time history of w-velocity at two center-plane locations

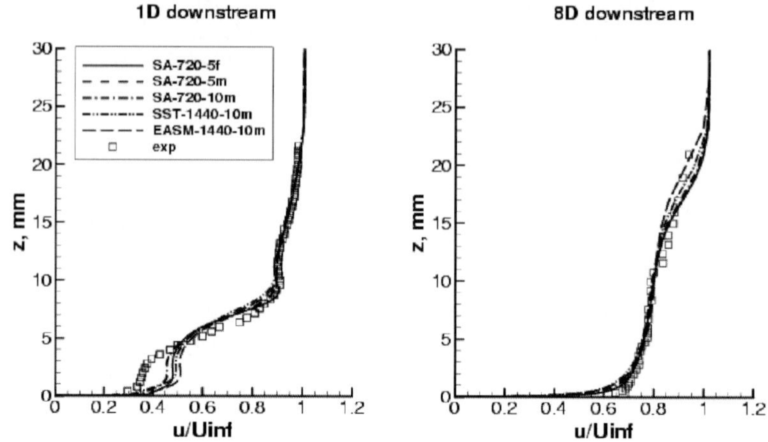

Fig. 2. Long-time-averaged u-velocity at two center-plane locations

various numerical parameters had little influence on the solution, and, except for a small difference in the inner part of the boundary layer 1D downstream, the CFD results were in good agreement with experiment. Phase-averaged results 1D downstream are shown in Fig. 3. Results were generally in very good agreement with experiment. However, at phase = 80°, the computations indicated a different response to the passing vortex than in the experiment: the reverse-flow region in the computations extended higher, and there was an overshoot in velocity near z=6 mm. The turbulence models had relatively little influence on the solution.

Contours of phase-averaged streamwise velocity at phase = 120° in the plane 1D downstream are shown in Fig. 4. Only the SA-720-5f computed result is shown; other results looked very similar. CFD predicted the overall structure of the passing vortex at this phase reasonably well.

4 Conclusions

A URANS CFD code was used to compute the unsteady flow of a synthetic jet issuing into a turbulent boundary layer through a circular orifice. The computation modeled the cavity and applied an assumed sinusoidal time-dependent boundary condition at the piston location. This boundary condition enabled the computation to capture the minimum and maximum vertical velocities at the orifice exit, but several other aspects of the experimental data

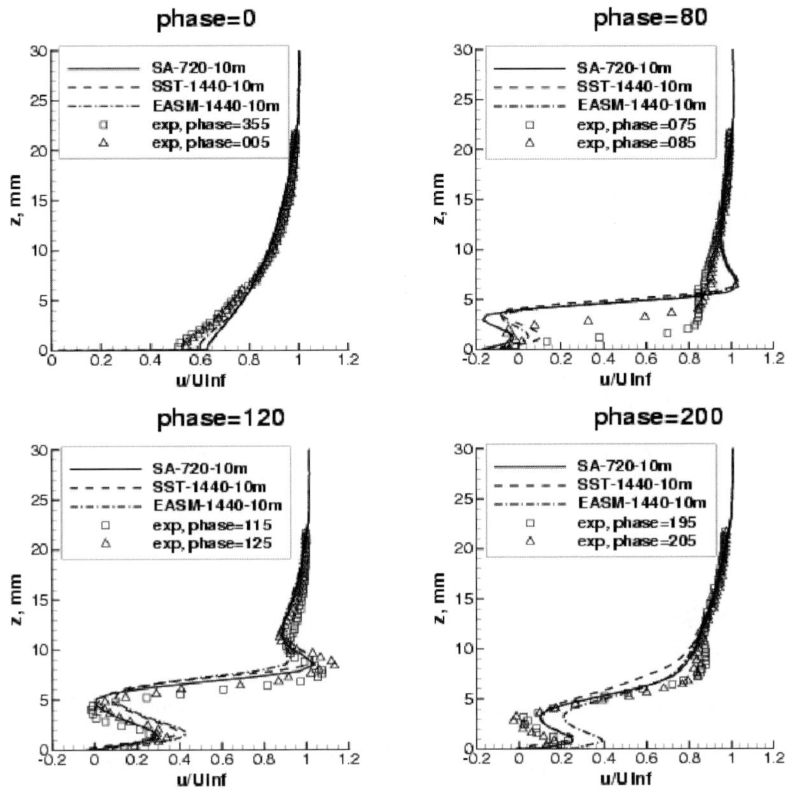

Fig. 3. Phase-averaged u-velocity at different phases 1D downstream (center-plane)

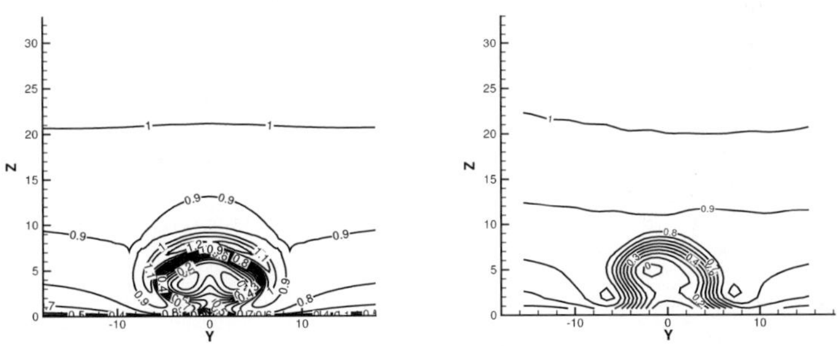

Fig. 4. Contours of u-velocity in the plane 1D downstream at phase=120; CFD on left, experiment on right

were missed. Although there were many differences evident in the quantitative comparisons, CFD was successful overall at predicting the global flow features – both long-time-averaged and phase-averaged – compared to experiment. Numerical parameters were investigated, and showed that the grid and time step were fine enough to capture the essential features of this flow field. The calculations did not exhibit a strong dependence on the type of turbulence model employed.

References

1. W. L. Sellers and C. L. Rumsey: http://cfdval2004.larc.nasa.gov (2004)
2. N. W. Schaeffler and L. N. Jenkins: The isolated synthetic jet in crossflow: a benchmark for flow control simulation. AIAA paper 2004-2219, June-July (2004)
3. S. D. Peterson and M. W. Plesniak: J. Fluid Mech. **503** (2004), pp 57–91
4. I. M. Milanovic and K. B. M. Q. Zaman: AIAA J. **42**, 5 (2004), pp 874–882
5. S. H. Smith and M. G. Mungal: J. Fluid Mech. **357** (1998), pp 83–122
6. R. M. Kelso, T. T. Lim, and A. E. Perry: J. Fluid Mech. **306** (1996), pp 111-144
7. J. Andreopoulos and W. Rodi: J. Fluid Mech. **138** (1984), pp 93–127
8. D. Crabb, D. F. G. Durao, and J. H. Whitelaw: J. Fluids Engng. **103** (1981), pp 142–153
9. L. L. Yuan, R. L. Street, and J. H. Ferziger: J. Fluid Mech. **379** (1999), pp 71–104
10. J. Cui, R. K. Agarwal, and A. W. Cary: Numerical simulation of the interaction of a synthetic jet with a turbulent boundary layer. AIAA paper 2003-3458, June (2003)
11. K. M. B. Gustafsson and T. G. Johansson: Turbulence and velocity fields of slanted jets in cross-flow – measurements and CFD simulations. In: *Turbulence, Heat and Mass Transfer 4*, ed by K. Hanjalic, Y. Nagano, and M. Tummers (Begell House, Inc., New York 2003) pp 763–770
12. R. Mittal, P. Rampunggoon, and H. S. Udaykumar: Interaction of a synthetic jet with a flat plate boundary layer. AIAA paper 2001-2773, June (2001)
13. R. Mittal and P. Rampunggoon: Phys. Fluids **14**, 4 (2002), pp 1533–1536
14. S. L. Krist, R. T. Biedron, and C. L. Rumsey: CFL3D user's manual. NASA/TM-1998-208444, June (1998)
15. C. L. Rumsey, T. B. Gatski, W. L. Sellers III, V. N. Vatsa, and S. A. Viken: Summary of the 2004 CFD validation workshop on synthetic jets and turbulent separation control. AIAA paper 2004-2217, June-July (2004)
16. P. R. Spalart and S. R. Allmaras: La Recherche Aerospatiale **1** (1994), pp 5–21
17. F. R. Menter: AIAA J. **32**, 8 (1994), pp 1598–1605
18. C. L. Rumsey and T. B. Gatski: Summary of EASM turbulence models in CFL3D with validation test cases. NASA/TM-2003-212431, June (2003)

Active Control of Shock/Boundary Layer Interaction in Transonic Flow Over Airfoils

Jose L. Vadillo[1], Ramesh K. Agarwal[1], and Ahmed A. Hassan[2]

[1] Department of Mechanical Enginering and Aerosapace Engineering, Washington University in St. Louis, MO 63130, USA, rka@me.wustl.edu
[2] The Boeing Company, Mesa, AZ

Summary. The objective of this paper is to evaluate, via numerical simulation, the feasibility of controlling the shock/boundary layer (SBL) interaction on airfoils in transonic flow thereby weakening the shock wave(s) and reducing the pockets of supersonic regions using synthetic jet actuators and thus achieving the desired changes in aerodynamic forces and moments. It is shown that by properly choosing the amplitude, frequency, momentum and location of the synthetic jet on the airfoil surface, desired modulation in lift, drag and pitching moment coefficients can be obtained for a given free-stream Mach number, Reynolds number and angle of attack. These calculations indicate the possibility of "hingeless" control of airplanes in the future using active flow control (AFC).

1 Introduction

In recent years, there has been a great deal of emphasis toward the development of advanced aerodynamic technologies based on the fluidic modification of aerodynamic and propulsive flowfieds forces, that can cover multiple flight conditions without the need for conventional control surfaces such as flaps, spoilers and variable wing sweep. The fluidic modification (or flow control) can be accomplished by employing micro-surface effectors or actuators operated by an intelligent control system. In subsonic flow, recently it has been shown experimentally by Amitay et al. [1] and computationally by the authors of this paper [2] that the pressure drag of an airfoil at low speeds and small angles of attack can be significantly reduced with a minimum change in lift by modification of the apparent aerodynamic shape of the airfoil using a synthetic jet actuator which displaces the local streamlines sufficiently to modify the local pressure distribution. These results have been very encouraging. However, although some computational and experimental work has been done in the application of AFC for virtual aeroshaping of airfoil to reduce drag in subsonic flow, hardly any work has been reported in the literature to date for active control of the strength of shock waves and their unsteadiness, the supersonic regions and shock-induced separation due to shock/boundary layer interaction. This problem is of great relevance for reducing the transonic cruise drag of a transport aircraft as well as for achieving "hingeless" control. This paper evaluates numerically the influence of a synthetic jet in modulating the forces and moments on an airfoil in transonic flow. By locally changing the surface pressures, one can affect the aerodynamic forces and moments experienced by the surface. At low subsonic speeds, it can be accomplished through the manipulation of the boundary layer flow (either by attaching an otherwise separated flow or, forcing the detachment of the flow such as in a spoiler or an air brake). At transonic and supersonic speeds, this can be accomplished through the manipulation of shock waves where a small disturbance in the vicinity of the shock can result in large changes in the aerodynamics of the configuration.

2 Results and Discussions

CFD simulations are performed by employing an extensively validated well known multi-zone Reynolds Averaged Navier-Stokes (RANS) flow solver WIND[3]. The solver has several turbulence models - one-equation Spalart-Allmaras model, two-equation Menter's Shear Stress Transport (SST) model and the Large Eddy Simulation (LES) model. Recently the code was modified to include the Detached Eddy Simulation (DES) formulation in the Spalart-Allmaras model as well as in the SST model. This extension allows for combining the RANS and LES models in a consistent manner in the flow field. For all the computations reported in this paper, the grid refinement study was performed to ensure that the computed solutions are grid-independent. The uncertainty in the computed solutions due to the numerics (size of computational domain, grid, numerical algorithm and boundary conditions) was minimized. The actual geometry of the control device (e.g. cavities of the synthetic jets) is included in the computations for accurate simulation of the flowfield; the simplified boundary conditions implemented on the airfoil surface can introduce significant errors.

The first part of the paper deals with the study of pressure-drag reduction of an airfoil in subsonic flow. To show the effect of virtual aeroshaping on pressure-drag reduction, Amitay et al. [1] performed experiments on a 24% thick Clark-Y airfoil wherein they deliberately introduced a small obstruction (bump) on the upper surface of the airfoil (Figure 1) to locally create a small stationary flow interaction domain adjacent to the surface to displace the local streamlines sufficiently to modify the local pressure distribution. Then they introduced a synthetic jet immediately downstream of the bump and experimentally measured the effect on drag and lift coefficients by varying the amplitude, frequency and momentum coefficient of the jet. The flow conditions were subsonic at Mach number 0.085, Reynolds number $R_{ec} = 381,000$ and angle of attack $= 3^o$. We have performed calculations to compare with the experimental data of Amitay et al. [1]. Figure 1 shows a non-scaled sketch of the airfoil with the bump. Figure 2 shows the zoomed-in view of the C-grid around the airfoil. Figure 3 shows the comparison of pressure distribution between numerical and experimental results for the baseline case without bump and synthetic jet. Figure 4 shows the computed Mach number contours on the baseline airfoil. Figures 5 and 6 show the computed pressure distributions on the airfoil in blowing and suction strokes of the synthetic jet of $C\mu = 1.2\text{e-}3$ at $f = 362Hz$ and $850Hz$ respectively and their comparisons with time-averaged experimental pressure distributions. A reasonable agreement between the computations and experimental data is obtained. Figures 7 and 8 show the lift and pressure-drag coefficients as a function of the actuation frequency of the synthetic jet. It is clear that the maximum benefit in drag reduction with minimal change in lift is achieved at higher frequencies of the synthetic jet.

The second part of the paper deals with the control of shock/boundary-layer interaction over a NACA64010 airfoil with the objective of achieving the desired modulation of aerodynamic forces and moments by weakening the shock wave(s) and reducing the pockets of supersonic regions. The goal is to achieve the "hingeless" control by using arrays of transverse oscillatory jets as shown in Figure 9. Figure 10 show the Schlieren photograph of the effect

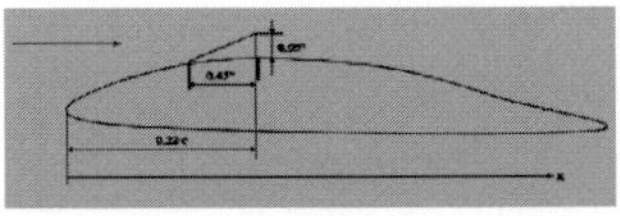

Fig. 1. Non-scaled sketch of the Clark-Y airfoil with bump

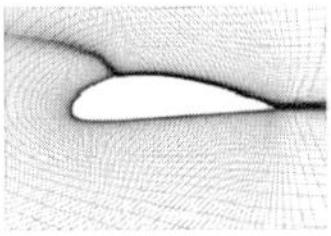

Fig. 2. Computational domain and C-mesh around the Clark-Y airfoil with a bump

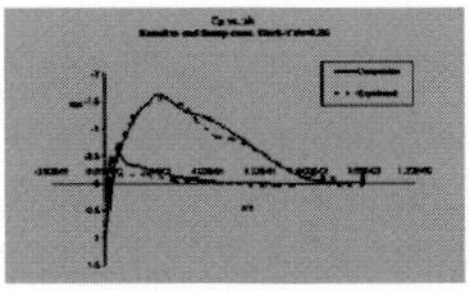

Fig. 3. Computed and experimental pressure distributions on the Clark-Y airfoil; $M_\infty = 0.085$, $Re_c = 381,000$, $\alpha = 3°$

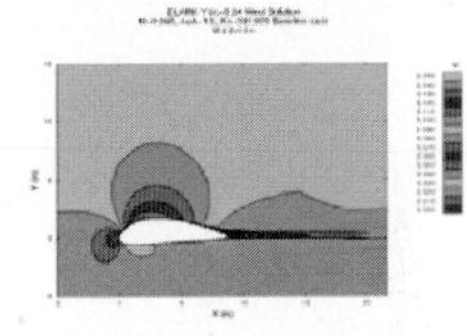

Fig. 4. Computed Mach number contours for flow past a 24% thick Clark-Y airfoil; $M_\infty = 0.085$, $Re_c = 381,000$, $\alpha = 3°$

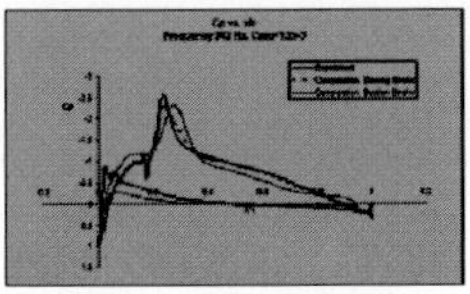

Fig. 5. Pressure distribution on Clark-Y airfoil with a bump and synthetic jet; $M_\infty = 0.085$, $Re_c = 381,000$, $\alpha = 3°$, $f = 362Hz$, $C\mu = 1.2\text{e-}3$

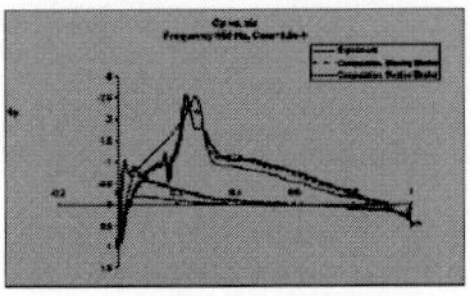

Fig. 6. Pressure distribution on Clark-Y airfoil with a bump and synthetic jet; $M_\infty = 0.085$, $Re_c = 381,000$, $\alpha = 3°$, $f = 850Hz$, $C\mu = 1.2\text{e-}3$

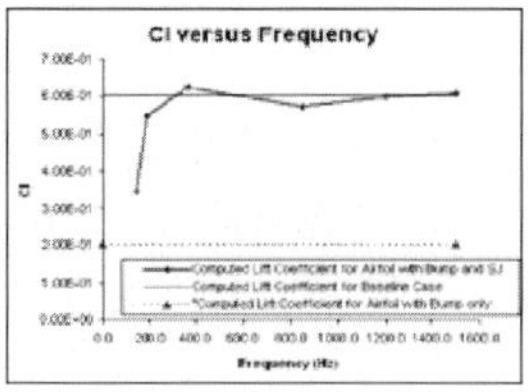

Fig. 7. Lift coefficient as a function of the actuation frequency for a Clark-Y airfoil with a bump and a synthetic jet

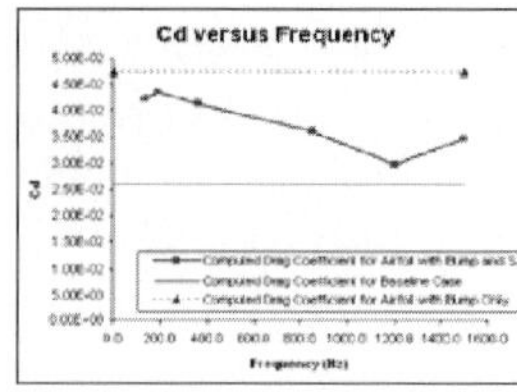

Fig. 8. Pressure drag coefficient as a function of the actuation frequency for a Clark-Y airfoil with a bump and a synthetic jet

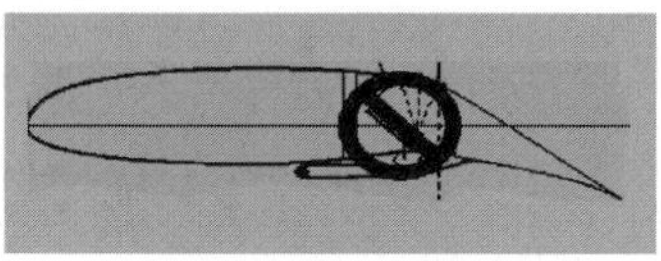

Fig. 9. Hingeless control: Trim forces can be effectively generated using arrays of transverse oscillatory jets

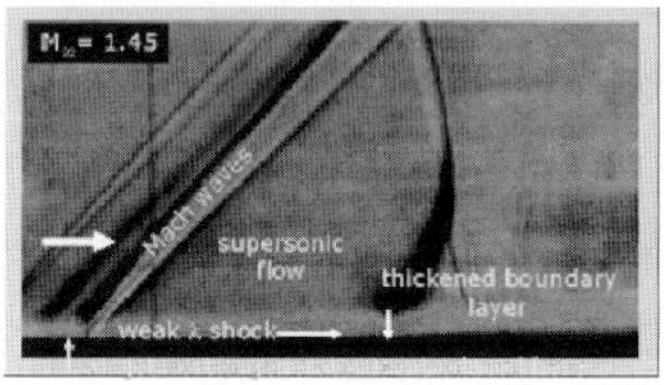

Fig. 10. Schlieren photograph of shock/boundary layer interaction on a flat plate, with active flow control using air-jet vortex-generators from the experiment of Doerffer et al. [4]; $M_\infty = 1.45$

Fig. 11. Computed Mach number contours for flow past a NACA64010 airfoil applying an oscillatory jet control; $M_\infty = 0.85$, $R_{ec} = 5.67e{+}06$, $\alpha = 1°$, $f = 150Hz$, $M_{jet} = 0.3$, $b/c = 0.0044$, $x_L/c = 0.53$

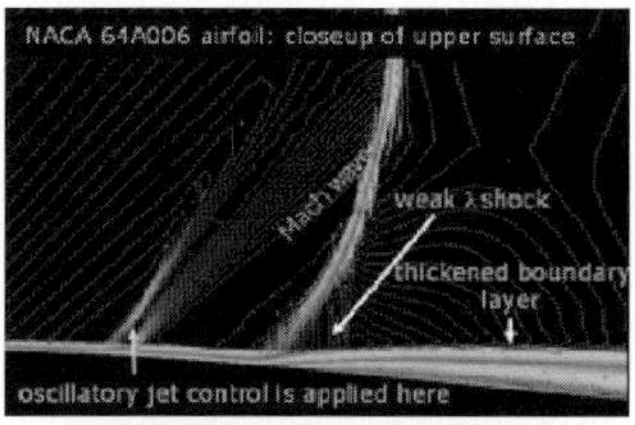

Fig. 12. Zoomed-in detail view of the weak shock and corresponding thickened boundary layer over NACA64010 airfoil with control of shock/boundary-layer interaction using a synthetic jet

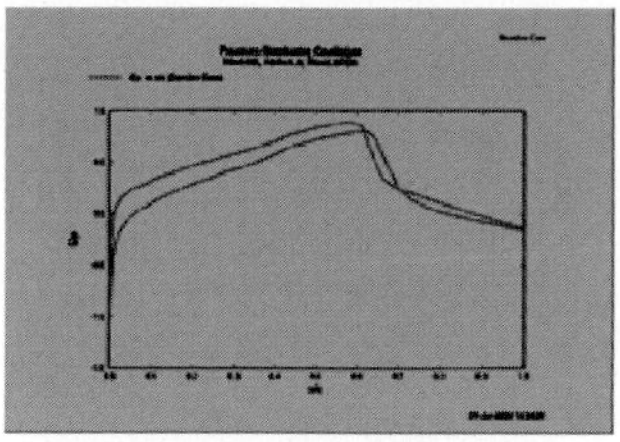

Fig. 13. Pressure Coefficient Distribution on a NACA64010 airfoil; $M_\infty = 0.85$, $R_{ec} = 5.67\text{e}{+}06$, $\alpha = 1°$, $f = 150Hz$, $M_{jet} = 0.3$, $b/c = 0.0044$, $x_L/c = 0.53$

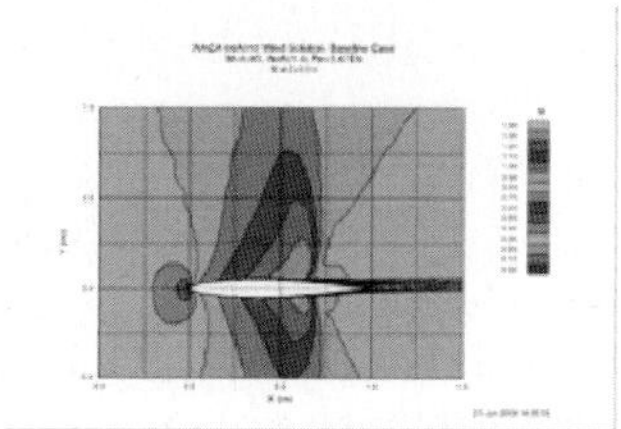

Fig. 14. Mach Contours for flow over a NACA64010 airfoil; $M_\infty = 0.85$, $R_{ec} = 5.67\text{e}{+}06$, $\alpha = 1°$, $f = 150Hz$, $M_{jet} = 0.3$, $b/c = 0.0044$, $x_L/c = 0.53$

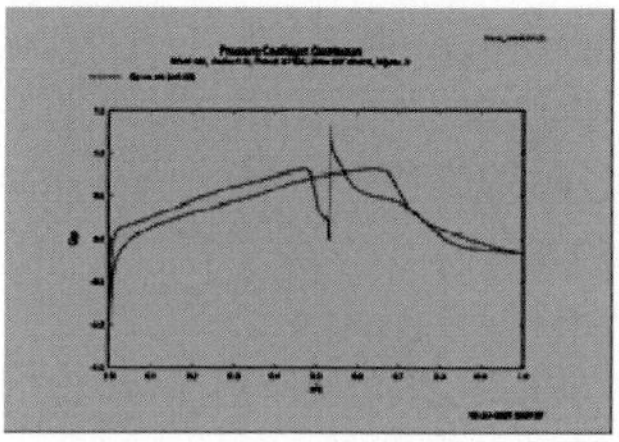

Fig. 15. Pressure Coefficient Distribution on a NACA64010 airfoil in maximum blowing; $M_\infty = 0.85$, $R_{ec} = 5.67\text{e}{+}06$, $\alpha = 1°$, $f = 150Hz$, $M_{jet} = 0.3$, $b/c = 0.0044$, $x_L/c = 0.53$

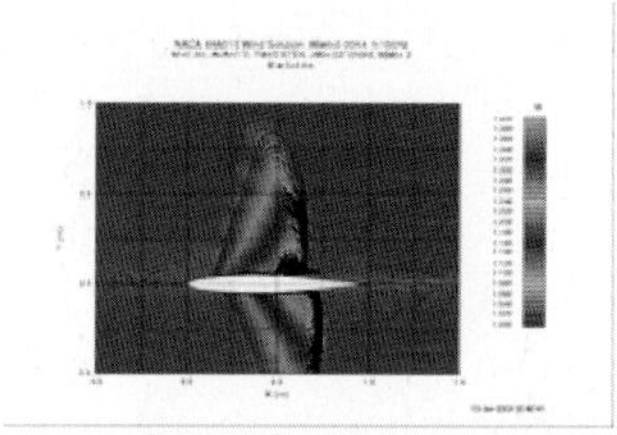

Fig. 16. Mach Contours for flow over a NACA64010 airfoil in maximum blowing; $M_\infty = 0.85$, $R_{ec} = 5.67\text{e}{+}06$, $\alpha = 1°$, $f = 150Hz$, $M_{jet} = 0.3$, $b/c = 0.0044$, $x_L/c = 0.53$

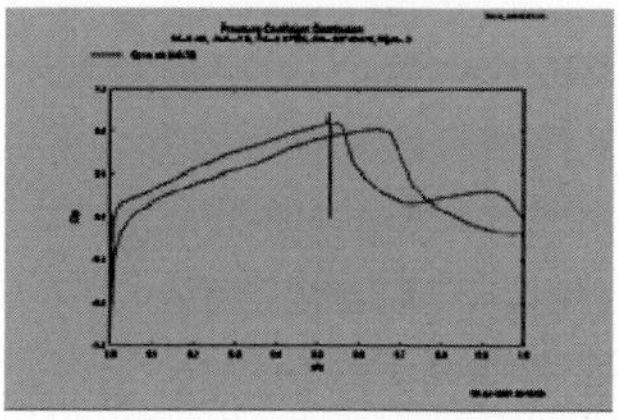

Fig. 17. Pressure Coefficient Distribution on a NACA64010 airfoil in maximum suction; $M_\infty = 0.85$, $R_{ec} = 5.67\text{e}{+}06$, $\alpha = 1°$, $f = 150Hz$, $M_{jet} = 0.3$, $b/c = 0.0044$, $x_L/c = 0.53$

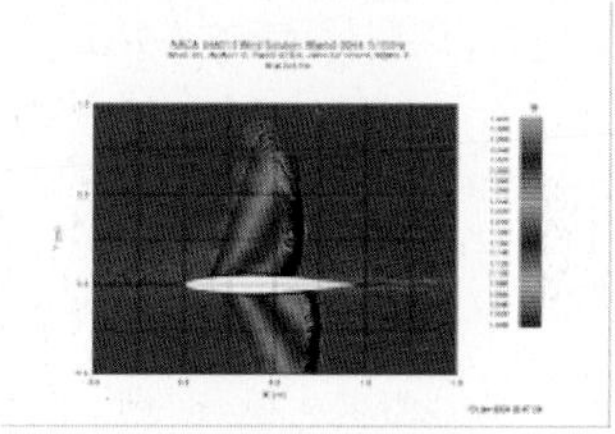

Fig. 18. Mach Contours for flow over a NACA64010 airfoil in maximum suction; $M_\infty = 0.85$, $R_{ec} = 5.67\text{e}{+}06$, $\alpha = 1°$, $f = 150Hz$, $M_{jet} = 0.3$, $b/c = 0.0044$, $x_L/c = 0.53$

of air-jets on SBL interaction on a flat plate obtained recently by Doerffer et al. [4]. Figure 11 presents a simulation of transonic flow over NACA64010 airfoil at Mach number = 0.85, angle of attack $\alpha = 3^o$, Reynolds number $R_{ec} = 5.67e{+}06$, $M_{jet} = 0.3$, $b/c = 0.0044$, $x_L/c = 0.53$, and $f = 150Hz$ (where b = jetwidth, c = airfoil chord, and x_L/c = jet location on airfoil surface) such that the flow in the supersonic and shocked region corresponds to the experiments of Doerffer et al. [4]. Figure 12 shows a zoomed-in view of this numerical simulation which shows a close match with the Schlieren photograph of Doerffer et al. [4].

A large number of numerical calculations were performed for flow over a NACA64010 airfoil, by varying the free-stream conditions and parameters of the synthetic jet, namely the jet width, the jet velocity and location. One numerical result is included here to show the influence of a synthetic jet in modulating the forces and moments on an airfoil in transonic flow, in particular the influence on pitching moment. Figures 13-18 show the pressure coefficient distribution and Mach contours for transonic flow over the airfoil at Mach number = 0.85, angle of attack $\alpha = 1^o$, Reynolds number $R_{ec} = 5.67e6$, $M_{jet} = 0.3$, $b/c = 0.0044$, $x_L/c = 0.53$, and $f = 150Hz$. SST model is employed in the computations. In particular, Figures 13 and 14 shown the results for the baseline airfoil without actuation. Figures 15 and 16 show the results for maximum blowing and Figures 17 and 18 show the results for maximum suction of the synthetic jet. In summary, the computations show that the forces and moments on an airfoil in transonic flow can be modulated by employing a synthetic jet. The desired values for C_L, C_D and C_M can be obtained by judiciously choosing the parameters of the synthetic jet for a given baseline flow.

References

1. M. Amitay, M. Horvath, M. Michaux, A. Glezer: *Virtual Aerodynamic Shape Modification at Low Angles of Attack Using Synthetic Jet Actuators*, AIAA Paper 01-2975, 2001
2. J.L. Vadillo, R.K. Agarwal, A.W. Cary, W.W. Bower: *Numerical Study of Virtual Aerodynamic Shape Modification of an Airfoil Using a Synthetic Jet Actuator*, AIAA Paper 03-4148, 2003
3. R.H. Bush: *The production Flow Solver of the NPARC Alliance*, AIAA Paper 88-0935, 1998
4. P. Doerffer, R. Szwaba: *Shock Wave Boundary Layer Interaction Control By Streamwise Vortices*, Proceedings of ICTAM, Warsaw, Poland, August 2004

Part IX

Fluid-Structure Interaction

Adaptive Solution of Some Steady-State Fluid-Structure Interactions

S. Étienne[1] and D. Pelletier[1]

École Polytechnique de Montréal, Montréal, Quebec, Canada H3C 3A7

Summary. This paper presents a monolithic formulation for the steady-state interaction of a viscous incompressible flow with an elastic structure undergoing large displacements. This is an initial step towards developing an unsteady analysis formulation for this class of problems. The entire problem is solved in an implicit manner via a Newton-Raphson pseudo-solid approach, using fluid velocity, pressure, and pseudo-structural displacements as unknowns in the flow domain and displacements in the structural components. The adaptive formulation is verified on a problem with a closed form solution and applied to a flexible hydrofoil.

1 Introduction

Interaction between solids and fluids has been a topic of engineering interest for many years. This paper presents a formulation which treats interactions between an incompressible flow and a structure undergoing large displacements.

Coupling between fluid and structure can be addressed in either of two ways. In *weakly-coupled* methods algorithms for the structure and the fluid are segregated. In *tightly-coupled* or *monolithic* methods [1, 2, 3] the formulation guarantees satisfaction of equilibrium of the interface between fluid and solid. This approach has been chosen for consistency reasons and to avoid instabilities stemming from segregated solutions. Following Sackinger *et al.*[4], a full Newton-Raphson pseudo-solid approach has been chosen to manage the geometry of the fluid region.

To cope with the deformations of both boundaries and fluid mesh, several techniques have been studied and discussed in the literature for unstructured meshes [4, 5]. We have opted for the pseudo-solid approach of Sackinger [4] because it allows for a fully coupled continuum formulation and is also compatible with existing mesh adaptation procedure.

The paper begins with a description of the steady state governing equations for laminar incompressible fluids, hyperelastic solid behavior, pseudo-solid material law and fluid-structure interface equilibrium. The following sections detail the solution procedure, the coupling of the fluid and solid domains, and the adaptive finite element procedure for the fluid-structure. Results are presented and the article ends with conclusions.

2 Governing Equations

The steady flow of an incompressible fluid is described by the continuity and momentum equations [6]:

$$\nabla \cdot \mathbf{u}_f = 0 \tag{1}$$
$$\rho_f \mathbf{u}_f \cdot \nabla \mathbf{u}_f = \nabla \cdot \boldsymbol{\sigma}_f + \mathbf{f} \tag{2}$$

where ρ_f is the fluid density, $\mathbf{u_f}$ is the fluid velocity, $\boldsymbol{\sigma}_f$ is the total fluid stress tensor, and $\mathbf{f}$ is a body force. We assume Newtonian fluid thus:

$$\boldsymbol{\sigma}_f = \mu_f(\nabla \mathbf{u}_f + (\nabla \mathbf{u}_f)^T) - p\mathbf{I}$$

where μ_f is the dynamic viscosity and p is the fluid pressure. The flow equations are closed with the following Dirichlet and Neumann boundary conditions,

$$\mathbf{u}_f = \overline{\mathbf{u}}_f \text{ on } \Gamma_D^f \qquad \boldsymbol{\sigma}_f \cdot \mathbf{n} = \overline{\mathbf{t}}_f \text{ on } \Gamma_N^f$$

where $\overline{\mathbf{u}}_f$ is the imposed velocity, and $\overline{\mathbf{t}}_f$ is a prescribed traction.

A Total Lagrangian formulation is ideally suited for the structural components [7]. In that context, the Piola-Kirchoff (or 2nd Piola-Kirchoff) stress tensor, $\boldsymbol{\sigma}_k$, corresponds to stresses on the undeformed configuration in terms of the undeformed surface; and the Piola-Lagrange (or 1st Piola-Kirchoff) stress tensor, $\boldsymbol{\sigma}_l$, corresponds to stresses on the deformed configuration in terms of the undeformed surface. We introduce the deformation gradient tensor $\mathbf{F} = \mathbf{I} + \mathbf{h}$ where $\mathbf{h} = \nabla \boldsymbol{\chi}$ represents the displacement gradient tensor expressed on the undeformed or initial configuration ($\boldsymbol{\chi} = \xi \mathbf{e}_x + \eta \mathbf{e}_y$ are the displacements). The Piola-Lagrange, the Piola-Kirchoff and the Cauchy stress tensors are related by $\boldsymbol{\sigma}_l = \mathbf{F}\boldsymbol{\sigma}_k$ and $\boldsymbol{\sigma}_c = \mathbf{F}\boldsymbol{\sigma}_k\mathbf{F}^T/J$ with $J = \det(\mathbf{F})$ the Jacobian of the transformation. The solid is assumed hyperelastic, isotropic and described by a St. Venant-Kirchoff material: $\boldsymbol{\sigma}_k = \lambda_s tr(\mathbf{E})\mathbf{I} + 2\mu_s \mathbf{E}$, λ_s and μ_s are the Lam constants, and $\mathbf{E} = \frac{1}{2}(\mathbf{h} + \mathbf{h}^T + \mathbf{h}.\mathbf{h}^T)$ is the nonlinear Green-Lagrange strain tensor. In the Total Lagrangian approach, equilibrium is expressed on the initial undeformed configuration:

$$\nabla \cdot \boldsymbol{\sigma}_l + \mathbf{f} = 0 \tag{5}$$

The structural equations are supplemented by the following Dirichlet and Neumann boundary conditions,

$$\boldsymbol{\chi}_s = \overline{\boldsymbol{\chi}_s} \text{ on } \Gamma_D^s \qquad \boldsymbol{\sigma}_l \cdot \mathbf{n} = \overline{\mathbf{t}}_s \text{ on } \Gamma_N^s$$

where $\overline{\boldsymbol{\chi}}_s$ a prescribed displacement, and $\overline{\mathbf{t}}_s$ is a prescribed traction.

A pseudo-solid formulation is used to manage the deforming geometry of the fluid domain:

$$\nabla \cdot \boldsymbol{\sigma}_l^{ps} = 0$$
$$\boldsymbol{\sigma}_k^{ps} = \lambda_{ps} tr(\mathbf{E}^{ps})\mathbf{I} + 2\mu_{ps}\mathbf{E}^{ps}$$

with $\mathbf{E}^{ps} = \frac{1}{2}(\mathbf{h}^{ps} + (\mathbf{h}^{ps})^T)$ and λ_{ps} and μ_{ps} the pseudo-solid Lam constants. The following boundary conditions are used:

$$\boldsymbol{\chi}_{ps} = 0 \text{ on } \Gamma - \Gamma_I \qquad \boldsymbol{\chi}_{ps} = \overline{\boldsymbol{\chi}_s} \text{ on } \Gamma_I$$

Their role is to provide physics based rules for moving the fluid mesh given the structural displacements at the solid-fluid interface.

The fluid and the solid are coupled by the no-slip condition and equilibrium of forces at the interface:

$$\mathbf{u}_f = \mathbf{u}_s = 0 \text{ on } \Gamma_I$$
$$\boldsymbol{\sigma}_c \cdot \mathbf{n}_s + \boldsymbol{\sigma}_f \cdot \mathbf{n}_f = 0 \text{ on } \Gamma_I$$

where $\mathbf{n}_s$ is the outward unit normal to the solid in the deformed configuration, $\mathbf{n}_f = -\mathbf{n}_s$, and $\mathbf{u}_s$ is the solid velocity. The stresses $\boldsymbol{\sigma}$ are Cauchy stresses, *i.e.* expressed on the deformed configuration in terms of the deformed surface.

3 Solution Strategy

We have chosen a monolithic solution strategy coupling all degrees of freedom: velocities, pressure, pseudo-solid and structural displacements. The monolithic approach also requires that all boundary conditions and interface loadings be treated implicitly. We proceed in 3 key steps:

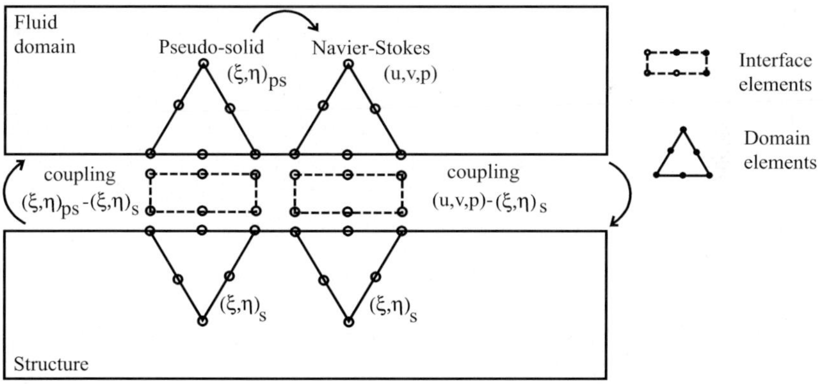

Fig. 1. Sketch of boundary and interface elements needed for the coupling.

- The boundary conditions of the pseudo-solid equations are of Dirichlet type: $\boldsymbol{\chi}_{ps} = \boldsymbol{\chi}_s$.
- Linearization of the pseudo-solid equations must account for all dependencies to achieve quadratic convergence of Newton's method [8].

– The fluid loads are applied to the structure through nodal reactions [8, 9].

These steps are implemented simply and in a straight forward manner through the introduction of interface elements, see fig. 1. Quadratic convergence of Newton's method is achieved at the cost of an increased number of unknowns compared to a decoupled approach. This is largely compensated by the significant reduction in the number of Newton iterations and CPU time.

4 Adaptive Finite Element Procedure

The velocity and displacement fields are discretized using 6-noded quadratic elements. Fluid pressure is discretized by piecewise linear continuous functions assembled in a skyline structure and solved by Gaussian elimination. Error estimates are obtained for all solution fields by a Zhu-Zienkewicz [10, 11] local projection error estimator. In our approach all variables are analyzed and contribute to the mesh adaptation process. For this, an error estimate is obtained separately for each dependent variable.

5 Numerical Results

First analytical solutions, in the solid and fluid regions are given by [8]:

$$\chi = \frac{r_e^2}{r}((\cos(\alpha) - 1)\mathbf{e_r} + \sin(\alpha)\mathbf{e}_\theta) \tag{8}$$

$$\mathbf{u} = (\tanh(a(r - r_i)) + \tanh(a(r_e - r)) - \tanh(a(r_e - r_i)))\mathbf{e}_\theta \tag{9}$$

with r_i the internal radius, r_e the interface radius. For given values of the parameter α and hydrostatic pressure, there exists one shape of the angular velocity u_θ.
Figure 2 compares the error estimates to the true error. This data confirms that the scheme is second order accurate for the solution derivatives. Furthermore, the estimator tends to the true error as the mesh is refined and the minimum size predicted by all fields is retained.

The proposed approach provides the tool to study the effect of flexibility on the drag and lift of an hydrofoil. Here, we have chosen an Eppler 817 hydrofoil. The Reynolds number is $R_e = \rho U_\infty C/\mu = 1000$ based on the profile chord C, free stream fluid velocity U_∞ and viscosity; the Young modulus is $E = 600$ and the Poisson coefficient is $\nu = 0.25$. The angle of attack is set to 6°. The computational domain extends a distance of $5C$ from the foil. The flow is horizontal from left to right. Figures 3(a) to 3(d) show the fluid and structural domains for the rigid and flexible foil configurations. Note the resulting deformations of the foil. A pressure trough causes the concavity observed on the suction side (Fig.3(d)). Second, the deformed geometry appears less aerodynamic than its rigid counterpart. Third, the foil is deformed to more or less align itself with the flow.

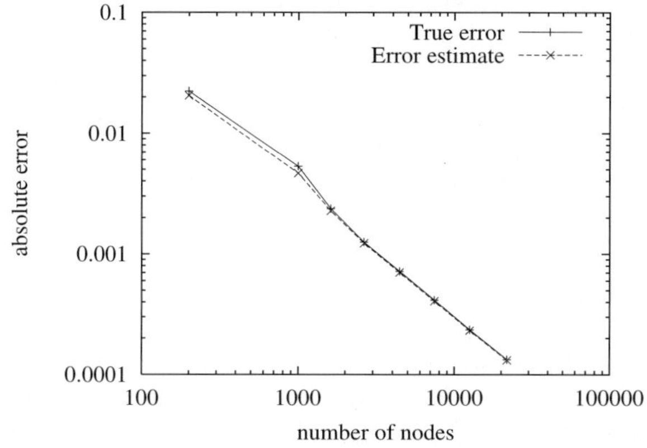

Fig. 2. Grid convergence for cylinder subject to viscous shear.

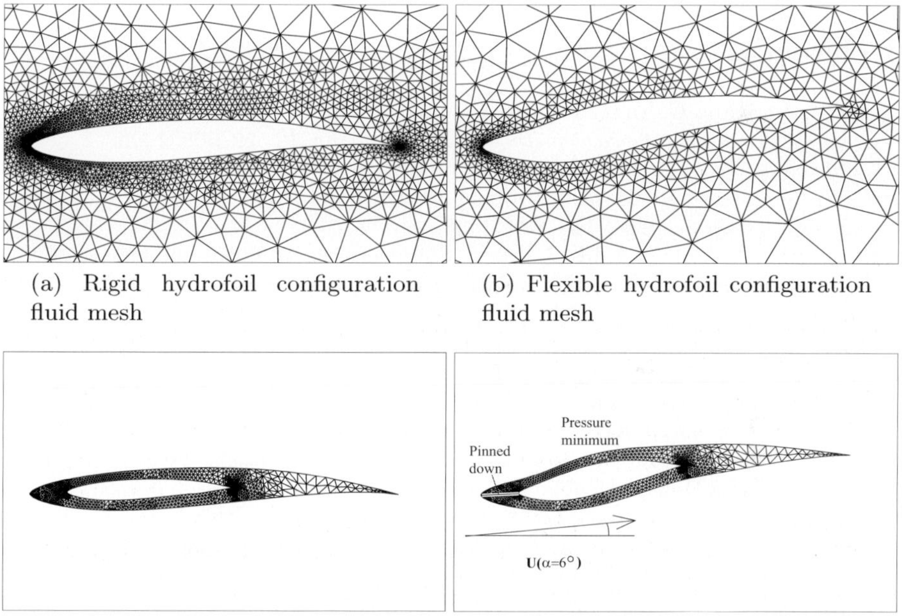

(a) Rigid hydrofoil configuration fluid mesh

(b) Flexible hydrofoil configuration fluid mesh

(c) Rigid hydrofoil configuration structural mesh

(d) Flexible hydrofoil configuration structural mesh

Fig. 3. Different meshes of the flow over a Eppler 817 hydrofoil.

6 Conclusions

We have developed a monolithic formulation for computing the interactions between an incompressible flow and a hyperelastic solid. The steady Navier-

Stokes equations in an Eulerian frame of reference are used to describe the fluid behavior, while a Lagrangian frame and large displacement-small strain theory is used for the solid. The pseudo-solid formulation provides control of mesh distortion during the computations. The formulation was verified on a problem with a closed form solution. Optimal rates of convergence are observed for all variables. Furthermore the error estimator is asymptotically exact. The methodology was successfully applied to the flow around a flexible hydrofoil.

Acknowledgements

This work was sponsored in part by NSERC (Government of Canada), the Canada Research Chair Program (Government of Canada), and by FQRNT (Government of Qubec).

References

1. R. Lohner and C. Yang. Improved ale mesh velocities for moving bodies. *communications in numerical methods in engineering*, 12:599–608, 1996.
2. O. Ghattas and X. Li. Domain decomposition methods for sensitivity analysis of a nonlinear aeroelasticity problem. *International Journal of Computational Fluid Dynamics*, pages 113–130, 1998.
3. E. Lund, H. Moller, and L. A. Lund. Shape design optimization of steady fluid-structure interaction problems with large displacements. In *AIAA 2001-1624*, 2001.
4. P.A. Sackinger, P.R. Schunk, and R.R. Rao. A newton-raphson pseudo-solid domain mapping technique for free and moving boundary problems: A finite element implementation. *Journal of Computational Physics*, 125:83–103, 1996.
5. C. Farhat, C. Degand, B. Koobus, and M. Lesoinne. Torsional springs for two-dimensional dynamic unstructured fluid meshes. *Comput. Methods Appl. Mech. Engrg.*, 163:231–245, 1998.
6. H. Schlichting. *Boundary-Layer Theory.* McGraw-Hill, 7th edition, 1979.
7. K.J. Bathe. *Finite element procedures in engineering analysis.* Prentice-Hall, Inc., Engelwood Cliffs, New Jersey 07632, 1982.
8. S. Etienne and D. Pelletier. A monolithic formulation for steady-state fluid-structure interaction problems. In *34th AIAA Fluid Dynamics Conference and Exhibit*, Portland, OR, June 2004. AIAA Paper 2004-2239.
9. G. Dhatt and G. Touzot. *Une présentation de la méthode des éléments finis.* Les Presses de l'Université Laval, Québec, Canada, 1981.
10. O. C. Zienkiewicz and J. Z. Zhu. The superconvergent patch recovery and *a posteriori* error estimates. Part 1: The recovery technique. *International Journal for Numerical Methods in Engineering*, 33:1331–1364, 1992.
11. O. C. Zienkiewicz and J. Z. Zhu. The superconvergent patch recovery and *a posteriori* error estimates. Part 2: Error estimates and adaptivity. *International Journal for Numerical Methods in Engineering*, 33:1365–1382, 1992.

Numerical Simulation for Impact of Elastic Deformable Body against Rigid Wall under Fluid Dynamic Force

Tomohisa Hashimoto[1], Koji Morinishi[2], and Nobuyuki Satofuka[3]

[1] Venture Laboratory, Kyoto Institute of Technology, Matsugasaki, Sakyo-ku, Kyoto 606-8585 Japan, hasimoto@ipc.kit.ac.jp
[2] Department of Mechanical and System Engineering, Kyoto Institute of Technology, Matsugasaki, Sakyo-ku, Kyoto 606-8585 Japan, morinishi@ipc.kit.ac.jp
[3] The University of Shiga Prefecture, Hassaka-cho, Hikone-shi, Shiga 522-8533 Japan, satofuka@office.usp.ac.jp

1 Introduction

The analysis of interactions between the elastic body and the surrounding flow field has been interested in engineering. It is important to reveal the mechanism in the deformation of a moving elastic body under the fluid dynamic force. There are few studies on the interaction between a largely deforming elastic body and the surrounding flow field. In this study, it is investigated that if a body can largely deform in the process of impact against the rigid wall under the fluid dynamic force, what shape does the body change into?

At a first step in this attempt, we proposed an elastic body model in 2D and 3D, with which a largely moving and deforming elastic body under the fluid dynamic force can be treated and developed a numerical method for simulating fluid-structure interaction, especially fluid-elastic body interaction problems. In this model, the elastic body whose inside is filled with fluid moves and deforms by the difference between inside and outside pressures of fluid and shear stress acting on the surface of body with interactive effects of tension and damping force of body. The model is verified for a test case where a cylindrical elastic body expands and contracts with damping effect in stationary fluid. The results are compared with those obtained analytically.

This model is applied to numerical simulations of incompressible unsteady flow around cylindrical and spherical elastic deformable bodies under the fluid dynamic force. In this approach, overset grid technique [1] is adopted for treating the moving and deforming body. The incompressible Navier-Stokes equations are solved by using the artificial compressibility method [2]. The unsteady incompressible flow field is obtained by solving the Navier-Stokes equations together with the equations of motion in this model through subiterations. We are successful in numerical simulations of flow around a largely moving and deforming elastic body under the fluid-dynamic force. The model is extended to treat the impact of an elastic body against the rigid

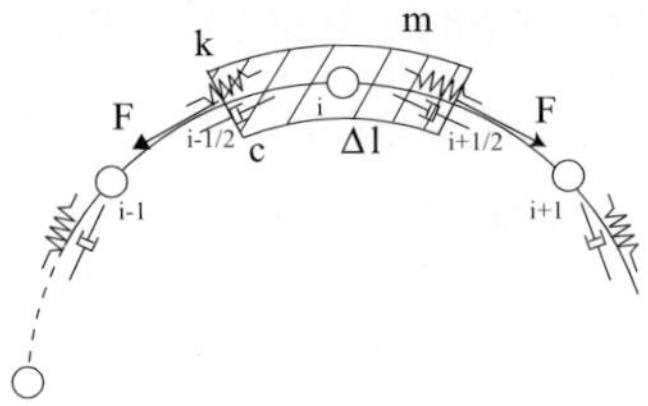

Fig. 1. Schematic representation of an elastic shell model.

wall by introducing bounce-back rule including the effect of friction between the elastic body and wall under the fluid dynamic force. Some numerical results for flow around a moving and deforming elastic body in the process of impact against the rigid wall under the fluid dynamic force are presented.

2 An Elastic Body Model

The equations of motion based on the wave equation with the term of damping effect are formulated for the elastic shell made of thin film, for example a balloon whose inside is filled with isothermal gas. The film model is composed of material particles connected with elastic springs and dashpots as shown in Fig. 1, so that the film can be expand and contract with the effect of damped free vibration. In this figure, the film shown in the slash portion corresponds to a material particle, where m, c, k, Δl and F denote the mass, the damping coefficient, the stiffness coefficient, the length of arc and the tension acting on the divided portion, respectively. The equations of motion for the material particle are given by

$$m\frac{d^2\boldsymbol{x}}{dt^2} = \boldsymbol{p} + \boldsymbol{\tau} + \boldsymbol{f} + \boldsymbol{f}' \tag{1}$$

$$\boldsymbol{p} = p\boldsymbol{n}\Delta l, \ \boldsymbol{\tau} = \tau\boldsymbol{t}\Delta l, \ p = p_{in} - p_{out} \tag{2}$$

where $\boldsymbol{x}$ is the coordinates of the material particle. The pressure $\boldsymbol{p}$ and shear stress $\boldsymbol{\tau}$ acting on the surface of body are obtained by solving the Navier-Stokes equations. p_{in} and p_{out} are the inside and outside pressures of fluid along the body respectively. $\boldsymbol{n}$ and $\boldsymbol{\tau}$ denote the outward unit normal and tangent vectors to the surface of body. The tension $\boldsymbol{f}$ and damping force $\boldsymbol{f}'$ of body are obtained as

$$\boldsymbol{f} = k(\boldsymbol{x}_{i+1,k} - 2\boldsymbol{x}_{i,k} + \boldsymbol{x}_{i-1,k}), \ \boldsymbol{f}' = c(\boldsymbol{u}_{i+1,k} - 2\boldsymbol{u}_{i,k} + \boldsymbol{u}_{i-1,k}) \tag{3}$$

where $\boldsymbol{u}$ is the velocity vector of the material particle.

3 Verification of the Elastic Body Model

The elastic body model in 2D is tested for free vibrations of a cylindrical elastic body in cases with and without damping effect in stationary fluid. In the starting condition for damped free vibration, the radius of body is expanded from $r = 0.5$ at the initial state to 0.6 and then begin vibrating. In determining the adjustable stiffness and damping coefficients in (3), notice that these coefficients in this system are different from those in simple harmonic vibration because there are additional effects of fluid and connection of material particles by tension and damping force of body. The analytical derivation for the period of vibration in this system is written in [3].

The damping effect is investigated by comparing the period of vibration for different damping factors; $\zeta = 0, 0.1, 0.5, 1$ and 1.5 in the case of $m = 100$. The results for $\zeta = 0$, 0.1 and 1 are shown in Fig. 2. The computed periods of vibration are $T = 129.1$ and 130.2 for $\zeta = 0$ and 0.1 respectively, while the corresponding analytical periods are $T = 129.5$ and 130.1. The comparisons between computed and analytical solutions are in good agreement. The case of $\zeta = 1$ is the fastest until the vibration is suppressed. It is found that the elastic body model is well working.

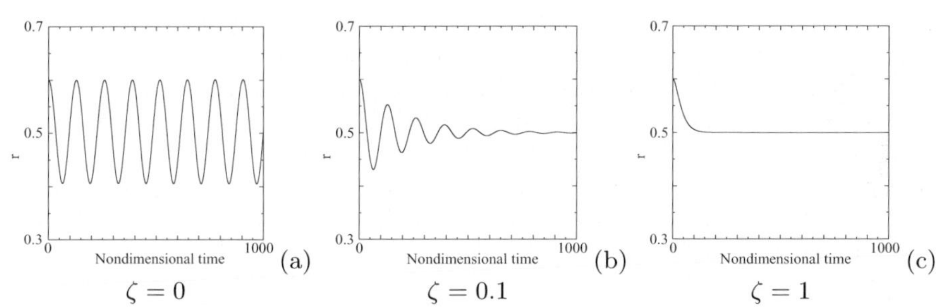

Fig. 2. Changes in the radius of elastic body for different damping factors; $\zeta = 0$, 0.1 and 1 ($m = 100$).

4 Flow around a Cylindrical Elastic Body

The elastic body model in 2D is applied to simulate flow around a moving and deforming cylindrical elastic body under the fluid dynamic force. In this simulation, the elastic body suddenly appears in fluid of uniform flow and then flows downstream with deformation. In the process of deformation, the inside fluid of body is specified as an ideal gas at the isothermal condition without solving the Navier-Stokes equations. For treating a largely moving and deforming elastic body, overset grid technique is adopted. The time history for changes in the shape of body from $t = 0$ to 750 in the nondimensional time at intervals of 5 for the Reynolds number of 40 is shown in Fig. 3. In this case, the deformation of elastic body is quite large in the process of flowing down. The cylindrical elastic body gradually becomes the slender shape like a jelly bean at t = 10, 20 and then changes into the shape like an egg at t = 30 and mushroom at t = 40. This mode is repeated several times until the noticeable deformation is suppressed. When the transitional speed of the elastic body finally reaches the uniform flow value, the fluid doesn't act any force to the body. At this moment, the elastic body shape returns to its initial state. In the present simulation, when the transitional speed of the elastic body reaches about 90% of uniform flow value, the elastic body shape almost returns to its initial state as shown in Fig. 4.

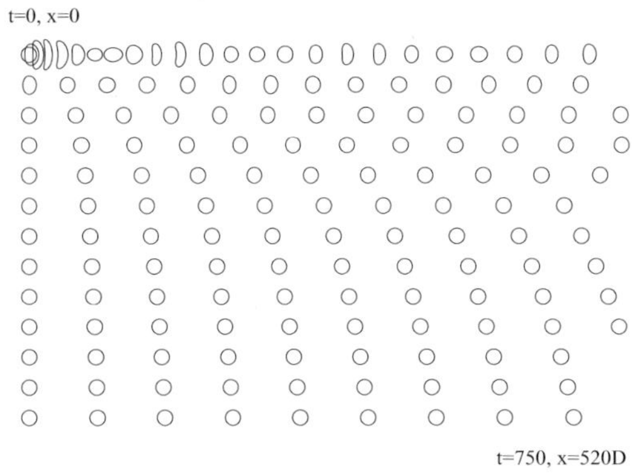

Fig. 3. Time history of the elastic body shape for $Re = 40$.

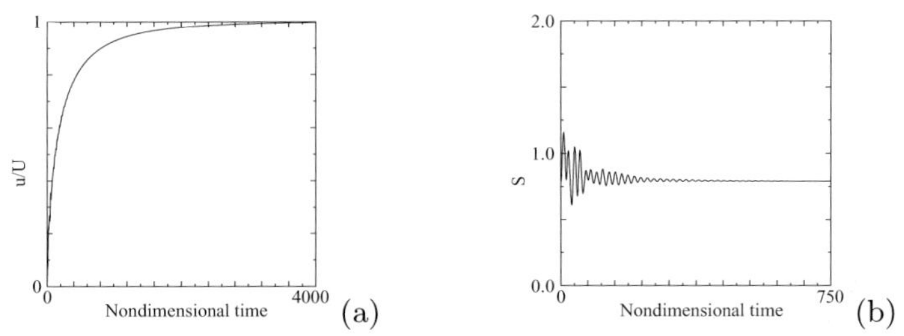

Fig. 4. (a) Velocity ratio of the elastic body to the uniform flow and (b) the change in the area of body.

5 Impact of a Cylindrical Elastic Body against Rigid Wall

In the starting condition, the elastic body moves with an initial velocity from a height at an oblique angle and collides with rigid wall including the effect of friction between the elastic body and wall in the gravitational field. For treating the impact of elastic body, bounce-back rule is introduced. The detailed treatment is referred to [3]. Figure 5 shows the time history for changes in the shape of body colliding with the rigid wall and rebounding in the cases with and without friction between the elastic body and wall in stationary fluid. Figure 6 shows the time history of pressure contours at the

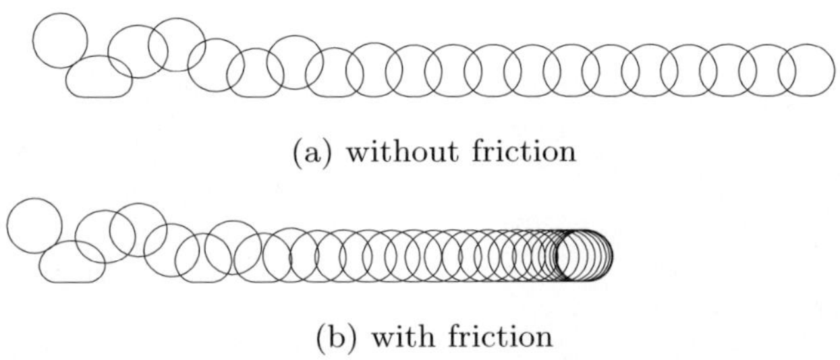

Fig. 5. Time history for changes in the shape of body for the impact against the rigid wall in the cases with and without friction between the elastic body and wall in stationary fluid.

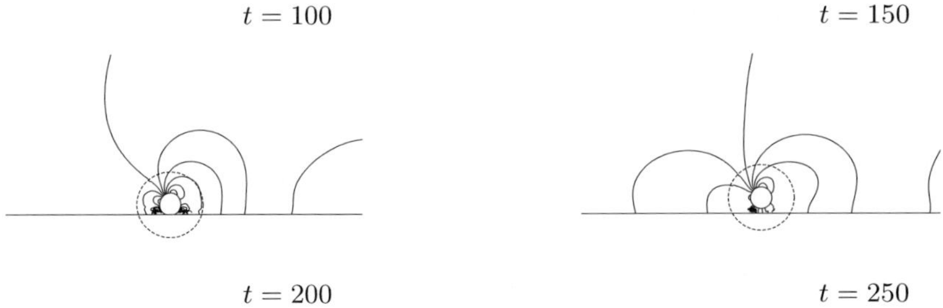

Fig. 6. Time history of pressure contours for the impact of elastic body against the rigid wall under the fluid dynamic force for $Re = 40$.

Reynolds number of 40 for the impact of elastic body against the rigid wall under the fluid dynamic force.

6 Flow around a Spherical Elastic Body

The elastic body model in 3D is also applied to simulate flow around a moving and deforming spherical elastic body under the fluid dynamic force. Figure 7 shows the computed instantaneous pressure contours for the Reynolds number of 100. Figure 8 shows the time history for changes in the shape of body from the side of view. In this case, the spherical body gradually becomes slender shape and then changes into the shape like a mushroom which has a sharp head in the front of body. Finally the shape returns to its initial state.

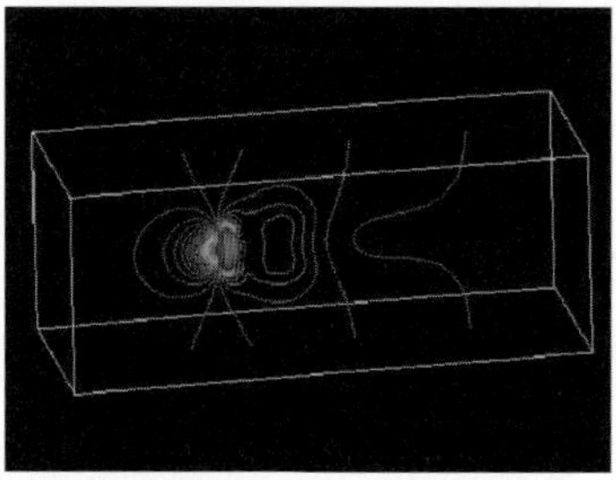

Fig. 7. Instantaneous pressure contours for $Re = 100$.

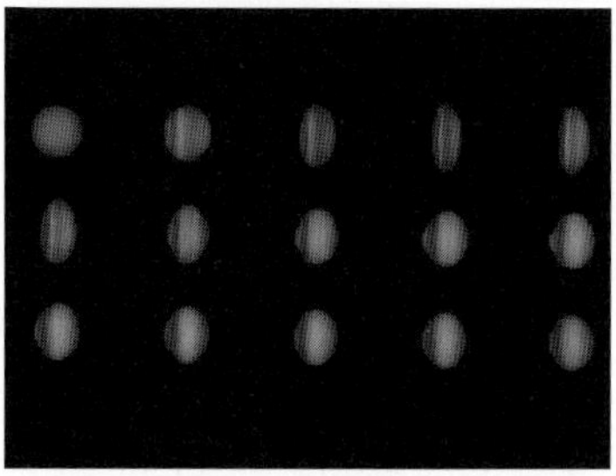

Fig. 8. Time history for changes in the shape of body for $Re = 100$.

7 Conclusions

Numerical simulations of fluid-elastic body interaction problem are presented. In the verification of elastic body model for damped free vibration in stationary fluid, the present solutions agree reasonably well the analytical ones. We can successfully simulate flows around a largely moving and deforming elastic body and collisions of the elastic body with rigid surface under the fluid dynamic force by using the elastic body model. In future work, the elastic body model should be validated through comparison between the numerical result and experiment.

References

1. J.L. Steger and J.A. Benek, Computer Methods in Applied Mechanics and Engineering, 64 (1987)
2. S.E. Rogers and D. Kwak, AIAA Journal, Vol. 28 (1990)
3. T. Hashimoto, K. Morinishi and N. Satofuka, ECCOMAS (2004)

Fluid Structure Interaction of a Hypersonic Generic Body-Flap Model

Andreas Mack[1], Roger Schäfer[2], Burkard Esser[3], and Ali Gülhan[3]

[1] German Aerospace Center, Institute of Aerodynamics and Flow Technology Lilienthalplatz 7, D-38108 Braunschweig
[2] Institute of Structures and Design, Pfaffenwaldring 38-40, D-70569 Stuttgart
[3] Institute of Aerodynamics and Flow Technology, Linder Höhe, D-51170 Köln

Summary. The purpose of the present investigation is the validation of a numerical coupling tool for fluid-thermal structure interaction for a generic body flap model by means of experimental data from an arc heated facility. Different geometrical configurations have been simulated and the main coupling phenomena are pointed out. The results show good agreement with the experimental data.

1 Introduction

Efficiency and safety of future space vehicles depend strongly on the design of the thermal protection system. Accurate and reliable predictions of heat flux and thermal loads on the surface would allow reducing costs and weight and significantly increasing safety due to new design concepts. Three different generic models have been tested experimentally in the DLR arc heated facility L3K and numerically rebuilt [1, 2, 3]. Particularly the numerical results for a generic body flap model presented here are focussing on the radiation effects between the body flap leeside and the opposite model surface. The model is designed resembling a realistic space vehicle body flap configuration. It consists of two flaps separated by a central gap. The deflection of the two flaps can be adjusted to 15 or 30°, symmetrically as well asymmetrically deflected. The actuators that drive the flaps in a real vehicle are also simulated. The model includes optional side plates that close the flap box in spanwise direction. The main effect to be investigated here is the temperature increasing at the leeside of the flaps due to heat conduction inside the model. The windward surfaces of the flaps face high heat fluxes being conducted through the structure to the lower sides and radiated from there to the opposite model surface. Therefore, the temperature on such surfaces is increased significantly. With that purpose the present investigation is devoted to validate a coupling environment for the prediction of thermal loads, consisting of the DLR's finite volume CFD-solver TAU [4] and the commercial finite element structural solver ANSYS [5].

2 Numerical Method

The CFD solutions are achieved with the unstructured DLR TAU-Code that has been validated in the past for different configurations at super- and hy-

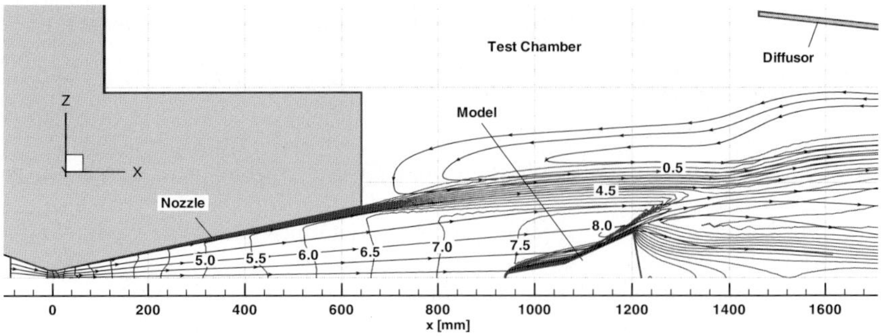

Fig. 1. Numerically rebuilt flowfield in arc heated facility L3K

personic flow conditions. The time-accurate three-dimensional Navier-Stokes equations are marched for steady or unsteady conditions by a three stage Runge-Kutta scheme. The upwind scheme employed is AUSMDV. Available gas models are Perfect gas formulation, equilibrium air and chemical nonequilibrium. State of the art acceleration techniques such as local time stepping, residual smoothing and multigrid are implemented. The numerical solutions for the thermal analysis of the structure have been carried out with the structural finite element solver ANSYS. Quasi structured meshes have been generated to increase the accuracy for the thermal solution. Surface radiation effects are taken into account by the structural solver. View factors of the different surface panels are calculated and reflections are taken into account, leading to a degraded surface radiation cooling in cavity areas. The material properties of the C/CSiC material are treated nonlinear, they are based on the analysis of the materials used during the experimental tests. The coupling algorithm follows the loose coupling approach for thermal coupling and uses Dirichlet-Neumann boundary conditions. Surface heat flux and temperature are calculated by the solvers and exchanged as boundary conditions. The surface values are interpolated from the fluid to the FEM mesh and vice versa with the commercial coupling routine MpCCI [6]. This routine allows an efficient neighborhood search also for large numbers of surface grid points.

3 Results

The study is carried out for conditions of DLR's high enthalpy facility L3K ($H_0 = 10.5 MJ/kg$, $Ma = 7.4$, $Re = 12.000$, $\alpha = 10°$) assuming laminar flow. Special attention is given to the model onflow condition in the facilitie's test chamber. The nozzle flow entering the test section is calculated including nonequilibruim effects and conical flow (Fig. 1). Due to the relatively large model dimensions strong flow gradients are present along the model, e.g. the pressure is reduced from the model nose to the trailing edge by a factor of two. The flow is fully frozen inside the test section and across the model. This means that only minor thermochemical effects take place in the flow field except surface catalytic effects.

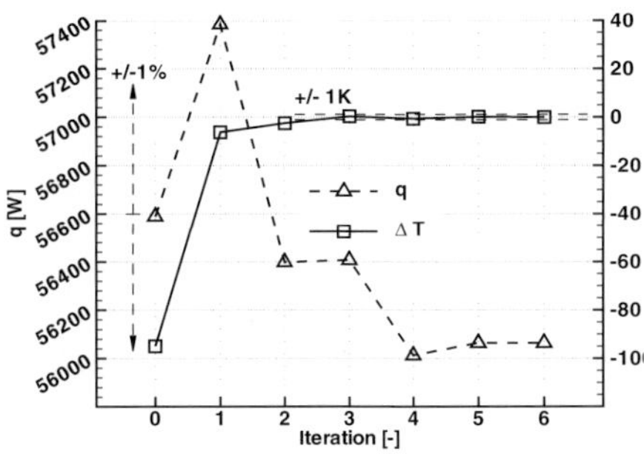

Fig. 2. Convergence of surface temperature and heat flux

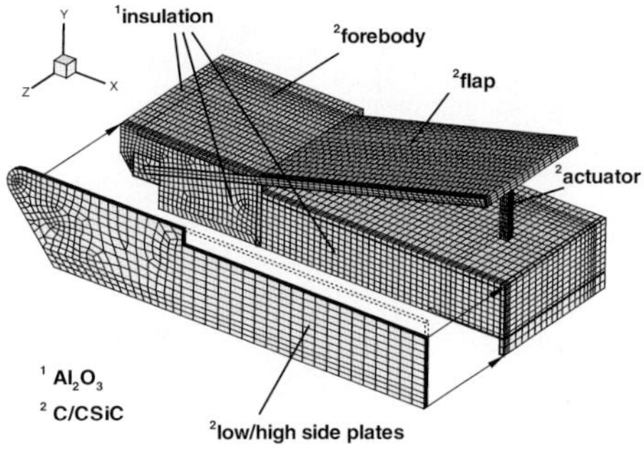

Fig. 3. Structural model for FEM simulation

The numerical coupling environment is investigated in order to estimate the numerical error including both the numerical errors of the solvers as well as the interpolation scheme. Global convergence of heatflux and temperature is achieved after 2 coupling iterations, compare Fig. 2. The heat flux step after the third iteration is due to a changed boundary treatment at the outflow. This does not affect the surface temperature convergence. The global heat flux does not change significantly due to the coupling effects. Grid refinement studies are performed partially on a flap quarter. The standard CFD mesh ($1.1 \cdot 10^6$ points) is refined twice by dividing the surface cells. The standard FEM mesh (70.000 nodes, Fig. 3) was also refined twice, the nodal distributions in the directions of high heat flux gradients e.g. at side edges are comparable for both CFD and FEM mesh. The heat flux at the flap side edge vicinity estimated by the CFD solver is plotted in Fig. 4. Grid refinement leads to a better resolution of the heat flux maximum whereas the integral heat flux is slightly reduced due to a reduced heat flux level apart of the edge, compare area A1 and A2 in Fig. 4. The integral heat flux is reduced by 1.6 % on the fine and another 0.3 % on the superfine mesh. The coarser CFD mesh therefore leads to an overprediction of the integral heatflux. The overall convergence of the heat flux can be seen clearly in Fig. 2. The interpolation error is calculated integrating the CFD surface heat flux on the superfine CFD mesh and the interpolated ones on the FEM meshes. The differences are 0.62 % (standard FEM mesh), 0.57 % (fine FEM mesh) and 0.51 % (superfine FEM mesh). The surface interpolation is non-conservative, in order to avoid local numerical peaks of the interpolated values, due to large differences in mesh resolution. Therefore a basic numerical error of 0.5 - 0.6 % can be derived from the interpolation scheme. The thermal analysis is not sensitive concerning the interpolated heat flux. A grid convergence study of the three grid resolutions is shown in Fig. 5. The surface heat flux is taken from the superfine CFD-mesh. The differences in temperature are 10 K at edges and 1 K on the surface which can be neglected. The standard calculation on standard-CFD and FEM mesh is also plotted and displays a

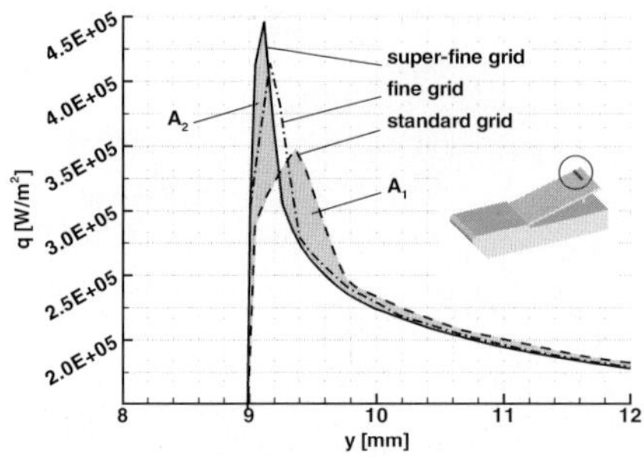

Fig. 4. CFD grid refinement study at inner flap section

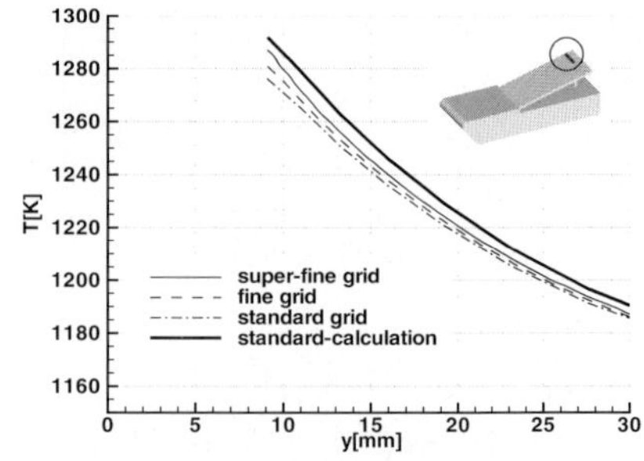

Fig. 5. FEM grid refinement study at inner flap section

difference of 5 K to the superfine solution due to the higher integral value of the heat flux.

3.1 Symmetric Flap Model

Fig. 6 and 7 show the coupling effects for nonequilibrium, noncatalytic flow, particularly the increasing temperature at the leeside of the flap due to the heat conduction of the C/CSiC material as well as the strong radiation effects in the flap region. Side edges that show high temperature for the CFD stand alone solution (Fig. 6) cool down due to heat conduction whereas temperature rises in cooler regions. The local heat peak on the actuator impinged by the supersonic gap flow is weakened by the structural heat conduction. The increasing temperature in the cold flow separation area below the flap leads to an incrasing pressure level. Therefore the separation disappears due to a pressure gradient from the cavity area in spanwise direction and in the main flow direction. Fully catalytic walls only lead to an almost constant temperature offset of 160 K without any further effects, whereas conical onflow due to the main pressure gradient along the model forces the separation to vanish and reduces the surface heatloads downstream and away from the symmetry axis of the flow core. The numerical data for conical, frozen flow inside the test chamber, simulating fully catalytic walls agree rather good with the experiments. Surface temperatures are shown in $y = 50mm$ in the flap middle section (Fig. 8). Temperatures from infrared (forebody and flap surface) and thermocouple experimental data (forebody and model surface) and the numerical results on the flap as well as the model surface show the same aerothermodynamic and physical coupling behaviour.

3.2 Asymmetric Flap Model

Deflecting the two flaps at different angles (15° and 30°) changes the flow field significantly. The 30° deflected flap interacts with the bow shock and rises

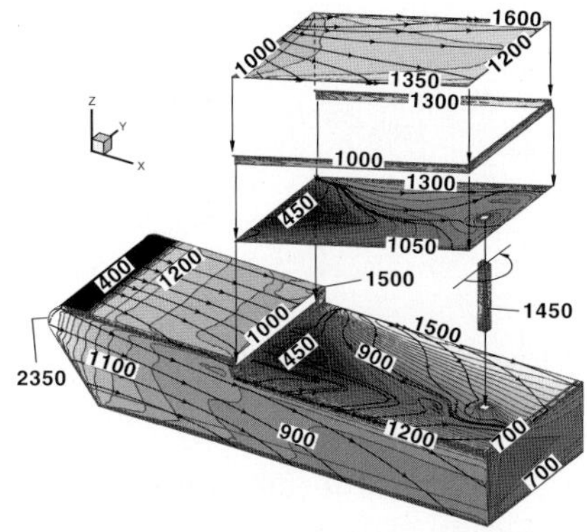

Fig. 6. Surface temperature and streamlines of radiation equilibrium CFD-solution (non-catalytic)

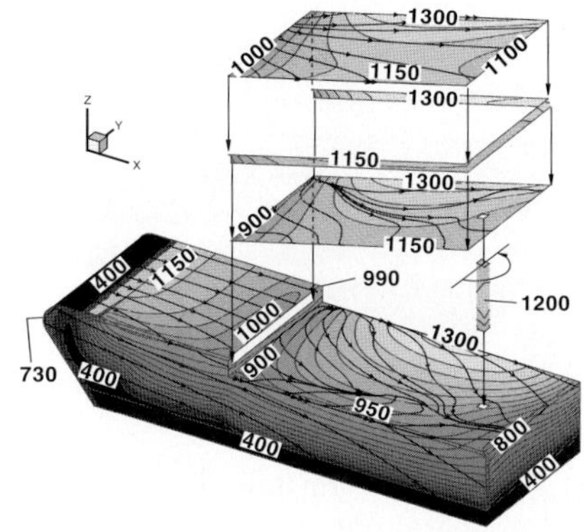

Fig. 7. Surface temperature and streamlines of coupled CFD-FEM solution (non-catalytic)

significantly the temperature while the 15° deflected flap now lies below the ramp shock of the other flap (Fig. 9, 10). Flow separation occurs at the hinge line of the 30° deflected flap. Flow is directed strongly below the 30° deflected flap impinging the actuator. The flowfield as well as the surface temperature distribution are asymmetric. The radiation effects below the 30° deflected flap are weakened due to better radiation cooling and less surface visibility. This flap radiates onto the 15° deflected flap, temperature there rises slightly. The crosswise sections at the $x = 190mm$ model position is shown in Fig. 11. The numerical data of the flap surfaces agree with the infrared data. Particularly at the symmetry plane the model surface temperature slightly increases. The heatflux in the symmetry plane with the asymmetric gap flow becomes more important. Due to the fully catalytic behaviour the heatflux might be slightly high. Another effect is the core diameter of the facility. The

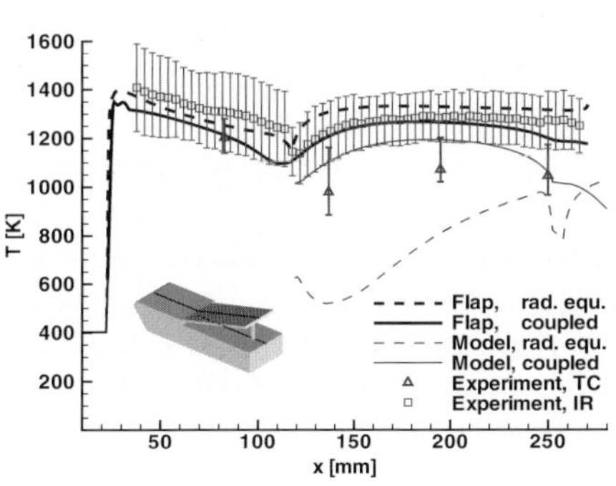

Fig. 8. Numerical (fully catalytic) and experimental data at spanwise section $y = 50mm$

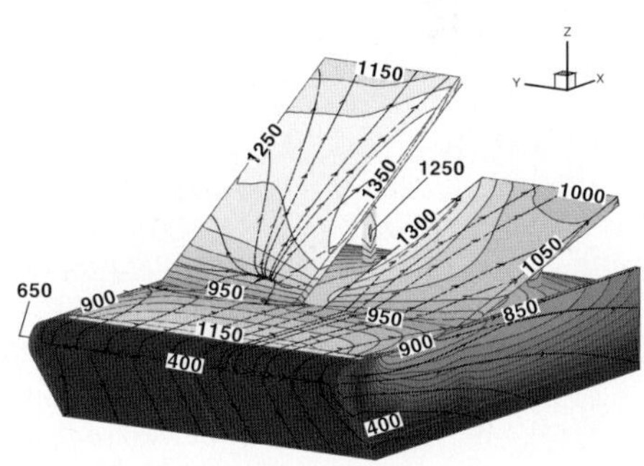

Fig. 9. Asymmetric laps: Coupled solution (non-catalytic, conical onflow), front view

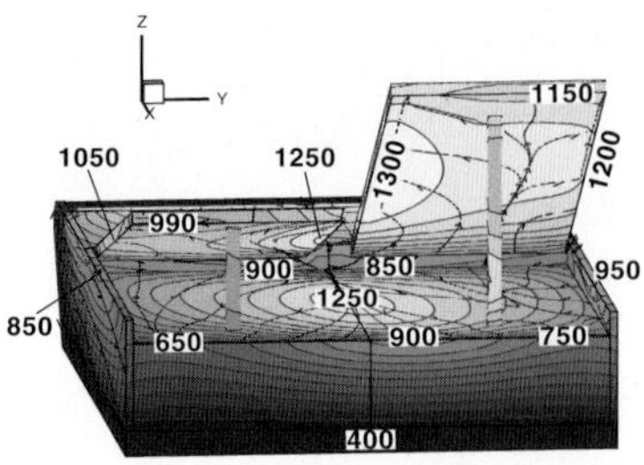

Fig. 10. Asymmetric flaps: Coupled solution (non-catalytic, conical onflow), rear view

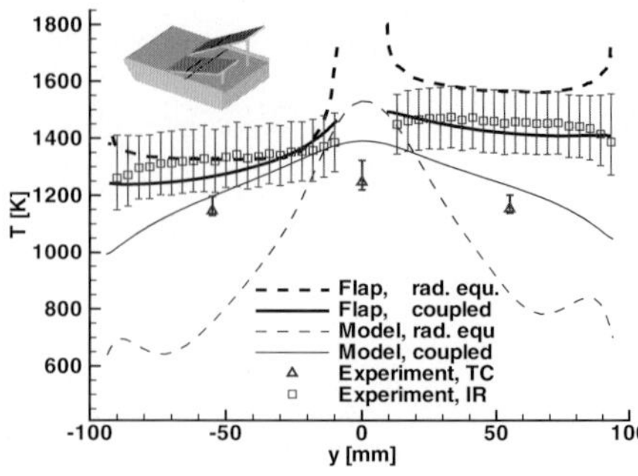

Fig. 11. Numerical (fully catalytic) and experimental data at crosswise section $x = 190mm$

trailing edge of the 30° deflected flap was outside the homogenous flow core. Because of this, the integral heat flux of the experimental setup was different from the numerical one, resulting in lower experimental values especially on the mainly passively heated model surface.

4 Summary

The present investigation shows that the coupled analysis of fluid and structure allows to increase the accuracy in the prediction of heat loads significantly, showing good agreement between numerical and experimental results for a generic body flap model in hypersonic flow. The main coupling effects such as radiation, heat conduction inside the model and the response of the fluid solution are reproduced well numerically for different configurations including asymmetric flap deflections. The validation of the present numerical coupling environment by experimental results allows to reduce safety margins due to the considered structural effects.

References

1. Gülhan, A., Esser, B., Koch, U.: *Experimental Investigation of Gap Flows on a Flap Model in the arc heated facility L3K*, DLR-IB-39113-99C01, 1999.
2. Schäfer, R., Mack, A., Esser, B., Gühan, A.: *Fluid Structure Interaction on a Generic Model of a Reentry Vehicle Noesecap.* 5th Inter. Congress on Thermal Stresses, Blacksburg, VA., 2003.
3. Mack, A., Schäfer, R.: *Flowfield topology changes due to fluid-structure interaction in hypersonic flow using ANSYS and TAU*, New Results in Numerical and Experimental Fluid Mechanics IV, Notes on Numerical Fluid Dynamics and Multidisciplinary Design, Vol. 87, Eds. C. Breitsamer *et al.*, pp. 196-203, 2002.
4. Mack, A., Hannemann, V.: *Validation of the unstructured DLR-TAU-Code for Hypersonic Flows* AIAA 2002-3111, 2002.
5. ANSYS, Inc.: *ANSYS User's Manual* Revision 5.2, August 31st, 1995.
6. Ahrem, R., Post, P., Wolf, K.: *A Communication Library to Couple Simulation Codes on Distributed Systems for Multi-Physics Computations* Proceedings ParCo99 Conference, Imperial College Press, 1999.

The Coupled Analysis of Pipe Burst and Multicomponents Fluid of Very High Pressured Natural Gas Pipeline

Tomoe Oda[1], Yoshiaki Tamura[2], Yoichiro Matsumoto[3], and Tetuya Kawamura[4]

[1] Tomoe Oda, Ochanomizu University, Otsuka 2-1-1, Bunkyo-ku, Tokyo 112-8610, Japan, tomoe@ns.is.ocha.ac.jp
[2] Yoshiaki Tamura, Toyo University, Kujirai 2100, Kawagoe, Saitama 350-8585, Japan, tamtam@eng.toyo.ac.jp
[3] Yoichiro Matsumoto, University of Tokyo, Hongo 7-3-1, Bunkyo-ku, Tokyo 113-8656, Japan, ymats@fel.t.u-tokyo.ac.jp
[4] Tetuya Kawamura, Ochanomizu University, Otsuka 2-1-1, Bunkyo-ku, Tokyo 112-8610, Japan, kawamura@is.ocha.ac.jp

1 Introduction

Nowadays, there are projects to construct natural gas pipelines that transport gases in higher pressure than conventional ones. The pressure inside is so high that an accidental eruption of gas due to a crack in the pipe may cause serious damage on the whole pipeline system. High-speed ductile fracture is a characteristic pattern of pipeline's fracture. Once a crack appear on the pipe, the crack will transmit on the pipe. Then the pipe wall will open. If the decompression of gas is fast, the transimission of crack will stop. If the decompression of gas is slow, the transimission of gas will continue. Although the pipe burst and the blow of gas depend on each other, the relationship between them has not been clear. It is necessary to investigate the characteristics of decompression of high-pressured gas. In this study, the characteristics of high-pressured natural gas expanding in the pipeline are simulated by two-dimensional coupled analysis of fluid and structure. The gas inside of the pipe is considered as natural gas and the gas outside is considered as air.

2 Numerical Method

2.1 Structure Analysis

The pipe-wall's opening is analyzed by finite element method with classical combined hardening elasto-plastic theory. The pipe wall is segmented for 3×36 cells and stress, which is obtained by CFD analysis, is distributed at 78 points.

2.2 Fluid Analysis

In this study, the gas inside of pipe is considered as natural gas and outside of pipe is considered as air. The governing equation of flowfield is two-dimensional compressible Euler equations that have two equations of conservation of mass.

$$\frac{\partial \boldsymbol{Q}}{\partial t}+\frac{\partial \boldsymbol{E}}{\partial x}+\frac{\partial \boldsymbol{F}}{\partial y}=0 \quad \boldsymbol{Q}=\begin{pmatrix}\rho_{\text{id}}\\ \rho\\ \rho u\\ \rho v\\ e\end{pmatrix} \boldsymbol{E}=\begin{pmatrix}\rho_{\text{id}}u\\ \rho u\\ \rho u^2+p\\ \rho uv\\ (e+p)u\end{pmatrix} \boldsymbol{F}=\begin{pmatrix}\rho_{\text{ideal}}v\\ \rho v\\ \rho uv\\ \rho v^2+p\\ (e+p)v\end{pmatrix} \quad (1)$$

One is for ideal gas and another is for mixed gas. In this study, C2-gas, which is one type of the natural gas, is considered as the gas inside of pipe. The BWRS state equation, which can treat natural gas as single component gas, is used as state equation of natural gas.

The BWRS (Benedict-Webb-Rubin equation of state modified by Starling) equation is given by

$$p=\rho RT+(B_0RT-A_0-\frac{C_0}{T^2}+\frac{D_0}{T^3}-\frac{E_0}{T^4})\rho^2+(bRT-a-\frac{d}{T})\rho^2$$

$$+\alpha(a+\frac{d}{T})\rho^6+\frac{c}{T^2}\rho^3(1+\gamma\rho^2)\exp(-\gamma\rho^2) \quad (2)$$

where the parameters take constant values, which depend on mixing ratio of components [1, 2, 3]. This equation can be applied for two-phase state of multi-components gas. The Density-Pressure curve in some temperature of C2 gas and ideal gas are plotted in Figure 1. The curves nearly equal to each other at high temperature, although the curves at low temperature are completely different. To save the calculation time, a set of data table is made by this equation before the simulation.

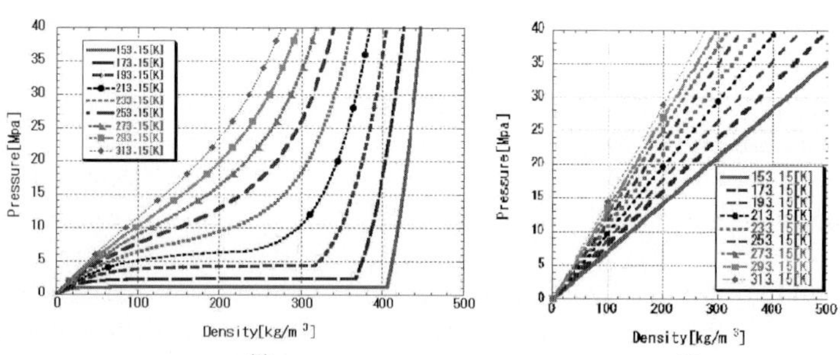

Fig. 1. Density-Pressure curve of BWRS equation and state equation of ideal gas.

The data files are referred at every grid point in every time step to get partial pressure value of natural gas. The temperatures of natural gas and air are considered to be the same at a grid point. The partial pressures and temperature are decided as follows. The subscript 1 is for natural gas and the subscript 2 is for ideal gas. The prime is indicates initial tentative value. The subscript n is for previous time step values and n+1 is present time step values. Asterisk indicates internal energy.

1. ${e_1}^{*\prime}$ and ${T_1}'$ are decided by data reference method with ${p_1}^n$, ${\rho_1}^{n+1}$.
2. ${p_2}'$ and ${e_2}^{*\prime}$ are calculated by state equation of ideal gas from $T = T_1 = T_2$.
3. ${e_1}^{*\prime}$ and ${e_2}^{*\prime}$ are substituted in conservation of energy equation. Total energy has been known.

$$e = {e_1}^* + {e_2}^* + \frac{1}{2}\rho(u^2 + v^2)$$

4. The temperature of gas is found by Newtonian method.
5. The partial pressure of natural gas is found by data reference method and the partial pressure of air is calculated from state equation of ideal gas.

2.3 Exchange of Boundary Conditions

Each of structural and fluid simulations needs an independent CPU. At every time step, boundary conditions are exchanged through interprocess communication. Fluid simulation process receives the position and velocity of pipe-wall from structural simulation process and then simulates the flowfield. Structural simulation process receives the pressure value at 78 points on the pipe-wall from fluid simulation process and then simulates the deformation of pipe-wall. This procedure is iterated at every time steps.

2.4 Pressure Data Reference Method

In this study, the partial pressure of natural gas is decided from data tables, which basis on the BWRS state equation, by the data reference method. Internal energy, sound velocity, and temperature values are calculated for density from $0.1[kg/m^3]$ to $400[kg/m^3]$ by every 0.01 $[kg/m^3]$. All data, which is in proportion to the pressure every 0.01[Mpa], are listed in a file. Firstry, the internal energy of natural gas ${e_2}^*$ is calculated from conservation of energy equation. Then, the 4 points combinationof ${e_2}^*$ and p_2 is serched. Figure 2 indicates the image of this combination. The p_A and p_B are calculated by Lagrangian interpolation. Then, the $left e_2{}^*, p_2 right)$ is calculated by Lagrangian interpolation, too.

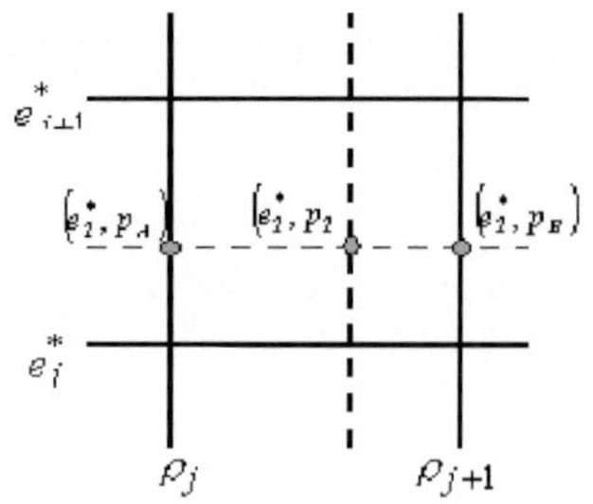

Fig. 2. Interpolating image of pressure data.

2.5 Initial and Boundary Conditions

The thickness of pipe wall is 20.0[mm]. The diameter of pipe is 36[inch] and the center of pipe is 1.0[m] above the floor. The 4[m]×4[m] area is considerd as flowfield. The flowfield can be considered as symmetry, the only half area is caluculated. The cartesian grid is used. The number of grid points is 201×401. Initial values of natural gas and air are indicated in Table 1. At the flowfield grid point which was inside of pipe wall in previous time step, the primitive values of density, velocity, energy and pressure, are given as the mean value of next gird points. At the flowfield grid point, which is next to pipe wall, the velocity of pipewall cell is added. The time interval is 10^{-7} [sec] by restriction of structure analysis.

Table 1. Initial values inside and outside of pipe.

	Pressure[MPa]	Density[kg/m^3]	Temperature[K]
Air(outside)	0.1	0.7945026	273.15[K]
C2(inside)	15.0	185.128074	273.15[K]

3 Simulation Results

3.1 One Dimensional Results

Firstly, the one-dimensional flowfield is simulated. In Figure 3, the left area is 0.1[MPa] and the right area is 15[MPa]. The pressure distribution at 1.0[sec] is showed in Figure 3. The pressure decompression of natural gas is slower than the one of ideal gas.

3.2 Two Dimensional Results

Figure 4 (a) and (b) are pressure distribution at 0.002[sec]. At the beginning of opening of the pipe wall, the high-pressured gas blow out above the pipe.

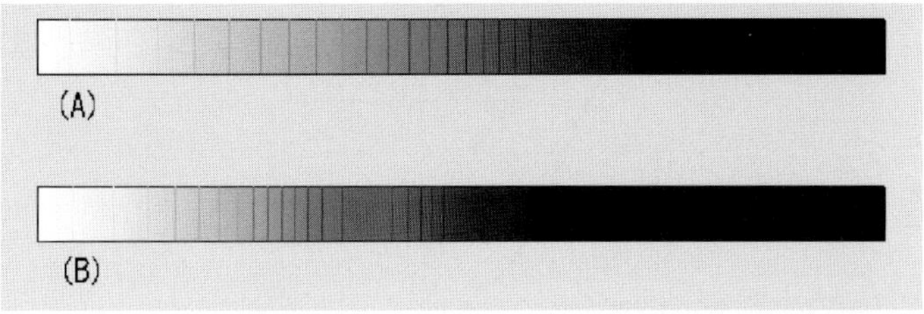

Fig. 3. Pressure distribution at 1.0[sec]. (A)Ideal gas and (B)C2 gas

The pressure inside of the pipe is drastically reduced. Figure 5 plots the time sequence of the pressure load. The pressure load is weakened drastically. But the deformation of pipe wall continues and deformation velocity increases more and more. It was suggested that the inertial force is ascendant for deformation of pipe wall. Figure 6 is temperature distribution at 0.002 [sec]. Natural gas spreads out from the crack, then the air compressed by natural gas and heated more than 350[K]. The temperature of expanded natural gas is less than 90[K]. From the curves in Fig 1, it is suggested that the liquefaction may be induced at this low temperature area.

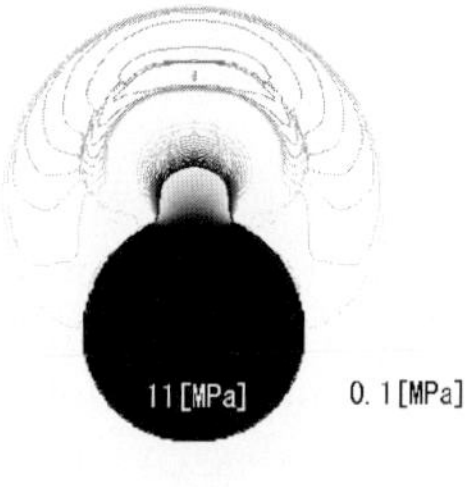

Fig. 4. Pressure distribution at 0.002[sec].

4 Discussions

The natural gas blows upward and gets around to the ground barely. So, it is considered that the natural gas diffuses widely when the gas explosion happens actually. The expanded gas is liquefied and the decompression of gas is observed. In one-dimensional simulation, the pressure decay of natural gas is slower than it of ideal gas. But, in this two-dimensional simulation, the inertia force influences the deformation of the pipe wall much more than the thermodynamic character of the natural gas.

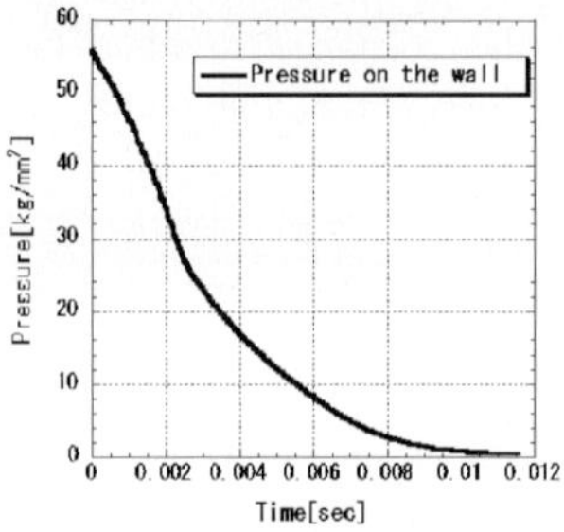

Fig. 5. Pressure load on the pipe wall.

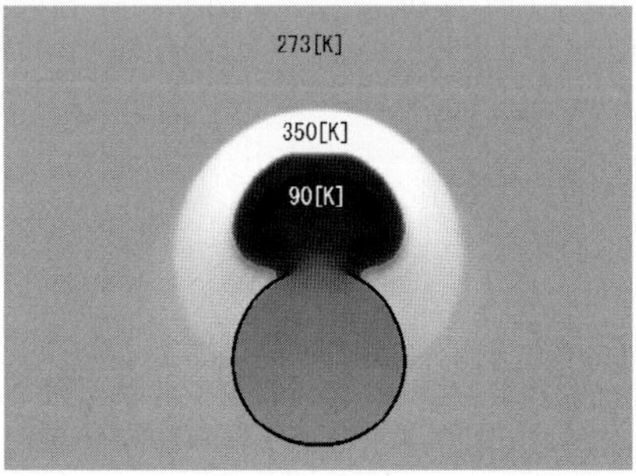

Fig. 6. Temperature distribution at 0.003[sec].

5 Conclusions

- The decompression of natural gas is slower than one of ideal gas.
- The two-dimensional coupled analysis of fluid and structure is developed.
- The large inertia force continues pipe deformation, although the pressure load on pipe wall is weakened drastically.
- At the boundary of air and natural gas, large gradient of temperature is observed.
- The liquefaction of expanded natural gas is observed.

References

1. Lin, C., Kwok Y. C., Starling, K. E., *Multiproperty analysis of PVT, Enthalpy and Phase Equilibrium Data*, Canadian Journal of Chemical Engineering, Vol.50, 1972.
2. Starling, E. K., Powers, J. E., *Enthalpy of Mixtures by Modified BWR Equation*, Industrial and Engineering Chemistry Fundamentals, Vol.9, pp.531-537, 1970.
3. Starling, K. E., Han, M. S., *Thermo data refined for LPG*, Hydrocarbon Processing, May, pp.129-132, 1972.

Part X

High-Order Schemes

High-Order Residual-Based Compact Schemes

Christophe Corre and Alain Lerat

Laboratoire SINUMEF, ENSAM, 151 Bd de l'Hopital, 75013 Paris, France
corre@paris.ensam.fr, lerat@paris.ensam.fr

Summary. A residual-based compact scheme, previously developed to compute d-dimensional inviscid compressible flows with 3rd-order accuracy on a 3^d-point stencil, is generalized to larger stencils allowing to reach a very high order of accuracy. Compactness is retained since for instance a 7th-order accurate dissipative approximation can be achieved on a 5^d stencil, without requiring the linear system solutions associated with usual compact schemes. The high-order generalization is also performed for unsteady and viscous flows.

1 Introduction

A residual-based compact (RBC) scheme has been recently proposed [1], [2], [3] for computing multidimensional, inviscid or viscous, steady or unsteady, compressible flows. While conventional high-order (HO) schemes rely on separate HO approximation of each space derivative when approximating the evolution problem $w_t + f_x + g_y = 0$, RBC schemes have been originally built on lower order approximation of f_x and g_y, combined to form a HO approximation of the residual $r = f_x + g_y$. The necessary dissipation for compressible flow computations is also built on the residual r and takes advantage of the residual vanishing to produce a HO dissipation at steady-state while ensuring robustness with a first-order dissipation in the transient phase. Starting from the well-established design principles of the RBC scheme, this paper proposes a systematic way to derive compact HO approximation of the discrete residuals on which the scheme is built: standard compact formulae based on Padé fractions [4] are used in a first step but taking advantage of the unique residual-vanishing property allows to avoid the linear algebra usually associated with Padé fractions.

2 HO RBC scheme for the Euler equations

In the sole purpose of simplifying the presentation, the $2D$ Euler equations $w_t + f_x + g_y = 0$ are considered and discretized by a conservative dissipative scheme of the form:

$$\frac{1}{\Delta t}(w^{n+1} - w^n) + \tilde{r}^n = \tilde{P}^n, \tag{1}$$

where w^n is the numerical solution at time-level $n\Delta t$, $\tilde{r}^n$ denotes some centered difference approximation to the steady-state residual $r = f_x + g_y$ while $\tilde{P}^n$ is a centered difference approximation to the following partial differential operator P built on the residual r:

$$P = \frac{\delta x}{2}(\Phi_1 r)_x + \frac{\delta y}{2}(\Phi_2 r)_y = \frac{\delta x}{2}[\Phi_1(f_x + g_y)]_x + \frac{\delta y}{2}[\Phi_2(f_x + g_y)]_y. \quad (2)$$

Matrices Φ_1 and Φ_2 are coefficients of order $O(1)$, designed once for all so as to ensure the dissipation of (2) - they are not tuning parameters nor limiters -; their expression depends on the eigensystems of the flux Jacobian matrices (see [1] for more details on the derivation of P). A fully-discrete expression for the RBC scheme is built by using the following basic difference and average operators:

$$(\delta_1 v)_{j+\frac{1}{2},k} = v_{j+1,k} - v_{j,k} \qquad (\delta_2 v)_{j,k+\frac{1}{2}} = v_{j,k+1} - v_{j,k}$$
$$(\mu_1 v)_{j+\frac{1}{2},k} = \tfrac{1}{2}(v_{j+1,k} + v_{j,k}) \qquad (\mu_2 v)_{j,k+\frac{1}{2}} = \tfrac{1}{2}(v_{j,k+1} + v_{j,k}),$$

where $v_{j,k}$ is defined on the Cartesian mesh $(x_j = j\,\delta x, y_k = k\,\delta y)$, with constant steps δx and δy of the same order $O(h)$. With these notations, RBC schemes applied at point (j,k) take the general form:

$$(w_t)_{j,k} + (\tilde{r}_0)_{j,k} = \frac{1}{2}[\delta_1(\Phi_1 \tilde{r}_1) + \delta_2(\Phi_2 \tilde{r}_2)]_{j,k}, \quad (3)$$

where $\tilde{r}_0$, $\tilde{r}_1$ and $\tilde{r}_2$ approximate r at point (j,k), $(j+\frac{1}{2},k)$ and $(j,k+\frac{1}{2})$.

In order to derive a HO expression for the residual at point (j,k) on a compact stencil, the following Padé fraction F based on the operator δ_l^2 $(l = 1,2)$ is used:

$$F(\delta_l^2) = \frac{\sum_{m=0}^{M} a_{2m}\delta_l^{2m}}{\sum_{m=0}^{N} b_{2m}\delta_l^{2m}}, \quad (4)$$

where $a_0 = b_0 = 1$ and δ^0 is the identity operator I. The Padé fraction which allows to approximate at high order, say $2q$, the first derivatives f_x and g_y at point (j,k) will be denoted $F^{(1)}$ and is such that:

$$[F^{(1)}(\delta_1^2)\frac{\delta_1\mu_1 f}{\delta x}]_{j,k} = (f_x)_{j,k} + O(h^{2q}) \ , \ [F^{(1)}(\delta_2^2)\frac{\delta_2\mu_2 g}{\delta y}]_{j,k} = (g_y)_{j,k} + O(h^{2q}).$$

Using $F^{(1)}$ to approximate $r = f_x + g_y$ at point (j,k) leads to:

$$\tilde{\tilde{r}}_0 = F^{(1)}(\delta_1^2)\frac{\delta_1\mu_1 f}{\delta x} + F^{(1)}(\delta_2^2)\frac{\delta_2\mu_2 g}{\delta y}. \quad (5)$$

with $\tilde{\tilde{r}}_0 = r + O(h^{2q})$. Since $F^{(1)}(\delta_l^2) = N^{(1)}(\delta_l^2)/D^{(1)}(\delta_l^2)$, a linear system solution is theoretically required to compute (5); this complexity is avoided if denominators are eliminated in (5), to yield:

$$\tilde{r}_0 = D^{(1)}(\delta_1^2)D^{(1)}(\delta_2^2)\tilde{\tilde{r}}_0 = N^{(1)}(\delta_1^2)D^{(1)}(\delta_2^2)\frac{\delta_1\mu_1 f}{\delta x} + N^{(1)}(\delta_2^2)D^{(1)}(\delta_1^2)\frac{\delta_2\mu_2 g}{\delta y}. \quad (6)$$

Since $\delta_l^2 = O(h^2)$ and $\tilde{\tilde{r}}_0 = r + O(h^{2q})$, it is clear that $\tilde{r}_0 = (I + O(h^2))(r + O(h^{2q}))$ so that, at steady-state where $r = 0$, $\tilde{r}_0 = O(h^{2q})$. Besides, it is easy to check formula (6) uses a compact 5×5 stencil iff $N = 1$ and $M = 2$, *i.e.* $N^{(1)}(\delta_l^2) = I + a_2\delta_l^2$ and $D^{(1)}(\delta_l^2) = I + b_2\delta_l^2 + b_4\delta_l^2$. As pointed out in [4] - and derived in [5] using the present operator notation - the following so-called supercompact formula:

$$\bar{F}^{(1)}(\delta_l^2) = \frac{\bar{N}^{(1)}(\delta_l^2)}{\bar{D}^{(1)}(\delta_l^2)} = \frac{I + \frac{5}{42}\delta_l^2}{I + \frac{2}{7}\delta_l^2 + \frac{1}{70}\delta_l^4} \tag{7}$$

yields a 8th-order accurate approximation of the first derivative in direction l on a 5-point stencil. Therefore, using this high-order Padé fraction in (6) leads to:

$$\tilde{r}_0 = (I+\frac{5}{42}\delta_1^2)(I+\frac{2}{7}\delta_2^2+\frac{1}{70}\delta_2^4)\frac{\delta_1\mu_1 f}{\delta x}+(I+\frac{5}{42}\delta_2^2)(I+\frac{2}{7}\delta_1^2+\frac{1}{70}\delta_1^4)\frac{\delta_2\mu_2 g}{\delta y}. \tag{8}$$

In the simple case where one looks for a 4th-order accurate approximation of r at point (j,k), choosing $N^{(1)} = I$ and $D^{(1)}(\delta_l^2) = I + b_2\delta_l^2$ preserves a 3×3-point stencil and, for $b_2 = \frac{1}{6}$, yields the following 4th-order expression:

$$\tilde{r}_0 = (I + \frac{1}{6}\delta_2^2)\frac{\delta_1\mu_1 f}{\delta x} + (I + \frac{1}{6}\delta_1^2)\frac{\delta_2\mu_2 g}{\delta y}. \tag{9}$$

Expression (9) was already proposed in [6]; however, it was not fully exploited since it was completed with HO non-compact dissipation. In the present work, the approach is not only extended at higher-order but the RB structure of the dissipation operator allows to achieve both high accuracy and compactness for the full dissipative scheme by following the same line of idea to build the mid-points estimates of the residuals defining the dissipation operator.

HO expressions of the mid-point residuals require to introduce the Padé fraction $F^{(\delta)}$, approximating at HO a first-derivative at point $(j+\frac{1}{2})$ or $(j,k+\frac{1}{2})$, as well as the Padé fraction $F^{(\mu)}$ which yields a HO approximation of fluxes f and g at these points; therefore, the following relations hold at midpoint $(j+\frac{1}{2},k)$:

$$(F^{(\delta)}(\delta_1^2)\frac{\delta_1 f}{\delta x})_{j+\frac{1}{2},k} = (f_x)_{j+\frac{1}{2},k} + O(h^{2q})\,,\ (F^{(\mu)}(\delta_1^2)\mu_1 g)_{j+\frac{1}{2},k} = g_{j+\frac{1}{2},k} + O(h^{2q}), \tag{10}$$

and similarly for $F^{(\delta)}(\delta_2^2)$ and $F^{(\mu)}(\delta_2^2)$ at midpoint $(j,k+\frac{1}{2})$. Using formula (10), r can be estimated at $(j+\frac{1}{2},k)$ as follows:

$$\tilde{\tilde{r}}_1 = F^{(\delta)}(\delta_1^2)\frac{\delta_1 f}{\delta x} + F^{(\mu)}(\delta_1^2)F^{(1)}(\delta_2^2)\frac{\delta_2\mu_2\mu_1 g}{\delta y} = r + O(h^{2q}). \tag{11}$$

A key point to preserve compactness and high-accuracy without resorting to any linear algebra is to find Padé fractions $F^{(\mu)}$ and $F^{(\delta)}$ sharing the same denominator, *i.e.* $F^{(\mu)} = N^{(\mu)}/D$ and $F^{(\delta)} = N^{(\delta)}/D$. It is possible in that case to eliminate the denominators $D(\delta_1^2)$ and $D^{(1)}(\delta_2^2)$ in (11) to obtain:

$$\tilde{r}_1 = D(\delta_1^2)D^{(1)}(\delta_2^2)\tilde{\tilde{r}}_1 = N^{(\delta)}(\delta_1^2)D^{(1)}(\delta_2^2)\frac{\delta_1 f}{\delta x} + N^{(\mu)}(\delta_1^2)N^{(1)}(\delta_2^2)\frac{\delta_2\mu_2\mu_1 g}{\delta y}. \quad (12)$$

Clearly, if $F^{(\mu)}$, $F^{(\delta)}$ and $F^{(1)}$ are designed so as to approximate respectively g, f_x and g_y at midpoint $(j+\frac{1}{2},k)$ at order $(4p-2)$ on a $2p \times (2p+1)$ stencil, the resulting approximation of $\delta_1(\phi_1\tilde{r}_1)$ is based on a $(2p+1)^2$ stencil and is of order $4p-1$; moreover similar developments can be performed to approximate $\delta_2(\phi_2\tilde{r}_2)$ at the same order on the same stencil so that the resulting RBC scheme is a $(4p-1)^{th}$-order approximation of the 2D Euler equations on a compact $(2p+1)^2$ stencil. In the trivial case $p=1$, $N^{(\mu)}$, $N^{(\delta)}$, $N^{(1)}$, $D^{(1)}$ (and D) are all necessarily equal to the identity operator I, yielding a 3rd-order dissipation on a 3×3 stencil [1]. For $p=2$, expressing $\tilde{r}_1$ on a 4×5 stencil supposes to choose:

$$\begin{aligned} N^{(\delta)}(\delta_1^2) &= I + a_2^{(\delta)}\delta_1^2 \ , \ N^{(\mu)}(\delta_1^2) = I + a_2^{(\mu)}\delta_1^2 \\ N^{(1)}(\delta_2^2) &= I + a_2^{(1)}\delta_2^2 \ , \ D^{(1)}(\delta_2^2) = I + b_2^{(1)}\delta_2^2 + b_4^{(1)}\delta_2^4 . \end{aligned} \quad (13)$$

Following the operator-based technique developed in [5], it can be easily found that choosing $a_2^{(\mu)} = \frac{1}{10}$, $a_2^{(\delta)} = \frac{11}{60}$ and $D(\delta_1^2) = I + \frac{9}{40}\delta_1^2 + \frac{3}{640}\delta_1^4$ ensures $F^{(\delta)}$ and $F^{(\mu)}$ approximate f_x and g at 6th-order. As for the Padé fraction $F^{(1)} = N^{(1)}/D^{(1)}$, it is given by a 6th-order formula derived in [5]:

$$F^{(1)}(\delta_l^2) = \frac{N^{(1)}(\delta_l^2)}{D^{(1)}(\delta_l^2)} = \frac{I + \frac{1}{30}\delta_l^2}{I + \frac{1}{5}\delta_l^2} \quad (14)$$

All the operators involved in expression (12) being defined, the 6th-order compact approximation of r at midpoint $(j+\frac{1}{2},k)$ may be explicitely written:

$$\tilde{r}_1 = (I + \frac{11}{60}\delta_1^2)(I + \frac{1}{5}\delta_2^2)\frac{\delta_1 f}{\delta x} + (I + \frac{1}{10}\delta_1^2)(I + \frac{1}{30}\delta_2^2)\frac{\delta_2\mu_2\mu_1 g}{\delta y}. \quad (15)$$

A similar process leads to the HO formula approximating r at midpoint $(j, k+\frac{1}{2})$:

$$\tilde{r}_2 = (I + \frac{1}{10}\delta_2^2)(I + \frac{1}{30}\delta_1^2)\frac{\delta_1\mu_1\mu_2 f}{\delta x} + (I + \frac{11}{60}\delta_2^2)(I + \frac{1}{5}\delta_1^2)\frac{\delta_2 g}{\delta y}. \quad (16)$$

Inserting the formulae (8), (15) and (16) into (3) completely defines a 7th-order (dissipative) approximation of the steady Euler equations on a 5×5-point stencil. With the same stencil size, conventional dissipative schemes are 3rd-order accurate only.

The previous ideas are readily extended to unsteady problems by taking advantage of a dual time framework: a time-accurate solution of the Euler equations is obtained as the steady solution of an evolution problem w.r.t. a dual time variable τ, namely $w_\tau + w_t + f_x + g_y = 0$, so that the steady residual to be considered when building the RBC scheme is $r = w_t + f_x + g_y$. The evolution problem w.r.t. τ is approximated by:

$$\frac{w_{j,k}^{n,m+1} - w_{j,k}^{n,m}}{\Delta\tau} + (\tilde{r}_0)_{j,k}^{n,m} = \frac{1}{2}[\delta_1(\Phi_1 \tilde{r}_1^{n,m}) + \delta_2(\Phi_2 \tilde{r}_2^{n,m})]_{j,k}, \quad (17)$$

where estimates of the residual r at point (j,k), midpoints $(j+\frac{1}{2},k)$, $(j,k+\frac{1}{2})$ and dual-time level (n,m) have been introduced. The simplest form of scheme (17) was derived in [3] using second-order centered expressions for the flux difference approximations and the three-level formula $T(w^{n,m},w^n,w^{n-1}) = \frac{3}{2}\frac{w^{n,m}-w^n}{\Delta t} - \frac{3}{2}\frac{w^n-w^{n-1}}{\Delta t}$ for the physical time-derivative, such that $T(w) = (w_t)^{n+1} + O(\Delta t^2)$ at steady-state on τ. This basic version, based on a 3×3-point stencil can be deduced from the following approximations of the residuals:

$$\begin{aligned}
\tilde{r}_0^{n,m} &= \bar{D}_1^{(1)}\bar{D}_2^{(1)}T(w^{n,m}) + \bar{N}_1^{(1)}\bar{D}_2^{(1)}\frac{\delta_1\mu_1 f^{n,m}}{\delta x} + \bar{N}_2^{(1)}\bar{D}_1^{(1)}\frac{\delta_2\mu_2 g^{n,m}}{\delta y}, \\
\tilde{r}_1^{n,m} &= N_1^{(\mu)}D_2^{(1)}\mu_1 T(w^{n,m}) + N_1^{(\delta)}D_2^{(1)}\frac{\delta_1 f}{\delta x} + N_1^{(\mu)}N_2^{(1)}\frac{\delta_2\mu_2\mu_1 g}{\delta y}, \\
\tilde{r}_2^{n,m} &= N_2^{(\mu)}D_1^{(1)}\mu_2 T(w^{n,m}) + N_2^{(\mu)}D_1^{(1)}\frac{\delta_1\mu_1\mu_2 f}{\delta x} + N_2^{(\delta)}D_1^{(1)}\frac{\delta_2 g}{\delta y},
\end{aligned} \quad (18)$$

by taking all the Padé fractions numerators and denominators equal to identity (where subscripts 1 and 2 indicate a dependence on δ_1^2 or δ_2^2). A HO version of the RBC scheme on a 5×5-point stencil is obtained when $\bar{F}^{(1)}$, $F^{(\mu)}$, $F^{(\delta)}$ and $F^{(1)}$ are those given in the previous section: it is then easy to check that, at steady-state on τ, the truncation error of (17)-(18) is $O(\Delta t^2, h^7)$.

3 HO RBC scheme for the Navier-Stokes equations

For the sake of presentation simplicity, the model equation $w_t + f(w)_x = \nu w_{yy}$ is considered instead of the full Navier-Stokes equations. As explained in [1], the expression of the RBC scheme for this viscous flow problem remains formally identical to its inviscid counterpart: it is still given by (3), but with a steady residual now equal to $r = f(w)_x - \nu w_{yy}$. The dissipation matrices are still based on the eigensystems of the inviscid Jacobian matrices and, since $g(w) = 0$ in the model problem considered, $\Phi_2 = 0$. Therefore, it remains only to derive HO approximation for $(r_0)_{j,k}$ and $(r_1)_{j+\frac{1}{2},k}$. The 5-point 8th-order Padé fraction approximating the second derivative w_{yy} at point (j,k) has been (re)derived in [5] and may be expressed as $\bar{F}^{(2)}(\delta_l^2) = \bar{N}^{(2)}(\delta_l^2)/\bar{D}^{(2)}(\delta_l^2)$ with $\bar{N}^{(2)} = I + \frac{31}{252}\delta_l^2$ and $\bar{D}^{(2)} = I + \frac{13}{63}\delta_l^2 + \frac{23}{3780}\delta_l^4$. Consequently, the following approximation of the residual, $\tilde{\tilde{r}}_0 = \bar{F}^{(1)}(\delta_1^2)\frac{\delta_1\mu_1 f}{\delta x} - \bar{F}^{(2)}(\delta_2^2)\nu\frac{\delta_2^2 w}{\delta y^2}$, is such that $\tilde{\tilde{r}}_0 = r + O(h^8)$, so that taking $\tilde{r}_0 = \bar{D}^{(1)}(\delta_1^2)\bar{D}^{(2)}(\delta_2^2)\tilde{\tilde{r}}_0$ leads to a 8th-order approximation of r using 5×5 points that reads:

$$\tilde{r}_0 = (I + \frac{13}{63}\delta_2^2 + \frac{23}{3780}\delta_2^4)(I + \frac{5}{42}\delta_1^2)\frac{\delta_1\mu_1 f}{\delta x} - (I + \frac{2}{7}\delta_1^2 + \frac{1}{70}\delta_1^4)(I + \frac{31}{252}\delta_1^2)\nu\frac{\delta_2^2 w}{\delta y^2}.$$

Following the same principle, the 6th-order approximation of the residual at midpoint $(j+\frac{1}{2},k)$ is obtained as:

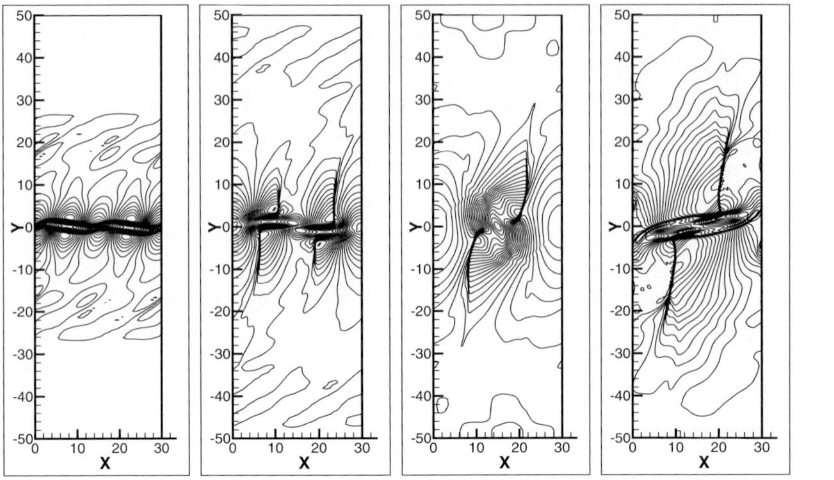

Fig. 1. Vortex pairing. Contours of temperature at time $t = 40, 80, 120$ and 160 (from left to right) on a 201×201 grid with $\Delta t = 0.5$. HO-RBC results.

$$\tilde{r}_1 = N^{(\delta)}(\delta_1^2)D^{(2)}(\delta_2^2)\frac{\delta_1 f}{\delta x} - N^{(\mu)}(\delta_1^2)N^{(2)}(\delta_2^2)\nu\frac{\delta_2^2\mu_1 w}{\delta y^2}$$

where $N^{(\delta)}$ and $N^{(\mu)}$ are unchanged w.r.t. the $2D$ inviscid case and $N^{(2)}$, $D^{(2)}$ are such that the Padé fraction $F^{(2)}$ approximates w_{yy} at order 6: $N^{(2)}(\delta_l^2) = I + \frac{1}{20}\delta_l^2$, $D^{(2)}(\delta_l^2) = I + \frac{2}{15}\delta_l^2$.

Application

Applying the same principles, a RBC scheme for the full $2D$ unsteady Navier-Stokes can be derived using compact (5×5) HO approximation of the residual $r = w_t + (f^E(w) - f^V(w, w_x, w_y))_x + (g^E(w) - g^V(w, w_x, w_y))_y$. The limited available space does not allow to explicit this scheme but its ability to accurately compute complex compressible flows involving shock waves is demonstrated on the figure below, which displays 7th-order results obtained for a vortex-pairing problem (see details of the problem set-up in [3]).

References

1. A. Lerat and C. Corre, A RBC Scheme for the Compressible Navier-Stokes Equations, *J. Comput. Phys.*, **170**, 642-675 (2001).
2. A. Lerat and C. Corre, RBC schemes for multiD hyperbolic system of conservation laws, *Computers & Fluids*, **31**, 639-661 (2002).
3. C. Corre, G. Hanss and A. Lerat, A RBC scheme for the unsteady compressible Navier-Stokes equations, to appear in *Computers & Fluids* (2004).
4. S.K. Lele, Compact finite difference schemes with spectral-like resolution, *J. Comput. Phys.*, **103**, p. 16-42 (1992).
5. A. Lerat et C. Corre, Approximations d'ordre élevé pour les écoulements compressibles, Ecole de Printemps de MFN, Fréjus (2003).
6. Abarbanel S. and Kumar A., Compact high-order schemes for the Euler equations, *J. Scientific Computing*, **3**, 275 (1988).

How Effective Are High-Order Approximations in Shock-Capturing Methods? Is There a Law of Diminishing Returns?

William J. Rider and James R. Kamm

Los Alamos National Laboratory, Continuum Dynamics Group, CCS-2, Los Alamos, NM 87545, USA, rider@lanl.gov, kammj@lanl.gov

1 Introduction

High-order approximations are effective in increasing the accuracy per unit mesh cell of finite difference methods under almost all circumstances. This is also appears to be true for flows containing discontinuities, prototypically shock waves. The nature of their effectiveness must include the increased cost incurred when invoking higher order approximations. Because formal order of accuracy is lost in a shock wave propagating in a coupled system of equations [9], the higher order methods do not accumulate the significant advantage available for smooth linear problems. Here, we examine whether shock-capturing methods are benefited with higher efficiency when blended with higher order approximations.

We will test the proposition in two separate situations: using forward-in-time Godunov methods with piecewise linear and parabolic reconstructions [16, 3, 4, 7], and method-of-lines with flux splitting using a weighted ENO method [8, 1, 15]. In both cases the presumption is that the high-order will provide greater accuracy. The question to ponder is whether the improvement in accuracy is worth the additional cost. To achieve our results we will examine several standard test problems [6] including Sod's shock tube [14], the interacting blast wave problem [17], and the Shu-Osher shock-entropy wave [13]. These problems include simple to complex structures in a standard one-dimensional setting. We will also add some discussion of the results achieved for a one-dimensional breaking wave problem that is analytic prior to the formation of a shock [5]. The evaluation will be done in the same manner as the study in [6].

2 Methods Studied

Below we briefly discuss the salient aspects of the methods studied here.

- **PLM**: The piecewise linear method was introduced by van Leer. It uses a linear interpolation to define its spatial differencing. Van Leer used second-order finite differences in this truly second-order method. Colella

introduced fourth-order differencing for the slopes in place of the second-order differencing that van Leer used. Here, we will examine the impact of this choice as well as even higher order approximations. Other details of the method we use are characteristic differencing [2, 15], a two-shock Riemann solver [10]. We also use classic monotone differencing as well as more recently nonlinear hybridizations developed by the authors [11, 12]
- **PPM** The piecewise parabolic method uses a parabolic interpolation to define its differencing. The original method employed fourth-order accurate estimates for the edge values used to define the interpolant. We will examine the increase in the accuracy of these estimate on the efficiency of the method. We also use classic monotone differencing as well as more recently nonlinear hybridizations developed by the authors [11, 12]
- **WENO** Weighted essentially non-oscillatory methods were introduced to remove potentially harmful pathological behavior with earlier ENO methods. This method produces a convex combination of high-order stencils into a higher order approximation (third to fifth-order, fourth to seventh-order and so on). These methods are usually implemented in a method of lines fashion and we follow this practice here.
- **AMP** This method is an earlier version of the approach we have applied above with PLM and PPM developed by Suresh and Huynh [15]. Here high-order differencing is applied and tested for suitability through the application of bounding estimates for the edge values predicted by the high-order differencing.

3 Method of Evaluation

For the purposes of this study all the methods are implemented into a common code framework. This is done to assure fairness of comparison. We provide an analysis using the same approach used in an earlier study by Greenough and Rider [6]. There we found that the higher order formal accuracy of WENO methods did not result in greater computational efficiency when compared with a monotone PLM method for the computation of shock physics. The study here follows that study closely in procedure and analysis.

The results will be displayed in terms of the efficiency of the method. We define efficiency as the relative computational effort that is required to achieve a certain solution quality. In the following results we will use the L1 norm error in density and pressure to describe the solution error. Solution errors can then be reduced through refining the computational mesh. This involves reduction in the spatial mesh spacing and concomitantly the time step size in keeping with the CFL condition. A doubling of the mesh resolution leads to four times the computational effort (mod cache effects on modern processors. Thus, in a one-dimensional problem the computational effort goes with the square of the number of zones, and the error roughly halves in keeping with first-order convergence in the presence of discontinuities. In two dimensions

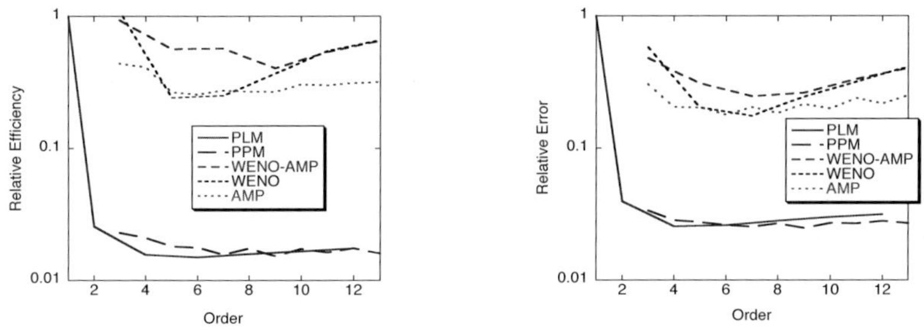

Fig. 1. The relative error for schemes on Sods shock tube. Both show that the error and efficiency saturates at 6th order.

the effort goes with the cube of the zone and with the fourth power in three dimensions. Thus the efficiency in one dimension is the computational cost multiplied by the square of the relative error.

4 Results

We summarize the results in the plots shown in Figures 1 and 2. The efficiency saturates beyond sixth or seventh order. With WENO schemes going to high order is quite expensive resulting in diminishing efficiency (although advances in accuracy). In three dimensions, should the trends in one dimension hold, the result would favor methods of perhaps one order higher due to the payoff for improvements in accuracy. For WENO methods the overall trend seems to indicate that seventh-order is best and with PPM sixth-order initial edge values. At the end of this study we left to ponder how much effort should be poured into high-order approximations for shocked problems (systems of equations). While it is clear that high-order approximations dramatically improve results and provide greater efficiency, the gains subside in going past sixth or seventh order in one dimension, and eighth-order in two and three dimensions. This seems to hold for Godunov-type methods and WENO methods.

5 Discussion

All of these results are dependent on the nature of computing hardware and implementation. With that said, the gain in accuracy in going from second-to-fourth order is unlikely to be overcome by the cost of moving to a slightly broader stencil. Parallel programming and computers may provide an even greater diminishing value for high-order differencing. As a result we expect that our results are relatively robust to the details.

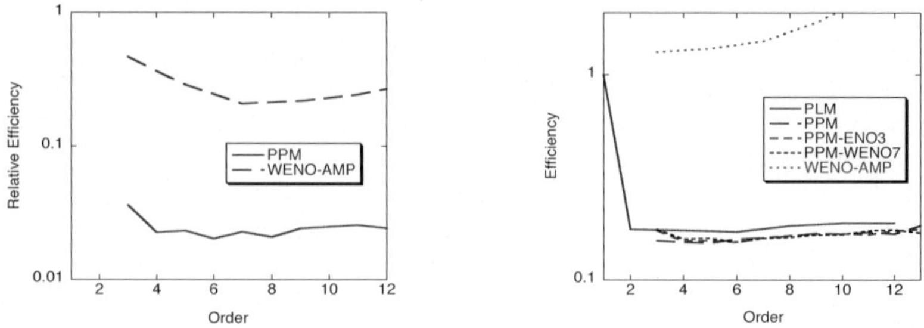

Fig. 2. The blast wave problem and Shu-Osher problem efficiency comparison. Both problems show a maximum efficiency occurring at sixth order

The important this to remember is that our results are predicated on the principle that the goal of computational physics is the best results for level of computational effort.

References

1. D. Balsara and C.-W. Shu. Monotonicity preserving weighted essentially non-oscillatory schemes with increasingly high order of accuracy. *Journal of Computational Physics*, 160(2):405–452, 2000.
2. D. S. Balsara. Total variation diminishing scheme for adiabatic and isothermal magnetohydrodynamics. *Astrophysical J. Supp. Ser.*, 116:133–153, 1998.
3. P. Colella. A direct Eulerian MUSCL scheme for gas dynamics. *SIAM Journal on Scientific and Statistical Computing*, 6:104–117, 1985.
4. P. Colella and P. Woodward. The piecewise parabolic method (PPM) for gas-dynamical simulations. *Journal of Computational Physics*, 54:174–201, 1984.
5. A. W. Cook and W. H. Cabot. A high-wavenumber viscosity for high-resolution numerical methods. *Journal of Computational Physics*, 195(2):594–601, April 2004.
6. J. Greenough and W. J. Rider. A quantitative comparison of numerical methods for the compressible Euler equations. fifth-order WENO and piecewise-linear Godunov. *Journal of Computational Physics*, 196(1):259–281, 2004. Los Alamos Unlimited Release Report, LA–UR–02–5640.
7. H. T. Huynh. Accurate upwind methods for the Euler equations. *SIAM Journal on Numerical Analysis*, 32:1565–1619, 1995.
8. G.-S. Jiang and C.-W. Shu. Efficient implementation of weighted ENO schemes. *Journal of Computational Physics*, 126:202–228, 1996.
9. A. Majda and S. Osher. Propogation of error into regions of smoothness for accuracte difference approximations to hyperbolic equations. *Communications on Pure and Applied Mathematics*, 30:671–705, 1977.
10. W. J. Rider. An adaptive Riemann solver using a two-shock approximation. *Computers and Fluids*, 28:741–777, 1999.
11. W. J. Rider, J. A. Greenough, and J. R. Kamm. Extrema, accuracy and monotonicity preserving methods for compressible flows. Technical Report AIAA–2003-4121, 16th AIAA CFD Conference June 23-26, 2003 Orlando FL, 2003.

12. W. J. Rider, J. A. Greenough, and J. R. Kamm. Extrema, accuracy and monotonicity preserving methods through adaptive nonlinear hybridizations. *jcp*, to be submitted, 2004.
13. C.-W. Shu and S. Osher. Efficient implementation of essentially non-oscillatory shock-capturing schemes. *Journal of Computational Physics*, 77:439–471, 1988.
14. G. Sod. A survey of several finite difference methods for systems of nonlinear hyperbolic conservation laws. *Journal of Computational Physics*, 27:1–31, 1978.
15. A. Suresh and H. T. Huynh. Accurate monotonicity-preserving schemes with Runge-Kutta time stepping. *Journal of Computational Physics*, 136:83–99, 1997.
16. B. van Leer. Towards the ultimate conservative difference scheme. V. A second-order sequel to Godunov's method. *Journal of Computational Physics*, 32:101–136, 1979.
17. P. Woodward and P. Colella. The numerical simulation of two-dimensional fluid flow with strong shocks. *Journal of Computational Physics*, 54:115–173, 1984.

Adaptive Numerical Dissipation Control in High Order Schemes for Multi-D Non-Ideal MHD

H. C. Yee[1] and B. Sjögreen[2]

[1] NASA Ames Research Center, Moffett Field, California, 94035, USA
[2] The Royal Institute of Technology, 100 44 Stockholm, Sweden

1 Motivation

The required type and amount of numerical dissipation/filter to accurately resolve all relevant multiscales of complex magnetohydrodynanics (MHD) unsteady high-speed shock/shear/turbulence/combustion problems are not only physical problem dependent, but also vary from one flow region to another. In addition, proper and efficient control of the divergence of the magnetic field (**Div(B)**) numerical error for high order shock-capturing methods poses extra requirements for the considered type of CPU intensive computations.

2 Objective

The goal is to extend our adaptive numerical dissipation control in high order filter schemes [3] [2] and our new divergence-free methods for ideal MHD [4, 5] to non-ideal MHD that include viscosity and resistivity. See [1] for the form of non-ideal MHD to be considered. The key idea in [3, 2] consists of automatic detection of different flow features as distinct sensors to signal the appropriate type and amount of numerical dissipation/filter where needed and leave the rest of the region free from numerical dissipation contamination. These scheme-independent detectors are capable of distinguishing shocks/shears, flame sheets, turbulent fluctuations and spurious high-frequency oscillations. The detection algorithm is based on an artificial compression method (ACM) [3] (for shocks/shears), and redundant multiresolution wavelets (WAV) [2] (for the above types of flow feature). These filters also provide a natural and efficient way for the minimization of **Div(B)** numerical error [4, 5].

The filter scheme consists of spatially sixth-order or higher non-dissipative spatial difference operators as the base scheme for the inviscid flux derivatives. If necessary, a small amount of high order linear dissipation is used to remove spurious high frequency oscillations. For example, an eighth-order centered linear dissipation (AD8) might be included in conjunction with a spatially sixth-order base scheme. The inviscid difference operator is applied twice for the viscous flux derivatives. After the completion of a full time step of the base scheme, the solution is adaptively filtered by the product of a "flow detector" and the "nonlinear dissipative portion" of a high-resolution shock-capturing scheme. It is noted that the nonlinear dissipative portion of a high-resolution shock-capturing scheme that is higher than second order is not unique. In addition, the scheme independent wavelet flow detector can be used in conjunction with spatially compact, spectral or spectral element type of base schemes. The ACM and wavelet filter schemes using the dissipative portion of a second-order shock-capturing scheme [5] with sixth-order spatial central base scheme for both the inviscid and viscous MHD flux derivatives and a fourth-order Runge-Kutta method are denoted by ACM66 (ACM66+AD8) and WAV66 (WAV66+AD8) respectively. Computation using the same temporal and spatial scheme for the viscous MHD flux derivatives, and the standard fifth-order WENO scheme for the inviscid flux derivatives is denoted by WENO5. Computations using a second-order MUSCL and the Harten-Yee [5] TVD scheme for the inviscid MHD flux with the second-order central scheme for the viscous flux and a second-order Runge-Kutta method are denoted by MUSCL and TVD22, respectively.

3 Numerical Examples

Extensive grid convergence studies using WAV66 and ACM66 for typical ideal MHD test cases were conducted in [4, 5]. More accurate solutions were obtained with WAV66 and ACM66 than with WENO5, which is more CPU intensive. Figures 1 and 2 show the comparison among MUSCL, WENO5 and WAV66 of a non-ideal MHD Orszag-Tang vortex problem with $Re = 1000$ and resistivity coefficient of 100 at time $T = 3.14$. Computations for the corresponding inviscid case are included for comparison.

The Orszag-Tang problem consists of periodic boundary conditions with smooth sinusoidal initial data [4, 5]. Five levels of grid refinement (101×101 to 1601×1601) indicate that all the three high order schemes exhibit a similar accuracy on their density, velocity and pressure variables. The accuracy of their three magnetic field variables are, however, different. Studies show that both ACM66 (ACM66+AD8) and WAV66 (WAV66+AD8) are divergence free preserving for the entire time evolution whereas WENO5 or TVD22 are only divergence free preserving at the very early stage of the time evolution. Their **Div(B)** numerical error quickly increases from a L^2-norm of 10^{-8} to be larger than 1, once the viscous shocks/shears have formed. The top left blank subfigure of Fig. 1 by WAV66 indicates divergence free preserving (10^{-14} to 10^{-11}), and the bottom left subfigure by WENO5 indicates its **Div(B)** numerical error. Figure 2 shows the time evolution of the L_2-norm of **Div(B)** numerical errors for the three methods. See [5] for the **Div(B)** contours and L_2-norm of **Div(B)** for each scheme.

Figures 3-5 show the comparison among the three methods for a supersonic shock interacting with a magnetic cloud with the same Re and resistivity coefficient as the first test case. Again the corresponding inviscid

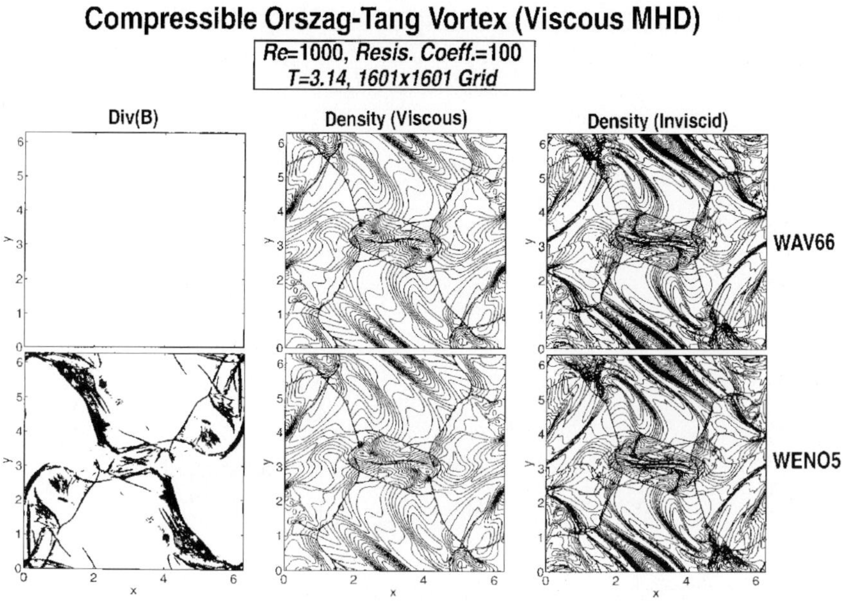

Fig. 1.

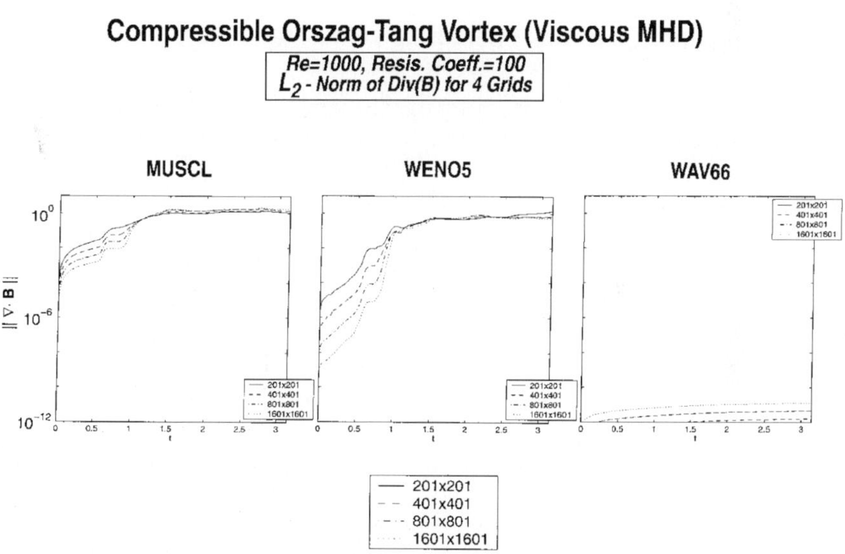

Fig. 2.

computations are included for comparison. For this test case, after the time evolution $T > 0.04$, very complex wave interactions occur at both physical boundaries. See [5] for the inviscid problem set up.

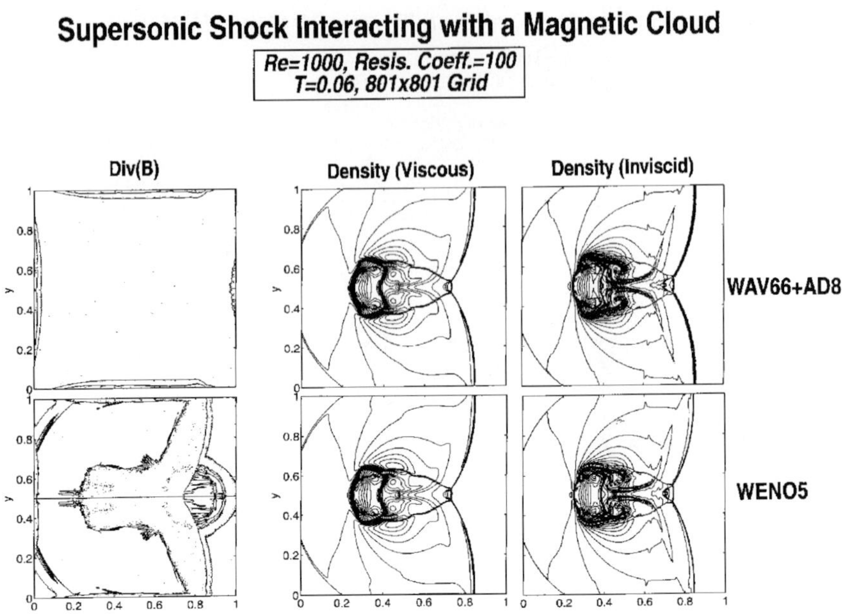

Fig. 3.

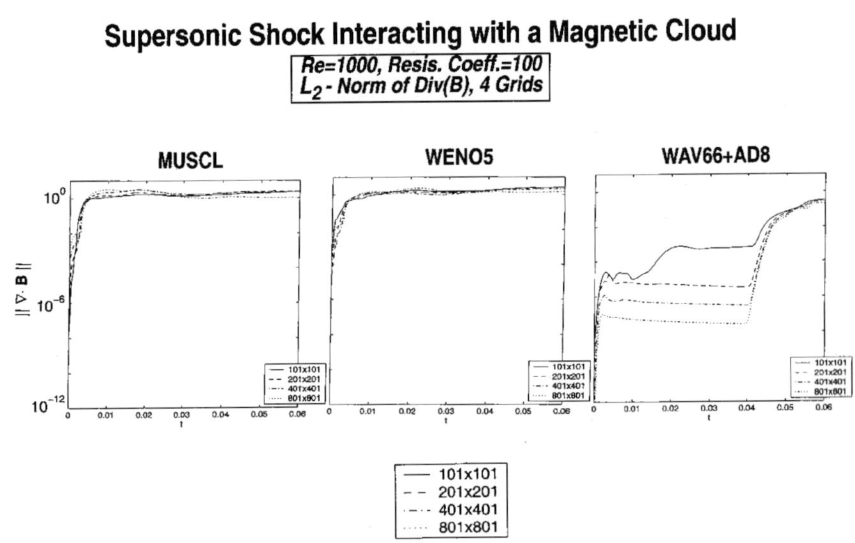

Fig. 4.

Figure 5 shows the advantage of our adaptive numerical dissipation control over the standard MUSCL and WENO5 schemes on coarse solutions. Aside from the small spurious oscillation, WAV66+AD8 exhibits the same resolution using a 101×101 grid as the WENO5 using a 201×201 grid. It

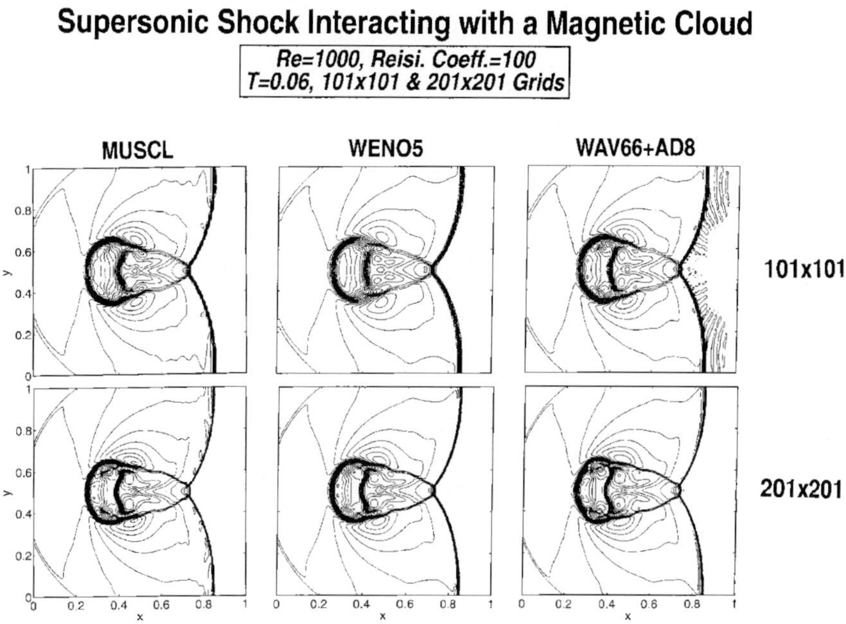

Fig. 5.

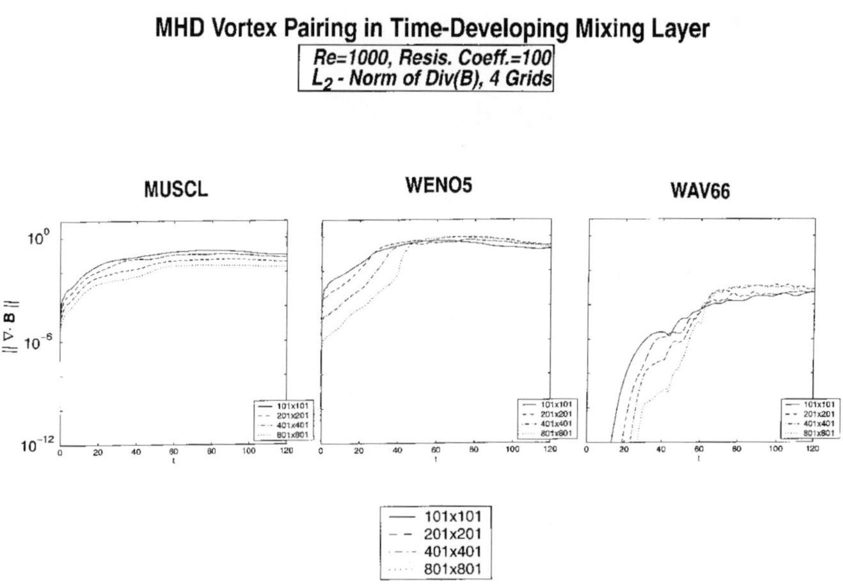

Fig. 6.

is interesting to note that the resolution of WAV66+AD8 using a 201×201 grid is comparable to WENO5 using a 801×801 grid.

Figures 6 and 7 show the computations of a MHD vortex pairing in a time-developing mixing-layer with the same *Re* and resistivity coefficient as the first test case. See [3] for the basic gas dynamic flow set up. Here the initial data for the three magnetic field components are 0.1 in the x-direction and zero in the other two directions. This time-developing mixing-layer problem consists of a periodic boundary condition in the x-direction and an open boundary in the y-direction. For the shock/cloud problem and the magnetized mixing-layer problem, although not all figures are shown, ACM66 (ACM66+AD8) and WAV66 (WAV66+AD8) give an overall improved accuracy over TVD22, MUSCL and WENO5. Machine zero **Div(B)** numerical error was not obtained for the entire time evolution due to the non-divergence-

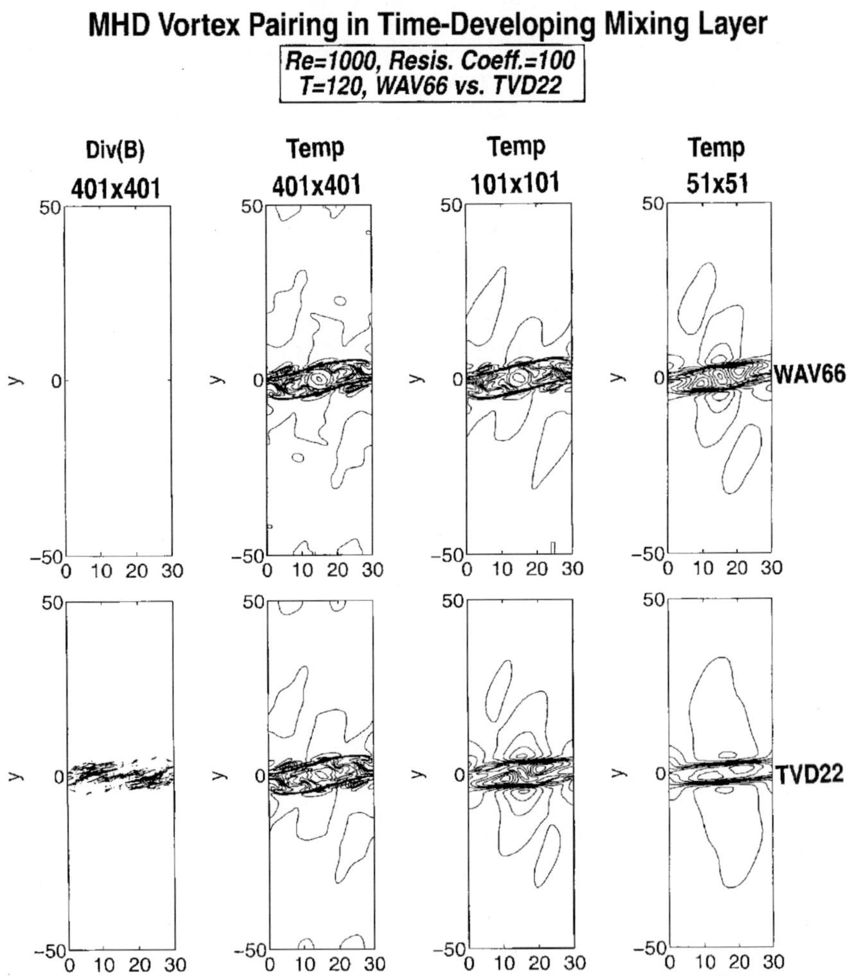

Fig. 7.

free numerical boundary treatment at the open boundaries (see Fig. 3 top left subfigure). For most of the time evolution, the ACM66(ACM66+AD8) and WAV66 (WAV66+AD8) exhibit orders of magnitude smaller **Div(B)** numerical error than TVD22, MUSCL and WENO5. See [6] for some issue on the stable numerical boundary condition treatment on high order base scheme and the importance of having a divergence-free numerical boundary treatment for complex wave interactions on open boundaries.

References

1. D.V. Gaitonde, *Development of a Solver for 3-D Non-Ideal Magnetogasdynamics* AIAA Paper 99-3610, 1999.
2. B. Sjögreen and H. C. Yee, *Multiresolution Wavelet Based Adaptive Numerical Dissipation Control for Shock-Turbulence Computations*, RIACS Report 01.01, Oct. 2000, NASA Ames Research Center, also, J. Scient. Computing, **20** (2), 211-255 (2004)
3. H. C. Yee, N. D. Sandham, and M. J. Djomehri, *Low Dissipative High Order Shock-Capturing Methods Using Characteristic-Based Filters*, J. Comput. Phys., **150**, 199 (1999).
4. B. Sjögreen and H. C. Yee, *Efficient Low Dissipative High Order Scheme for Multiscale MHD Flows, I: Basic Theory*, AIAA 2003-4118 (2003).
5. H.C. Yee and B. Sjögreen, *Efficient Low Dissipative High Order Scheme for Multiscale MHD Flows, II: Minimization of Div(B) Numerical Error*, RIACS Technical Report TR03.10, July 2003, NASA Ames Research Center.
6. H.C. Yee and B. Sjögreen, *On Divergence-Free Low Dissipative High Order Scheme for non-ideal MHD*, proceedings of the ICOSAHOM, Brown University, Providence, RI, June 21-25, 2004.

A NURBS-Based Shape Optimization Method for Hydraulic Turbine Stay Vane

Didier Poueymirou-Bouchet, Christophe Tribes, and Jean-Yves Trépanier

Départment of mechanical engineering, École Polytechnique de Montréal, C.P. 6079, Succursale Centre Ville, Montréal, QC, Canada, H3C 3A7,
didier.poueymirou-bouchet@polymtl.ca, christophe.tribes@polymtl.ca
jean-yves.trepanier@polymtl.ca

Summary. This paper presents an optimization methodology adapted to rehabilitation of the stay vanes within a Francis type hydraulic turbine. The proposed methodology enables to take into account the constraints and objectives specific to such a design problem. Results obtained with a multi-island genetic algorithm and a gradient-based algorithm are compared.

1 Introduction

Leveraging the increasing usage and confidence in CFD predictions, computational-based design is the next technology to be deployed in industrial settings. The paper presents results from a feasibility study concerning the development of an automated design optimization procedure for the shape design of Francis type hydraulic turbine stay vanes to be refurbished. This kind of design problem is a challenge for engineers because it is necessary to manage the interactions between new, old and renovated components of the hydropower plant under the constraints of construction feasibility. Moreover, the time limitation constraint on the design cycle and the expected improvement on the performance are of crucial importance for obtaining profitable contracts. Thus, efficient optimization methodologies appear as powerful tools that can help engineers to deal with some of the issues of rehabilitation projects [1, 2].

The stay vanes are the first series of blades that the flow enters after leaving the spiral case. The original design has put the emphasis on the principal functions of the stay vanes, which are to preserve the structural integrity of the spiral case and to pre-turn the flow from the tangential to the radial direction. The hydrodynamic performances of these design are far from being optimal as the blades often generate considerable energetic loss due to separation of the boundary layer. The rehabilitation of stay vanes consists in modifying a part of the profile to maximize the ratio between the rehabilitation costs and the improvements of the turbine output. As a consequence, the refurbishment will be focused on the leading and trailing edge areas of the stay vane profile by adding or removing material from the original profile, leaving the central portion unchanged, and thus maintaining the primary structural function.

2 Vane Shape parameterization and Constraints

The shape of the stay vane is parameterized in 2D using Non-Uniform Rational B-Splines (NURBS) with special features to account for the particular constraints of the application.

The coordinate of a point on the NURBS curve is represented as a vector-valued explicit function of an independant parameter u:

$$C(u) = \frac{\sum_{i=0}^{n} N_{i,p}(u)\omega_i P_i}{\sum_{i=0}^{n} N_{i,p}(u)\omega_i} \tag{1}$$

where $n+1$ is the number of control point, P_i is the vector position of the ith control point, ω_i is the weight associated with the ith control point, $N_{i,p}(u)$ is the ith B-spline basis function of p-degree (order $p+1$) defined on the nondecreasing knot vector $U = \{u_0, \ldots, u_{n+p+1}\}$ (see Ref. [3] for the expression of the B-spline basis functions and details about the NURBS). The degree of the basis functions determines the geometric regularity of the curve; the continuity of the curve, its tangents and its curvature are desired in our case. A few properties related to the regularity of the NURBS can be outlined:

- The first and the last control points are the end points of the curve.
- The curve is tangent to the end segments of the control points polygon.
- The curve is C^∞ continuous for u on any interval $]u_i, u_{i+1}[$ of the knot vector, and C^{p-k} continuous at $u = u_i$, k being the multiplicity of a knot in the knot vector.

We have considered a cubic NURBS ($p = 3$) and a knot vector such that $u_i = 0$ for $i = 0, \ldots, p$ (i.e., multiplicity $k = p+1$), $u_i < u_{i+1}$ for $i = p+1, \ldots, n-1$ ($k = 1$) and $u_i = 1$ for $i = n, \ldots, n+p+1$ ($k = p+1$). Thus, except at the end points, the selected parameterization produces a curve that is at least C^2 continuous.

The domain of influence of the control point P_i is limited to the portion of curve on the interval $[u_i, u_{i+p+1}[$ where the basis function $N_{i,p}$ is nonzero. Therefore, for any $u \in [u_i, u_{i+1}[$, there is at most $p+1$ control points influencing the position of the point $C(u)$: $P_{i-p}, \ldots, P_i$. The position of a point corresponding to a knot of multiplicity one depends at most on p control points: $P_{i-p}, \ldots, P_{i-1}$. For example, with $p = 3$, a point $C(u_k)$ (u_k in the knot vector) is invariant as long as the control points P_{k-3}, P_{k-2}, P_{k-1} verify the following vectorial relation:

$$Q_f = C(u_k) = \frac{\sum_{i=k-1}^{k-3} N_{i,3}(u_k)\omega_i P_i}{\sum_{i=k-1}^{k-3} N_{i,3}(u_k)\omega_i} \tag{2}$$

This corresponds to two scalar relations with 9 unknowns for a 2D curve (x, y coordinates of each control points and the weights ω), i.e., 7 degrees of freedom. By fixing the weight of the control point, the vectorial relation (2)

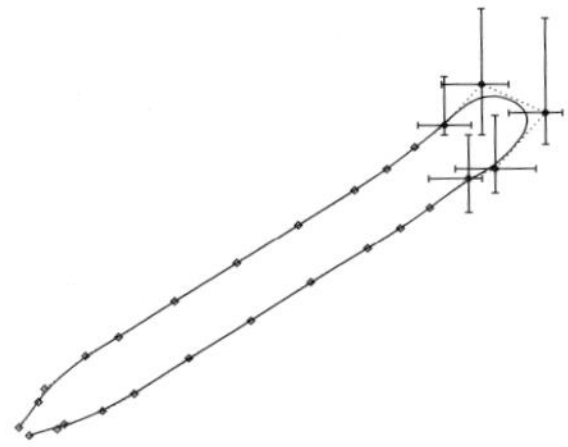

Fig. 1. NURBS control points and design variables of the initial shape

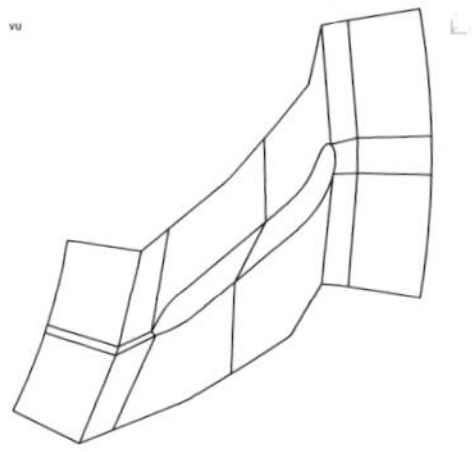

Fig. 2. Multi-zone subdivision of the domain

becomes two scalar linear relations and we have 4 degrees of freedom left. Thus, the curve will pass through the point Q_f by specifying two control points a priori whereas the third control point is obtained from the solution of the system of equations. Once P_{k-1}, P_{k-2}, P_{k-3} are properly selected, the curve $C(u)$ will always pass through the point P_f even if the rest of the control points are moved. Moreover, the curve $C(u)$ is "separated" in two parts, this means that shifting a control point before P_{k-3} changes the portion of curve downstream $u = u_k$ and shifting a control point after P_{k-1} changes the portion of curve upstream $u = u_k$.

The proposed parameterization of the stay vane uses this property of local influence of each control points to define several zones on the 2D profile delimited by fixed points. This parameterization enables local optimization of portions of the stay vane. The NURBS properties insure that the curve is C^2 continuous at the frontiers of the zones.

The initial shape of the stay vane is obtained by a B-spline (i.e., a NURBS with the $\omega_i = 1$) interpolation of a selection of points Q_j including the desired fixed points Q_f and some points spread out according to the curvature. First, we assign a parameter value $\overline{u}_j$ to each of the selected points $(C(\overline{u}_j) = Q_j)$ by using the centripetal method. Second, the knot vector is obtained by an "averaging" that reflects the distribution of $\overline{u}_j$ (see Ref. [3]). After that, the control points of the B-spline are obtained by solving a system of linear equations obtained from the vectorial relations with $u = \overline{u}_j$ in Eq. (1). Finally, additional control points and knots must be inserted to ensure that the fixed points are influenced only by specific control points—each fixed point and the corresponding additional control points must verify equation (2).

The requirements from the industry were to keep untouched the mid portion of the vanes, to add a longer trailing edge and to principally allow to modify the shape of the leading edge to better match the inflow conditions. The resulting parameterization of the stay vane profile is presented on Figure 1. Five control points have been used at the leading edge and their position has been allowed to vary according to the range illustrated on Figure 1.

A total of 15 design variables are used, 10 for the x and y positions of the control points and 5 for their weight.

Other constraints from structural requirements cannot be verified implicitly by the parameterization. One constraint that has been identified is the maximal curvature on the leading edge of the stay vane. Indeed, in this region, a profile too sharp would make the manufacturing process more complex, and would be too fragile because of its thinness. Therefore, if we consider $C_{le}(X)$ as the curvature on the leading edge for the current design and C_{lim} the maximal acceptable curvature, this constraint can be expressed as follows:

$$g_1(X) = \frac{C_{lim} - C_{le}(X)}{C_{lim}} \geq 0 \tag{3}$$

In addition, considering that the stay vane is a component of a whole installation, and that some of the components such as the spiral case can not be modified, the modifications applied to the stay vane must not interfere with the other components. In particular, the length of the stay vane must be limited. This constraint can be expressed as follows:

$$g_2(X) = l_{max} - \frac{L(X) - L_0}{L_0} \geq 0 \tag{4}$$

where $L(X)$ and L_0 are respectively the current and initial length of the profile, and l_{max} is the maximum allowed elongation.

3 Flow Solution

A commercial CFD package is used to obtain a measure of the performance of the vanes in terms of total pressure loss. The automation of the total pressure loss calculation is performed in four steps. The first step constructs a multi-zone subdivision of the computational domain and the second step generates a computational structured grid. The grid is clustered close to the vanes to respect the flow solver requirements in terms of Y^+ for the turbulence model. The third step is the computation of the flow solution using the CFX-TASCFlow solver and the last step involves extracting the average total pressure loss between the inlet and outlet surfaces from the solution.

The first and the second step use several methods for geometric modelling, mesh generation and adaptation developped for a large range of turbomachinery components such as runner blades, stay vanes and wicket gates. As illustrated on Figure 2, the computational domain and topology is automatically adapted to the shape and the periodicity of the blade in the stay ring (number of blades, external and internal diameters).

A study of the influence of the mesh has permitted to find that a grid made up of 27588 elements constitutes a satisfying compromise between the duration and the accuracy of the calculation for optimization purposes.

Table 1. Optimization results and off-design performances

	$f(X)$ for different inflow angles			
Design	15 deg	20 deg	25 deg	30 deg
Original	0.3993	0.3014	0.2062	0.1531
Initial	0.3089	0.1815	0.0849	0.0620
Optimization at 15 deg	**0.1775**	0.0910	0.09175	0.1293
Optimization at 20 deg	0.1822	**0.0778**	0.06008	0.0709
Optimization at 25 deg	0.2104	0.0849	**0.0533**	0.0471
Optimization at 30 deg	0.2591	0.11657	0.0580	**0.0440**

4 Design Optimization Results

The optimization aims to improve hydrodynamic behaviour with the objective to reduce the loss of total pressure through the stage subject to the geometric constraint given in Section 2. The optimization problem statement is:

$$\begin{aligned} \min_{X} f(X) &= \frac{P_{tot}^{in} - P_{tot}^{out}}{P_{tot}^{in}} \\ \text{s.t.:} \quad & g_1(X) \geq 0 \\ & g_2(X) \geq 0 \end{aligned} \tag{5}$$

where P_{tot}^{in} and P_{tot}^{out} are respectively the inflow and outflow average total pressure, X is the vector of the design variables.

Two optimization algorithms have been tested. The first one is a multi-islands Genetic Algorithm (GA) which is known to globally explore the design space, thus increasing the chance to find a global optimum. The second one is a gradient-based algorithm (DONLP based on Sequential Quadratic Programming method, see Ref. [4]) which has the advantage to rapidly converge to a local optimum. To compare these two approaches, we have performed a single point optimization for an inflow angle of 20 degrees using these two algorithms implemented in the iSIGHT 7.0 environment.

To increase the efficiency of the DONLP algorithm and to reduce the risk of convergence to a local optimum, we have choosen a coarse step size initially to overcome some local optima and the step size value is progressively reduced as the optimizer gets closer to the optima area. The final value of the step size is obviously fixed by the level of numerical noise, previously determined by a sensitivity analysis (see Ref. [5] for details). The profile used to start all the optimizations is shown on Figure 1.

After 500 evaluations of the objective function (limit imposed to the GA), the design obtained by the Genetic Algorithm corresponds to $f(X) = 0.0930$. The DONLP algorithm obtained a better design ($f(X) = 0.0780$) after 345 evaluations. This test made us choose the DONLP sequence to solve the optimization problem.

Figure 3 illustrates the stay vane shape optimized for different inflow. As could have been expected, the mean camber line in the leading edge region

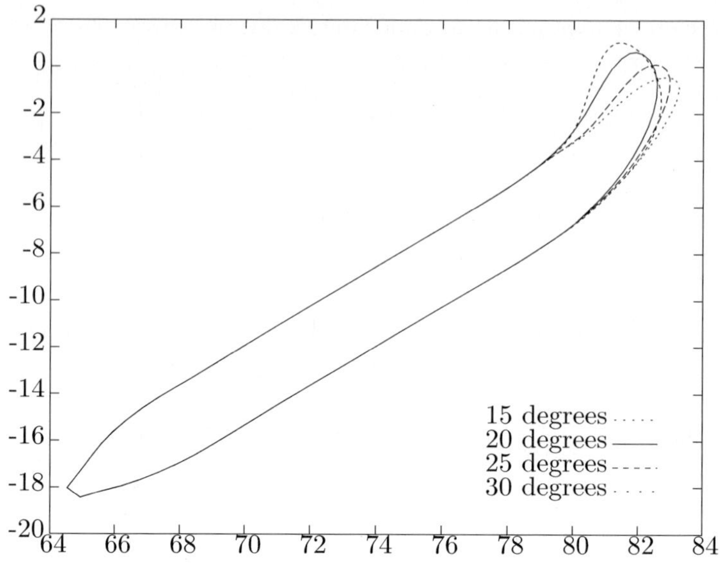

Fig. 3. Optimized shape vs inflow angles

tend to align with the inflow direction. As shown on Figure 4, the shape of the leading edge is also modified to permit a progressive change in the direction of the flow from the inflow angle to a direction parallel with the stay vane mid portion, thus avoiding the separation of the boundary layer on the suction side. The comparison of the streamlines given in Figure 4 reveals that the optimization considerably reduces flow separation. The sharp leading edge also helps to reduce the deceleration in this region of the flow whithin the respect of the constraints $g_1(X) \geq 0$ and $g_2(X) \geq 0$.

We have considered two references for evaluating the improvements brought by the optimization: the original blunt profile and the initial profile used as a starting point of the optimization with a smoother leading edge and a streamlined trailing edge (see Figure 1). Table 1 shows that the initial

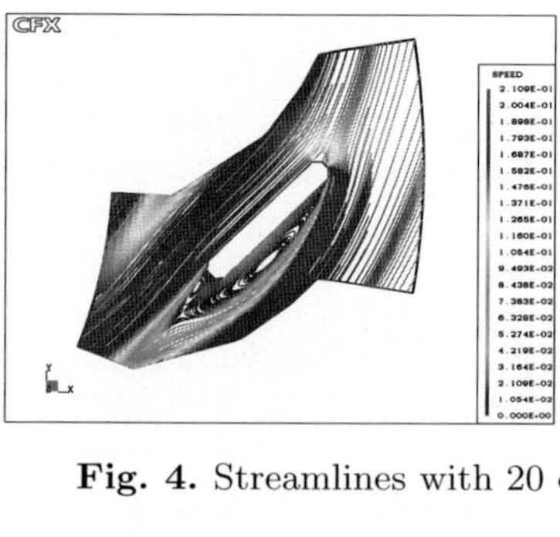

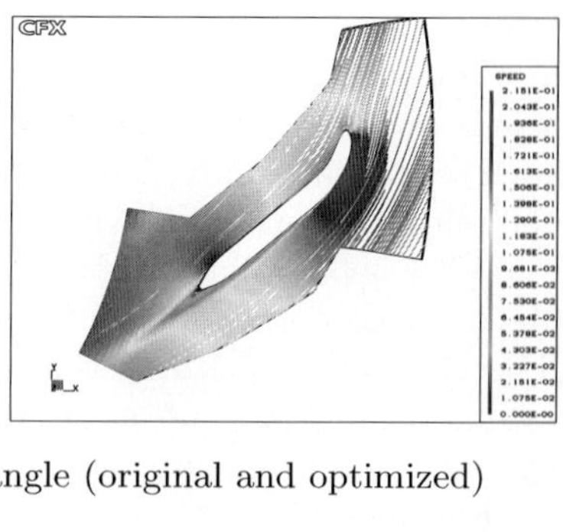

Fig. 4. Streamlines with 20 degrees inflow angle (original and optimized)

profile produces considerable improvements over the original profile; after that, the optimizations conducted for each inflow angle reduce even more significantly the objective function (bold values).

The optimizations produce a different profile depending on the inflow angle. The off-design performances for each profile are also given in table 1. For the optimized profile, the objective function is larger for a different inflow angle than the one used for the optimization.

5 Conclusion

A methodology to design hydraulic turbine stay ring for rehabilitation projects is presented. The methodology includes a NURBS parameterization of the shape that includes geometrical constraints and the use of optimization algorithms to get a correct solution faster.

Two optimization algorithms have been tested: a gradient-based algorithm and a multi-island genetic algorithm. The comparison of the two approaches shows that the gradiant-based algorithm is able to find a better design more rapidly.

The shape optimizations of the leading edge portion of the stay vane for different inflow angles lead to significant reductions of energetic losses compared with the original shape. Also, we have observed that the performance of the optimized shapes is slightly impaired for off-design conditions, which is typical of single point optimizations.

Acknowledgments

The present research work was carried out under a collaborative R&D project, between École Polytechnique de Montréal and GE Hydro, entitled "Geometric modelling and mesh generation and adaptation for turbo machinery components" and partially funded by NSERC of Canada.

References

1. Tomas, L., Pedretti, C., Chippa, T. and Stoll, P., "Automated design of a Francis turbine runner using global optimization algorithms", *Proceedings of the XXIst IAHR Symp. on Hydraulic Machinery and Systems,* Lausanne, 2002.
2. Sallaberger, M., Eisele, K. and Casey, M., "Accelerated design procedure for Francis runners in rehabilitation projects", *Hydropower & Dams,* No. 1, pp. 87-92, 2001.
3. Piegl, L. and Tiller, W.: *The NURBS Book*, Springer-Verlag, 1995.
4. Spelluci, P., "SQP method for general nonlinear programs using only equality constrained subproblems", Math. Prog., 82:413–448, 1998.
5. Poueymirou-Bouchet, D., "Optimisation hydrodynamique de profils 2D", Mémoire de Maîtrise, École Polytechnique de Montréal, Montréal (2003).

Super Compact Spatial Differencing for the Linear and Nonlinear Geophysical Fluid Dynamics Problems

V. Esfahanian[1], S. Ghader[2], and A.R. Mohebalhojeh[3]

[1] Mechanical Engineering Department, University of Tehran, Campus#2, North Karegar Ave., Tehran, Iran evahid@ut.ac.ir

[2] Atmospheric Science& Meteorological Research Center, Karaj Highway, Tehran, Iran sghader@ut.ac.ir

[3] Institute of Geophysics, University of Tehran, North Karegar Ave., Tehran, Iran amoheb@ut.ac.ir

Summary. The Super Compact Finite Difference Method (SCFDM) is applied to spatial differencing of some prototype linear and nonlinear geophysical fluid dynamics problems. For the frequency of linear inertia-gravity waves on different numerical grids (Arakawa's A-E and Randall's Z), the sixth-order SCFDM shows a substantial improvement on the conventional methods. For the Jacobians involved in vorticity advection by nondivergent flow and in Bolin–Charney balance equation, as nonlinear problems, it is found that the sixth-order SCFDM provides a noticeably more accurate representation of the wavenumber distribution of the Jacobians, when compared with the conventional methods. In addition, computation of a normalized global error at different horizontal resolutions in longitude and latitude directions for the Rossby–Haurwitz wave on a sphere shows that sixth-order SCFDM can markedly improve on the fourth-order compact.

1 Introduction

The compact finite difference schemes, introduced as far back as the 1930s, have been found as simple, yet powerful ways of reaching the objectives of high accuracy and low computational cost [14, 12]. A recent development in compact schemes has been the introduction of a general procedure to generate highly accurate schemes of arbitrary order with minimal stencil size by Fu Dexun and Ma Yanwen [4], who called it Super Compact Finite Difference Method (SCFDM). The derivation and the application of the method in uniform grid have been presented in [16, 5]. In addition, some aspects of the scheme in nonuniform grid have been investigated by Ghader [9] and Esfahanian *et al.* [7].

This paper is devoted to the assessment of the accuracy of SCFDM in the atmosphere–ocean dynamics context. We apply the sixth order SCFDM to spatial differencing of some prototype linear and nonlinear geophysical fluid dynamics problems.

2 Super Compact Method

The super compact comprises of a basic equation and a set of auxiliary equations. The basic equation is obtained by truncating the Taylor series which relates a function (ϕ) to its derivatives. To arrive at a closed system of equations, a set of $(n-1)$ additional independent equations called auxiliary equations, is needed. The SCFDM relations can be written in a vector form:

$$-\frac{1}{2}\mathbf{L}(\frac{1}{\sigma_j})\mathbf{F}_{j-1}+(\mathbf{A}+\mathbf{L})(\sigma_j)\mathbf{F}_j-\frac{1}{2}\mathbf{L}(\sigma_{j+1})\mathbf{F}_{j+1}=\frac{1}{2}(\alpha\Delta_x^+ +\beta\Delta_x^-)\phi_j\mathbf{E} \quad (1)$$

in which α and β are free parameters, $\Delta_x^+\phi_j=\phi_{j+1}-\phi_j$, $\Delta_x^-\phi_j=\phi_j-\phi_{j-1}$ and the definition of the vectors and matrices are given in [4, 7]. At the boundaries, forward and backward relations are used [16, 9].

The derivation of a more convenient form of the SCFDM relations is also possible [7]. The alternative form of the sixth-order approximations of the first and second derivatives are obtained as

$$\phi_j^{<1>}=\left(\frac{10(12+\Delta_x^2)\Delta_x^\circ}{120+30\Delta_x^2+\Delta_x^4}\right)\phi_j, \qquad \phi_j^{<2>}=\left(\frac{30(12+\Delta_x^2)\Delta_x^2}{360+60\Delta_x^2+\Delta_x^4}\right)\phi_j \quad (2)$$

where $\Delta_x^\circ=(\Delta_x^+ +\Delta_x^-)/2$, $\Delta_x^2=\Delta_x^+\Delta_x^-=\Delta_x^-\Delta_x^+$. The sixth-order approximation of the first derivative for the cell-centered can be found as

$$\phi_j^{<1>}=\left(\frac{320(12+\Delta_x^2)}{1920+240\Delta_x^2+\Delta_x^4}\right)(\phi_{j+\frac{1}{2}}-\phi_{j-\frac{1}{2}})/2 \quad (3)$$

3 Spatial Differencing of the Linear Problems

For the spatial differencing of the linear problems using SCFDM, the linearized shallow water equations on an f-plane are considered. The shallow water equations will be discretized on the well known Arakawa A-E grids [1], and on the Randall's Z grid [13].

By assuming wave solutions, for the continuous linearized shallow water equations [1, 13] the dispersion relation $(\omega/f_\circ)^2=1+\lambda^2(\kappa^2+\ell^2)$ is obtained. In this equation ω is the frequency, $\lambda=\sqrt{gH}/f_\circ$ is the Rossby radius of deformation, and κ and ℓ are the wavenumbers in the x and y directions, respectively.

To represent the discrete dispersion relationships for the inertia-gravity waves on Arakawa A-E grids, the generic expressions of the discrete dispersion relationships derived by Blayo [2] are used. This idea, here, can be extended in a similar way to obtain the discrete dispersion relation $(\omega/f_\circ)^2=1-\lambda^2\left[T_2(\kappa)+T_2(\ell)\right]$ for the Z grid. We note that the transfer function, T of a scheme S is defined by $S(e^{i\kappa x})=T(\kappa)e^{i\kappa x}$.

The results of the discrete relations of the different methods on the numerical A-E and Z grids are presented for a resolved ($\lambda/d=2$) and for an

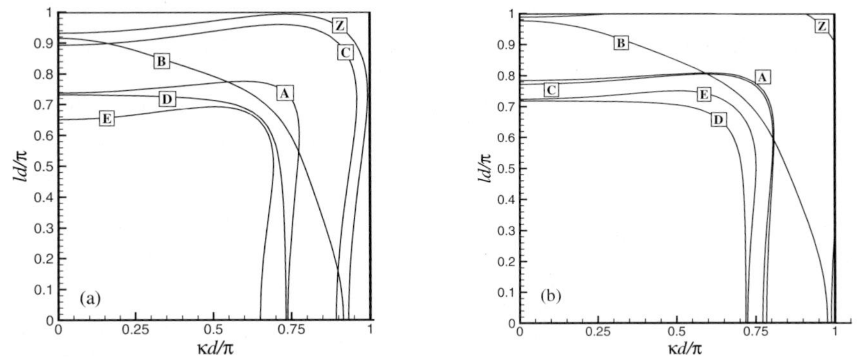

Fig. 1. Contour lines of 10% error in frequency for the dispersion relationship of the inertia-gravity wave on A-E and Z grids obtained from the sixth-order super compact method, (a) $\lambda/d = 2$ and (b) $\lambda/d = 1/2$.

under-resolved ($\lambda/d = 1/2$) cases. Figure 1 show the 10% relative error contour for the sixth-order SCFDM. The relative error is defined as $|\omega_e - \omega_n|/|\omega_e|$, where ω_e is the exact frequency and ω_n is the grid frequency which is computed for different methods on A-E and Z grids. By comparing these results with those obtained from the second-order centered [6] and the fourth-order compact [2] methods, it can be seen that the sixth-order SCFDM leads to a very clear increase in the accuracy of the frequency of inertia-gravity waves.

4 Spatial Differencing of Nonlinear Problems

In this part, we study the two Jacobians appearing in the vorticity and divergence equations as the dominant nonlinear terms. The Jacobians are $J(\psi, \zeta)$ and $J(u_\psi, v_\psi)$, where ψ is the streamfunction, $u_\psi = -\partial\psi/\partial y$, $v = \partial\psi/\partial x$, and $J(X, Y) = (\partial X/\partial x)(\partial Y/\partial y) - (\partial X/\partial y)(\partial Y/\partial x)$ for any two arbitrary quantities X and Y. It is assumed that the streamfunction ψ, and vorticity ζ, are on an unstaggered grid. The different, though identical in continuous case, forms of the Jacobian operator for vorticity advection are given in [11, 10]. It should be emphasized that the Jacobian operators are different when discretized by numerical methods. Following Chang and Shirer [3], for a doubly periodic domain, the continuous ψ is expanded in a complex Fourier series to find the analytical and discrete forms of two Jacobians.

In this paper, only the results for the amplitude errors of $J_A(\psi, \zeta)$ and $J_1(u_\psi, v_\psi)$ are presented. We assume that ψ is composed of two components having different wanenumbers m_1 and m_2 in the x-direction and wavenumber n in the y-direction [3] i.e., $\psi = A_{m_1,n} \exp[i(m_1 x + ny)] + A_{m_2,n} \exp[i(m_2 x + ny)]$.

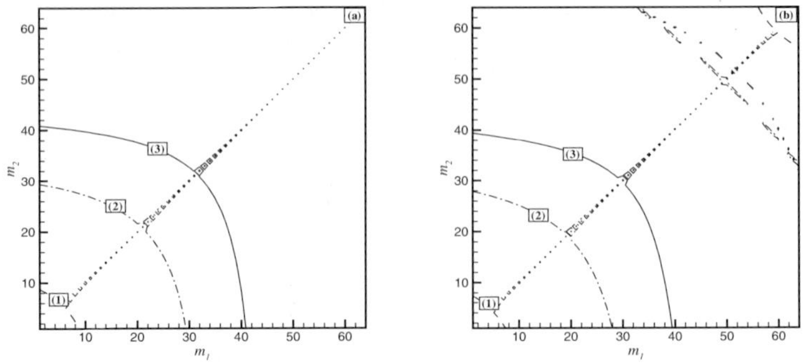

Fig. 2. The contours represent the 5% relative error for (a) $J_{\mathrm{A}}(\psi,\zeta)$ and (b) $J_1(u_\psi, v_\psi)$. The dashed, dashed-dotted, and solid lines are for, respectively, (1) the second-order centered, (2) the fourth-order compact, and (3) the sixth-order SCFDM.

To evaluate the exact and approximate Jacobians, it is appropriate to make a physically meaningful model for the amplitude $A_{m,n}$. To recover the amplitude for streamfunction, we use quasigeostrophic relations in the potentio-vortical energy E in spectral space given in [17], and obtain $A_{m,n} = \left(2E/(H(m^2+n^2+1/\lambda^2))\right)^{1/2}$.

Figure 2 presents the 5% relative errors for $J_{\mathrm{A}}(\psi,\zeta)$ and $J_1(u_\psi, v_\psi)$ and the three numerical schemes [amplitude errors for the wave $(m_1+m_2, 2n)$]. Notice that the error increases from small to large values of m_1 and m_2 almost isotropically, except for a reduction at large values of the wavenumbers at the upper right-hand corner for $J_1(u_\psi, v_\psi)$. It is clear that the fourth-order compact method gives a significant improvement on the second-order centered. Further, the sixth-order SCFDM enjoys a noticeably larger region with relative error less than 5% compared with the fourth-order compact. It is expected that the use of sixth-order SCFDM will further improve the representation of wave-wave interactions.

5 Spherical Geometry

In this section, the problem of reconstruction of the streamfunction field from the vorticity on the sphere is considered. A well-known analytical solution is used to compare the results of the three schemes. For the analytical case the Rossby–Haurwitz wave, which is one of the standard tests proposed by Williamson *et al.* [15] is used. To find the streamfunction from the known vorticity field a Poisson equation on the sphere must be solved. The Poisson equation is solved on a doubly periodic longitude-latitude grid system. The longitude-latitude grid system of Fornberg [8] and Nihei and Ishii [12] is used for the computation.

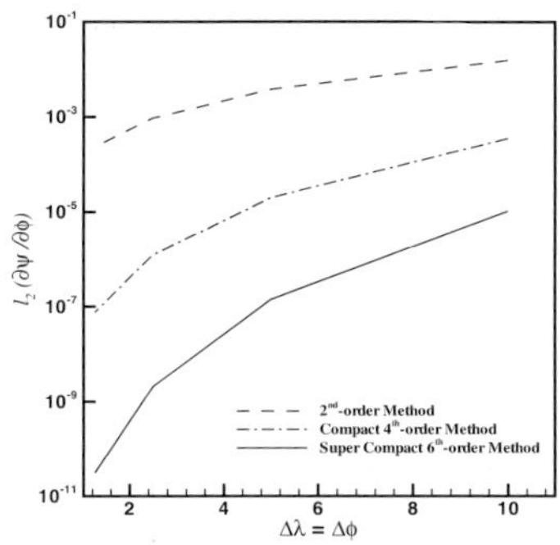

Fig. 3. Normalized l_2 error of $\partial\psi/\partial\varphi$ field as a function of resolution obtained using the second-order centered, fourth-order compact, and sixth-order SCFDM for the Rossby–Haurwitz wave. A uniform resolution, shown in degrees in the figure, is used in both λ and φ directions.

The second-order centered, fourth-order compact and sixth-order SCFDM schemes are used to discretize the spatial derivatives to find the numerical streamfunction filed of the Rossby–Haurwitz wave. Spatial discretization of the derivatives are performed in both longitude and latitude directions which leads to solve a cyclic block tridiagonal system in each direction for the fourth-order compact and the sixth-order SCFDM and a cyclic tridiagonal system for the second order centered. These solutions are compared with the analytical streamfunction of the Rossby–Haurwitz wave to find the error of each scheme. Figure 3 shows the normalized global error l_2 [15] for $\partial\psi/\partial\varphi$ at different horizontal resolutions in longitude and latitude directions, calculated for the second-order centered, fourth-order compact, and sixth-order SCFDM. It can be seen that the sixth-order SCFDM gives a marked improvement in accuracy on the fourth-order compact, almost the same as the improvement shown by the fourth-order compact on the scend-order centered. It is worth mentioning that the convergence rates shown by the three Poisson solutions are consistent with their order of accuracy.

6 Concluding Remarks

For the linear propagation of inertia-gravity waves on the f-plane, the sixth-order SCFDM shows a significant improvement on the conventional second-order centered and the fourth-order compact. As a remark, we expect the same for the linear propagation of Rossby waves on the β-plane and on the sphere. For the nonlinear problems represented by the Jacobians involved in vorticity advection by nondivergent flow and in Bolin–Charney balance equation, the improvement by the sixth-order SCFDM on the fourth-order compact in wavenumber distribution of the Jacobians is noticeable, amounting to nearly a 25% reduction in the global error measures calculated. For the reconstruction of the streamfunction of the vorticity field of the Rossby–Haurwitz wave, the sixth-order SCFDM improves markedly on the fourth-order compact.

Acknowledgment

Authors would like to thank university of Tehran and Atmospheric Science & Meteorological Research Center (ASMERC) for supporting this research.

References

1. A. Arakawa, R. Lamb: Methods in Computational Physics **17**, pp 174–265 (1977)
2. E. Blayo: J. Comput. Phys. **164**, pp 241–257 (2000)
3. H. Chang, H. N. Shirer: Mon. Wea. Rev. **113**, pp 409–423 (1985)
4. Fu Dexun, Ma Yanwen: High resolution schemes. In: *Computational Fluid Dynamic Review*, ed by M. Hafez and K. Oshima (John Wiley & Sons, 1995) pp 234–250
5. Fu Dexun, Ma Yanwen: Inter. J. for Numerical Methods In Fluids **36**, pp 773–805 (2001)
6. J. K. Dukowicz: J. Comput. Phys. **119**, pp 188–194 (1995)
7. V. Esfahanian, S. Ghader, Kh. Ashrafi: Inter. J. for Numerical Methods In Fluids, in press (2003)
8. B. Fornberg: SIAM J. Sci. Comput. **16 (5)**, pp 1071–1081 (1995)
9. S. Ghader: Numerical simulation of the stability of the laminar boundary layer flow over a flat plate by PSE equations using super compact method. M.Sc. Thesis, University of Tehran, Tehran (2000)
10. G. J. Haltiner, R. T. Williams:*Numerical Prediction and Dynamic Meteorology*, 2nd edn, (Wiley & Sons, 1980)
11. D. K. Lilly: Mon. Wea. Rev. **93**, pp 11–26 (1965)
12. T. Nihei, K. Ishii: J. Comput. Phys. **187**, pp 639–659 (2003)
13. D. A. Randall: Mon. Wea. Rev. **122**, pp 1371-1377 (1994)
14. W. F. Spotz, M. A. Taylor, P. N. Swarztrauber: J. Comput. Phys. **145**, pp 432–444 (1998)
15. D. L. Williamson, J. B. Drake, J. J. Hack, R. Jakob-Chien, P. N. Swarztrauber: J. Comput. Phys. **102**, pp 211–224 (1992)
16. Ma Yanwen, Fu Dexun: Computational Fluid Dyn. J. **5(2)**, pp 259-276 (1996)
17. L. Yuan, K. Hamilton: J. Fluid Mech. **280**, pp 369–394 (1994)

A New Discretization Method of Governing Equations for High Order Accuracy

Dehee Kim and Jang Hyuk Kwon

Korea Advanced Institute of Science and Technology,
373-1 Guseong-dong, Yuseong-gu, Daejeon 305-701, Korea

Summary. To obtain high order accurate numerical solutions with the finite volume method, the Euler or the Navier-Stokes equations are discretized with the second order accuracy and cell-face numerical fluxes are calculated by high order accurate shock-capturing schemes such as ENO/WENO. In that approach, numerical dissipations are added due to built-in second order discretization errors. In this work, a discretization method with fourth order spatial accuracy is proposed. We investigate the effect of discretization order and numerical flux order on the solution accuracy using the third order MUSCL and fifth order WENO schemes for numerical fluxes.

1 Introduction

High order accurate schemes are used widely to enhance the accuracy of numerical solutions and reduce the number of grid points, simultaneously. The accuracy of numerical solutions is affected seriously by numerical dissipations. Since numerical dissipations could smear or contaminate the physical phenomena of flows, complex flows such as vortical, aeroacoustic and turbulent flows are usually simulated with low dissipative schemes.

Numerical dissipations arise from the discretization of governing equations and the calculation of numerical fluxes. Therefore, to obtain high order accurate numerical solutions, it is necessary to discretize governing equations with a high order accuracy and calculate cell-face numerical fluxes accurately as well. High order discretization methods are developed by several researchers [1, 2, 3]. Yee [1] formulated two families of explicit and implicit high order discretization methods for the multidimensional compressible Euler equations and proposed many variations of these two families of methods. Deng et al. [2] developed a high order discretization method for governing equations and proposed a high order interpolation method for flow variables at cell face. Ekaterinaris [3] used the compact schemes [4] for the discretized governing equations in dimension by dimension and implemented the developed method to solve linear aeroacoustic problems and subsonic flows around an airfoil. It is our intention to discretize the governing equations with 4th order spatial accuracy using similar strategy as Ekaterinaris's [3]. In our work, three cells in each coordinate direction are involved for the discretization of the governing equations at a cell, $(i,\ j)$, namely 5 cells for 2 dimensions,

$(i-1,\ j)$, $(i,\ j)$, $(i+1,\ j)$, $(i,\ j-1)$, and $(i,\ j+1)$. The derived discretized equations are 1st order accurate temporally and 4th order spatially.

Then, the discretized equations are time-marching implicitly using the DADI method and residuals are calculated using the MUSCL and WENO schemes at the cell face. The effect of the discretization order and the numerical flux order on solution accuracy is investigated through several numerical experiments.

2 Numerical Methods

2.1 Governing Equations

We consider the conservative form of the two dimensional unsteady compressible Navier-Stokes equations as

$$\frac{\partial \mathbf{Q}}{\partial t} + \frac{\partial \mathbf{F}}{\partial \xi} + \frac{\partial \mathbf{G}}{\partial \eta} = \frac{\partial \mathbf{F}_v}{\partial \xi} + \frac{\partial \mathbf{G}_v}{\partial \eta} \tag{1}$$

in curvilinear coordinates, where $\mathbf{Q}$, $\mathbf{F}$, $\mathbf{G}$, $\mathbf{F}_v$, and $\mathbf{G}_v$ denote conservative variables, inviscid fluxes, and viscous fluxes, respectively. A high order discretization method proposed in this work is derived for the Euler equations and the viscous terms of Navier-Stokes equations are discretized using the usual 2nd order centeral difference.

2.2 High Order Discretization Method

Euler equations can be written in generalized coordinates as

$$\frac{\partial \mathbf{Q}}{\partial t} + \frac{\partial \mathbf{F}}{\partial \xi} + \frac{\partial \mathbf{G}}{\partial \eta} = 0 \tag{2}$$

Following equations for four adjacent cells around $(i,\ j)$ cell are used for the derivation of a high order discretization method.

$$\alpha \left.\frac{\partial \mathbf{Q}}{\partial t}\right|_{i-1,j} + \alpha \left.\frac{\partial \mathbf{F}}{\partial \xi}\right|_{i-1,j} + \alpha \left.\frac{\partial \mathbf{G}}{\partial \eta}\right|_{i-1,j} = 0 \tag{3}$$

$$\alpha \left.\frac{\partial \mathbf{Q}}{\partial t}\right|_{i+1,j} + \alpha \left.\frac{\partial \mathbf{F}}{\partial \xi}\right|_{i+1,j} + \alpha \left.\frac{\partial \mathbf{G}}{\partial \eta}\right|_{i+1,j} = 0 \tag{4}$$

$$\alpha \left.\frac{\partial \mathbf{Q}}{\partial t}\right|_{i,j-1} + \alpha \left.\frac{\partial \mathbf{F}}{\partial \xi}\right|_{i,j-1} + \alpha \left.\frac{\partial \mathbf{G}}{\partial \eta}\right|_{i,j-1} = 0 \tag{5}$$

$$\alpha \left.\frac{\partial \mathbf{Q}}{\partial t}\right|_{i,j+1} + \alpha \left.\frac{\partial \mathbf{F}}{\partial \xi}\right|_{i,j+1} + \alpha \left.\frac{\partial \mathbf{G}}{\partial \eta}\right|_{i,j+1} = 0 \tag{6}$$

Six equations (three equations in each direction), namely, Eq.(3) for the cell $(i-1,\ j)$, Eq.(2) for the cell $(i,\ j)$, Eq.(4) for the cell $(i+1,\ j)$, Eq.(5) for the cell $(i,\ j-1)$, Eq.(2) for the cell $(i,\ j)$, and Eq.(6) for the cell $(i,\ j+1)$ are summed. Then using the Taylor series and the 4th order compact scheme as

$$\alpha \left.\frac{\partial \mathbf{F}}{\partial \xi}\right|_{i-1,j} + \left.\frac{\partial \mathbf{F}}{\partial \xi}\right|_{i,j} + \alpha \left.\frac{\partial \mathbf{F}}{\partial \xi}\right|_{i+1,j} = a\left(\mathbf{F}_{i+1/2,j} - \mathbf{F}_{i-1/2,j}\right) \tag{7}$$

where $a = 2\alpha + 1$, the discretized equations are derived as follows:

$$\begin{aligned}
2(2\alpha+1)\Delta\mathbf{Q}^n_{i,j} &+ a\Delta t\left(\mathbf{F}^{n+1}_{i+1/2,j} - \mathbf{F}^{n+1}_{i-1/2,j}\right) + a\Delta t\left(\mathbf{G}^{n+1}_{i,j+1/2} - \mathbf{G}^{n+1}_{i,j-1/2}\right)\\
&- \frac{2\alpha+1}{24}\Delta t\left(\mathbf{F}^{n+1}_{i+3/2,j} - 27\mathbf{F}^{n+1}_{i+1/2,j} + 27\mathbf{F}^{n+1}_{i-1/2,j} - \mathbf{F}^{n+1}_{i-3/2,j}\right)\\
&- \frac{2\alpha+1}{24}\Delta t\left(\mathbf{G}^{n+1}_{i,j+3/2} - 27\mathbf{G}^{n+1}_{i,j+1/2} + 27\mathbf{G}^{n+1}_{i,j-1/2} - \mathbf{G}^{n+1}_{i,j-3/2}\right)\\
&+ \alpha\Delta t\left[\left(\mathbf{F}^{n+1}_{i+1/2,j+1} - 2\mathbf{F}^{n+1}_{i+1/2,j} + \mathbf{F}^{n+1}_{i+1/2,j-1}\right)\right.\\
&- \left.\left(\mathbf{F}^{n+1}_{i-1/2,j+1} - 2\mathbf{F}^{n+1}_{i-1/2,j} + \mathbf{F}^{n+1}_{i-1/2,j-1}\right)\right]\\
&+ \alpha\Delta t\left[\left(\mathbf{G}^{n+1}_{i+1,j+1/2} - 2\mathbf{G}^{n+1}_{i,j+1/2} + \mathbf{G}^{n+1}_{i-1,j+1/2}\right)\right.\\
&- \left.\left(\mathbf{G}^{n+1}_{i+1,j-1/2} - 2\mathbf{G}^{n+1}_{i,j-1/2} + \mathbf{G}^{n+1}_{i-1,j-1/2}\right)\right] = 0
\end{aligned}$$

where $a/24 = \alpha$ and $\alpha = 1/22$. Dividing both sides by $2a$ and rearranging in the ADI form, final equations can be written as

$$\begin{aligned}
&\left[c\bar{\mathbf{A}}_{i-1} + \left\{\mathbf{I} + c\bar{\mathbf{A}}_i\right\} + c\bar{\mathbf{A}}_{i+1}\right]\\
&\left[c\bar{\mathbf{B}}_{j-1} + \left\{\mathbf{I} + c\bar{\mathbf{B}}_j\right\} + c\bar{\mathbf{B}}_{j+1}\right]\Delta\mathbf{Q}^n = \Delta t\mathbf{R}^n
\end{aligned} \tag{8}$$

$$\begin{aligned}
\mathbf{R}^n = &- \frac{1}{48}s\left(\hat{\mathbf{F}}^n_{i+1/2,j+1} - \hat{\mathbf{F}}^n_{i-1/2,j+1} + \hat{\mathbf{F}}^n_{i+1/2,j-1} - \hat{\mathbf{F}}^n_{i-1/2,j-1}\right)\\
&- \frac{1}{48}s\left(\hat{\mathbf{G}}^n_{i+1,j+1/2} - \hat{\mathbf{G}}^n_{i+1,j-1/2} + \hat{\mathbf{G}}^n_{i-1,j+1/2} - \hat{\mathbf{G}}^n_{i-1,j-1/2}\right)\\
&+ \frac{1}{48}s\left(\hat{\mathbf{F}}^n_{i+3/2,j} - \hat{\mathbf{F}}^n_{i-3/2,j} + \hat{\mathbf{G}}^n_{i,j+3/2} - \hat{\mathbf{G}}^n_{i,j-3/2}\right)\\
&- c\left[\hat{\mathbf{F}}^n_{i+1/2,j} - \hat{\mathbf{F}}^n_{i-1/2,j} + \hat{\mathbf{G}}^n_{i,j+1/2} - \hat{\mathbf{G}}^n_{i,j-1/2}\right]
\end{aligned} \tag{9}$$

where, $\bar{\mathbf{A}},\ \bar{\mathbf{B}}$ are the Jacobian matrices of residual, and $\hat{\mathbf{F}},\ \hat{\mathbf{G}}$ are the numerical fluxes at cell face, and $c = 49/48,\ s = 1$ for interior cells, and $c = 1,\ s = 0$ for boundary cells.

For practical applications, the implicit time-marching method is used, but explicit methods can be used for high order temporal accuracy. Because above equations are used for steady calculations, a dual-time stepping method is applied for unsteady calculations to obtain 2nd order accuracy in time.

Using conventional discretized equations with 2nd order accuracy (denoted by "D2") and the derived discretized equations with 4th order accuracy (denoted by "D4"), we simulated several test problems and compared the solution accuracies. The 3rd order MUSCL and the 5th order WENO [5] (denoted by "WENO5") are implemented to calculate numerical fluxes at cell face.

3 Numerical Experiments

A convecting vortex [6], double mach reflection [7], weak shock/vortex interaction [8], and turbulent flowfields around a RAE2822 airfoil are simulated.

3.1 Advection of an isentropic vortex

This problem was chosen in Yee et al. [6] to assess the performance of numerical schemes. An isentropic vortex convects in an inviscid free stream. Periodic boundary conditions in both directions are implemented. Detailed problem discription can be found in [6]. Figs. 1 and 2 show the density distributions along the center line($y = 0$). Fig. 1 shows the results of grid refinement study with 41×40, 81×80, and 161×160 grid systems, respectively. The unsteady time step size was set to 0.04. The result using the WENO scheme is superior to that using the MUSCL scheme. For the coarse grid system, the high order discretization method gives a little improved solutions. Fig. 2 shows the effect of time step size with $\Delta t = 0.1,\ 0.04,\ 0.02$, respectively. A 81×80 grid system was used and all results are obtained at $t = 100$.

3.2 Double Mach reflection of a strong shock

This is a famous problem for high resolution schemes [7]. The problem involves a Mach 10 strong shock in air ($\gamma = 1.4$) which makes a 60° angle with a reflecting wall. Fig. 3 shows the density contours using "MUSCL-D2" and "WENO5-D2" at $t = 0.2$. Unsteady time step size was set to 0.0002. A 481×121 grid system was used. The WENO scheme gives more accurate solutions for the roll-up region. Since results using "-D2" and "-D4" are nearly same each other, those results using "MUSCL-D4" and "WENO5-D4" are not shown. This is because the accuracy of numerical fluxes decides overall solution accuracy for strong shock problems.

3.3 Weak shock/vortex interaction

This test case was tailored to investigate the shock-capturing ability of numerical schemes to predict the generation and the transport of acoustic waves during a shock/vortex interation [8]. Fig. 4 shows the density distributions

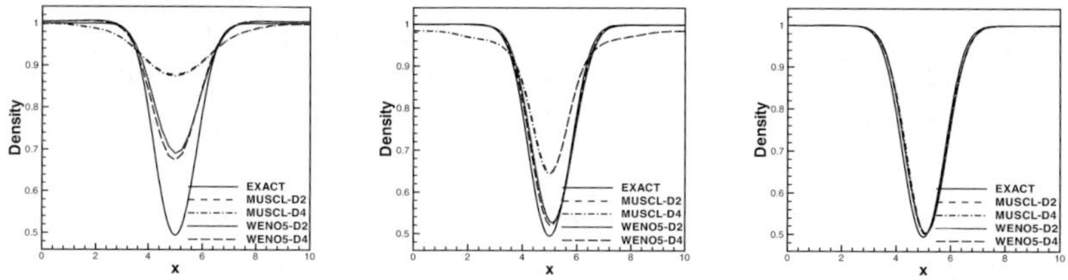

Fig. 1. Density distributions of a convecting vortex along center line with 41×41, 81×80, and 161×160 grid systems (left to right) at $t = 100$.

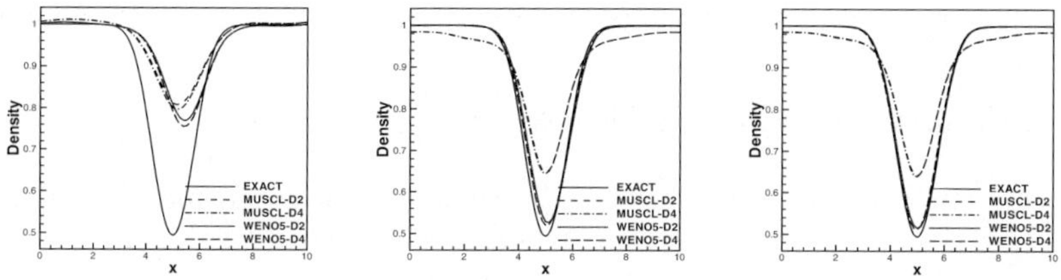

Fig. 2. Density distributions of a convecting vortex along center line with $\Delta t = 0.1,\ 0.04,\ 0.02$ (left to right) at $t = 100$.

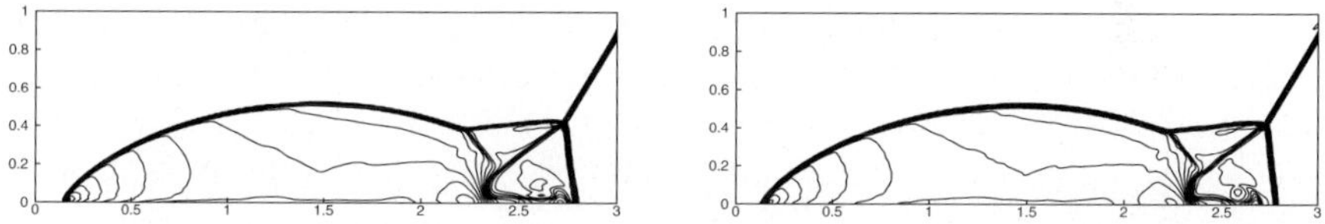

Fig. 3. Density contours of double Mach reflection using "MUSCL-D2"(left) and "WENO5-D2"(right) at $t = 0.2$.

along center line at $t = 0.7$ using 101×101 and 201×201 grid systems (numbers in parentheses, "101" and "201" denote grid systems). The unsteady time step size was set to 0.002. High order discretization method gives more accurate solutions for the acoustic waves.

3.4 Turbulent flowfields around a RAE2822 airfoil

A RAE2822 airfoil in transonic turbulent flowfields was simulated. The Mach number in the free stream and the angle of attack are 0.75, 3.19^o, respectively and the Reynolds number based on the airfoil chord length is 6.2×10^6. The $k - w$ WD+ turbulence model was used [9] with 385×65 grid system. As shown in Fig. 5, shock resolution is improved slightly by using high order discretization method ("-D4").

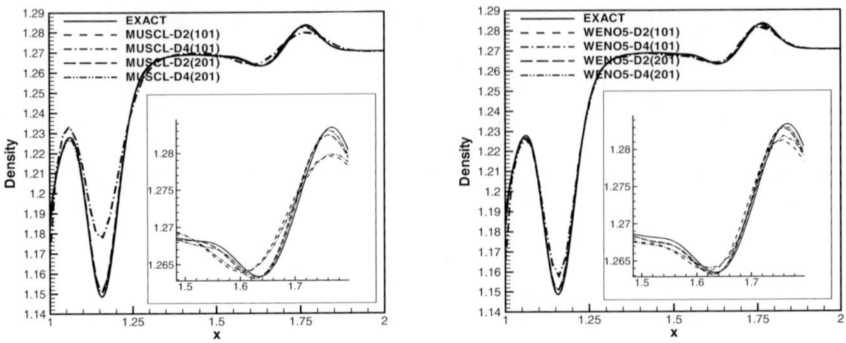

Fig. 4. Density distributions along the center line at $t = 0.7$.

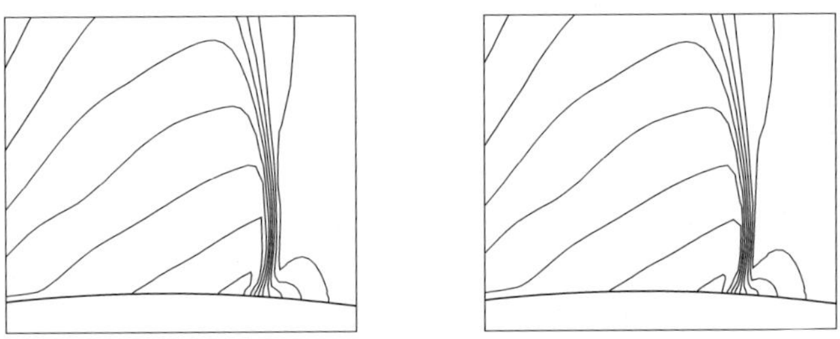

Fig. 5. Pressure contours of a RAE2822 airfoil using "WENO5-D2"(left) and "WENO5-D4"(right).

4 Conclusions

A 4th order discretization method for the Euler equations has been proposed and the effect of the discretization order and the numerical flux order at cell face on solution accuracy has been reviewed. It is found that the order of numerical flux plays an important role for solution accuracy, but the accuracy of numerical solutions can be improved by using high order discretization methods.

References

1. H.C. Yee: J. Comp. Phys. **131**, 216 (1997)
2. X. Deng, M. Mao: AIAA Paper 2001-2626, June, 2001
3. J.A. Ekaterinaris: J. Comp. Phys. **156**, 272 (1999)
4. S.K. Lele: J. Comp. Phys. **103**, 16 (1992)
5. G.S. Jiang, C.W. Shu: J. Comp. Phys. **126**, 202 (1996)
6. H.C. Yee, N.D. Sandham, M.J. Djomehri: J. Comp. Phys. **150**, 199 (1999)
7. P. Woodward, P. Colella: J. Comp. Phys. **54**, 115 (1984)
8. C. Tenaud, E. Garnier, P. Sagaut: Int. J. Numer. Meth. Fluids **33**, 249 (2000)
9. S.H. Park, J.H. Kwon: AIAA J. in revision (2004)

A High-Order Accurate Unstructured GMRES Solver for the Compressible Euler Equations

Amir Nejat and Carl Ollivier-Gooch

Department of Mechanical Engineering, University of British Columbia, BC V6T 1Z4, Canada. nejat@mech.ubc.ca, cfog@mech.ubc.ca

Summary. A preconditioned matrix-free GMRES method is presented for unstructured higher-order computation of 2D Euler equations. Lower-Upper Symmetric Gauss-Seidel (LU-SGS) has been used as the preconditioning strategy, and a first-order Jacobian as the preconditioner. The numerical algorithm demonstrates promising convergence performance for both the second and the third order discretization methods in supersonic flow.

1 Introduction

Recent research on structured mesh flow solver for aerodynamics shows that for practical levels of accuracy using a higher order accurate method can reduce computation time and memory usage and improve robustness compared to a second-order scheme [7, 15]. To take advantage of the high flexibility in mesh generation and adaptation for unstructured meshes, we want to apply higher-order accurate methods for unstructured meshes. This approach combines benefits of higher-order methods and unstructured meshes. Although high-order accurate methods for unstructured meshes are reasonably well established [1, 2, 6, 10], application of these methods for physically complicated flows is still a challenge due to very slow convergence. This eliminates the efficiency benefits of higher-order unstructured discretization. Consequently, convergence acceleration becomes the key issue for the practical usage of higher-order unstructured solvers. Recent results of an unstructured mesh solver for Poisson's equation [9] clearly showed the possibility of reducing computational cost required for a given level of solution accuracy using higher-order methods and matrix free GMRES [13] as a convergence acceleration technique. As Poisson's equation is simple and has a linear flux, the resultant Jacobian matrix is well-conditioned, and the convergence rate for GMRES technique is good. However, as GMRES convergence is sensitive to the condition number of Jacobian matrix, in the case of complex equations (such as Euler) which normally have an ill-conditioned Jacobian, employing an effective preconditioner for GMRES technique becomes a key necessity. For structured meshes, Pueyo and Zingg [11] presented an efficient matrix-free Newton-GMRES solver using multi-grid for steady aerodynamic flows. Barth and Linton [3] successfully applied both full-matrix and matrix-free

GMRES iteration for computing compressible fluid flow over unstructured meshes. Blanco and Zingg [4] compared the use of two variations on quasi- and full-Newton method for solution of the Euler equations on unstructured grids. Recently Manzano et. al. [8] developed a preconditioned matrix-free GMRES method to solve 2-D and 3-D Euler equations using ILU preconditioner based on an approximation to the flow Jacobian.

Our objective in this research is to develop an efficient and accurate unstructured solver for inviscid compressible fluid flow. This paper presents preliminary results of using LU-SGS for preconditioning of the matrix-free GMRES to compute the higher-order solution of Euler equations.

2 Governing Equations

The unsteady (2D) Euler equations which model compressible inviscid fluid flows, are conservation equations for mass, momentum, and energy. The finite-volume formulation of Euler equations for an arbitrary control volume can be written in the following form of a volume and a surface integral.

$$\frac{d}{dt}\int_{cv} \boldsymbol{U}\,dv + \oint_{cs} \boldsymbol{F}\,dA=0 \tag{1}$$

where

$$\boldsymbol{U} = \begin{bmatrix} \rho \\ \rho u \\ \rho v \\ E \end{bmatrix}, \qquad \boldsymbol{F} = \begin{bmatrix} \rho u_n \\ \rho u u_n + P\hat{n}_x \\ \rho v u_n + P\hat{n}_y \\ \rho(E+P)u_n \end{bmatrix} \tag{2}$$

In (2), $u_n = u\hat{n}_x + v\hat{n}_y$ and $(\rho \quad \rho u \quad \rho v \quad E)^T$ are the densities of mass, x-momentum, y-momentum, and energy, respectively. The energy is related to the pressure by the perfect gas equation of state: $E = P/(\gamma-1)+\rho(u^2+v^2)/2$, with γ the ratio of specific heats for the gas.

3 Numerical Algorithm

Linearization of the governing equations in time and applying implicit time integration leads to implicit time advance formula:

$$\left(\frac{I}{\Delta t} + \frac{\partial R}{\partial U}\right)\delta U = -R_i \quad , \quad \delta U_i = U_i^{n+1} - U_i^n \tag{3}$$

where $\partial R/\partial U$ is the Jacobian matrix. To advance the solution from one time step to the next and eventually converging to the steady-state solution, a large system of linear equations must be solved. The linear system is asymmetric both in fill and values making GMRES the obvious first choice in iterative solvers. The linear system arising from high-order discretization has four-five times as many non-zero entries as a second-order scheme. Because of the size of these matrices and the difficulty in computing their entries analytically (even for the second order), we have chosen to use an entirely matrix-free implementation of GMRES. Since the GMRES algorithm only needs matrix-vector products, and these products can be computed by matrix-free approach [3], matrix-free GMRES is a very attractive technique for dealing with the complicated Jacobian matrices resulting from higher-order discretization. However, as GMRES convergence is highly dependent on the condition number of Jacobian matrix, in the case of Euler with nonlinear flux function and possible discontinuities in the solution, using a preconditioner for GMRES becomes necessary for practical purposes. At the same time, using high-order discretization makes the Jacobian matrix more off-diagonally dominant and quite ill-conditioned. Therefore, we use flexible GMRES known as FGMRES [14] with LU-GSG [5] as a preconditioning strategy. To build the preconditioner, we form a first-order Jacobian matrix, reducing the size and complexity of the matrix.

In our discretization scheme, a higher-order accurate least-square reconstruction procedure [10] has been used in the interior of the domain in order to compute fluxes using Roe's difference-splitting scheme [12] at control volume boundaries up to the desired accuracy. Having computed the fluxes, we integrate along each control volume boundary using Gauss quadrature integration technique with the proper number of points to get the required order of accuracy for the flux integral. Imposing the boundary conditions also has been done to high-order accuracy.

4 Results

To investigate the convergence performance and robustness of the purposed GMRES solver with a higher order unstructured discretization, different supersonic flow cases have been studied. Here for brevity, we only present one of them which shows the general convergence behavior of our numerical experiments. Our test case (Fig.1) is a 5m supersonic duct with an expansion corner located 1m after inlet at the lower wall and the corner angle is 9.23°. The height of the inlet and outlet are 0.75m and 1.4m respectively. 2100 control-volumes are used in this test case with a proper refinement at expansion corner, and inlet Mach number is equal to 2. As convergence performance and stability of GMRES technique especially for compressible flows are quite sensitive to the start-up process and the solution at initial iterations, we have to do several implicit iterations before switching to GMRES iteration. For both the second and third order discretization, 20 implicit iterations have been performed before GMRES iteration. The iterations start with $CFL = 1.0$ and CFL number is increased gradually to 100. For the fourth order discretization, we do 275 pre-iterations with similar CFL number increasing trend.

The residual convergence history in terms of number of iterations and residual evaluations for the test case has been shown in Fig.2 and Fig. 3. The figures show considerable drop in residual as GMRES iteration starts. GMRES-LUSGS effectively decreases the residual (10 orders of magnitude) by about 15 iterations for the second order and about 35 iterations for the third order discretization. However, for the fourth order discretization, GMRES-LUSGS loses its quadratic convergence behavior after 15 iterations and the convergence slows considerably. The convergence result shows 10 order drop in L2-norm (density residual) by about 1200 and 2700 residual evaluations for the second and the third order cases. Also 9 order residual reduction has been achieved by 13000 residual evaluations for 4th order discretization by GMRES-LUSGS(Fig.3) . To demonstrate the effect of (LU-SGS) preconditioning on GMRES, the residual convergence histories for the simple GMRES without preconditioner have been shown in similar graphs

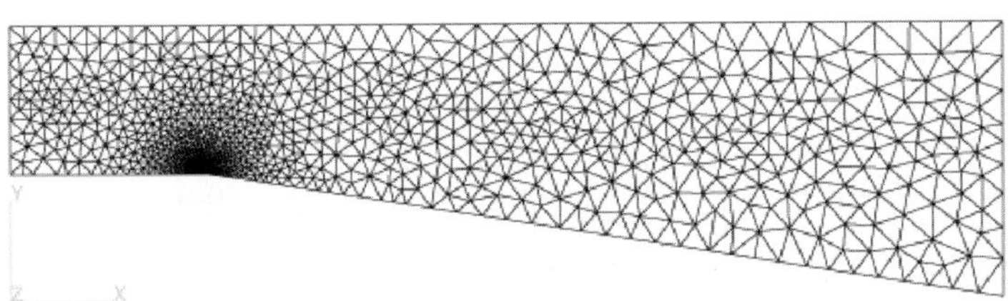

Fig. 1. Domain and mesh for the supersonic duct

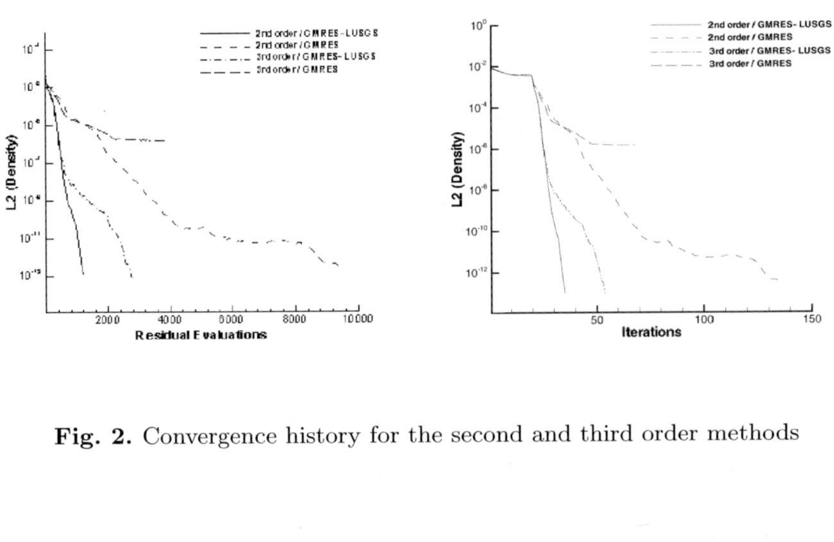

Fig. 2. Convergence history for the second and third order methods

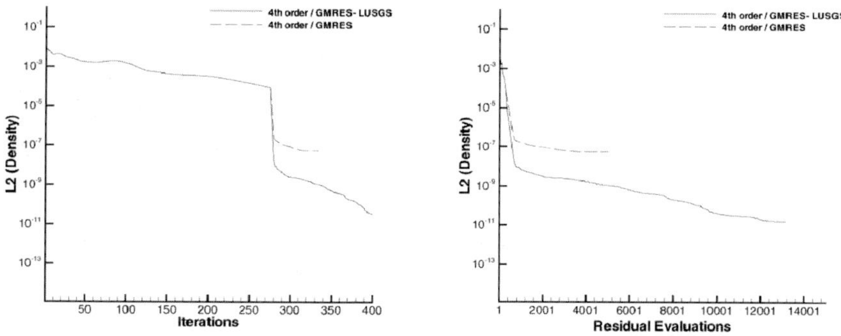

Fig. 3. Convergence history for the fourth order method

for all different orders of accuracy, and as shown using a preconditioner has improved the performance and robustness of GMRES technique noticeably.

5 Concluding Remarks

A preconditioned matrix-free LUSGS-GMRES algorithm has been presented for higher-order computation of solution of 2D Euler equations. The results show that LUSGS-GMRES works almost as efficiently for the third order discretization as for the second order one. In the case of fourth order discretization, still there are challenges ahead, and as a future work we will concentrate on improving current approach especially for the 4th order accurate unstructured mesh discretization.

6 Acknowledgment

This research was supported by the Canadian Natural Science and Engineering Research Council under Grant OPG-0194467.

References

1. T. J. Barth, P. O. Fredrickson, M. Stuke et al: Higher-Order Solution of the Euler Equations on Unstructured Grids Using Quadratic Reconstruction, AIAA conference paper 90-0013, (1990)
2. T. J. Barth: Recent Development in High-Order K-Exact Reconstruction on Unstructured Meshes, AIAA conference paper 93-0668, (1994)

3. T. J. Barth, S. W. Linton: An Unstructured Mesh Newton Solver for Compressible Fluid Flow and Its Parallel Implementation, AIAA conference paper 95-0221, (1995)
4. M. Blanco, D. W. Zingg: A Fast Solver for the Euler Equations on Unstructured Grids Using a Newton-GMRES Method, AIAA conference paper 97-0331, (1997)
5. R. F. Chen, Z. J. Wang: Fast, Block Lower-Upper Symmetric Gauss-Seidel Scheme for Arbitrary Grids, AIAA Journal, Vol. 38, no. 12, pp. 2238-2245 (2000)
6. M. Delanaye, J. A. Essers: Quadratic-Reconstruction Finite-Volume Scheme for Compressible Flows on Unstructured Adaptive Grids, AIAA Journal, Vol. 35, pp. 631-639, (1997)
7. S. De Rango, D. W. Zingg: Aerodynamic Computations Using a Higher-Order Algorithm, AIAA conference paper 99-0167, (1999)
8. L. M. Manzano, J. V. Lassaline, P. Wong, D. W. Zingg: A Newton-Krylov Algorithm for the Euler Equations Using Unstructured Grids, AIAA paper 2003-0274, (2003)
9. A. Nejat, C. Ollivier-Gooch: A High-Order Accurate Unstructured GMRES Solver for Poisson's Equation, CFD 2003 conference proceeding, pp. 344-349, (2003)
10. C. Ollivier-Gooch: Quasi-ENO Schemes for Unstructured Meshes Based on Unlimited Data-Dependent Least Squares Reconstruction., Journal of Computational Physics, Vol. 133, pp. 6-17, (1997)
11. A. Pueyo, D. W. Zingg: Improvement to a Newton-Krylov Solver for Aerodynamic Flows, AIAA conference paper 98-0619, (1998)
12. P. L. Roe: Approximate Riemann Solvers, Parameter vectors, and difference schemes, Journal of Computational Physics, Vol. 43, pp. 357-372, (1981)
13. Y. Saad, M. H. Schultz: A Generalized Minimal Residual Algorithm for Solving Non-Symmetric Linear Systems, SIAM J. Sci., Stat. Comp. Vol. 7, pp. 856-869, (1986)
14. Y. Saad: A Flexible Inner-Outer Preconditioned GMRES Algorithm, SIAM J. Sci., Stat. Comp. Vol. 14, pp. 461-469, (1993)
15. D. W. Zingg, S. De Rango, M. Nemec, T. H. Pulliam: Comparison of Several Spatial Discretizations for the Navier-Stokes Equations, Journal of Computational Physics, Vol.160, pp.683-704, (2000)

Computation of Aeroacoustic Waves with High Order Spectral Volume Method

Z.J. Wang

Michigan State University, East Lansing, MI 48824, U.S.A

Summary. A recently developed high-order finite volume method named the spectral volume (SV) method is employed to solve several benchmark problems from the Fourth Computational Aeroacoustics (CAA) Workshop on Benchmark Problems. In particular, the computational results for multiple cylinder wave scattering problems are presented, and compared with analytical solutions. It is shown that a 4th order SV scheme with an equivalent 13 points-per-wave is capable of producing numerical solutions which agree well with analytical solutions.

1 Introduction

A new high-order finite volume (FV) method named the spectral volume (SV) method has been developed recently for hyperbolic conservations laws, and successfully demonstrated for both scalar and system conservation laws [1-4]. The SV method is a Godunov-type finite volume method [5], which has been under development for over three decades, and has become the-state-of-the-art for the numerical solution of hyperbolic conservation laws. The SV method is also related to the discontinuous Galerkin (DG) method [6] and the k-exact FV method [7]. For a review of the literature on the Godunov-type FV and DG methods, refer to [1], and the references therein.

In this study, the SV method is evaluated for several problems in the Fourth Computational Aeroacoustics (CAA) Workshop on Benchmark Problems. As pointed out by Tam in [8], acoustic waves have their own characteristics which make their computation unique and challenging. Acoustic waves are inherently unsteady. Their amplitudes are several orders smaller than the magnitudes of the mean flow and their frequencies are generally very high and broad ranging. Computational methods with high order accuracy are required to capture the acoustic portion of the solution.

The paper is organized as follows. In the next section, the basic formulation of the SV method is reviewed. In Section 3, solutions for the benchmark problems are presented. Conclusions are summarized in Section 4.

2 Spectral (Finite) Volume Method

In the present study, the non-linear Euler equations are employed, which can be expressed as

$$\frac{\partial Q}{\partial t} + \frac{\partial E}{\partial x} + \frac{\partial F}{\partial y} = S, \tag{1a}$$

$$S = \begin{bmatrix} 0 \\ 0 \\ 0 \\ e^{-ln2\frac{x^2+y^2}{0.2^2}} Sin(\omega t) \end{bmatrix}. \tag{1b}$$

In (1a) Q is the vector of conserved variables, E and F are the inviscid flux vectors in x and y directions respectively and S is the time dependent source term. The governing equation can be cast into the following divergence form

$$\frac{\partial Q}{\partial t} + \nabla . f = S, \tag{2}$$

where f = (E,F). Assume that equations (1a) and (1b) are solved in the computational domain Ω subject to proper initial and boundary conditions. The domain is discretized into N non-overlapping simplex elements (triangular elements in 2D) called spectral volumes (SVs) $\Omega = \cup_i^N S_i$. In order to support a degree k polynomial reconstruction within each SV, the SV is further partitioned into m subcells, with m = (k+1)(k+2)/2 . Note that m is also the dimension of P^k, the space of polynomials of degree at most k. Many candidate partitions are evaluated in [2]. It was found that the partitions for various k shown in Figure 1 perform satisfactorily. Denote the jth CV of S_i by C_{ij}. In the SV method, the degrees-of-freedom (DOFs) or unknowns are the cell-averaged conservative variable Q at the subcells, or the CVs. Given the DOFs $\overline{Q}_{ij}$, a polynomial $p_i \in P^k$ can be reconstructed such that it is a (k+1)-th order accurate approximation to the solution Q inside S_j. The reconstruction can be conveniently expressed as

$$p_i(\mathbf{r}) = \Sigma_{j=1}^m L_j(\mathbf{r}) \overline{Q}_{ij}, \tag{3}$$

where $L_j(\mathbf{r}) \in P^k$ are the "shape" functions shape functions, which are given in [4] for 2D reconstructions. The high-order reconstruction is then used to generate high-order updates for the DOFs using the usual FV method. Integrating (2) in C_{ij}, we obtain the following integral equation for the DOFs

$$\frac{d\overline{Q}_{ij}}{dt} + \frac{1}{V_{ij}} \Sigma_{r=1}^K \int_{A_r} (\mathbf{f}.\mathbf{n}) dA = \frac{1}{V_{ij}} \int_{C_{ij}} S dV, \tag{4}$$

where K is the number of faces in C_{ij} and A_r represents the rth face of C_{ij}. The surface and volume integrals on each face are performed with Gauss quadrature formulas which are exact for degree k polynomials. The flux vectors at the quadrature points $\mathbf{r}_{rq}$ are computed with a Riemann solver

$$\mathbf{f}(Q(\mathbf{r}_{rq})).\mathbf{n} \approx \hat{f}(Q_L(\mathbf{r}_{rq}), Q_R(\mathbf{r}_{rq}), \mathbf{n}), \tag{5}$$

where Q_L and Q_R are the vector of conserved variables just to the left and right of the interface. A third-order TVD Runge-Kutta scheme is employed for time integration [9].

3 Numerical Results

3.1 Category 2 Problem 1 Two Cylinder Wave Diffraction Problem

The scattering of a periodic acoustic source from two circular cylinders is considered. The acoustic source used in this case has a transient term expressed in the following form

$$S = e^{-ln2\frac{x^2+y^2}{0.2^2}} Sin(\omega t) f(t),$$
$$f(t) = min(1, \frac{t}{t_0}^3).$$

The following parameters are chosen in the present study: $\omega = 8\pi$, $t_0 = 4$. Since the configuration is symmetric, only the upper half of the physical domain is used in the computation. The computational mesh is displayed in Figure 2a. The entire computational domain extends to r = 15. The grid within r = 9 is nearly uniform with a mesh size = 0.06. Since each triangle is further partitioned into 10 subcells, the grid has an equivalent points-per-wave (PPW) of about 13 (sqrt(10)*0.25/0.06). The mesh is coarsened from r = 9 to r = 15 with an expansion factor of 1.1 to serve as the buffer zone. A constant time step of 0.002 was used in the computation with a total of 20,000 time steps. The 4th order SV scheme was employed in the computation. The rms pressure is computed in the last 2000 time steps. The computed pressure field at a certain time is shown in Figure 2b. Note that the outgoing waves are significantly damped in the buffer zone without visible reflections. The computed rms pressure along the center line is compared with the analytical solution [10] in Figure 3. The rms pressure on the two cylinder surfaces is also compared with the analytical solution in Figure 4. Note that the agreement between the computational and analytical solutions is good with the current grid resolution.

3.2 Problem 2 Three Cylinder Wave Diffraction Problem

The flow conditions are similar to the last problem. The only difference is the geometry. The computational grid is shown in Figure 5a. The mesh has an inner zone inside r = 9, and an outer buffer zone from r = 9 to r = 14. The inner zone has a grid size of 0.06. The mesh is coarsened from r = 9 to r = 14 with an expansion factor of 1.1 to serve as the buffer zone. The computed pressure field is shown in Figure 5b. The computed rms pressure along the center line is compared with the analytical solution in Figure 6. The rms pressure on the two cylinder surfaces is also compared with the analytical

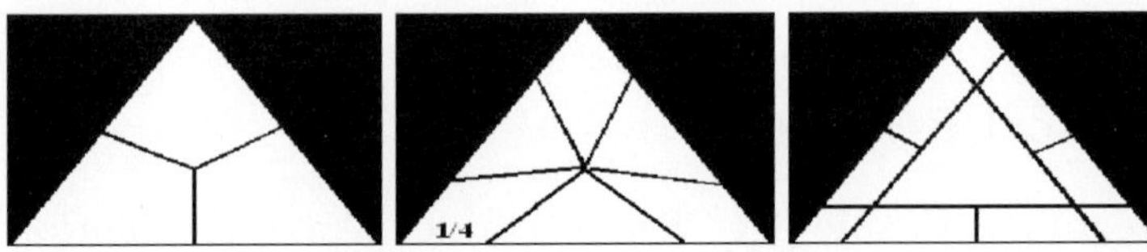

(a) Linear SV (b) Quadratic SV (c) Cubic SV

Fig. 1. Spectral volumes of various degrees

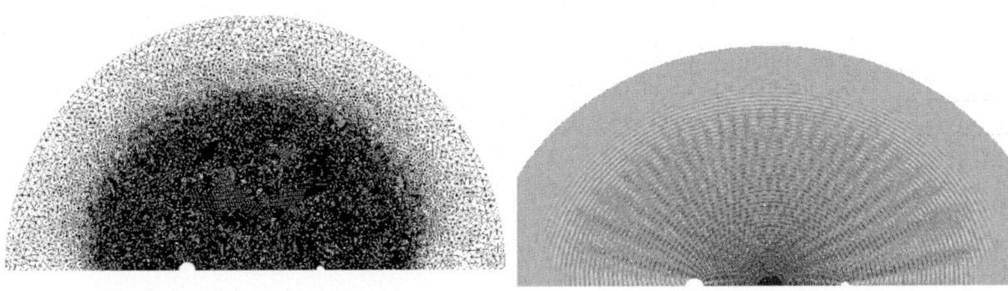

(a)Grid (b)Pressure Distribution

Fig. 2. Computational grid and Pressure Respectively for the two-cylinder diffraction problem

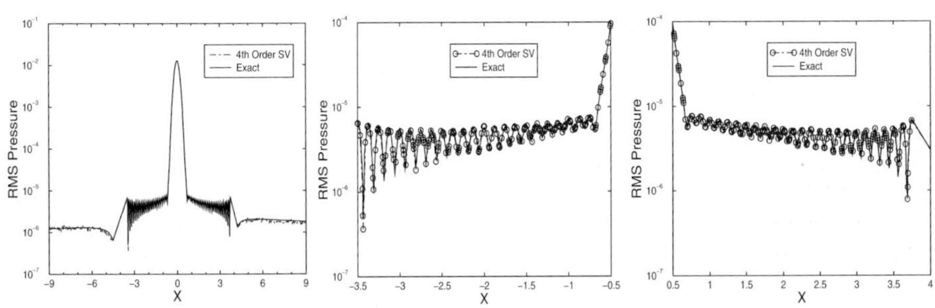

Fig. 3. Comparison of the computational and analytical RMS pressure along the center line

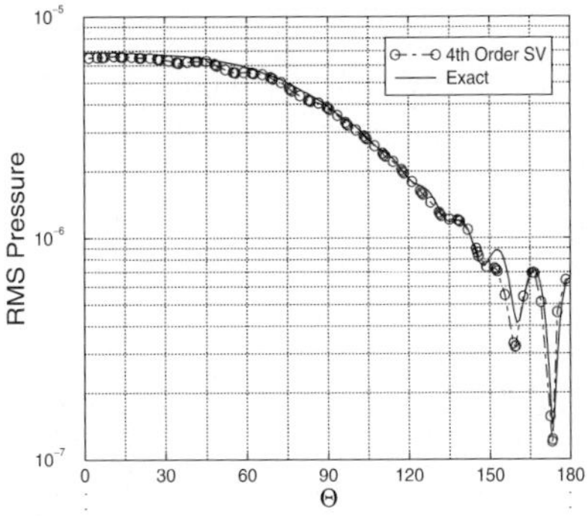

(a) Left Cylinder

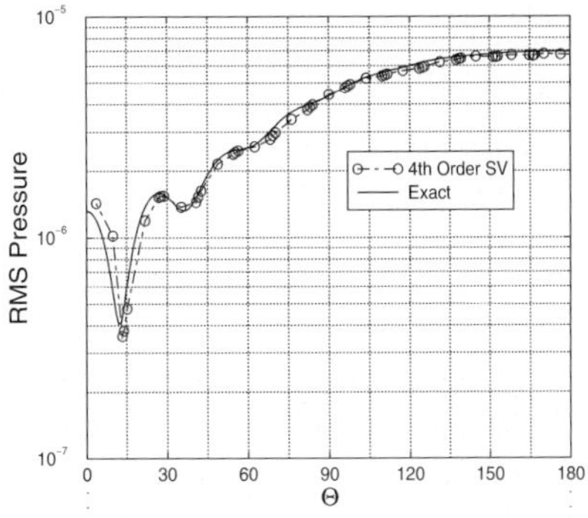

(b) Right Cylinder

Fig. 4. Comparison of the computational and analytical RMS pressure along the two cylinders

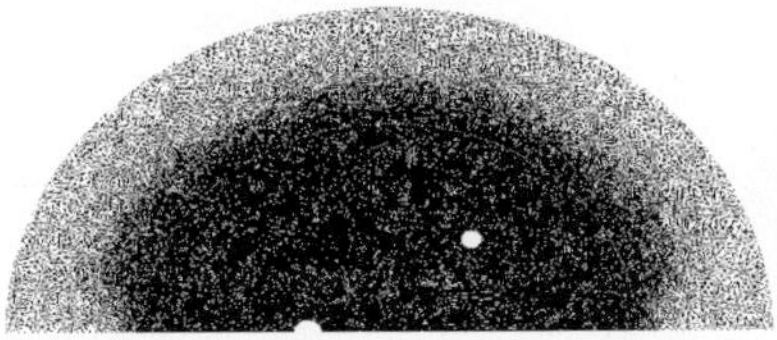

(a)Grid

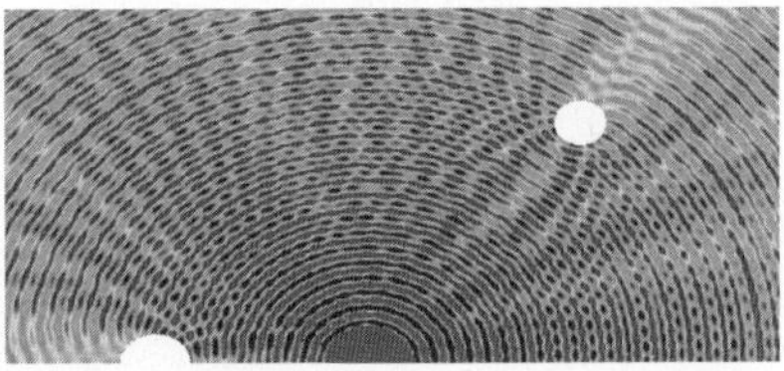

(b)Pressure Distribution

Fig. 5. Pressure distribution computed with the 4th order SV scheme

solution in Figure 7. Note again that there is a good agreement between the computational and analytical solutions.

4 Conclusions

The SV method has been tested on two benchmark problems in the Fourth Computational Aeroacoustics (CAA) Workshop on Benchmark Problems. For the Category 2 problems, the 4th SV scheme was capable of producing reasonable results with a grid resolution of about 13 PPW. The advantage of the SV method lies in its capability of handling complex geometries in a very flexible manner. The problems can all be set up in minutes.

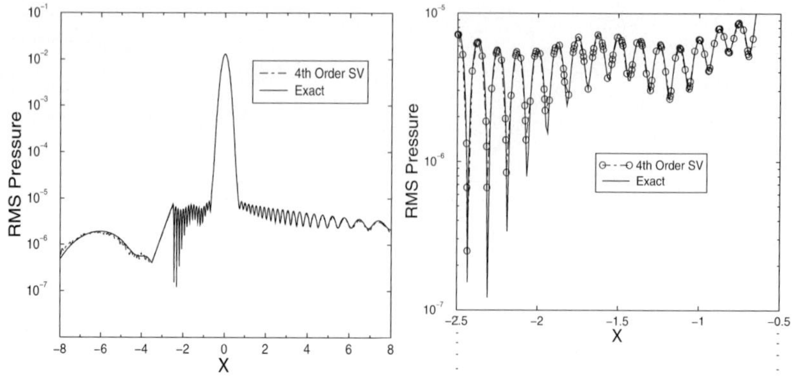

Fig. 6. Comparison of the computational and analytical RMS pressure along the center line

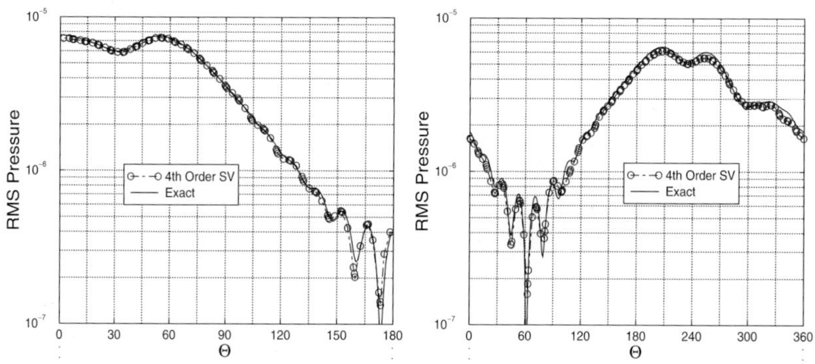

(a) Left Cylinder (b) Right Cylinder

Fig. 7. Comparison of the computational and analytical RMS pressure along the two cylinders

Acknowledgements

Some of the work presented in the paper was conducted while the author was visiting the Air Force Research Laboratory, Dayton with support from the AFRLs summer faculty program. The author thanks Drs. Miguel Visbal and Scott Sherer for many helpful discussions and for providing the analytical solutions for the Category 2 problems.

References

1. Wang, Z.J.: *J. Comput. Phys.*, **178**, 210(2002).
2. Wang, Z.J. and Liu, Y.: *J. Comput. Phys.*, **179**, 665(2002).
3. Wang Z.J. and Liu, Y.: *J. of Sci. Comp.*, **20**, 137(2004).
4. Wang, Z.J., Zhang, L. and Liu, Y.: *J. Comput. Phys.*, **194** 716(2004).
5. Godunov, S.K.: *Mat. Sb.* **47**, 271(1959)
6. Cockburn, B., Hou S. and Shu, C.-W.: *Math. of Comp.* **54**, 545(1990).
7. Barth, T.J. and Frederickson, P.O. : AIAA Paper No. 90-0013, 1990.
8. Tam, C.K.W.: AIAA J., **33**, 1788(1995).
9. Shu, C.-W.: *SIAM J. on Sci. and Stat. Comput.* **9**, 1073(1988).
10. Sherer, S. and Visbal, M.: AIAA Paper No. 2003-3203.

Discontinuous Spectral Difference Method for Conservation Laws on Unstructured Grids

Yen Liu[1], Marcel Vinokur[2], and Z.J. Wang[3]

[1] NASA Ames Research Center, Moffett Field, CA 94035, USA, Yen.Liu@nasa.gov
[2] Eloret Corp., Sunnyvale, CA 94087, USA, vinokur@nas.nasa.gov
[3] Michigan State University, East Lansing, MI 48824, USA, zjw@egr.msu.edu

1 Summary

A new, high-order, conservative, and efficient method for conservation laws on unstructured grids is developed. The concept of discontinuous and high-order local representations to achieve conservation and high accuracy is utilized in a manner similar to the Discontinuous Galerkin (DG)[1] and the Spectral Volume (SV)[2] methods, but while these methods are based on the integrated forms of the equations, the new method is based on the differential form to attain a simpler formulation and higher efficiency. Conventional unstructured finite-difference (FD)[3] and finite-volume (FV)[4] methods require data reconstruction based on the least-squares formulation using neighboring point or cell data. Since each unknown employs a different stencil, one must repeat the least-squares inversion for every point or cell at each time step, or store the inversion coefficients. In a high-order, three-dimensional computation, the former would involve impractically large CPU time, while for the latter the memory requirement becomes prohibitive. In addition, the finite-difference method does not satisfy the integral conservation in general. By contrast, the DG and SV methods employ a local, universal reconstruction of a given order of accuracy in each cell in terms of internally defined conservative unknowns. Since the solution is discontinuous across cell boundaries, a Riemann solver is necessary to evaluate boundary flux terms and maintain conservation. In the DG method, a Galerkin finite-element method is employed to update the nodal unknowns within each cell. This requires the inversion of a mass matrix, and the use of quadratures of twice the order of accuracy of the reconstruction to evaluate the surface integrals and additional volume integrals for non-linear flux functions. In the SV method, the integral conservation law is used to update volume averages over subcells defined by a geometrically similar partition of each grid cell. As the order of accuracy increases, the partitioning for 3D requires the introduction of a large number of parameters, whose optimization to achieve convergence becomes increasingly more difficult. Also, the number of interior facets required to subdivide non-planar faces, and the additional increase in the number of quadrature points for each facet, increases the computational cost greatly.

In the spectral difference (SD) method, the conservative unknowns in each cell are the number of nodal values required to support a reconstruction of a given order of accuracy. Their locations are chosen so that a quadrature approximation for the volume integral exists at least to the same order of accuracy. The fluxes are calculated at a different set of nodes, whose number will support a reconstruction of one order higher accuracy, since the flux derivatives are used to update the conservative unknowns. They are located so that quadrature approximations for surface integrals over the cell boundaries exist to a required order of accuracy. In addition, the locations of the conservative nodes and the flux nodes must be such that the integral conservation law is satisfied for the cell to the desired order of accuracy. If the nodes are distributed in a geometrically similar manner for all cells, the discretizations become universal, and can be expressed as the same weighted sums of the products of the local metrics and fluxes. These metrics are constants for the line, triangle, and tetrahedron elements, and can be computed analytically for curved elements. We can also show that the number of flux nodes is far less than the number of quadrature points in the SV method. Since all unknowns are decoupled, no mass matrix inversion is required. The SD formulation for the line element is a generalization of the multidomain spectral method[5]. Its tensor products can be used for quadrilateral and hexahedral elements.

2 The Spectral Difference Method

The most general form of a conservation law can be written as

$$\frac{\partial u}{\partial t} + \nabla * F = 0, \tag{1}$$

where the conservative variable u can be a scalar or a vector, and the generalized flux F can be a vector or tensor. The term $\nabla * F$ represents the divergence or curl of F, depending on the physical definition of u. Integrating (1) over cell i, we obtain

$$\frac{d}{dt}\int_{V_i} u\, dV + \sum_{l=1}^{M} \int_{S_{l,i}} d\mathbf{S} * F = 0, \tag{2}$$

where V_i is the volume of the cell i , and $S_{l,i}$ is the area of face l for cell i. (In 2D, each face is actually a line.) Here M is the number of faces, which is one more than the dimension D of the domain.

In each cell, the discrete unknowns are the values of u at quadrature points for the volume integral over the cell. We denote these points, some of which may lie on the cell faces, as $\mathbf{r}_{j,i}$, and define $u_{j,i}$ as $u(\mathbf{r}_{j,i})$. The expansion of u in the cell can be written in the cardinal form

$$u_i(\mathbf{r}) = \sum_{j=1}^{N_n} L_{j,i}(\mathbf{r}) u_{j,i}, \tag{3}$$

where $L_{j,i}(\mathbf{r})$ are the cardinal basis functions and N_n is the number of basis functions required to support a desired degree of precision n of the reconstruction. We will use polynomials as an independent basis. The locations of $\mathbf{r}_{j,i}$ then uniquely define the $L_{j,i}(\mathbf{r})$. In order to evaluate the surface integrals in (2) efficiently, we discretize F at points $\mathbf{r}_{k,i}$, most or all of which are located at quadrature points for those integrals. The expansion of F in the cell can also be written in the cardinal form

$$F_i(\mathbf{r}) = \sum_{k=1}^{N_{n+1}} M_{k,i}(\mathbf{r}) F_{k,i}, \tag{4}$$

where $M_{k,i}(\mathbf{r})$ are now the set of cardinal basis functions defined by $\mathbf{r}_{k,i}$ and $F_{k,i} = F(\mathbf{r}_{k,i})$. We can satisfy (1) at points $\mathbf{r}_{j,i}$ by evaluating

$$\nabla * F(\mathbf{r}_{j,i}) = \sum_{k=1}^{N_{n+1}} \nabla M_{k,i}(\mathbf{r}_{j,i}) * F_{k,i}. \tag{5}$$

In order to evaluate $F_{k,i}$, $u_{k,i}$ is required, which can be obtained directly from (3) as

$$u_{k,i} = \sum_{j=1}^{N_n} L_{j,i}(\mathbf{r}_{k,i}) u_{j,i}. \tag{6}$$

To reduce the cost of this interpolation, some of points $\mathbf{r}_{k,i}$ may be chosen to coincide with $\mathbf{r}_{j,i}$. If the points $\mathbf{r}_{j,i}$ and $\mathbf{r}_{k,i}$ are distributed in a geometrically similar manner for all cells and within each cell i the gradient of a function is expressed in terms of its area vectors $\mathbf{S}_i^l$ and volume V_i, the coefficients in (5) and (6) become universal, independent of cell i. There are only a few of these coefficients, which can be calculated exactly and stored in advance. For points $\mathbf{r}_{k,i}$ in the interior of cell, $F_{k,i} = F(u_{k,i})$. For those points $\mathbf{r}_{k,i}$ located on the cell faces, since u may be discontinuous, we must replace the flux F by a Riemann flux.

In order to check the integral conservation law for the cell, we must show that

$$V_i \sum_{j=1}^{N_n} w_j \nabla * F(\mathbf{r}_{j,i}) = \sum_{l=1}^{M} \mathbf{S}_i^l * \sum_k w_k F_{k,i}. \tag{7}$$

Here w_j are the volume quadrature weights at the points $\mathbf{r}_{j,i}$ and w_k are surface quadrature weights for face l, and for each face the summation is over those points $\mathbf{r}_{k,i}$ located on that face. The total contribution from any interior point $F_{k,i}$ to the volume integral in Eq. (7)(lefthandside term) must vanish.

From Eqs. (5) and (6), we also see that the SD formulation is very similar to that of the FD method for structured grids. The SD method thus retains the simplicity and computational efficiency of the structured FD method.

However, the metric terms in the latter are evaluated by numerically differencing the grid point coordinates. Since numerical grid generators are mostly only second-order accurate, the overall accuracy of the solution can be severely degraded if the grid is not sufficiently smooth. In contrast, the metric terms in the SD method are computed exactly from the geometry of the grid, no matter how it was generated. It thus retains its formal accuracy, even for very unsmooth unstructured grids. Furthermore, in contrast to the FD method, the integral conservation law is satisfied to the desired accuracy.

3 Locations of the Unknown and Flux Points

The critical part of the SD method is the locations of the u points $\mathbf{r}_{j,i}$ and F points $\mathbf{r}_{k,i}$, which are determined by symmetry groups associated with the cell centroids, vertices, edges, and faces. All but the first contain arbitrary parameters that can be varied to obtain optimum solutions. The number of points required for a reconstruction with a specified degree of precision is greater than the minimum number of Gaussian quadrature points for that precision. One can obtain greater efficiency by locating some u points to coincide with F points. For F points on the vertices in 2D and edges and vertices in 3D, more than one Riemann solver is necessary. For those formulations with expensive Riemann solvers, these points should be minimized. Another criterion for the placement of u and F points is that the reconstruction matrix is nonsingular. The final criterion is that integral conservation is satisfied within the desired degree of precision. We can show that the number of F points is far less than the number of flux quadrature points in the SV method with the same accuracy.

We first show some representative placements of u points (circles) and F points (squares) that satisfy integral conservation, for various orders of accuracy (which are one more than the degree of precision). Figures 1a and 1b show the placements for line element with second and fifth order of accuracy, and Figures 2a, 2b, and 2c show the placements for the triangular element with first, second, and third order of accuracy, respectively. Here the u points are placed at the Gaussian quadrature points, while the F points on the surface are placed at Gauss-Lobatto points, except in Fig. 2a where they are located at the Gaussian points.

(a) 2nd order (b) 5th order

Fig. 1. Placement of unknown and flux points for line element.

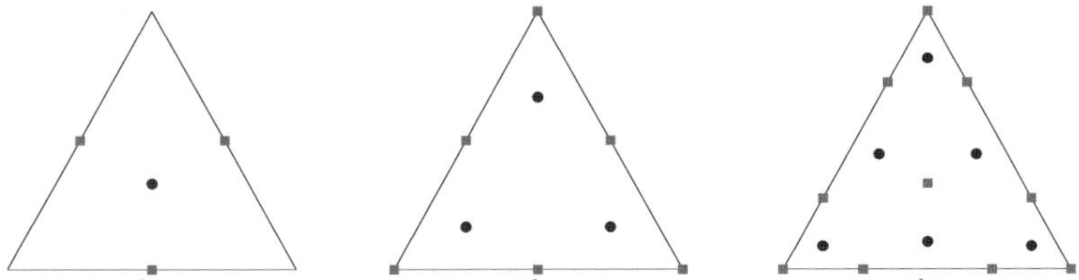

(a) 1st order (b) 2nd order (c) 3rd order

Fig. 2. Placement of unknown and flux points for triangular element.

4 Numerical Results

We first show some one-dimensional convergence studies for the linear convection equation. Plotted are L_1 (solid lines) and L_∞ (dash lines) error norms as functions of grid size for second to fifth order accurate methods. Figure 3a (1D-G) shows the plots for the u points at the Gaussian quadrature points, while Figure 3b (1D-GL) shows the plots for the u points at the Gauss-Lobatto quadrature points. Both exhibit the expected orders of accuracy. In Fig. 3c (2D) we show analogous plots for a plane wave propagating at 45 degree through a square domain for second and third order accurate methods, using the point placements in Figs. 2b and 2c. The third example is the scattering of a TM wave incident on a perfectly conducting circular cylinder. Figure 4 shows the unstructured grid consisting of 2024 triangular cells. The wave propagates from the left to the right with the wave number equal to 5, based on the radius of the cylinder. This gives approximately 6 cells per wavelength. Contour plots for E_z with the exact solution (solid lines) are shown in Figs. 5a, 5b, and 5c for first, second, and third order accurate methods, respectively. It is seen that the first order solution is very dissipative with this grid resolution, while the second and third order solutions show an excellent agreement with the exact solution.

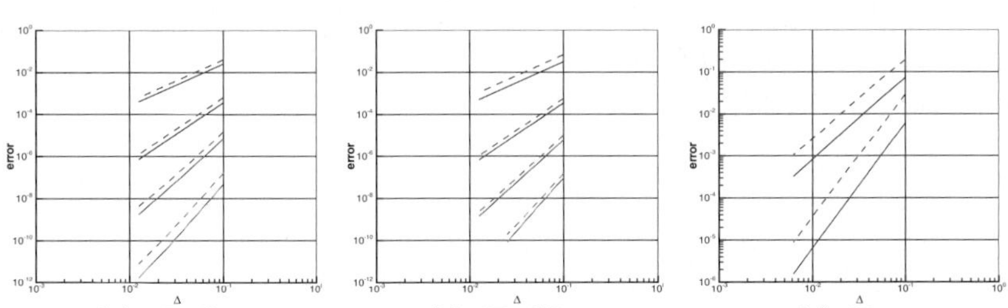

(a) 1D-G (b) 1D-GL (c) 2D

Fig. 3. Error norms of various order of accuracy for line and triangular elements.

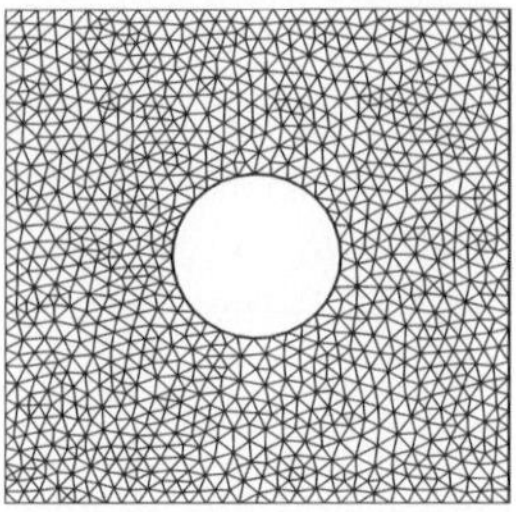

Fig. 4. Grid for a region exterior to a circular cylinder.

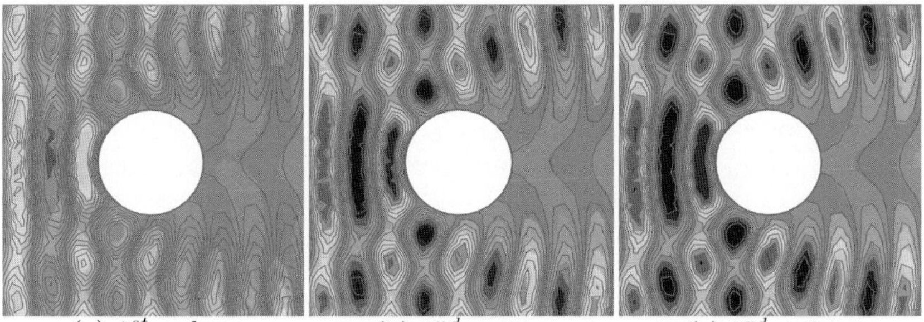

(a) 1st order (b) 2nd order (c) 3rd order

Fig. 5. Contour plots of $\mathbf{E}_z$ for a plane wave incident on a perfectly conducting cylinder. (numerical solutions: color contours, exact solution: solid lines)

References

1. B. Cockburn and C.W. Shu, *J. Comput. Phys.*, **141**, 199-227 (1998).
2. Z.J. Wang and Yen Liu, *J. Comput. Phys.*, **179**, 665-697 (2002).
3. D. Sridar and N. Balakrishnan, *J. Comput. Phys.*, **189**, 1-29 (2003).
4. T.J. Barth and P.O. Frederichson, AIAA Paper No. 90-0013, 1990.
5. D.A. Kopriva and J.H Kolias, *J. Comput. Phys.*, **125**, 244-261 (1996).

Multigrid Solution for High-Order Discontinuous Galerkin Discretizations of the Compressible Navier-Stokes Equations

Todd A. Oliver[1], Krzysztof J. Fidkowski[2], and David L. Darmofal[3]

[1] Department of Aeronautics and Astronautics, Massachusetts Institute of Technology, Cambridge, MA 02139. toliver@mit.edu

[2] kfid@mit.edu

[3] darmofal@mit.edu

1 Introduction

While CFD has achieved significant maturity, computational costs are extremely large for aerodynamic simulations of aerospace vehicles. In this applied aerodynamics context, the discretization of the Euler or Navier-Stokes equations is performed predominantly by finite volume algorithms. The accuracy of many of these methods is at best second order; i.e. the error decreases as $O(h^2)$ where h is a measure of the grid spacing. Recent studies have shown that these existing algorithms may not be adequate to achieve desired engineering accuracy for modern aerodynamic applications [7, 10]. The development of a practical higher-order solution method could alleviate this problem by significantly decreasing the computational time required to achieve an acceptable error level.

An attractive approach for achieving higher-order accuracy is the discontinuous Galerkin (DG) formulation in which element-to-element coupling exists only through the fluxes at the shared boundaries between elements. This limited coupling is an enabling feature which permits the development of an efficient higher-order solver. In this paper, we present a p-multigrid algorithm with an elemental line smoother for the solution of higher-order DG discretizations of the compressible Navier-Stokes equations. Using Fourier analysis, we demonstrate that higher-order DG discretizations can be stably marched with locally-implicit schemes without the use of multi-stage iterations. Finally, computational results for a laminar airfoil show that higher-order solutions offer significant reductions in the degrees of freedom required to achieve engineering accuracy.

2 Discontinuous Galerkin Discretization

The two-dimensional, compressible Navier-Stokes equations in strong, conservation form are given by:

$$\boldsymbol{u}_t + \nabla \cdot \mathcal{F}_i(\boldsymbol{u}) - \nabla \cdot \mathcal{F}_v(\boldsymbol{u}, \nabla \boldsymbol{u}) = 0, \tag{1}$$

where $\boldsymbol{u} = (\rho, \rho u, \rho v, \rho E)^T$ is the conservative state vector, $\mathcal{F}_i = (\boldsymbol{F}_i^x, \boldsymbol{F}_i^y)$ is the inviscid flux, and $\mathcal{F}_v = \mathcal{A}_v \nabla \boldsymbol{u} = (\boldsymbol{F}_v^x, \boldsymbol{F}_v^y)$ is the viscous flux.

Eqn. 1 is discretized using the second form of Bassi and Rebay (BR2) [2]. This discretization was chosen because it achieves optimal order of accuracy with a compact stencil without relying on an ad hoc penalty parameter [6]. To write the discretization in a concise form, jump, $[\![\cdot]\!]$, and average, $\{\cdot\}$, operators are defined as follows,

$$[\![s]\!] = s^+\hat{\boldsymbol{n}}^+ + s^-\hat{\boldsymbol{n}}^-, \qquad \{\boldsymbol{\varphi}\} = \frac{1}{2}(\boldsymbol{\varphi}^+ + \boldsymbol{\varphi}^-),$$

where s is a spatial scalar quantity and $\boldsymbol{\varphi}$ is a spatial vector quantity. The $()^+$ and $()^-$ notation indicates the trace value taken from opposite sides of a face. Note that both operators produce spatial vectors.

Denote by $\mathcal{V}_h^p$ the space of discontinuous vector-valued polynomials of degree p on a subdivision T_h of the domain Ω into elements such that $\Omega = \bigcup_{\kappa \in T_h} \kappa$. The BR2 discretization is given by the following form: find $\boldsymbol{u}_h \in \mathcal{V}_h^p$ such that $\forall \boldsymbol{v}_h \in \mathcal{V}_h^p$,

$$\begin{aligned}
&\mathbb{E} + \sum_{\kappa \in T_h} \int_\kappa \nabla \boldsymbol{v}_h^T \cdot (\mathcal{A}_v \nabla \boldsymbol{u}_h) d\boldsymbol{x} - \int_{\Gamma_i} \Big([\![\boldsymbol{u}_h]\!]^T \cdot \{\mathcal{A}_v^T \nabla \boldsymbol{v}_h\} + [\![\boldsymbol{v}_h]\!]^T \cdot \{\mathcal{A}_v \nabla \boldsymbol{u}_h\} \Big) ds \\
&+ \int_{\Gamma_i} \eta_f [\![\boldsymbol{v}_h]\!]^T \cdot \{\boldsymbol{\delta}_f\} ds + \int_{\partial\Omega} (\boldsymbol{u}^b - \boldsymbol{u}_h^+)^T (\mathcal{A}_v^T \nabla \boldsymbol{v}_h)^+ \cdot \hat{\boldsymbol{n}} ds \\
&- \int_{\partial\Omega} \boldsymbol{v}_h^{+T} (\mathcal{A}_v \nabla \boldsymbol{u}_h)^b \cdot \hat{\boldsymbol{n}} ds + \int_{\partial\Omega} \eta_f \boldsymbol{v}_h^{+T} \boldsymbol{\delta}_f^b \cdot \hat{\boldsymbol{n}} ds = 0. \qquad (2)
\end{aligned}$$

where $\mathbb{E}$ is the discretization of the Euler equations [5], Γ_i is the union of interior faces, and $\boldsymbol{\delta}_f$ and $\boldsymbol{\delta}_f^b$ are auxiliary variables. The auxiliary variables are given by the following weak statement: find $\boldsymbol{\delta}_f \in [\mathcal{V}_h^p]^2$ such that $\forall \boldsymbol{\psi}_h \in [\mathcal{V}_h^p]^2$,

$$\int_{\kappa^\pm} \boldsymbol{\psi}_h^T \cdot \boldsymbol{\delta}_f^\pm d\boldsymbol{x} = -\frac{1}{2} \int_{\sigma_f} [(\mathcal{A}_v^T \boldsymbol{\psi}_h) \cdot \hat{\boldsymbol{n}}]^{\pm T} (\boldsymbol{u}_h^\mp - \boldsymbol{u}_h^\pm) ds \qquad (3)$$

for interior faces, and

$$\int_{\kappa^+} \boldsymbol{\psi}_h^T \cdot \boldsymbol{\delta}_f^b d\boldsymbol{x} = - \int_{\sigma_f} [(\mathcal{A}_v^T \boldsymbol{\psi}_h) \cdot \hat{\boldsymbol{n}}]^{+T} (\boldsymbol{u}_h^b - \boldsymbol{u}_h^+) ds \qquad (4)$$

for boundary faces, where σ_f denotes the face indexed by f. The auxiliary variables are locally eliminated from Eqn. 2 by solving Eqn. 3 or 4 at each face in the domain.

The boundary conditions on $\partial\Omega$ are imposed weakly by constructing an exterior boundary state, $\boldsymbol{u}_h^b$, and a normal derivative of the state, $(\nabla \boldsymbol{u} \cdot \hat{\boldsymbol{n}})^b$, that are functions of the interior quantities and boundary condition data. These quantities are then used to construct the inviscid and viscous fluxes at the domain boundaries.

3 Solution Method

The DG discretization given by Eqns. 2 through 4 can be written as a vector of nonlinear equations $\boldsymbol{R}(\boldsymbol{u}) = 0$. To solve the steady-state, nonlinear system, a p-multigrid scheme with an iterative smoother is used. A generic iterative scheme can be written as,

$$\boldsymbol{u}^{n+1} = \boldsymbol{u}^{n} - \boldsymbol{P}^{-1}\boldsymbol{R}(\boldsymbol{u}^{n}), \tag{5}$$

where the preconditioner, $\boldsymbol{P}$, is an approximation to $\frac{\partial \boldsymbol{R}}{\partial \boldsymbol{u}}$. Two preconditioners were considered: elemental block-Jacobi, in which the unknowns on each element are solved simultaneously, and elemental line block-Jacobi, in which the unknowns on each line of elements are solved simultaneously. This work focuses on the details of the line smoother and the multigrid solver based on it.

In strongly convective systems, transport of information proceeds along characteristic directions. By solving implicitly on lines of elements connected along these directions, one can alleviate the stiffness associated with strong convection. Furthermore, for viscous flows, the line solver is an important ingredient in removing the stiffness associated with regions of high grid anisotropy frequently required in viscous layers [1, 11]. In such cases, the lines are formed between elements which exhibit the strongest coupling. In this work, the coupling is defined by the Jacobian of the $p = 0$ discretization of the scalar transport equation,

$$\nabla \cdot (\rho \boldsymbol{u} \phi) - \nabla \cdot (\mu \nabla \phi) = 0$$

where $\rho \boldsymbol{u}$ and μ are taken from the solution at the current iteration. More specifically, the coupling between two elements j and k that share a face is given by

$$\boldsymbol{C}_{j,k} = \max\left(\left| \frac{\partial R_j}{\partial \phi_k} \right|, \left| \frac{\partial R_k}{\partial \phi_j} \right| \right). \tag{6}$$

Using this definition of coupling, lines of elements are formed using the line creation algorithm of Fidkowski and Darmofal [8, 7].

The initial preconditioner for line smoothing consists of subsystems, $\boldsymbol{M}^l$, formed from $\frac{\partial \boldsymbol{R}}{\partial \boldsymbol{u}}$, for each line l. $\boldsymbol{M}$ denotes the assembled $\boldsymbol{M}^l$ matrices. To improve the robustness during the initial iterations, $\boldsymbol{M}$ is augmented by the addition of an unsteady term, yielding the final preconditioner, $\boldsymbol{P} = \boldsymbol{M} + \mathcal{M}/\Delta t$, where $\mathcal{M}$ is the mass matrix. As the solution begins to converge, $\Delta t \rightarrow \infty$ [7].

The inversion of $\boldsymbol{P}$ uses a block-tridiagonal algorithm in which the block diagonal is LU decomposed. As the dominant cost of the line solver (especially for higher-order schemes) is the LU decomposition of the diagonal, the computational cost of the line smoother scales as that of the simpler elemental block-Jacobi. However, the performance of the line smoother is significantly better due to the increased implicitness along strongly coupled directions.

p-multigrid is used in conjunction with the line smoother to improve the performance of the solver. In standard multigrid techniques, solutions on spatially coarser grids are used to correct solutions on the fine grid. In p-multigrid, the idea is similar except that lower-order approximations serve as the "coarse" grids[13, 9].

To solve the nonlinear system in question, the Full Approximation Scheme (FAS) [3, 4] was chosen as the multigrid method. The different p levels are sequenced in a V-cycle using ν_1 pre-smoothing and ν_2 post-smoothing steps. Performance and robustness are further improved using full multigrid (FMG), in which the calculation is started on the coarsest level rather than the finest [3]. For more details on the multigrid implementation, see [8, 7].

4 Stability Analysis

For higher-order discretizations based on extended stencils, stable iterative methods are often developed using multi-stage or multi-step methods. In contrast, higher-order DG discretizations can be iterated stably using an element-implicit scheme in which all of the degrees of freedom associated with an element are solved simultaneously[7]. Recently, we have analyzed the stability of element and line element methods for convection-diffusion and found these to be stable independent of order, mesh characteristics, and flow conditions; in this paper, we briefly report on the line relaxation.

To determine the stability of the line relaxation applied to the DG discretization of the Navier-Stokes equations, Fourier analysis is performed for two-dimensional convection-diffusion with periodic boundary conditions. The convection-diffusion problem is given by

$$au_x + bu_y - \nu(u_{xx} + u_{yy}) = f(x, y) \quad \text{on} \quad [-1, 1] \times [-1, 1]. \tag{7}$$

Triangulating the domain $[-1, 1] \times [-1, 1]$ into $N_x N_y$ rectangular elements, the discretized form of Eqn. 7 can be written as the system $\mathbf{A}\boldsymbol{u} = \mathbf{f}$. The relaxation footprint consists of the eigenvalues of $-\mathbf{P}^{-1}\mathbf{A}$, where $\mathbf{P}$ is the line-implicit preconditioner defined in Section 3. These eigenvalues are a function of p, $Re \equiv a\Delta x/\nu$, $\tan\alpha \equiv b/a$, and $\text{AR} \equiv \Delta x/\Delta y$, where Δx and Δy are the grid spacing in the x and y directions, respectively. While the footprints cannot be computed analytically, numerical results show that, over all combinations of Re, α, AR, and p considered, the eigenvalues are stable. Fig. 1 shows the footprints for $p = 0$ and $p = 3$ for two cases.

5 Results

The test case considered is that of laminar flow over a NACA 0012 airfoil at $\alpha = 0^o$, $Re = 5000$, $M = 0.5$, with a no-slip, adiabatic wall [12]. The multigrid V-cycles used here consist of 4 pre- and post-smoothing steps with 100

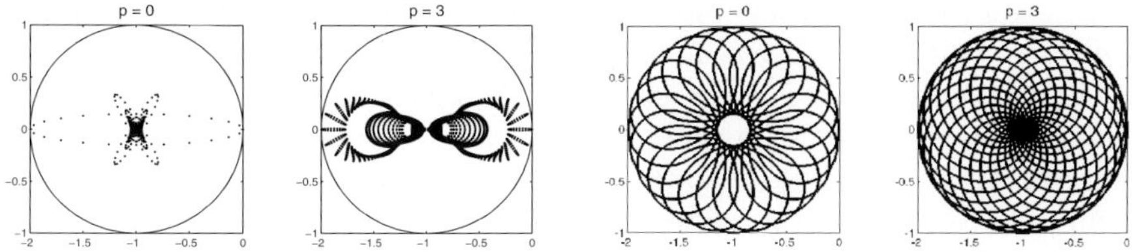

(a) $Re = 10$, AR = 1, $\alpha = 1^o$ (b) $Re = 10^4$, AR = 100, $\alpha = 1^o$

Fig. 1. Line relaxation footprints

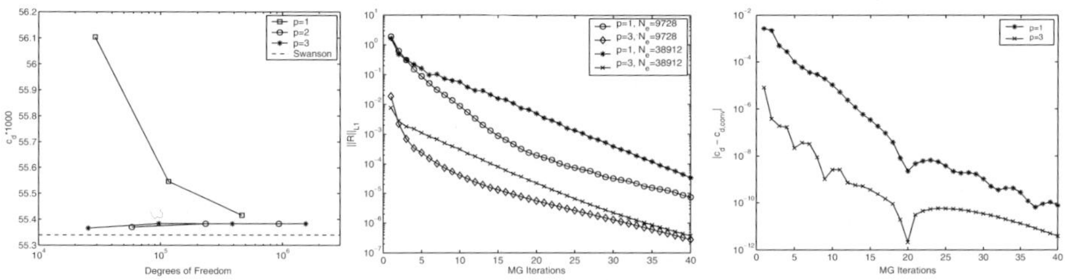

(a) Drag vs. DOF (b) Residual History (c) Drag Error History

Fig. 2. Results for NACA 0012, $\alpha = 0^o$, $Re = 5000$, $M = 0.5$

smoothing steps on the coarsest level. Fig. 2(a) shows the dramatic benefits of high-order solutions in terms of degrees of freedom (DOF) required to attain a desired error level. Defining the error for a given case as the difference between the computed drag and the fine grid, $p = 3$ drag, the coarsest grid, $p = 3$ solution (2.6e4 DOF, 640 elements) has an error of approximately 0.17 drag counts compared to 0.31 drag counts for the finest, $p = 1$ solution (4.7e5 DOF, 38912 elements).

The residual histories for $p = 1$ and $p = 3$ are plotted in Fig. 2(b). The asymptotic convergence rates show only a slight dependence on p and h. While p independence is expected, the residual convergence rate has been shown to have stronger h dependence for other problems [7]. Fig. 2(c) shows the drag error convergence histories for a 9728 element grid, where the drag error for iteration i is the difference between the drag computed at iteration i and the fully converged drag for that case. We see that the the drag converges to within 0.1 drag counts in far fewer iterations than required to converge the residual. For $p = 3$, after performing ten $p = 1$ and $p = 2$ V-cycles, only one $p = 3$ V-cycle is necessary to drive the drag error to 0.08 drag counts.

6 Conclusions

While higher-order DG discretizations and the p-multigrid algorithm with elemental-line smoothing have significant potential to reduce computational requirements to achieve engineering-required accuracy, many opportunities for future work remain. Specifically, no problems with shocks, have been considered here. In cases with shocks, limiting is required to stabilize the oscillatory behavior of high-order approximations near discontinuities. Furthermore, only laminar flows have been considered. A turbulence model must be implemented to allow consideration of a wider class of flows of engineering interest.

References

1. S.R. Allmaras. Analysis of semi-implicit preconditioners for multigrid solution of the 2-d compressible Navier-Stokes equations, 1995.
2. F. Bassi and S. Rebay. GMRES discontinuous Galerkin solution of the compressible Navier-Stokes equations. In Karniadakis Cockburn and Shu, editors, *Discontinuous Galerkin Methods: Theory, Computation and Applications*, pages 197–208. Springer, Berlin, 2000.
3. Achi Brandt. *Guide to Multigrid Development.* Springer-Verlag, 1982.
4. William Briggs, Van Emden Henson, and Steve F. McCormick. *A Multigrid Tutorial, 2nd Ed.* SIAM, 2000.
5. Bernardo Cockburn and Chi-Wang Shu. Runge-Kutta discontinuous Galerkin methods for convection-dominated problems. *Journal of Scientific Computing*, pages 173–261, 2001.
6. D. N. Arnold; F. Brezzi; B. Cockburn; and L. D. Marini. Unified analysis of discontinuous Galerkin methods for elliptic problems. *SIAM J. Numer. Anal.*, 39(5):1749–1779, 2002.
7. K. Fidkowski. A high-order discontinuous Galerkin multigrid solver for aerodynamic applications. Master's thesis, Massachusetts Institute of Technology, 2004.
8. K. Fidkowski and D. Darmofal. Development of a higher-order solver for aerodynamic applications. *42nd AIAA Aerospace Sciences Meeting.*, AIAA Paper Number 2004-0436, 2004.
9. B.T. Helenbrook, D.J. Mavriplis, and H.A. Atkins. Analysis of p-multigrid for continuous and discontinuous finite element discretizations. AIAA Paper 2003-3989, 2003.
10. David W. Levy, Thomas Zickuhr, John Vassberg, Shreekant Agrawal, Richard A. Wahls, Shahyar Pirzadeh, and Michael J. Hemsch. Data summary from the First AIAA Computational Fluid Dynamics Drag Prediction Workshop. *Journal of Aircraft*, 40(5):875–882, 2003.
11. D.J. Mavriplis. Multigrid strategies for viscous flow solvers on anisotropic unstructured meshes. *Journal of Computational Physics*, 145:141–165, 1998.
12. R. Radespiel and R.C. Swanson. An investigation of cell centered and cell vertex multigrid schemes for the Navier-Stokes equations. AIAA Paper Number 89-0453, 1989.
13. E. M. Rønquist and A. T. Patera. Spectral element multigrid. I. formulation and numerical results. *J. Sci. Comput.*, 2(4):389–406, 1987.

Part XI

Incompressible Flow

Accurate Solution of Corner Singularities in Axisymmetric and Plane Flows Using Adjusted Mesh of Finite Elements

Pavel Burda[1], Jaroslav Novotný[2], and Jakub Šístek[1]

[1] Department of Mathematics, Faculty of Mechanical Engineering, Czech University of Technology, Karlovo nám. 13, CZ-121 35 Praha 2, Czech Republic
burda@fsik.cvut.cz, sistek@seznam.cz

[2] Institute of Thermomechanics AS CR, Dolejškova 5, CZ-182 00 Praha 8, Czech Republic, novotny@bivoj.it.cas.cz

Summary. We investigate the finite element solution of the flow of incompressible fluid in channels or in axisymmetric tubes with abrupt changes of diameters. The problems of singularities near the nonconvex corners are solved on a priori refined meshes. The refinement is based on a priori error estimate and on the asymptotic behaviour of the solution near the corners. In case of "geometric" singularities this is an efficient alternative to the adaptive mesh refinement. We present numerical results of flows in a channel with sharp obstacle, and in a channel with sharp cavity.

1 Introduction

In the papers [2], [3] we solved the problem with corner singularities by means of adaptive refinement strategy based on a posteriori estimates. In this paper we present an alternative approach. For problems on domains with corners we use the a priori error estimates and asymptotic behaviour of the solution near the corner, and derive an algorithm for adjusting the mesh near the corner.

2 Steady Navier-Stokes Equations and FEM Solution

We consider the flow of an incompressible fluid governed by the Navier-Stokes equation. In primitive variables V (velocity vector) and p (pressure) it reads

$$(V \cdot \nabla)V - \nu \Delta V + \nabla p = 0 \quad , \quad \nabla \cdot V = 0 \quad \text{in } \Omega, \tag{1}$$

where ν is the viscosity.

In [1] we studied the problem of axisymmetric flow in a tube with sudden changes of diameter, as in Fig. 1. To investigate the asymptotic behaviour of the solution near the corners we can proceed as follows. The rotational symmetry allows to restrict the problem into two dimensions using the cylindrical coordinates. By means of the stream function-vorticity formulation of the Navier-Stokes equations we obtain fourth order partial differential equation for the stream function. Using the Fourier transform technique we come to ordinary differential equation which enables finally to obtain the asymptotic behavior of the solution near the corners. E.g. for the internal angle

$\alpha = \frac{3}{2}\pi$ (as in Fig. 1), we proved in [1] that the velocity components depend on the distance ρ from the corner approximately as

$$v_l(\rho, \vartheta) \approx \rho^{0.54448374} \varphi_l(\vartheta), \quad l = 1, 2. \tag{2}$$

The same asymptotics has been known for the plane flow, cf. e.g. [4]. In what follows we restrict ourselves to the plane flow, and Fig. 1 is to be viewed as a channel, the proportions being in milimeters.

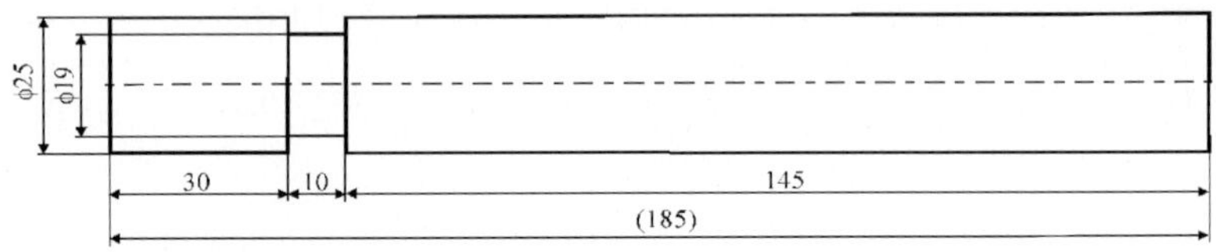

Fig. 1. Geometry of the channel or tube

To fix ideas we take as the solution domain Ω a polygon in $\mathbb{R}^2$, corresponding (due to symmetry) to the upper half of the domain of Fig. 1. We use Hood-Taylor finite elements, where velocity values are given in vertices - nodes and in midside nodes of the quadrilateral or the triangle, and pressure values only in vertices - nodes, so the Babuška-Brezzi condition is satisfied. For the case of the Stokes problem (first term missing in (1)) the following a priori error estimate can be proved:

$$\begin{aligned} &\|\nabla(V - V_h)\|_0 + \|p - p_h\|_0 \leq \\ &\leq C\Big[\Big(\sum_T h_T^{2k} \mid V \mid^2_{H^{k+1}(T)}\Big)^{1/2} + \Big(\sum_T h_T^{2k} \mid p \mid^2_{H^k(T)}\Big)^{1/2}\Big], \end{aligned} \tag{3}$$

where h_T is the diameter of the triangle T of a triangulation $\mathcal{T}_h$, $k = 2$ for Hood-Taylor elements. By means of (2) we derive the estimate

$$\mid V \mid^2_{H^{k+1}(T)} \approx C \int_{r_T - h_T}^{r_T} \rho^{2(\gamma-k-1)} \rho \, d\rho \approx C \left[-r_T^{2(\gamma-k)} + (r_T - h_T)^{2(\gamma-k)} \right], \tag{4}$$

where r_T is the distance of the element T from the corner. In order to equidistribute the errors on the elements and get the error estimate of the order $O(h^k)$, we need

$$h_T^{2k} \left[-r_T^{2(\gamma-k)} + (r_T - h_T)^{2(\gamma-k)} \right] \approx h^{2k} \tag{5}$$

Thus we have the algorithm that near the corners generates the mesh which is adjusted to singularity: the proportions of elements are computed by (5) using Newton's method. We start with $r_1 = 0,25\ mm$, and take $\gamma = 0,5444837$, $h = 0,1732\ mm$. This corresponds to cca 3 % of relative error equidistributed on the elements. The resulting elements are in Table 1.

3 Adjusted Mesh Refinement and Numerical Results

Using the above algorithm, J. Šístek [5] suggested several variants of the mesh refinement near the corner that corresponds to the "forward step" in Fig. 1. One of these fits very well to polar coordinate nature of the asymptotics near the corner (cf. Fig. 6). We used this adjusted mesh to solve the incompressible flow problem governed by the Navier-Stokes equations (1) on the domain corresponding to Fig. 1. Data: kinematic viscosity $\nu = 0.000025$ m^2/s and $v_{in\ max} = 1$ m/s, which give maximum Reynolds number around 760.

The results of plane flow obtained on this mesh adjusted to singularity are in good agreement with those obtained by the adaptive mesh refinement in [3]. We show some results on Figs. 2 - 6. The strength of singularity (both for velocity and pressure) is well captured, as seen on Figs. 4, 5. Due to refinement adjusted to singularity the results on Figs. 4, 5 are much sharper (more precise) than in [3]. To evaluate the error on elements we use the modified absolute error based on a posteriori error estimates, defined as

$$\mathcal{A}_m^2(v_1^h, v_2^h, p^h, \Omega_l) = \frac{|\Omega|\mathcal{E}^2(v_1^h, v_2^h, p^h, \Omega_l)}{|\overline{\Omega}_l| \|(v_1^h, v_2^h, p^h)\|_{X,\Omega}^2}, \tag{6}$$

where $\mathcal{E}^2(v_1^h, v_2^h, p^h, \Omega_l)$ is the estimate of the error on element l, $|\Omega|$ is the area of the whole domain and $|\overline{\Omega}_l|$ is the mean area of elements obtained as $|\overline{\Omega}_l| = \frac{|\Omega|}{n}$, $X = (H^1(\Omega))^2 \times L_2(\Omega)$. Here n means the number of all elements in the domain. More about the evaluation of the error can be found in [2].

On Fig. 6 we show the errors on individual elements: we can see that the distribution of the error on elements is quite uniform. Fig. 3 shows the location of the maximum of the velocity component v_y. Comparing it with Figs. 4 and 5 we can conclude that it is pressure who is responsible for needing so fine mesh near the corner.

Table 1. Resulting refinement

i	$r_i(mm)$	$h_i(mm)$
1	0.25000	0.0600369
2	0.18996	0.0480779
3	0.14189	0.0379492
4	0.10394	0.0294666
5	0.07447	0.0224535
6	0.05202	0.0167410
7	0.03527	0.0121680
8	0.02311	0.0085813
9	0.01453	0.0058366
10	0.00869	0.0037980
11	0.00489	0.0023392
12	0.00255	0.0013434
13	0.00121	0.0007042
14	0.00050	0.0005042

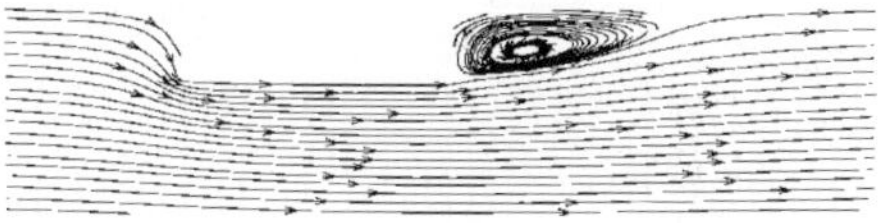

Fig. 2. Streamlines near the corners

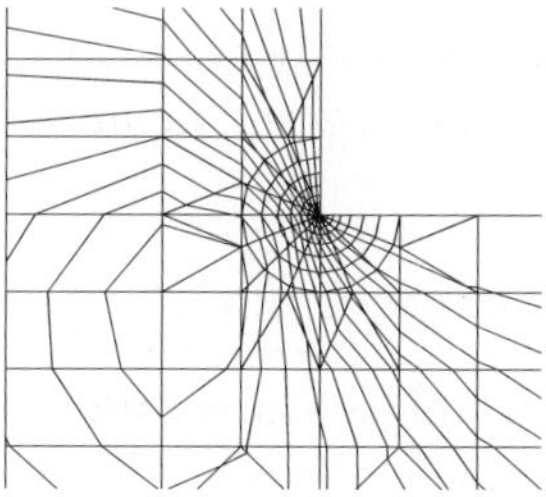

Fig. 3. Isolines of v_y

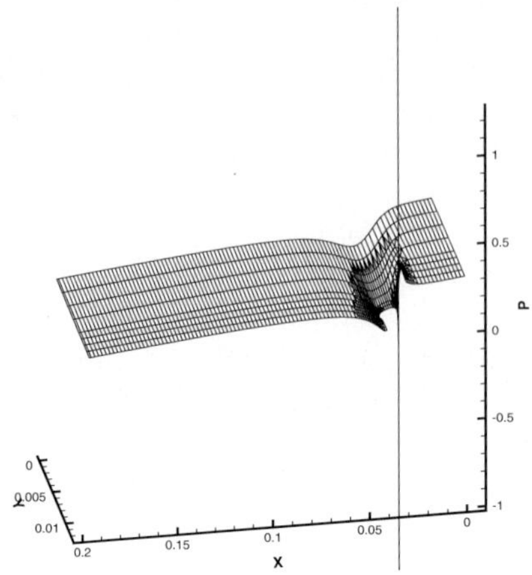

Fig. 4. Pressure p

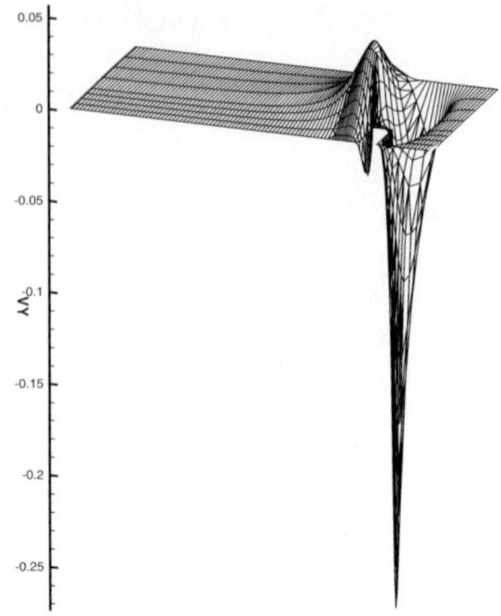

Fig. 5. Velocity component v_y

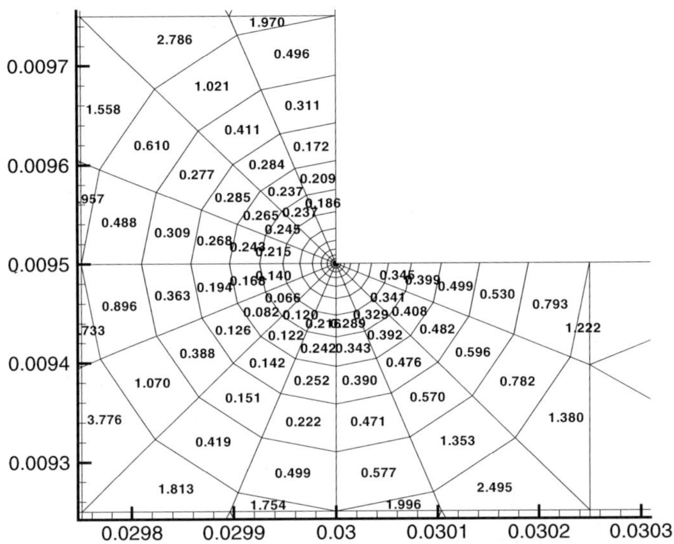

Fig. 6. Errors on elements near the corner (forward step)

Flow in a Channel with Sharp Cavity

We applied the algorithm of adjusted mesh refinement also to the problem of flow in a channel with a kind of cavity, cf. Fig. 7, where the proportions are now in meters. Again due to symmetry the solution domain Ω is only one half of the channel. At the inflow parabolic velocity profile is taken, on the walls (upper part) zero velocity. Data: $v_{in\ max} = 1$ m/s, Re=400. The mesh is refined near both nonconvex corners in the same manner as in the channel on Fig. 1. On Figures 8, 9, 10, 11 we show in turn the velocity component v_y, the streamlines, the error distribution on the refined elements, and the pressure.

In case of axisymmetric flow in tubes the same asymptotics apply, (cf. [1]). Our algorithm for adjusted meshes is applicable also in this case.

The algorithm is in fact applicable to problems with any nonconvex corners, the refinement will be different when the exponent in (2) is different due to different angle α.

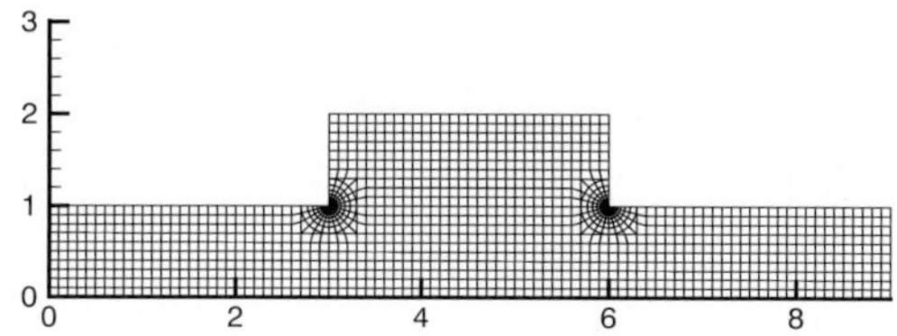

Fig. 7. Adjusted mesh for cavity problem

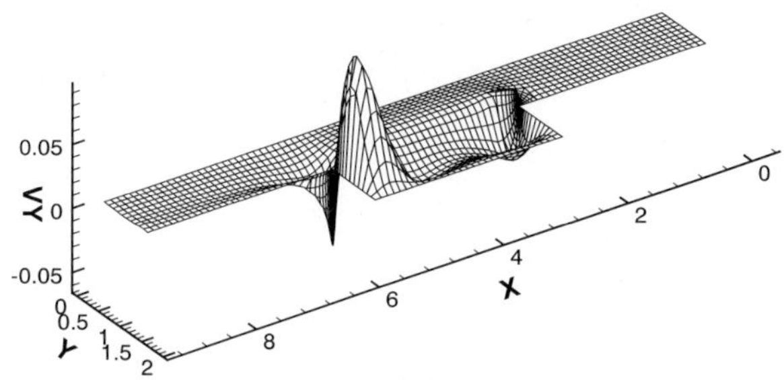

Fig. 8. Velocity component v_y

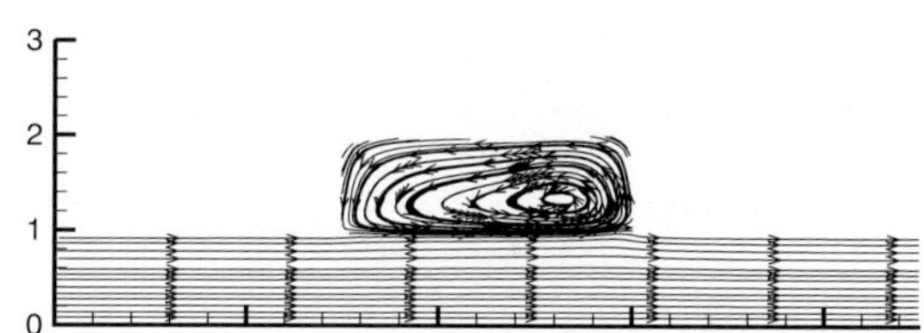

Fig. 9. Streamlines

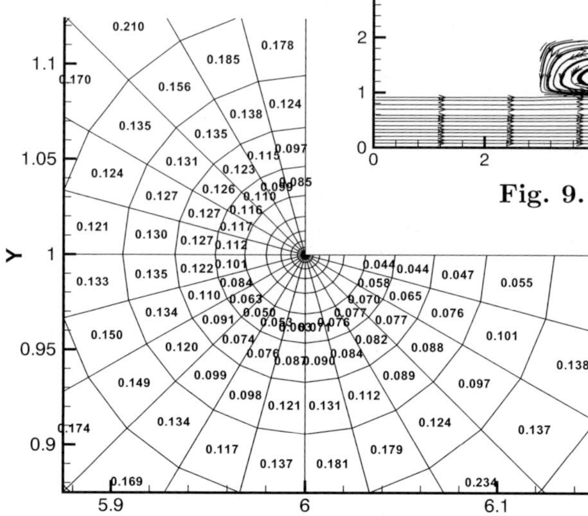

Fig. 10. Errors on elements near the second corner

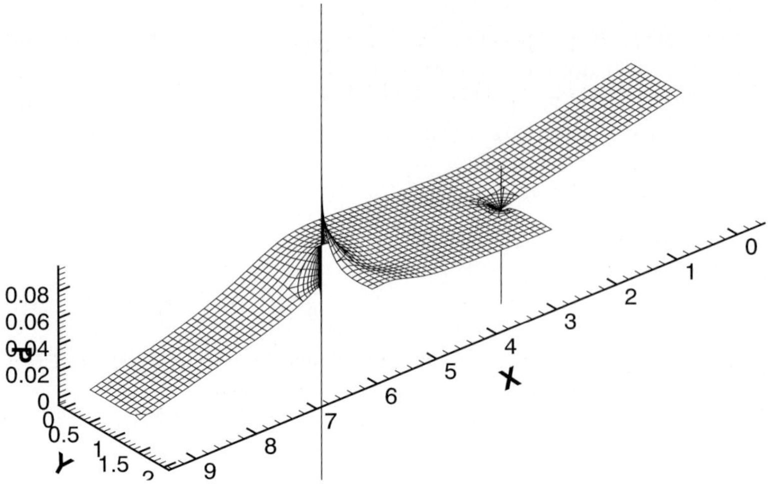

Fig. 11. Pressure p

4 Conclusions

The algorithm based on the asymptotic behaviour of the solution near the corners allows to get more accurate results in a shorter time than the adaptive refinement. The efficiency of the algorithm is high: desired accuracy requires to run the computation only ones (compared with adaptive approach - the same precission would need approx. 10 successive refinements). But, of course, the adaptive refinement based on a posteriori error estimates is more robust.

Acknowledgements

This research has been supported partly by the grant of the Czech Acad. Sci. No. IAA2120201/02, and partly by the State Research Project No. J04/98/210000010.

References

1. P. Burda: On the F.E.M. for the Navier-Stokes equations in domains with corner singularities. In: *Finite Element Methods, Supeconvergence, Post-Processing and A Posteriori Estimates,* ed. by M. Křížek et al. (Marcel Dekker, New York 1998) pp 41-52
2. P. Burda, J. Novotný, B. Sousedík: A posteriori error estimates applied to flow in a channel with corners, Mathematics and Computers in Simulation, **61**, Nos. 3-6, pp. 375-383 (2003)
3. P. Burda, J. Novotný, B. Sousedík: A posteriori error estimates and adaptive meshing applied to flow in channels with corners. In: *Computational Fluid Dynamics 2002,* ed. by S. Armfield et al. (Springer, Berlin) pp. 465 - 470.
4. V. A. Kondratiev: Asimptotika rešenija uravnienija Nav'je-Stoksa v okrestnosti uglovoj točki granicy, Prikl. Mat. i Mech., **1**, 119-123 (1967)
5. J. Šístek: Solution of the fluid flow in a tube with singularity using suitable mesh refinement (in Czech), Student Conference STČ, ČVUT Praha (2003)

Sensitivity Analysis of Transient Incompressible Laminar Flows

H. Hristova[1], S. Etienne[1], D. Pelletier[1], and J. Borggaard[2]

[1] École Polytechnique de Montréal, Montréal, Canada H3C 3A7, Dominique.Pelletier@Polymtl.ca
[2] ICAM Virginia Tech, Blacksburg, VA 24061, USA, borggajt@icam.vt.edu

Summary. This paper presents a general sensitivity equation method (SEM) for time dependent incompressible laminar flows. The methodology is applied to pulsatile flow around a square cylinder. A numerical study indicates that sensitivities predict changes in the flow structure sooner than the velocity fields.

1 Introduction

Sensitivities of an incompressible flow are defined as the derivatives of the velocity and pressure with respect to parameters controlling the response of flow. Sensitivities thus provide an enriched base of information on which to develop an understanding of complex flow problems. A continuous Sensitivity Equation Method (SEM) was developed by the authors for laminar isothermal flows and laminar heat transfer [1, 2, 3]. The work presented here is an extension, to time-dependent flows, of the methodology developed previously for steady state flows.

2 Flow and Sensitivity Equations

The flow regime of interest is modeled by the momentum and continuity equations:

$$\rho \frac{\partial \boldsymbol{u}}{\partial t} + \rho \boldsymbol{u} \cdot \nabla \boldsymbol{u} = -\nabla p + \boldsymbol{f} + \nabla \cdot \left[\mu \left(\nabla \boldsymbol{u} + (\nabla \boldsymbol{u})^T \right) \right] \tag{1}$$

$$\nabla \cdot \boldsymbol{u} = 0 \tag{2}$$

where ρ is the density, $\boldsymbol{u}$ is the velocity, p is the pressure, μ is the viscosity, t represents time and $\boldsymbol{f}$ is a body force. The state of the fluid, for $t > 0$ and $\boldsymbol{x}$ in the flow domain Ω, is determined by specifying the initial and boundary conditions:

$$\boldsymbol{u}(\boldsymbol{x}, t = 0) = \boldsymbol{U}_0(\boldsymbol{x}) \qquad \text{in } \Omega \tag{3}$$

$$\boldsymbol{u}(\boldsymbol{x}, t) = \boldsymbol{U}_b(\boldsymbol{x}, t) \qquad \text{on } \Gamma_D \text{ and } t > 0, \tag{4}$$

$$\boldsymbol{t} = [-p\mathbb{I} + \mu(\nabla \boldsymbol{u} + \nabla \boldsymbol{u}^T)] \cdot \hat{\boldsymbol{n}} = \boldsymbol{F}^N \qquad \text{on } \Gamma_N \text{ and } t > 0, \tag{5}$$

where $\mathbb{I}$ is the identity tensor and $\boldsymbol{F}^N$ is the imposed boundary value of the surface traction force $\boldsymbol{t}$.

The continuous sensitivity equations (CSE) are derived formally by implicit differentiation of the flow Eqs. 1 to 5 with respect to parameter a. Thus, not only do we treat the variable $\boldsymbol{u}$ as a function of space and time, but also as a function of the parameter a. This dependence is denoted as $\boldsymbol{u}(\boldsymbol{x},t;a)$. Defining the flow sensitivities as the partial derivatives $\boldsymbol{s}_u = \frac{\partial \boldsymbol{u}}{\partial a}$ and $s_p = \frac{\partial p}{\partial a}$, and the derivatives of the fluid properties and other flow parameters by a $(')$, we obtain

$$\rho' \frac{\partial \boldsymbol{u}}{\partial t} + \rho \frac{\partial \boldsymbol{s}_{\boldsymbol{u}}}{\partial t} + \rho' \boldsymbol{u} \cdot \nabla \boldsymbol{u} + \rho \boldsymbol{s}_{\boldsymbol{u}} \cdot \nabla \boldsymbol{u} + \rho \boldsymbol{u} \cdot \nabla \boldsymbol{s}_{\boldsymbol{u}} = -\nabla s_p + \boldsymbol{f}' + \nabla \cdot \left[\mu' \left(\nabla \boldsymbol{u} + (\nabla \boldsymbol{u})^T \right) + \mu \left(\nabla \boldsymbol{s}_{\boldsymbol{u}} + (\nabla \boldsymbol{s}_{\boldsymbol{u}})^T \right) \right]$$

$$\nabla \cdot \boldsymbol{s}_{\boldsymbol{u}} = 0.$$

Initial and boundary conditions for the sensitivity equations are obtained, for example, by implicit differentiation of Eqs. (3) and (4). This yields

$$\boldsymbol{s}_{\boldsymbol{u}}(\boldsymbol{x}, t=0) = \frac{d\boldsymbol{U}_0}{da}(\boldsymbol{x}) \quad \text{in } \Omega \tag{6}$$

$$\boldsymbol{s}_{\boldsymbol{u}}(\boldsymbol{x}, t; a) = \frac{d\boldsymbol{U}_b}{da}(\boldsymbol{x}, t; a) - \frac{\partial \boldsymbol{u}}{\partial x}\frac{\partial x_b}{\partial a} - \frac{\partial \boldsymbol{u}}{\partial y}\frac{\partial y_b}{\partial a} \quad \text{on } \Gamma_D. \tag{7}$$

The first term on the right hand side results from the possible dependence of the boundary condition $\boldsymbol{U}_b$ on a, while $\frac{\partial x_b}{\partial a}$ and $\frac{\partial y_b}{\partial a}$ are the geometric sensitivities of the boundary. For a value parameter the geometry of the domain does not depend on a, $\frac{\partial x_b}{\partial a}$ and $\frac{\partial y_b}{\partial a}$ vanish, and the boundary condition reduces to

$$\boldsymbol{s}_{\boldsymbol{u}}(\boldsymbol{x}, t; a) = \frac{\partial \boldsymbol{U}_b}{\partial a}(\boldsymbol{x}, t; a) \quad \text{on } \Gamma_D. \tag{8}$$

Neumann boundary conditions are obtained similarly by differentiation of Eq. (5). The flow equations and the CSE are solved by a finite element method. Time is discretized by either an implicit Crank-Nicholson or Euler scheme. Element matrices are constructed using a numerical Jacobian technique.

3 Numerical Results

First we perform a grid refinement study to verify the temporal and spatial accuracy of the flow and sensitivity solvers. The SEM is then applied to pulsatile flow around a square cylinder.

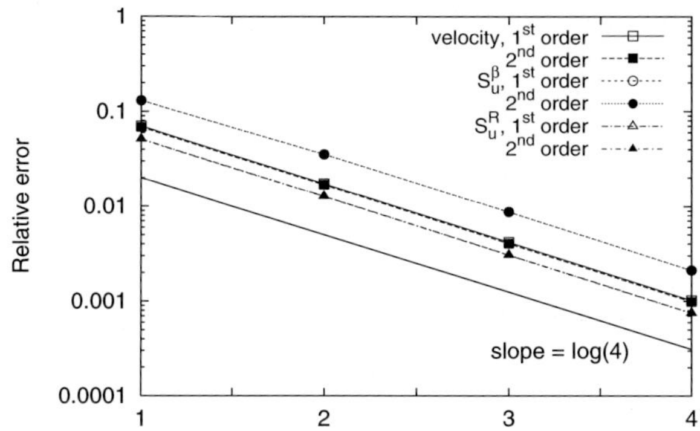

Fig. 1. Convergence of u and two sensitivity variables

3.1 Verification

We use the following expressions taken from the Stommel ocean flow model [4]

$$u(x, y) = \frac{\mathcal{F}}{R} \cos\left(\frac{\pi y}{b}\right) \left[C_1 e^{k_1 x} + C_2 e^{k_2 x} - 1\right] \tag{9}$$

$$v(x, y) = -\frac{b}{\pi} \frac{\mathcal{F}}{R} \sin\left(\frac{\pi y}{b}\right) \left[C_1 k_1 e^{k_1 x} + C_2 k_2 e^{k_2 x}\right] \tag{10}$$

$$p(x, y) = -2\frac{\mathcal{F}}{R} \cos\left(\frac{\pi y}{b}\right) \left[C_1 k_1 e^{k_1 x} + C_2 k_2 e^{k_2 x}\right], \tag{11}$$

with

$$k_{1,2} = -\frac{D\beta}{2R} \pm \sqrt{\left(\frac{D\beta}{2R}\right)^2 + \left(\frac{\pi}{b}\right)^2}, \tag{12}$$

and

$$\mathcal{F} = F \sin(\pi t). \tag{13}$$

A grid and time step convergence study is carried out for the flow and its sensitivities with respect to the Coriolis coefficient β and the friction coefficient R. The mesh and time step sequences are chosen so that the true error at the final time of the simulation to be reduced by a factor of 4 from one mesh to the next.

Fig. 1 shows the results for the velocity and its sensitivities with respect to β and R. The the flow and sensitivity solvers exhibit their theoretical rates of convergence: first-order in time for implicit Euler, and second-order accuracy for the implicit Crank-Nicholson scheme.

3.2 Pulsatile Flow Around a Square Cylinder

The computational domain and boundary conditions for this problem are described in Fig. 2. The inflow velocity varies in time according to:

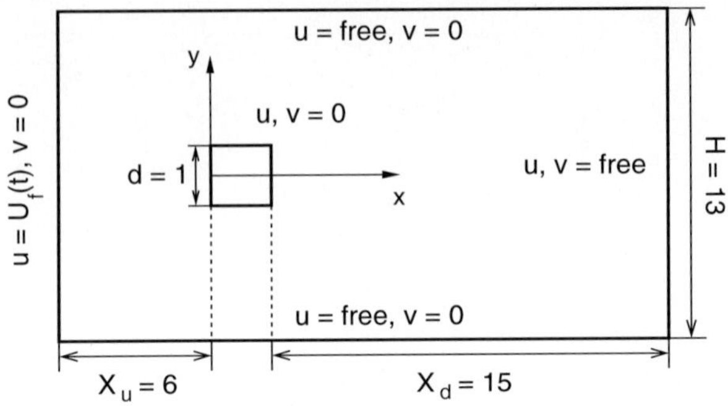

Fig. 2. Description of pulsed flow around a square cylinder

$$U_f(t) = U_0 \left(1 + \alpha \sin \frac{2\pi t}{T_\alpha}\right), \tag{14}$$

where U_0 is the time mean value of the free-stream velocity, α the amplitude of the sinusoidal variation with period T_α. These parameters are set to $U_0 = 1, \alpha = 0.4, T_\alpha = 4$. The initial conditions are obtained from a steady state solution of the flow and sensitivity equations. The Reynolds number $Re = \rho U_0 \, d/\mu$ is set to 100. Sensitivities are computed with respect to U_0, α, and T_α. The only non-zero boundary conditions for the sensitivities are those at the inlet.

The time response of the flow at a point one diameter downstream of the cylinder is shown in Fig. 3. The top curve is the free-stream variation while the bottom curves show the velocity and pressure signals. Initially symmetrical

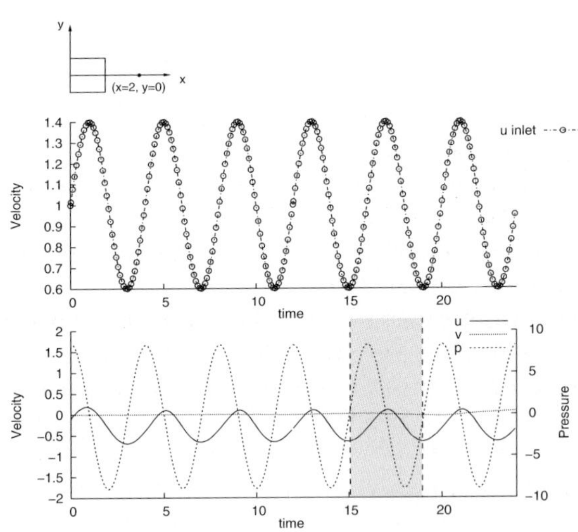

Fig. 3. Flow response at $(x = 2, y = 0)$

Fig. 4. Symmetric and non-symmetric vortex shedding

vortex shedding occurs. At time $t = 20$ the flow transitions to the usual Kaman vortex street is shown on Fig. 4.

The time signals of the flow sensitivities are shown in Fig. 5. Note that while the flow shows a harmonic response at nearly all times, sensitivities show deviation from this behavior at much earlier times. In this example, sensitivities foretell the transition from symmetric to asymmetric vortex shedding before it is visible in the flow signal.

4 Conclusion

A general sensitivity equation formulation was developed for two dimensional time-dependent incompressible laminar flows. The method was verified on a problem with a closed form solution to confirm the temporal and spatial rates of convergence of the solver. Results indicate optimal convergence rates for both the flow and its sensitivities.

The method was applied to pulsatile flow around a square cylinder. The flow starts with symmetrical vortex shedding and then goes through a transition phase leading to the usual Karman vortex street characterized by non-symmetric and alternate vortex shedding. The flow sensitivities are especially useful in predicting this flow transition. Indeed, signs of transition to asymmetrical vortex shedding appear much earlier in the sensitivity solution than in the flow variable themselves.

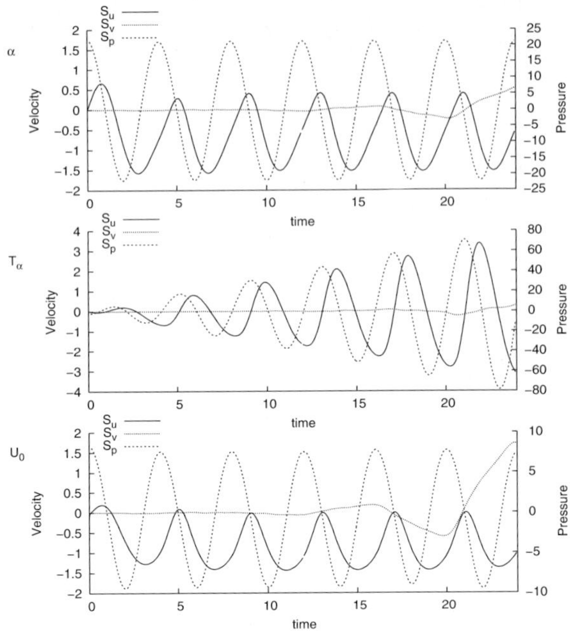

Fig. 5. Sensitivities at $(x = 2.0, y = 0.0)$

5 Acknowledgments

This work was sponsored in part by NSERC (Government of Canada), FCAR (Government of Québec), the Canada Research Chair Program and by the Air Force Office of Scientific Research under grants AFOSR F49620-00-1-0299 and F49620-03-1-0243.

References

1. É. Turgeon, D. Pelletier, and J. Borggaard. A general purpose sensitivity equation formulation for complex flows. In *Proceedings of the 8th Annual Conference of the Computational Fluid Dynamics Society of Canada*, volume 2, pages 697–704, June 11-13, 2000 / Montréal, Canada.
2. É. Turgeon, D. Pelletier, and J. Borggaard. A general continuous sensitivity equation formulation for complex flows. In *8th AIAA/NASA/USAF/ISSMO Symposium on Multidisciplinary Analysis and Optimization*, Long Beach, CA, Sep. 2000. AIAA Paper 2000-4732.
3. É. Turgeon, D. Pelletier, and J. Borggaard. A continuous sensitivity equation method for flows with temperature dependent properties. In *8th AIAA/NASA/USAF/ISSMO Symposium on Multidisciplinary Analysis and Optimization*, Long Beach, CA, Sep. 2000. AIAA Paper 2000-4821.
4. J. J. Von Schwind. *Geophysical Fluid Dynamics for Oceanographers.* Prentice-Hall, 1980.

Comparison of Artificial Compressibility Methods

Cetin Kiris[1], Jeffrey Housman[2], and Dochan Kwak[3]

[1] NASA Ames Research Center, Moffet Field, CA, ckiris@mail.arc.nasa.gov
[2] UC Davis, Davis, CA, jhousman@math.ucdavis.edu
[3] NASA Ames Research Center, Moffet Field, CA, dkwak@mail.arc.nasa.gov

Summary. Various artificial compressibility methods for calculating the three-dimensional incompressible Navier-Stokes equations are compared. Each method is described and numerical solutions to test problems are conducted. A comparison based on convergence behavior, accuracy, and robustness is given.

1 Introduction

The difficulty in computing solutions to the incompressible Navier-Stokes system of PDEs lies in satisfying the divergence-free velocity condition. Artificial compressibility methods, developed by A. Chorin [1], provide a mechanism to march in pseudo-time towards the divergence-free velocity field such that mass and momentum are conserved in the pseudo steady-state.

The classical artificial compressibility method transforms the mixed elliptic/parabolic type equations into a system of hyperbolic or parabolic equations in pseudo-time, which can be numerically integrated. The method has been generalized to curvilinear coordinates and used for various applications, i.e. Kiris et. al. [2].

Since the publication of Chorin's original paper many alternative forms of artificial compressibility have been developed. These methods include a generalized preconditioning matrix to equalize the wave speeds and the use of differential preconditioning, Turkel and Radespiel [3], as well as the addition of artificial viscosity such as an artificial Laplacian of pressure term in the continuity equation, Shen [4]. We present a direct comparison of four different versions of the artificial compressibility method on a series of test problems.

2 Artificial Compressibility Methods

The general form of the preconditioned incompressible, constant density and constant viscosity flow equations written in conservative form in generalized coordinates with the density absorbed by the non-dimensionalization of the pressure term are

$$\Gamma \frac{\partial}{\partial \tau} \left[I - \epsilon_\tau I_1 \Delta \right] \hat{Q} + I_m \frac{\partial \hat{Q}}{\partial t} + \frac{\partial(\hat{E}_1 - \hat{E}_{v1})}{\partial \xi_1} + \frac{\partial(\hat{E}_2 - \hat{E}_{v2})}{\partial \xi_2} + \frac{\partial(\hat{E}_3 - \hat{E}_{v3})}{\partial \xi_3} = \epsilon_x I_1 \Delta \hat{Q}, \quad (1)$$

where,

$$\hat{Q} = \frac{1}{J} \begin{bmatrix} p \\ u \\ v \\ w \end{bmatrix}, \quad I_m = \begin{bmatrix} 0000 \\ 0100 \\ 0010 \\ 0001 \end{bmatrix}, \quad I_1 = \begin{bmatrix} 1000 \\ 0000 \\ 0000 \\ 0000 \end{bmatrix}.$$

The advection and viscous terms, $\hat{E}_i$ and $\hat{E}_{vi}$, are defined in [2]. The preconditioning matrix Γ is introduced at the psuedo-time level to help advect the artificial pressure waves out of the domain. The constants ϵ_τ and ϵ_x control the amount of artificial dissipation in pseudo-time and space respectively. The parameters descibed above are chosen such that four distinct methods of artificial compressibility are tested.

2.1 Classical and Generalized Artificial Compressibility

The classical and generalized artificial compressibility methods are defined by setting $\epsilon_\tau = \epsilon_x = 0$ and $\Gamma = \Gamma_c$ for the classical method, and $\Gamma = \Gamma_p$ for the generalized method. Where,

$$\Gamma_c = \begin{bmatrix} \frac{1}{\beta}000 \\ 0100 \\ 0010 \\ 0001 \end{bmatrix}, \quad \Gamma_p = \begin{bmatrix} \frac{1}{\beta}000 \\ \frac{u}{\beta}100 \\ \frac{v}{\beta}010 \\ \frac{w}{\beta}001 \end{bmatrix},$$

and $\beta > 0$ is the artificial compressibility parameter.

2.2 Artificial Dissipation

To assist in the dissipation of spurious pressure waves introduced by the classical method, i.e. choosing $\Gamma = \Gamma_c$, we may choose $\epsilon_x > 0$ as in [4]. This term has the affect of adding a second difference artificial dissipation term to the continuity equation.

2.3 Differential Preconditioning

The artificial dissipation described above manipulates the physical dissipation properties of the PDE system. Alternatively we can manipulate the convective properties of the system. Following [3], we choose $\epsilon_\tau > 0$. This term will have an effect of propagating the low-frequency components of error more quickly than the high-frequency components which will be dissipated by the discretization scheme.

3 Numerical Results

The INS3D code [5] and [6] has been adapted to include each of the artificial compressibility methods described. An implicit line symmetric Gauss-Seidel relaxation scheme is used with fully-implicit boundary conditions. Iterations are performed until the residual of the nonlinear system has been reduced nine orders of magnitude in the l^2 norm. Results for $\beta = 1, 10, 100$ and $CFL = 1$ and $CFL = 1000$ are provided. For the artificial dissipation method ϵ_x values of $1.0^{-2}, 1.0^{-1}, 1.0^{+0}$ are used to scale the Laplacian term. The differential preconditioned method uses values of $\epsilon_\tau = 1.0^{-1}, 1.0^{+0}, 1.0^{+1}$. For each test-case the inlet velocity is specified and a constant pressure is enforced at the outlet. The classical artificial compressibility method is denoted by $P = 1$, the generalized version by $P = 2$, the artificial dissipation by $P = 3$, and the differential preconditioned by $P = 4$. Plots of residual convergence use the following symbols; $\circ$ $P = 1$; $\square$ $P = 2$; $\times$ $P = 3, \epsilon_x = 1.0^{-2}$; $+$ $P = 3, \epsilon_x = 1.0^{-1}$; $*$ $P = 3, \epsilon_x = 1.0^{+0}$; $\diamond$ $P = 4, \epsilon_\tau = 1.0^{-1}$; $\triangle$ $P = 4, \epsilon_\tau = 1.0^{+0}$; $\star$ $P = 4, \epsilon_\tau = 1.0^{+1}$.

Test 1: Inviscid Flow in a Square Duct

Each method is used to calculate the inviscid flow in a square duct with dimensions $10 \times 1 \times 1$ non-dimensional units. The exact solution is $Q = (0, 1, 0, 0)^T$. A grid of dimension $33 \times 9 \times 9$ is used. Table 1 displays the number of iterations required. Each of the methods computed the correct solution up to double precision. For $\beta = 1$ the generalized preconditioned method has the best convergence. For $\beta > 1$, the classical has the best convergence rate with the exception of $CFL = 1000$ where the differential preconditioning scheme is slightly better.

Table 1. Inviscid square duct: Number of iterations for residual reduction of 9 orders of magnitude in the discrete L^2 norm.

		P=1	P=2		P=3			P=4		
CFL	β			$\epsilon =$	0.01	0.10	1.00	0.10	1.00	10.0
1	1	220	209		325	1876	>9000	269	875	6992
	10	151	160		158	266	1874	165	309	2290
	100	242	261		243	250	328	246	291	1093
1000	1	212	202		314	1867	>9000	212	213	219
	10	137	137		145	255	1863	137	137	131
	100	238	255		239	242	304	238	238	238

Test 2: Viscous Flow in a Circular Pipe

A simple viscous flow in a circular pipe of radius one and length ten is computed. A Reynolds number of 1000 is used for which an exact solution is derived. Grids of dimension $17 \times 9 \times 9$, $33 \times 17 \times 17$, and $65 \times 33 \times 33$ are used.

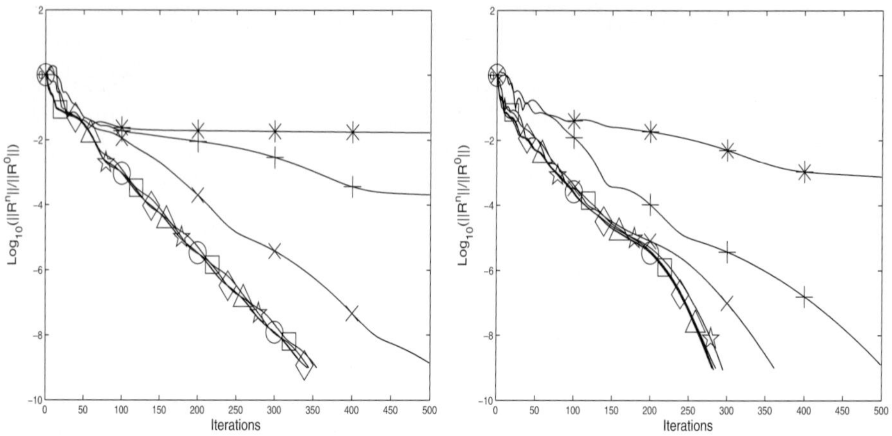

Fig. 1. Viscous Circular Pipe: Comparison of the convergence between preconditioning methods for $\beta = 1$(left) and $\beta = 10$ (right).

Each method is verified to produce second order accurate results for $\beta = 10$. Figure 1 plots the normalized l^2 residual for varying β and $CFL = 1000$. Table 2 displays the number of iterations required to converge on the finest mesh. The third and fourth methods fail to converge for low CFL numbers and the artificial dissipation scheme is especially sensitive to the ϵ parameter. For high CFL and high β the differential preconditioning becomes effective, but a moderate β must be used for accuracy purposes.

Table 2. Viscous Circular Pipe: Number of iterations for residual reduction of 9 orders of magnitude in the discrete L^2 norm.

		P=1	P=2		P=3			P=4		
CFL	β			$\epsilon =$	0.01	0.10	1.00	0.10	1.00	10.0
1	1	352	363		518	>2500	>2500	399	1233	>2500
	10	290	301		364	500	>2500	374	1577	>2500
	100	734	743		753	1044	1242	922	1130	>2500
1000	1	343	355		509	>2500	>2500	343	343	345
	10	281	286		362	503	>2500	281	283	295
	100	744	754		748	1030	1210	743	736	688

Test 3: Viscous Flow in an Square Duct with a 90° Bend

A more complicated viscous flow in a square duct with a 90° bend is used for the final test. The geometry used is described in Humphrey [7], where experimental results were obtained for Reynolds number 790. A grid of dimension $65 \times 33 \times 33$ is used. Figure 2 plot the residual for varying β and $CFL = 1000$. Table 3 displays the number of iterations required. The symbol $****$ denotes

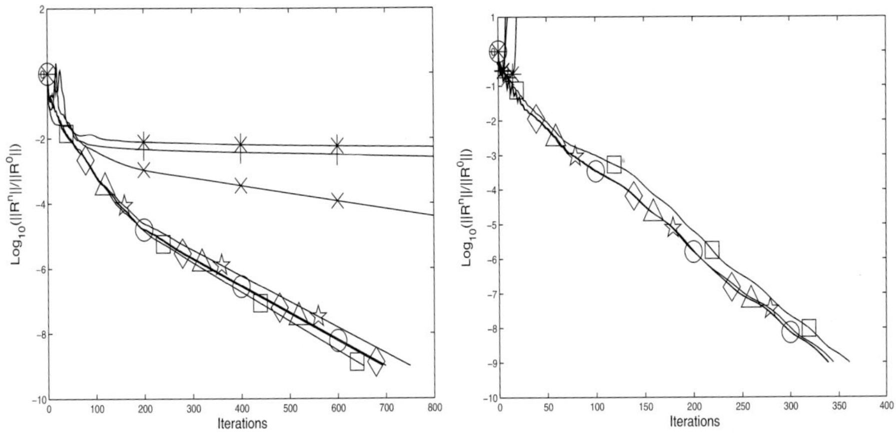

Fig. 2. Viscous Square Duct with 90° bend: Comparison of the convergence between preconditioning methods for $\beta = 1$ (left) and $\beta = 10$ (right).

the method failed to converge. Figure 3 displays a comparison of the different computed solutions with the experimental data for $\beta = 1, 10, 100$ at $\theta = 90^o$ plane of the curved duct. Robustness is an issue for the third and fourth schemes. Comparing the experimental data with the computed solutions we observe that using $\beta > 10$ leads to poor solution accuracy and $\beta = 1$ is the most accurate.

Table 3. Viscous Square Duct with 90° bend: Number of iterations for residual reduction of 9 orders of magnitude in the discrete L^2 norm.

		P=1	P=2		P=3			P=4		
CFL	β			$\epsilon =$	0.01	0.10	1.00	0.10	1.00	10.0
1	1	697	655		2764	5481	9999	1141	****	****
	10	341	365		****	****	****	388	****	****
	100	524	594		****	****	****	****	****	****
1000	1	693	653		2760	5483	9999	693	699	750
	10	338	361		****	****	****	338	339	344
	100	534	****		****	****	****	534	535	543

4 Conclusion

Four variations of the artificial compressibility method have been implemented and compared on a series of test problems. The classical and generalized artificial compressibility methods have similar convergence rates on each test case for each combination of the CFL and β parameters. The artificial dissipation and differential preconditioning methods lack of robustness

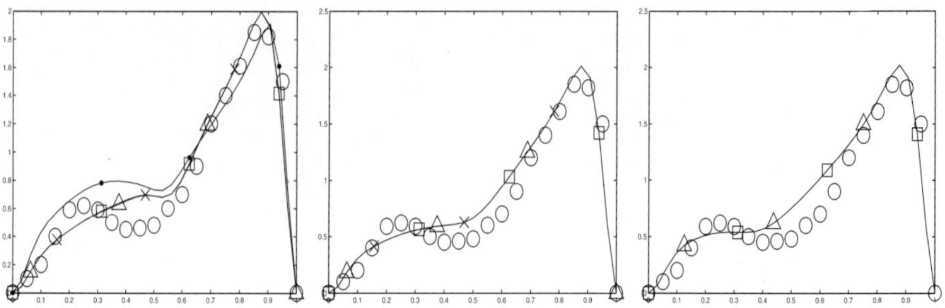

Fig. 3. Comparison of the solution at $\theta = 90^o$ and $\beta = 1$ (left), $\beta = 10$ (center), and $\beta = 100$ (right). $\circ$ *Exper.*; $\square$ $P = 1$; $\triangle$ $P = 2$; $\times$ $P = 3$; $\bullet$ $P = 4$.

and do not converge for certain parameter values. High values of β lead to poor accuracy for all the methods considered. For moderate values of β the classical and generalized methods appear to be the most accurate. These two versions will be evaluated for more complicated engineering applications.

References

1. A.J. Chorin: J. Comp. Phys. **2**, 12 (1967)
2. C. Kiris, D. Kwak and S. Rogers: Incompressible Navier-Stokes Solvers in Primitive Variables and Their Applications to Steady and Unsteady Flow. In: *Numerical Simulations of Incompressible Flows*, ed by M.M. Hafez (World Scientific Publishing, 2003) pp 3-34
3. E. Turkel and R. Radespiel: Preconditioning Methods for Multidimensional Aerodynamics. In: *VKI Lecture Notes*, 1997
4. Jie Shen: Pseudo-Compressibility Methods for the Unsteady Incompressible Navier-Stokes Equations. In *Beijing Symposium on Nonlinear Evolution Equations and Infinite Dynamical Systems*, ed by B. Guo (2003)
5. D. Kwak, J.L.C. Chang, S.P. Shanks and S.R. Chakravarthy: AIAA J. **24**, 3 (1986)
6. S.E. Rogers, D. Kwak and C. Kiris: AIAA J. **29**, 4 (1991)
7. J.A. Humphrey, A.M. Taylor and J. Whitelaw: J. Fluid Mech. **83**, 3 (1977)

Part XII

Magnetohydrodynamics

A Central, Diamond-Staggered Dual Cell, Finite Volume Method for Ideal Magnetohydrodynamics

P. Arminjon[1] and R. Touma[2]

[1] Universit de Montral arminjon@crm.umontreal.ca
[2] Universit de Montral touma@dms.umontreal.ca

1 Introduction

The adaptation of shock capturing numerical methods to Magnetohydrodynamics (MHD) has been a very dynamic and continuous process since the early eighties; Since MHD plays an important role in astrophysical flows, which are highly compressible, it was soon observed that Godunov-type methods might be a useful approach to solve these problems. Among many important contributions let us mention Brio and Wu [3], Dai and Woodward [2].

In this paper we extend central Nessyahu-Tadmor-type one-dimensional [6] or multi-dimensional finite volume schemes to the resolution of some problems in ideal (inviscid and non-resistive) compressible Magnetohydrodynamics (MHD). In MHD, the interaction between magnetic and fluid dynamic phenomena is described by a set of eight equations [1]: one mass conservation law, three momentum conservation laws, one energy conservation law and Faraday's (three-dimensional) law for the magnetic field $\mathbf{B}$.

$$\frac{\partial}{\partial t}\begin{bmatrix}\rho\\ \rho\mathbf{u}\\ \rho e\\ \mathbf{B}\end{bmatrix} + \nabla\cdot\begin{bmatrix}\rho\mathbf{u}\\ \rho\mathbf{u}\mathbf{u} + I(p + \frac{\mathbf{B}\cdot\mathbf{B}}{2\mu_o}) - \frac{\mathbf{B}\mathbf{B}}{\mu_o}\\ (\rho e + p + \frac{\mathbf{B}\cdot\mathbf{B}}{2\mu_o})\mathbf{u} - \frac{1}{\mu_o}(\mathbf{u}\cdot\mathbf{B})\mathbf{B}\\ \mathbf{u}\mathbf{B} - \mathbf{B}\mathbf{u}\end{bmatrix} = 0 \tag{1}$$

where $\rho, \mathbf{u}, p, \mathbf{B}$ and e are the mass density, velocity field vector, thermal pressure, magnetic field vector and the specific total energy. μ_o is the permeability of the vacuum and I is the (3×3) identity matrix. This system of equations is completed by the equation of state $p = (\gamma - 1)\rho\epsilon$, where γ is the ratio of specific heats and ϵ denotes the specific internal energy.

2 Two-Dimensional Cartesian Diamond-Staggered Scheme

We consider the two-dimensional hyperbolic system of conservation laws:

$$\vec{U}_t + \nabla\cdot\vec{F}(\vec{U}) \equiv \vec{U}_t + \vec{f}_x + \vec{g}_y = 0 \tag{2}$$

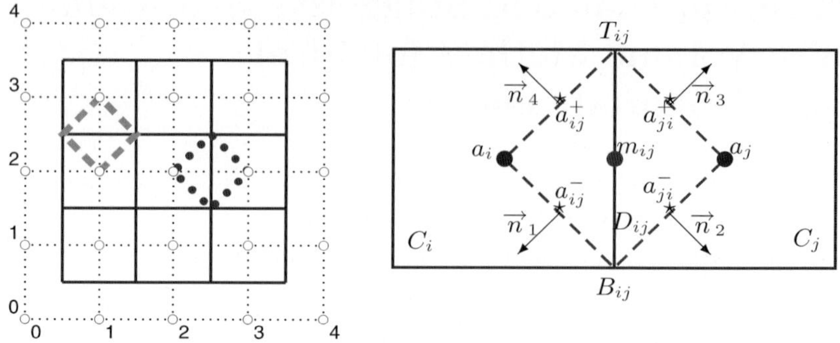

Fig. 1. (Left) Square cell centers denoted by 'o', dual oblique cells in red (dotted) and green (dashed); (Right) Original cells C_i, C_j and the dual cell D_{ij}; case when $a_i a_j$ is parallel to the x-axis

with the initial condition $\overrightarrow{U}(x, y, 0) = \overrightarrow{U_o}(x, y)$. We consider for our computational domain a uniform rectangular grid with M^2 square cells. Starting from the original Cartesian grid with cells C_i at time t^n, we alternate to the diamond dual cell D_{ij} at time t^{n+1}, and return back to the original cell C_i of the original structured grid as shown in Fig.1. Let $\Delta x = \Delta y = h = x_{i+1/2} - x_{i-1/2}$ denote the mesh size, $(x_i, y_j) = (ih, jh), 0 \le i, j \le M$, denote the nodes of the first grid. We consider a numbering of the nodes "a_i", $1 \le i \le (M+1)(M+1)$; for any arbitrary node a_i we consider the corresponding finite volume cell C_i for the first grid to be the square centered at a_i with sides parallel to the axes as in Fig.1. The diamond dual cells D_{ij} are shown in Fig.1. There are 2 cases depending on whether the axis joining the two nodes of the original grid used to define the diamond cell is parallel to the $x-$axis (the "i-j" or the "x-direction" case) or to the $y-$axis (the "i-k" or the "$y-$direction" case). In Fig.1 the dual cell (in the $x-$ *direction*) is the quadrilateral $a_i T_{ij} a_j B_{ij}$. Nodes of the staggered grid are the centroids of the diamond cells. Let $\overrightarrow{U}_i^n \cong \overrightarrow{U}(a_i, t^n)$ and $\overrightarrow{U}_{ij}^{n+1} \cong \overrightarrow{U}(m_{ij}, t^{n+1})$ denote respectively the average values in the first and second grid at time t^n and t^{n+1}. Performing the first time step gives $\overrightarrow{U}_{ij}^{n+1}$ (n even), while the cell values $\{\overrightarrow{U}_i^{n+2}\}$ are obtained at the end of the second time step.

As can be seen in Fig.1, the first time step should be performed by successively scanning in the x and y directions, in order to cover the whole computational domain. We consider here two adjacent nodes a_i and a_j, where $a_i a_j$ is parallel to the $x-$axis, and D_{ij} denotes the corresponding dual cell. Integrating (2) on $D_{ij} \times [t^n, t^{n+1}]$:

$$\int_{t^n}^{t^{n+1}} \iint_{D_{ij}} \overrightarrow{U}_t dA dt = - \int_{t^n}^{t^{n+1}} \iint_{D_{ij}} \nabla \cdot \overrightarrow{F} dA dt \tag{3}$$

and applying Green's theorem gives,

$$\iint_{D_{ij}} \vec{U}(x,y,t^{n+1})dA = \iint_{D_{ij}} \vec{U}(x,y,t^{n})dA - \int_{t^n}^{t^{n+1}} \oint_{\partial D_{ij}} (\vec{f}n_x + \vec{g}n_y)d\sigma dt \quad (4)$$

where $\vec{n} = (n_x, n_y)$ is the unit outward normal vector to the boundary of the diamond cell ∂D_{ij}. The left-hand side of (4) defines the value $A(D_{ij})\vec{U}_{ij}^{n+1}$.

The first integral in the RHS of (4) is approximated to second order by

$$\iint_{D_{ij}} \vec{U}(x,y,t^{n})dA \cong \vec{U}(x_i + \frac{\Delta x}{3}, y_i, t^n)A(D_{ij} \cap C_i) + \vec{U}(x_j - \frac{\Delta x}{3}, y_j, t^n)A(D_{ij} \cap C_j) \quad (5)$$

Applying van Leer's piecewise linear interpolants will guarantee second-order accuracy and preserve the monotonicity. Using (5) then leads to

$$\begin{aligned}\iint_{D_{ij}} \vec{U}(x,y,t^{n})dA &\cong \vec{U}_i^n A(D_{ij} \cap C_i) + \vec{U}_j^n A(D_{ij} \cap C_j) \\ &+ \frac{1}{3}\vec{U}_{i,x}^{lim} A(D_{ij} \cap C_i) - \frac{1}{3}\vec{U}_{j,x}^{lim} A(D_{ij} \cap C_j) \\ &= \frac{h^2}{4}(\vec{U}_i^n + \vec{U}_j^n + \frac{1}{3}\vec{U}_{i,x}^{lim} - \frac{1}{3}\vec{U}_{j,x}^{lim}) \quad (6)\end{aligned}$$

where $\nabla\vec{U}^{lim} = (\vec{U}_x^{lim}/\Delta x, \vec{U}_y^{lim}/\Delta y)$ is a limited numerical gradient. Using (6) and applying the midpoint rule to the flux-integral in (4) both in time and space finally leads to:

$$\begin{aligned}\vec{U}_{D_{ij}}^{n+1} &= \frac{1}{2}(\vec{U}_i^n + \vec{U}_j^n) + \frac{1}{6}(\vec{U}_{i,x}^{lim} - \vec{U}_{j,x}^{lim}) - \frac{\Delta t}{h}[(-\vec{f}_{a_{ij}^-}^{n+1/2} - \vec{g}_{a_{ij}^-}^{n+1/2}) \\ &+ (\vec{f}_{a_{ji}^-}^{n+1/2} - \vec{g}_{a_{ji}^-}^{n+1/2}) + (\vec{f}_{a_{ji}^+}^{n+1/2} + \vec{g}_{a_{ji}^+}^{n+1/2}) + (-\vec{f}_{a_{ij}^+}^{n+1/2} + \vec{g}_{a_{ij}^+}^{n+1/2})]\end{aligned}$$

where $(\vec{f}_{a_{ij}^\pm}^{n+1/2})$ are values predicted at time $t^{n+1/2}$, using (2) and an explicit Euler time discretization. For a description of the second time step, see [7].

3 The divB=0 Constraint and Numerical Experiments.

It is shown in electromagnetic theory that the magnetic field vector **B** must be solenoidal, and thus satisfy Maxwell's equation $\nabla \cdot \mathbf{B} = 0$. If the initial magnetic field satisfies the corresponding divergence-free constraint ($\nabla \cdot \mathbf{B}|_{t=0} = 0$), Faraday's law ensures that it remains divergence-free for all subsequent time. On the other hand, due to the accumulation of errors, the

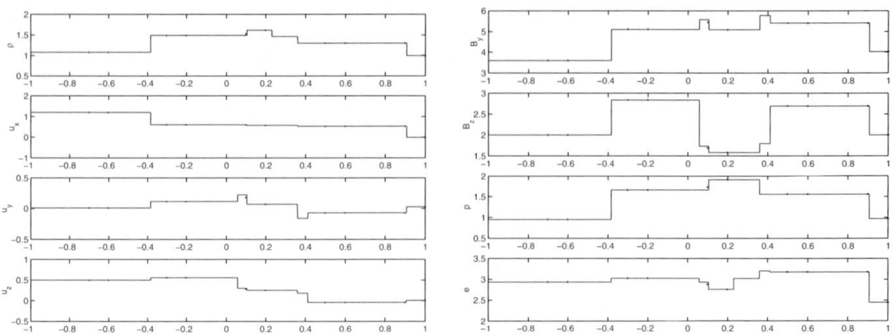

Fig. 2. One-dimensional MHD shock tube problem, MC-($\theta = 2$) limiter

numerical solution will tend to lose track of this constraint, leading to a non-solenoidal magnetic field, which in turn can lead to serious losses in accuracy, non-physical waves, and to the production of negative pressure and density in the ideal MHD case. Inspired from [4], we have constructed a method to enforce the $\nabla \cdot \mathbf{B} = 0$ constraint in the case of staggered Central Schemes, which we call the CTCS method ("Constrained Transport for Central Schemes"). See [7] for details. The CTCS applies indifferently to central schemes with Cartesian grids, 2 and 3-D diamond dual cells and Unstructured triangular or tetrahedric grids.

We first solved the 1-D MHD shock tube problem. We use the interval [-1,1] of the $x-$axis, let $\gamma = 5/3, B_x = 2$ and consider the initial data for the Riemann problem at $x = 0$, $U_r = [0.989112, -0.013123, 0.026933, 0.010037,$ $4.024421,$ $2.002600, 0.971588]$ and $U_l = [1.08, 1.2, 0.01, 0.5, 3.6, 2.0, 0.95]$ with $U = [\rho, u_x, u_y, u_z, B_y, B_z, p]$. Numerical results are compared with the exact solution of this problem presented in [3].This first test case features seven discontinuities. We have considered a grid with 1000 meshpoints. The solution is computed at time $t = 0.25$ along with a CFL condition of 0.485. Fig.2 shows a very good agreement between the numerical and exact solutions. We then solved a 2D-adaptation of the one-dimensional MHD shock tube problem, involving a compound wave, which consists of a shock and, directly attached, a rarefaction wave. The computational domain is the rectangle $-1 \leq x \leq 1, 0 \leq y \leq 1$. The initial conditions feature a shock along the axis $x = 0$ with the following data: $U_l = [1, 0, 0, 0, \sqrt{4\pi}, 0, 1]$, $Ur = [0.125, 0, 0, 0, -\sqrt{4\pi}, 0, 0.1]$, and $B_x = 0.75\sqrt{4\pi}$. The solution is computed at time $t = 0.25$ with a CFL condition of 0.485. The computations were performed with an MC-θ ($\theta = 1.5$) limiter. Fig.3 shows a good agreement of the numerical solution (400 data points dotted in blue) with the solution of the corresponding one-dimensional problem obtained from a Riemann solver (10000 data points, solid red line). For our second 2D problem, we consider a Riemann problem where the four initial states (ρ, p, u_x, u_y) are

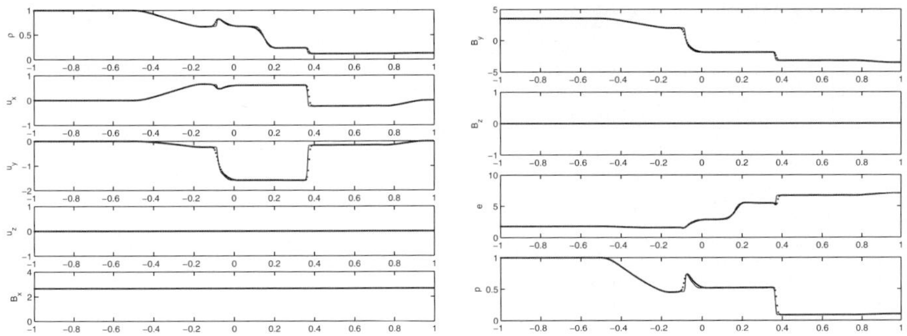

Fig. 3. Two-dimensional MHD shock tube problem, MC-($\theta = 1.5$) limiter

(1, 1, 0.75, 0.5) for $x > 0$ and $y > 0$, (2,1,0.75,0.5) for $x < 0$ and $y > 0$, (1,1,-0.75,0.5) for $x < 0$ and $y < 0$ and (3,1,-0.75,-0.5) for $x > 0$ and $y < 0$. We consider a uniform initial magnetic field $\mathbf{B} = (2, 0, 1)$. The solution is computed at time $t = 0.6175$ on a 300×300 grid. Fig.4 shows the contour lines for mass density and energy, respectively. Here again our numerical results are similar to those presented in [2]. For our final 2D test, we consider the Orszag-Tang MHD vortex problem. The initial conditions for our example are: $\rho(x, y) = \rho_0, p(x, y) = p_0\mathbf{u}(x, y) = -\sin(2\pi y)\mathbf{i} + \sin(\mathbf{2\pi x})\mathbf{j}, \mathbf{B(x, y)} = -\sin(\mathbf{2\pi y})\mathbf{i} + \sin(\mathbf{4\pi x})\mathbf{j}$ with $\rho_0 = 25/(36\pi)$ and $p_0 = 5/(12\pi)$. $\mathbf{i}$ and $\mathbf{j}$ are unit vectors in the $x-$ and $y-$ directions. Fig.5 shows the mass density contours at time $t = 0.5$ (left) and $t = 2$ (right).

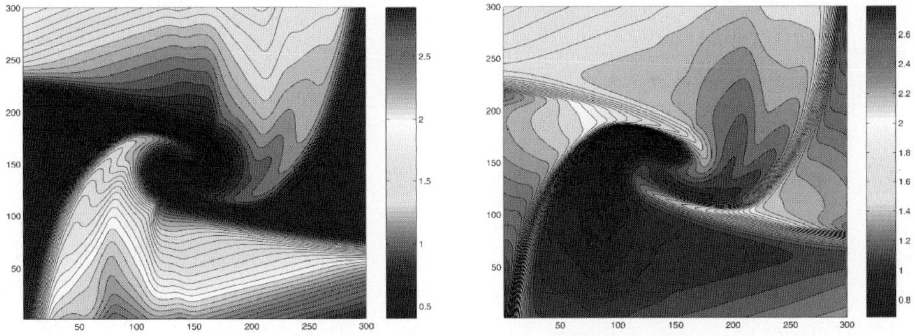

Fig. 4. Two-dimensional MHD Riemann problem

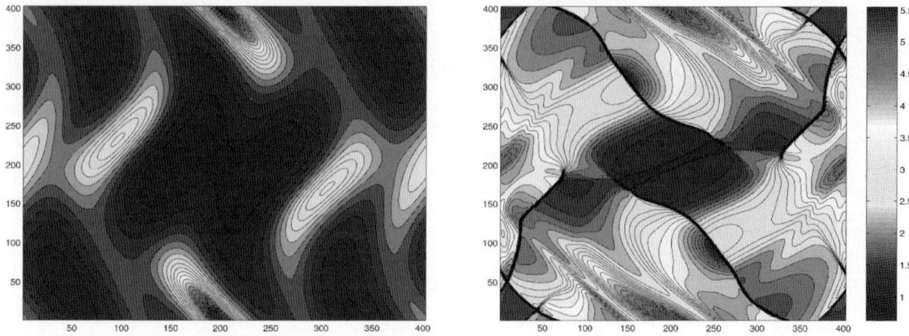

Fig. 5. Two-dimensional Orszag-Tang MHD turbulence problem, MC-(θ = 1.5) limiter

4 Conclusion

In this paper, we have extended our Nessyahu-Tadmor-type central finite volume methods to one and two-dimensional MHD problems, using Cartesian cells for the original grid, and our previously introduced diamond cells for the dual grid. The resolution of the Riemann problems at the cell interfaces is by-passed by central schemes, leading to computing time reductions. Our results show the high potential of the method which was robust enough to allow us to avoid, in a first attempt, the use of a strategy to maintain the div·**B**=0 constraint. For more elaborate numerical tests, we found it necessary to construct a method to satisfy this constraint, suitably designed to be applied in the context of our staggered central schemes. Using this method the constraint was satisfied to 10e-14 accuracy and we obtain results (Fig.5) identical with those appearing in the literature. For a description of our method to satisfy the div·**B**=0 constraint and more numerical tests see [7].

References

1. L. D. Landau, E. Lifshitz: *Electrodynamics of Continuous Media*, (Pergamon, New York, 1960)
2. W. Dai and P.R. Woodward: J. Comp. Phys. **142**, 331 (1998)
3. M. Brio, C.C. Wu: J. Comp. Phys. **75**, 400 (1988)
4. C. R. Evans and J. F. Hawley: Appl. Phys. Astrophys. J. **332**, 659 (1988)
5. G. Tóth: J. Comp. Phys. **161**, 605 (2000)
6. H. Nessyahu, E. Tadmor: J. Comp. Phys. 2 **87**, 408 (1990)
7. P. Arminjon and R. Touma: preprint submitted to App. Num. Math. (2004)
8. P. Arminjon and R. Touma: Central Finite Volume Methods for one and two-dimensional ideal magnetohydrodynamics. In: *12th Annual Conference of the CFD Society of Canada*, ed by S. Chen, S. Mcllwain, 585 (2004)

Simulation of Supersonic Flows in Inductively Coupled Plasma Tunnels

James R. Diebel[1], Thierry E. Magin[1], Marco Panesi[1], Pietro Rini[2], David Vanden Abeele[2], and Gérard Degrez[2]

[1] von Karman Institute for Fluid Dynamics, Belgium, magin@vki.ac.be
[2] Université Libre de Bruxelles, Belgium, gdegrez@ulb.ac.be

1 Introduction

High pressure inductively coupled plasma (ICP) sources are used in numerous applications, in particular for the testing of thermal protection materials used to protect space (re-)entry vehicles. The planetary gas is injected in a swirling, annular way into a heat-resistant quartz tube, surrounded by an inductor. A radio-frequency electric current runs through the inductor and induces a secondary current through the gas inside the tube, which heats up by means of Ohmic dissipation. Such an ICP torch (Fig. 1) typically produces a low

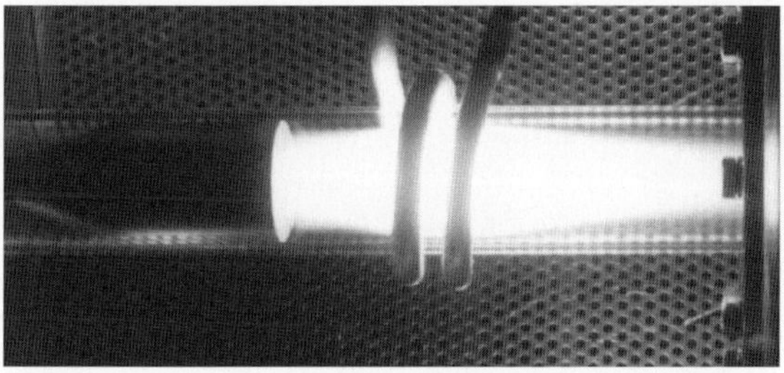

Fig. 1. VKI pilot ICP facility in operation

Mach number plasma jet, which can possibly be accelerated to supersonic flow conditions by a nozzle. Whereas numerical models for low Mach number inductively coupled plasma flows have been available for a long time, it is only recently that an integrated numerical model for a supersonic ICP facility has been proposed by Utyuzhnikov *et al.* [6]. A similar model is presented in the present contribution, which differs by some physical modelling as well as numerical aspects.

2 Governing Equations and Physical Models

The facility is modeled by a fully axisymmetric configuration. The equations of resistive magneto-hydrodynamics at low magnetic Reynolds number are solved assuming local thermodynamic equilibrium [7]. Although this is not rigorously correct [5], it is assumed that the plasma elemental composition is uniform. The temporal variations of the electromagnetic field can be represented by harmonic oscillations at the torch frequency. The associated oscillations of the gasdynamic quantities are small, whence the (time averaged) flow is governed by the steady axisymmetric Navier-Stokes equations (where x, y and z represent the axial, radial, and azimuthal coordinates respectively):

$$\frac{\partial(y\mathbf{F}^c)}{\partial x} + \frac{\partial(y\mathbf{G}^c)}{\partial y} = \frac{\partial(y\mathbf{F}^d)}{\partial x} + \frac{\partial(y\mathbf{G}^d)}{\partial y} + \mathbf{S} \tag{1}$$

in which

$$\mathbf{F}^c = (\rho u, \rho u^2 + p, \rho uv, \rho uw, \rho uH)^t \tag{2}$$

$$\mathbf{G}^c = (\rho v, \rho uv, \rho v^2 + p, \rho vw, \rho vH)^t \tag{3}$$

$$\mathbf{F}^d = (0, \tau_{xx}, \tau_{xy}, \tau_{xz}, u\tau_{xx} + v\tau_{xy} + w\tau_{xz} - \sum_{s=1}^{N_s} J_{xs}h_s - q_x)^t \tag{4}$$

$$\mathbf{G}^d = (0, \tau_{xy}, \tau_{yy}, \tau_{yz}, u\tau_{xy} + v\tau_{yy} + w\tau_{yz} - \sum_{s=1}^{N_s} J_{ys}h_s - q_y)^t \tag{5}$$

and the sources flux vector

$$\mathbf{S} = (0, yF_x, (p + \rho w^2 - \tau_{zz}) + yF_y, (\rho vw + \tau_{yz}), y\dot{Q})^t \tag{6}$$

includes time-averaged Lorentz forces (F_x, F_y) and Joule heating ($\dot{Q}$), as well as axisymmetric source terms. Because of the high temperatures in the facility, the Reynolds number is small and the flow is therefore assumed to be laminar.

Unlike Utyuzhnikov *et al.* [6] who use a quasi-1D approximation, in the present work, the EM field is computed using the more accurate 2D form of the MHD induction equation

$$\frac{\partial^2 E}{\partial x^2} + \frac{1}{y}\frac{\partial}{\partial y}\left(y\frac{\partial E}{\partial y}\right) - \frac{E}{y^2} - i2\pi\mu_0\sigma f E = -i\mu_0 2\pi f \delta(\boldsymbol{r} - \boldsymbol{r}_C) I_C \tag{7}$$

both inside the torch and on a far field domain covering the space surrounding the torch. The Dirac distribution in the latter equation represents the singular current density in the inductor rings assumed to be infinitely thin. To account for phase differences in the plasma, the electric field amplitude stands for a complex variable.

Thermodynamic properties of the mixture are computed using a semi-classical statistical mechanics formulation whereas the equilibrium mixture composition is derived from mass and charge conservation and equilibrium conditions of a minimal chemical reaction set [1].

Transport properties, including the electrical conductivity σ, are obtained by means of a perturbative Chapman-Enskog solution of the Boltzmann equation valid for partially ionized plasmas [4]. The resulting linear systems for the mixture shear viscosity and thermal conductivity, and the Stefan-Maxwell systems for the diffusive fluxes are solved using efficient transport algorithms based on iterative methods [4] .

3 Numerical method

3.1 Space discretization

The governing equations (1–7) are discretized using a structured multiblock cell-centered finite-volume method. The induction equation (7) is spatially discretized using a central scheme on a mesh that covers the inductive torch and surrounding space including the coil to an extent such that a vanishing electric field boundary condition can be applied on the outer boundary. The main advantage of this far-field approach is that it produces a sparse linear system that can be efficiently solved using modern iterative linear solvers, in contrast with the integral boundary conditions used in most ICP models [7].

The MHD induction equation (7) being linear, it is found convenient to decompose the electric field between the contribution of the coil rings E_V, which can be computed analytically using the Biot-Savart law, and the plasma-induced part E_P [7]. Substracting the equation for the contribution of the coil rings from Eqn. 7, one then obtains the induction equation for the plasma contribution

$$\frac{\partial^2 E_P}{\partial x^2} + \frac{1}{y}\frac{\partial}{\partial y}\left(y\frac{\partial E_P}{\partial y}\right) - \frac{E_P}{y^2} - i2\pi\mu_0\sigma f(E_P + E_V) = 0 \tag{8}$$

which is regular over the entire space despite the singularity in E_V at the location of the coil rings, because the electrical conductivity σ vanishes outside of the torch.

As far as the gas dynamics equations are concerned, the diffusive fluxes and source terms are centrally discretized, whereas the convective fluxes are computed using an upwind flux formula together with a one-dimensional MUSCL reconstruction for second order accuracy. In order to cope with the low flow Mach number in the torch, the preconditioned AUMS+(P) scheme proposed by Liou [3, 2] is selected in the present work, with modifications for an appropriate scaling of the dissipation in the low Reynolds number limit as proposed by Venkateswaran and Merkle [8]. A nice feature of this scheme is that it reduces to the pressure-stabilized incompressible scheme used by Vanden Abeele and Degrez [7] in the low Mach number limit.

3.2 Iterative Solution Strategy

In order to reduce the number of (costly) evaluations of thermodynamic and transport properties to a minimum, the discretized induction and flow field equations are solved in a loosely coupled manner using a backward Euler time stepping:

$$\sum_{j \in \mathcal{M}_{EM}} \left(\frac{\partial R_{EM,i}}{\partial E_{P,j}} \right)^n \delta E_{P,j} = -R^n_{EM,i} \tag{9}$$

$$\sum_{j \in \mathcal{M}_{NS}} \left[\frac{1}{\Delta t} \left(\frac{\partial Q}{\partial U} \right)^n_i \delta_{ij} + \left(\frac{\partial R_{NS,i}}{\partial U_j} \right)^n \right] \delta U_j = -R^n_{NS,i} \tag{10}$$

where $\partial Q/\partial U$ is the jacobian of the conservative variables $(\rho, \rho u, \rho v, \rho w, \rho E)$ with respect to the set of primitive variables $(\Delta p, \rho u, \rho v, \rho w, T)$ which are solved for.

The linear systems (9–10) are solved using the GMRES algorithm with BILU(k) preconditioning. The Navier-Stokes equations are solved coupled on all the blocks in order to achieve rapid convergence.

4 Numerical Results

To illustrate the capabilities of the code, simulations of the flow in the VKI pilot ICP facility equipped with a converging-diverging nozzle are presented. The torch diameter is 3 cm and the nozzle outlet diameter is 0.9 cm. The flow conditions are as follows: a 0.55 g/s air mass flow is injected at 10 kPa and ambient temperature in the torch. The coil current frequency is 27 MHz, and the electrical power dissipated in the plasma is 10 kW. Two back pressures corresponding respectively to underexpanded (1.5 kPa) and overexpanded (4.5 kPa) conditions have been considered.

The flow mesh, composed of three blocks corresponding respectively to the torch, nozzle and exhaust jet is shown on Fig. 2. The temperature and Mach

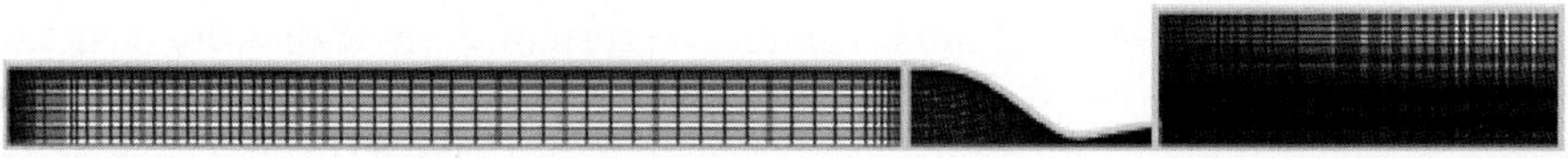

Fig. 2. Hydrodynamic multiblock mesh ($81 \times 31 + 110 \times 31 + 126 \times 89$)

number fields in the whole facility are shown in Fig 3 for the underexpanded case. The peak temperature exceeds 10000 K in the coil area. The flow is seen to be very slow in the torch ($M < 0.03$), then strongly accelerates in the nozzle to reach a Mach number around 2.4 at the nozzle exit. A close-up view

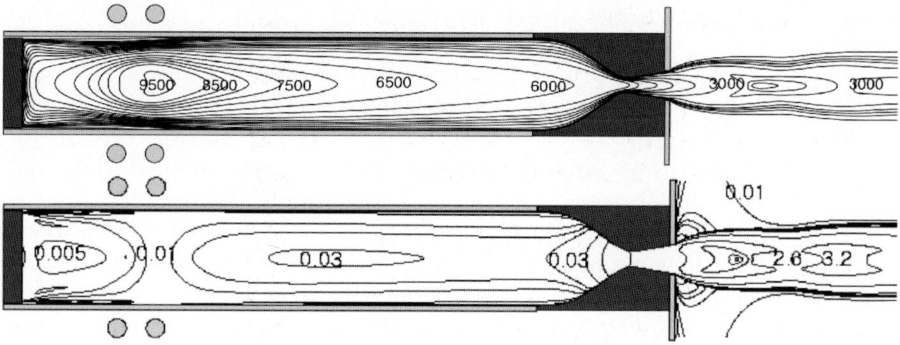

Fig. 3. Temperature [K] (above) and Mach number (below) fields in the torch, nozzle, and jet.

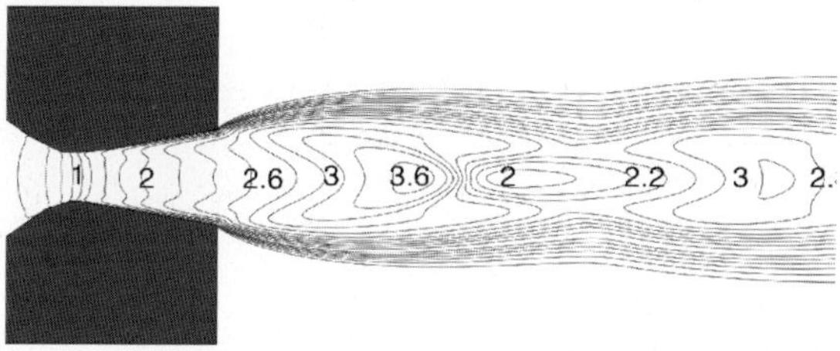

Fig. 4. Mach number in the jet downstream of the nozzle ($p_{\mathrm{back}} = 1.5$ kPa).

of the Mach number field in the nozzle and jet regions is shown on Fig. 4, in which the typical expansion-compression patterns are clearly visible.

The same close-up view for the overexpanded case is shown on Fig. 5, in which the expansion-compression patterns are seen to be inverted.

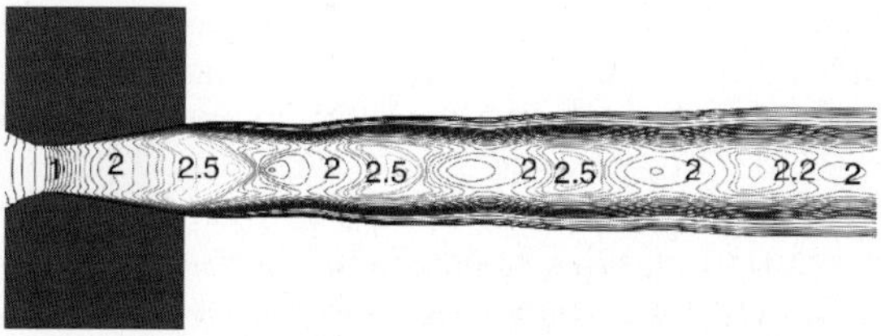

Fig. 5. Mach number in the jet downstream of the nozzle ($p_{\mathrm{back}} = 4.5$ kPa).

5 Conclusions

A numerical model for all speed ICP flows has been presented and applied to the VKI pilot ICP facility equipped with a converging-diverging nozzle producing a Mach 2.5 plasma jet. The main features of the model are the following:

- thermodynamic properties computed using a statistical mechanics formulation,
- transport properties evaluated from a perturbative Chapman-Enskog analyis, and computed using modern linear algebra algorithms,
- multiblock cell-centered finite volume formulation, with preconditioned AUSM+ flux formula for the convective terms in the Navier-Stokes equations,
- implicit time stepping/damped quasi-Newton iterative solution strategy with loose coupling of the electromagnetic and flow problems.

The numerical simulations demonstrate the ability of the solver to handle plasma flows with a combination of (extended) low speed regions and supersonic regions. It now remains to validate these encouraging preliminary results against other numerical simulations and experiments.

References

1. B. Bottin, D. Vanden Abeele, M. Carbonaro, G. Degrez, and G. S. R. Sarma. Thermodynamic and transport properties for inductive plasma modeling. *Journal of Thermophysics and Heat Transfer*, 13(3):343–3550, 1999.
2. M.-S. Liou. A further development of the AUSM+ scheme towards robust and accurate solutions for all speeds. In *16th AIAA Computational Fluid Dynamics Conference*, Orlando, Florida, 2003. AIAA Paper 2003-4116.
3. M.-S. Liou and J. R. Edwards. AUSM schemes and extensions for low Mach and multiphase flows. In *30th Computational Fluid Dynamics*, Rhode-St.-Genèse, Belgium, 1999. von Karman Institute for Fluid Dynamics. VKI LS 1999-03.
4. T. E. Magin and G. Degrez. Transport properties of partially ionized and unmagnetized plasmas. *Phys. Review E*, 2004. to be published.
5. P. Rini, D. Vanden Abeele, and G. Degrez. Elemental demixing in inductively coupled air plasmas at high pressures. In *37th Thermophysics and Heat Transfer Conference*, Portland, Oregon, June 2004. AIAA 2004-2742.
6. S. Utyuzhnikov, A. Konyukhov, D. Rudenko, S. Vasil'evskii, A. Kolesnikov, and O. Chazot. Simulation of sub- & supersonic flows in inductive plasmatrons. In *34th AIAA Plasmadynamics and Lasers Conference*, Orlando, Florida, 2003. AIAA Paper 2003-3866.
7. D. Vanden Abeele and G. Degrez. Efficient computational model for inductive plasma flows. *AIAA Journal*, 38(2):234–242, 2000.
8. S. Venkateswaran and L. Merkle. Analysis of preconditioning methods for the Euler and Navier-Stokes equations. In *30th Computational Fluid Dynamics*, Rhode-St.-Genèse, Belgium, 1999. von Karman Institute for Fluid Dynamics. VKI LS 1999-03.

Drift-Diffusion Model for Magneto-Fluid-Dynamics Interaction

J.S. Shang[1] and S.T. Surzhikov[2]

[1] Wright State University, Dayton Ohio, USA, jshang@cs.wright.edu
[2] Russian Academy of Science, Moscow, Russia, surg@ipmnet.ru

1 Introduction

The aerodynamic performance of most aerospace vehicles has reached a level of sophistication in which any additional improvement must be derived from new physical dimensions beyond traditional fluid dynamics [1-3]. An effective physical mechanism in interdisciplinary aerodynamics is the electromagnetic force. However, this environment requires an electrically conducting medium where the Lorentz force and Joule heating are present. In most hypersonic flights, the bow shock compresses and bounds air mixture in the shock layer to attain a weakly ionized state. Therefore, the magneto-fluid-dynamics (MFD) is closely associated with hypersonic flows.

In hypersonic flow, a well-known viscous-inviscid interaction always occurs at the leading edge of a flight vehicle. A shock wave will be generated by the outward displacement of boundary layer over a solid surface. The induced pressure is proportional to the pressure interaction parameter $\overline{\chi} = M^3(C/Rx)^{1/2}$ [4]. When an electromagnetic force is exerted to modify the boundary-layer structure, this dynamic event may lead to a revolutionary magneto-fluid-dynamics interaction for flow control. The MFD interaction together with a nonlinear aerodynamic amplification is equally applicable either as a virtual control surface or a variable-geometry hypersonic inlet in a combined cycle propulsive system. In principle, a series of segmented, gas discharge electrodes in a hypersonic inlet can produce a nearly isentropic compression that is impossible to achieve by pure aerodynamic mechanism alone [1].

In the present effort, the magneto-aerodynamic interaction over a sharp leading edge plate is demonstrated to be a plasma-based flow control mechanism. The displacement thickness of the boundary layer is modified by a glow discharge emitted from embedded electrodes. The ensuing viscous-inviscid interaction generates a pressure rise equivalent to a movable control surface. The numerical simulation is accomplished by solving the coupled MFD and electro-dynamic equations and has been verified by experimental observations.

2 Approach

In computational magnetohydrodynamics, complicate eight eigenvalues were considered and led to a very complex formulation and computational procedure [5]. For most aerospace applications the magnetic Reynolds number, $Re_m = \mu_m \sigma u L$, is always much less than unity the Lorentz force and the Joule heating can be treated as source terms in the Navier-Stokes equations; the formulation is significantly simplified [6]. However, the most complex and challenging issue in MFD simulation is the determination of the transport property of the plasma [7]. Since the electrical conductivity of the fluid medium is mostly by the electron collision ionization, the formulation of high-temperature plasma does not apply. From the most recent research result, a drift-diffusion glow discharge model has successfully simulated gas discharge phenomena and even included an externally applied magnetic field [8,9]. In the present approach, the charge conservation, magneto-aerodynamics, electrodynamics, and external circuit equations are coupled to ensure the self-consistent electromagnetic simulations.

The basic formulation of drift-diffusion theory is based on the two dominant mechanisms in plasma to describe the electro-dynamic motion of charged particles [7,8]. For the present purpose, the charge conservation equation is adopted to describe the electrically conducting medium to appear as:

$$\frac{\partial n_e}{\partial t} - \nabla \cdot (D_e \nabla n_e + n_e \mu_e E) = \alpha(E,p)|D_e \nabla n_e + n_e \mu_e E| - \beta n_e n_+ \quad (1)$$

$$\frac{\partial n_+}{\partial t} + \nabla \cdot (D_+ \nabla n_+ + n_+ \mu_+ E) = \alpha(E,p)|D_e \nabla n_e + n_e \mu_e E| - \beta n_e n_+ \quad (2)$$

$$\nabla^2 \phi = 4\pi e(n_e - n_i) \quad (3)$$

where α and β are the Townsend discharge and recombination coefficients. D_e, D_+ are the electron and ion diffusion coefficients, and μ_e, μ_+ are the mobility of the electron and ion respectively [7]. Further simplification of the species conservation equations (1.1)- (1.3) is possible for the quasi-neutral plasma ($n_e = n_i$) [9]. The complete formulation of the MFD equations at the low magnetic Reynolds number can be given as [6]:

$$\frac{\partial \rho}{\partial t} + \nabla \cdot (\rho U) = 0 \quad (4)$$

$$\frac{\partial \rho U}{\partial t} + \nabla \cdot \left(\rho \overline{\overline{UU}} + p\overline{\overline{I}} - \frac{1}{Re}\overline{\overline{\tau}} \right) = S Re_m \left(J \times B \right) \quad (5)$$

$$\frac{\partial \rho e}{\partial t} + \nabla \cdot \left[(\rho e + p) U - \frac{1}{Re} \left(U \cdot \overline{\overline{\tau}} \right) - \frac{Q}{(\gamma - 1) Pr Re M^2} \right] = S Re_m \left(E \cdot J \right) \quad (6)$$

where the MFD interaction parameter is defined as S $= \sigma B^2 L / \rho u$ for the low Hall parameter codition. Both the MFD and the electro-dynamic equations can be cast into the flux vector form [8,9].

Since the wave speeds of the governing equations system cover an extraordinary range from the sonic speed to speed of light, a loosely coupled solving scheme was developed for the magneto-aerodynamics and electro-dynamics equations. The energy equation in terms of temperature, the Gauss's law for electrical potential, and the species concentration equations are decoupled from the rest and solved by the five-point successive line over-relaxation (SLOR) scheme:

$$A_{i.j}U^{n+1}_{i-1,j} + B_{i,j}U^{n+1}_{i+1,j} + A^{*}_{i,j}U^{n+1}_{i,j-1} + B^{*}_{i,j}U^{n+1}_{i,j+1} - C_{i,j}U^{n+1}_{i,j} = R_{i,j} \quad (7)$$

where $U = U(n_e, n_+, \phi, T)$, all coefficients of polynomials can be found in references 8 and 9.

The initial values boundary conditions for the governing equation system are straightforward. For the fluid dynamic variables, the no-slip condition for the velocity components, the vanishing pressure pressure gradient condition, as well as the prescribed surface temperature are imposed on the solid surface. For the electromagnetic variables, the normal component of the electric field must be balanced by the surface charge density at the interface. To meet these conditions on cathode and anode surfaces, the invariant gradients of charge number density are approximated for both. The full electric potential is imposed on the anode in reference to the cathode.

3 Discussion of Results

The present numerical simulation duplicates the experiment in a Mach 5.15 plasma channel. The flow field is characterized by a static pressure of 78.4 Pa, a temperature of 43 K, and streamwise velocity of 675.5 m/s. Under these condition, the Reynolds number per meter is 1.625x10^5. The surface plasma is generated by a Universal Voltronic DC power supply that can deliver a current of 800 mA up to 10 kV.

The relatively small model has the physical dimension of 38x67 mm and the electrodes dimension of 30.5x0.6 mm. The cathode is placed parallel to the leading edge and at a distance of 4.8 mm downstream. The anode is embedded farther downstream and the distance between electrode is 37.5 mm. A series of grid refinements for the viscous-inviscid interaction was performed. It was found that the numerical results were essentially grid independent with a judiciously mesh clustering in the high gradient domains. All numerical results were generated on a (500x70) grid system with high grid density at the leading edges of the plate and electrodes.

The DC glow discharge over the plate is displayed in Figure 1. The surface plasma after the initial breakdown is maintained by an electric potential of 1.2 kV and a current of 50 mA. The total power required is merely 60 Watts. The dominant and visible emission is over the cathode region. This distinctive feature is the consequence of a layer of positive space charge at the cathode with a substantial drop of electrical potential.

The ion number density contour based on the drift-diffusion theory is depicted in Figure 2. In the present side-by-side electrode arrangement, the positive column between electrodes is not clearly identifiable. However, the high concentration of ions is noted over the electrodes and correctly reflects the existence of plasma sheath.

The verification of numerical results with experimental data is presented in Figure 3. The data of the charge particle number density over the electrodes was measured by a double Langmuir probe. Based on the channel flow condition, the Debye length is estimated around 7 μm and the probe radius is 250 μm, which means the collision process is significant. The data were collected over a vertical distances from 0.32 cm to 1.27 cm above the plate and the number density spanned a range from 7.6x10^9 to 1.8x10^{12} per cc [10]. In general, the data over the cathode is nearly an order of magnitude greater than the computed values and reaches a good agreement over the anode. In view of the current measuring and computing uncertainty in hypersonic flow condition, The disparity between results reflects the state-of-the-art in plasma modeling.

Figure 4 gives three surface pressure distributions over the plate by the pressure interaction, electrode and Joule heating. The electrode heating is a component of the resistive heating by the surface discharge. To illustrate the fact that the additional induced surface pressure rise is truly a consequence of magneto-aerodynamic interaction, the separated electrode heating is simulated by increasing the electrode surface temperature identical to that of surface discharge, $T_e = 600K$. Indeed, the Joule heating consists of both the volumetric heating and conductive heat transfer from electrodes. The Joule heating shifts the velocity outward and decreases the density in the boundary layer to produce a greater growth rate of the displacement thickness. The ensuing viscous-inviscid interaction produces a pressure rise, as if the plate has deflected one degree from the Mach 5 free stream. This control effectiveness can be scaled by the power requirement as 20 W/cm per degree.

The Lorentz force also exerts a strong effect to the magneto-aerodynamic interaction. An applied magnetic field transverse to the electrical current accelerates the charged particles toward or away from the electrodes. The momentum exchanging collision between the heavy ion and neutral particles has shown to further intensify or suppress the viscous-inviscid interaction. These phenomena have been observed by both the present computations and experiments [9,10].

4 Concluding Remarks

The drift-diffusion model for the low-temperature, nonequilibrium plasma has captured all essential features of a glow discharge. The numerical results duplicate the global behavior of a hypersonic magneto-aerodynamic interaction. The basic model is equally applicable to all flow regions.

The magneto-aerodynamic interaction initiated by a surface plasma and ensued by viscous-inviscid interaction has been demonstrated to be a potential plasma-based flow control mechanism. The control effectiveness is illustrated by the low power input to achieve positive response without any moving components.

Fig. 1. Gas discharge over wedge in hypersonic stream

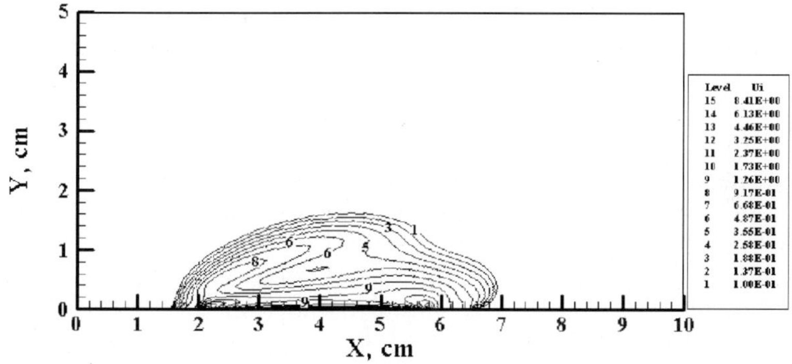

Fig. 2. Ion number density contour over plate, $10^{10}/cm^{-3}$

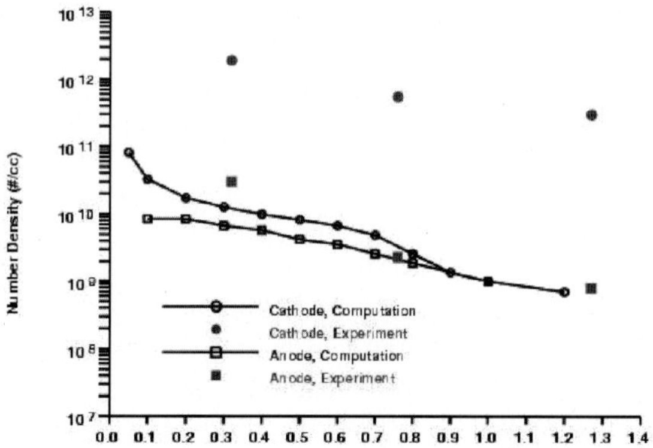

Fig. 3. Ion number density distributions above electrodes

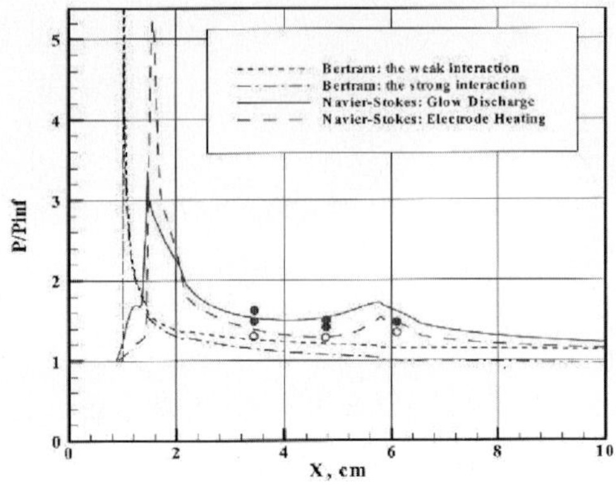

Fig. 4. Surface pressure by magneto-aerodynamic interaction

Acknowledgment

Authors gratefully acknowledge the support of the U.S. Air Force Research Laboratory, Russian Academy of Science, and Wright State University.

References

1. E.L. Resler, W.R. Sears, The Prospect for Magneto-aerodynamics, J. Aero. Science 1958, Vol. 25, 1958, pp. 235-245 and 258.

2. J.S.Shang, Recent Research in Magneto-Aerodynamics, Progress in Aerospace Sciences, Vol. 31, 2001, pp. 1-20.

3. J.S. Shang, Historical Perspective of Magneto-Fluid-Dynamics, Introduction to Magneto-Fluid-Dynamics for Aerospace Applications, Lecture Series 2004-01, von Karman Institute for Fluid Dynamics, Rhode-ST-Genese Belgium, Oct. 2003.

4. W. Hayes and R. Probstein, Hypersonic Flow Theory, Academic Press, New York, 1959, pp. 333-374.

5. M. Brio and C.C. Wu, An Upwind Differencing Scheme for the Equations of Ideal Magnetohydrodynamics, JCP Vol. 75, 1988, pp. 400-422.

6. M. Mitchner and C. Kruger, Partial Ionized Gases, John Wiley & Sons, New York, 1973.

7. Yu. P. Raizer and S.T. Surzhikov, Two-dimensional Structure of the Normal Glow Discharge and the Role of Diffusion in Forming of Cathode and Anode Current Spots, High Temperatures, No. 3, Vol. 26, 1988.

8. S.T. Surzhikov and J.S. Shang, Glow Discharge in Magnetic Field with Heating of Neutral Gas, AIAA 2003-3654, Orlando FL., June 2003.

9. J.S. Shang and S.T. Surzhikov, Magneto-Fluid-Dynamics Interaction for Hypersonic Flow Control, AIAA 2004-0508, Reno NV, 4-8 January 2004.

10. R. Kimmel, J. Hayes, J. Menart, and J.S. Shang, Effect of Surface Plasma Discharges on Boundary Layers at Mach 5, AIAA 2004-0509, Reno NV.,January 2004.

Part XIII

Meshless Methods

Gridless Computation Using the Unified Coordinates

W.H. Hui, J.J. Hu, and G.P. Zhao

Hong Kong University of Science & Technology

1 Introduction

A central problem in CFD is flow past a body, e.g. an airplane. Most CFD work use Eulerian coordinates, which require generating a body-fitted grid prior to flow filed computation. Despite three decades of research, grid generation is still a bottleneck of CFD, as it is time consuming, tedious and requires specialized training. It will be shown in this paper that using the Unified Coordinates introduced by Hui *et. al* [1, 2], there is no need for grid generation prior to flow computations; the grid is automatically generated while computing the flow. This greatly saves computing time. For simplicity, we consider only 2-D and axisymmetric flow.

2 Steady Flow

2.1 Special case: Supersonic/Hypersonic Flow – Space Marching Method

In this case the Euler equations of motion for a γ-law gas in the Eulerian coordinates (x, y) are ($\alpha = 0$ for 2-D and $\alpha = 1$ for axisymmetric flow):

$$\frac{\partial F_e}{\partial x} + \frac{\partial G_e}{\partial y} + H_e = 0 \tag{1}$$

where $F_e = \left[\rho u, \rho u^2 + p, \rho uv, u(\rho e + p)\right]^T$, $G_e = \left[\rho v, \rho uv, \rho v^2 + p, v(\rho e + p)\right]^T$, $H_e = \alpha\left[\rho v, \rho uv, \rho v^2, v(\rho e + p)\right]^T / y$. In (1), p is pressure, ρ density, u and v are the x- and y-component of fluid velocity $\mathbf{q}$ (or the axial- and transverse component of $\mathbf{q}$ in the axisymmetric flow case), and $e = \frac{1}{2}(u^2 + v^2) + \frac{1}{\gamma - 1}\frac{p}{\rho}$.

The unified coordinates (λ, ξ) are given via the transformation: $dx = hud\lambda + Ad\xi$ and $dy = hvd\lambda + Bd\xi$, where ξ is a stream function and the arbitrary function $h(\lambda, \xi)$ is chosen so that the grid $\lambda\xi$ is orthogonal[3], yielding $hq = \exp\int^{\xi} \frac{1}{\rho q^2}\frac{\partial p}{\partial \xi} d\xi$. These coordinates are optimal.

Equation (1) is transformed to the unified coordinates $\lambda\xi$, and we use the shock-adaptive Godunov scheme [3, 4] to get

$$\frac{\partial E}{\partial \lambda} + \frac{\partial F}{\partial \xi} + S = 0 \tag{2}$$

where $E = [K, H, p/\rho^\gamma, Ku + pB, Kv + vpB/u]^T$, $F = [0, 0, 0, -pv, pu]^T$, and $S = \frac{\alpha hKv}{y} [1, 0, 0, u, v]^T$ with $K = \rho(uB - vA)$, $H = e + \frac{p}{\rho}$, $uA + vB = 0$.

For supersonic/hypersonic steady flow, Eq. (2) is hyperbolic and is solved by marching in the space coordinate λ with the fractional step method to account for the source term S. The computing procedure is as follows: To compute a steady flow $\mathbf{q}_\infty$ past a body, we start at $\lambda = 0$ with a column of cells in the η direction. This initial column of cells will move with velocity $h\mathbf{q}_\infty$. After m time-steps when the distance it travels is equal to the grid size, we add a new column of initial cells on the left. We repeat this process. When the leading column of cells reach the body, the boundary condition of zero normal velocity is imposed there. This changes the grid but it remains orthogonal as guaranteed by the choice of h given above.

2.2 General Case – Time Marching Method

In this case the 2-D Euler equations in the Eulerian coordinates are

$$\frac{\partial E_e}{\partial t} + \frac{\partial F_e}{\partial x} + \frac{\partial G_e}{\partial y} = 0 \tag{3}$$

where $E_e = [\rho, \rho u, \rho v, \rho e]^T$. The unified coordinates (λ, ξ, η) are given via the transformation $dt = d\lambda, dx = hud\lambda + Ad\xi + Ld\eta$, and $dy = hvd\lambda + Bd\xi + Md\eta$.

It is shown [1] that coordinates ξ and η move with velocity $h\mathbf{q}$, hence in the steady state one set of coordinates, $\eta =$ const. say, coincide with streamlines. In particular, the body surface coincides with a coordinate line of the unified coordinates. So the unified grid, which is flow generated, is a body-fitted grid.

Eq. (3) is transformed to the unified coordinates [1]. For general steady flow that may have subsonic region, the transformed equations are solved by marching in the time variable λ until a steady state is reached. At each time step, the arbitrary function h is determined so that the grid angles are preserved during the λ-marching [1]. In particular, if we start with an orthogonal grid, it remains orthogonal for all time.

3 Unsteady Flow

In this case the unified coordinates (λ, ξ, η) are given via the transformation $dt = d\lambda, dx = hud\lambda + Ad\xi + Ld\eta$, and $dy = kvd\lambda + Bd\xi + Md\eta$. Under this transformation, Eq. (3) become

$$\frac{\partial E}{\partial \lambda} + \frac{\partial F}{\partial \xi} + \frac{\partial G}{\partial \eta} = 0 \tag{4}$$

where $E = [\rho\Delta, \rho\Delta u, \rho\Delta v, \rho\Delta e, A, B, L, M]^T$, $F = [\rho X, \rho Xu + pM, \rho X - pL, \rho Xe + p(uM - vL), -hu, -kv, 0, 0]^T$ and $G = [\rho Y, \rho Yu - pB, \rho Y + pA, \rho Ye +$

$p(vA-uB), 0, 0, -hu, -kv]^T$ with $\Delta = AM - BL, X = (1-h)uM - (1-k)vL$ and $Y = (1-k)vA - (1-h)uB$.

To compute time-accurate unsteady flow, such as that around an oscillating airfoil, Eq. (4) is solved by marching in time λ. At each time step the arbitrary functions h and k are determined to satisfy two requirements:

(a) the coordinate line $\eta =$ const. shall coincide with a material line of a fluid particle, yielding $(1-k)vA = (1-h)uB$. As the body surface is a material line, this condition guarantees that the unified grid, which is generated by the flow, is a body-fitted grid at all time.

(b) the grid angles shall be preserved, yielding $\frac{\partial h}{\partial \eta} + P(\eta; \lambda, \eta)h = Q(\eta; \lambda, \xi)$, where $P(\eta; \lambda, \xi) = \frac{S^2}{T^2\Delta}\left(A\frac{\partial B}{\partial \xi} - B\frac{\partial A}{\partial \xi}\right) - \frac{L}{A\Delta}\left(A\frac{\partial B}{\partial \eta} - B\frac{\partial A}{\partial \eta}\right)$, $Q(\eta; \lambda, \xi) = -\frac{S^2 A}{T^2\Delta}\left(A\frac{\partial v}{\partial \xi} - B\frac{\partial u}{\partial \xi}\right)\frac{1}{u} - \frac{S^2}{T^2\Delta}\left(A\frac{\partial B}{\partial \xi} - B\frac{\partial A}{\partial \xi}\right) + \frac{L}{\Delta}\left(A\frac{\partial v}{\partial \eta} - B\frac{\partial u}{\partial \eta}\right)\frac{1}{u} - \frac{L}{A\Delta}\left(A\frac{\partial B}{\partial \eta} - B\frac{\partial A}{\partial \eta}\right)$, $S^2 = L^2 + M^2, \quad T^2 = A^2 + B^2$.

4 Examples

Example 1. Steady supersonic flow, $M_\infty = 1.745$, past a diamond-shape airfoil with 15° vertex angle placed at an angle of attach of 10° (Fig. 1a). To solve Eq. (2), a space-marching method is used. The flow-generated grids at different λ are plotted in Figs. 1b to 1f. The computed surface Mach numbers are plotted in Fig. 1g which are in complete agreements with the exact solution. The whole computation takes 1.8 second on a(P4, 2.8GHz) PC. In Eulerian computation, it takes 2,393 seconds after grid generation,

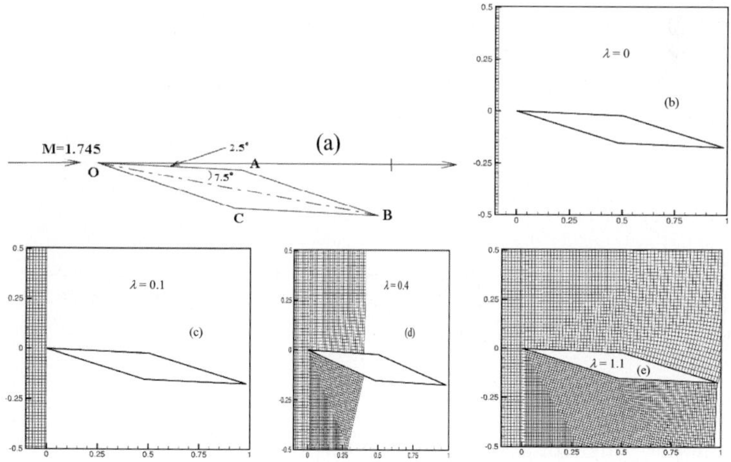

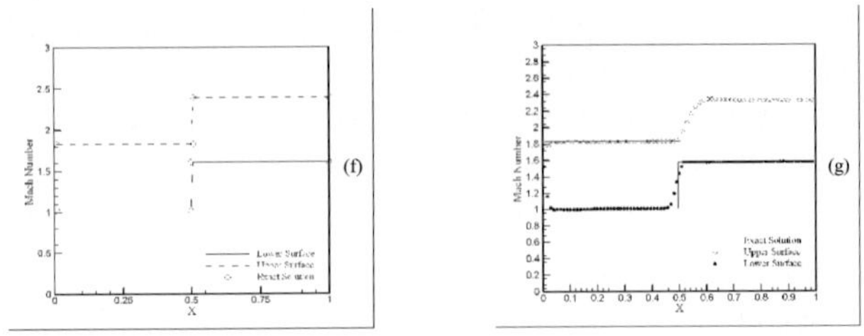

Fig. 1. Steady supersonic flow past a diamond-shape airfoil by space-marching method. (a) sketch, (b) – (e) flow generated grids, (f) surface Mach number, 120 cells. Computing time: 1.8s (P4, 2.8 GHz) (g) Eulerian computation (5th order WEND scheme), 100×200 cells, $2,395$s

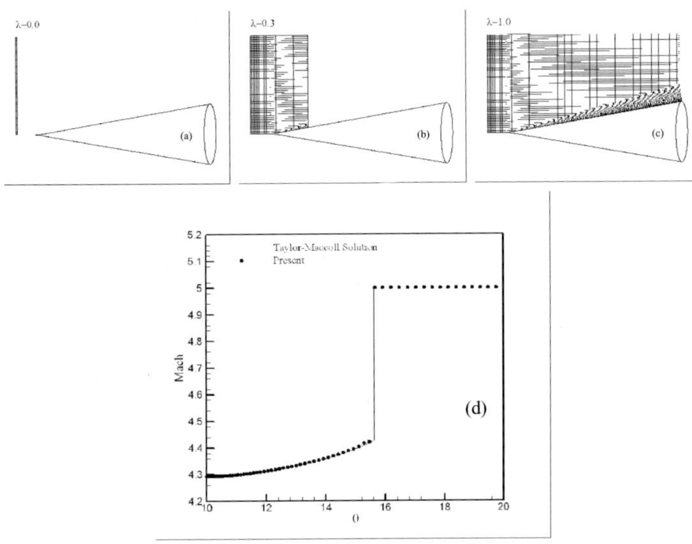

Fig. 2. Hypersonic flow past a cone. (a) – (c), flow generated grids, (d) Mach number, 120 cells, computing time 7.21s (P4, 2.8GHz)

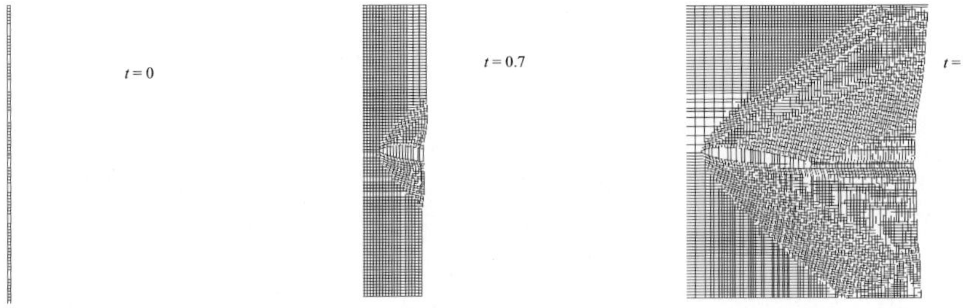

Fig. 3. Flow-generated grids for steady supersonic flow, $M = 2.2$, past a NACA 0012 airfoil at 8° angle of attack, at $t = 0, 0.7, 4$

yet the computed solutions are smeared near the leading edge and the mid-section of the airfoil (Fig. 1h).

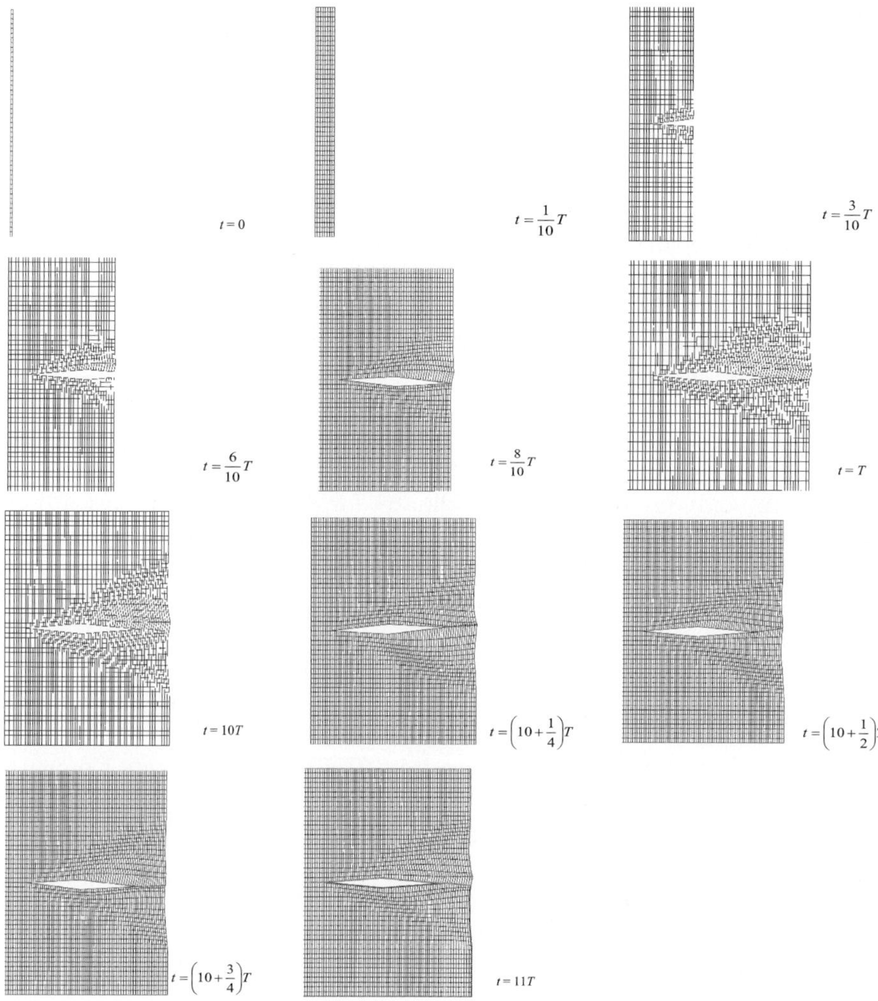

Fig. 4. Flow-generated grids for oscillating diamond-shape airfoil. Apex angle = 10°, pitching motion about the apex: $\theta(t) = 2^\circ \sin t/T, T = 1/30$

Example 2. Steady hypersonic flow, $M_\infty = 5$, past a circular cone with 20° vertex angle placed at zero incidence. Again, Eq. (2) is solved by a space-marching method with an adaptive Godunov scheme [4]. The flow-generated grids at different λ are plotted in Figs 2a to 2c, whereas the computed Mach number distribution is plotted in Fig 2d, compared excellently to the Taylor-Maccoll solution. Our computation takes 7.21 seconds, while the Taylor-Maccoll solution, which needs iteration due to the shock position being unknown, takes more than 2,000 seconds.

Example 3. Steady supersonic flow, $M_\infty = 2.2$, past a NACA 0012 airfoil at an angle of attack equal to 8°. Since the flow near the nose of the airfoil is subsonic, we compute the steady flow solution marching in time λ long enough until a steady state is reached. The computed flow-generated grids a different times are plotted in Figs. 3a to 3c.

Example 4. Supersonic flow, $M_\infty = 3.0$, past a diamond-shape airfoil with 10° vertex angle which is oscillating about its vertex according to $\theta = 2° \sin 30t$, where θ is the instantaneous pitching angle. The 2-D unsteady equation (4) is solved by splitting it into two 1-D equations in $\lambda\xi$ and $\lambda\eta$, each of them is solved by using the standard Godunov/MUSCL scheme. The computed flow-generated grids at different times are plotted in Figs. 4.

5 Conclusions

In the commonly-used Eulerian coordinates for computing flow past a body, it is necessary to generate a body-fitted grid prior to computing the flow. With the use of the unified coordinates, the grid-generation stage is rendered redundant; the grid is generated automatically by the flow. This greatly saves computing time.

Acknowledgements

This research was funded by a grant from the Research Grants Council of Hong Kong.

References

1. Hui, W.H., Li, P.Y., and Li, Z.W., **J. Comput. Phys.**, 153 pp.507, (1999).
2. Hui, W.H. and Kudriakov, S., **J. Comput. Phys.**, 172, pp.235, (2001).
3. Hui, W.H. and Chu, D.L., **Comput. Fluid Dyn.**, 4, pp.403 (1996).
4. Lepage, C.Y. and Hui, W.H., **J. Comput. Phys.**, 22, pp.291 (1995).

Viscous Flow Computations Using a Meshless Solver, LSFD-U

Anup Ninawe[1], N. Munikrishna[2], and N. Balakrishnan[3]

[1] Indian Institute of Science, Bangalore-12 anupsn@aero.iisc.ernet.in
[2] Indian Institute of Science, Bangalore-12 kishan@aero.iisc.ernet.in
[3] Indian Institute of Science, Bangalore-12 nbalak@aero.iisc.ernet.in

Summary. The Upwind-Least Squares Finite Difference (LSFD-U) scheme has been successfully applied for inviscid flow computations. In the present work, we extend the procedure for computing viscous flows. Different ways of discretizing the viscous fluxes are analysed for the positivity, which determines the robustness of the solution procedure. The scheme which is found to be more positive is employed for viscous flux computation. The numerical results for validating the procedure are presented.

1 Introduction

The meshless methods requiring only a cloud of grid points for solution update are competing with unstructured data based finite volume solvers for inviscid flows involving complex industrial configurations. A great deal of research interest has been generated in the use of these methods for industrial computations. The Upwind-Least Squares Finite Difference scheme [1, 4] is one such procedure. Several theoretical details of this procedure and its application to inviscid flows have been discussed in reference [1]. Here we discuss its extension to solve viscous flows. Two important requirements of viscous flux discretization(VFD) procedure are: 1. consistency 2. positivity. The "consistency" requirement is fundamental and is ignored by some of the meshless solvers [2]. The robustness of the procedure is linked to the second requirement. Use of the upwind schemes with limiters for the convective fluxes results in a monotonicity preserving procedure for inviscid flows. On the other hand, for viscous flows, the solution procedure fails due to lack of positivity of numerical scheme employed for VFD. The ultimate objective of the present work is to arrive at a consistent and a positive VFD procedure. In this paper, different variants resulting in a consistent VFD procedure are discussed. The positivity of these variants is studied by analysing the discrete Laplacian [5]. The variant which is observed to be most positive is employed for VFD. The procedure is validated for standard laminar and turbulent flow test cases. The LSFD-U procedure as applied to inviscid flows is presented in section 2. The details and analysis of the proposed variants for VFD are presented in section 3. The numerical results are presented in section 4 with concluding remarks in section 5.

2 LSFD-U

The compressible Navier-Stokes equation in divergence form reads,

$$U_t + (f_i + f_v)_x + (g_i + g_v)_y \; = \; 0, \tag{1}$$

where, U is vector of Conserved variables, f_i & g_i are inviscid fluxes and f_v & g_v represent the viscous fluxes. The LSFD-U framework operates on a cloud of points used in solution update at 'o' as shown in figure 1a. The fundamental idea here is to use the Taylor series expansion for obtaining the discrete approximation to the flux derivatives. The function value, say ϕ in the neighbourhood of 'o' (say at node 'j') can be estimated using truncated Taylor series,

$$\phi_j \; = \; \phi_o \; + \; \sum_{q=1}^{l} \sum_{m=0}^{q} \binom{q}{m} \frac{\Delta x_j^{q-m} \, \Delta y_j^m}{q!} \frac{\partial^q \phi}{\partial x^{q-m} \partial y^m}. \tag{2}$$

Using the method of least squares, discrete approximation to the derivatives at node'o' can be obtained. This procedure when applied to estimate the derivatives of the inviscid fluxes does not take into account the hyperbolicity of Euler equations and therefore the scheme is expected to be unstable. One of the ways to overcome this problem is to enforce upwinding in the numerical procedure. For this purpose, we introduce a fictitious interface 'J' associated with neighbouring node 'j'. Inviscid flux along 'oj' at the fictitious interface, $F_{i\uparrow J}$ is given by

$$F_{iJ} = \frac{1}{\Delta r_J} (f_{iJ} \Delta x_J + g_{iJ} \Delta y_J), \tag{3}$$

where $\Delta x_J = x_J - x_o$, $\Delta y_J = y_J - y_o$. Expanding f_{iJ} and g_{iJ} in Taylor series around 'o', we have,

$$\begin{aligned} \Delta F_{iJ}^e = & \frac{1}{\Delta r_J} \sum_{q=1}^{l} \left[\frac{\partial^q f_i}{\partial x^q} \frac{\Delta x_J^{q+1}}{q!} + \sum_{m=0}^{q-1} \left\{ \binom{q}{m+1} \frac{\partial^q f_i}{\partial x^{q-(m+1)} \partial y^{(m+1)}} \right. \right. \\ & \left. \left. + \binom{q}{m} \frac{\partial^q g_i}{\partial x^{q-m} \partial y^m} \right\} \frac{\Delta x_J^{q-m} \, \Delta y_J^{m+1}}{q!} + \frac{\partial^q g_i}{\partial y^q} \frac{\Delta y_J^{q+1}}{q!} \right] \end{aligned} \tag{4}$$

Employing a suitable reconstruction procedure, the interfacial flux F_{iJ} and thus the flux difference ΔF_{iJ} can be computed using any upwind flux formula. The resulting over determined system of equations, represented by equation 4, can be solved for the flux derivatives using the method of least squares. These flux derivatives are used along with suitable time integration procedure for solution update at node 'o'. The order of accuracy of this procedure and the means to construct schemes with specified order of accuracy are discussed in reference [1]. For a consistent discretization, LSFD-U requires F_J accurate atleast to $O(h^2)$.

3 Viscous Flux Discretization (VFD)

The different variants arising out of VFD are presented here. The least squares finite difference method (equation 2) approximates the nth derivative for $n \leq l$ to $O(h^{l-n+1})$ [1]. Therefore, a quadratic least squares (QLS) procedure (l=2) will determine the gradient $\nabla\phi$ to $O(h^2)$ and Hessian $\mathcal{H}\phi$ to $O(h)$. The two possibilities for VFD are: (1) use the inviscid framework itself, demanding the definition of F_{vJ} to $O(h^2)$ (2) use of $\nabla\phi_o$ and $\mathcal{H}\phi_o$ resulting from QLS procedure. Both the procedures should employ atleast a QLS procedure and are consistent. The resulting variants are:
Variant 1a: $\nabla\phi_J = \frac{\nabla\phi_o+\nabla\phi_j}{2}$, $\nabla\phi_o$ and $\nabla\phi_j$ are nodal values obtained to $O(h^2)$ using QLS procedure,
Variant 1b: $\nabla\phi_J = \frac{\nabla\phi_J^o+\nabla\phi_J^j}{2}$, $\nabla\phi_J^o$ and $\nabla\phi_J^j$ are projected values at J,
Variant 2: the viscous flux derivatives are determined using $\nabla\phi_o$ and $\mathcal{H}\phi_o$.

3.1 Positivity Studies

The robustness of the scheme is linked to the positivity of the VFD procedure. It is an usual practice to test the basic requirements of VFD procedure such as consistency and positivity by analysing the discrete laplacian. For a discrete Laplacian given by $\phi_o = \sum_{j=1}^{N} a_j\phi_j$, the following positivity parameter is defined [5]:

$$\alpha_{min} = \frac{min(a_j, 0)}{\sqrt{\sum_{j=1}^{N} \frac{a_j^2}{N}}}, \tag{5}$$

where N is the number of neighbors in the stencil. The Cartesian like grid shown in figure 1b, which is a representative grid commonly used for viscous flow computations, is considered. The positivity of the aforesaid variants is studied with respect to the stretching factor ϵ and aspect ratio AR.

The results for the positivity analysis are shown in figures 1c to 1f. The variants 1a and 1b are always non-positive. Even for a uniform grid(ϵ=1 and AR=1) these variants do not preserve positivity. Variant 2 results in a positive stencil for uniform grid and for grids with unit AR and small values of $\epsilon \sim 1.5$. For other AR's the variant 2 is less negative when compared to variants 1a and 1b and is shown in figure 1f. It is interesting to observe that the positivity parameter becomes insensitive to AR for values of AR$\geq$10. Based on the positivity analysis, variant 2 is employed for VFD.

4 Numerical Results

The procedure is validated for standard laminar and turbulent flow test cases. In conjunction wth QLS based reconstruction, the inviscid fluxes are computed using Roe scheme [7]. The convergence acceleration to steady state is obtained by employing SGS implicit relaxation procedure [6]. Baldwin-Lomax turbulence model is used to compute eddy viscosity. No slip and adiabatic conditions are imposed on the wall boundary nodes. The pressure is obtained by setting normal pressure gradient to zero. The characteristic Riemann boundary condition is applied for nodes on the farfield boundary. The results obtained are compared with those obtained using a cell vertex finite volume flow solver on the same grids.

<u>Case 1:</u>Laminar flow, NACA 0012: Re_∞=500, M_∞=0.85, α=0°

The point distribution (8000 points) used here is shown in figure 2a. The surface coefficient distributions are shown in figures 2b and 2c.

<u>Case 2:</u>Laminar flow, NACA 0012: Re_∞=500, M_∞=0.8, α=10°

The point distribution used is same as that in Case1. The surface coefficient distributions are shown in figures 3a and 3b. Figure 3c shows the Mach num-

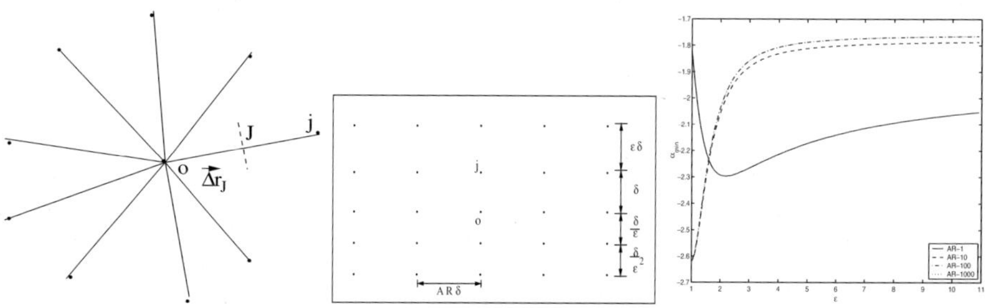

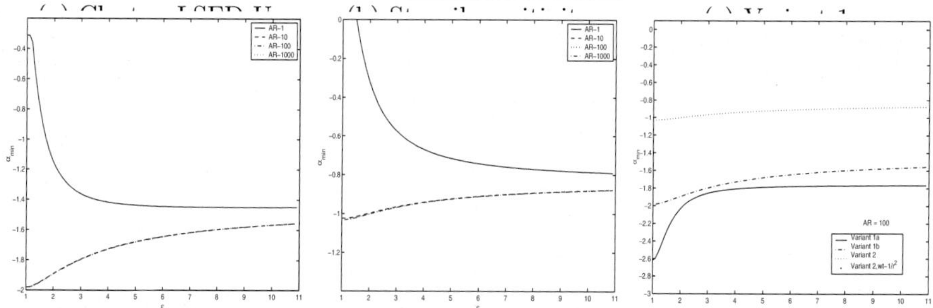

(d) Variant 1b (e) Variant 2 (f) All Variants

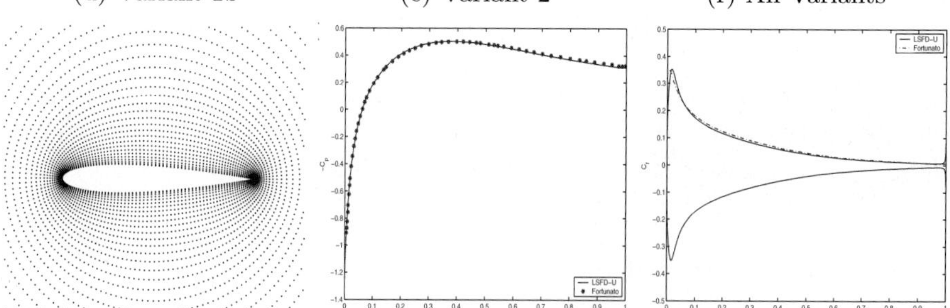

(a) Point Distribution (b) $-C_p$ Distribution (c) C_f Distribution

Fig. 2. Laminar Flow, NACA 0012: $Re_\infty = 500, M_\infty = 0.85, \alpha = 0^o$

ber contours. The surface coefficients predicted by the LSFD-U is in good agreement with the values given in reference [8].

Case 3:Laminar flow, NACA 0012: Re_∞=5000, M_∞=0.5, α=3°

The point distribution (9885 points) used here is shown in figure 4a. The surface coefficient distributions are compared with the values given in reference [9] in figure 4b. Figure 4c shows the Mach number contours.

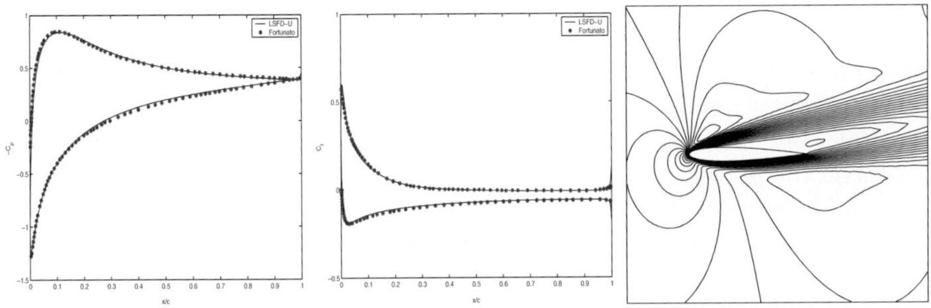

(a) $-C_p$ Distribution (b) C_f Distribution (c) Mach Contours

Fig. 3. Laminar Flow, NACA 0012: $Re_\infty = 500, M_\infty = 0.8, \alpha = 10^o$

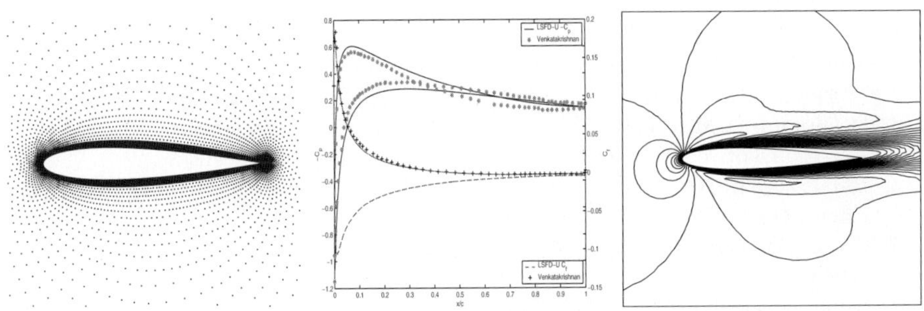

(a) Point Distribution (b) $-C_p$&C_f Distribution (c) Mach Contours

Fig. 4. Laminar Flow, NACA 0012: $Re_\infty = 5000, M_\infty = 0.5, \alpha = 3^o$

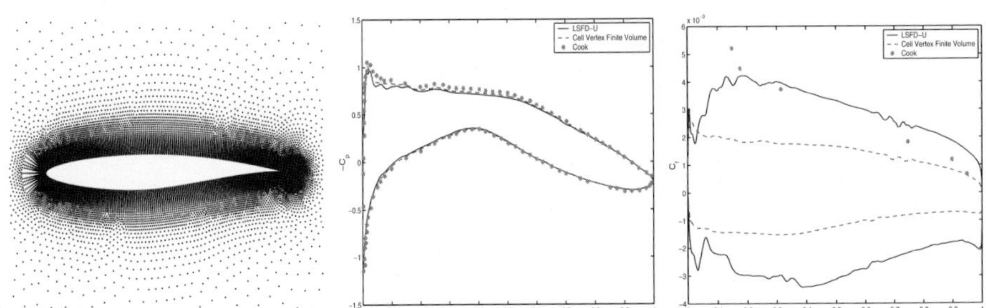

(a) Point Distribution (b) $-C_p$ Distribution (c) C_f Distribution

Fig. 5. Turbulent Flow, RAE 2822: $Re_\infty = 5.7\text{x}10^6, M_\infty = 0.676, \alpha = 1.92^o$

Case 4:Turbulent flow, RAE 2822: Re_∞=5.7x10^6, M_∞=0.676, α=1.92^o
The point distribution (28817 points) used here is shown in figure 5a. The pressure distribution on the airfoil is compared with the experimental values [10] in figure 5b and the agreement is good. The computed friction coefficients are compared with the experimental values in figure 5c. This clearly suggests the use of finer grid for turbulent flow computations.

5 Conclusion

In the present work, different possibilities that results in consistent formulation for viscous flux discretization are analysed for the positivity. The variant that fairs best in terms of positivity is employed and the solver is validated. Based on the numerical results, it can be concluded that the LSFD-U procedure can be employed for solving viscous flows. The present effort involves the use of Cartesian like grids for obtaining the point distribution. The sensitivity of the solution procedure to the grid asymmetry, neighbourhood selection and the boundary treatment are also being analysed.

References

1. Sridar D., Balakrishnan N., JCP, Vol. 189, pp 1-29 (2003).
2. Morinishi K., Computational Fluid Dynamics Jl, 9 (1) (2000).
3. Deshpande S.M., Ghosh A.K., Mandal J.C., *Fluid Mechanics Report No. 89 FM 04*, Dept. of Aerospace Engg., IISc., Bangalore 12 (1989).
4. Balakrishnan N., *Fluid Mechanics Report No. 99 FM 09*, Dept. of Aerospace Engg., IISc., Bangalore 12 (1999).
5. Coirier W.J.:*Ph.D. Thesis*, Dept. of Aerospace Engg., The Univ. of Michigan, (1994).
6. Sridar D., Balakrishnan N., *Fluid Mechanics Report No. 2004 FM 13* , Dept. of Aerospace Engg., IISc., Bangalore 12 (2004).
7. Roe P.L., JCP, Vol. 43, pp 357-372 (1981).
8. Fortunato B., Vinicio M.: In *Proceedings of $14^t h$ ICNMFD, Lecture Notes in Physics*, ed by S.M.Deshpande, S.S.Desai and R.Narasimha (Springer-Verlag Berlin 1995), p 259.
9. Venkatakrishnan V., Computers and Fluids, Vol. 18, (2), pp 191-204 (1990).
10. Cook P.H., Report No. AR-138, AGARD, (May, 1979).

Part XIV

Microscale Flows

Preconditioning Method for Compressible Near-critical Fluids in Micro-Channel

Satoru Yamamoto

Dept. of Aeronautics and Space Engineering, Tohoku University,
Sendai 980-8579, Japan
yamamoto@caero.mech.tohoku.ac.jp

1 Introduction

A preconditioned flux-splitting scheme for simulating condensate flows has been proposed by our group in the last ICCFD conference at Sydney [1]. This method is based on the numerical method for transonic condensate flows [2] and the preconditioning method developed by Weiss and Smith [3]. Since the flux-splitting formulation can be easily applied to existing compressible flow solvers and the present method can remove the stiffness completely from the calculation for flows at very low Mach number, critical flow problems such as natural convections with condensation have been successfully resolved [1] [4].

In this study, we apply this method to supercritical fluids including the near-critical region, especially, compressible near-critical fluids in a micro-channel and around a micro-scale body. Supercritical fluids are being so important in energy, environment, and material studies. Those flows have compressibility even if the flow speed is very low and the Reynolds number is very small. Existing compressible flow solvers may be confronted by the stiffness if those flows are calculated by them. The Peng-Robinson(P-R) equation of state(EOS) [5] is newly introduced in this study to calculate physical properties for supercritical fluids. Also an accurate formulation for speed of sound governed by the P-R equation is derived and installed in the present method. In this paper as numerical examples, a supercritical water flow in a micro-channel with a cavity region and a supercritical carbon-dioxide flow around a very small airfoil are calculated using the present preconditioning method. Finally, the calculated physical properties using the P-R equation are compared with those using the EOS for ideal gas.

2 Numerical Methods

2.1 Fundamental equations

The preconditioning method developed by Weiss and Smith [3] is applied to the 3-D compressible Navier-Stokes equations. The unknown variables in the unknown vector $\hat{Q}$ for the preconditioned equations are defined as $\hat{Q} =$

$J[p \ \ u_1 \ \ u_2 \ \ u_3 \ \ T]^T$. Then the set of preconditioned equations in general curvilinear coordinates is derived by using a preconditioning matrix Γ as

$$\Gamma \partial \hat{Q}/\partial t + L(\hat{Q}) = \Gamma \frac{\partial \hat{Q}}{\partial t} + \frac{\partial F_i}{\partial \xi_i} + \frac{\partial F_{vi}}{\partial \xi_i} + S = 0 \tag{1}$$

where, F_i, F_{vi}, and S are the vectors of inviscid flux, viscous flux and the source term.

2.2 Preconditioning flux-splitting form

The numerical flux $(F_i)_{\ell+1/2}$ for F_i in Eq.(1) defined at the interface between the control volume ℓ and $\ell+1$ in each coordinate $i(i=1,2,3)$ can be written by a flux-vector splitting form as

$$\begin{aligned}(F_i)_{\ell+1/2} &= (F_i^+)_{\ell+1/2} + (F_i^-)_{\ell+1/2} \\ &= (\hat{A}_i^+)_{\ell+1/2}\hat{Q}^L_{\ell+1/2} + (\hat{A}_i^-)_{\ell+1/2}\hat{Q}^R_{\ell+1/2}\end{aligned} \tag{2}$$

The superscripts $\pm$ indicate the sign of characteristic speeds. $F_i^{\pm}$ and $\hat{A}_i^{\pm}$ are the numerical flux vectors and the preconditioned Jacobian matrices divided by positive or negative characteristic speeds, respectively. $\hat{Q}^L$ and $\hat{Q}^R$ are the unknown vectors extrapolated by the compact MUSCL [6] from the left and the right directions. The preconditioned flux-vector splitting form for $(\hat{A}_i^{\pm})_{\ell+1/2}\hat{Q}^M_{\ell+1/2}$ is given by

$$\begin{aligned}(\hat{A}_i^{\pm})_{\ell+1/2}\hat{Q}^M &= (\Gamma L_i^{-1}\Lambda_i^{\pm}L_i)_{\ell+1/2}\hat{Q}^M \\ &= \hat{\lambda}_{i1}^{\pm}\Gamma\hat{Q}^M + \frac{\hat{\lambda}_{ia}^{\pm}}{\hat{c}_i\sqrt{g_{ii}}}\hat{Q}_{ia} + \frac{\hat{\lambda}_{ib}^{\pm}}{\hat{c}_i^2}\hat{Q}_{ib}\end{aligned} \tag{3}$$

where, $g_{ii} = \nabla\xi_i \cdot \nabla\xi_i$. The superscript M is replaced by L or R. L_i and Λ_i are the matrices composed of preconditioned left eigenvectors and preconditioned characteristic speeds(eigenvalues). $\hat{\lambda}_{ia}^{\pm}$ and $\hat{\lambda}_{ib}^{\pm}$ are defined by

$$\hat{\lambda}_{ia}^{\pm} = (\hat{\lambda}_{i4}^{\pm} - \hat{\lambda}_{i5}^{\pm})/2, \quad \hat{\lambda}_{ib}^{\pm} = (\ell_i^-\hat{\lambda}_{i4}^{\pm} - \ell_i^+\hat{\lambda}_{i5}^{\pm})/(\ell_i^- - \ell_i^+) - \hat{\lambda}_{i1}^{\pm}$$

where $\hat{\lambda}_{ij}^{\pm}(j=1,4,5)$ and $\ell_i^{\pm}$ are calculated by

$$\hat{\lambda}_{ij}^{\pm} = (\hat{\lambda}_{ij} \pm |\hat{\lambda}_{ij}|)/2, \quad \ell_i^{\pm} = \rho U_r^2/(U_i(1-\alpha)/2 \pm \hat{c}_i\sqrt{g_{ii}})$$

$\hat{\lambda}_{ij}(j=1,4,5)$ are the preconditioned characteristic speeds derived as

$$\hat{\lambda}_{i1} = U_i, \quad \hat{\lambda}_{i4} = (1+\alpha)U_i/2 + \hat{c}_i\sqrt{g_{ii}}, \quad \hat{\lambda}_{i5} = (1+\alpha)U_i/2 - \hat{c}_i\sqrt{g_{ii}}$$

where, $\hat{c}_i$ is the numerical speed of sound. It is defined by

$$\hat{c}_i = \sqrt{U_i^2(1-\alpha)^2/g_{ii} + 4U_r^2}/2$$

$\alpha = U_r^2(\rho_p + \rho_T/\rho C_p)$, where $\rho_p = \partial\rho/\partial p$ and $\rho_T = \partial\rho/\partial T$. If U_r equals the physical speed of sound, α is reduced to unit and characteristic speeds and physical speed of sound for compressible flows are recovered in Eq.(1). $\hat{Q}_{ia}$ and $\hat{Q}_{ib}$ are the subvectors defined by

$$\hat{Q}_{ia} = \hat{q}_1^M Q_{ic} + \rho\hat{U}_i Q_d, \quad \hat{Q}_{ib} = (\rho\hat{U}_i\hat{c}_i^2/g_{ii})Q_{ic} + (\hat{q}_1^M\hat{c}_i^2/U_r^2)Q_d$$

$\hat{q}_j^M$ and $\hat{U}_i[= (\partial\xi_i/\partial x_j)\hat{q}_{j+1}^M (j = 1,2,3)]$ are the j-th element of $\hat{Q}$ and the contravariant velocities extrapolated by the compact MUSCL [6] in which $M = L$ or R, respectively. Q_{ic} and Q_d are the subvectors given by

$$\hat{Q}_{ic} = [0 \ \ \partial\xi_i/\partial x_1 \ \ \partial\xi_i/\partial x_2 \ \ \partial\xi_i/\partial x_3 \ \ U_i]^T, \quad \hat{Q}_d = [1 \ \ u_1 \ \ u_2 \ \ u_3 \ \ H]^T$$

This flux-vector splitting form is applied to the LU-SGS scheme [7]. We developed the preconditioning LU-SGS scheme [1] to accelerate the convergence of solution. This form is further extended to the form for supercritical fluids in this study.

2.3 Peng-Robinson equation

Supercritical fluids have been studied in several fields in engineering such as chemical, material, and mechanical engineerings. In the supercritical region, some anomalous properties for the fluids are observed. Especially, a near-critical point, the thermal properties, such as the heat conductivity, the viscosity, and the density have an anomalous value, especially the speed of sound is rapidly decreased near the critical point. The EOS for ideal gas usually used in the compressible Navier-Stokes solver calculates them inaccurately. In this study, we employ one of cubic equations to solve real gas flows in high pressure and high temperature. The equation is defined in a general form by

$$(p+\beta)(1-b\rho) = \rho RT \tag{4}$$

where, if $\beta = a\rho^2$ or $\beta = a\rho^2/(1+2b\rho - b^2\rho^2)$, Eq.(4) is the van der Waals equation(W-A) or the Peng-Robinson equation(P-R). α used in Eq.(3) defined in the previous study assuming ideal gas is alternatively derived as $\alpha = U_r^2(\rho_p + \rho_T(1-\rho h_p)/\rho h_T)$ for supercritical fluids, and the square of the sound speed is defined by $dp/d\rho = \rho h_T/(\rho_T(1-\rho h_p) + \rho\rho_p h_T)$, where h_p and h_T are the partial derivatives of enthalpy by the pressure and the temperature. Also the preconditioning matrix Γ is slightly modified according to this assumption.

3 Results

We calculated the speed of sound in supercritical water and carbon dioxide theoretically using the speed of sound derived from the P-R equation. The pressure for the supercritical water is specified to 22.12[MPa] and that for the carbon dioxide is to 7.377[MPa]. Figures 1 and 2 show the calculated speeds of sound compared with the experimental data and those calculated by the EOS for ideal gas. The ideal gas equation can not trace the experimental data. The speed of sound calculated from the P-R equation can evaluate the curve of the experimental data in the supercritical region accurately. The calculated values at the critical point and its temperature are also quite in good agreement with the experiments. However, the calculated values at the subcritical region may be inaccurate in both water and carbon dioxide cases. This result suggests that the P-R equation has a limitation when it is used in the subcritical region. We should use the P-R equation only in the supercritical region without crossing the critical point.

Next, a supercritical water flow at very low Mach number through a micro-channel with a cavity region is calculated using the present method with the P-R equation. Figure 3 shows the computational grid which has 201x41 grid points in the channel and 41x21 grid points in the cavity. The inlet height is 1E-4[m]. The channel length is 9E-4[m], and the edge length of the cavity is 1E-4[m]. The flow conditions are specified: the inlet temperature is 750[K], the inlet pressure is 22.18 [MPa], the wall temperature is fixed to 650[K], and the inlet Mach number is 0.001. Figure 4 shows the calculated temperature contours obtained by the present method using the P-R equation. The calculated speed of sound using the P-R equation is shown in Fig.5 and this result can be compared with that calculated using the EOS of ideal gas in Fig.6. The minimum value of the speed of sound in Fig.5 is far smaller than that in Fig.6 and the value may approach to the minimum value at the critical point shown in Fig.1. The minimum value in Fig.6 is at least 200[m/s] higher evaluated than that in Fig.1. Despite the flow is very slow and the Reynolds number is very small(Re=120), the speed of sound is quite variable in the supercritical flow field. It indicates that the compressibility may be very important for supercritical fluids even if they are in a micro-channel. Table 1 shows that calculated minimum and maximum densities obtained by using the P-R equation and the EOS of ideal gas. This table also suggests the importance of the compressibility in supercritical fluids.

The present method is further applied to a supercritical carbon-dioxide flow around a very small NACA0012 airfoil. The chord length is 1E-4[m]. Uniform Mach number is 0.001. The Reynolds number is 147. The uniform pressure is 7.377[MPa]. Three cases in Table 2 changing the uniform temperature, the wall temperature and the EOS are calculated. Figure 7 shows the calculated temperature contours in the CASE A1. A thick boundary layer for temperature is formed around the airfoil. Figure 8 shows the calculated density contours in the CASE A1. The difference between the minimum and maximum values are 80[kg/m^2]. Also the calculated minimum and maximum densities in three cases are shown in Table 2. The density difference in the CASE A1 where the temperature is close to that at the critical point is relatively large compared with other two cases assuming a higher temperature(CASE A2) and assuming the ideal gas(CASE B).

Table 1. Calculated minmum and maximum densities for supercritical water

	P-R equation	Ideal gas
Min. Density[kg/m^2]	82.2	64.1
Max. Density [kg/m^2]	179.9	74.0

Table 2. Calculated minmum and maximum densities for supercritical carbon dioxide

	CASE A1	CASE A2	CASE B
EOS	P-R equation	P-R equation	Ideal gas
Uniform temp.[K]	450	650	450
Wall temp.[K]	330	530	330
Min. density[kg/m^2]	98.0	95.5	57.1
Max. density [kg/m^2]	178.0	122.2	77.9

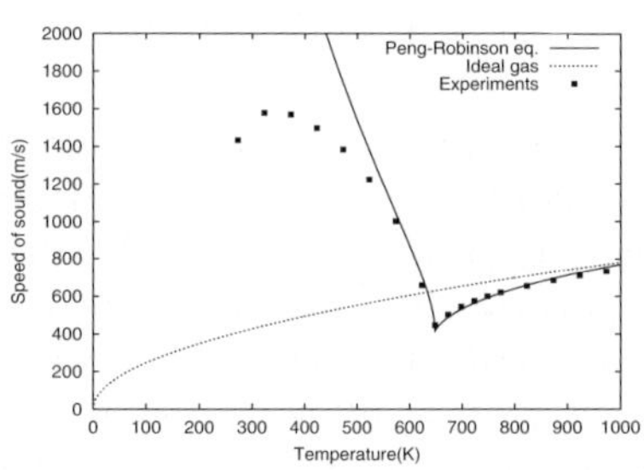

Fig.1 Speed of sound in H_2O

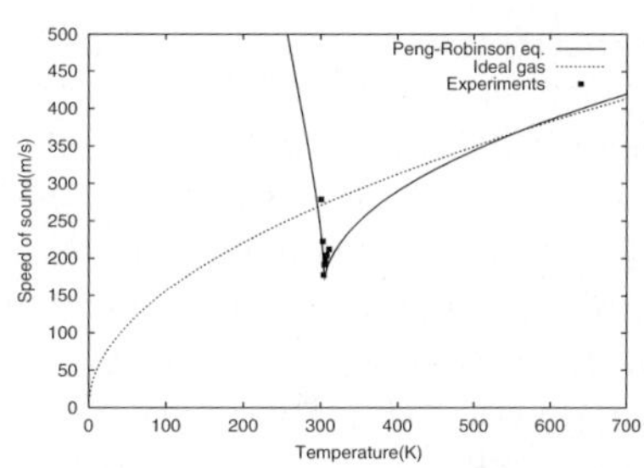

Fig.2 Speed of sound in CO_2

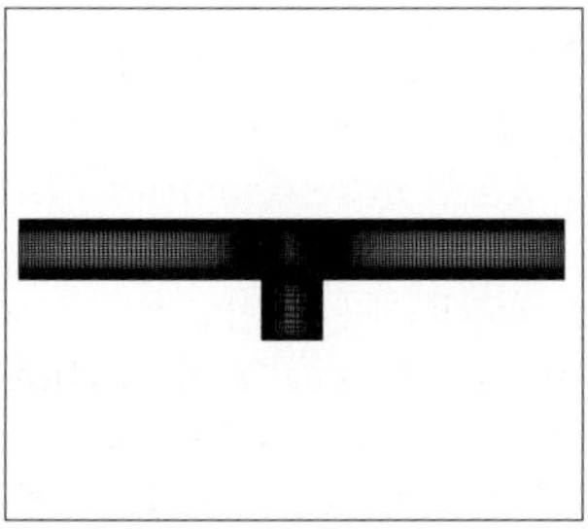
Fig.3 Computational grid

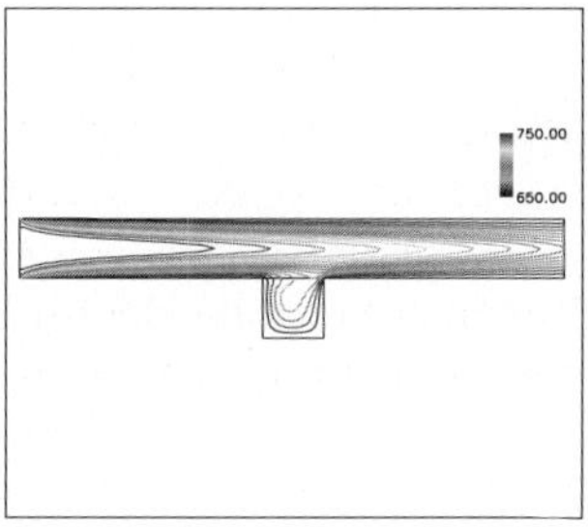

Fig.4 Temperature (H_2O, P-R)

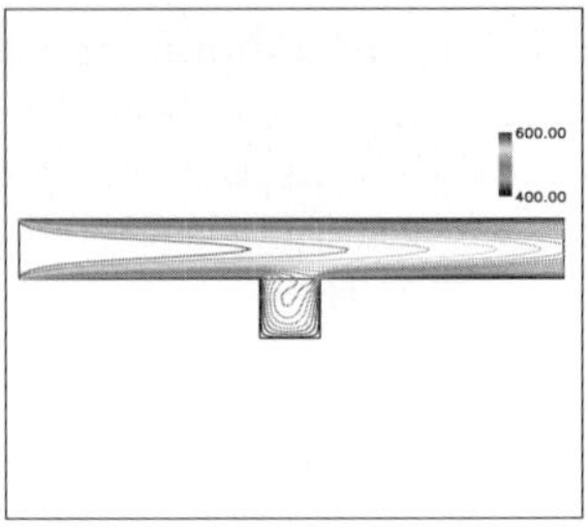

Fig.5 Speed of sound(H_2O, P-R)

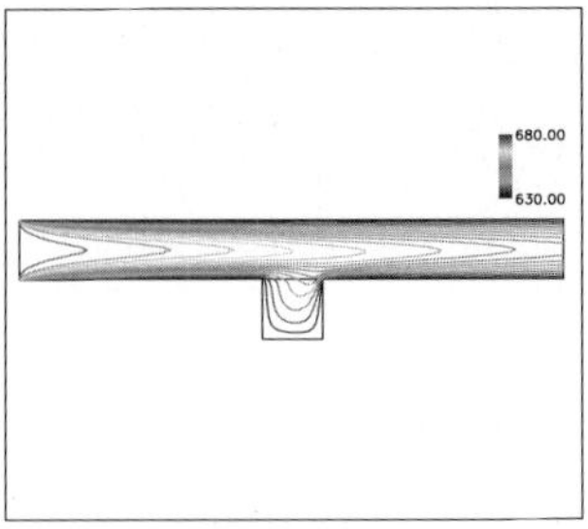

Fig.6 Speed of sound(H_2O, Ideal gas)

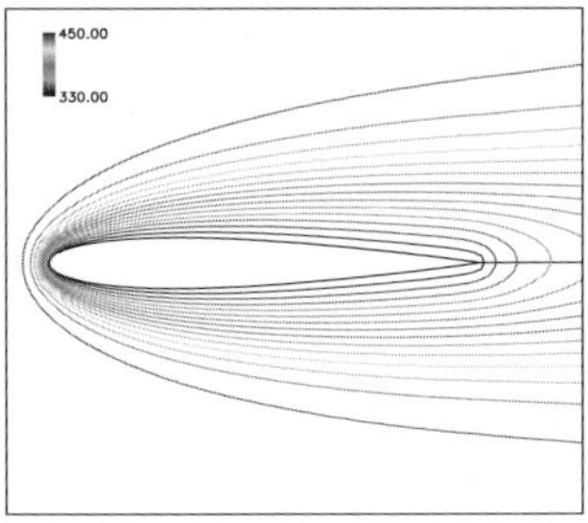

Fig.7 Temperature(CO_2, P-R)

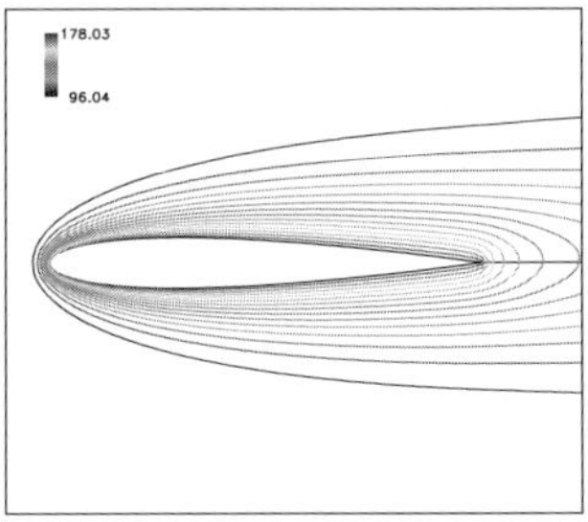

Fig.8 Density(CO_2, P-R)

4 Conclusion

A supercritical water flow in a micro-channel and a supercritical carbon-dioxide flow around a very small airfoil were calculated using the preconditioning method coupled with the Peng-Robinson equation of state. All calculated results indicate that the compressibility should be taken in the supercritical fluids into consideration even if the flow region is very narrow and the flow speed is very low. I believe that the present method based on the preconditioning method is the best method for simulating supercritical fluids near the critical point as an engineering tool.

References

1. S. Yamamoto and B.-R. Shin, Proc. of the Second Int. Conf. on Computational Fluid Dynamics, (2002), 112-117, Springer.
2. S. Yamamoto, H. Hagari and M. Murayama, Trans. JSASS, **42**(2000), pp.182-189.
3. J.M. Weiss and W.A. Smith, AIAA J., **33**(1995), pp.2050-2056.
4. S.Yamamoto, D. Niiyama and B.-R. Shin, Proc. FEDSM'03, 4th ASME_JSME Joint Fluids Engineering Conference, FEDSM2003-45390, (2003).
5. D.-Y. Peng and D.B.Robinson, Ind. Eng. Chem. Fundam, **15**(1976), pp.59-64.
6. S. Yamamoto and H. Daiguji, Computers and Fluids, **22**(1993), pp.259-270.
7. S. Yoon and A. Jameson, AIAA J., **26**(1988), pp.1025-1026.

Comparison of Kinetic and Navier-Stokes Solutions for Rarefied Gas Flows in Micro-channels

Nobuyuki Satofuka[1], Koji Morinishi[2], and Keigo Kamitsuji[3]

[1] The University of Shiga Prefecture, Hassaka-cho, Hikone-shi, Shiga 522-8533 Japan, satofuka@office.usp.ac.jp
[2] Department of Mechanical and System Engineering, Kyoto Institute of Technology, Matsugasaki, Sakyo-ku, Kyoto 606-8585 Japan, morinishi@ipc.kit.ac.jp
[3] DIHATSU Motor Co., Ltd.

1 Introduction

Until recently, non-continuum slip flows were only encountered in low-density (rarefied gas) applications such as vacuum or space-vehicle technology. However, the rapid progress in fabricating and utilizing Micro Electro Mechanical Systems (MEMS) over the last decade has led to considerable interest in non-continuum flow-processes. The small length scales commonly encountered in MEMS imply that rarefaction effects occur in micron-sized channels at normal operating pressures. In these cases it is necessary to solve the Boltzmann equation. Needless to say, the direct numerical integration of the equation is computationally too costly due to the complexity of the collision term. So far most of the attempts to solve the Boltzmann equations have been carried out using the Direct Simulation Monte Carlo (DSMC) method proposed by Bird [1]. Most gaseous micro flows related to MEMS correspond to a low Mach number and low Reynolds number in contrast to their aeronautical counterparts. In these cases, the DSMC method needs an enormous number of test particles to isolate the small macroscopic flow velocity from the statistical noise. Therefore, development of more efficient method is strongly needed.

Many years ago we proposed a time dependent finite difference method for solving kinetic model equations with BGK type simplified collision terms and applied to simulate both supersonic and hypersonic flow problems [2].

In this paper we apply our method of lines approach to the solution of model Boltzmann equations for pressure-driven plane Poiseuille flows and compare the results with those of the DSMC method and the Navier-Stokes (NS) solutions with slip boundary conditions. For low Mach number cases, comparison is also made with that of the linearized Boltzmann equations. The present method is parallelized by using domain decomposition and implemented on a parallel computer using MPI.

2 Kinetic Approach

2.1 BGK Model Equation

The Boltzmann equation with BGK collision model [3] for diatomic molecules without any external force may be written in the Cartesian coordinates as

$$\frac{\partial f}{\partial t} + \mathbf{c_x} \cdot \frac{\partial f}{\partial \mathbf{x}} = \nu(f_e - f) \tag{1}$$

where $f(t, \mathbf{x}, \mathbf{c})$ is the distribution function, $\mathbf{c_x}$ the translational molecular velocity (c_x, c_y, c_z), $\mathbf{x}$ the spatial coordinates (x, y, z), and ν the collision frequency. The Maxwellian equilibrium distribution function f_e is given by

$$f_e = \frac{\rho}{(2\pi RT)^{5/2}} \exp\left(-\frac{\mathbf{C}^2}{2RT}\right) \tag{2}$$

where R is the gas constant.

The density ρ, macroscopic flow velocity $\mathbf{u}(u, v, w)$, and temperature T are obtained from the distribution function f as:

$$\rho = \int f d\mathbf{c}\,, \quad \rho\mathbf{u} = \int \mathbf{c_x} f d\mathbf{c}\,, \quad \frac{5}{2}\rho RT = \int \frac{1}{2}\mathbf{C}^2 f d\mathbf{c} \tag{3}$$

The molecular velocity $\mathbf{c}$ and the thermal velocity $\mathbf{C}$ are each composed of the three translational components and two rotational components as

$$\mathbf{c} = (c_x, c_y, c_z, c_{r1}, c_{r2})\;, \quad \mathbf{C} = (c_x - u, c_y - v, c_z - w, c_{r1}, c_{r2}) \tag{4}$$

where c_{r1} and c_{r2} are the molecular rotational velocity components around two axes of symmetry. The pressure p is obtained from the equation of state as

$$p = \rho RT \tag{5}$$

2.2 Non-dimensionalization

All quantities are normalized by the equilibrium values at a infinite upstream $(\rho_\infty, T_\infty, C_\infty, \mu_\infty)$ and the reference length L_∞. The most probable molecular thermal speed is chosen as the reference speed C_∞, that is,

$$C_\infty = \sqrt{2RT_\infty} \tag{6}$$

Then (1) to (3) can be rewritten in the dimensionless forms as

$$\frac{\partial f}{\partial t} + \mathbf{c_x} \cdot \frac{\partial f}{\partial \mathbf{x}} = \nu(f_e - f) \tag{7}$$

$$f_e = \frac{\rho}{(\pi T)^{5/2}} \exp\left(-\frac{\mathbf{C}^2}{T}\right) \tag{8}$$

$$\rho = \int f d\mathbf{c}\,, \quad \rho\mathbf{u} = \int \mathbf{c_x} f d\mathbf{c}\,, \quad \frac{5}{2}\rho T = \int \mathbf{C}^2 f d\mathbf{c} \tag{9}$$

For inverse power law molecules, the dimensionless collision frequency ν is obtained as

$$\nu = \frac{8\rho T^{1-S}}{5\sqrt{\pi}Kn} \tag{10}$$

where Kn is the Knudsen number which is defined as

$$Kn = \frac{l_\infty}{L_\infty} \tag{11}$$

Here l_∞ is the following mean free path at the infinite upstream.

$$l_\infty = \frac{16\mu_\infty}{5\sqrt{\pi}\rho_\infty C_\infty} \tag{12}$$

2.3 Reduced Distribution Functions

Since our present concern is the analyses of two-dimensional flows, the following reduced distribution functions G and H are conveniently introduced in order to reduce computing operational count and storage.

$$\begin{aligned} G(t,x,y,c_x,c_y) &= \int\int\int f dc_z dc_{r1} dc_{r2} \\ H(t,x,y,c_x,c_y) &= \int\int\int \left(c_z^2 + c_{r1}^2 + c_{r2}^2\right) f dc_z dc_{r1} dc_{r2} \end{aligned} \tag{13}$$

Now let us introduce the following 2-dimensional vector field.

$$\mathbf{x} = (x,y)\,, \quad \mathbf{u} = (u,v)\,, \quad \mathbf{c} = (c_x,c_y)\,, \quad \mathbf{C} = (c_x - u, c_y - v)$$

Then by integrating (7) over c_z, c_{r1} and c_{r2} space using the weights of 1 and $c_z^2 + c_{r1}^2 + c_{r2}^2$, we have the following equations of G and H.

$$\frac{\partial G}{\partial t} + \mathbf{c}\cdot\frac{\partial G}{\partial \mathbf{x}} = \nu(G_e - G)\,, \quad \frac{\partial H}{\partial t} + \mathbf{c}\cdot\frac{\partial H}{\partial \mathbf{x}} = \nu(H_e - H) \tag{14}$$

The reduced equilibrium distribution functions G_e and H_e become as:

$$G_e = \frac{\rho}{(\pi T)}\exp\left(-\frac{\mathbf{C}^2}{T}\right), \quad H_e = \frac{3}{2}TG_e \tag{15}$$

where the density ρ, macroscopic flow velocity $\mathbf{u}$, and temperature T are obtained from the reduced distribution functions G and H as:

$$\rho = \int\int G dc_x dc_y, \quad \rho\mathbf{u} = \int\int \mathbf{c} G dc_x dc_y, \quad \frac{5}{2}\rho T = \int\int \left(\mathbf{C}^2 G + H\right) dc_x dc_y \tag{16}$$

In the case of higher moment model, G_e and H_e in (14) should be replaced as

$$\begin{aligned} G_0 &= G_e \left[1 + \frac{4}{5}(1 - Pr)\, \mathbf{C}\cdot\mathbf{q}\, (\frac{C^2}{T} - 2) \,/\, nT \right] \\ H_0 &= H_e \left[1 + \frac{4}{5}(1 - Pr)\, \mathbf{C}\cdot\mathbf{q}\, (\frac{C^2}{T} - 1) \,/\, nT \right] \end{aligned} \tag{17}$$

2.4 Numerical Procedure

For obtaining numerical solution of the BGK equation using conventional finite difference methods on an arbitrary flow domain, curvilinear coordinate system (ξ, η), of which coordinates exactly fit the physical boundaries and smoothly spread over the domain, are introduced. Then (14) may be rewritten in the coordinate as:

$$\frac{\partial G}{\partial t} + c_\xi \frac{\partial G}{\partial \xi} + c_\eta \frac{\partial G}{\partial \eta} = \nu(G_e - G)\,, \quad \frac{\partial H}{\partial t} + c_\xi \frac{\partial H}{\partial \xi} + c_\eta \frac{\partial H}{\partial \eta} = \nu(H_e - H) \tag{18}$$

where c_ξ and c_η are the following contravariant molecular velocity components.

$$c_\xi = \frac{1}{J}\left(c_x y_\eta - c_y x_\eta\right), \quad c_\eta = \frac{1}{J}\left(c_y x_\xi - c_x y_\xi\right) \tag{19}$$

The spatial derivatives in (18) are first discretized with the second order upwind differential approximations. For example ξ derivative of G at a grid point (i, j) is approximated as:

$$\left.\frac{\partial G}{\partial \xi}\right|_{i,j} = \begin{cases} \dfrac{3G_{i,j} - 4G_{i-1,j} + G_{i-2,j}}{2\Delta\xi} & c_\xi \geq 0 \\[2ex] \dfrac{-3G_{i,j} + 4G_{i+1,j} - G_{i+2,j}}{2\Delta\xi} & c_\xi < 0 \end{cases} \tag{20}$$

where $\Delta\xi$ is the grid spacing in ξ-direction. Evaluation of the collision terms is rather simple. The density, mean flow velocity and temperature are first obtained with (16) using trapezoidal rule. Then the collision frequency and the reduced equilibrium distribution functions are easily evaluated with (10) and (15), respectively.

Evaluating the spatial derivatives and collision terms in (18) leads to the following system of ordinary differential equations in time.

$$\frac{d\mathbf{F}}{dt} = Q(\mathbf{F}) \tag{21}$$

Here $\mathbf{F}$ denotes $(G,H)^T$. The system of ordinary differential equations can be integrated with any conventional method. For example, the Euler method can be written for (21) as

$$\mathbf{F}^{n+1} = \mathbf{F}^n + \Delta t Q(\mathbf{F}^n) \tag{22}$$

For steady state problems, time step size Δt can be determined locally so that the local Courant number becomes a constant for each grid point.

3 Micro-channel Flow

As a basic element of MEMS device, micro-channel gas flows have been studied extensively. In this study, numerical simulations under rarefied conditions are carried out for the pressure-driven plane Poiseuille flows using the BGK and higher moment model equations. The geometric form of the micro-channel is quite simple as shown in Fig. 1, but it is able to reveal the distinguishing feature of the rarefied gas micro-scale internal flows. The height of the micro-channel is $10l_\infty$ and the aspect ratio of the channel is taken to 3. The simulated fluid is a hard sphere gas with inlet Mach number, $M_a = 0.5$ and Knudsen number, $Kn = 0.1$. For the purpose of comparison, the inflow and outflow boundary conditions are the same as those used in the DSMC simulation by Zheng et al [4]. Figure 2 shows comparison of cross sectional profile of velocity, pressure and temperature among those of the BGK, higher moment, DSMC and NS with slip boundary condition. The agreement in velocity and pressure profile is good among the BGK, higher moment, and DSMC solution. The NS pressure profile has the opposite curvature to the others. Temperature profile obtained by the higher moment model agrees excellently with the DSMC one. Figure 3 shows comparison of stream-wise distribution of velocity, pressure and temperature. Agreement is generally satisfactory except temperature distribution in which significant difference is observed between the BGK and higher moment model. For low Mach number cases, $M_a = 0.1$ and 0.01, the results are also compared with those of the NS with slip boundary condition and of the linearized Boltzmann equations [5].

4 Conclusions

Micro-channel flows in slip flow regime are simulated by using kinetic model equations for various Mach and Knudsen numbers. BGK solutions are generally in good agreement with the DSMC solutions except the temperature distribution. Better agreement in the temperature distribution is obtained by using a higher moment model rather than the BGK model. Significant difference in the flow field is observed between the Navier-Stokes and kinetic model solutions.

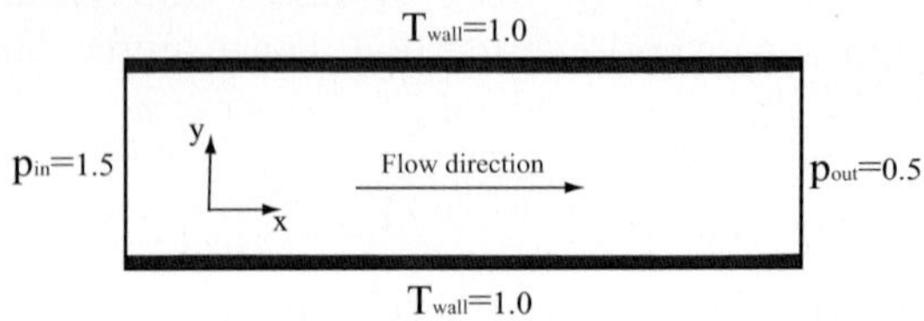

Fig. 1. The geometric form of the micro-channel.

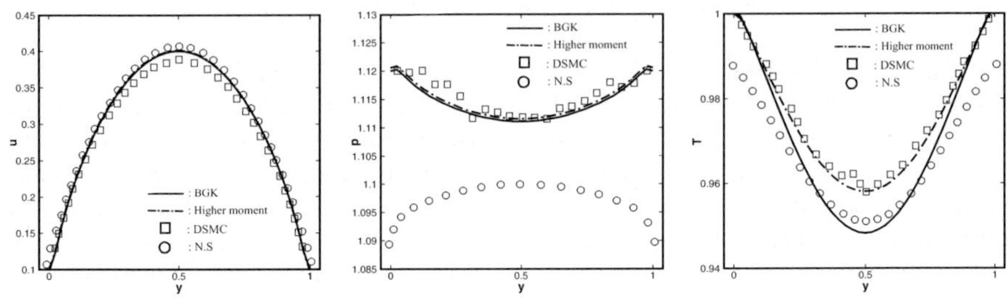

Fig. 2. Comparison of cross sectional profile of velocity, pressure and temperature.

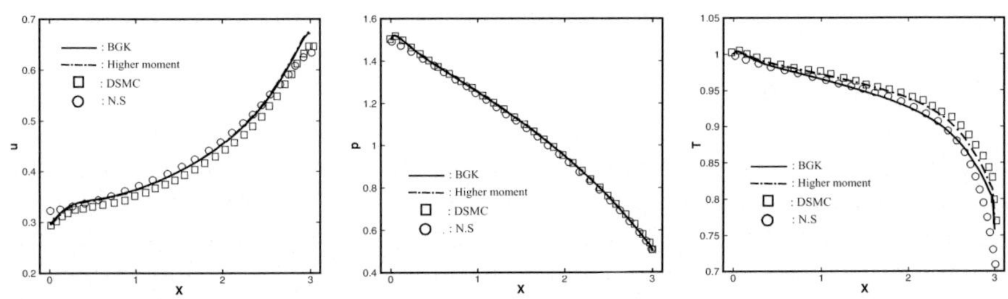

Fig. 3. Comparison of stream-wise distribution of velocity, pressure and temperature.

References

1. G.A. Bird, *Molecular gas dynamics*, Clarendon Press, (1976)
2. N. Satofuka, K. Morinishi and T. Oishi, *Computational Mechanics*, **11**, pp 452–464 (1993)
3. P.L. Bhatnagar, E.P. Gross, and M. Krook, *Physical Review* **94**, pp 511–525 (1954)
4. Y. Zheng, A.L. Garcia and B.J. Alder, *Proceedings of 23rd International Symposium on Rarefied Gas Dynamics*, (2002)
5. T. Ohwada, Y. Sone and K. Aoki, Physics of Fluids A, 1–12, pp 2042–2049 (1989)

Application of the 10-Moment Model to MEMS Flows

Yoshifumi Suzuki[1], Shintaro Yamamoto, Bram van Leer[2], Quanhua Sun, and Iain D. Boyd

[1] The University of Michigan, Ann Arbor, MI 48109 fumifumi@umich.edu
[2] The University of Michigan, Ann Arbor, MI 48109 bram@umich.edu

Summary. In this paper we use the 10-moment description of fluid flow to compute external flow around a micro-airfoil in the continuum-transition regime. The numerical method used is Hancock's scheme, a second-order non-oscillatory Godunov-type scheme; the numerical flux function incorporated is of the Harten-Lax-van Leer (HLL) type and due to Linde (HLLL). The numerical solutions are validated by a comparison to results obtained with a Navier-Stokes code, a hybrid DSMC/Navier-Stokes method and an experiment. The 10-moment solution is close to the experimental results and obtained more efficiently than the DSMC solution, with considerable speed-up still in store.

1 Introduction

Flow in or around micro-electro-mechanical systems (MEMS) typically is in the so-called transition regime between continuum and free-molecular flow, with Knudsen numbers in the range $0.1 \leq \mathrm{Kn} \leq 10$. In this regime the Navier-Stokes equations, even if allowing for slip at a solid boundary, do not describe the flow with sufficient accuracy. An alternative is to use a model suitable for a higher Kn range, but this comes at a computational penalty. In particular, the Direct-Simulation Monte-Carlo (DSMC) method, a particle-based method, is required for the highest Knudsen numbers, but in the transition regime it has competition from extended-hydrodynamics methods based either on higher-order PDE's or on large sets of first-order moment equations. The DSMC method, stochastic in nature, gives statistical scatter in the produced solutions, and requires a cell size of the order of the molecular mean free path; the PDE-based methods do not. The multi-moment approach has the additional advantage that the equations only include first derivatives, allowing discretization on the narrowest stencils.

In order to avoid using DSMC where it is not required, Sun and Boyd [1] computed the flow over a micro-airfoil using the particle method only in the vicinity of the airfoil, and a Navier-Stokes code farther away. This calls for blending of the two methods in a buffer zone. The present work attempts to solve the same class of flow problems using a multi-moment model everywhere.

2 10-Moment Model

The description chosen is the 10-moment model, which is the best known and most studied among models that use multiple moments of the Boltzmann equation. This model is based on a Gaussian velocity distribution (Gaussian closure) [2]. The general form of the Gaussian velocity distribution $\mathcal{G}$ is as follows.

$$\mathcal{G}(\boldsymbol{x}, \boldsymbol{v}, t) = \frac{n(\boldsymbol{x}, t)}{(2\pi)^{3/2}(\det \Theta)^{1/2}} \exp\left(-\frac{1}{2}\Theta_{ij}^{-1} c_i c_j\right), \tag{1}$$

where

$$\Theta_{ij} = \frac{P_{ij}}{\rho}, \tag{2}$$

$n(\boldsymbol{x}, t)$ is the number density, $\boldsymbol{c}(\boldsymbol{x}, t)$ the random velocity and P_{ij} the generalized stress tensor. The model is equivalent to the Navier-Stokes equations without heat conduction; this is sufficiently accurate for the flow problem studied, which has an almost isothermal solution.

The 10-moment model is derived as follows. Assume the velocity distribution function used with the Boltzmann equation is Gaussian, $\mathcal{G}$, and integrate over all particle velocities. The Gaussian velocity distribution has the mathematical property that third-order velocity moments are zero (leading to zero heat flux), which leads to closure of the set of moment equations. Using the BGK approximation for the collision operator and expressing the equations in vector form in a 3-D Cartesian coordinate system, the 10-moment transport equations assume the form

$$\frac{\partial \boldsymbol{U}^{10}}{\partial t} + \frac{\partial \boldsymbol{F}^{10}}{\partial x} + \frac{\partial \boldsymbol{G}^{10}}{\partial y} + \frac{\partial \boldsymbol{H}^{10}}{\partial z} = \boldsymbol{S}_c^{10}, \tag{3}$$

where $\boldsymbol{U}^{10}$ is the vector of conserved quantities, $(\rho,\ \rho u_x,\ \rho u_y,\ \rho u_z,\ \rho u_x^2 + P_{xx},\ \rho u_x u_y + P_{xy},\ \rho u_x u_z + P_{xz},\ \rho u_y^2 + P_{yy},\ \rho u_y u_z + P_{yz},\ \rho u_z^2 + P_{zz})^T$; $\boldsymbol{F}^{10}, \boldsymbol{G}^{10}$ and $\boldsymbol{H}^{10}$ are the flux vectors, and $\boldsymbol{S}_c^{10}$ is the source vector which contains a factor $1/\tau$, where τ is a characteristic relaxation time related to viscosity and hydrostatic pressure:

$$\tau = \frac{\mu}{p}. \tag{4}$$

3 Numerical Scheme

Among numerical methods for hyperbolic systems, those of the Godunov type have been most successful; these require an algorithm for solving the Riemann problem arising at each cell interface, either exactly or approximately. For large systems of equations it is practical to use an approximate Riemann solver that does not attempt to account for all separate waves through which the cells interact, but lumps the information. Harten, Lax and Van Leer [3] described two families of such methods; the latest member is due to Linde [4]. The HLLL solver uses three waves to cover the domain of influence of the cell interface; it requires only the following knowledge:

- The PDE system is hyperbolic and possesses a convex entropy function;
- maximum and minimum wave speeds are known.

A detailed derivation of the HLLL flux function for the 10-moment equations is presented in [5]. For the integration of the 10-moment equations, the HLLL flux is inserted into Hancock's predictor-corrector version of the MUSCL scheme [6]. Second-order accuracy in space and time is achieved by introducing linear subcell distributions and evaluating flux and source terms halfway during the time step. The half-time (predictor) step which includes gradient-limiting, is done with primitive variables instead of conservative variables $\boldsymbol{U}$ to prevent non-physical values such as negative pressures. The gradients of the primitive variables are obtained by solving least-square problems involving data from all adjacent cells.

4 Numerical Results

4.1 Resolving 1-D shock structure

We present some 1-D results from validation studies in which we tried to produce steady shock profiles for various inflow Mach numbers. Assuming a steady state leads to a system of ordinary differential equations (ODE), which can be solved by a standard fourth-order Runge-Kutta method [7]. The resulting ODE solutions are compared with the solution of the PDE's obtained by the finite-volume method described in the previous section. Upstream and downstream boundary conditions are assumed to be in equilibrium; given the upstream Mach number, density and velocity; downstream conditions are determined from the jump-equations. To avoid constraints by upstream and downstream values, a sufficiently wide computational domain is taken. The cell size is the same as the upstream local mean free path and we assume the monatomic gas is Argon.

In Figures 1 and 2, density distributions are shown for two different upstream Mach numbers ($M_\infty = 1.1, 5.0$), superimposed on ODE results. The PDE-based solutions (symbols) agree well with the solutions obtained by integrating the ODE's describing the steady structures (solid line). Using the ODE-based solutions as a benchmark ($M = 1.1$), the maximum density error was computed for a sequence of grids; its convergence rate demonstrates the second-order spatial accuracy of the method used (Figure 3).

4.2 NACA0012 Micro-Airfoil flow

Next, the external flow around a NACA0012 micro-airfoil is computed using the 10-moment model. The free-stream initial conditions are given in Table

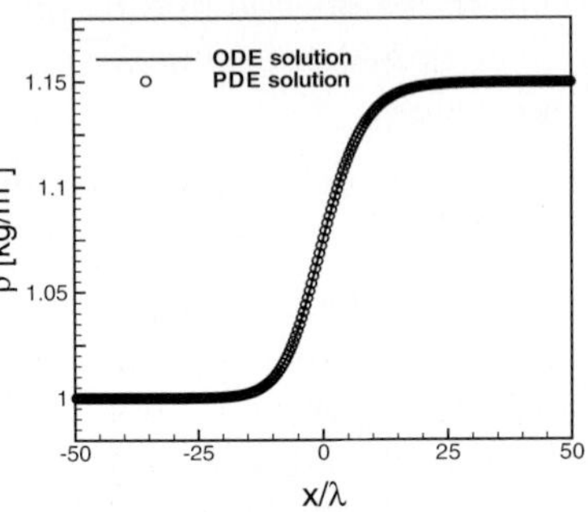

Fig. 1. Density distribution in steady shock structure for $M_{inflow} = 1.1$. The space coordinate is normalized by the upstream mean free path λ

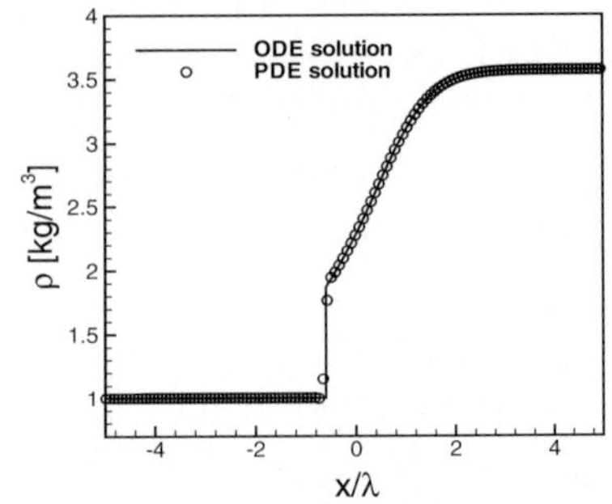

Fig. 2. Density distribution in steady shock structure for $M_{inflow} = 5.0$. A "frozen" shock is followed by a relaxation zone

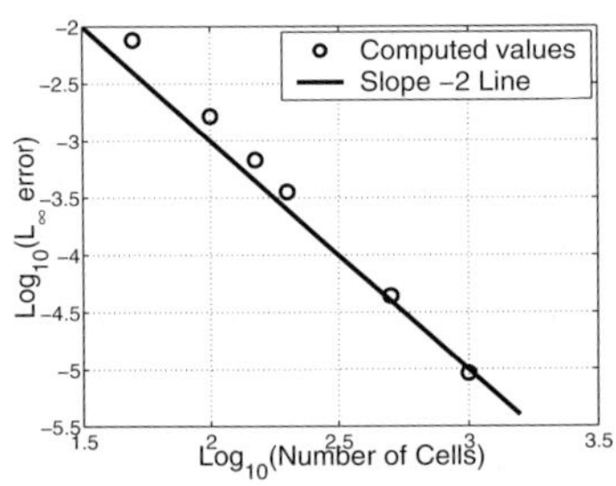

Fig. 3. Grid convergence study of L_∞ density error shows that steady shock solutions ($M = 1.1$) are second-order accurate

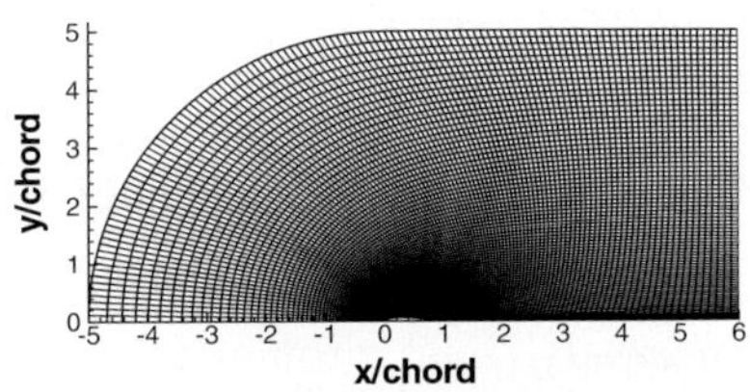

Fig. 4. Computational grid around NACA0012 micro-airfoil. The coordinate is normalized by the cord length. The number of cells is 120 × 76

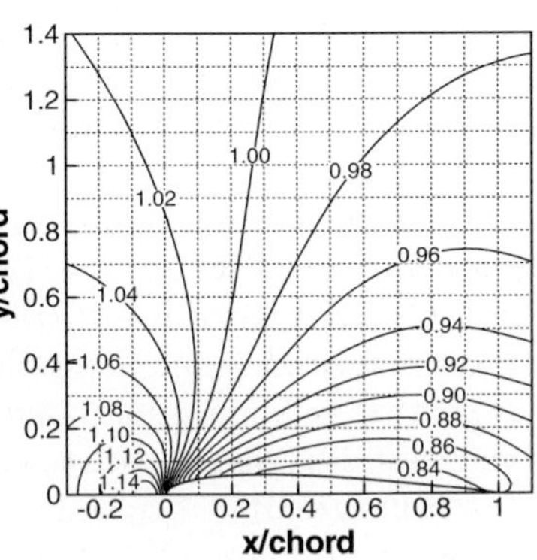

Fig. 5. Density distribution (ρ/ρ_0) around NACA 0012 airfoil by the 10-moment model

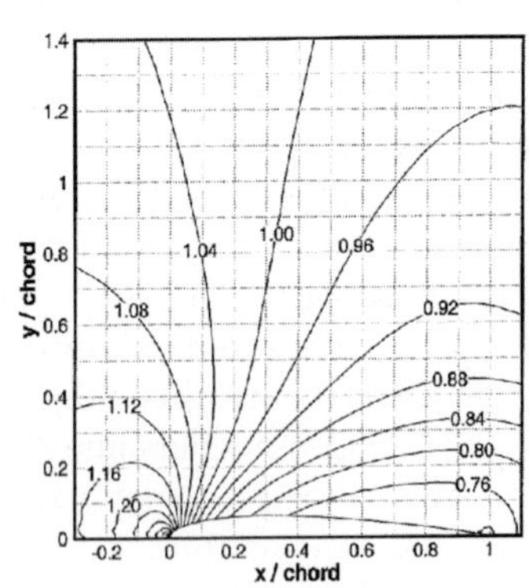

Fig. 6. Density distribution (ρ/ρ_0) around NACA 0012 airfoil by the IP method

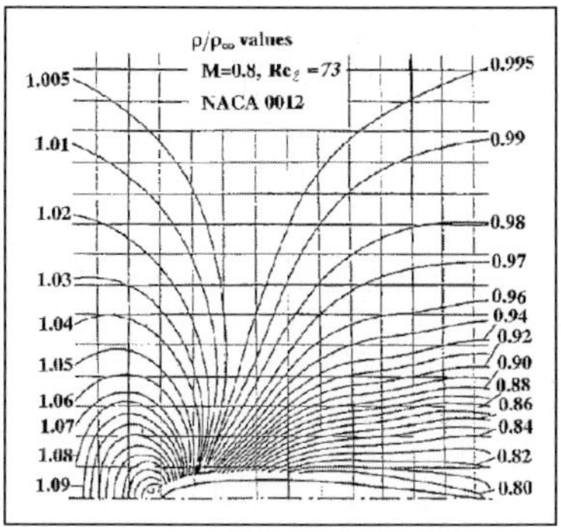

Fig. 7. Density distribution (ρ/ρ_0) around NACA 0012 airfoil by the experiment

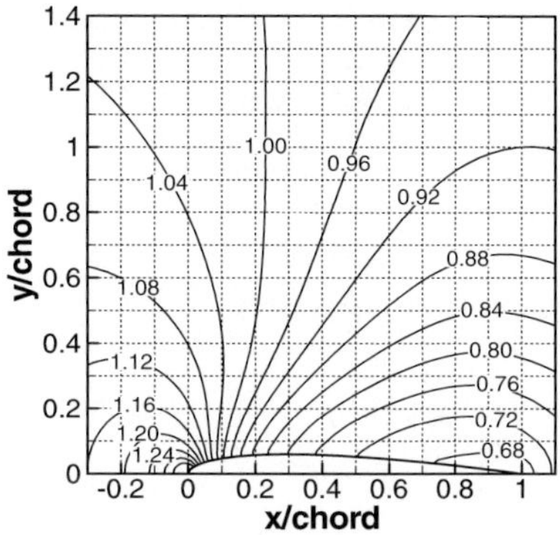

Fig. 8. Density distribution (ρ/ρ_0) around NACA 0012 airfoil by the Navier-Stokes equations

1. The chord length of the airfoil is 0.04 m and a C-type grid is used. The grid geometry is shown in Figure 4. The 10-moment result is shown in Figure 5 with the corresponding Sun-Boyd result reproduced in Figure 6. In this continuum-transition regime, the flow on the wall has a finite velocity. This slip velocity is given by Maxwell's first-order slip boundary condition

$$u_{\text{gas}} - u_{\text{wall}} = \frac{2-\sigma}{\sigma}\lambda \left.\frac{\partial u_t}{\partial n}\right|_w , \tag{5}$$

where σ is an accommodation coefficient and u_t is the tangential component of velocity at the wall. At the airfoil, completely diffuse molecular reflection is assumed in formulating the boundary condition for both methods (achieved by setting $\sigma = 1$). There are clear differences between the solutions, especially upstream of the airfoil. Near the stagnation point the 10-moment approach gives significantly lower density values than the DSMC/NS approach, with the former values expected to be the more accurate ones. This is borne out by the experimental results reproduced from [8] in Figure 7.

Despite the good agreement with the experiment near the leading edge, the normalized density near the trailing edge is slightly higher than the experimental value. This might be improved upon by developing a new set of boundary conditions for the 10-moment model, more accurate than Maxwell's first-order slip model. Navier-Stokes results using the first-order slip boundary condition are shown in Figure 8; they are closer to the DSMC/NS results than to the 10-moment or experimental results. The 10-moment results are obtained in about one third the time it takes the DSMC/NS result to converge; a much greater efficiency gain is still in store.

Table 1. Free stream condition for micro-airfoil flow

M_∞	Re_∞	Kn_∞	$\rho_\infty[kg/m^3]$	$U_\infty[m/s]$	$T_\infty[K]$	$L_{\mathrm{chord}}[m]$
0.8	73	0.017	1.161×10^{-4}	257	257	0.04

Acknowledgement

This work was supported by Grant Nr. F49620-03-1-0226 from the Air Force Office of Scientific Research.

References

1. Quanhua Sun, Iain D. Boyd, and Graham V. Candler. A hybrid continuum/particle approach for modeling subsonic, rarefied gas flows. *Journal of Computational Physics*, 194:256–277, 2004.
2. C. D. Levermore. Moment closure hierarchies for kinetic theories. *Journal of Statistical Physics*, 83(5-6):1021–1065, 1996.
3. Amiram Harten, Peter D. Lax, and Bram Van Leer. On upstream differencing and Godunov-type schemes for hyperbolic conservation laws. *SIAM Review*, 25(1):35–61, Jan. 1983.
4. Timur Linde. A practical, general-purpose, two-state HLL Riemann solver for hyperbolic conservation laws. *International Journal for Numerical Methods in Fluids*, 40(391-402), 2002.
5. Yoshifumi Suzuki and Bram Van Leer. Application of the 10-moment model to MEMS flows. to be presented at 43rd AIAA Aerospace Sciences Meeting and Exhibit, January 2005.
6. Bram van Leer. Upwind and high-resolution methods for compressible flow: From donor cell to residual-distribution schemes. *16th AIAA Computational Fluid Dyanmics*, 2003-3559, June 2003.
7. S. Brown. *Approximate Riemann Solvers for Moment Models of Dilute Gases*. PhD thesis, University of Michigan, 1996.
8. J. Allegre, M. Raffin, and J. C. Lengrand. Experimental flowfields around NACA 0012 airfoils located in subsonic and supersonic rarefied air streams. In M. O. Bristeau, R. Glowinski, J. Periaux, and H. Viviand, editors, *Numerical Simulation of Compressible Navier-Stokes Flows*, volume 18 of *Notes on Numerical Fluid Mechanics*, pages 59–68, Germany, 1987. Friedr. Vieweg and Sohn, Braunschweig.

Part XV

Modelling and Simulation of Turbulence

Calculation of Static and Dynamic Stability Derivatives of the F/A-18E in Abrupt Wing Stall Using RANS and DES

James R. Forsythe[1], Charles M. Fremaux[2], and Robert M. Hall[3]

[1] Cobalt Solutions, LLC, Springfield, OH forsythe@cobaltcfd.com
[2] NASA Langley Research Center, Hampton, VA, c.m.fremaux@larc.nasa.gov
[3] NASA Langley Research Center, Hampton, VA, r.m.hall@larc.nasa.gov

1 Introduction

During envelope expansion flights of the preproduction F/A-18E in the Engineering and Manufacturing Development phase, the aircraft encountered uncommanded lateral activity, which was labeled "wing drop" . An extensive resolution process was undertaken to resolve this issue. A production solution was developed, which included revising the flight control laws and the incorporation of a porous wing fold fairing to eliminate the wing drop tendencies of the pre-production F/A-18E/F. The wing drop events were traced to an abrupt wing stall (AWS) on one side of the wing causing a sudden and severe roll-off in the direction of the stalled wing. Development of a reliable computational tool for prediction of abrupt wing stall would enable designers to screen configurations prior to building the first prototype, reducing costs and limiting risks.

The F/A-18E provides an excellent testing ground for simulation tools due to the large amount of experimental data obtained [1, 2]. Previous computational research [3] focused on predicting the zero sideslip characteristics of the aircraft, including the break in the lift curve slope characteristic of AWS. It was found that by applying Detached-Eddy Simulation (DES) to this problem to predict the unsteady shock motion seen experimentally, a better mean flow prediction could be obtained compared to industry standard Reynolds-averaged (RANS) models [4]. Detached-Eddy Simulation is a hybrid RANS and LES model that for natural applications (i.e. applied as intended) uses RANS in the attached boundary layer, and LES elsewhere [5, 6].

The current work seeks to extend the past computational successes to predicting stability derivatives (both static and dynamic) in the AWS regime. Both Menter's SST RANS model and Spalart-Allmaras based Detached-Eddy Simulation were applied. To assess the accuracy of the simulations, comparisons are made against experiments.

In order to obtain approval for releasing this paper to the public, quantitative information has been removed from most vertical scales as per guidelines from the Department of Defense.

2 Governing Equations and Flow Solver

The commercial unstructured flow solver *Cobalt* was used [7]. The numerical method is a cell-centered finite-volume approach applicable to arbitrary cell topologies. The spatial operator uses an exact Riemann Solver, least squares gradient calculations using QR factorization to provide second-order accuracy in space, and TVD flux limiters to limit extremes at cell faces. A point implicit method using analytic first-order inviscid and viscous Jacobians is used for advancement of the discretized system. For time-accurate computations, a Newton sub-iteration scheme is employed, and the method is second-order accurate in time. The compressible Navier-Stokes equations were solved with an arbitrary Lagrangian Eulerian method to handle rigid body motion.

2.1 Menter's Shear Stress Transport RANS Model

In order to provide a baseline for comparison, computations were performed with the Menter's Shear Stress Transport (SST) model [8]. The method is a blend of the $k-\epsilon$ and $k-\omega$ models, using a parameter F_1 to switch from $k-\omega$ to $k-\epsilon$ in the wake region.

2.2 Detached-Eddy Simulation

The original DES formulation is based on a modification to the Spalart-Allmaras RANS model [9] such that the model reduces to its RANS formulation near solid surfaces and to a subgrid model away from the wall [5]. The DES formulation replaces in the S-A model the distance to the nearest wall, d, by $\widetilde{d}$, where $\widetilde{d} \equiv \min(d, C_{DES}\Delta)$. In "natural" applications of DES, the wall-parallel grid spacings are at least on the order of the boundary layer thickness and the S-A RANS model is retained throughout the boundary layer, i.e., $\widetilde{d} = d$. Consequently, prediction of boundary layer separation is determined in the 'RANS mode' of DES. Away from solid boundaries, the closure is a one-equation model for the sub-grid-scale (SGS) eddy viscosity. When the production and destruction terms of the model are balanced, the length scale $\widetilde{d} = C_{DES}\Delta$ in the LES region yields a Smagorinsky eddy viscosity $\widetilde{\nu} \propto S\Delta^2$. The additional model constant $C_{DES} = 0.65$ was set in homogeneous turbulence, and was used in the following calculations.

For the current runs, the grid contained tight clustering around the wing fold fairing and on the surface that reduced Δ low enough to make the RANS-LES interface occur in the boundary layer. To remedy this situation, the DES length scale was modified according to the equation $\widetilde{d} = \min(C_{des}\max(n^2 C_{des}\Delta^2/d, \Delta), d)$ where n is the ratio of the new RANS-LES interface height to the original height. Above the interface, the length scale draws down smoothly to $C_{des}\Delta$. For the current simulations $n = 3$ was chosen to push the interface outside of the boundary layer. Methods that detect the edge of the boundary layer would be far preferable to reduce the burden on the users.

3 Results

The configuration examined was an 8% scale pre-production F/A-18E with 6°/8°/4° flaps set. The Mach number for all cases was 0.9, with a freestream static pressure and temperature of 8.575 psi and 518.5° R respectively, leading to a chord based Reynolds number of 3.9×10^6. The wind tunnel comparisons are from the model tested in NASA Langley's 16 ft Transonic Tunnel (16TT).

The grids used were unstructured grids created using the tetrahedral grid generator VGRIDns [10]. The *Cobalt* utility *blacksmith* was used to recombine the high aspect ratio tetrahedra in the boundary layer into prisms. The grid was 8.4×10^6 cells for both sides of the aircraft. The average first y^+ for the grid was < 0.7. The grid was created by performing a solution based adaptation using the time averaged results of a run at 9° angle of attack, using NASA Langley grid adaption code RefineMesh and the process described in reference [4]. Although the absolute number of cells would normally be considered quite coarse for a DES of a full aircraft, the grid cells were tightly clustered in the separation bubble because of the method of grid adaption.

3.1 Static Lateral Stability

Calculations were performed with various bank (ϕ) and pitch (θ) angles. RANS runs were performed using large timesteps to converge to steady state, while DES runs were run at a non-dimensional timestep (by chord and freestream velocity) of 0.01 (determined in reference [4]). Time-averages for the DES were taken over 120 non-dimensional time-units, which increased the cost of DES over RANS by about a factor of eight. For the experiments and computations, the pitch angle was held fixed, and the model rolled around the longitudinal axis of the aircraft. This leads to a reduction in alpha, and an increase in beta. Thus the calculations do not strictly give derivatives with respect to beta.

Figure 1 shows rolling moment vs. bank angle for four different pitch angles around the AWS regime. Note that for the experiments all four cases exhibit lateral static stability about $\phi = 0°$. However, for the pitch angles greater than 7°, there are asymmetries in the data, and there appear to be significant rolling moments at zero bank. For the 9° case, model dynamics due to unsteady shock motion were strong enough to prevent taking a full set of data. The asymmetries and gaps in the experimental data make comparison difficult. All DES results predicted strong shock oscillations due to separation as was seen previously in [4] to give more realistic smeared mean pressure distributions. At $\theta = 7°$ (Figure 1a) the SST RANS results (which show separation near the trailing edge at zero bank) are in good agreement for the low bank angles, but reverse sign at $\phi = 30°$. Examination of flow visualizations showed this reversal to be caused by separation from the leading edge of the upwind wing. DES results for this angle showed a reduced slope, likely caused by the fact that at this angle DES was separated further forward on the wing, in the mean. For the 8.5° and 10° cases there was a significant rolling moment at near zero bank for both the experiments and the DES. Running time averages of the DES rolling moment for 10° were examined and tended to stabilize at this non-zero moment despite averaging over one second of full scale time. So even if longer time-averages would tend towards zero, there is a rolling moment over a time significantly long enough to affect the aircraft motion. The RANS runs did not show this zero bank rolling moment, probably because the moment seems to be caused by low frequency shock oscillation, which was not captured in the RANS. Although not shown, yawing moment and side force were also plotted and compared to the experiments, with excellent agreement for both the RANS and DES. This is likely due to the fact that these coefficients are dominated by the flow around the vertical stabilizers, which is attached and easily predicted.

3.2 Dynamic Lateral Stability

Dynamic lateral stability was also examined by performing forced sinusoidal roll oscillations around the body axis. A roll rate was picked in order to stay low enough to be in the linear range as much as possible, yet have a large enough effect on the moments to be distinguishable from the effects of unsteady shock oscillation. An amplitude (5°) and frequency were chosen to give a peak non-dimensional roll rate (when passing through wings level) that would induce an effective angle of attack change of 1° at the wingtip, i.e. $(\partial\phi/\partial t)b/2U_\infty = tan(1°) = 0.0175$. For both the RANS and DES a nondimensional timestep of 0.02 was used, which was verified by a timestep study to be small enough to adequately resolve the motion. Since DES predicted strong shock oscillations, it was necessary to phase average over multiple cycles to obtain roll damping information. Most RANS cases gave the same (or nearly the same) results for two subsequent oscillations, so phase averaging was not required. So although the cost of the RANS and DES was the same per cycle (since both were run time-accurate), the cost of DES was higher due to the need to phase-average over multiple cycles.

Figure 2 plots the rolling moment vs. roll rate. The left and right hand sides of the loops correspond to when the aircraft is passing through wings level rolling either left or right respectively. Rolling moments of opposite sign to the roll rate yield stable roll damping (i.e. negative slope of a linear fit through the loops). The SST results exhibited stable roll damping in all cases, but for the 7° case had a significant rolling moment offset (i.e. a rolling moment with zero roll rate). This was seen from flow visualizations to be caused by the flow separating on one wing at the leading edge, and the other at the trailing edge. This is likely caused by the starting procedure which was used to begin the simulation by rolling from wings level towards one side. A hysterisis effect could then allow that wing to stay separated while the other remained attached. The reduction in roll damping for angles above 7° is caused by the drop in the lift curve slope once the flow separates from the leading edge of the wing. DES results showed highly non-linear rolling moments over each cycle and near zero roll damping (in a linear derivative sense) for 9°, although it would be useful to run more cycles to ensure that there are su cient samples in the phase averages. At this angle (and the angles close to it) the variations in rolling moment due to shock unsteadiness are far stronger than the effect of the motion.

4 Conclusions

RANS and DES calculations were performed for the F/A-18E to predict static and dynamic stability derivatives in the challenging abrupt wing stall regime. Comparison to experiments for static stability derivatives demonstrated that DES picked up zero bank rolling moment offsets which could be a trigger for wing drop. DES also predicted a strong reduction of roll damping in the AWS

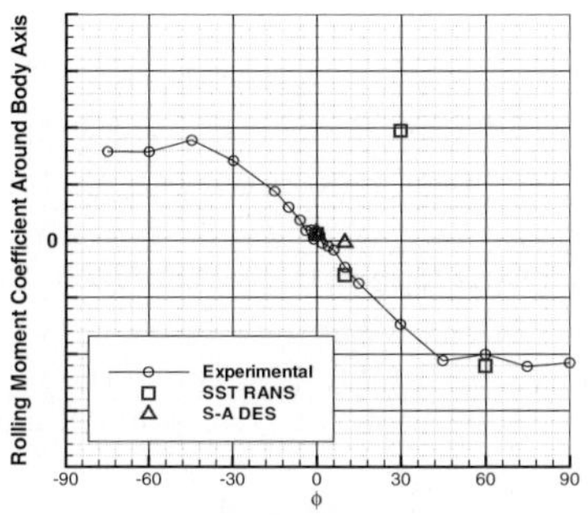

(a) Run 247, $\theta \approx 7.0°$

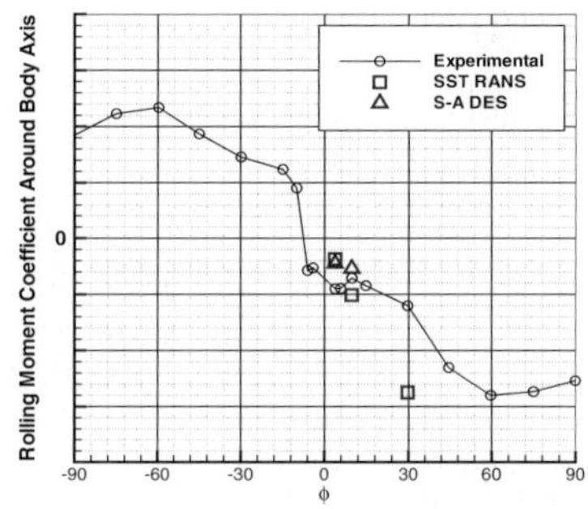

(b) Run 240, $\theta \approx 8.5°$

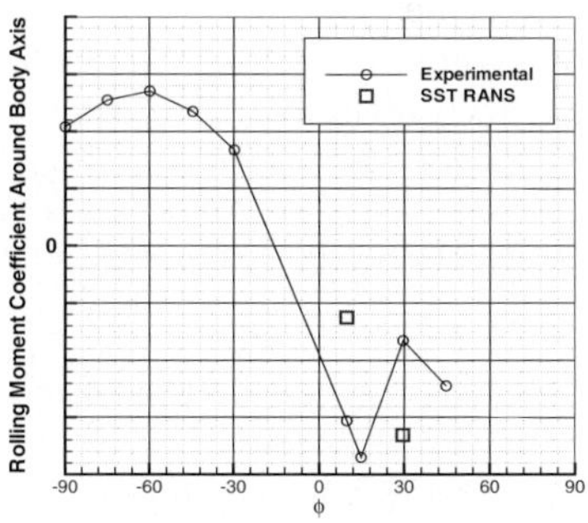

(c) Run 242, $\theta \approx 9.0°$

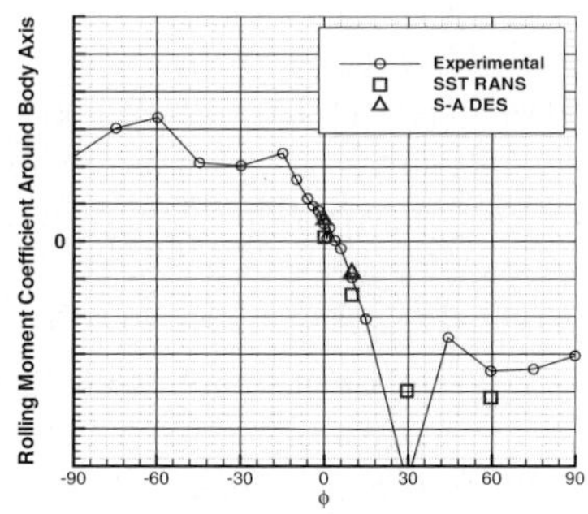

(d) Run 244, $\theta \approx 10.0°$

Fig. 1. Rolling moment coefficient vs. bank angle

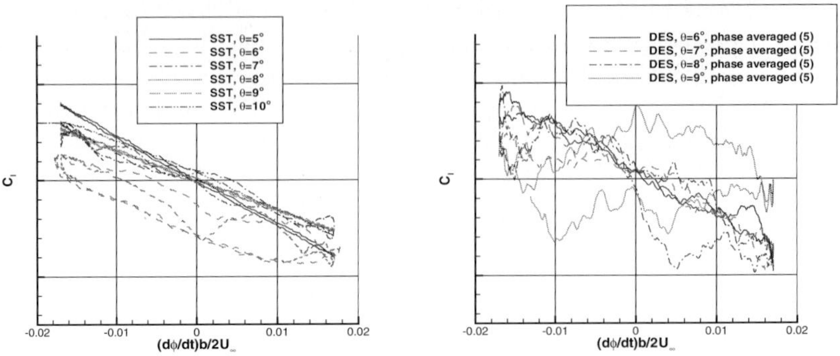

(a) SST results

(b) DES results phased averaged over five cycles

Fig. 2. Forced oscillation runs, Rolling moment coefficient vs. roll rate

regime that could also contribute to wing drop. More detailed comparisons to experiments is ongoing.

References

1. Lamar, J., Hall, R., "AWS Figures of Merit Developed Parameters from Static Transonic Model Tests," emphAIAA 03-0745, Jan 2003.
2. Owens, B., Brandon, J., Capone, F., Hall, R., Cunningham, K., "Free-to-Roll Analysis of Abrupt Wing Stall on Military Aircraft at Transonic Speeds," *AIAA 03-0750*, Jan 2003.
3. Woodson, S.H., Green, B.E., Chung, J.J., Grove, D.V., Parikh, P.C., Forsythe, J.R., "Recommendations for CFD Procedures for Predicting Abrupt Wing Stall," *AIAA 2003-0923*, Jan 2003.
4. Forsythe, J.R., Woodson, S.H., "Unsteady CFD Calculations of Abrupt Wing Stall using Detached-Eddy Simulation," *AIAA 2003-0594*, Jan 2003.
5. Spalart, P. R. , Jou W-H. , Strelets M. , and Allmaras, S. R., "Comments on the Feasibility of LES for Wings, and on a Hybrid RANS/LES Approach," *Advances in DNS/LES, 1st AFOSR Int. Conf. on DNS/LES*, Aug 4-8, 1997, Greyden Press, Columbus Oh.
6. Strelets, M., "Detached-Eddy Simulation of Massively Separated Flows," *AIAA 01-0879*, Jan 2001.
7. `www.cobaltcfd.com`
8. Menter, F.R., 1994, "Two-Equation Eddy-Viscosity Turbulence Models for Engineering Applications," *AIAA Journal*, 32(8), pp 1598–1605.
9. Spalart, P.R. and Allmaras, S.R., 1994, "A One-Equation Turbulence Model for Aerodynamic Flows," *La Recherche Aerospatiale* **1**, pp. 5-21.
10. Pirzadeh, S., "Three-dimensional Unstructured Viscous Grids by the Advancing Layers Method," *AIAA Journal*, V. 34, pp. 43-49.

Large Eddy Simulation of Flow Around a Slat with a Blunt Trailing Edge

Saloua Ben Khelil

ONERA, Applied Aerodynamics Department. BP 72, 92322 Chtillon Cedex
France. khelil@onera.fr

Abstract

A numerical prediction of the aerodynamic noise generated by high-lift devices (HLD) is performed using Large Eddy Simulation (LES) approach. Emphasis is placed on accurate simulation of the local flow field of a slat with a blunt trailing edge. Therefore, the LES zone is limited to the slat region thanks to an innovative grid technique computation using selective 2D/3D domains.

1 Introduction

The aerodynamic noise generated by the high-lift devices (slats and flaps) of transport aircraft is an important contributor to the total radiated airframe noise, especially in the approach phase. In particular, the leading edge slat and the side edges of the flaps have been identified as dominant noise sources [1]. The present computational study focuses on the relevant aeroacoustic flow features associated with a slat. For several years, ONERA has been pursuing the objective of developing an hybrid Computational AeroAcoustics (CAA) process with a capability to fully simulate the aeroacoustic mechanism. This process includes the noise source generation by turbulent flow around airfoils and the mid- and far field noise propagation. It combines CFD techniques and acoustic numerical methods, each one being adapted to a particular domain in which specific physical mechanisms are simulated solving an adequate set of equations. In previous studies, the CAA hybrid method has been successfully applied to the configuration of a lifting NACA0012 airfoil with a blunt trailing edge [2]. The objective behind this work is to apply the same process to simulate the aerodynamic noise generated by a high-lift configuration. The first step of this hybrid method is to identify and capture the unsteady aerodynamic noise sources using a 3D unsteady simulation of the flow near the wing profile. Generally, URANS methods are used to compute unsteady flows around airfoils for aeroacoustic predictions, but Large Eddy Simulation (LES) is often considered as one of the most promising method in the context of CAA applied to high-lift noise. However, the LES simulation of the entire high-lift configuration is still beyond the capabilities of available supercomputers. We thus have chosen to perform only the unsteady three dimensional simulation of the flow in the slat cove of a high-lift thanks to an innovative grid technique computation using selective 2D/3D domains [3]. In this paper, the main features of the LES is reminded first. Then, the high-lift geometry and the computational grid are presented. Finally some results are given revealing the existence of several flow features responsible for noise generation.

2 Large Eddy Simulation

The LES method has been implemented in the FLU3M solver, an industrial CFD software which has been developed at ONERA for several years [4]. This solver is based on the discretization of the compressible Navier-Stokes equation on multiblock structured meshes by a cell-centered finite volume technique. The viscous fluxes are discretized by a second order accurate centered scheme. A second order accurate implicit temporal integration is achieved thanks to an approximate Newton method. Usually LES requires high-order centered scheme for the Euler fluxes discretization (with spectral-like resolution) in order to minimize dissipative numerical errors. However, such scheme cannot be applied easily on complex geometry. As several works (see for instance [5]) have shown that LES can be carried out with low-order centered scheme in case of sufficient mesh resolution, only second-order accurate scheme is employed in this study but a special effort has been carried out to minimize the intrinsic dissipation of the scheme [3]. For the sub-grid scale modelling the selective mixed scale model, developed by Sagaut and Lenormand [6], has been retained, because it realized a good compromise between accuracy, stability and computational cost. More particularly, the use of a selective function allows to handle transitional flows [7]. Otherwise, in order to reduce the total number of grid, a local mesh refinement technique is used: a 2D/3D coupling strategy [3]. In such simulations, the flow is laminar and two-dimensional in a large part of the computational domain. Thus three-dimensional computation can be limited to a zone close to the airfoil where the flow is turbulent, whereas two-dimensional simulations are used for the far field. This strategy which does not alter the accuracy of the simulation, allows the use of a mesh with LES resolution, while keeping the total number of grid points at a reasonable level.

3 High-Lift Configuration

The configuration considered here is two-dimensional (constant section in the spanwise direction) and includes three elements, a leading edge slat with a blunt trailing edge and with 23° deflection, the main element and a trailing edge flap with 17° deflection. Such configurations generate very complex flow-field with the presence of separation, wakes and boundary layer mixing. The computation is performed at Mach number of 0.19 and angle of attack of 4°. It corresponds to a Reynolds number of 2.5 10^6 based on the stowed chord length (0.61 m). A steady, 2D RANS computation is first carried out over the full high-lift configuration using the Spalart-Allmaras model [8], on a 413,890 grid points. This computation provides an initial flow solution to the LES computations. For the LES simulation the 2D/3D coupling strategy is used. The three-dimensional computation is limited to the slat cove region (colored in black in figure 1 (left)) whereas the two-dimensional simulation is limited to the regions of the main element, the flap and the far field. The spanwise extent is about 36% of the slat chord with 52 mesh points in the three-dimensional domain. Figure 1 (right) presents an overview of the mesh in the vicinity of the slat. The total number of the grid is about 7.3 million points. The mesh resolution in wall unit, usually required by LES of wall bounded flow ($\Delta y^+ \leq 2$ (normal direction), $\Delta x^+ \leq 20$ (streamwise direction) and $\Delta z^+ \leq 100$ (spanwise direction)) is almost reached everywhere (figure 2). A non-slip condition is applied at the high-lift surface and a periodic condition is imposed in the spanwise direction. Non-reflective characteristic boundary conditions are applied for the far field. In order to ensure the time-accuracy of the results, the physical time step is fixed at 0.2 μs, meaning a sampling frequency of 5 MHz. After an initial phase, the useful computation was performed with a total duration of 20 ms, representing the convection of the flow around 2.14 airfoil chord, corresponding to 490 CPU Hours on a single processor of NEC SX5.

4 Results and Discussion

Figure 3 compares the mean pressure coefficient obtained with the LES and 2D steady RANS computations. The agreement is globally good except on the slat and the leading edge of the main element. These differences are confirmed when comparing the mean flow streamlines inside the slat cove between the two computations (figure 4). Indeed, an extra separation zone at the leading edge of the main element is present in the LES result. These results highlight additional structures which are captured by LES computation in comparison with the 2D steady RANS approach. Figure 5 (left) shows an instantaneous Schlieren-like view of the field. It highlights several phenomena. The iso-contours clearly display the spatial location of the free shear layer. A mixing layer is developed from the cusp of the slat in the shear region and borders the main separation bubble on the lower side of the slat. The mixing layer is convected through the slot, then it slightly impacts on the lower side of the main element and in a more important way on the lower side of the slat. We can also notice the formation of discrete vortices which roll-up the shear layer. This vortices are convected and amalgamate as they approach the reattachment point. At the reattachment point, some of the vortices turn upstream and get trapped in the cove recirculation flow field, while some of them go downstream and interact with the slat wake . Moreover, the bubble in the leading edge of the main element interacts with the slat wake, creating thus a very turbulent zone close to the wing, after the reattachment point. The slat wake vortex shedding behind the blunted trailing edge is also clearly visible in the figure. Figure 5 (right) displays an instantaneous pressure fluctuations fields near the slat trailing-edge (right). At every point of LES grid, pressure fluctuations are computed by subtracting the time-average pressure to the instantaneous pressure. It shows the propagating waves on the upper side of the slat. Concentric waves are observed near the trailing edge with a wave length corresponding to the vortex shedding frequency of 30 kHz. This wave pattern vanishes within a short spatial distance from the airfoil because of the radial stretching in the 2D zones of the LES grid, which acts on the noise field as a low-pass filter. Three pressure spectrum are shown in figure 6 on different points (figure 5). On the first point on the cusp of the slat, we distinguish three peaks around 4 kHz, 7.8 kHz and 30 kHz. The pressure spectrum on the recirculation bubble (point 2) appears to be a full spectrum due to the high level of turbulence observed in the slot. A main frequency of 30 kHz is detected on the third point on the slat trailing edge corresponding to the frequency of the vortex shedding.

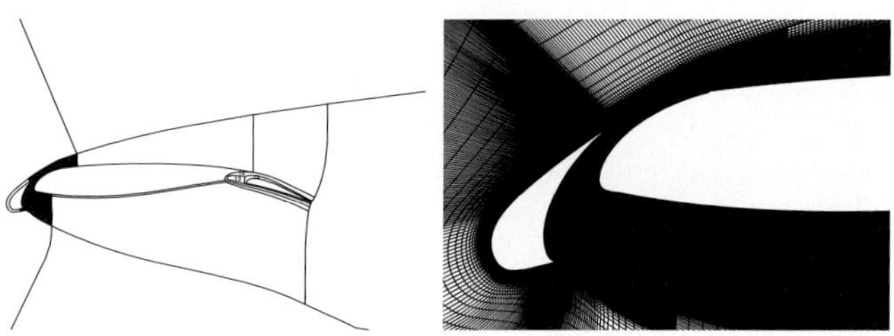

Fig. 1. Left:2D/3D location. Right:Grid distribution

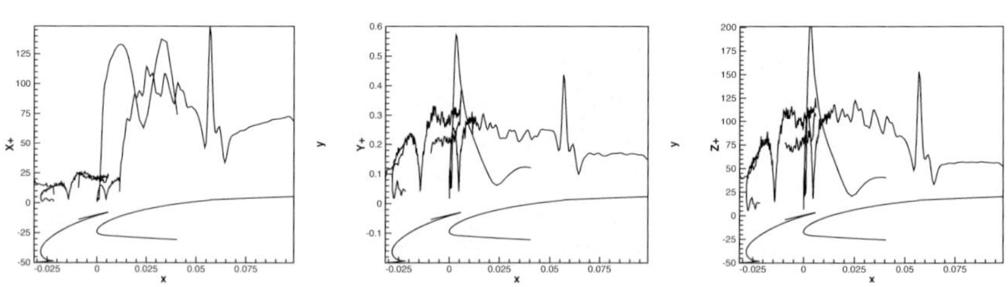

Fig. 2. Evolution of the mesh resolution on the 3D zone in wall unit

5 Conclusion

Accurate LES simulation was carried out to study the turbulent 3D unsteady flow around a slat cove of a high-lift wing. Thanks to the use of local mesh refinement and adapted discretization of the convective fluxes, the LES computational resolution was adequate to capture the details of the local flow field that are relevant to airframe noise generation. Analysis of the unsteady data revealed the presence of several flow features, which are associated to aerodynamic noise emission. The next step of this work is, firstly, a more precise analysis of the flow field in the slat zone in order to get a better

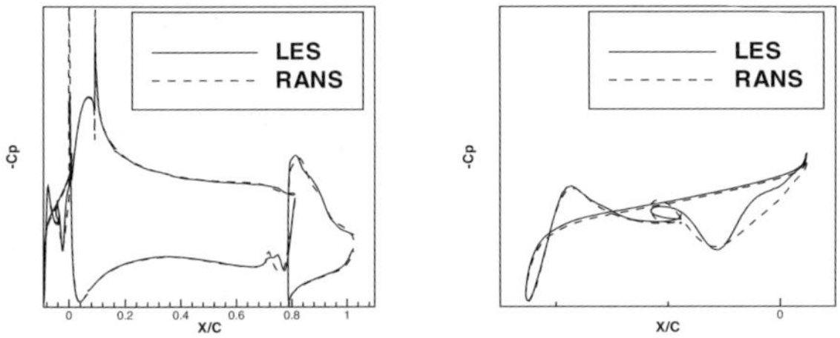

Fig. 3. Mean pressure coefficient. Right: slat

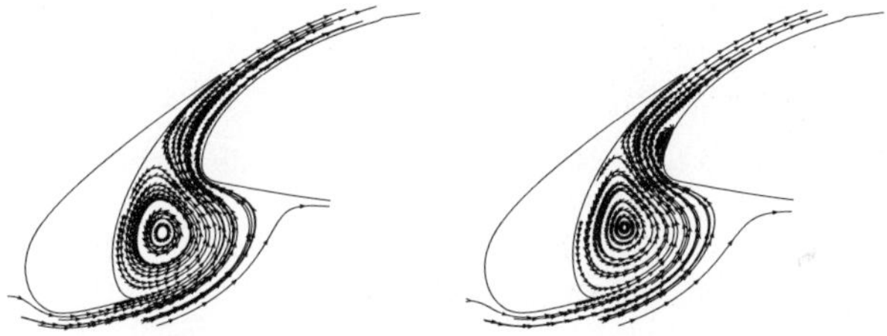

Fig. 4. Mean flow streamlines. Left: RANS computation. Right: LES computation

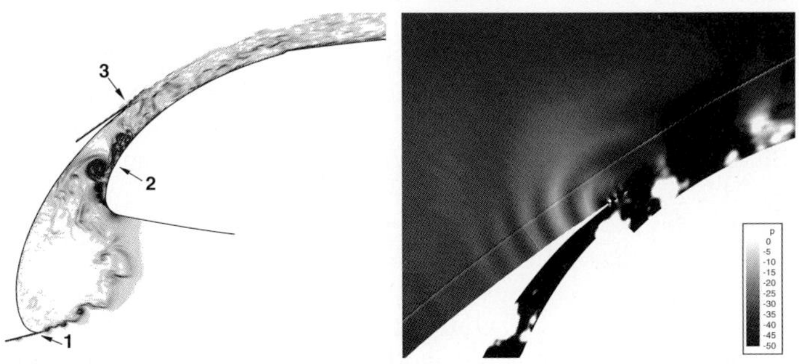

Fig. 5. Left: instantaneous Schlieren-like view. Right: instantaneous iso-values of pressure fluctuations

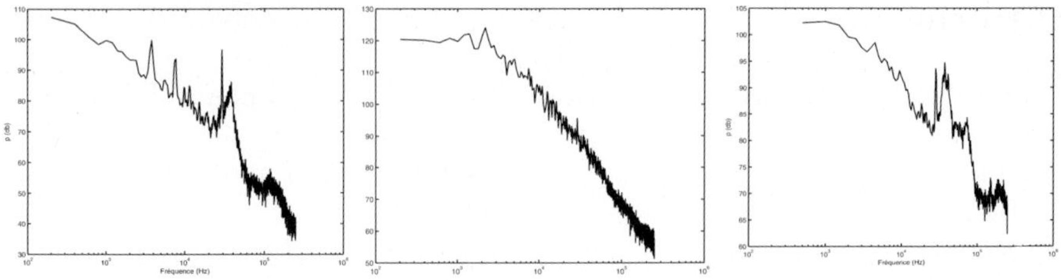

Fig. 6. Pressure spectrum. Left: at the cusp of the slat(point 1) , middle: on the leading-edge of the main element(point 2), right: at the trailing-edge (point 3)

understanding of the flow physics. Then, within the framework of the CAA process, acoustic techniques will be used for the prediction of the mid- and far field noise propagation.

6 Acknowledgements

This work was funded by the SPA (French Agency for Aeronautical Programs). The author is grateful to I. Mary for his help to use the LES, F. Moens for the choice of the high-lift configuration and Ch. Franois who has generated the mesh.

References

1. R. Davy, H. Remy. Airframe noise characteristics of a 1/11 scale Airbus model. *AIAA paper 98-2335*, 1998.
2. E. Manoha, C. Delahay, P. Sagaut, I. Mary, S. Ben Khelil, Ph. Guillen. Numerical prediction of the unsteady flow and radiated noise from a 3D lifting airfoil. 7th *AIAA/CEAS Aeroacoustics Conference. AIAA 2001-2133*, May 28-29 2001.
3. I. Mary, P. Sagaut. Large Eddy Simulation of Flow around an airfoil. *AIAA paper 2001-2559. 15th CFD Conference, Anaheim (Ca)*, 11014 June 2001.
4. M. Pchier, Ph. Guillen, R. Gayzac. Magnus Effect over Finned Projectiles. *J. of Aircraft and Rockets*, 2001.
5. X. Wu, R. Jacobs, J. Hunf, P. Durbin. Simulation of Boundary Layer Transition Induced by Periodically Passing Wake. *J. Fluid Mech,, col 398, 109-153*, 1999.
6. E. Lenormand, P. Sagaut, L. Ta Phuoc, and P. Comte . Subgrid-Scale Models for Large Eddy Simulations of Compressible Wall Bounded Flows at Moderate Reynolds Number. *AIAA Journal*, 38, 2000.
7. E. David . *Modlisation des coulements compressibles et hypersoniques : une approche instationnaire.* PhD thesis, Institut National Polytechnique de Grenoble, 1999.
8. S. Deck, Ph. Duveau, P. d'Espiney, Ph. Guillen. Development and application of Spalart Allmaras one equation turbulence model to three-dimensionnal supersonic complex configurations. *Aerospace Science and Technology*, 6(3):171–183, March 2002.

Implicit Large Eddy Simulation of a Flow Around a Subsonic Airfoil Near its Stall Angle

Satoko Komurasaki[1] and Kunio Kuwahara[2]

[1] Nihon University, Kanda-Surugadai, Chiyoda-ku, Tokyo 101-8308, Japan
[2] Institute of Space and Astronautical Science, Yoshinodai, Sagamihara, Kanagawa 229-8510, Japan

Summary. In the present paper, a simulation of a subsonic flow around an airfoil near its stall angle, at Reynolds number of 10^6, is investigated. In this computation, the 3D time-dependent incompressible Navier-Stokes equations are solved by the multi-directional finite-difference method. A curvilinear coordinate system of O-type topology is used. No explicit turbulence models are employed, but a third-order upwind scheme is adopted. In the present computation, Cl agrees well with the experimental values near the stall angle. At the angle of attack just before the stall, the value of Cl fluctuates, and a flow near the leading edge is unsteady, even after the flow is developed fully.

1 Introduction

A flow around an airfoil is one of the most fundamental problems in aerodynamics. Many simulations have been done but some important problems still remain unsolved.

In the present paper, as one of those unsolved problems, a simulation of a subsonic flow over an airfoil near the stall angle, is attacked. In the computation, the time-dependent incompressible Navier-Stokes equations are solved by finite-difference approximation using O-grid without any turbulence model. An approach similar in philosophy but different in method is adopted by Boris et al. (1992). Most successful simulations of this kind at high Reynolds numbers are based on the third-order upwind formulation (Kawamura and Kuwahara, 1984). To increase the accuracy, we employ the multi-directional finite-difference method (Suito et al., 1995, Kuwahara, 1999).

Kuwahara et al. simulated the flow around a subsonic airfoil in 2D (2000) and 3D (2001). The 3D computations (2001) were carried out with free-slip boundary conditions in the spanwise direction, and their results agreed with experiments (Abbott et al., 1959).

In the present article, the periodic boundary condition is employed in the spanwise direction. Under these conditions, the flow over NACA0012 airfoil is simulated near the stall angle.

2 Computational Method

The governing equations are the 3-d incompressible Navier-Stokes equations and the equation of continuity as follows:

$$\mathrm{div}\boldsymbol{u} = 0 \tag{1}$$

$$\frac{\partial \boldsymbol{u}}{\partial t} + \boldsymbol{u} \cdot \mathrm{grad}\boldsymbol{u} = -\mathrm{grad}p + \frac{1}{Re}\triangle \boldsymbol{u}. \tag{2}$$

where $\boldsymbol{u}, p, t$ and Re denote the velocity vector, pressure, time and the Reynolds number respectively. For high-Reynolds-number flows, time-dependent computations are required owing to the strong unsteadiness.

The numerical procedure is based on the projection method. The pressure field is obtained by solving the following Poisson's equation:

$$\triangle p = -\mathrm{div}(\boldsymbol{u} \cdot \mathrm{grad}\boldsymbol{u}) + \frac{D^n}{\delta t} \tag{3}$$

$$D = \mathrm{div}\boldsymbol{u},$$

where n is the time step and δt is the time increment. D^{n+1} is assumed to be zero, but D^n is retained as a corrective term.

The equations are discretized based on the multi-directional finite-difference method. In case of 2D computations, when structured grid points are given, the black points in Fig.1(a) are usually used to approximate the derivatives at the central point (system A). If we introduce another 45° rotated grid system, the white points in Fig.1(b) can be used to approximate the derivative at the central point (system B). In order to improve the derivative value at the central point, the values from both systems are used. If a ratio $A : B = 2/3 : 1/3$ is adopted, the resulting finite-difference scheme for the Laplacian coincides with well-known 9 point formula with fourth-order accuracy. This method improves the rotational invariance of the coordinate system, then those cases where flow direction is not parallel to the grid direction, are better simulated. In three-dimensions, three different grid systems are used. Each grid system is obtained by rotating a perpendicular plane 45° with respect to each coordinate axis. One of such systems is shown in Fig.1(c).

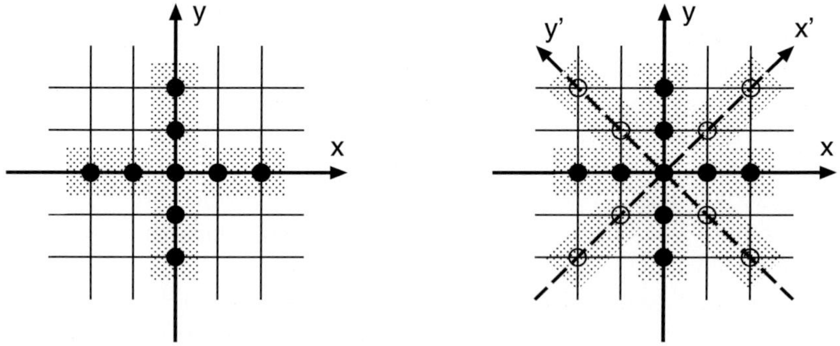

(a) System A. (b) System B.

Fig. 1. (a), (b) Grid for multi-directional scheme for 2D case

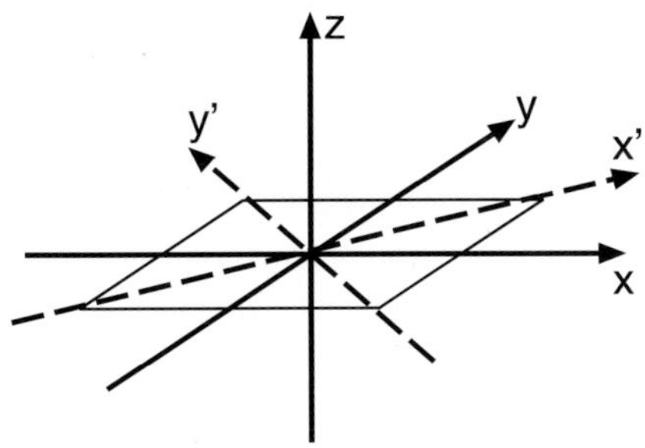

(c) System Z: $x' - y' - z$.

Fig. 1. (c) Grid for multi-directional scheme for 3D case

For the convective terms, a third-order upwind scheme is used to stabilize the computation. This is the most important point for high-Reynolds-number computations and the detail is given below.

Strong numerical instability caused by the aliasing error occurs at high Reynolds numbers, owing to the non-linear convection terms, if enough grid points are not used to resolve the small-scale structures. When digitizing a continuous function into a finite number of the values, it is very important to filter out the high-frequency part of the original function which can not be resolved by the digital system. If not, aliasing error makes the approximation meaningless.

Usually a turbulence model or a large eddy simulation is used to get rid of this instability. The diffusion coefficients increased by the added turbulent viscosity reduce the aliasing error and suppress the numerical instability. In most of the models, this diffusion has the same form as the viscous diffusion and the diffusion coefficient is usually much larger than that of the viscous diffusion. Therefore, the effect of physical diffusion is concealed, resulting no dependency of the flow on the Reynolds number is captured.

Another way to stabilize the computation is to use an upwind scheme. A third-order upwind scheme has been found to be most suitable for high-Reynolds-number-flow computation, and it is given by,

$$u\frac{\partial u}{\partial x} \approx u_k \cdot \frac{-u_{k+2} + 8(u_{k+1} - u_{k-1}) + u_{k-2}}{12\Delta x} + |u_k| \cdot \frac{u_{k+2} - 4u_{k+1} + 6u_k - 4u_{k-1} + u_{k-2}}{4\Delta x}. \tag{4}$$

If the first and second terms on the right-hand side of Eq. (4) are developed into Taylor series, and following terms are derived respectively,

$$\frac{\partial u}{\partial x} + O(\Delta x^5) \qquad \text{and} \qquad \frac{\Delta x^3}{4} \cdot \frac{\partial^4 u}{\partial x^4} + O(\Delta x^5).$$

The leading numerical error term of Eq. (4) contains a fourth-order derivative, where the effects of the second numerical diffusions are carefully removed. The

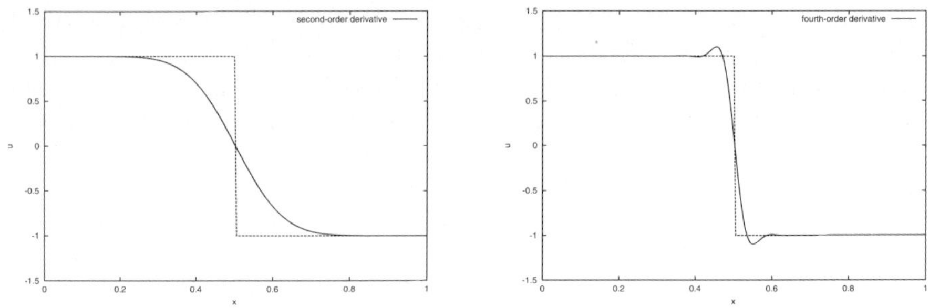

(a) Second-order diffusion. (b) Fourth-order diffusion.

Fig. 2. Two kinds of diffusion

numerical diffusion of fourth-order derivatives is of short range and does not conceal the effect of molecular diffusion, but at the same time stabilizes the computation.

The second-order Crank-Nicolson implicit scheme is used for time integration. The equations are iteratively solved at each time step by SOR method. For Poisson's equation, a multi-grid method is utilized. The computation is done with the periodic boundary condition in the spanwise direction.

A generalized coordinates system is employed, and grid points can be concentrated near the body surface. For the airfoil simulation, O-grid is adopted here with a grid size of 129*65*65 (Fig.3).

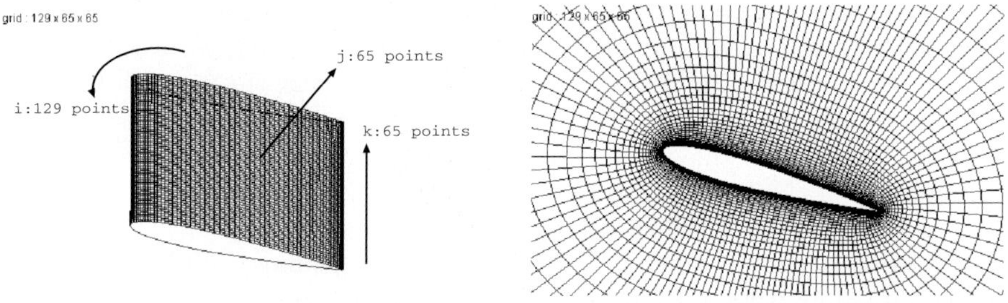

Fig. 3. Computational grid

3 Computational Results

Three-dimensional flows around NACA0012 airfoil are simulated at Reynolds number of 10^6. Fully developed 2-d flow is used as an initial condition for 3-d computation to save the computation time.

The flow pattern is visualized using Fig.4-8. Section lift coefficients Cl are given in Fig.4. They are computational and experimental values at each angle of attack. It is captured that the value of Cl decreases abruptly after the stall angle 16°. Figure 5 shows pressure distribution on the body surface Cp at angles of attack 14°, 16° and 18°. Three kinds of the temporal variation of the Cl at angles of attack 16° (the stall angle), 14° and 12°, are illustrated in Fig.6. At the stall angle, Cl fluctuates unsteadily. Figures 7 and 8 show instantaneous flow field at angles of attack 8° and 16°. In Fig.7, pressure field is visualized by shading on the body, contour lines and a contour surface.

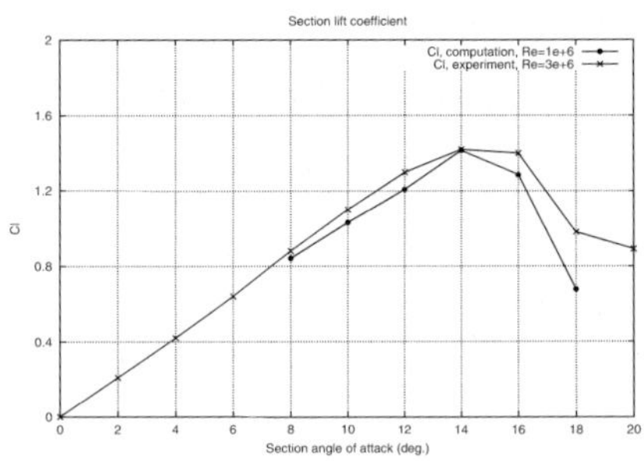

Fig. 4. Section lift coefficient

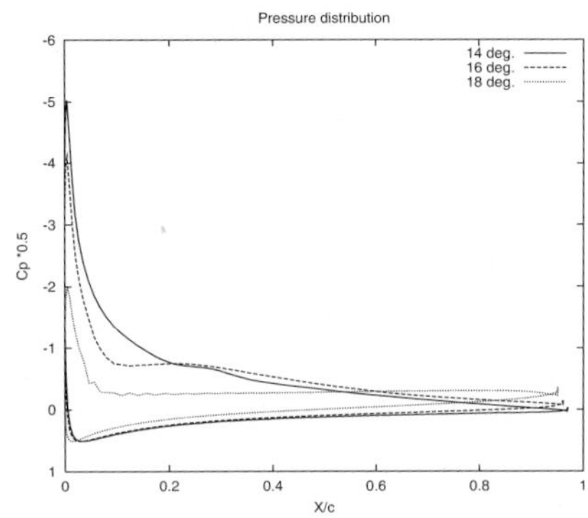

Fig. 5. Pressure distribution

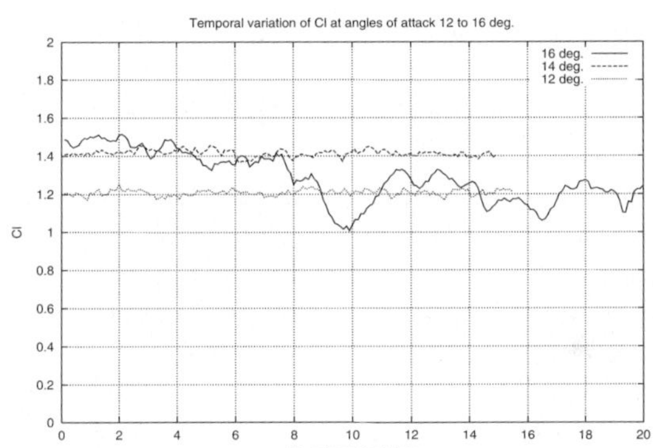

Fig. 6. Temporal variation of Cl at angles of attack 12° – 16°

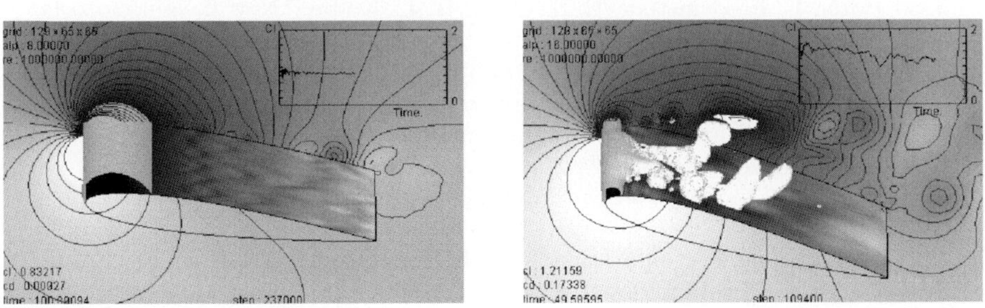

(a) $\alpha = 8°$.

(b) $\alpha = 16°$.

Fig. 7. Instantaneous pressure field at angles of attack 8° and 16°

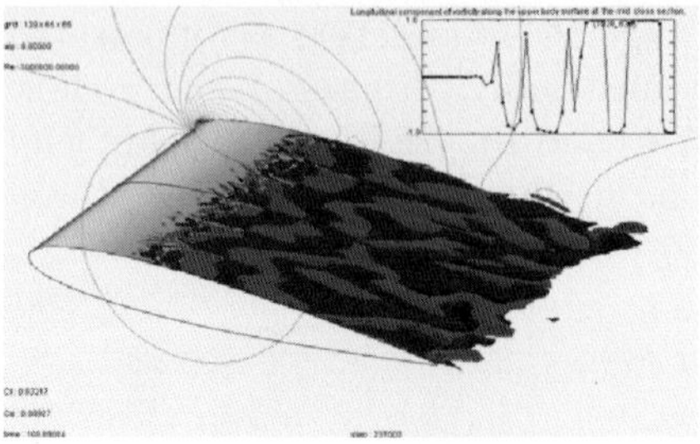

(a) $\alpha = 8°$.

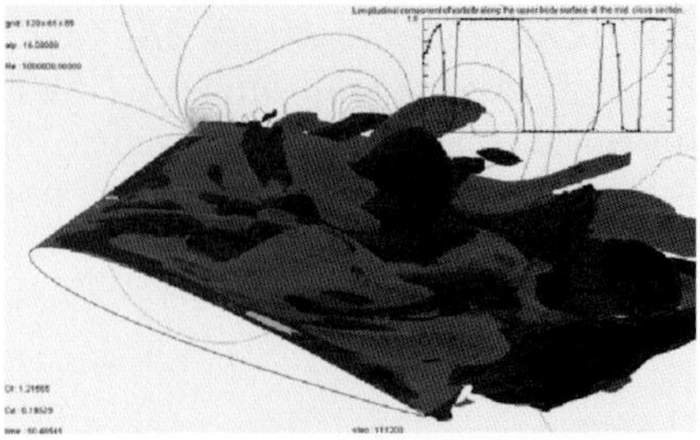

(b) $\alpha = 16°$.

Fig. 8. Longitudinal component of vorticity at angles of attack 8° and 16°

Longitudinal component of vorticity is given using two equi-surfaces of a positive and a negative values in Fig.8. These figures exhibit flow pattern changes clearly between the stall angle and lower angle of attack.

4 Conclusion

In the present 3D computation, Cl agrees well with the experimental values at near the stall angle.

At the angle of attack just before the stall (16°), the value of Cl fluctuates, and also the flow near the leading edge is unsteady, even after the flow is developed fully. Although the flow near the leading edge is unstable just before the stall angle, a low pressure region near the leading edge is kept well, and a pressure increase is not found there.

In the present paper, computational results agreed well with the experimental values without explicit turbulence models; the flow structure to cause the stall was clarified with separation near the leading edge.

References

1. J. P. Boris, F. F. Grinstein, E. S. Oran, R. L. Kolbe: "New insights into large eddy simulation". Fluid Dynamics Research **10**, 1992, pp. 199-228.
2. T. Kawamura, K. Kuwahara: "Computation of high Reynolds number flow around a circular cylinder with surface roughness". AIAA Paper 84-0340, 1984.
3. H. Suito, K. Ishii and K. Kuwahara: "Simulation of Dynamic Stall by Multi-Directional Finite-Difference Method". AIAA paper 95-2264, 1995.
4. K. Kuwahara: "Unsteady Flow Simulation and Its Visualization". AIAA Paper 99-3405, 1999.
5. K. Kuwahara and S. Komurasaki: "Semi-Direct Simulation of a Flow around a Subsonic Airfoil". AIAA Paper 2000-2656, 2000.
6. K. Kuwahara and S. Komurasaki: "Direct Simulation of a Flow around a Subsonic Airfoil". AIAA Paper 2001-2545, 2001.
7. I. H. Abbott, A. E. Von Doenhoff: "Theory of Wing Sections". Dover Pub., 1959.

DNS of Compressible Turbulent Boundary Layer Over a Blunt Wedge

Xinliang Li and Dexun Fu

LNM, Institute of Mechanics, CAS, Beijing China. lixl@lnm.imech.ac.cn

Wall turbulence is more complicated than free turbulence, and the direct numerical simulation (DNS) of wall turbulence is more difficult. In recent years, most of DNS cases for wall turbulence are simplified by using time mode, where streamwise periodic boundary condition is imposed. In time mode, spatial transition will be considered as an analogue of time-evolving transition. For the channel turbulence, an equivalent body force can substitute for the mean gradient of pressure, and then the streamwise periodic boundary condition can be used. Because of evolving of the boundary layer, there is no streamwise periodic property, but the evolving of the boundary layer can be replaced by an equivalent body force and then the periodic condition can be used. The time mode can only be used in relatively simple flow, such as channel flow or flat plate boundary layer, but cannot be used in the flow with complex shape, such as flow over a blunt body. Compared with the time mode, there is no artificial assumption in spatial mode, and the result seems more reliable. Because of the high cost of computation for spatial mode, only a few papers on DNS of spatially evolving boundary layer have been published.

Compared with incompressible turbulence, DNS for compressible turbulence started much later. Guarini et al.[1] performed DNS of a flat-plate boundary layer with free stream Mach number 2.5, and Maeder et al.[2] performed the same DNS with free stream Mach number 3, 4.5 and 6. Compressibility effects are studied in those two papers. According to Maeder et al.[2], the maximum turbulent Mach number for the boundary layer with $M_\infty = 6$ is 0.45, and Morkovin's theory is not valid. Both Guarini and Maeder used time mode in their numerical work. Rai et al.[3] performed DNS for spatially evolving flat-plate boundary at free stream Mach number 2.5, and studied the statistics of turbulence.

Most of the DNS work conducted is of flat-plate boundary layer. But the boundary layer of a blunt body is more interesting in aeronautics and astronautics. The prediction of laminar-turbulent transition in hypersonic boundary layers is a critical part of the aerodynamic design and control of hypersonic vehicles[4]. Zhong et al.[4] performed numerical simulation of leading-edge receptivity to freestream disturbance waves for hypersonic flow over a body with parabolic shape, but the simulation was limited to two-dimensional. Laurien[5] performed analysis and DNS of boundary layer of a re-entry capsule. But the DNS in Laurien's work is on time evolving flat-plate boundary layer. There is still no DNS work on the turbulent boundary layer of a blunt body.

Direct numerical simulation of spatially evolving compressible boundary layer over a blunt wedge with the free stream Mach number 6 is performed in this paper. The disturbance source produced by blowing and suction is located at the wall down stream of the sound-speed point. Statistics are studied and compared with the result for incompressible flat-plate boundary layer. The mean pressure gradient effects on the vortex structure are studied. Because of existence of bow shock, the turbulent Mach number in turbulent boundary layer is not high, and the compressible effects are not significant. It is also found that near-wall disturbance is strongly suppressed in the near nose region of the blunt wedge, but the suppression of far-from-wall disturbance is not significant.

1 Numerical simulation

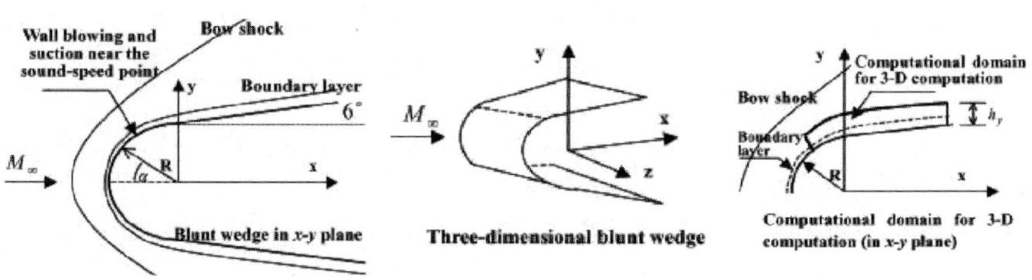

Fig. 1. Supersonic flow over a blunt wedge

A schematic diagram of supersonic flow over a blunt wedge is shown in figure 1. The wedge angle is θ , and free stream Mach number is defined as $M_\infty = u_\infty / c_{infty}$, and the Reynolds number is $Re = \rho_\infty u_\infty R / \mu_\infty$. We suppose that the temperature on the wall is constant. The flow parameters are given in table 1.

Table 1. Flow parameters

M_∞	θ	Re	T_w
6.0	6°	2.0×10^6	5.0

At the first step, numerical simulation of two-dimensional steady laminar flow is performed, and the numerical results are used in the initial condition and the inlet and outer boundary condition of three-dimensional computation.

The 3-D computational domain (see figure 1) is shown in Table 2. The governing equations are Jocabian-transformed Navier-Stokes equations, and Sutherland equation is used to calculate the viscous coefficient. The viscous terms of the Navier-Stokes equations are discretized with a 6th order central difference. The flux vector splitting is used for the inviscous terms, and 7th order accurate upwind-biasing difference approximations are used to discretize the inviscous terms. Three order Range-Kutta method is used for time advancing.

Table 2. Computational domain of three-dimensional numerical simulation

Streamwise	Vertical	Spanwise
$-0.628 \le x \le 1.224$	$h_y = 0.134$	$h_z = 0.1$

Non-slip and isothermal boundary conditions are used at the wall. The density, velocity and temperature are invariable at the inlet section and outer boundary, and the values are obtained from two- dimensional simulation. Supersonic non-reflecting outflow boundary condition is used at the outlet section.

The blow and suction disturbance is imposed at the wall of $-0.5735 \le x \le -0.5062$, and the form of the disturbance is shown in Ref[3],The base frequency of disturbance is $\beta = 2.5$, and the grid number is $800 \times 160 \times 80$.

Programs are coded by using MPI Fortran, and the DNS results are computed with computer LSSC2 of LSEC (State Key Laboratory of Scientific and Engineering Computing). Averaged performance is 8.3 seconds/time-step (40 CPUs in 20 nodes).

2 Result analysis

The two-dimensional steady flow is used as the initial condition of three-dimensional computation, and three-dimensional computation is performed until the turbulence reached equilibrium.

Figure 2 shows the skin friction in the streamwise direction. It can be seen that skin friction increases fast in the transition region.

Figure 3 shows the spanwise energy spectra $\hat{E}(k,x)$ as a function of x at $y_0 = 0.0002(y_0^+ = 4.2)$, and k=4,8,16 and 24. Where $\hat{E}(k,x) = \frac{1}{h_z}\int E(x,y_0,z)e^{-ikz}dz, E = \frac{1}{2}(u^2+v^2+w^2)$

Figure 2 and Figure 3 show that transition occurs soon after the disturbance zone, and then transition is depressed at the head of the blunt-wedge, and finished at the body of the blunt-wedge.

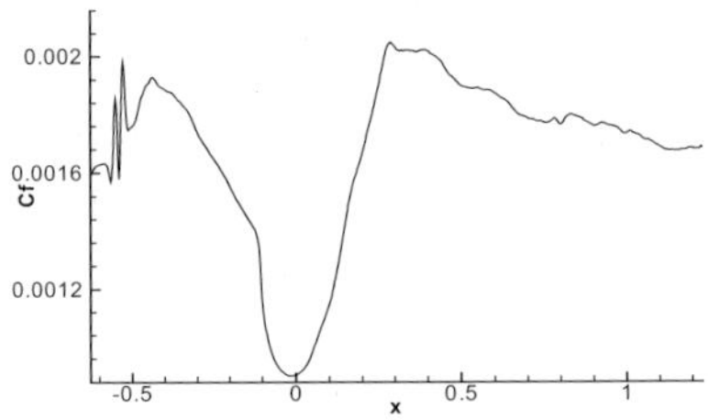

Fig. 2. Skin friction in the streamwise direction

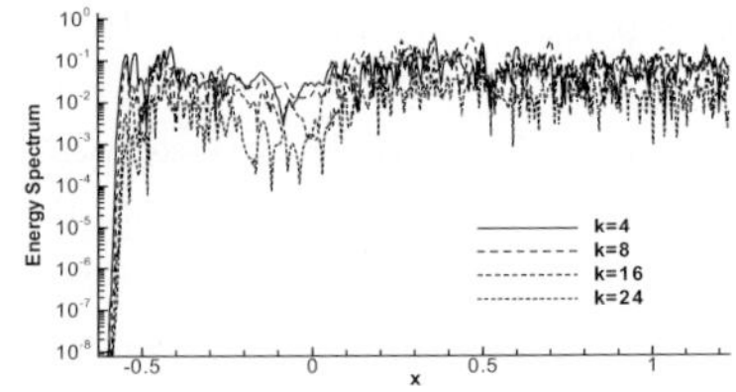

Fig. 3. Spanwise energy spectra vary as a function of x.

Figure 4 shows the mean Van Driest velocity profile normalized by wall-shear velocity at x=1.0. The computed results obey with the wall law and log law. The log law is expressed as $u^+ = 1/\kappa \ln y^+ + C$where $\kappa = 0.4, C = 7.8$. Because of existence of stream pressure gradient, the coefficient C is different from that in the incompressible flat-plate boundary layer where C=5.5.

Figure 5 shows the RMS of velocity fluctuation $u'_{rms}, v'_{rms}, w'_{rms}$ at x=1.0 as a function of the vertical distance to the wall. We can see that the peak RMS is located at $y^+ = 20$. In the near-wall zone $u'_{rms} > w'_{rms} > v'_{rms}$, which means that the turbulence are strong anisotropic near the wall. In the far-from-wall zone, difference between and is not large, which means that the turbulence in the far-from-wall zone is nearly isotropic.

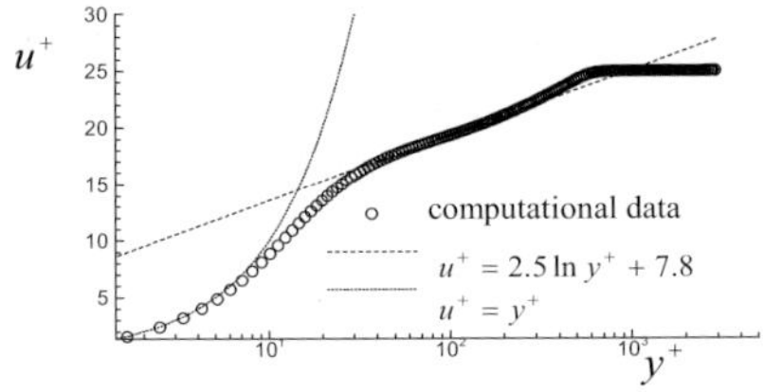

Fig. 4. Mean Van Driest velocity

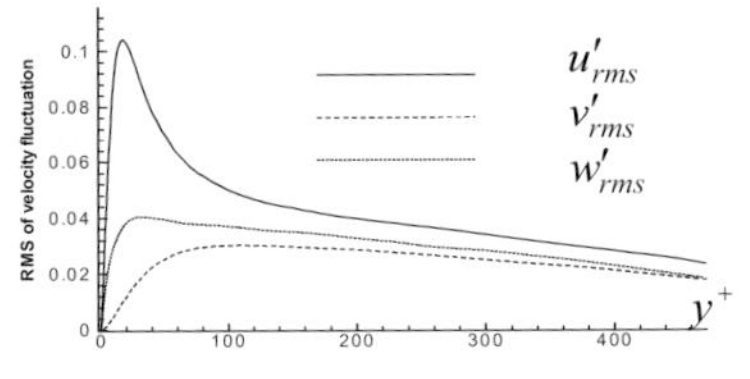

Fig. 5. RMS of velocity fluctuation at x=1.0

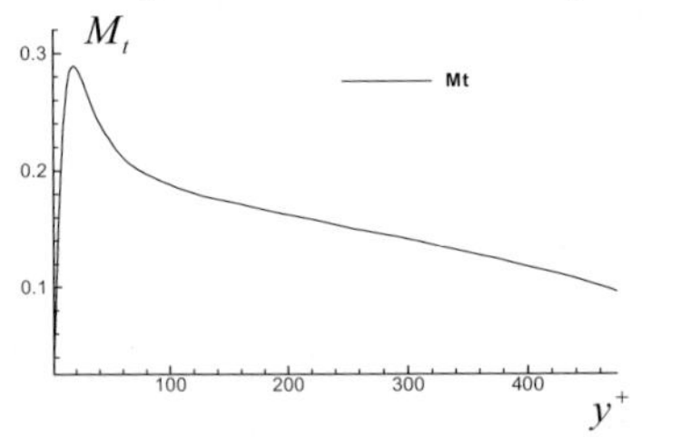

Fig. 6 Distribution of turbulent Mach number at x=1.0

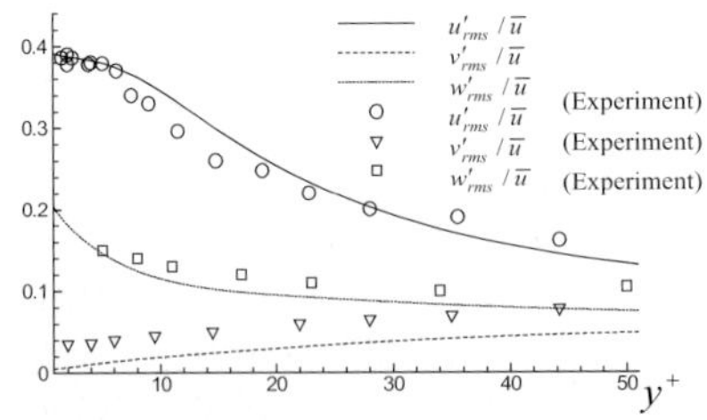

Fig. 7. Normalized RMS velocity fluctuations at x=1.0

Figure 6 shows the turbulent Mach number defined by $M_t = \frac{V'_{rms}}{\bar{c}}$ at the location x=1.0 as a function of vertical distance to the wall y. It can be seen that the peak of the turbulent Mach number equal to 0.29 is located at

$y^+ = 20$, which is much less than 0.45 in the flat-plate boundary layer for the same free stream Mach number $M_\infty = 6$. This is because of existence of strong bow shock out of the boundary layer. Figure 7 and 8 show the normalized turbulence intensities and normalized Reynolds shear stress at x=1.0, where the symbols denote the experiment data for incompressible flat-plate boundary. Because the turbulent Mach number is not big, the difference between those two data is not significant.

Figure 9 shows turbulent intensity V'_{rms} at y=0.00020, 0.00043,0.00102and 0.00180 (y^+ =4.2,9.1,21.2 and 37.6) as a function of streamwise coordinate x. We can see that the turbulent intensity increases fast after the disturbance zone, and it means that transition occurs, and then the turbulent intensity is depressed at the head of the blunt-wedge ($x < -0.1$), and the transition is finished at the body of the blunt-wedge. This figure shows that that the near-wall disturbance is strongly suppressed at the nose of the blunt wedge, but the suppression of disturbance at the nose of the blunt wedge far from the wall is not significant.

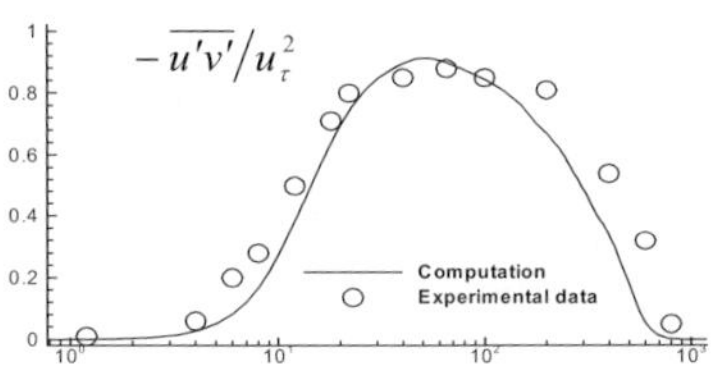

Figure 8. Normalized Reynolds shear stress

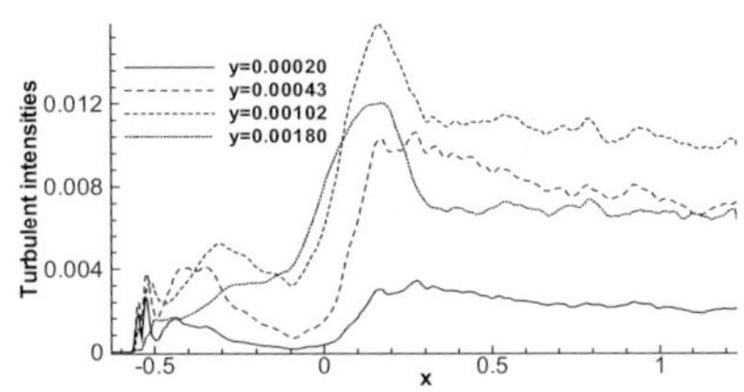

Figure 9. Turbulent intensity as a function of x

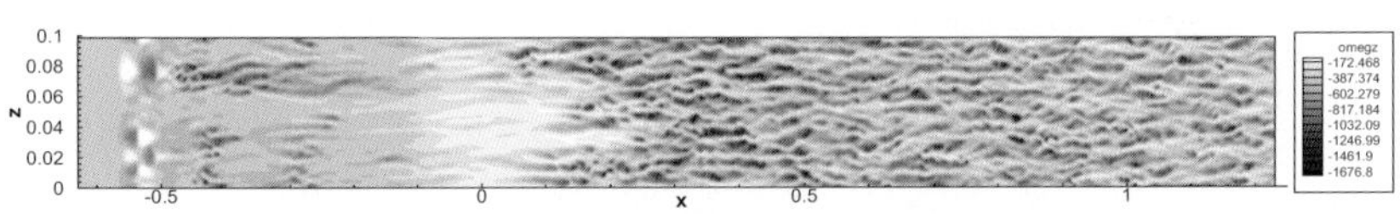

Spanwise vorticity contours on the plane y=0.0002are shown in figure 10, in which the near-wall streaky structures can be seen clearly. From this figure we can see that at x=-0.1, the near-wall streaky structures are depressed, and this depression is not strong in the zone far from wall.

3 Conclusion

In this paper, DNS of turbulent boundary layer of supersonic flow over a blunt wedge is performed, and the following conclusions are drawn.

1. Because of the bow shock out of the boundary layer, the turbulent Mach number in the turbulent boundary layer is not very high, and the turbulent Mach number is much less than that in the flat plate boundary layer.

2. Because the turbulent Mach number is not high, the compressible effects are not significant.

3. In this flow, transition begins at the head of blunt wedge, and then the transition is depressed. Transition is finished at the body of the wedge. The near-wall disturbance is depressed strongly.

Acknowledgements

This work is supported by NKBRSF (CG199032805) and NNSFC (90205025). The authors thank Professor Zhang Linbo for his help in programming, and LSEC for providing computer time.

References

1. Guarini, S.E., Moser, R.D., Shariff, K., et al., J. Fluid Mech.,414:1-33 (2000).
2. Maeder, T., Adams, N.A., Kleiser, J. Fluid Mech., 429:187-216 (2001).
3. Rai, M.M., Gatski, T.B., Erlebacher, G., AIAA paper, 95-0583 (1995).
4. Zhong, X.L., J. Fluid Mech. 441:315-367 (2001).
5. Laurien, E., Journal of spacecraft and rockets. 33 (3):313-318 (1996).

Computation of the Turbulent Boundary Layer on a Long Circular Cylinder in Axial Flow with a Vorticity Boundary Condition

Milton Woods and Max Bull

Department of Mechanical Engineering, University of Adelaide, Australia, 5005.
milton.woods@adelaide.edu.au

1 Introduction

The present work is part of a programme of direct numerical simulation of axial and near-axial turbulent flow over long cylinders. Here attention is concentrated on axial flow. Previous DNS calculations for axial flow are reported by Neves, Moin and Moser [1]. In their procedure, the boundary conditions at the outer radial limit of the computational domain were imposed on the time-dependent velocity: the radial component U_r and the radial derivatives of the azimuthal component U_θ and axial component U_z were set to zero. In the present work, vorticity boundary conditions have been used: the radial component Ω_r is set to zero and the remaining components are calculated to minimise the magnitude of the vorticity vector at the outer boundary of the computational domain. In order to identify the effects of the boundary conditions, the results of the present calculations are compared with those of Neves et al. and the experimental data presented by Luxton, Bull and Rajagopalan [2] for Reynolds numbers $Re_a = U_\infty a/\nu$ (based on cylinder radius a) between 311 and 674 and boundary layer thickness in the range 5 to 40 cylinder radii.

2 Numerical Method

The numerical method developed for the present study is similar in many respects to the simulation procedure described by Neves [3]. Most of the differences between the two methods are associated with the formulation of the boundary conditions at the radial limits of the computational domain. The boundary conditions employed here are designed to accommodate both axial and off-axial freestream flow.

The boundary layer is confined to a cylindrical region aligned with the cylinder axis. The outer limit of the computational domain is at radius b, and the boundary layer thickness at $0.99U_\infty$ is $\delta < b$. Flow outside the boundary layer is not excluded from the model but is assumed to be a time-dependent potential flow. The velocity field is required to be continuous across the interface at radius b. The flow is truncated to an axial length L_z and periodic axial boundary conditions are imposed, allowing simulations to be advanced to a statistically steady state without advance knowledge of the time-varying inflow and outflow.

As the external flow is assumed irrotational, the vorticity $\boldsymbol{\Omega} = \nabla \times \boldsymbol{U}$ is confined to the boundary layer region. Continuity of the velocity field across the interface between the regions requires that the radial component of vorticity, Ω_r, is set to zero at the outer boundary. A smooth transition between boundary layer and potential flow is achieved when the remaining components of vorticity, Ω_θ and Ω_z, are set to the smallest possible values. Vorticity boundary conditions can be imposed on the flow by solution of the vorticity transport equation,

$$\frac{\partial \boldsymbol{\Omega}}{\partial t} = \nabla \times (\boldsymbol{U} \times \boldsymbol{\Omega}) + \frac{1}{Re} \nabla^2 \boldsymbol{\Omega}. \tag{1}$$

Further constraints on the solution are

$$\nabla \cdot \boldsymbol{U} = \nabla \cdot \boldsymbol{\Omega} = 0. \tag{2}$$

The vorticity field is advanced in time by integration of the vorticity transport equation using a stiffly-stable scheme of third order accuracy (see [4]). The cylindrical components of the equation are coupled by both the viscous and nonlinear terms. Time advancement of the coupled terms is most conveniently achieved with an explicit formulation, which uses previously computed values of the flow variables. The uncoupled parts of the viscous terms must be treated implicitly to avoid numerical stability problems. The instability reported by Woods and Bull [5] for a third order scheme has been found to be the result of a programming error.

Fourier expansion of the flow variables in the azimuthal and axial directions allows the flow equations to be expressed as a set of ordinary differential equations involving functions of radial position. The equations are solved by use of the Galerkin method with Chebyshev expansions in the radial direction. The equations for the Fourier modes are independent, with the exception of the nonlinear term of the vorticity transport equation which is most efficiently evaluated in the space domain.

While vorticity is the primary quantity in the numerical method, the velocity field is also needed at each timestep of the simulation. The equations relating vorticity and velocity are solved over the boundary layer with arbitrary boundary conditions, after which the proper boundary conditions are enforced by addition of a potential flow. The potential flow added within the boundary layer and the potential flow outside the layer are determined so that there is zero radial velocity at the cylinder surface, velocity is continuous across the interface at radius b, and the freestream velocity U_∞ occurs at infinite radius. The Fourier modes of the required potential flows can be expressed analytically and partially precalculated for efficiency.

Boundary conditions for the vorticity field are imposed on the solution of the time-integrated vorticity transport equation. The radial component of vorticity is fixed to zero both at the cylinder surface, where the no-slip condition applies, and at the outer limit of the boundary layer, where the velocity is assumed to be continuous across the interface to the potential flow. After solution for the radial component of vorticity, the outer boundary values of the azimuthal and axial vorticity components are determined so that (2) is satisfied and the magnitude of the vorticity vector is minimised at the boundary. The inner boundary values of the azimuthal and axial vorticity are initially set to zero, and the surface slip velocity is removed from the solution by addition of a preprocessed solution with inhomogeneous inner boundary conditions.

The pressure field may be calculated as a post-processing step or at arbitrary intervals during the simulation. The divergence of the Navier-Stokes equation gives a Poisson equation for pressure,

$$\nabla^2 P = -\nabla \cdot (\boldsymbol{U} \cdot \nabla \boldsymbol{U}). \tag{3}$$

Boundary conditions for the pressure are derived from the Navier-Stokes equation for the radial direction. The radial derivative of pressure is evaluated at the cylinder surface, while the pressure at the outer limit of the boundary layer is determined analytically by integration of the total pressure from infinite radius through the potential flow region.

3 Results

The mean and fluctuating vorticity calculated for $Re_a = 674$ and $b/a = 6$ are shown in Fig. 1. The mean vorticity Ω_θ decays smoothly with increasing radius. The root-mean-square vorticity fluctuations ω' vary smoothly over most of the computational domain, but there is some local distortion in ω'_θ and ω'_z close to the outer boundary. The distortion is a consequence of the simultaneous requirements for zero radial vorticity at the outer boundary and divergence-free vorticity everywhere in the flow. The intensity of the velocity fluctuations u' for the same flow is shown in Fig. 2. The variation of the velocity fluctuations is smooth throughout the boundary layer. The turbulence intensity decays but does not vanish at the outer limit of the computational domain.

Calculated mean-velocity profiles, in terms of outer layer parameters, for $Re_a = 311$ with b/a varying from 12 to 41 are shown in Fig. 3. They are compared with the experimental results of Luxton et al. [2] for $Re_a = 325$ at $\delta/a = 39$. The simulated data tend towards the experimental profile as δ/a approaches the experimental value. The mean velocity computed by Neves et al. [1] for $Re_a = 311$ and $b/a = 12$ is consistent with the present calculation.

Mean-velocity profiles in terms of inner layer parameters are shown in Fig. 4. The calculations show that as the curvature increases, that is as a^+ decreases, the curves deviate more and more from the flat plate case ($a^+ =$

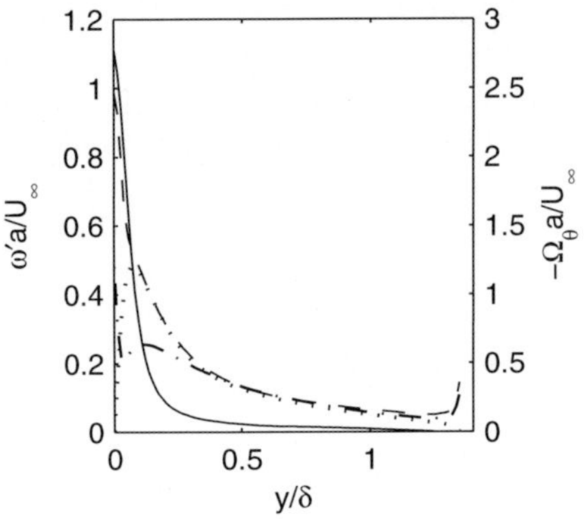

Fig. 1. Calculated vorticity profiles for $Re_a = 674$, $b/a = 6$. Mean flow: — Ω_θ. RMS fluctuations: $\cdots\ \omega_r'$; $--\ \omega_\theta'$; $-\cdot\ \omega_z'$.

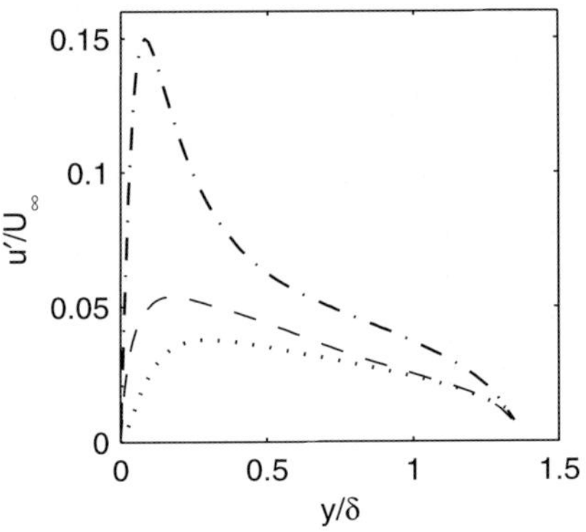

Fig. 2. Calculated turbulence intensity for $Re_a = 674$, $b/a = 6$: $\cdots\ u_r'$; $--\ u_\theta'$; $-\cdot\ u_z'$.

∞). The calculated values are essentially indistinguishable from the numerical results of Neves et al. (not shown in the figure). The experimental values of Willmarth et al. [6] show similar trends with a^+, but there is a noticeable difference between the calculated and experimental values, for reasons at present not understood.

The intensity of axial velocity fluctuations normalised by the friction velocity u_τ is shown in Fig. 5. The results of the present calculations are virtually identical to those of Neves et al. for $Re_a = 311$ and 674 with $b/a = 12$ and 6 respectively. The calculated results are in good agreement near the cylinder with the experimental data of Luxton et al. for $Re_a = 325$ and 620 with $\delta/a = 39$ and 27 respectively. The departure of the calculated profiles from the measurements at large radial distances from the cylinder wall is to be expected given the relatively small δ/a values of the simulations.

The root-mean-square vorticity fluctuations computed with the present method are compared in Fig. 6 with the results of Neves et al. for $Re_a = 674$ and $b/a = 6$. There is close agreement for all three components of vorticity, indicating that the difference in outer boundary condition does not play a significant role near the cylinder wall.

Contours of the instantaneous axial velocity fluctuations in a plane through the cylinder axis are shown in Fig. 7. The entire axial and radial extent of the calculation domain is shown for $Re_a = 311$ and $b/a = 31$. Some of the turbulence structures are significantly larger than the cylinder, and their motion in the radial direction is unlikely to be restrained by the cylinder. At some axial positions, high speed fluid on one side of the cylinder corresponds to low speed fluid on the other side, which may indicate the presence of a local cross-flow. Further investigation of this phenomenon is in progress.

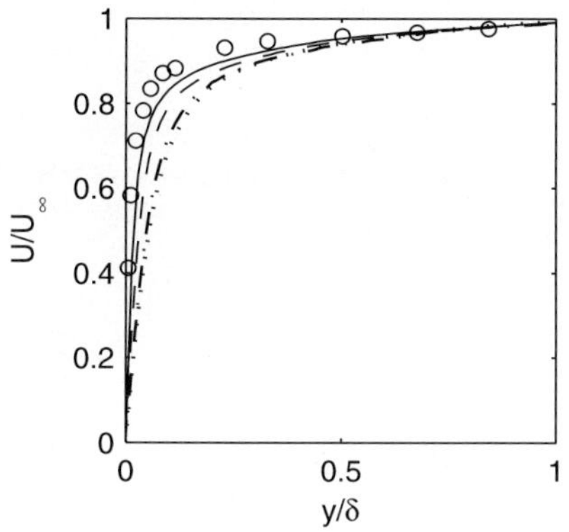

Fig. 3. Mean velocity profile in outer units. Luxton et al. [2] (experiment): ∘ $Re_a = 325$, $\delta/a = 39$. Neves et al. [1] (simulation): $-\cdot$ $Re_a = 311$, $b/a = 12$, $\delta/a \approx 9.2$ (estimated for present purposes). Present study, $Re_a = 311$: $\cdots$ $b/a = 12$, $\delta/a = 7.6$; $--$ $b/a = 21$, $\delta/a = 15.1$; — $b/a = 41$, $\delta/a = 28.4$.

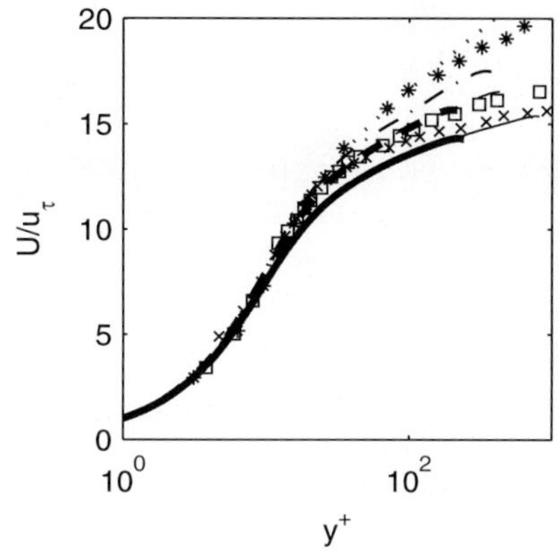

Fig. 4. Mean velocity profile in wall units. $\cdots$ Flat plate. Willmarth et al. [6] (experiment): × $\delta/a = 37.5$, $a^+ = 46.1$; □ $\delta/a = 27.0$, $a^+ = 83.4$; ∗ $\delta/a = 16.0$, $a^+ = 198$. Present study: **—** $\delta/a = 7.6$, $a^+ = 22.0$; — $\delta/a = 28.4$, $a^+ = 20.2$; **- -** $\delta/a = 3.7$, $a^+ = 43.0$; $--$ $\delta/a = 8.1$, $a^+ = 40.8$; $-\cdot$ $\delta/a = 3.7$, $a^+ = 74.3$.

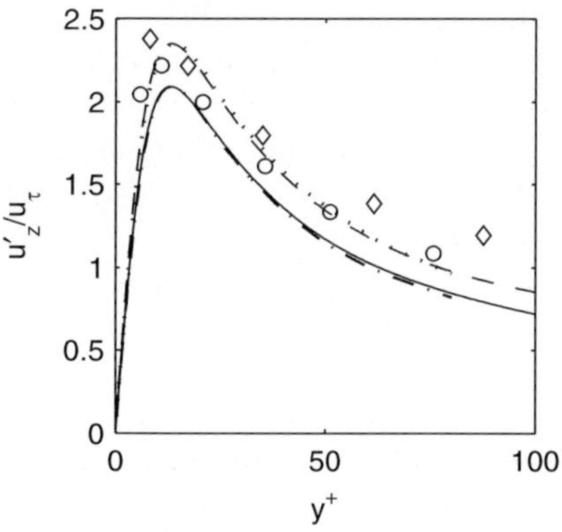

Fig. 5. Axial turbulence intensity. Luxton et al. [2] (experiment): ∘ $Re_a = 325$, $\delta/a = 39$; ⋄ $Re_a = 620$, $\delta/a = 27$. Neves et al. [1] (simulation): $-\cdot$ $Re_a = 311$, $b/a = 12$; $\cdots$ $Re_a = 674$, $b/a = 6$. Present study: — $Re_a = 311$, $b/a = 12$; $--$ $Re_a = 674$, $b/a = 6$.

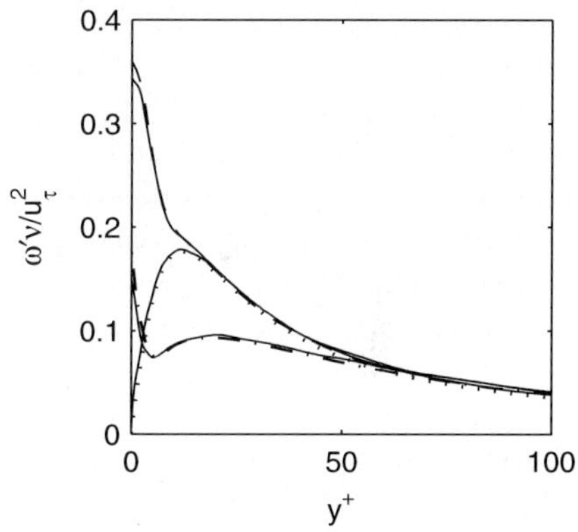

Fig. 6. RMS vorticity fluctuations for $Re_a = 674$, $b/a = 6$. Neves et al. [1] (simulation): — $\omega'_r, \omega'_\theta, \omega'_z$. Present study: $\cdots$ ω'_r; $--$ ω'_θ; $-\cdot$ ω'_z.

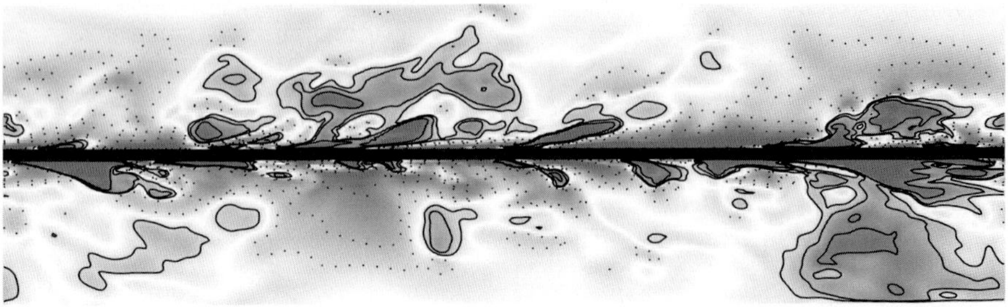

Fig. 7. Contours of calculated instantaneous axial velocity fluctuations in a plane through the cylinder axis for $Re_a = 311$, $b/a = 31$. Mean flow is from left to right. Contour levels at $\pm u_z/U_\infty = 0.02, 0.04, 0.08$: — $u_z < 0$; ⋯ $u_z > 0$.

4 Conclusions

From the calculations that have been made and comparison of them with experiment, the following conclusions can be drawn.

1. The DNS calculation procedure produces turbulent boundary layer properties that are, in general, in good agreement with or consistent with previous numerical simulations and existing experimental data.
2. The simulation results for comparable Re_a and δ/a are mildly dependent on the choice of velocity or vorticity boundary conditions. At the present stage of the investigation, there is no strong evidence to indicate superiority of either boundary condition.
3. The numerical simulations extending to $\delta/a \approx 30$ show that, as in the experimental case, wall similarity in the mean velocity profiles in the inner layer deviates increasingly from flat-plate similarity as the curvature (measured by $1/a^+$) increases. In the outer layer, the fullness of the mean velocity profile increases with δ/a, in accordance with experiment.

References

1. J.C. Neves, P. Moin, R.D. Moser: J. Fluid Mech. **272**, 349 (1994)
2. R.E. Luxton, M.K. Bull, S. Rajagopalan: Aeronautical Journal **88**, 186 (1984)
3. J.C. Neves: Numerical study of axial turbulent flow over long cylinders. PhD Thesis, Stanford University, Ann Arbor MI (1992)
4. G.E. Karniadakis, M. Israeli, S.A. Orszag: J. Comput. Phys. **97**, 414 (1991)
5. M.J. Woods, M.K. Bull: Spectral vorticity method for computation of axial and near-axial flow over long circular cylinders. In: *Computational Fluid Dynamics*, ed by S. Armfield, P. Morgan, K. Srinivas (Springer, Berlin Heidelberg New York 2002) pp 387–392
6. W.W. Willmarth, R.E. Winkel, L.K. Sharma, T.J. Bogar: J. Fluid Mech. **76**, 35 (1976)

LES of Combined Forced and Natural Turbulent Convection in a Vertical Slot

J. Yin[1] and D.J. Bergstrom[2]

[1] Department of Mechanical Engineering, University of Saskatchewan Saskatoon, Saskatchewan S7N 5A9, CANADA, jiy080@mail.usask.ca
[2] Department of Mechanical Engineering, University of Saskatchewan Saskatoon, Saskatchewan S7N 5A9, CANADA, Don_Bergstrom@engr.usask.ca

Summary. The present paper reports a Large Eddy Simulation (LES) of turbulent combined forced and natural convection between two vertical parallel plates maintained at two different temperatures. For the subgrid-scale (SGS) model, both a Smagorinsky eddy viscosity model (SM) and a dynamic eddy viscosity model (DM) with a constant SGS Prandtl number are investigated. The present predictions indicate that LES using both models capture many important features of this flow, such as the effect of buoyancy on the mean velocity profile. However, the dynamic model gives a better prediction for the temperature field.

1 Introduction

Given that many turbulent buoyant flows are inherently unsteady and feature important large-scale motions, often they are not well predicted by Reynolds Average Navier-Stokes (RANS) methods. Instead a time dependent simulation, such as a Direct Numerical Simulation (DNS) or LES, is required. In LES, the numerical solution only resolves motions whose scales are larger than the filter scale, and the small scale motions are represented by a SGS model. To correctly predict the detailed flow structure and wall heat transfer rates, both the SGS stress and SGS heat flux terms need to be properly modeled. Eddy viscosity type models, which are based on an assumption that the SGS flow structure is isotropic, are widely used for many LES applications. However, their application to thermally buoyant flows has not been extensively investigated. Unlike the advecting velocity field, the large and small scales in a scalar field, such as temperature, are strongly coupled. Even at high Reynolds number, the scalar field is anisotropic both at the dissipation and inertial scales [1, 2]. This anisotropy of a scalar field presents a challenge for SGS heat flux modeling.

The present paper considers turbulent combined forced and natural convection between two vertical parallel plates maintained at two different temperatures. The pressure gradient drives the mean flow upward, while the buoyancy force acts in the upward (aiding flow) and downward (opposing flow) direction near the high- and low-temperature walls, respectively. The turbulent transport is enhanced in the opposing flow and suppressed in the aiding flow. Compared to the vertical channel flow, the mean velocity profile becomes more asymmetric as the Grashof number (or buoyancy effect)is increased. The shape of the velocity profile in the aiding flow is very different from that in the opposing flow, and a classic logarithmic region no longer exists [3]. This particular feature enables us to investigate the SGS models in terms of the response of the velocity field to the effect of temperature both in the aiding and opposing flow. Previous calculations have explored the SGS model in the context of thermal transport in turbulent Couette flow [4].

The focus of the present study relates to the SGS heat flux modeling problem. Since the modeling of the SGS heat flux is not as well developed as for the SGS stress, in many LES applications for turbulent buoyant flows, the SGS heat flux is modeled by combining a SGS eddy viscosity model for momentum transport with the assumption of a constant SGS Prandtl number. In the present study, we test both a Smagorinsky model [5] and a dynamic model [6, 7], with a constant SGS Prandtl number of 0.5 for the SGS heat flux. The LES results are compared to the DNS data of Kasagi and Nishimura [3].

2 Mathematical Model

In the present calculations, the filtered continuity equation, Navier-Stokes equations with the Boussinesq approximation, and energy equation are used. They are expressed in the following forms:

$$\frac{\partial \overline{u}_i}{\partial x_i} = 0 \tag{1}$$

$$\frac{D\overline{u}_i}{Dt} = -\frac{1}{\rho}\frac{\partial \overline{p}}{\partial x_i} + \nu \frac{\partial^2 \overline{u_i}}{\partial x_j \partial x_j} + \frac{\partial \tau_{ij}}{\partial x_j} + \beta g_i \left(\overline{\theta} - \Theta_r\right) \tag{2}$$

$$\frac{\partial \overline{\theta}}{\partial t} + \frac{\partial}{\partial x_j}\left(\overline{u}_j \overline{\theta}\right) = \alpha \frac{\partial^2 \overline{\theta}}{\partial x_j \partial x_j} + \frac{\partial h_j}{\partial x_j} \tag{3}$$

The additional SGS terms, τ_{ij} and h_j need to be modeled. For the SGS stress, we let

$$\tau_{ij} - \frac{\delta_{ij}}{3}\tau_{kk} = -2C\overline{\Delta}^2 \left|\overline{S}\right| \overline{S}_{ij} \tag{4}$$

where $\left|\overline{S}\right| = (2\overline{S}_{ij}\overline{S}_{ij})^{1/2}$, and $\overline{S}_{ij} = (\overline{u}_{i,j} + \overline{u}_{j,i})/2$ is the total strain rate. In the present calculation, the Smagorinsky model [5] assumes the coefficient C to be a constant so that a wall damping function is needed. The dynamic model of Lilly [6] obtains C automatically and locally by applying a novel dynamic procedure. Germano et al [7] originally introduced a test grid which is coarser than the grid used in the simulation, and assumed similarity between the test-grid scale (TGS) and SGS flow structures to obtain a relation for C. In the numerical implementation of a dynamic model, a large negative value of C can result in stability problems. One approach used to avoid large negative values of the coefficient is to average it over a homogeneous plane. However, the coefficient thus obtained is no longer related to the local flow structure. Furthermore for an inhomogeneous flow, this approach is inapplicable. To avoid this problem, the negative coefficients are simply *clipped* in many LES applications. Here, we adopt a somewhat different strategy: we implement local volume averaging for C and then bound the magnitude of the coefficient such that $|C| \leq 0.3$. We also limit the minimum value of the eddy viscosity, ν_{SGS} to -1.5ν, which still allows the possibility of a small negative value for the total viscosity.

A SGS heat flux model is also required to solve the energy equation. In this case, a simple eddy diffusivity relation was adopted,

$$h_j = \alpha_{SGS}\overline{\theta}_{,j} \tag{5}$$

where the SGS eddy diffusivity, α_{SGS}, was evaluated from the SGS eddy viscosity using a constant SGS Prandtl number Pr_{SGS}, $\alpha_{SGS} = \nu_{SGS}/\mathrm{Pr}_{SGS}$. In the present study, it was set as 0.5.

NOTATION

Gr	Grashof number, $g\beta\Delta\theta(2\delta)^3/\nu^2$	θ	temperature
g_i	gravitational acceleration	θ_τ	friction temperature on, each wall $q_w/(\rho c_p u_\tau)$
Pr	Prandtl number, ν/α	Θ_r	reference temperature
$\overline{p}$	filtered pressure	τ_w	wall shear stress
q_w	heat flux		
Re_τ	Reynolds number, $u_\tau\delta/\nu$		
$\overline{u}_i$	filtered velocity component	**Superscripts**	
u_i'	unresolved velocity fluctuations	$+$	normalized with wall variables on each wall
u_τ	friction velocity		
α	viscous diffusivity	$*$	normalized with wall variables averaged on the two walls
β	thermal expansion coefficient		
Δ	size of grid filter	$-$	grid filter

3 Numerical Method

The governing equations written above were discretized by using a second-order central difference scheme on a collocated grid. The momentum equations were solved using a fractional-step method; the velocity field is advanced in time using a semi-implicit Crank-Nicolson method; the pressure correction equation is solved by using a multi-grid method with a control strategy. To solve the energy equation, a fourth-order Runge-Kutta method was used to advance the temperature field over a single time step.

The flow geometry and computational domain in the present simulation were selected to mimic those in a DNS of this flow [3]. The flow is driven upward through a vertical slot with two different wall temperatures. In the present study, the flow is set to be air; the Reynolds number Re_τ, based on the friction velocity u_τ, is 150; and the Grashof number is 9.6×10^5. A no-slip and no-penetration boundary condition was imposed on the velocity components at the walls. The temperatures on the walls are set to fixed values. Periodic boundary conditions were imposed in the streamwise and spanwise directions. The computational domain is $5\pi\delta \times 2\delta \times 2\pi\delta$ and the grid size is $48 \times 32 \times 48$ in the streamwise, wall-normal and spanwise directions, respectively. The grid is uniform in the streamwise and spanwise directions, but made non-uniform in the wall normal direction by applying a hyperbolic-tangent function.

A laminar velocity profile, with a random perturbation, was input as the initial condition. Both the temperature and velocity field were allowed to develop in time before data were collected for staticical analysis. The time increment in the present simulation is 0.001s. The statistics were obtained from averaging over space and time based on more than 15,000 time steps for each simulation.

4 Simulation Results

Figure 1 and 2 shows the mean velocity profiles predicted by LES using the Smagorinsky and dynamic model. Compared to the DNS data [3], both SGS models give satisfactory results, except that they slightly over predict the opposing flow. Figure 3 shows the instantaneous secondary flow vectors in a transverse plane. It is evident that the opposing flow is much more *turbulent* than the aiding flow. Large vortical structures are observed in the opposing flow. Figure 4 is an iso-contour of the instantaneous temperature field in the same plane. We observe that the effect of the secondary flow is to mix a quantity of cold fluid into the center region.

In this paper, we were specially interested in the effect of SGS modeling on the mean and fluctuating temperature field. Figure 5 shows the mean temperature profile in inner coordinates. As indicated by Kasagi and Nishimura[3], the mean temperature profile shifts upward in the aiding flow and downward in the opposing flow similar to the velocity profile. The present LES predictions are consistent with their observation. Figure 5 also indicates that the dynamic model better predicts the opposing flow. The resolved temperature fluctuation is shown in Figure 6; again the prediction of the dynamic model is slightly better than that of the Smagorinsky model. Although the predictions for the temperature field by the two models show a difference, predictions for the mean and fluctuating velocity field are almost the same (the results are not shown here). Since the coefficient of the dynamic model is obtained by applying a similarity assumption that responds to the local flow structure, the dynamic model should better predict the anisotropic temperature field.

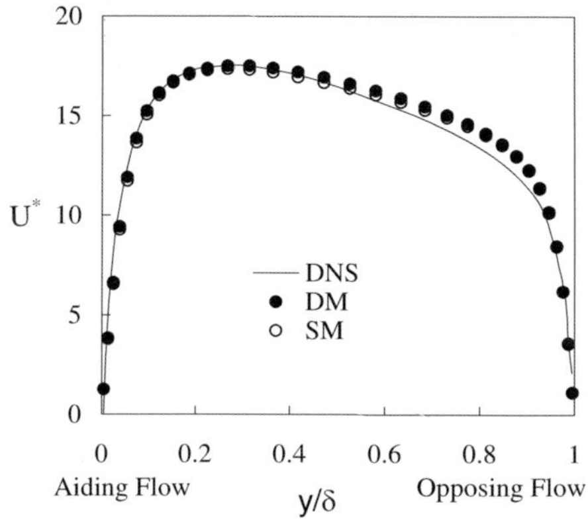

Fig. 1. Mean velocity profiles using global coordinates.

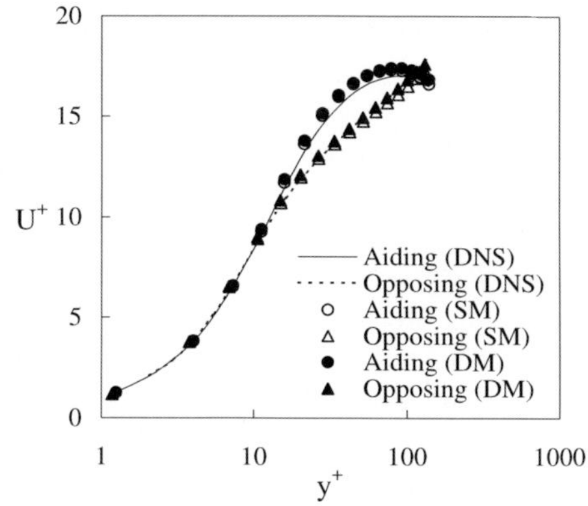

Fig. 2. Mean velocity profiles using wall coordinates.

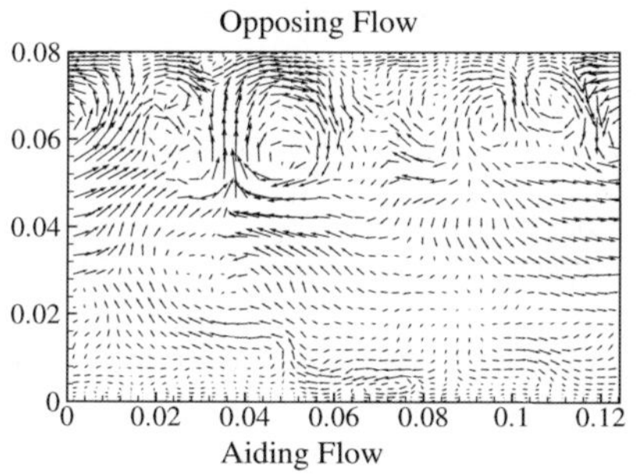

Fig. 3. Instantaneous secondary flow vector in a transverse plane.

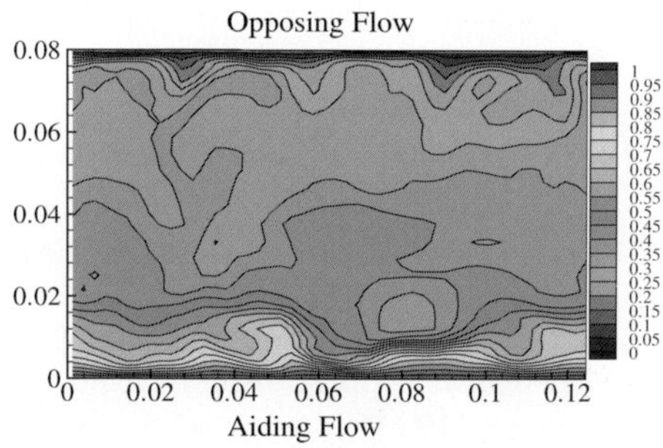

Fig. 4. Iso-contours of instantaneous temperature in a transverse plane.

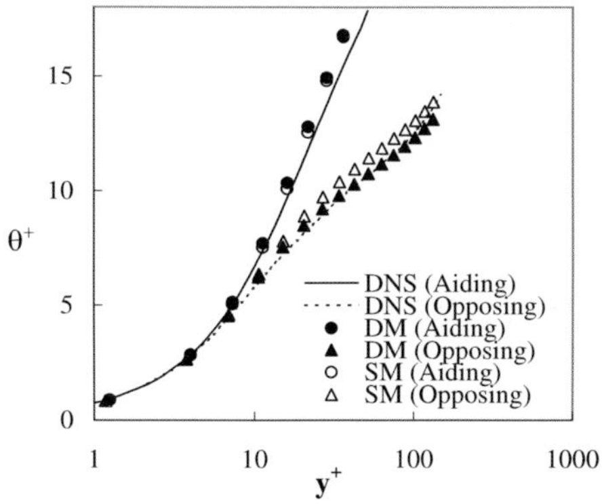

Fig. 5. Mean temperature profile in wall coordinates.

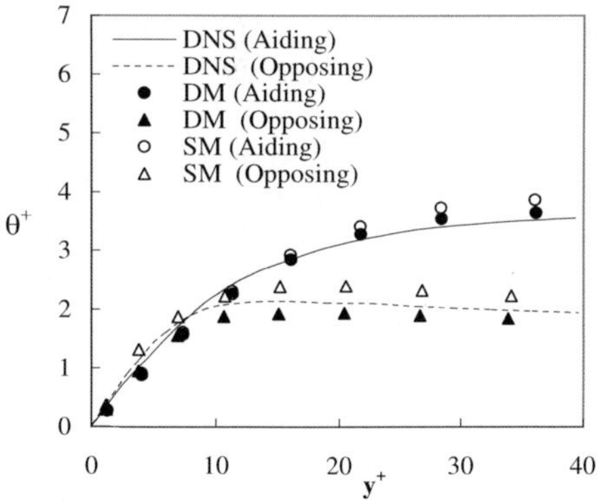

Fig. 6. Profile of normalized temperature fluctuation.

5 Conclusions

The present paper reports the results of LES of turbulent combined forced and natural convection between two vertical parallel plates. Even thought it used a relatively coarse grid, the LES gave satisfactory results. With respect to the prediction of the velocity field, the difference between the Smagorinsky and the dynamic model using a constant SGS Prandtl number is insignificant. However, the dynamic model yields better predictions for the temperature field. This substantiates the conclusions of previous research that the scalar field is more sensitive to the choice of SGS model.

6 Acknowledgements

The financial support provided by a NSERC research grant to DJB and a University of Saskatchewan Scholarship to JY are gratefully acknowledged.

References

1. Warhaft Z.: Passive scalars in turbulent flows, Annu. Rew. Fluid Mech. **32**, pp. 203-240 (2000)
2. Liu S., Meneveau C. and Katz J.: On the properties of similarity subgrid-scale models as deduced from measurements in a turbulent jet, J. Fluid Mech. **276**, pp. 83-119 (1994)
3. Kasagi N. and Nishimura M.: Direct numerical simulation of combined forced and natural turbulent convection in a vertical plane channel, Int. J. Heat and Fluid Flow **18**, pp. 88-99 (1997)
4. Bergstrom, D.J. Yin, J. and Wang B.-C.: LES of near-wall turbulent heat transfer using dynamic SGS models. In: *Turulence, Heat and Mass Transfer 4*, 2003, pp. 483-490
5. Smagorinsky J.: General circulation experiments with the primitive equation. I. The basic experiment. Mon. Weather Rev. 91:99 , 91:99 (1963)
6. Lilly D. K.: A proposed modification of the Germano subgrid-scale closure method, Phys.Fluids A4 **3**, pp. 633-635 (1992)
7. Germano M., Piomelli U., Moin P., and Cabot W.H.: A dynamic subgrid-scale eddy viscosity model, Phys. Fluids A3 **7**, pp. 1760-1765 (1991)

Numerical Study on $k - \omega$ Turbulence Models for Supersonic Impinging Jet Flow Field

Eugene Kim, Soo Hyung Park, and Jang Hyuk Kwon

Department of Aerospace Engineering, Korea Advanced Institute of Science and Technology, Daejeon, 305-701, Republic of KOREA

Summary. The result of a computational investigation for an underexpanded jet and impingement on a wall with various distances from the jet exit is presented. The three-dimensional Navier-Stokes equations and a number of two-equation turbulence models based on the ω equation are computed. It is shown that the pressures on the impinging plate with the weakly nonlinear eddy viscosity model is in good agreement with the experimental data, particularly, in the case of occurrence of the stagnation bubble.

1 Introduction

A great deal of effort has been made on experimental studies of the stagnation bubble and peak pressure positions when a jet impinges on the wall. Gubanova et al.[1], Ginzburg et al.[2], and Kalghatgi and Hunt[3] reported investigations of this flow structure, particularly in the impingement zone. They also found that the flow pattern varies with nozzle-to-plate distances(h/R_e) and the exit Mach number. Ginzburg et al. proposed that stagnation bubbles occur at distances in which a free jet shocks interact with the stand-off shock resulting in the triple-shock structure. A slip-line emanates from the triple-point divides flow close to the plate into two regions, the outer has the higher total pressure. Mixing then occurs along the slip-line which exist between these two regions, and the interaction of the high pressure shear layer formed with the inner boundary layer close to the plate causes separation and the formation of a bubble of recirculating fluid. Recently, jet impinging problems can be simulated by solving the compressible Navier-Stokes equations. Furthermore, some investigations of turbulence models capable of capturing these complex flow fields above the impinging wall are presented. Cumber et al.[4] tested the standard $k - \epsilon$ model and Alvi et al.[5] verified the accuracy of SARC(Spalart-Allmaras with Rotation and Curvature correction) one equation model and SST(Shear-Stress Transport) two-equation model. The overall goal of this study is to simulate a vertical launching missile. For the first step, the present work will provide analysis of this complex flow and test performance of $k - \omega$ turbulence models through comparison between experiments and simulations.

2 Governing Equations and Numerical Methods

In the present work, the compressible Navier-Stokes equations and $k-\omega$ turbulence equations are considered. The Reynolds-Averaged Navier-Stokes equations:

$$\frac{\partial \mathbf{q}}{\partial t} + \frac{(\partial \mathbf{f}_j - \partial \mathbf{f}_{vj})}{\partial x_j} = 0 \tag{1}$$

where $\mathbf{q}$ is the flow variable, $\mathbf{f}_j$ and $\mathbf{f}_{vj}$ are the inviscid and viscous fluxes in each direction,

$$\mathbf{q} = \begin{bmatrix} \rho \\ \rho u_i \\ \rho E \end{bmatrix}, \ \mathbf{f}_j = \begin{bmatrix} \rho u_j \\ \rho u_j u_i + p\delta_{ij} \\ \rho u_j H \end{bmatrix}, \ \mathbf{f}_{vj} = \begin{bmatrix} 0 \\ \tau_{ij} \\ u_j \tau_{ij} - q_j \end{bmatrix} \tag{2}$$

Here, ρ and p are the density and pressure, and u_j are the Cartesian velocity components. E is the total energy and $H = E + p/\rho$ is total enthalpy. q_j are the heat fluxes in each direction, and τ_{ij} are the viscous stresses, which are defined as

$$\tau_{ij} = 2(\mu_l + \mu_t)\left(S_{ij} - \frac{1}{3}S_{kk}\delta_{ij}\right) \tag{3}$$

where $S_{ij} = \frac{1}{2}\left(\frac{\partial u_i}{\partial x_j} + \frac{\partial u_j}{\partial x_i}\right)$ are the velocity strain rates, μ_l is the molecular viscosity determined by the Sutherland law and μ_t is the eddy viscosity provided by a turbulence model.

The $k-\omega$ equations can be written as

$$\frac{\partial \mathbf{q_T}}{\partial t} + \frac{(\partial \mathbf{f_T}_j - \partial \mathbf{f_T}_{vj})}{\partial x_j} = \mathbf{S}_{k\omega} \tag{4}$$

where $\mathbf{q_T} = [\rho k, \rho\omega]^T$ and the convection and diffusion terms are expressed as

$$\mathbf{f_T}_j = \begin{bmatrix} \rho u_j k \\ \rho u_j \omega \end{bmatrix}, \ \mathbf{f_T}_{vj} = \begin{bmatrix} (\mu_l + \sigma_k \mu_t)\frac{\partial k}{\partial x_j} \\ (\mu_l + \sigma_\omega \mu_t)\frac{\partial \omega}{\partial x_j} \end{bmatrix} \tag{5}$$

and $\mathbf{S}_{k\omega}$ is the turbulent source vector consist of the production rates of k and ω, and destruction rates.

$$\mathbf{S}_{k\omega} = \begin{bmatrix} P_k - D_k \\ P_\omega - D_\omega \end{bmatrix} = \begin{bmatrix} \mu_t S^2 - \frac{2}{3}\rho k S_{kk} - \beta' \rho k \omega \\ \alpha \frac{\omega}{k}\left(\mu_t \frac{S^2}{\alpha_\nu} - \frac{2}{3}\rho k S_{kk}\right) - \beta \rho \omega^2 \end{bmatrix} \tag{6}$$

where S is the mean strain rate, defined as $S^2 = 2S_{ij}S_{ji} - \frac{2}{3}S_{kk}^2$. The cross-diffusion term is added to the ω equation for the $k-\omega$ SST model. The turbulence eddy viscosity in terms of k and ω is defined as

$$\mu_t = \rho c_\mu \frac{k}{\omega} \tag{7}$$

The $k - \omega$ model is known that it behaves well for wall-bounded problems. Moreover, this model is simple and has faster convergence than the $k - \epsilon$ model. Despite favorable numerical aspects, however, it suffers from free-stream dependence. Menter therefore combined the $k - \omega$ model in the near wall region and the $k - \epsilon$ model outside of the boundary layer into a single model. This model is known as the SST model. A weakly non-linear eddy viscosity model, the WD+ model is known to improve the performance of the turbulence models for flows in the presence of adverse pressure gradients, particularly in Shock-Wave/Boundary-Layer Interactions(SWBLI). In this study, three turbulence models based on the ω equation are applied to the impinging jet problem. Original $k - \omega$ model of Wilcox[6], Wilcox-Durbin+ model[7] and Menter's $k - \omega$ SST[8] model are tested. The closure coefficients and detailed functions of models are given in References[6-8]. The finite volume technique is used as the discretization of the governing equations. To avoid numerical instability such as the carbuncle that sometimes accompanies the approximate Riemann solver, the HLLE+ scheme is employed for inviscid flux at the cell interfaces[9]. This switching scheme can remove some instability by changing the dissipation mechanisms without degrading accuracy. Simple central differencing is applied to obtain variable gradients of the viscous flux. The Diagonalized Alternating Direction Implicit(DADI) method is applied to find steady-state solutions. When the variation of pressure on the wall is below 10^{-4}, the simulation is considered to be in the steady state.

3 Grids and Boundary Conditions

The nozzle geometry used to produce numerical results is a conical-type nozzle with the exit diameter of 30mm and 5-deg half-angle following the experiment of Ginzburg et al.[2] To avoid reconstruction of grid varies with nozzle-to-plate distances, the chimera grid technique is applied. The distance to the first cell centers from the impinging wall is 5×10^6 so that $y+$ is set below 1. Fig. 1(a) shows axisymmetric structured and chimera grids. The full 3-dimensional grid of nozzle is based on an axisymmetric chimera grid which is the O-type structure, whereas an H-type grid is used at the impinging plate to remove the singularity at the centerline. At the solid wall the no-slip boundary condition for velocities is applied and the density and energy are extrapolated from the interior cells. The adiabatic wall condition is applied since that heat flux is not considered in the present work. The kinetic energy k is set to zero at the wall and ω has asymptotic value as

$$\omega_{wall} = \frac{19}{9} \frac{6\mu_{wall}}{\beta \rho d_1^2} \tag{8}$$

where d_1 is the distance to first cell centers from the wall. The boundary conditions at the nozzle throat are calculated from isentropic relations and the perfect gas law. A specific head ratio of 1.4 is used.

4 Numerical Results and Discussion

The conditions at nozzle exit are $M_e = 2.185$, $n_e = p_e/p_a = 1.9$, underexpanded nozzle flow. The Reynolds number at throat is 5.2×10^6. The distance ratio, $\bar{h} = h/R_e$, between the nozzle exit and the center of plate is chosen as 2.8, 4.0, and 6.8 to investigate the presence of stagnation bubble and the structure of jet flow. Ginzburg et al. stated that the off-center peak pressure on impinging plate caused by separation can occur when $\bar{h} > 3.6$.

To investigate effects of the chimera grid, three solutions of each grid system with same conditions are shown in Fig. 1(b). The distributions of pressure are almost same, but higher resolution of the structured grid at stand-off shock position than the chimera grid makes a small difference. Fig. 2 shows solutions using different flux schemes. In the Fig. 2(a), the HLLE+ scheme keeps the symmetry, whereas the Roe scheme makes some instability in spite of large enough dissipation. Furthermore, this large numerical dissipation of Roe scheme gives lower peak pressure than the solutions with the HLLE+ scheme as can be seen in Fig. 2(b).

In Fig. 3, the pressure distributions in the impinging zone given by three turbulence models are compared with the experimental data. All three models produce nearly the same flow pattern and the peak pressures are shown with these models. The recirculating flow near the wall caused by separation is shown as off-center peak pressure, and $k-\omega$ WD+ model predicts better solutions than other models in all three cases. In the second case, the pressures at the center and the separation position are almost same and the velocity above the wall is accelerated again between two positions. Therefore the stagnation bubble created at outer peak position does not reach centerline. Fig. 4 shows a closed bubble shape by computation with the WD+ model.

The Mach number distributions with the WD+ model in the plane of symmetry are shown in Fig. 5. The results keep symmetry and show the flow structures of impinging jet. The WD+ model using the weakly non-linear eddy viscosity is in better overall agreement with the experimental data, especially the separation caused by adverse pressure gradients in the boundary layer.

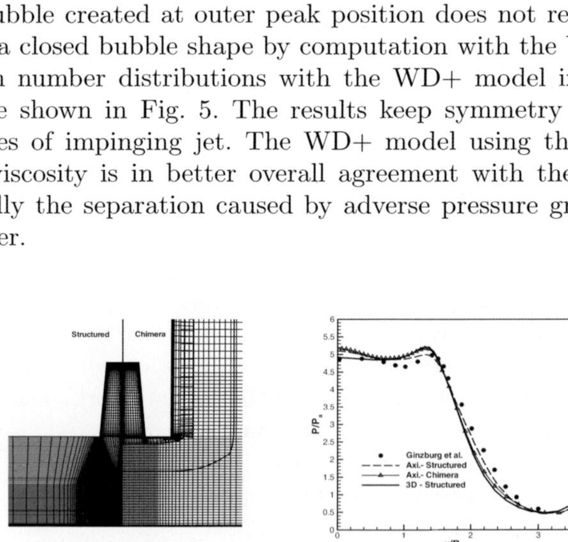

(a) Axisymmetric Grids (b) Pressure distribution

Fig. 1. Grids and Pressure on the wall($\bar{h} = 4.0$)

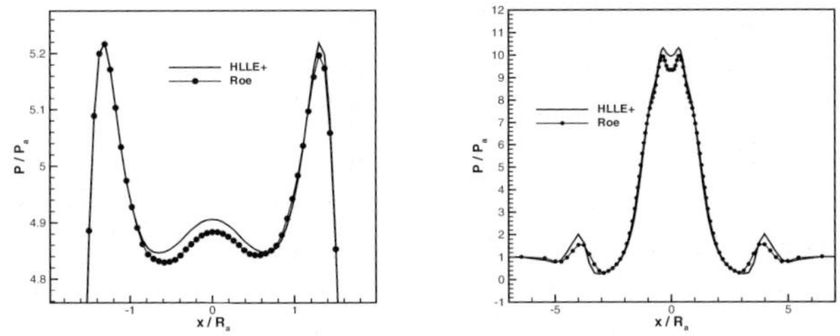

(a) $\bar{h} = 4.0$ (b) $\bar{h} = 6.8$

Fig. 2. Numerical instabilities in stagnation region on the wall

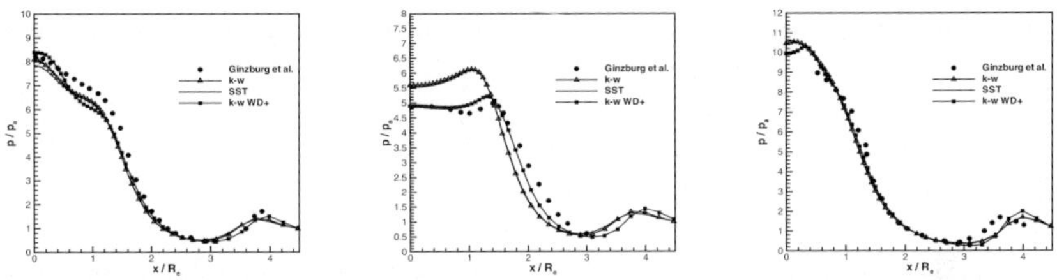

(a) $\bar{h} = 2.8$ (b) $\bar{h} = 4.0$ (c) $\bar{h} = 6.8$

Fig. 3. Pressure distributions on the wall for various $\bar{h}$

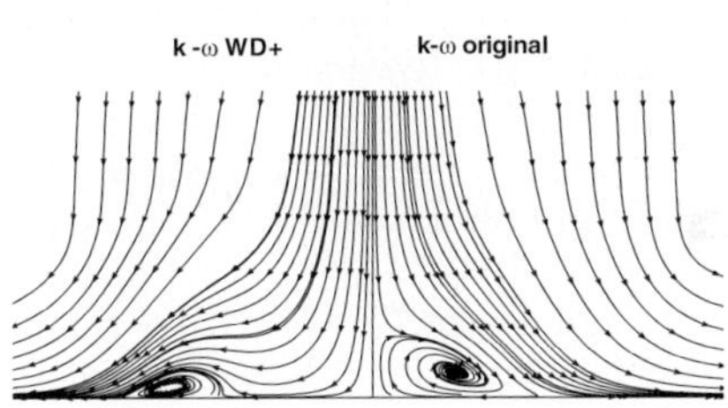

Fig. 4. Streamlines near the wall($\bar{h} = 4.0$)

5 Concluding Remarks

The Wilcox $k - \omega$, $k - \omega$ SST and WD+ models were tested in supersonic impinging jet flow fields by comparing of pressure on the impinging wall with experimental data. As a result, the overall agreement is fairly good. Three models revealed the presence of stagnation bubble which contains a low-velocity recirculating flow. But the pressure on the impinging wall using the weakly nonlinear eddy viscosity model, the WD+ model is in better agreement with the experimental data than other models in the adverse pressure gradient region. The ability to predict accurately the impinging jet behavior,

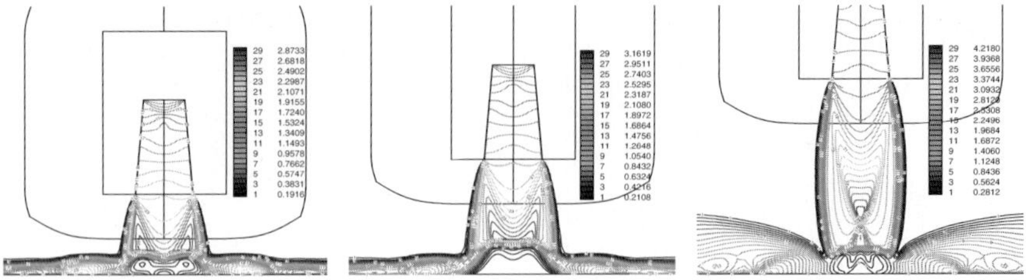

(a) $\bar{h} = 2.8$ (b) $\bar{h} = 4.0$ (c) $\bar{h} = 6.8$

Fig. 5. Mach contours in symmetric plane for various $\bar{h}$

especially near the ground plane can be applied to missile problems during launch.

References

1. Gubanova, O. I.,Lunev, V. V.m.,and Plastinina, L. N., "The Central Breakaway Zone with Interaction between a Supersonic Under expanded Jet and a Barrier," *Fluid Dynamics*, Vol. 6, July 1973, pp. 298-301.
2. Ginzburg. I. P., Semiletenko, B. G., Terpigor'ev, V. S., and Uskov, V. N., "Some Singularities of Supersonic Underexpanded Jet Interaction with a Plane Obstacle," *Journal of Engineering Physics*, Vol. 19, 1973, pp. 1081–1084.
3. Kalghatgi, G .T., and Hunt, B. L., "The occurrence of stagnation bubbles in supersonic jet impingement flows," *Aeronautical Quarterly*, Vol. 27, 1976, pp. 169–185.
4. Cumber, P. S., Fairweather, M., Falle, S.A.E.G., and Giddings, J. R., "Predictions of Impacting Sonic and Supersonic Jets," *Journal of Fluids Engineering*, Vol. 119, 1997, pp. 83–89.
5. Alvi, F. S., Ladd, J. A., and Bower, W. W., "Experimental and Computational Investigation of Supersonic Impinging Jets," *AIAA Journal*, Vol. 40, No. 4, 2002, pp. 599–609.
6. Wilcox, D. C., "Reassessment of the scale-determining equation for advanced turbulence models," *AIAA Journal*, Vol. 26, No. 11, 1988, pp. 1299–1310.
7. Thivet, F., "Lessons learned from RANS simulations of shock-wave/boundary-layer interactions," AIAA paper 2002-0583, Jan. 2002.
8. Menter, F. R., "Two-equation eddy-viscosity turbulence models for engineering applications," *AIAA Journal*, Vol. 32, No. 8, 1994, pp. 1598–1605.
9. Park, S. H., and Kwon, J. H., "On the dissipation mechanism of Godunov-type schemes," *Journal of Computational Physics*, Vol. 188, Issue 2, 2003, pp. 524–542.

Comparative Study of Reynolds Stress Turbulence Models in Free-Shear and Wall-Bounded Flows

Valerio Viti, George Huang[1] and Peter Bradshaw[2]

[1] University of Kentucky, Lexington, KY 40502, USA vviti@uky.edu
[2] Stanford University, Stanford, CA, USA

1 Introduction and Approach

The present work focuses on the performance and validation of some recent Reynolds-Stress models in simple wall-bounded and free-shear flows. In addition to this, the study addresses some of the difficulties that were encountered in the process of extracting the correct form of the models from the open literature and implementing them in a numerical code. The latter issue was totally unexpected and originally it was not supposed to be part of the present work. However, as work progressed, it appeared that the difficulties associated with the implementation of the different turbulence models were numerous and, in some cases, troublesome. These difficulties include typographical errors in the equations, inconsistencies between the forms of the model published in different journals and/or at different times, differences between the published models and the version implemented in the numerical code of the model developer. As a consequence of this unexpected difficulty, the study of more complex flowfields (separated flows, rotational flows) cannot be performed unless the above issues are first solved. Therefore, the present work encompasses only those turbulence benchmark flows such as free-shear flows (plane and round jets, mixing layer) and simple wall-bounded flows (flat plate with and without adverse pressure gradient). Future work will focuse on the extension of this work to more complex flowfields such as those studied by Hsu et al. [1] and Bardina et al. [2] and Aspley et al. [3]

The boundary layer code uses second-order upwind spatial differentiation for the convective fluxes, second-order central differencing for the diffusive terms and the SIMPLE scheme for the mass continuity equation. The models here considered are the Launder-Shima [4], the Jakirlic-Hanjalic [5, 6], Craft model [7, 8, 9], Manceau-Hanjalic [10, 11],and the Perot models [12, 13, 14]. The transport equations for the RS tensor is different for each second order model. The only exception is the Perot Turbulence Potential Model (TPM) in which the divergence of the RS tensor is modeles rather than the RS tensor itself.

The cases chosen represent those "benchmark" flowfields usually used as starting point for the validation of new turbulence models. The flowfields encompass solid-bounded flows such as the flat plate with and without adverse pressure gradient, and free-shear flows such as the mixing layer, plane and round jets. The numerical results were then compared to either proven correlations (the Von Karman-Schoenherr skin friction correlation for the flat plate or the self-similar analytical solutions for the free-shear flow), or the experimental results of Samuel and Joubert [15].

2 Results and Conclusions

For the flat plate calculations a constant inlet velocity profile was used as the inlet condition. For the mixing layer and jet cases, a step profile was used as inlet velocity. For all the cases a constant freestream turbulence intensity of 1% was assumed, as well a as a turbulent to laminar viscosity ratio of 1. For the wall bounded flows, a no slip condition was enforced at the wall. At freestream, a zero-gradient boundary was used for the flat plate without pressure gradient while a pressure profile taken from the experimental data of Samuel and Joubert [15] was imposed at the freestream for the case with pressure gradient. For the free shear flows, a zero-gradient boundary conditions was implemented at the freestream. For the plane and round jets, a symmetrical boundary condition is imposed at the jet center.

The skin friction results for the flat plate with zero pressure gradient are shown in Fig. 1. It is clear that all of the Reynolds Stress models perform quite well in the prediction of the skin friction with the exception of the Launder-Shima model. Notice also that the Manceau-Hanjalic model slightly over-predicts the results of the other models.

The skin friction results for the flat plate case with an adverse pressure gradient are shown in Fig. 2.expect for the Launder-Shima model, all models predict this case quite well.

The results for the free shear flows show the similarity in performance of the tested Reynolds Stress models. Figure 3 shows the velocity profile and the spreading rate for the mixing layer case. Apart from Perot's model, all models predict well against the experimental spreading rate while miss the profile near the high velocity side.

Figure 4 shows the velocity profiles for the plane jet. With the expection of Perot's model, all the models predict a lower spreading rate than the experimental value.

In the case of the round jet, Fig. 5, although Launder-shima and Perot's models seemed to predict the round/plane jet anomaly, the spreading rates predicted are lower than experimental data.

From the results obtained in this study and the experience gained in the validation effort we can summarize the following points.

The more complex and computationally more expensive Reynolds-Stress models perform as well as the simpler eddy viscosity models in the simple wall bounded and shear free flows. However, it expected to see an advantage in accuracy of Reynolds Stress models for more complex flows.

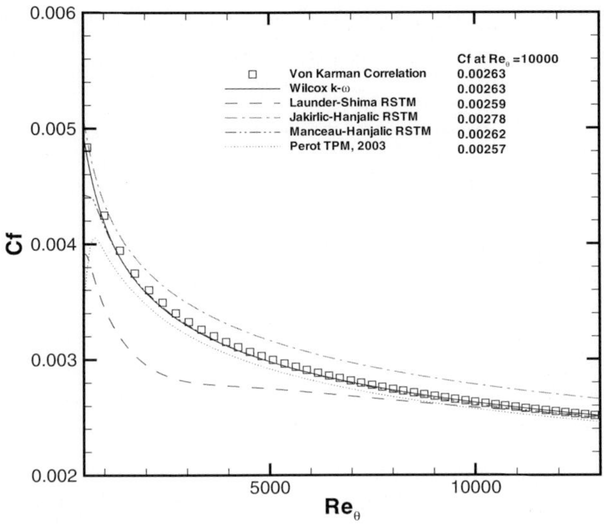

Fig. 1. Velocity Profile for the flat plate without pressure gradient.

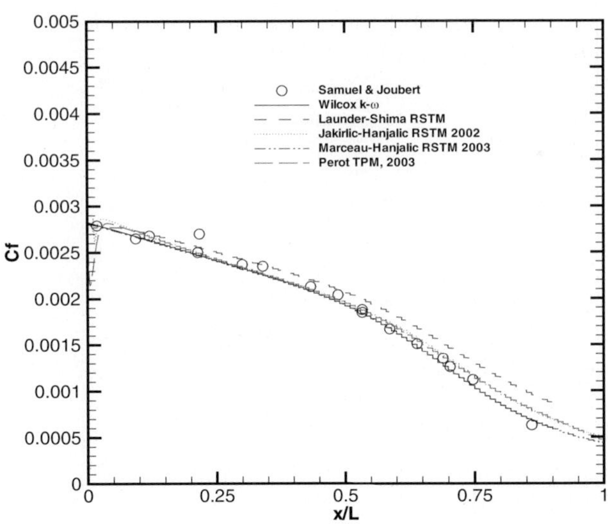

Fig. 2. Velocity Profile for the Samuel-Joubert flat plate with adverse pressure gradient.

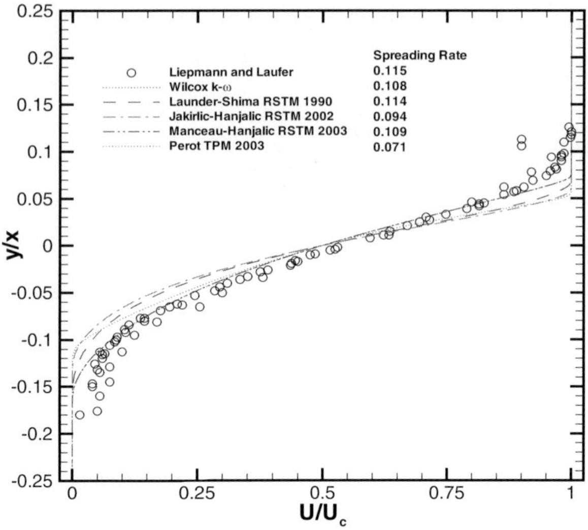

Fig. 3. Velocity Profile for the Mixing Layer.

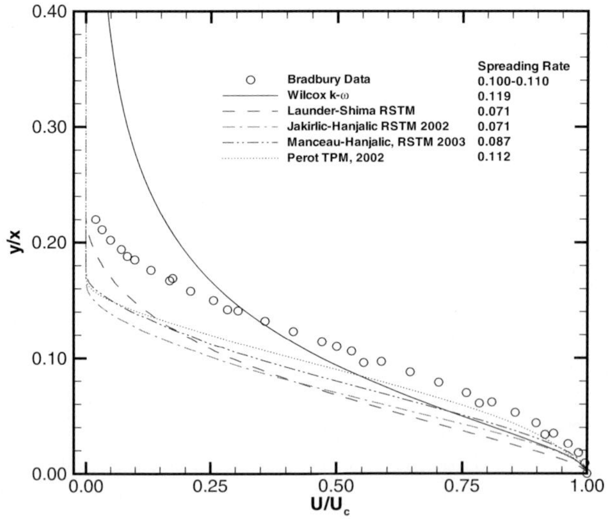

Fig. 4. Velocity Profiles for the Plane Jet.

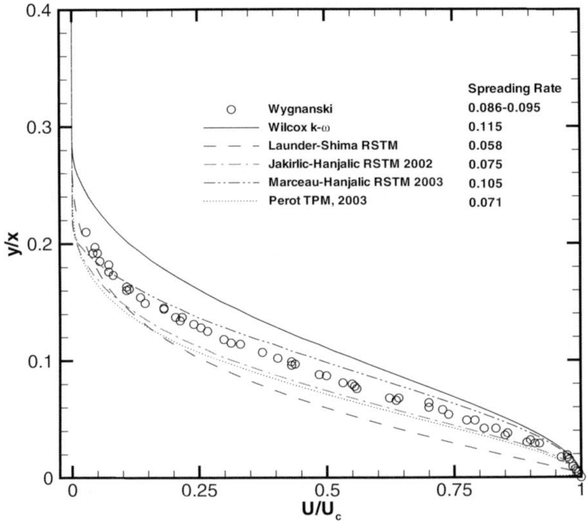

Fig. 5. Velocity Profile for the Round Jet.

The Launder-Shima model does not match the experimental skin friction or the results obtained from the other turbulence models in wall-bounded flows. However, it performs as well as the other models in the free shear flows.

The Perot Turbulent Potential Model, which solves scalar and vector equations for the divergence of the Reynolds Stress instead of the Reynolds Stress itself, gave good results, matching the performance of the other Reynolds Stress models. The computational advantages produced by this approach were not studied in this work.

Turbulence models that have already been validated by the model developer are seldom well-behaved and generally they do not produce a converged solution when implemented as published in the open literature.

Often, even though not always, the form of the model equations published in the open literature does not correspond to the actual version implemented in the numerical code used by the model developer. This problem is probably due to the nature of turbulence research itself (and probably also that of academic research), in which there is a continuous evolution of the methods, approaches and findings that does not match the tracking effort.

Turbulence research seems to lack a general structure that facilitates and organize communication between the model developers and the actual users.

References

1. M. C. Hsu, K. Vogiatzis, P.G. Huang, Validation and Implementation of Advanced Turbulence Models in Swirling and Separated Flows, AIAA 2003-766, January 2003.
2. Bardina, J. E., Huang, P. G. and Coakley, T. J., 1997, Turbulence Modeling Validation, Testing and Development, NASA Technical Memorandum 110446
3. Aspley D. D., M. A. Leschziner Investigation of Advanced Turbulence Models for the Flow in a Generic wing-Body Junction Flow:Turbulence and Combustion, Vol. 67, pp. 25-55, 2001.
4. Launder B.E., Shima N., Second Moment Closure for the Near-Wall Sublayer: Development and Application, AIAA Journal, Vol. 27, No.10, October 1989.
5. K. Hanjalic, S. Jakirlic, Contribution Towards the Second-Moment Closure Modeling of Separating Turbulent Flows, Computers and Fluids, Vol. 27, No. 2, pp. 137-156, 1998.
6. S. Jakirlic, K. Hanjalic, A New Approach to Modeling Near Wall Turbulence Energy and Stress Dissipation, Journal of Fluid Mechanics, Vol. 459, No. 2, pp. 139-166, 2002.
7. T. J. Craft, B.E. Launder, A Reynolds Stress Closure Designed for Complex Geometries, International Journal of Heat and Fluid Flow, Vol. 17, pp. 245-254, 1996.
8. T. J. Craft, Developments in a Low-Reynolds-Number Second-Moment Closure and Its Application to Separating and Reattaching Flows, International Journal of Heat and Fluid Flow, Vol. 19, pp. 541-548, 1998.
9. P. Batten, T. J. Craft, M. A. Leschziner, H. Loyau, Reynolds-Stress Transport Modeling for Compressible Aerodynamics Applications, AIAA Journal, Vol. 37, No. 7, pp.785-793, 1999.
10. R. Manceau, K. Hanjalic, Elliptic Blending Model: a New Near-Wall Reynolds-Stress Turbulence Closure, Physics of Fluids, Vol. 14, No. 2, pp. 744-754, February 2002.
11. R. Manceau (University of Poitiers, France), Private Communications with V. Viti, September-October 2003.
12. B. Perot, Turbulence Modeling Using Body Force Potentials, Physics of Fluids, Vol. 11, No. 9, pp. 2645-2656, 1999.
13. B. Perot, A. Sasanka, C. Wang, Application of the Turbulent Potential Model to Complex Flows, Proceedings of the 5th International Conference on Turbulence Modeling and Measurements, Mallorca, Spain, 2002.
14. B. Perot, J. Taupier, Modeling Three-Dimensional Boundary Layers Using the Turbulent Potential Model, AIAA 2000-0914, 2000.
15. S. J. Kline, B. J. Cantwell, G. M. Lilley, 1980-81 AFOSR-HTTM-Stanford Conference on Complex turbulent Flows, Thermo-sciences Division, Stanford University, CA, USA, 1981.

Part XVI

Multifluid and Multiphase Flows

Lattice Boltzmann Simulations in Chemical Engineering

D. Hänel, U. Lantermann, and R. Kaiser

Institute of Combustion and Gasdynamics, University of Duisburg-Essen
D-47048 Duisburg, Germany, E-mail: haenel@ivg.uni-duisburg.de

Summary. Lattice Boltzmann approaches are presented for solving flow problems in chemical engineering. The lattice BGK (LBGK) method has been adapted to different prescribed model problems based on Taylor and Chapman-Enskog analysis and is demonstrated in two examples. A LBGK method on Cartesian and axisymmetric coordinates is presented for calculating gas phase reactions in the low Mach number limit with strongly changing density. Further, a LBGK method has been derived for solving equations of Poisson-type, essentially to utilize the favorable, gas kinetic treatment of boundaries for potential fields of electric or magnetic forces.

Key words: Lattice Boltzmann, low Mach number flow, potential fields

1. Introduction

Chemical engineering offers a wide field for CFD, as well for challenging applications as for developments of suited methods. The present paper deals with developments of methods but connected with simulations of the generation process of nano-particles by high temperature gas phase reactions. Among the different problems arising in the generation process, two important partial processes are picked up in this paper, where new approaches of lattice Boltzmann methods were successfully introduced. These are the process of gas phase reactions in an axisymmetric reactor vessel and the transport of nano-particles to surfaces under the influences of fluid flow, random diffusion and external forces. The lattice Boltzmann methods in form of the so called lattice BGK (LBGK) methods are used in our working group in different applications, in addition a number of algorithmic improvement were developed which essentially have increased the applicability and efficiency of the LBGK method and are partially included in the present calculations. Examples for algorithmic developments are e.g. the boundary fitting concept, [1], the local grid refinement, [2], and acceleration strategies, [3].

2. Gas phase reactions at low Mach numbers (LMNA)

Gas phase reactions take place in an axisymmetric reactor vessel with a mean, slow flow of low Mach number, but remarkable changes of density due to external wall heating and due to chemical reactions. Neglecting acoustic modes, such a situation is modeled with advantage by the low Mach number approximation (LMNA) of the Navier-Stokes equations, [4], which essentially is based on the effect that the thermodynamic pressure is nearly constant in space. The complete set of equations describing the gas phase reaction, is written in Cartesian coordinates if $\theta = 0$ with $\alpha, \beta = x, y, z$ and holds for an axisymmetric coordinate system with $\alpha, \beta = x, r$, if $\theta = 1$.

$$\partial_t \rho_{mix} + \partial_\alpha(\rho_{mix} u_\alpha) = -\frac{\theta}{r}\, \rho_{mix}\, u_r \tag{1}$$

$$\partial_t \rho_{mix} u_\alpha + \partial_\beta(\rho_{mix} u_\beta u_\alpha) + \partial_\alpha p^{(1)} - \partial_\beta \mu(\partial_\beta u_\alpha + \partial_\alpha u_\beta - \frac{2}{3}\delta_{\alpha\beta}\partial_\gamma u_\gamma) = \frac{\theta}{r}\left(-\rho_{mix}\, u_r\, u_\alpha + \mu\left(\delta_{\alpha x}(\partial_r u_x + \partial_x u_r) + \delta_{\alpha r}\, 2\cdot\partial_r(\frac{u_r}{r})\right)\right) \tag{2}$$

$$\rho_{mix}\, c_p\,(\partial_t T + u_\beta \partial_\beta T) - \partial_\gamma\, \lambda\, \partial_\gamma T - \rho_{mix} \sum_k h_k w_k - \frac{d\,p_{th}(t)}{dt} = \frac{\theta\,\lambda}{r}\delta_r T \tag{3}$$

$$\rho_{mix}\,(\partial_t \xi_k + \, u_\beta \partial_\beta \xi_k) - \partial_\gamma(\rho D)_{kl}\, \partial_\gamma \xi_k + \rho_{mix}\, w_k = \frac{\theta\, \rho D_{kl}}{r}\delta_r \xi_k \quad k = 1, ..., N \tag{4}$$

The reactive flow at low Mach numbers is modeled by the continuity and momentum equations Eq. (1) and Eq. (2), by the energy equation for the temperature (3) and by Eq. (4) for the mass fraction $\xi_k = \rho_k/\rho_{mix}$ of the k-th species, where w_k and h_k are the production rate and the heat of formation. The pressure $p = p_{th} + p^{(1)}$ splits in the thermodynamic pressure $p_{th}(t)$, which is a function of time only, and in the hydrodynamic pressure $p^{(1)}$, ensuring mass conservation. The density $\rho_{mix}(T, \xi_i)$ is a function of temperature T and of species concentrations ξ_k given as $\rho_{mix} = \frac{p_{th}}{RT}$ with $R = \sum_k \frac{\xi_k}{W_k}$.

A LBGK method for solving the LMNA of the Navier-Stokes equations with variable density was developed by Filippova et al. in [5] for Cartesian-like lattices. The equations for temperature and species concentration are coupled with the LBGK method versus density and are solved either by finite difference approximations or by a similar LBGK method for convection-diffusion equations with source terms and variable density. The LBGK method in [5] has shown to be a reliable and efficient method for computing reactive flows in particular due to the inherent small time steps which reduce essentially the stiffness of the problem. However if problems can be treated axisymmetrically, then 3-D Cartesian calculations become to expensive. To compensate this drawback, the LBGK method for the LMNA has been extended here to axisymmetric coordinates, as shown in the following.

The LBGK equation for a distribution function $f_i(t, \boldsymbol{r})$ in direction of the discrete molecular speed i reads

$$f_i(t + \delta_t, \boldsymbol{r} + \boldsymbol{c}_i \delta_t) = f_i(t, \boldsymbol{r}) + \Omega\, [F_i(t, \boldsymbol{r}) - f_i(t, \boldsymbol{r})] + \delta t\, s_i^{(k)} \tag{5}$$

The additional source terms $s_i^{(k)} = s_i^{(1)} + \delta t\, s_i^{(2)}$ are determined later. The dimensionless collision frequency Ω is connected with the kinematic viscosity by $\nu = \mu/\rho_{mix} = \delta t\, c_s^2(\frac{1}{\Omega} - \frac{1}{2})$, where c_s^2 is $c_0^2/3$. The equilibrium distribution F_i with variable density ρ_{mix} is defined by

$$F_i = t_p \left[\frac{P}{c_s^2} + \frac{\rho_{mix}\, u_\alpha\, c_{i\alpha}}{c_s^2} + \frac{\rho_{mix}\, u_\alpha\, u_\beta}{2c_s^2} \left(\frac{c_{i\alpha} c_{i\beta}}{{c_s}^2} - \delta_{\alpha\beta} \right) \right] + F_{i,corr} \quad (6)$$

where $F_{i,corr}$ is a correction term, too. The pseudo-pressure $P(t, \boldsymbol{r})$ collects diagonal elements of the stress tensor which act like the hydrodynamic pressure $p^{(1)}$. All the additional quantities, the source terms $s_i^{(k)}$, the distribution $F_{i,corr}$ and the "pressure" $P(t, \boldsymbol{r})$ are used to adapt the LBGK scheme to the requirements of the governing equations, here to the LMNA.

The analysis using Taylor series and Chapman-Enskog expansion connects the discrete LBGK approach with macroscopic moment equations. To give an idea how to determine the unknown quantities $s_i^{(k)}$, $F_{i,corr}$ and $P(t, \boldsymbol{r})$, this analysis is presented in the following but in a very short form neglecting many details. The expanded LBGK-equation Eq. (5) with a perturbation distribution $f_i^{(1)}$ from a Chapman-Enskog expansion $f_i = F_i + \delta t\, f_i^{(1)} + \delta t^2\, f_i^{(2)} + \cdots$ reads

$$D_\alpha\, F_i = \delta t\, \left(\frac{1}{\Omega} - \frac{1}{2}\right) D_\alpha D_\beta F_i - \Omega\, f_i^{(1)} - \frac{\delta t}{\Omega}\, D_\alpha s_i^{(k)} + s_i^{(k)} \qquad D_\alpha = \partial_t + c_{i,\alpha} \partial_\alpha \quad (7)$$

The invariant, discrete moments of f_i satisfy $\sum_i f_i^{(1)} = \sum_i f_i^{(1)} c_{i,\alpha} = 0$ and

$$\sum_i f_i = \sum_i F_i = P \quad \text{and} \quad \sum_i f_i c_{i,\alpha} = \sum_i F_i\, c_{i,\alpha} = \rho_{mix} u_\alpha \quad (8)$$

The zeroth moment M_0 of Eq. (7) describes the mass conservation in the LBGK frame

$$\partial_t (P/c_s^2) + \partial_\alpha (\rho_{mix}\, u_\alpha) - \sum_i s_i^{(1)} = O(\delta t) \quad (9)$$

and correspondingly the momentum conservation by multiplication with $c_{i,\alpha}$ and summation over i

$$\begin{aligned} &\partial_t \rho_{mix} u_\alpha + \partial_\beta (\rho_{mix} u_\beta u_\alpha) + \partial_\alpha P \\ &\qquad - \partial_\beta\, \nu\, [\partial_\beta (\rho_{mix} u_\alpha) + \partial_\alpha (\rho_{mix} u_\beta) + \delta_{\alpha\beta} \partial_\gamma (\rho_{mix} u_\gamma)] \\ &= -\partial_\beta \sum_i c_\alpha c_\beta\, F_{i,corr} - \frac{\delta t}{\Omega} \sum_i D_\alpha c_\alpha\, s_i^{(1)} + \delta t \sum_i c_\alpha\, s_i^{(2)} + O(\delta t^2) \end{aligned} \quad (10)$$

The macroscopic moments Eq. (9) and Eq. (10) of the LBGK approach are unclosed through the quantities s_i, $F_{i,corr}$ and P. These quantities are determined by comparisons with the governing equations of the actual problem.

A LBGK method for the LMNA in Cartesian frames, i.e. for Eq. (1) and Eq. (2) with $\theta = 0$, is constructed by using the following quantities:

$$s_i^{(1)} = -\partial_t\, \rho_{mix} \quad \text{if} \quad c_{i,\alpha} = 0 \quad \text{else} \quad s_i^{(1)} = 0 \quad \text{and} \quad s_i^{(2)} = 0 \quad (11)$$

$$F_{i,corr} = t_p\, \nu\, u_\gamma\, \partial_\delta \rho_{mix} \left(\frac{c_{i\gamma} c_{i\delta}}{{c_s}^2} - \delta_{\gamma\delta} \right) \quad P = p^{(1)} + 2/3\, \mu\, \partial_\gamma u_\gamma + \nu \partial_\gamma (\rho_{mix} u_\gamma)$$

The P-term is to evaluate at boundaries only. The resulting LBGK method agrees exactly with the method derived in [5].

An axisymmetric LBGK method for the LMNA of the Navier-Stokes equations, i.e. Eq. (1) and Eq. (2) with $\theta = 1$, is realized by changing the source terms in

$$s_i^{(1)} = -\partial_t \rho_{mix} - \rho_{mix} u_r/r \quad \text{if} \quad c_{i,\alpha} = 0 \quad \text{else} \quad s_i^{(1)} = 0 \qquad (12)$$

$$s_i^{(2)} = \frac{t_p}{c_s^2} \left[\frac{1}{r}(\partial_r u_x + \partial_x u_r)\, c_{i,x} + 2\, \partial_r (u_r/r)\, c_{i,r}\right] \qquad (13)$$

but with unchanged quantities P and $F_{i,corr}$ as in Eq. (12). Corresponding LBGK transformations in axisymmetric systems can be performed in similar way for convection-diffusion equations with source terms and variable density, as e.g. for Eq. (3) or Eq. (4).

An axisymmetric chemical reactor, based on the experimental arrangement, was designed to validate and to compare the new LBGK developments with results of a pressure relaxation method using finite differences (FD) for the axisymmetric system of equations Eq. (1) to Eq. (4). The same data and computational grids are used in both methods. The reactor, as sketched in Fig. 2, consists of a cylindrical vessel with heated outer walls. The reacting precursor gas, here $Fe(CO)_5$, is feed in through an inner pipe together with an inert gas N_2. The precursor decomposes under external heat to Fe-atoms, which grow along their paths by coagulation to nano-particles. Fig. 3 shows three typical LBGK results for colors of constant values of axial velocity, temperature and reaction rate versus radius at a wall temperature of 1073 K.

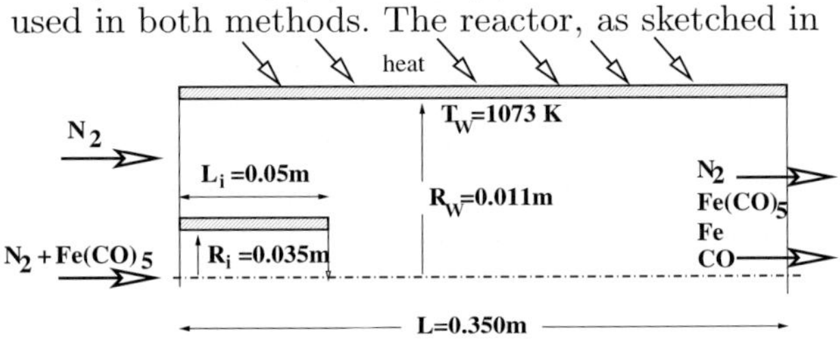

Fig. 2. Model of axisymmetric, chemical reactor

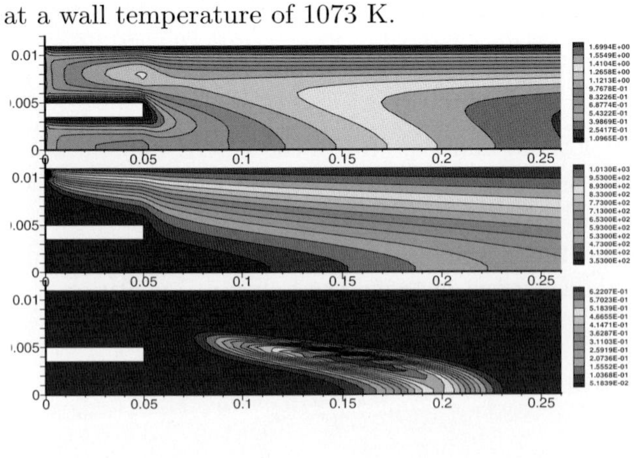

Fig. 3. LBGK solutions for axisymmetric, reactive flow. Colors of constant axial velocity, temperature and reaction rate

A quantitative comparison between LBGK solutions and pressure relaxation finite difference (FD) solutions is shown in Fig. 4 for the axisymmetric flow.

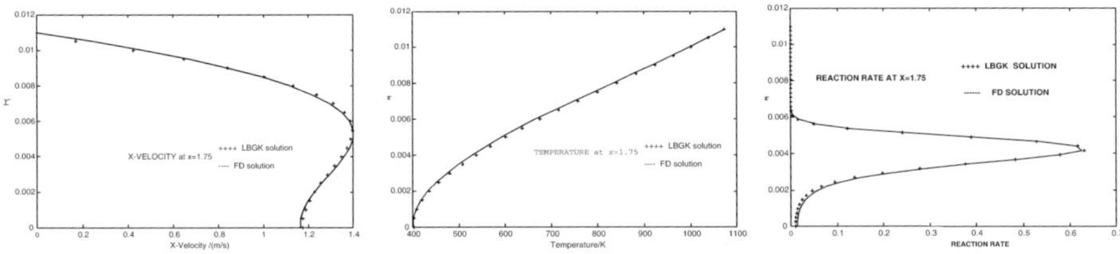

Fig. 4. Comparisons between a LBGK solution (crosses) and a FD solution (full) in a cut $x = .175\ m$ for axial velocity, temperature and reaction rate of $Fe(CO)_5$ versus radius in the axisymmetric reactor of Fig. 2 and Fig. 3

3. Particle transport and deposition

The particle Monte-Carlo method for computing the translatory motion discrete particles with drag, external and random forces due to Brownian diffusion reads in short form

$$m_p \frac{d\boldsymbol{v}_p}{dt} = \boldsymbol{F}_{drag} + \boldsymbol{F}_{ext} + \boldsymbol{F}_{random} \quad \text{and} \quad \frac{d\boldsymbol{r}_p}{dt} = \boldsymbol{v}_p \tag{14}$$

A set of three additional equations for the angular momentum with corresponding external and random torques has to be considered in addition, if magnetic particles and interaction are taken into account.

The solution of the problem requires the knowledge of the flow field and of the potential fields for determining the external, electrical or magnetic forces $\boldsymbol{F}_{ext}$ or torques $\boldsymbol{T}_{ext}$. The geometrical boundaries are relatively complex in both cases due to complex filter geometries but also due particles already deposited on the surfaces. Complex boundaries are well fitted by LBGK methods, [1, 2]. The incompressible flow and the potential fields are computed here by corresponding LBGK methods. The LBGK algorithm for 3-D incompressible fluid flows is directly based on the LBGK algorithm Eq. (5), Eq. (6) and Eq. (8) with $s_i = F_{i,corr} = 0$ and $P = p^{(1)}$.

A LBGK algorithm for equations of Poisson or Laplace-type for electric or magnetic potentials,

$$\frac{\partial^2 \phi_\alpha}{\partial x_\beta^2} = -RS_\alpha \tag{15}$$

has been developed from the basic LBGK concept Eq. (5) and a linear, equilibrium distribution

$$F_i(t, \boldsymbol{x}) = t_p \left[1 + \frac{M_{1\alpha} c_{pi\alpha}}{c_s^2}\right]. \tag{16}$$

The source terms in Eq. (5) are chosen as

$$s_i^{(1)} = 0 \qquad \delta t\, s_i^{(2)} = t_p\,(1 - \frac{\Omega}{2}) \left(2\,\frac{\partial M_0}{\partial x_\alpha} + \frac{\delta t}{\Omega}\, RS_\alpha(t, \boldsymbol{x})\right)$$

The term $\sim \frac{\partial M_0}{\partial x_\alpha}$ corrects non-equilibrium effects, since $M_0 = \sum_{p,i} f_{pi} \neq \sum_{p,i} F_i$. The resulting potential ϕ_α is defined by the components of the first moment M_1

$$\phi_\alpha = M_{1\alpha} = \sum_{p,i} f_{pi} c_{pi\alpha}, \quad \alpha = 1, 2, 3 \tag{17}$$

The LBGK solution results in three potentials ϕ_α with $\alpha = 1, 2, 3$, which look like vector potentials. However they are not independent through the coupling over boundary conditions. A demonstrative example for particle transport to filters under electrical forces using LBGK solutions for the flow field and the electrical field is shown in Fig. 5.

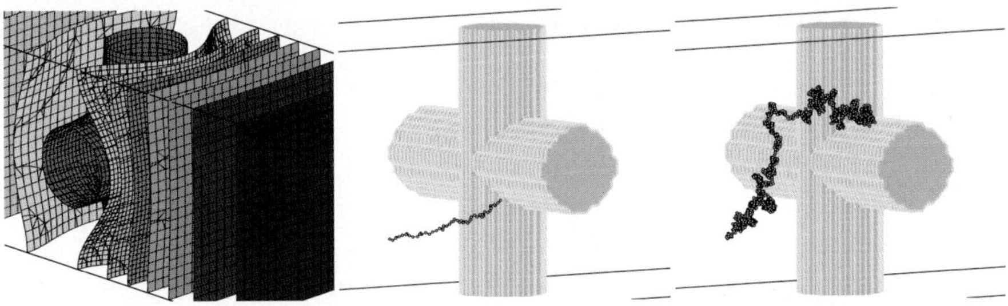

Fig. 5. Planes of constant electrical potential around a periodical, crossing filter fibers (left), trajectory of a charged nano-particle (10 nm) in electrical field and fluid drag and random diffusion (middle), trajectory of the same particle without electrical field (right)

Conclusions

The range of applications of LBGK methods has been extended by introducing additional free parameters to be adapted to prescribed macroscopic equations using Taylor and Chapman-Enskog analysis. This way is demonstrated here by means of axissymmetric, reactive flow and by solutions of electric or magnetic potentials over complex boundaries, where LBGK methods offer advantages against conventional solution methods.

References

1. O. Filippova, D. Hänel: Boundary Fitting and Local Grid Refinement for Lattice-BGK Models. Int. Journal of Modern Physics C, vol. 9, N 8, (1998).
2. O. Filippova, D. Hänel, Grid refinement for lattice-BGK models, *J. Comp. Phys.* vol 147, 219 (1998).
3. O. Filippova, D. Hänel: Acceleration of Lattice-BGK Schemes with Grid Refinement J. Comp. Physics, vol. 165, pp. 407-427 (2000).
4. Y.H. Choi, C.L. Merkle: The application of preconditioning in viscous flows. J. Comp. Physics, vol. 105, pp. 207-223, (1993).
5. O. Filippova, D. Hänel, A novel lattice-BGK approach for low Mach number combustion, J. Comp. Phys. vol 158, 139 (2000).
6. D. Hänel, U. Lantermann: Simulation of Particle Transport and Deposition. To be published in: Proc. of Traffic Granular Flow 2003, Springer-Verlag Berlin, Heidelberg, (2004).

A Numerical Scheme for Compressible Multiphase Flows

Rémi Abgrall[1,2] and Vincent Perrier[1,3]

[1] Mathmatiques Appliques de Bordeaux, Universit Bordeaux I
33 405 Talence Cedex, France
[2] Institut Universitaire de France
[3] Centre des Lasers Intenses et Applications, Universit Bordeaux I
33 405 Talence Cedex, France
abgrall,perrier@math.u-bordeaux1.fr

Summary. We start from a scheme that is able to represent many physical situations where the flow is composed of several non miscible phases. The topology of bubbles is introduced into the scheme via relaxation parameters. In several situations, we may consider that at least one of these relaxation parameters is very large. For example when they are all infinite, this means that the interfacial area between phases becomes large. This leads to many numerical difficulties. We study the behavior of the scheme in the limit of infinite relaxation parameters, and provide numerical illustrations.

1 Introduction

The modelisation of multiphase flows is an extremely complex subject. Their numerical simulation is at least as complex because besides the modelisation difficulties coexist numerical ones. In [7, 3], we have proposed a numerical model which is in principle able to simulate many situations. Assuming two non miscible phases, the flow is described by the volume fraction of the phases α_k, their densities ρ_k, their velocities $\mathbf{u}_k$ and their total energy E_k via

$$\begin{aligned}
&\frac{\partial \alpha_k}{\partial t} + \mathbf{u}_I \cdot \nabla \alpha_k &&= \mu(P_k - P_{\bar{k}})\\
&\frac{\partial(\alpha_k\rho_k)}{\partial t} + \text{div }\big(\alpha_k\rho_k\mathbf{u}_k\big) &&= 0\\
&\frac{\partial(\alpha_k\rho_k\mathbf{u}_k)}{\partial t} + \text{div }\big(\alpha_k\rho_k\mathbf{u}_k\otimes\mathbf{u}_k\big) + \nabla(\alpha_k P_k) &&= P_I\nabla\alpha_k + \lambda(\mathbf{u}_{\bar{k}} - \mathbf{u}_k)\\
&\frac{\partial(\alpha_k\rho_k E_k)}{\partial t} + \text{div }\big(\alpha_k(\rho_k E_k + P_k)\mathbf{u}_k\big) &&= P_I\mathbf{u}_I\cdot\nabla\alpha_k - \mu P_I(P_k - P_{\bar{k}})\\
& &&\quad +\lambda\mathbf{u}_I(\mathbf{u}_{\bar{k}} - \mathbf{u}_k)
\end{aligned} \tag{1}$$

In (1), P_k is the pressure of phase k, the total energy is related to the pressure by $E_k = \varepsilon_k(P_k, \rho_k) + \frac{1}{2}\mathbf{u}_k^2$ where the internal energy is a concave function of its arguments. We define $\bar{k} = 1$ (resp. $= 2$) when $k = 2$ (resp. $k = 1$). The interface velocity and pressure $\mathbf{u}_I$ and P_I are modelised, this is one of the hard points, see e.g. Drew and Passman [4].

The variables μ and λ are relaxation parameters that depend on the interfacial area, and can be considered large when the two phases are well mixed, as for alloys, but also when interface problems are computed (in which case the computational domain is splitted in two subdomains where one of the volume fractions is set to machine zero).

The advantage of (1) is that the system is *always* hyperbolic, whatever the choices of the interface variable. But it also leads to several numerical difficulties : the system cannot be cast into conservative form (hence what is a shock ?, which form should possess a numerical scheme ?), the relaxation terms have to be splitted for simplicity reasons, but the meaning of this becomes questionable when λ or μ tends to infinity.

In the limit $\lambda, \mu \to +\infty$ while μ/λ remains bounded, it is possible to derive the asymptotic system satisfied in the limit, namely

$$\begin{aligned}
\frac{\partial \alpha_k}{\partial t} + \mathbf{u}\cdot\nabla\alpha_k &= \frac{\rho_{\bar{k}} a_{\bar{k}}^2 - \rho_k a_k^2}{\frac{\rho_k a_k^2}{\alpha_k} + \frac{\rho_{\bar{k}} a_{\bar{k}}^2}{\alpha_{\bar{k}}}} \text{div } \mathbf{u} \\
\frac{\partial(\alpha_k\rho_k)}{\partial t} + \text{div }(\alpha_k\rho_k\mathbf{u}) &= 0 \\
\frac{\partial(\rho\mathbf{u})}{\partial t} + \text{div }(\rho\mathbf{u}\otimes\mathbf{u}) + \nabla(P) &= 0 \\
\frac{\partial(\rho E)}{\partial t} + \text{div }(\rho E + P\mathbf{u}) &= 0,
\end{aligned} \tag{2}$$

where ρ and E are mixture variables, and a_k and $a_{\bar{k}}$ are the sound speed of each phase, see [6] for details. In (2) appears a new term which indicates that the volume fraction should be modified accross shock waves, but once more the mathematical meaning of $\frac{\rho_{\bar{k}} a_{\bar{k}}^2 - \rho_k a_k^2}{\frac{\rho_k a_k^2}{\alpha_k} + \frac{\rho_{\bar{k}} a_{\bar{k}}}{\alpha_{\bar{k}}}} \text{div } \mathbf{u}$ is far from being clear.

The purpose of this note is to sketch a method and a scheme that clarify these questions.

2 A scheme for multiphase flows

For the sake of simplicity, we consider only one dimensional problems. We are aware that the physical meaning of one dimensional multiphase flows is very questionable, since many important effects are indeed multidimensionals. See [1] for an extension to more than one dimension.

The derivation of the scheme uses the set averaging ideas of Drew and Passman [4] combined with the discretisation principle introduced by Godunov. At each time step, the flow is described in each computional cell by the average $W_j = (\alpha_{1,j},\ \alpha_{1,j}\rho_{1,j},\ \alpha_{1,j}\rho_{1,j}u_{1,j},\ \alpha_{1,j}\rho_{1,j}E_{1,j},\ \alpha_{2,j},\ \alpha_{2,j}\rho_{2,j},\ \alpha_{2,j}\rho_{2,j}u_{2,j},\ \alpha_{2,j}\rho_{2,j}E_{2,j})$. Following [3], we consider a family of random subdivison of the cell $\mathcal{C}_j =]x_{j-1/2}, x_{j+1/2}[$. In each of the subcells of $\mathcal{C}_j$, we randomly set the variables $W_{1,j} = (\rho_{1,j},\ \rho_{1,j}u_{1,j},\ \rho_{1,j}E_{1,j})$ or $W_{2,j} =$

$(\rho_{2,j}, \rho_{2,j}u_{2,j}, \rho_{2,j}E_{2,j})$, and the random process is done so that the average is W_j. In [3], we notice that it is sufficient that the average length of the phase Σ_1 (resp. Σ_2) in $\mathcal{C}_j$ is $\alpha_{1,j}(x_{j+1/2}-x_{j-1/2})$ (resp. $\alpha_{2,j}(x_{j+1/2}-x_{j-1/2})$). This means that if X is the indicator function of Σ_1, the average $\mathcal{E}\left(\int_{x_{j-1/2}}^{x_{j+1/2}} X dx\right)$ is precisely $\alpha_{1,j}(x_{j+1/2} - x_{j-1/2})$.

Given any realisation, we have splitted the computational domain in sub(–sub) domains filled with pure fluids (either Σ_1 or Σ_2) with the states $W_{1,j}$ or $W_{2,j}$. We can apply Godunov method to this situation and ensemble–average in $\mathcal{C}_j$ to get an evolution equation for W_j. In the sequel, we denote by $F(U,U')$ the Godunov flux between the two (pure fluids) states $U$and U', $\sigma(U,U')$ is the contact discontinuity velocity for the Riemann problem between U and U', and $F^{lag}(U,U') = F(U,U') - \sigma(U,U')U^*$ where U^* is the state in the Riemann problem at the right or the left of the contact (no matter of the choice, because of the Rankine Hugoniot relations). Thanks to these notations, we get

$$\begin{aligned}\frac{d}{dt}(\alpha_{1,i}W_{1,i}) + &\frac{\mathcal{E}(XF)_{i+1/2} - \mathcal{E}(XF)_{i-1/2}}{\Delta x} \\ &= \frac{\mathcal{E}\left(F^{lag}\dfrac{\partial X}{\partial x}\right)_{i,\mathrm{bound}}}{\Delta x} + \mathcal{E}\left(F^{lag}\frac{\partial X}{\partial x}\right)_{i,\mathrm{relax}}\end{aligned} \tag{3}$$

Defining $\beta_{i+1/2}^{(l,p)} = \mathrm{sign}(\sigma(W_{l,i}, W_{p,i-1}), \beta^+ = \max(0,\beta), \beta^- = \min(0,\beta)$ and $P_{i+1/2}^{l,l} = \min(\alpha_{l,i}, \alpha_{l,i+1})$, $P_{i+1/2}^{l,q} = \max(\alpha_{l,i} - \alpha_{q,i+1}, 0)$ for $l \neq q$, we have

$$\begin{aligned}\mathcal{E}(XF)_{i+1/2} = &P_{i+1/2}^{1,2}\left(\beta_{i+1/2}^{(1,2)}\right)^+ F(W_{1,i}, W_{2,i+1}) + P_{i+1/2}^{1,1} F(W_{1,i}, W_{1,i+1}) \\ &- P_{i+1/2}^{2,1}\left(\beta_{i+1/2}^{(2,1)}\right)^- F(W_{2,i}, W_{1,i+1})\end{aligned} \tag{4}$$

and

$$\begin{aligned}\mathcal{E}\left(F^{lag}\frac{\partial X}{\partial x}\right)_{i,\mathrm{bound}} = &P_{i+1/2}^{1,2}\left(\beta_{i+1/2}^{(1,2)}\right)^- F^{lag}(W_{1,i}, W_{2,i+1}) \\ &- P_{i+1/2}^{2,2}\left(\beta_{i+1/2}^{(2,1)}\right)^- F^{lag}(W_{2,i}, W_{1,i+1}) \\ &- P_{i-1/2}^{1,2}\left(\beta_{i-1/2}^{(1,2)}\right)^+ F^{lag}(W_{1,i-1}, W_{2,i}) \\ &+ P_{i-1/2}^{2,2}\left(\beta_{i-1/2}^{(2,1)}\right)^+ F^{lag}(W_{2,i-1}, W_{1,i}) \\ \mathcal{E}\left(F^{lag}\frac{\partial X}{\partial x}\right)_{i,\mathrm{relax}} = &\lambda_i\left(F^{lag}(W_{2,i}, W_{1,i}) - F(W_{1,i}, W_{2,i})\right)\end{aligned} \tag{5}$$

where λ_j is the average number of internal subdivisions in $\mathcal{C}_j$ (i.e. the interfacial area in 1D). Similar expressions for the fluid Σ_2 can be got. Last, the volume fraction evolution equation is given by

$$\begin{aligned}\frac{d\alpha_{1,j}}{dt}=\frac{1}{\Delta x}\Bigg(& P^{1,2}_{j-1/2}\left(\beta^{(1,2)}_{j-1/2}\right)^{+}\sigma(W_{1,j-1},W_{2,j})\\ & -P^{2,1}_{j-1/2}\left(\beta^{(2,1)}_{j-1/2}\right)^{-}\sigma(W_{2,j-1},W_{1,j})\\ & +P^{2,1}_{j+1/2}\left(\beta^{(1,2)}_{j+1/2}\right)^{+}\sigma(W_{1,j},W_{2,j+1})\\ & -P^{2,1}_{j+1/2}\left(\beta^{(2,1)}_{j+1/2}\right)^{-}\sigma(W_{2,j},W_{1,j+1})\Bigg)\\ & +\lambda_j\left(\sigma(W_{2,j},W_{1,j})-\sigma(W_{1,j},W_{2,j})\right).\end{aligned} \tag{6}$$

Details can be found in [3]. In practical applications, the discretisation is achieved via a splitting method. First, we integrate

$$\begin{aligned}\frac{(\alpha_{1,i}W_{1,i})^{n+1/2}-(\alpha_{1,i}W_{1,i})^{n}}{\Delta t}&+\frac{\mathcal{E}\left(XF\right)_{i+1/2}-\mathcal{E}\left(XF\right)_{i-1/2}}{\Delta x}\\ &=\frac{\mathcal{E}\left(F^{lag}\dfrac{\partial X}{\partial x}\right)_{i,\mathrm{bound}}}{\Delta x}\end{aligned} \tag{7}$$

and then a relaxation step

$$\frac{(\alpha_{1,i}W_{1,i})^{n+1}-(\alpha_{1,i}W_{1,i})^{n+1/2}}{\Delta t}=\mathcal{E}\left(F^{lag}\frac{\partial X}{\partial x}\right)_{i,\mathrm{relax}} \tag{8}$$

The flux are computed at time t_n. The relaxation step is carried out implicitely.

3 Derivation of the relaxed scheme, example of Godunov scheme

Assuming that λ_j is large, the scheme (3)–(6) can be rewritten formally as

$$\frac{\partial W}{\partial t}+\frac{G}{\Delta x}=\frac{R(W)}{\varepsilon_j} \tag{9}$$

where $\varepsilon_j=\frac{1}{\lambda_j}$, G is the vector composed of the flux variations $\mathcal{E}\left(XF\right)_{i+1/2}-\mathcal{E}\left(XF\right)_{i-1/2}$ and of the interaction between fluids at the boundary of the cell $\mathcal{E}\left(F^{lag}\frac{\partial X}{\partial x}\right)_{i,\mathrm{bound}}$, except for the first and fourth component of W. In that case, G is set to the second term of (6).

We are interested in the limit situation when $\lambda_j\to+\infty$. To this end, we look for an expansion of the type

$$W=W^0+\varepsilon_jW^1+o(\varepsilon_j)$$

which is plugged into (9). This leads to

$$R(W^0) = 0 \qquad \frac{\partial W^0}{\partial t} + \frac{G(W^0)}{\Delta x} = R'(W_0)W^1.$$

This shows that the interesting term W^0 lies in the equilibrium set $\mathcal{V} = \{W, R(W) = 0\}$ which depends on the base scheme (Godunov' in the example we develop). We assume (and this is true for practical schemes) that $\mathcal{V}$ is of dimension $n = 6$ and can be parametrized by a mapping $W \mapsto M(W)$. It can be shown that the columns of $\frac{\partial M}{\partial W}$ form a basis of $\mathcal{K}$ the null space of $\frac{\partial R}{\partial W}(M(W))$. We denote by P the projection onto $\mathcal{K}$. When $\lambda_j \to +\infty$, the limit scheme is obtained by projecting (9) onto $\mathcal{K}$, i.e.

$$P\frac{\partial W^0}{\partial t} + P\frac{G(W^0)}{\Delta x} = 0.$$

Let us give some details in the case of the Godunov flux. The variety $\mathcal{V}$ is defined by

$$\mathcal{V} = \{W = (\alpha_1 W_1, \alpha_2 W_2) \text{ such that } F^{lag}(W_1, W_2) = F^{lag}(W_2, W_1)\}$$

and a simple analysis shows that in that case $P^1 = P^2 := P$, $u^1 = u^2 := u$. The set $\mathcal{V}$ is the one for which we have a single pressure and a single velocity.

Using this, rewriting (9) as

$$\frac{\partial \alpha_{k,j}}{\partial t} + VF_{k,j} = 0, \qquad \frac{\partial \alpha_{k,j}\rho_{k,j}}{\partial t} + M_a F_{k,j} = 0$$
$$\frac{\partial \alpha_{k,j}\rho_{k,j}u_{k,j}}{\partial t} + M_o F_{k,j} = 0, \frac{\partial \alpha_{k,j}\rho_{j,k}E_{j,k}}{\partial t} + EF_{k,j} = 0$$

We get, after having transformed the system in primitive variables (α_k,$\alpha_k\rho_k$, $\alpha_k\rho_k u_k$, $\alpha_k\rho_k E_k$), and defining $\beta_k = \left(\frac{\partial \varepsilon_k}{\partial P_k}\right)_{\rho_k}$ and $\kappa_k = \left(\frac{\partial \varepsilon_k}{\partial \rho_k}\right)_{P_k}$, we find the following set of PDEs

$$\begin{aligned}\frac{\partial \alpha_2}{\partial t} &+ VF_2 + \frac{\alpha_1\alpha_2}{\alpha_2\rho_1 a_1^2 + \alpha_1\rho_2 a_2^2}\left(\frac{EF_2}{\alpha_2\rho_2\beta_2} - \frac{u_2 M_o F_2}{\alpha_2\rho_2\beta_2}\right.\\ &+ \frac{(\frac{u_2^2}{2} - \varepsilon_2 - \rho_2\kappa_2)}{\alpha_2\rho_2\beta_2} M_a F_2 + \frac{\rho_2^2\kappa_2 VF_2}{\alpha_2\rho_2\beta_2} - \frac{EF_1}{\alpha_1\rho_1\beta_1} + \frac{u_1 M_o F_1}{\alpha_1\rho_1\beta_1}\\ &\left.- \frac{(\frac{u_1^2}{2} - \varepsilon_1 - \rho_1\kappa_1)}{\alpha_1\rho_1\beta_1} M_a F_1 - \frac{\rho_1^2\kappa_1 VF_1}{\alpha_1\rho_1\beta_1}\right) = 0\end{aligned}$$

$$\frac{\partial(\alpha_k\rho_k)}{\partial t} + M_a F_k = 0, \quad \frac{\partial(\rho u)}{\partial t} = M_o F_1 + M_o F_2, \quad \frac{\partial(\rho E)}{\partial t} = EF_1 + EF_2.$$

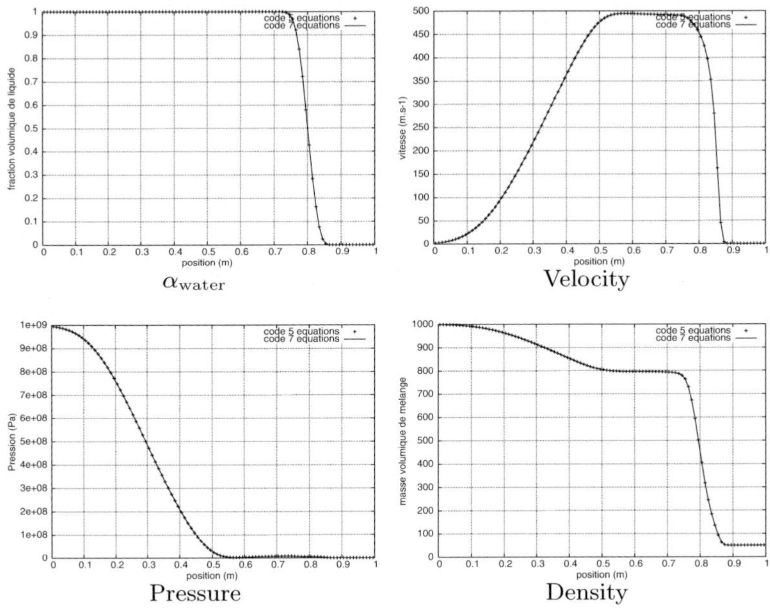

Fig. 1. Liquid–gas shock tube, $t = 229\mu$s

4 Numerical tests

All the tests are done using stiffened equation of state for pure fluids, $\varepsilon(P,\rho) = \frac{P+\gamma P^\infty}{(\gamma-1)\rho}$; We have used the following parameters

	air	water	epoxy	spinel
γ	1.4	4.4	2.94	1.62
P^∞(Pa)	0.	6.10^8	3.2×10^9	$141. \times 10^9$

In some cases, pure fluids are set in one of the chambers. In that case, we impose residual volume fractions $\alpha = 10^{-8}$ for the phase that does not exist.

4.1 Liquid–gas shock tube

At the initial time, we set on the left high pressure water ($P = 10^9$ Pa), and on the right air at atmospheric conditions ($P = 10^5$ Pa). The density of air is 50 Kg/m^3, the density of water is 1000 Kg/m^3. The two fluids are at rest. We compare the original scheme and the present scheme on Figures 1 on a 100 points mesh. The results are undistinguishable. This is a very stiff test case.

4.2 Multiphase problem

The right and left chambers are filled with water and air and $\alpha = 0.5$ in both cases. The pressure is 10^9 Pa on the left and 10^5 Pa on the right. The

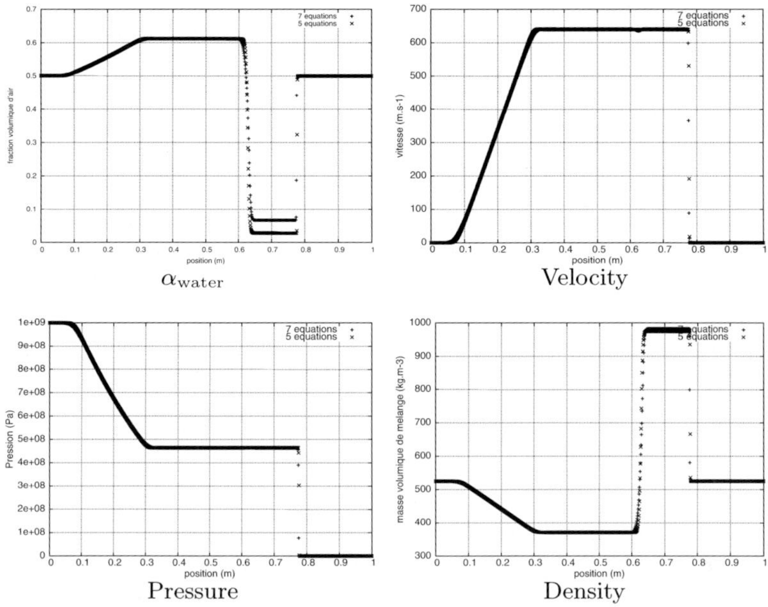

Fig. 2. Multiphase problem, $t = 200\mu$s.

density of water is 1000 Kg/m^3, that of air is 50 Kg/m^3. The test is done with 1000 mesh points, we compare the original splitted (seven equations) and the present scheme (five equations), see Figure 2. The main effect is on the volume fraction and the density.

4.3 Propagation of a shock in a solid media

We make a simulation on an alloy composed of epoxy and spinel for which experimental data exist. The tube is impacted by a piston, so that a shock wave propagates . There exists a linear relation between the shock speed u_c and the impact velocity u_i : $u_c = a_0 + su_i$.

Initially, the two solids are at atmospheric pressure, their densities are $\rho_1 = 1185\,\text{kg.m}^{-3}$, $\rho_2 = 3622\,\text{kg.m}^{-3}$. The volume fractions are $\alpha_1 = 0.595$ and $\alpha_2 = 0,405$. The experimental set–up is such that the shock speed is measured. On Figure 3, we compare our results with the experimental measures and taken from [5], those of [6], those obtained by the original scheme and the present scheme. We see that our results are very close to those of the original scheme, and in good agreement with the experimental results.

5 Conclusion

We have developped a new scheme for multiphase flow problems starting from the scheme decribed in [3]. The advantage of the scheme is that less variables

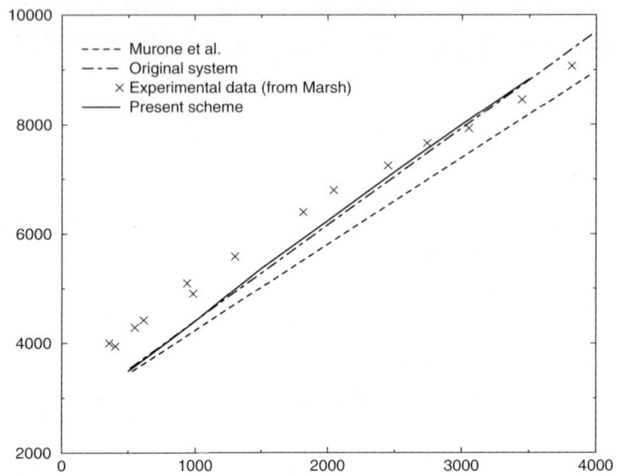

Fig. 3. Shock velocity, comparison between several approaches.

are needed, thus the CPU cost is reduced by a non negligible factor, while the results are very similar to those of the original scheme. More details, in particular about the second order extension of the scheme, can be found in [2].

References

1. R. Abgrall, M. Papin, and L. Hallo. A scheme for compressible two-phases flows and interface problems. In Roland Glowinski Zhangxin (John) Chen and Kaitai Li, editors, *ICM2002-Beijing Satellite Conference on Scientific Computing*, American Mathematical Society in the Contemporary Mathematics (CONM) book series. American Mathematical Society, 2003.
2. R. Abgrall and V. Perrier. Construction of numerical scheme for multiphase problem by asymptotic expansion. *in preparation*, 2004.
3. R. Abgrall and R. Saurel. Discrete equations for physical and numerical compressible multiphase mixtures. *J. Comput. Phys.*, 186(2):361–396, 2003.
4. D. A. Drew and S. L. Passman. *Theory of multicomponent fluids*, volume 135 of *Applied Mathematical Sciences*. Springer-Verlag, New York, 1999.
5. S.P. Marsh. *LASL Shock Hugoniot data.* Los Alamos series on dynamic material properties. University of California Press, 1980.
6. Angelo Murrone and Hervé Guillard. A five equations reduced model for compressible two phase flow problems. *J. Comput. Phys.*, 2004. in revision.
7. R. Saurel and R. Abgrall. Some models and methods for compressible multifluid and multiphase flows. *J. Comp. Phys.*, 150(2):425–467, 1999.

A New Accurate Method for Simulating Polydispersed Two-Phase Flows

G. Dufour[1]

Laboratoire MIP, Université Paul Sabatier Toulouse III, 118 Route de Narbonne, 31062 Toulouse Cedex dufour@mip.ups-tlse.fr

Dispersed two-phase flows are encountered in numerous industrial applications ranging from combustion in aeronautical engines, solid propellant rocket boosters, aerosols in chemical engineering, up to meteorology or planet formation in solar nebulae. Our study is mainly concerned with sprays where the dispersed phase is constituted of droplets of various sizes carried by a gaseous flow, which have their own inertia, can evaporate, break-up and coalesce. Such sprays can be described "at the kinetic level" by an evolution equation of Boltzmann-type in the phase space [4]:

$$\partial_t f + v.\nabla_x f + \nabla_v.(Af) + \partial_s(Rf) = \Gamma(f) + Q(f) \tag{1}$$

Where f is the droplet number density depending on the time t, the position x, the velocity v and the size s. The term A denotes the acceleration of the particles (due to the drag force), R denotes the effect of evaporation and the terms $\Gamma(f)$ and $Q(f)$ correspond respectively to breakup and coalescence.

There are several strategies in order to solve such an equation. A first choice is to approximate f by a sample of discrete numerical parcels of particles of various sizes through a Lagrangian–Monte-Carlo approach. This approach has been shown to be efficient but for unsteady computations may require large memory and CPU cost.

A second one is to use a Eulerian description. However, the full dynamics in the phase space can not be simulated using deterministic classical numerical methods and one has to reduce these dynamics and derive, at least formally, a set of "fluid" conservation equations on some moments of the distribution function f. This approach is numerically very competitive but suffers from severe physical limitations due to the intrinsic poor description of the droplet size distribution. In particular, breakup and coalescence cannot be taken into account in the standard two-fluid approach where only some moments in size of the distribution are resolved. Recently, Beck and Watkins proposed a method based on droplet size moments without using classes yielding the development of spray submodels, but also needing to presuppose a shape for the whole distribution function in size.

Another attempt at deriving a Eulerian model for sprays polydispersed in size was developed by Tambour et al. [8]; considering the dispersed phase as a set of continuous media : "fluids", each "fluid" corresponding to a statistical average between two fixed droplet sizes : the section. More recently, Laurent and Massot provided a rigorous "kinetic" framework for dilute sprays in laminar flames [2], as well as for dense sprays with coalescence in [3]. However, this approach was shown to be only first order accurate in the size phase space discretization [4] thus resulting in strong numerical diffusion when too few sections are to be used.

We present here a new approach combining both ideas of *Multi-Fluid* and *Moment Methods* and also provide numerical simulations using a Eulerian formalism in configurations coupling droplet dynamics and evaporation.

1 Multi-Fluid Approach

1.1 Classic Multi-Fluid model

In all the following, the drag force will be assumed to follow a Stokes law : $A(t,x,s,v) = \frac{U_g(t,x)-v}{\gamma . s}$ where U_g is the gas velocity and γ a constant. Moreover, we take the *surface of the droplets* as the variable of size and we will consider the d^2-law for evaporation, assuming that $R = -K$ is a negative constant. Starting from the kinetic evolution equation we first deduce evolution equations on the zero-th and first moment of f in velocity:

$$n(t,x,s) = \int_v f(t,x,s,v)dv \quad ; \quad \overline{v}(t,x,s) = \frac{1}{n(t,x,s)} \int_v v f(t,x,s,v)dv \quad (2)$$

We then discretize the obtained equations : N being a fixed positive integer, choose a finite increasing serie of size of droplets $(s_i)_{0\leq i\leq N}$ defining N finite intervals of size $I_i = [s_{i-1}, s_i[$ and one infinite interval $I_{N+1} = [s_N, +\infty[$. Following the original notations of Tambour ([7], [8]) we will denote by the term "section" any interval of size. In order to close our model, we assume that all droplets in a same section have the same velocity. This hypothesis is equivalent to suppose that there is no velocity dispersion at a given size. Moreover, we assume that the profile of n in a given section is fixed as a function only of the size of the droplet and does not depend on t and x :

$$\begin{cases} n(t,x,s) = n_i(t,x)\kappa_i(s) \\ \overline{v}(t,x,s) = \overline{v}_i(t,x) \end{cases} \quad \text{for } s_{i-1} \leq s < s_i \quad (3)$$

Such an assumption is quite natural to do since the principle of the sectional approach is to take some moments at a given size of the distribution function with velocity [2]. In the classic Multi-Fluid Method, κ_i is assumed to be a constant in the radius variable. This choice for the size variable and the one for the size discretization are discussed in [2] or [9]. Moreover, the mass is preserved, defining the value of the constant and yielding 2 equations on mass density and momentum in any section. Finally, we obtain a set of $2N+2$ equations on mass density and momentum which is the classic Multi-Fluid Model.

1.2 Improved Multi-Fluid method

The main idea of the improved method is to take into account within any section the information on two moments in size of the number density function f and to approximate the density function using one function with an *a priori* form which preserves these two moments of f. one may consider as long as one is able to find a reliable family of positive functions $\mathcal{F}$ fulfilling the following *realisability condition* : for any acceptable density function f there exists one and only one function $\tilde{f}$ in $\mathcal{F}$ such that f and $\tilde{f}$ have the same moments m_1 and m_2 in any section. In all the following, In this paper, we choose for κ_i an exponential function (in the surface variable) under mass and number conservation constraints, whereas the classical Multi-Fluid Model uses a constant (in the radius variable) function and only preserves the mass. Such a choice is justified by the following results :

Proposition 1. *Let I be any (finite or not) interval of $[0,+\infty[$ and $f \in L^1(I,(1+s^{3/2})ds)$ be a positive function not everywhere equal to zero. Then there exists a unique couple $(a,b) \in \mathbb{R}_+ \times \mathbb{R}$ such that the function defined by $\tilde{f}(s) = a\exp(-bs)$ has the same moments of order 0 and $3/2$ on I than f. Moreover if the interval I is infinite we have $b > 0$.*
We note $\Pi_I(f)$ the corresponding exponential function.

Theorem 1. *Let I be any interval of length h and s_0 be its middle point. Let f not everywhere equal to 0 be a function of class C^{p+3} where p is the smallest integer (eventually 0) such as $f^{(p)}(s_0) \neq 0$. We then have the following estimate :*

$$\|\Pi_I(f) - f\|_{L^\infty(I)} = O(h^2)$$

In all the following, we will note $\Pi_{[s_{i-1},s_i[}(f)(s) = a_i\exp(-b_i s)$. The process is then quite the same as for the classic Multi-Fluid Method, the main difference being the number of preserved quantities since we obtain a set of $3N+3$ evolution equations on the number density n, the mass density m and the momentum $m\overline{v}$. On any section i we have :

$$\begin{cases} \partial_t n_i + \partial_x(n_i\overline{v}_i) - K.(F_{i+1} - F_i) = 0 \\ \partial_t m_i + \partial_x(m_i\overline{v}_i) + (E_i^{(1)} + E_i^{(2)})m_i - E_{i+1}^{(1)}m_{i+1} = 0 \\ \partial_t(m_i\overline{v}_i) + \partial_x(m_i\overline{v}_i \otimes \overline{v}_i) + (E_i^{(1)} + E_i^{(2)})m_i\overline{v}_i - E_{i+1}^{(1)}m_{i+1}\overline{v}_{i+1} = m_i\overline{A_i} \end{cases}$$

$$E_i^{(1)} = \frac{\rho_l K}{3\sqrt{4\pi}} s_{i-1}^{3/2} a_i \exp(-b_i s_{i-1}) \; ; \; E_i^{(2)} = \int_{s_{i-1}}^{s_i} \frac{\rho_l K}{3\sqrt{4\pi}} s^{1/2} a_i \exp(-b_i s) ds$$

$$\overline{A_i} = \int_{s_{i-1}}^{s_i} \frac{\rho_l s^{3/2}}{3\sqrt{4\pi}} A(t,x,s,\overline{v}_i) a_i \exp(-b_i s) ds \; ; \; F_i = a_i(t,x)\exp(-b_i(t,x)s_{i-1})$$

2 Numerical schemes

Since the numerical methods used for the transport part (Finite Volume method) and for the evaporation part (Improved Sectional method) are completely different, we will compute these effects separetely and then use a fractional step in time method to solve the global system though the Strang Splitting method [5].

2.1 Transport in space scheme

Concerning the transport in space scheme, relation between m and n has to be preserved . Indeed, in any section i, since n_i and m_i denote respectively the number and mass density in the section, their ratio m_i/n_i has therefore to remain in the interval $]\frac{\rho_L}{3\sqrt{4\pi}}s_{i-1}^{3/2}, \frac{\rho_L}{3\sqrt{4\pi}}s_i^{3/2}[$. Thus the transport scheme has to preserve this property. The section been fixed, we will omit to note the indices related to the size and note $\mu = m/n$ so that we are solving :

$$\begin{cases} \partial_t n + \nabla_x.(nv) = 0 \\ \partial_t(n\mu) + \nabla_x.(n\mu v) = 0 \\ \partial_t(n\mu v) + \nabla_x.(n\mu v \otimes v) = 0 \end{cases} \tag{4}$$

Performing a classic finite volume approach with affine reconstruction on n, μ and v and slope limitations in order to have a second-order accurate scheme, the theory ensures us that we can find limiters which preserve the positivity of n and therefore a maximum principle on μ. Then we have also the positivity of $n\mu$ and obviously a maximum principle on v. Second-order in time is then achieved thanks to a Runge-Kutta method. [6].

2.2 Evaporation scheme

Choosing the d^2-law for evaporation, the kinetic equation on evaporation written in the surface variable s reads $\partial_t f - K\partial_s f = 0$ where K is assumed to be a constant. Given an initial condition f_0, the exact solution to this equation for any time $t \geq 0$ and surface $s \geq 0$ then reads $f(t,s) = f_0(s+Kt)$.

Let $(s_i)_{0 \leq i \leq N}$ be a discretization of the size space in the surface variable. A time step in our evaporation scheme may be decomposed into three elementary steps. The first one consists in rebuilding the exponential approximate number density function under the constraint of preserving the number and the mass density in every section. Then the rebuild density is exactly transported during a time step Δt and finally we compute the new number and mass density in every section. If p is related to the time discretization and i is related to the size discretization we have, denoting $U = (n, m, m\overline{v})^T$:

$$\frac{U_i^{p+1} - U_i^p}{\Delta t} - K\frac{F_{i+1/2}^{p+1/2} - F_{i-1/2}^{p+1/2}}{\Delta s_i} = S_i \tag{5}$$

$$F_{i+1/2}^{p+1/2} = \frac{a_{i+1}}{\Delta t} \int_0^{\Delta t} \begin{pmatrix} 1 \\ (s_{i+1}+Kt)^{3/2} \\ (s_{i+1}+Kt)^{3/2}\overline{v}_{i+1}^{p} \end{pmatrix} e^{-b_{i+1}(s_{i+1}+Kt)} dt$$

$$S_i = \frac{1}{\Delta s_i \Delta t} \int_{s_{i-1}}^{s_i} \begin{pmatrix} 0 \\ 1 \\ \overline{v}(u) \end{pmatrix} \left(u^{3/2} - (u+K\Delta t)^{3/2}\right) a(u) e^{-b(u)(K\Delta t+u)} du$$

Theorem 2. *Under the CFL condition $K\Delta t \leq \min_i \Delta s_i$, the scheme (5) preserves the positivity of n and m and fulfills a maximum principle on the velocity. Moreover, this scheme can be proved to be consistent of order 2 with the evaporation equation if the size step is assumed to be constant.*

3 Numerical results

We aim here at comparing our exponential method with the classic Multi-Fluid Method and with a classic MUSCL method performed on the number density. We first present computations on evaporation only with a regular bimodal initial condition yielding to a non-trivial evolution of the Sauter Mean radius (Fig. 1). Figure 2 presents an estimation based on both errors on remaining mass and Sauter mean radius. The MUSCL method is very accurate for a fine discretization (100 sections) but does not have better results than the first order multi-fluid method for a rough one whereas the improved multi-fluid method remains accurate due to its property of mass preserving.

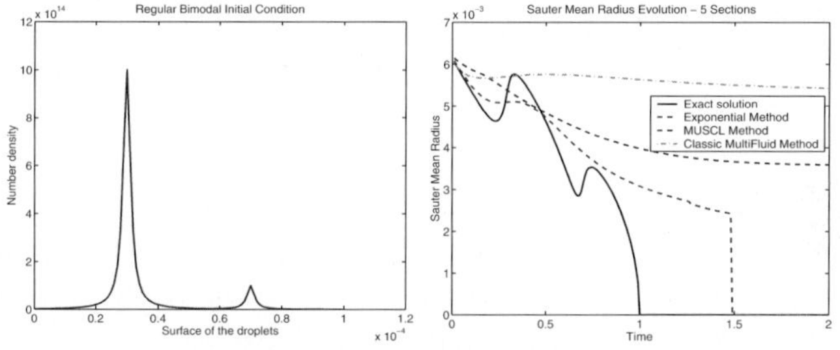

Fig. 1. Initial Condition and Evolution of the Sauter mean Radius - 5 Sections

Figures 3 and 4 are related to 1-Dimensional computations on the stationary solution of the kinetic equation (1). Both Improved multi-fluid methods and MUSCL Methods are compared to the exact solution computed assuming that all droplets have the same velocity at the point $x = 0$. This shows the importance of having information on the mean radius, *within every section*.

4 Conclusion

We have developed an hybrid model using both ideas of Multi-Fluid and Moment Methods which is able to simulate quite accurately the polydisperse

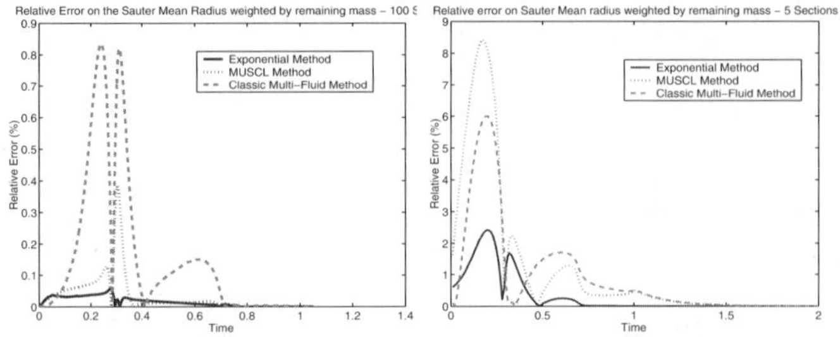

Fig. 2. Reduced error on the Sauter Mean Radius - 100 and 5 Sections

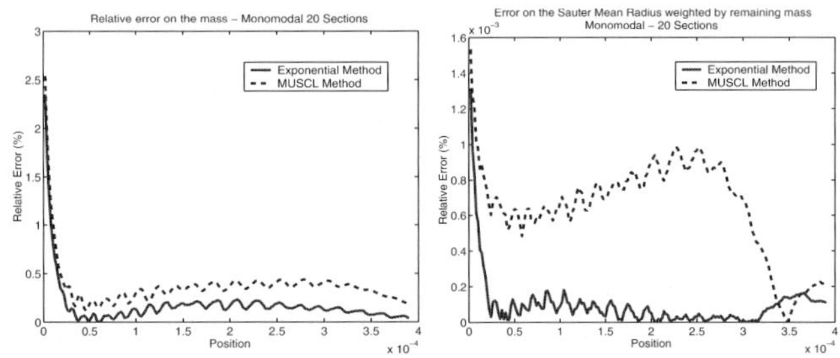

Fig. 3. Error on mass and Sauter Mean Radius - 20 Sections

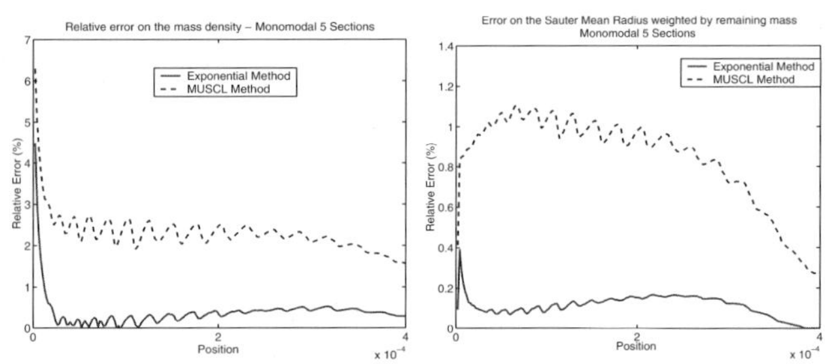

Fig. 4. Error on mass and Sauter Mean Radius - 5 Sections

aspect of the spray even with a few number of sections. This is an important feature since the coupling of size-dependant phenomena (e.g : drag force, break-up, coalescence) requires a good estimate of the mean size and using a great number of sections would not be competitive.

References

1. J.C. Beck, A.P. Watkins : J. Comp. Physics, **182**, 586-621 (2002).
2. F. Laurent, M. Massot: Comb. Theory and Model., **Vol. 5**, 4, 537-572 (2001)
3. F. Laurent M. Massot, P. Villedieu : J. Comp. Physics, **194**, 505-543 (2004)
4. F. Laurent : C.R. Acad. Sci. Paris, Ser. I **334**, 417-422 (2002).
5. A. V. Bobylev, T. Ohwada : Applied Mathematics Letters, **14**, 45-48 (2001).
6. G. Dufour, P. Villedieu : (In preparation)
7. Y. Tambour : Comb. and Flame, **60**, 15-28 (1985).
8. J.B Greenberg, I. Silverman, Y. Tambour : Comb. and Flame, **93**, 90-96 (1993).
9. K. Domelevo : Atomization and Sprays, **11**, 291-303 (2001).

The Characteristics-Based Matching Method (CBM) for High-Speed Fluid-Fluid Flows

Nourgaliev, R.R.[1], Dinh, T.N.[2], Liou, M.-S.[3], and Theofanous T.G.[4]

[1] Center for Risk Studies and Safety, UCSB, robert@engr.ucsb.edu
[2] Center for Risk Studies and Safety, UCSB, nam@engr.ucsb.edu
[3] NASA Glenn Research Center, Cleveland, Meng-Sing.Liou-1@nasa.gov
[4] Center for Risk Studies and Safety, UCSB, theo@engr.ucsb.edu

1 Introduction

This paper is concerned with direct numerical simulations of multimaterial free interfaces in high-speed flows involving propagating shock waves. Of particular interest are high acoustic impedance mismatch interfaces, such as those between a gas and a liquid. Therein lie the intense multiphase interactions, including interfacial breakup, involved in the injection of liquids in supersonic gas flows (atmospheric dissemination of liquid agents) or in explosion waves supported by, or propagating through multiphase gas-liquid media (inertia confinement fusion first wall, pulse detonation engines, etc.). We expect that such simulations will aid understanding and development of models needed for treatment at the effective field level, which more often than not becomes necessary due to the scale/complexity in engineering applications.

The purpose of this paper is to introduce [8] and demonstrate a method that contributes in the above context The basic guidance in this development is to respect information transport in inherently transient and complex flow and pressure fields across multi-material domains. We achieve this objective through a consistent use of numerical schemes based on Riemann problem solutions in both single-fluid interior domains and two-fluid interfacial regions. We bring to bear best features of front-tracking ([5] and [3]) and front-capturing ([1], [4] and [7]) algorithms, namely the accuracy of the interface boundary condition treatment in the former and the algorithmic simplicity for the interface geometry treatment in the later. The resulting method, named the Characteristics-Based Matching (CBM) [8], is Eulerian-grid-based, utilizes the high-order-accurate Godunov schemes for capturing of discontinuities in compressible flows, and uses a level-set technique to capture the multi-dimensional complex topology of evolving interfaces, including breakup and coalescence. The CBM method is shown to produce accurate and robust solutions for a number of challenging compressible multiphase flow problems, including one-dimensional stiff shock tubes with gas/liquid interfaces, interaction of shock waves with two-dimensional gas bubble, and collapse of a three-dimensional air bubble under strong shock wave in water pool. The method is favorably compared to the Ghost Fluid Method (GFM)

([4] and [7]) for multi-gas flows, and it is found to perform more accurately and robustly under extremely strong shock waves at gas-liquid interfaces.

2 Characteristics-Based Matching

We employ a level set method [10]. The level set is a signed distance function which separates computational domain on "positive" and "negative" fluids. A zero-level of the level set function represents an interface. Compressible fluid dynamics inside bulk of each fluid is governed by Euler equations.

Basic numerical scheme. Computational domain is discretized using a uniform Cartesian grid. To advance solution on each side of the interface, we apply the high-order-accurate Godunov-based conservative finite-difference approach. Specifically, we employ the third-order-accurate Runge-Kutta Total Variation Diminishing (RK-TVD) scheme for time advancement; the third-order-accurate limited MUSCL scheme for polynomial reconstruction; and a variety of flux treatment methods (Godunov-based) to construct numerical fluxes at the edges of computation nodes.

Interface boundary treatment [8]. Near the interface (within a stencil range of the numerical scheme for space polynomial reconstruction), the numerical fluxes are modified in order to effectively couple (match) numerical solutions on each side of the interface, respecting a proper propagation of information flow (transmission and reflection of shock and rarefaction waves). In difference to the closely-related Ghost Fluid Method [4], we use a *characteristics-based method* for interface boundary treatment, which is borrowed from front-tracking methods ([5] and [3]).

More specifically, at the begining of each time step, based on the currently-available level set function, we identify subcell position of the interface as all intersections (denoted as *markers*) of the zero level with grid lines. During current time step, these markers are allowed to "slide" along grid lines (*"discrete front tracking"*), based on the grid line normal velocity available from the pseudo-multidimensional two-fluid exact Riemann solution, obtained at the begining of each level of the RK time discretization procedure. In addition to grid line normal velocities, the exact Riemann solution at marker positions provides the wave structure at the interface, which is used for correction of numerical fluxes at the near-interface cells within a stencil range of the MUSCL scheme. Also, the interface normal velocity available from the exact Riemann solution at markers is utilized to construct the speed function, needed to advance the level set equation.

3 Numerical Examples

In the present section, we present three examples, demonstrating performance of the above-described Characteristics-Based-Matching in one, two and three dimensions, comparing the results with analytical solution, performance of the closely-related Ghost Fluid Method, and experimental visualization data.

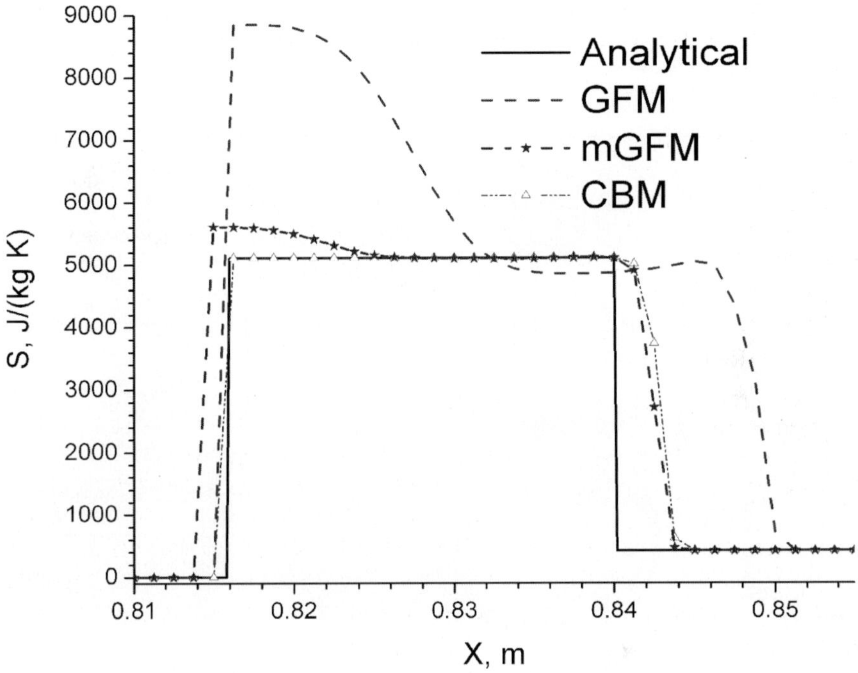

Fig. 1. Comparison of GFM, mGFM and CBM for water-air shock-tube problem.

3.1 Strong Shock for Gas-Liquid Interface in 1D

As our first numerical example we consider a gas-liquid (stiff) shock-tube test introduced by Saurel and Abgrall in [11]. We use a generalized γ-gas equation of state for for both gas and liquid. The conditions of this test involve an interface located at $x = 0.7$ m, and initial Left and Right states specified as

$$\begin{aligned}(\rho, u, p, \gamma, \Pi)_{\mathrm{L}} &= \left(1000,\ 0,\ 10^9,\ 4.4,\ 6 \cdot 10^8\right)\\ (\rho, u, p, \gamma, \Pi)_{\mathrm{R}} &= \left(50,\ 0,\ 10^5,\ 1.4,\ 0\right)\end{aligned} \tag{1}$$

The analytical and computed profiles of entropy for $t = 240$ μsec are plotted in Fig.1. The original GFM coupling procedure [4] performs poorly for this numerical test. Even though there are no pressure oscillations near the interface, the numerical solutions for density and entropy are unsatisfactory due to well-known overheating problems, Fig.1. Apparently, the constant-entropy isobaric fix employed here is not sufficient at stiff conditions of this problem. Moreover, there is an error in the prediction of the transmitted M_{sh}=2.35 shock wave position. As one can see from Fig.1, the modified GFM approach [7] is a significant improvement in terms of overheating and position of the transmitted shock. At the same time, the CBM approach is found to produce very accurate results, having no problems with both overheating and transmitted shock.

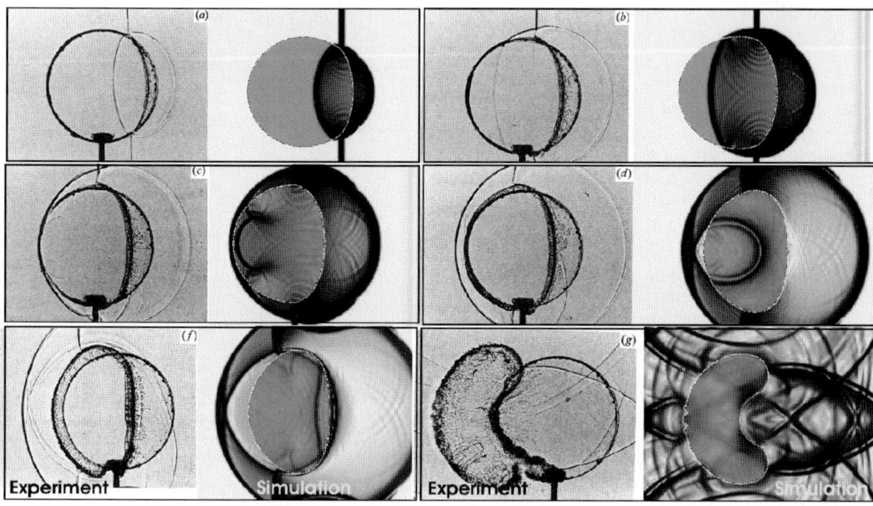

Fig. 2. Interaction of weak air shock wave with Helium cylinder: Experimental shadow-photographs by Haas and Sturtevant (1987) [6] vs. idealized numerical Schlieren from present study.

3.2 Interaction of Shock Wave with Multimaterial Gaseous Media

As our next numerical test, we consider a weak shock in air interacting with a cylindrical "bubble" of Helium. This problem was studied experimentally by Haas and Sturtevant in [6]. The results of the simulation (Fig.2) are in excellent agreement with experimental data for general shape and position of the Helium bubble. Even though our uniform grid resolution is way below the one used in the previous AMR simulation by Quirk and Karni [9], we are able to predict correctly all major flow structures.

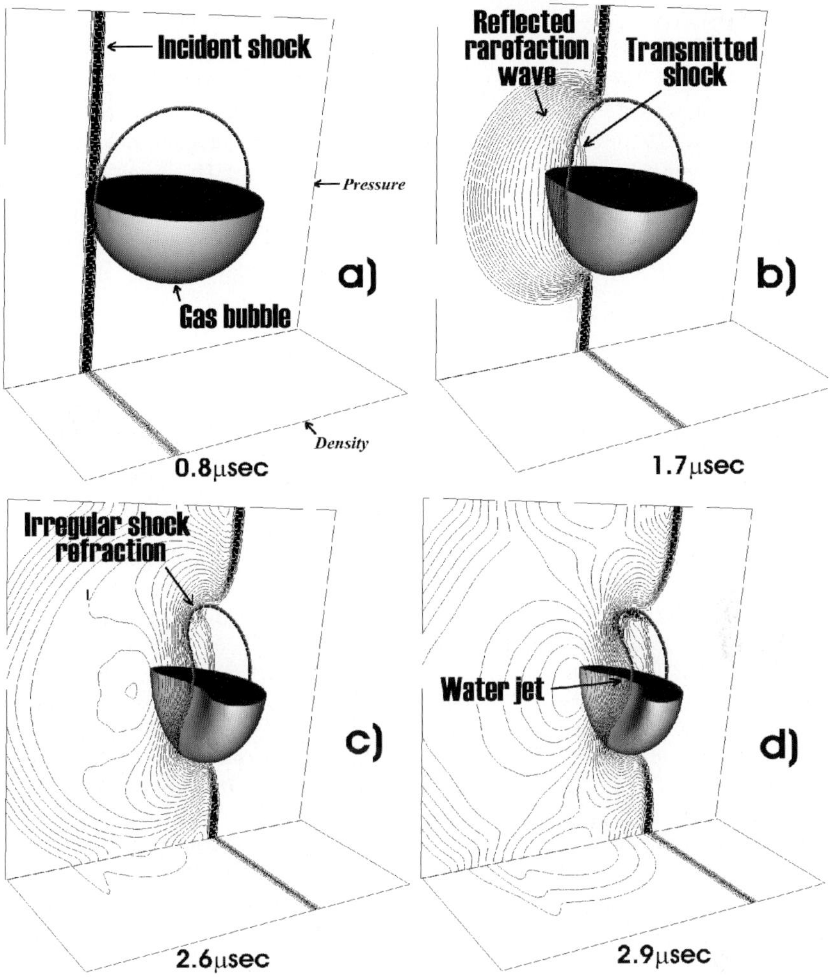

Fig. 3. Dynamics of pressure and density fields and shape of bubble for shock/cavity interaction, under incident M=1.7 shock wave in water.

3.3 Collapse of an Air Bubble in Water

As our final numerical example we consider collapse of an air bubble in water. Similar problem was studied previously in [3], [11] and [2] in two dimensions. Here, we will consider the three-dimensional case, the first of such simulations to our knowledge. (This is for a variety of reasons in each case, for example in front tracking a 3D treatment becomes extremely complicated). While the main features (indicated in Fig.3) are similar to the 2D case, the quantitative aspects are clearly different.

4 Conclusion

As a conclusion, we would like to emphasize the importance of the Characteristics-(Riemann solver)-based treatment of interface boundary conditions in compressible multimaterial fluid flows (as in the present paper), which provides accuracy and, more importantly, robustness of the overall numerical procedure. These are especially crucial for direct numerical simulation of high-acoustic impedance mismatch (gas-liquid and gas-solid) interfaces.

Acknowledgment:

This work is sponsored by the Lawrence Livermore National Laboratory ("ALPHA" and "MIX" projects). Support of Drs. Frank Handler, Glen Nakafuji and Dan Klem are gratefully acknowledged.

References

1. R. Abgrall, and S. Karni, Computations of Compressible Multifluids, *Journal of Computational Physics*, **169**, 594-623 (2001).
2. G.J. Ball, B.P. Howell, T.G. Leighton, and M.J. Schofield, Shock-Induced Collapse of a Cylindrical Air Cavity in Water: a Free-Lagrange Simulation, *Shock Waves*, **10**, 265-276 (2000).
3. J.-P. Cocchi, and R. Saurel, A Riemann Problem Based Method for the Resolution of Compressible Multimaterial Flows, *Journal of Computational Physics*, **137**, 265-298 (1997).
4. R.P. Fedkiw, T. Aslam, B. Merriman, and S. Osher, A Non-oscillatory Eulerian Approach to Interfaces in Multimaterial Flows (the Ghost Fluid Method), *Journal of Computational Physics*, **152**, 457-492 (1999).
5. I.L. Chern, J. Glimm, O. McBryan, B. Plohr, and S. Yaniv, Front Tracking for Gas Dynamics, *Journal of Computational Physics*, **62**, 83-110 (1985).
6. J.-F. Haas, and B. Sturtevant, Interaction of Weak Shock Waves with Cylindrical and Spherical Gas Inhomogeneties, *Journal of Fluid Mechanics*, **181**, 41-76 (1987).
7. T.G. Liu, B.C. Khoo, and K.S. Yeo, Ghost Fluid Method for Strong Shock Impacting on Material Interface, *Journal of Computational Physics*, **190**, 651-681 (2003).
8. R.R. Nourgaliev, T.N. Dinh, and T.G. Theofanous, On Treatment of High Acoustic Impedance Mismatch Interfaces in Compressible Multimaterial Flows, *Intern. Journal of Multiphase Flow*, submitted, (2004).
9. J.J. Quirk, and S. Karni, On the Dynamics of a Shock-Bubble Interaction, *Journal of Fluid Mechanics*, **318**, 129-163 (1996).
10. S. Osher, and J.A. Sethian, Fronts Propagating with Curvature-Dependent Speed: Algorithms Based on Hamilton-Jacobi Formulations, *Journal of Computational Physics*, **79**, 12-49 (1988).
11. R. Saurel, and R. Abgrall, A Simple Method for Compressible Multifluid Flows, *SIAM Journal on Scientific Computing*, **21**(3), 1115-1145 (1999).

Simulation of Multifluid Multiphase Flows with AUSM$^+$-up Scheme

Chih-Hao Chang[1] and Meng-Sing Liou[2]

[1] NASA Glenn Research Center at Lewis Field, Cleveland, OH 44135, USA, chih_hao_chang@hotmail.com
[2] NASA Glenn Research Center at Lewis Field, Cleveland, OH 44135, USA, MENG-SING.LIOU-1@nasa.gov

Summary. An accurate and robust method is proposed for the simulation of two dimensional compressible multifluid flows. Our method incorporates the AUSM$^+$-up scheme and the stratified flow model to discretize compressible single-pressure multifluid equations. This method satisfies the pressure-undisturbed condition and can capture the moving discontinuities accurately. It's also able to compute a very strong shock wave propagating across the gas-liquid interface. Two test problems, including the shock-bubble interaction problem and the shock-water column interaction problem, will be demonstrated in this paper.

1 Introduction

Assuming the fluids to be compressible, immiscible and inter-penetrating to each other, the multifluid system can be formulated into a compressible single-pressure multifluid model, which includes two sets of the Navier-Stokes equations for each phase. The multifluid model introduced by Ishii [1] has been employed in many applications, for example hydrodynamics of the cooling system within the nuclear reactor, the formation and collapse of the cavities in the underwater propellor, the dynamics of the fuel injector, etc. However, the multifluid model also has presented challenges to the numerical simulation for many years.

Unlike the conventional conservation laws, the multifluid model is not in conservation form and is not necessarily of hyperbolic type. This results in an ill-posed problem, and the solution may not depend continuously on the initial data. Since there is no weak solution for the multifluid model, it is difficult to capture the shock wave or the contact discontinuity well. Several amendments to the model, including the interfacial pressure correction term [2], the two-pressure model [3] or the virtual mass term [4], were proposed to make it hyperbolic or more hyperbolic. Even so, an implicit operator or additional numerical dissipation was still necessary to make the calculation stable. As a result, excessive smearing was usually found in the solution.

We found that part of the instability of the multifluid model came from the non-linear terms which represent the interaction between different phases. As pointed out by Saurel and Abgrall [5], to exactly capture a contact discontinuity, the pressure non-disturbed condition must be satisfied, which leads to the following condition in our formulation:

$$\nabla(\alpha_i p_i) = p_i \nabla(\alpha_i) \tag{1}$$

where the subscript "i" can be either "g" or "l" representing for the gas or liquid phase respectively. As a result, the discretization of the term $\nabla(\alpha_i)$ strictly depends on the scheme used for computing the numerical flux. Hence the discretization may not be general and the numerical result will be affected by how the term $\nabla(\alpha_i)$ is handled. To completely avoid this requirement, we propose another viewpoint to describe pressure interactions between like and different phases.

Considering the physics behind the inter-phasic interaction term, we found that all of the previous articles on the multifluid flows dealing with the non-conservative terms assumed that the force and the energy only took place within the same phase. It means the non-conservative terms associated with the gas and liquid phases in the same cell will cancel with each other. That is, only the inter-phasic terms marked by **A** in Figure (1) are considered. This is correct when the void fraction function is continuous. However, when a contact discontinuity exists in the flow field, we find that the inter-phasic term between neighboring cell (marked by **B** in Figure (1)) must be taken into consideration in order to correctly represent the physics.

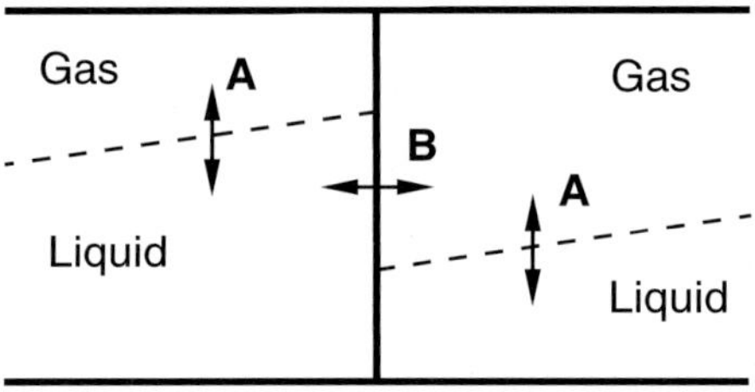

Fig. 1. Illustration of the inter-phasic terms between different phases.

To do so, we introduced the stratified flow model [6] to describe each discretized cell in which each fluid is confined within a separated control volume. The interfaces between the same or different phases are defined on the control surface. Consequently, the numerical flux on the interface between different phases, either within the same cell or between neighboring cells can be taken into account.

The stratified flow model has been applied on several one dimensional cases in our early work [6], including the air-water shock tube problem, Ransom's faucet problem and the phase separation problem, etc. In this paper, we will extend the stratified flow model to solve two dimensional flows.

2 Numerical Method

In this paper, we can only give a brief description of how we extend the stratified flow model to higher space dimension. A more detailed discussion of the 1D stratified flow model and the AUSM$^+$ scheme can be found in our previous work [6]. Based on the stratified flow model, the governing equations for each fluid within the control volume can be written as:

$$\begin{aligned}
&\frac{\partial}{\partial t}\int_{V_i} \rho_i \,\mathrm{d}V_i + \oint_{S_i} (\rho_i \mathbf{v}_i)\cdot \boldsymbol{n}\,\mathrm{d}S_i = 0\\
&\frac{\partial}{\partial t}\int_{V_i} \rho_i \mathbf{v}_i \,\mathrm{d}V_i + \oint_{S_i} (\rho_i \mathbf{v}_i \mathbf{v}_i)\cdot \boldsymbol{n}\,\mathrm{d}S_i + \oint_{S_i} p\,\boldsymbol{n}\,\mathrm{d}S_i = 0\\
&\frac{\partial}{\partial t}\int_{V_i} \rho_i E_i \,\mathrm{d}V_i + p\frac{\partial}{\partial t}\int_{V_i} \mathrm{d}V_i + \oint_{S_i} (\rho_i H_i \mathbf{v}_i)\cdot \boldsymbol{n}\,\mathrm{d}S_i = 0
\end{aligned} \tag{2}$$

where $V_i = \alpha_i V$ and S_i are the control volume and control surface of fluids. The above equations are equivalent to the original multifluid model when the void fraction function is continuous. Then Equation (2) can be discretized and written as (see details in [6]):

$$\frac{V\,\delta_t}{\Delta t}\begin{pmatrix}\alpha_i\rho_i\\ \alpha_i\rho_i u_i\\ \alpha_i\rho_i v_i\\ \alpha_i E_i\end{pmatrix} + p\frac{V\,\delta_t}{\Delta t}\begin{pmatrix}0\\0\\0\\ \alpha_i\end{pmatrix} + \sum_k (\mathbf{f}_{i-i} + \mathbf{f}_{i-i'})_k\, S_k + p\sum_k \begin{pmatrix}0\\ (\alpha_i)_L\, n_x\\ (\alpha_i)_L\, n_y\\ 0\end{pmatrix}_k S_k = 0 \tag{3}$$

where δ_t is the time difference operator. The subscript k is for every cell boundary around the cell, and $(\alpha_i)_L$ is the reconstructed void fraction function on the cell boundary. The pressure work and force exchanged between different phases within the same cell is represented by the 2nd and 4th terms in the above equation. The usual inviscid fluxes have two part, $\mathbf{f}_{i-i}$ and $\mathbf{f}_{i-i'}$, respectively representing contributions on the cell boundaries between the same and different phases. The introduction of $\mathbf{f}_{i-i'}$ is vital for making Equation (3) satisfy the pressure-undisturbed condition (although not explicitly needed in our discretization) and capture the contact discontinuity exactly.

In this paper, the stiffened gas model is used as the equation of state for the water. To overcome the stiffness accompanied with the stiffened gas model, dissipation terms based on the pressure and velocity field are added to the AUSM$^+$ scheme [6]. The new AUSM$^+$-up scheme [6, 7] is used to compute the numerical flux $\mathbf{f}_{i-i}$. For the numerical flux $\mathbf{f}_{i-i'}$, the exact Riemann solver, under the condition that different fluids are separated by a contact discontinuity, is used here because to our knowledge there is no approximate Riemann solver designed for such situation.

3 Simulation Results

3.1 Shock-bubble interaction problem

The interaction of a moving underwater shock with an air bubble is studied. The initial condition is basically same as the case used by Hankin [8]. An air bubble (diameter 6.0 mm) is immersed in the water with its center at the origin. The incoming shock is initially located at $x = -4.0$mm. The fluid states before the shock are $p = 1.013250 \times 10^5$Pa, $u = 0.0$m/s and $T = 292.98$K, and the fluid states behind the shock are $p = 1.6 \times 10^9$Pa, $u = 661.81$m/s, and $T = 595.14$K. The Mach number of the shock wave is $\mathrm{M} = 1.509$. The simulation is computed on a mesh of about 149,000 cells.

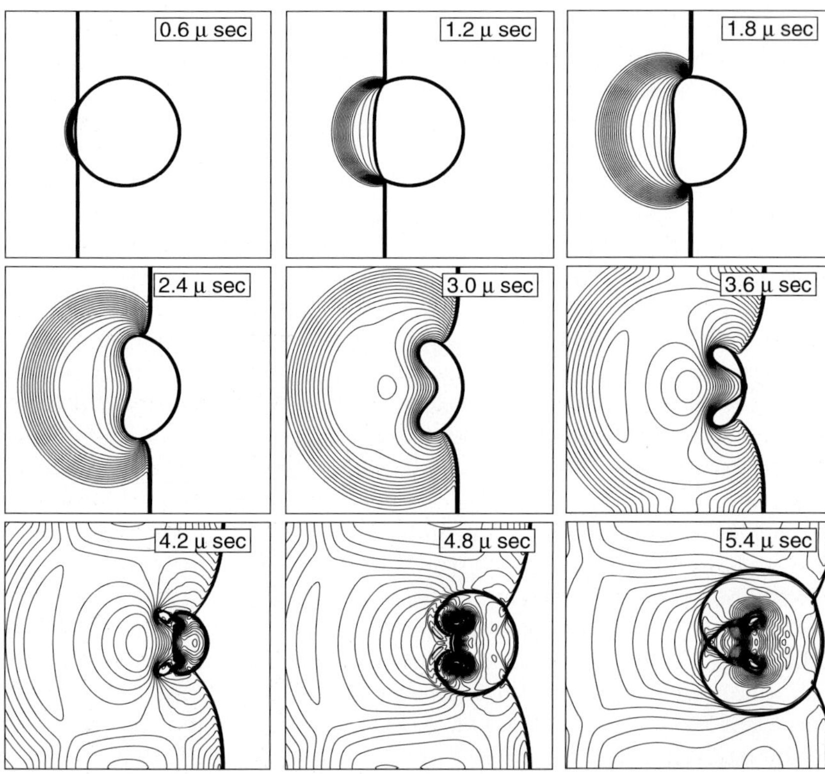

Fig. 2. Pressure and void fraction contours for the shock-bubble interaction problem.

The time evolution of the simulation result is presented in Figure (2). We observe that, after the water shock wave hits the bubble, a strong reflection rarefaction wave is developed and a relatively weak shock is transmitted into the air bubble ($t = 0.6 \sim 3.0\mu$ sec). The strength of the shock in the air bubble is relatively weak (no more than 0.1% the strength of the incoming shock), making it difficult to identify the pressure contours within the bubble. However, the shock wave can be clearly seen in the Mach contours which is not shown here due to the space limit. A water jet generated by the rarefaction wave continuously pushes the bubble into a crescent shape. The bubble finally breakups and the water jet collides with the still water ahead of the bubble ($t = 3.6\mu$ sec). The collision generates several shock waves propagating in all directions radially. At the same time, the separated air bubbles are compressed into a very small volume due to the extremely high pressure imposed on it ($t = 5.4\mu$ sec). This case clear demonstrates the capability of our method in capturing a complex interface and shock wave system, without serious smearing found in the result.

3.2 Shock-water column interaction problem

In this case, we have a water column (diameter 6.4 mm) at the origin and an incoming air shock wave at the position $x = -4.0$mm initially. The initial condition is based on the case in the paper by R. Nourgaliev, et al. [9]. The fluid states before the shock are $p = 1.0 \times 10^5$Pa, $u = 0.0$m/sec, $T = 347.0$K, and after the shock are $p = 2.3544 \times 10^5$Pa, $u = 246.24$m/sec, $T = 451.2$K. A mesh of about 186,000 cells is used in the simulation.

Figure (3) shows the time evolution of the simulation result. We find that when the incident shock hits the water column, part of the shock wave is transmitted into the water and part of it is reflected. Since the incompressibility of the water is much larger than the air, the shock wave within the water column moves much faster than the shock wave in air ($t = 4.0\mu$sec). Also, the pressure wave in the air can be transmitted into the water easily, but it's very difficult for the pressure wave to be transmitted from water to air, as in the previous case. As a result, the pressure wave transmitted into the water column is basically confined within the water. We find the waves bounce back and forth within the water quickly, and the pressure field of air is essentially not influenced before the two branched incoming shock waves merge again ($t \geq 18.5\mu$sec).

4 Conclusion

In this paper, we extend the stratified flow model and the AUSM$^+$-up scheme to solve the two dimensional compressible single-pressure multifluid model. Two critical problems, namely the water shock-bubble interaction problem and the shock-water column interaction problems, are simulated. We show that our method can capture the deformation and breakup of the fluid interface automatically and sharply, and allow the shock wave to pass across the

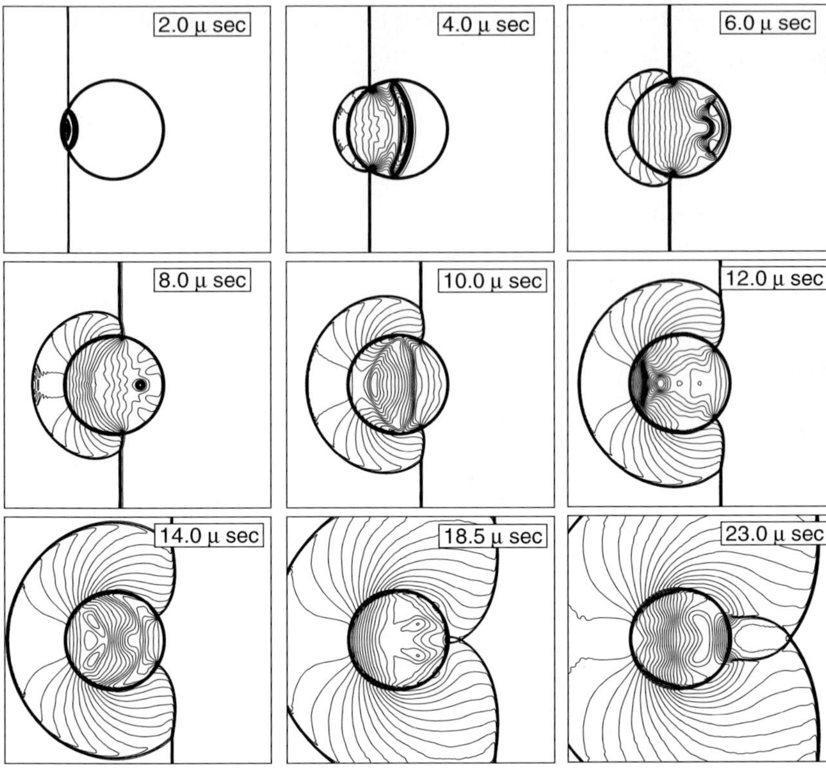

Fig. 3. The pressure and void fraction contours for the shock-water column interaction problem.

interface smoothly, demonstrating that our method is capable of providing accurate and robust solution for the multifluid flow simulation.

References

1. M. Ishii: *Thermo-fluid dynamic theory of two-phase flow*, (Eyrolles, Paris, 1975)
2. D. Bestion: Nuclear Engineering and Design **124**, 229 (1990)
3. V.H. Ranson and D.L. Hicks: Journal of Computational Physics **75**, 498 (1988)
4. H. Städtke and G. Franchello and B. Worth: Nuclear Engineering and Design **17**, 199 (1997)
5. R. Saurel and R. Abgrall: Journal of Computational Physics **150**, 425 (1999)
6. C.-H. Chang and M.-S. Liou: AIAA paper 03–4107, Orlando, USA (2003)
7. M.-S. Liou: AIAA paper 03–4116, Orlando, USA (2003)
8. R. K. S. Hankin: Journal of Computational Physics **172**, 808 (1999)
9. R. Nourgaliev, N. Dinh and T. Theofanous: International Conference on Multiphase Flow, Yokohama, Japan (2004)

Computational Framework for Complex Fluid Physics Applications

Ding Li, Guoping Xia, Venkateswaran Sankaran, and Charles L. Merkle

School of Mechanical Engineering
Chaffee Hall, 500 Allison Road
Purdue University, West Lafayette, IN 47907
dli@purdue.edu

1 Introduction

Computer capabilities have progressed to a level that problems involving simultaneous solution of multiple physical phenomena are currently within reach. High fidelity simulations of complex engineering problems, however, require an algorithm that combines the coupling between the various physical phenomena in an efficient manner and a computational implementation of this algorithm that will function efficiently on current computer hardware. Our multi-physics interests lie in problems that center on fluid mechanics but involve additional physics such as combustion, thermal and structural effects in adjacent solids, and MHD/electromagnetic phenomena in the fluid, solid and surrounding media. In the present paper, we describe a method for dealing with such multi-physics problems with emphasis on the computational implementation. Because fluid mechanics is central in most of our applications, the formulation is focused on a generalized implementation of the fluids equations that extends to additional complex physics as well.

The equations of motion are written in a generalized form that allows solids, liquids, gases, supercritical fluids and multi-phase or multi-component mixtures to be treated in a common manner. The partial differential equations are complemented by arbitrary thermodynamic and caloric equations of state and constitutive equations for the solid. Their solution is obtained by adding a pseudo-time derivative to the space-time conservation form of the equations to enable standard finite volume procedures. The pseudo-time allows a numerical flux and a convergence algorithm to be defined, and in addition introduces an artificial property procedure to ensure proper conditioning of the eigenvalues [1]. The computational implementation of the general solver for the system uses an arbitrary, structured/unstructured grid capability to enable applications to complex geometry while the general conservation formulation allows simulations of arbitrary material. Data and code structures are implemented in a fashion that mimics conventional mathematical notation and operation for tensors, vectors and partial differential equations. To allow for multiple species, phases and number of dimensions, the number of equations is chosen at input. These general equations, arbitrary mesh and multiple physics are implemented in our in-house General Equation and Mesh Solver (GEMS) code [2].

In the following sections we first present the model formulation and outline the numerical solution procedure. Following this we describe the multi-physics zone method and our parallel computing approach. Representative results are presented for a variety of applications including a trapped vortex combustor, an Argon condensation flow in a hypersonic nozzle [3], analysis of an MHD power generator [4] and conjugate heat transfer in an ultrahigh pressure hypersonic nozzle. For brevity, only the last case is discussed in detail.

2 Generalized Equations

It is straightforward to write the conservation equations in a form that applies to all types of fluids. The primary issues are to retain density inside the differentials even for the incompressible fluid case and to retain either enthalpy or internal energy (as opposed to the specific heats or the specific heat ratio) in the energy equation. Here we choose enthalpy and write the equations of motion in the vector form,

$$\frac{\partial Q}{\partial t} + \nabla \cdot \left(\boldsymbol{F} - \boldsymbol{F}_v \right) = S \tag{1}$$

The conservative variables, inviscid and viscous flux vectors in Eq. 1 are given in their standard conservative form,

$$Q = \begin{pmatrix} \rho \\ \rho \boldsymbol{Y} \\ \rho \boldsymbol{V} \\ \rho h^0 - p \\ \rho \boldsymbol{K} \\ \boldsymbol{B} \end{pmatrix}, F = \begin{pmatrix} \rho \boldsymbol{V}^T \\ \rho \boldsymbol{Y} \boldsymbol{V}^T \\ \rho \boldsymbol{V} \boldsymbol{V}^T + p \bar{\boldsymbol{I}} \\ \rho h^0 \boldsymbol{V}^T \\ \rho \boldsymbol{K} \boldsymbol{V}^T \\ \boldsymbol{B} \boldsymbol{V}^T - \boldsymbol{V} \boldsymbol{B}^T \end{pmatrix}, F_v = \begin{pmatrix} 0 \\ \rho \boldsymbol{D} \nabla \boldsymbol{Y} \\ \bar{\boldsymbol{\tau}} \\ \bar{\boldsymbol{\tau}} \cdot \boldsymbol{V} - \boldsymbol{q} \\ \mu_K \nabla \boldsymbol{K} \\ \bar{\boldsymbol{\pi}} \end{pmatrix}, S = \begin{pmatrix} 0 \\ \dot{\omega} \\ \rho \boldsymbol{g} + \boldsymbol{f}_B \\ \dot{q}_B \\ \boldsymbol{S}_K \\ 0 \end{pmatrix} \tag{2}$$

where ρ, $\boldsymbol{V}$ and p are density, velocity vector and pressure, respectively; $\boldsymbol{Y}$ is a species mass fraction vector which is also used to distinguish phases for applications with multiple phases; h^0 is the total enthalpy and K represents the turbulence variables (ν_T for one-equation turbulence model and $(k,\epsilon)^T$ or $(k,\omega)^T$ for two-equation models); $\boldsymbol{B}$ is the magnetic field for MHD problems. In the viscous flux, $\boldsymbol{D}$ is the molecular diffusion coefficient; $\bar{\boldsymbol{\tau}}$ is the stress tensor and $\boldsymbol{q}$ is the heat flux. The resistivity tensor in the magnetic equation is represented by $\bar{\boldsymbol{\pi}}$. For generality, a source term, S, has been included in Eq. 1 to allow for transformation between species or phases and the addition of external body forces or heat sources. An example of the combination of these source terms is given in Eq. 2 with the reaction rate, $\dot{\omega}$, the gravity force, $\boldsymbol{g}$, the Lorentz force and the electric power dissipation. Further, the turbulence production and dissipation are also included in the source term of the turbulence transport equations.

The equations of motion given in Eq. 1 are not closed and must be augmented by an equation of state, an enthalpy relation and pertinent transport property relations that describe the fluid of interest. In order to make our formulation apply to arbitrary fluids, we introduce an arbitrary equation of state and an arbitrary enthalpy relation in the general form,

$$\rho = \rho(p, T, \boldsymbol{Y}^T) \qquad h = h(p, T, \boldsymbol{Y}^T) \tag{3}$$

with analogous relations for the transport variables,

$$\mu = \mu(p, T, \boldsymbol{Y}^T) \qquad K = K(p, T, \boldsymbol{Y}^T) \qquad \boldsymbol{D} = \boldsymbol{D}(p, T, \boldsymbol{Y}^T) \tag{4}$$

and the state-dependent electrical conductivity and electron number density.

3 Numerical Approach

To obtain the numerical solution of the above system, we embed Eq. 1 in a pseudo-time that can be transformed to the more convenient primitive variables, $Q_p = (p, \boldsymbol{Y}, \boldsymbol{V}, T, \boldsymbol{K}, \boldsymbol{B})^T$ using the chain rule:

$$\frac{\partial Q}{\partial \tau} = \frac{\partial Q}{\partial Q_p} \frac{\partial Q_p}{\partial \tau} \tag{5}$$

We then replace the Jacobian by the matrix, Γ_p, which we designate as the preconditioning matrix. The equations then take the following form:

$$\Gamma_p \frac{\partial Q_p}{\partial \tau} + \frac{\partial Q}{\partial t} + \nabla \cdot \left(\boldsymbol{F} - \boldsymbol{F}_v \right) = S \tag{6}$$

where the "preconditioning" matrix is given by:

$$\Gamma_p = \begin{pmatrix} \rho'_p & \rho_Y & 0 & \rho_T & 0 & 0 \\ \rho'_p \boldsymbol{Y} & \rho_Y \boldsymbol{Y} + \rho & 0 & \rho_T \boldsymbol{Y}^T & 0 & 0 \\ \rho'_p \boldsymbol{V} & \rho_Y \boldsymbol{V} & \rho \bar{\boldsymbol{I}} & \rho_T \boldsymbol{V}^T & 0 & 0 \\ \rho'_p h^0 - (1 - \rho h_p) & \rho_Y h^0 + \rho h_Y & \rho \boldsymbol{V}^T & \rho_T h^0 + \rho h_T & 0 & 0 \\ \rho'_p \boldsymbol{K} & \rho_Y \boldsymbol{K} & 0 & \rho_T \boldsymbol{K} & \rho & 0 \\ 0 & 0 & 0 & 0 & 0 & I \end{pmatrix} \tag{7}$$

This matrix contains the density, enthalpy, velocity and species mass fractions along with the four thermodynamic properties, ρ_p, ρ_T, h_p and h_T, that are obtained by taking derivatives of the state relations. We note that the preconditioning matrix differs from the physical Jacobian in that the physical property, ρ_p, is replaced by an artifical property, ρ'_p, which is defined so as to ensure that the time-marching system remains well conditioned for all equations of state, including the incompressible limit [1].

We discretize the equations by integrating Eq. 6 over a control volume,

$$\bar{\Gamma}_p \frac{\partial \bar{Q}_p}{\partial \tau} \Omega + \frac{\partial \bar{Q}}{\partial t} \Omega + \sum_{k=1}^{K} \left(\boldsymbol{F}_k^{inv} - \boldsymbol{F}_k^{vis} \right) \cdot \boldsymbol{A}_k = \bar{S} \Omega \tag{8}$$

where $\boldsymbol{F}_k^{inv}$ and $\boldsymbol{F}_k^{vis}$, represent the numerical invisid and viscous flux tensors on the kth face, and $\boldsymbol{A}_k$ is the face area normal vector. An approximate Riemann solver is employed to define the invicid fluxes on the face,

$$\left(\boldsymbol{F}^{inv} \cdot \boldsymbol{A} \right)_k = \frac{1}{2} \left(F_{nR} + F_{nL} \right)_k - \frac{1}{2} \Gamma_p \left| \Gamma_P^{-1} \frac{\partial F_n}{\partial Q_p} \right|_k \left(Q_{pR} - Q_{pL} \right)_k \tag{9}$$

where the subscripts R and L represent values on the right and left side of the kth face.

4 Multi-Physics Zone Method

Practical applications often involve multiple media having different physical behavior. In our computational framework, we therefore identify discrete physics zones, each of which are governed by a unique set of conservation equations. For example, in a conjugate heat transfer problem, there are two physics zone. The continuity, momentum and energy equations are solved in the fluid zone, while only the energy equation is solved in the solid zone. Similarly, MHD problems can be divided into three physical zones, the fluid, solid and vacuum zones. The continuity, momentum, energy, species and magnetic diffusion equations are solved in the fluid zone; the energy and magnetic diffusion equations are solved in the solid zone, and only the magnetic diffusion equation is solved in the vacuum zone. Interfaces between multiple physics zones are treated as a mixture of internal and external boundary conditions.

The division into zones is incorporated within a parallel solution framework to optimize the load-balancing between processors. Each physics zone may then be subdivided into several sub-zones, each of which is computed on a separate processor. For example, in Fig. 1, Zone 1 is processed in a 4-processor sub-cluster, while Zone 2 is processed in a 3-processor sub-cluster, and Zone 3 is processed in a 2-processor sub-cluster. The optimization is carried out by taking into account the number of equations and the number of grid points in each sub-zone.

5 Numerical Results

The multi-zone GEMS code [2] has been successfully applied to a variety of applications including a trapped vortex combustor (TVC) with a liquid fuel spray, an MHD power generator with plasma channel, electric conductor, dielectric and surrounding air zones [4], a condensing Argon flow in a supersonic nozzle [3] and conjugate heat transfer in a hypersonic nozzle flow.

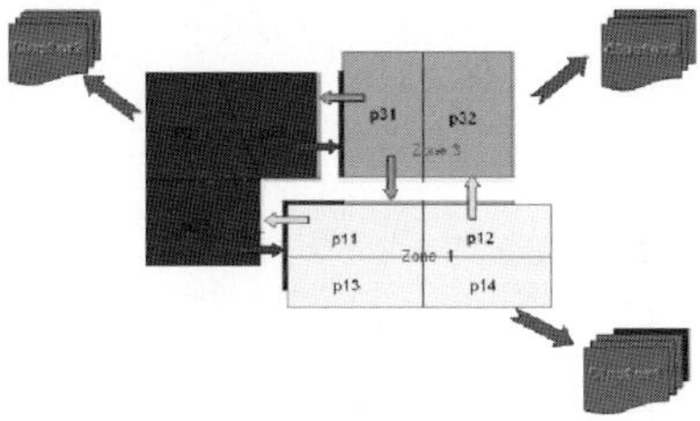

Fig. 1. Sketch of multi-physical zones with their interfaces.

In the present paper, the conjugate heat transfer with real gas air property effect is presented.

For the hypersonic nozzle flow, the state and enthalpy relations are obtained from the REFPROP fluid, thermodynamic and transport property database [5] that is converted into a look-up table format. The effects of the real gas properties under these extreme conditions (P=1700 MPa, T=750K) are demonstrated in Fig. 2, which shows a comparison of the temperature in the nozzle for perfect and real gas properties. The left half (where the axial coordinate runs from left to right) shows the temperature for the perfect gas case while the right half (where the axial coordinate runs from right to left) shows the results for the real gas case. At any given axial location the maximum temperature will generally occur on the wall where the fluid reaches the recovery temperature while the minimum temperature will lie in the inviscid portion of the stream where the velocity is fastest. Note that the maximum temperature in the real gas case is much higher while the minimum temperature is slightly lower. In the real gas case, the recovery temperature on the wall increases rapidly through the throat from the 750 K stagnation temperature at the inlet to a final value of approximately 1750 K. The reason for this dramatic increase in the recovery temperature is that the enthalpy in the real-gas case is a strong function of pressure, whereas for the perfect gas case it is a function of temperature only.

The extremely high recovery temperature in turn implies the nozzle wall will undergo very high heat loads, which we next estimate by studying the conjugate heat transfer between nozzle flow and nozzle wall. In this case, only the energy equation is solved in the solid zone while there are six equations describing the three-dimensional turbulent flow in the fluid zone. The converged temperature field for both the fluid and the solid is shown in Fig. 3 along with the Mach number contours in the fluid zone. The thermal boundary layer and real gas effects in the flow zone are easily seen in these distributions. The temperature in the steel shell increases dramatically up to the gas temperature at the fluid-wall interface and decreases to the outside wall temperature (500K). These results provide good understanding of the interaction between the fluid and the steel shell thermal fields and showcase the capabilities of the generalized framework utilized in this study.

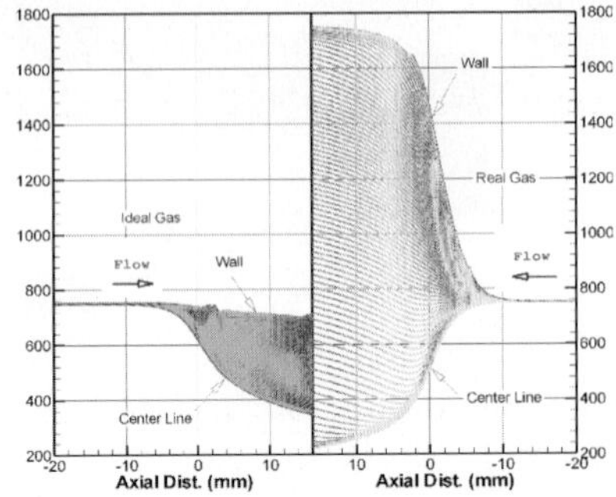

Fig. 2. Comparison of temperature distribution for real gas and ideal gas. Left half is for ideal gas and right half for real gas.

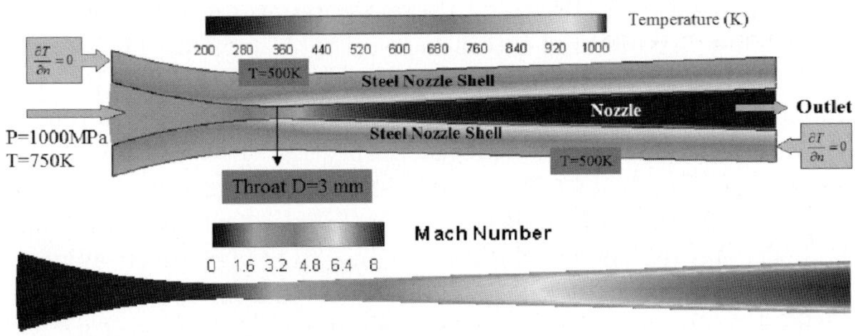

Fig. 3. Temperature contours in solid and fluid zone (top) and the Mach number contour in fluid nozzle zone (bottom).

References

1. V. Sankaran and C. L. Merkle, "Analysis of Preconditioning Methods for Euler and Navier-Stokes Computations," Von Karman Institute Lecture Series on Computational Fluid Dynamics, March, 1999.
2. D. Li, F. Hakhari, V. Sankaran and Charles L. Merkle, Convergence Assessment of General Fluid Equations on Unstructured Hybrid Grids, AIAA 2001-2557, Anaheim, CA, June 11-14, 2001.
3. D. Li and C. L. Merkle, Analysis of Real Fluid Flows in Converging Diverging Nozzles, AIAA-2003-4132, July 2003.
4. D. Li, D. Keefer, R. Rhodes, C. L. Merkle, K. Kolokolnikov and R. Thibodeaux, Analysis of MHD Generator Power Generation, AIAA2003-5050, July 2003.
5. E. Lemmon, M. McLinden, and M. Huber, NIST Reference Fluid Thermodynamic and Transport Properties-REFPROP, Users Guide, NIST-Report, 2002.

A Novel Physical Model and Computational Method for Non-Isentropic, Compressible Two-Fluid Flow

Jeroen Wackers and Barry Koren

CWI, P.O. Box 94079, 1090 GB Amsterdam, The Netherlands.
Jeroen.Wackers@cwi.nl, Barry.Koren@cwi.nl

Summary. This paper describes a five-equation model for compressible two-fluid flow, based on physical flow equations only. The model is conservative and pressure-oscillation free. Equations for continuous flow and jump conditions for discontinuities are given, as well as a discretisation of the equations and an adaptation of the HLL Riemann solver to two-fluid flow. Numerical tests in 1D and 2D show the accuracy of the method.

1 Introduction

Interface-capturing methods for compressible two-fluid flows are based on mixture-fluid models. The interface between the fluids appears as a numerically smeared transition from fluid 1 to fluid 2. Many conservative formulations of such models produce large pressure errors. This problem can be solved by using locally non-conservative methods [1] or by solving the full two-phase flow model [2].

Here, an intermediate approach is presented: a two-fluid method that is fully conservative and pressure-oscillation free. It is an extension of the work by Van Brummelen and Koren [3], it will be described in detail in a future paper. A similar method is derived in a different way by Kapila et al. [4].

The present method has two major advantages. First, the conservative formulation gives good capturing of shocks and interfaces, also for problems with very strong shocks. And second, the model strongly resembles a single-fluid model: it does not require a complex interface-tracking algorithm. It can thus be solved with existing techniques, even on complex, irregular grids.

2 Flow model

The physical model used here for two-fluid flow is based on a mixture model. However, the fluids are not fully mixed: the 'mixture' may be thought to consist of very small bits of the two pure fluids, arranged in an arbitrary pattern. Each fluid has its own pure-fluid equation of state and the fluids interact only by exerting forces on each other. In the model, the pressure and the velocity of the fluids are equal, but each fluid has its own density. The volume fraction of fluid 1, α, is used to denote the relative amounts of the two fluids. Thus, in 1D, we have five independent state variables (p, u, ρ_1, ρ_2 and α), so we need five differential equations to solve the flow.

The two-fluid bulk flow satisfies the standard Euler equations:

$$(\rho)_t + (\rho u)_x = 0 \,, \tag{1a}$$

$$(\rho u)_t + \left(\rho u^2 + p\right)_x = 0 \,, \tag{1b}$$

$$(\rho E)_t + (\rho E u + p u)_x = 0 \,. \tag{1c}$$

In these equations, the bulk quantities are $\rho = \alpha\rho_1 + (1-\alpha)\rho_2$ and $\rho E = \alpha\rho_1 E_1 + (1-\alpha)\rho_2 E_2$, with the total energy for each fluid $j = 1, 2$ defined as $E_j = e_j + \frac{1}{2}u^2$. Here e_j is the internal energy of fluid j.

Two more flow equations are needed to close the system. The first one is the conservation of mass for fluid 1:

$$(\rho_1\alpha)_t + (\rho_1 u\alpha)_x = 0 \,. \tag{2a}$$

Together with equation (1a), this equation gives mass conservation for both fluids. For the last equation, the energy balance of fluid 1 is used. As the fluids exert forces on each other, they exchange energy, which appears as a source term in the equation:

$$(\rho_1 E_1 \alpha)_t + (\rho_1 E_1 u\alpha + p u\alpha)_x = S \,. \tag{2b}$$

An expression for this source term is derived in the next section.

To close the system, equations of state (EOS) are needed for the two fluids. A possible EOS is the ideal gas law,

$$p = (\gamma_1 - 1)\rho_1 e_1 = (\gamma_2 - 1)\rho_2 e_2 \,, \tag{3}$$

with constant γ's. For this equation, it is easy to compute the primitive variables p and α from the total energies.

3 The source term

3.1 Derivation of the source term

The source term S in equation (2b) models the exchange of energy between fluid 2 and fluid 1. Euler flow has no heat conduction, so the only energy exchanged is the work done by the force between the fluids. This force is found from a momentum analysis.

Consider a fluid element in a smooth 1D flow (Fig. 1). The element contains fluid 1 and fluid 2 (the interface is drawn schematically). The force on the entire fluid element is $p(x) - p(x+dx)$ and its bulk mass is ρdx. The force on fluid 1 in the element is $(p\alpha)(x) - (p\alpha)(x+dx) + S_M dx$. Its mass is

$\rho_1 \alpha dx$ and its acceleration is *equal* to the acceleration of the entire element (because both fluids have the same velocity). Therefore

$$\frac{p(x) - p(x+dx)}{\rho dx} = \frac{(p\alpha)(x) - (p\alpha)(x+dx) + S_M dx}{\rho_1 \alpha dx} .$$

The force S_M follows from this expression (using the mass fraction $\beta = \frac{\rho_1 \alpha}{\rho}$):

$$S_M = p\alpha_x + (\alpha - \beta)\, p_x . \tag{4}$$

The energy source term S is the work done by the force S_M:

$$S = uS_M = pu\alpha_x + (\alpha - \beta)\, up_x . \tag{5}$$

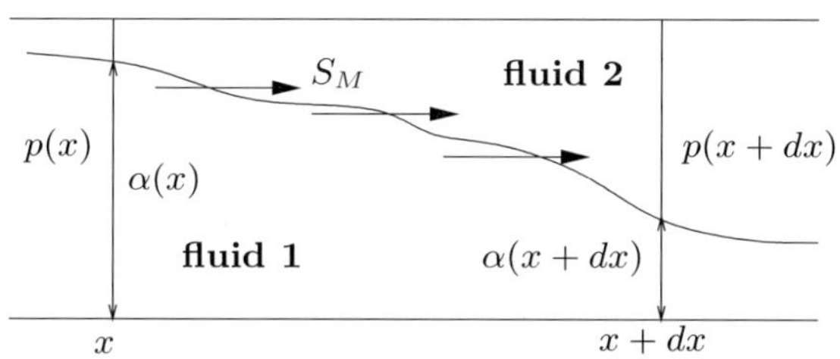

Fig. 1. Two-fluid element in smooth 1D flow.

3.2 Characteristic analysis of the system for ideal gas

The source term (5) is valid for any EOS. Substitution of the ideal gas law (3) allows a characteristic analysis of the flow equations. This results in five wave speeds,

$$\lambda_1 = u - c,\ \lambda_{2,3,4} = u,\ \lambda_5 = u + c, \quad \text{with } c = \sqrt{(\alpha\gamma_1 + (1-\alpha)\gamma_2)p/\rho} . \tag{6}$$

This combination of wave speeds is physically correct. It can be proved that (5) is the only possible source term that gives such a combination.

3.3 Source term in discontinuities

To allow weak solutions with discontinuities of the two-fluid flow equations, we need a proper definition of the flow across a discontinuity. The first four equations, (1a) – (1c) and (2a), satisfy the Rankine-Hugoniot condition $[\mathbf{f}] = c_s[\mathbf{q}]$, with c_s the speed of the discontinuity. For the fifth equation, (2b), this condition becomes

$$[\mathbf{f}] = c_s[\mathbf{q}] + \int_{x_L}^{x_R} S dx , \tag{7}$$

where x_L and x_R denote the coordinates of the left and right side of the discontinuity. The integral must be evaluated across the discontinuity, which is impossible. However, if we assume that the discontinuity is the inviscid limit of a viscous layer and thus has a continuous internal structure (the precise shape is unimportant), then we can write the state variables as continuous functions of p and integrate the source term:

$$\int_{x_L}^{x_R} S dx = [pu\alpha] + \tfrac{1}{2}\beta_L \rho_L (u_L - c_s)[u]^2 + \frac{1}{\rho_L(u_L - c_s)} \int_{p_L}^{p_R} p\alpha dp \,. \quad (8)$$

A derivation of this expression will be given in a future paper. The last integral can be evaluated exactly, but it requires an EOS. So there is a unique jump condition for the present two-fluid model, but, unlike the single-fluid jump condition, it depends on the material properties of the fluids.

4 Numerical method

4.1 Second-order accurate discretisation

The flow equations are discretised with a second-order accurate finite-volume scheme. Fluxes are computed with an improved version of Linde's three-wave HLL approximate Riemann solver [5], combined with a limited reconstruction of the cell interface states. The limiter is applied to the primitive variables ρ, u, p, α and β. Time stepping is done with a two-step scheme (see [7]).

4.2 Numerical source term

A discretisation of the source term is needed in two places. First, an approximation of the source term in a discontinuity is needed to compute the HLL flux. The HLL solver models a Riemann problem with three discontinuous waves. The easiest way to incorporate the source term in these waves is to compute only one approximate solution of (8), using the left and right cell interface state, and to divide this source term proportionally over the three waves. This procedure causes some small inaccuracies, but it is fast and straightforward.

Secondly, the source term for the time integration is computed. It consists of two parts:

i) sources in the discontinuities at the cell faces, that are summed over all HLL waves on interfaces $i - \frac{1}{2}$ and $i + \frac{1}{2}$, that actually run into cell i,
ii) sources in the continuous flow in the cell, which are integrated over the piecewise linear approximations to the primitive variables, that follow from the use of the limiter.

These two sources are summed per cell.

5 Numerical results

5.1 Shock-tube test

The method is tested first on a 1D Riemann problem for ideal gases: a two-fluid variant of Sod's problem, with a ten times higher left pressure and density, giving it a pressure ratio of 100:1. Figure 2 shows that the discontinuities (shock and two-fluid interface) are in the proper locations. The pressure is constant over the contact discontinuity and the volume fraction is constant over both the shock and the expansion fan. A convergence study for this problem shows that the L^1-errors in ρ, u and p converge approximately with the power 0.96 of the mesh width. The volume fraction converges with the power 0.78 of the mesh width. This rate of convergence is comparable to that for similar single-fluid solutions.

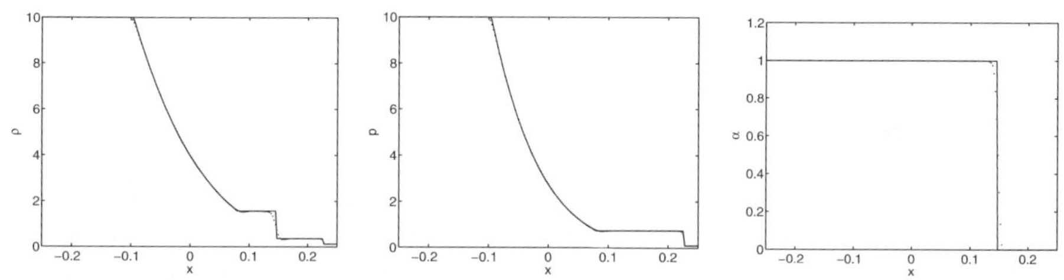

Fig. 2. High-pressure, two-fluid Sod problem. $(\rho, u, p)_L = (10,\ 0,\ 10)$, $(\rho, u, p)_R = (0.125,\ 0,\ 0.1)$, $\gamma_L = 1.4$ and $\gamma_R = 1.6$. The grid has 200 cells, 160 time steps, $\Delta t/\Delta x = 0.2$ (CFL $= |\lambda_{\max}|\Delta t/\Delta x = 0.56$). Solid lines: exact solution.

5.2 Shock hitting helium bubble in air

This 2D test case has been taken from literature [6]. It consists of a cylindrical helium bubble in air, which is hit by an incoming shock wave. The problem is solved on a grid of 200×400 cells, with $\Delta t = 1.25 \times 10^{-5}$. Figure 3 shows the solution at two times. The (half) bubble is visible between $x = -0.025$ and $x = 0.025$. The incident shock, coming from the right, can be seen in the air above the bubble, the curved shock in the bubble runs ahead of this shock. The rightmost wave is an expansion wave, reflected into the air behind the shock. At the later time, a complicated λ-shock structure has developed above the bubble. Figure 4 shows the pressure and the volume fraction for this time. Of the waves appearing in the density plot, the shock waves and expansions are visible in the pressure plot only and the interface in the volume fraction plot only, as it should be. The pressure is continuous over the interface.

The speeds of the shocks and the interface at the centerline ($y = 0$) have been compared with results from Quirk and Karni [6] (obtained on a very fine, adapted grid). The difference is between 0.7% and 2.2%.

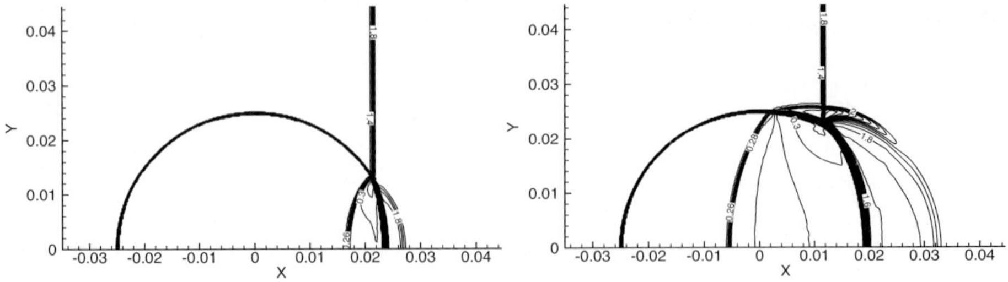

Fig. 3. Shock hitting helium bubble, density at $t = 2.74 \times 10^{-3}$ and $t = 10.74 \times 10^{-3}$.

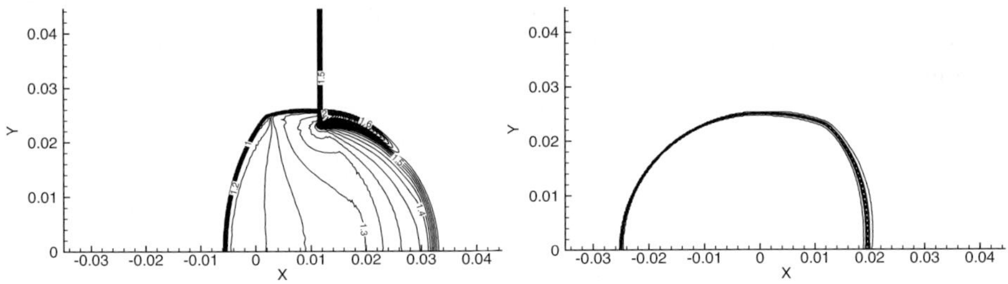

Fig. 4. Shock hitting helium bubble, pressure (left) and volume fraction (right) at $t = 10.74 \times 10^{-3}$.

6 Conclusions

A model for compressible two-fluid flow is proposed, that is conservative and pressure-oscillation free. 1D tests show that the model resolves contact discontinuities without creating pressure errors and that it accurately handles problems with strong shocks. A 2D test shows that the method properly resolves curved shocks and interfaces too.

Acknowledgement: This work was supported by the Dutch government through the national program BSIK (ICT project BRICKS, theme MSV1).

References

1. R. Abgrall, S. Karni: J. Comp. Phys. **169**, 594 (2001)
2. R. Abgrall, R. Saurel: J. Comp. Phys. **186**, 361 (2003)
3. E.H. van Brummelen, B. Koren: J. Comp. Phys. **185**, 289 (2002)
4. A.K. Kapila, J.B. Bdzil, R. Menikoff, S.F. Son, D.S. Steward: Phys. Fluids **13**, 3002 (2001)
5. T. Linde: Int. J. Num. Meth. Fluids **40**, 391 (2002)
6. J.J. Quirk, S. Karni: On the dynamics of a shock-bubble interaction. ICASE Report 94-75, NASA Langley Research Center, Hampton, VA (1994)
7. J. Wackers, B. Koren: A simple and efficient space-time adaptive grid technique for unsteady compressible flows. AIAA Paper 2003-3825 (2003)

Modeling Turbulent Interfacial Flows

Ali Jafari[1] and Nasser Ashgriz[2]

[1] Departemt of Mechanical and Industrial Engineering, University of Toronto, Toronto, Canada. ajafari@mie.utoronto.ca

[2] Departemt of Mechanical and Industrial Engineering, University of Toronto, Toronto, Canada. ashgriz@mie.utoronto.ca

Summary. Spreading of a turbulent liquid jet as a first step to model the primary atomization is presented. It is shown that a high speed turbulent liquid jet issuing in a gaseous surrounding spreads due to the reduction in its velocity. In addition, turbulence in the jet and its surrounding fluid may cause large deformations at the liquid surface. These surface deformations may be the main cause of droplet formation or the primary atomization. New models for the turbulence-interface interaction are presented. A set of governing equations for the description of the turbulent interfacial flows, based on the Reynolds Averaged Navier-Stokes (RANS) equations for two fluids with interfaces, is developed. As a result of averaging process, some new unclosed correlations (that are specific to interfacial flows) between turbulent flow characteristics and the interface properties (fluctuations of the interface position and curvature) appear that need to be closed. Some physical ideas are presented and appropriate models for these correlations are introduced. The new turbulence models are used in simulating the spreading of a high-speed water jet in air.

1 Introduction

For high-velocity liquid jet discharging into a quiescent or a moving gaseous surrounding, it is believed that the main cause of atomization is the action of the surrounding gas. In addition, jet turbulence contributes to the ruffling of the jet surface, making it more responsive to aerodynamic effects. Fully turbulent liquid jets may breakup by themselves if the turbulence intensities are sufficiently large without any need for the aerodynamic forces. In some cases, a turbulent liquid jet injected into a vacuum disintegrates solely under the influence of its own turbulence. In order to model the primary atomization of a liquid jet, we need to model very complex liquid-gas interface topologies as well as the effect of turbulence on the interface. The shape of the interface (as a sharp discontinuity) plays an important role in dynamics of the problem. More information on properties of turbulent interfacial flows can be found in [1]. The accurate prediction of position, curvature and topology of the interface is essential in simulating the problem. Numerous methods have been developed to capture the interface motion. Among these are the front tracking and volume tracking methods which are capable of capturing large interface deformations. Almost all of these methods are designed for laminar flows and no consideration is given for the turbulence effects. The main objective of this work is to introduce new ideas and models of turbulence necessary to simulate interfacial flows, especially those which may result in the production of droplets. A new model for correlation of the mean pressure with fluctuations of the interface location is developed in this work. In addition, the correlation of the volume fraction fluctuations with velocity is modeled and the model is used in the volume fraction equation. Although these models are used in RANS formalism in this study, they can be applied to LES formalism as well.

2 Governing Equations

The mathematical formulation describes the transient flow of two immiscible, incompressible fluids, each having a constant viscosity and including surface tension. It comprises a single set of conservation equations for the whole flow field even though material properties are discontinuous across the fluid boundaries. The time average of mass continuity and momentum equations read as follows:

$$\frac{\partial \overline{u}_i}{\partial x_i} = 0 \tag{1}$$

$$\frac{\partial \overline{u}_i}{\partial t} + \frac{\partial (\overline{u}_i \overline{u}_j)}{\partial x_j} + \frac{\partial \overline{u'_i u'_j}}{\partial x_j} = -\frac{1}{\overline{\rho}} \frac{\partial \overline{p}}{\partial x_i} + \frac{\mu}{\overline{\rho}} \frac{\partial^2 \overline{u}_i}{\partial x_j \partial x_j} + \frac{1}{\overline{\rho}} \overline{B_i^{st}} \tag{2}$$

Here u_i's are the velocity components, p is pressure, ρ the density, overbar denotes the time average, and prime denotes the fluctuating quantities. In general, B_i^{st} can be written as $\sigma \kappa \delta_s \mathbf{n}$ [2], where σ is the the surface tension coefficient, κ the curvature of the liquid surface, δ_s is delta function, and $\mathbf{n}$ is the unit vector normal to the interface. The time-averaged equation for volume fraction, which is needed to construct and capture the interface motion in the VOF method [3], is the following equation:

$$\frac{\partial \overline{F}}{\partial t} + \frac{\partial (\overline{F} \overline{u}_i)}{\partial x_i} + \frac{\partial \overline{F' u'_i}}{\partial x_i} = 0 \tag{3}$$

where F is the volume fraction that is equal to 1 if the computational cell contains liquid, 0 if it contains gas, and between 0 and 1 if the cell contains a portion of the free-surface. The local density ρ and viscosity μ can be written in terms of the liquid volume fraction using the following equation:

$$\rho = F \rho_l + (1 - F) \rho_g \tag{4}$$

$$\mu = F \mu_l + (1 - F) \mu_g \tag{5}$$

where the subscripts l and g refer to liquid and gas respectively. The pressure in the interfacial region can be written as [4]:

$$p = p_2 + H(p_1 - p_2) \tag{6}$$

where H is a non-dimensional parameter corresponding to the location of the interface. The time average of pressure equation is:

$$\overline{p} = \overline{p_2} + (\overline{p_1} - \overline{p_2})\overline{H} + \overline{(p_1' - p_2')H'} \tag{7}$$

3 Proposed Models

In order to close the Reynolds averaged equations, the average of fluctuating quantities appearing in Eqs. (1) to (3) along with Eqs. (4),(5) and (7), must be modeled. The terms that require modeling are:

$$\overline{u_i'u_j'}, \overline{F'u_i'}, \overline{(p_1' - p_2')H'} \tag{8}$$

3.1 Reynolds stress term

To model the fist term (Reynolds stress term), a $k-\epsilon$ based turbulence model which consists of a realizable Reynolds stress algebraic equation model [5] has been used.

3.2 Volume fraction-velocity term

For flow near the interface, where the turbulence is inhomogeneous, a model introduced by Lumley [6] which consists of a gradient transport and a convective transport term which vanishes in homogeneous flows is used. This model is presented for passive scalar admixture in inhomogeneous turbulent flows. The model is:

$$-\overline{F'u_i'} = C_F[\nu_\tau \frac{\partial \overline{F}}{\partial x_i} + \frac{1}{2}\overline{F}\frac{\partial \nu_\tau}{\partial x_i}] \tag{9}$$

where C_F is the correlation coefficient between the fluctuation of volume fraction and velocity components and should be determined (it is inversely proportional to the turbulent Schmidt number). The resulting differential equation for F, after including the turbulence model, may be obtained from inserting Eq. (9) into Eq. (10). The result is:

$$\frac{\partial \overline{F}}{\partial t} + \frac{\partial(\overline{F}\overline{u}_i)}{\partial x_i} = C_F[\nu_\tau \frac{\partial \overline{F}}{\partial x_i} + \frac{1}{2}\overline{F}\frac{\partial \nu_\tau}{\partial x_i}] \tag{10}$$

3.3 Pressure term

For modeling pressure, we correlate the fluctuations of curvature κ and non-dimensional interface location H to the turbulent flow characteristics. More details can be found in [1]. The final form of pressure model is:

$$\overline{(p_1' - p_2')H'} = \frac{C_P\sigma}{L}[\frac{\nu_\tau}{\nu}]^{1/2} \tag{11}$$

where L is a physical length scale and ν_τ is the kinematic eddy-viscoisty.

4 Results and Discussion

The models introduced in this paper are implemented into a Navier-Stokes flow solver which is capable of tracking interface motion. More information about this flow solver can be found in [1]. As a test case, a plane turbulent liquid jet of water which is injected into still air, is simulated. The experimental work for this case is done by Sallam et al. [8]. The mean jet exit velocity and the jet diameter in the simulation are taken from the experiments. However, since the turbulence intensities were not measured, we have assumed our own values for the turbulence intensities. Due to large computational resources necessary for a three-dimensional simulation and also because of the symmetry, only half of the flow is solved using two-dimensional equations of motion. A 128×64 grid (domain size is 12b×6b where b is the nozzle half width) is employed. This grid resolution is not enough to capture the small scale interfacial waves and, thus, it would only give an average view of the interface deformation and its diffusion. This study can give an overall picture of the flow and some spray quantities such as amount of jet spreading and diffusion. Two cases are considered here. In the first case, the mean jet exit average velocity ($\overline{u}_0$) is 15.8m/s and in the second case it is 28.2m/s. The turbulence intensity at the inlet is assumed to be 1%. Figures (1) and (2) show the experimental shadowgraphs taken from Sallam et al. [8], as well as the results of numerical simulation. The simulations show the surface of the liquid jet along with the velocity vectors (only one out of each three vectors are shown and vectors are scaled by $\overline{u}_0$) in the flow for cases 1 and 2, respectively. The boundaries are inflow at the top, no gradient outlet at the bottom, symmetry at the right (as we are solving for half of the domain) and free at the left boundary. The boundary conditions of turbulent quantities for this case are similar to [5]. For this type of flow, different values of C_F and C_P coefficients were tested. The values of C_F=0.1 and C_P=1 provide satisfactory results. Increasing C_F causes more diffusion and spreading of the jet, but increasing C_P increases the rate of breakup or surface instability. Therefore, these two parameters act in the opposite of each other. The comparison between preliminary numerical simulations and shadowgraph pictures of the liquid jet shows satisfactory agreement. Comparison of the experimental results provided in Figs. (1) and (2) shows that the jet with a lower velocity ($\overline{u}_0$=15.8m/s) has longer surface waves than the one with a higher velocity ($\overline{u}_0$=28.2m/s). The jet with higher velocity has high frequency fine surface fluctuations. Therefore, the numerical simulations could not capture such small fluctuations. Much higher grid resolution is needed to capture such fluctuations, which is under progress at this time. However, the computation has been able to capture the smaller frequency fluctuations of the lower speed jet, as shown in Fig. (1). Also, the variation of the non-dimensional axial centerline velocity (centerline axial velocity u_{CL} scaled by the average jet exit velocity $\overline{u}_0$) and turbulent kinetic energy per unit mass along the jet axis are shown in Fig. (3). As can be seen from Fig. (3), for case 2 (higher velocity), the rate of decrease of this non-dimensional velocity is higher than the case 1, since the spreading of the jet is higher in this case, thus lowering the axial velocity. In the case of turbulent kinetic energy (T.K.E.), we see that the rate of decay of T.K.E. is faster for lower Reynolds number flows (case 1 here). This is due to the longer times required for the lower Reynolds number flow to reach a particular axial position and is in agreement with the experimental observations of Chigier et al. [9] for turbulent liquid jets.

5 Conclusion

The spreading of a high speed water jet in air as a first step in modeling the primary atomization process is simulated numerically. The effect of pressure fluctuations on the interface fluctuations are modeled for the first time. Two different models are developed to include the turbulence-interface effects in standard turbulence models. The results show that turbulence in the flow can generate significant interface fluctuations. The flow pattern is sensitive

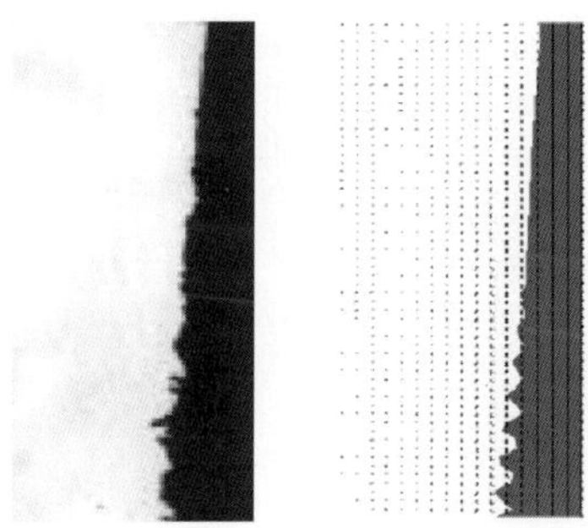

Fig. 1. (left) Shadowgraph of a plane turbulent liquid jet [8] and (right) its numerical simuation ($\overline{u}_0$=15.8m/s). Vectors are scaled by $\overline{u}_0$

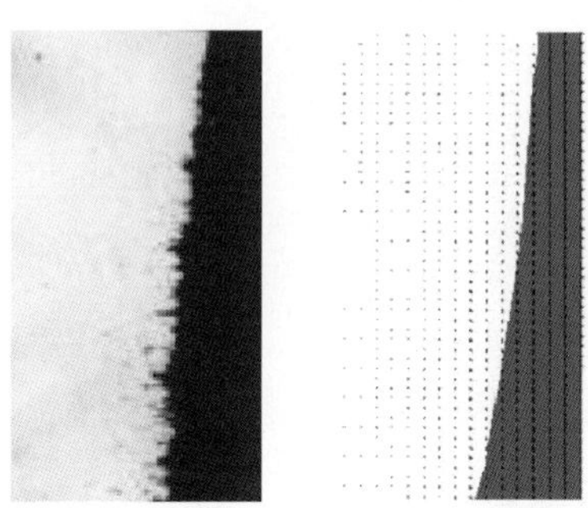

Fig. 2. (left) Shadowgraph of a plane turbulent liquid jet [8] and (right) its numerical simuation ($\overline{u}_0$=28.2m/s). Vectors are scaled by $\overline{u}_0$

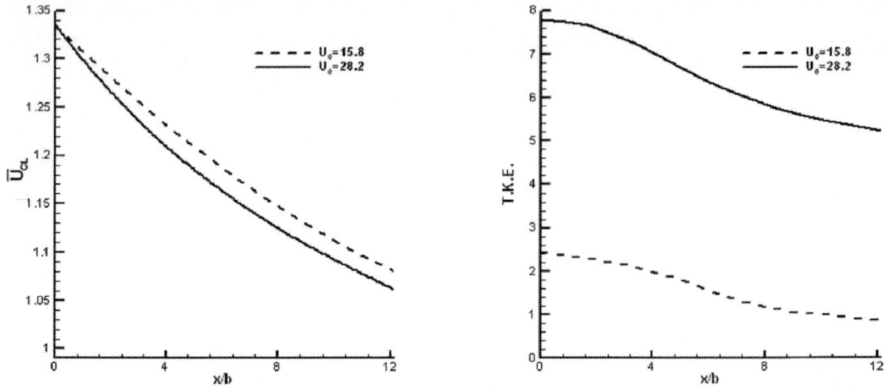

Fig. 3. (left) Variation of non-dimensional axial centerline velocity and (right) variation of T.K.E. per unit mass along the jet axis for $\overline{u}_0$=15.8 and 28.2m/s.

to the values of the model coefficients. A set of coefficients are determined which provide good results as compared to the experiments on the spreading of a high-speed turbulent liquid jet. These are: C_F=0.1 and C_P=1. However, determination of the global model coefficients requires further tests and validation including examining effects of density ratio and other parameters on the flow characteristics. Once a set of global coefficients are determined, the present model can be used to predict the spray angle. In order to develop more reliable models for atomization, incorporation and testing of these models in an LES code would be the next step.

References

1. E. Shirani, A. Jafari, N. Ashgriz: Turbulence models for flows with free surfaces and interfaces. 42nd AIAA Aerospace Sciences Meeting and Exhibit, Reno, NV (2004)
2. J.U. Brackbill, D.B. Kothe, C. Zemach: J. Comp. Phys. **100**, 335 (1992)
3. C.W. Hirt, B.D. Nichols: J. Comp. Phys. **39**, 201 (1981)
4. A. Jafari, E. Shirani, N. Ashgriz: An improved model for interface pressure calculations in free surface flows. 12th Annual Conference on Computational Fluid Dynamics, Ottawa, Canada (2004)
5. T.H. Shih, J. Zhu, J.L. Lumley: Comput. Methods Appl. Mech. Engrg. **125**, 287 (1995)
6. J.L. Lumley: Phys. Fluids **18**, 619 (1975)
7. B Lafaurie, C. Nardone, R. Scardovelli, S. Zaleski, G. Zanetti: J. Comp. Phys. **113**, 134 (1994)
8. K.A. Sallam, Z. Dai, G.M. Faeth: Int. J. Multiphase Flows **25**, 1161 (1999)
9. N Chigier, A. Mansour, U. Shavit: The separate influences of air and liquid turbulence on atomization. In: *Mechanics and Combustion of Droplets and Sprays*, ed by Chiu, Chigier (Begell House 1995) pp 307–21

On Modeling of Collisions in Direct Numerical Simulation of High-Speed Multiphase Flows

Nourgaliev, R.R.[1], Dinh, T.N.[2], and Theofanous T.G.[3]

[1] Center for Risk Studies and Safety, UCSB, robert@engr.ucsb.edu
[2] Center for Risk Studies and Safety, UCSB, nam@engr.ucsb.edu
[3] Center for Risk Studies and Safety, UCSB, theo@engr.ucsb.edu

1 Introduction

In this paper, we are concerned with numerical modeling of particle-particle collisions within a framework of direct numerical simulations of compressible multiphase fluid-solid flows. Such flows are present in applications relevant to explosive dispersal of particulate matter (intended or accidental), shock-induced powder compaction and fluidization, protection of structures against explosions using particle layers or foams, etc. It has been argued that particle-particle collisions affect global dynamics of the particulate multiphase flows, and, therefore, appropriate modeling of collisions is of paramount importance.

In the case of low-speed particle flows, particle-particle collisions are implemented in terms of (i) lubrication theory [4]; (ii) "hard sphere collision" models [3]; and (iii) the "effective-collision-force" models [2]. Application of these models to high-speed flows is hampered by severe stability issues (i.e., instantaneous change of particle velocities due to "hard-sphere"-based collision approach is prohibited by the CFL stability requirements) and inability to represent inelastic collisions, which happen under high impact velocity conditions.

In this work, we present a different approach that is appropriate for high-speed flows. The objective is to be able to represent both elastic and viscoelastic behaviors robustly under high-speed flow and collision conditions. The central idea is to use an integral relation (in time and space) to represent singular, subgrid scale collision forces in a continuum description of fluid mechanics. The model is demonstrated on examples of a shock-induced two-particle collision and a shock-induced dispersal of particle cloud.

2 High-Speed Particle-Particle Collision Model

Our particle collision model is based on the work of Brilliantov et al. [1]. They developed a first-principles description of (spherical) particle collisions so as to predict the restitution coefficient in terms of material properties and particle velocities, and showed it to compare favorably with experiments. The basic idea can be conveyed by the interparticle force

$$\begin{aligned} \mathbf{F}^{(\mathrm{Col})} &= F^N \mathbf{n}; & F^N &= C\varsigma^{\frac{1}{2}}\left(\varsigma + A\dot{\varsigma}\right) \\ \mathbf{g}_{ij} &= \mathbf{V}_i - \mathbf{V}_j; \varsigma & &= \tilde{R}_i + \tilde{R}_j - \left|\mathbf{r}_i(t) - \mathbf{r}_j(t)\right| \\ \mathbf{n} &= \frac{\mathbf{r}_i - \mathbf{r}_j}{\left|\mathbf{r}_i - \mathbf{r}_j\right|}; & m_{ij}^{\mathrm{eff}} &= \frac{m_i m_j}{m_i + m_j} \end{aligned} \tag{1}$$

which features in Newton's law written for a two-particle system:

$$\dot{\mathbf{g}}_{ij} = \frac{\mathbf{F}_i^{(\mathrm{HD})}}{m_i} - \frac{\mathbf{F}_j^{(\mathrm{HD})}}{m_j} + \frac{\mathbf{F}^{(\mathrm{Col})}}{m_{ij}^{\mathrm{eff}}}; \ \dot{\mathbf{r}}_i = \mathbf{V}_i \tag{2}$$

where $\mathbf{g}_{ij}$ is the relative velocity of the colliding particles of mass m_i and m_j; m_{ij}^{eff} is the effective mass; $\tilde{R}_i$ and $\tilde{R}_j$ are the effective particle radii (see below);$\mathbf{V}$ and $\mathbf{r}$ are the velocity and position vectors of particles; ς is a deformation; and $\mathbf{F}^{(\mathrm{HD})}$ is a hydrodynamical force (see below). Brilliantov et al. [1] was able to express coefficients A and C in eq.(1) in terms of solid-mechanic interactions (force-deformation-dissipation) and reduce them simply to functions depending only on material properties. In this work, we are able to embed this basic idea in a CFD framework by smearing the collision (in space-time, as described immediately below) in such a way that coefficient A and C are adjusted so as to preserve desirable (real) collision properties. In particular, we can recover Brilliantov's restitution coefficient, or we can impose a restitution coefficient at will. Thus, in our treatment the coefficients A and C depend, in addition to material properties, on the collision process itself.

In order to numerically implement the above-described collision forces, we numerically "smear" a collision event over N_c time steps, introducing the "collision smearing" length scales, δ_c and δ_{min} and the "smearing" collision time $t_c = N_c \Delta t$. The first length scale δ_c defines when the collision forces become active. Typically, this parameter was set to $\delta_c = 2\Delta h$, where Δh is the grid size. The second "smearing length scale" δ_{min} is introduced to define the closest distance two particles are allowed to approach during the collision. In most calculations presented here we use the value $\delta_{\mathrm{min}} = \Delta h$. The "smearing" collision time is needed to eliminate stability issues due to sudden change of the fluid/solid boundary velocity, caused by collisions. The effective radia of ("smeared") particles i and j are defined as $\tilde{R}_i = R_i + \frac{\delta_c}{2}$ and $\tilde{R}_j = R_j + \frac{\delta_c}{2}$. A collision between two particles is activated when the "smeared" particles overlap, i.e. $\varsigma > 0$. Next, we apply a two-step "predictor-corrector" procedure (a) to determine the coefficients C and A (predictor); (b) to compute collision force and advance particles (corrector). The outcome of the predictor step is values of A and C, which i) ensure a proper conservation/loss of energy due to elastic/viscoelastic collision; and ii) prevent particle-particle overlap. Detailed description of this procedure is given in [6].

3 Level-Set-Based Cartesian Grid Method (LSCG)

Fluid Dynamics. For description of complex-geometry moving fluid-solid boundaries in compressible flows, we employ the Level-Set-Based Cartesian Grid (LSCG) approach [5]. The basic idea of the LSCG is the following. The physical time-space is discretized using a uniform mesh. The fluid-solid boundaries in particulate flows are represented by the level set function φ, defined in all computational nodes as a signed distance to the boundary. Zero-level of this function represents an interface which separates fluid ($\varphi > 0$) and solid ($\varphi < 0$). We trace particle' centers of inertia using particle logics algorithm (see below). After each time step of the numerical solution, based on the current positions of all particles, the level set is reconstructed analytically. Next, based on the current value of the level set function, all computational nodes are "tagged" into four different groups: *i)* REAL nodes (RN); *ii)* BOUNDARY nodes (BN); *iii)* GHOST nodes (GN); and *iv)* UNUSED nodes (for details - see [5]). Solution in REAL nodes is obtained using the 'Characteristics-Based Conservative Finite Difference (CBCFD)' approach (RK_3-TVD for time discretization and $MUSCL_3$-, approximate Riemann solver-based schemes for space discretization). Numerical solution in BNs is computed using the 'Characteristics-Based Matching (CBM)'. GNs are populated using the 1st-order-accurate level-set-PDE-based extrapolation technique.

Particle Logics. Equations of motion for particles (Newton' laws) are discretized using the first-order-accurate Euler backward differencing scheme. The hydrodynamical forces $\mathbf{F}^{(\mathrm{HD})}$ and torques acting on each particle are computed using the bi-cubic spline interpolation technique together with the level-set-based body-fitting transformation, needed to reconstruct shear and normal stresses at the particle surfaces (see for details [6]).

4 Numerical Examples

In the computational domain of size $L_x \times L_y$, periodical in y-direction, we initialize two or more circular particles, with diameter of 1mm. Initially, these particles are motionless in a γ-gas. From the left, there is an incoming $M_{\mathrm{sh}} = 5$ shock wave, defined by the following pre- and post-shock conditions:

$$\textit{Pre-shock:} \begin{cases} P{=}10^5 \text{ Pa} \\ T{=}293.1 \text{ K} \\ \mathbf{u}{=}(0;0) \text{ m/s} \end{cases} \quad \textit{Post-shock:} \begin{cases} P{=}2.9 \cdot 10^6 \text{ Pa} \\ T{=}1,700 \text{ K} \\ \mathbf{u}{=}(0;1,372.71) \text{ m/s} \end{cases}$$

Driven by the impact from the transmitted shock, particles start to move to the right, colliding with each other under different impact configurations.

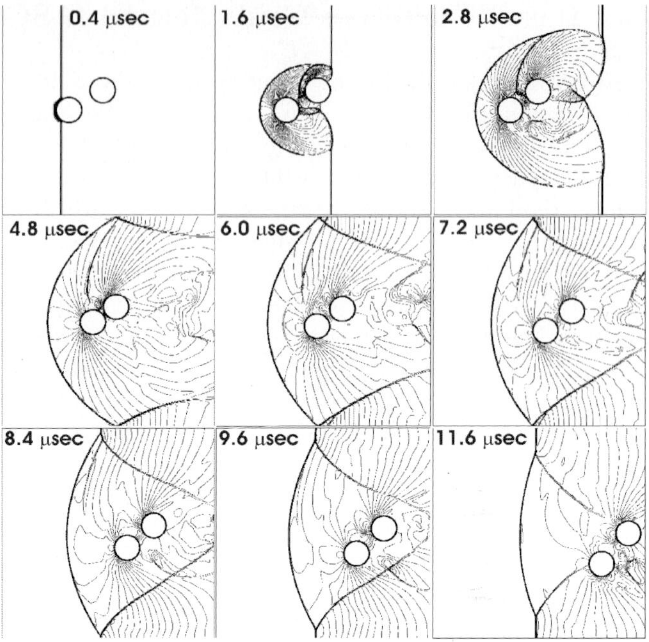

Fig. 1. Dynamics of an elastic collision. Pressure field is rendered with 50 isolines uniformly distributed in the range from 0 to 178 bar.

4.1 Two-Particle Collision

As a first test, we will consider offset collision of two isolated particles. The size of the computational domain is (8×8)mm. Initial positions of particles are $\mathbf{r}_1 = (2.5; 4.0)$mm and $\mathbf{r}_2 = (3.75; 3.25)$mm. To speed up the dynamics, we consider artificially light particles with density of $\rho_p = 100\frac{\text{kg}}{\text{m}^3}$.

The dynamics of the pressure field in the case of elastic collision is shown in Fig. 1. The particle-particle collision occurred at the simulation time of $t_c = 4.72$ μsec, causing a deflection of particle's trajectories, Fig. 2a. As it can be seen in Figs. 2b and 2c, the effect of the viscoelasticity is to create an effective particle clustering. Furthermore, while in the case of the elastic collision the relative velocity of particles is increased, in the case of the viscoelastic collision, it remains the same, Fig.2d. Apparently, this effect should impact on the global behavior of dense multi-particle disperse system, which is considered next.

4.2 Shock-Induced Dispersal of Particle Cloud

In our second example, we consider interaction of an incident shock wave with a cloud of solid ceramic particles ($\rho_p = 2,000\frac{\text{kg}}{\text{m}^3}$), initially arranged in a cluster shown in Fig.3, and subject to the same shock as above.

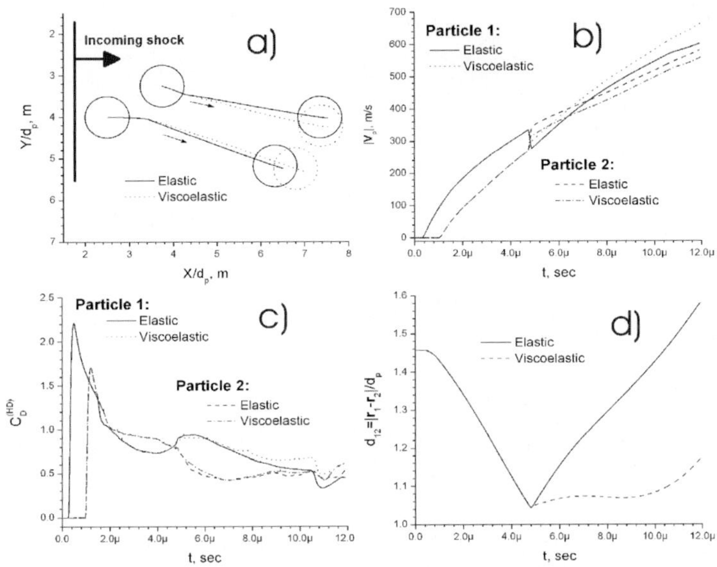

Fig. 2. Particle's trajectory (a), absolute velocity (b), drag coefficient (c) and relative distance (d) for elastic and viscoelastic two-particle offset collisions.

The dynamics of the shock/particle cloud interaction is shown in Fig.3, for the case of elastic (left) and viscoelastic (right) collisions. As one can see, the dispersal of the particle cloud under viscoelastic collisions is quite different from the case of elastic ones. First, there is no lateral spreading of the material. Second, the total number of collisions is significantly higher. This is because the collisions are dissipative and the particles do not run away from each other due to the 'particle pressure', but rather they prefer to 'cluster'. An interesting result is that in terms of an overall cloud-average drag, the viscoelastic case is of lower magnitude.

5 Conclusion

High-speed particle-particle collisions are central to collective behavior, including pattern formation, in particulate clouds under dispersive forces. The new computational framework introduced in this paper allows singular interaction forces be incorporated in the continuum fluid dynamics. We demonstrate that a proper representation of dissipation is of paramount importance to the prediction of cloud global dynamics.

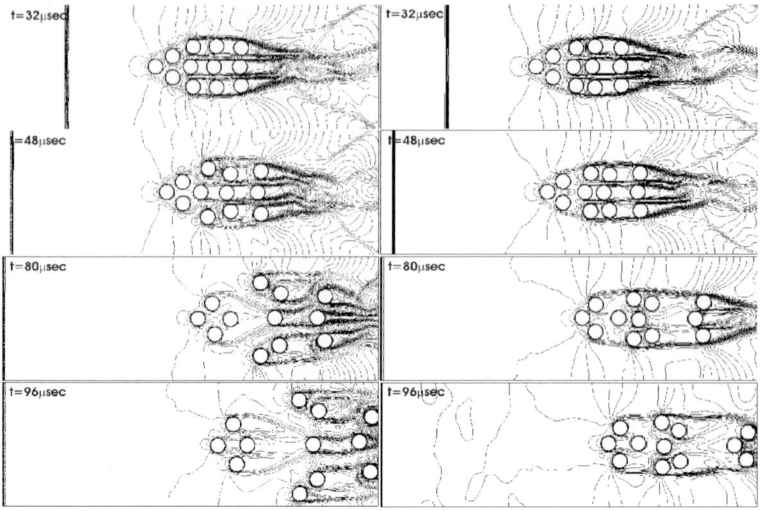

Fig. 3. Dynamics of the Mach number for shock-induced dispersal of solid particle cloud. Elastic (left) vs. viscoelastic (right) collisions. Mach number is rendered with 30 isolines uniformly distributed in the range from 0 to 4.6.

Acknowledgment

This work is sponsored by the Lawrence Livermore National Laboratory ("ALPHA" and "MIX" projects). Support of Drs. Frank Handler, Glen Nakafuji and Dan Klem are gratefully acknowledged.

References

1. N.V. Brilliantov, F. Spahn, J.-M. Hertzsch, and T. Poschel, Model for Collisions in Granular Gases, *Physical Review E*, **53**(5), pp.5382-5392, May 1996.
2. R. Glowinski, T.-W. Pan, T.I. Hesla, and D.D. Joseph, A Distributed Lagrange Multiplier/Fictitious Domain Method for Particulate Flows, *International Journal of Multiphase Flow*, **25**, pp.755-794, 1999.
3. A.A. Johnson, and T.E. Tezduaer, 3D Simulation of Fluid-Particle Interactions with the Number of Particles Reaching 100, *Comput. Methods. Appl. Mech. Engrg.*, **145**, pp.301-321, 1997.
4. N.Q. Nguyen, and A.J.C. Ladd, Lubrication Corrections for Lattice-Boltzmann Simulations of Particle Suspensions, *Physical Review E*, **66**, 046708, 2002.
5. R.R. Nourgaliev, T.N. Dinh, and T.G. Theofanous, The 'Characteristics-Based Matching' (CBM) Method for Compressible Flow with Moving Boundaries and Interfaces, *ASME Journal of Fluids Engineering* (in press), 2004.
6. R.R. Nourgaliev, T.N. Dinh, and T.G. Theofanous, On Modeling of Elastic and Viscoelastic Collisions in Direct Numerical Simulation (DNS) of High-Speed Multiphase Compressible Fluid-Particle Flows. Shock-Induced Dispersal of Solid Material, *CRSS Research Report 11/04-1* November 10, 2003.

A Second-Order Adaptive Sharp-Interface Method for Incompressible Multiphase Flow

M. Sussman[1], M.Y. Hussaini[2], K. M. Smith[2], Ren Zhi-Wei[1], and V. Mihalef[3]

[1] Department of Mathematics, Florida State University, Tallahassee, FL 32306
[2] School of Computational Science and Information Technology, Florida State University, Tallahassee, FL 32306
[3] CBIM, Rutgers, The State University of New Jersey, Piscataway, NJ 08854

Summary. The present method combines the advantages of the Cartesian-grid approach, second-order coupled level set and volume-of-fluid approach, and the dynamic adaptive mesh refinement(AMR) procedure to treat complex three-dimensional problems with sharp interfaces. Its accuracy and robustness are demonstrated by its application to bubble dynamics, ship hydrodynamics, and "water crown" computation. It is shown to capture or preserve sharp slip-lines, phase boundaries remarkably well.

1 Introduction

Discretization strategies for the treatment of problems containing slip-lines and phase boundaries in incompressible flow generally fall into the following distinct categories: continuum approach including level set and volume-of-fluid methods(e.g., Brackbill, et al. [3, 12, 11]), Cartesian-grid approach (e.g., Ye, et al. [13]) and ghost fluid or"dummy cell" method (e.g., Nguyen, et al. [6]; Helenbrook, et al. [5]). The continuum approach is robust and formally second-order accurate. However, it can excessively smear jumps in tangential velocity and its overall accuracy is affected by the low-order treatment of moving boundary conditions (see e.g., [9]). In the work of Nguyen, et al., the convergence of the interface location is first-order accurate that will consequently affect the overall accuracy of the method. Ye, et al. [13] employ front-tracking on a two-dimensional Cartesian grid, and it is not clear if the method can handle complex three-dimensional geometries with large density ratios. Helenbrook, et al. [5] developed a second-order method for two-phase flows, but the applications did not include complex geometries. It is unlikely that a straightforward application of their methods is possible to flow configurations with such wide parameter ranges and complex geometries as in the present study. The present effort develops an accurate and robust methodology that employs semi-implicit Crank-Nicolson temporal discretization and a spatial discretization strategy based on the second-order coupled level set and volume-of-fluid approach, the Cartesian-grid approach, and dynamic adaptive mesh refinement. The present methodology retains the advantages of these methods while precluding their weaknesses.

2 Governing Equations and Solution Technique

The governing equations are the well-known three-dimensional incompressible unsteady Navier-Stokes equations for multiphase flows with appropriate jump relations at interface boundaries [4]. The solution technique is described below on a single uniform rectangular mesh. Details on the extension to an adaptive hierarchy of rectangular grids are given in [8]. Simply stated, adaptive grids are dynamically added and deleted over the course of a simulation depending on resolution requirements, and in the present computations, they cluster near the interface between fluids.

The variables for the two fluids are computed and stored *separately*, and then the solutions in the two fluids are coupled during the implicit computation of the new pressure and the viscous forces. Figure 1 illustrates the locations of the discrete variables. We provide here a brief outline of the method.

1. Advance the location of the interface using the coupled level set and volume-of-fluid method ([9, 7]).
2. Calculate nonlinear advective force terms in each fluid *separately* using high order, upwind, slope limited discretization.
3. Implicit Crank-Nicolson computation of viscous forces [1]. The new velocity field is continuous across the phase boundary (zero level set). Surface tension is included as a body force as in [9] except with zero thickness. Curvature is calculated directly from the volume fractions as described in [7].
4. Implicit calculation of updated pressure and solenoidal velocity field (projection step):

$$\nabla \cdot \frac{\nabla p}{\rho} = \nabla \cdot \boldsymbol{V} \qquad \boldsymbol{u} = \boldsymbol{V} - \frac{\nabla p}{\rho} \tag{1}$$

   The new velocity field satisfies the continuity condition and the new pressure satisfies the appropriate jump relations. The Cartesian-grid approach is used to approximate the divergence operator. We resort to an additional "correction" pressure solve to accurately enforce the continuity condition at the phase interface. The resulting matrix systems are symmetric and are solved by the multigrid preconditioned conjugate gradient method.
5. Extend liquid variables into the gas, and extend gas variables into the liquid (see [7]).

3 Results

3.1 Bubble Dynamics

We present two examples. For the first example, the density ratio is 1000:1, viscosity is zero, and the surface tension coefficient is 1/200. For the second

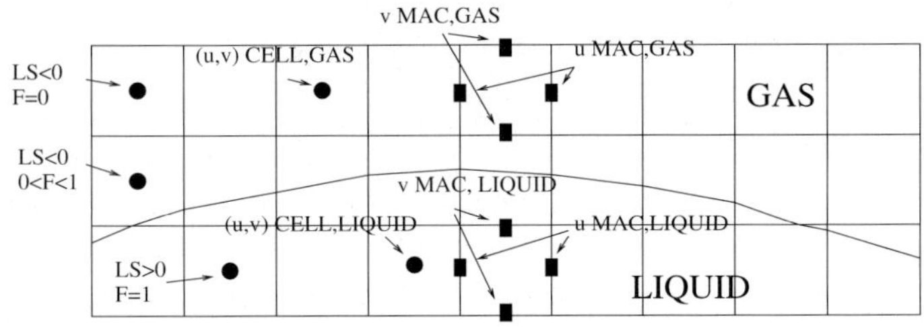

Fig. 1. Location of level set function, LS, volume-of-fluid function F, MAC velocity $\boldsymbol{U}^{MAC}$ and CELL velocity $\boldsymbol{U}^{CELL}$ in relation to the computational grid.

example, the density ratio is 1050:1, the viscosity ratio is 31765:1, the liquid viscosity is 0.54 and the surface tension coefficient is 0.078. Results for the inviscid bubble are found to be similar to those of [10] which were obtained by the boundary integral method and "continuum" approach. Comparison of the results for the viscous bubble with the experimental observations of Bhaga and Weber[2] indicate good agreement (not shown owing to limited space).

The rate of convergence of the present method (Figure 2) is studied by refining the grid by a factor of two, and checking the difference or error between the level set function values on the coarse- and fine-grids. The error is defined as

$$E = \frac{\sum_{\Gamma_{ij}} |\phi_{ij}^c - \phi_{ij}^f|}{\sum_{\Gamma_{ij}} 1}, \qquad \Gamma_{ij} = \left\{(i,j)|\phi_{ij}^c \text{or} \phi_{ij}^f \text{changes sign}\right\} \tag{2}$$

where ϕ^c and ϕ^f correspond to the coarse-grid and fine-grid level set functions respectively. The error for varying grid resolutions is given in Table 1, which clearly demonstrates second-order convergence.

Table 1. Grid refinement study; left: inviscid air bubble rising in liquid, $t = 1.25$, $r = 1$; right: viscous air bubble rising in liquid, $t = 0.25$, $r = 0.013$.

coarse grid size	fine grid size	E
32x64	64x128	0.0102
64x128	128x256	0.0028

coarse grid size	fine grid size	E
32x64	64x128	0.00069
64x128	128x256	0.00017

3.2 Three-Dimensional Ship Hydrodynamics

Figure 3 shows wave overturning at the bow of a DDG5415 navy ship (details of ship geometry provided at http://www50.dt.navy.mil/5415/). The density ratio of water to air is 1000:1. The ship runs lengthwise from $x = 0.0$ to

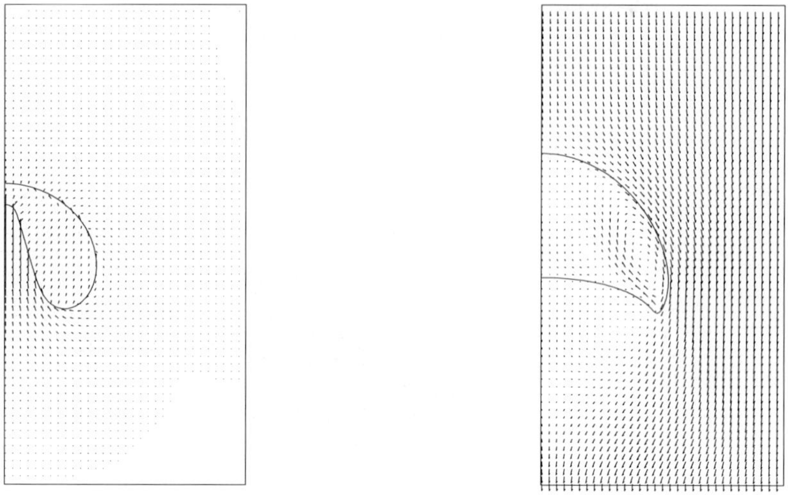

Fig. 2. Axisymmetric gas bubbles rising in a liquid medium; 128x256 grid; left: inviscid bubble, right: viscous bubble.

$x = -1.0$. Figure 4 compares the numerical predictions (continuous line) of wave height above the quiescent level with the whisker-probe measurements (jagged line) at various positions along the x-axis. In general, the agreement between predictions and measurements is very good. Table 2 presents the details of adaptive mesh refinement: the first column gives the number of levels of refinement, the second column the number of grid blocks together with the number of blocks added at each refinement level, the third column the total number of cells, the fourth column the number of processors, the fifth column the minimum mesh width and the last column the cpu time in seconds per timestep. The speed-up due to increasing the processors or the levels of adaptivity is found to slightly exceed 70%.

Table 2. CPU time per time step for varying levels of adaptivity; 3D ship hydrodynamics.

levels	grids	cells	δ	processors	CPU time/step
1	64	2097152	1/128	32	376
2	64+148	5324800	1/256	32	1300
3	64+116+475	17940480	1/512	64	3000

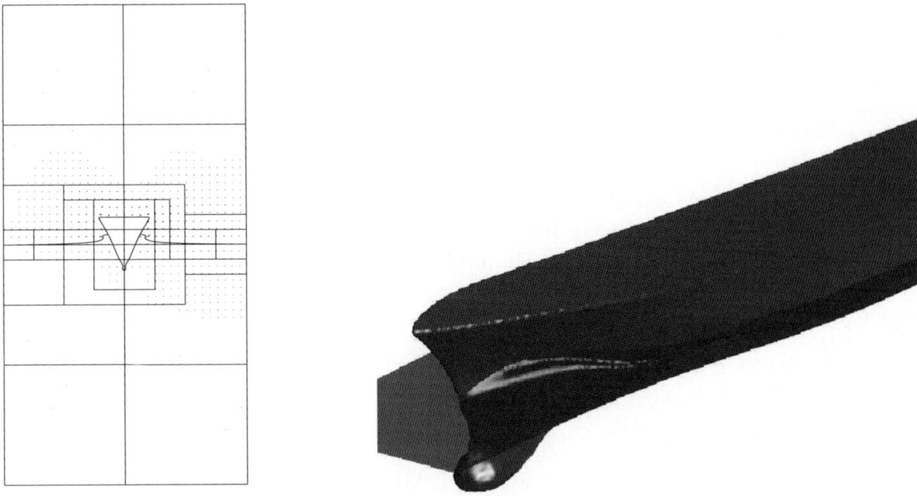

Fig. 3. Left: AMR grid for DDG5415 at $x = -0.1$; right: bow view of DDG5415.

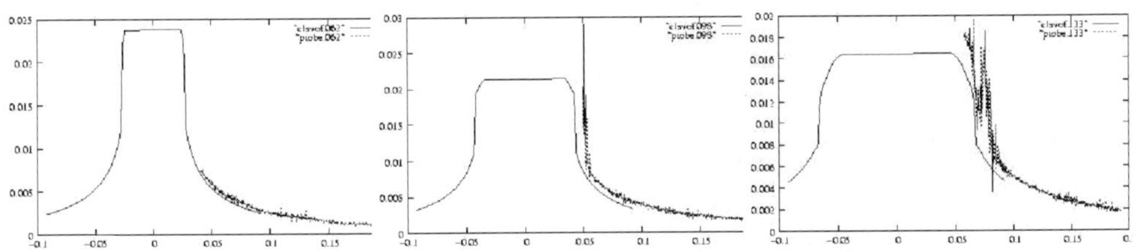

Fig. 4. Wave height predictions compared to whisker-probe measurements for the DDG5415; left to right: $x = -0.062$, $x = -0.098$, $x = -0.133$.

3.3 Three-dimensional "water crown" computation

Figure 5 illustrates a spherical water drop falling on the surface of water and the consequent formation of what is known as "water-crown". The present method sharply captures the evolution of the fluid interface merging and breaking on a relatively coarse mesh of dimensions 32x32x64.

Fig. 5. "Water crown" computation; 32x32x64 grid

4 Conclusions

The accuracy and efficiency of the present method is demonstrated by the solutions to the problems of bubble dynamics, ship hydrodynamics and the "water crown." These computational tests indicate that the results previously obtained on fine meshes can now be obtained on a relatively coarse mesh, resulting in a speed-up of 8:1. We also show that additional speed-up results from the use of adaptive mesh refinement and parallelization. Evidently, the present method robustly handles the merging and breaking of fluid interfaces involving density ratios on the order of 1000:1.

Acknowledgements

This work is partially supported by NSF grant 0108672.

References

1. J. B. Bell, P. Colella, and H. M. Glaz. A second-order projection method for viscous, incompressible flow. In *8th AIAA Computational Fluid Dynamics Conference*, Honolulu, June 9–11, 1987.
2. D. Bhaga and M.E. Weber. Bubbles in viscous liquids: Shapes, wakes and velocities. *J. Fluid Mech.*, 105:61–85, 1981.
3. J. U. Brackbill, D. B. Kothe, and C. Zemach. A continuum method for modeling surface tension. *J. Comput. Phys.*, 100:335–353, 1992.
4. Y.C. Chang, T.Y. Hou, B. Merriman, and S. Osher. Eulerian capturing methods based on a level set formulation for incompressible fluid interfaces. *J. Comput. Phys.*, 124:449–464, 1996.
5. B. Helenbrook, L. Martinelli, and C. Law. A numerical method for solving incompressible flow problems with a surface of discontinuity. *Journal of Computational Physics*, 148:366–396, 1999.
6. D. Nguyen, R. Fedkiw, and M. Kang. A boundary condition capturing method for incompressible flame discontinuities. *J. Comput. Phys.*, 172:71–98, 2001.
7. M. Sussman. A second order coupled levelset and volume of fluid method for computing growth and collapse of vapor bubbles. *Journal of Computational Physics*, 187:110–136, 2003.
8. M. Sussman. A parallelized, adaptive algorithm for multiphase flows in general geometries. *International Journal of Computers and Structures*, 2004. to appear.
9. M. Sussman and E.G. Puckett. A coupled level set and volume of fluid method for computing 3d and axisymmetric incompressible two-phase flows. *J. Comp. Phys.*, 162:301–337, 2000.
10. M. Sussman and P. Smereka. Axisymmetric free boundary problems. *J. Fluid Mech.*, 341:269–294, 1997.
11. M. Sussman, P. Smereka, and S.J. Osher. A level set approach for computing solutions to incompressible two-phase flow. *J. Comput. Phys.*, 114:146–159, 1994.
12. S. O. Unverdi and G. Tryggvason. A front-tracking method for viscous, incompressible, multi-fluid flows. *J. Comput. Phys.*, 100:25–37, 1992.
13. T. Ye, W. Shyy, and J.N. Chung. A fixed-grid, sharp interface method for bubble dynamics and phase change. *J. Comp. Phys.*, 174:781–815, 2001.

Large-Scale Direct Simulation of Two-Phase Flow Structure Around a Spacer in a Tight-Lattice Nuclear Fuel Bundle

Kazuyuki Takase[1], Hiroyuki Yoshida[2], Yasuo Ose[3], and Hajime Akimoto[4]

[1] Japan Atomic Energy Research Institute, Tokai-mura, 319-1195, Japan
takase@popsvr.tokai.jaeri.go.jp
[2] Japan Atomic Energy Research Institute, Tokai-mura, 319-1195, Japan
yoshida@hflwing.tokai.jaeri.go.jp
[3] Yamato System Engineer Co. Ltd., Hitachi-shi, 317-0063, Japan
ose@atom.tokai.jaeri.go.jp
[4] Japan Atomic Energy Research Institute, Tokai-mura, 319-1195, Japan
akimoto@hflwing.tokai.jaeri.go.jp

1 Introduction

Subchannel codes [1, 2, 3] are generally used for the thermal-hydraulic analysis of fuel bundles in nuclear reactors, however, many composition equations and empirical correlations based on experimental results regarding the water-vapor two-phase flow behavior are needed. When there are no experimental data such as the reduced-moderation light water reactor (RMWR) [4] which is currently studied by the Japan Atomic Energy Research Institute, therefore, it is very difficult to obtain highly precise predictions. The RMWR core has remarkably narrow gap spacing between furl rods (i.e., around 1 mm) which are arranged at a triangular tight-lattice configuration in order to reduce the moderation of the neutron. In such a tight-lattice core, there is no sufficient information about the effects of the gap spacing and the spacer configuration on the two-phase flow characteristics. Then, the authors tried to analyze the water-vapor two-phase flow configurations in a tight-lattice fuel bundle with spacers, using a large-scale simulation with the earth simulator of Japan [5]. The earth simulator is the highest parallel vector supercomputer system in the world and consists of 5120 arithmetic processors with 10 TB of main memory. Its theoretical performance becomes 40 Tflops. This paper describes the predicted results of the two-phase flow structure around a spacer in the tight-lattice fuel bundle.

2 Specification of RMWR

A tight-lattice fuel bundle is required to the RMWR core to attain a higher conversion ratio more than unity. Although the coolant is 100% water at the core inlet, it changes water and vapor along the flow direction, and then the vapor occupies 90% or more at the core outlet. Therefore, the RMWR has very severe cooling conditions on the viewpoint of the thermal engineering.

Figure 1 shows a bird-eye view of the actual RMWR design. It consists of a core, control rod, separator and dryer region, and a pressure vessel. The core is composed of 282 fuel bundles. Each fuel bundle has a hexagonal shape horizontally. A length of one side of the hexagonal shape is about 0.13 m and the axial length of a fuel bundle is about 2.9 m. A heating section in the core consists of two seed and three blanket regions and its length is about 1.3 m. The mixed oxide fuel is used to the seed region and then the depleted UO_2 is used to the blanket region.

Four spacers are installed in order to keep the gap spacing between fuel rods and to restrict the movement of a fuel rod to the radial and circumferential directions. The shape is shown in Fig. 3: the fuel rod diameter is 13 mm; the gap spacing is 1.3 mm; and, thickness of the spacer is 0.3 mm.

3 Numerical Analysis

3.1 Analytical Method

A new two-phase flow analysis code, TPFIT, was developed [6]. It uses the interface tracking method [7] to predict precisely an interface between liquid and gas phase. Its method was improved to simulate unstable behavior of water-vapor liquid film flow around the fuel rods with high accuracy. The surface tension of water is calculated using the continuum surface force (CSF) model [8].

3.2 Governing Equations

- `Mass conservation,` $$\frac{D\rho}{Dt} = -\rho\,\frac{\partial u_i}{\partial x_i} \tag{1}$$

$\rho = \rho_l\phi_l + \rho_g\phi_g\,,\quad \phi_g = 1 - \phi_l$

- `Density function,` $$\frac{D\rho_l\phi_l}{Dt} = -\rho_l\phi_l\,\frac{\partial u_i}{\partial x_i},\quad \frac{D\rho_g\phi_g}{Dt} = -\rho_g\phi_g\,\frac{\partial u_i}{\partial x_i} \tag{2}$$

- `Momentum conservation,` $$\frac{Du_i}{Dt} = -\frac{1}{\rho}\frac{\partial p}{\partial x_i} + \frac{1}{\rho}\frac{\partial \tau_{ij}}{\partial x_i} + g_i + \sigma_i \tag{3}$$

- `Energy conservation,` $$\frac{De}{Dt} = -\frac{p}{\rho}\frac{\partial u_i}{\partial x_i} + \frac{1}{\rho}\frac{\partial}{\partial x_i}\left[\lambda\frac{\partial T}{\partial x_i}\right] + q \tag{4}$$

- `Volume fraction,` $$\frac{D\phi_l}{Dt} = 0\,,\quad \frac{D\phi_g}{Dt} = 0 \tag{5}$$

Here, D, e, g, p, q, T, t, u, x, ϕ, λ, ρ, σ, τ denote the substantial time derivative, internal energy, gravity acceleration, pressure, heat source, temperature, time, velocity, coordinate, volume fraction, viscosity, density, surface tension, stress tensor, respectively. Moreover, subscripts of g, i, j and l represent the gas phase, species in multi-component system, and liquid phase.

3.3 Analytical Model and Boundary Conditions

Figure 2 shows the analytical geometry. It simulates the RMWR fuel bundle with 37 fuel rods and four spacers. The fuel rod outer diameter is 13 mm and the gap spacing between each rod is 1.3 mm. Water flows upward from the bottom of the fuel bundle. Analytical conditions of water at the fuel bundle inlet are 283° C, 7.2 MPa, and 400 kg/m^{2}s (i.e., the RMWR design values). Three-dimensional computations were carried out for single-phase and two-phase flow conditions under the non-heated isothermal flow. It is set to remove the effect of heat transfer by the fuel rods. The maximum and minimum mesh divisions were 0.2 and 0.01mm.

4 Results and Discussion

Predicted axial velocity distributions in the horizontal direction are shown in Fig. 4. Here, Fig. 4(a) and (b) are the results at the position A and B in Fig. 1. The velocity is indicated using color gradation; blue and red mean 0 and 18 m/s. In Fig. 4 (a) the axial velocity distribution is almost uniform. It is accelerated by reduction of the channel cross-section due to existence of a spacer as can be seen in Fig. 5 (b).

Figure 5 shows the void fraction distributions around fuel rods in the horizontal direction. Figure 5 (a) is the predicted result. Here, the void fraction of blue and red indicate 100% water and 100% vapor, respectively. Each fuel rod surface is enclosed with very thin water film, and vapor flows the outside. In the region where the gap spacing between fuel rods is small, the bridge formation in which adjacent fuel rods are connected by water film is confirmed. On the other hand, the vapor flows through the center area of three fuel rods arranged by the triangular pitch. It is easy for vapor to flow into the triangular region in which the spacing is large because the local flow resistance is lower than that of the small spacing region.

Figure 5 (b) is an example of the experimental result of the void fraction distribution around the fuel rods. This figure was obtained by an advanced neutron radiography technique which was developed by Kureta [9]. Although the experiment condition differs from the calculated condition of Fig. 5 (a) slightly, a tendency of the water and vapor distribution of Fig. 5 (a) is in good agreement with that of Fig. 5 (b).

Figure 6 shows the two-phase flow configuration in the perpendicular direction of just a spacer position. Here, blue represents a region where the void fraction is 0.1 or less and it is occupied with water of about 100%; Moreover, red represents a region where the void fraction is 0.9 or more and it is occupied with vapor of about 100%. Water widely exists on the wall as the liquid film and vapor partially exists in the water region.

Figure 7 is an example of the predicted vapor region around the fuel rods. A flow configuration of vapor shows the streak structure in the vertical direction.

5 Conclusions

Two-phase flow structure around a spacer in a RMWR fuel bundle was clarified quantitatively with the large-scale simulations. From the results of the present study the following summaries are derived:

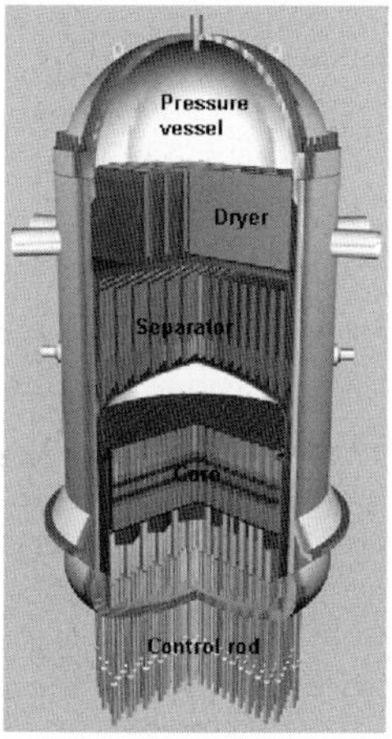

Fig. 1. Structural design of the RMWR.

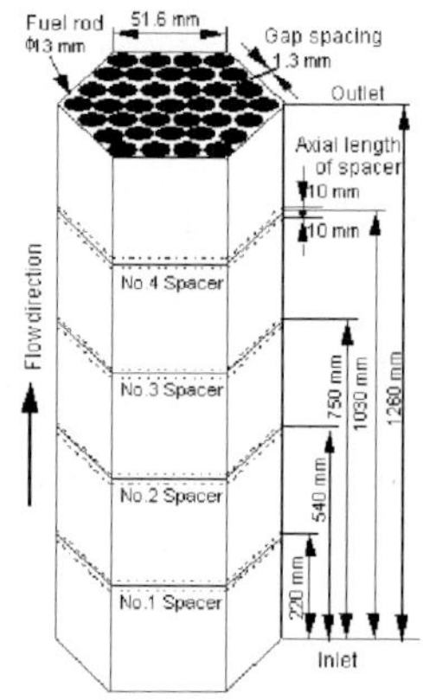

Fig. 2. Three-dimensional analytical geometry of a tight-lattice fuel bundle.

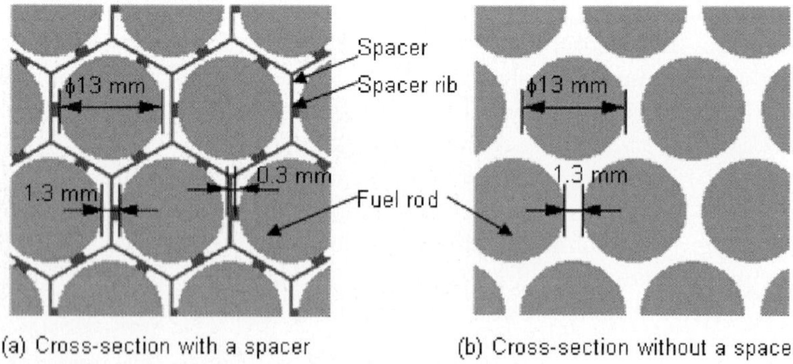

Fig. 3. Cross-sectional views of a fuel bundle with and without a spacer.

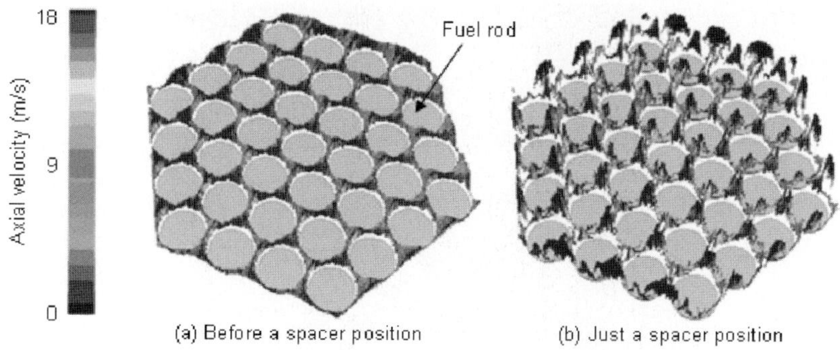

Fig. 4. Predicted axial velocity distributions.

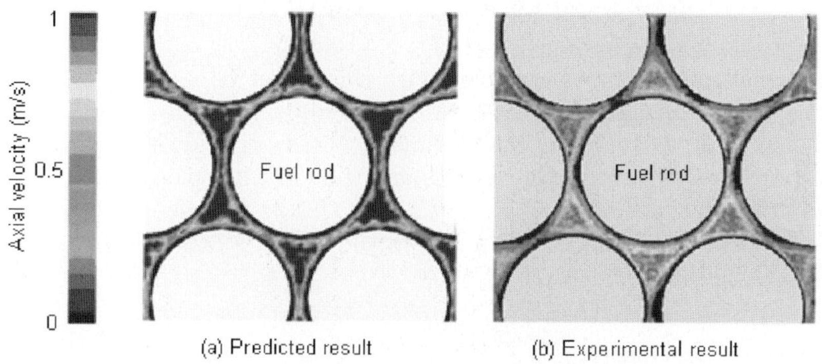

Fig. 5. Predicted and experimental void fraction distributions.

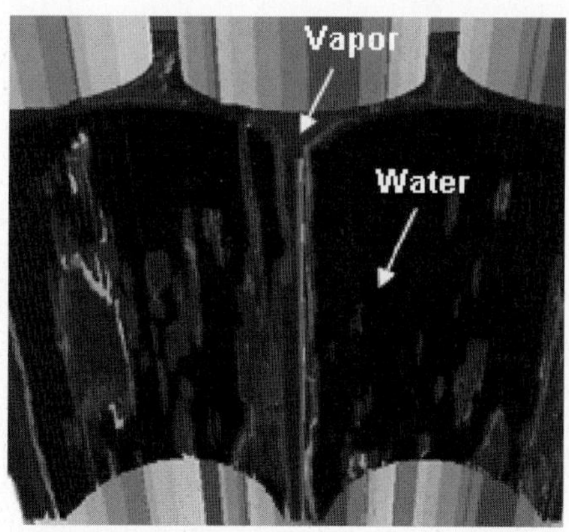

Fig. 6. Predicted water and vapor distribution around a spacer vertically (blue and red correspond to water and vapor).

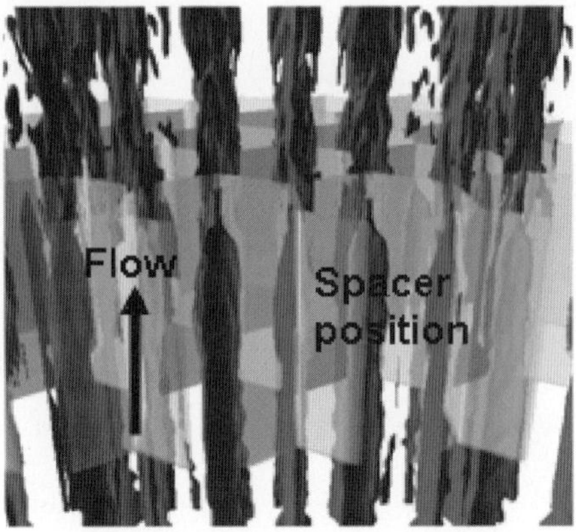

Fig. 7. Axial flow image of vapor through a spacer.

1) The fuel rod surface is encircled with thin water film;
2) The bridge formation by the water film appears in the region in which the spacing between fuel rods is small;
3) Vapor flows downward the triangular region in which the spacing between fuel rods is large; and,
4) A flow configuration of vapor shows the streak structure in the vertical direction.

Acknowledgements

The present study was conducted under collaboration among JAERI, JAPC, Hitachi Ltd., Toshiba Corp., Mitsubishi Ltd., Univ. of Tokyo and Univ. of Osaka with the governmental funding from Publicly Invited Research Projects for Development of Innovative Nuclear Technologies by the Ministry of Education, Culture, Sports, Science and Technology (MEXT).

References

1. J. K. Kelly, S. P. Kao., M. S. Kazimi: MIT-EL-81-014, (1981).
2. M. J. Thurgood, M. J.: NURREG/CR-3046, PNL-4385, Vol. 1, R4, (1983).
3. S. Sugawara, Y. Miyamoto: Nuclear Engineering and Design, 129, (1990) pp.146-161.
4. T. Iwamura, et al., Proc. of 13th Pacific Basin Nuclear Conference (PBNC 2002), Shenzhen, China, (2002) pp.1631-1637.
5. Earth Simulator Center: Annual report of the earth simulator center (April 2002- March 2003), (2003).
6. H. Yoshida, et al.: Proc. of GENES4/ANP2003, No.1111, Kyoto, Japan. (2003).
7. D. L. Youngs, (K. W. Morton and M. J. Baine, Edi.): Numerical methods for fluid dynamics (Academic Press, 1982) pp.273-285.
8. J. U. Brackbill: J. Computational Physics, 100, 2, (1992) pp.335-354.
9. M. Kureta, et al.: Nuclear Technology, 136, (2001) pp.241-254.

Part XVII

Optimization

Surface Mesh Movement for Aerodynamic Design of Body-Installation Junction

Hyoung-Jin Kim[1] and Kazuhiro Nakahashi[2]

[1] Tohoku University, Sendai 980-8579, Japan, hjkim@ad.mech.tohoku.ac.jp
[2] Tohoku University, Sendai 980-8579, Japan, naka@ad.mech.tohoku.ac.jp

1 Introduction

In aerodynamic design of complete aircraft configurations having body installation junctions such as wing-fuselage or nacelle-pylon-fuselage junctions, the design of wing, fuselage or nacelle-pylon geometry requires redefinition of the junction lines and regeneration or movement of surface mesh points on both the wing and body.

For automated aerodynamic shape design of the wing-body junction region, overset-grid methods can easily be applied [1, 2]. Another approach is direct link of flow solver and CAD system [3], which requires regeneration of surface and volume mesh for every design geometry.

As an alternative way, Murayama *et al.* [4] developed a surface mesh movement method adopting a surface mapping method in which a three-dimensional surface mesh region to be moved is partially cut out to make a patch and mapped onto a two-dimensional parameter domain. Then meshes on the parameter domain are moved appropriately and transformed back to the physical domain. For design problems the patch map and a formulation relating the surface mesh and the patch mesh should be generated for every design geometry. This may cause inefficiency for the whole design procedure.

In this study, an efficient and robust method for surface mesh movement is presented for aerodynamic design optimization of body-installation juncture. Instead of mapping the surface mesh to a parametric patch mesh, we move the surface mesh points directly on the curved body surface by using a spring analogy based on the geodesic distance. New mesh locations are restricted to be on the initial surface by a surface recovery method.

2 Methodology

2.1 General Procedure

Figure 1 shows a schematic diagram for a wing-body configuration in which both of the wing and fuselage geometry are modified. Solid lines represent an initial geometry and dotted lines a modified geometry. For a wing-fuselage

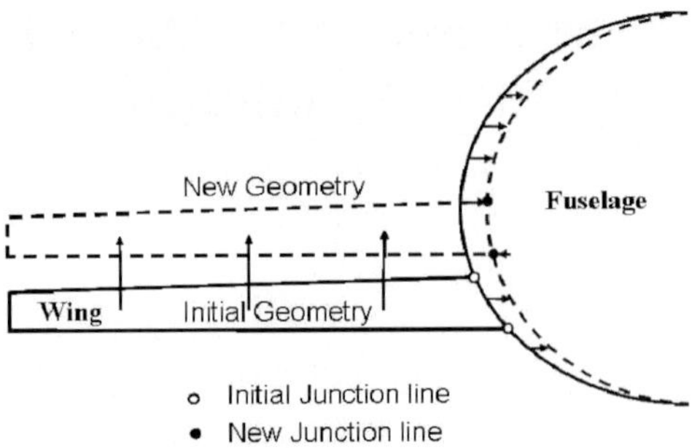

Fig. 1. Change of wing-body junction location

design problem as an instance, the proposed procedure can be briefly described as follows:

1) Modify wing and fuselage geometry respectively.
2) For the modified geometry, find the new location of the wing-fuselage junction nodes. And calculate the geodesic distances from initial to new junction nodes along the faceted surface.
3) With the body-fuselage junction node movements as boundary conditions, move the surface mesh on the fuselage surface using a two-dimensional spring analogy based on the geodesic distance.
4) The same as step 3 is done for mesh points on the wing surface.
5) The new surface mesh points on the initial faceted surfaces are projected to the body surface by a surface recovery method [5] with the initial grid as the background grid.
6) Modify the volume grid according to the surface mesh movement using a three-dimensional volume mesh modification method.

2.2 Finding New Junction Points

In order to find the new junction node positions, the wing surface geometry needs to be extrapolated spanwisely when undefined wing geometry inside the fuselage is exposed by the geometric modification. For structured grids, the spanwise grid lines on wing surfaces are usually along the local sweep angles. Thus spanwise extrapolation can be easily done along the grid lines. Since unstructured grids do not have aligned grid lines, at every junction nodes we calculate an extrapolating vector which is along local sweep angle and tangential to wing surface cell adjacent to the juncture.

The next step is to find an intersection point where the extrapolation vector meets the fuselage surface, which is conducted using a neighbor-to-neighbor jump search method. See Fig. 2. Although the neighbor-to-neighbor search enables an efficient one-dimensional search, it may fail to find the target cell if there is a hole between the initial and target node points as

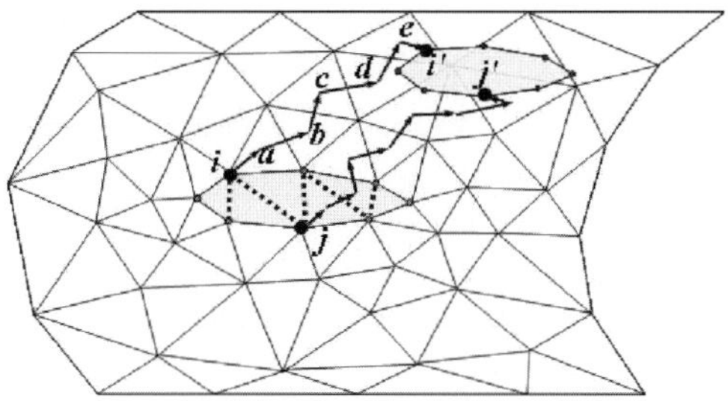

Fig. 2. Neighbor-to-neighbor search from initial to new junction location (Subsidiary cells are added by the edges in dotted lines)

the wing installation position on the fuselage surface. This is the case for the node j on the lower juncture line. In order to circumvent this problem subsidiary cells are added in the junction hole as can be seen in dotted lines. The subsidiary cells also help to interpolate fuselage geometry on the hole efficiently by a surface recovery method.

2.3 Surface Mesh Movement

When the new junction point is found by the neighbor-to-neighbor search method one need to calculate the geodesic distance between the initial and new junction points so that it can be used as the bounddary condition for surface mesh movement. The geodesic distance can be calculated by the following procedure (see Fig. 2):

1) Set the node i (initial junction point) as a starting point, and the node i' (new junction point) as a target point.

2) Select a face cell having node i as one of its three vertices and lying on the way from the starting point to the target point (a in Fig. 2 for instance).

3) Calculate $\overrightarrow{\boldsymbol{V}}$ from the starting point to the target point as a direction vector.

4) Project the direction vector to the cell plane and keep the vector length unchanged as follows:

$$\overrightarrow{\boldsymbol{R}} = |\overrightarrow{\boldsymbol{V}}| \frac{\overrightarrow{\boldsymbol{V}} - (\overrightarrow{\boldsymbol{V}} \cdot \boldsymbol{n})\boldsymbol{n}}{|\overrightarrow{\boldsymbol{V}} - (\overrightarrow{\boldsymbol{V}} \cdot \boldsymbol{n})\boldsymbol{n}|}, \tag{1}$$

where $\overrightarrow{\boldsymbol{R}}$ is the resultant vector, and $\boldsymbol{n}$ is a unit normal vector of the current surface cell.

5) Find an intersection between $\overrightarrow{\boldsymbol{R}}$ and three edges of the cell.

6) If $\overrightarrow{\boldsymbol{R}}$ and the edges have an intersection except the starting point, calculate the distance from the starting point to the intersection point, and add it to the total geodesic distance. And then move to the cell sharing the edge on which the intersection point lies. Set the intersection point as the new starting

point and go to step 3.
7) If $\overrightarrow{\boldsymbol{R}}$ and the edges have no intersection except the starting point, the target point is inside the current cell. In this case calculate the distance from the starting point to the target point, and add it to the total geodesic distance.
8) Repeat step 1–7 for all junction points.

With the junction nodes movemet as boundary conditions, surface mesh points are moved using a spring analogy. A spring force on node i by edge connecting nodes i and j can be represented as follows:

$$\boldsymbol{F}_{ij} = k_{ij} \Delta \boldsymbol{x}_{ij} \tag{2}$$

where $\Delta \boldsymbol{x}_{ij}$ is defined by a direction vector and geodesic distance between nodes i and j as follows.

$$\Delta \boldsymbol{x}_{ij} = G \overrightarrow{\boldsymbol{V}}_{ij} / |\overrightarrow{\boldsymbol{V}}_{ij}|, \tag{3}$$

where G is a geodesic distance, which is a distance on the curved surface in contrast to Euclidian distance. The geodesic distance should be used because the traveled distance by the boundary junction node is the geodesic distance which is larger than the Euclidian distance. If the Euclidian distance is used for the surface mesh movement problem, one would obtain a flipped surface mesh by comparatively larger junction movement than inner surface nodes.

Stiffness coefficient of edge spring k_{ij} is defined as inversely proportional to edge length and distance from the junction line:

$$k_{ij} = \frac{1}{l_{ij} \cdot (ndist_j + 1)^2}, \tag{4}$$

where l_{ij} is the edge length and $ndist_j$ is minimum number of edges from junction nodes to node j. The new node position can be obtained by the force equilibrium for neighboring nodes. Form the above force equilibrium equation, one can obtain the required movement vector $\Delta \boldsymbol{x}$. The surface mesh are then moved along the faceted surface using $\Delta \boldsymbol{x}$. This procedure is similar to the procedure for the geodesic distance calculation of junction point in section 2.2 except that now the movement vector $\Delta \boldsymbol{x}$ is given instead of the target point. The vector $\Delta \boldsymbol{x}$ is rotated to be tangential to the cell and the traveled distance is summed up until the traveled distance reaches the given geodesic distance for the node.

New mesh points are now located on the faceted background mesh surface. In order to reposition the new mesh point on initial surface geometry, we adopted a surface recovery algorithm [5].

3 Results

The present method is applied to ONERA M5 wing-body configuration. Figure 3 shows surface mesh of the configuration and Fig. 4 represents the fuselage with surface mesh cells and additional cells on the junction hole region.

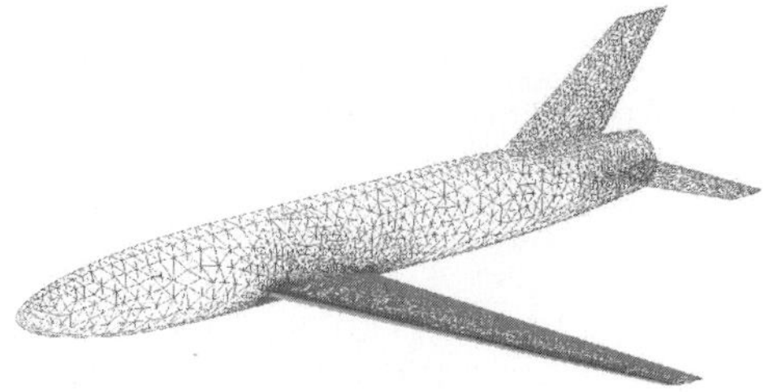

Fig. 3. Surface mesh for ONERA M5 wing-body configuration

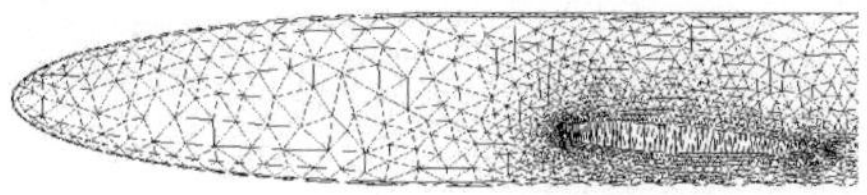

Fig. 4. Fuselage surface mesh with subsidiary cells

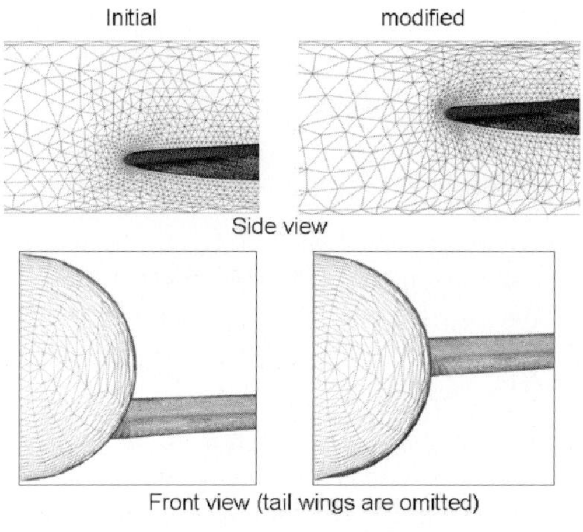

Fig. 5. Vertical movement of ONERA M5 main wing

The first example is a vertical movement of the main wing (Fig. 5) and the second example is slimmed fuselage (Fig. 6).

4 Conclusion

A surface mesh movement method has been developed for aerodynamic design optimization of body-installation juncture. Instead of mapping the surface mesh to a parametric patch mesh, we moved the surface mesh points directly on the curved surface using a spring analogy based on geodesic distance. The initial mesh is used as background mesh. New mesh locations are restricted

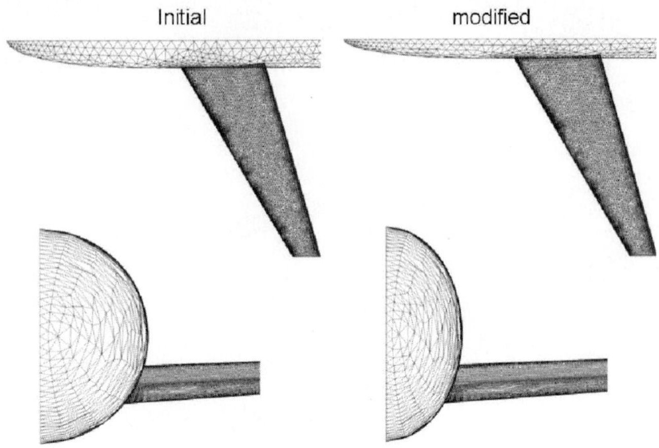

Fig. 6. ONERA M6 fuselage modification

to be on the initial surface by a surface recovery method. Examples verify the efficiency and robustness of the present method for wing-body geometry.

References

1. Steger, J.L., Dougherty, F. C., Benek, J.A.: A Chimera Grid Scheme, ASME Mini-Symposium on Advanced Grid Generation, 1982
2. Nakahashi, K., Togashi, F., and Sharov, D.: Integrated-Boundary Definition Method for Overset Unstructured Grid Approach, *AIAA J.*, Vol.38, No.11, 2000, pp.2077-2084.
3. Jameson, A. Martinelli, L. and Haimes, B.: Aerodynamic Shape Optimization of Complete Aircraft Configurations using Unstructured Grids, AIAA 2004-0533. 2004
4. Murayama, M., Nakahashi, K., and Matsushima, K.: A Robust Method for Unstructured Volume/Surface Mesh Movement, *Trans. Japan Soc. Aero. Space Sci.* Vol.46, No.152, pp.104–112, Aug. 2003
5. Löhner, R.: Regridding Surface Triangulations, *J. Comp. Phys.* 126, pp.1–10, 1996

Formulation and Multigrid Solution of the Discrete Adjoint Problem on Unstructured Meshes

Dimitri J. Mavriplis

Department of Mechanical Engineering
University of Wyoming
Laramie, WY USA
mavripl@uwyo.edu

1 Introduction

The use of the adjoint equations is now well established for design-optimization problems in computational fluid dynamics (CFD) for the Euler and Navier-Stokes equations. In the continuous adjoint method approach, the original continuous governing equations are first linearized, and then discretized, while in the discrete adjoint approach, these two steps are performed in reverse order. Because the continuous approach affords more flexibility at the discretization stage, formulations involving reduced memory and cpu overheads have generally been achieved with this approach [1]. On the other hand, the discrete approach reproduces the exact sensitivity derivatives of the original discretization of the governing equations, which provides a verifiable consistency check on the final gradients produced by the adjoint solution [2]. If $f(w^*, D)$ represents a cost functional to be minimized in an optimization process, where D is a design variable, and w^* represents flow-field values which result from the solution of the discretized flow equations, given in residual form as $R(w, D) = 0$, the sensitivity derivative of f with respect to D is given by:

$$\frac{df}{dD} = \frac{\partial f}{\partial w}\frac{\partial w}{\partial D} + \frac{\partial f}{\partial D} \tag{1}$$

where the change in (converged) flow variables with respect to the design variable is given by differentiating the statement $R(w, D) = 0$, which yields:

$$\left[\frac{\partial R}{\partial w}\right]\frac{\partial w}{\partial D} = -\frac{\partial R}{\partial D} \tag{2}$$

Because each sensitivity derivative requires a new solution of equation (2) with a different right-hand side, a more efficient strategy for large numbers of design variables is to use the adjoint-based sensitivity equation:

$$\frac{df}{dD} = -\Lambda^T\frac{\partial R}{\partial D} + \frac{\partial f}{\partial D} \tag{3}$$

where the co-state variables Λ are given by the solution of the adjoint problem [1, 2]:

$$\left[\frac{\partial R}{\partial w}\right]^T \Lambda = \left(\frac{\partial f}{\partial w}\right)^T \tag{4}$$

The discrete adjoint approach described above benefits from the relative simplicity of implementation. In fact, if a linearization of the discrete equations is already available in the form of a Jacobian for use in a fully implicit scheme, the adjoint equations (including boundary conditions) are obtained by simply transposing the Jacobian matrix. Additionally, exact duality preserving iteration strategies can be constructed for the solution of equations (2) and (4), thus ensuring similar convergence rates between governing and adjoint equations [3, 4]. In order to take advantage of these properties, the formulation and solution efficiency of the discrete adjoint equations should be comparable to that achieved in current state-of-the-art continuous adjoint formulations. In this paper, we propose an efficient discrete-adjoint formulation and demonstrate a duality preserving agglomeration multigrid approach for solving the resulting equations.

2 Formulation of Discrete Adjoint

We seek the solution of the steady-state Euler or Navier-Stokes equations in two dimensions, which we denote as: $\mathbf{R}(w) = 0$. If this system is solved using a fully implicit (Newton) scheme, the iteration strategy may be written as: $\left[\frac{\partial R}{\partial w}\right] \Delta w = -\mathbf{R}(w)$. In this case, the adjoint equations are obtained simply by transposing the Jacobian as: $\left[\frac{\partial R}{\partial w}\right]^T \Lambda = \left(\frac{\partial f}{\partial w}\right)^T$, where f represents the cost function to be minimized, and Λ represents the co-state variables. In practice, storage of the exact Jacobian for a second-order accurate discretization involves an extended stencil and is therefore not practical. However, second order-accurate discretizations are most-often constructed in two nearest-neighbor passes, suggesting a similar approach for the construction of the discrete adjoint. For example, the artificial dissipation approach achieves higher-order accuracy by replacing differences of neighboring flow variables $w_k - w_i$ by differences in undivided Laplacians i.e. $q_i = \sum_{k=1}^{neighbors} w_k - w_i$. Therefore, rewriting the Jacobian using the chain rule,

$$\left[\frac{dR}{dw}\right] = \frac{\partial R}{\partial w} + \frac{\partial R}{\partial q}\frac{\partial q}{\partial w} \tag{5}$$

the discrete adjoint can be obtained as

$$\left[\frac{dR}{dw}\right]^T = \left[\frac{\partial R}{\partial w}\right]^T + \left[\frac{\partial q}{\partial w}\right]^T \left[\frac{\partial R}{\partial q}\right]^T \tag{6}$$

The first term represents contributions from convective terms, and in the case of the Navier-Stokes equations, physical dissipation terms (which operate on a nearest neighbor stencil for triangular elements). The last term is equivalent to contributions from a first-order artificial dissipation (applied to the q variables), thus constituting a symmetric matrix (neglecting variations in the interface coefficient matrix, which is evaluated at the Roe state [5]), while the third term represents the linearization (transposed) of the undivided Laplacian, which is a trivial matrix in this case containing unity entries for each edge of the mesh. Assuming this matrix need not be stored due to its simplicity, the evaluation of $\left[\frac{dR}{dw}\right]^T \Lambda$ makes use of existing (edge-based) flow-solver data-structures, and requires the storage of two matrices: $\left[\frac{\partial R}{\partial w}\right]^T$ which is equivalent to a nearest-neighbor stencil-based sparse matrix, and $\left[\frac{\partial R}{\partial q}\right]^T$ of which only half the matrix need be stored, thus resulting in 1.5 times the storage of a nearest neighbor 1st order Jacobian.

For a MUSCL type reconstruction scheme, the linearization can be cast in a similar form, where the residual $\mathbf{R}(w)$ now depends on the flow variables w, as previously, and on the computed gradients of these variables, which take the place of the q variables. The discrete adjoint equations are obtained in a similar fashion, where $\frac{\partial q}{\partial w}$ represents the linearization of the gradient formulation, which is again a matrix which need not be stored, since the matrix entries are grid metrics which are already stored for the gradient computations in the residual evaluation [6].

3 Duality Preserving Iterative Strategy

From equations (1) and (3), we obtain the duality relation:

$$\frac{\partial f}{\partial w}\frac{\partial w}{\partial D} = -\Lambda^T \frac{\partial R}{\partial D} \tag{7}$$

which must hold for the finally converged values of $\frac{\partial w}{\partial D}$ and Λ. However, iterative solution strategies for equations (2) and (4) which preserve relation (7) at each iteration for partially converged values of $\frac{\partial w}{\partial D}$ and Λ have been demonstrated [3, 4]. If equation (2) is solved as:

$$[P]\left(\frac{\partial w}{\partial D}^{n+1} - \frac{\partial w}{\partial D}^{n}\right) = -\frac{\partial R}{\partial D} - \left[\frac{\partial R}{\partial w}\right]\frac{\partial w}{\partial D}^{n} \tag{8}$$

where $[P]$ is the iteration matrix to be inverted (or preconditioner), the corresponding duality preserving iteration strategy for equation (4) is given by:

$$[P]^T\left(\Lambda^{n+1} - \Lambda^n\right) = \frac{\partial f}{\partial w} - \left[\frac{\partial R}{\partial w}\right]^T \Lambda^n \tag{9}$$

The right-hand side of equation (9) requires the evaluation of the adjoint residual, which is performed as a matrix-vector product, using the two-pass construction given by equation (6). A common strategy is to take $[P]$ as the first-order Jacobian of the flow residual. The $[P]$ matrix is then approximately inverted by splitting it into diagonal blocks $[D]$, and off-diagonal blocks $[O]$, and employing a Jacobi or Gauss-Seidel iteration, following:

$$[D]^T \left(\Lambda^{n+1} - \Lambda^n\right)^{k+1} = \frac{\partial f}{\partial w} - \left[\frac{\partial R}{\partial w}\right]^T \Lambda^n - [O]^T \left(\Lambda^{n+1} - \Lambda^n\right)^k \quad (10)$$

in the case of the adjoint problem, where $[D]^T$ refers to the block-diagonals of the $[P]^T$ matrix, $[O]^T$ refers to the off-diagonal blocks, and k denotes the Jacobi or Gauss-Seidel iteration counter. In order to conserve memory space, the individual blocks of the $[P]^T$ matrix need not be stored separately. Rather, they can be quickly reconstructed from the matrices used to evaluate the full second-order adjoint residual as:

$$\left[\frac{\partial R}{\partial w}\right]^T_{first-order} = \left[\frac{\partial R}{\partial w}\right]^T + \left[\frac{\partial R}{\partial q}\right]^T \quad (11)$$

where the matrices on the right-hand side of equation (11) correspond to those defined in equation (6). In practice, a separate copy of the $[D]^T$ block diagonals are factorized and stored in LU form, but the off-diagonal blocks $[O]^T$ are reconstructed as per equation (11) at each iteration. The Gauss-Seidel scheme provides faster convergence due to the ordered sweep across the mesh which employs latest available information. Additionally, both methods enable multiple sweeps before updating the adjoint residual, which incurs an additional expense, since this is evaluated in a two-pass strategy as described above. These Jacobi and Gauss-Seidel schemes may also be viewed as defect-correction approaches [7].

These iterative schemes are then used as the driver for an agglomeration multigrid strategy to accelerate the solution of the adjoint problem. The use of agglomeration multigrid for solving the Euler and Navier-Stokes equations on unstructured meshes is now well established [8]. A duality preserving agglomeration multigrid algorithm has been developed for solving the discrete adjoint equations on unstructured meshes. To preserve exact duality, the order of operations in the multigrid cycle must be reversed when applied to the adjoint equations [4]. For example, a multigrid cycle which employs one pre-smoothing and no post-smoothing iterations must be replaced by a cycle which uses no pre-smoothing and one post-smoothing iteration for the adjoint problem. Furthermore, the restriction and prolongation operators must be exact transposes of one-another.

4 Results

Figures 1 illustrates the unstructured mesh (7884 points) and the convergence of the discrete adjoint problem (c.f. equation (4)) on this mesh using a duality preserving agglomeration multigrid algorithm, compared with the original multigrid algorithm applied to the linearized flow equations (c.f. equation (2)). The freestream Mach number is 0.6 and the incidence is 1.25 degrees for this test case. Five multigrid levels are used, with 4 smoothing iterations on each grid level (two pre-smoothing, and two post-smoothing), using either a Jacobi or Gauss-Seidel iteration for the smoother. Rapid convergence to machine accuracy is observed and the Gauss-Seidel smoother is seen to deliver better overall multigrid convergence than the Jacobi smoother. In both cases, primal and adjoint problems converge at similar rates, which mirror the convergence of the agglomeration algorithm for the non-linear solution of the flow equations. The third part of the figure shows the difference between the sensitivity derivatives calculated by equation (2) and equation (4) for the lift with respect to vertical displacement of a surface grid point (at x=63% chord on the lower surface), as a function of the iteration number, for the duality preserving algorithm, and for a non-duality preserving algorithm where three smoothing passes were performed in the multigrid coarsening phase, and one pass in the refinement phase, for both primal and dual solvers. In this case, the values computed by the two formulations initially differ by approximately 1.e-02, and the difference decreases to machine zero as the respective problems converge, while the difference remains at machine zero for the duration of the iteration procedure for the duality preserving schemes. For this case, the final value of the sensitivity derivative was 0.2530, which is close to the value of 0.2534 computed using a finite-difference approach. Figure 2 depicts

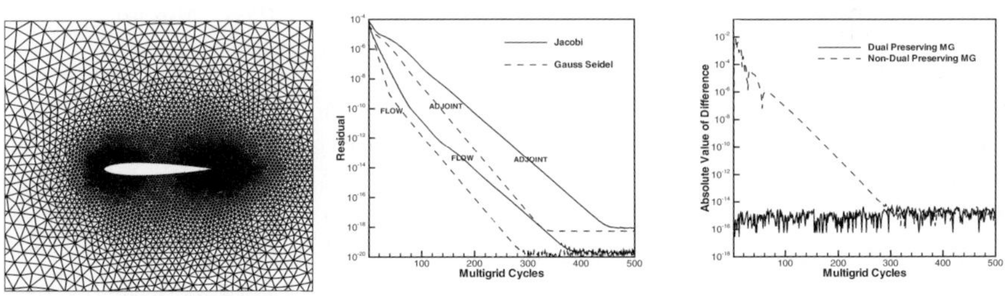

Fig. 1. Unstructured mesh (left), multigrid convergence of primal and adjoint equations (center), convergence of difference between both formulations for dual-preserving and non-preserving multigrid algorithms (right).

a similar experiment for the solution of the laminar Navier-Stokes equations on the same grid, at a Reynolds number of 500 with the same incidence, and a Mach number of 0.8. Using the Jacobi driven multigrid scheme, primal and adjoint problems both display similar rapid convergence rates. The convergence of the computed lift sensitivity derivative (as defined previously) is depicted in the third part of the figure for both formulations using the non-duality preserving multigrid scheme, showing only minor deviations between formulations, and good final agreement with the value predicted by finite-difference. In future work, this technique will be extended to three dimensions and the effect of turbulence modeling will be incorporated.

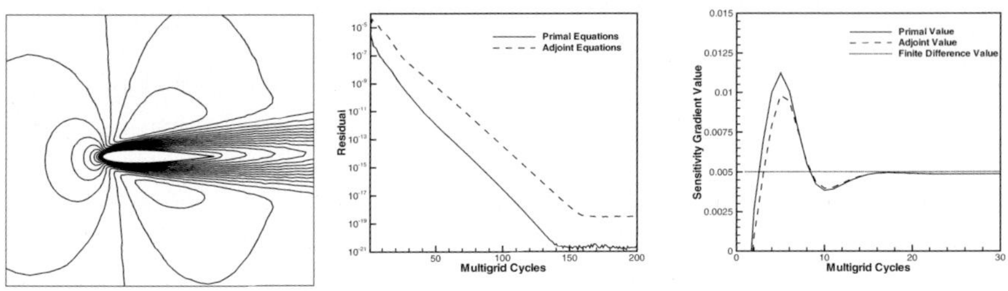

Fig. 2. Viscous flow field (left), convergence of Jacobi multigrid for primal and dual equations (center), convergence of sensitivity derivative for both methods (non-dual preserving algorithm)(right).

References

1. A. Jameson, J. J. Alonso, J. J. Reuther, L. Martinelli, and J. C. Vassberg. Aerodynamic shape optimization techniques based on control theory. AIAA Paper 98-2538, June 1998.
2. E. J. Nielsen and W. K. Anderson. Recent improvements in aerodynamic design optimization on unstructured meshes. AIAA Paper 2001-0596, January 2001.
3. M. B. Giles. Adjoint code developments using the exact discrete approach. AIAA Paper 2001-2596, June 2001.
4. E. J. Nielsen, J. Lu, M. A. Park, and D. L. Darmofal. An exact dual adjoint solution method for turbulent flows on unstructured grids. AIAA Paper 2003-0272, January 2003.
5. C. Hirsch. *Numerical Computation of Internal and External Flows, Volume II: Computational Methods for Inviscid and Viscous Flows.* Wiley, New York, NY, 1988.
6. T. J. Barth and S. W. Linton. An unstructured mesh Newton solver for compressible fluid flow and its parallel implementation. AIAA paper 95-0221, January 1995.
7. D. J. Mavriplis. On convergence acceleration techniques for unstructured meshes. AIAA paper 98-2966, presented at the 29th AIAA Fluid Dynamics Conference, Albuquerque, NM, June 1998.
8. D. J. Mavriplis and S. Pirzadeh. Large-scale parallel unstructured mesh computations for 3D high-lift analysis. *AIAA Journal of Aircraft*, 36(6):987–998, December 1999.

Aerodynamic Design of Gas Turbine Cascades Using Global Optimizers and Artificial Neural Networks

Temesgen Mengistu[1] and Wahid Ghaly[2]

[1] Department of Mechanical and Industrial Engineering, Concordia University, Montreal, Quebec, CANADA. mengistu@me.concordia.ca
[2] Department of Mechanical and Industrial Engineering, Concordia University, Montreal, Quebec, CANADA. ghaly@alcor.concordia.ca

Summary. A simulation-based optimization scheme for gas turbine cascades is developed and is implemented using Computational Fluid Dynamics (CFD) for the flow simulation coupled with Genetic Algorithm (GA), Simulated Annealing (SA) and Artificial Neural Networks (ANN) for the optimization process. It is parallelized and tested for a cascade of compressor blades with the goal of improving their aerodynamic performance over their entire operating range by properly reshaping their profile, which is parameterized using a NURBS approximation. The validity and effectiveness of the developed optimization method is demonstrated on the redesign of a compressor rotor.

1 Introduction

As CFD matured over the last decade and as computing technology has greatly improved and has become more affordable, simulation-based optimization is becoming affordable and more popular than ever. One possible design objective is to minimize flow losses, which can be achieved by properly reshaping the blade profile. Automated aerodynamic design is accomplished by coupling a CFD flow simulation code with numerical optimization methods. As the aerodynamic shape optimization problem is a complex one with possibly many local minima, heuristic/evolutionary global algorithms such as GA and SA are used to ensure reaching the global minimum and have been recently applied in turbomachinery design problems, e.g. Refs. [1, 2, 3]. To reduce computation time, Artificial Neural Networks, ANN can be used to approximate the objective function over the design space [4, 5].

In this work, a simulation-based optimization scheme for gas turbine cascades is developed and is implemented using CFD for the flow simulation coupled with GA, SA and ANN for the optimization process, and is tested for a cascade of compressor blades with the goal of improving their aerodynamic performance over their entire operating range by properly reshaping their profile, which is parameterized using a NURBS approximation. Parallel processing is implemented for the GA and SA. The validity and effectiveness of the developed optimization method is demonstrated on the redesign of a compressor rotor.

2 The Design Methodology

2.1 Problem Definition and Objective Function

A common aerodynamic shape design problem is to achieve a better performance (e.g. maximizing the adiabatic efficiency) of gas turbine cascades by varying the blade camber line, thickness distribution and blade spacing while fixing the inlet and outlet boundary conditions as well as constraining the minimum blade thickness and cross-sectional area for structural reasons. Considering design and off-design points, the objective function can take the following form:

$$F_{obj}(X) = Min \left[C_1 \sum_{i=0}^{n} (1 - \eta_i) + C_2 \sum_{i=0}^{n} \sum_{j=0}^{n} |\eta_j - \eta_i| + PT \right] \tag{1}$$

Where X is the vector of design variables, which include the back pressure as well as the shape parameters that control the blade profile. The latter is defined by the mean camber line and a thickness distribution; each of which was parameterized by a NURBS function with a number of control points ranging from nine to eleven [6]. The appropriate number and value of the NURBS parameters describing the blade profile can be obtained as described in [7]. η is the adiabatic efficiency and PT is a Penalty Term that accounts for the mechanical and aerodynamic constraints imposed on the optimization process. The constraints include the exit flow angle, the spacing to chord ratio, and the stall margin. The summation is carried out over n pre-selected operating points, namely the design point and three off-design points, i.e. $n = 4$. The weights C_1 and C_2 are prescribed by the designer to have the desired optimization objective.

2.2 Flow Simulation Method

The two-dimensional turbulent flow in a linear cascade is simulated using a time-marching cell-vertex finite volume method on an unstructured triangular mesh, where the Reynolds-averaged Navier-Stokes equations are integrated in pseudo-time using an explicit five-stage Runge-Kutta scheme. The discretization of the Euler equations, given in [8], was modified[3] to solve the Reynolds-averaged Navier-Stokes equations where turbulence was modeled using the Baldwin-Lomax model. The flow at the inlet and exit planes,

[3]This modification was implemented by Mr. Daneshkhah, a Ph.D. student working with the second author

which are placed at about one chord upstream and downstream of the cascade, is subsonic. The prescribed quantities at inlet are the total pressure, total temperature and the tangential velocity; at exit, the exit static pressure (referred to as the back pressure) is specified. This set of boundary conditions allows for computing the flow at any point on a given speed line.

3 Optimization Algorithms

3.1 Genetic Algorithm

GA combines selection, crossover, mutation, and elitism operators with the goal of finding the best solution to a problem. In this work, a real coded genetic algorithm is used; selection is based on the Roulette method (rank biased) so that over time, individuals with the higher fitness will have more offspring than those with lower fitness [9]. In crossover operation, two "better" parents will be selected and combined to create two new offspring. During mutation the offspring is changed randomly, to prevent the solutions in a given population from falling into a local optimum. For a generational GA, elitism makes a few (e.g. two copies) of the best performer in the old pool and places them in the new pool, thus ensuring the fittest chromosome survives. GA iterates over several number of generations, as the algorithm runs from generation to generation, the population become saturated with "better" candidate design until it finds the "best" solution. The parallel nature of GA allows for using parallel computation to reduce the overall computation time. Using n-processors in parallel, a speed-up close to n-folds can be achieved. The parallelization is run on an SGI Origin 2000 server using Message Passing Interface (MPI).

3.2 Simulated Annealing

The Simulated Annealing algorithm used in this task for continuous variables is described in Corana *et al.* [10]. The algorithm starts with a user supplied initial guess and a step size and it then employs a random search, which not only accepts decreases in the objective function E, but also some increases. The latter are accepted with a probability $\exp(-\delta E/T)$, advanced by Metropolis *et al.* [11], where δE is the increase in the objective function E and T is given by the cooling schedule. Thus, the control space is randomly examined among local minima of depth less than about T. As T is lowered, the number of such minima qualifying for frequent visits is gradually reduced. The major advantage of this algorithm over other methods is the ability to circumvent local minima.

4 Artificial Neural Networks

ANN is used as a low order approximation of the objective function. The construction of an ANN model involves training and testing. The successful ANN design should predict the behavior of the original objective function for any combination of physically plausible inputs [13].

The ANN training involves determining the type of ANN network and its architecture, and choosing a training strategy. These choices depend on e.g. the presence of local minima, high dimensionality, disparity in input scales in the function being approximated. Back propagation with a single hidden layer with 41 nodes that are connected to the input and output layers through weights [13], was chosen as the ANN network. The training strategy is based on a combination of GA, SA, and a gradient descent method. GA and SA are used in the initial phase of weights estimation, which is followed by the gradient descent method. The ANN-based aerodynamic optimization process involves: 1) selecting the blade geometry candidates, 2) generating the CFD solution for these candidates, 3) training the ANN model (RSM construction), and 4) applying ANN in the aerodynamic optimization. These four tasks take 7%, 75%, 17% and 1% of the model development time. There is an optimum point where the training is stopped although it can still reduce the error so as to avoid memorization behavior [13]. This optimum point is determined by the test data that are set aside for the ANN testing.

5 Results and Discussion

The optimization algorithms, presented in the previous sections, were used to design a NACA 65 subsonic axial compressor [12] for best sustained efficiency. The weights in Eq. (1) are taken as $C_1 = 1$ and $C_2 = 0.5$, respectively. The flow is assumed to be turbulent, at a Reynolds number of $2.45 * 10^5$.

The ANN model was used to construct the response surface using a family of 50 blade profiles, 35 were used in training the ANN model and 15 were used in testing it. The flow was simulated at the design point and at three off-design points; thus the total number of CFD runs amounts to 200 simulations that cover a reasonably large range of the design space, see Figs. 1 and 2. The CFD flow simulations were carried out in parallel (one CFD simulation per node) on four nodes of an SGI 2000 server using Message Passing Interface (MPI), which required about 60 hours of wall-clock time. The ANN training took about five hours on a Pentium IV running at 2.2GHz clock speed and the average error was found to be less than $\pm 5\%$, as seen in Fig. 3.

Figure 2 shows the performance map for the candidates used in building the ANN model as well as the optimal blade; the latter has shown more than 4% gain in adiabatic efficiency. The optimum blade profile shows a better performance in terms of adiabatic efficiency as well as pressure ratio (although the latter was not part of the optimization function) over the full range of operation; this behavior was also observed by Oyama *et al.* [3]. Figure 4 shows that the optimized blade camber has significantly changed near the trailing edge, and the maximum thickness has increased by 3.7% and has slightly moved downstream. The high camber near the trailing edge resulted in a higher flow turning at the cost of an increase in profile loss. Figures 5 and 6, where the efficiency is plotted vs. mass flow rate and vs. pressure ratio, show a 7% improvement in efficiency and about 1% increase in total pressure ratio for the same blade speed and mass flow rate.

6 Conclusion

A flexible and robust simulation-based optimization scheme for gas turbine cascades is developed and is implemented using CFD for the flow simulation coupled with GA, SA and ANN for the optimization process, and is

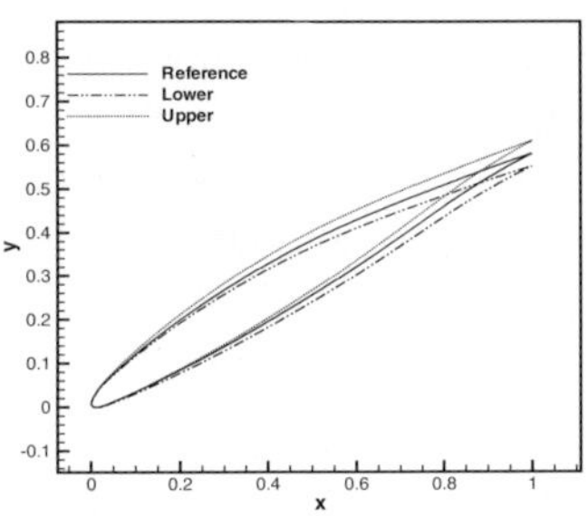

Fig. 1. The range of geometry explored for the design space.

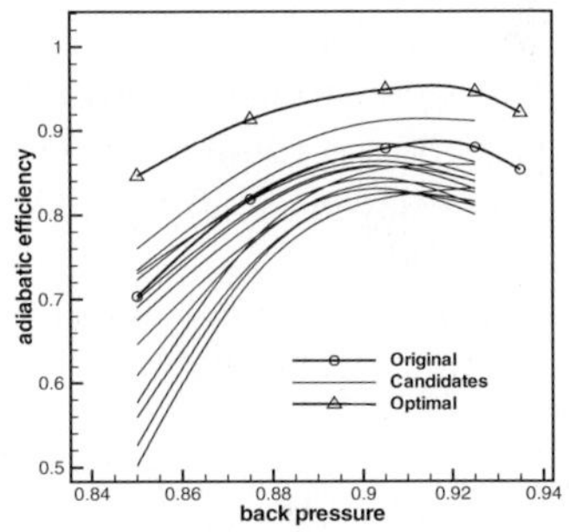

Fig. 2. Speed lines of blade profiles used in training/testing the ANN.

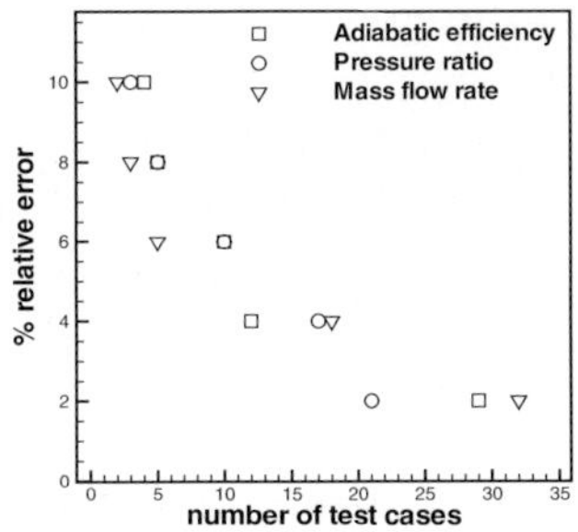

Fig. 3. The ANN model validation.

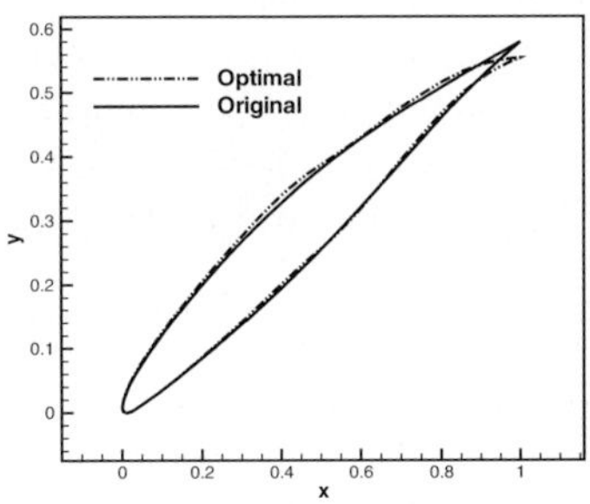

Fig. 4. The original and optimized blade profiles.

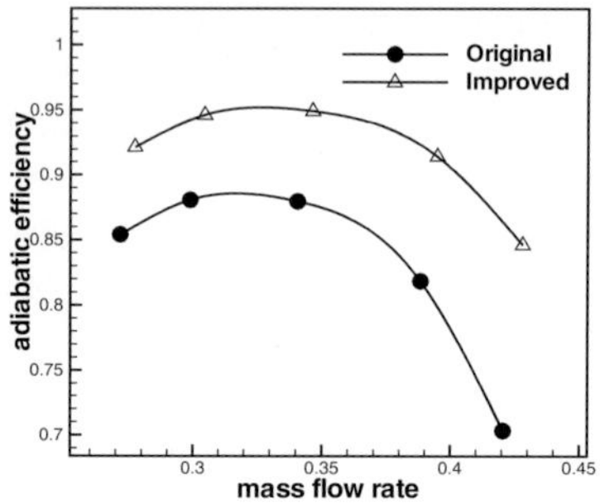

Fig. 5. The performance map for the original and improved designs at the four preselected optimization points.

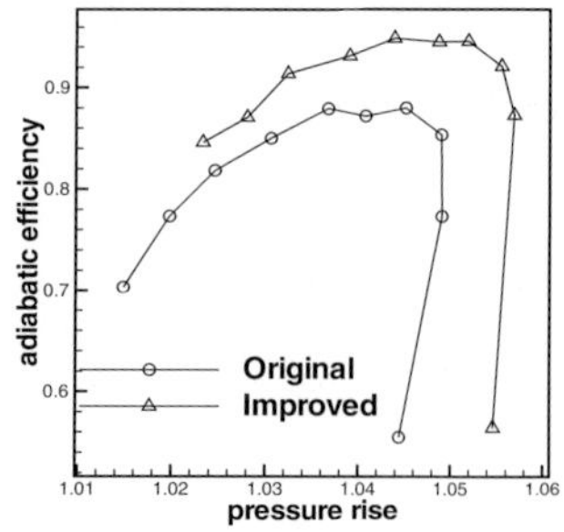

Fig. 6. The efficiency vs pressure ratio for the original and improved designs.

parallelized and tested for a cascade of compressor blades with the goal of improving its aerodynamic performance by properly reshaping the blades. The NURBS functions used to parameterize the blades profile were found to be flexible, smooth and accurate. The ANN reduced the computing time by ten folds while approximating the objective function to within $\pm 5\%$. The validity and effectiveness of the developed optimization method was demonstrated on the redesign of a compressor rotor where the adiabatic efficiency was increased in a sustained manner over the full operating range.

References

1. Dennis, B.H., Dulikravich, G.S., and Han, Z. X., ASME Paper No. 99-GT-441, 1999.
2. Wang, X., and Damodaran, M., AIAA-2000-4847, 2000.
3. Oyama, A., Liou, M.-S., and Obayashi, S., AIAA 2002-5642, 2002.
4. Pierret, S., Demeulenaere, A., et al., AIAA-2000-4879, 2000.
5. Rai, M.M., and Madavan, N., AIAA-98-4928, 1998.
6. Piegl, L., and Tiller, W., "The NURBS Book," Springer, Germany, 1995.
7. Ghaly, W.S., and Mengistu, T.T., J. of Inverse Problems in Engineering, Vol. 11, No. 5, pp. 359-373, October 2003.
8. Ahamdi, M., and Ghaly, W., CASI Journal ,Vol. 44, No. 3, pp. 175-181,1998.
9. Gen,M., Cheng, R., "Genetic Algorithm and Engineering Design," Wiley, 1997.
10. Corana, A., Marchesi, et al., ACM Transaction on Mathematical Software, Vol. 13, No. 3, pp. 262-280, 1987.
11. Metropolis, N., Rosenbluth, A., et al., J. Chem. Phys., Vol. 21, No. 6, pp. 1087-1092, 1953.
12. Emery, J.C., Herrig, L.J., et al., NACA Report 1368, 1957.
13. Schalkoff, R.J., "Artifical Nerual Networks",MIT press and McGraw-Hill, 1997.

An Analysis of Bodies Having Minimum Pressure Drag in Supersonic Flow: Exploring the Nonlinear Domain

Karthik Palaniappan[1] and Antony Jameson[2]

[1] Stanford University, Stanford, CA 94305, USA. karthikp@stanford.edu
[2] Stanford University, Stanford, CA 94305, USA. jameson@baboon.stanford.edu

Summary. One of the most interesting problems in supersonic aerodynamics has been to find profile shapes that have minimum pressure drag, subject to a set of geometric constraints. Considerable work was done in this area in the fifties, all of which relied on a linearized flow model. We revisit this problem using a more sophisticated flow model. It is logical to expect the optimum profile shape to look different. We confirm this, but also note that the differences are very small. We then examine the equations of fluid flow and try to see why the linearized flow model works well for this problem and where the differences come from.

1 Problem Definition

The problem that we study in this paper is that of finding the two-dimensional (airfoils) and axisymmetric (bodies of revolution) profile sections that have minimum pressure drag in supersonic flow. We assume that the flow is inviscid, modelled by the nonlinear Euler equations. We also enforce the constraints that the ends are pointed and the enclosed area/volume is fixed.

2 Results from Classical Theory

Analytical solutions for the problem being studied have been obtained, assuming a linearized flow model. For the $2-d$ case the optimum profile is parabolic.

$$y(x) = 3Ax(1-x)\ , \quad \tau = \frac{3A}{2}\ , \tag{1}$$

where A is the area enclosed and τ is the thickness-chord ratio. The drag coefficient is given by

$$C_d = \frac{12A^2}{\sqrt{M^2-1}}\ . \tag{2}$$

For the axisymmetric case, the profile shapes that solve this problem are the well known Sears-Haack profiles, discovered independently by Sears(1947) and Haack(1947). The derivation of the Sears-Haack profiles is outlined in the book by Ashley and Landahl [1] and also in an article by Carlo Ferrari [2]. The Sears-Haack profile is given by

$$y(x) = \sqrt{\frac{16V}{3\pi^2}} [4x(1-x)]^{\frac{3}{4}} \ , \tau = \sqrt{\frac{64V}{3\pi^2}} \ , \tag{3}$$

where V is the enclosed volume and τ is the fineness ratio. The drag coefficient is given by

$$C_D = 24V \ . \tag{4}$$

As can be observed, these profile shapes have some interesting properties. Firstly, they are unique solutions to the optimization problem. Moreover, they are just a function of the enclosed area/volume and not the Mach Number.

3 Nonlinear Optimization via Control Theory

In this work we apply the adjoint method developed by Jameson and his associates during the last 15 years. ([3, 4, 5, 6]) The aerodynamic shape optimization problem involves minimizing(or maximizing) a given cost function, with parameters that define the shape of the body as the design variables, usually of the form

$$I = \int_{\mathcal{B}_\xi} \mathcal{M}(w, S) \, d\mathcal{B}_\xi \ , \tag{5}$$

where w is the vector of flow state variables and S_{ij} are the coefficients of the Jacobian matrix of the transformation from physical space to computational space. $\mathcal{M}(w, S)$ in our case is just C_p, the pressure coefficient. We also have the constraint that the state variables at the computational points have to satisfy the flow equations, irrespective of the shape of the boundary.

$$\int_{\mathcal{B}} n_i \phi^T f_i(w) d\mathcal{B} = \int_{\mathcal{D}} \frac{\partial \phi^T}{\partial x_i} f_i(w) d\mathcal{D} \ , \tag{6}$$

or, when transformed to computational space

$$\int_{\mathcal{B}_\xi} n_i \phi^T S_{ij} f_j(w) d\mathcal{B}_\xi = \int_{\mathcal{D}_\xi} \frac{\partial \phi^T}{\partial \xi_i} S_{ij} f_j(w) d\mathcal{D}_\xi \ , \tag{7}$$

where ϕ is any arbitrary test function.

Since equation (7) is true for any test function ϕ, we can choose ϕ to be the adjoint variable ψ. We can then add equation (7) to the cost function defined in (5) to form the following augmented cost function.

$$I = \int_{\mathcal{B}_\xi} \mathcal{M}(w, S) \, d\mathcal{B}_\xi + \int_{\mathcal{B}_\xi} n_i \psi^T S_{ij} f_j(w) d\mathcal{B}_\xi - \int_{\mathcal{D}_\xi} \frac{\partial \psi^T}{\partial \xi_i} S_{ij} f_j(w) d\mathcal{D}_\xi \ . \tag{8}$$

We then take a variation of the cost function described in (8)

$$
\begin{aligned}
\delta I= & \int\limits_{\mathcal{B}_\xi} \left(\frac{\partial \mathcal{M}}{\partial w} \delta w + \delta \mathcal{M}_{II} \right) d\mathcal{B}_\xi \\
& + \int_{\mathcal{B}_\xi} n_i \psi^T \left(S_{ij} \frac{\partial f_j(w)}{\partial w} \delta w + \delta S_{ij} f_j(w) \right) d\mathcal{B}_\xi \\
& - \int_{\mathcal{D}_\xi} \frac{\partial \psi^T}{\partial \xi_i} \left(S_{ij} \frac{\partial f_j(w)}{\partial w} \delta w + \delta S_{ij} f_j(w) \right) d\mathcal{D}_\xi \ .
\end{aligned} \tag{9}
$$

We choose ψ such that the variation in the cost function δI does not depend on the variation of the solution δw. ψ is then a solution of the adjoint equations

$$
\begin{aligned}
\frac{\partial \mathcal{M}}{\partial w} &= -n_i \psi^T S_{ij} \frac{\partial f_j(w)}{\partial w} \ , \text{on } \mathcal{B}_\xi \ , \\
\left(S_{ij} \frac{\partial f_j(w)}{\partial w} \right)^T \frac{\partial \psi}{\partial \xi_i} &= 0, \quad \text{on } \mathcal{D}_\xi \ .
\end{aligned} \tag{10}
$$

One thus obtains an expression for the change in the cost function of the form

$$
\delta I = \int_{\mathcal{B}_\xi} \mathcal{G} \delta \mathcal{F} d\mathcal{B}_\xi \ , \tag{11}
$$

where $\mathcal{F}(\xi)$ is a function defining the shape and $\mathcal{G}$ is the required gradient.

The gradient with respect to the design variables is obtained from the solutions to the adjoint equations by a reduced gradient formulation ([5]). This is modified to account for the area/volume constraints. In order to preserve the smoothness of the profile the gradient is smoothed by an implicit smoothing formula. This corresponds to redefining the gradient with respect to a weighted Sobolev inner product ([4]). The optimum is then found by a sequential procedure in which the shape is modified in a descent direction defined by the smoothed gradient at each step, and the flow solution and the gradient are recalculated after each shape change.

4 Results and Discussions

Optimum Profile Shapes

The results of the 2D optimization can be seen in Fig. 1 and the results of the axisymmetric optimization can be seen in Fig. 2. As can be observed, the nonlinear optimum profiles are slightly different from the classical optimum profiles. They have a more rearward point of maximum thickness. The primary difference between a linearized flow model and a nonlinear model is the appearance of shocks at the leading edge in the case of the nonlinear flow model. Reducing the included angle at the leading edge and moving the point of maximum thickness backward is consistent with reducing the magnitude of the leading edge shock. This results in a lower drag and at the same time brings the flow closer to the linear regime.

The difference is hardly noticable for small thickness-chord/fineness ratios. This is just an indicator to the fact that linear theory is a very good approximation for small fineness ratios. Moreover, the nonlinear optimum profiles for axisymmetric flow are a lot closer to their corresponding classical profiles than for 2D flow. This is because of the three-dimensional relieving effect experienced in axisymmetric flow.

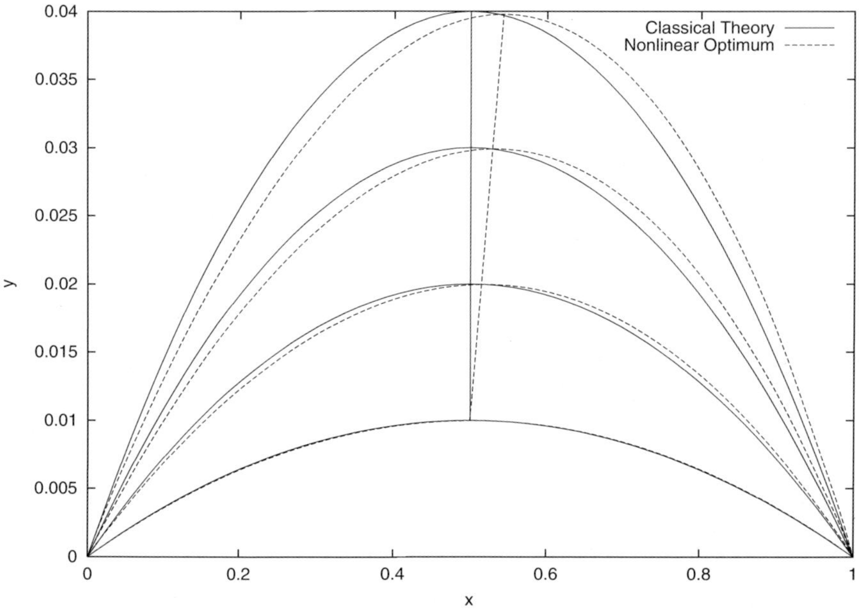

Fig. 1. Classical and Nonlinear Optimum Profiles for 2-D flow

Variation with Mach Number

The optimum profile for 2D flow changes with Mach number. The optimum shape for two Mach numbers is shown in Fig. 3. It is seen that the point of mazimum thickness is more backward for the higher Mach number. This again is consistent with our earlier argument that the main goal of the nonlinear optimization is to reduce the magnitude of the leading edge shock. Such a variation is not observed for axisymmetric optimum profiles. This is due to the fact that the drag coefficient is not sensitive to changes in Mach number in this case. This can also be seen from Eqn. 4.

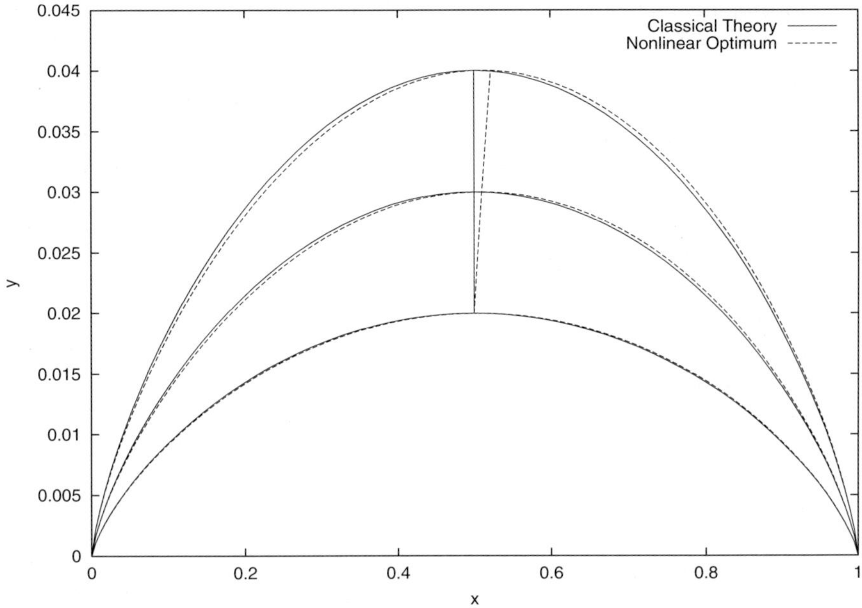

Fig. 2. Classical and Nonlinear Optimum Profiles for Axisymmetric flow

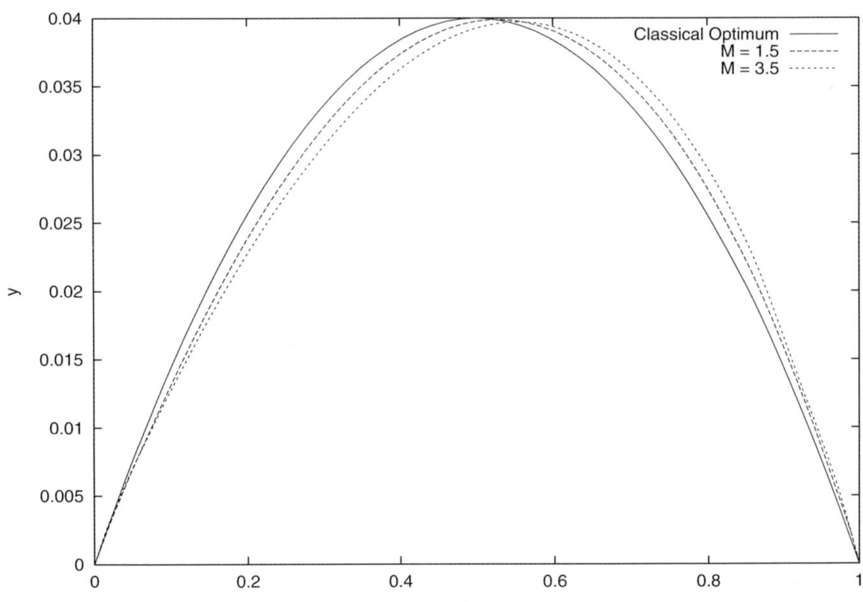

Fig. 3. Variation of 2D Optimum Profiles with Mach Number

Discussion of Results

The main assumptions of linear theory are that the flow is isentropic, irrotational, and the perturbation velocities in the axial and normal directions are very small compared to the free-stream velocity. We observe that these are valid assumptions except in the vicinity of the leading and the trailing edges, where we have stagnation points. Here we have shocks that cause entropy jumps. Moreover, the perturbation velocities are no longer small enough. Thus we expect to see the biggest change at these points. This is found to be true.

5 Conclusions

The minimum pressure drag problem was solved using a nonlinear flow model. It is observed that the optimum shapes thus obtained are slightly different from the classical results. The difference is even smaller for the case of axisymmetric flow. It can be concluded that linearized theory provides a very good approximation of the flow field for the regimes considered and thus the optimum shapes obtained using a nonlinear flow model will be very close to the classical results based on linear theory.

6 Acknowledgements

This work has been carried out under the support of the NASA Langley Research Center, under contract NAG-1-03046.

References

1. Ashley and Landahl: *Aerodynamics of Wings and Bodies*, (Dover Publications Inc New York 1985)
2. Carlo Ferrari: Bodies of Revolution having Minimum Pressure Drag. In: *Theory of Optimum Aerodynamic Shapes*, ed by Angelo Miele (Academic Press Inc, New York 1965) pp 103–124
3. Antony Jameson: Aerodynamic Design via Control Theory, *Journal of Scientific Computing*, pp 3:233–260 (1988)
4. Antony Jameson: A Perspective on Computational Algorithms for Aerodynamic Analysis and Design. In: *Progress in Aerospace Sciences*, (2001) pp 37:197–243
5. Antony Jameson and Sangho Kim: Reduction of the Adjoint Gradient Formula in the Continuous Limit, *AIAA Paper 2003-0040*. (41st AIAA Aerospace Sciences Meeting and Exhibit, Reno, NV, January 2003)
6. Siva Nadarajah: The Discrete Adjoint Approach to Aerodynamic Shape Optimization: PhD Dissertation, Stanford University, Stanford (2003)

Optimum Multidisciplinary and Multi-Objective Wing Design in CFD Using Evolutionary Techniques

L. González[1], E. Whitney[1], K. Srinivas[1], and J. Périaux[2]

[1] School of Aerospace, Mechanical and Mechatronic Engineering, The University of Sydney, NSW 2006, Australia.
[2] Dassault Aviation, 78 Quai Marcel Dassault, 99214 Saint-Cloud, France.

Summary. This paper details some current extensions and applications of hierarchical asynchronous parallel evolutionary algorithms (HAPEA) for multidisciplinary and multi-objective wing design and optimisation problems. In this work the search for the solution takes place in separate hierarchical layers comprising different CFD solvers or resolutions. The performance and advantages of the algorithm are compared to that of a classical EA which would normally use only a single complex model and involve larger computational expense. The formulation and implementation of the algorithm are described and a test case for a multidisciplinary transonic wing design in structures and aerodynamics is presented. The trade–off between the objective functions produced a set of compromise designs represented in an optimal Pareto front. Results indicate that the algorithm is fast and robust for multi-objective and multidisciplinary optimisation problems and as designed produces classical as well as alternative wing configurations.

1 Introduction

Aerospace design and optimisation is a challenge which is subject to many variables and constraints. The use of CFD within aerospace design optimisation is in continuous development and flow solvers have reached a point where a confident application for aerospace design and MDO problems is possible [5]. There is an increasing interest in the application of Evolutionary Algorithms (EAs) for aerospace design problems. The benefits of using EAs are that in contrast to traditional optimisation methods, they require no derivatives or gradients of the objective function, have the capability of finding globally optimum solutions amongst many local optima, are easily executed in parallel and can be adapted to arbitrary solver codes without major modification. Another advantage is that EAs can be adapted suitably for multi-objective problems. Comprehensive studies on the coupling of EAs with CFD for aerospace applications have been developed by different researchers[3] but the continuing challenge in EA research has been to lower the computational expense and the number of function evaluations. Towards this end several approaches have been proposed, these include a combination of variable fidelity models, parallelization strategies and hybridization techniques [1]. In this paper we explore the coupling of one of the strategies; Hierarchical Asynchronous Parallel Evolutionary Algorithms (HAPEA), for three-dimensional problems and multi-objective, multidisciplinary design optimisation. The paper gives an introduction to the features of the method and describes the multi–objective and multidisciplinary optimisation problem, then the paper discusses and presents results for a wing multidisciplinary design optimisation problem.

2 Multidisciplinary and Multi-Objective Evolutionary Algorithms

In aerospace engineering design and optimisation the designer is usually presented with a problem which involves considering not only one objective but numerous objectives and multi-physics. Hence a systematic approach, which is regarded as Multidisciplinary Design Optimisation (MDO) that accounts for multi-physics and coupling between the variables is required. Wing design is an example of a multi-objective MDO problem as there is a strong interaction between aerodynamics and structures. Wing optimisation methods have relied on traditional techniques that use aggregating functions for the solution. But recent developments have explored the benefits of using EAs and Multi-objective EAs (MOEAs) and have shown to be robust and efficient to find global solutions [1].

EAs are based on Darwinian evolution, whereby populations of individuals evolve over a search space and adapt to the environment by the use of different mechanisms such as mutation, crossover and selection. An attractive feature of EAs is that they evaluate multiple populations of points and are capable of finding a number of solutions in a Pareto set. EAs have been successfully applied to different aeronautical design problems and there have been various efforts to explore the capabilities of EAs for MDO, aircraft, wing, aerofoil and rotor blade design [3]. Deb [1] describes and analyses some problem features that might cause a multi-objective genetic algorithm to fail to converge to the true Pareto front. The algorithm used in this research has been tested for some of these cases and has been proven to be robust and efficient in finding optimal Pareto fronts [6]. Another concern associated when solving an optimisation problem involving CFD problems is that the time for the residual of the solution to converge to a specified level (either machine zero or an arbitrarily selected higher value) can vary over a significant range. The traditional genetic algorithm and evolution strategy both use this approach. A problem with generational models is that they create an unnecessary bottleneck when used on parallel computers. If the population size is approximately equal to the number of processors, and most of the candidate offsprings sent for solution can be successfully evaluated, then some processors will complete their task quickly with the remainder taking more time. With a generational approach, those processors that have already completed their solutions will remain idle until all processors have completed their work.

3 Asynchronous Evaluation and Hierarchical Topology

The approach used in by the algorithm, is to ignore any concept of a generation based solution. This is similar other non-generational approaches, however the selection operator is quite different, as it couples one-by-one (steady-state) function evaluation with a direct multi-objective fitness criterion. Whilst a parent population exists, offspring are not sent as a complete 'block' to the parallel slaves for solution. Instead one candidate is generated at a time, and is sent to any idle processor where it is evaluated at its own speed. When candidates have been evaluated, they are returned to the optimiser and either accepted for insertion into the main population or rejected. A hierarchical population topology, when integrated into an evolution algorithm, means that a number of separate populations are established in a hierarchical layout to solve the given problem, rather than a single 'cure-all' type population layout. This method was first proposed by Sefrioui [4]. The bottom layer can be entirely devoted to exploration, the intermediate layer is a compromise between exploitation and exploration and the top layer concentrates on refining solutions. In this work we also use the benefits of parallel computing wherein the master carries on the optimisation process while remote nodes compute the solver code.

4 Formulation of the MDO Problem

We consider a multidisciplinary, multi-objective conceptual design optimisation of a swept forward wing design for an Unmanned Aerial Vehicle (UAV). The two objectives are minimisation of wave drag and wing weight. The cruise Mach number and altitude are 0.69 and 10000 ft. The wing area is set to 2.94 m^2 and the corresponding C_L is fixed at 0.19. For the solution we initially compute the pressure distribution over the wing using a potential flow solver to obtain the wing aerodynamics characteristics which include the span-wise pressure distribution, C_L and total drag coefficients C_{D_w}. The lift distribution is replaced by concentrated loads and the spar cap area is calculated to resist the bending moment. The weight is then approximated as the sum of the span-wise cap weight. The interaction between the aerodynamic pressure distribution and the structural deflections is ignored.

4.1 Design Variables and Constraints.

The wing geometry is represented by three aerofoil sections and nine plan-form variables. Each aerofoil section is represented by the combination of a

Table 1. Design variables for a UAV wing design.

Description	Variable	*Lower bounds*	*Upper bounds*
Wing Aspect Ratio	AR	3.50	7.00
Break to root Taper	λ_{rb}	0.65	0.80
Break to tip Taper	λ_{bt}	0.20	0.45
Wing 1/4 Chord inboard Sweep, deg	Λ_{inb}	10.00	20.00
Wing 1/4 Chord outboard Sweep, deg	Λ_{outb}	-20.00	0.00
Twist at Root, deg	ϵ_r	0.00	3.00
Twist at Break, deg	ϵ_b	-1.00	0.00
Twist at Tip, deg	ϵ_t	-1.00	0.00
Break Location, %	b_l	0.20	0.35

mean line and thickness distribution, which is very common in classical aerodynamics. Both the lines are represented by a Bézier curves. In this case for each aerofoil we take six free control points on the mean line and ten free control points on the thickness distribution. The wing planform design variables and its upper and lower bounds are represented in Table 1. In total fifty nine design variables are used for the optimisation. Constraints are imposed on minimum thickness ($t/c\ \geq 14\ \%root$), $\geq 12\ \%\ intermediate$, $\geq 11\%\ tip$) and position of maximum thickness $\left(20\% \leq x_{t/c} \leq 55\%\right)$. If any of these constraints is violated both fitness are linearly penalised to ensure an unbiased Pareto set.

4.2 Fitness Functions

The two fitness functions to be optimised are defined as:

$$f_1 = min(Cd_w) \quad , \ f_2 = min\left(TotalSparCap_{weight}\right) \tag{1}$$

4.3 Aerodynamics and Weight Analysis

The aerodynamic characteristcs of the wing configurations are evaluated using a three dimensional full potential wing analysis code which uses sheared parabolic coordinates and accounts for wave drag [2]. The lift can be satisfied by performing an extra two function evaluations by varying the angle of attack at the wing root and assuming a linear variation of the lift coefficient. The lift distribution is replaced by concentrated loads. The wing weight is estimated from the wing spar cap area designed to resist the bending moment. The local stress has to be less than the ultimate shear stress in this case for Aluminum Alloy 2024-T6 $\sigma < \sigma_{ult}$.

4.4 Implementation

The solution to this problem has been implemented using two approaches, the first approach uses a traditional EA with a single population model and

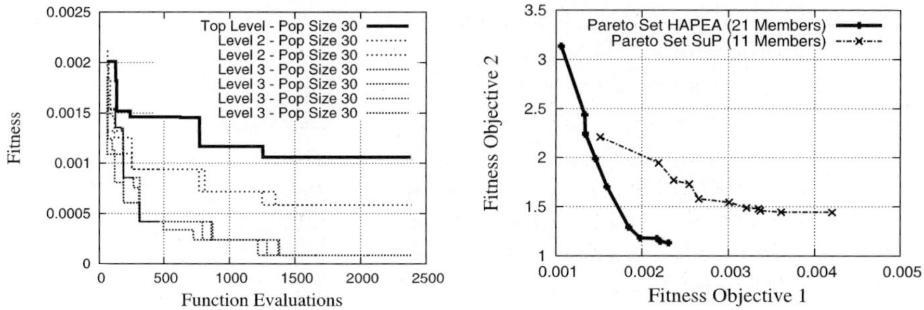

Fig. 1. a) Convergence history for objective one and b) Pareto fronts after 2000 function evaluations and

computational grid of $96 \times 12 \times 16$. The second approach uses a hierarchical topology of CFD resolutions with a computational grid of $96 \times 12 \times 16$, $72 \times 9 \times 12$ and $48 \times 6 \times 8$ on the top, intermediate and bottom layers respectively. Six machines were used in the calculation. The population size for this problem was set to 30 in all levels.

4.5 Results

The algorithm was run five times for 2000 function evaluations and took in average six hours to compute. Figure 1a shows convergence history for objective one and Figure 1b shows the Pareto fronts obtained by using the two approaches. It can be seen how the optimisation technique gives a uniformly distributed front in both cases. By inspection we can see that the use of a multi-fidelity approach gives an overall lower front as compared to a single model approach. A compromise design, Pareto member ten (PM10), taken from the middle of the Pareto set is taken for evaluation. Figure 2a shows the root, break and tip aerofoils and Figure 2b shows the wing geometry and pressure coefficient distribution.

5 Conclusion

In this paper a robust hierarchical evolution algorithm has been studied and applied to realistic practical multi criteria and multidisciplinary design problems. The benefits of using a hierarchical topology on a MDO problem have been investigated. The results show a computational gain on using a hierarchical topology of fidelity models as compared to a single model during the optimisation. The algorithm was capable of identifying the trade-off between the multi-physics involved and provide classical aerodynamic shapes as well as alternative configurations from which the designer can choose. Research is now underway to include higher fidelity models and a third objective for Radar Cross Section (RCS) minimisation.

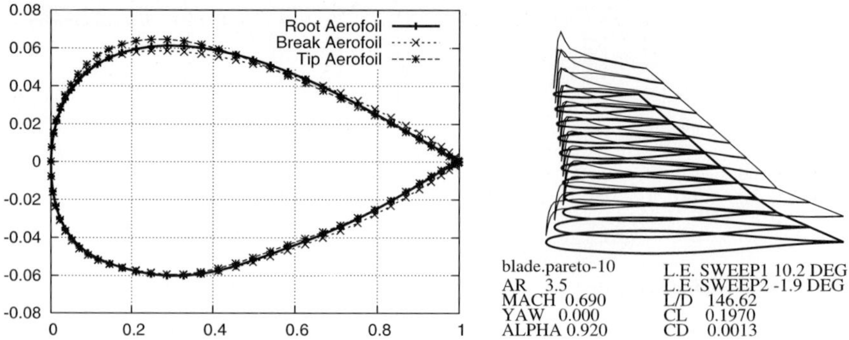

Fig. 2. a) Aerofoil sections and b) Wing span pressure coefficient distribution for Pareto member ten.

Acknowledgment

The authors would like to thank Dr M. Sefrioui at Dassault Aviation for contributions on the hierarcical topology model and NASA Langley for providing the use of the solver code. We also acknowledge Professor Steve Armfield and Dr Patrick Morgan at The University of Sydney for providing the facilities on using the cluster of computers.

References

1. K. Deb. *Multi-Objective Optimization Using Evolutionary Algorithms.* Wiley, 2003.
2. A. Jameson, D. Caughey, P. Newman, and R. Davis. A Brief Description of the Jameson Caughey NYU Transonic Swept-Wing Computer Program FLO22. Technical Report NASA TM X-73996, NASA/American Institute of Aeronotics and Astronautics, 1976.
3. A. Oyama, M. Liou, and S. Obayashi. Transonic axial-flow Blade Shape Optimization Using Evolutionary Algorithm and Three-Dimensional Navier- Stoke Solver. *AIAA Paper* 2002-5642, April 2002.
4. M. Sefrioui and J. Périaux. A Hierarchical Genetic Algorithm Using Multiple Models for Optimization. In M. Schoenauer, K. Deb, G. Rudolph, X. Yao, E. Lutton, J.J. Merelo, and H-P. Schwefel, editors, *Parallel Problem Solving from Nature, PPSN VI*, pages 879–888. Springer, 2000.
5. Z. Thomas and A. Green. Multidisciplinary Design Optimization Techniques: Implications and Opportunities for Fluid Dynamics Research. *AIAA Paper* 1999-3798, June 1999.
6. E.J. Whitney, M. Sefrioui, K. Srinivas, and J. Périaux. Advances in Hierarchical Parallel evolutionary Algorithms for Aerodynamic Shape Optimisation. *JSME International Journal*, 45(1):23–28, 2002.

Advances in Aerodynamic Shape Optimization

Antony Jameson

Stanford University, Stanford, CA 94305-4035
jameson@baboon.stanford.edu

1 Introduction

The focus of CFD applications has shifted to aerodynamic design. This shift has been mainly motivated by the availability of high performance computing platforms and by the development of new and efficient analysis and design algorithms. In particular automatic design procedures, which use CFD combined with gradient-based optimization techniques, have had a significant impact on the design process by removing difficulties in the decision making process faced by the aerodynamicist.

A fast way of calculating the accurate gradient information is essential since the gradient calculation can be the most time consuming portion of the design algorithm. The computational cost of gradient calculation can be dramatically reduced by the control theory approach since the computational expense incurred in the calculation of the complete gradient is effectively *independent* of the number of design variables. The foundation of control theory for systems governed by partial differential equations was laid by J.L. Lions [1]. The method was first used for aerodynamic design by Jameson in 1988 [2, 3]. Since then, the method has even been successfully used for the aerodynamic design of complete aircraft configurations [4].

In the present work a continuous adjoint formulation has been used to derive the adjoint system of equations, in which the adjoint equations are derived directly from the governing equations and then discretized. This approach has the advantage over the discrete adjoint formulation in that the resulting adjoint equations are independent of the form of discretized flow equations. The adjoint system of equations has a similar form to the governing equations of the flow, and hence the numerical methods developed for the flow equations [5, 6, 7] can be reused for the adjoint equations. Moreover, the gradient can be derived directly from the adjoint solution and the surface motion, independent of the mesh modification.

In order to accelerate the convergence of the descent process the gradient is then smoothed implicitly via a second order differential equation. This is equivalent to redefining the gradient in a Sobolve space. The resulting procedure is very efficient, often yielding the optimum in 10-20 design cycles.

Recently wing planform parameters have been included as design variables and the Aerospace Computing Laboratory at Stanford University has successfully designed a wing which produces a specified lift with minimum drag, while meeting other criteria such as low structure weight, sufficient fuel volume, and stability and control [8]. Based on the promising results from our wing planform optimization strategy applied to inviscid flow and from our viscous aerodynamic design techniques [9, 10], we are now applying wing shape and planform optimization methods to viscous flow in order to take into account the viscous effects such as shock/boundary layer interaction, flow separation, and skin friction and eventually produce more realistic designs [11].

Additionally, the design method, which is greatly accelerated by the use of control theory, has been further enhanced by the use of a new continuous adjoint method, which reduce the volume integral part of the adjoint gradient formula to a surface integral [12], thus eliminating the dependence of the gradient formulas on the mesh perturbation. The computational savings in the gradient calculation are particularly significant for three-dimensional aerodynamic shape optimization problems on general unstructured and overset meshes. The use of unstructured grid techniques hold considerable promise for aerodynamic design by facilitating the treatment of complex configurations without incurring a prohibitive cost and bottleneck in mesh generation. The computational feasibility of using unstructured meshes for design is essentially enabled by the use of the continuous adjoint approach and the reduced gradient formulas [13]. Representative calculations are displayed in figures 2 through 6.

2 Adjoint and Gradient Formulations for the Equations of Transonic Flow

The adjoint method may be applied directly to the partial differential equations to derive a continuous adjoint equation, which must then be discretized to obtain a numerical solution. Alternatively one may derive a discrete adjoint equation directly after first discretizing the flow equations. In this work the first procedure has been adopted because it allows more flexibility in the formulation of the gradient.

The procedure is illustrated here for the Euler equations. These are represented in transformed coordinates ξ_i on a fixed computational domain. Let

$$S = JK^{-1}$$

where

$$K_{ij} = \frac{\partial x_i}{\partial \xi_j}, J = det(K)$$

Then the transformed equations are

$$\frac{\partial F_i}{\partial \xi_i} = \frac{\partial (S_{ij} f_j)}{\partial \xi_i} = 0$$

As an example, consider the case of an inverse problem where one wishes to find the shape which brings the pressure as close as possible to the specified target pressure, p_t. Hence we try to minimize the cost function

$$\mathcal{I} = \frac{1}{2}\int_{\mathcal{B}} (p - p_t)^2 dS$$

over the design surface $\mathcal{B}$, which for convenience is assumed to be the surface $\psi_2 = 0$. Now a shape modification induces a change δp in the pressure and consequently

$$\delta\mathcal{I} = \int_{\mathcal{B}} (p - p_t)\delta p dS + \frac{1}{2}\int_{\mathcal{B}} (p - p_t)^2 d\delta S$$

Also the change in the solution is given by

$$\frac{\partial}{\partial\psi_i}\left(\delta F_i(w)\right) = 0$$

Here the flux changes are

$$\delta F_i = \delta S_{ij} f_j + C_i \delta w$$

where

$$C_i = S_{ij}\frac{\partial f_j}{\partial w}$$

Consequently one can augment the cost variation by

$$\int_{\mathcal{D}} \psi^T \frac{\partial \delta F_i}{\partial \xi_i} d\xi = \int_{\mathcal{B}} n_i \psi^T \delta F_i d\xi_{\mathcal{B}} - \int_{\mathcal{D}} \frac{\partial \psi^T}{\partial \xi}\delta F_i d\xi$$

Now choose ψ to satisfy the adjoint equation

$$C_i^T \frac{\partial \psi}{\partial \xi_i} = 0$$

with the boundary condition

$$\psi_2 \eta_x + \psi_3 \eta_y + \psi_4 \eta_z = p - p_t$$

where η_x, η_y, η_z are the components of the surface normal. Then the boundary integrals involving δp and the field integral involving δw are eliminated and the gradient is reduced to

$$\frac{1}{2}\int_{\mathcal{B}} (p - p_t)^2 d\delta S - \int\int_{\mathcal{B}} \left(\delta S_{21}\psi_2 + \delta S_{22}\psi_3 + \delta S_{23}\psi_4\right) p d\xi_1 d\xi_3$$
$$- \int_{\mathcal{D}} \frac{\partial \psi^T}{\partial \xi}(\delta S_{ij} f_j) d\xi$$

where typically the first term is negligible and can be dropped. Other cost functions, such as the drag coefficient, lead to different adjoint boundary conditions. When a mesh perturbation procedure is defined the perturbations δS_{ij} can be directly related to the surface perturbation. Finally one obtains the cost variation as an inner product

$$\delta I = (\mathcal{G}, \delta\mathcal{F}) = \int\int \mathcal{G}\delta\mathcal{F} d\xi_1 d\xi_2$$

where $\mathcal{G}$ is the pointwise gradient, and $\mathcal{F}$ defines the surface shape.

3 Optimization Procedure

Another key issue for successful implementation of the continuous adjoint method is the choice of an appropriate inner product for the definition of the gradient. It turns out that there is an enormous benefit from the use of a modified Sobolev gradient, which enables the generation of a sequence of smooth shapes.

The gradient $\mathcal{G}$ is generally in a lower smoothness class than the initial shape $\mathcal{F}$. Then a sequence of steps

$$\delta\mathcal{F} = -\lambda\mathcal{G}$$

progressively reduces the smoothness, leading to instability.

In order to prevent this we can introduce a weighted Sobolev inner product [14] of the form

$$\langle u, v\rangle = \int (uv + \epsilon u' v')dx$$

in one dimension, where the parameter ϵ controls the smoothness. Correspondingly we define a modified gradient $\bar{\mathcal{G}}$ such that

$$\delta I =< \bar{\mathcal{G}}, \delta\mathcal{F} > .$$

In the one dimensional case $\bar{\mathcal{G}}$ is obtained by solving the smoothing equation

$$\bar{\mathcal{G}} - \frac{\partial}{\partial\xi_1}\epsilon\frac{\partial}{\partial\xi_1}\bar{\mathcal{G}} = \mathcal{G}. \tag{1}$$

In the multi-dimensional case the smoothing is applied in product form. Finally we set

$$\delta\mathcal{F} = -\lambda\bar{\mathcal{G}} \tag{2}$$

with the result that

$$\delta I = -\lambda < \bar{\mathcal{G}}, \bar{\mathcal{G}} > \quad < 0,$$

unless $\bar{\mathcal{G}} = 0$, and correspondingly $\mathcal{G} = 0$.

The implicit smoothing procedure acts as a preconditioner which allows the use of much larger steps for the search procedure and leads to a large reduction in the number of design iterations needed for convergence.

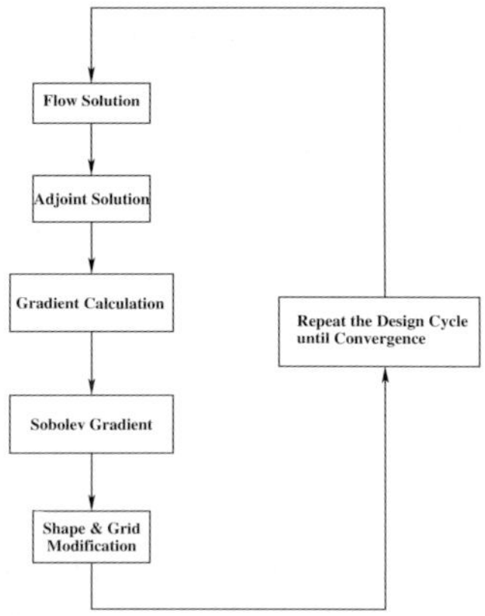

Fig. 1. Design cycle

3.1 Outline of the design procedure

The design procedure can finally be summarized as follows:

1. Solve the flow equations for ρ, u_1, u_2, u_3, p.
2. Solve the adjoint equations for ψ subject to appropriate boundary conditions.
3. Evaluate $\mathcal{G}$ and calculate the corresponding Sobolev gradient $\bar{\mathcal{G}}$.
4. Project $\bar{\mathcal{G}}$ into an allowable subspace that satisfies any geometric constraints.
5. Update the shape based on the direction of steepest descent.
6. Return to 1 until convergence is reached.

Practical implementation of the design method relies heavily upon fast and accurate solvers for both the state (w) and co-state (ψ) systems. The result obtained in Section 4 have been obtained using well-validated software for the solution of the Euler and Navier-Stokes equations developed over the course of many years [5, 15, 16]. For inverse design the lift is fixed by the target pressure. In drag minimization it is also appropriate to fix the lift coefficient, because the induced drag is a major fraction of the total drag, and this could be reduced simply by reducing the lift. Therefore the angle of attack is adjusted during each flow solution to force a specified lift coefficient to be attained, and the influence of variations of the angle of attack is included in the calculation of the gradient. The vortex drag also depends on the span loading, which may be constrained by other considerations such as structural loading or buffet onset. Consequently, the option is provided to force the span loading by adjusting the twist distribution as well as the angle of attack during the flow solution.

The design procedure has been implemented in the computer programs Syn88 and Syn107 for three-dimensional wing-fuselage design using the Euler and RANS equations respectively on structured meshes. Both codes include automatic mesh generation. The method has also been implemented for unstructured meshes in the computer program Synplane, which treats complete aircraft.

4 Case Studies

4.1 B747 Euler planform result

The shape changes in the section needed to improve the transonic wing design are quite small. However, in order to obtain a true optimum design larger scale changes such as changes in the wing planform (sweepback, span, chord, section thickness, and taper) should be considered. Because these directly affect the structure weight, a meaningful result can only be obtained by considering a cost function that accounts for both the aerodynamic characteristics and the weight.

In references [8, 11, 17] the cost function is defined as

$$I = \alpha_1 C_D + \alpha_2 \frac{1}{2} \int_{\mathcal{B}} (p - p_d)^2 dS + \alpha_3 C_W,$$

where $C_W \equiv \frac{W}{q_\infty S_{ref}}$ is a dimensionless measure of the wing weight, which can be estimated either from statistical formulas, or from a simple analysis of a representative structure, allowing for failure modes such as panel buckling. The coefficient α_2 is introduced to provide the designer some control over the pressure distribution, while the relative importance of drag and weight are represented by the coefficients α_1 and α_3. By varying these it is possible to calculate the Pareto front of designs which have the least weight for a given drag coefficient, or the least drag coefficient for a given weight. The relative importance of these can be estimated from the Breguet range equation;

$$\begin{aligned} \frac{\delta R}{R} &= - \left(\frac{\delta C_D}{C_D} + \frac{1}{log \frac{W_1}{W_2}} \frac{\delta W_2}{W_2} \right) \\ &= - \left(\frac{\delta C_D}{C_D} + \frac{1}{log \frac{W_1}{W_2}} \frac{\delta C_W}{\frac{W_2}{q_\infty S_{ref}}} \right). \end{aligned}$$

Figure 2 shows the Pareto front obtained from a study of the Boeing 747 wing [17], in which the flow was modeled by the Euler equations. The wing planform and section were varied simultaneously, with the planform defined by six parameters; sweepback, span, the chord at three span stations, and wing thickness. The weight was estimated from an analysis of the section thickness required in the structural box. The figure also shows the point on the Pareto front when $\frac{\alpha_3}{\alpha_1}$ is chosen such that the range of the aircraft is

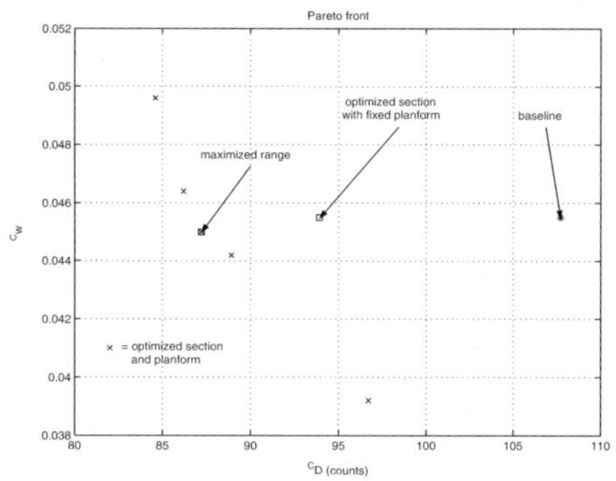

Fig. 2. Pareto front of section and planform modifications

maximized. The optimum wing, as illustrated in figure 3, has a larger span, a lower sweep angle, and a thicker wing section in the inboard part of the wing. The increase in span leads to a reduction in the induced drag, while the section shape changes keep the shock drag low. At the same time the lower sweep angle and thicker wing section reduce the structural weight. Overall, the optimum wing improves both aerodynamic performance and structural weight. The drag coefficient is reduced from 108 counts to 87 counts (19%), while the weight factor C_W is reduced from 455 counts to 450 counts (1%).

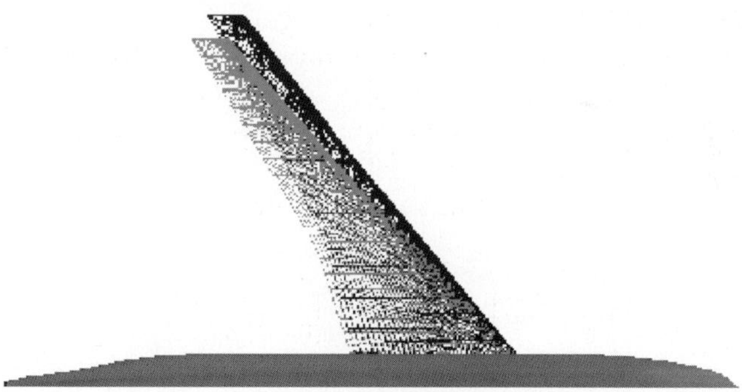

Fig. 3. Superposition of the baseline (green) and the optimized section-and-planform (blue) geometries of Boeing 747. The redesigned geometry has a longer span, a lower sweep angle, and thicker wing sections, improving both aerodynamic and structural performances. The optimization is performed at Mach .87 and fixed C_L .42, where $\frac{\alpha_3}{\alpha_1}$ is chosen to maximize the range of the aircraft.

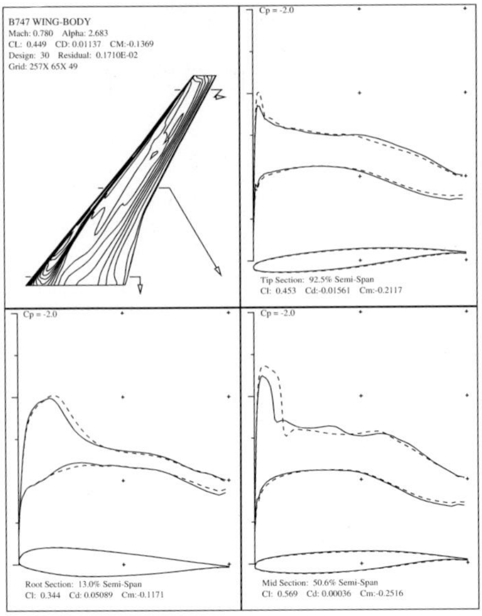

(a) Mach .78

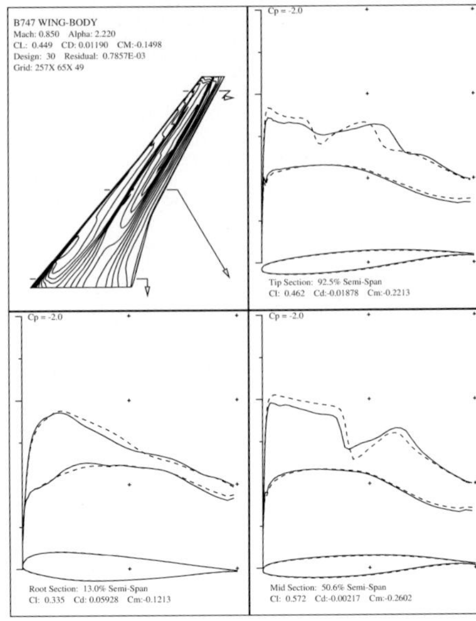

(b) Mach .85

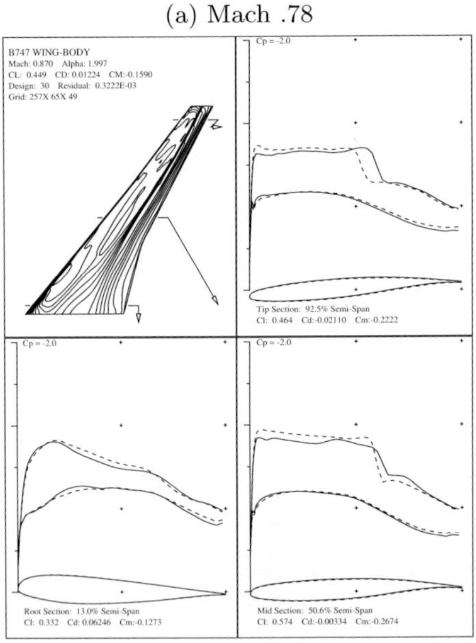

(c) Mach .87

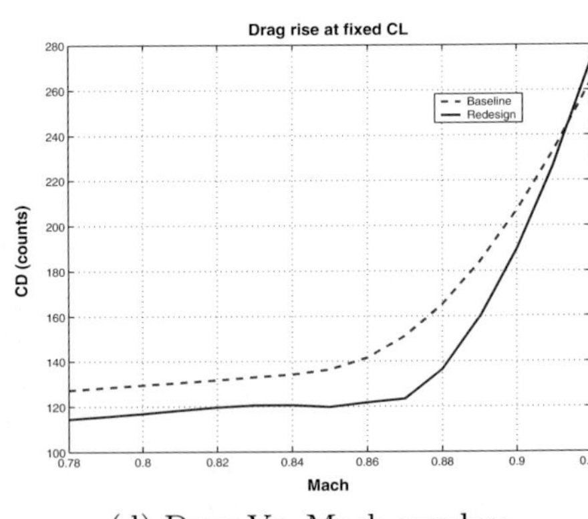

(d) Drag Vs. Mach number

Fig. 4. (a)-(c): Super B747 at Mach .78, .85, and .87 respectively. Dash line represents shape and pressure distribution of the initial configuration. Solid line represents those of the redesigned configuration. **(d)**: Drag Vs. Mach number of Super B747.

Table 1. Comparison between Baseline B747 and Super B747 at Mach .86

	C_L	C_D counts	C_W counts
Boeing 747	.45	141.3 (107.0 pressure, 34.3 viscous)	499 (82,550 lbs)
Super B747	.50	141.9 (104.8 pressure, 37.1 viscous)	427 (70,620 lbs)

4.2 Super B747

In order to explore the limits of attainable performance the B747 wing has been replaced by a completely new wing to produce a "Super B747". An initial design was created by blending supercritical wing sections obtained from other optimizations to the optimum planform which was found in the planform study described in the previous section. Then the RANS optimization code Syn107 was used to obtain minimize drag over 3 design points at Mach .78, .85, and .87, shown in figures 4 (a)-(c) with a fixed lift coefficient of .45 for the exposed wing, corresponding to a lift coefficient of about .52 when the fuselage lift is included. Because the new wing sections are significantly thicker, the new wing is estimated to be 12,000 pounds lighter than the baseline B747 wing as shown in table 1. At the same time the drag is reduced over the entire range from Mach .78 to .90 with a maximum benefit of 25 counts at Mach .87, as shown in figure 4 (d). Figure 5 and table 2 display the lift-drag polar at Mach .86. The drag coefficient of the Super B747 is 142 counts at a lift coefficient of .5, whereas the baseline B747 has the same drag at a lift coefficient of .45. This represents improvement in L/D of more than

Table 2. Comparison of drag polar; B747 Vs. Super B747

Boeing 747		Super B747	
C_L	C_D	C_L	C_D
0.0045	94.3970	0.0009	76.9489
0.0500	82.2739	0.0505	67.8010
0.1000	74.6195	0.1005	64.6147
0.1501	72.1087	0.1506	65.5073
0.2002	73.9661	0.2006	69.4840
0.2503	79.6424	0.2507	76.0041
0.3005	88.7551	0.3008	84.9889
0.3507	101.5293	0.3509	95.6117
0.4009	118.0487	0.4010	106.9625
0.4512	**141.2927**	0.4510	121.7183
0.5014	177.0959	**0.5010**	**141.8675**
0.5516	228.1786	0.5512	175.2569
0.6016	298.0458	0.6014	222.5459

(C_D in counts)

Note equal drag of the baseline B747 at C_L .45 and the Super B747 at C_L .5.

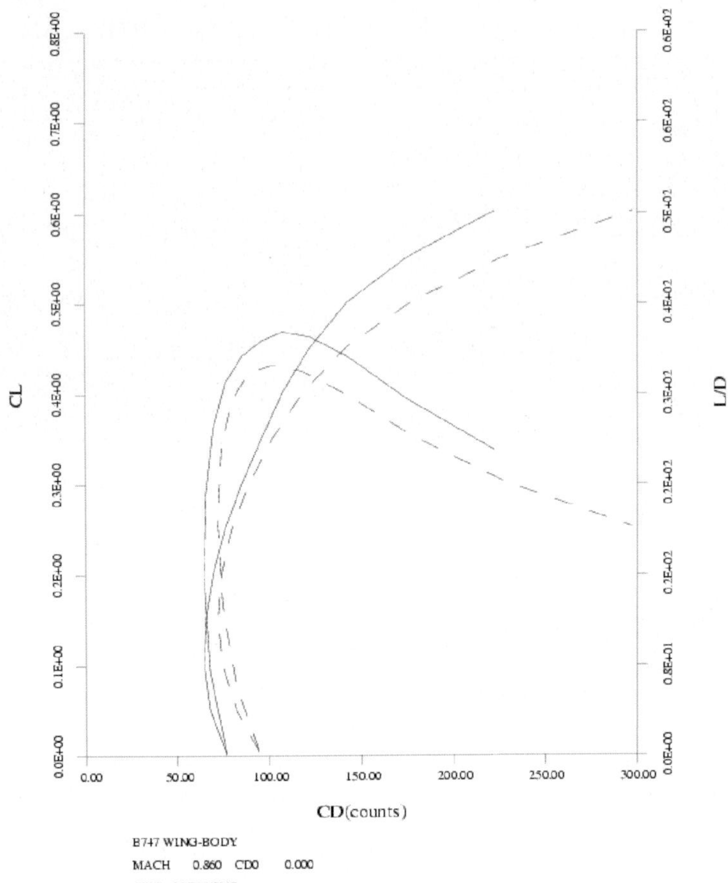

Fig. 5. Drag Polars of Baseline and Super B747 at Mach .86. (Solid-line represents Super B747. Dash-line represents Baseline B747.)

10 percent. In combination with the reduction in wing weight and an increase in fuel volume due to the thicker wing section, this should lead to an increase in range which is substantially more than 10 percent.

4.3 Shape optimization for a transonic business Jet

The unstructured design method has also been applied to several complete aircraft configurations. The results for a business jet are shown in figures 6 (a) and (b). There is a strong shock over the out board wing sections of the initial configuration, which is essentially eliminated by the redesign. The drag was reduced from 235 counts to 215 counts in about 8 design cycles. The lift was constrained at 0.4 by perturbing the angle of attack. Further, the original thickness of the wing was maintained during the design process

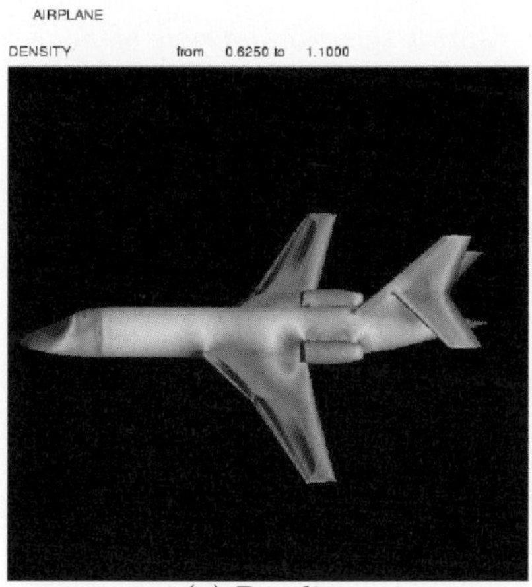

(a) Baseline

(b) Redesign

Fig. 6. Density contours for a business jet at $M = 0.8$, $\alpha = 2^o$

ensuring that fuel volume and structural integrity will be maintained by the redesigned shape. Thickness constraints on the wing were imposed on cutting planes along the span of the wing and by transferring the constrained shape movement back to the nodes of the surface triangulation.

5 Conclusion

The accumulated experience of the last decade suggests that most existing aircraft which cruise at transonic speeds are amenable to a drag reduction of the order of 3 to 5 percent, or an increase in the drag rise Mach number of at least .02. These improvements can be achieved by very small shape modifications, which are too subtle to allow their determination by trial and error methods. When larger scale modifications such as planform variations or new wing sections are allowed, larger gains in the range of 5-10 percent are attainable. The potential economic benefits are substantial, considering the fuel costs of the entire airline fleet. Moreover, if one were to take full advantage of the increase in the lift to drag ratio during the design process, a smaller aircraft could be designed to perform the same task, with consequent further cost reductions. Methods of this type will provide a basis for aerodynamic designs of the future.

References

1. J. L. Lions. *Optimal Control of Systems Governed by Partial Differential Equations.* Springer-Verlag, New York, 1971. Translated by S.K. Mitter.

2. A. Jameson. Aerodynamic design via control theory. *Journal of Scientific Computing*, 3:233–260, 1988.
3. A. Jameson. Computational aerodynamics for aircraft design. *Science*, 245:361–371, July 1989.
4. J. Reuther, A. Jameson, J. Farmer, L. Martinelli, and D. Saunders. Aerodynamic shape optimization of complex aircraft configurations via an adjoint formulation. *AIAA paper 96-0094*, 34th Aerospace Sciences Meeting and Exhibit, Reno, Nevada, January 1996.
5. A. Jameson, W. Schmidt, and E. Turkel. Numerical solutions of the Euler equations by finite volume methods with Runge-Kutta time stepping schemes. *AIAA paper 81-1259*, January 1981.
6. A. Jameson and T.J. Baker. Improvements to the aircraft Euler method. *AIAA paper 87-0452*, AIAA 25th Aerospace Sciences Meeting, Reno, Nevada, January 1987.
7. T. J. Barth. Apects of unstructured grids and finite volume solvers for the Euler and Navier Stokes equations. *AIAA paper 91-0237*, AIAA Aerospace Sciences Meeting, Reno, NV, January 1991.
8. K. Leoviriyakit and A. Jameson. Aerodynamic shape optimization of wings including planform variations. *AIAA paper 2003-0210*, 41st Aerospace Sciences Meeting & Exhibit, Reno, Nevada, January 2003.
9. A. Jameson. A perspective on computational algorithms for aerodynamic analysis and design. *Progress in Aerospace Sciences*, 37:197–243, 2001.
10. S. Kim, J. J. Alonso, and A. Jameson. Design optimization of high-lift configurations using a viscous continuous adjoint method. *AIAA paper 2002-0844*, AIAA 40th Aerospace Sciences Meeting & Exhibit, Reno, NV, January 2002.
11. K. Leoviriyakit, S. Kim, and A. Jameson. Viscous aerodynamic shape optimization of wings including planform variables. *AIAA paper 2003-3498*, 21st Applied Aerodynamics Conference, Orlando, Florida, June 23-26 2003.
12. A. Jameson and S. Kim. Reduction of the adjoint gradient formula in the continuous limit. *AIAA paper 2003-0844*, AIAA 41st Aerospace Sciences Meeting & Exhibit, Reno, NV, January 2003.
13. A. Jameson, Sriram, and L. Martinelli. An unstructured adjoint method for transonic flows. *AIAA paper*, 16th AIAA CFD Conference, Orlando, FL, June 2003.
14. A. Jameson, L. Martinelli, and J. Vassberg. Reduction of the adjoint gradient formula in the continuous limit. *AIAA paper*, 41st AIAA Aerospace Sciences Meeting, Reno, NV, January 2003.
15. L. Martinelli and A. Jameson. Validation of a multigrid method for the Reynolds averaged equations. *AIAA paper 88-0414*, 1988.
16. S. Tatsumi, L. Martinelli, and A. Jameson. A new high resolution scheme for compressible viscous flows with shocks. *AIAA paper* To Appear, AIAA 33nd Aerospace Sciences Meeting, Reno, Nevada, January 1995.
17. K. Leoviriyakit and A. Jameson. Aero-structural wing planform optimization. Reno, Nevada, January 2004. Proceedings of the 42st Aerospace Sciences Meeting & Exhibit.

On the Use of Parametric-CAD Systems and Cartesian Methods for Aerodynamic Design

Marian Nemec[1], Michael J. Aftosmis[2], and Thomas H. Pulliam[3]

[1] NASA Ames Research Center, Moffett Field, CA, USA *nemec@nas.nasa.gov*
[2] *aftosmis@nas.nasa.gov*
[3] *tpulliam@mail.arc.nasa.gov*

1 Introduction

Automated, high-fidelity tools for aerodynamic design face critical issues in attempting to optimize real-life geometry and in permitting radical design changes. Success in these areas promises not only significantly shorter design-cycle times, but also superior and unconventional designs. To address these issues, we investigate the use of a parametric-CAD system in conjunction with an embedded-boundary Cartesian method. Our goal is to combine the modeling capabilities of feature-based CAD with the robustness and flexibility of component-based Cartesian volume-mesh generation for complex geometry problems. We present the development of an automated optimization framework with a focus on the deployment of such a CAD-based design approach in a heterogeneous parallel computing environment.

2 Problem Formulation

The aerodynamic optimization problem consists of determining values of design variables such that a given objective function is minimized, while satisfying the governing flow equations and any other side constraints. Examples of objectives and constraints include performance functionals, such as lift and drag, and geometric quantities, such as volumes and thicknesses. The flow equations are the three-dimensional Euler equations of a perfect gas.

A modular framework is constructed to solve the optimization problem based on two optimizers: 1) a genetic algorithm (GA), and 2) a BFGS quasi-Newton algorithm, where the objective function gradient is evaluated via central-differences. At the core of the framework is the analysis module that consists of a CAD-system interface and the *Cart3D* flow-analysis package. The primary components of *Cart3D* include a Cartesian grid generator [1], and a flow solver [2]. Below, we provide a description of the CAD interface, followed by the optimization framework. See [3] for additional details.

3 CAD Interface: CAPRI

The CAD interface controls the regeneration of a CAD model in response to a design change and provides a corresponding watertight surface triangulation. This is accomplished by the use of the Computational Analysis PRogramming Interface (CAPRI) [4, 5]. CAPRI exposes the master-model feature tree of the CAD model and allows direct modification of parameters within that tree. Most design variables are associated directly with values exposed in the feature tree. Depending on the CAD-face topology of the model, CAPRI triangulates the faces by using either a quality triangulation algorithm [6, 4], or by decomposing quadrilaterals generated with an auto-blocking algorithm [7].

4 Optimization Framework

The synthesis of CAPRI with the optimization process is shown in Fig. 1. At each iteration of the optimization, CAD geometry requests are generated for different design-variable values and these are placed in a central repository. Independent of the optimization runs, a distributed geometry server is initiated that consists of multiple CAD nodes. The nodes process the geometry requests by retrieving the required parts and assemblies, regenerating the CAD models, and providing surface triangulations for the optimization processes. Since the geometry requests are independent, the geometry server achieves nearly linear scalability.

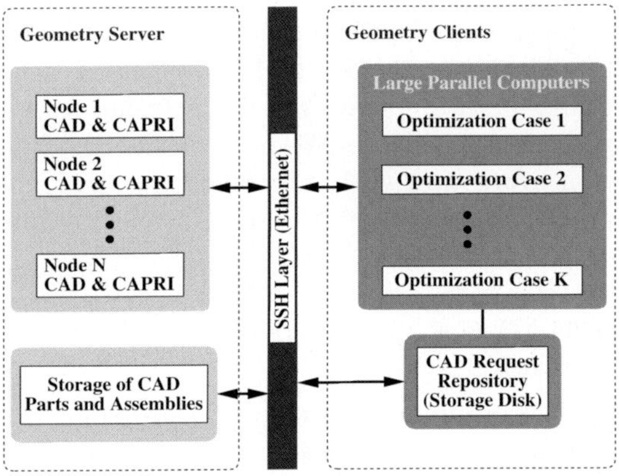

Fig. 1. Layout of the interface between optimization processes and geometry server

The number of CAD nodes is limited by the number of available CAD licenses, as each node consumes one license. An immediate concern is that the CAD nodes may become the bottleneck of the optimization process, idling the processors of the compute engines. To avoid such bottlenecks, we dynamically allocate the processors of the optimization to the number of completed surface triangulations. Figure 2 illustrates this on an example with 64 processors. At the start of each design iteration, all processors are dedicated to the solution of the first returned surface triangulation from the CAD nodes. This is the base state of the gradient method and the first chromosome of the GA, denoted as “Geometry 1” in Fig. 2. Upon completion of the first geometry analysis, we check the number of completed surface triangulations. These are processed by the CAD nodes while the analysis of the first geometry is performed on the compute engine, denoted as "Geometries $2\dots K$" in Fig. 2. The number of processors is distributed among the completed surface triangulations and multiple analysis modules are executed on subsets of the available processors. This dynamic, coarse-grained parallelism provides not only concurrent execution of serial tasks, but also ensures high parallel efficiency of the flow solver by limiting the number of processors available to each analysis module.

The worst case scenario occurs when the wall-clock time required for the processing of a geometry request exceeds the time for completion of the flow solution. If only one CAD node is available, then this CAD node would not be able to feed the compute engine with geometries without processor idle time. This situation is unlikely, since CAD model regeneration and tessellation tasks have computational complexity of $\mathcal{O}(N^2)$, while volume mesh generation and flow solution tasks are $\mathcal{O}(N^3)$.

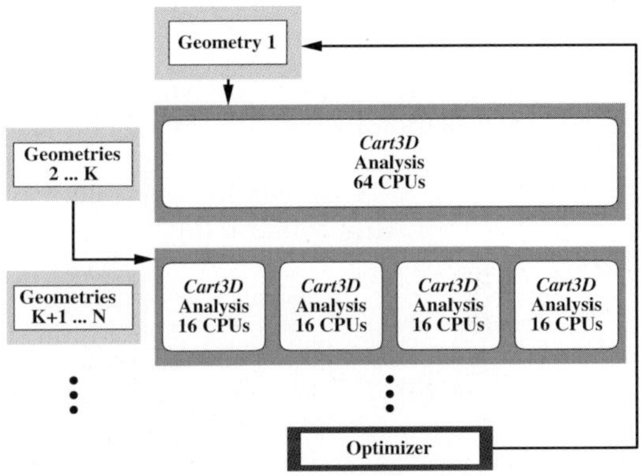

Fig. 2. Dynamic allocation of processors to mask the latency of CAD geometry processing

5 Design Example

We investigate the performance of the optimization framework for a design example based on the configuration shown in Fig. 3. This generic model is a CAD assembly of five parts, where the wing and canard are "attached" to the fuselage via two parameters, their horizontal and vertical locations, respectively. These parameters are constrained to intersect the projection of the fuselage on the symmetry plane within the CAD system. This simple construct avoids non-physical configurations, for example wings that detach from the fuselage during the optimization, even if the fuselage

Fig. 3. CAD model configuration (before component intersection)

shape and dimensions change. The CAD model is constructed using the Pro/ENGINEER CAD system.

Before presenting optimization results, we characterize the performance of the optimization framework. Table 1 presents average CPU timing results for the CAD model regeneration and surface triangulation using CAPRI. It is clear from Table 1 that CAD-model regeneration times are not a significant expense even for problems with many design variables. While the time required for surface triangulation is not prohibitive, it is important to avoid all unnecessary re-triangulations during the optimization. This is accomplished by caching an associated baseline triangulation for each part prior to the optimization.

Table 1. Average CPU time for CAD-model regeneration and tessellation (600 MHz R14000 SGI Octane Workstation)

Part	CAD-Model Regeneration (s)	Tessellation (s)	Number of Triangles	Tessellation Algorithm
Fuselage	2.0[a]	93.3	$\approx 41,000$	Quality-based
Wing	3.0[b]	16.5	$\approx 50,000$	Right-triangle

[a] No shape-section change, only global parameter modifications
[b] Shape-section change and planform parameter modifications

Table 2 presents average timing results for individual components within the analysis module. The volume mesh contains roughly 1.5 million cells for a half-span model of the configuration. The time for the mesh-solution transfer algorithm used to "warm-start" finite-difference gradient computations is also shown.

Table 2. Wallclock times for individual components of the analysis module (600 MHz R14000 SGI Origin 3000)

Component	Time (s)	Algorithm
Mesh Generation[a]	132.0	Serial
Flow Solution (flowCart 1.2)	455.0	Parallel (64 processors)
Mesh Solution Transfer	26.0	Serial

[a] Also includes component intersection, domain decomposition, and multigrid coarse-mesh generation

Comparison of Tables 1 and 2 indicates that the time required to complete a CAD-model regeneration and surface triangulation is a factor of six smaller than the time required for a flow solution. This means that by the time the analysis module completes the flow solution of the first chromosome of the

GA or the base-state of the gradient method, six new surface triangulations are ready for analysis, ensuring good CPU efficiency.

We consider the optimization problem of attaining a nearly zero pitching moment coefficient by optimizing the canard/tail control surface. The design variables are the control surface aspect ratio (AR), twist (tip relative to root section), and axial position along the center line of the fuselage. The problem has two local optima, the tail or canard configuration, with the canard configuration as the global optimum due to an aft location of the center of gravity. For optimization using the GA, the canard area is also a design variable. This introduces the possibility of a topology change in the design space. We use 16 chromosomes for each generation of the GA. For the quasi-Newton algorithm, the control surface area is kept constant (60.0). The freestream Mach number is 0.85, and the angle of attack is 1.0 deg.

The objective function uses a quadratic-penalty formulation, with a target lift coefficient of 0.222 and a target pitching moment coefficient of 0.001. The initial pitching moment is −0.0714. Figure 4(a) shows the convergence history of the objective function. The label "Design Iteration" refers to the number of generations evaluated by the GA, and the number of objective function and gradient evaluations by the quasi-Newton algorithm. The GA converges within 5 design iterations, requiring only 80 function evaluations. The quasi-Newton algorithm requires 56 function evaluations to trim the configuration.

Figures 4(b) and 4(c) show the initial and final designs for the quasi-Newton algorithm. The control surface converged to the minimum allowable forward location on the fuselage (8% of fuselage length), a twist angle of 2.98 deg., and AR of 6.03. Figure 4(d) shows the final design using the GA. For this case, the optimization converged to the upper bound of the control surface area, which is 60.0, a forward location of 8.2% of fuselage length, a twist angle of 3.41 deg., and AR of 4.36. The small differences in the two designs indicate that there may be many control surfaces that trim this configuration and further constraints are required to define a unique problem.

6 Conclusions

A modular framework for the aerodynamic optimization of complex geometries has been developed. The framework includes a direct interface to a parametric-CAD system that allowed an efficient manipulation and surface tessellation of generic-CAD models. Furthermore, the use of a component-based Cartesian method reduced the demands on the CAD system by reusing cached component triangulations, and improved the robustness of the framework due to the decoupling of the surface mesh form the volume mesh. Parallel efficiency of the framework was maintained even when subject to limited CAD resources by dynamically re-allocating the processors of the flow solver, thereby using the scalability of the solver to mask the latency of the geometry server.

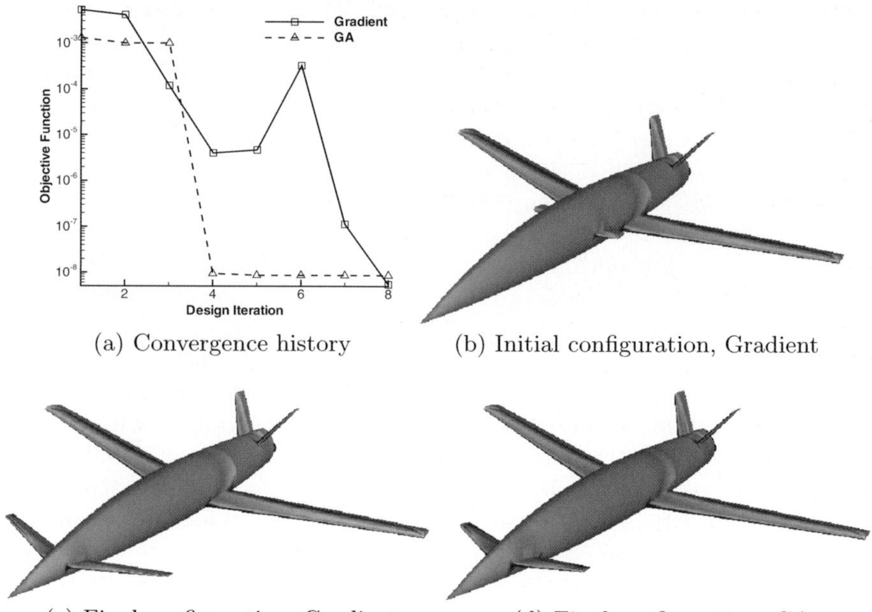

(a) Convergence history (b) Initial configuration, Gradient

(c) Final configuration, Gradient (d) Final configuration, GA

Fig. 4. Design example summary (Contour plots denote surface Mach number)

Acknowledgments

The authors gratefully acknowledge Robert Haimes (MIT). This work was performed while the first author held an NRC Research Associateship Award.

References

1. Aftosmis, M. J., Berger, M. J., and Melton, J. E., "Robust and Efficient Cartesian Mesh Generation for Componenet-Based Geometry," *AIAA J.*, **36**, 1998.
2. Aftosmis, M. J., Berger, M. J., and Adomavicius, G., "A Parallel Multilevel Method for Adaptively Refined Cartesian Grids with Embedded Boundaries," AIAA Paper 2000–0808, Reno, NV, Jan. 2000.
3. Nemec, M., Aftosmis, M. J., and Pulliam, T. H., "CAD-Based Aerodynamic Design of Complex Configurations Using a Cartesian Method," AIAA Paper 2004–0113, Reno, NV, Jan. 2004.
4. Haimes, R. and Aftosmis, M. J., "On Generating High Quality "Water-tight" Triangulations Directly from CAD," ISGG, June 2002.
5. Haimes, R. and Crawford, C., "Unified Geometry Access for Analysis and Design," 12th International Meshing Roundtable, Sept. 2003.
6. Aftosmis, M. J., "On the Use of CAD-Native Predicates and Geometry in Surface Meshing," NASA TM 1999–208782, Aug. 1999.
7. Haimes, R. and Aftosmis, M. J., "Watertight Anisotropic Surface Meshing Using Quadrilateral Patches," 13th International Meshing Roundtable, Sept. 2004.

Improvement of the Optimization Method of the TSTO Configuration – Application of Accurate Aerodynamics

Koji Shimoyama[1], Kozo Fujii[2], and Hiroaki Kobayashi[3]

[1] University of Tokyo, 3-1-1 Yoshinodai, Sagamihara, Kanagawa, 229-8510, Japan, simoyama@flab.eng.isas.jaxa.jp
[2] Institute of Space and Astronautical Science, Japan Aerospace Exploration Agency, 3-1-1 Yoshinodai, Sagamihara, Kanagawa, 229-8510, Japan, fujii@flab.eng.isas.jaxa.jp
[3] Institute of Space Technology and Aeronautics, Japan Aerospace Exploration Agency, 7-44-1 Jindaiji-Higashimachi, Chofu, Tokyo, 182-8522, Japan, kobayashi.hiroaki@jaxa.jp

1 Introduction

At present in Japan, Two-Stage-To-Orbit (TSTO) spaceplane is considered as one of space transportation systems in the next generation. One of the advantages of TSTO is that the operation cost of TSTO is lower than that of Single-Stage-To-Orbit (SSTO) as TSTO needs less fuel and lighter gross take-off weight than SSTO. On the other hand, the design cost of TSTO is higher than that of SSTO as the TSTO system consists of more components than SSTO. Design of complicated total systems such as TSTO requires a comprehensive optimization to satisfy design requirements, in which each component of the systems is optimized simultaneously with other components taken into account.

The conventional comprehensive optimization method of TSTO spaceplane, [1] however, did not have high fidelity for each component model (propulsion, aerodynamics, trajectory and structure weight) as shown in Fig. 1. The aerodynamics model, for example, was based on the existing experimental aerodynamic data of a configuration different from what was used in the structure weight model. In addition, aerodynamic performance of the TSTO configuration was estimated by a simple summation of the two separate airframes, assuming that the Booster and the Orbiter have the same aerodynamic performance. Therefore it is necessary to improve the aerodynamics model for the reliable TSTO design.

In this study, a new aerodynamics model is constructed using high-fidelity CFD simulations of TSTO configuration, and the effect of the aerodynamics model on the optimization result is investigated.

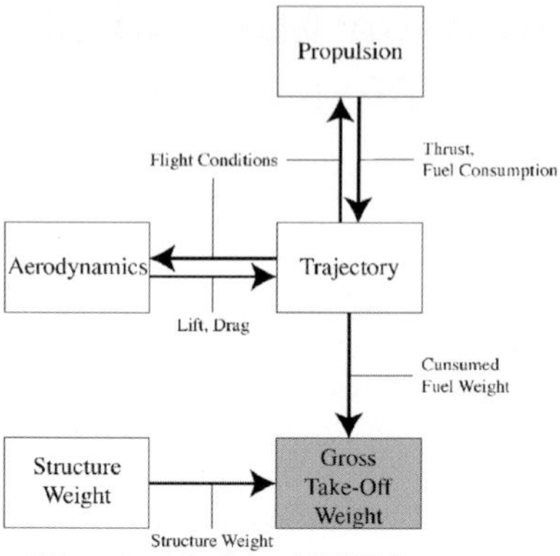

Fig. 1. Model of TSTO system.

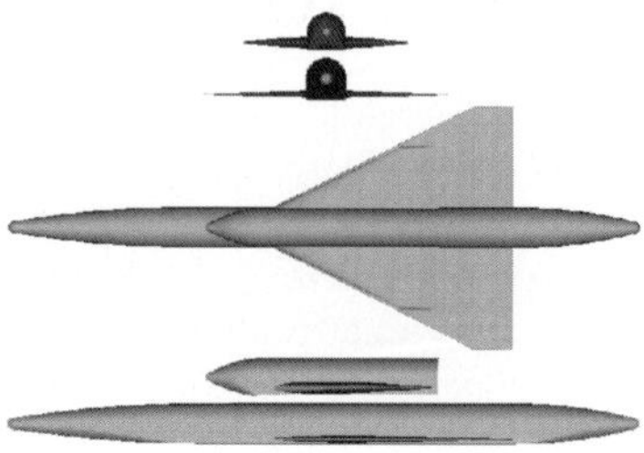

Fig. 2. TSTO configuration.

2 CFD Simulations of the TSTO Spaceplane

2.1 Geometry, Numerical Method and Flow Conditions

Figure 2 shows the TSTO configuration used in this study. This configuration is based on the baseline configuration used in the structure weight model and the optimum solution obtained in the former study. [1]

The governing equations for the CFD simulations are the three-dimensional Favre-averaged compressible thin-layer Navier-Stokes equations. The convective terms are evaluated by the SHUS scheme [2] which is a family of the AUSM-type schemes. In the convective terms, furthermore, high-order accuracy is obtained by the 3rd-order upwind biased MUSCL interpolation [3] based on the primitive variables. The viscous terms are evaluated by the central differencing and the eddy viscosity is modeled by the Baldwin-Lomax turbulence model. [4] Only the stady state flow fields are considered, the LU-ADI factorized implicit algorithm [5] is used for the time integration.

The zonal method is used to handle the TSTO configuration. The zonal grids are generated for each component (body/wing of the Booster/Orbiter) of the TSTO configuration, respectively, and the information of the physical values of each zonal grid is transferred to the other zones by the fortified solution algorithm (FSA). [6] The total number of grid points is about 1.8 million.

The Reynolds number based on the Booster body length is fixed at $Re = 1 \times 10^6$. The freestream Mach number is set as $M_\infty = 0.3 \sim 3.0$ because the non-linearity exists in this range. The angle of attack is set as $\alpha = 0 \sim 10$[degs] considering the TSTO flight control range.

2.2 Results and Discussion

Aerodynamic performance of the three configurations (Booster-only, Orbiter-only and Booster/Orbiter-integrated configurations) is estimated using the

CFD simulations. Figure 3 shows the aerodynamic coefficients (C_L and C_D) of the Booster/Orbiter-integrated configuration against the angle of attack α. Here, "lin. sum." (linear summation) in the legend denotes the data obtained by the conventional method [1] derived from the following equations.

$$\begin{aligned} C_{D\text{tsto}} &= 1.05 \left(S_{ref_{\text{bst}}} C_{D\text{bst}} + S_{ref_{\text{obt}}} C_{D\text{obt}} \right) / S_{ref_{\text{tsto}}} \\ C_{L\text{tsto}} &= S_{ref_{\text{bst}}} C_{L\text{bst}} / S_{ref_{\text{tsto}}} \end{aligned} \tag{1}$$

where S_{ref} denotes the reference area of aerodynamic coefficients. In this model, aerodynamics are assumed to be the same for the Booster and the Orbiter and only scaling of each configuration is considered assuming $C_{D\text{obt}} = C_{D\text{bst}}$ and $C_{L\text{obt}} = C_{L\text{bst}}$. Figure 3(a) shows that the total lift evaluated by the present CFD simulations of the TSTO is larger than that evaluated by the conventional method. Figure 4 shows the pressure contour plots on the airframe surface and symmetric plane at $M_\infty = 0.95$, $\alpha = 5$[degs] of the Booster-only and Booster/Orbiter-integrated configurations. These results show that shock wave occurs at the upper side of the Orbiter due to the flow interaction of the Booster and the Orbiter, resulting in the lift increase. Figure 3(b) shows that the total drag evaluated by the present CFD simulations of the TSTO is larger than that evaluated by the conventional method at low angles of attack. The drag increases as there exists a low pressure region in the Orbiter's base, which is not considered in the conventional method. In the actual TSTO spaceplane, effect of engine plume should be considered for still-higher accuracy.

3 Comprehensive Optimization of the TSTO Spaceplane

3.1 Aerodynamics Model

The C_{D0} and $C_{L\alpha}$ curves at each flight Mach number obtained by the present high-fidelity CFD simulations are implemented in the aerodynamics model and used in the comprehensive optimization. In the present optimization, the aerodynamic performance at arbitrary flight condition (M_∞, α) is estimated by the following equations.

$$\begin{aligned} C_D &= C_{D0} + C_{L\alpha}\alpha^2 \\ C_L &= C_{L\alpha}\alpha \end{aligned} \tag{2}$$

Figure 5 compares the present CFD-based and the conventional [1] aerodynamics models. In this study, the CFD simulations are carried out only in the range of $M_\infty \leq 3.0$ as non-linearity exists in this Mach number range. The data of the aerodynamics model in the range of $M_\infty > 3.0$ is simply extrapolated from the data in the range of $1.0 < M_\infty \leq 3.0$, considering that the aerodynamic performance at the high Mach number can be evaluated easily by the Newtonian impact theory.

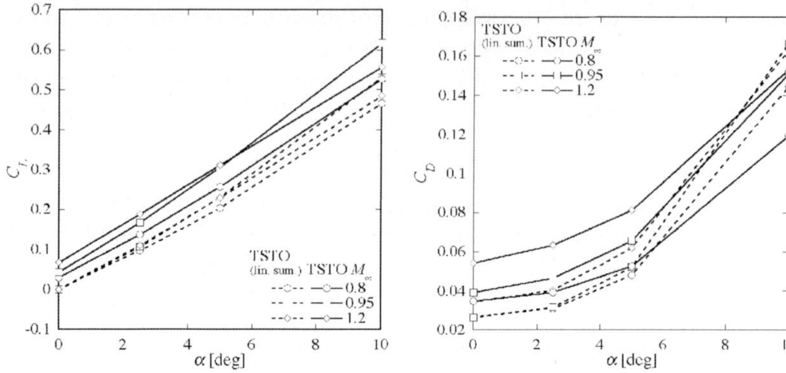

(a) Lift coefficients. (b) Drag coefficients.

Fig. 3. Aerodynamic coefficients of the Booster/ Orbiter-integrated configuration against α.

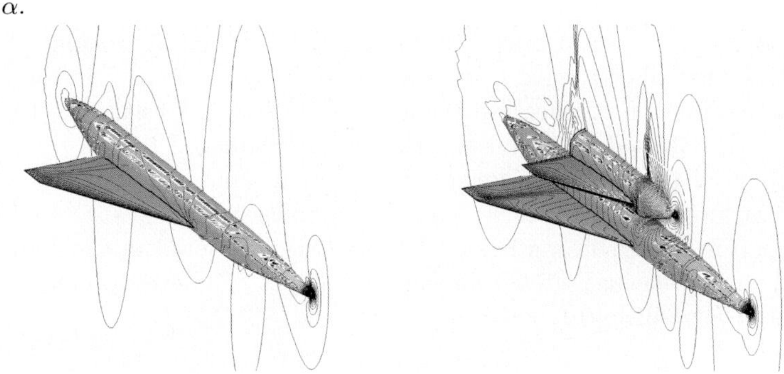

(a) Booster-only. (b) Booster/Orbiter-integrated.

Fig. 4. Pressure contour plots on the airframe surface and symmetric plane ($M_\infty = 0.95$, $\alpha = 5$[degs]).

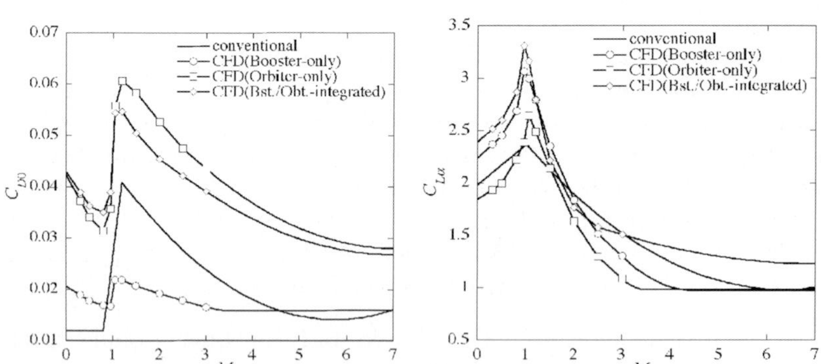

(a) C_{D0} distributions against M_∞. (b) $C_{L\alpha}$ distributions against M_∞.

Fig. 5. Aerodynamics models.

3.2 Definition of Optimization Problem

The present TSTO mission is to put a payload of 10t into the equatorial orbit at the altitude of 400km. For simplicity, the take-off and landing sites are assumed to be on the equator. The engine of the Booster is assumed to be the air-turbo-ramjet engine with expander cycle (ATREX), [7] which is under development in Japan. Propulsion, trajectory and airframe configuration parameters (total fourteen) are considered as design variables and the gross take-off weight is minimized as an objective function.

3.3 Optimization Method and Conditions

A genetic algorithm (GA) is used as the optimization method. In general, the GA has capability of finding global optimum solutions, but this characteristics leads to the deterioration in local convergency and the vast computation load. To improve the local convergency, the unimodal normal distribution crossover (UNDX) [8] is used for crossover. In addition, to preserve variety of solutions, the minimal generation gap (MGG) [8] is used for the alternation of generation. The optimization conditions are set as the population size of 20, the number of generations of 300, the number of crossover of 10 and mutation rate of 5%.

3.4 Results and Discussion

Both the conventional and CFD-based aerodynamics models are applied to the optimizations. The optimum gross take-off weight evaluated by using the CFD-based aerodynamics model obtained by the optimization is heavier than that evaluated by using the conventional aerodynamics model as can be observed in Table 1.

Figure 6 shows the histories of the computed trajectory control variables (lift L, drag D and engine thrust T) of the optimum solutions against the flight Mach number M. The larger drag at transonic flight condition evaluated by the CFD-based aerodynamics model requires the larger engine thrust and thus the larger engine size. Besides, the increased engine thrust increases the consumed fuel weight and enlarges the fuel tank size, which leads to a large body size. The increased drag, then, increases the lift, resulting in the large wing size. The gross take-off weight eventually increases.

Table 1. Optimum gross take-off weight [t].

conventional	CFD-based
331.46	368.77

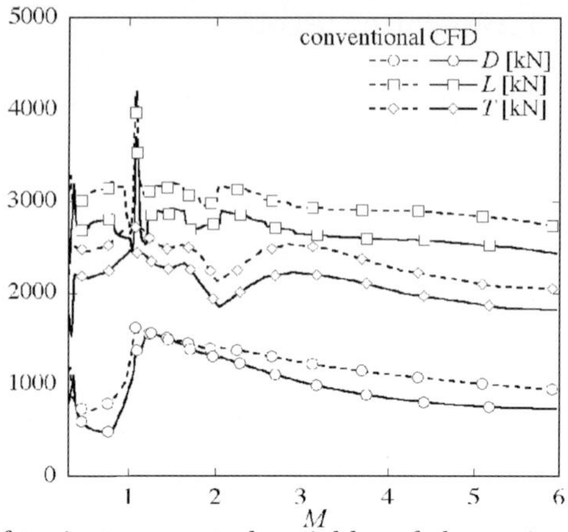

Fig. 6. Histories of trajectory control variables of the optimum solutions against M.

4 Conclusions

A new aerodynamics model was constructed using the data obtained by the contemporary CFD simulations of the TSTO spaceplane, and the effect of the aerodynamics model on the comprehensive optimization of TSTO was investigated. The comparison of the aerodynamic coefficients by the CFD simulations and those by the conventional method shows that both the lift and drag of the TSTO configuration increases at transonic to supersonic flow region by the CFD simulations. There are two keys fot these increases: the interaction of the flow fields over the Booster and the Orbiter, and low pressure region in the Orbiter's base. Therefore, these keys should be considered for an accurate optimization. The CFD-based aerodynamics model evaluated aerodynamic performance of the TSTO more severely than the conventional model and the optimization using the CFD-based aerodynamics model results in the heavier gross take-off weight compared to the result using the conventional model.

References

1. H. Kobayashi, N. Tanatsugu: AIAA Paper 2001–3965 (2001)
2. E. Shima, T. Jounouchi: Proceedings of 14th NAL Symposium on Aircraft Computation Aerodynamics, NAL SP–34 (1996) pp 7–12
3. B. van Leer: Journal of Computational Physics, **32** (1979) pp 101–136
4. B. S. Baldwin, H. Lomax: AIAA Paper 78–257 (1978)
5. K. Fujii, S. Obayashi: AIAA Paper 86–0513 (1986)
6. K. Fujii: Journal of Computational Physics, **118** (1995) pp 92–108
7. N. Tanatsugu, P. Carrick: AIAA Paper 2003–2586 (2003)
8. I. Ono, S. Kobayashi: Proceedings of the 7th International Conference on Genetic Algorithms (1997) pp 246–253.

Part XVIII

Parallel Algorithms

Parallel Simulation for Strong Blast Wave from TNT Explosion on Large-scale PC-Cluster

Takayuki Aoki[1], Kaori Kato[1], Tei Saburi[1], and Masatake Yoshida[2]

[1] Global Scientific Information and Computing Center, Tokyo Institute of Technology, O-okayama, Muguro-ku, Tokyo 152-8550,Japan, taoki@gsic.titech.ac.jp

[2] Research Center for Explosion Safety, National Institute of Advanced Industrial Science and Technology, Azuma 1-1-1, Tsukuba-shi Ibaraki 305-8565, Japan

1 INTRODUCTION

Explosives are very dangerous materials because enormous energy is released at the moment through detonation process. When the detonation process treminated in solid explosives, the pressure reaches tens of thousands times the air and the solid density still remains. The detonation product becomes the gas and expands rapidly driving strong shock wave. They are called blast waves. From the viewpoint of safety study, it is very important to estimate of damages by blast waves. For many years, the numerical schemes and the basic research on blast waves [1] $\sim$ [7] have been developed. However, few schemes can give quantitative estimation for the wide calculation scale. In the early stage of blast wave propagation, the large density jump between the detonation product gas and the air makes the numerical simulation difficult. A negligible numerical error for high-density detonation product gas has serious effect on the air and degrades the shock waves propagation. In order to study such blast waves, a stable and accurate numerical scheme have to be developed.

2 NUMERICAL METHOD

We develop a stable and accurate Eulerian scheme for propagation of strong blast waves. The governing equations (1)-(4) are spilt into the advection phase and the non-advection phase in the same way as the original CIP scheme [8]. The fractional step procedure is introduced for the stability of the scheme. The time advance from the n-th step to the $n+1$-th step consists of the following procedure. The dependent variables of the n-th time step is integrated to the intermediate step by solving the advection equation. Then the variables are advanced to the $n+1$-th step by solving the non-advection equation. Because we need better coupling between the pressure and the velocity for the phenomena with a steep gradient profile in the non-advection phase, we adapt staggered grid.

3 Two-dimensional simulation for the blast wave propagation around the magazine driven by the 0.1-*kg* pentolite explosion in Cylindrical coordinate

The simulation for the blast wave driven by the 0.1-kg pentolite is performed in the two-dimensional cylindrical coordinate where the z-axis is chosen to be the axial direction of the magazine and the radial direction is denoted as r. In the initial condition, the gas of the state just after the detonation is set as the energy source in the region of $-0.95m \leq z \leq -0.0908m$ and $0.0m \leq r \leq 0.022m$. The gas has the density of $1640.0kg/m^3$, internal energy of $5.225 \times 10^6 J/kg$ and the pressure of $7.67 \times 10^9 Pa$, and employs the JWL equation of state[9]. The total grid number of $N_r \times N_z = 600 \times 1120$ is assigned to the area of $1.5m \times 2.8m$ and the mesh size is $\Delta x = \Delta z = 0.0025m$. The magazine of $0.1m$ in radius, $1.0m$ in length and $0.11m$ in thickness is implemented as the rigid wall boundary in the simulation.

The expansion of the detonation product gas in the confined space makes the flow turbulent due to multiple interactions with the magazine wall. In the numerical result, the front of the blast wave reaches the outlet at the time $0.39ms$. The contour plots of the density and pressure profiles at the time of $0.45ms$, $1.12ms$ and $2.31ms$ are shown in Fig.1, respectively. When the detonation gas is erupted from the magazine outlet, the front velocity of the blast wave rapidly decreases due to switching to three-dimensional expansion from one-dimension. The shock wave in the air is detached from the detonation product gas and starts to propagate spherically as shown in Fig.1(a1) and Fig.1(b1). The spherically expanding front of the primary shock wave into the air is described well in all the contour plots. The contact discontinuity between the air and the detonation product gas becomes clear in the density profiles of the later time Fig.1(a2) and Fig.1(a3). The Kelvin-Helmholtz instability drives the turbulence of the detonation product gas. At the time of $1.12ms$, the low-density the void appears due to the rapid expansion in Fig.1(a2), and the secondary shock waves are generated in closing void region. The complex shock waves are found in the middle of the pressure profile in Fig.1(b3). The axial eruption of the detonation product gas keeps faster velocity than the sound velocity of the air and pushes the shock wave ahead. The shock front is swelled in the axial direction as shown in Fig.1(a2).

4 Three-dimensional simulation for the blast wave propagation around the magazine driven by the 32-*kg* TNT explosion in Cartesian coordinate

The three-dimensional blast wave simulation is performed on the same condition with the open-air mock experiment of the 32-kg TNT explosion of the underground magazine which was done in the exercises field of Japan Ground

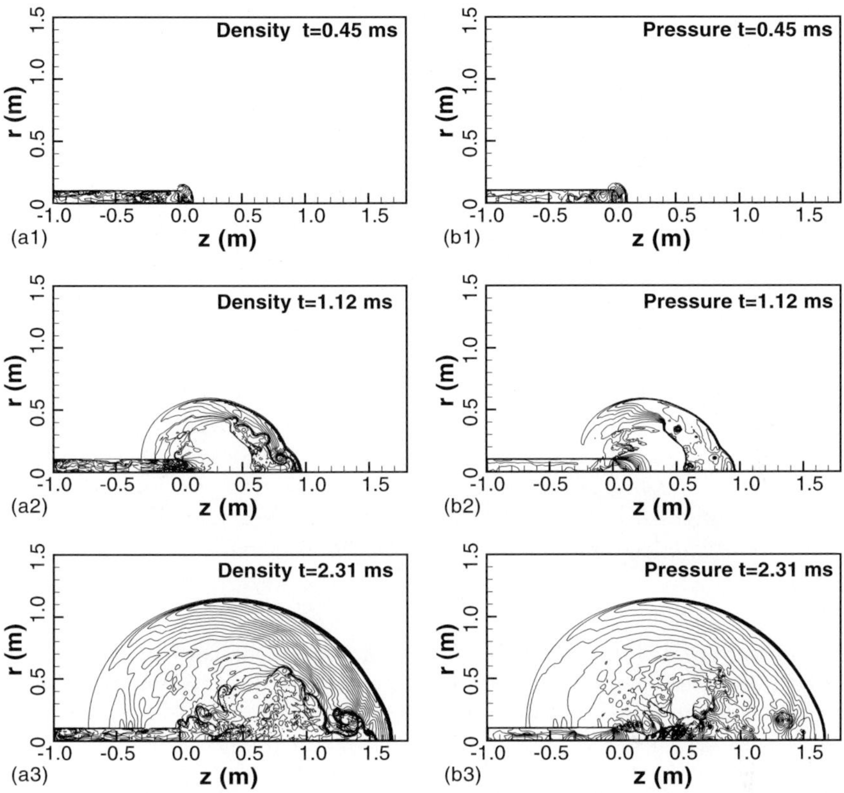

Fig. 1. Contours of blast wave driven by 0.1-kg pentolite explosion at the magazine: (a1) the density at $0.45ms$ with the range of $1.0\sim10.0kg/m^3$ and the interval of $1.0kg/m^3$; (b1) the pressure at $0.45ms$ with the range of $0.5\sim3.5MPa$ and the interval of $0.5MPa$; (a2) the density at $1.12ms$ with the range of $0.5\sim4.0kg/m^3$ and the interval of $0.25kg/m^3$; (b2) the pressure at $1.12ms$ with the range of $0.6\sim8.0MPa$ and the interval of $0.2MPa$; (a3) the density at $2.31ms$ with the density range of $0.1\sim2.0kg/m^3$ and the interval of $0.05kg/m^3$; and (b3) the pressure at $2.31ms$ with the range of $0.4\sim2.0MPa$ and the interval of $0.1MPa$.

Self Defense Force in 2003 [10]. The heap with an earth was constructed and the magazine was horizontally embedded. The pyramid shape has the base area of $10m \times 10m$ and the height of $4.4m$ and the magazine size is $5.0m$ in length and $1.8m$ in diameter. The schematic of them is shown in Fig.2(a). We set the x-z plane on the ground and the y-axis in the vertical direction and the origin of the coordinate system is located at the magazine outlet. It is assumed that the phenomena are symmetric due to the symmetric geometries of all the objects, so that we calculate only a half of the computational domain in the x-direction. The grid number of $N_x \times N_y \times N_z = 180 \times 180 \times 450$ is assigned to the domain $0m \leq x \leq 40m$, $-20m \leq z \leq 45m$, $0m \leq y \leq 60m$ in the Cartesian grid with non-uniform intervals. Fine meshes of $\Delta x = \Delta y = \Delta z = 0.1m$ are arranged inside the magazine and around the outlet, and distant meshes have larger size to $\Delta x = \Delta y = \Delta z = 0.4m$ as shown in Fig.2(b).

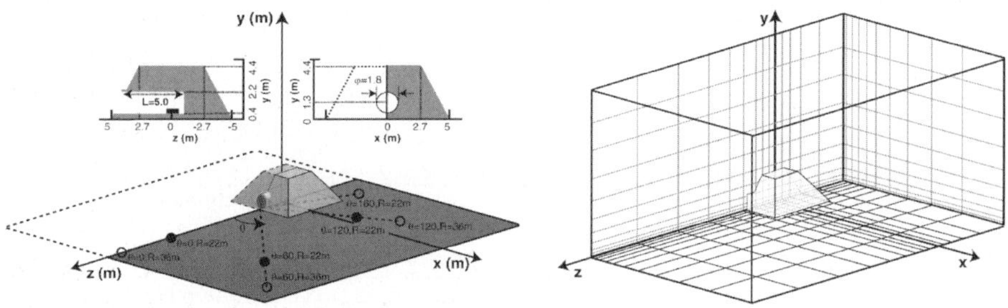

(a) (b)

Fig. 2. Configuration of the comutational domain for three-dimensional simulation: (a) a schematic of magazine and measurement point; and (b) a schematic of mesh covering the computational domain.

The simulation was executed on the PC-Cluster of Intel Xeon 2.2GHz×40 with Myrinet2000 interconnection at Research Center for Explosion Safety, National Institute of Advanced Industrial Science and Technology (AIST). It makes parallel computation possible that our numerical scheme is easily implemented to domain decomposition. The mesh resolution of $\Delta x = \Delta y = \Delta z = 0.1m$ inside the magazine is too large for the size of 32-kg TNT, so that we calculate the propagation of the blast wave by using two-dimensional cylindrical simulation. When the front of the blast wave reaches the outlet, the two-dimensional result is planted in the three-dimensional Cartesian grid as the initial condition. The rigid bodies are assumed for the heap with the magazine and the ground.

The density contour and the pressure iso-surface of the simulation result at the time of $51.28ms$ are displayed in Fig.3(a) and (b). Almost spherical shock wave is described also in the three-dimensional simulation and it is qualitatively understood that the shock is stronger in the forward direction. The distribution of the detonation product gas is found in front of the magazine outlet and the interface to the air evolves unstably. In the three-dimensional simulation, the geometrical shape of the heap is accurately taken into consideration and unstable behavior of the detonation product gas is different from two-dimensional simulation as shown in Fig.4.

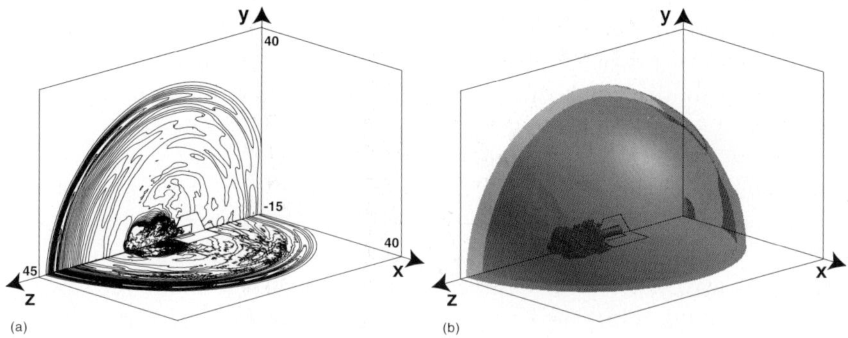

Fig. 3. Blast wave propagation driven by 32-kg TNT explosion at the magazine at t=51.28ms: (a)density contour on z-x and z-y plane with the range of 0.5∼1.35kg/m^3 and the interval of $\Delta\rho$=0.01kg/m^3; and (b)iso-surface for the volume fraction of α=0.1 and the pressure of p=1.05HPa in the range of 63.2∼115.6kPa.

Fig. 4. Photo-realistic visualization of the unstable detonation product gas:

5 SUMMARY

For the simulation of strong blast waves, a high-accurate numerical scheme was presented. The fractional step method splits the compressible fluid equation into the advection phase and non-advection phase. The monotone preserving RCIP scheme is used for the density and the internal energy in the advection phase to avoid negative values, and the IDO scheme is adapted with 3-stage Runge-Kutta time integration in the non-advection phase to describe the sound wave accurately. It is noticed that in both the phases, the schemes are based on the cubic Hermite interpolations constructed by the value and the gradient, and the overall numerical accuracy of the scheme keeps 3rd order in space.

It is concluded that the scheme proposed in this paper is quit available to solve the compressible fluid equation and applicable to the simulation of strong blast waves. Possible damage estimations of the blast wave will be given by the simulation based on our numerical scheme.

References

1. Harold L. Brode. Numerical Solution of spherical Blast waves *Journal of Applied physics* 1955; **26**:766–775.
2. Harold L. Brode. Blast wave from a spherical charge *Phisics Of Fluid.* 1959; **2**:217–229.
3. J.W.Kury, H.C.Hornig, E.L.Lee, J.W.McDonnel, D.L.Ornellas, M.Finger, F.M.Starnge, M.L.Wilkins Metal Acceleration by Chemical explosives, *Fourth Symposium on detonation* 1965: 3–13.
4. D. L. Jones. Blast Waves and Scaling Laws *Phys. Fluid* 1970; **13**:1398–1399.
5. D. Kh. Ofengeim, D. Drikakis. Simulation of blast wave propagation over a cylinder *Shock Waves* 1997; **7**:305–317.
6. S. Cieslak, S. Ben Khelil, I. Choquet, A. Merlen.Cut cell strategy for 3-D blast waves numerical simulations *Shock Waves* 2001; **10**:241–249.
7. S. M. Liang, J. S. Wang, H. Chen. Numerical study of spherical blast-wave propagation and reflection *Shock Waves* 2002; **12**:59–68.
8. T.Yabe, T.Aoki. A unuversal solver for hyperbolic equations by cubic-polynomial interpolation I. One-dimensional solver *Computer Physics Communications* 1991; **66**:219–232.
9. B.M.Dobratz *Properties of Chemical Explosives and Explosive Simulants* LLNL Explosives Handbook UCRL-52997 Distribution Category UC-45.
10. *Bakuhatsu-eikyou-hyouka-iinkai-houkoku-sho All Japan Association for security of explosives*, 2003;

A Parallel Implicit Adaptive Mesh Refinement Algorithm for Body-Fitted Multi-Block Mesh

Clinton P. T. Groth

University of Toronto Institute for Aerospace Studies, 4925 Dufferin Street, Toronto, Ontario, Canada M3H 5T6 groth@utias.utoronto.ca

Summary. A parallel implicit adaptive mesh refinement (AMR) algorithm is described for the system of partial-differential equations governing steady two-dimensional compressible gaseous flows. The AMR algorithm uses an upwind finite-volume spatial discretization procedure in conjunction with limited linear solution reconstruction and Riemann-solver based flux functions to solve the governing equations on multi-block mesh composed of structured curvilinear blocks with quadrilateral computational cells. A flexible block-based hierarchical data structure is used to facilitate automatic solution-directed mesh adaptation according to physics-based refinement criteria. A Jacobian-free inexact Newton method is used to solve the system of nonlinear equations arising from this finite-volume spatial discretization procedure and a preconditioned generalized minimal residual (GMRES) method is used to solve the resulting non-symmetric system of linear equations at each step of the Newton algorithm. An additive Schwarz global preconditioner with variable overlap is used in conjunction with block-fill incomplete lower-upper (BFILU) type preconditioners based on the Jacobian of the first-order upwind scheme for each sub-domain. The Schwarz preconditioning and block-based data structure readily allow efficient and scalable parallel implementations of the implicit AMR approach on distributed-memory multi-processor architectures. Numerical results will be described for several flow cases, demonstrating both the effectiveness of the mesh adaptation and algorithm parallel performance. The proposed parallel implicit AMR method allows for anisotropic mesh refinement and is well suited for predicting complex flows with disparate spatial and temporal scales.

1 Introduction

The development of an effective solution strategy for physically complex steady flows with disparate spatial and temporal scales is considered. A parallel implicit method is proposed for multi-block body-fitted mesh which combines a block-based adaptive mesh refinement (AMR) approach with a Newton-Krylov-Schwarz (NKS) solution of the discretized nonlinear equations governing compressible gaseous flows. The proposed AMR algorithm allows for anisotropic refinement on body-fitted meshes. Future research will focus on the further development of the algorithm for unsteady and three-dimensional flows and for more complex systems of equations.

2 Equations of Compressible Gas Dynamics

Solutions of the equations governing compressible flows of polytropic gases are considered. For two-dimensional planar flows, the conservative form of the Euler equations of compressible gas dynamics reflecting the conservation of mass, momentum, and energy can be summarized as follows:

$$\frac{\partial \mathbf{U}}{\partial t} + \frac{\partial \mathbf{F}}{\partial x} + \frac{\partial \mathbf{G}}{\partial y} = 0 \tag{1}$$

where $\mathbf{U} = [\rho, \rho, \rho v, \rho e]^{\mathrm{T}}$ is the conserved variable solution vector, x and y are the spatial coordinates of the frame, t is time, ρ is the gas density, u and v are the velocity components in the x- and y-coordinate directions, $e = p/(\rho(\gamma-1)) + (u^2+v^2)/2$ is the specific total energy, $p = \rho RT$ is the pressure, T is the gas temperature, R is the gas constant, γ is the specific heat ratio, and $\mathbf{F}$ and $\mathbf{G}$ are x- and y-direction solution flux vectors.

3 Parallel Implicit AMR Algorithm

3.1 Finite-Volume Scheme

The proposed AMR algorithm uses an upwind finite-volume spatial discretization procedure in conjunction with limited linear solution reconstruction and Riemann-solver based flux functions to solve the preceding equations on multi-block mesh composed of quadrilateral computational cells. The semi-discrete form of the scheme for cell (i, j) is given by

$$\frac{d\mathbf{U}_{i,j}}{dt} = -\frac{1}{A_{i,j}} \sum_{k} (\vec{\mathbf{F}} \cdot \vec{\mathbf{n}}\, \Delta\ell)_{i,j,k} = \mathbf{R}_{i,j}(\mathbf{U}) \,, \tag{2}$$

where $\mathbf{U}_{i,j}$ is the conserved solution state for cell (i, j), $\vec{\mathbf{F}} = (\mathbf{F}, \mathbf{G})$ is the flux dyad, A_{ij} is the area of the cell, and $\Delta\ell$ and $\vec{\mathbf{n}}$ are the length of the cell face and unit vector normal to the cell face or edge, respectively. The vector, $\mathbf{R}$, is referred to as the residual vector. The numerical fluxes at the faces, k, of each cell, $\vec{\mathbf{F}} \cdot \vec{\mathbf{n}}$, are determined from the solution of a Riemann problem. The left and right solution states are determined via a least-squares piece-wise limited linear solution reconstruction procedure in conjunction with either the Barth-Jesperson or Venkatakrishnan limiters [1, 2]. In the present algorithm, both exact and approximate Riemann solvers can be used to solve the Riemann problem and evaluate the numerical flux. The Roe linearized Riemann solver [3], HLLE and modified HLLE flux function due to Linde [4, 5], the HLLC flux function [6], and the exact Riemann solver of Gottlieb and Groth [7] have all been implemented.

3.2 Block-Based Adaptive Mesh Refinement

Following the approach developed by Groth *et al.* for computational magnetohydrodynamics [8, 9], a flexible block-based hierarchical data structure has been developed and is used in conjunction with the finite-volume scheme described above to facilitate automatic solution-directed mesh adaptation on multi-block body-fitted quadrilateral mesh according to physics-based refinement criteria. The approach readily permits local refinement of the mesh for complex flow geometries and also allows for anisotropic mesh refinement to resolve thin solution layers, such as boundary and free shear layers. The proposed AMR formulation borrows from previous work by Berger and coworkers [10, 11, 12, 13], Quirk [14, 15], and De Zeeuw and Powell [16] for Cartesian mesh and has similarities with the block-based approaches described by Quirk and Hanebutte [15] and Berger and Saltzman [12]. Note that other researchers have considered the extension of Cartesian mesh adaptation procedures to more arbitrary quadrilateral and hexagonal mesh. See for example the work by Davis and Dannenhoffer [17] and Sun and Takayama [18]. Note that although the proposed block-based AMR approach is somewhat less flexible and incurs some inefficiencies in solution resolution as compared to a cell-based approaches (i.e., for the same solution accuracy, generally more computational cells are introduced in the adapted grid), the block-based method can offer many advantages over cell-based techniques when computational performance and parallel implementation of the solution algorithm is considered.

3.3 Inexact Newton Method

The semi-discrete form of the governing equations given in Eq. (2) form a coupled set of non-linear ordinary differential equations. For steady flows, time-invariant solutions Eq. (2) are sought satisfying $\mathbf{R}(\mathbf{U}) = 0$, requiring the solution of a large coupled nonlinear system of algebraic equations. Starting with an initial estimate, $\mathbf{U}^0$, a solution is obtained using Newton's method:

$$\left(\frac{\partial \mathbf{R}}{\partial \mathbf{U}}\right)^n \Delta\mathbf{U}^n = \mathbf{J}^n \Delta\mathbf{U}^n = -\mathbf{R}(\mathbf{U}^n)\,, \tag{3}$$

where at each Newton step, n, and an improved approximation for the solution is given by $\mathbf{U}^{n+1} = \mathbf{U}^n + \Delta\mathbf{U}^n$ Here $\mathbf{J} = \partial\mathbf{R}/\partial\mathbf{U}$ is the residual Jacobian. The iterative procedure is repeated until an appropriate norm of the solution residual is sufficiently small, i.e., $||\mathbf{R}(\mathbf{U}^n)||_2 < \epsilon||\mathbf{R}(\mathbf{U}^0)||_2$ where ϵ is some small parameter (typically, $\epsilon \approx 10^{-12}$–$10^{-10}$). As discussed by Dembo *et al.* [19], an exact solution of the linear system is not necessary for rapid convergence of Newton's method and an inexact approach is adopted here.

3.4 Parallel GMRES with Schwarz Preconditioning

Each step of the Newton method requires the solution of a system of linear equations which are solved here using restarted version of the generalized minimal residual (GMRES) algorithm of Saad [20]. Right preconditioning of linear system is used to improve performance of the GMRES method. A combination of global and local preconditioning techniques is applied where an additive Schwarz global preconditioner with variable domain overlap is used in conjunction with block-fill incomplete lower-upper (BFILU) local preconditioning based on the Jacobian of the first-order upwind scheme for each subdomain. The GMRES algorithm does not explicitly require the evaluation of the complete global Jacobian matrix, $\mathbf{J} = \partial\mathbf{R}/\partial\mathbf{U}$, The required matrix-vector products $\mathbf{JM}^{-1}\mathbf{x}$ are approximated using numerical differentiation, thereby avoiding the storage of the Jacobian matrix. The proposed "Jacobian-free" inexact Newton method with additive Schwarz/BFILU preconditioning and restarted GMRES linear solveris a variant of the NKS algorithm of Gropp *et al.* [21].

In order to increase the radius of convergence and ensure global convergence of the NKS method for almost any initial data, a good startup startup algorithm is invariably required. A parallel full multigrid algorithm with both V- and W-cycles is used as a startup procedure for the NKS algorithm to obtain a good initial estimate of the solution.

3.5 Parallel Implementation

By design, the multi-block AMR scheme and NKS algorithm are well suited to parallel implementation on distributed-memory multi-processor architectures. A parallel implementation has been developed using the MPI (message passing interface) library. For homogeneous architectures with multiple processors all of equal speed, the self-similar nature of the solution blocks is exploited and parallel implementation is carried out via domain decomposition where the solution blocks are simply distributed equally among the available processors, with more than one block permitted on each processor. A Morton-ordered space filling curve is used to provide nearest-neighbour ordering of blocks for more efficient load balancing [22].

4 Numerical Results

The numerical results are now described for supersonic flow past a two-dimensional circular cylinder with its axis of symmetry perpendicular to the free-stream flow direction such that a stationary bow shock forms about the body. The free-stream Mach number is $\mathrm{M}_\infty = 2.5$ for the case of interest. Numerical solutions are calculated on a 128×128 mesh consisting of 16,384 computational cells and several different blockings (partitionings) of the computational domain are examined in order to investigate the influence of the Schwarz preconditioning. A single-block grid consisting of one 128×128 solution block is considered, along with 4-, 16-, and 64-block grids composed of 4 64×64, 16 32×32, and 64 16×16 solutions blocks. The single- and 16-block

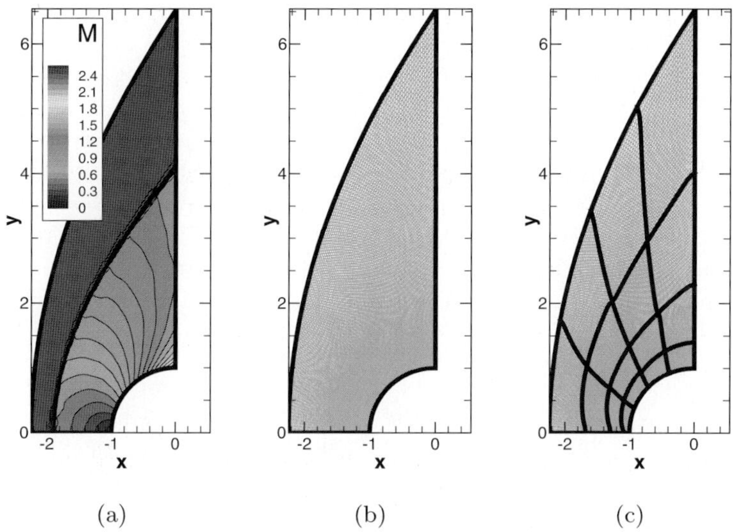

Fig. 1. Computed solution of parallel NKS algorithm for supersonic flow past a circular cylinder; Mach number $M_\infty = 2.5$; showing (a) Mach number distribution and (b) single (128 × 128) and (c) 16 (32 × 32) block, 16,384 cell, meshes used in computations.

computational mesh are shown in Figures 1(b) and 1(c) for comparison. Mesh refinement is not considered for this problem.

The computed Mach number distribution for the blunt-body flow is shown in Figure 1(a) and the convergence of the parallel NKS algorithm is given in Figure 2(a), where the 2-norm of density residual is depicted as a function of the number of equivalent residual evaluations and total processor time residual evaluation. The latter are related to one another by the CPU time required to evaluate the residual vector, **R**, on the finest mesh. Convergence results are shown for the 1-, 4-, and 16-block cases in Figure 2(a). In each of these cases, a 4-level, V-cycle full multigrid procedure, with 100 multigrid cycles on each successively finer mesh, was used to provide a good estimate for the solution before initiating the NKS iterations. No or zero domain overlap was employed in the Schwarz preconditioning. he influence of subdomain overlap was experimented with here and it was also found that for fill levels of 0, 1, and 2, overlap did not provide any significant benefits. In addition, the values of the slope limiters were "frozen" and held constant after the norm of the solution residual was reduced by four orders in magnitude.

It is also evident from the convergence histories Figure 2(a) that the multigrid startup algorithm is very effective in providing a good initial estimate of the steady solution for the parallel NKS algorithm. Furthermore, the matrix-free inexact Newton method rapidly converges to the steady-state solution.

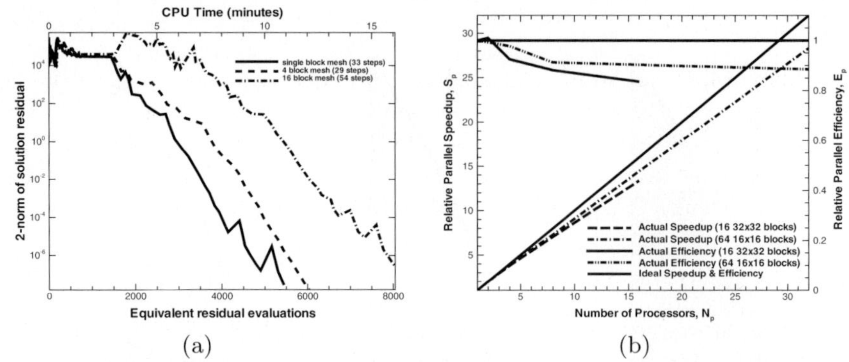

Fig. 2. Performance of parallel NKS algorithm for supersonic flow past a circular cylinder; Mach number $M = 2.5$; (a) convergence history showing 2-norm of density residual as a function of the number of equivalent residual evaluations and total processor time and (b) relative parallel speed-up, $S_p = (t_1/t_p)p$, and parallel efficiency, E_p, as a function of the number of processors used in the calculation.

Just 33 Newton steps are required to reduce the solution residual by 12 orders in magnitude on the single block 128×128 mesh. This number drops slightly to 29 for the 4-block mesh and only increases to 54 for the 16-block mesh. It is clear that there is some deterioration of the Newton method performance produced by the increased partitioning used in the additive Schwarz preconditioning. The results for this case indicate that the 16-block partitioning leads to approximately a 40% increase in the overall computational costs of the Newton method as compared to the single partition case. Note that in all cases, the number of GMRES iterations required to reduce the residual for the linear problem by a factor of $\zeta = 0.2$ is typically between 10 and 30. In many cases the number of search directions can be fewer than 5.

Estimates of the parallel performance and scalability of the proposed method on this parallel architecture for the blunt-body flow problem are shown in Figure 2(b). The figure depicts both the relative parallel speed-up, $S_p = \frac{t_1}{t_p} p$ and the relative parallel efficiency, $E_p = \frac{S_p}{p}$ for a fixed-size problem (fixed total computational work) as a function of the number of processors, p, where t_p is the total processor time required to solve the problem using p processors and t_1 is the processor time required to solve the problem using a single processor. Two separate performance curves are shown and compared to the idealized values of each performance measure for up to 32 processors. The results correspond to two different domain decomposition scenarios for the blunt-body flow problem: the 16- and 64-block partitionings of the computational domain. It is evident that the relative parallel speedup is nearly linear in both cases and the parallel efficiency remains above 85%.

5 Conclusions

A highly parallelized implicit AMR algorithm has been developed for predicting two-dimensional compressible gaseous flows on body-fitted multi-block mesh. The parallel implicit method allows for anisotropic mesh refinement and appears to be well suited for predicting complex flows with disparate spatial and temporal scales in a reliable and efficient fashion.

Acknowledgements

This research was supported by a Premier's Research Excellence Award from the Ontario Ministry of Energy, Science, and Technology and by the Natural Sciences and Engineering Research Council of Canada. Funding for the parallel computing facility used to perform the computations described herein was obtained from the Canadian Foundation for Innovation and Ontario Innovation Trust (CFI Project No. 2169).

References

1. T. J. Barth. Paper 93-0668, AIAA, January 1993.
2. V. Venkatakrishnan. Paper 93-0880, AIAA, January 1993.
3. P. L. Roe. *J. Comput. Phys.*, 43:357–372, 1981.
4. B. Einfeldt. *SIAM J. Numer. Anal.*, 25:294–318, 1988.
5. T. J. Linde. PhD thesis, University of Michigan, May 1998.
6. E. F. Toro, M. Spruce, and W. Speares. *Shock Waves*, 4(1):25–34, 1994.
7. J. J. Gottlieb and C. P. T. Groth. *J. Comput. Phys.*, 78:437–458, 1988.
8. C. P. T. Groth, D. L. De Zeeuw, K. G. Powell, T. I. Gombosi, and Q. F. Stout. Paper 99-3273, AIAA, June 1999.
9. C. P. T. Groth, D. L. De Zeeuw, T. I. Gombosi, and K. G. Powell. *J. Geophys. Res.*, 105(A11):25,053–25,078, 2000.
10. M. J. Berger. *J. Comput. Phys.*, 53:484–512, 1984.
11. M. J. Berger and P. Colella. Local adaptive mesh refinement for shock hydrodynamics. *J. Comput. Phys.*, 82:67–84, 1989.
12. M. J. Berger and J. S. Saltzman. *Appl. Numer. Math.*, 14:239–253, 1994.
13. M. J. Aftomis, M. J. Berger, and J. E. Melton. *AIAA J.*, 36(6):952–960, 1998.
14. J. J. Quirk. PhD thesis, Cranfield Institute of Technology, January 1991.
15. J. J. Quirk and U. R. Hanebutte. Report 93-63, ICASE, August 1993.
16. D. De Zeeuw and K. G. Powell. *J. Comput. Phys.*, 104:56–68, 1993.
17. R. L. Davis and J. F. Dannenhoffer. *Int. J. Comput. Fluid Dyn.*, 1:79–93, 1993.
18. M. Sun and K. Takayama. *J. Comput. Phys.*, 150:143–180, 1999.
19. R. S. Dembo, S. C. Eisenstat, and T. Steihaug. *SIAM J. Numer. Anal.*, 19(2):400–408, 1982.
20. Y. Saad. PWS Publishing Company, Boston, 1996.
21. W. D. Gropp, D. K. Kaushik, D. E. Keyes, and B. F. Smith. *Parallel Computing*, 27:337–362, 2001.
22. M. J. Aftomis, M. J. Berger, and S. M. Murman. Paper 2004-1232, AIAA, January 2004.

MPI Parallelization of Unstructured Mesh Adaptation

C.Y. Lepage[1], A. St-Cyr[2], and W.G. Habashi[3]

[1] Newmerical Technologies International, 680 Sherbrooke Street, West, Montreal, QC, Canada H3A 2M7 Claude.Lepage@newmerical.com
[2] National Center for Atmospheric Research, 1850 Table Mesa Drive, Boulder, CO 80305, USA amik@ucar.edu
[3] Computational Fluid Dynamics Laboratory, McGill University, 688 Sherbrooke Street, West, Montreal, QC, Canada H3A 2S6 Wagdi.Habashi@mcgill.ca

Summary. A 3D hybrid unstructured anisotropic mesh adaptation module is parallelized using MPI on distributed-memory architectures. Attention is given to the parallelization issues of the adaptation algorithm, with focus on domain decomposition, dynamic load-balancing, and treatment of the nodes on the frozen interfaces between partitions to achieve a uniformly adapted high-quality converged mesh. A cell agglomeration technique is proposed for the treatment of mesh periodicity, internal faces, and Y^+ adaptation in columns of prisms.

1 Introduction

With the increasing role of mesh adaptation in CFD, recent trends now focus on its parallelization. In many cases, the mesh adaptation module is embedded in a particular flow solver and parallelization of the latter has left the former jeopardizing the overall scalability of the calculations.

Parallelization of mesh adaptation codes is reported in a few instances for hierarchical mesh adaptation on shared-memory architectures [1] and on distributed-memory systems [2, 3]. In hierarchical mesh adaptation, the subdivision of cells is recursively performed, up to a few refinement levels, and coalescence of cells is only permitted for child cells. Node movement is prohibited, as well as the coalescence of parent cells. The adaptation process may also introduce hanging nodes for certain classes of CFD solvers.

The current work presents the parallelization, using MPI, of a fully unstructured mesh adaptation technology which features node movement, edge refinement, coarsening, and swapping [4]. The unstructured implementation aims at equidistributing the error at a given level, as calculated a posteriori from the Hessian of a selected scalar field of the solution. The adapted grids produced are optimal in the sense that they maximize accuracy for a given number of grid points, leading to cost-effective CFD solutions. On the other hand, the hierarchical approach ensures that the error threshold is satisfied, without equidistributing it, thus yielding meshes with a much larger number of grid points.

2 Parallelization Issues

In a distributed-memory environment, the domain is decomposed into as many partitions as there are processors, with the goal being to balance the workload within each partition while minimizing the size of the interface between partitions. Here, the mesh decomposition is performed by applying the multilevel k-way graph partitioning algorithm on the dual graph for the cell-to-cell connectivity of the mesh, using ParMeTiS [5]. The main advantage of the distributed-memory environment over the shared-memory environment is the reduction in the peak memory per processor, despite the added complexities in the implementation for the management of the data structures.

For the partitioning algorithm, a weight is assigned to each cell, based on the estimated workload which is proportional to the solution error. A large error indicates local refinement, while a small error indicates local coarsening. Moreover, CAD projections must be accounted for during the node movement iterations for cells adjacent to walls.

In the parallel adaptation algorithm, the original mesh is partitioned and each partition is adapted for one iteration. An *iteration* of adaptation refers to the following sequence of operations: node movement, edge refinement and coarsening, edge swapping, and node movement. The cells inside each partition are adapted, while the nodes and the faces on interfaces between partitions remain momentarily frozen. This is the minimal overlap region possible between partitions in order to avoid inter-domain communications. It is noted that no volumetric cell overlap is created between partitions.

At the end of each iteration, the mesh is "mixed" and repartitioned, again using the multilevel k-way graph partitioning algorithm in ParMeTiS, and mesh data are migrated across partitions. It is noted that the adaptive repartitioning algorithm offered in ParMeTiS for adaptively refined meshes, `AdaptiveRepart`, yields partitions which are not simply connected. A full repartitioning is thus performed, despite the fact that more cells need to be migrated – which proves beneficial to the quality of the adaptation as more cell "mixing" is performed. The cost of repartitioning is about 20% of the cost of an iteration. This cost also factors in the time to rebuild the cell-to-cell dual graph, to migrate the nodes and the cells, and to regenerate the mesh edges on the new partitions.

The role of the repartitioning step is two-fold: to re-balance the workload and to allow for the adaptation of previously frozen nodes. For the former, an imbalance in the workload soon emerges when refinement and coarsening are allowed and a new estimate of the workload is needed based on the anticipated levels of refinement and coarsening at the next iteration. For the latter, a frontal algorithm is developed that mixes up the cells on new interfaces to make sure that any two volume cells on opposite sides of an old frozen face are assigned to the same processor (see figure 1). In 3-D, this algorithm prevents repeated frozen faces (2-D), although repeated frozen edges (1-D) and nodes (0-D) are possible from one iteration to the next (intersections between surfaces and curves). The fact that several iterations are needed to reach convergence allows for the complete adaptation of the frozen nodes on the interfaces, provided that these nodes do not remain frozen from one iteration to the next. By the nature of this strategy, inter-process communications can be limited to take place only at the end of iterations.

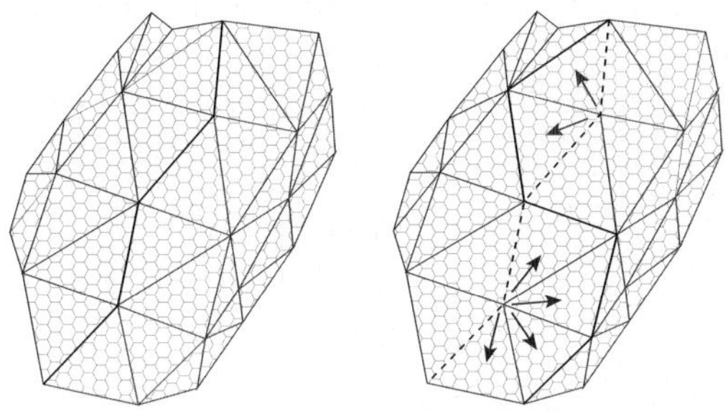

Fig. 1. Treatment of repeated frozen nodes at partition boundaries.

Despite efforts to predict an accurate cell weight factor, the synchronization times are still significant. First, the cost estimate is very crude, and second, ParMeTiS cannot always provide a perfect balance of the workload on all partitions. In order to reduce the elapsed idle time, the node movement iterations after refinement/coarsening/swapping are terminated on all processors as soon as one processor has completed its adaptation cycle. Such a strategy does not lead to a degradation in the quality of the adapted mesh, but does improve scalability significantly.

The parallel adaptation algorithm is extended to include special mesh topologies, namely: periodicity, internal faces, and columns of prisms on no-slip walls. See figure 2. Each of these configurations requires that the involved cells be kept together on the same processor to avoid inter-process communications during the adaptation phase. Such a requirement is satisfied by agglomerating such cell pairings into super cells, via the dual graph of the mesh. For periodicity, translational or rotational, any operation effected on a periodic node or edge must be duplicated on its periodic counterpart. The cell-to-cell connectivity must also "wrap around" the periodic plane to prevent the periodic plane from becoming a permanent partition boundary.

Internal faces are treated in a similar way, with both cells adjacent to an internal face glued together. Finally, a column of prisms is lumped as one super-prism, since any edge operation (refinement, coarsening, or swapping) or node movement operation must be performed on the column of prisms as a whole. Columns of prisms on no-slip walls provide an efficient means by which to satisfy mesh orthogonality and Y^+ control for turbulent flows, while benefiting from the advantages of unstructured meshes. It is noted that

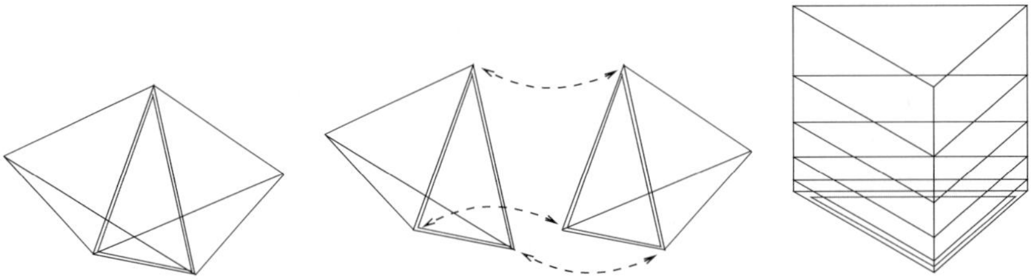

Fig. 2. Special mesh configurations: periodicity, internal faces, columns of prisms.

the proposed cell-agglomeration technique can support any configuration of meshes combining periodicity, internal faces, and/or periodicity.

3 Results

The first example consists of adapting the mesh for a low speed air flow past a ship. Such simulations are useful for training pilots on landing helicopters on the rear deck. This example illustrates the parallelization of the adaptation technology (see figure 3) for tetrahedral meshes. The speed-ups are plotted in figure 4 up to 8 processors. The *adaptation* curve refers to the parallel operations only and excludes overhead processing costs such as loading the CAD geometry, reading and writing the mesh and the solution, and calculating the adaptation error. The *overall* curve includes all costs.

The second example displays geometric mesh smoothing of a Boeing 737 nacelle with 10 layers of prisms. Figure 5 shows the initial and final partitions using 4 processors. Figure 6 shows zooms of the initial and adapted meshes on the symmetry plane, confirming the proper adaptation of the prisms in parallel and the clustering of tetrahedral cells in the boundary layer.

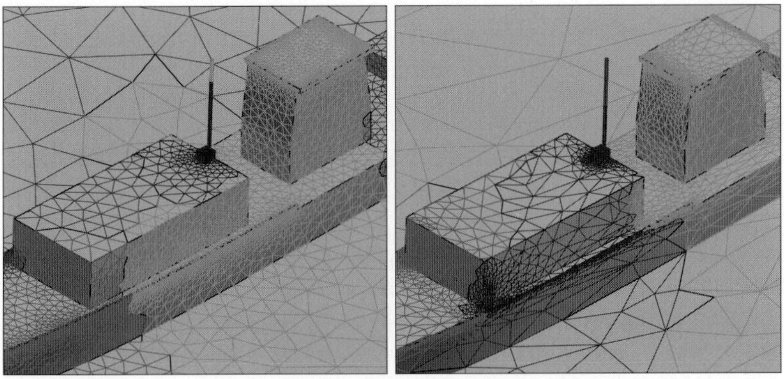

Fig. 3. Parallel Adaptation of a Ship (4 cpus): Initial Partitions (left) and Adapted Partitions (right).

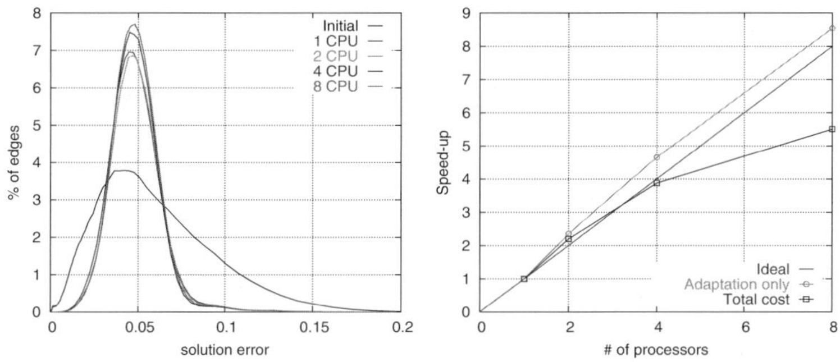

Fig. 4. Observed Error Distribution and Speed-Ups for the Parallel Adaptation of a Ship.

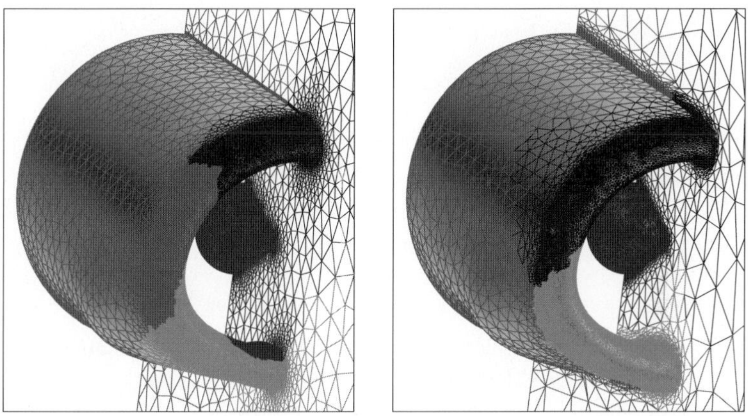

Fig. 5. Parallel Adaptation of a Boeing 737 Nacelle (4 cpus): Initial Partitions (left) and Adapted Partitions (right).

4 Conclusions

Demonstrated results for tetrahedral meshes show excellent scalability, without any degradation in the quality of the adapted mesh due to the parallelization of the algorithm. Parallelization of the pre-processing steps, in particular the evaluation of the error metric, will yield a much improved overall scalability. A cell mixing treatment at repeated identical partition boundaries prevents repeated frozen nodes from iteration to iteration and ensures mesh quality uniformly throughout the mesh without traces of incomplete adaptation on interfaces between partitions. Generalizations to the dual graph of the cell connectivity, used in the partitioning algorithm, allows the treatment of complex mesh topologies such as periodicity, internal faces, and columns of prisms for Y^+ adaptation.

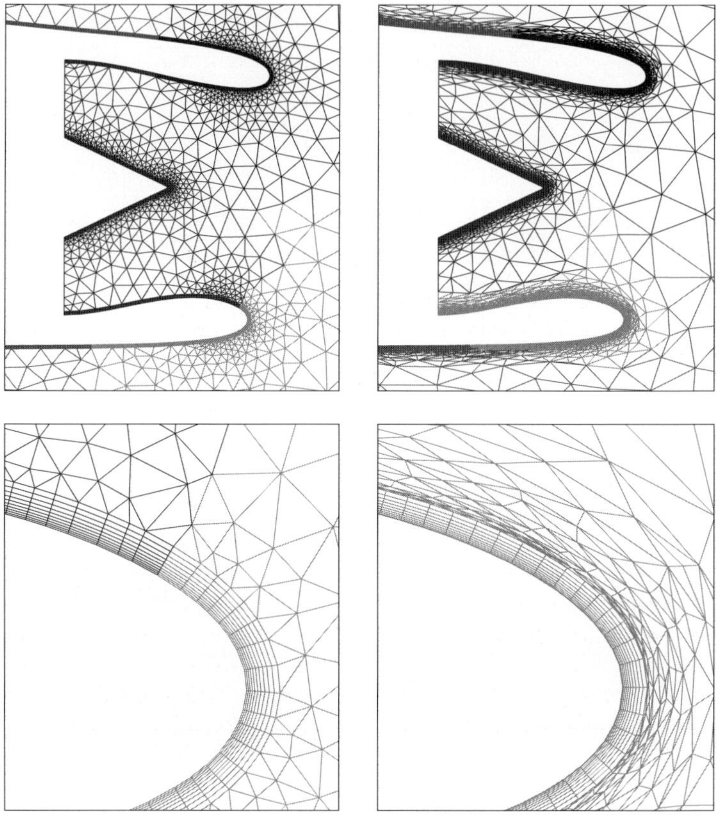

Fig. 6. Parallel Adaptation of the Layers of Prisms for a Boeing 737 Nacelle (4 cpus): Initial Partitions (left) and Adapted Partitions (right).

Acknowledgments

The authors thank CLUMEQ Supercomputing Center for the usage of the computing facilities in the development of this work (`www.clumeq.mcgill.ca`).

References

1. J. Waltz: AIAA Paper 2003-1115 (2003)
2. L. Oliker, R. Biswas, and H.N. Gabow: Parallel Computing, **26**, (2000)
3. X. Zhao et al.: AIAA Paper 2003-0955 (2003)
4. C.Y. Lepage, A. St-Cyr, and W.G. Habashi: AIAA Paper 2004-2532 (2004)
5. G. Karypis, K. Schloegel, and V. Kumar: ParMeTiS: *Parallel Graph Partitioning and Sparse Matrix Ordering Library, Version 3.0* (University of Minnesota, March 2002)

Parallel Implementation of a Dynamic Overset Unstructured Grid Approach

A. Madrane

German Aerospace Center,DLR, Lilienthalplatz 7, 38108 Braunschweig, Germany, aziz.madrane@dlr.de

Summary. The purpose of this study is to develop an efficient parallel computational algorithm for complex geometries using overset unstructured grid technique. The strategy adopted in the parallelization of the overset grids method including the use of **hierarchical** data structure and communication, is described. Numerical results are presented to demonstrate the efficiency of the resulting parallel overset grid method.

1 Introduction

The overlapping grid technique has gradually become one of the most popular algorithms in the field of CFD, allowing flows with arbitrarily shaped flow geometies to be analysed, such as flow around aeroplanes or in solving moving body problems. Benek et al. [1] developed overset grids method in two and three dimensions for Euler Equations. This method is quite powerful for dealing with the moving bodies because remeshing is not required. To apply this method to the unstructured mesh, three major steps are needed:
1. The creation of an overlapping region, this step is part of the grid generator, the user should specify the analytic shapes, such as cubes,spheres, cones, etc. as hole cutting geometries.
2. The cell surrounding each interpolation point is searched from the other grids, using Alternating Digital Tree ADT.
3. The conservative variables are transferred by linear or trilinear interpolation [2].
The purpose of this study is to develop an efficient parallel computation algorithm for analysing the flowfield of complex geometries using **hierarchical** data structure and communication.

This paper is divided into three sections. The second section describes the parallel implementation of the overset unstructured grid method. The last section presents some recent industrial results for unsteady flows.

2 Parallel Implementation

In this section we explain the algorithm and the implementation of a parallel chimera that has been developed for the DLR TAU-Code. At the beginning we have several grid blocks where each grid block is a physical grid with own volume triangulation and boundary description. These grid blocks are not connected. The only communication between these blocks is done by a

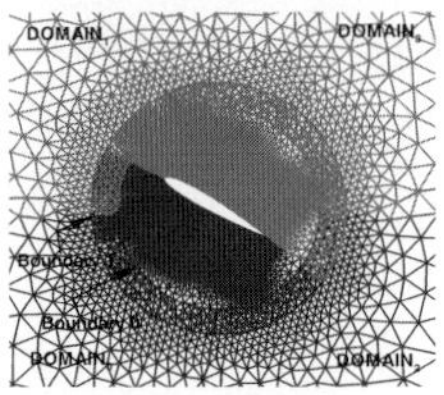

Fig. 1. Overset grid system and partitioned grid

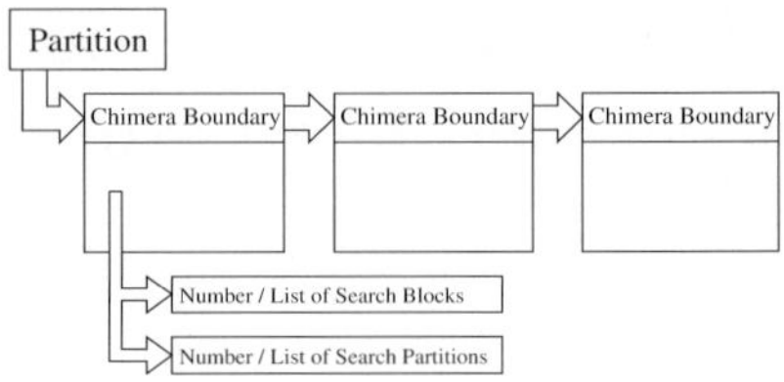

Fig. 2. Input lists for Chimera search

chimera interpolation. The grid blocks are written into one data structure and afterwards these data structure is partitioned. Each partition may have a different set of blocks. There exists a local numbering of the grid blocks for the partitions and a global numbring for the non partitioned grid. The parallel algorithm is explained for the rotating profile see Fig. 1. The grid consists of two blocks and four partitions. The algorithm is explained for partition 0. This partition is denoted the "own" domain. Each partition contains a list of chimera boundaries. Each chimera boundary needs a search partition list and a search block list which is the donor cell block list. The lists are shown in Fig. 2. If no knowledge and no further algorithm is provided the list of search partitions contains all partitions and the list of search blocks contains all blocks excluding the chimera block of the boundary. The list of search blocks should be specified by user input. The search block list has to be determined only once at the beginning whereas the list of search partitiones must be updated each physical time step. **All the information which is sent from the own partition for a chimera search on another partition is prefixed with send. All information which is received from another partition for a chimera search on the own partition is prefixed with recv.** For a parallel chimera search from the list of search partitions and the list of search blocks for each chimera boundary on the domain the following information has to be prepared.

- the number of send partitions and the list of send partitions.
- the number of chimera boundaries per send partition and the list of chimera boundaries per send partition

The lists are generated by setting up a list of pairs of each combination chimera boundary - search partition. The list is reordered to the search partitions in increasing size. The number of send partitions is determined running through the list of pairs and comparing each element with its successor. If it is unequal to the successor and unequal to own domain the number of send partitions is increased. If it is equal to own domain the counter of own domain is set to one. After the size is determined the list of send partitions is built in a second loop. The number of chimera boundaries per send partition is implicitly stored in an index list and the chimera boundaries per search partition are stored in an associated data list. An index list contains the start

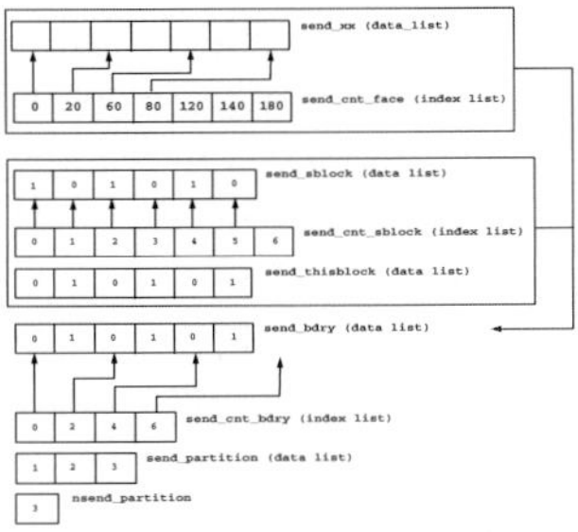

Fig. 3. Data sets to be sent to other domains (hierarchical send lists)

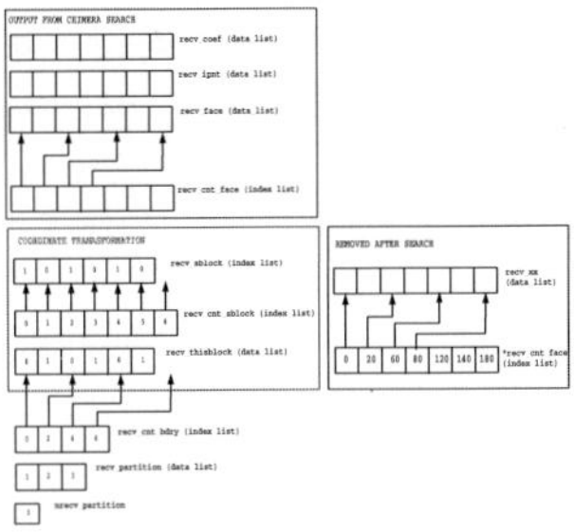

Fig. 4. Receive partition (hierarchical receive lists)

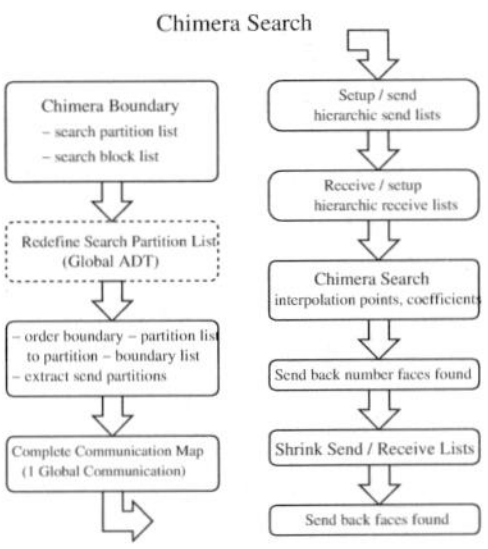

Fig. 5. Steps for parallel overset unstructured grids

index of each entry of an associated data list for each entry of a given data list. The range of values of the associated data list for one entry of the given data list is the difference of entries **i+1** and **i** where **i** is the position of the entry of the given data list. The size of the index list is one more than the size of the given data list. For our example the lists are shown in Fig. 3. The number of send partitions is denotiated with **nsend_partition**, the list of send partitions with **send_partition**, the index list with **send_cnt_bdry** and its associated data list with **send_bdry**.

The index list is built by looping through the list of pairs. First a counter is initialized with zero. Then the partition number of each pair is compared to its successor and each time it is equal and unequal to the own domain the counter is increased. If this test fails or the end of the list is reached the counter is stored at position **i+1** for loop index **i**. The loop stops at the end of the list. To complete the index list the first entry is initialized with zero and successively each element is added up to its successor.

The associated list of chimera boundaries are the chimera boundary entries of the pair list after the reordering. For our example these are the even entries of the ordered list of pairs The list of chimera boundaries to be searched on the own domain similarly can be extracted from the list of pairs. The complete set of data to be sent to the other partitions (of the send partition list) is also shown in Fig. 3. The coordinates of the target points are stored in **send_xx** and are associated with the list of chimera boundaries via the index list **send_cnt_face**. For moving bodies the blocks of the chimera boundaries and the search block list for each chimera boundary has to be communicated. The search block list **send_sblock** is associated with the list of chimera blocks **send_thisblock** via the index list **send_cnt_sblock**. These additional sets are only needed on the receiving partitions. After the communication they can be removed. For the communication the start and end index of each data list which has to be sent to partition **i** is known via the underlying index lists. For example the set of target points to be sent to partition **i** is **send_xx[send_cnt_face[send_cnt_bdry[i+1]]]**−

send_xx[send_cnt_face[send_cnt_bdry[i]]] . Index lists cannot be communicated in this form. They have to be rewritten as lists of differences of consecutive entries and can then be communicated in the same way like a data list. The new composed list of differences after the communication can then be added up the new index list.

For the receiving partitions it is unknown from which partitions data has to be received. To get the number and list of receive partitions communication to all other partitions has to be done. A signal of 1 is sent in case that the corresponding partition is in the send partition list otherwise a signal of zero is sent. Because there is no global synchronisation the receivings have to be stored in a list of the size **ndomains - 1**. After the communication the number of receive partitions is the number of 1 entries in this list and the list of receive partitions are the indices with 1 entries of this list.

The data set which is set up after the communication and which is used for the chimera search is illustrated in Fig. 4. The output of the chimera search are the interpolation points of the donor cells stored in the list **recv_ipnt** and their weights stored in the list **recv_coef**. The local numbers of target points found per search block are stored in **recv_face**. The local numbers refer the numbering of boundary points within each chimera boundary. The index list **recv_cnt_face** setup by communication and containing the number of target points per chimera boundary has been prefixed with a star (***recv_cnt_face**). This list is only used as input to the chimera search and is not used anymore afterwards.

At the beginning for the search of the target points of a chimera boundary the list **recv_face** is initialized from zero up to the number of target points of this boundary. A counter (nifaces) with the number of target points found is set to zero. Then each global block of the search block list (**recv_sblock**) is mapped to the local block number of the own domain. If a block is not found the search skipps to the next search block. If a block is found the coordinates are transformed from the chimera block (**recv_thisblock**) to the search block. The search goes from the number of target points found from the previous search blocks to the number of target points of the boundary. The loop counter is i. If a target point is found nifaces is increased. The interpolation points are written consecutively to **recv_ipnt** and the corresponding weights to **recv_coef**. If the target point is not found position **i** is exchanged with the last position in the list not exchanged before. This position is **number of target points - (nifaces - i)**. After the search the entries of **recv_cnt_face** have to be added up. The sums are the number of target points found per chimera boundary. These values have to be communicated back to the sending partitions. Then both the receive and send lists downwards from **send_cnt_face** and **recv_cnt_face** are shrinked independently see [3]. The intension is to keep only non zero entries in the lists except the first entry of index lists. As soon as the detailed information of the chimera search is available the dependencies and the communication map is adapted to the minimum necessary size. The exchange of data and the communication is optimal for the interpolation of the conservative variables. Within a dual time stepping scheme the interpolation has to be done every inner time step and the chimera search only every physical time step. After all lists have been shrinked the list **recv_face** with the local point numbers of the chimera boundary for each target point found boundary has to be communicated back to the sending partitions.

2.1 Interpolation of Conservative Variables

The interpolation of the conservative variables is done by looping over all target points in one loop. The conservative variables for each entry **recv_ipnt** have to be weighted with the corresponding entry of **recv_coef** and each consecutive set of four variables has to be summed up corresponding to one entry of **recv_face** and written to a list. For interpolations with different types of donor cells another index list has to be introduced which relates **recv_ipnt** with **recv_face**.

For moving grids the interpolated velocity vectors and gradients have to be transformed back from the search block to the chimera block. For each chimera block **i** and each search block **j** in the range **recv_cnt_sblock[i+1]** − **recv_cnt_sblock[i]** one transformation matrix has to be calculated and each interpolated value in the range **recv_cnt_face[j+1]** − **recv_cnt_face[j]** has to be transformed with this transformation matrix. As a last step the interpolated values are communicated back to the sending partitions. The range of interpolated values for partition **i** is

recv_cnt_face[recv_cnt_sblock[recv_cnt_bdry[i+1]]]−
recv_cnt_face[recv_cnt_sblock[recv_cnt_bdry[i]]]

On the sending partition an interpolated value in the range

send_cnt_face[send_cnt_bdry[i+1]]−
send_cnt_face[send_cnt_bdry[i]]

with **recv_face** = **k** has to be written to variable **bpoint[k]** where **bpoint[k]** is the list of boundary points of chimera boundary i of the list **send_bdry** see Fig. 3. The parallel algorithm for overset unstructured grids is summarized in Fig. 5.

3 Numerical Experiments

3.1 Simulation of an Isolated Propeller

An important application area for chimera grids in relative motion is the modeling of the complex aerodynamics of propeller flows. The DLR TAU-Code was used to investigate the low-speed aerodynamics of a modern highly

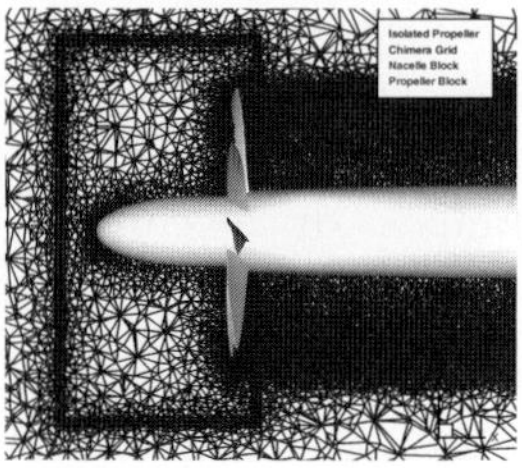

Fig. 6. Isolated propeller geometry, chimera grid and a detail view at the intermesh boundary

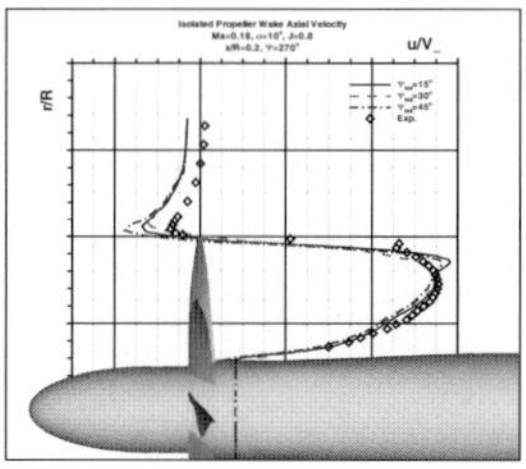

Fig. 7. Comparison of computed axial wake velocity profiles at $x/R = 0.2$ with experiments

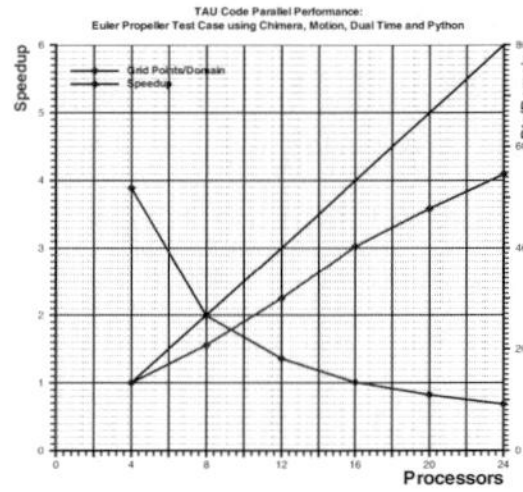

Fig. 8. Parallel speedup - 3D isolated propeller

loaded 8-bladed propeller designed to be used for turboprop transports travelling at high cruise speeds of around $Ma_\infty = 0.7$.

Figure 6 shows the geometry of the propeller-nacelle configuration and a cut of the grid along the symmetry plane. This grid block features a cylindrical hole at the position of the rotating propeller grid block and has a total of 993.549 nodes. For the second block, a grid was generated around the rotating parts of the geometry, i.e. the spinner, hub and the eight propeller blades. The final propeller block grid consists of 1.020.214 nodes so that the total point number for the two grid blocks is 2.013.763.

The flow parameters set for the CFD simulation emulating the scale-model propulsion system used in the experiment are representative for the low-speed take-off conditions and correspond to settings used in the wind tunnel tests. An onflow Mach number of $Ma_\infty = 0.18$ and an angle of attack of $\alpha = 10°$ were specified, the advance ratio was set at $J = 0.8$ and a fine blade pitch angle was selected, corresponding to a thrust coefficient of $c_T = 0.1$. As the intent of the tunnel campaign was to replicate the full-scale advance ratio, Mach number and thrust coefficients, the necessary angular velocity of the model propeller needed to be on the order of 14000 rpm. Initially a steady state computation was run with both grid blocks rotating at the same angular velocity. The solution obtained after two full revolutions served as the starting solution for an unsteady computation with only the propeller block rotating relative to the stationary nacelle block. One propeller revolution was resolved with 360 dual time steps with 40 inner iterations each. All in all seven full propeller rotations were computed using four 1.6GHz processors on a Linux-PC Cluster resulting in a wallclock run-time of 490 hours.

Figure 7 shows a comparison of the computed and experimentally determined axial velocity profiles in the propeller wake at a position of $x/R = 0.2$ behind the propeller along a ray extending out to the right of the nacelle. The figure shows the axial velocity profiles as computed for every 15° of propeller rotation for a 45° period. Agreement with the experimental results is quite favorable.

In the performance of parallel computing, the speedup ratio is plotted in Fig. 8.The speedup ratio of 2 processors over 24 processors is about 8 without multigrid.

References

1. Benek, J.A., Buning, P.G, Steger, J.L., *A 3-D Chimera Grid Embedding Technique*, AIAA 85-1523, 1985.
2. Madrane, A., Heinrich R. and Gerhold T., *Implemetation of the Chimera method in the unstructured hybrid DLR finite volume Tau-Code*, 6th Overset Composite Grid and Solution Technology Symposuim. Ft.Walton Beach, Florida, USA, October 8-10, 2002. http://www.arl.hpc.mil/Overset2002/
3. Madrane, A., Raichle, A., Stuermer, A., *Parallel Implemetation of a Dynamic Unstructured Chimera Method in the DLR Finite Volume TAU-Code*, 12th Annual Conference of the CFD Society of CANADA, Ottawa, Mai 9–11, 2004, pp.524-534.

A Parallel Multi-Block Method for the Unsteady Vorticity-Velocity Navier-Stokes Equations

A. Grimaldi, G. Pascazio, and M. Napolitano

DIMeG, Sez. Macchine ed Energetica, & CEMeC, Politecnico di Bari, Via Re David 200, 70125 Bari, Italy

1 Introduction

Vorticity plays a fundamental role in the physics of vortex dominated flows [1]; thus, it is convenient to use it as one of the dependent variables when developing a new method for the time-dependent incompressible Navier–Stokes equations. The authors have developed numerical methods for solving the vorticity-velocity formulation of the Navier–Stokes equations in two-dimensional general curvilinear geometries for both steady and unsteady problems [2, 3]. The scalar vorticity transport equation is coupled with a second-order equation for the velocity vector, obtained by combining the vorticity definition and the incompressibility condition. The equations are discretized in space by a staggered-grid second-order-accurate finite-volume method and the problem is closed by computing the vorticity at the boundary by means of its definition. For steady flows, the three equations are solved by an alternating direction line-Gauss–Seidel relaxation scheme, using a false-transient approach [2]. For unsteady flows, the vorticity transport equation is discretized in time by means of an implicit three-level scheme and the solution at each time level is obtained through a dual time-stepping technique [4]: a multigrid line-Gauss–Seidel relaxation scheme is applied to the coupled vorticity-velocity system, the vorticity time-derivative being a source term in the vorticity transport equation.
The aim of the present paper is to further reduce the overall CPU time needed for large scale computations either to reach the steady-state solution or to achieve the solution at each time level in unsteady computations using the dual time-stepping technique. Therefore, on one hand more efficient relaxation schemes have been used and, on the other hand, a domain decomposition of the physical space has been employed: the multi-block algorithm allows to handle multiply-connected domains and complex configurations and, more importantly, a straightforward parallelization of the code by solving each grid-block on a single processor.

In the following, the numerical method is described with reference to the two-dimensional case and with particular emphasis on the multi-block algorithm. The standard two-dimensional driven cavity flow problem is considered to verify the effectiveness of the proposed methodology; then, the unsteady two-dimensional flow past a square cylinder is solved to assess the capability of the proposed method for unsteady flow problems.

2 Numerical Method

The non-dimensional incompressible vorticity–velocity Navier–Stokes equations are written in conservative form as:

$$\frac{\partial \omega}{\partial t} + \boldsymbol{\nabla} \cdot (\boldsymbol{u}\omega) - \frac{1}{Re}\nabla^2 \omega = 0, \tag{1}$$

$$\boldsymbol{\nabla} \cdot \boldsymbol{u} = 0, \tag{2}$$

$$\boldsymbol{\nabla} \times \boldsymbol{u} = \omega \boldsymbol{k}, \tag{3}$$

where $\boldsymbol{u}$ is the velocity vector, ω is the only component of $\boldsymbol{\omega}$, $\boldsymbol{k}$ is the unit vector in the direction normal to the plane of the flow, t is the time, and Re is the Reynolds number. It is convenient to obtain a second-order elliptic PDE for the velocity vector in place of the two first-order scalar equations (2) and (3), so as to use standard relaxation methods. Following [2], one subtracts the gradient of equation (2) from the *curl* of equation (3), to obtain:

$$\boldsymbol{\nabla} \times \boldsymbol{\nabla} \times \boldsymbol{u} - \boldsymbol{\nabla}(\boldsymbol{\nabla} \cdot \boldsymbol{u}) = \boldsymbol{\nabla} \times \omega \boldsymbol{k}. \tag{4}$$

By virtue of Gauss' theorem, equations (1) and (4) are discretized in space as:

$$\mathcal{V}\frac{d\omega}{dt} + \sum_l \left(\boldsymbol{u}\omega - \frac{1}{Re}\boldsymbol{\nabla}\omega\right) \cdot \boldsymbol{S}^l = 0, \tag{5}$$

$$\sum_l (\boldsymbol{\nabla} \times \boldsymbol{u} - \omega \boldsymbol{k}) \times \boldsymbol{S}^l + \sum_l (\boldsymbol{\nabla} \cdot \boldsymbol{u})\boldsymbol{S}^l = 0, \tag{6}$$

where $\mathcal{V}$ and $\boldsymbol{S}^l$ indicate the cell control volume and its face area vectors, respectively, the summation being extended to all faces of $\mathcal{V}$. In the present staggered-grid finite-volume discretization, the vorticity is defined at the nodes, whereas the discrete velocity field is approximated by means of the flow rates across the faces of the cells:

$$U = \boldsymbol{u} \cdot \boldsymbol{S}^\xi, \quad V = \boldsymbol{u} \cdot \boldsymbol{S}^\eta. \tag{7}$$

The vorticity transport equation, equation (5), is rewritten as:

$$\mathcal{V}_\omega \frac{d\omega_{i,j}}{dt} = R_{i,j} = -\mathcal{C}_{i,j}(U, V, \omega) + \frac{1}{Re}\mathcal{L}_{i,j}(\omega), \tag{8}$$

where $\mathcal{V}_\omega$ is the vorticity control volume, whereas $\mathcal{C}_{i,j}$ and $\mathcal{L}_{i,j}$ indicate the centred second-order accurate discrete convective and diffusive operators, as given in [2]. The vorticity time derivative is approximated using a three-points backward difference,

$$\frac{d\omega_{i,j}}{dt} = \frac{3\,\omega_{i,j}^{n+1} - 4\,\omega_{i,j}^{n} + \omega_{i,j}^{n-1}}{2\,\Delta t}.$$

Then, the dot product of the vector equation (6) times each face area vector, $\boldsymbol{S}^{\xi}$, $\boldsymbol{S}^{\eta}$, provides two scalar equations for the dependent variables U and V, respectively. Adding a pseudo-time derivative to each governing equation and discretizing it in pseudo-time using a two-level implicit scheme and the delta form of Beam and Warming [5], one obtains:

$$\mathcal{V}_{\omega}\frac{\Delta\omega_{i,j}}{\Delta t} + \mathcal{C}_{i,j}^{u}(U,V,\Delta\omega) - \frac{1}{Re}\mathcal{L}_{i,j}(\Delta\omega) = -\mathcal{V}_{\omega}\frac{d\omega_{i,j}}{dt} + R_{i,j}. \quad (9)$$

$$\mathcal{V}_{U}\frac{\Delta U_{i,j+\frac{1}{2}}}{\Delta\tau} - \mathcal{L}^{\xi}_{i,j+\frac{1}{2}}(\Delta U,V) = \mathcal{L}^{\xi}_{i,j+\frac{1}{2}}(U,V) - \mathcal{D}^{\xi}_{i,j+\frac{1}{2}}(\omega), \quad (10)$$

$$\mathcal{V}_{V}\frac{\Delta V_{i+\frac{1}{2},j}}{\Delta\tau} - \mathcal{L}^{\eta}_{i+\frac{1}{2},j}(U,\Delta V) = \mathcal{L}^{\eta}_{i+\frac{1}{2},j}(U,V) - \mathcal{D}^{\eta}_{i+\frac{1}{2},j}(\omega). \quad (11)$$

In the equations above, $\mathcal{V}_U$ and $\mathcal{V}_V$ are the control volumes for the velocity components U and V, respectively, $\Delta\tau$ is the pseudo-time step, $\Delta\omega$, ΔU, ΔV are the unknowns, namely, the variations of the dependent variables at the new physical time level $(n+1)$ from the old pseudo-time level (k) to the new one $(k+1)$, and the right-hand-sides (*rhs*) are the unsteady residuals, evaluated at $(n+1,k)$. Moreover, in equation (9), $\mathcal{C}^{u}_{i,j}$ indicates the first-order-accurate upwind operator, according to the deferred correction strategy [6], whereas $\mathcal{L}^{\xi}$, $\mathcal{D}^{\xi}$, $\mathcal{L}^{\eta}$ and $\mathcal{D}^{\eta}$ are shorthand notations for the diffusion-like operators and the vorticity coupling terms for the U and V equations, respectively. The operators in equations (10) and (11) are discretized using a second-order finite volume centered scheme, so as to produce solutions satisfying conditions (2) and (3) exactly in the discrete [2]. The vorticity at the boundary is computed using the discrete form of its definition, equation (3).

The three uncoupled equations (9), (10), and (11) are solved sequentially by a single application of one of the following relaxation schemes: forward-backward line-Gauss–Seidel relaxation procedure combined with a FAS multigrid scheme [7] (LGS); the incomplete LU decomposition of Stone [8] accelerated by a FAS multigrid scheme (SIP); a stabilized conjugate gradient squared method, see, e.g. [9], with incomplete Cholesky preconditioning (CGS); a combination of the CGS and SIP methods obtained by performing sequentially the SIP and CGS scheme (CGS-SIP), thus achieving a more effective preconditioning of the CGS scheme. For the CGS and CGS-SIP methods, a restart equal to five and four has been used, respectively.

Concerning the multigrid strategy, full weighting collection and bilinear prolongation operators have been used, see [10]. After driving the *rhs* unsteady residuals to zero, the solution is advanced to the new physical time step.

In this paper multi-block grids are used to discretize the space domain, which allow to handle multiply-connected domains and complex configurations and, more importantly, a straightforward parallelization of the code by solving each grid-block on a single processor. In such an approach, the main concern when applying the aforementioned relaxation methods on multi-block meshes is the treatment of intra-block boundaries, which affects the rate of convergence. In this paper a Schwarz additive algorithm has been used, see [11], characterized by matching overlapping grids. The influence of the number of cells in the overlap region on the convergence rate is crucial and has been investigated in details by numerical experiments in the following section.

3 Results

In order to verify the effectiveness of the multi-block parallel strategy, the steady two-dimensional driven cavity flow at $Re = 1000$ has been computed on a uniform Cartesian mesh with 512×512 cells. All of the four relaxation strategy have been employed to solve the problem using 2, 4, and 9 equal-size blocks and, for each computation, three different numbers of cells in the overlapping region have been used, namely, 0 (non-overlapping blocks), 8, and 16. The single-block computation has also been carried out, for comparison. The computations have been carried out on an IBM eServer 1300 cluster

Table 1. Single-block computations

	LGS	SIP	CGS	CGS-SIP
N_{it}	23803	7349	3867	1827
CPU time [s]	156700	28775	21475	11143

consisting of 16 two-processor nodes (Intel Pentium III, 1.4 GHz). Table 1 provides the number of iterations (N_{it}) and the CPU time required by the single-block computations to drive the residual to 10^{-12}, whereas tables 2 and 3 provide, for each relaxation method and number of cells in the overlapping region, the number of iterations and the speedup, S, evaluated as the ratio between the corresponding CPU time and that of the single-block calculation. It appears that the efficiency increases going from the LGS method to the

Table 2. Multi-block computations with the LGS and SIP methods

	LGS						SIP					
	0		8		16		0		8		16	
Blocks	N_{it}	S	N_{it}	S	N_{it}	S	N_{it}	S	N_{it}	S	N_{it}	S
2	26072	1.83	24320	1.97	24245	1.97	8423	1.86	7382	1.99	7352	2.00
4	29885	3.06	27872	3.68	26710	3.75	10857	2.78	7630	3.96	7374	3.99
9	33408	6.33	26898	7.33	25023	7.94	11473	5.79	8528	7.55	7519	8.70

Table 3. Multi-block computations with the CGS and CGS-SIP methods

	CGS						CGS-SIP					
	0		8		16		0		8		16	
Blocks	N_{it}	S	N_{it}	S	N_{it}	S	N_{it}	S	N_{it}	S	N_{it}	S
2	8427	0.91	5328	1.48	4410	1.63	4110	0.93	2319	1.57	2008	1.80
4	13573	1.01	6822	2.11	4474	3.21	5245	1.32	2782	2.58	2699	2.61
9	11812	2.96	6969	4.97	5735	5.96	5697	2.90	3619	4.62	3377	4.92

CGS-SIP one and, more importantly, that the modification of the preconditioning procedure by means of the SIP method is indeed effective. Concerning the performance of the multi-block parallel procedure, the number of cells in the overlap region plays a fundamental role, the speed-up increasing with the number of cells. In particular, using 16 cells, the CPU times of the LGS and SIP methods scale linearly with the number of processors, whereas the *more implicit* CGS and CGS-SIP methods do not achieve the theoretical linear scaling.

Then, the unsteady flow past a square cylinder at $Re = 150$ has been computed using the mesh employed in [12], namely, a grid with eight blocks and a total of 168×120 cells, see figure 1, and $\Delta t = 0.025$. At each time

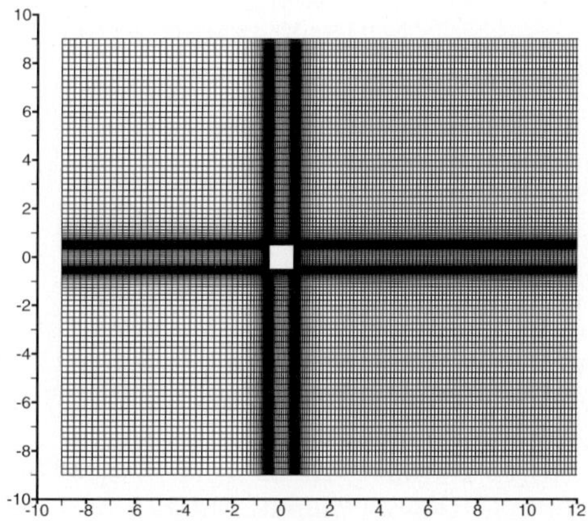

Fig. 1. Square cylinder: Computational grid.

step, the unsteady residual has been reduced to 10^{-7}, requiring about 350 iterations using the CGS-SIP method, which amounts to about 40 CPU seconds on the aforementioned computer. The computed Strouhal number is equal 0.165, the reference value provided in [12]. Figures 2 and 3 provide the horizontal and vertical velocity components at points (4,0) and (7,0), respectively, the origin of the Cartesian axes coinciding with the centre of the square. Notwithstanding the agreement with the results of [12], the same

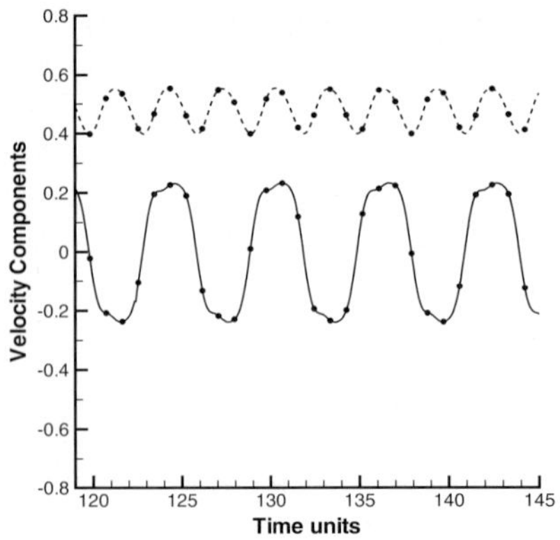

Fig. 2. Time history of the horizontal (dashed line) and vertical (solid line) velocity components at $(4, 0)$.

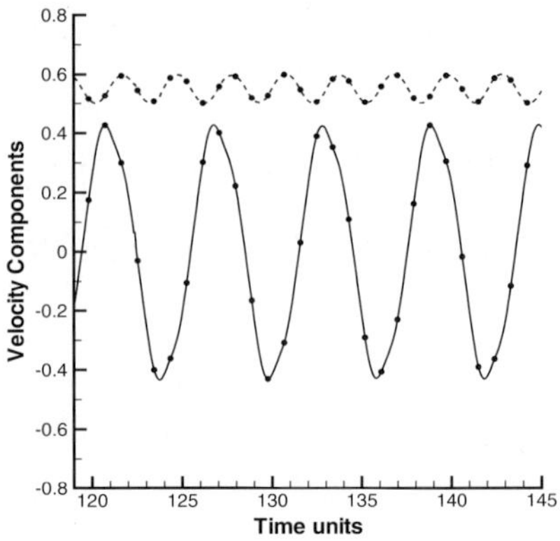

Fig. 3. Time history of the horizontal (dashed line) and vertical (solid line) velocity components at $(7, 0)$.

problem has been computed using a grid with a total of 336×240 cells and $\Delta t = 0.0125$. The corresponding results are provided as symbols in figures 2 and 3, which clearly indicate grid convergence.

References

1. H. Fasel: J. Fluid Mech. **78**, 355 (1976).
2. G. Pascazio and M. Napolitano: Comput. Fluids **25**, 433 (1996).
3. P. De Palma, G. Pascazio, and M. Napolitano: Int. J. Num. Meth. Heat and Fluid Flow **11**, 286 (2001).
4. Pascazio G., Grimaldi A., and Napolitano M., 'An accurate and efficient technique for unsteady viscous flows in two dimensions'. In *Computational Fluid Dynamics 2002*, ed by S. Armfield, P. Morgan, and K. Srinivas (Springer, New York 2003) pp. 725-730.
5. R. M. Beam and R. F. Warming: AIAA J. **16**, 393 (1978).
6. P. K. Khosla and S. G. Rubin: Comput. Fluids **2**, 207 (1974).
7. A. Brandt: 'Guide to multigrid development'. *Lecture Notes in Mathematics* **960**, pp. 220-312 (1982).
8. H.L. Stone: SIAM J. Numer. Anal. **5**, 530 (1968).
9. J. H. Ferziger and M. Perić: *Computational methods for fluid dynamics.* (Springer, Berlin Heidelberg 1997)
10. M. Napolitano and G. Pascazio: Comput. Fluids **19**, 489 (1991).
11. B. Smith, P. Bjorstad, and W. Gropp: *Domain decomposition: parallel multilevel methods for elliptic partial differential equations.* (Cambridge University Press, Cambridge 1996)
12. Ahmad Sohankar, C. Norberg, and L. Davidson: Phys. Fluids **11**, 288 (1999).

Parallelization of an Unstructured Data Based Cell Centre Finite Volume Code, HIFUN-3D

Gopal N. Shinde[1], Nikhil V. Shende[2], and N. Balakrishnan[3]

[1] Indian Institute of Science, Bangalore-12 gopal@aero.iitb.ac.in
[2] Indian Institute of Science, Bangalore-12 nikvijay@aero.iisc.ernet.in
[3] Indian Institute of Science, Bangalore-12 nbalak@aero.iisc.ernet.in

Summary. This work describes the parallelization of High Resolution flow solver on unstructured meshes, HIFUN-3D, an unstructured data based finite volume solver for 3-D Euler equations. For mesh partitioning, we use METIS , a software based on multilevel graph partitioning. The unstructured graph used for partitioning is associated with weights both on its vertices and edges. The data residing on every processor is split into four layers. Such a novel procedure of handling data helps in maintaining the effectiveness of the serial code. The communication of data across the processors is achieved by explicit message passing using the standard blocking mode feature of *Message Passing Interface* (MPI). The parallel code is tested on PACE++128 available in CFD Center (Department of Aerospace Engg.) of Indian Institute of Science (IISc), Bangalore.

1 Introduction

The code HIFUN-3D [1] is a general purpose flow solver for Euler equations. It uses an unstructured data employing a face based connectivity. The code can handle grids with basic elements like hexahedra, tetrahedra, prisms, pyramids or any combination of these elements. The code is equipped with embedded grid refinement allowing for hanging nodes [2] which leads to the formation of grids with arbitrary polyhedral volumes. The code is provided with a variety of numerical schemes for interfacial flux computations and a variety of boundary conditions. The code makes use of least squares based linear reconstruction procedure [3, 4, 5] with Venkatakrishnan limiter [6]. It is equipped with implicit relaxation procedures like Point Jacobi [7], Symmetric Gauss Seidel [7, 8] and Lower-Upper Symmetric Gauss Seidel [7, 8, 9, 10] in the framework of matrix free algorithms, to accelerate convergence to steady state. Due to these features, the code HIFUN-3D is capable of handling geometric complexities and intricate flow physics frequently arising in the industrial computations. But more often, the size of problems in terms of computational resources and time, involved in the industrial computations is so large that one needs to go for parallel computing in order to provide meaningful results, in a turnaround time small enough to impact design. Thus, the present work aims at parallelization of the code HIFUN-3D in such a fashion that the generality of the sequential code is also carried forward to parallel HIFUN-3D.

2 Algorithm

Our objective is to parallelize the code HIFUN-3D satisfying the constraints: (1) The computational load on each of the processors should be balanced. (2) The data to be communicated between the processors should be minimized. (3) The performance of the serial version of the code should not degenerate because of parallelization for both explicit and implicit procedures. (4) The data availability on the processor should ensure minimal or no change in the serial code.

The first two objectives are associated with domain decomposition procedure. The requirements three and four which greatly depend on preprocessing of the data after domain decomposition are related and are very important for an unstructured data based code.

2.1 Domain Decomposition

For the purpose of domain decomposition, we have employed METIS [11], a software for multi-constraint partitioning of unstructured graphs. This feature of METIS is very important for codes like HIFUN-3D, which allow for arbitrary polyhedral volumes and employ least squares based reconstruction procedure. The first step involves generating the graph file to be input to METIS from the grid file. The generation of the unstructured graph is pictorially depicted for a 2D-example in figure 1-(a). While converting the mesh into a graph, the volumes form the vertices of the graph and the edges are formed based on the connectivity employed in the reconstruction procedure. It is important to note that a face based neighbour (which is most commonly used for generating the edges on the graph) is necessarily included in the connectivity set used for reconstruction. Also, even one way connectivity is represented as an edge on the graph. By this we mean that vertex i may have been included in the connectivity set of vertex j but vice versa may not be true. This kind of situation may be encountered in tetrahedral mesh, with a large number of node based neighbours. The use of this kind of enlarged connectivity can substantially increase the communication between the processors. At the same time, this problem is not specific to least squares based reconstruction procedure alone; for example, use of Green Gauss reconstruction will require a connectivity set including node sharing neighbours. The next issue is associating weights with the vertices and the edges of the graph.

The weights associated with the vertices dictate the resulting load balance and the weights associated with the edges dictate the communication overheads. For a code like HIFUN-3D, there are two important tasks which are compute intensive. They are: (1) Solution reconstruction. (2) Implicit/explicit state update involving flux integration on the volume interface. This immediately suggests use of a multi constraint partitioning software like METIS , which can ensure load balancing on both these counts. The computational effort involved in state update is proportional to the number of faces the volume has; this suggests a vertex weight proportional to the number of faces the corresponding volume has. Similarly, for the purpose of reconstruction, the weight associated with a vertex is proportional to the size of its connectivity set. If an edge on the graph, represents two way connection, it is given a unity weight, otherwise, a weight of $\frac{1}{2}$ is associated with the edge. In our view, the success of the partitioning strategy critically depends on the choice of weights. Presently efforts are on to understand the aforesaid partitioning strategy using METIS .

2.2 Data Classification

In order to satisfy the third and fourth constraints, it is just not sufficient to ensure that any processor is not starved of required data, but also the update of all the cells on a given domain should proceed as it would on a serial machine. To achieve this goal we employ a “multi-layered” approach to solution update on a given domain.

The data required for solution update of the cells belonging to any subdomain may be available on the corresponding processor itself or may have to be communicated from the neighbouring domains. In the present procedure, if D_i represents the set of cells belonging to domain i and A_{ij} represent the set of auxiliary cells belonging to domain j, whose data is required for update on i, then the entire data associated with $D_i \cup A_{ij}$ is contiguously stored on i. The set A_{ij} in turn can be considered to comprise of two sets of cells, flux cells F_{ij} and state cells S_{ij}, i.e., $A_{ij} = F_{ij} \cup S_{ij}$. A cell $k \in D_j$ is called a flux cell if at least one of its volume interface is shared with any cell belonging to D_i. If R_m represents the connectivity set (for reconstruction) of any cell $m \in D_i$, then a cell $k \in (D_j - F_{ij})$ is called a state cell if $k \in R_m \forall m \in D_i$. From these definitions, it is amply clear that,
(1) if cell $k \in D_i$, computations associated with k involves both reconstruction and solution update.
(2) if cell $k \in F_{ij}$, not only the cell averaged states of this cell are required in subdomain i but also the states extrapolated on the face shared by k with subdomain i are required.
(3) if cell $k \in S_{ij}$, the data corresponding to cell averaged states of this cell need to be just made available for this cell.
Thus, this approach can be considered as a three layered approach, with an inner core D_i and two layers of flux cells and state cells. Further distinction of the core cells may also be drawn for effecting non-blocking communication of MPI [12]. For this purpose, we define a peripheral set of cells $P_i \in D_i$ requiring data from outside the subdomain i for state update. This results in a four layered approach, where the state update of any cell $k \in (D_i - P_i)$ may be performed without having to wait for data from other subdomains. While these computations are being performed, the data associated with P_i may

be transfered. Thus, we can visualize any subdomain of the computational domain to be made up of four layers of cells as shown in figure 1-(b).

3 Results and Discussion

Two test cases have been considered to demonstrate the performance of parallel HIFUN-3D. For both the cases, interfacial flux computation is using van Leer scheme [13] and Point Jacobi implicit relaxation procedure is used for time integration. The first test case is transonic flow past ONERA M6 wing with $M_\infty = 0.84$ and $\alpha = 3.06^o$. For this case, two grids have been made use of. Grid-1 is the hexahedral grid with number of points, cells and faces as 418200, 400000 and 1216400 respectively. Grid-2 is the tetrahedral grid with number of points, cells and faces as 223383, 1184118 and 2405853 respectively. Figure 2 shows the pressure distribution at different stations along the wing span compared with the experimental results using Grid-1 along with the pressure contours obtained on wing and the symmetry plane using Grid-2. The second test case is supersonic flow past a cruciform wing–body [14] with $M_\infty = 1.6$ and $\alpha = 10^o$. The grid-3 used for this case consists of 363762 points, 1991252 cells and 4040979 faces. Figure 3 shows the surface grid and pressure contours on the wing–body. The normal force coefficient obtained numerically is 5.532 while that reported experimentally is 5.6. Figure 4 shows the speed up ratio curves for all the three grids. If the *reference time* is the total computational time required for reaching steady state using minimum number of processors, then the speed up ratio for a particular run with given number of processors is given by

$$\text{Speed up ratio} = \frac{\text{Reference time}}{\text{Time req. for particular run}} \tag{1}$$

The slope of the linear part of speed up curve for Grid-1 is 0.8013 (ideal slope = 1), that for Grid-2 is 0.2798 (ideal slope $= \frac{1}{3} = 0.3333$) and that for Grid-3 is 0.18 (ideal slope $= \frac{1}{4} = 0.25$). Also, the optimal number of processors for Grid-1 are 14, that for Grid-2 are 32 and that for Grid-3 are 48.

4 Conclusion

In the present work, an attempt has been made to develop parallel version of HIFUN-3D, a general purpose code based on unstructured finite volume methodology for 3-D Euler equations. The parallelization of the code is achieved using standard blocking mode message passing feature of MPI. The code is tested on PACE++128 machine. The parallel code is validated using two test cases mentioned in the previous section. From the results, it can be concluded that the present code can be used for performing the distributed memory parallel computations for the flow past realistic configurations.

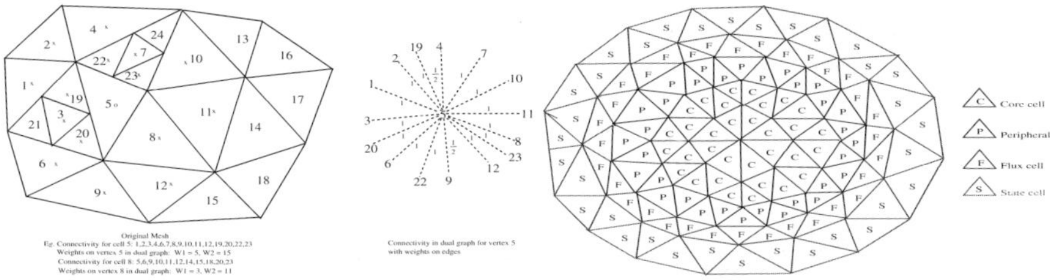

(a) Mesh to Dual Graph conversion (b) Subdomain with four cell layers

Fig. 1. Graph input to METIS and structure of a subdomain

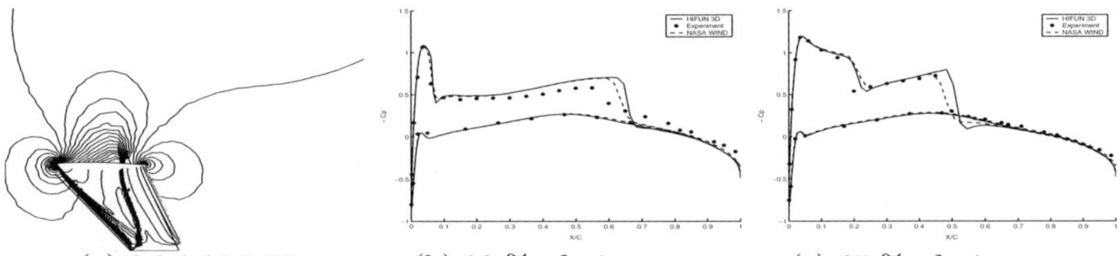

(a) 0.3:0.02:1.52 (b) 20 % of wing span (c) 65 % of wing span

Fig. 2. Pressure contours (pmin:Δp:pmax) and Cp distribution for ONERA M6

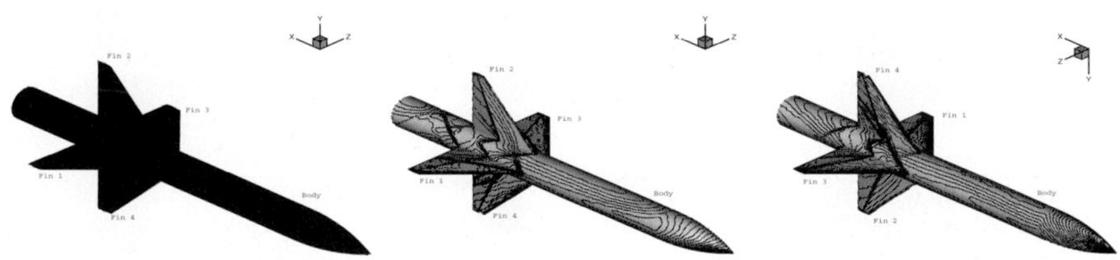

(a) Surface Mesh (b) 0.01:0.013:0.64 (c) 0.01:0.013:0.64

Fig. 3. Surface Mesh and pressure contours (pmin:Δp:pmax) on leeward and windward surfaces for slender body with fins

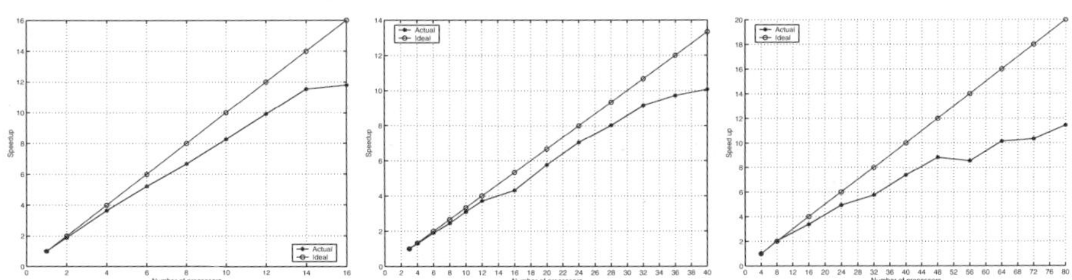

(a) Grid 1 (b) Grid 2 (c) Grid 3

Fig. 4. Speedup achieved on various grids

Acknowledgement

Authors would like to thank Mr. Sundar, Mr. Ramkumar and Mr. Natarajan of REALTIME TECHSSOLUTIONS Pvt. Ltd., Bangalore, for helping us with making runs of parallel HIFUN-3D on PACE++128. Authors are also thankful to Mr. Suresh, Research Scholar, Dept. of Aerospace Engg., IISc, for his valuable comments and suggestions while including the parallelization modules in HIFUN-3D. Finally, authors are thankful to Fluent India Pvt. Ltd. for their support in grid generation.

References

1. Shende N., Balakrishnan N.: Fluid Mechanics Rept. 2004 FM 10, Dept. of Aerospace Engg., Indian Inst. of Science, Bangalore (2004).
2. Shende N., Garg U., Karthikeyan D., Balakrishnan N.: In: *Computational Fluid Dynamics 2002*, ed by F. Armfield, P. Morgan, K. Srinivas (Springer, Berlin Heidelberg New York 2002) p 94.
3. Barth T. J.: In: *Proceedings of 13th ICNMFD, Lecture Notes in Physics*, 414, ed by M. Napolitano and F. Sabetta (Springer-Verlag, Berlin 1992) p 240.
4. Ashford G. A., Powell K. G.: CFD Lecture Series: 1996-06, von Karman Institute, Belgium (1996).
5. Balakrishnan N., Deshpande S. M.: In: *Proceedings of the First Asian CFD conference*, Vol. 1, The Hong Kong University of Science & Technology, Hong Kong , pp. 359-364 (1995).
6. Venkatakrishnan, V., JCP, Vol. 118, pp 120–130 (1995).
7. Nikhil Shende, Balakrishnan N.: To appear in AIAAJ (2004).
8. Hong Luo, Dimtri Sharov, Joseph D Baum, Rainald Lohner: In: *Computational Fluid Dynamics 2000*, ed by N. Satofuka (Springer, Berlin Heidelberg New York 2000) p 93.
9. Hong Luo, Joseph D. Baum, Rainald Lohner, JCP, Vol. 146, pp 664–690 (1998).
10. Yoon, S., Jameson, A., AIAA Journal, Vol.26, No.9, pp 1025–1026 (1988).
11. George Karypis, Vipin Kumar: Metis–version 4.0, University of Minnesota, Department of Computer Science/Army HPC Research Center, Minneapolis, MN 55455 (1998).
12. Marc Snir, Steve W. Otto, Steven Huss-Lederman, David W. Walker, Jack Dongarra: MPI– The Complete Reference. *Scientific and Engineering Computation*, ed by Janusz Kowalik (1996).
13. van Leer B., Report No. 82-30, NASA Langley Research Centre , Hampton, Virginia (1982).
14. Robert L. Stallings Jr., Milton Lamb, Carolyn B. Watson, NASA Technical Paper 1683, Langley Research Centre, Hampton, Virginia (1980).

Parallel Turbulent Flow Computations Using a Hybrid Spectral/Finite-Element Method on Beowulf Clusters

David Vanden-Abeele[1], Gérard Degrez[1], and Deryl Owen Snyder[2]

[1] Université Libre de Bruxelles, Belgium
[2] Brigham Young University, Provo, Utah 84602, USA,
david.vanden.abeele@ulb.ac.be, gdegrez@ulb.ac.be

1 Introduction

Direct or large eddy simulations (DNS/LES) of incompressible turbulent flows are often restricted to academic configurations with trivial boundary conditions, for which the full power of pseudo-spectral methods can be harnessed: superior accuracy combined with high computational speed and very low memory overhead. The use of higher-order finite-difference (-volume) methods on multi-block structured meshes allows to consider more challenging flows [1]; nevertheless, geometries remain fairly simple. During the past decade, turbulence researchers have started to explore the use of finite-element methods (FEMs) on unstructured meshes [2], which in principle allow to handle entirely general geometries.

Considerable work remains to be done before FEMs will become accepted by the wider turbulence community. The feasibility of FEMs has not yet been sufficiently demonstrated for the standard DNS and LES test cases. Also, the dispersive and dissipative behavior of finite-element schemes requires deeper investigation. Last but not least, unlike for finite-differences, classical FEMs inherently require the solution of a large spare linear system at each time step; it is not obvious to do this in acceptable computation times.

Rather than to directly create an unstructured finite-element solver for flows over fully 3D geometries, our team has chosen to first address these open problems using a hybrid spectral/finite-element research code, which allows to simulate fully 3D flows around 2D geometries of arbitrary complexity (SFELES, see Ref. [3]). The present paper provides a description of this code, with emphasis on issues of parallelization. The performance of the code is analyzed by means of the direct numerical simulation of the flow around a square cylinder at moderate Reynolds numbers on a cluster of LINUX PCs.

2 Discretization of the Governing Equations

We advance the incompressible Navier-Stokes equations in time in a second-order accurate manner, by setting the following continuity and momentum residuals to zero at each new time step:

$$
\begin{aligned}
R_{\text{Cont}}^{n+1} &= \nabla\cdot(\boldsymbol{u}^{n+1}) = 0, \\
\boldsymbol{R}_{\text{Mom}}^{n+1} &= (\boldsymbol{u}^{n+1} - \boldsymbol{u}^{n})/\Delta t \\
&\quad + \nabla p^{n+1/2} - \frac{1}{2Re}\nabla^2(\boldsymbol{u}^{n+1} + \boldsymbol{u}^{n}) \\
&\quad + (3/2)\nabla\cdot(\boldsymbol{u}^{n}\otimes\boldsymbol{u}^{n}) - (1/2)\nabla\cdot(\boldsymbol{u}^{n-1}\otimes\boldsymbol{u}^{n-1}) = 0.
\end{aligned}
\tag{1}
$$

Herein, the pressure and viscous terms are treated in an implicit manner using the second-order accurate Crank-Nicolson scheme, whereas the convective terms are treated explicitly using the Adams-Bashforth method.

We restrict our attention to flows which are periodic in the span-wise direction z. In planes $z = cst$, we represent the unknowns $q = u, v, w, p$ by means of linear, triangular elements:

$$
q^n(x,y,z) = \Sigma_{j=1}^{N_{\text{Nod}}} q_j^n(z,t) N_j(x,y), \tag{2}
$$

where the index j runs over all N_{Nod} nodes in the mesh. Each basis function $N_j(x,y)$ takes a value of 1 in the node j and decreases to zero in a piecewise linear manner over the neighboring elements.

We discretize Eqs. (1) in a second-order accurate manner in space by means of the following weighted residual statement:

$$
\int N_i R_{\text{Cont}}^{n+1}\,\mathrm{d}\Omega + \int \tilde{\nabla} N_i\cdot(\tau_{\text{pspg}}\boldsymbol{R}_{\text{Mom}}^{n+1})\,\mathrm{d}\Omega = 0, \tag{3}
$$

$$
\int N_i \boldsymbol{R}_{\text{Mom}}^{n+1}\,\mathrm{d}\Omega - \int \tilde{\nabla} N_i\cdot(\nu_{\text{jst}}\tilde{\nabla}^2\boldsymbol{u}^n)\,\mathrm{d}\Omega = 0. \tag{4}
$$

where $\tilde{\nabla} = (\partial/\partial x, \partial/\partial y, 0)$. Readers familiar with CFD will identify the stabilization term in Eq. (3) as the well known Pressure-Stabilized Petrov-Galerkin dissipation [4], whereas the term added to the momentum residual closely resembles the higher-order dissipation used in the finite-volume scheme of Jameson, Schmidt and Turkel [5].

The time scale τ_{pspg} and hyper-diffusivity coefficient ν_{jst} appearing in the stabilization terms are constant over each element and are given by

$$
\tau_{\text{pspg}} = \min(1, Re_e/3)\frac{h_e}{2\beta}; \qquad \nu_{\text{jst}} = \alpha\,\min(1, Re_e/3)\beta h_e^3. \tag{5}
$$

Herein, h_e represents the length of the shortest edge of an element e, β stands for an estimate of the maximal velocity in the solution domain and the element Reynolds number is defined to be $Re_e = h_e\beta/\nu$. To avoid adding excessive amounts of artificial dissipation, the coefficient α in ν_{jst} is set to a small value of 0.1. The velocity Laplacians that appear in the stabilization of the momentum equation (4) are computed in each node using a lumped mass-matrix:

$$
(\tilde{\nabla}^2\boldsymbol{u}^n)_i = -\int \tilde{\nabla} N_i\cdot\tilde{\nabla}\boldsymbol{u}^n\,\mathrm{d}\Omega \Big/ \int N_i\,\mathrm{d}\Omega. \tag{6}
$$

3 Time Integration Scheme

We cast the system of discretized equations (3) and (4) under a matrix form:

$$\left[M_1 + M_2\frac{\partial}{\partial z} + M_3\frac{\partial^2}{\partial z^2}\right]\begin{bmatrix}\Delta p\\ \Delta \boldsymbol{u}\end{bmatrix} = \begin{bmatrix}R^n_{\text{Cont}}\\ \boldsymbol{R}^n_{\text{Mom}}\end{bmatrix}. \tag{7}$$

where $\Delta p(z) = [(p_i^{n+1} - p_i^n); i = 1, ..., N_{\text{Nod}}]$ and $\Delta \boldsymbol{u}(z) = [(\boldsymbol{u}_i^{n+1} - \boldsymbol{u}_i^n); i = 1, ..., N_{\text{Nod}}]$. The physical space unknowns in this linear system are still a function of the transverse coordinate z. Coupling in the transverse direction occurs explicitly through the nonlinear terms in the RHS and implicitly through the z-derivatives in the LHS.

To take into account variations in the z-direction, we represent the nodal unknowns $q_j^n(z)$ by means of a discrete sum of N_{mod} Fourier modes:

$$q_j^n(z) = \frac{1}{N_{\text{mod}}}\Sigma_{k=0}^{N_{\text{mod}}-1}\hat{q}_{k,i}\exp\left(2\pi Ikz/L\right), \tag{8}$$

where L stands for the periodic length. If we now apply the fast Fourier transform (FFT) algorithm to transform Eqs. (7) to Fourier-space, then the above 3D problem reduces to a set of loosely coupled 2D problems for the Fourier mode unknowns $\hat{q}_{k,i}$:

$$\left[M_1 + M_2(2\pi Ik/L) + M_3(2\pi Ik/L)^2\right]\begin{bmatrix}\Delta \hat{p}_k\\ \Delta \hat{\boldsymbol{u}}_k\end{bmatrix} = \begin{bmatrix}\hat{R}^n_{\text{Cont},k}\\ \hat{\boldsymbol{R}}^n_{\text{Mom},k}\end{bmatrix}. \tag{9}$$

We remark that:

- Coupling between the different Fourier mode happens in a purely explicit manner, via the RHS terms $\hat{R}^n_{\text{Cont},k}$ and $\hat{\boldsymbol{R}}^n_{\text{Mom},k}$.
- For each mode k, a 2D linear system needs to be solved at each time step; the system matrix $A_k = [M_1 + M_2(2\pi Ik/L) + M_3(2\pi Ik/L)^2]$ is different for each Fourier mode, but constant in time.
- We solve for both the real and complex parts of the Fourier coefficients $\hat{q}_{i,k}^{n+1}$. As the physical variables themselves are real-valued, however, it can be shown that
$$\hat{q}_{i,k}^{n+1} = (\hat{q}_{i,N_{\text{Mod}}-k}^{n+1})^*. \tag{10}$$
This implies that we must only solve for modes $k = 0$ up to $N_{\text{Mod}}/2$; the remaining modes follow directly from Eq. (10). In addition, since modes 0 and $N_{\text{Mod}}/2$ are both purely real, we can combine them in a single complex-valued linear system, such that in practice, only $N_{\text{Mod}}/2$ linear systems are solved.

4 Parallel Implementation

In Fig. 1, we show a typical spectral/finite-element mesh used by the algorithm. The mesh has been partitioned into 16 subdomains in FEM space, using METIS2004. The parallel 'pseudo-spectral' time-integration method functions as follows:

(1) The nonlinear terms are first computed in physical space on the different partitions, using stabilized finite-elements in the (x, y) directions and central finite-differences in the z-direction.
(2) FFTs are performed on each sub-domain, to obtain the residual terms in Fourier space.
(3) **Transformation from physical to Fourier space:** for each Fourier mode k ($k = 0, N_{\mathrm{Mod}}/2$), the residuals $\hat{R}^n_{\mathrm{Cont},k}$ and $\hat{\boldsymbol{R}}^n_{\mathrm{Mom},k}$ are gathered from all partitions onto process k.[3]
(4) The solution update is obtained in Fourier space by solving $N_{\mathrm{Mod}}/2$ 2D linear problems using a sequential linear solver.
(5) **Transformation from Fourier to physical space:** the solution update is scattered back to the respective subdomains. By means of an inverse FFT, we retrieve the updated solution q_i^{n+1} in physical space, after which we return to step (1).

The transformations between physical and Fourier space are equivalent to a parallel matrix transposition and are achieved using the ALLTOALL command of the MPI library.

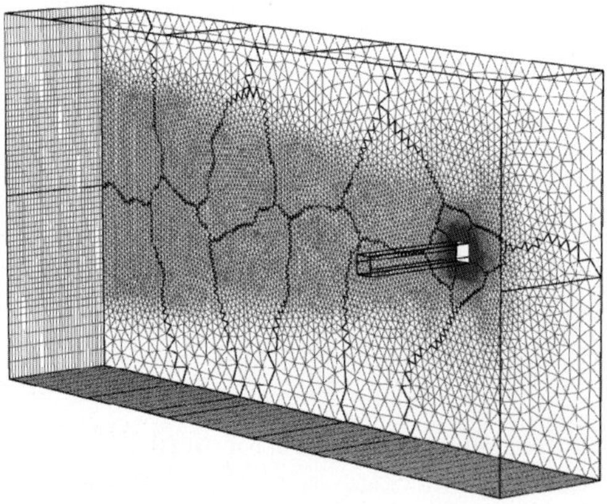

Fig. 1. Hybrid spectral/finite-element mesh around a square cylinder ($\sim$ 8000 nodes; 32 Fourier modes; 16 subdomains).

[3]With the exception of the real-valued mode $N_{\mathrm{Mod}}/2$, which is treated as if it were the complex part of mode 0, on process 0.

5 Results

To evaluate the performance of the parallel algorithm, we consider the DNS of the incompressible flow over a square cylinder with edges of length D at $\mathrm{Re} = 300$, using the mesh shown in Fig. (1). The sequential linear solver of the SUPERLU library was used to solve the linear systems (9) [8]. At time step 1, the $N_{\mathrm{Mod}}/2$ complex-valued system matrices are factorized. Since they remain constant in time, the factorizations are kept stored in memory, such that linear systems can be solved by means of rapid LU back-substitution for all subsequent iterations.

Simulations were run on the LINUX cluster ANIC-2 built by our research team [7], which consists of 32 PCs (1.8 GHz clock speed, 1.5 GB RAM each), put together in a private network with 100 Mb/s switch. The solution was advanced in time by 85 dimensionless time units using a time step $\Delta t \beta / D$ of 0.0015, requiring 56 hours of overall computation time on 16 processors. Due to the large volume of data that needs to be communicated at each iteration, ($8000 \times 32 \times 8$ double precision numbers), a time step using 32 modes and 16 processors was found to be 1.8 times slower than an iteration using the same FEM mesh with 2 modes on a single processor.

In Fig. (2), we show an instantaneous iso-surface of u-velocity; the von Kármán vortex street is clearly visible in the cylinder wake. In Fig. (3), we show iso-surfaces of positive and negative w-velocity. Stream-wise vortices with diameter $\sim D$ are detected; these are thought to correspond to the 'mode-B' instabilities found in the wake of a circular cylinder for the same Reynolds number [3].

6 Concluding Remarks

The solution algorithm presented in this paper does not scale well with increasing number of Fourier modes/processors, because it requires excessive amounts of communication. This is not a major source of concern to us, as the current parallel implementation is only an intermediate step towards the development of a more efficient version based upon a parallel algebraic multi-grid solver. Our numerical results nevertheless show that the obtained speed-up is already sufficient to achieve acceptable performance on a self-assembled LINUX cluster with relatively slow connectivity.

Acknowledgements

The distributed memory parallel version of SFELES has been developed with the help of Dr. J.-M. Cela at the European Center for Parallelism of Barcelona (CEPBA). The authors also indebted to O. Debliquy, who built and maintains the LINUX cluster ANIC-2.

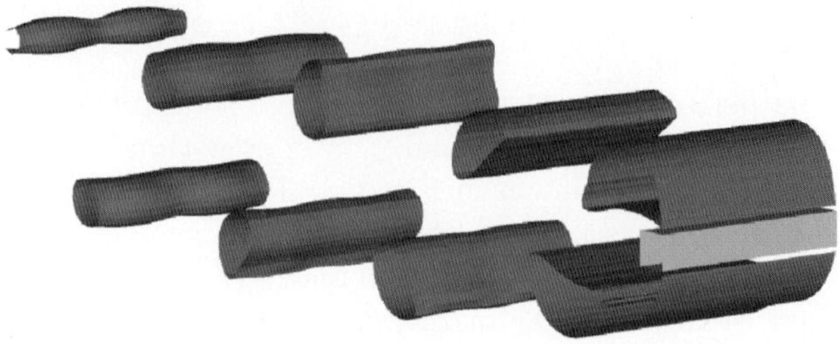

Fig. 2. Computed iso-contours of u-velocity magnitude in the wake behind the cylinder (Re = 300).

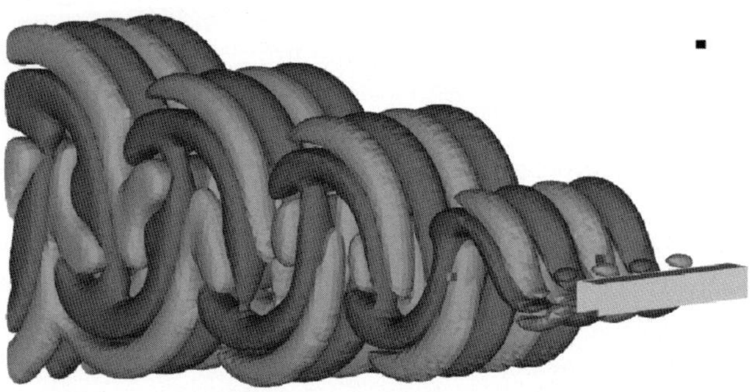

Fig. 3. Computed iso-contours of positive and negative w-velocity magnitude in the wake behind the cylinder (Re = 300).

References

1. E. Simons: An efficient multi-domain approach to large-eddy simulation of incompressible turbulent flows in complex geometries. PhD Thesis, von Karman Institute for Fluid Dynamics / Catholic University of Leuven, Belgium (2000)
2. K. Jansen: Comput. Methods Appl. Mech. Engrg. **174** (1999)
3. D. O. Snyder and G. Degrez: Large-eddy simulation with complex 2D geometries using a parallel finite-element/spectral algorithm. Int. J. Num. Meth. Fluids **41**(10) (2003)
4. T. J. R. Hughes *et al.*: Comput. Methods Appl. Mech. Engrg. **59** (1986)
5. A. Jameson *et al.*: AIAA Technical Paper 81-1259 (1981)
6. G. Karypis: METIS: a family of multilevel partitioning algorithms, www-users.cs.umn.edu/~karypis/metis/metis/index.html/ (2004)
7. O. Debliquy *et al.*: A network of inexpensive computers (ANIC-2), http://mach.ulb.ac.be/~dcarati/RECHERCHES/ANIC/anic2.html (2004)
8. J.W. Demmel *et al.*: A supernodal approach to sparse partial pivoting, SIMAX J. **20**(3) (1999)

Part XIX

Upwind Schemes

An Upwind Moment Scheme for Conservation Laws

H. T. Huynh

NASA Glenn Research Center, Cleveland, OH 44135, USA; huynh@grc.nasa.gov

Summary. A new piecewise linear scheme called upwind moment scheme is presented. Similar to the standard upwind scheme, the cell average values are updated by using a linear function in each cell, a march to the half time level by a Taylor series expansion, and a flux calculation that takes into account the wind direction. The slopes, however, are updated by evaluating the moment about the cell center in a manner similar to that of the discontinuous Galerkin method. The resulting scheme is third-order accurate in both space and time; in addition, the leading error, which is a dissipative error, is only 1/9 that of the standard piecewise-parabolic upwind method.

1 Introduction

Upwind schemes are popular due to their shock-capturing capability. The upwind scheme employed here is among the simplest. It consists of a piecewise linear reconstruction [6], a march to the half time level by a Taylor series expansion (Hancock's observation) [5], and a flux calculation that takes into account the wind direction [3].

The discontinuous Galerkin (DG) methods are also popular [1]. These schemes were introduced by Reed and Hill for the neutron transport equation (steady state linear hyperbolic) using a triangular mesh in 1973. They were extended to the case of unsteady nonlinear conservation laws by Cockburn and Shu, where a Runge-Kutta time stepping and an upwind flux were employed.

In this paper, a new scheme called the upwind moment scheme is presented. Here, both the cell average values and the slopes must be stored. The cell average values are updated in a manner identical to that of the upwind scheme: a march to the half time level via a Taylor series and an upwind flux. The slopes are updated by evaluating the moment about the cell center using an integration by parts; the semi-discrete form of this update is identical to that of the DG method, but the fully discrete form is different.

The resulting scheme is more accurate than the standard piecewise-linear upwind scheme. For the case of convection with a constant speed, the upwind moment scheme reduces to Van Leer's scheme III [6]. Thus, although the scheme employs a linear function in each cell, it is third-order accurate in both space and time; in addition, the leading error (dissipative type) is only 1/9 that of the standard piecewise-parabolic method. Note that Van Leer's scheme III, which consists of an exact time evolution and a least square projection, was extended to the case of systems of equations in multiple dimensions in [2]. The upwind moment scheme here replaces the exact time evolution by a Taylor series and the least square projection by an integration by parts; as a result, the extension is considerably simpler.

Compared with the linear element DG method, the upwind moment scheme is more efficient since each time step involves roughly one and a half stages while that for DG, three stages (for third-order accuracy).

To capture shocks without oscillations, we need a limiting procedure. While limiting typically causes a loss of accuracy near extrema, the procedure employed here is designed to preserve both monotonicity and accuracy.

2 Upwind moment scheme for a scalar conservation law

Consider the scalar conservation law

$$u_t + f_x = 0, \tag{2.1}$$

where t is time, x spatial coordinate, and $f = f(u) = f(u(x,t))$ the flux. The wave speed is $a(u) = df/du$. Applying the chain rule to f_x, (2.1) implies

$$u_t + a u_x = 0. \tag{2.2}$$

Let h be the cell width of a uniform mesh, $x_j = jh$ be the cell centers, $x_{j+1/2} = (j+1/2)h$ be the cell interfaces, and $I_j = [x_{j-1/2}, x_{j+1/2}]$ be the cell j for $j = 0, 1, 2, \ldots$ Let the test functions on I_j be $\phi_0(x) = 1$ and $\phi_1(x) = x - x_j$. Formally, the solution in I_j is required to satisfy, for $k = 0$ and $k = 1$,

$$\int_{I_j} u_t(x,t)\phi_k(x) + f_x(x,t)\phi_k(x)\,dx = 0. \tag{2.3}$$

Using integration by parts on the second term, the result is

$$\frac{\partial}{\partial t}\int_{I_j} u(x,t)\phi_k(x)\,dx + [\,f(x,t)\phi_k(x)\,]_{x_{j-1/2}}^{x_{j+1/2}} + \int_{I_j} f(x,t)\,(\phi_k)_x\,dx = 0. \tag{2.4}$$

Since the solution u will be approximated by piecewise linear functions, which are discontinuous across the cell interfaces, for the square bracket term above, the fluxes $f(x_{j-1/2},t)$ and $f(x_{j+1/2},t)$ are the *upwind fluxes*, whereas $\phi_k(x_{j-1/2})$ and $\phi_k(x_{j+1/2})$ are evaluated by the ϕ_k of I_j (i.e., no upwinding). Integrating the above from t^n to t^{n+1}, we obtain

$$\int_{I_j} u(x,t^{n+1})\phi_k(x)\,dx - \int_{I_j} u(x,t^n)\phi_k(x)\,dx = \mathrm{BI} - \mathrm{VI}, \tag{2.5}$$

where the boundary integral BI and the volume integral VI are

$$\mathrm{BI} = \int_{t^n}^{t^{n+1}} f(x_{j-1/2}, t)\phi_k(x_{j-1/2})\, dt - \int_{t^n}^{t^{n+1}} f(x_{j+1/2}, t)\phi_k(x_{j+1/2})\, dt, \tag{2.6}$$

$$\mathrm{VI} = \int_{t^n}^{t^{n+1}} \int_{I_j} f(x,t)\, (\phi_k)_x\, dx\, dt. \tag{2.7}$$

Note that with $\phi_1(x) = x - x_j$, the quantity $\int_{I_j} u(x, t^n)\phi_1(x)\, dx$ in (2.5) is the *moment* of u about x_j in I_j. If a function v is expanded as a Taylor series:

$$v(x) = v(x_j) + v'(x_j)(x - x_j) + \tfrac{1}{2}v''(x_j)(x - x_j)^2 + ...$$

then, the moment calculation relates to the slope estimate:

$$\frac{12}{h^3} \int_{I_j} v(x)\phi_1(x)\, dx = v'(x_j) + O(h^2). \tag{2.8}$$

Thus, (2.5) with $k = 1$ yields the slope update and, with $k = 0$, the cell average update. The problem, therefore, reduces to evaluating BI and VI.

At time $t^n = n\Delta t$, suppose u_j^n and $(u_x)_j^n$, which respectively approximate the average and the slope of the solution u in cell j, are known and stored for all j. We wish to calculate u_j^{n+1} and $(u_x)_j^{n+1}$. For simplicity, the superscript n in u_j^n and $(u_x)_j^n$ is omitted and the data are denoted by u_j and $(u_x)_j$. Other superscripts such as $n+1/2$, however, are retained. As is routine, with $\sigma = a\Delta t/h$, the time step is assumed to satisfy the CFL condition $|\sigma| \le 1$.

First, we calculate u_j^{n+1}. Discretizing (2.5–2.7) with $k = 0$ at the point $(x_j, t^{n+1/2})$ via central differences, we obtain

$$u_j^{n+1} = u_j + \tfrac{\Delta t}{h}\, (f_{j-1/2}^{n+1/2} - f_{j+1/2}^{n+1/2}). \tag{2.9}$$

To march to the half time level where the flux $f_{j+1/2}^{n+1/2}$ is evaluated, for each cell j, calculate

$$(u_t)_j = -a_j(u_x)_j. \tag{2.10}$$

Next, approximate the solution on the domain $[x_{j-1/2}, x_{j+1/2}] \times [t^n, t^{n+1}]$ by

$$r_j(x,t) = u_j + (u_x)_j(x - x_j) + (u_t)_j(t - t^n). \tag{2.11}$$

For each cell j, the values at the two interfaces at time level $n + 1/2$ are $r_j(x_{j-1/2}, t^{n+1/2})$ and $r_j(x_{j+1/2}, t^{n+1/2})$. Therefore, at each interface $x_{j+1/2}$ and time $t^{n+1/2}$, we have two values: one from cell j denoted by u_L, and the other from cell $j + 1$ denoted by u_R

$$u_L = u_j + \tfrac{h}{2}(u_x)_j + \tfrac{\Delta t}{2}(u_t)_j \quad \text{and} \quad u_R = u_{j+1} - \tfrac{h}{2}(u_x)_{j+1} + \tfrac{\Delta t}{2}(u_t)_{j+1}. \tag{2.12}$$

With u_L and u_R as above, the flux $f_{j+1/2}^{n+1/2}$ is obtained by upwinding [3],

$$f_{j+1/2}^{n+1/2} = f_{\mathrm{Roe}}(u_L, u_R). \tag{2.13}$$

This flux and (2.9) complete the cell average update.

Next, we update the slope. It is here where the new scheme differs from the standard upwind scheme. Instead of interpolating, the slope is calculated by (2.5–2.7) with $k = 1$. The boundary term (2.6) is evaluated at time level $n + 1/2$ by again the upwind flux (2.13). The slope update, after a little algebra, with VI defined by (2.7), is given by

$$(u_x)_j^{n+1} = (u_x)_j - \Delta t \left\{ \frac{6}{h^2} (f_{j-1/2}^{n+1/2} + f_{j+1/2}^{n+1/2} - 2\frac{\text{VI}}{h\Delta t}) \right\}. \tag{2.15}$$

Since $(\phi_1)_x = 1$,

$$\text{VI} = \int_{t^n}^{t^{n+1}} \int_{I_j} f(x,t)\, dx\, dt. \tag{2.16}$$

The quantity $\text{VI}/(h\Delta t)$ is the average of f over the domain $[x_{j-1/2}, x_{j+1/2}] \times [t^n, t^{n+1}]$. Therefore, the quantity in the curly bracket in (2.15) represents an approximation to f_{xx}. This fact is consistent with the fact that the evolution of u_x can be obtained by differentiating (2.1) with respect to x: $(u_x)_t = -f_{xx}$.

How should we calculate VI in (2.16)? If we use $\frac{1}{2}(f_{j-1/2}^{n+1/2} + f_{j+1/2}^{n+1/2})$ to estimate $\text{VI}/(h\Delta t)$, the result is a slope that is constant in time, i.e., the inaccurate update of $(u_x)_j^{n+1} = (u_x)_j$. Thus, VI must be evaluated with care.

It appears reasonable to require that for the case of advection with a constant speed, the slope update yields a scheme identical to Van Leer's third [6], which is third-order accurate. To satisfy this requirement, first, calculate the cell average value at time $t^{n+1/2}$ by balancing fluxes in a manner similar to (2.9):

$$u_j^{n+1/2} = u_j + \tfrac{\Delta t}{h} (f_{j-1/2}^{n+1/4} - f_{j+1/2}^{n+1/4}); \tag{2.17}$$

here $f_{j+1/2}^{n+1/4}$ is the upwind flux at time level $n + 1/4$ given by expressions similar to (2.12) and (2.13). Next, the average of u on the space-time cell $[x_{j-1/2}, x_{j+1/2}] \times [t^n, t^{n+1}]$ is denoted by u_A and obtained by

$$u_A = \tfrac{1}{6}(u_j + 4u_j^{n+1/2} + u_j^{n+1}). \tag{2.18}$$

The time level associated with u_A is $n + 1/2$. As for the slope, we can use

$$(u_x)_j^{n+1/2} = (u_x)_j. \tag{2.19}$$

(For nonlinear equations, the above can be employed as a first guess in an iteration process where $(u_x)_j^{n+1/2} = \frac{1}{2}[(u_x)_j + (u_x)_j^{n+1}]$; numerical experiments show no improvement over (2.19), however.) Finally, set

$$r_A(x) = u_A + (u_x)_j^{n+1/2}(x - x_j), \quad \text{and}$$

$$\mathrm{VI} = \tfrac{1}{6}\left[f(r_A(x_{j-1/2})) + 4f(u_A) + f(r_A(x_{j+1/2}))\right]. \tag{2.20}$$

These expressions complete the description of the scheme.

The above upwind moment scheme can be easily extended to the two dimensional Euler equations. To preserve third-order accuracy for 2D convection, however, the flux across each edge must be evaluated at the two Gauss points. The reason is that for the boundary integral, along each edge, both u and the test function ϕ are linear, and the product $u\phi$ is parabolic.

3 Numerical results

The first test has both shocks and local extrema [4]. In the interval $-1 \le x \le 1$, a moving Mach 3 shock interacts with sine waves in density described by

$$(\rho, u, p) = \begin{cases} (3.857, 2.629, 10.333) & \text{if} -1 \le x < -.8, \\ (1 + 0.2\sin 5\pi x, 0, 1) & \text{if} -.8 \le x \le 1. \end{cases}$$

The final time is $t = 0.36$. Fig. 1 shows the results with 300 cells by the upwind moment and the standard upwind (with Van Albada limiter) schemes. The solid lines represent the solution with 3600 cells by the standard upwind scheme. Here, the upwind moment solution is highly accurate.

The last test is the (2D) oblique shock problem [7]. The domain is $[0, 4] \times [0, 1]$. The boundary conditions are: left, inflow with $(\rho, u, v, p) = (1., 2.9, 0., 1./1.4)$; top, $(1.7, 2.6193, -.5063, 1.5282)$; bottom, solid wall; right, outflow. The rectangular mesh is 80×20. Fig. 2 shows pressures along $y = .475$ by the upwind moment and the standard upwind schemes. The solid line represents the exact solution. Again, the upwind moment solution is more accurate.

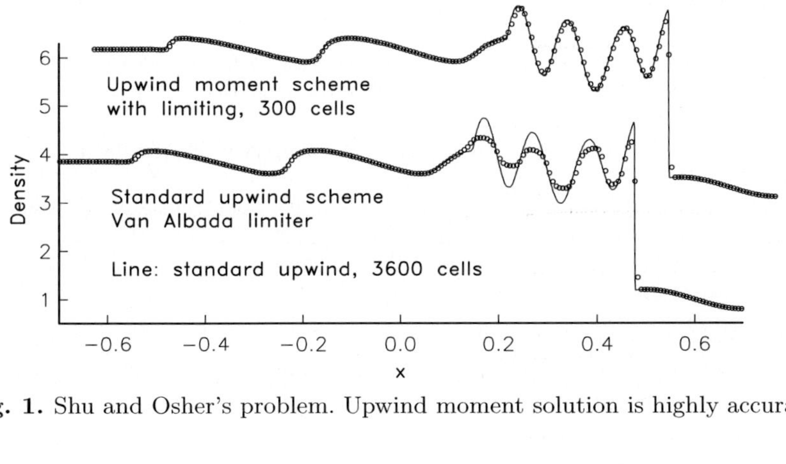

Fig. 1. Shu and Osher's problem. Upwind moment solution is highly accurate.

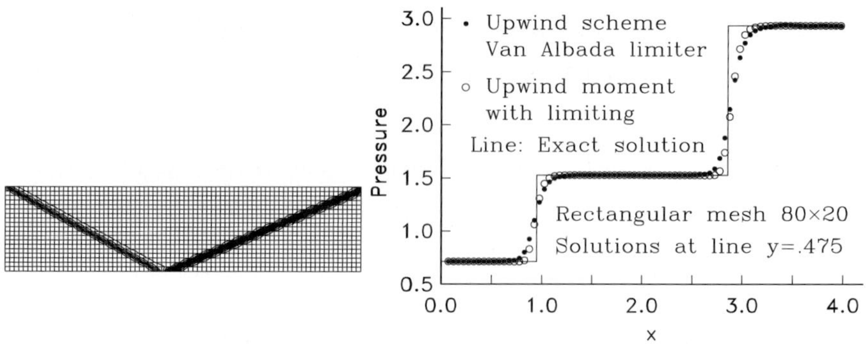

Fig. 2. Oblique shock. (Left) Mesh and pressure contours by upwind moment scheme. (Right) Solutions along $y = .475$; upwind moment solution is more accurate.

Conclusions

In conclusion, a new scheme called upwind moment scheme, which combines the cell average update of the standard upwind scheme with the slope update via a moment calculation similar to that of the DG method, is introduced. The scheme is more accurate than the standard piecewise-linear upwind method; for the two-dimensional Euler equations, computing cost increases by roughly a factor of five per time step. Additional results will be presented elsewhere.

Acknowledgements. This work was supported by the Engine System Noise Reduction project, which is part of the Quiet Aircraft Technology Program at the NASA Glenn Research Center.

References

1. B. Cockburn, G. Karniadakis, and C.-W. Shu, 'The development of discontinuous Galerkin methods', Springer (2000), pp 3–50.
2. R. B. Lowrie, P. L. Roe, B. van Leer, 'A space-time discontinuous Galerkin method for the time accurate numerical solution of hyperbolic conservation laws', (1995) AIAA 95-1658.
3. P. L. Roe, 'Characteristic-based schemes', Ann. Rev. Fluid Mech., 18 (1986), pp 337–365.
4. C-W. Shu and S. Osher, 'Efficient implementation of essentially non-oscillatory shock-capturing schemes', J. Comp. Phys. 83 (1989), pp 32–78.
5. G. D. van Albada, B. van Leer, and W. W. Roberts, Jr., 'A comparative study of computational methods in cosmic gas dynamics' Astronom. and Astrophys., 108 (1982), pp 76–84.
6. B. van Leer, 'Towards the ultimate conservative difference scheme. IV. A new approach to numerical convection', J. Comp. Phys. 23 (1977), pp 276–298.
7. H. C. Yee, R. F. Warming, and A. Harten, 'Implicit Total Variation Diminishing (TVD) Schemes for Steady-State Calculations', (1983) AIAA 83-1902.

Accurate and Efficient Re-evaluation of Cell-interface Convective Fluxes

Sung-Hwan Yoon, Kyu-Hong Kim, Chongam Kim, and Oh-Hyun Rho

Department of Aerospace Engineering, Seoul National University, Seoul 151-742, Korea. E-mail: chongam@snu.ac.kr

In order to reduce the excessive numerical dissipation which is induced when a grid system is not aligned with a discontinuity, a new spatial treatment of cell-interface fluxes is introduced. The M-AUSMPW+ in this paper possesses an additional procedure of re-defining transferred properties at a cell-interface based on AUSMPW+ [1], and it shows a complete monotonic property regardless of sonic point location. The newly defined transferred property could eliminate numerical dissipation effectively in non-flow aligned grid system of multi-dimensional flows.

1 Re-evaluation of Cell-interface Convective Fluxes

Upwind scheme may produce excessive numerical dissipation in a continuous region. If an upwind flux could distinguish a continuous region from a continuous region and it is modified in a continuous region appropriately, it is expected to be low dissipative and to yield the more accurate results. The convective flux of AUSM-type scheme is written as follows.

$$\boldsymbol{F}_{\frac{1}{2}} = m_{\frac{1}{2}} c_{\frac{1}{2}} \boldsymbol{\Phi}_{LorR}, \tag{1}$$

where $m_{\frac{1}{2}}$ is the Mach number that is estimated at a cell-interface and $\boldsymbol{\Phi}$ is the transferred property vector. For the improvement of Eq.(1), a flexible flux form is adopted to distinguish the region of continuity from discontinuity.

$$\boldsymbol{F}_{\frac{1}{2}} = m_{\frac{1}{2}} c_{\frac{1}{2}} \boldsymbol{\Phi}_{LorR,\frac{1}{2}}, \tag{2}$$

where the subscript $\frac{1}{2}$ represents the property defined at a cell-interface. The transferred property $\boldsymbol{\Phi}_{L,R,\frac{1}{2}}$ is determined to satisfy the following requirements.

r1. The transferred property should be able to distinguish the region of continuity from discontinuity to increase accuracy in a continuous region.
r2. The transferred property should satisfy the monotonic condition.
r3. The transferred property should maintain the upwind characteristics in a supersonic flow.

The requirement 1 would be the major objective of this paper. If the requirement 2 is not satisfied, the newly defined transferred property leads to oscillatory behaviors in a discontinuity. The requirement 3 is necessary to represent physical phenomena rightly in a supersonic flow.

Firstly, the characteristics of continuous and discontinuous regions are examined to establish the criterion that divides the two regions. The convective quantity at a cell-interface is then re-evaluated according to the criterion. In order to obtain the information on the distribution of the physical value, TVD interpolation is analyzed. It can be shown that Minmod limiter always assumes a gently varied region to give the interpolated values of

$$\boldsymbol{\Phi}_i < \boldsymbol{\Phi}_L < \boldsymbol{\Phi}_R < \boldsymbol{\Phi}_{i+1}, \tag{3}$$

And superbee limiter always produces a rapidly varied region to yield the interpolated values of

$$\boldsymbol{\Phi}_i < \boldsymbol{\Phi}_R < \boldsymbol{\Phi}_L < \boldsymbol{\Phi}_{i+1}. \tag{4}$$

Thus, it is observed that a continuous region, or a gently varied region, can be identified as the region where TVD interpolation satisfies Eq.(3), and Eq.(4) for region of discontinuity, or a rapidly varied region. Therefore, the following criterion is proposed to guarantee the accuracy improvement.

(Proposition) If the interpolated value of $\boldsymbol{\Phi}_{LorR}$ satisfies the condition of Eq.(3), the physical state at a cell-interface is considered to be in a continuous region and the convective quantity at a cell-interface is modified as $\boldsymbol{\Phi}_{L,i+\frac{1}{2}} = \boldsymbol{\Phi}_{R,i+\frac{1}{2}} = 0.5(\boldsymbol{\Phi}_L + \boldsymbol{\Phi}_R)$. If $\boldsymbol{\Phi}_{LorR}$ satisfies the condition of Eq.(4), it is considered to be in a discontinuous region and the interpolated value is not re-evaluated. Then, the re-evaluated convective quantity is always closer to the physical value than the interpolated value and thus it increase solution accuracy both in continuous and in discontinuous region.

Eq.(3) can be rewritten as follows.

$$(0.5(\boldsymbol{\Phi}_L + \boldsymbol{\Phi}_R) - \boldsymbol{\Phi}_L)(\boldsymbol{\Phi}_{L,superbee} - \boldsymbol{\Phi}_L) \geq 0. \tag{5}$$

Here, $\boldsymbol{\Phi}_L$ lies between $\boldsymbol{\Phi}_{L,minmod}$ and $\boldsymbol{\Phi}_{L,superbee}$, and the convective quantity is determined as follows.
when $0.5(\boldsymbol{\Phi}_R - \boldsymbol{\Phi}_L)(\boldsymbol{\Phi}_{L,superbee} - \boldsymbol{\Phi}_L) \geq 0$,

$$\boldsymbol{\Phi}_{L,i+\frac{1}{2}} = 0.5(\boldsymbol{\Phi}_L + \boldsymbol{\Phi}_R). \tag{6}$$

when $0.5(\boldsymbol{\Phi}_R - \boldsymbol{\Phi}_L)(\boldsymbol{\Phi}_{L,superbee} - \boldsymbol{\Phi}_L) < 0$,

$$\boldsymbol{\Phi}_{L,i+\frac{1}{2}} = \boldsymbol{\Phi}_L. \tag{7}$$

As a consequence, $\boldsymbol{\Phi}_{L,i+\frac{1}{2}}$ always gives a greater variation than the original value and thus accuracy is always enhanced.

For the requirement of r2 for the monotonic characteristics, the calculated property at a cell-interface should be between the value by minmod limiter and the value by superbee limiter.

$$\boldsymbol{\Phi}_i < \boldsymbol{\Phi}_{L,minmod} < \boldsymbol{\Phi}_{L,i+\frac{1}{2}} = 0.5(\boldsymbol{\Phi}_L + \boldsymbol{\Phi}_R) < \boldsymbol{\Phi}_{L,superbee} < \boldsymbol{\Phi}_{i+1}. \quad (8)$$

When the monotonic condition of Eq.(8) is applied to Eq.(6),(7), we have

if $(\boldsymbol{\Phi}_R - \boldsymbol{\Phi}_L)(\boldsymbol{\Phi}_{L,superbee} - \boldsymbol{\Phi}_L) \geq 0$,

$$\boldsymbol{\Phi}_{L,\frac{1}{2}} = \boldsymbol{\Phi}_L + sign(\boldsymbol{\Phi}_{L,superbee} - \boldsymbol{\Phi}_L)min(|0.5(\boldsymbol{\Phi}_R - \boldsymbol{\Phi}_L)|\ ,\ |\boldsymbol{\Phi}_{L,superbee} - \boldsymbol{\Phi}_L|). \quad (9)$$

if $(\boldsymbol{\Phi}_R - \boldsymbol{\Phi}_L)(\boldsymbol{\Phi}_{L,superbee} - \boldsymbol{\Phi}_L) < 0$,

$$\boldsymbol{\Phi}_{L,\frac{1}{2}} = \boldsymbol{\Phi}_L. \quad (10)$$

In brief, the property at a cell-interface can be rewritten as follows.

$$\boldsymbol{\Phi}_{L,\frac{1}{2}} = \boldsymbol{\Phi}_L + \frac{max[0,(\boldsymbol{\Phi}_R - \boldsymbol{\Phi}_L)(\boldsymbol{\Phi}_{L,superbee} - \boldsymbol{\Phi}_L)]}{|\boldsymbol{\Phi}_R - \boldsymbol{\Phi}_L|(\boldsymbol{\Phi}_{L,superbee} - \boldsymbol{\Phi}_L)} min\left[\frac{|\boldsymbol{\Phi}_R - \boldsymbol{\Phi}_L|}{2}, |\boldsymbol{\Phi}_{L,superbee} - \boldsymbol{\Phi}_L|\right]. \quad (11)$$

$$\boldsymbol{\Phi}_{R,\frac{1}{2}} = \boldsymbol{\Phi}_R + \frac{max[0,(\boldsymbol{\Phi}_L - \boldsymbol{\Phi}_R)(\boldsymbol{\Phi}_{R,superbee} - \boldsymbol{\Phi}_R)]}{|\boldsymbol{\Phi}_R - \boldsymbol{\Phi}_L|(\boldsymbol{\Phi}_{R,superbee} - \boldsymbol{\Phi}_R)} min\left[\frac{|\boldsymbol{\Phi}_R - \boldsymbol{\Phi}_L|}{2}, |\boldsymbol{\Phi}_{R,superbee} - \boldsymbol{\Phi}_R|\right]. \quad (12)$$

The flow information is transferred elliptically in a subsonic region, and in a supersonic region, it is transferred hyperbolically. While $\boldsymbol{\Phi}_{\frac{1}{2}} = 0.5(\boldsymbol{\Phi}_L + \boldsymbol{\Phi}_R)$ is physically appropriate for subsonic flows, it is not suitable for supersonic flows. Since it is essential that the flux should maintain the upwind characteristics in a supersonic region, the property at a cell-interface $\boldsymbol{\Phi}_{\frac{1}{2}}$ should be determined by considering whether the cell-interface is located on a subsonic or a supersonic region, which is realized by introducing a Mach number-based function, a. Finally, the property at a cell-interface of the present method satisfying the conditions (r1, r2, r3) is determined as follows.

$$\boldsymbol{\Phi}_{L,\frac{1}{2}} = \boldsymbol{\Phi}_L + \frac{max[0,(\boldsymbol{\Phi}_R - \boldsymbol{\Phi}_L)(\boldsymbol{\Phi}_{L,superbee} - \boldsymbol{\Phi}_L)]}{|\boldsymbol{\Phi}_R - \boldsymbol{\Phi}_L|(\boldsymbol{\Phi}_{L,superbee} - \boldsymbol{\Phi}_L)} min\left[a\frac{|\boldsymbol{\Phi}_R - \boldsymbol{\Phi}_L|}{2}, |\boldsymbol{\Phi}_{L,superbee} - \boldsymbol{\Phi}_L|\right]. \quad (13)$$

$$\boldsymbol{\Phi}_{R,\frac{1}{2}} = \boldsymbol{\Phi}_R + \frac{max[0,(\boldsymbol{\Phi}_L - \boldsymbol{\Phi}_R)(\boldsymbol{\Phi}_{R,superbee} - \boldsymbol{\Phi}_R)]}{|\boldsymbol{\Phi}_R - \boldsymbol{\Phi}_L|(\boldsymbol{\Phi}_{R,superbee} - \boldsymbol{\Phi}_R)} min\left[a\frac{|\boldsymbol{\Phi}_R - \boldsymbol{\Phi}_L|}{2}, |\boldsymbol{\Phi}_{R,superbee} - \boldsymbol{\Phi}_R|\right]. \quad (14)$$

where $a = 1 - min(1, max(|M_L|, |M_R|))^2$. a is chosen as a quadratic function because its derivative is continuous when the Mach number becomes zero. In supersonic flows, the function a has the value of zero. Although the analysis in this section is carried out using TVD limiters, the *proposition* is general in the sense that the present approach is still available when other monotonic interpolation schemes are adopted, such as ENO [2].

2 Complete Monotonicity

In order to reduce the grid dependency and improve the convergence characteristics in a steady shock discontinuity, the pressure splitting function of AUSM-type scheme is modified. Most useful relations in realizing a numerical shock profile are the Rankine-Hugoniot relation and the Prandtl relation. Roe's FDS exploits the Rankine-Hugoniot relation and AUSM-type schemes use the Prandtl relation. In AUSM-type schemes, the Rankine-Hugoniot relation, which includes the relation among thermodynamic variables, e.g., density, pressure and temperature, cannot be exploited any more due to the split of convective and pressure flux. Instead, they notice a shock through the Prandtl relation. Unlike the Rankine-Hugoniot relation, the Prandtl relation does not possess the information between thermodynamic variables. This lack of information yields non-monotonic overshoots or slightly diffusive results of a discrete shock profile. In M-AUSMPW+, the defect is cured by the information on pressure jump across a shock, directly derived from the governing equations. The results can be written as follows.
If $M_{L,s} > 1, 0 < M_{R,s} < 1, M_{L,s}M_{R,s} < 1$,

$$P_R^- = max\left(0, min\left(0.5, 1 - \frac{\rho_L U_L(U_L - U_R) + p_L}{p_R}\right)\right), \tag{15}$$

If $-1 < M_{L,s} > 0, -1 < M_{R,s}, M_{L,s}M_{R,s} < 1$,

$$P_R^+ = max\left(0, min\left(0.5, 1 - \frac{\rho_R U_R(U_R - U_L) + p_R}{p_L}\right)\right). \tag{16}$$

These pressure splitting function can achieve the complete monotonic characteristic and reduce grid dependency.

3 Numerical Results

3.1 Vortex flow

This test case is the weak vortex flow and initial conditions are that the core radius of vortex is 0.2m and the angular velocity is 2. Figure 1 shows density distributions along the line AB. It is the results after 1000 iterations. Roe's FDS and AUSM+ show the similar and diffusive results. The present method yields much more improved results, which show that the present method is very useful in a multi-dimensional flow. The present method seems to show the two times accuracy improvement compared with conventional appraoches.

3.2 Viscous shock tube problem

The test problem is the same as the study by Daru and Tenaud. It is the shock tube problem in 2-D square box with a unit side length. Fig.2(a) is the comparison of density contour of AUSMPW+ and the present method with van Leer limiter. The primary vortex of AUSMPW+ is shown to be slower in the growth and less rotated compared with that of the present method. Figure 2(b) shows that the results by the present method on 250 x 125 (=31250) grids is almost the same as that by AUSMPW+ on 350 x 175 (=61250) grids which consists of two times grid points. It is reasonable that the present method provides the two times efficiency improvement, i.e., the present method can provide the same accuracy in the half grid points compared with existing upwind schemes such as AUSMPW+ or Roe's FDS and so on.

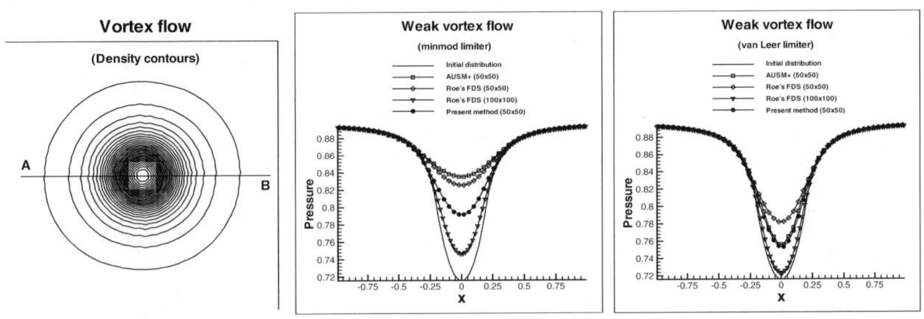

Fig. 1. Comparison of pressure distribution of Roe's FDS and the present method along the line AB

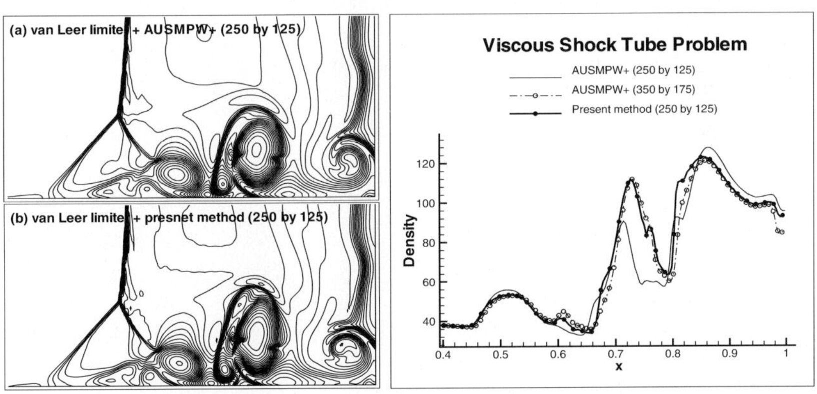

(a) (b)

Fig. 2. Comparison of density contours and density distribution along the wall

4 Conclusion

A new treatment of cell-interface convective fluxes in AUSM-type methods is introduced to substantially reduce the numerical dissipation in a smooth region, including a contact discontinuity, without compromising accuracy in shock regions. The core idea of the new scheme is the modification of the transferred property at a cell-interface under the consideration of physical and multi-dimensional phenomena. The transferred property of the M-AUSMPW+ is revealed in two aspects. One is that the newly defined cell-interface value is closer to the real physical value. The other is that it can eliminate numerical dissipation effectively in a non-flow aligned grid system.

Another desirable characteristic of M-AUSMPW+ is the complete monotonicity in capturing a steady shock wave, regardless of the location of sonic transition position. As a result, convergence characteristics and grid dependency of AUSM-type methods are remarkably enhanced.

Through numerous test cases, M-AUSMPW+ is proved to be equally efficient but twice more accurate than previous schemes. If M-AUSMPW+ would be applied to three-dimensional problems, accuracy and efficiency is expected to be improved further. In addition, if present idea would be applied to E-CUSP type scheme, we may also expect accuracy and efficiency enhancement.

References

1. K. H. Kim, C. Kim and O. H. Rho, "Methods for the Accurate Computations of Hypersonic Flows, Part I: AUSMPW+ Scheme,": J. Computational Physics. A **174**, 38-80 (2001)
2. Harten, B.Engquist, S.Osher, and Chakravarthy," Uniformly high order accurate essentially non-oscillatory schemes,": J. of Computational Physics. A **71**, 231-303 (1987)

Computation of The Flow Around a Bluff Body By Multi-Directional Finite Difference Method

Mi Young Lee[1], Tetuya Kawamura[2], and Kunio Kuwahara[3]

[1] Ochanomizu University, 2-1-1 Ootsuka Bunkyo-Ku Tokyo 112-8610 JAPAN
lmy@ns.is.ocha.ac.jp

[2] Ochanomizu University, 2-1-1 Ootsuka Bunkyo-Ku Tokyo 112-8610 JAPAN
kawamura@is.ocha.ac.jp

[3] The Institute of Space and Astronautical Science, 3-1-1 Yoshinodai Sagamihara-Shi Kanagawa 229-8510 JAPAN kuwahara@pub.isas.jaxa.jp

Summary. We present calculation of compressible flow based on solving the Navier-Stokes equations at equal intervals in Cartesian coordinate system. For a fundamental understanding of unsteady flow, we apply the multi-directional upwind finite difference method to the compressible flow past bluff bodies of various shapes. In order to catch the complicated nature of flows around bluff bodies, it is effective to use the multi-directional finite difference method for all the space derivative terms.

1 Introduction

A fundamental understanding of compressible fluid flow with shock waves is important to the flow around bluff bodies of various shapes in high speed range. It has not been widely studied even though it has wide range of aerodynamic applications[1].

In this paper, for understanding the flow past bluff bodies[2], we present simulation of compressible flow based on solving the compressible Navier-Stokes equations in high-speed range. We apply the MSUCL TVD cheme[3,4] to the convection terms and 2nd-order centered difference scheme for the viscosity terms.

We use a Cartesian coordinate system at equal intervals. Adopting Cartesian grid makes computation time of grid generation short and the process simple even for complex two and three-dimensional configurations. On the other hand, in general, when we use Cartesian coordinate system, flow direction is not always parallel to a coordinate line. In order to improve this problem, we propose to use the multi-directional finite difference method[5] for all the space derivative terms. In simulations of incompressible flow, more accurate results are achieved by the present method. Since the flow behind the bluff body gets fluctuated, it is difficult to catch the information of flow including turbulence and vortices. We show the effectiveness of our method by calculating the compressible flow with shock waves and vortices in Cartesian coordinate system. Therefore, we aim to catch the information of the flow more clearly and to prevent non-physical oscillation using our method. No explicit turbulence model is incorporated into the present study.

2 Numerical Methods

2.1 Numerical method

The governing equations are the compressible Navier-Stokes equations. A differential form of the compressible Navier-Stokes equations can be obtained by applying conservation principles to an infinitesimally small fluid element fixed in space. In the computation, all the variables in equations are replaced to non-dimensional counterparts.

For the solution of the governing equations, we have applied a 3rd-order accurate interpolation scheme called the MUSCL TVD scheme to interpolate the flow properties stored at cell centers to cell faces for solving convection terms. In order to simplify the analysis of the equations, it computed using roe's averages of Q at adjacent points. For higher-order schemes, the values assigned to each cell are the averages over each cell. In this case, non-physical oscillations are produced, which must be prevented. Unwanted oscillations can be avoided if the slopes of the variables used in the extrapolations are limited in such a way that the end point values do not create a new maximum or minimum. For this specification of the extrapolated values of the dependent variables, and easy control is devised by the introduction of slope limiter and min mod limiter. The min mod limiter is used extensively in TVD numerical methods. The viscosity terms are discretized by the 2nd-order centered difference scheme.

2.2 Multi-directional difference finite method

When we use Cartesian grids, the decrease in accuracy occurs near the boundary because flow direction is not always parallel to a coordinate line. For improving this problem, the multi-directional finite difference method is used in the computation process. This method is summarized as follows:

Figure 1 shows the coordinate system. We consider nine grid points. In brief, the essential ingredients can be easily described in the form of the two-dimensional unsteady Euler equations on a Cartesian grid system at equal interval.

$$\frac{\partial Q}{\partial t} + \frac{\partial e}{\partial x} + \frac{\partial f}{\partial y} = 0 \tag{1}$$

Here, Q is the flow properties (state) vector, and e, f contain the flux terms. We suppose that the interval of lattice is the length h(=△x=△y). When structured grid points are given, the black point in fig. 1(a) are usually used to approximate the derivatives at the central point (System A). If we use

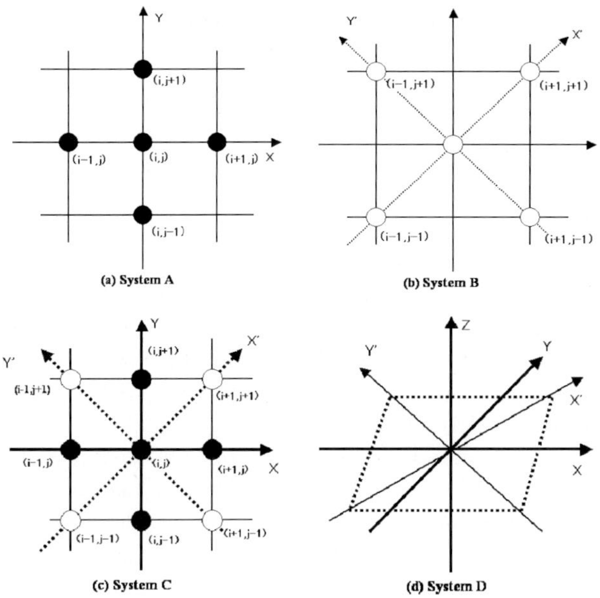

Fig. 1. The grid for multi-directional method.

the 2nd-order centered difference scheme for eq. (1) in system A, it can be approximated as follow

$$\frac{Q^{n+1}-Q^{n}}{\triangle t}+\frac{e_{i+1,j}-e_{i-1,j}}{2h}+\frac{f_{i,j+1}-f_{i,j-1}}{2h}=0 \tag{2}$$

Then, we introduce the other 45 degrees-rotated local grid points, the white ones in fig. 1(b), it can be used to approximate the derivatives at the central point (System B). The interval of lattice is the length$\sqrt{2}$h(=△X=△Y). In system B, eq. (1) is changed as follow,

$$\frac{\partial Q}{\partial t}+\frac{\partial E}{\partial x}+\frac{\partial F}{\partial y}=0 \tag{3}$$

where

$$E=\frac{e+f}{\sqrt{2}},F=\frac{-e+f}{\sqrt{2}},X=\frac{x+y}{\sqrt{2}},Y=\frac{-x+y}{\sqrt{2}} \tag{4}$$

And we use the 2nd-order centered difference scheme for eq. (3) in system B, it can be showed that as follow,

$$\frac{Q^{n+1}-Q^{n}}{\triangle t}+\frac{E_{i+1,j+1}-E_{i-1,j-1}}{2\sqrt{2}h}+\frac{F_{i-1,j+1}-F_{i+1,j-1}}{2\sqrt{2}h}=0 \tag{5}$$

Eq. (5) can be changed to eq. (6) by adopting eq. (4).

$$\frac{Q^{n+1}-Q^{n}}{\triangle t}+\frac{e_{i+1,j+1}+f_{i+1,j+1}-e_{i-1,j-1}-f_{i-1,j-1}}{4h}$$
$$+\frac{-e_{i-1,j+1}+f_{i-1,j+1}+e_{i+1,j-1}-f_{i+1,j-1}}{4h}=0 \quad (6)$$

In order to improve the derivative value at the central point, we mix the derivative values calculated from both system A and B at a proper ratio (System C). We adopt the ratio A:B=2/3:1/3. Adopting this ratio, the resulting finite-difference scheme for the Laplacian coincides with well-known 9 points formula with forth-order accuracy. This method improves the rotational invariance of the finite difference system. Then those flows where flow direction is not parallel to the grid direction are better simulated. For 3-dimensional computation, we introduce three different grid systems. Each grid system is obtained by rotating a plane (system A) with respect to its perpendicular axis. For example, one of those systems, system D (x'-y'-z) is shown in fig. 1(d). In the same way as 2-dimensional computation, white circle points are used instead of black ones on the x-y plane, and ordinary ones in the z-direction. Other gird systems are also introduced similarly. Thus we can obtain three different values at the same point, and they are averaged, since any physical phenomena are equivalent in each grid system.

3 Numerical Results

Flow condition used in the present study is supersonic flow. The grid used for the computation is at equal intervals in Cartesian coordinate system. Bluff bodies are placed in an initially uniform flow. On the solid boundary, the non-slip conditions are enforced, i.e. u=v=0 in 2D, u=v=w=0 in 3D. The flow condition is M_{∞}=1.4 and the number of grid points are 200×200 in fig. 2. Fig.3 is shown the grid points of 128×64×64 and M_{∞}=1.4. Along the upper and lower boundaries, slip wall condition is adopted. By this condition, we see shock wave reflection near the boundaries. The calculations are normalized with density=1.4, pressure =1.0 at infinity. The Reynolds number is 1.0×10^{6}.

In Fig. 2 (b), the flow behind the bluff body is fluctuated and it is difficult to catch the information of flow including turbulence and vortices. Results by multi-directional method (fig. 2(a)) show that vortices are caught more clearly and numerical oscillation is prevented in comparison with results by standard method (fig. 2(b)). It shows disturbance behind the cylinder exactly. It also clearly shows the difference between two computational methods. Fig. 3 shows result of density distribution of a square pillar using the multi-directional method. The value is non-dimensional.

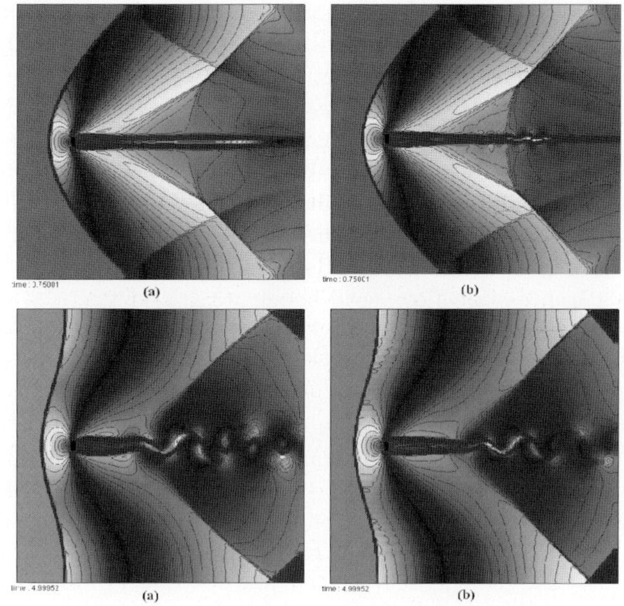

Fig. 2. Time development of Mach number distribution; (a) multi-directional method; and (b) standard method (M_∞=1.4, Re=1,000,000).

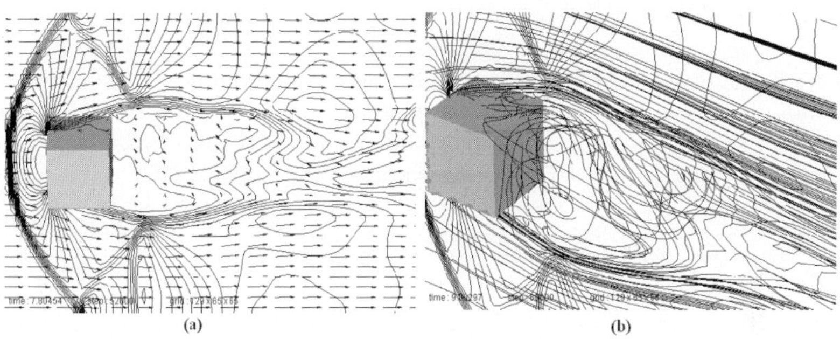

Fig. 3. Density distribution of a square pillar using the multi-directional method; (a) velocity vector and denstiy distribution; and (b) Streamlines and denstiy distribution.

4 Conclusions

In this paper, we present computation of compressible fluid flow around bluff bodies using multi-directional method. The computational grid is generated based on the Cartesian coordinate system at equal intervals. Adopting Cartesian grid makes computation time of grid generation short and the process simple even for complex two and three-dimensional configurations. On the other hand, in general, when we use Cartesian coordinate system, flow direction is not always parallel to a coordinate line. In order to improve this problem, we proposed to use the multi-directional upwind finite difference method for all the space derivative terms.

In calculation, by the standard method, the flow is fluctuated and due to the low accuracy, it is difficult to catch the information of flow including turbulence and vortices. For flow that is difficult to predict, the use of multi-directional method enables to reproduce complicated flow without the expense of the simplicity using Cartesian coordinate system. Using the present method, we have solved this problem successfully and observed accurate information in compressible flow with shock wave and vortices in Cartesian coordinate system. Also, we prevented numerical oscillation.

The present computations clearly captured the features of the unsteady flow fields.

In future study, applying the multi-directional method to other cases, we hope to get more accurate information on compressible fluid flow. We will compute the flow around the bodies of more complicated shape. For example, we try to simulate the phenomena which will take place when a satellite reentries in the atmosphere using Cartesian coordinate system.

References

1. T. V. Karman., "Supersonic Aerodynamics-Principles and Applications", Journal of the Aeronautical Sciences, **14**, No. 7, 1947, pp.373-402.
2. Ishii,K. and Kuwahara,K., 1984, "Computation of Compressible Flow Around a Circular Cylinder", AIAA Paper 84-1631
3. Paul-Henry C.,Christophe D. and Alain D., "A Positive MUSCL Scheme for Triangulations", Rapport de recherche no 3465-Juillet 1998.
4. Bram Van Leer, 1979, "Towards the Ultimate Conservative Difference Scheme. V. A Second-Order Sequel to Godunov's Scheme", Journal of Computational Physics **32**, 101-136
5. Kuwahara,K. and Komurasaki,S., 2002, "Simulation of High Reynolds Number Flows Using Multi-Directional Upwind Scheme",AIAA Paper 2002-0133

A One Point Shock Capturing Kinetic Scheme for Hyperbolic Conservation Laws

Dominic D.J. Chandar, S.V. Raghurama Rao, and S.M. Deshpande

Indian Institute Of Science, Bangalore, India
dominic@aero.iisc.ernet.in

1 Introduction

Kinetic Schemes [1],[2] have been demonstrated to be very robust and efficient in solving practical problems of interest from subsonic to hypersonic flows. The main drawback of most of the Kinetic Schemes is the presence of a large numerical dissipation in their first order accurate versions, as compared to the schemes of Roe [4] and Jameson [3]. While the approximate Riemann solver of Roe can capture steady discontinuities exactly without any numerical dissipation, the Convective upwind and Split Pressure (CUSP) scheme of Jameson can capture the shocks with at the most a single interior point. Both these schemes are based on enforcing some form of the Rankine-Hugoniot condition at the discrete level. Another method based on enforcing the Rankine-Hugoniot condition is the recently developed Accurate Shock Capturing Algorithm with a Relaxation System (ASCARS) of Raghurama Rao and Balakrishna [5]. Since the Rankine-Hugoniot (jump) condition is an exact solution of the conservation equations for the case of a single steady shock, these schemes can capture the shocks very accurately by forcing the jump condition at the discrete level. In this work, using a similar strategy, a single point shock capturing Kinetic Scheme is developed. The basic idea is to introduce a new type of molecular velocity splitting (modifying the Courant splitting) by introducing a free parameter which can be fixed in such a way that a cell containing a shock is in equilibrium with the left and right states.

2 Kinetic Flux Vector Splitting with Modified Courant Splitting

Consider the 1-D (collisionless) Boltzmann equation

$$\frac{\partial f}{\partial t} + v\frac{\partial f}{\partial x} = 0 \tag{1}$$

In an upwind discretization for the above equation, the molecular velocity is split as

$$v = \frac{v + |v|}{2} + \frac{v - |v|}{2} \tag{2}$$

leading to the modified equation

$$\frac{\partial f}{\partial t} + v\frac{\partial f}{\partial x} = |v|\frac{\Delta x}{2}\frac{\partial^2 f}{\partial x^2} \tag{3}$$

where the right hand side represents the numerical diffusion. We introduce a new splitting as

$$v = \frac{v + |v|\phi}{2} + \frac{v - |v|\phi}{2} \text{ where } \phi \text{ is a free parameter of choice.} \tag{4}$$

This results in the modified equation

$$\frac{\partial f}{\partial t} + v\frac{\partial f}{\partial x} = \phi|v|\frac{\Delta x}{2}\frac{\partial^2 f}{\partial x^2} \tag{5}$$

We infer from the above modified equation that ϕ can be used to reduce dissipation. The modified split fluxes resulting from taking moments of the Boltzmann equation are,

$$G^{\pm} = \int\int \frac{\psi}{2}(v \pm |v|\phi) F dv dI \tag{6}$$

where F is the Maxwellian distribution function. Treating ϕ as a quantity that is independent of the molecular velocity, the above expression can be simplified to

$$G^{\pm} = \frac{1}{2}\left[G \pm \phi(\tilde{G}^{+} - \tilde{G}^{-})\right] \tag{7}$$

Where $\tilde{G}^{\pm}$ denote KFVS split fluxes.

Our basic motive as mentioned previously, is to choose ϕ appropriately such that the cell containing the shock is in equilibrium with the left and the right states. In the following section, we formulate this condition based on splitting of the fluxes.

3 One point Shock Capturing and Conditions for Splitting the Fluxes in One Dimension

Jameson [3] has analyzed the discrete shock structure and and the requirements for obtaining a single point shock capturing macroscopic scheme. Using a similar strategy, we derive in this chapter the conditions for a flux splitting scheme to capture a steady shock with a single interior point and use these conditions in our new Kinetic Scheme. Consider a set of finite volumes as shown in figure 1 in which a stationary shock is located in a single cell, j, with left and right states denoted by $j-1$ and $j+1$ respectively. Let us also denote the left state by L, the right state by R and the intermediate cell containing the shock as A.

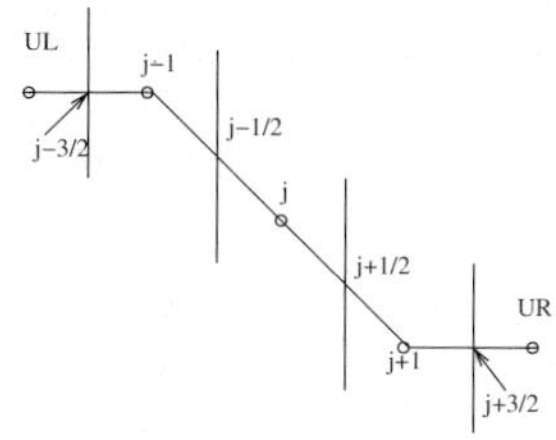

Fig. 1. Shock structure for a single point

Consider a one dimensional conservation law

$$\frac{\partial U}{\partial t} + \frac{\partial G}{\partial x} = 0 \tag{8}$$

Let us discretize the spatial derivative in the above conservation equation using a finite volume method as

$$\frac{\partial U}{\partial t} + \frac{1}{\Delta x}\left[G_{j+\frac{1}{2}} - G_{j-\frac{1}{2}}\right] = 0 \tag{9}$$

Here $G_{j+\frac{1}{2}}$ and $G_{j-\frac{1}{2}}$ are the cell interface fluxes and are given by

$$G_{j+\frac{1}{2}} = G_j^+ + G_{j+1}^-,\ G_{j-\frac{1}{2}} = G_{j-1}^+ + G_j^- \tag{10}$$

Let us now apply the update formula of the finite volume method with the above interface fluxes at the grid points j-1, j, j+1.

Semi-discrete update formula at grid point j-1

$$\frac{\partial u}{\partial t} + \frac{1}{\Delta x}\left[G_{j-\frac{1}{2}} - G_{j-\frac{3}{2}}\right] = 0 \tag{11}$$

Using expressions for the cell interface fluxes $G_{j-\frac{1}{2}} = G_{j-1}^+ + G_j^-, G_{j-\frac{3}{2}} = G_{j-2}^+ + G_{j-1}^-$ in the above equation, we obtain

$$\frac{\partial u}{\partial t} + \frac{G_{j-1}^+ + G_j^-}{\Delta x} - \frac{G_{j-2}^+ + G_{j-1}^-}{\Delta x} = 0 \tag{12}$$

At steady state, the transient term vanishes and we then obtain

$$G_{j-1}^+ - G_{j-2}^+ = G_{j-1}^- - G_j^- \tag{13}$$

In the case of a shock being located in a single cell j, the states j-1 and j-2 are not different and they belong to the same left state. Thus $G_{j-1}^+ = G_{j-2}^+$, and we obtain a simpler equation as

$$G_{j-1}^- = G_j^- \text{ or } G_L^- = G_A^- \tag{14}$$

Similarly, the condition at grid points j+1 and j will result in $G_R^+ = G_A^+$, $G_R = G_L$ (which is the Jump condition) respectively.

Requirements for Single Point Shock Capturing Kinetic Scheme

In the previous section, we derived the modified Split fluxes with the free parameter ϕ as

$$G^{\pm} = \int_0^{\infty}\int_{-\infty}^{\infty} \psi \frac{(v \pm |v|\phi)}{2} F dv dI \tag{15}$$

which is equivalent to

$$G^{\pm} = \frac{1}{2}\left[G \pm \phi(\tilde{G}^{+} - \tilde{G}^{-})\right] \tag{16}$$

Therefore, using the conditions derived above for single point shock capturing, we obtain

$$G_L^- = G_A^- \text{ or } G_L - \phi_{LA}(\tilde{G}_L^+ - \tilde{G}_L^-) = G_A - \phi_{LA}(\tilde{G}_A^+ - \tilde{G}_A^-) \tag{17}$$

$$G_R^+ = G_A^+ \text{ or } G_R + \phi_{AR}(\tilde{G}_R^+ - \tilde{G}_R^-) = G_A + \phi_{AR}(\tilde{G}_A^+ - \tilde{G}_A^-) \tag{18}$$

From (17 & 18) we determine ϕ_{LA}, ϕ_{AR}, where the subscripts LA, AR define the interface between the left and intermediate state, right and intermediate state respectively.

$$\phi_{LA} = \frac{G_A - G_L}{(\tilde{G}_A^+ - \tilde{G}_A^-) - (\tilde{G}_L^+ - \tilde{G}_L^-)} \tag{19}$$

$$\phi_{AR} = \frac{G_A - G_R}{(\tilde{G}_R^+ - \tilde{G}_R^-) - (\tilde{G}_A^+ - \tilde{G}_A^-)} \tag{20}$$

We apply the above conditions at high gradients of density so that shocks are captured accurately. Such high gradients can be sensed by calculating numerically $\frac{|\nabla \rho|}{max(|\nabla \rho|)}$ which is near unity in regions of high gradients. In regions where the flow is smooth without appreciable gradients, ϕ is set to unity.

For two dimensions, we consider a structured two dimensional finite volume and apply (17 & 18) across the interface between two cells by considering the flow locally one dimensional and treating the fluxes G as fluxes normal to any surface.

4 Numerical Results

The performance of this new kinetic scheme is tested for supersonic and transonic internal and external flows. For internal flow cases, the following test cases were solved:

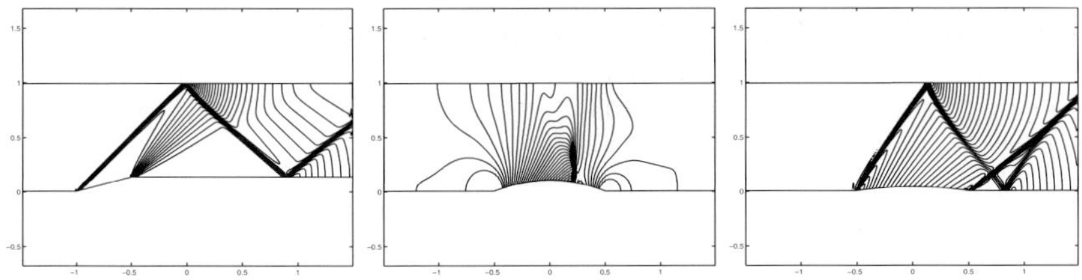

Fig. 2. Mach contours for test cases 1,2,3 respectively

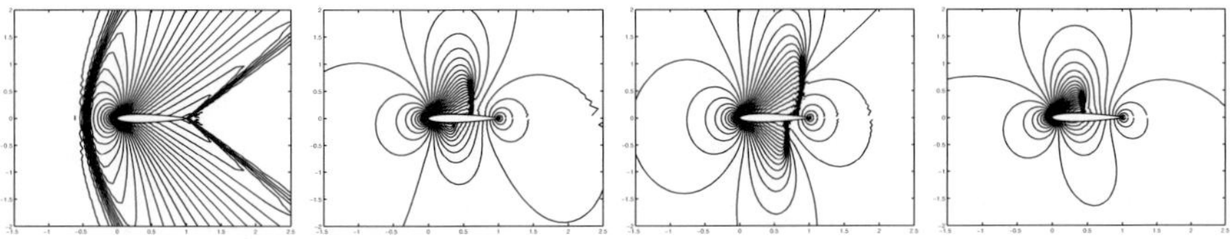

Fig. 3. Mach contours for test cases 4,5,6,7 respectively

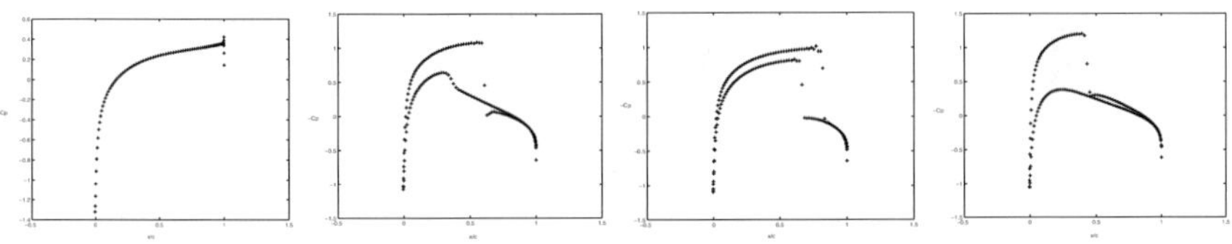

Fig. 4. Cp distribution for test cases 4,5,6,7 respectively

1. Steady supersonic flow at $Mach$ 2.0 **across a ramp of included angle** 15°.
2. Steady transonic flow at $Mach$ 0.675 **over a 10% circular arc bump**.
3. Steady supersonic flow at $Mach$ 1.4 **over a 4% circular arc bump**.

For external flow configurations we consider (i) Flows past NACA 0012 airfoil with the following conditions:

4. $Mach$ 1.2**, at an angle of attack** 0°.
5. $Mach$ 0.8**, at an angle of attack** 1.25^o.
6. $Mach$ 0.85**, at an angle of attack** 1.00^o.
7. $Mach$ 0.75**, at an angle of attack** 2.00^o.

and (ii) Steady Supersonic flow past a full cylinder at $Mach$ 3.00.

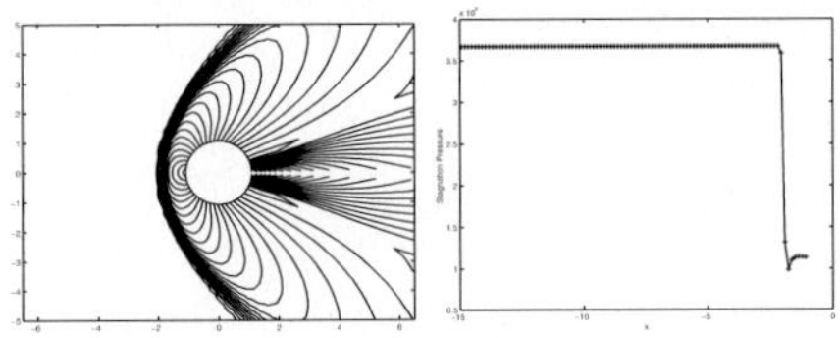

Fig. 5. Mach contours and Stagnation Pressure along y=0 for test case(ii)

5 Concluding Remarks

We have presented a new low dissipative kinetic scheme based on a *modified Courant Splitting* of the molecular velocity through a parameter ϕ. Conditions for the split fluxes derived based on equilibrium determine ϕ for a one point shock. It turns out that ϕ is a function of the Left and Right states to the shock and that these states should satisfy the Rankine-Hugoniot Jump condition. Hence ϕ is utilized in regions where the gradients are sufficiently high, and is switched to unity in smooth regions. Numerical results confirm a discrete shock structure with a single interior point when the shock is aligned with the grid.

References

1. Mandal,J.C and Deshpande,S.M, Kinetic Upwind methods for Inviscid Compressible Flows, Computers and Fluids, **23**, 2,447-478, (1993).
2. Deshpande S M, Kinetic Flux Splitting schemes , Computational Fluid Dynamics Review (1995), John Wiley and Sons. pp. 161.
3. Antony Jameson. Analysis and Design of Numerical Schemes foir Gas Dynamics 2 Artificial Discussion and Discrete Shock structure,International Journal of Computational Fluid Dynamics, Vol **5** pp1-38, (1995).
4. Roe,P.L Approximate Riemann solvers , Parameter Vectors, and Finite Difference Schemes, Journal of Computational Physics,Vol.**43**, p.357, (1981).
5. S.V. Raghurama Rao and K. Balakrishna, An Accurate Shock Capturing Algorithm with a Relaxation system for Hyperbolic Conservation Laws. AIAA-2003-4115 (2003).

Accurate and Efficient Multi-dimensional TVD Interpolation

Sung-soo Kim, Kyu-Hong Kim, and Chongam Kim

Department of Aerospace Engineering, Seoul National University, Seoul 151-742, Korea. E-mail:chongam@plaza.snu.ac.kr

As an accurate and efficient multi-dimensional higher-order interpolation, Higher-order TVD interpolation in multi-dimension is developed. The major advantage of MHTVD (Multi-dimensional Higher-order TVD) is the complete multi-dimensional monotonic characteristics and it is possible to be more than 3rd order spatial accuracy. In addition, MHTVD shows good convergence characteristics and is computationally efficient. Through test cases, MHTVD is verified to present very accurate and robust solution in regions of a shock discontinuity as well as a boundary layer.

1 Baseline Spatial Discretization

The baseline spatial discretization is M-AUSMPW+, which can be briefly written as follows.

$$F_{\frac{1}{2}} = \overline{M}_L^+ c_{\frac{1}{2}} \Phi_{L,\frac{1}{2}} + \overline{M}_R^- c_{\frac{1}{2}} \Phi_{R,\frac{1}{2}} + (P_L^+ \mathbf{P}_L + P_R^- \mathbf{P}_R), \tag{1}$$

where Φ is the transferred property defined at a cell-interface. The basic idea of this scheme is to modify the transferred property, Φ, appropriately to represent physical distribution more accurately and to yield a complete monotonic profile regardless of sonic point location [1].

2 MHTVD (Multi-dimensional Higher-order TVD)

Monotonic characteristics in multi-space dimension are directly related to accuracy and convergence. Especially, when hypersonic flows including strong shock discontinuity is calculated, monotonicity in multi-dimension is crucial for successful computations. Moreover, oscillations in a shock region interferes stable convergence. The monotonicity condition for multi-space dimension was studied by Spekreijse [2]. He extended the one-dimensional TVD condition to two-dimensional case and presented the two-dimensional TVD condition. However, the derived condition is so loose that the results are very diffusive since the condition was obtained by five-point stencil. In addition, it has undetermined coefficients. In this paper, we derive the accurate multi-dimensional TVD condition without any undetermined parameter and with nine point values.

2.1 Multi-dimensional TVD condition

The interpolated value at a grid point in a multi-dimensional case should be within the maximum and minimum of neighboring cells.

$$\Phi^{min}_{neighbor} < \Phi < \Phi^{max}_{neighbor}. \tag{2}$$

In a two-dimensional problem, if the discontinuity is inclined by θ_j and $0 < \theta_j < 90°$ as in Fig.1, the values at grid points should satisfy Eq.(3).

$$\Phi_{i,j-1} < \Phi_1 < \Phi_{i+1,j}, \Phi_{i,j-1} < \Phi_1 < \Phi_{i+1,j},$$
$$\Phi_{i,j-1} < \Phi_1 < \Phi_{i+1,j}, \Phi_{i,j-1} < \Phi_1 < \Phi_{i+1,j}. \tag{3}$$

The values at four vertex points are given as

$$\Phi_1 = \Phi_{i,j} + \Delta\Phi^-_\eta + \Delta\Phi^+_\xi, \Phi_2 = \Phi_{i,j} + (1 + \tan\tilde{\theta_j})\Delta\Phi^+_\eta,$$
$$\Phi_3 = \Phi_{i,j} + \Delta\Phi^+_\eta + \Delta\Phi^-_\xi, \Phi_4 = \Phi_{i,j} + (1 + \tan\tilde{\theta_j})\Delta\Phi^-_\eta, \tag{4}$$

where $\tan\tilde{\theta_j} = \frac{\Delta\Phi^+_\xi}{\Delta\Phi^+_\eta}$. Since $\Delta\Phi^\pm_{\xi,\eta}$ is determined by the TVD MUSCL limiter, Φ_1 and Φ_3 always satisfy the monotonic condition of Eq.(4). Now, let us check whether Φ_2 satisfies the condition of Eq.(3). When $r_L = \frac{\Delta\Phi_{j+1/2}}{\Delta\Phi_{j-1/2}} > 1$, the maximum allowable variation is $\Delta\Phi^+_\eta = 0.5\min(\Delta\Phi_{j+1/2}, 2\Delta\Phi_{j-1/2}) \leq 0.5\Delta\Phi_{j+1/2}$ and the following should be satisfied for the monotonic condition of Eq.(3).

$$\Phi_2 = \Phi_{i,j} + (1+\tan\tilde{\theta_j})\Delta\Phi^+_\eta \leq \Phi_{i,j} + 0.5(1+\tan\tilde{\theta_j})\Delta\Phi_{j+1/2} < \Phi_{i+1,j+1}. \tag{5}$$

When $0 < r_L < 1$, the maximum allowable variation is now $\Delta\Phi^+_\eta = 0.5\min(\Delta\Phi_{j-1/2}, 2\Delta\Phi_{j+1/2}) \leq \Delta\Phi_{j+1/2}$ and Eq.(6) should be satisfied.

$$\Phi_2 = \Phi_{i,j} + (1 + \tan\tilde{\theta_j})\Delta\Phi^+_\eta \leq \Phi_{i,j} + (1 + \tan\tilde{\theta_j})\Delta\Phi_{j+1/2} < \Phi_{i+1,j+1}. \tag{6}$$

By limiting the maximum variation of $\Delta\Phi^\pm_{\xi,\eta}$, multi-dimensional TVD condition can be obtained. From Eq.(6),

$$\Phi_{i,j} + (1 + \tan\tilde{\theta_j})\Delta\Phi^+_\eta \leq \Phi_{i+1,j+1} = \Phi_{i,j} + \frac{\tan\bar{\theta}_{j+1}}{r_{R,j+1}}\Delta\Phi_{j+1/2}. \tag{7}$$

When $r_L > 1$, the maximum variation is modified as $\Delta\Phi_\eta^+ = 0.5\min(\Delta\Phi_{j+1/2}, \alpha\Delta\Phi_{j-1/2})$. From $\Delta\Phi_\eta^+ = 0.5\min(\Delta\Phi_{j+1/2}, \alpha\Delta\Phi_{j-1/2}) \leq 0.5 \frac{\alpha}{r_L}\Delta\Phi_{j+1/2}$, we have

$$1 \leq \alpha \leq \min[2, \frac{2r_L(1 + \frac{\tan\bar{\theta}_{j+1}}{r_{R,j+1}})}{1 + \tan\tilde{\theta}_j}]. \quad (8)$$

Similarly, when $0 < r_L < 1$, the maximum variation is modified again as $\Delta\Phi_\eta^+ = 0.5\min(\Delta\Phi_{j-1/2}, \alpha\Delta\Phi_{j+1/2})$ and we obtain

$$1 \leq \alpha \leq \min[2, \frac{2(1 + \frac{\tan\bar{\theta}_{j+1}}{r_{R,j+1}})}{1 + \tan\tilde{\theta}_j}]. \quad (9)$$

Equations (8) and (9) always make Φ_2 less than $\Phi_{i+1,j+1}$ and the monotonic condition of Eq.(3) is realized.
When $\tan\bar{\theta}_{j+1}/r_{R,j+1} < 0$, we require

$$\Phi_2 = \Phi_{i,j} + (1 + \tan\tilde{\theta_j})\Delta\Phi_\eta^+ \leq \Phi_{i,j+1} = \Phi_{i,j} + \Delta\Phi_{j+1/2}. \quad (10)$$

By applying monotonicity condition, finally we obtain the multi-dimensional TVD condition as follows.

$$1 \leq \alpha \leq \min[2, \frac{2\max(1, r_L)(1 + \max(0, \frac{\tan\bar{\theta}_{j+1}}{r_{R,j+1}}))}{1 + \tan\tilde{\theta}_j}]. \quad (11)$$

α contains two angles related to property variation, $\tilde{\theta}$ and $\bar{\theta}$, defined at each cell. For computational efficiency we define the representative angle,θ , which indicates the angle between the maximum gradient angle of property at each cell. Obtaining the property gradient by

$$\theta_{max} = \tan^{-1}(\frac{\nabla\Phi_{\eta,i,j}}{\nabla\Phi_{\xi,i,j}}). \quad (12)$$

$\theta_i = \theta_{max} - 0°$ and $\theta_j = 90° - \theta_{max}$ where θ_i is the angle of the ξ-direction calculation and θ_j is the angle of the η-direction calculation. (See Fig.1)

2.2 General form of MHTVD

The MHTVD can be written as follows.

$$\Phi_{i+1/2,L} = \Phi_i + 0.5\max(0, \min(2, \alpha_L r_L, \beta_L))\Delta\Phi_{i-1/2}, \quad (13)$$

The value of θ using the representative angle does not have any information on the shape of cell. Thus, additional information on cell shape such as cell aspect ration has to be added. From Fig. 1, we have $\Delta\Phi_\xi = \Delta\Phi'_\xi$ and

$\frac{a}{b}\Delta\Phi_\eta = \Delta\Phi'_\eta$. And the equivalent angle of θ_e for a high aspect ratio cell is defined as

$$\tan\theta_{e,j} = \frac{\Delta\Phi_\eta}{\Delta\Phi_\xi} = \frac{b}{a}\tan\theta_j, \tan\theta_{e,i} = \frac{\Delta\Phi_\xi}{\Delta\Phi_\eta} = \frac{a}{b}\tan\theta_i. \tag{14}$$

Followings are the β for the 3rd and 5th order form of MHTVD respectively.

3rd order MHTVD $\beta_L = \frac{1+2r_L}{3}$, $\beta_R = \frac{1+2r_R}{3}$.

5th order MHTVD

$\beta_L = \frac{-4/\max[0.5,\min(2,r_{L,i-1})]+14+24r_{L,i}-4r_{L,i}\max[0.5,\min(2,r_{L,i+1})]}{30}$,

$\beta_R = \frac{-4/\max[0.5,\min(2,r_{R,i+2})]+14+24r_{R,i+1}-4r_{R,i+1}\max[0.5,\min(2,r_{R,i})]}{30}$.

Efficiency is the one of key advantages compared with other higher order interpolation scheme such as ENO or WENO [3] and so on. Detailed analysis about MHTVD can be found in Ref. [4]

Table 1. Comparison of efficiency of MHTVD

Interpolation scheme	van Leer limiter Roe's FDS	3rd MHTVD Roe's FDS	5th MHTVD Roe's FDS	5th MHTVD M-AUSMPW+
Time/Time$_{MUSCL}$	1	1.075	1.125	1.075

3 Multi-dimensional Interpolation

Truly higher-order accuracy can be achieved in multi-dimension using dimension by dimension interpolation. In 2-D flow, we can achieve kth order accuracy using $Q(x,y) = q(x,y) + O(\Delta x^k, \Delta y^k)$ with the condition $\int\int Qdxdy = \int\int qdxdy$. Firstly, we interpolate $\int Q(x,y)$ dy with k-1th order polynomial of x as follows:

$$\int Q(x,y)dy = p(x) + O(\Delta x^k \Delta y), \tag{15}$$

where $\int\int Qdydx = \int p(x)dy$.
Then we can obtain the cell interface values as follows:

$$p(x_{i+1/2}) = \int Q(x_{i+1/2},y)dy + O(\Delta x^k \Delta y). \tag{16}$$

Since Eq.(16) gives single value at call interface in smooth region, exact fluxes rather than numerical fluxes are used. Interpolating the values $p_j(x_{i+1/2})$ with k-1th order polynomial of y, we can obtain approximate solution with kth order accuracy as follows:

$$Q(x_{i+1/2}, y) = q(x_{i+1/2}, y) + O(\Delta x^k, \Delta y^k), \tag{17}$$

where $\int q(x_{i+1/2}, y)dy = p(x_{i+1/2})$.

With Gauss quadrature point values which can be obtained from Eq. (17), the surface flux integration on each face can be performed with a kth order accurate Gauss quadrature formula.

4 Numerical Results

The results of shock/boundary interaction and double shock reflection are presented. The free stream conditions in shock boundary interaction are that Mach number is two and shock impinging angle is 32.585 degree angle. Reynolds number is 2.96×10^5. MHTVD with M-AUSMPW+ in Fig. 2 give more accurate results than MUSCL with van Leer limiter. In Fig.3 , the 3rd order and the 5th order MHTVD show a good convergence property, while superbee and multi-dimensional superbee show a convergence stall. The free stream Mach number is two and the deflection angle of the lower wall is 15 degree angle in Double Mach reflection problem. Even if the test case is very simple, superbee limiter and van Leer limiter are shown not to be converged because they do not satisfy the multi-dimensional TVD condition. On the other hand, the multi-dimensional superbee limiter, the 5th order and the 3rd order MHTVD show good convergence characteristics. It is because the monotonic condition in a multi-dimension removes an unwanted local maximum or minimum distribution.

5 Conclusions

Through numerical analysis and tests, MHTVD is derived and shown to enhance accuracy and robustness simultaneously in regions of a shock dis-

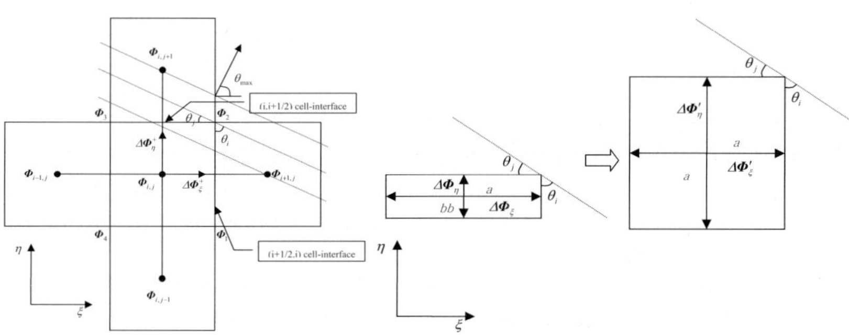

Fig. 1. Determination of angle θ

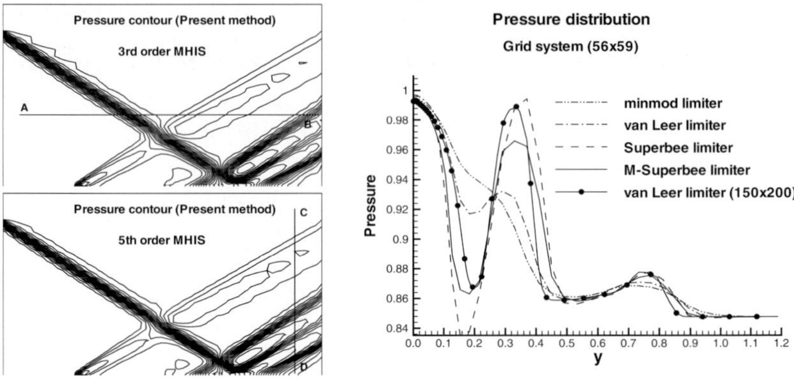

Fig. 2. Comparison of pressure distribution along the line CD

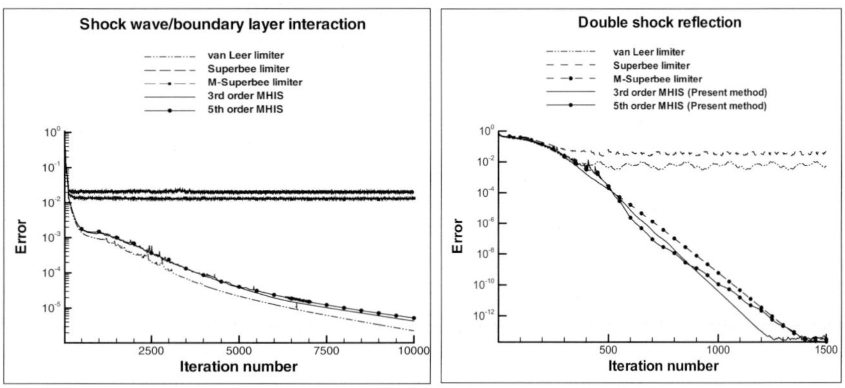

Fig. 3. Error history

continuity as well as a boundary-layer. In addition, it is also very efficient and has the good convergence characteristics.

References

1. K. H. Kim, S. S. Kim, C. Kim: Accurate and Efficient Re-evaluation of Cell-interface Convective Fluxes *presented at ICCFD3*,(2004)
2. S. P. Spekreijse: Mathematics of Computation, **49**, 135 (1987)
3. X-D. Liu, S. Osher, and T. Chan: J. of Computational Physics, **115**, 200 (1994)
4. K. H. Kim. and C. Kim: Accurate, Efficient and Monotonic Numerical Methods for Multi-dimensional Compressible Flows *submitted to J. of Computational Physics*, (2004)

Exact Flux Linearization for Convergence Improvement in the Implicit Godunov Method

Igor Men'shov and Yoshiaki Nakamura

Nagoya University, Chikusa-ku, Nagoya 464-8603, Japan

Summary. The present paper addressees the implicit Godunov method. A common way to handle the implicitness is to apply Newton's iterations that require exact linearization of the residual. However, because the Godunov method adopts the exact, rather complicated, solution of the initial-value Riemann problem for the numerical flux, the linearization looks cumbersome. We show how to settle this problem and obtain the Godunov linearized flux in a compact analytical form. The linearization is based on the variational Riemann problem that is succeeded to solve analytically. The linear system of the Newton's iteration is solved in two steps with the LU-SGS factorization method. Comparison with the method that uses conventional approximate flux linearization of Turkel and Jameson reveals certain advantages of the exact linearization in the rate of convergence.

1 Introduction

The difficulty of using implicit Godunov schemes is that the state parameter vector to be solved is defined by a system of discretized equations, which have a rather complicated non-linear form due to an intricate non-linear relationship between the numerical flux and the solution vector [1]. This difficulty may be overcome by using the standard Newton's approach that can efficiently solve the equations with fast, quadratic convergence providing the linearization of the equations is virtually exact.

However, until recently, it has been a widespread opinion that the exact linearization (EL) of the Godunov's flux is extremely expensive, if not impossible, not only for exact, but also for approximate RP solutions (as say the Roe's solution [2]). Therefore, a common approach has been proposed that consists in the replacement of the EL with an approximate linearization (AL) in the Newton's approach.

Jameson and Turkel suggested one way for such approximation [3], which is rather popular among CFD researches. The basic idea of this approximation is to use not the exact, but an approximate Godunov's flux for linearization, namely the Roe's flux [2] with the Jacobian matrix to be substituted by its spectral radius. The numerical flux is much simplified under this substitution, and the linearization can be readily performed. However, a penalty arises due to the flux modification that the quadratic convergence of Newton's iterations can no longer be achieved in so far as a mismatch and inconsistency between the right- and left-hand sides arise in the linearized equations.

The purpose of the present paper is to obtain the genuinely EL of the Godunov flux and accomplish it in the implicit time-integration scheme with the aim of improving the convergence rate of Newton's iterations. To do this, first we consider an auxiliary problem on variation of the exact RP solution with respect to small change in initial data. This problem is shown to admit the unique solution that can be expressed by explicit analytical formulas. Then, by means of this solution, the exact linearization of the Godunov's flux-function is readily fulfilled.

The linearized equations are solved at each iteration with the LU-SGS approximate factorization method of Yoon and Jameson [4]. Numerical experiments are carried out for both steady and unsteady compressible flows under different flow conditions ranging from low subsonic to strongly supersonic speeds. Comparisons are made for the convergence rate of the implicit LU-SGS Godunov method between the EL and AL of the numerical flux-function.

2 Variational Riemann problem

The Variational Riemann Problem (VRP) can be set forward as follows. Let the exact solution to the RP be known for initial data given by vectors $\mathbf{Q}_l$ and $\mathbf{Q}_r$: $\mathbf{Q}^R = \mathbf{Q}^R(\lambda, \mathbf{Q}_l, \mathbf{Q}_r)$, where λ is the similarity parameter, $\lambda = x/t$, $\mathbf{Q}$ is the vector of state parameters, subs l and r denote the left- and right-hand side of the initial discontinuity, respectively. Considering small variations of the initial state vectors

$$\mathbf{Q}_l \to \mathbf{Q}_l + \delta\mathbf{Q}_l, \;\; \mathbf{Q}_r \to \mathbf{Q}_r + \delta\mathbf{Q}_r \tag{1}$$

the first variation of the baseline solution is introduced by

$$\delta\mathbf{Q}^R = M_l \delta\mathbf{Q}_l + M_r \delta\mathbf{Q}_r \tag{2}$$

where M_l and M_r are referred to as variation matrices (VM) associated with the left- and right-hand side of the initial discontinuity, respectively, and defined by

$$M_i = M_i(\lambda, \mathbf{Q}^R) = \frac{\partial \mathbf{Q}^R}{\partial \mathbf{Q}_i}, \;\; i = l, r \tag{3}$$

The VRP is to find these VMs for arbitrary initial data of the baseline RP.

Matrices M_l and M_r are piecewise analytical matrix-functions of λ with the domains of analyticity to be the same as those of the basic solution $\mathbf{Q}^R(\lambda)$. These domains are defined by a set of waves originating from the initial discontinuity break-up, or in other words, by the wave pattern proceeded in the non-varied RP.

Moreover, VMs are obviously constant with respect to λ in each analyticity domain except for the fan of centered expansion waves, within which they vary with λ. In unperturbed zones, we have $M_l = I$ and $M_r = O$ on the left-hand side and $M_l = O$ and $M_r = I$ on the right-hand side, where I is the unit-matrix and O is the null-matrix. The VM inside the expansion fan can be found analytically by varying Riemann invariants and integrating the resultant system of differential equations [5].

In contact zones adjacent to the contact discontinuity, the VMs are constant with λ, and should be defined by considering conjugate relations at the rear characteristic of the expansion fan, shock wave, or weak discontinuity, depending on which wave pattern appears in the basic RP. Analyzing these relations for all possible cases, one can show that the variational vector $\delta\mathbf{Q}^i$ in each contact zone is determined for all three cases in a unique form that includes one arbitrary constant c_i $(i = l, r)$ as follows:

$$\delta\mathbf{Q}^i = N^i \delta\mathbf{Q}_i + c_i \mathbf{m}^i \tag{4}$$

Here the matrix N^i and the vector $\mathbf{m}^i$ can be analytically expressed in terms of the solution to the basic RP. The corresponding rather tedious formulas are not given here, but can be found in [5].

To complete the solution, it is necessary to determine two constants c_l and c_r in eq.(4). There are just two conjugate relations for this. They link the variations of pressure and velocity at the contact discontinuity: $[\delta u] = 0$, $[\delta p] = 0$.

With the use of these relations, the constants c_l and c_r are determined in a linear form with respect to the initial variations $\delta\mathbf{Q}_i$, $i = l, r$, and the variational matrices are defined throughout the flow. The resultant formulas can be cast in a rather simple and compact form (for details the reader is referred to [5, 6]).

3 Time Marching Scheme

The Navier-Stokes equations

$$\partial_t \mathbf{q} + \partial_k \mathbf{f}_k = \partial_k \mathbf{g}_k \tag{5}$$

where $\mathbf{q}$ is the conservative vector of density, Cartesian components of momentum and total energy, $\mathbf{f}_k$ and $\mathbf{g}_k$ are inviscid and viscous fluxes are discretized in space with the finite-volume method and in time with the implicit backward scheme. The resultant system of discretized equations is solved iteratively by Newton's method that leads to a system of linear equations with respect to the iterative increment of the solution vector $\Delta^s\mathbf{q} = \mathbf{q}^{n+1,s+1} - \mathbf{q}^{n+1,s}$:

$$\left[\frac{V_i}{\Delta t} + D_i^{n+1,s}\right] \Delta^s \mathbf{q}_i = \mathbf{R}_i^{n+1,s} - \sum_\sigma \mathbf{G}\left(\Delta^s \mathbf{q}_\sigma\right) \tag{6}$$

where superscripts n and s indicate time level and Newton's iteration index, respectively, Δt is the time step, V is the cell volume, and the summation over σ on the right-hand side is fulfilled over all cells surrounding the cell i .

In eq.(6), $\mathbf{R}$ represents the iterative unsteady residual,

$$\mathbf{R}_i^{n+1,s} = \sum_{\sigma} \left(-\mathbf{f}_\sigma^{n+1,s} + \mathbf{g}_\sigma^{n+1,s}\right) - V_i \left(\mathbf{q}_i^{n+1,s} - \mathbf{q}_i^n\right) / \Delta t \tag{7}$$

where the inviscid flux is approximated by the Godunov flux-function based on the exact RP solution: $\mathbf{f}_\sigma = \mathbf{f}(\mathbf{q}^R)$, and $\mathbf{q}^R = \mathbf{q}^R(\mathbf{q}_i, \mathbf{q}_\sigma)$.

The matrix D and the vector $\mathbf{G}$ in eq.(6) result from linearization of the numerical flux, and have the following form:

$$\begin{aligned} D_i^{n+1,s} &= \textstyle\sum_\sigma \left(M_1^{n+1,s} + \rho_{dis}^{n+1,s}\right) \\ \mathbf{G}\left(\Delta^s \mathbf{q}_\sigma\right) &= \left(M_2^{n+1,s} - \rho_{dis}^{n+1,s} I\right) \Delta^s \mathbf{q}_\sigma \end{aligned} \tag{8}$$

where matrices M_1 and M_2 define linearization of the inviscid numerical flux,

$$\delta \mathbf{f} = M_1 \delta \mathbf{q}_i + M_2 \delta \mathbf{q}_\sigma \tag{9}$$

and ρ_{dis} is a majorant of the spectral radius of the viscous Jacobian.

Exact linearization corresponds to the matrices $M_{1,2}$ that are evaluated through the VMs of the exact RP solution:

$$M_{1(2)} = A\left(\mathbf{q}^R\right) \frac{\partial \mathbf{q}^R}{\partial \mathbf{q}_{i(\sigma)}} \tag{10}$$

For comparison, we also consider the conventional approximate linearization that was proposed by Turkel and Jameson [3], in which the matrices M_1 and M_2 are approximated by the inviscid Jacobian A and its spectral radius ρ_{inv} as

$$M_1 = 0.5[A(\mathbf{Q}_i) + \rho_{inv} I]; \; M_2 = 0.5[A(\mathbf{Q}_i) - \rho_{inv} I] \tag{11}$$

The foregoing two approaches are referred to as the exact linearization (EL) scheme and the approximate linearization (AL) scheme respectively, hereinafter.

Eq.(6) is solved at each s iteration with the LU-SGS method [4] performed in two (forward and backward) sweeps. Details of the solution can be found in [6, 7].

4 Results

Several calculations are presented in this section. The results will be shown that concern the convergence behavior of the implicit Godunov method with the exactly linearized numerical flux (the EL scheme) in comparison with that of the AL scheme.

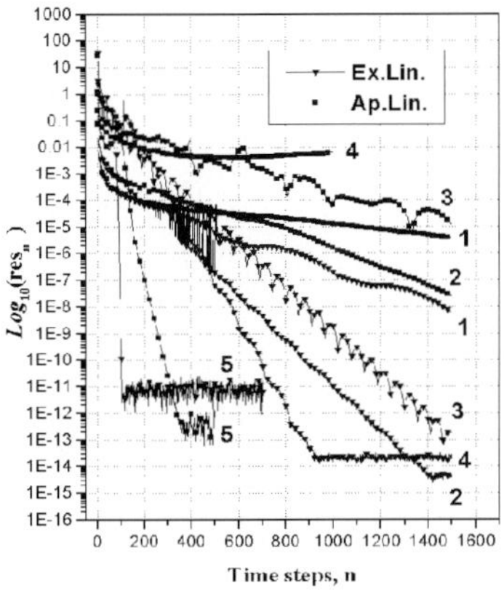

Fig. 1. Convergence histories: 1 – $M_{\text{inf}} = 0.085$, 2–$M_{\text{inf}} = 0.25$, 3–$M_{\text{inf}} = 0.85$, 4 – $M_{\text{inf}} = 1.1$, 5 – $M_{\text{inf}} = 8.5$

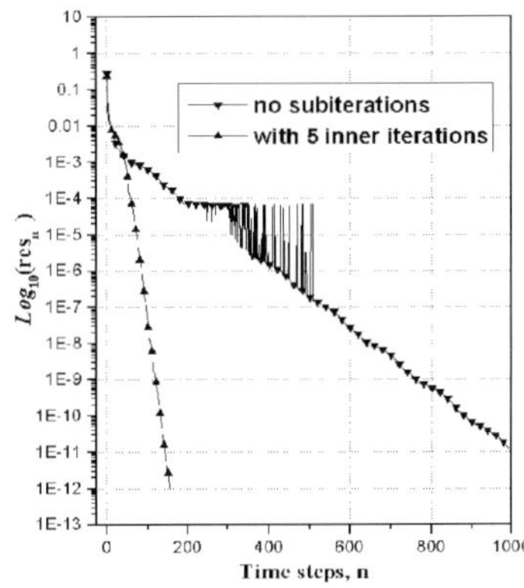

Fig. 2. Effect of inner iterations on the residual in the EL scheme for $M_{\text{inf}} = 0.25$ flow

4.1 Inviscid Steady Flow

An inviscid flow past a Zhukovskii airfoil at an attack angle of 5° under different freestream conditions is considered. Calculations have been carried out with a C-type grid of 272 × 100 mesh points (128 points are placed on the body surface, and 100 points between the body and an outer boundary, which is placed about 15 airfoil chord lengths away from the body). Only one inner iteration (iteration in s) was used to solve eq.(6) at each time step. The time step Δt defined in terms of the CFL number was taken to be of the order $O(1)$ at the beginning of calculations; then it is gradually increased up to a value of 10^4 by the 50th time step.

Figure 1 shows the convergence history of the EL and AL schemes versus the time step for several values of the freestream Mach number. One can see that the EL scheme converges much faster than the AL scheme. However, the quadratic convergence is not achieved as only one iteration is performed, and therefore Newton's iterative process is not in fact realized.

The convergence rate is much improved by implementing a number of inner iterations (the quadratic convergence can be achieved in the limit of full convergence of inner iterations at each time step). This improvement can be seen in Fig. 2, where convergence histories are shown for the EL scheme executed with 1 and 5 inner iterations, respectively.

4.2 Viscous Flow

The convergence for viscous steady flows is illustrated by Fig. 3 that displays results for the forebody flow of $M_\infty = 6$ and $Re = 1.3 \cdot 10^6$ around a cylinder. A 120 × 80 grid is used with a minimal grid spacing of $10^{-5} \times R_{cyl}$. One can

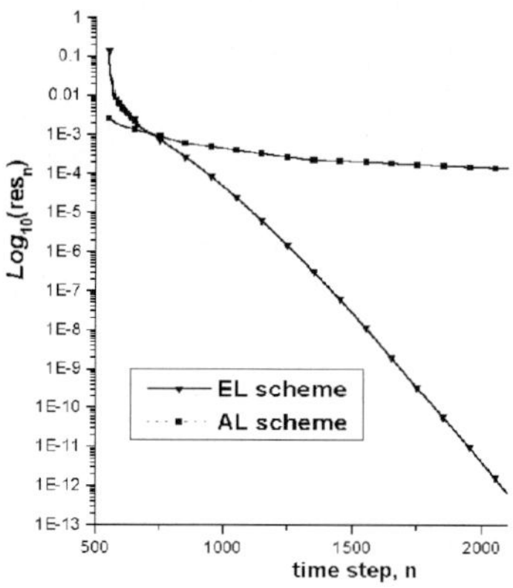

Fig. 3. Residual convergence histories for supersonic laminar flow around a circular cylinder forebody

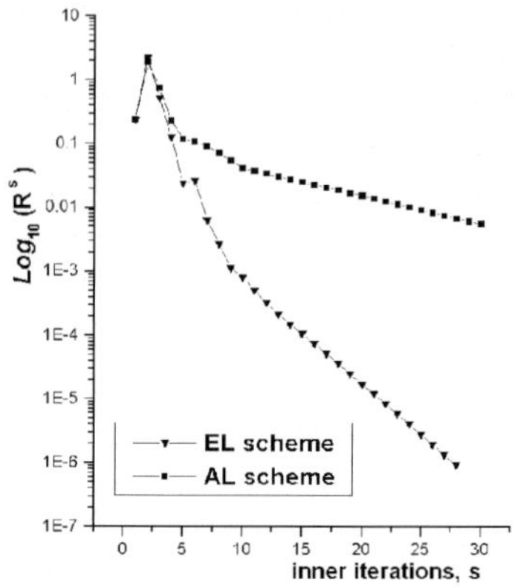

Fig. 4. Convergence of inner iterations for unsteady flow past a Zhukovskii airfoil

see again that the convergence of the EL scheme is much faster than that of the AL scheme. It should also be noted that this calculation is executed with single inner iteration at each time step, and therefore the strict Newton iterative process is far from it fulfilled, even if the exact linearization is performed.

For unsteady flow calculations, it is important to drive the unsteady residual, eq. (7), to zero for better resolution of unsteady phenomena. This process is accomplished by implementing inner iterations. The convergence of inner iterations in the EL scheme is examined for the foregoing flow past the Zhukovskii airfoil, which is now studied with the viscous model and grid that clusters toward the airfoil with a minimal normal mesh spacing of $10^{-3} \times L_{chord}$. The freestream Mach number is 0.5, the attack angle is 5°, and the Reynolds number based on the chord length is 23000. Under these conditions the flow is unsteady due to boundary layer separation accompanied by the formation of large vortical structures that travel downstream in the wake.

Comparison between convergence rates of inner iterations of the EL and AL schemes is given in Fig. 4, which clearly displays the superiority of the EL scheme over the AL scheme. Only 5 inner iterations need to reduce the unsteady residual by two orders in this scheme, while more than 20 to do this with the AL scheme.

5 Conclusions

The exact linearization of the Godunov flux-function has been obtained on the base of the solution to the Variational Riemann problem and applied in the Newton's iterative method for the implicit time integration scheme. Comparative analysis has been performed for the convergence rate of the proposed approach and the conventional method of approximate flux linearization of Turkel and Jameson. Substantial acceleration of the residual convergence has been found when the exact linearization was used. Moreover, the convergence acceleration was obtained even with truncated inner iterations.

References

1. S. Godunov: Math. Sb. **47**, 3 (1959)
2. P. Roe: J. Comput. Phys. **43**, 357 (1981)
3. A. Jameson, E. Turkel: Mathematics of Computation **37**, 385 (1981)
4. A. Jameson, S. Yoon: AIAA Journal **25**, 929 (1987)
5. I. Men'shov and Y. Nakamura: J. Comput. Phys., **181**, 118 (2002)
6. I. Men'shov and Y. Nakamura: Computers and Fluids., **29**, 595 (2000)
7. I. Men'shov and Y. Nakamura: AIAA J., **42**, 551 (2004)

On High-Order Fluctuation-Splitting Schemes for Navier-Stokes Equations

Hiroaki Nishikawa and Philip Roe*

Department of Aerospace Engineering,
University of Michigan, Ann Arbor, MI 48109, USA

1 Introduction

The Navier-Stokes equations express a balance of advection and diffusive terms; the former are always dominant far from the solid surfaces, and the latter dominate close to the surface. One needs numerical methods that are accurate in both of these limiting situations, and everywhere between. In the purely diffusive limit, it is appropriate to use central differencing, and to appeal to minimization principles. For the opposite limit, upwinding or artificial dissipation must be used to provide stability, and minimum principles do not apply. There is a lack of uniformly applicable design principles, and some dissatisfaction with current practice [7].

Here we will examine a simpler problem that raises the same issues, scalar advection-diffusion, surveyed comprehensively by Morton [6]

$$u_t + a\,u_x + b\,u_y = \nu\,(u_{xx} + u_{yy}), \tag{1}$$

We define Reynolds numbers $Re_{(L,h)} = \sqrt{a^2+b^2}(L,h)/\nu$, where the characteristic length is either a typical length scale L in the problem or the mesh spacing h. We focus on schemes based on multidimensional upwinding, or fluctuation-splitting, because of their reduced sensitivity to mesh quality for inviscid flow [1, 2]. This derives in part from the fact that the advection schemes preserve polynomial solutions of certain order on arbitrary grids.

These schemes are based on nodal variables, leading to cell-based residuals that are distributed to the nodes. For purely elliptic problems, there is a superficially similar method that distributes gradients, rather than residuals. It is equivalent to the classical Galerkin discretization, and is second-order accurate. However, simply adding this to a successful advection scheme may yield only a first-order method for (1) at finite Reynolds number.

2 Losing $\mathcal{O}(h^2)$

It has been common to construct fluctuation-splitting Navier-Stokes codes by adding a Galerkin discretization of the viscous term to a second-order Euler code. [3, 10, 11, 12]. This leads to the following discretization of (1),

*This work was supported under NASA URETI award NCC 3989 funded by NASA Marshall and Glenn centers and by the USAF.

$$\frac{du_j}{dt} = \sum_{T\in\{T_j\}} \left[\beta_j^T \phi_{\text{adv}}^T - \frac{\nu}{2}(\text{grad}\, u)^T \cdot \mathbf{n}_T\right], \tag{2}$$

where the right hand side collects updates contributed from the triangles $\{T_j\}$ around the node j. The first term distributes, according to some advection scheme with weights β_j^T, the advective residuals ϕ_{adv}^T, and the second term evaluates the standard Galerkin discretization of the Laplacian.

For the above scheme on a grid obtained by inserting diagonals into a uniform Cartesian grid with the spacing h, we present the Taylor expansion of the RHS for an arbitrary function. The result is, after some manipulation,

$$\begin{aligned}\mathcal{TE} =& r - \frac{\nu h}{2a}(a\partial_x + b\partial_y)(u_{xx} + u_{yy}) \\ &- \frac{h^2}{6}(au_{xxx} + bu_{yyy} + 3bu_{xyy} + 3bu_{xxy}) + \frac{\nu h^2}{12}(u_{xxxx} + u_{yyyy}) + \mathcal{O}(h^3)\end{aligned}$$

where $r = -(au_x + bu_y) + \nu(u_{xx} + u_{yy})$ which vanishes for exact steady solutions. The last term on the first line arises from interaction between the advective and diffusive effects, and is generally of order h, except in the limiting cases of very large or very small Re_L.

To confirm this analysis, we set up a test problem in a square domain $(0 < x < 1, 0 < y < 1)$ with the exact solution

$$u = -\cos 2\pi\eta \, \exp\left[0.5\xi(1 - \sqrt{1 + 16\pi^2\nu^2})/\nu\right)]$$

with $\xi = ax + by$, $\eta = bx - ay$. Solutions for $\nu = 0.1$ are computed on a series of uniform grids with spacings $h = 0.1 \times 2^{-n}, n = 0 \dots 5$. We also analysed and tested a version of this scheme, with third-order gradient correction [9, 8]. This also gave a first-order error at low Reynolds number.

Figure 1 (a) and (b) show the convergence of the numerical errors in the L_2 norm. The 1st-order behavior is obvious for fine grids; the sudden transition between orders of accuracy is striking. The third example (c) is 2nd-order throughout; this scheme is the outcome of the next section.

3 Preserving $\mathcal{O}(h^2)$

To achieve uniform 2nd-order accuracy for the advection-diffusion equation, we follow the philosophy **[2]** that leads to high-order advection schemes: evaluate the measure of the error (*fluctuation*) over the element and make this drive changes to each nodal solution (*distribution*). That is, the fluctuation for the whole equation should be distributed:

$$\frac{du_j}{dt} = \sum_{T\in\{T_j\}} \alpha_j^T \phi^T \tag{3}$$

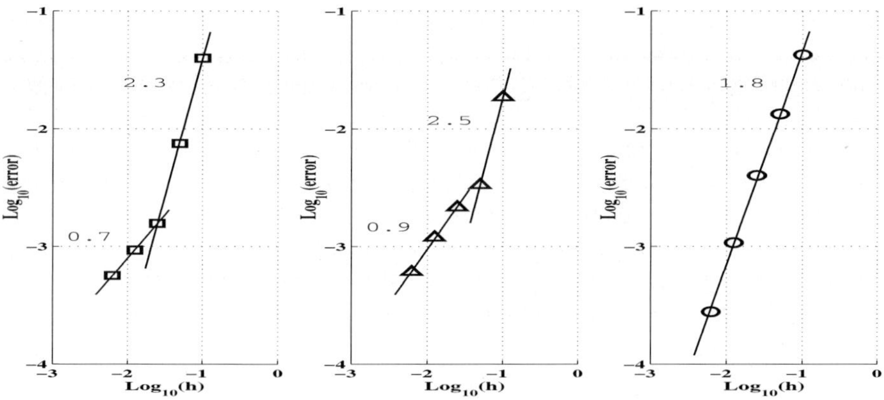

Fig. 1. Error Convergence: (left) LDA advection plus Galerkin diffusion; (center) The same but with gradient corrections applied to both terms; (right) unified LDA distribution of both residuals.

where

$$\phi^T = \int_T \left[-(a\, u_x + b\, u_y) + \nu\, (u_{xx} + u_{yy})\right]\, dxdy. \tag{4}$$

Eqn (2) is not of this form. This way, we can make sure that nothing happens if the fluctuation is zero. The viscous contribution to the residual needs to be evaluated cell-by-cell, combined with the advective residual, and then the combination is distributed to the nodes. In the previous section, the viscous residual was evaluated directly at the vertices (the second term of (2)).

In order to carry out this strategy, the viscous stresses (gradients) need to be available at the vertices. Either they can be computed there from the adjacent cells (for example by the Green-Gauss formula, or by least-squares fitting) or stored there as part of a Hermitian representation of the solution. Caraeni and Fuchs[4] developed a scheme employing reconstruction for which they claim third-order accuracy, but their distribution is purely advective, and is not appropriate in the diffusive limit. To remedy this we may use

$$\alpha_j^T = \left(\beta_j^T + \frac{k}{3\, Re_h}\right) / (1 + \frac{k}{Re_h}) \tag{5}$$

where k is some constant. This becomes, for diffusion-dominated flows, a simple isotropic diffusion. The truncation error of this scheme with the Green-Gauss reconstruction is

$$\mathcal{TE} {=} r + \frac{h}{2a(1+\frac{k}{Re_h})}(a\partial_x + b\partial_y)\, r + \mathcal{O}(h^2) \tag{6}$$

There is still a contribution proportional to h, but this now multiplies the total residual, which vanishes for an exact steady-state solution, making this method uniformly 2nd-order accurate even with $k = 0$. It is this scheme that produced the result in Figure 1 (c). However, we found that obtaining uniformly higher-order accuracy, as in the next section, required us to take $k \neq 0$.

In fact, for pure diffusion, the method is then fourth-order on the regular triangularization employed. Knowing this, we attempted to construct a 3rd-order scheme (improving on Caraeni and Fuchs[3]) by upgrading the advective part to 3rd-order with the high-order correction[9]. But this again resulted in the same form of the truncation error as (6), i.e. only 2nd-order. It became apparent that we were still missing a guiding principle.

4 Beyond $\mathcal{O}(h^2)$

To see what goes wrong, let $r = r_a + r_d$ where $r_a = -(a\, u_x + b\, u_y)$, $r_d = \nu\, (u_{xx} + u_{yy})$. A useful property of many higher-order residual distribution schemes is that their truncation error contains terms proportional to the residual. Thus, if we evaluate the residuals to third order (using corrections obtained from estimated vertex gradients), and then distribute these with central weights ($\alpha_j^T \equiv 1/3$) we obtain the following truncation errors, in the limits of pure advection and pure diffusion.

$$\mathcal{TE}_{\text{adv}} = r_a + \frac{5h^2}{36}(\partial_{xx} + \partial_{xy} + \partial_{yy})\, r_a + \mathcal{O}(h^4) \tag{7}$$

$$\mathcal{TE}_{\text{diff}} = r_d + \frac{h^2}{3}(\partial_{xx} + \partial_{xy} + \partial_{yy})\, r_d + \mathcal{O}(h^4) \tag{8}$$

both of which are 4th-order accurate at their own steady-states ($r_a = 0$ or $r_d = 0$). However the truncation error for their sum is what is relevant in solving the advection-diffusion equation, and that is

$$\mathcal{TE} = r_a + r_d + \frac{h^2}{3}(\partial_{xx} + \partial_{xy} + \partial_{yy})\,(\frac{5}{12}\, r_a + r_d) + \mathcal{O}(h^4) \tag{9}$$

which does not vanish for $r = 0$. This leads only to a 2nd-order method, due to incompatible discretizations of advective and diffusive fluctuations. It is noteworthy that accuracy in the two limits does not guarantee accuracy in intermediate cases. A discretization method that does have this property has been developed by Lerat and Corre [5] for structured grids, but at the present time it is not clear to us how to construct an unstructured version.

The approach we have taken here may not be necessary, and may be overelaborate. We avoided the problem of finding compatible discretizations for first- and second-order derivatives by writing the governing equation as a first-order system. We introduce $p - u_x$ and $q = u_y$ as additional unknowns and consider the first-order version of (1)

$$u_t + a\, u_x + b\, u_y = \nu\, (p_x + q_y) \tag{10}$$

$$p - u_x = 0 \tag{11}$$

$$q - u_y = 0. \tag{12}$$

Assume first a piecewise linear variation of u, p, q. By integrating equation (10) over a triangle, we obtain a 2nd-order residual, distributed with coefficients (5) to update u. The residuals of (11), (12), are distributed isotropically ($\alpha_j^T \equiv 1/3$) to update p, q; this minimizes those residuals in an L_2 norm.

To go to higher order, we simply follow the previous works: the fluctuations are made more accurate by adding high-order correction [9, 8], (or equivalently reconstruct gradients $\{u_x, u_y, p_x, p_y, q_x, q_y\}$ at nodes, construct the Hermite interpolation over each edge, integrate the equations as a contour integral to obtain fluctuations). The modified fluctuations are then distributed with the same coefficients as the 2nd-order version.

The truncation error of the fourth-order version is, for (10)

$$\mathcal{TE}=r + c_1 h + c_2 h^2 + c_3 h^3 + \mathcal{O}(h^4) \tag{13}$$

where

$$c_1=\frac{a r_x + b r_y}{2a}$$

$$c_2=-\frac{1}{36a}\left\{(6a+5\nu a)r_{xx} - (3a+3b+5\nu a)r_{yy} - (5\nu a + 6b + 3a)r_{xy}\right\}$$

$$c_3=\frac{1}{24a}\left\{a r_{xxx} + (a+b)(r_{xyy}+r_{xxy}) + b r_{yyy}\right\},$$

The truncation error for (11) is

$$\mathcal{TE}_p=r^p + \frac{5}{36}(r^p_{xx} + r^p_{xy} + r^p_{yy})h^2 + \mathcal{O}(h^4), \tag{14}$$

and similarly for (12). Therefore the method is 4th-order accurate for u, p, and q at steady-state ($r=0,\ r^p=0,\ r^q=0$).

The schemes just described were applied for the test problem in Section 2, and the results are shown in Figure 3. Clearly we have achieved 2nd-order with the base scheme, 3rd-order with the Green-Gauss gradient reconstruction, 4th-order with a quadratic gradient reconstruction. We remark that the 3rd-order accuracy is only due to poor performance of the Green-Gauss reconstruction near boundaries.

Finally, we obtained results from the fourth-order method on a sequence of fully unstructured grids. As hoped, the convergence was rapid, with an experimental order of accuracy of about 3.5 (see Figure 5).

5 Concluding Remarks

This paper has shown that uniformly second-order fluctuation-splitting schemes can be developed for the advection-diffusion equation by distributing the full residual with proper distribution coefficients. We were able to go beyond second-order by computing the slopes as independent unknowns. All of the ideas should be readily extendable to the Navier-Stokes system.

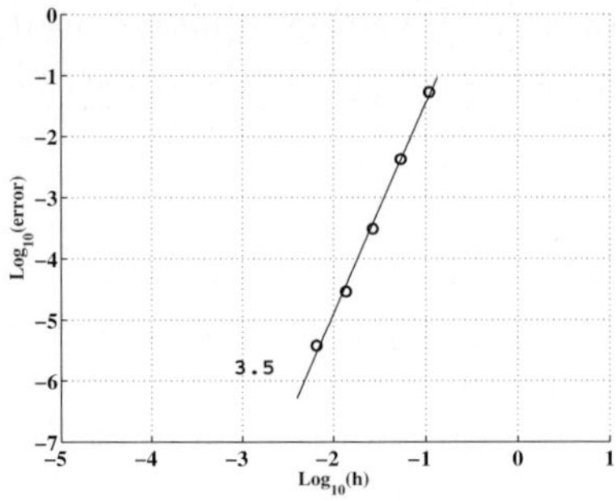

Fig. 2. Error Convergence: the numbers indicate slopes determined by the least-squares fit

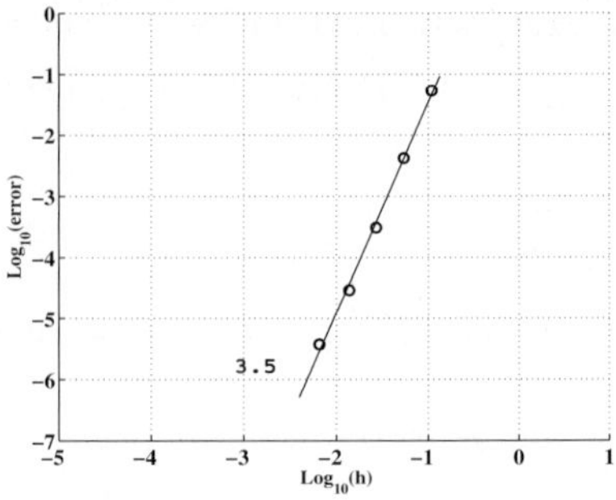

Fig. 3. Results for unstructured grids, each with the same number of nodes as in the regular counterpart

In future work we need to study how these ideas can be combined with the nonlinear distribution schemes needed for monotone shock-capturing, It is likely that a successful distribution coefficient involves the (perhaps local) Reynolds number as a parameter. Second, instead of the high-order correction method to solve the first-order system, we may introduce additional degrees of freedom at the midpoint of each edge to achieve 4th-order accuracy without a need for gradient reconstruction. This type of high-order extension was studied in [2] for the advection equation.

References

1. R. Abgrall, **JCP**, 167, pp. 277-315, 2001.
2. R. Abgrall and P. L. Roe, **JSC**, 19, Number 1-3, pp. 3-36, 2002.
3. A. Bonfiglioli, **IJCFD**,14, pp. 21-39, 2000
4. D. Caraeni and L. Fuchs, Theoret. and Comput. Fluid Dynamics, 15:6, 2002.
5. A. Lerat and C. Corre, **JCP**, 170, pp 642–675, 2001
6. K. W. Morton, *Numerical Solution of Convection-Diffusion Problems,* Chapman and Hall, 1996.
7. High Energy Flow Solver Synthesis (LAURA Benchmarks: Cylinder) at `http://hefss.larc.nasa.gov/index.cfm?selection=laura&laura_type=benchmarks`
8. H. Nishikawa, ICCFD Proceedings, Toronto. Springer-Verlag, 2004.
9. H. Nishikawa, M. Rad, P. L. Roe, AIAA Paper 2001-2595, Anaheim, June 2001
10. G. T. Tomaich, *PhD Thesis*, Aerospace Engineering, Univ. of Michigan, 1995.
11. E. v.d. Weide, H. Deconinck, E. Issman, G, Degrez, **Comp Mech.**, 23:2, 1999.
12. W. A. Wood and W. L. Kleb, AIAA 2001-2630, Anaheim, June 2001.

Computation of Weakly Ionized Atmospheric Entry Flows Using an Extended Roe Scheme

Tristan Soubrié, Olivier Rouzaud, and Jouke Hylkema

ONERA,DMAE, BP 4025, 31055 Toulouse, France, rouzaud@onecert.fr

1 Introduction

To accurately model atmospheric entry aerothermodynamics is challenging because of the complex chemical and physical processes involved in such hyperenthalpic hypersonic flows, and in particular ionization. Coquel and Marmignon [1] have developed a numerical method based on the Roe scheme to deal with weakly ionized flows. Soubrié *et al.* [2] have extended its application to flows containing several ionized species. This report presents further results, especially the influence of the mesh and of the chemical model.

2 Ionized Flows in Thermochemical Nonequilibrium

The partitioning across the energy modes is described using a $(T,\ T_{v,m},\ T_e)$ temperature model. Heavy particles translational and rotational modes are equilibrated at temperature T. The number density of vibrational levels obeys a Boltzmann distribution, characterized by a specific vibrational temperature $T_{v,m}$ for O_2 and N_2 species and by the temperature T for other molecules. T_e is the electron temperature. Heavy particles electronic excitation is neglected. We assume local charge neutrality and no conduction current. The governing equations, derived from [1], are

$$\partial_t \rho_s + \partial_{x_j}(\rho_s u_j) = \partial_{x_j}\left(\rho D_s \partial_{x_j} y_s\right) + \dot{w}_s \tag{1}$$

$$\partial_t \rho u_i + \partial_{x_j}(\rho u_i u_j + p) = \partial_{x_j}[\mu\left(\partial_{x_j} u_i + \partial_{x_i} u_j\right) - \frac{2}{3}\mu \partial_{x_k} u_k \delta_{ij}] \tag{2}$$

$$\begin{aligned}\partial_t \rho E + \partial_{x_j} \rho H u_j &= \partial_{x_j}[\eta_{eq}\partial_{x_j} T + \sum\left(\eta_{v,m}\partial_{x_j} T_{v,m}\right)] \\ + \partial_{x_j}(\rho \sum h_s D_s \partial_{x_j} y_s) &+ \partial_{x_j}[u^i \mu\left(\partial_{x_j} u_i + \partial_{x_i} u_j\right) - \frac{2}{3}\mu u_i \partial_{x_k} u_k \delta_{ij}]\end{aligned} \tag{3}$$

$$\begin{aligned}\partial_t \rho_m e_{v,m} + \partial_{x_j}\left(\rho_m e_{v,m} u_j\right) &= \partial_{x_j}\left(\eta_{v,m}\partial_{x_j} T_{v,m}\right) + \partial_{x_j}\left(\rho h_{v,m} D_m \partial_{x_j} y_m\right) \\ &+ \rho_m(e_{v,m}(T) - e_{v,m}(T_{v,m}))/\tau_{t-v,m} + \dot{w}_m e_{v,m}\end{aligned} \tag{4}$$

$$\partial_t s_e + \partial_{x_j} s_e u_j = (\gamma_e - 1)\frac{S_{e-t}}{y_e \rho^{\gamma_e - 1}}\, g'\!\left(\frac{p_e}{y_e \rho^{\gamma_e}}\right) \tag{5}$$

$$\frac{\rho_e}{M_e} = \sum_{s=ion} \frac{\rho_s}{M_s} \tag{6}$$

Subscripts s, m and e respectively refer to chemical species, molecules in vibration thermal nonequilibrium and electrons. ρ is the density, p the total pressure, M_s the molecular weight, R the universal gas constant and u_i the velocity in direction i. The total energy E and enthalpy H are linked by the relation $H = E + p/\rho$.
Considering the electron temperature T_e implies to write the electron energy balance equation in which a term of electric field due to charge separation appears. In the framework of our assumptions, this term is proportional to the electron pressure gradient, leading to a non conservative convective system and thus complicating the application of numerical schemes for the shock capture. Moreover, solving a non conservative system does not mathematically ensure that the numerical converged solution will be the correct one. Ignoring dissipative phenomena and inertial effects within the electron gas, Coquel and Marmignon [1] have replaced the electron energy equation by the electronic entropy equation (5) written under conservative form, using the following definition (y_e mass fraction, $\gamma_e = 5/3$ isentropic coefficient):

$$s_e = \rho\, g\left(\frac{p_e}{y_e \rho^{\gamma_e}}\right) = \left(\frac{p_e}{(\gamma_e - 1) y_e}\right)^{1/\gamma_e} \tag{7}$$

3 Physico-Chemical Modelling

Chemical Kinetic Model: The Park's [3] and Dunn-Kang's [4] 11-species air model (O_2, N_2, NO, O_2^+, N_2^+, NO^+, N, O, N^+, O^+, e^-) have been used and enable to calculate the species mass production rate $\dot{w}_s$.

Transport Properties: The **viscosity** for a pure gas is given by the Blöttner interpolation law [5] using Blöttner's [5] and Gupta's [6] parameter values. The mixture viscosity μ is then obtained using Armaly-Sutton's model [7]. The definition of **thermal conductivity** is splitted in two parts. For all the modes in thermal equilibrium, we define a global thermal conductivity η_{eq} based on a fixed Prandtl number $Pr = 0.72$. For molecules in vibration thermal nonequilibrium, $\eta_{v,m}$ follows Armaly-Sutton's model [8]. The **diffusion coefficient** D_s is supposed to be equal for all neutral species and is based on a constant Lewis number $Le = 1.2$. Ion and electron diffusion is modeled by the ambipolar diffusion approach [9].

Relaxation Processes: The vibrational-translational energy exchange between a molecule and heavy particles is written in the Landau-Teller form in equation (4) and Millikan-White's model as reviewed by Park [3] enables to evaluate the characteristic relaxation time $\tau_{t-v,m}$. The term S_{e-t} in equation (5) represents the energy exchange resulting from elastic collisions between electrons and heavy particles. Appleton-Bray's [10] model has been used, the collision section between electron and neutral species being written following Matsuzaki *et al.* [11]. All other energy exchanges have been neglected.

4 Numerical Method

The spatial discretization of the inviscid fluxes is treated by using an extension of the Roe scheme, proposed by Coquel and Marmignon [1] [12]. Minmod and van Albada limiters are available and ensure a second-order accuracy. The discretization of the viscous fluxes is based on a central differencing. Harten's entropy correction has been applied to locally avoid that the eigenvalues magnitude reaches zero. Time stepping is performed with an implicit Euler method. Van Leer's Flux Vector Splitting method [13] provides the implicit matrix of the linear system of which the solution is obtained by a Krylov-GMRES method with an ILU(0) preconditioning. Wall boundary conditions (no slip, isothermal wall) are partially implicit to enforce the stability of the method.
To adapt the Roe scheme to weakly ionized flows, consider a cell interface separating a left state L and a right state R. The Roe numerical flux writes

$$F^{Roe}(\mathbf{u}_L, \mathbf{u}_R) = \frac{1}{2}\left(\mathbf{f}(\mathbf{u}_L) + \mathbf{f}(\mathbf{u}_R)\right) - \frac{1}{2}|\overline{A}\left(\mathbf{u}_L, \mathbf{u}_R\right)|(\mathbf{u}_R - \mathbf{u}_L) \quad (8)$$

where $\mathbf{f}$ represents the exact flux and $\mathbf{u}$ the vector of conservative variables. The matrix $\overline{A}\left(\mathbf{u}_L, \mathbf{u}_R\right)$ has to satisfy the three following conditions

$$\bullet \overline{A}(\mathbf{u}_L, \mathbf{u}_R) \text{ is diagonalizable with real eigenvalues,} \quad (9)$$
$$\bullet \quad \overline{A}(\mathbf{u}_L, \mathbf{u}_R).(\mathbf{u}_R - \mathbf{u}_L) = \mathbf{f}(\mathbf{u}_R) - \mathbf{f}(\mathbf{u}_L) \quad (10)$$
$$\bullet \quad \overline{A}(\mathbf{u}, \mathbf{u}) = \nabla_{\mathbf{u}}\mathbf{f}(\mathbf{u}) \quad (11)$$

It is tightly connected to the jacobian matrix and can be written using a set of averages between the left and the right states as

$$\overline{A}\left(\mathbf{u}_L, \mathbf{u}_R\right) = A(\overline{y_s}, \overline{u}, \overline{v}, \overline{H}, \overline{y_m e_{v,m}}, \overline{s_e/\rho}, \widetilde{\nabla_{\mathbf{u}} p}, \widetilde{\nabla_{\mathbf{u}} p_e}) \quad (12)$$

The $\bar{.}$ average is simply the Roe average. The $\widetilde{\nabla_u p}$ average shares also the same definition as for neutral gas mixtures [14]. For the term $\nabla_u p_e$, Coquel and Marmignon have derived an average form which is consistent with the three requirements (9)-(11). The numerical flux can be found in [1].

5 Numerical Results

The Radio Attenuation Measurement RAM-C II experiment [15] [16] was carried out in the sixties to study phenomena such as the communication blackout during the reentry phase. The electron number density was measured in the flowfield around the probe and the flight data are among the few experimental results that are currently available for code validation.
The body is a spherical cap-cone configuration with a $R = 0.1524m$ nose radius, $9°$ cone half angle and an overall length of $1.295m$. Computations

have been performed to replicate the RAM-C II flight experiment at altitudes $61km$ and $71km$ (Tab.1). Wall temperature T_w is fixed at $1500K$ and the wall is supposed to be non-catalytic. The meshes contain 50×50 (resp. 100×100) cells, the first cell height is 10^{-5} (resp. $5 \cdot 10^{-6}$). Computation on a finer mesh with first cell height of $1 \cdot 10^{-6}$ is in progress.

Table 1. RAM-C II flow conditions

Altitude [km]	T_∞ $[K]$	P_∞ $[Pa]$	u_∞ $[m/s]$	Mach	Re_∞
61	254	19.7	7650	23.9	19500
71	216	4.8	7659	25.9	6287

RAM-C II 61km : As pointed out by Walpot [17], using a finer mesh resolution sharpens the shock and increases the peak of stagnation line translational temperature (Fig.1). In addition, since first cell size has decreased, viscous effects are better described. The influence of the chemical model is weak on temperature distributions but significant on the peak of electron density around the body (Fig.3) since ion mass fractions are not equally predicted (Fig.2). Dunn-Kang's model predicts a large amount of O^+ species and thus does not fit to experimental electron density in the rear flow. In contrary, Park's model gives results in good agreement with flight data.

RAM-C II 71km : The flow is further from equilibrium than at altitude $61km$. Therefore, the shock is larger and the temperatures equilibrate closer to the wall (Fig.4). Prediction of the electron density is fairly good compared to flight measurements (Fig.5). Heat flux at the wall is presented in Fig.6 and values at the stagnation point are compared to other numerical computations in Tab.2. Dispersion in the results is large.

Table 2. Heat flux at the stagnation point of RAM-C $71km$. (D-K: Dunn-Kang)

Source	Present		Tchuen [18]	Walpot [17]			Candler *et al.* [9]
Catalytic wall	no		no	no	no	yes	no
Chemical model	D-K	Park	Park	D-K	Park	D-K	Park-Wray
Heat flux $[MW/m^2]$	0.99	0.96	2.4	0.90	0.74	2.2	2.5

6 Conclusion

The results prove the validity of the numerical method employed since comparison with other numerical computations and flight data is fair, except on the heat flux values. However, note that the authors consulted (Tab.2) deal with a governing system written under a non-conservative form. In addition, Walpot [17] and Tchuen [18] have considered the heavy species electronic excitation to be at thermal equilibrium with the electron gas energy. Therefore they have solved an electron-electronic excitation energy equation, whereas with additional assumptions we have solved an electron entropy equation under conservative form and have neglected electronic excitation.

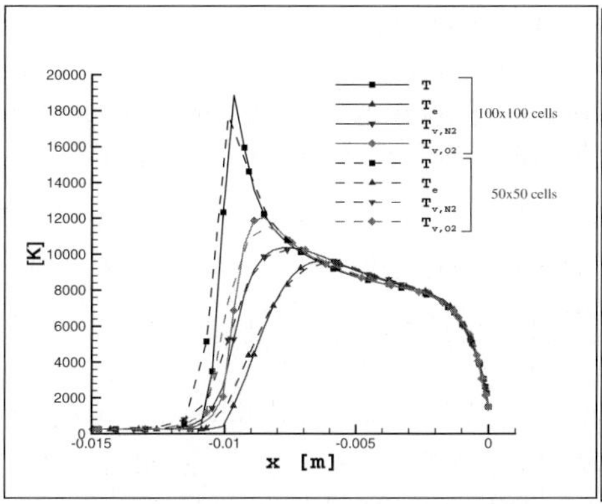

Fig. 1. Grid convergence on stagnation line temperature at altitude 61km using Dunn-Kang's model.

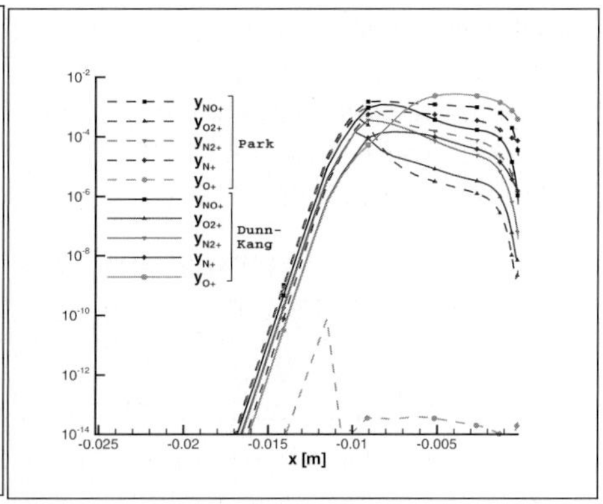

Fig. 2. Influence of chemical model on stagnation line ion mass fractions at altitude 61km.

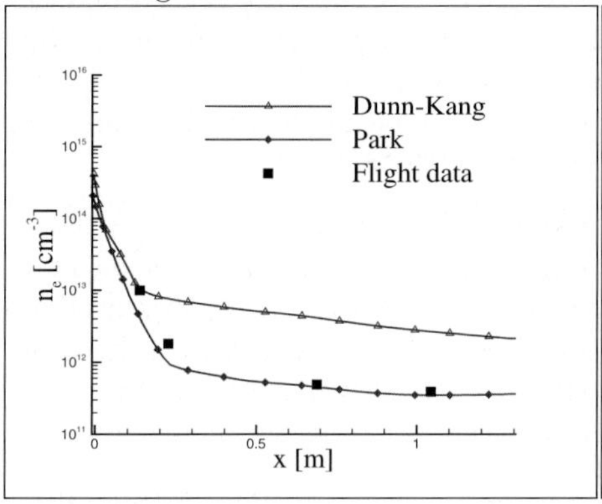

Fig. 3. Influence of chemical model on electron density peak at altitude 61km.

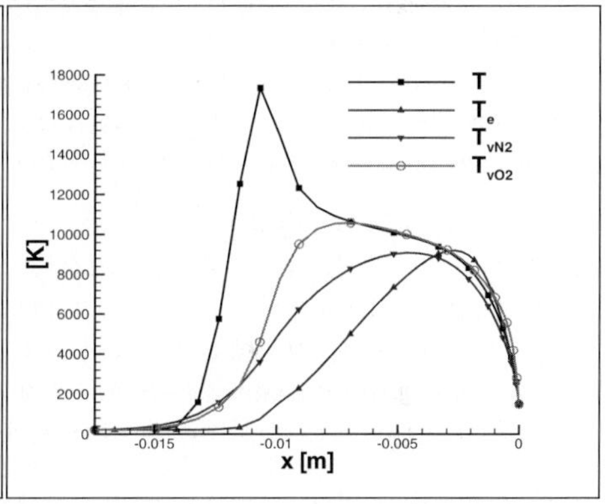

Fig. 4. Stagnation line temperature at altitude 71km using Park's model.

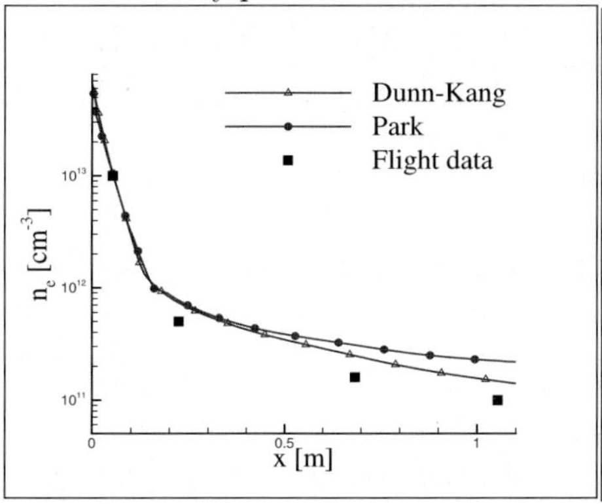

Fig. 5. Influence of chemical model on electron density peak at altitude 71km.

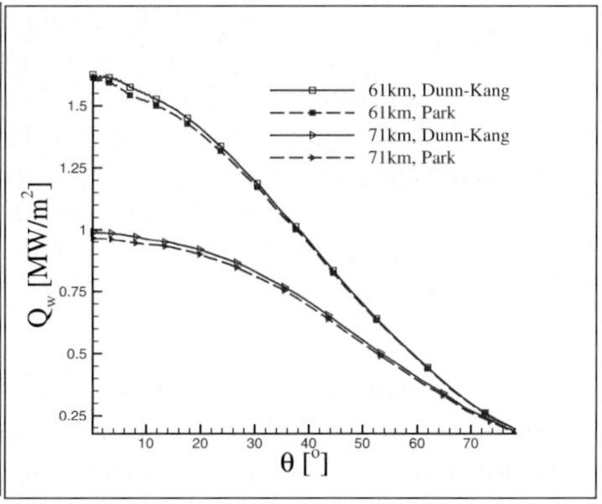

Fig. 6. Influence of chemical model on wall heat flux at both altitudes.

References

1. F. Coquel, C. Marmignon: A Roe-Type Linearization for the Euler Equations for Weakly Ionized Multi-Component and Multi-Temperature Gas. AIAA Paper 95-1675 (1995)
2. T. Soubrié, O. Rouzaud, D.E. Zeitoun: Computation of Weakly Multi-ionized Gases for Atmospheric Entry Using an Extended Roe Scheme. In: *Proc. ECCOMAS* (2004)
3. C. Park. Review of Chemical-Kinetic Problems of Future NASA Missions, I: Earth Entries. In: *Journal of Thermophysics and Heat Transfer*, 7(3) (1993).
4. M.G. Dunn, S.W. Kang: Theoretical and Experimental Studies of Re-entry Plasmas. NASA CR-2232 (1973)
5. F.G. Blöttner, M. Johnson, M. Ellis: Chemically Reacting Viscous Flow Program for Multicomponent Gas Mixtures. Sandia Laboratories, Technical Report Sc-RR-70-754 (1971)
6. R.N. Gupta, J.M. Yos, R.A. Thompson, K.P. Lee: A Review of Reaction Rates and Thermodynamic and Transport Properties for an 11-species Air Model for Chemical and Thermal Nonequilibrium Calculations to 30000K. NASA Reference Publication 1232 (1990)
7. B.F. Armaly, K. Sutton: Viscosity of Multicomponent Partially Ionized Gaz Mixtures. AIAA Paper 80-1495 (1980)
8. B.F. Armaly, K. Sutton: Thermal Conductivity of Partially Ionized Gas. AIAA Paper 81-1174, (1981)
9. G.V. Candler, R.W. MacCormack: The Computation of Hypersonic Ionized Flows in Chemical and Thermal Nonequilibrium. AIAA Paper 88-0511 (1988)
10. J.P. Appleton, K.N.C. Bray: The Conservation Equations for a Non-equilibrium Plasma. In: *J. Fluid Mech.*, 20(4), pp 659–672 (1964)
11. R. Matsuzaki, N. Hirabayashi: Study of Chemical Reaction Models for High Enthalpy Flow of Air. NAL TR-982 (1988)
12. C. Marmignon, F. Coquel: Simulation numérique d'écoulements en déséquilibre thermochimique et faiblement ionisés. ONERA TRS 14/1929 AY (1997)
13. M.-S. Liou, B. van Leer, J.-S. Shuen: Splitting of Inviscid Fluxes for Real Gases. In: *Journal of Computational Physics*, 87, pp 1–24 (1990)
14. R. Abgrall: Extension of Roe's Approximate Riemann Solver to Equilibrium and Nonequilibrium Flows. In: *Proc. 8th GAMM Conference on Num. Method in Fluids Mech.*, 29 (P.Wesseling Ed. Vieweg 1990)
15. C.J. Schexnayder, J.S. Evans. P. Huber: Comparison of Theoretical and Experimental Electron Density for Ram C Flights. NASA, SP-252 (1970)
16. A.E. Cross, W. L. Jones, Jr.: Electrostatic Probe Measurements of Plasma Parameters for Two Reentry Flight Experiments at 25,000 Feet per Second. NASA, TN-D-6617 (1972)
17. L.M. Walpot: Development and Application of a Hypersonic Flow Solver. PhD Thesis, TU Delft (2002)
18. G. Tchuen: Modélisation et Simulation Numérique des Ecoulements à Haute Enthalpie: Influence du Déséquilibre Electronique. PhD Thesis, Université d'Aix-Marseille I (2003)

Part XX

Technical Notes

3D Prediction of Developing Turbulent Flow in a 90° Duct of Rectangular Cross-Section

H. Alemi and M. Raisee

Department of Mechanical Engineering, Faculty of Engineering, University of Tehran, P. O. Box: 11365/4563, Tehran, Iran. mraisee@ut.ac.ir

Summary

Turbulent flow in curved ducts occurs in many engineering applications such as centrifugal pumps, aircraft intake and in the cooling passages of gas turbine blades. An important characteristic which distinguishes such flows from those in straight ducts is the generation of strong secondary motions. In this work, a numerical investigation has been undertaken to study incompressible, developing turbulent flow through a 90° duct of rectangular cross-section [1]. The main objective has been to assess the effectiveness of the Launder and Sharma [2] low-Re $k-\epsilon$ model and the nonlinear low-Re $k-\epsilon$ model of Craft et al. [3] in predicting turbulent flow in curved ducts. In this study, the governing equations of mean flow are solved using an extended version of STREAM-3D [4] finite volume code. The pressure field is obtained using the SIMPLE algorithm and the convection terms in all transport equations are approximated using the QUICK scheme. The predicted velocity vectors, in Figure 1(a), indicates that the duct curvature causes flow acceleration and deceleration along the convex surface, and reverse along the concave surface. Furthermore, Figure 1(b) shows that the curvature induces a strong secondary motion which as a result of that the core fluid displaces towards the convex surface.

It is noted that both the linear and nonlinear $k-\epsilon$ models return similar axial velocity contours and secondary flow motions. Comparisons of the predicted axial velocity, turbulent kinetic energy, and turbulent shear-stress with measurements of Kim and Patel [1], shown in Figure 2, indicates that while both turbulence models successfullly reproduce the measured axial velocity, the nonlinear $k-\epsilon$ model returns more accurate turbulence field predictions. As shown in Figure 3(a), both turbulent models correctly reproduce the variation of pressure coefficient. Note, however, that as shown in Figure 3(b) the distribution of friction coefficient is more faithfully reproduced by the nonlinear $k-\epsilon$ model. In general, it appears that the nonlinear low-Re $k-\epsilon$ model does offer the better route for the reliable computation of turbulent flow in curved ducts.

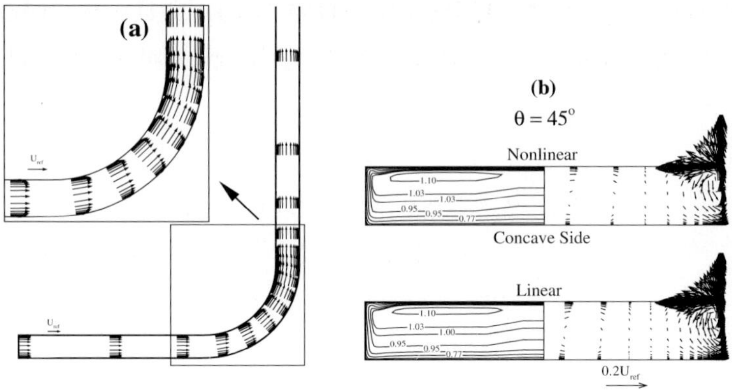

(a) me (b)

Fig. 1. (a) Predicted flow field in mid-plane of the duct using the nonlinear low-Re $k - \epsilon$ model; and (b) predicted axial velocity contours and secondary flow vectors.

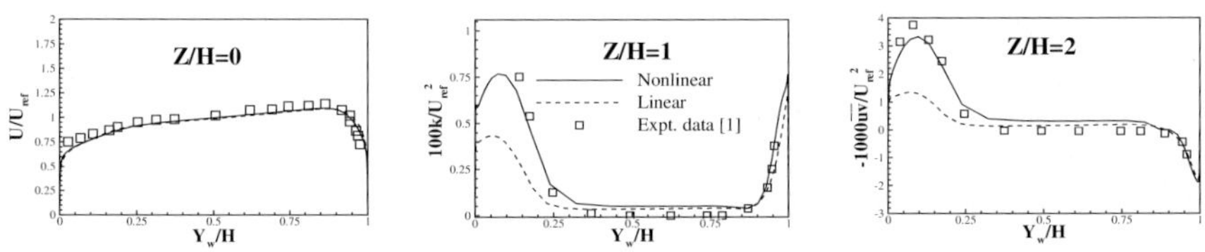

Fig. 2. Comparisons between predicted and measured streamwise velocity component, turbulent kinetic energy, and turbulent shear-stress at $\theta = 45°$.

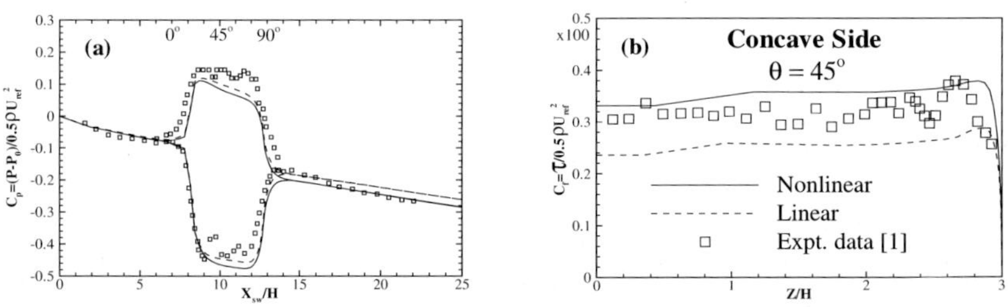

Fig. 3. (a) Local pressure coefficient comparisons; and (b) friction factor comparisons.

References

1. Wu J. Kim and V. C. Patel: J. of Fluids Engineering **116**, 45 (1994)
2. B. E. Launder and B. I. Sharma: Letters in Heat Mass Transfer **1**, 131 (1974)
3. T. J. Craft, H. Iacovides and J. H. Yoon: Turbulence Flow and Combustion **63**, 59 (1999)
4. F.-S. Lien and M. A. Leschziner: Comp. Meth. Appl. Mech. Eng. **114**, 123 (1994)

Numerical Simulation of Steady Newtonian and Non-Newtonian Flow Through Vascular Stenoses

Geoffrey R. Behrens and Ramesh K. Agarwal

Department of Mechanical and Aerospace Engineering, Washington University in St. Louis, MO 63130, USA

1 Introduction

The high cost and risk factors associated with surgical procedures on vascular diseases such as arterial aneurysms and stenoses make accurate diagnosis of such conditions critical. Phase Contrast Magnetic Resonance Imaging (PC-MRI) has become widely used to measure blood velocities as a fast, non-invasive technique to diagnose vascular diseases. The primary purpose of this paper is to demonstrate the ability of a commercially available Computational Fluid Dynamics (CFD) solver, FLUENT, to validate results obtained from in vitro PC-MRI experiments on vascular stenoses and to investigate the effects of non-Newtonian flow through the same geometry.

2 Results and Discussions

The PC-MRI results shown in this paper in Figure 1 were collected by colleagues in the CVIA lab in the Washington University School of Medicine [1] using a 1.5 Tesla MR imaging system on an in-vitro axi-symmetric flow phantom. The flow at the inlet was laminar for all cases, and a laminar solver was employed for all 50% area reduction cases as well as the 75% stenosis cases with low flow rates. For the most severe stenosis (90% area reduction) and the 75% stenosis at high flow rates, however, turbulence effects became significant and thus the standard k-epsilon turbulence model with some modifications was applied. Computations for steady, Newtonian flow were performed for three stenosis models with area reductions of 50%, 75%, and 90%; each at flow rates of 10, 15, and 20 ml/s. A direct comparison of centerline static pressures from FLUENT and PC-MRI data for Newtonian flow is shown in Figure 1. Good agreement is seen for most of the cases, however noticeable error is seen in the case of 90% stenosis wherein the turbulence effects became significant. Figure 2 displays comparisons of centerline pressure from FLUENT for Newtonian and non-Newtonian flow (using Casson's model for blood viscosity). The most significant result seen in these plots is the increased pressure gradient required to maintain the same flow rate for non-Newtonian flow.

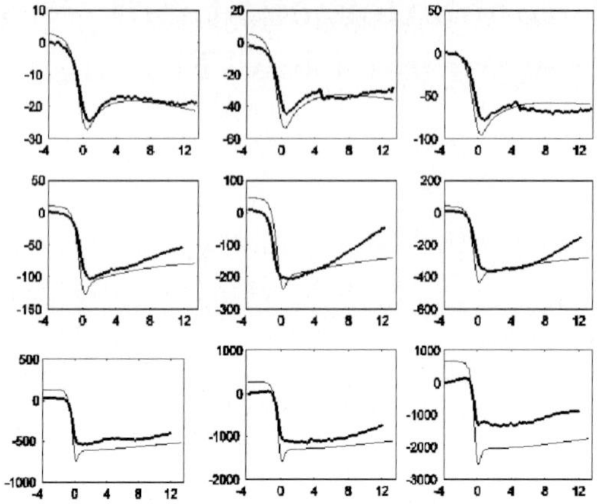

Fig. 1. Centerline Pressures computed from FLUENT (thin lines) and PC-MRI data (thick lines). Rows from top to bottom represent 50, 75, and 90% stenosis models; and columns from left to right represent flow rates of 10, 15, and 20 ml/s, respectively. Horizontal axes are measured in cm and vertical axes are in dynes/cm^2.

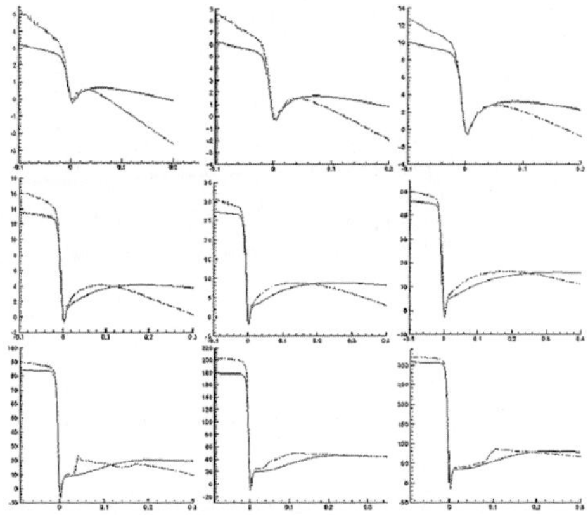

Fig. 2. Comparison of computed centerline pressures for Newtonian (solid lines) and non-Newtonian (dashed lines) flow from FLUENT. Rows from top to bottom represent 50%, 75%, and 90% stenoses, and columns from left to right represent flow rates of 10, 15, and 20 ml/s, respectively.

References

1. A.N. Moghaddam, G. Behrens, R.K. Agarwal, E.T. Choi, A.A Amini: *Affecting Accuracy in Measurement of Intravascular Pressure from Phase-Contrast MRI*, accepted for publication in MRM journal, 2004.

Finite Volume Methods for Fluid Flow Through Elastic Tubes

Marek Brandner

Centre of Applied Mathematics, University of West Bohemia, 22 Univerzitní Street, Pilsen, Czech Republic, PO BOX 314 06. brandner@kma.zcu.cz

The fluid flow through elastic tubes (in 1D) can be described by the non-autonomous system of conservation laws in the form

$$\begin{aligned} \mathbf{q}_t + [\mathbf{g}(\mathbf{q}, x)]_x &= \mathbf{0}, \;\; x \in R, \;\; t > 0, \\ \mathbf{q}(x, 0) &= \mathbf{q}_0(x), \;\; x \in R. \end{aligned} \tag{1}$$

This system can be rewritten to

$$\begin{aligned} \mathbf{u}_t + [\mathbf{f}(\mathbf{u})]_x &= \mathbf{0}, \;\; x \in R, \;\; t > 0, \\ \mathbf{u} = \left[\mathbf{q}^T, w\right]^T, \;\; \mathbf{f} &= \left[\mathbf{g}^T(\mathbf{q}, w), 0\right]^T, \\ \left[\mathbf{q}^T(x, 0), w(x, 0)\right]^T &= \left[\mathbf{q}_0^T(x), x\right]^T, \;\; x \in R. \end{aligned} \tag{2}$$

There are two basic approaches towards solving this augmented system. The first one is based on componentwise methods – for example, the central schemes. We propose to use the semidiscrete version of these algorithms where the degenerate conservation law $w_t = 0$ is solved exactly. The second approach is based on upwind schemes. If we use a finite volume method based on Roe's linearization $\mathbf{f}(\mathbf{U}_{j+1}^n) - \mathbf{f}(\mathbf{U}_j^n) = \bar{\mathbf{A}}_{j+1/2}^n(\mathbf{U}_{j+1}^n - \mathbf{U}_j^n)$ where

$$\mathbf{f}(\mathbf{U}_{j+1}^n) - \mathbf{f}(\mathbf{U}_j^n) = \begin{bmatrix} \mathbf{g}(\mathbf{Q}_{j+1}^n, x_{j+1}) - \mathbf{g}(\mathbf{Q}_j^n, x_j) \\ 0 \end{bmatrix}$$

we obtain the approximation of the Jacobi matrix in the following form

$$\bar{\mathbf{A}}_{j+1/2}^n = \begin{bmatrix} (\mathbf{A}_{11})_{j+1/2}^n & (\mathbf{a}_{12})_{j+1/2}^n \\ \mathbf{0} & 0 \end{bmatrix} \tag{3}$$

where

$$[\mathbf{A}_{11}(x_j)]_{j+1/2}^n(\mathbf{Q}_{j+1}^n - \mathbf{Q}_j^n) = \mathbf{g}(\mathbf{Q}_{j+1}^n, x_j) - \mathbf{g}(\mathbf{Q}_j^n, x_j),$$

$$(\mathbf{A}_{11})_{j+1/2}^n = \frac{1}{2}[\mathbf{A}_{11}(x_j)]_{j+1/2}^n + \frac{1}{2}[\mathbf{A}_{11}(x_{j+1})]_{j+1/2}^n,$$

$$(\mathbf{a}_{12})_{j+1/2}^n = \frac{1}{2}\left[\frac{\mathbf{g}(\mathbf{Q}_{j+1}^n, x_{j+1}) - \mathbf{g}(\mathbf{Q}_{j+1}^n, x_j)}{x_{j+1} - x_j} + \frac{\mathbf{g}(\mathbf{Q}_j^n, x_{j+1}) - \mathbf{g}(\mathbf{Q}_j^n, x_j)}{x_{j+1} - x_j}\right].$$

The decomposition (used in the upwind schemes) has the form

$$\sum_{p=1}^{m+1} |\bar{\lambda}^{n,p}_{j+1/2}| \Upsilon^{n,p}_{j+1/2} \mathbf{r}^{n,p}_{j+1/2} = \begin{bmatrix} \sum_{p=1}^{m} |\bar{\lambda}^{n,p}_{j+1/2}| \Upsilon^{n,p}_{j+1/2} (\mathbf{r}_{11})^{n,p}_{j+1/2} \\ 0 \end{bmatrix} \tag{4}$$

where

$$\mathbf{\Upsilon}^n_{j+1/2} = (\mathbf{R}^n_{j+1/2})^{-1} (\mathbf{U}^n_{j+1} - \mathbf{U}^n_j), \tag{5}$$

$\mathbf{r}^{n,p}_{j+1/2}$ are the eigenvectors and $\bar{\lambda}^{n,p}_{j+1/2}$ are the eigenvalues of the matrix $\bar{\mathbf{A}}^n_{j+1/2}$ (the eigenvectors are columns of $\mathbf{R}^n_{j+1/2}$); $(\mathbf{r}_{11})^{n,p}_{j+1/2}$ are the eigenvectors of the matrix $[\mathbf{A}_{11}]^n_{j+1/2}$ There are also two possibilities. The first approach is based on the direct application of the conservative upwind scheme improved by an entropy fix procedure. If we apply Roe's linearization we suppose that Riemann problems at cell interfaces (typically) have a large jump at most one wave family (i.e., the solution consists of a single wave and $||\Upsilon^{n,p}_{j+1/2} \mathbf{r}^{n,p}_{j+1/2}|| = O(\Delta x)$ for all other waves). In our case we obtain two wave families but the second one (for the function w) with $O(\Delta x)$. In fact this scheme is equivalent to the approach described in [2]. The system of Eq. (2) fulfils the simplified version of the jump condition

$$[\mathbf{g}] = s[\mathbf{q}], \;\; [w](0 - s) = 0$$

discussed in the mentioned article. This observation is used in the second approach. We solve the non-augmented system of Eq. (1) but in order to obtain the correct projection into the characteristics field, we have to use the full left eigenvectors. In the case we have a weakly hyperbolic system we use the approach based on the complementary projection method [1]. The Godunov type method can also be used but the more general structure of waves for weakly hyperbolic systems has to be respected. If the function $\mathbf{g}(\mathbf{q}, x)$ is sufficiently smooth in $\mathbf{q}$ and x it is possible to use any standard entropy fix procedure. In other cases a special entropy fix procedure has to be applied (and it is better to use the formulation in the form $\mathbf{q}_t + [\mathbf{g}(\mathbf{q}, \mathbf{a}(x))]_x = \mathbf{0}$ where $\mathbf{a} = \mathbf{a}(x)$ is a non-smooth function).

References

1. R. P. Fedkiw, B. Merriman, and S. Osher. Efficient characteristic projection in upwind difference schemes for hyperbolic systems. *J. Comput. Phys.*, 141(1):22–36, 1998.
2. R. P. Fedkiw, B. Merriman, and S. Osher. Simplified discretization of systems of hyperbolic conservation laws containing advection equations. *J. Comput. Phys.*, 157(1):302–326, 2000.
3. G. S. Jiang and E. Tadmor. Nonoscillatory central schemes for multidimensional hyperbolic conservation laws. *SIAM J. Sci. Comput.*, 19(6):1892–1917, 1998.
4. R. J. LeVeque. *Finite volume methods for hyperbolic problems.* Cambridge University Press, 2002.

Parallel 2D/3D Unsteady Incompressible Viscous Flow Computations Using an Unstructured CFD Code

H. Chen[1], P.G. Huang[2], and R.P. LeBeau[3]

[1] University of Kentucky, 151 RGAN Building, Lexington, KY, 40506,
hchen@engr.uky.edu
[2] University of Kentucky, 151 RGAN Bldg, Lexington, KY 40506,
ghuang@engr.uky.edu
[3] University of Kentucky, 151 RGAN Bldg, Lexington, KY 40506,
rplebeau@engr.uky.edu

This paper presents a cell-centered pressure-based CFD method implemented in the parallel unstructured code UNCLE. The code is a two/three-dimensional finite volume unsteady incompressible Navier-Stokes solver based on the SIMPLE algorithm with second order accuracy in both time and space. UNCLE can use a variety of grid types, such as triangular, quadrilateral, tetrahedral, prisms, and hexahedral meshes, generating essentially identical solutions for different mesh types.

UNCLE makes use of METIS [1] for domain decomposition. The key features of METIS include extremely fast partition, high quality partitions, and low fill orderings. Our test results using METIS in UNCLE exhibit excellent load balance in two/three-dimensional grids with various types of meshes, where the average load balance is up to 97% for all our test cases.

The parallel construction of UNCLE is based on MPI protocols and has worked successfully on systems ranging from commodity PC clusters up to traditional supercomputers. Performance tests of UNCLE including speedup and cache miss rate have been undertaken on multiple architectures of Linux clusters [2]. These tests show that UNCLE exhibits excellent parallel and good single-node speedup and steady cache performance.

Several test cases are presented for validation of UNCLE including steady two/three-dimensional driven cavity flow [3] [4], and unsteady two/three-dimensional flow over a circular cylinder for low Reynolds numbers [5]. Figure 1 shows the streamlines of two-dimensional driven cavity flow for Re=400. Figure 2 shows the contours of vorticity components from a three-dimensional simulation of cylinder flow for Re=200 at y=0 plane (top view). All the test cases investigated yielded good agreement in comparison with previous benchmark computational and experimental results.

Support for this research was provided by Kentucky NASA EPSCoR, grant WKU 516140-02-06.

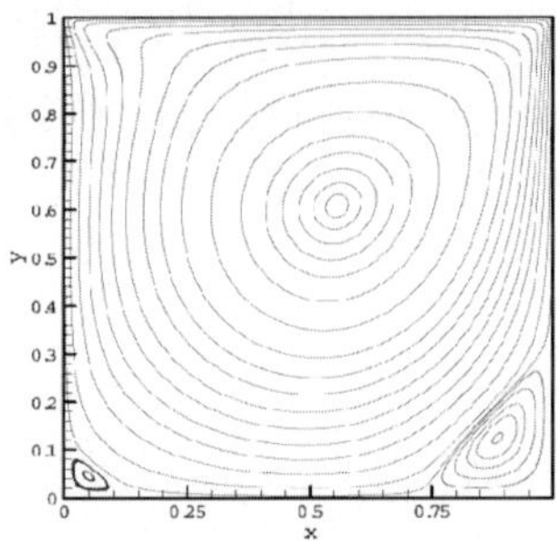

Fig. 1. The streamline plot for 2D driven cavity flow for Re = 400.

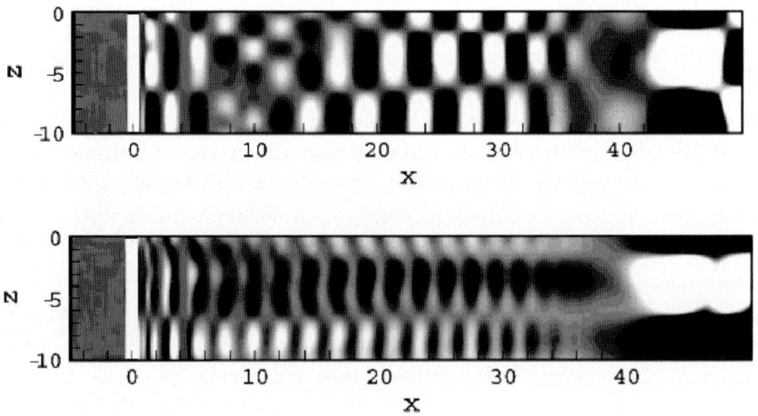

Fig. 2. Vorticity component contours for three-dimensional simulation at y=0 plane (top view), where the top is ω_x and the bottom is ω_y for Re = 200.

References

1. G. Karypis, V. Kumar: A Software Package for Partitioning Unstructured Graphs, Partitioning Meshes, and computing Fill-Reducing Orderings of Sparse Matrices Version 4.0, http://www.cs.umn.edu/ karypis (1998)
2. R.P. LeBeau, H. Chen, P. Kristipati, S. Gupta, P.G. Huang: Joint Performance Optimization of Two CFD Codes on Commodity Clusters. Submitted to: *AIAA 43rd Aerospace Sciences Meeting and Exhibit* (2005)
3. U. Ghia, K.N. Ghia, C.T. Shin: J. Comp. Phys. **48**, 387 (1982)
4. H.C. Ku, R.S. Hirsh, T.D. Taylor: J. Comp. Phys. **70**, 439 (1987)
5. C.H.K. Williamson: Ann. Rev. Fluid Mech. **28**, 477 (1996)

Adaptive Bounds to Outputs of the Three Dimensional Steady Incompressible Navier-Stokes Equations

Hae-Won Choi[1] and Marius Paraschivoiu[2]

[1] Department of Mechanical and Industrial Engineering, University of Toronto, 5 King's College Road, Toronto, ON M5S 3G8, Canada; haewon@mie.utoronto.ca

[2] Department of Mechanical and Industrial Engineering, Concordia University, H-549, 1455 De Maisonneuve Blvd. W., Montreal, PQ H3G 1M8, Canada; paraschi@ME.Concordia.CA

1 Introduction and Motivation

The simulation based engineering design requires repetitive procedures for testing prototypes to meet an acceptable design goal. Typically the design with high fidelity leads to computationally expensive numerical analysis for the prediction of design quantities termed outputs. To circumvent the high cost of CFD for designs, an implicit *a-posteriori* finite element technique – termed the bound method – is utilized to permit an inexpensive design without sacrificing numerical accuracy. This novel technique based on the augmented Lagrangian by appealing to the dual *max-min* (or *inf-sup*) theory provides lower and upper bounds to the output of the underlying partial differential equations (PDEs). The bound method is applied herein to address a realistic design scenario applied to the coupled steady incompressible Navier-Stokes and convection-diffusion (i.e., energy) equations in three space dimensions. To handle more complex applications, an adaptive subdomain refinement strategy leading to enhancement of the bound method is adopted. For the bound method, the methodology developed in [1] is used. The inter-subdomain boundary data (i.e., hybrid-fluxes), required to be self-equilibrated, are calculated by utilizing the directly equilibrated hybrid-flux technique recently proposed in [2]. The spatial discretization uses the $\mathbf{P}_2^+ - \mathbf{P}_1$ (Crouzeix-Raviart) element. The $\mathbf{P}_1$ approximation for the pressure renders the divergence-free subdomain thanks to the discontinuous pressure approximation. The implicit nature of this framework leads to two global coarse mesh calculations and a multitude of elemental subdomain computations. Subdomains used in this work are in fact the coarse mesh tetrahedra further refined into self-similar tetrahedra leading to discrete local Neumann subproblems. By refining subdomains (i.e., global tetrahedra) sharper bounds can be obtained.

To motivate the bound method we consider a series of electronic chip devices. In a realistic design scenario, engineers need to determine the interested design output such as a mass flow rate, mean temperature, and drag/lift force in a specific output region. One then verifies that the output quantities are within an acceptable design range and a specific design condition is accepted or rejected accordingly. In practice, a number of design situations must be tested in a given simulation environment. For a single design process, due to no initial information regarding the 'ideal' mesh for simulation, many different types of meshes should be tested. Hence engineers face the issue of design efficiency. To reduce computational efforts for iterative design procedures, the use of the adaptive bound method exploiting the direct equilibration technique is applied herein to concept designs – provide bounds to the specific output at the fastest rate with reliable numerical accuracy.

2 Numerical Results

To validate the proposed technique, a convective heat transfer for a pressure driven flow around two obstacles for $Re = 100$, $Pe = 100$, and $q = \frac{1}{Pe}\frac{\partial \Theta}{\partial n} = 1$ is considered as a model problem. The output selected is the mean temperature in the specific output region Γ^O. To provide bounds to the mean temperature in the specific output region, two subdomain mesh refinement strategies are considered herein. An uniformly refined subdomain mesh exploiting 19,200 tetrahedra gives an output as $s_h = 13.33 \pm 0.928\,(6.96\%)$ whereas an adaptively refined subdomain mesh having 18,161 tetrahedra, as shown in Fig. 1(left), gives bounds to the output $s_h = 13.39 \pm 0.818\,(6.11\%)$. Under similar situations, clearly the adaptive refinement results in sharper output bounds than the uniform refinement. Furthermore, the adaptively refinement mesh properly captures vortex and temperature contours behind the obstacles as displayed in Fig. 1(right). Hence, the technique described herein can serve as an efficient design tool for engineering applications.

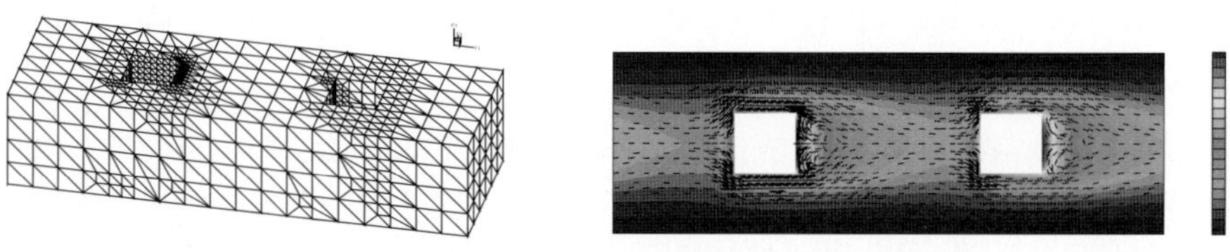

Fig. 1. Adaptively refined subdomain mesh with 18,161 tetrahedra (left); Its uniform velocity vector and temperature contour at $x_3 = 1/3$ (right).

References

1. Z. Cheng, M. Paraschivoiu: Int. J. Numer. Methods in Engrg. Accepted in April (2004).
2. H.-W. Choi, M. Paraschivoiu: Int. J. Compute. Fluid Dyn. In Press (2004).

A Comparative Study of Three Composite Schemes: Lax-Wendroff/Lax-Friedrichs, Mac-Cormack/Lax-Friedrichs and Corrected Lax-Friedrichs Lax-FriedrichS Schemes, Based on Conservation Laws

M.Z. Dauhoo and A.R. Appadu

Dept. of Mathematics, Faculty of Science, University of Mauritius, Mauritius.
m.dauhoo@uom.ac.mu

Summary. Liska and Wendroff[4] blended the Lax-Wendroff scheme (LWS) and Lax-Friedrichs scheme (LFS) in order to reduce oscillations and damping from dispersive and diffusive schemes. In this work, we combine the MacCormack Scheme (MCS) and LFS to obtain a new composite scheme which we have formulated as MacCormack/Lax-Friedrichs (MCLF) scheme. Based on the numerical results for conservation laws in 1-D and in 2-D, we observe that the MCLF scheme is more efficient than LWLF to capture shocks. We show analytically that MCLF scheme is less diffusive than LWLF scheme in 1-D and 2-D. Also, we have used another composite scheme in 2-D [3] consisting of Corrected Lax-Friedrichs scheme (CFS) and of Lax-Friedrichs scheme, developed by Liska and Wendroff to solve a 2-D contact discontinuity problem given in [2]. Lastly , we present some preliminary work on a composite scheme made up of MCS and MCLF scheme.

1 LWLF, MCLF and MCMCLF in 1-D

Figure 1 shows the diffusion error of the schemes MC and MCMCLF for the 1-D linear advection equation. Figure 2 shows the numerical results of the 1-D shallow water problem described in [3].

2 Composite Schemes in 2-D

We study the diffusion errors of the 3 composite schemes: LWLF, MCLF and CFLF using the 2-D scalar advection equation [1] which is given by:

$$\frac{\partial u}{\partial t} + A\frac{\partial u}{\partial x} + B\frac{\partial u}{\partial y} = 0. \tag{1}$$

Figure 3 show the corresponding diffusion error of the schemes. Figure 4 shows the numerical results for a 2-D contact discontinuity problem.

3 Conclusion

In 1-D and 2-D, MCLF scheme captures shocks better than LWLF scheme as confirmed by our numerical results. A composite scheme of MCS with MCLF scheme improves results in 1-D but not in 2-D. A low dissipative and low dispersive scheme in 1-D is MCMCLF scheme and in 2-D, the CFLF scheme.

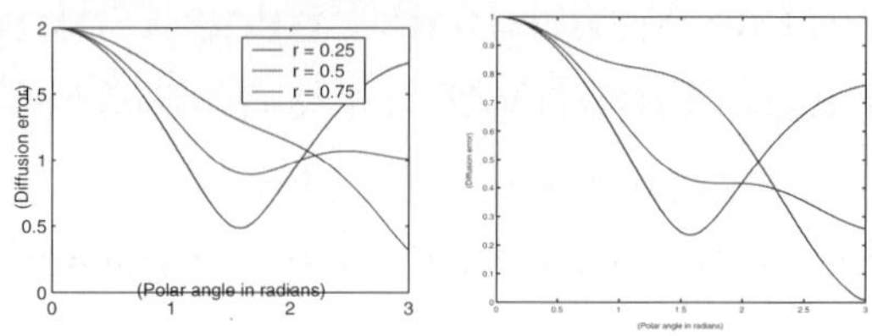

(a) MCLF (b) MCMCLF

Fig. 1. Diffusion error for each scheme

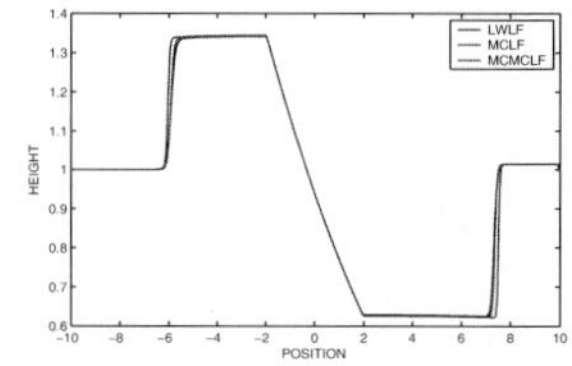

Fig. 2. Comparison of height profile for the 1-D shallow water problem calculated by LWLF, MCLF and MCMCLF.

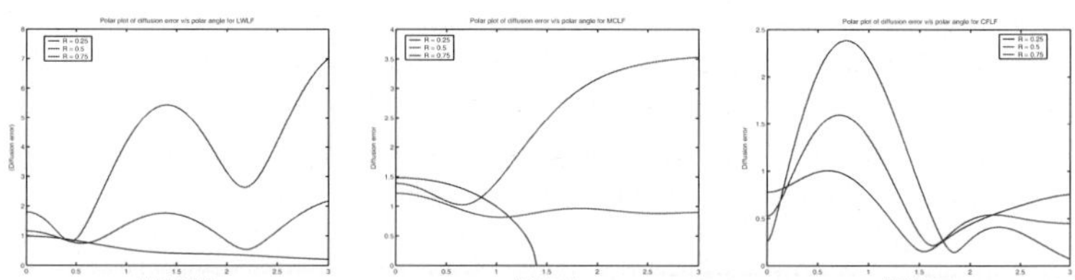

(a) LWLF (b) MCLF (c) CFLF

Fig. 3. Diffusion error for each scheme

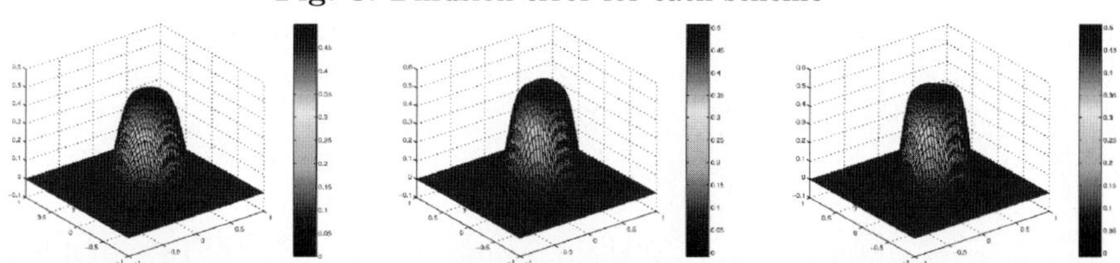

(a) LWLF (Ratio of LWS to LFS is 1:1) (b) LWLF (Ratio of LWS to LFS is 2:1) (c) MCLF (Ratio of MCS to LFS is 1:1)

Fig. 4. Numerical result for the 2-D contact discontinuity problem (I)

References

1. C. Hirsch: Numerical Computation of Internal and External Flows, Volume 2, John Wiley and Sons, 1990.
2. S.M. Ould Kaber: J. Computat. Phys. 128, 165-180 (1996).
3. R. Liska and B. Wendroff: Technical Report LA-UR 96-3589, LANL, Los Alamos, 1996.
4. R. Liska and B. Wendroff: Czech Technical University in Prague, 1997.

Evaluation of Reynolds Number Effects on the CFD Simulation of Downwind Sails

G. Delussu[1], N. Erriu[2], R.G.J. Flay[3], M. Mulas[1], P. Puddu[2], and M. Talice[1]

[1] CFD group CRS4, Italy talice@crs4.it
[2] Mechanical Engineering Department, University of Cagliari, Italy
[3] Mechanical Engineering Department, University of Auckland, New Zealand

1 Introduction

Experimental approach is still widely used in the design of sails for high performance yacht. Even though the testing sections of some wind tunnel could reach remarkable dimensions, yachts are too big to be experimentally tested into a wind tunnel and scaled model are used instead. In order to enforce consistency between wind tunnel and real operating conditions, Reynolds similarity should be respected, thus both geometric and flow similarity conditions must held. Enforcing Reynolds similarity on a scaled model would result in values of the incoming flow velocity, unbearable for the model structure, so that *de facto* experiments in wind tunnel cannot respect Reynolds similarity. The actual relevance of this approximation is still an open question and in the present paper CFD will be used as a tool to investigate the matter.

2 Results and Discussion

The available experimental data were collected in the Twisted Flow Wind Tunnel (TFWT) of the University of Auckland, where a model of a Volvo Ocean 60 hull and rig scaled 12.5 times, was positioned on a six component force balance [2]. Tests were performed with a Code 4 asymmetric gennaker and a main sail at 90° AWA and 0° heel. The onset flow in the TFWT reproduced the apparent wind twist that occurs during sail conditions, the velocity profiles of which can be found in [2]. For being the description of the sails under loaded conditions the only available geometrical data, only the two sails have been taken into account in the CFD model, whereas the presence of the hull has been neglected. The computational grid was of about $1,000,000$ hexahedral cells, stretched at solid walls to resolve boundary layers. The Spalart and Allmaras turbulence model [1] has been used through the computations. The experimental twisted velocity profile was imposed at the domain boundaries and computations have been repeated for Reynolds numbers ranging from TFWT conditions to typical full scale values and above. Moreover, an inviscid calculation has been carried out to represent numerically infinite Reynolds number. In the Figure 1 the computed force coefficients for different values of the ratio Re/Re_{wt} are shown, together with the corresponding experimental data. Results of the inviscid computation have been reported in error bar form, due to lack of numerical convergence. Table 1 reports the quantities $(q - q_{ref})/q_{ref}\,\%$, with q = ($C_x$, C_y, C_z, C_{mx}, C_{my}, C_{mz}, experimental data) and $q_{ref} = q$ at $Re/Re_{wt} = 30$ (first line) and q_{ref} = experimental data (second line). Analysis of the Table 1, shows a not negligible influence of the Reynolds number on the computed results. Comparison with experimental values should take into account the fact that in the numerical model the presence of the hull was not considered. Differences with experimental values in fact reduce to about 15% in the sail-boat axial direction, in which the portion of hull surface directly exposed to the flow is smaller.

Table 1.

C_x	C_y	C_z	C_{mx}	C_{my}	C_{mz}	
6.57	32.37	16.95	16.76	25.00	6.71	Re/Re_{wt} 1 *vs.* Re/Re_{wt} 30
36.21	46.29	14.64	17.84	93.87	37.98	*Experiment vs.* Re/Re_{wt} 1

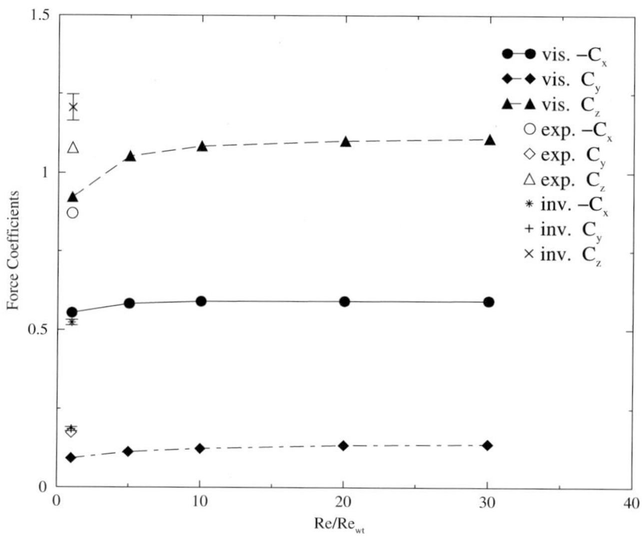

Fig. 1. Force coefficients

References

1. P. R. Spalart and S.R. Allmaras: A one-equation turbulence model for aerodynamic flows. In: *AIAA 30th Aerospace Sciences Meeting*, vol 92, pp 439
2. Le Pelley D., Ekblom P., Flay R.: Wind Tunnel Testing of Downwind Sails. In: *High Performance Yacht Design Proceedings*, 2002, pp 155–161

Assessment of the Immersed Boundary Technique for Compressible CFD codes

G. Delussu, M.Mulas and M. Talice

CFD group, CRS4, Italy, mulas@crs4.it

1 Introduction

In the recent years there has been a renewed interest in the Immersed Boundary (IB) technique where the presence of the body, on an underlying regular grid (non conformal with the geometrical definition of the body), is accounted for by means of suitable forcing terms added to the Navier-Stokes equation. The IB method was originally developed in the 60's mainly by Peskin [1]. A recent review article [2] well documents the regained interest of the method, particularly for the typical industrial applications in complex geometries, occasionally interested, but not necessarily, by moving boundaries.

The basic reason of the renewed interest in the IB technique stems from the fact that the mesh generation process stilll represents a difficult and time-consuming task. Moreover, another important motivation comes from the demand of software for aero-dynamic optimization which requires re-meshing steps. In the framework of the IB technique, such a re-meshing step would only require the recalculation of the forcing terms on the unchanged underlying regular mesh, taking negligible computing time, without the need of any interactive user's activity.

The most recent developments of the IB method have been carried out for fractional-step methods for the Large Eddy Simulation of incompressible turbulent flows [3]. In the framework of the RANS model, the use of the IB technique presents more difficulties with respect to LES: near wall meshes with aspect ratios up to 1000, depending on the Reynolds number. One possibility is to resort to adaptive mesh refinement, as described in ref. [4].

The present implementation of the IB technique into a 2D inviscid and compressible flow solver is still based on the use of forcing functions and on the desire to keep the simple and straightforward treatment of structured meshes.

The presence of the body, whose surface cuts the underlying mesh, does not break the structured data-base: the external cells are normally treated, the internal ones are also normally treated, however never used, and only the boundary, or cut, cells need some sort of special treatment. With reference to the figure 1, showing a cut cell, the fluxes through the four faces (six in 3D) are scaled by the coefficient ω no matter if a cell face is completely ($\omega = 0$), or only partially, immersed inside the body. What is missing is only the wall pressure flux due to the surface of the body, which is introduced as a special additional term. The wall pressure flux is in fact the only non-zero portion of the overall flux through a solid boundary, which contributes to the cell momentum budget. A pre-processing stage is needed to classify all cells (external, internal or cut) and to compute the ω coefficients and the cell volumes. Preliminary test cases of internal channel flow with area change give correct results, validated against computations on standard body fitted meshes.

2 Conclusions and Future Work

A simple extension of the IB technique to 2D Euler flow solver is presented. Preliminary results show the feasibility of the approach. Issues concerned with the accuracy of the method have yet to be addressed. Extension to 3D solvers is straightforward. The treatment of bluff bodies with zero thickness (for instance, sails of racing boats) presents problems related to the break down of the structured data-base (each cut cell has to be duplicated). Finally, the extension to viscous and turbulent simulation will also present problems related to the necessary requirements on mesh stretching, without the need to resort to local mesh refinement (which would also break the block-structure database).

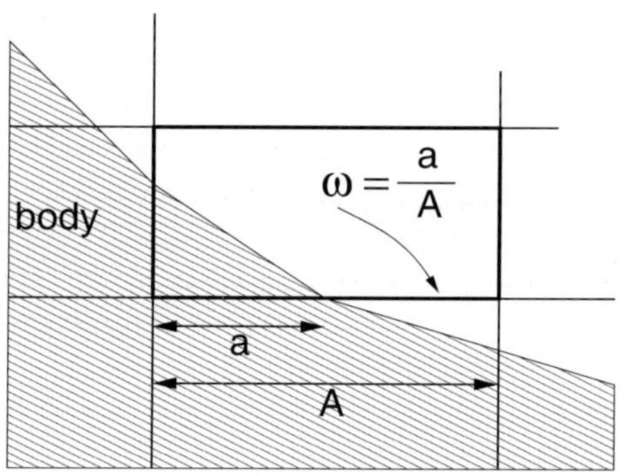

Fig. 1. A boundary, or cut, cell: the fluxes through the four faces are scaled by the coefficient ω; the forcing term should introduce the missing wall pressure flux

References

1. Peskin C.S.: J. of Comput. Phys. **25**, 220-252, (1977)
2. Iaccarino G. and Verzicco R.: Appl. Mech. Rev. **56**, 331-347, (2003)
3. Verzicco R., Mohd-Yusof J., Orlandi P. and Haworth D.: AIAA J. **38**, 427-433, (2000)
4. Majumdar S., Iaccarino G., and Durbin P.A.: RANS solvers with adaptive structured boundary non-conforming grids. In: *Annual Briefs, Center for Turbulence Research, 2001*

Spectral Solution of High Speed Flows Over Blunt Bodies with Improved Boundary Treatment

V. Esfahanian[1], M. Boroomand[2], and M. Najafi[2]

[1] Dept. of Mechanical Engineering, University of Tehran, Tehran, Iran
evahid@ut.ac.ir

[2] Dept. of Aerospace Engineering, Amirkabir University of Technology, Tehran, Iran

Summery

In some fluid dynamics problems, like flow stability and transition, accurate solution of mean flow is required. While such an accurate solution cannot be obtained by means of ordinary finite difference approximations, spectral approximations are able to surmount. Furthermore, the low inherent dissipation of spectral methods is not enough to stabilize the method in the presence of shocks [1]. The spectral methods with shock-fitting have been shown to solve the inviscid blunt body problem accurately and efficiently. In addition, errors decay exponentially fast with the increase in the grid points [2].

In this study, the chebyshev spectral collocation method is used to solve Euler equations over a blunt body in a supersonic flow fields. The axisymmetric and non-conservative form of Euler equations is considered as the governing equations. The initial shock shape is taken from an empirical formula developed for flow over spherically blunted cones experimentally [3]. The Rankine-Hugoniot relations are used for shock boundary condition in accordance with a proper compatibility relation, which carry information from the flow field to the shock. The shock acceleration is derived similar to the method found in [2] but with different formulation to treat the shock boundary condition. Zero normal velocity along with compatibility relations in a matrix form is used as a boundary condition for solid body, which simplifies the implementation. Since the outflow is supersonic (all four characteristics are leaving the flow field), no explicit boundary conditions are necessary; therefore the governing equations are used to obtain the boundary variables. At the symmetry line, the symmetry conditions are applied using spectral differentiation by means of the procedure found in [4].

The sensitivity of spectral methods to boundary conditions, which can not be overstated, makes the boundary treatment the most difficult part of the solution. In contrast to [2] which uses different treatments for all types of boundaries, a uniform approach is used in the present study by extending the matrix method proposed by [5] for finite difference method to the spectral one. This procedure allows a more robust systematic solver. The solution is advanced in time using 4th order Runge-Kutta method. As a famous check on the validity of the spectral code, supersonic flow over a sphere is considered and the results for shock shape and body surface quantities are compared with the results of [5]; and the spectral accuracy is observed.

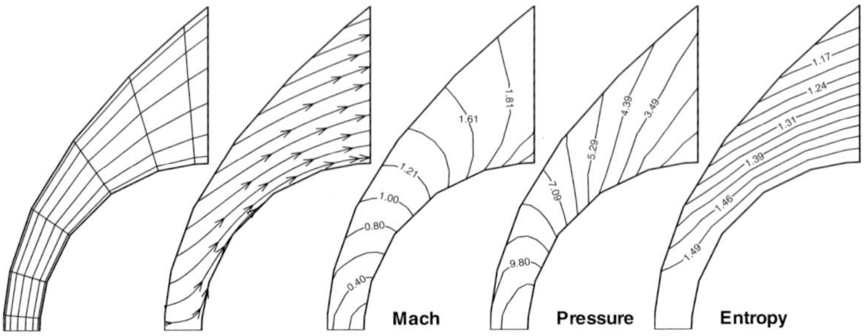

Fig. 1. Grid, streamlines and flow variables for flow over a sphere, M =2.94

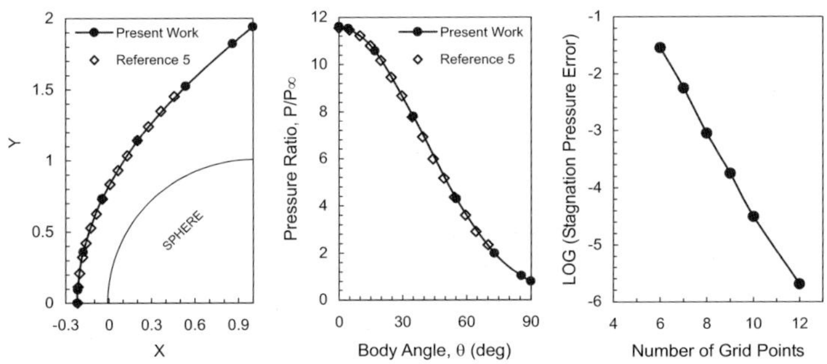

Fig. 2. Shock shape, pressure ratio and error decay for flow over a sphere, M =2.94

References

1. C. Canuto, M. Y. Hussaini, A. Quarteroni and T. A. Zang: *Spectral Methods in Fluid Dynamics*, (Springer, Berlin Heidelberg New York 1987) pp 241
2. D. A. Kopriva, T. A. Zang and M. Y. Hussaini: AIAA J. **29**(9) pp 1458–1462 (1991)
3. F. S. Billig: J. of Spacecraft and Rockects **4** pp 822–823 (1967)
4. D. Funaro: *Polynomial Approximation of Differential Equations*, (Springer, Berlin Heidelberg New York 1992) pp 125–150
5. S. R. Chakravarthy : AIAA J. 21(5) pp 699–706 (1983)

Verification of Mathematical Model of the Shock Wave/Dust Layer Interaction Problem

A.V. Fedorov, N.N. Fedorova, and I.A. Fedorchenko

Institute of Theoretical and Applied Mechanics, Novosibirsk, Russia

1 Introduction

The problem of dust lifting due to the unstability of a layer situated on a rigid wall under the action of travelling shock wave is of considerable applied and theoretical interest. The practical aspect relates with phenomenon of initial phase of a "layer detonation" process. The theoretical one is a possibility to get a new information about a mechanism of heterogeneous medium and air mixing phenomenon at contact surface. In the paper [1], the review of several works made from beginning of 60th and devoted to the physical and mathematical investigations of the phenomenon has been given. The main goal of our paper is to verify the mathematical model and numerical method by means of comparing with an available experimental data.

2 Mathematical Model and Problem Statement

The problem statement is as follows. The planar SW travels over the dust layer in a still gas. A particle layer with thickness h is situated on the rigid wall. Behind the SW the flow is forming in an air which parameters is derived from planar shock relations.

The problem has been investigated in the frame of two mathematical models. The first one is the 2-D Euler equations for the mixture with additional equation for the volume concentration of solid phase m_2 and special equation of state for the mixture. The higher density of the layer is caused by the presence of a dispersed phase. The description of the mathematical model and details of CIP-scheme implemented in the work are presented in [2]. The second approach is based on two-dimensional Favre-averaged Navier–Stokes equations closed by $k - \omega$ turbulence model of Wilcox [3]. The dust layer in this case has been simulated by a lower temperature gas. Detailed description of the model and numerical method is presented in [4].

3 Results and Discussion

The calculations have shown that the flow scheme is similar in both of these cases. In the near-wall dense layer the front of the leading shock wave becomes curved. The curved wave falls onto the rigid wall and reflects from it. It has been shown that either regular and irregular regimes of reflection of the curved SW from the surface can be observed. The reflected wave arrives at the contact surface between clean and dusted gases and is reflecting from it as an expansion wave, which again comes onto the wall. Then the process is repeated.

Three possible mechanisms of dust lifting behind the travelling SW have been found. The first one is associated with the formation of an area with a rather high positive vertical velocity of the mixture behind the curved SW, which may lead to upward ejection of large particles. The second mechanism is conditioned by an unsteady vortex structure into which the dusty gas jet is "swirled". The third possible mechanism of particle lifting is related to the Kelvim-Helmholtz instability of the shear layer, which develops in the stratified layer under the action of internal waves and external pertuberations.

The verification of the model has been also carried out with the help of experimental data [5, 6]. A good agreement between experimental and calculated values of the first peak of the wall pressure distribution has been observed. Moreover, the numerical solution demonstrates exactly the steady pressure value far downstream. In a case of a low dust volume concentration, the waves period inside the layer decreases while an amplitude of the waves increases.

Acknowledgements

The work was supported by the Russian Foundation of the Basic Research (Grant No 03-01-00453) and Ministry of Education of the Russian Federation (Grant No 03-2.10-544).

References

1. A.V. Fedorov: Combustion, Explosion, and Shock Waves **40**, 1 (2004)
2. A.V. Fedorov, N.N. Fedorova, I.A. Fedorchenko, V.M. Fomin: Journal of Applied Mechanics and Technical Physics **43**, 6 (2002)
3. D.C. Wilcox: AIAA Paper. N.Y., N 584 (1974)
4. A.V. Borisov, N.N. Fedorova: Thermophys. Aeromech. **4**, 1 (1996)
5. H. Matsui: Structure and Propagation Mechanism of the Soot Layer Detonation. In: *Proceedings of Research on the Processes of Combustions and Modelling of Fires*, (Khabarovsk 1992) pp 57–62
6. B.E. Gelfand, S.P. Medvedev, A.A. Borisov et al: Archivum combustionis **9** 1/4 (1989)

Automated Unstructured Mesh Generation for Objects in Direct Contact

Dr. F. Fortin[1] and T. Zoric[2]

[1] Institute for Aerospace Research(IAR), National Research Council(NRC), Ottawa, Ontario, Canada, K1A 0R6, francois.fortin@nrc-cnrc.gc.ca
[2] Carleton University, Department of Mechanical and Aerospace Engineering, Ottawa, Ontario, Canada, K1S 5B6, tanya-zoric@canada.com

The Institute for Aerospace Research (IAR), at the National Research Council of Canada (NRC), undertook the initiative in 1994 to develop a Canadian Computational Fluid Dynamics (CFD) capability for the numerical simulation of external stores carriage and release. The methodologies developed as an outcome of this initiative, were mainly sponsored by the Canadian Department of National Defence (DND), and have now been transferred to Bombardier Aerospace.

Following evaluation by the industry, it was agreed that automated procedures were required to accomplish repetitive tasks and thus increase efficiency. The mesh generation process was identified as the most time-consuming procedure and has thus been fully automated. This process consists of combining the initial basic shape of the releasing aircraft along with its different components (stores, fuel tanks, etc.) and then obtaining an unstructured mesh around the loaded aircraft. The components are positioned interactively through the use of a developed Graphical User Interface, which runs on a PC and is capable of providing a 3D visual representation of the actual loaded configuration.

The various components, defined inside a dynamic database, are accessed through the Graphical User Interface. Various settings, defining an adequate surface mesh as well as the CAD representation of each component at its inital position, are contained in a specific file. This file is a TETIN file, originally generated by the mesh generation package ICEM CFD. When the component is moved to a new position, the application of the basic transformation matrices (rotation and translation) reevaluates the new setting values.

After completion of the aircraft loading, the independent files of the various components are concatenated in a single file. For the components having a multiplicity greater than one, the boudaries names are altered, in order to be able to distinguish the same component type at different locations. An outer shell, representing the far-field, is added to bound the extent of the three-dimensional domain to be meshed.

A script file, written in TCL (Tool Command Language), is then run with the commercial mesh generation package ICEM CFD to get an initial mesh. The script file reproduces all the commands interactively given by a user

a) b) c)

Fig. 1. Automated intersection procedure: (a) Breakdown into 8 components, (b) Resulting components together, (c) Intersection cockpit with vertical fin.

using the keyboard or the mouse. Some parts of this technique have been presented in [1].

Direct surface contact will exist for missiles mounted at the wing tips and for stores carried under the belly of the aircraft. It was then required, for any series of objects put together, to check for the possibility of existence of intersection. When this is happening, the automated procedure will create a 3D curve defining this intersection. The mesh size selected, in order to adequately represent this intersection, will be selected as the smallest of the two intersecting component sizes.

The concept is demonstrated in Figure 1. A simple geometry made up of eight components (Figure 1a), i.e. a cone for the nose, an upper canopy, a seat, a lower canopy, a rear fuselage, left and right wings as well as a vertical fin, was created (Figure 1b) with automatic intersection estimations. The inner space bounded by the nose, the upper canopy, the lower canopy and the rear fuselage was not meshed. The upper canopy and the seat were then given new positions and orientations, as seen in Figure 1c. The intersection between the upper canopy and the vertical fin was correctly computed, using the automated procedure.

This automated mesh generation procedure is also used during store release trajectory predictions, when needed.

Acknowledgements

The authors would like to express their gratitude to the Canadian Department of National Defence for their financial support.

References

1. A. Tahi, F. Fortin and A. Benmeddour: Modular and Automated Unstructured Grid Generation Using ICEM CFD, Proceedings of the CASI 8th Aerodynamics Symposium, Toronto, Canada, 2001.

Complex Flow Patterns in Realistic Geometry of Human Aorta

R. Gårdhagen, J. Svensson, and M. Karlsson

Departments of Mechanical and Biomedical Engineering Linköping University, S-581 83 Linköping, SWEDEN, {rolga,johsv,matka}@ikp.liu.se

1 Introduction

Magnetic resonance (MR) images are used to construct a realistic geometry for CFD simulations of blood flow in a human aorta, Fig. 1. The patient specific data used here come from a juvenile with a coarctation of the aorta (CoA) and an aneurysm. This work focuses on the development of an efficient tool for diagnoses and intervention planning for vascular surgery capable of presenting parameters such as velocity, pressure and wall shear stress (WSS).

Normally, the aortic blood flow is rotating clockwise (CW) and here we use the tool to examine how the parameters mentioned above are affected by different rotations.

Fig. 1. Geometry of human aorta.

2 Method

Six mass flow rates (MFRs) (from 0.01 to 0.06 kg/s) and five rotations (high and low CW, high and low counter-clockwise (CCW) and no rotation) are modelled for laminar, steady state flow conditions. The geometry is meshed in Gambit and the simulations are performed in Fluent. We study pressure, velocity and WSS in the CoA region. To study the WSS (and other quantities at the wall) we propose the use of colour coded maps (here presented in black and white). These show the vessel wall in one single 2D image.

3 Results

The overall WSS pattern is significantly affected by the different rotations. Regions with high WSS change due to the varying flow field causing the jet developed after the CoA to change both position and shape. The local WSS maximum as well as average WSS in the entire vessel wall are, however, relatively independent of the rotation. A parameter estimation is used in order to condense the large amount of data from a simulation to a few informative parameters, see Fig. 2.

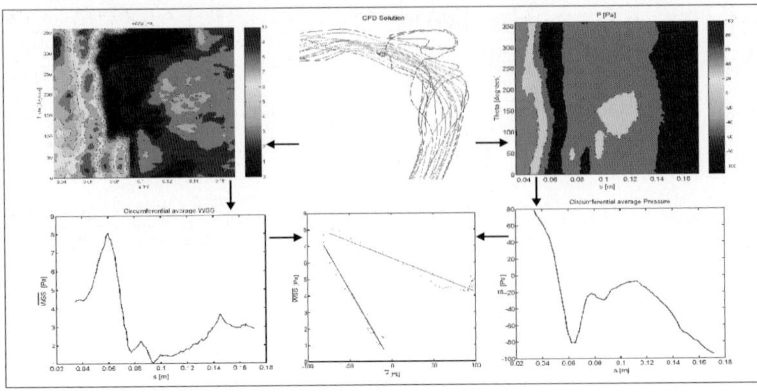

Fig. 2. WSS and wall pressure are determined from the simulation, top left and top right, respectively. Circumferential average values of these quantities are calculated, bottom left and bottom right, respectively. Average WSS magnitude is plotted against average pressure, bottom centre. Two lines are approximated to the points using a least square technique.

Circumferential average WSS magnitude is plotted against circumferential average wall pressure in a number of cross sections, Fig. 2 bottom-centre. There is a tendency to two lines, one before and one after the CoA and by means of a least square technique straight lines are approximated to the points. Direction coefficients for similar lines for all mass flow rates and rotations are shown in Fig. 3. Significant differences upstream and downstream to the CoA are seen. The procedure of reducing the large amount of data may have implications for further clinical usefulness.

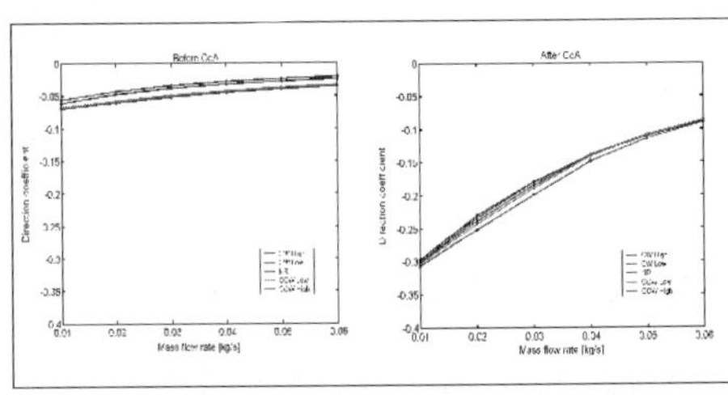

Fig. 3. Direction coefficients for WSS as linear function of pressure before and after the CoA for different mass flow rates and rotations

4 Conclusions

The overall WSS pattern is significantly affected by the rotation of the blood flow in the aortic arch, whilst local maximum values and average values remain almost constant when the rotation is varied. Dependency of mass flow rate and rotation may differ between different regions of the geometry. Linear relations between WSS and pressure are found upstream and downstream to the CoA. A small but clear dependency of both rotation and mass flow rate is found upstream, while downstream the dependency of rotation has disappeared whilst the dependency of mass flow rate has strongly increased.

Interplay Between Inertia and Gravity in Transient Thin-Jet Flow

Radoslav German and Roger E. Khayat

Department of Mechanical & Materials Engineering, The University of Western Ontario, London, Canada, N6A 5B9, rgerman@uwo.ca, rkhayat@eng.uwo.ca

1 Introduction

The interplay between inertia and gravity is examined in this theoretical study for the transient two-dimensional flow of a thin Newtonian jet. The fluid emerges from a channel and is driven by both a pressure gradient maintained inside the channel and/or gravity. The problem consists of obtaining the shape of the free surface and the flow field inside the domain, as the fluid emerges from the channel. Given the small thickness of the fluid jet compared to jet length and local curvature, the thin-film approximation will be used. In this study a unified spectral method is proposed to model the problem. The proposed methodology becomes particularly suited for highly non-linear thin film flow.

2 Problem Formulation

Consider the two-dimensional flow of an incompressible Newtonian fluid of density, ρ, and viscosity, μ. L and H are taken as reference length and thickness in streamwise and depthwise directions, respectively with x and z being the corresponding dimensionless coordinates. The streamwise and depthwise velocities u and w are respectively scaled by V and HV/L. After scaling, three important dimensionless groups emerge, namely, the Reynolds number, $Re = \rho V H^2/(\mu L)$, Froude number, $Fr = V/\sqrt{gL}$, and the aspect ratio, $\varepsilon = H/L$. The conservation equations are obtained in the dimensionless form, with terms of O(ϵ^2) and higher being excluded [1].

$$u_x + w_z = 0, \qquad Re(u_t + uu_x + wu_z) = \frac{Re}{Fr^2}, \tag{1}$$

where subscripts denote partial differentiation. The shape of the free surface is dictated by the kinematic condition:

$$h_t + u(x, z = h, t)h_x = w(x, z = h, t) \tag{2}$$

The relevant dynamic condition at the surface and symmetry condition at the centerline are applied. Flow conditions at the channel exit and initial condition over the whole domain are assumed to be known. The flow equations are first mapped onto the rectangular domain and then the formal expansion of the velocity field is introduced in terms of orthonormal basis functions. The Galerkin projection method is used to generate the equations that govern the expansion coefficients. The equations are solved by 6-th order Runge-Kutta integration in streamwise direction, combined with implicit finite difference scheme in time.

3 Results

The method predicts the shape of the free surface, as well as the flow field. It is found that a fluid with low inertia tends to contract at the channel exit. In this case the depthwise flow can be significant, contrary to what is usually believed in the literature on thin films.

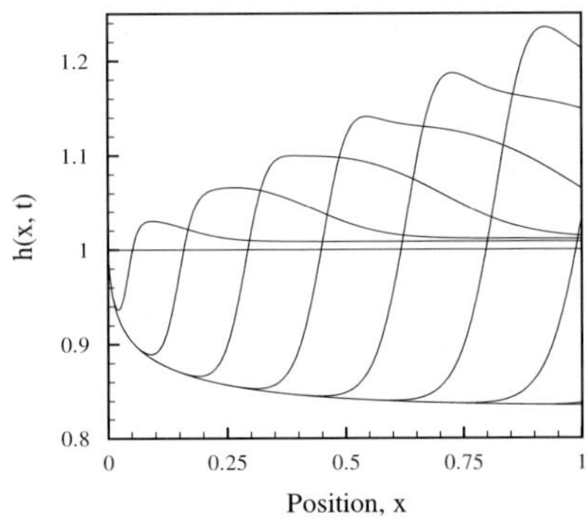

Fig. 1. Evolution of free surface for Re = 10, in the absence of gravity.

Sequence of transient flow is given in Fig. 1. In the early stages, the film thickness exhibits a contraction near the channel exit, with the film reaching already the steady state near $x = 0$. The contraction, however, coincides with the occurrence of a minimum at some $x > 0$. The film height rises further downstream, exhibiting a maximum (forming a wave), and asymptotically embracing the initial height for large x. With time, the disturbance propagates downstream with increase in amplitude. The film may or may not be unstable, depending on the flow parameters and initial conditions.

References

1. R.E. Khayat, S. Welke: Influence of inertia, gravity and substrate topography on the two-dimensional transient coating flow of a thin Newtonian fluid film. Phys. Fluids **13**, 355 (2001)

Numerical Modeling of Cell Deformation Passing Through a Nozzle to Determine its Viscosity and Surface Tension

Amirreza Golpaygan and Nasser Ashgriz

University of Toronto, 5 King's college Road, Toronto, ON M5S 3G8
ashgriz@mie.utoronto.ca

One important aspect during the process of leukocyte migration through endothelial cell in human body is the deformation of the cell [1]. Deformation and migration of leukocyte through endothelial cells depend on the cell rheological properties [2]. A new method is proposed to determine the cell rheologies, including its apparent viscosity, as well as its surface tension. In this method, a cell has been modeled as a viscous liquid drop with constant surface tension, which has been forced to migrate through a converging nozzle. The full nonlinear, transient Navier-Stokes equations subject to appropriate initial and boundary conditions are solved in three dimensions to analyze the dynamics of the deformation and recovery of liquid drop with different viscosity and surface tension coefficient. The motion through the nozzle involves very large deformation which has been captured using Volume-Of-Fluid (VOF). A comprehensive set of result including shapes of the drop after being released from the nozzle based on a wide range of viscosity and surface tension coefficient has been determined. Once the shape information is available, the inverse problem, namely, determining the cell properties based on the cell shape and time can be solved. In this presentation we have demonstrated the existance of correlationes between the cell shape and cell properties.

Figure 1 shows various shapes of a droplet with different properties just after existing a small converging nozzle. We present the results for the motion of the droplet with the initial velocity of 1 m/s moving through a nozzle which has the conic angle of 35°. The length of the droplet after exiting the nozzle is chosen as the characteristic parameter to compare deformation of droplet with different properties. By increasing the viscosity of the fluid, the droplet tends to exit with a longer length. At the same time, higher viscosity corresponds to higher dissipation of the energy for the droplet. Surface tension tends to keep the droplet in its spherical shape. After the droplet exits the nozzle, it starts to oscillate to its initial spherical shape. Figure 2 shows the changes in the deformation of the droplet versus surface tension and viscosity. The characteristic length of the droplet increases when surface tension decreases. On the other hand, for the amount of deformation versus viscosity, there is a maximum point. By incresing viscosity, first the length of the droplet increases, then starts to decrease. Higher dissipation which occures by increasing viscosity is the main reason for this behavior in the length of the droplet.

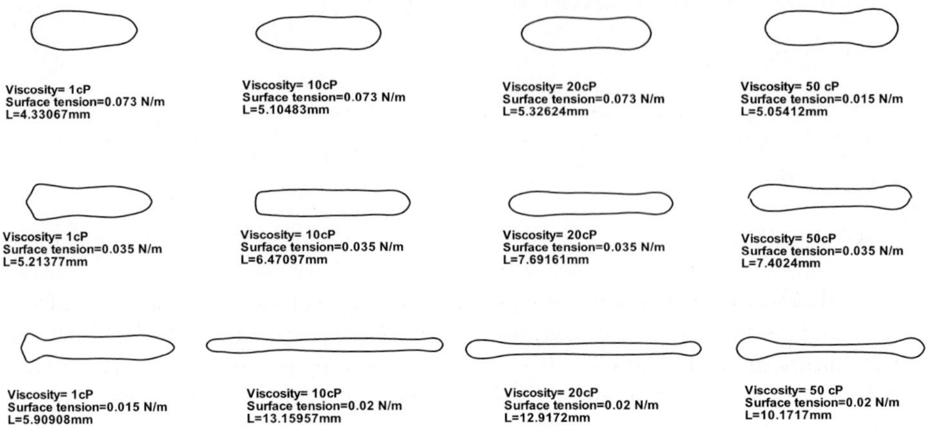

Fig. 1. Shapes of a droplet just after existing the nozzle based on its viscosity and surface tension.

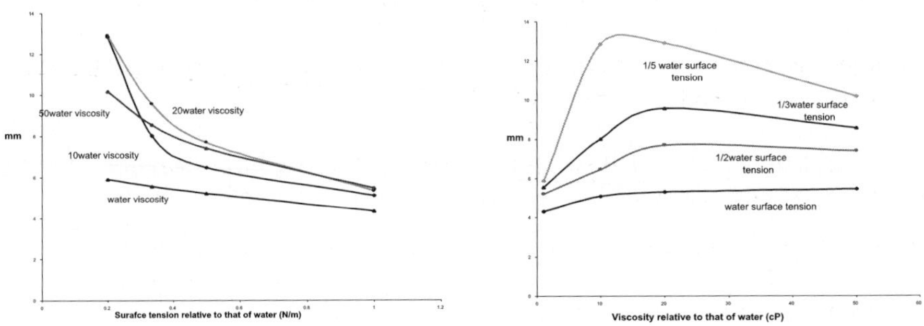

Fig. 2. Changes in the droplet characteristic length versus viscosity and surface tension.

References

1. A. Lauffenburger, A.F. Horwitz: cell **359**, 84 (1996)
2. R.D. Kamm: Annu. Rev. Fluid Mech. **34**, 211 (2002)

Gradient Computation for Variational Assimilation of Meteorological Observations

Y. Horibata

Faculty of Engineering, Hosei University, 3-7-2, Kajino-cho, Koganei, Tokyo 184-8584, Japan, horibata@k.hosei.ac.jp

1 Introduction

A single Doppler radar observes spatial and temporal distribution of one component of wind velocity and rainwater. From the Doppler radar observations, the complete atmosphere field is estimated at a given time. The assimilation is formulated as a constrained nonlinear programming problem. The gradient method iterates search for the minimum point. Each iteration requires the gradient of the objective function with respect to its variables. This paper develops a method for gradient computation.

2 Constrained Nonlinear Programming Problem

An objective function is defined such that it measures the distance between a model solution and the observations:

$$J(u, q_r) = \int_x \int_y \int_z \int_t [w_u(u - u^{ob})^2 + w_q(q_r - q_r^{ob})^2] dx\, dy\, dz\, dt \qquad (1)$$

where u and q_r are x component of the wind velocity and the mixing ratio of rainwater, respectively, and u^{ob} and q_r^{ob} are their observations. The coefficients w_u and w_q are weights.

The constraint is the model equations for the atmosphere field.

The field variables defined at the grid points at the given time are taken to be the initial data for the model equations; the finite difference scheme integrates the model equations by marching outward from the inital data [1]. Thus, the objective function is reagarded as a function of the initial data.

3 Gradient Computation

If the initial data are perturbed, the model solution is also perturbed. From the discretized model equations, the equations are derived for the perturbations [2]. From the perturabation equations and the gradient perturbation, the algorithm for gradient computation is developed.

4 Numerical Experiment

A numerical experiment is presented for a three-dimensional downburst. A downburst is a strong downdraft which induces an outburst of damaging winds on or near the ground. A downburst is simulated by integrating the model equations. The simulation domain is 20km in both horizontal directions and 10km in the vertical. The origin of the coordinate system is located at the center of the base. Out of the simulation result, x component of the wind velocity and the rainwater are assumed to be the observations from a Doppler radar. For an initial guess of the field variables at the initial time, the gradient of the objective function is computed, and is compared with one computed by the finite difference method. Figure 1 shows the gradient with respect to the rainwater computed by the present method in the $x - z$ cross section at $y = 0$. Figure 2 shows the corresponding one computed by the finite difference method.

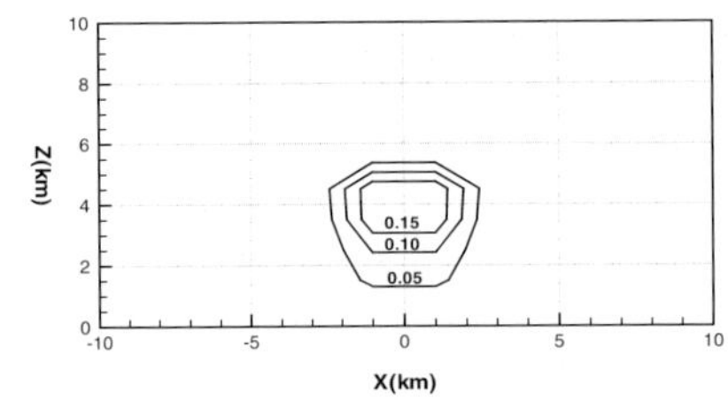

Fig. 1. Gradient by the present method.

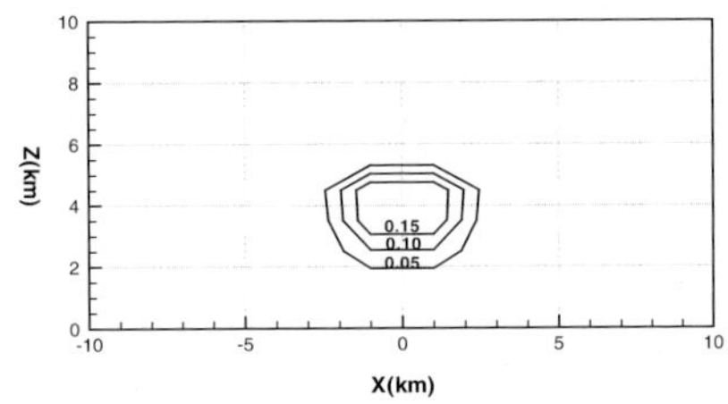

Fig. 2. Gradient by the finite difference method.

5 Conclusions

The two gradients agree well. The present method reduces CPU time by a factor of about 10 compared with the finite difference mrethod.

References

1. Y. Horibata, H. Oikawa: Numerical simulation of downbursts using a terrain-following coordinate tJournalransformation. In: *Computational Fluid Dynamics '94*, ed by S. Wagner, J. Périaux, E.H. Hirschel (Wiley, Chichester New York 1994) pp 1002–1009
2. O. Talagrand, P. Courtier: Q.J.R. Meteorol. Soc. **113**, 1311 (1987)

3-D Heat and Fluid Flow Analysis of Successively Variable Louver Angle of Louver Fin Geometry in Compact Heat Exchangers

C. T. Hsieh[1] and J. Y. Jang[2]

[1] Department of Mechanical Engineering Southern Taiwan University of Technology, Tainan, Taiwan, also Ph.D graduate student of NCKU
[2] Department of Mechanical Engineering National Cheng Kung University, Tainan, Taiwan, 701 E-mail:jangjim@mail.ncku.edu.tw

1 Introduction

Louver fin geometry is used widely for heat transfer to air in automotive and room air conditioning heat exchangers such as radiators, condensers, and oil coolers. The louvers act to interrupt the airflow and create a series of thin boundary layers that have lower thermal resistance.

Although many studies on louver fin have been reported, they are based on the assumption that the louver angle is uniform. In this study, a 3-D investigation into the thermal-hydraulic characteristics of successively variable louver angle of louver fin geometry in compact heat exchangers is presented. Numerical simulations of the turbulent are employed.

2 Numerical Results

Figure 1 designates the computation domain and the grid system for the louvered- fin heat exchanger. It consists of an entrance and exit louver with three louvers on either side of the center or redirection louver. Table 1 shows the five different cases of louver angle ($+2°, +4°, -2°, -4°$, and,$u20°$), used in this study. A grid system of 239 x 46x 11 grid points is adopted typically in the computational domain.

The effects of different louver angle patterns are investigated in detail for inlet air velocity ranging from 1m/s to 5m/s. The numerical results for Nusselt number and pressure drop coefficient are shown and compared with the plain fin configuration. Fig. 2 shows the Colburn factor ratio j/j_o and friction factor ratio f/f_o, respectively, between various successively variable louver angle and uniform pattern. For the pattern of case 2($+4°$), the maximum heat transfer improvement interpreted by j/j_o is 119% and corresponding friction factor ratio f/f_o is 119%. For the other cases of case 4($-4°$), case 1($+2°$), case 3($-2°$), the maximum j/j_o is increase up to 115%, 107%,105% and corresponding f/f_o is 116%, 110%, 108%. Fig. 3 shows the maximum area reduction for case 2($+4°$) is the largest and could reach 25.5%, and the other three cases (case 4($-4°$), case 1 ($+2°$), case 3 ($-2°$)) are can be up to 23.4%, 20.2%, 17.5% respectively.

3 Conclusions

The results indicated the successively variable louver angle pattern applied in heat exchangers could effectively enhance heat transfer performance.

Table 1. The louver pattern for different louver angle

Louver pattern								
Case 1 (+ 2°)	20	22	24	26	26	24	22	20
Case 2 (+ 4°)	20	24	28	32	32	28	24	20
Case 3 (- 2°)	26	24	22	20	20	22	24	26
Case 4 (- 4°)	32	28	24	20	20	24	28	32
Case 5 (u20 °)	20	20	20	20	20	20	20	20

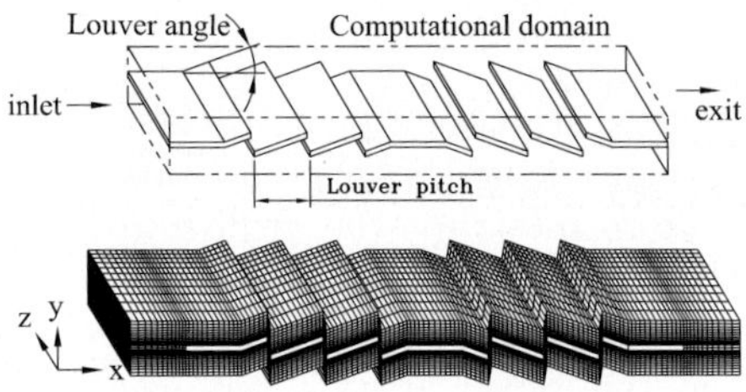

Fig. 1. The physical model, computational domain and grid system

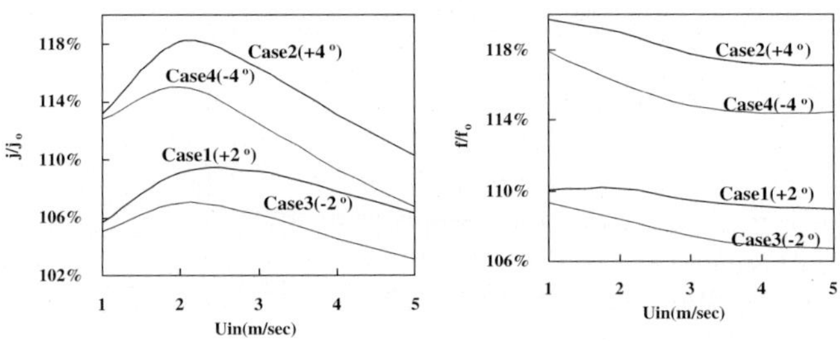

Fig. 2. The (a)j/j_o (b)f/f_o vs. frontal velocity for different louver angle

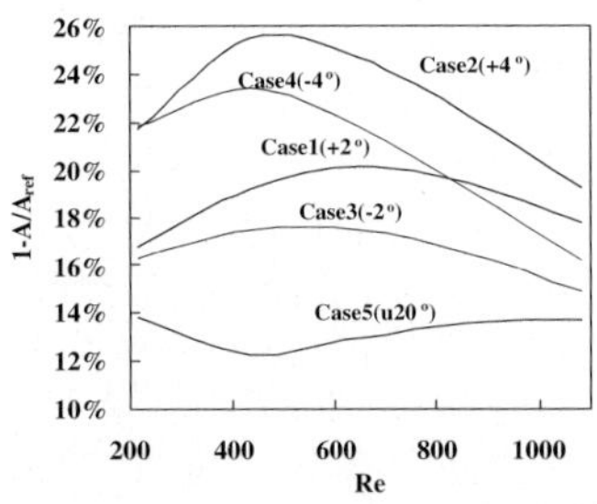

Fig. 3. The Area reduction ratio for different louver angle

Pattern Formation in Viscoelastic Thermal Convection

Zhenyu Li and Roger Khayat

Department of Mechanical and Materials Engineering, the University of Western Ontario, London, Ontario, Canada N5X 2L8

1 Introduction

Convection motions in a shallow fluid layer heated from below or cooled from above by evaporation occur naturally under many circumstances. The spatial variation of convection cells is often referred to as pattern. Pattern formation is determined by nonlinear aspects of the system under study. As such, the elucidation of pattern formation is a challenging problem in fluid mechanics. The objective of this study is to understand the fundamental aspects of patterns and their instabilities, especially in the presence of viscoelastic fluids such as biofluids, polymeric solutions, suspensions, and granular materials.

2 Problem Formulation and Discussion of Results

Consider a thin layer of an incompressible viscoelastic fluid is confined between two infinite and flat plates. The fluid is heated from below and the temperature difference between two plates is δT. The Boussinesq's approximation is assumed to hold. The governing equations to be solved are the conservation of mass, momentum, and energy equations [1], and the constitutive equation [2]. In this work, the fluid is assumed to obey the Oldroyd-B constitutive model [3]. There are four important dimensionless groups in the problem, namely the Rayleigh number, Ra, the Prandtl number, Pr, the elasticity number, E, and the solvent-to-solute viscosity ratio, Rv:

$$Ra = \frac{g\alpha D^3 \delta T}{\nu\kappa}, Pr\frac{\nu}{\kappa}, E = \frac{\lambda\kappa}{D^2}, Rv = \frac{\eta_s}{\eta_p} \tag{1}$$

where α is the coefficient of volumetric expansion, ν the kinematic viscosity, κ the thermal diffusivity, λ the relaxation time, η_s Newtonian solvent viscosity, and η_p polymer solute viscosity. In this study, only the stress free boundary condition is considered. An amplitude equation method[4] is used to examine parameter ranges of three-dimensional stationary convection.

The mostly commonly observed convection patterns between two horizontal layers are: rolls, hexagonal and squares cells. Square pattern is found to be always unstable under the conditions studied here. Figure 1 shows typically the regions of existence of roll and hexagonal patterns for a fluid with

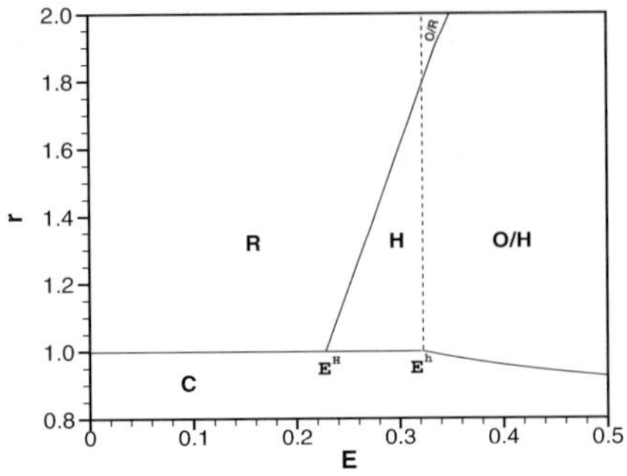

Fig. 1. Stability picture for roll and hexagonal convective patterns.

$Rv = 3.75$ and $Pr = 1000$. In the figure, H and R indicate stable hexagonal and roll regions, respectively. For relatively small E, the conductive state, C, is lost to two-dimensional stationary convection (rolls) when Ra exceeds a critical value Ra_c ($r = Ra/Ra_c > 1$). When the level of elasticity exceeds a critical value E^H, only the hexagonal pattern is stable at the onset of stationary convection. It is found that E^H increases linearly with r, which seems to be the case for other values of Rv and Pr. Thus, rolls are expected to be more stable at higher Ra. The figure also indicates that stationary roll and oscillatory convection can coexist. This region is indicated by O/R in the figure. Similarly, there is a region marked by O/H, where both stationary hexagons and oscillatory convection are possible.

3 Conclusion

Stationary hexagonal patterns are predicted to be stable for certain range of fluid elasticity, which is in contrast to the Newtonian case, where only rolls are predicted to be stable. It is found that the viscosity ratio plays a more important role in determining the likelihood of the two- or three-dimensional patterns for typical polymeric solutions.

References

1. P. G. Drazin, W. H. Reid *Hydrodynamic Stability*, (Cambridge University Press 1981)
2. R. E. Khayat: J. Non-Newtonian Fluid Mech **53**, 227 (1994)
3. R. B. Bird, R. C. Armstrong, O. Hassager *Dynamics of Polymeric Liquids*, vol 1, 2nd edn (John Wiley and Sons, New York 1987)
4. P. Parmentier, G. Lebon, V. Regnier: J. Non-Newtonian Fluid Mech. **89**, 63 (2000)

Turbulent Transport of Passive Scalar Emitted from Line Sources in an Open Channel Flow

Chun-Ho Liu[1] and Dennis Y.C. Leung[2]

[1] Department of Mechanical Engineering, The University of Hong Kong, liuchunho@graduate.hku.hk
[2] Department of Mechanical Engineering, The University of Hong Kong, ycleung@hku.hk

This paper discusses the fluid turbulence structure and transport behavior of passive scalar (emitted from line sources aligned in the spanwise direction) in an open channel calculated by direct numerical simulation (DNS).

The mathematical model consists of the unsteady Navier-Stokes, continuity and scalar transport equations. The implicit coupling between velocity and pressure is decoupled by the 2nd-order accurate fractional-step method. The spatial domain is discretized by the Galerkin finite element method with trilinear interpolating polynomials for velocity, pressure and scalar mixing ratio. The advection and diffusion terms are integrated in time by the 3rd-order accurate Runge-Kutta and the 2nd-order accurate Crank-Nicolson schemes, respectively. The Conjugate Gradient method is used to solve the linear equation systems. The detailed methodology of the calculation is discussed elsewhere [1, 2].

The spatial domain is a channel with 12H (streamwise x) $\times$ 4H (spanwise y) $\times$ H (vertical z), where H is the height of the channel. The flow is pressure-driven in the x direction and periodicity is assumed in the horizontal domain. The Reynolds number Re_τ (based on wall properties) and Schmidt number Sc are 180 and 0.72, respectively. A wall boundary condition (BC) $u_i = 0$ is applied at $z = 0$ while a shear-free BC i. e. $\partial u/\partial z = \partial v/\partial z = w = 0$, where u, v and w are velocity components in the x, y and z directions, respectively, is applied at $z = 1$. Line sources of constant scalar mixing ratio are placed at $x = 0$, $z_s = 0$, 0.25, 0.5, 0.75 and 1 where z_s is the dimensionless emission height. For the scalar, $c = 0$ and a non-reflective open BC $\partial c/\partial t + u\partial c/\partial x = 0$ are assumed, respectively, at the upstream inlet and downstream outlet. Periodic BC is assumed in the y direction and zero Neumann BC is assumed elsewhere.

The vertical profile of the calculated mean streamwise velocity expressed in wall coordinates u^+ exhibits the characteristics surface sublayer, buffer and logarithmic region as observed in most turbulent boundary layers (Figures 1a). The calculated u^+ in the surface sublayer and buffer region is close to that of turbulent channel or plane Couette flow while u^+ is higher in the logarithmic region. As shown in Figure 1b, the structure of the vertical profiles of root-mean-square (RMS) streamwise velocity fluctuation $< u^{+\prime\prime} u^{+\prime\prime} >^{1/2}$ is close to that of a forced channel flow in which the maximum $< u^{+\prime\prime} u^{+\prime\prime} >^{1/2} = 2.7$ at $z^+ = 18$. The calculated RMS spanwise velocity fluctuation $< v^{+\prime\prime} v^{+\prime\prime} >^{1/2}$ is fairly constant except at $z^+ \leq 20$. The calculated RMS vertical velocity fluctuation $< w^{+\prime\prime} w^{+\prime\prime} >^{1/2}$ is fairly constant at the center core of channel while its maximum shifts slightly downward at $z^+ = 70$. The mean plume heights $\overline{z}$ converge toward the channel center as well mixing is being developed downstream (Figure 2). w'', which directly account for the vertical turbulent scalar transport, is large at the center core of the channel and so is the vertical scalar dispersion coefficient σ_z for the source at $z_s = 0.5$ (Figure 2b). In fact, because of the broad constant w'', σ_z is comparable for the sources at $z_s = 0.25$, 0.5 and 0.75. Whilst, w'' is close to zero at the channel bottom and top, thus, σ_z for the sources at $z_s = 0$ and $z_s = 1$is small.

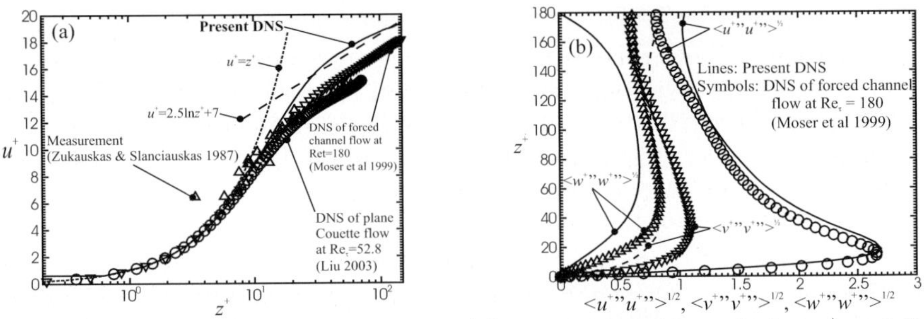

Fig. 1. Vertical profiles of dimensionless (a) mean streamwise velocity u^+ and (b) root-mean-square velocity fluctuations $< u_i^{+\prime\prime} u_i^{+\prime\prime} >^{1/2}$ in wall coordinates z^+.

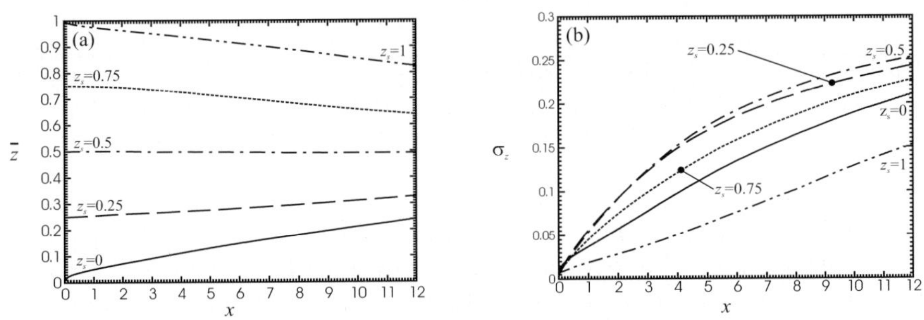

Fig. 2. Dimensionless (a) mean plume height $\overline{z}$, and (b) vertical scalar dispersion coefficient σ_z expressed as functions of dimensionless downstream distance x.

References

1. C.-H. Liu, D.Y.C. Leung: Comput. Meth. Appl. Mech. Eng. **190**, 4301 (2001)
2. C.-H. Liu, D.Y.C. Leung, C.-M. Woo: Computational Mechanics **32**, 185 (2003)
3. C.-H. Liu: J. Heat Transfer **125**, 988 (2003)
4. R.D. Moser, J. Kim, N.N. Mansour: Phys. Fluids **11**, 943 (1999)
5. A. Žukauskas, A. Šlančiauskas: *Heat Transfer in Turbulent Fluid Flows*, (Hemisphere Publishing Corporation 1987)

Mass Transpiration Cooling Analysis at Hypersonic Mach Numbers Using CFD Tools

P.S. Kulkarni[1], V.N. Kulkarni[2], K.P.J. Reddy[2], T. Saito[1], and K. Takayama[1]

[1] ISWRC, Institute of Fluid Science, Tohoku University,Sendai, Japan 980-8577, psk@rainbow.ifs.tohoku.ac.jp
[2] Department of Aerospace Engineering, Indian Institute of Science, Bangalore, India 560012, psk@aero.iisc.ernet.iisc.in

Summary

Navier-Stokes equations coupled with species continuity equation which govern the transpiration cooling are solved numerically using CFD tools in the conditions prevailing over the earth's atmosphere and that of the planet Mars. In the mass transfer cooling technique a cold fluid is injected into the boundary layer through a porous media such that coolant issues forth as a continuous mass. The CFD code developed is based on the finite volume method [1,2]. It uses Roe's upwind method for computations of inviscid fluxes. The modified non-dimensional N-S equations in the conservative integral form can be written as:

$$\frac{\partial \bar{U}}{\partial t} + \frac{1}{A} \oint_S \boldsymbol{F} \cdot \boldsymbol{n} dS = 0 \tag{1}$$

Where, $\bar{U} = \frac{1}{A} \int \int$ UdA is the average value of the column vector of conserved variables U= $[\rho,\ \rho u,\ \rho v,\ \rho E,\ \rho m_l]^T$ in the control volume and $\boldsymbol{F}$ is the flux vector, m_1 is the mass fraction of the species l and other variables have their usual meaning. Earlier numerical investigations were carried out to study transpiration cooling at $M_\infty = 8.5$ over the earth's atmosphere [3]. The modified code is further used compute the flow fields over a flat plate for Mach numbers 6 and 7 and for Reynolds numbers of 1 and 2 million under the conditions prevaling over the planet Mars for different blowing parameters. The variation of skin friction over the flat plate in the earths atmosphere in presence of transpiration cooling for various blowing parameters is shown in Fig. 1. The effect of mass injection on the wall temperature and on the temperature normal to the direction of the flat plate are shown in Figs. 2 and 3. The analysis using CFD tools shows that the transpiration cooling has significant effect on the flow field profiles which results in the reduction of the heat transfer and skin friction at the wall and increase in the boundary layer thickness. Helium and carbon dioxide are found to be relatively more efficient coolants in the atmospheres of earth and Mars respectively. Preliminary investigations accounting for real gas effects in computations show promising results. Experimental investigations which are being planned in ISWRC shok tunnel will become very useful for code validation.

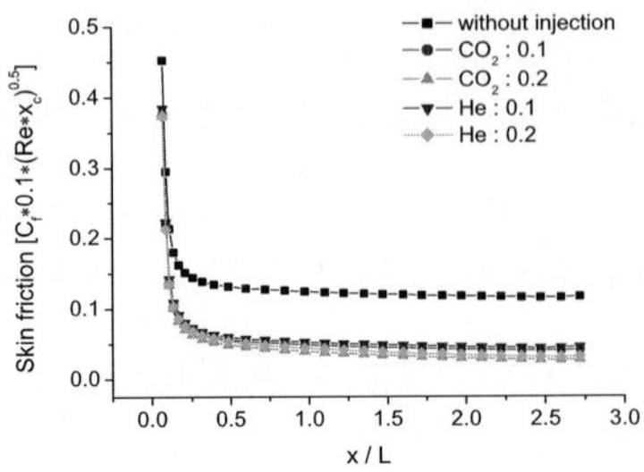

Fig. 1. Skin friction for $M_\infty = 8.5, Re = 5 \times 10^6$ with air as test gas

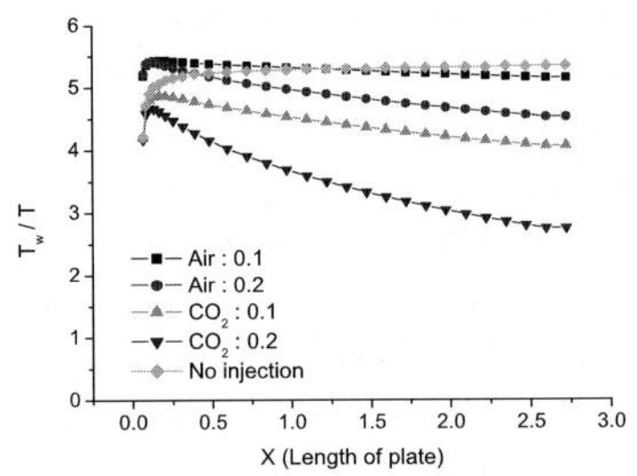

Fig. 2. Wall temperature for $M_\infty = 6.0, Re = 10^6$ with co2 as test gas

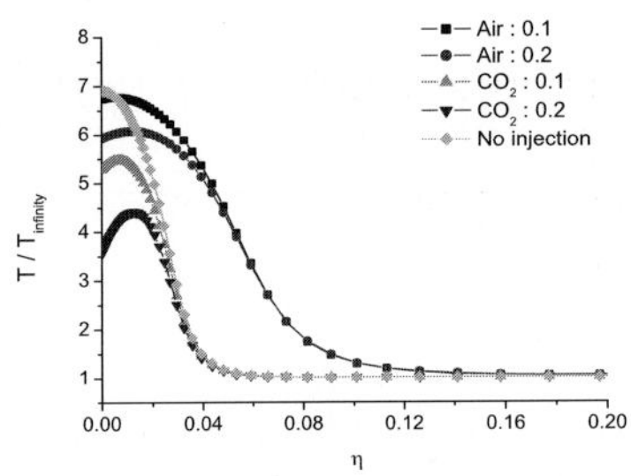

Fig. 3. Temperature normal to the plate $M_\infty = 7.0, Re = 10^6$ with co2 as test gas

References

1. K. Sreekanth,N.M. Reddy:Transpiration cooling analysis at hypersonic Mach numbers using Navier-Stokes equations. AIAA paper 94-2075, (1994)
2. B.R. Ravi:Transpiration cooling analysis including binary diffusion using 2-D Navier-Stokes equations at hypersonic Mach numbers. MSc thesis, 1997, Dept. of Aero Engg, IISc, Bangalore, India
3. P.S. Kulkarni,B.R. Ravi,K.P.J. Reddy:Two dimensional Navier-Stokes solutions for transpiration cooling at hypersonic Mach numbers.Shock Waves,2004

FEM in Domain Decomposition for Fluid-Structure Interaction Problems

Pavel Moses[1], Jaroslav Novotný[2], and Pavel Burda[1]

[1] Department of Mathematics, Faculty of Mechanical Engineering, Karlovo nám. 13, CZ-121 35 Praha 2, Czech Republic, wingco@seznam.cz, burda@fsik.cvut.cz

[2] Institute of Thermomechanics AS CR, Dolejškova 5, CZ-182 00 Praha 8, Czech Republic, novotny@bivoj.it.cas.cz

1 Domain Decomposition Algorithm for Interaction

As a typical example of fluid-structure interaction we can consider a profile in a stream of fluid (air, water). Motion of the fluid exerts forces on the profile, which change its attitude and shape (elastic body), thus forming new geometrical conditions influencing the flow itself.

To model such problem we combine mathematical description of the fluid (Navier-Stokes equations) on the fluid domain Ω_f and the structure (force-equilibrium equation) on the structural domain Ω_s and add a set of conditions on the fluid-structure interface Γ_I to account for their mutual influence.

Now, we will be solving for the fluid velocity and pressure $\mathbf{v} = \mathbf{v}(\mathbf{x}, t)$, $p = p(\mathbf{x}, t)$ and structural displacements $\mathbf{u} = \mathbf{u}(\mathbf{x}, t)$ on discrete time levels.

Schematically, assuming we know the solution at certain time level, solution in the next time step will be found in several steps [1, 2]:

1. find approximation of new interface Γ_I
2. compute velocity v_I^{n+1} on the approximate Γ_I
3. On Ω_f compute $\mathbf{v}, p$ satisfying t^{n+1}

$$\frac{\partial \mathbf{v}}{\partial t} + (\mathbf{v} \cdot \nabla)\mathbf{v} - \nu \Delta \mathbf{v} + \nabla p = \mathbf{f},$$

$$\text{div } \mathbf{v} = 0 \quad \text{on } \Omega_t^f,$$

   with initial conditon $\mathbf{v}(x, t^n) = \mathbf{v}^n(x), x \in \Omega_f$,
   and boundary condition $\mathbf{v}(x, t^{n+1}) = \mathbf{v}_I^{n+1}(x)$ na Γ_I.

4. On Γ_I compute fluid stress

$$\tau_{ij}(v) = \rho_f \left[-p\delta_{ij} + \nu \left(\frac{\partial v_i^{n+1}}{\partial x_j} + \frac{\partial v_j^{n+1}}{\partial x_i} \right) \right]$$

5. On Ω_s find $\mathbf{u}(x, t^{n+1})$ by solving

$$\rho_s \frac{\partial^2 u_i}{\partial t^2} = \frac{\partial \sigma_{ij}(u)}{\partial x_j}$$

with initial and boundary conditions

$$u_i(x,t^n) = u_i^n(x),$$
$$\dot{u}_i(x,t^n) = \dot{u}_i^n(x), \ x \in \Omega_s,$$
$$\sigma_{ij}(u(x,t^{n+1}))n_j^s = \tau_{ij}(v(x,t^{n+1}))n_j^f \text{ on } \Gamma_I.$$

6. On Ω_s find $\dot{u}_i(t^{n+1})$. Then on Γ_I define new velocity approximation e.g. according to

$$x_i^{NEW} := \gamma x_i^f + (1-\gamma)\dot{x}_i^s \text{ for } \gamma \in [0,1)$$

2 Numerical Results

Included is an example of fluid solution for NACA0012 profile at Re=5000.

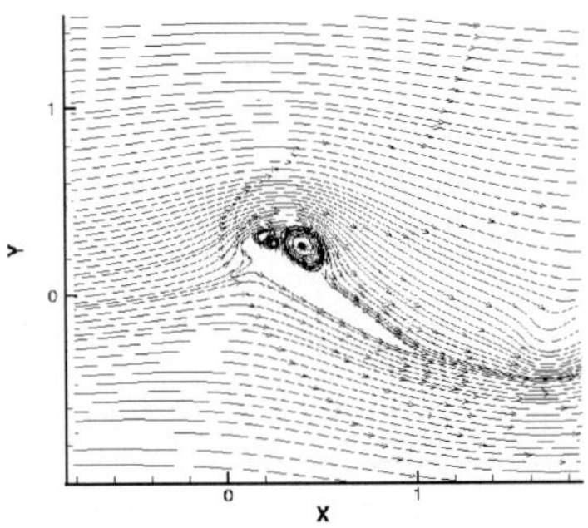

Fig. 1. Streamlines Re 5000

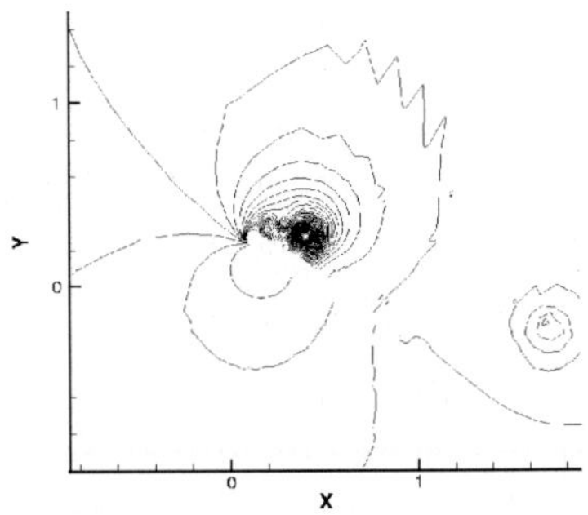

Fig. 2. Pressure isolines Re 5000

Acknowledgements

This research has been supported partly by the GAČR grant No. GA101/02/0391, and partly by the State Research Project No. J04/98/210000010.

References

1. P. Le Tallec, J. Mouro: Fluid structure interaction with large structural displacements. In *ECCOMAS 98* ed by D. Kyriacos at al (John Wiley, New York, 1998) pp 1032–1039
2. J. Novotný: Mathematical modelling of fluid and elastic body interaction (in Czech). PhD Thesis, Czech Technical University, Prague (1998)

LES of Turbulent Flow Around a Simplified Railway Vehicle Model Under Cross Winds

Koji Nakade[1] and Masahiro Suzuki[2]

[1] Environmental Engineering Division, Railway Technical Research Institute 2-8-38, Hikari-cho, Kokubunji-shi, Tokyo 185-8540 JAPAN, nakade@rtri.or.jp
[2] Environmental Engineering Division suzuki@rtri.or.jp

1 Introduction

In order to investigate the aerodynamic characteristics of a railway vehicle under strong cross winds, a number of wind tunnel experiments are performed (for example [1]). In this study, we perfomed LES to study unsteady turbulent flows and simulate vehicle motion against the ground.

2 Numerical Procedure

Three-dimensional filtered incompressible Navier-Stokes equations are used as governing equations. The dynamic subgrid scale model is used with the least square technique. The conservative higher order finite difference schemes [2, 3] are used to discretize the governing equations, which are advanced in time by the fractional-step method. In this study, the authors used the fourth order schemes for space discretization, and the third order Adams-Bashforth method and Crank-Nicolson method for time discretization.

3 Simulation Models and Main Results

Three simulations performed are summarized in Table 1. The physical configuration of the 3D model is shown in Fig. 1. The length of train of the 2D model is 5H. An inflow turbulent boundary layer simulation [4] (Fig. 2) is performed in parallel with the simulation on the 2D model. Figure 3 shows the time history of the lift coefficient. Large periodic oscilation is observed

Table 1. Conditions of simulation

	3D model1	3D model2	2D model
Running speed	$0.5U_\infty$	$0.1U_\infty$	0
Inflow condition	laminar ($\delta = 6H$)		turbulence ($\delta \approx 8H$)
Reynolds number ($U_\infty H/\nu$)		8000	

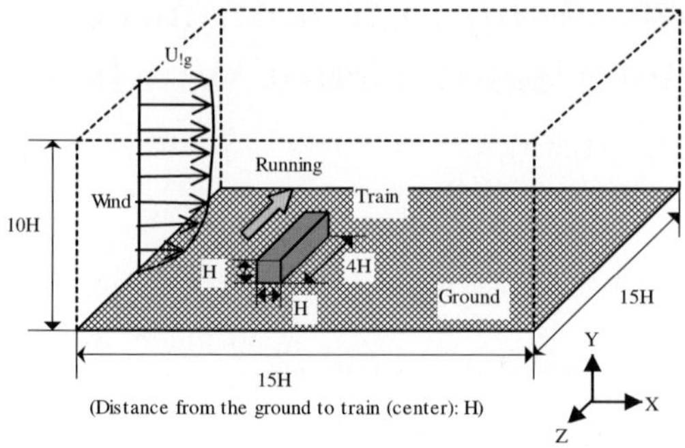

Fig. 1. 3D model

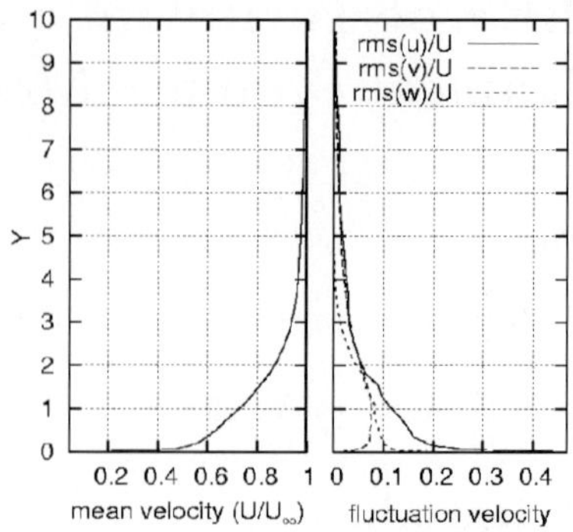

Fig. 2. Inflow for 2D model

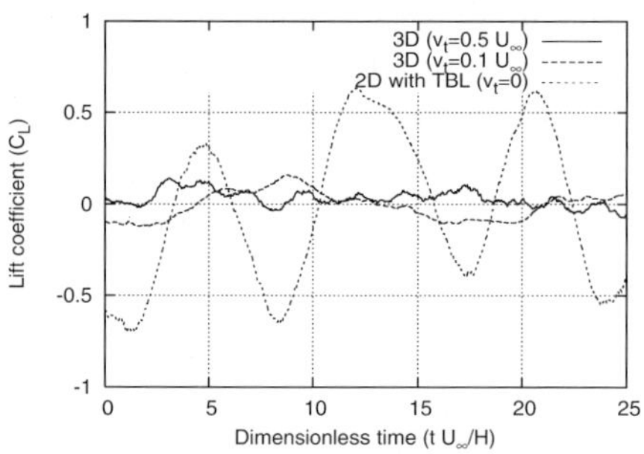

Fig. 3. Time history of C_L

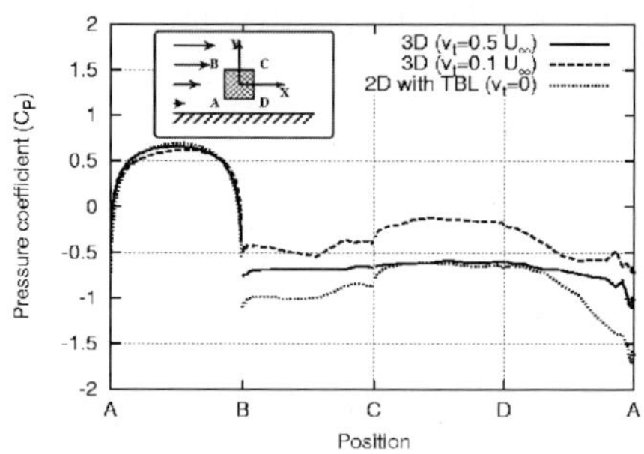

Fig. 4. Mean pressure coefficient

only on the 2D model, whose Strouhal number is 0.11 ∼0.15. Figure 4 shows the mean pressure coefficient on the middle surface of the car. We can see that the side force depends on the 3D effect and running speed. In each case, similar ground effects are observed at the positions D to A in Fig. 4.

References

1. M. Suzuki, K. Tanemoto and T. Maeda: Aerodynamic Characteristics of Train/Vehicles under Cross Winds. J. of Wind Eng. Ind. Aerodyn. 91 pp. 209–218 (2003).
2. Y. Morinishi, T.S. Lund, O.V. Vasilyev and P. Moin: Fully Conservative Higher Order Finite Difference Schemes for Incompressible Flow J. of Computational Physics 143, pp. 90–124 (1998).
3. T. Ohta, Y. Miyake and T. Kajishima: Direct Numerical Simulation of Turbulent Flow in a Wavy Channel. JSME International Journal, B 41 (2), pp. 447–453 (1998).
4. T.S. Lund, X. Wu and K.D. Squires: Generation of Turbulent Inflow Data for Spatially-Developing Boundary Layer Simulations. J. of Computational Physics 140, pp. 233–258 (1998).

Development of Compressible Navier-Stokes Equations into Higher Order DNS of Incompressible Turbulence

Hidetoshi Nishida[1] and Motoyoshi Hatta[2]

[1] Department of Mechanical and System Engineering, Kyoto Institute of Technology,
Matsugasaki, Sakyo-ku, Kyoto 606-8585, JAPAN, nishida@kit.ac.jp
[2] DENSO Corporation, Showacho 1-1, Kariya 448-8661, JAPAN

1 Introduction

In order to simurate the incompressible flows, we usually use the incompressible Navier-Stokes equations as the governing equations. Then, we have to solve not only the momentum equations but also the elliptic partial differential (Poisson) equation, e.g., the pressure equation for MAC[1] approach. However, the computational time for solving this Poisson equation occupies the large part of total compuational time. On the other hand, we consider to use the compressible Navier-Stokes equations for solving the incompressible flow. In the very low speed flow, however, the compressible Navier-Stokes equations have the property of instability and stiffness. In order to overcome these properties, the preconditioning method[2] has proposed, but the time accuracy is not always guaranteed.

In this paper, we will develop the compressible Navier-Stokes equations to solve the incompressible flows with higher order of accuracy in space and time. We adopt the higher order method of lines approach[3] to solve these equations. The spatial derivatives are discretized by the modified differential quadrature (MDQ) method[4]. The MDQ method gives the arbitrary order of accuracy by changing only one parameter. Then, the partial differential equations are reduced to the system of ordinary differential equations (ODEs) in time. The system of ODEs is integrated by the 4th order Runge-Kutta scheme.

In the very low speed and high Reynolds number flow simulation, the unphysical oscillations will appear. In order to improve this unphysical solution, the artificial viscosity terms are usually added to the governing equations. However, the artificial viscosity terms include the intentional parameter. This is not desirable for more accurate simulations, e.g., simulation of turbulence. Then, instead of adding the artificial viscosity, we introduce here the shifted flux approach for discretizing the convective terms. The inviscid fluxes are discretized at cell center and these fluxes at usual grid point are specified by using the interpolation. We consider the direct numerical simulation (DNS) of homogeneous isotropic turbulence.

2 Computational Results

The computational conditions are the uniform grid with 128^3 grid points, the Reynolds number $Re = 1/0.0045$ ($\nu = 4.5 \times 10^{-3}$, ν is the kinematic viscosity), the Mach number $Ma = 0.05$ and 10th order of spatial accuracy. In order to compare the flow field, Fig.1 shows the iso-surface of enstrophy of the present and spectral solutions at $t = 2.5$ and $t = 10.0$. At $t = 2.5$, the flow field is more complicated, and the difussion is dominant at $t = 10.0$. We can obtain the smooth solution without the unphysical oscillations at any time. The present enstrophy field is in more excellent agreement with solution obtained by the spectral method. Figure 2 shows the comparison of statistical quantities, i.e., the total energy, energy dissipation and enstrophy dissipation, with the spectral solution. The present method gives very good results, too.

Then, it is concluded that the present method is very promising for the DNS of incompressible flows, especially incompressible turbulence.

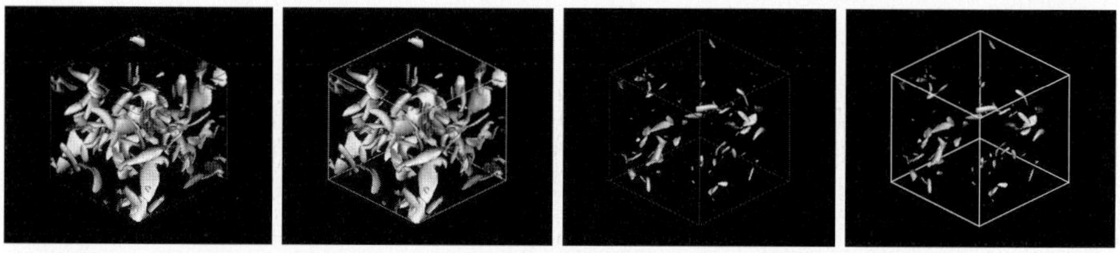

(a)Present solution ($t = 2.5$, $\Omega = 30$) (b)Spectral solution ($t = 2.5$, $\Omega = 30$) (c)Present solution ($t = 10.0$, $\Omega = 15$) (d)Spectral solution ($t = 10.0$, $\Omega = 15$)

Fig. 1. Comparison of enstrophy field

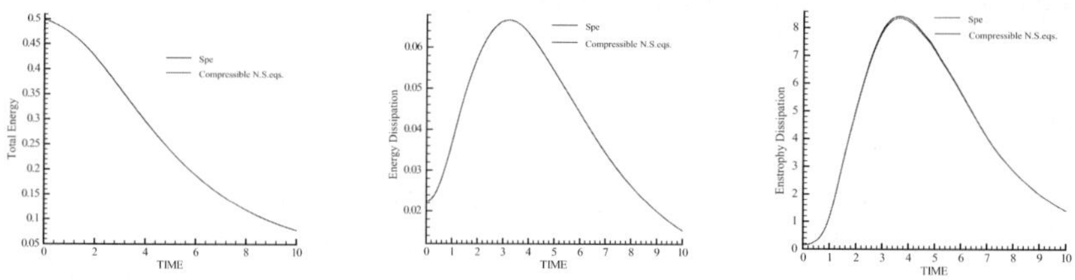

(a)Total energy (b)Energy dissipation (c)Enstrophy dissipation

Fig. 2. Comparison of statistical quantities

References

1. F.H.Harlow and J.E.Welch, Phys. Fluids, **8**, pp 2182–2189 (1965)
2. E.Turkel, J. Comp. Phys., **72-2**, pp 277–298 (1987).
3. H.Nishida and N.Satofuka, Computational Mechanics '95 (Springer 1995), Vol.1, pp 941–946
4. N.Satofuka and K.Morinishi, *NASA TM-81339*, (1982)

Direct Numerical Simulation of Mixed Convection in Horizontal Pipe Flow

M. Piller

Università degli Studi di Trieste, Dipartimento di Ingegneria Navale, del Mare e per l'Ambiente, via A. Valerio, 10 – 34127 Trieste, ITALY, piller@units.it

1 Introdution

Direct numerical simulations (DNS) of fully-developed mixed convection in horizontal pipes were carried out to investigate the interactive shear and buoyancy effects on the turbulent momentum and heat transport. The friction Reynolds number, based on the pipe diameter, and the Prandtl number are assumed $\mathrm{Re}^* = 360$ and $\mathrm{Pr} = 0.71$, respectively. The intensity of buoyancy is defined by the friction Richardson number: $\mathrm{Ri}^* = g\beta T^* D/{u^*}^2 = 0.0, 0.1, 0.3, 0.7, 1.0$. The flow is driven by a constant mean pressure gradient and *ideal* isoflux conditions are imposed on the pipe wall, resulting in a mixed stable/unstable stratification with a complex secondary flow pattern. The ill-posedness of the resulting differential problem for the temperature field is inherently avoided, due to the integral formulation of the finite volume method, used for the present simulations.

Previous investigations on turbulent mixed convection were carried out for the channel flow [3, 4] and for the vertical pipe flow [2, 1] while, to the author's best knowledge, no DNS investigation has been reported so far for the turbulent mixed convection in horizontal tubes.

2 Results

Previous experimental and numerical studies [7, 6, 5] have shown that the buoyancy forces generate a secondary flow, symmetric with respect to the vertical midplane, with the warmer fluid moving upward close to the pipe wall and, by continuity, flowing downward through the core region. While in the laminar and transitional regimes the increased advection, due to the secondary motions, causes an augmentation of both the momentum and the heat transfer [7, 2], the present results show the opposite trend. This circumstance is related to two different phenomena. First, the buoyancy tends to reduce and, finally, annihilate the near-wall streamwise vortices, as already verified by Iida *et al.* [3] for the unstably stratified turbulent channel flow. In addition, a region of stable stratification extends from the upper portion of the pipe to the core region, as the Richardson number increases, causing a significant reduction of the turbulent mixing. The Fanning friction factor

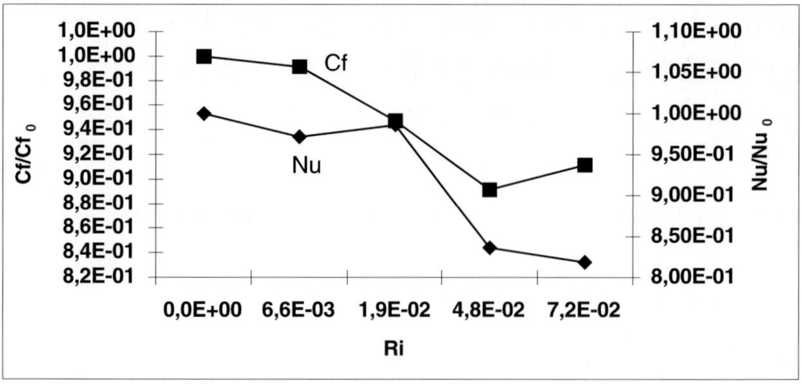

Fig. 1. Trend of the Fanning friction factor and the Nusselt number, based on the pipe diameter and the difference between the (azimuthally-averaged) wall and bulk temperatures.

and the Nusselt number, reported in Figure 1, give a synthetic indication of the hydraulic and thermal efficiency, respectively.

References

1. G. Celata, F. D'Annibale, A. Chiaradia, and M. Cumo, *Upflow turbulent mixed convection heat transfer in vertical pipes*, Int. J. Heat Mass Transfer, 41 (1998), pp. 4037–4054.
2. A. Ghajar and L. Tam, *Flow regime map for a horizontal pipe with uniform wall heat flux and three inlet configurations*, Exp. Thermal Fluid Science, 10 (1995), pp. 287–297.
3. O. Iida and N. Kasagi, *Direct numerical simulation of unstably stratified turbulent channel flow*, Int. J. Heat Transfer, 119 (1997), pp. 53–61.
4. N. Kasagi and M. Nishimura, *Direct numerical simulation of combined forced and natural convection in a vertical plane channel*, Int. J. Heat and Fluid Flow, 18 (1997), pp. 88–99.
5. L. Li, C. Lin, and M. Ebadian, *Turbulent mixed convective heat transfer in the entrance region of a curved pipe with uniform wall-temperature*, Int. J. Heat Mass Transfer, 41 (1998), pp. 3793–3805.
6. S. Morcos and A. Bergles, *Experimental investigation of combined forced and free laminar convection in horizontal tubes*, ASME J. Heat Transfer, 97 (1975), pp. 212–219.
7. P. Newell and A. Bergles, *Analysis of combined free and forced convection for fully developed laminar flow in horizontal tubes*, J. Heat Transfer, Trans. ASME, Series C, 92 (1970), pp. 83–89.

Simulation of Inviscid, Unsteady Flows in Hypersonic Air Inlets Using an Adaptive, Unstructured, Multi-Block Method

Råbi Bin Tahir[1], Eugene Timofeev[1], Peter Voinovich[2], and Sannu Mölder[3]

[1] McGill University. 817 Sherbrooke St. West, Montreal, QC, H3A 2K6, Canada. rabi.tahir@rogers.com; evgeny.timofeev@mcgill.ca

[2] St. Petersburg Branch of the Joint Supercomputer Center. Polytekhnicheskaya 26, St. Petersburg 194021, Russia. vpeter@scc.ioffe.ru

[3] Ryerson University, 89 Joseph Street, Port Carling, ON, P0B 1J0, Canada. smolder@ryerson.ca

The present work is devoted to application of a high-resolution, unstructured, solution-adaptive, multi-block method to simulate high Mach number flows in the inlet portion of a hypersonic air-breathing engine. The numerical method features a MUSCL-type, globally second-order, finite-volume, TVD scheme which has proven to be well-suited to such flows (unsteady, inviscid flow cases), particularly when high temporal accuracy is desired.

In the context of high speed internal flows, simulations involving removal of a diaphragm or opening and closing of a wall perforation require the presence of internal boundaries within the computational domain; this is applicable whenever two or more sub-domains can not be separated by use of initial conditions alone. A drastic change in flow conditions may occur at these internal boundaries, *e.g.*, when the diaphragm is ruptured or when a wall perforation is opened or closed. A simple strategy based on twin-nodes situated along the internal boundaries [3] is suitable for use in multi-block discretizations based on unstructured meshes. A generalization of this concept to gasdynamic flows (see Fig. 1) and to arbitrary number of neighbouring sub-domains has been implemented [1, 4], and results shown below demonstrate its application to both pseudo-steady as well as to fully unsteady flows.

Consider, for instance, the geometry of Fig. 2, which is designed for simulation of flow through slits in the side walls of the inlet. In general, the flow properties are different across the wall. Thus, if the wall slits are to be opened and/or closed during the simulation, then this geometry can be easily represented by placing twin-nodes along the wall.

Another convenient application of the twin-nodes concept is for diaphragm rupture. Consider, as an example, the geometry of a classical shock-tube problem as shown in Fig. 3, where two zones are initially separated by a diaphragm. In our inlet starting studies, we "install" diaphragms of different geometries at various locations inside high-contraction inlets. It turns out[2] that under certain conditions, instantaneous rupture of such diaphragms can be used to achieve started flow through the inlets.

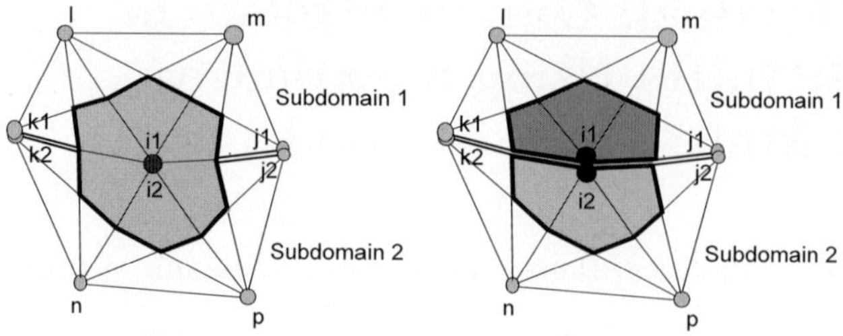

Fig. 1. Twin nodes at an internal boundary and control volumes for different types of boundary conditions ("invisible boundary" – left; a wall – right).

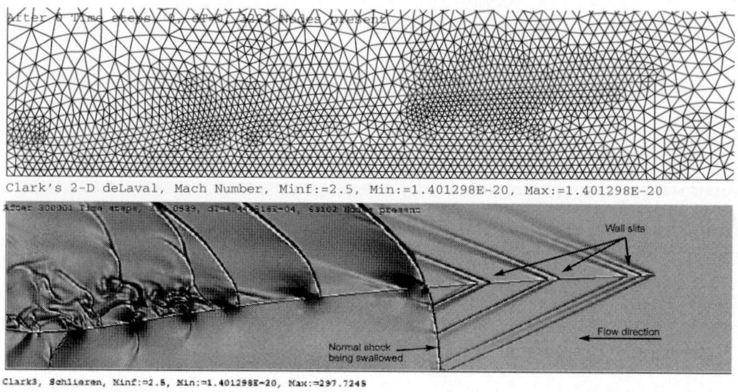

Fig. 2. Inlet with a perforated wall: initial basis mesh with closed slits (top); mass spillage through open slits and shock-swallowing during the starting process (bottom).

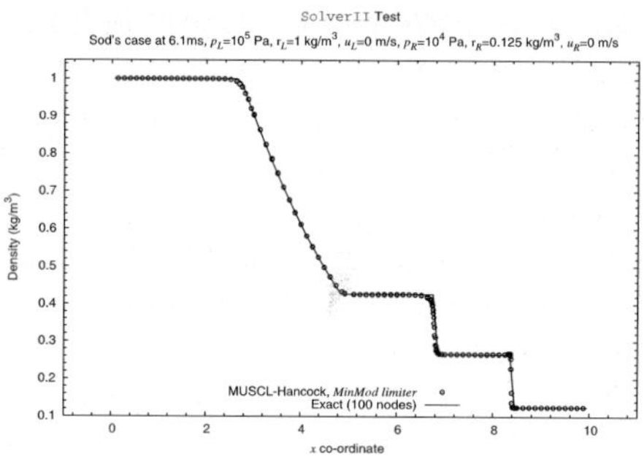

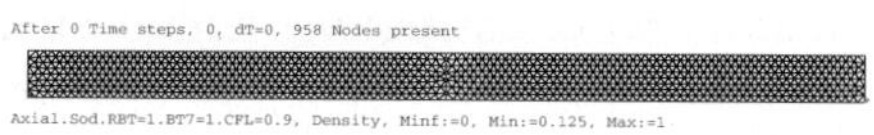

Fig. 3. Axisymmetric shock tube: initial basis mesh with intact diaphragm (right), solution after diaphragm rupture (left).

References

1. R. Tahir, P. Voinovich, and E. Timofeev. `SolverII`—Two-dimensional, unstructured, adaptive, multi-block, time-depenedent CFD applications for MS Windows®, 1999–2003. `e-mails: rabi.tahir@rogers.com; timofeev@sympatico.ca; vpeter@scc.ioffe.ru.`
2. R. B. Tahir, S. Molder, and E. V. Timofeev. Unsteady starting of high Mach number air inlets—a CFD study. AIAA Paper 2003-5191, 2003.
3. P. Voinovich and A. Merlen. Two-dimensional unstructured elastic model for acoustic pulse scattering at solid-liquid interfaces. *Shock Waves*, 12(5):421–429, 2003.
4. P. A. Voinovich and A. O. Galyukov. UGG. Two and three-dimensional, triangular and tetrahedral, unstructured grid generators. 1993–1999.

Fluid Structure Interaction for Strongly Coupled Problems Based on a Sensitivity Analysis

Jan Vierendeels[1], Kris Dumont[2], Erik Dick[1], and Pascal Verdonck[2]

[1] Ghent University, Department of Flow, Heat and Combustion Mechanics, Sint-Pietersnieuwstraat 41, 9000 Gent, Belgium. Jan.Vierendeels@UGent.be
[2] Ghent University, Institute Biomedical Technology, Hydraulics Laboratory, same address. Pascal.Verdonck@UGent.be

A fluid-structure interaction (FSI) procedure is presented that allows the calculation of strongly coupled FSI problems. The application that is addressed is the opening and closing of a stiff valve with frictionless hinge. The coupling method consists of an iteration procedure for each time step in which alternatively solutions of the fluid and the structure problem are obtained until convergence is reached. Such an iteration procedure can be unstable depending on the density of the material and fluid, the gap size and the area of the valve, even when substantial underrelaxation is used. Therefore the method uses a sensitivity analysis to predict the behaviour of the fluid system. With this information, the structural problem can be solved and implicit coupling is achieved. The equation of motion for the stiff valve leaflet is given by $M = I\ddot{\theta}$ with M the moment resulting from the forces acting on the surface of the leaflet, I the moment of inertia and θ the angle that determines the position of the leaflet. The moment of inertia I with the rotational axis at the end of the leaflet is given by $I = (1/3)ml^2$ with $m = \rho lt$ the mass of the leaflet per unit length, ρ the density of the leaflet (1100 kg/m^3), t the thickness of the leaflet (0.5 mm) and l the length of the leaflet (22 mm). A reference length of 1 m is used in the third dimension, corresponding to the reference length used in the CFD package to calculate the moments. The moment of inertia $I = 1.9545$ kg mm^2. We consider a class of time integration schemes for the valve motion given by $\dot{\theta}^{n+1} = \dot{\theta}^n + (1-\beta)\,\ddot{\theta}^n \Delta t + \beta\ddot{\theta}^{n+1}\Delta t$, $\theta^{n+1} = \theta^n + \dot{\theta}^n \Delta t + \gamma\ddot{\theta}^n \Delta t^2 + \alpha\ddot{\theta}^{n+1}\Delta t^2$. For γ=1/2-α this corresponds to the class of Newmark schemes. For γ=0, $\alpha = \beta = 1$ this corresponds to the backward Euler scheme. The valve motion is coupled with the fluid motion which is computed with the commercial CFD software package Fluent (Fluent Inc.). The dynamic mesh model of Fluent 6.1, with backward Euler time integration, has been used. Only the position of the solid is exchanged between both solvers. Figure 1 shows snapshots of the flow pattern in a 2D geometry representing a stiff aortic valve rotating about a frictionless hinge. Blood (ρ_f=1000 kg/m^3, dynamic viscosity μ=4 mPa s) is used as fluid.

As a generic 1D model for the kind of FSI problem treated in this paper, a moving cylinder in a cylindrical tube is considered (figure 2). Between the cylinder and the tube there exists a small gap which connects the fluid at the front and the back of the cylinder. With a backward Euler discretization for the fluid motion, $F^{n+1}/m = \frac{\rho_f}{\rho_s}\left(\frac{1}{a}\frac{u^{n+1}-u^n}{\Delta t} - \frac{1-a}{a}\frac{\dot{x}_{NS}^{n+1}-\dot{x}_{NS}^{n}}{\Delta t}\right)$. Here $\dot{x}_{NS}$ denotes the velocity of the solid in the fluid solver. The resulting equations for this generic 1D problem can be written in matrix form: $\begin{bmatrix} 0 & 1 & -\alpha \\ 1 & 0 & -\beta \\ 0 & K & 1 \end{bmatrix} \begin{bmatrix} \dot{x}\Delta t \\ \dot{x}_{NS}\Delta t \\ \ddot{x}\Delta t^2 \end{bmatrix}^{n+1} = \begin{bmatrix} 1 & 0 & -\gamma \\ 1 & 0 & 1-\beta \\ 0 & K & 0 \end{bmatrix} \begin{bmatrix} \dot{x}\Delta t \\ \dot{x}_{NS}\Delta t \\ \ddot{x}\Delta t^2 \end{bmatrix}^{n} + \begin{bmatrix} 0 \\ 0 \\ \frac{K\Delta u \Delta t}{1-a} \end{bmatrix}$ where $K = \frac{\rho_f}{\rho_s}\frac{1-a}{a}$, with $a = A_g/A$, $A_g = A - A_b$. The eigenvalues λ_i are $\lambda_1 = 1$, $\lambda_{2,3} = \frac{(\alpha-\beta-\gamma)K \pm \sqrt{(\alpha+\beta+\gamma)^2K^2 - 4K(1+\alpha K-\beta-\gamma)}}{2(1+\alpha K)}$. Only for the backward Euler scheme $\lambda_{2,3} = 0$ and perfect damping is obtained. Figure 3 shows the comparison of the evolution of the angular acceleration of the valve for the first cycle computed with the backward Euler scheme and the second order Newmark scheme ($\alpha = 0.25, \beta = 0.5$).

From the study it can be concluded that it is very important to match the time integration schemes of fluid and structural solvers if strong coupling is present.

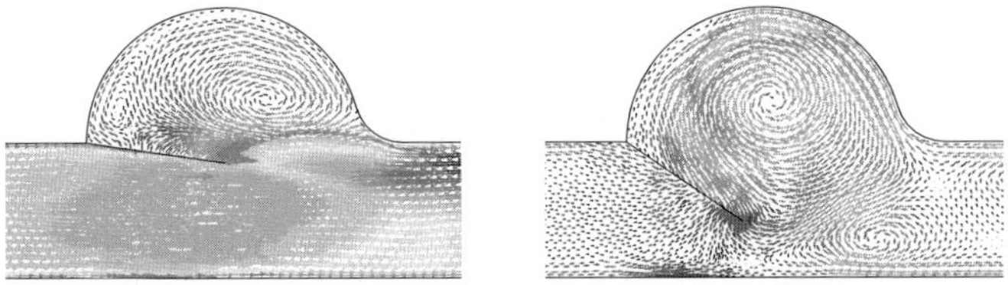

Fig. 1. Flow patterns during opening and closing of the valve

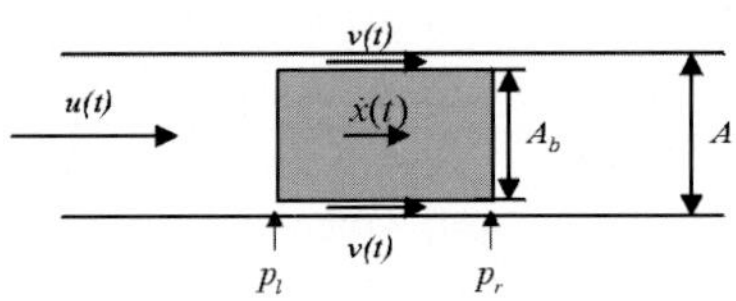

Fig. 2. Motion of a rigid body in a tube. $u(t)$ and $\dot{x}$ are the velocities of the oncoming fluid and the rigid body. $v(t)$ is the fluid velocity in the gaps.

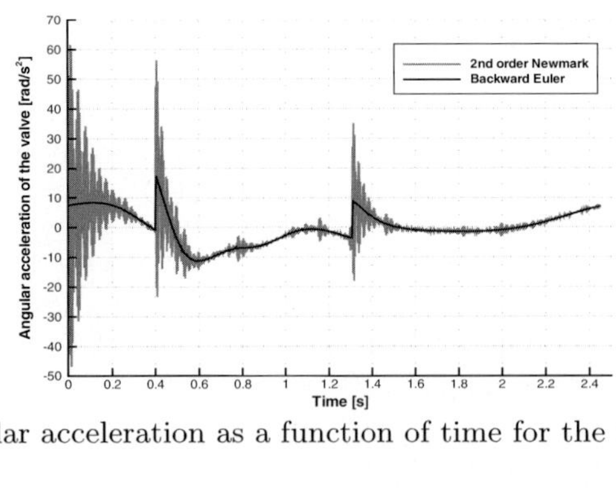

Fig. 3. Angular acceleration as a function of time for the valve calculation.

Numerical Simulation of Flow Conditioners Used for Flow Meter Calibration

E. von Lavante[1], G. Moniz Pereira[1], U. Banaszak[1], and B. Mickan[2]

[1] University of Duisburg- Essen, Institute of Turbomachinery, Essen, Germany, ernst3.vonlavante@tigger.turbo.uni-essen.de
[2] Physikalisch- Technische Bundesanstalt, Department of Gas, Braunschweig, Germany, bodo.mickan@ptb.de

It is a well known and recognized fact that the behavior of mass and volumetric flow rate measuring devices can be strongly affected by the flow conditions prevailing upstream of their location. Disturbed velocity profiles caused by pipe configurations such as bends, headers, pressure regulators and convergent or divergent pipe sections in front of a flow meter can lead to errors of several percent in the meter reading. It is therefore necessary to apply flow conditioning before the flow enters the meter. Flow conditioning means in this case the generation of fully developed symmetric flows in the inlet pipe section of meters to avoid installation effects. Since this is in practice not always possible, it is necessary to investigate the influence of flow perturbations on the behavior of flow meters, so that such investigations became a part of pattern approval in legal metrology. In the present investigation, the attempt was made to simplify the above procedure by introducing a reproducible disturbance in the inflow to the meter. Hence, flow conditioning means here the generation of disturbed flows with a definite and reproducible level of perturbation. The International Organization of Legal Metrology (OIML) defined therefore standard pipe configurations (OIML Recommendation R 32) to perform perturbations test within pattern approvals.

The test configurations recommended by OIML were defined for metering applications with the metered media being air at atmospheric conditions. The main reason for this somewhat restrictive scope of implementation was the otherwise huge effort necessary to perform investigations in natural gas under high pressure conditions. The idea behind the present work is to generate a prescribed disturbance of the inflow velocity profile by introducing swirl and non-symmetric velocity distribution at the inflow by employing a plate with asymmetric hole configuration. The model flow conditioner, shown in Fig. 1, consisted of small pipes of different diameters, discharging into a larger pipe of 200 mm diameter. The smaller pipes are arranged off-center and are tilted under different angles from the symmetry line. The sudden expansion formed by the transition from two smaller pipes to one larger pipe is shown in Fig. 1.

The flow in the above configuration was investigated numerically as well as experimentally, allowing a direct comparison and, therefore, validation of the numerical results. The numerical work used the commercial product FLUENT version 6.x. The segregated solver that is appropriate for weakly

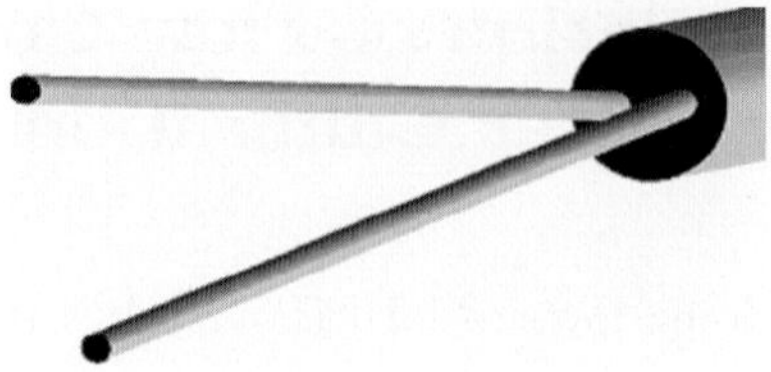

Fig. 1. Model flow conditioner (40mm inlet pipes into 200mm pipe)

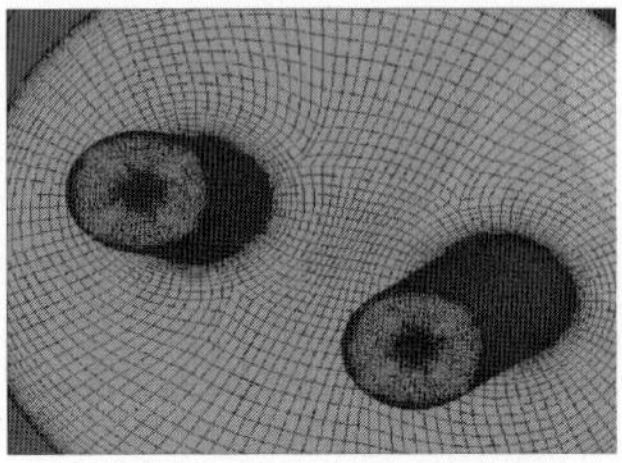

Fig. 2. Computational grid at the inlet of the flow conditioner

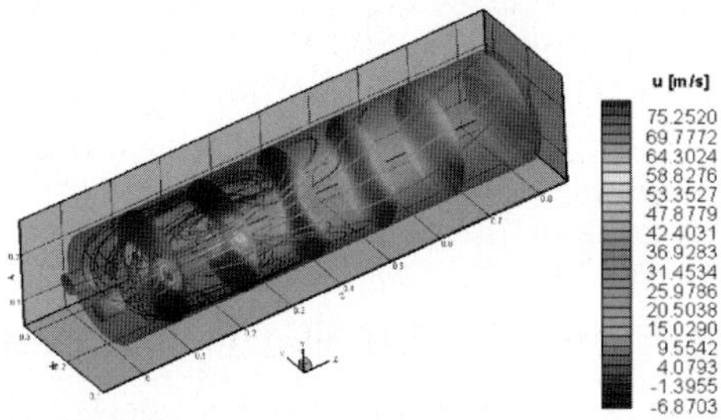

Fig. 3. Computed axial velocity and stream lines

compressible flow was used. In Fig. 2, the surface grid is displayed at a location where the two smaller pipes merge with the large one. The grid consisted of 536, 000 cells arranged in 22 blocks, representing a reasonable compromise between required resolution and numerical speed. Different turbulence models were tested and validated. These were the widely used $k-\epsilon$ realizable model, the $k-\omega$ model and the rather computationally intensive full Reynolds Stress Model (RSM), as implemented in the Fluent program. A representative mass flow rate of $\dot{m}=0.2136$ kg/s was selected for the present work, resulting in a Reynolds number based on the larger pipe diameter of $Re=200,000$.

The computed axial velocity in the flowfield downstream of the sudden expansion can be seen in Fig. 3. It was found that the resulting flowfield and, therefore also the characteristic flow parameters depend critically on turbulence models used. In particular, numerical simulations using the RSM agreed rather well with the experimental data. The recirculation zone, reattachment and pressure drop as well as momentum and swirl numbers were well predicted.

Large Eddy Simulation Using Tetrahedral Elements

Tao Xu[1], German Cardenas[1], and Marius Paraschivoiu[2]

[1] Dept. of Mech. and Ind. Eng., Univ. of Toronto, Toronto, Canada, M5S 3G8. xutao@mie.utoronto.ca, german@mie.utoronto.ca

[2] Dept. of Mech. and Ind. Eng., Concordia Univ., Montreal, Canada, H3G 1M8. paraschi@ME.concordia.ca

1 Introduction

We present a parallel finite element method for large eddy simulation (LES). The objectives is to apply the LES to solve practical engineering problems, that is, on complex geometries and large Reynolds numbers flows. Only preliminary work is presented here where an in-house finite element code was modified to perform LES. The accuracy and efficiency of this tool for turbulence simulations are examined in detail for isotropic decaying turbulence. The Smagarinsky model is implemented by the scheme proposed by Mullen [1]. Taylor-Hood tetrahedral elements are implemented for the spatial discretization. The algebraic system arising from a splitting scheme is solved using a matrix-free conjugate gradient iterative method. Domain decomposition and parallel computation are implemented.

2 Numerical Framework and Parallel Computation

The governing equation is the filtered Navier-Stokes equation. The Smagorinsky model is used as the subgrid-scale model. Implementing the Galerkin finite element approach, we employ the variable viscosity formulation introduced by Mullen [1] to discretize the viscous term. Substituting Taylor-Hood basis functions into other terms of the variational forms and take twice $diag(\mathbf{A})$ results in the following semi-discrete equations:

$$\mathbf{M}\frac{\partial \bar{\mathbf{u}}_\mathbf{i}^\mathbf{h}}{\partial \mathbf{t}} + \mathbf{C}\left(\bar{\mathbf{u}}^{\mathbf{h(n)}}\right)\bar{\mathbf{u}}_\mathbf{i}^{\mathbf{h(n)}} + \mathbf{2A}\bar{\mathbf{u}}_\mathbf{i}^{\mathbf{h(n+1)}} - \mathbf{D}_\mathbf{i}^\mathbf{T}\bar{\mathbf{P}}^{\mathbf{h(n)}} = \underline{\mathbf{D}}_{ij}\bar{u}_j^{h(n)} + \mathbf{A}\bar{\mathbf{u}}_\mathbf{i}^{\mathbf{h(n)}}, \quad \mathbf{i} = \mathbf{1,2,3}, \qquad (1)$$

and $\mathbf{D_i}\bar{\mathbf{u}}_\mathbf{i}^\mathbf{h} = \mathbf{0}$ where $\mathbf{n}$ is the current time step. The above system has been proved to be unconditioned stable for the Stokes part of the system [1]. The strategy for temporal discretization is a semi-implicit, fractional time-stepping scheme. In this case, a 3rd- order accurate, explicit Adams-Bashforth $AB3$ time integration scheme is used for the nonlinear convection terms, while a 2nd-order accurate, implicit backward differentiation $BD2$ scheme is pursued for the remainder Stokes subproblem. The FEM based LES is implemented on two different Beowulf clusters. MPI is employed to do the gathering and scattering operations. The computational domain is pre-processed

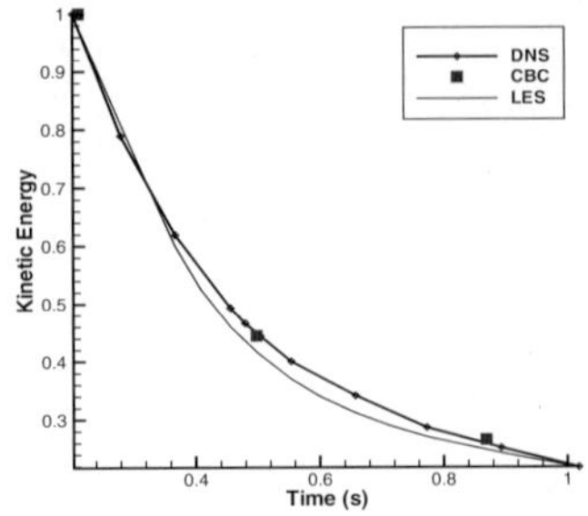

Fig. 1. Decay of turbulence kinetic energy: comparison between LES, DNS, and experimental results (CBC).

and broken into subdomains using METIS. We employ Jacobi preconditioner in this reaserch, but more effective preconditioning technique will be explored in the future.

3 Numerical results

We validate the parallelized LES code by simulating isotropic decaying turbulence. A 48^3 cubic mesh is used in this test case. There are 857,375 nodes and 622,938 Taylor-Hood elements employed to address the large eddy simulation. The initial conditions for the code are based on the AGARD database [2] which includes a restart file containing a velocity field in physical space at $Re_\lambda = 96$. A domain decomposition is also necessary to apply the initial condition into each subdomains.

We compare the numerical LES solution with the experimental data from Comte-Bellot & Corrsin (CBC) [3]. DNS results by A. Wray [2] are also compared. Figure 1 illustrates the comparison among the turbulence kinetic energy normalized by the initial value for CBC, LES and DNS data, which shows that the LES data have a good agreement with the DNS and CBC data for the decaying rate of turbulence kinetic energy. We have tested the parallelized LES code on two different clusters: a HPCAF Alpha Beowulf Cluster and Unix PC based cluster. Results show that the HPCAF supercomputer is faster and more efficient due to its faster communication network. For the HPACF supercompeter, the parallel speed-up of running the code on 12 processors with respect to 4 processors is 2.7.

References

1. J. S. Mullen. Development of a Parallel Spectral Element Based Large Eddy Simulation Model for the Flow of Icompressible Fluids in Complex Geometries. PhD Thesis, Brown University, May 1999.
2. J. Jimenez, O. Leuchter, S. Tavoularis. A Selection of Test Cases for the Large Eddy Simulation of Turbulent Flows. *Agard Advisory Report*, No. 345, 1998.
3. G. Compte-Bellot, S. Corrsin. The Use of a Contraction to Improve the Isotropy of Grid Generated Turbulence. *J. Fluid. Mech.*, 48, 273-337.

Printing: Krips bv, Meppel
Binding: Stürtz, Würzburg